3rd International Conference on Geoscience and Environmental Chemistry (ICGEC 2021)

E3S Web of Conferences Volume 290 (2021)

Xining, China
18-20 June 2021

Editors:

Jun Sun
Zhaojian Li
Xianbin Liu

ISBN: 978-1-7138-3481-6

Printed from e-media with permission by:

Curran Associates, Inc.
57 Morehouse Lane
Red Hook, NY 12571

CURRAN ASSOCIATES INC. proceedings.com

Some format issues inherent in the e-media version may also appear in this print version.

For additional information, please contact EDP Sciences – Web of Conferences at the address below.

EDP Sciences – Web of Conferences
17, Avenue du Hoggar
Parc d'Activité de Courtabœuf
BP 112
F-91944 Les Ulis Cedex A
France

Phone: +33 (0) 1 69 18 75 75
Fax: +33 (0) 1 69 28 84 91

contact-edps@webofconferences.org

Additional copies of this publication are available from:

Curran Associates, Inc.
57 Morehouse Lane
Red Hook, NY 12571 USA
Phone: 845-758-0400
Fax: 845-758-2633
Email: curran@proceedings.com
Web: www.proceedings.com

TABLE OF CONTENTS

ENVIRONMENTAL CHEMISTRY AND CHEMICAL TECHNOLOGY APPLICATION

ENVIRONMENTAL ENERGY PROTECTION AND ENERGY-SAVING SUSTAINABILITY

Preface

Due to recent pandemic, the 2021 3rd International Conference on Geoscience and Environmental Chemistry (ICGEC 2021) which was planned to be held in Xining, China, was held virtually online during June 18-20, 2021. The decision to hold the virtual conference was made in compliance with many restrictions and regulations that were imposed by countries around the globe. Such restrictions were made to minimize the risk of people contracting or spreading the COVID-19 through physical contact. There were 120 individuals who attended this online conference, represented many countries including Algeria, Cameroon and China.

The primary goal of the symposium is to promote research and developmental activities in geoscience and environmental chemistry, and another goal is to promote scientific information interchange between researchers, developers, engineers, students, and practitioners working all around the world. An ideal platform for seeking global partners had been established. We sincerely invited you to attend the forum to share views and experiences in geoscience, environmental chemistry and related areas.

During the conference, the conference model was divided into three sessions, including oral presentations, keynote speeches, and online Q&A discussion. In the first part, some scholars, whose submissions were selected as the excellent papers, were given about 5-10 minutes to perform their oral presentations one by one. Then in the second part, keynote speakers were each allocated 30-45 minutes to hold their speeches. In the second part, we invited three professors as our keynote speakers. Prof. Xinyong Li, from Dalian University of Technology, performed a speech on *In Situ Spectroscopy Investigation of the Surface-Interface Charge-Transfer Process over Catalysis Reactions.* And then we had Prof. Guoan Wang, from China Agricultural University. He talked about *Effects of Arbuscular Mycorrhizal Fungi on Plant Nitrogen Uptake Strategy Under Different Soil Water Conditions.* Lastly, we were glad to invite Prof. Lei Chen, from Shandong University of Science and Technology as our finale keynote speakers. He shared a speech on *An assemblage of macroscopic and diversified carbonaceous compression fossils in the Tonian Period.* Their insightful speeches had triggered heated discussion in the third session of the conference. Every participant praised this conference for disseminating useful and insightful knowledge.

The proceedings are a compilation of the accepted papers and represent an interesting outcome of the conference. Topics include but are not limited to the following areas: Geophysics & Geochemistry, Environmental Chemistry, Geography & Geology, Hydrology & Ocean Science, Geodesy & Cartography, Atmospheric Sciences, Earth Science History and more related topics. All the papers have been through rigorous review and process to meet the requirements of international publication standard.

We would like to acknowledge all of those who supported ICGEC 2021. The help and contribution of each individual and institution was instrumental in the success of the conference. In particular, we would like to thank the organizing committee for its valuable inputs in shaping the conference program and reviewing the submitted papers.

We sincerely hope that the ICGEC 2021 turned out to be a forum for excellent discussions that enabled new ideas to come about, promoting collaborative research. We are sure that the proceedings will serve as an important research source of references and knowledge, which will lead to not only scientific and engineering findings but also new products and technologies.

Committee of ICGEC 2021

Committee member

Conference Chair
A. Prof. Ahmad Safuan Bin A Rashid, Universiti Teknologi Malaysia, Malaysia

Program Committees
Prof. Wei Gao, Hohai University, China
Prof. Xingong Tang, Yangtze University, China
Prof. Shiqi Huang, Xijing University, China
Prof. Shanzhong Qi, Shandong Normal University, China
Prof. Weifeng Liang, Guilin University of Technology, China
Prof. Keren Dai, Chengdu University of Technology, China
Prof. Xianhong Chu, Qingdao Institute of Marine Geology, China
A. Prof. Bobo Li, Guizhou University, China
A. Prof. Huan Yu, Chengdu University of Technology, China
A. Prof. Yang Wang, Guangzhou Institute of Geography, China
Dr. Yong Zhang, East China University of Technology, China
Dr. Deyu Yin, Huaiyin Institute Of Technology, China
Dr. Long Yuan, China Petroleum Logging Co., Ltd., China
Dr. Huairen Cao, Lanzhou University, China
Dr. Xuliang Feng, Xi'an Shiyou University, China
Dr. He Xiao, Guilin University of Technology, China
Dr. Lei Liu, Wuhan Center, China Geological Survey, China

Technical Program Committees
Prof. Hua Wang, China Coal Research Institute CCRI, China
Prof. Huisheng Zhuang, Shanghai Jiaotong University, China
Prof. Luping Li, Changsha University of Science & Technology, China
Prof. Pingping Liu, Beijing Wuzi University, China
Prof. Yang-Ming Lu, National University of Tainan, Taiwan, China
Dr. Zhenxing Chi, Harbin Institute of Technology, China
Dr. Liang Deng, Shanghai University of Electric Power, China

edp sciences

Web of Conferences

Statement of Peer review

In submitting conference proceedings to *Web of Conferences*, the editors of the proceedings certify to the Publisher that

1. They adhere to its **Policy on Publishing Integrity** in order to safeguard good scientific practice in publishing.
2. All articles have been subjected to peer review administered by the proceedings editors.
3. Reviews have been conducted by expert referees, who have been requested to provide unbiased and constructive comments aimed, whenever possible, at improving the work.
4. Proceedings editors have taken all reasonable steps to ensure the quality of the materials they publish and their decision to accept or reject a paper for publication has been based only on the merits of the work and the relevance to the journal.

Title, date and place of the conference

2021 3rd International Conference on Geoscience and Environmental Chemistry
June 18-20, 2021
Xining, China

Proceedings editor(s):

Jun Sun, Tianjin University of Science & Technology, China
Zhaojiang Li, Geographical Society of Tianjin, China
Xianbin Liu, Tianjin University of Science & Technology, China

Date and editor's signature

Jun Sun Zhaojiang Li Xianbin Liu

17, avenue du Hoggar - PA de Courtabœuf – BP 112 - 91944 Les Ulis Cedex A (France)
Tél. : 33 (0)1 69 18 75 75 - Fax : 33(0)1 69 07 45 17 – www.edpsciences.org

https://doi.org/10.1051/e3sconf/202129001001

Theoretical Study on the Mechanisms of Catalytic Hydration of Diiodine Trioxide in Marine Regions

Yan Liang[1], Xiuhui Zhang[1,*], and Wenguo Xu[1,*]

[1]School of Chemistry and Chemical Engineering, Beijing Institute of Technology, Beijing 100081, China.

Abstract. Diiodine trioxide (I_2O_3) is one of the most common iodine oxides in the marine boundary layer (MBL). Both theoretical and experimental studies have confirmed that they can be quickly formed and are relatively stable under dry conditions. However, there is no report on the field observation of I_2O_3, which means that I_2O_3 is likely to be lost in the actual marine atmosphere. But the specific loss pathways and mechanisms are still unclear. Considering that the humidity in the marine regions is generally high and the loss of I_2O_3 will be affected by some substances in the marine atmosphere, water (H_2O, **W**) and iodic acid (HIO_3, **IA**) were selected as a catalyst to investigate the catalytic hydration mechanisms of I_2O_3 at DLPNO-CCSD(T)//ωB97X-D/aug-cc-pVTZ + aug-cc-pVTZ -PP (for iodine) level of theory. The results show that hydration of I_2O_3 presents a high energy barrier, but **IA** can reduce it to 3.76 kcal/mol. Therefore, in the marine atmosphere, I_2O_3 can be hydrolyzed under the catalysis of **IA**, and cannot directly participate in the new particle formation process.

1 Introduction

Atmospheric aerosols have great impacts on human health and the atmospheric environment ([1]). Ultrafine aerosols (particle matter with a size of less than 2.5μm) are key components of local and regional air pollution. There is increasing evidence that exposure to ultrafine aerosols can cause some acute and chronic diseases ([2]). In addition, aerosols can also affect the cloud formation process by increasing the total amount of cloud condensation nuclei (CNN), which in turn has an important impact on the climate ([3]). Marine covers 71% of the total surface of the earth, thus, marine aerosols have a non-negligible position in the global aerosol system. In recent years, the outbreaks of ultrafine particles have been frequently observed in marine areas ([4-7]). Researchers have confirmed that the formation of new particles in the marine regions is linked to the iodine-containing substances in the marine atmosphere.

Diiodine trioxide (I_2O_3) is a common iodine oxide in the marine boundary layer (MBL) ([8]). A laboratory study has shown that the polymerization between IO and OIO radicals will generate I_2O_3 ([9]). Lifetime calculation showed that the time for I_2O_3 to resist pyrolysis is 1.67×10^{11} s ([10]), indicating that I_2O_3 can therefore remain in the marine atmosphere for a certain period. However, so far, there is no report on the field observation of I_2O_3, which means that I_2O_3 is likely to be lost due to certain reactions in the actual marine atmosphere. But the specific loss pathways and mechanisms are still unclear, which greatly affects people's understanding of the biogeochemical cycle of iodine. Considering that the humidity in the marine regions is generally high and the loss of I_2O_3 will be affected by some substances in the marine atmosphere, it is speculated that I_2O_3 may be hydrolyzed in the gas phase. Thus, common substances in the marine atmosphere, water (H_2O, **W**) and iodic acid (HIO_3, **IA**), were selected as a catalyst to investigate the catalytic hydration reaction mechanisms of I_2O_3.

In the present paper, the hydrations of I_2O_3 without catalyst and catalyzed by **W** as well as **IA** were studied at DLPNO-CCSD(T)//ωB97X-D/aug-cc-pVTZ + aug-cc-pVTZ-PP level of theory to explore whether hydration is one of the loss pathways of I_2O_3.

2 Methods

The method of quantum chemistry calculation is used to study the hydration reactions of I_2O_3 when common substances **W** and **IA** in the marine atmosphere act as catalysts. The configurations of the stationary points and transition states are optimized by using the Gaussian-09 ([11]) software package at 298 K and 1 atm. All calculations are performed at DLPNO-CCSD(T)//ωB97X-D/aug-cc-pVTZ + aug-cc-pVTZ-PP level of theory ([12]). The aug-cc-pVTZ-PP basis set with effective core potential ECP28MDF ([13,14]) is used for the iodine atom and the conventional aug-cc-pVTZ ([15]) basis set for the other atoms. And intrinsic reaction coordinate (IRC) calculations were performed to ensure the translation states were correctly connected to the corresponding reactants and products. And the Gibbs free energies (G) of compounds were calculated as:

$$G = E_{SPE} + G_{thermal} \quad (1)$$

* Corresponding author: zhangxiuhui@bit.edu.cn, xuwg60@bit.edu.cn

where E_{SPE} is the single-point energy calculated by the DLPNO-CCSD(T) method, and $G_{thermal}$ is the thermal corrections to Gibbs free energy calculated by the ωB97X-D method.

In order to predict the reaction sites of I_2O_3 hydration, Multiwfn 3.6 ([16]) and Visual Molecular Dynamics (VMD) ([17]) software were also used to analyze the electrostatic potential (ESP) ([18]) on the molecular surface and draw ESP-mapped of molecular van der Waals (vdW) surfaces of reactants.

3 Results and discussion

3.1. The hydration of I_2O_3 without catalysts

In order to explore the reaction sites of I_2O_3 hydration, the electrostatic potential (ESP) on the surfaces of I_2O_3 and **W** molecules was first analyzed. As shown in **Fig. 1**, the two hydrogen atoms of **W** have a relatively positive electrostatic potential (red area), and the two oxygen atoms of I_2O_3 have a relatively negative electrostatic potential (blue area). Therefore, the complex I_2O_3-H_2O (**I_2O_3-W**) can be formed through electrostatic attraction between **W** and I_2O_3.

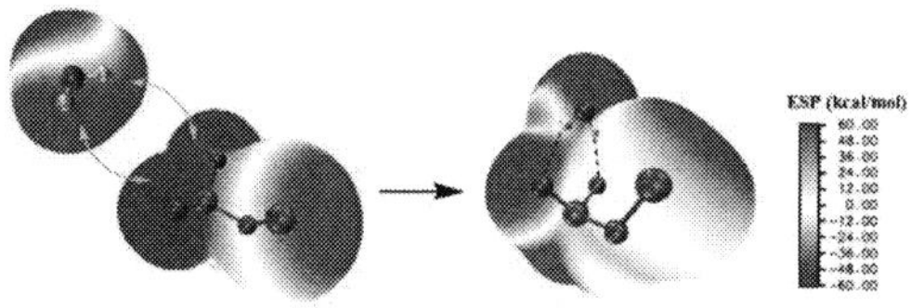

Fig. 1. The ESP-mapped molecular vdW surfaces of reactants I_2O_3 and **W** as well as the pre-reaction complex **I_2O_3-W**. The purple, red, and grey spheres represent **I**, **O**, and **H** atoms, respectively. The black and grey dashed line refers to the hydrogen bond between I_2O_3 and **W**, respectively.

Subsequently, the I_2O_3 hydration reaction was calculated at the DLPNO-CCSD(T)//ωB97X-D/ aug-cc-pVTZ (H, C, N, O, S) + aug-cc-pVTZ-PP (for iodine) level of theory. Take the sum of Gibbs free energies (G) of the reactants as the zero point of the potential energy surface. **Fig. 2** (a) is the potential energy surface of the I_2O_3 hydration reaction, and **Fig. 2** (b) is the corresponding energy profile. The hydration of I_2O_3 starts with the collision between I_2O_3 and **W** molecules to form the pre-reaction complex **I_2O_3-W** with a relative Gibbs free energy of 4.31 kcal/mol. Then, **I_2O_3-W** overcomes a relatively high Gibbs free energy barrier of 32.60 kcal/mol to generate the product HIO_3-HOI (**IA-HA**). The energy barrier of the reaction is as high as 32.60 kcal/mol, and it is an endothermic reaction, indicating that without the participation of catalysts, the hydration reaction of I_2O_3 is difficult to occur in the actual marine atmosphere.

On this basis, the mechanism of I_2O_3 hydration was further analyzed. As shown in **Fig. 2** (a), a proton of **W** transfers to the **O** atom of I_2O_3 while the **OH** moiety binds to the **I** atom of I_2O_3 to form the product. Therefore, the nature of the I_2O_3 hydration is the transfer of a hydrogen proton.

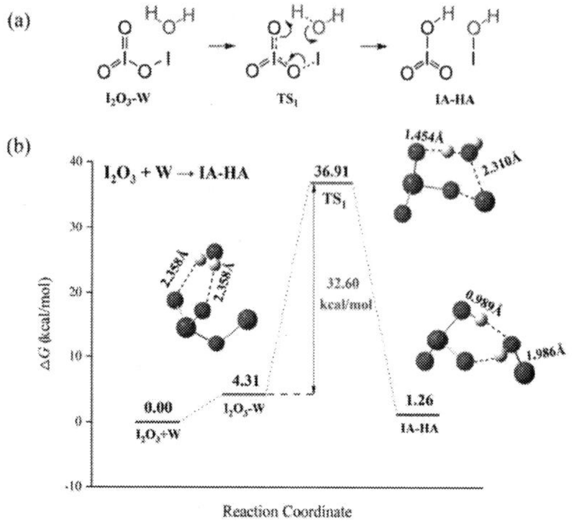

Fig. 2. (a) Reaction mechanism for the gas-phase hydration of I_2O_3 with no catalyst. (b) Potential energy profiles (kcal/mol) for gas-phase hydration of I_2O_3 with no catalyst calculated at DLPNO-CCSD(T)//ωB97X-D/aug-cc-pVTZ + aug-cc-pVTZ-PP (for iodine) level of theory at 298 K and 1 atm. The purple, red, and white spheres represent I, O, and H atoms, respectively.

3.2 The hydration of I_2O_3 catalyzed by W

The marine atmosphere is very humid, so the concentration of **W** is generally high. **W** can act as a proton acceptor and donor at the same time, thereby accelerating the transfer of protons between molecules, so **W** may be used as a catalyst. In this section, the hydration reaction of I_2O_3 when **W** acts as a catalyst was studied. **Fig. 3** (a) shows the I_2O_3 hydration mechanism catalyzed by **W**, and **Fig. 3** (b) shows the potential energy surface of the corresponding reaction.

As shown in **Fig. 3** (b), the catalytic hydration of I_2O_3 has two pathways: I_2O_3 first collides with **W** to form **I_2O_3-W** and two **W** molecules collide first to form **W-W** dimer. The pathway that first forms the **W-W** dimer requires the lowest energy and is the main pathway. Then, **W-W** collides with I_2O_3 to form the pre-reaction complex **I_2O_3-W-W**, and **I_2O_3-W-W** overcomes a relatively Gibbs free energy barrier of 16.91 kcal/mol to generate the product HIO_3-HOI-H_2O (**IA-HA-W**). The catalytic reaction is endothermic and there is a certain energy barrier. Therefore, when **W** acts as a catalyst, the hydration reaction of I_2O_3 is unfavorable in both kinetics and thermodynamics.

It is worth noting that, compared with the reaction of I_2O_3 without a catalyst, **W** as a catalyst effectively reduces the energy barrier from 32.60 kcal/mol to 16.91 kcal/mol. This is because, with the participation of the catalyst **W**, the transition state changes from the original six-membered ring structure to an eight-membered ring structure, and the ring strain of the transition state is therefore reduced, so the energy barrier becomes lower. When **W** acts as a catalyst, it assists the transfer of protons between molecules by accepting and donating protons simultaneously. In summary, **W** as a catalyst assists the transfer of protons between molecules, thereby lowering the energy barrier.

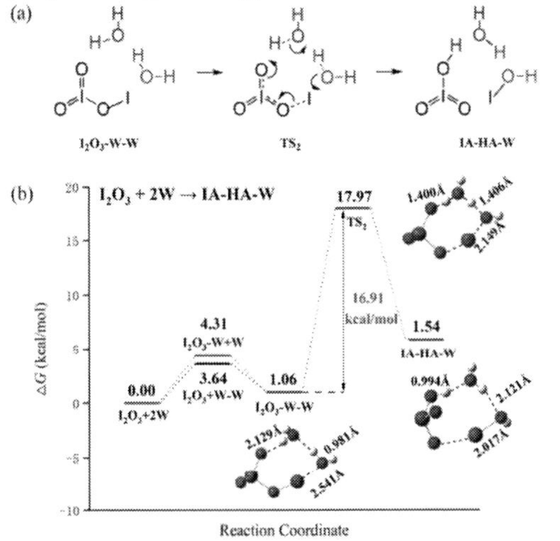

Fig. 3. (a) Reaction mechanism for the gas-phase hydration of I_2O_3 catalyzed by **W**. (b) Potential energy profiles (kcal/mol) for gas-phase hydration of I_2O_3 catalyzed by **W** calculated at DLPNO-CCSD(T)//ωB97X-D/aug-cc-pVTZ + aug-cc-pVTZ-PP (for iodine) level of theory at 298 K and 1 atm. The purple, red, and white spheres represent I, O, and H atoms, respectively.

3.3 The hydration of I_2O_3 catalyzed by IA

As discussed above, the nature of the I_2O_3 hydration is the transfer of protons between I_2O_3 and **W**. As one of the hydration products of I_2O_3, **IA** is very common in marine atmosphere and has a relatively high concentration ([21]), and it can act as both an acceptor and a donor of protons. Thus, it is necessary to study whether **IA** can catalyze the hydration of I_2O_3 by assisting the transfer of protons. **Fig. 3** (a) shows the I_2O_3 hydration mechanism catalyzed by **IA**, and **Fig. 3** (b) shows the potential energy surface of the corresponding reaction.

There are three different bi-molecular collision modes between I_2O_3, **IA**, and **W** (**I_2O_3-W**, **I_2O_3-IA**, and **IA-W**) so there are three pathways for the catalytic reaction. The formation of **I_2O_3-W** and **I_2O_3-IA** is endothermic, so these two pathways are not likely to occur in the actual marine atmosphere. The pathway that first forms the **IA-W** is exothermic, thus it is the main pathway. And **IA-W** collides with I_2O_3 to form pre-reaction complex **I_2O_3-IA-W**, which then forms **$(IA)_2$-HA** by a stepwise mechanism. This reaction has a very low energy barrier (3.76 kcal/mol) and is exothermic, so it is kinetically and thermodynamically favorable. In addition, the energy released by the formation of the reaction complex **I_2O_3-IA-W** is higher than the energy barrier, so the reaction can occur spontaneously in the common marine atmosphere. It should be noted that the Gibbs free energy of the Intermediate (**Int**) is slightly higher than that of Transition State 2 (**TS_2**) because of the calculation error, which is caused by the approximate electron density used in the DFT calculations, and it always occurs in systems involving proton transfer ([18,19]).

As shown in **Fig. 4** (a), the participation of the catalyst **IA** changed the transition state into a ten-membered ring structure, which further reduces the ring strain of the transition states and makes the reaction energy barrier lower (3.76 kcal/mol). The specific catalytic mechanism is that **IA** first gives a proton to I_2O_3, thus, forming the intermediate (**Int**) HIO_3-IH_2O^+-IO_3^-, and then IO_3^- accepts the proton from **W** to form the product **$(IA)_2$-HA**. The mechanisms of **IA** catalytic and **W** catalytic are similar, the **IA** and **W** both assist the transfer of protons between I_2O_3 and **W** by accepting and donating protons. The difference is that when **IA** acts as a catalyst, the transfer of protons is step by step, but the transfer of protons is simultaneous when **W** acts as a catalyst. Compared with **W**, **IA** is a stronger proton donor and is easier to give protons, so the energy barrier is lower than when **W** catalyzing.

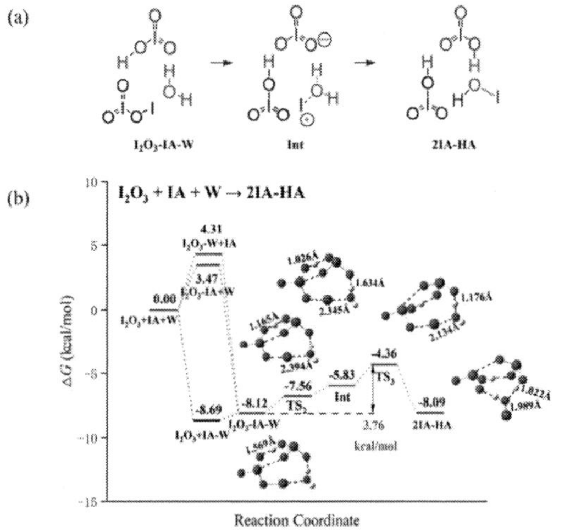

Fig. 4. (a) Reaction mechanism for the gas-phase hydration of I_2O_3 catalyzed by **IA**. (b) Potential energy profiles (kcal/mol) for gas-phase hydration of I_2O_3 catalyzed by **IA** calculated at DLPNO-CCSD(T)//ωB97X-D/aug-cc-pVTZ + aug-cc-pVTZ-PP (for iodine) level of theory at 298 K and 1 atm. The purple, red, and white spheres represent I, O, and H atoms, respectively.

4 Conclusions

The gas-phase hydrations of I_2O_3 with no catalyst and with common atmospheric substances **W** and **IA** as catalysts were calculated at DLPNO-CCSD(T)//ωB97X-D/aug-cc-pVTZ + aug-cc-pVTZ-PP (for iodine) level of theory. The results showed that under the catalysis of **W** and **IA**, the energy barrier is reduced. Especially under the catalysis of **IA**, the energy barrier can be reduced to 3.76 kcal/mol, and the catalytic reaction spontaneously occurred. **IA** is both one of the products of the I_2O_3 hydration reaction and a catalyst, so the reaction can be autocatalyzed. The concentration of **IA** gradually increases as the reaction proceeds, so the role of **IA** in the I_2O_3 hydration gradually increases. Thus, I_2O_3 is unstable and short-lived, and much likely to undergo hydration to generate **IA** in the marine regions, which is one of the loss pathways of I_2O_3. I_2O_3 cannot directly participate in the formation of new particles in the marine atmosphere, but the products **IA** may be involved in the nucleation process, therefore catalytic hydration reaction may trigger the formation of new particles.

This research can provide a comprehensive understanding of the biogeochemical cycle of iodine and provide theoretical guidance for the formation of new particles in marine and coastal regions.

References

1. Murphy D.M. Ravishankara A.R. Trends and patterns in the contributions to cumulative radiative forcing from different regions of the world. Proc Natl Acad Sci U S A. 115,52 (2018)
2. Gong J.C. Zhu T. Kipen H. et al. Comparisons of Ultrafine and Fine Particles in Their Associations with Biomarkers Reflecting Physiological Pathways. Environ Sci Technol. 48, 9 (2014)
3. Glasow R. Seaweed, Iodine, New Particles and Atmospheric Chemistry - The Current State of Play. Environ Chem. 2, 4 (2005)
4. Dall'Osto M. Simo R. Harrison R.M. et al. Abiotic and biotic sources influencing spring new particle formation in North East Greenland. Atmos Environ. 190 (2018).
5. McFiggans G. Coe H. Burgess R. et al. Direct evidence for coastal iodine particles from Laminaria macroalgae - linkage to emissions of molecular iodine. Atmos Chem Phys. 4 (2004)
6. Jimenez J.L Bahreini R. Cocker III D.R. et al. New particle formation from photooxidation of diiodomethane (CH_2I_2). J Geophys Res. 108, D10 (2003)
7. Saiz-Lopez A. Plane J.M.C. Novel iodine chemistry in the marine boundary layer. Geophys Res Lett. 31, 4 (2004)
8. Saiz-Lopez A. Plane J.M. Baker A.R. et al. Atmospheric Chemistry of Iodine. Chem Rev. 112, 3 (2012)
9. Galvez O. Gomez Martin J.C. Gomez P.C. et al. A theoretical study on the formation of iodine oxide aggregates and monohydrates. Phys Chem Chem Phys. 15, 3 (2013)
10. Gomez Martin J.C. Galvez O. Baeza-Romero M.T. et al. On the mechanism of iodine oxide particle formation. Phys Chem Chem Phys. 15, 37 (2013)
11. Frisch M.J. Trucks G.W. Schlegel H.B. et al. Gaussian 09, Revision A.1, Gaussian, Inc. Wallingford, CT. (2009)
12. Peterson K.A. Systematically Convergent Basis Sets with Relativistic Pseudopotentials. I. Correlation Consistent Basis Sets for the Post-D Group 13–15 Elements. J Chem Phys. 119, 21 (2003)
13. Feller D. The role of databases in support of computational chemistry calculations. J Comput Chem. 17, 13 (1996)
14. Schuchardt K.L. Didier B.T. Elsethagen T. et al. Basis set exchange: a community database for computational sciences. J Chem Inf Model. 47, 3 (2007)
15. Peterson K.A. Systematically Convergent Basis Sets with Relativistic Pseudopotentials. I. Correlation Consistent Basis Sets for the Post-D Group 13–15 Elements. J Chem Phys. 119, 21 (2003)
16. Lu T. Chen F. Multiwfn: a multifunctional wavefunction analyzer. J Comput Chem, 33, 5 (2012)
17. Humphrey W. Dalke A. Schulten K. VMD: visual molecular dynamics. J Mol Graph. 14, 1 (1996)
18. Mohan N. Suresh C.H. A molecular electrostatic potential analysis of hydrogen, halogen, and dihydrogen bonds. J Phys Chem A. 118, 9 (2014)
19. You Z.S.; Guo C. Pan Y.J. An experimental and theoretical study on fragmentation of protonated N-(2-pyridinylmethyl) indole in electrospray ionization mass spectrometry. Rapid Commun Mass Spectrom. 26, (2012)
20. Kim M.C. Sim E. Burke K. Understanding and reducing errors in density functional calculations. Phys Rev Lett. 111,7 (2013)
21. Sipila M. Sarnela N. Jokinen T. et al. Molecular-scale evidence of aerosol particle formation via sequential addition of HIO_3. Nature. 537, 7621 (2016)

Capacity Limitations of brine Contamination on High Density Oil-Based Drilling Fluid

Shuanggui Li[1,2], Sheng Fan[1,2], Jiaxue Li [3*], Yingting Li[1,2], Biao Qi[1,2] and Junwei Fan[1,2]

[1]Key Laboratory of Enhanced Recovery for Fracture-Cave Oil Reservoir, Xinjiang, China
[2]Sinopec Northwest Oilfield company, Xinjiang, China
[3]China University of Petroleum (Beijing) @ Karamay, Xinjiang, China

Abstract. Aimed on the problem that the Drilling fluids were easily intruded by the saline water in high-pressure salt-gypsum formations, and the properties of the drilling fluid were damaged and accidents easily occur. The mechanism of drilling fluid contamination by the formation brine was analyzed and the capacity limits of the contamination were found out. The results show that small amount of brine invasion causes emulsification while large amount of brine invasion lead to demulsification. Two capacity limitations were proposed. The first limitation is 10% of saline water concentration, below which the performance of the drilling fluid remains relatively good. The second limitation is 30% of saline water concentration, below which the drilling fluid needs to be systematically maintained, and after which the fluid has no potential for maintenance. This proposal has been successfully used in 8 wells in Kuqa foreland area, and the drilling cycle was saved by 35 days on average.

1 Introduction

In DB, KS, BZ and other blocks of Tarim basin, a large number of salt gypsum layers are developed, in which high-pressure salt water are developed between the layers. During drilling, a lot of salt water and other salts will invade or dissolve into the drilling fluid due to the high-pressure salt water layer, which will damage the coalescence stability of the drilling fluid itself, deteriorate the performance of the drilling fluid, and even cause accidents such as sticking, seriously affecting the drilling safety [1,2]. Table 1 Statistics of salt water encountered during drilling in DB, KS, BZ and other blocks of Tarim basin in recent five years.

Table 1. Brine overflow in Tarim basin from 2011 to 2015.

well	Lost circulation times	Brine overflow times	Blocking times	Other truble times
DB5	13	1	42	
DB206	6	3	17	
DB303	6	2		
DB304	22	1	28	2
DB306	17	4	11	
KS7	11	1	40	1
KS2-1-4	1	2	11	
KS2-1-11	4	1	30	
KS2-1-18	17	4	15	
BZ101	14	1	44	12

High density oil-based drilling fluid (2.45-2.60g/cm3) is generally used in this block due to large oil-bearing salt gypsum layer. When drilling with high-pressure brine, the invasion performance of drilling fluid will become worse, which will lead to complex drilling truble such as blocking and sticking. However, the mechanism of brine invasion and the capacity limit of high-density oil-based drilling fluid affected by brine invasion are not clear, which can not effectively guide drilling operation. Therefore, the laboratory experiments and mechanism analysis are used to improve the understanding of the mechanism and capacity limit of brine contamination of drilling fluid, and to guide the field operation.

2 Mechanism of brine invasion

In order to study the mechanism of oil-based drilling fluid contaminated by salt water, a certain amount of salt water is added to the oil-based drilling fluid commonly used in the field, and its performance is measured. According to the change of performance, the influence mechanism of the oil-based drilling fluid on the drilling fluid is analyzed. The experimental oil-based drilling fluid system was taken from ks2-1-4 well with a sampling depth of 5546m. The drilling fluid is: diesel oil + 2% - 3% main emulsion + 2% - 3% secondary emulsion + 0.5% - 0.8% organic soil + 0.3% - 0.6% suspending agent + 1.5% - 2.5% duratone + barite.

2.1. Performance effect of salt water addition on into oil based drilling fluid

The oil-based drilling fluid was soaked with 5%, 10%, 15%, 20%, 30%, 40%, 50% and 60% brine by volume to test its influence on the rheology property. The experimental results are shown in Fig. 1 and Fig. 2.

* Corresponding author: liji3211979@163.com

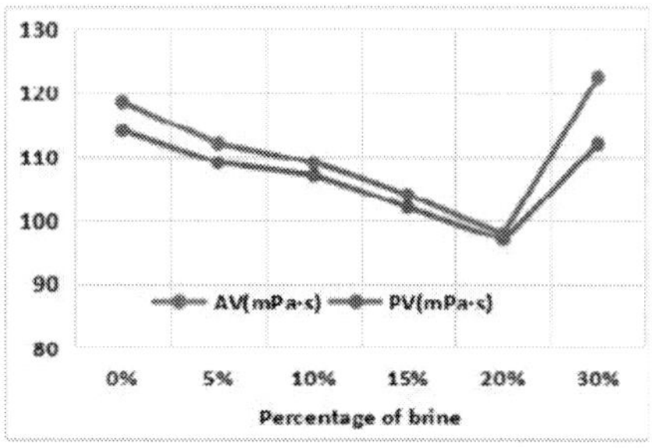

Fig. 1 Effect of different percentage of brine on viscosity of oil-based drilling fluid

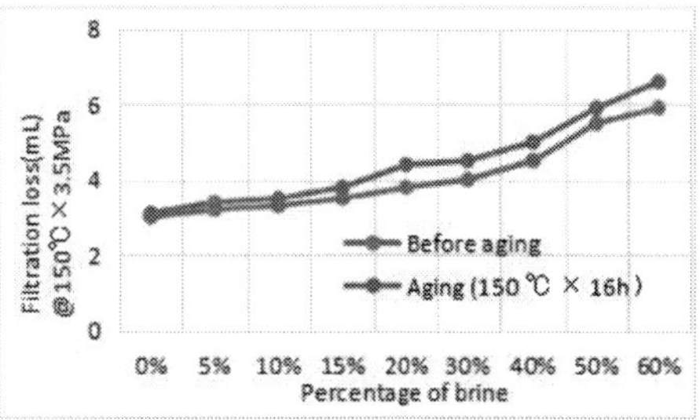

Fig. 2 Effect of different percentage of brine on filtration loss of oil-based drilling fluid

The results show that when the brine percentage is less than 20%, the viscosity decreases with the increase of brine percentage; When the percentage of brine is more than 30%, the viscosity increases, and the percentage of brine is more than 50%, the viscosity of the system has exceeded the measurement range of the instrument.The High temperature and high pressure filtration loss increased slowly with little rate.

2.2 Effect of brine on settlement stability of oil-based drilling fluid system

Due to the contamination of brine, the settlement stability of high-density oil-based drilling fluid is bound to be affected, and the settlement stability is one of the important parameters to be concerned after the drilling fluid is contaminated by brine [3,4]. In order to determine the effect of brine contamination on the settling stability of drilling fluid, a steel cylinder with inner diameter of 6.35cm, outer diameter of 7.6cm, inner clear height of 35cm, total height of 39cm and volume of 1107ml was used to simulate the wellbore. The contaminated drilling fluid settles in the steel cylinder, and the steel cylinder is heated by the incubator to keep the drilling fluid at a certain temperature. After a certain settling time, the upper and lower parts drilling fluid of the steel cylinder are taken respectively to measure their density. Then the density difference between the upper and lower parts is used to reflect the settling stability of the drilling fluid at different temperatures. The settling stability of oil-based drilling fluid at temperature of 120 ℃ and 150 ℃ after adding 5%, 10%, 15%, 20%, 30%, 40%, 50% and 60% brine was tested. The test results are shown in Fig. 3.

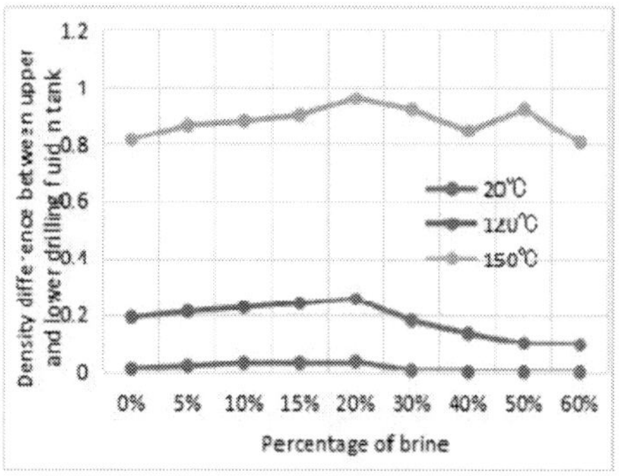

Fig. 3 Effect of different percentage of brine on settling stability of oil-based drilling fluid

The results in Fig. 3 show that the density difference of drilling fluid first increases and then decreases with the increase of brine percentage; With the increase of temperature, the settling stability of the drilling fluid decreases, the performance of the oil-based drilling fluid has become worse at 150 ℃, and more clear fluid is precipitated from the upper part

2.3 Eeffect of brine on demulsification voltage

Demulsification voltage is one of the indexes of oil-based drilling fluid stability [5]. The demulsification voltage of 0%, 5%, 10%, 15%, 20%, 30%, 40%, 50% and 60% brine was tested. The test results are shown in Table 2.

Table 2 Relationship between percentage of brine and demulsification voltage of oil based drilling fluid

percentage of brine(%)	Demulsification voltage of drilling fluid after ageing(v)	Demulsification voltage of drilling fluid after ageing (v)
0	468	445
5	396	410
10	340	380
15	225	305
20	180	295
30	175	263
40	150	225
50	121	180
60	90	143

The experimental results in Table 2 show that the demulsification voltage decreases gradually after adding brine, and it slightly increased after ageing (150 °C ,16 h).

2.4 influence of brine on lubricity and filter cake of oil-based drilling fluid

The lubrication coefficient was tested when the salt water dosage was 5%, 10%, 15%, 20%, 30%, 40%, 50% and 60%,the test results are shown in Table 3. The mud cake measured in the experiment is shown in Figure 4.

Table 3 Relationship between brine percentage and lubricity of oil based drilling fluid system

percentage of brine (%)	0	10	20	30	40	50	60
Lubricity of drilling fluid (KF)	0.075	0.094	0.138	0.152	0.170	0.191	0.217

https://doi.org/10.1051/e3sconf/202129001002

10% brine 30% brine 50% brine

Fig.4 Mud cake of drilling fluid with different brine percentage

The data in Table 3 and Figure 4 show that with the increase of salt water dosage, the filter cake tends to thicken, resulting in the increase of lubrication coefficient, but the overall lubricity is still good.

2.5 Mechanism of brine invasion into oil based drilling fluid

The effects of brine addition on the rheological properties, settling stability, filtration loss, demulsification voltage, filter cake and lubricity of oil-based drilling fluid system are analyzed. The pollution mechanism of brine to oil-based drilling fluid is as follows:

1)When a small amount (≤ 20%) of brine is invaded into oil-based drilling fluid, the viscosity decreases slowly and the filtration loss increases slowly. The system still has a high demulsification voltage, which indicates that the immersed brine and the excessive emulsifier in the system form a relatively stable emulsion.

2)When a large amount (> 20%) of brine is invaded into the oil-based drilling fluid system , the stability of the emulsion is broken, and the viscosity and filtration loss continue to increase, which indicates that the brine enters the oil-based drilling fluid in the form of water phase, reduces the oil-water ratio of the system, increases the viscosity of the oil-based drilling fluid, and worsens the mud cake.

3)When the volume of brine in the system increases to more than 30%, the viscosity and filtration loss increase a lot, but it is still water in oil emulsion. When the actual experiment reaches 60%, the system is demulsified. Therefore, the possibility of wetting reversal is relatively small in actual drilling.

3 Research on the capacity limit of drilling fluid and maintenance

According to the mechanism of brine contamination on drilling fluid, combined with the actual requirements of drilling in KS, DB and BZ blocks of Tarim basin, the key performance indexes of oil-based drilling fluid are selected. According to the performance changes and maintenance after brine contamination, the capacity limit of drilling fluid contaminated by brine is determined to guide the field drilling fluid performance maintenance.

3.1 index of drilling fluid contaminated by brine

Combined with the influence and mechanism of brine on the rheological properties, filtration rate, temperature resistance and sedimentation stability of drilling fluid system, the first capacity limit is determined by the fact that the drilling fluid can be used in the well after simple treatment, and the second capacity limit is the end point of maintenance treatment, which means that the performance of drilling fluid has been damaged over that, the amount of drilling fluid by maintenance treatment has increased a lot, the treatment cost is large. According to the actual drilling performance of oil-based drilling fluid, demulsification voltage, Gel strength of 10 minute and apparent viscosity are selected as indexes to divide the capacity limit of brine invasion. The results are shown in Table 4.

Table 4 Invasion capacity limit index

index	demulsification voltage (V)	Gel strength of 10 minute (Pa)	apparent viscosity (mPaS)
first capacity limit	>400	≤ 8	100~120
second capacity limit	400 ~ 260	8 ~ 15	<100

3.2 Invasion capacity limit of oil-based drilling fluid

According to the capacity limit index, the key performance of drilling fluid is divided. If the index value of drilling fluid is still within the design range after contamination, it is recorded as the safe area (blue). There is not much beyond the design range, the index value can be adjusted to the design range as the warning area (yellow) after simple treatment. Those that may lead to drilling accidents and complications are classified as dangerous areas (red), from which the salt water invasion amount and drilling fluid safety chart are drawn, as shown in Table 5.

Table 5 Safety chart of drilling fluid by brine Invasion

percentage of brine(%)	Gel strength of 10 minute(Pa)	apparent viscosity (mPaS)	demulsification voltage (V)	Lubricity of drilling fluid KF
0%	5.5	118.5	445	0.075
5%	6	112	410	0.081
10%	7	109	380	0.094
15%	5.5	104	305	0.125
20%	6	98	295	0.138
30%	6	122	263	0.152
40%	9	133	225	0.170

According to the chart of oil-based drilling fluid, the capacity limit of oil-based drilling fluid polluted by brine is determined, as shown in Table 6.

Table 6 Invasion capacity limit of oil based drilling fluid

capacity limit	percentage of brine (%)	Key points of maintenance
First	<10%	Add weighting materials and adjust density
Second	<30%	Adding treatment agent and maintaining performance

3.3 Maintenance and treatment of drilling fluid polluted by brine

In the experiment, 30% brine was Invased, diesel oil, emulsifier, organic soil, suspending agent and other treatment agents were added, and their properties were determined, as shown in Table 7. The formula of drilling fluid is oil base drilling fluid + 30% brine + 100% diesel oil + 1.5% main emulsion + 1.5% secondary emulsion + 0.9% organic soil + 0.3% SC + 1.5% fluid loss additive + barite in Table 7.

Table 7 maintenance of fluid polluted by 30% brine

condition	AV/ mPa.s	PV/ mPa.s	YP/ Pa	Gel Pa	HTHP 150°C mL/mm	demulsification voltage (V)
Before aging	113	102	11	6/7	3.2/2	/
After aging	117	107	10	6.5/8	3.0/2	565

For the oil-based drilling fluid system with 30% salt water, 100% diesel oil, 1.5% main emulsion, 1.5% secondary emulsion, 0.9% organic soil, 0.3% suspending agent and 1.5% fluid loss additive are used for maintenance. The pH value is adjusted to 9 and the original density is increased. The performance of the system is restored to the same level as that of the field drilling fluid.

4 Conclusions

1)The main mechanism of salt water immersion in dirty oil-based drilling fluid is as follows: when a small amount of salt water is immersed in dirty oil-based drilling fluid (≤ 20%), the viscosity decreases slowly, and the system still has a high demulsification voltage, indicating that the immersed salt water forms a relatively stable emulsion; When a large amount of brine intrudes into the oil-based drilling fluid system, the brine enters into the drilling fluid in the form of water phase to reduce the oil-water ratio.

2)According to the performance of drilling fluid, the demulsification voltage, final shear value, apparent viscosity and lubricity of oil-based drilling fluid are determined as the key indexes. According to the influence of brine on the key indexes, the first and second capacity limits of high-density oil-based drilling fluid are determined to be 10% and 30%, within the first capacity limit, simple treatment (weighting) is enough, In the second capacity limit, the treatment agent needs to be added for treatment and maintenance. If it exceeds the second capacity limit, the treatment cost is too large, and it is of great significance to treat the polluted water.

3)For the drilling fluid within the second capacity limit, diesel oil, emulsifier, organic soil, suspending agent and other treatment agents should be added in proportion and increased to the original density, so that the performance of the drilling fluid can be restored to the same level as that of the original field drilling fluid.

Acknowledgments:

This paper is one of the phased achievements of Karamay science and technology project "quantitative evaluation system of drilling fluid plugging based on wellbore stability of hard and brittle shale formation (No. 2020CXRC0019)" . I would like to express my sincere thanks to the Karamay science and Technology Bureau, who funded me in the course of writing this paper.

References

1. Yin D., Ye Y., Li L.,et al. High pressure salt water treatment technology of well Keshen7 in foothill structural Zone of Tarim [J].Drilling fluid and completion fluid. 2012,29(5):6-8.
2. Li J., Zhao J., Yang H.,et al. The research and application of safe controlled drilling in evaporite salt layer of Missan oilfield [J]. Journal of Yangtze university(Nat sci edit). 2013,10(16):92-94.
3. Shen W., Yin D., Wang Z., et al. Sedimentation Stability of Organic Salt Drilling Fluid [J]. Chemistry and Bioengineering.2016,33(5):56-58.
4. Wang J., Peng F., Xu T., et al .Research Progress on Testing and Predicting Methods for Settling Stability of Drilling Fluid [J].Drilling fluid and completion fluid. 2016,29(5):79-63.
5. Zhang Y., Xiong H., Ding F., al .Impact of water/oil ratio on the rheological of oil based drilling fluid [J]. Science technology and Engineering. 2016,16(12):238-242.
6. Li J., Huang J., Luo P., et al. Technology of controlling loss by drilling in slugging for fracture formation [J]. Journal of Basic Science and Engineering. 2011, 19(5):1–6.
7. Wang G., Pu X., Discrete element simulation of granular lost circulation material plugging a fracture [J]. Particulate Science and Technology. 2014, 32(2):112–117.
8. Zhu B., Tang H., Wang X., et al. Coupled CFD–DEM simulation of granular LCM bridging in a fracture. [J] Particulate Science and Technology. 2019, 38(03):371–380.

laboratory simulation on drifting of hazardous chemical substance

Xing Wang[1,2], Lai Wei[1,*], Xiuren Li[3],Fan Yang[1],Ying Liu[1], Chunhui Wang[1], Yunlong Liu[1]

[1]North China Sea Environmental Monitoring Center, State Oceanic Administration, Qingdao, China
[2]Shandong Provincial Key Laboratory of Marine Ecology and Environment & Disaster Prevention and Mitigation, Qingdao, China
[3]College of Oceanic and Atmospheric Sciences,Ocean University of China,Qingdao, China

Abstract. The continuously increase of hazardous chemicals transportation leads to a high risk of chemicals leakage. Researches on drifting of chemical substances are of vital importance in damage reducing. Laboratory simulation on drifting of hazardous chemical substance carried out inside a wave tank at the Shandong Provincial Key Laboratory. Different environmental conditions (wind, wave, etc.) were simulated in the wave tank to find out the influence of these factors on substance drifting and diffusion. To identify the difference between hazardous substance, floating ball and dyed petroleum oil were used to simulate solid and liquid floating hazardous chemical substance. The result revealed that wave can improve diffusion, the diffusion speed varies with wave height. Wind can drive surface substance, the drifting coefficient ranges from 2.1% to 3.0%, while liquid drifting coefficient is relatively larger. The laboratory results provide a basis for the study on the drifting and diffusion of hazardous chemicals at sea. Meanwhile, the coefficient could be applied as a correction in numerical models to improve prediction accuracy.

1 Introduction

With the development of chemical engineering industry, the growing sea transportation of hazardous chemical substance leads to a higher frequency of leaking accidents. These accidents had severe influence on ocean environment as well as financial loss. On Jan 6, 2018, tanker "Sanchi" had a collision in the East China Sea. Eight days later, it sank with 111 kilotons of condensate oil. The poisonous leaked oil led to the deaths of countless sea animals[1]. Pollution treatment is becoming one of the research focuses currently.

To reduce pollution damage, determine the location of leaking substance is necessary. Numerical model is a low-cost method to predict the position of leaking substance. Existing oil spilling prediction model can provide comparatively accurate calculation result[2,3], while the research on other types of hazardous chemicals is far from satisfactory. According to the "RCRA Hazardous Wastes"[4], more than 1300 kinds of hazardous wastes have high frequency of leakage. Therefore, it is necessary to exploit new models on other kinds of hazardous wastes.

A key to build up a numeric model is to determine the critical parameter, like wind-induced and current-induced drifting coefficient as well as wave induced mixing. Temperature could speed up molecular movement, thus contribute to diffusion[5]. Wave-induced turbulence will introduce mixing[6], which will dramatically increase chemical substance diffusion. Wind and current can drive substance drifting[7]. Laboratory simulation is an effective method to figure out the effect of different factors.

Recent researches mainly focus on numerical simulation, only minor studies are related to laboratory work. Cong and Zhu studied sediment transport under the influence of wave and tide, providing a new sediment concentration vertical distribution function[8]. Zhao et al used salt water as a simulation of hazardous wastes with same density to study the horizontal diffusion distribution characteristics, offering parameters for environmental pollution mathematical model[9]. Song utilized an ecological environment simulated wave tank to seek out the attenuation coefficient of hazardous wastes[10].

In this work, laboratory research were carried out to increase the accuracy of numerical prediction, thus reducing the environmental damage. The drifting and diffusion of hazardous chemicals (or its simulant) under different environmental conditions were observed. the main parameters were extracted from observed result could be used in improve numerical models while the transportation characteristic could provide basis for drifting and diffusion researches at sea.

2 Experiment design

The experiments were performed inside a wave tank at the Shandong Provincial Key Laboratory of Marine Ecology and Environment & Disaster Prevention and Mitigation, Qingdao. The wave tank was 32 m in length, 0.8 m in width, and 2 m in height and was made of

* Corresponding author: weilai198911@163.com

 https://doi.org/10.1051/e3sconf/202129001003

corrosion-resistant steel with a glass inspection window on the side. The tank was equipped with a mechanical wave maker that can generate regular waves or breaking waves with a height of 0–50 cm and a period of 0.5–5 s. The matched current simulation system can simulate a uniform current field with a velocity of 0–15 cm/s. Additionally, a wind simulation system consist of fan and plastic cover was added. The plastic cover fixed the top of wave tank to form a stable wind channel and reduce wind energy loss. A fan was placed at the east end of wave tank to blow wind, the practical maximum wind speed can reach 5m/s.

Considering the specialty of hazardous chemical substance, floating objects and dyed petroleum oil were used to simulate solid and liquid floating hazardous chemical substance, respectively.

Fig. 1. Part of the wave tank

Fig. 2. Newly installed cover

Various wind speed and current speed were tested to find out the coefficient, the practical wind speed decided by the frequency of fan and section from water surface to plastic cover. In the experiment, wind speed was measured with anemometer, hazardous chemical substance simulant was introduced into wave tank after wind field stabled (about 3 minute).

Wave generated with a mechanical wave maker, 3 conditions of wave with different wave height(10cm, 20cm, 30cm) were measure with one condition without wave as a contrast. Wave height and period could be controlled by computer connected to wave maker, wave condition measured with wave gauge to make sure the practical wave match the setting.

For better simulation, the tank was filled with 1.3 m of seawater obtained from Jiaozhou Bay. Cameras were used to record the movement of hazardous chemical substance, several reference gauges were placed at the side of the wave tank to help calculate the velocity of hazardous chemical substance. To get better contrast, the petroleum oil was dyed with oil-soluble red stain (Juents B664789) of 1 g per 200 ml of oil, the floating wooden objects were also painted red.

3 Results & Discussion

Amount of factors could influence the spread of hazardous chemical substance. Higher temperature can speed up the diffusion, wave can increase the mixing of chemicals and water. Wind and current can drive chemicals drift swiftly. To identify the location of leaking hazardous chemical substance, the drifting coefficient of wind and current is necessary.

3.1 Effect of wind field

Floating objects and dyed petroleum oil were used to simulate solid and liquid float hazardous chemical substance. Surface wind-driving current present under the influence of wind, the speed of wind-driving current was identically measured with ADV. In actual experiments, floating objects usually transported to the side of wave tank, the interaction between objects and wave tank will decrease drifting speed. Only result of spherical object was chosen as spherical object had minor interaction.

Fig. 3. Oil movement under influence of wind

Hazardous chemical substance moved downstream with wind. The velocity of hazardous chemical substance is larger than wind-driving current but less than wind speed. Under same wind speed, liquid drifting velocity was larger than solid drifting velocity, and both velocity increase with wind speed. The result showed in table 1.

The result illustrated that the average solid drifting coefficient is 2.2%, the average liquid drifting coefficient is 2.9%. In the past studies, Kuang et al analyzed the field drifting trajectory data from south China sea joint search-and-rescue exercises, regarding that for dummy model (solid substance), the drifting coefficient is 1.7%[11]. Li et al studied drifting coefficient of green tide alga, giving the drifting coefficient of 0.015[12]. Based on the experiment on the

Table 1. drifting coefficient of solid and liquid substance

Fan frequency	Wind speed (m/s)	Current speed (cm/s)	Ball Velocity (cm/s)	Oil velocity (cm/s)	Solid drifting coefficient (%)	Liquid drifting coefficient (%)
50	3.90	0.82	9.13	12.76	2.1	3.1
50	3.90	0.82	9.54	12.91	2.2	3.1
50	3.90	0.82	9.02	12.65	2.1	3.0
40	3.32	0.68	7.78	10.30	2.1	2.9
40	3.32	0.68	7.73	10.06	2.1	2.8
40	3.32	0.68	7.76	10.46	2.1	2.9
30	2.44	0.46	5.97	7.34	2.3	2.8
30	2.44	0.46	6.26	7.42	2.4	2.9
30	2.44	0.46	5.75	7.37	2.2	2.8

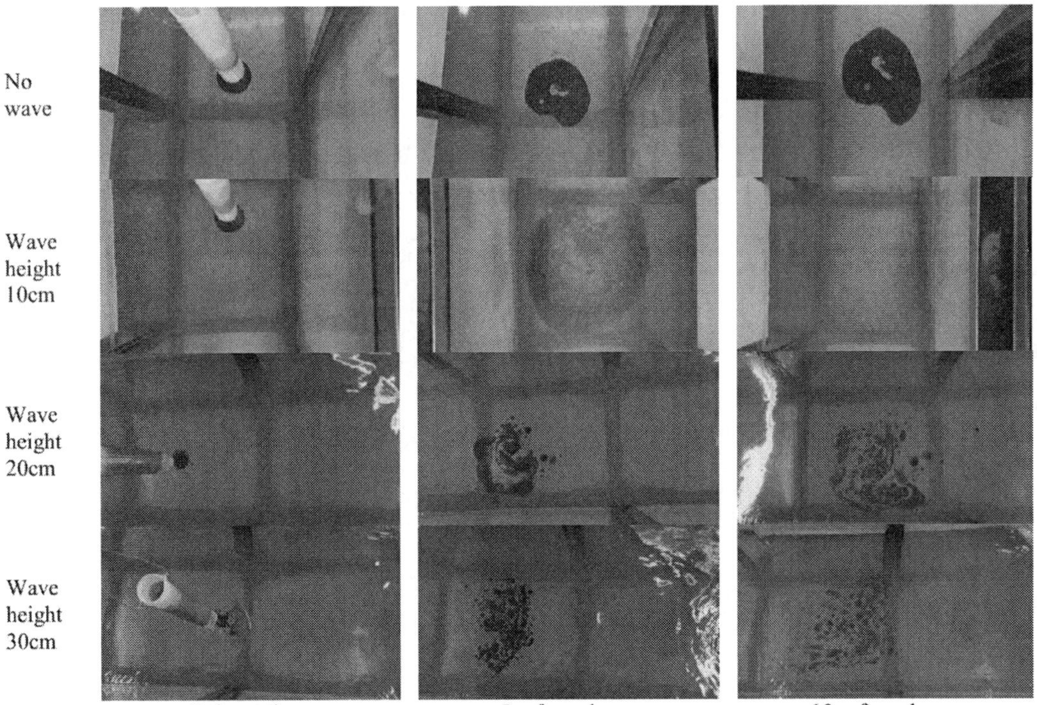

Fig. 4. Oil dispersion under different wave conditions

sea, Du et al calculated that the drifting coefficient for oil is 0.027[13]. Ji and Chen calculated a result of 0.02[7]. Considering those studies, the drifting coefficient should ranging from 2.1% to 3.0%, varies with the density and characteristic of substance.

3.2 Effect of wave

Wave can introduce mixing, thus accelerate chemical diffusion. Dyed petroleum oil used to identify the effect of wave on hazardous chemical substance, the oil was restricted in a plastic cylinder located in the middle of wave tank, 0.4m from wave tank boundary. A camera placed right above release spot to record oil movement.

Under the influence of wave, oil dispersed into small oil droplets, and will become difficult to observe as they spread. This experiment record the time when the oil reached the boundary of wave tank(0.4m) as the diffusion speed under the influence of wave. The result illustrated that wave can accurate diffusion speed, but with the increasing of wave height, the diffusion speed showed a decrease, but still faster than no wave condition.

Table 2. oil diffusion speed under different wave conditions

Wave height (cm)	0	10	20	30
diffusion time (s)	23	6	11	12
diffusion speed (cm/s)	1.74	6.67	3.64	3.33

4 Conclusions

Laboratory simulation can reappear the influence of wind and wave on hazardous substance, but with the restrict of size, the results might have a few differences since the boundary effect of wave tank. Compared with past work, the wind related drifting coefficient of this experiment is dependable.

The results of this experiment illustrated that wave can improve diffusion, while larger wave will decrease substance size and diffusion speed. Wind can drive surface substance, the drifting coefficient ranges from 2.1% to 3.0%, the average solid drifting coefficient is 2.2%, the average liquid drifting coefficient is 2.9%.

The results provide a basis for the study on the drifting and diffusion of hazardous chemicals at sea. Meanwhile, the coefficient could be applied as a correction in numerical models to improve prediction accuracy. Further research could focus on the difference of drifting coefficient.

Acknowledgments

This work was supported by National Key R&D Program of China (No.2016YFC1402404-02).

References

1. Yin L., Zhang M., Zhang Y., Qiao F., (2018) The long-term prediction of the oil-contaminated water from the Sanchi collision in the East China Sea. Acta Oceanologica Sinica.37(03):1-4
2. Mackay D, Leinonen P J.(1977) Mathematical model of the behavior of oil spills on water with natural and chemical dispersion[R]. Dispersers,
3. Lehr W J. (1996) Modeling the benzene inhalation hazard from spilled oil[J]. Spill Science & Technology Bulletin, 3(4):199-202.
4. Crowell, Lip M. (2001) RCRA hazardous wastes handbook[J]. Landolt-Börnstein-Group III Condensed Matter, 72(2):235-235.
5. Ye T., Hang J., Zhou J., Lin T., Zong L., Luo C., (2016) Experimental study on characteristics of sewage dilution and dispersion in estuarine and coastal areas[J].Water Resources Protection, 4:131-135
6. Wei, L., Guan, C.,Troitskaya, Y., (2018) Laboratory Experiment on Wave Induced Turbulence. J. Ocean Univ. China, 17(04):721-726
7. Ji,R., Chen G., Hu Y., Li Y., (1996) Laboratory model simulation of the oil spill drifting with parameters of current, wind and wave. J. Ocean Univ. China, 26(3):353-359
8. Cong X., Zhu W., Research on sediment transportation pattern under the influence of wave and tide. Zhou Shan, 18th China ocean (shore) engineering academic seminar.
9. Zhao X., LiX., Xu X., wang X., (2012) Experimental Research on Transverse Diffusion Coefficients of Typical Pollutants. Proceedings of Conference on Environmental Pollution and Public Health
10. Song Z., Lu J., Ren X., Wang Z., (2018) Experimental Study on Pollutant Attenuation Law and Its Application in Songyuan River. Water Conservancy Science and Technology and Economy. 24(12)
11. Kuang F., Jing C., Zhang J., (2017) Study of wind-induced drift coefficients based on observation and numerical model. Journal of Applied Oceanography 1:41-48
12. Li Y., Pan L., Xiao W., Hu S., Yang H., (2014) Effect of wind on the drifting of green macroalgae in the Yellow Sea. Marine Environmental Science 5:772-776
13. Du W., Liao S., Zhang G., (1986) Offshore experiment on oil drift diffusion . Marine Environmental Science 5(3)53-55

Study of the diffusion of metals and metals-EDTA complexes in water by capillary method

Zilong Geng[1], Hui Chen[1], Chenyu Yang[1], Qiang Shen[1], Tao Wu[*1]

[1]Department of Engineering, Huzhou University, 313000 Huzhou, P. R. China

Abstract. EDTA can complex with radionuclides (RNs) to form negatively charged complexes, making it difficult for clay minerals to retard the diffusion of RNs waste. The diffusion coefficient of RNs in water (D_w) is an important parameter for the safety assessment of the repository. In this study, the effectsof EDTA on the diffusion of metal ions (Cu^{2+}, Sm^{3+}, Nd^{3+}, Lu^{3+} and Zn^{2+}) were investigated by a capillary method. The experimental results showed that $[Cu\text{-}EDTA]^{2-}$, $[Sm\text{-}EDTA]^-$ and $[La\text{-}EDTA]^-$ have higher D_w thanthe Mn+. Whereas, $[Nd\text{-}EDTA]^-$ and $[Zn\text{-}EDTA]^{2-}$ have lower D_w than Nd^{3+} and Zn^{2+} cations. The D_w is consistent with the literatures, indicating the validity of the capillary method to determine the diffusion coefficients. According to Stokes-Einstein relation, the ionic radius and ionic potential of the ion are in disproportional to the D_w value. Cu-, Sm- and La-EDTA complexes have smaller molecular size than the uncomplexed metal ions, indicating that the M^{n+} ions might be associated with many water molecules to form hydrated ions with larger ionic radius. Whereas the $[Nd\text{-}EDTA]^-$ and $[Zn\text{-}EDTA]^{2-}$ have larger molecular size than Nd^{3+} and Zn^{2+} cations.

1. Introduction

Ethylenediaminetetraacetate (EDTA) is often used as a chelate for agricultural and industrial applications, resulting in an increase of EDTA concentrations in various water sources. It may complex with free heavy metal ions to form solublecomplexation, thus enhancing the transportation of $M\text{-}EDTA^{(Z-4)+}$ complexes in water and clays barrier systems [1-3]. The effect of EDTA on the transportation of metal ions has been attracted a lot of attention, such as $^{85}Sr\text{-}EDTA^{2-}$ in sediments [4], $Cd/Co\text{-}EDTA^{2-}$ in saprolite [5], and $Eu\text{-}EDTA^-$ in hard clay rock [6]. EDTA can decrease the adsorption of metal ions on minerals and enhance their transportation not only due to the formation of negatively charge complexes[7, 8], but also due to the reducing of the clay specific surface area [9]. For example, equimolar EDTA inhibited the sorption of U(VI) on montmorillonite due to the formation of negatively charged UO_2HEDTA^-, $(UO_2)_2OHEDTA^-$ and $(UO_2)_2EDTA_2^{4-}$ [7]. The transportation of Th^{4+} through sands was enhanced by EDTA due to the formation of $Th(OH)(EDTA)_2^{2-}$ and $Th(OH)(EDTA)^-$ [2].

In order to retard the transportation of radionuclidewaste, it often uses porous materials with low permeability such as clay rock and compacted bentonite as backfill material for the repository. Diffusion is the predominant behavior of radionulide ions. Diffusion coefficient of RNs in water (D_w) is one of the most important parameters to predict the diffusion behavior of RNs in backfill material and the surrounding rock. The diffusion cell method is often used in determining the diffusion coefficient of RNs in water [1, 10] and in minerals [6, 11]. EDTA can alter diffusion of RNs by complexaion reaction or by the modification of the properties of minerals. Due to the large molceular size of EDTA, the diffusion coefficient of $M\text{-}EDTA^{(Z-4)+}$ complexes in water was found to be similar among various metal ions [1]. Descostes et al. (2017) reported that EDTA increase the diffusion of RNs in clays due to the formation of negatively charged complexes. Anionic exclusion of $[Eu\text{-}EDTA]^-$ was found in the rock clay. However, to the best of our knowledge, only a few studies related to the effect of EDTA on the diffusion of RNs were reported due to the long experimental period [1, 6, 9]

Tri-valent lanthanides were often used as the surrogates of tri-valent actinides to avoid the operation of radioactivity experiments. In this study, the effect of EDTA on metals ions (Cu^{2+}, Sm^{3+}, Nd^{3+}, Lu^{3+} and Zn^{2+}) were investigated by a capillary method. The aim is to verifiy this method to the classis conductivity method and diffusion cell method. Some new diffusion coefficients of ions were also provided. We wish to have more knowledge of the diffusion properties of $M\text{-}EDTA^{(Z-4)+}$ complexes before the diffusion experiment in clay starts.

* Corresponding author: twu@zjhu.edu.cn

 https://doi.org/10.1051/e3sconf/202129001004

2. Materials and Methods

2.1 Experiments

2.1.1 Materials

The diffusion experiments were conducted at aerobic condition. Reagent-grade $CuCl_2{\cdot}H_2O$, $SmCl_3{\cdot}6H_2O$, $LaCl_3{\cdot}6H_2O$, $NdCl_3{\cdot}6H_2O$, $ZnCl_2$, $LuCl_3{\cdot}6H_2O$ (Aaladdin) and ethylenediaminetetraacetic acid disodium salt (Na_2EDTA, Aaladdin). The solution was prepared in Milli-Q (18 MΩ/cm) water.

The aqueous chemical speciation of M-$EDTA^{(Z-4)+}$ complexes were calculated by using HySS (Hyperquad Simulation and Speciation) computer program. The stablity constants from literature were used for calculations [12]. All M^{n+} cations was mainly formed with EDTA as soluable M-$EDTA^{(Z-4)+}$ complexes with almost 100% in 0.5 M NaCl solution due to the large stability constantes of M-$EDTA^{(Z-4)+}$ complexes.

2.1.2 Diffusion experiment

The diffusion behavior of M^{n+} and M-$EDTA^{(Z-4)+}$ complexes in water will be investigated by a capillary method with single element solution (Fig.1). A working solution of Cu^{2+} was prepared by dissolving solid metal chlorides in 0.5 M NaCl solution. 0.1 mol/L of Cu^{2+} stock solution was prepared by adding some Cu^{2+} working solution in 0.5 mol/L of NaCl solution. The pH was adjusted to 7.0 ± 0.2 by adding minor HCl or NaOH. The 5.0 ml vials were filled with 4.0 ml of tracer free NaCl solution. After a capillary was fit into an opening in the cap of the vial, the glass open-ended capillary (diameter ca. 0.85 mm) was filled with 25 μL Cu^{2+} at a length of ca. 4.7 cm, and then a parafilm membrane sealed the upper end of the capillary. The bottom end of the capillary immerged into 0.5 mol/L of NaCl solution (Fig. 1A). The out-diffusion of Cu^{2+} from the capillary into the solution started. After a certain time interval, the capillary was removed from the solution. The concentration of copper in vials was measured by an inductively coupled plasma-optical emission spectrometer (ICP-OES, Perkin Elmer, Optima 7000 DV). The same experimental procedures were conducted for the other metal ions (Sm^{3+}, La^{3+}, Nd^{3+}, Zn^{3+} and Lu^{3+}) and M-$EDTA^{(Z-4)+}$ complexes.

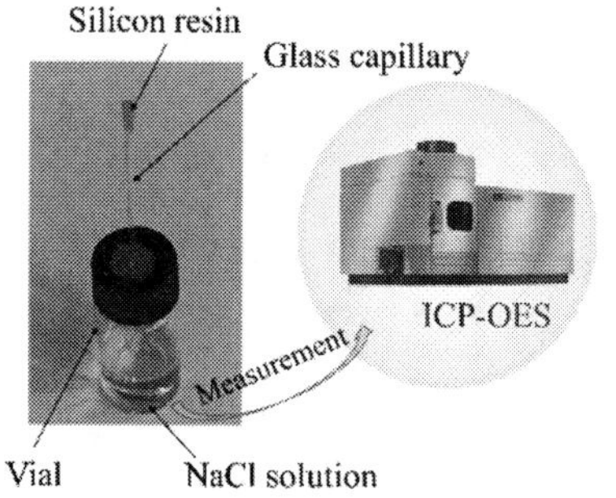

Fig.1 The diffusion set-up for the capillary method.

2.2 Computational methods

2.2.1 Determination of the diffusion coefficients of free metal ions

The diffusion coefficient (D_w) for a metal ion in NaCl solution was obtained by fitting the concentration of the metal ion in vials as a function of time by the following equation [13]:

$$\frac{C}{C_0} = \frac{8}{\pi^2}\sum_{n=0}^{\infty}\frac{1}{(2n+1)^2}\exp\left[-(2n+1)^2\cdot\pi^2\cdot D_w\cdot t/4L^2\right] \quad (1)$$

where C (mol/L) is the concentration of the metal ion in the vials. C_0 (mol/L) is the initial concentration of the metal ion. L (m) is the filled length of metal ions solution in the capillary. It was 0.047 m in this study.

D_w is related to the diffusion coefficients of M^{n+} and M-$EDTA^{(Z-4)+}$ complexes ($D_{w,0}$) and NaCl ion ($D_{w,e}$) in diluted solution as follows [14]:

$$D_w = \frac{Z_1^2c_1 + Z_2^2c_2}{Z_1^2c_1/D_{w,e} + Z_2^2c_2/D_{w,0}} \quad (2)$$

where $D_{w,Na+}$ is 1.33×10^{-9} m^2/s [14].

2.2.2 Calculation of ionic radius of M^{n+} and M-$EDTA^{(Z-4)+}$ complexes

The ionic radius of the diffusing species (r, Å) is calculated by the classic Stokes-Einstein relation as follows:

$$r = \frac{kT}{6\pi\eta D_w} \quad (3)$$

where k (1.380649×10^{-23} $Pa{\cdot}m^3/K$) is the Boltzmann constant, T (K) is the temperature and η (Pa·s) is the viscosity of the medium, which is 0.9365×10^{-3} Pa·s in 0.5 mol/L of NaCl solution [15].

The ionic potential (I_p) is defined as:

$$I_p = |z|/r \quad (4)$$

where z is the charge of ion and r is the ionic radius (Å).

3. Results & Discussion

Fig. 2 shows the diffusion profile of M^{n+} and M-$EDTA^{(Z-4)+}$ complexes in 0.5 M NaCl solution. Due to the large stability constants of M-$EDTA^{(Z-4)+}$ complexes, almost 100% complexes were formed in 0.5 M NaCl solution. The concentration ratio of M^{n+} and M-$EDTA^{(Z-4)+}$ complexes increased with increasing time. The concentration ratio of M-$EDTA^{(Z-4)+}$ (M = Cu^{2+}, Sm^{3+} and La^{3+}) complexes are higher than that of M^{n+}, indicating that the complexes diffuse faster than the metal ions. Whereas, the ratio of M-$EDTA^{(Z-4)+}$ (M = Nd^{3+} and Zn^{2+}) complexes are lower than that of M^{n+}, implying a higher diffusion velocity of metal ions than that of the complexes.

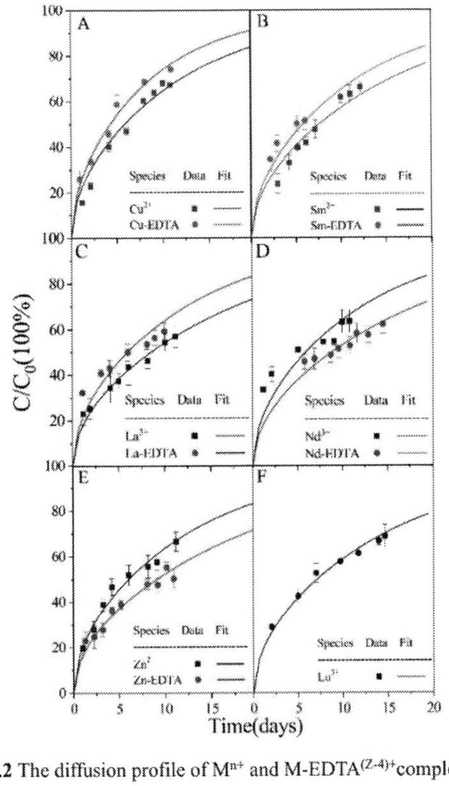

Fig.2 The diffusion profile of M^{n+} and M-$EDTA^{(Z-4)+}$complexes in 0.5 M NaCl solution. T = 20 °C, $C_0(M^{n+})$ = 1.0 × 10^{-3} mol/L.

The diffusion coefficients, the ionic radii and the ionic potentials (I_p) of M^{n+} and M-$EDTA^{(Z-4)+}$ complexes were summerized in Table 1. The results were compared with the D_w measuremed by diffusion cell method and the limiting molar ionic conductivites using the Nernst-Einstein equation [1, 10, 14, 16]. The diffusion coefficient was obtained by fitting the experimental data as shown in Fig.1 by Eq.(1). The ionic radius was calculated by Eq.(3) in this work and for the literatures. I_p was calculated by Eq.(4). The I_p from literatures are listed in the round bracket in Table 1. The diffusion coefficients of M^{n+} ions in diluted water are calculated by Eq.(2), which are consistent with the D_w from literatures, indicating the validity of the capillary method to determine the diffusion coefficients. However, the D_w of M-$EDTA^{(Z-4)+}$ complexes were higher in this work than in literature. Since multi-element solution was employed to determine the diffusion coefficient by diffusion cell method as reported by Furukawa et al. (2007), the descrepancy could be explained by the co-complexation reaction of ions and EDTA.

According to Stokes-Einstein relation as shown in Eq.(1), the diffusion coefficient of ion is in dispropotional to the ionic radius. For Cu^{2+}, Sm^{3+} and La^{3+}, the ionic radius of M-$EDTA^{(Z-4)+}$ complexes are smaller than that of M^{n+}, indicating that the M^{n+} ions might be associated with many water molecules to form hydrated ions with larger ionic radius. Whereas, the ionic radius of $[Nd\text{-}EDTA]^-$ and $[Zn\text{-}EDTA]^-$ are larger than that of Nd^{3+} and Zn^{2+}. The D_w of $[Nd\text{-}EDTA]^-$ and $[Zn\text{-}EDTA]^-$ are in good agreement with that of Furukawa, et al. (2008), who reported that the D_w of Ln-$EDTA^{(Z-4)+}$ complexes were in the range of 5.43 × 10^{-10} – 5.76 × 10^{-10} m^2/s, which is close to that of H_2EDTA^{2-}. Since the molceular size of EDTA is much larger tha thato fo metal ions, the diffusion of metal ions lose their characteristics by the complexation. However, the diffusion coefficient of $[Cu\text{-}EDTA]^{2-}$, $[Sm\text{-}EDTA]^-$ and $[La\text{-}EDTA]^-$ were higher. More experiments will be conducted to clarifed the descrepancy of the diffusion coefficient due to the multi-element solution and single element solution.

Table 1 The diffusion coefficients and molecular radii and of M^{n+} and M-$EDTA^{(Z-4)+}$ complexes in 0.5 M NaCl solution. T = 20 °C, $C_0(M^{n+})$ = 1.0 × 10^{-3} mol/L.

Species	D_w(×10^{-10} m^2/s)	RSD%	$D_{w,0}$(×10^{-10} m^2/s)	r(Å)	I_p	Reference
Cu^{2+}	7.88	5.24	7.84	2.91	0.69 (2.74*)	This work
Cu^{2+}			7.14	3.21	0.62 (2.95)	Li, et al. (1974)
Cu^{2+}			7.33	3.13	0.64	Vanysek (2000)
Sm^{3+}	6.57	5.47	6.53	3.49	0.86	This work
Sm^{3+}			6.08	3.77	0.80	Vanysek (2000)
La^{3+}	5.87	5.43	5.84	3.90	0.77 (2.78*)	This work
La^{3+}			6.19	3.70	0.81 (2.47)	Vanysek (2000)
Nd^{3+}	7.39	5.44	7.35	3.10	0.97	This work

https://doi.org/10.1051/e3sconf/202129001004

Nd^{3+}			6.16	3.72	0.81	Vanysek (2000)
La^{3+}			6.17	3.71	0.81	Li, et al. (1974)
Zn^{2+}	8.34	5.46	8.3	2.75	0.73 (2.70*)	This work
Zn^{2+}			8.85	2.68	0.75	Li, et al. (1974)
Zn^{2+}			7.03	3.26	0.61	Vanysek (2000)
Lu^{3+}	7.03	5.44	6.99	3.26	0.92	This work
Lu^{3+}			5.9	3.88	0.77	Furukawa (2008)
$[Cu\text{-}EDTA]^{2-}$	11.7	5.47	11.7	1.77	1.13	This work
$[Sm\text{-}EDTA]^{-}$	8.16	5.51	8.1	2.54	0.39	This work
$[Sm\text{-}EDTA]^{-}$		4.50	5.53	3.75	0.27 (1.6)	Furukawa (2007)
$[La\text{-}EDTA]^{-}$	8.35	5.58	8.29	2.48	0.40	This work
$[La\text{-}EDTA]^{-}$			5.47	3.79	0.26 (1.6)	Furukawa (2007)
$[Nd\text{-}EDTA]^{-}$	5.63	5.44	5.57	3.68	0.27	This work
$[Nd\text{-}EDTA]^{-}$			5.53	3.75	0.27	Furukawa (2007)
$[Zn\text{-}EDTA]^{2-}$	5.54	5.56	5.48	3.74	0.53	This work

* Data from Nordstrom (1997)

Fig.3 shows the diffusion coefficients of M^{n+} and M-$EDTA^{(Z-4)+}$ complexes as a function of ionic potentials. It is defined as $|z|/r$, where z is the charge of ion and r is the ionic radius (Å). It shows a linear relationsip between the diffusion coefficients and ionic potentials. The intercepts decrease with increasing the charge of ions. The ionic potentials of are lower than 1.2, which is inaccordance with the anions reported by Li and Gergory (1974). However, cations have the lower ionic potentials in this work than in literatures [1, 10, 16]. It can be explained that the ionic radii of M^{n+} ions are larger due to the formation with water molecules to form hydrated ions in this work. The I_p is in proprtional to the D_w when combined Eq.(2) and Eq. (3), indicating that the ions with high ionic potentials diffuse faster in the water.

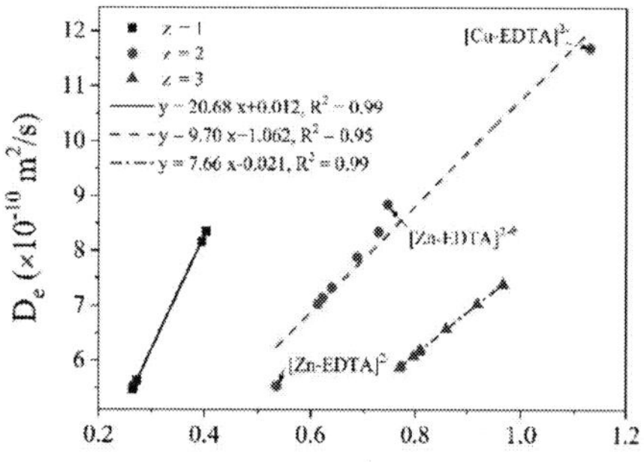

Fig.3 The diffusion coefficients of M^{n+} and M-$EDTA^{(Z-4)+}$ complexes in 0.5 M NaCl solution as a function of ionic potentials.

4.Conclusions

The results of this work indicate that EDTA can alter the diffusion behavior of the metal ions. The diffusion coefficients of ions is in disappropotional to the molecular size and ionic potential of metal ions. The ions with smaller molecular sizes and ionic potentials have higher diffuion ability. Most of M-$EDTA^{(Z-4)+}$ complexes have similar diffusion velocities with the D_w values of ~5.5 $\times 10^{-10}$ m^2/s. Due to the heterogeneity of RNs, the M-$EDTA^{(Z-4)+}$ complexes do not show similar diffusion behavior. This work provides a simple method to determine the diffusion coefficients of RNs.

Acknowledgments

This study was supported by National Science Foundation of Zhejiang Province (LY18B070006 and LQ20E040001). We would like to thank Ms. Yudie Wang and Ms. Jiatong Zhao their help for checking the diffusion experimental results.

References

1 K. Furukawa, Y. Takahashi, H. Sato, Geochimica et Cosmochimica Acta 2007, 71, 4416-4424 https://doi.org/10.1016/j.gca.2007.07.009.

2 C. C. May, L. Young, P. J. Worsfold, S. Heath, N. D. Bryan, M. J. Keith-Roach, Water Research 2012, 46, 4870-4882.

3 E. Reinoso-Maset, P. J. Worsfold, M. J. Keith-Roach, Environmental Pollution 2012, 162, 399-405 https://doi.org/10.1016/j.envpol.2011.11.038.

4 M. N. Pace, M. A. Mayes, P. M. Jardine, L. D. Mckay, X. L. Yin, T. L. Mehlhorn, Q. Liu, H. Guerleyuek, Journal of Contaminant Hydrology 2007, 91, 267-287.

5 M. A. Mayes, P. M. Jardine, I. L. Larsen, S. C. Brooks, S. E. Fendorf, Journal of Contaminant Hydrology 2000, 45, 243-265.

6 M. Descostes, I. Pointeau, J. Radwan, J. Poonoosamy, J.-L. Lacour, D. Menut, T. Vercouter, R. V. H. Dagnelie, Journal of Hydrology 2017, 544, 125-132 https://doi.org/10.1016/j.jhydrol.2016.11.014.

7 W. Q. Zhou, J. Y. Wang, J. G. He, X. Y. Yang, Y. L.

Shi, X. Y. Wang, C. L. Liu, Colloids and Surfaces a-Physicochemical and Engineering Aspects 2019, 583.
8 D. Read, D. Ross, R. J. Sims, Journal of Contaminant Hydrology 1998, 35, 235-248 https://doi.org/10.1016/S0169-7722(98)00136-3.
9 X. J. You, S. G. Liu, C. M. Dai, G. H. Zhong, Y. P. Duan, Y. J. Tu, Science of the Total Environment 2020, 706.
10 K. Furukawa, Y. Takahashi, Chemosphere 2008, 73, 1272-1278 https://doi.org/10.1016/j.chemosphere.2008.07.017.
11 R. V. H. Dagnelie, P. Arnoux, J. Radwan, D. Lebeau, P. Nerfie, C. Beaucaire, Applied Clay Science 2015, 105, 142-149 10.1016/j.clay.2014.12.004.
12 M. Zabiszak, M. Nowak, K. Taras-Goslinska, M. T. Kaczmarek, Z. Hnatejko, R. Jastrzab, Journal of Inorganic Biochemistry 2018, 182, 37-47 https://doi.org/10.1016/j.jinorgbio.2018.01.017.
13 T. Wu, S. Amayri, J. Drebert, L. R. Van Loon, T. Reich, Environmental Science & Technology 2009, 43, 6567-6571.
14 P. Vanysek, CRC Handbook of Chemistry and Physics 2000, 83, 76-78.
15 F. ., A., Gonalves, J., Kestin, Berichte der Bunsengesellschaft für physikalische Chemie 1977, 81, 1156-1161.
16 Y. H. Li, S. Gregory, Geochimica et Cosmochimica Acta 1974, 38, 703-714 https://doi.org/10.1016/0016-7037(74)90145-8.

https://doi.org/10.1051/e3sconf/202129001005

Design of an acrylic wastewater treatment method based on deep oxidation technology

Wei Liang[1]

[1]Acrylic fiber factory of Daqing Petrochemical Company in Heilongjiang Province, Daqing, China

Abstract. Acrylic fiber is one of the earliest developed chemical fiber species in China. Acrylic products are fluffy, warm, soft to the touch, and have good weather resistance and anti-mildew and anti-moth properties. Acrylic fiber has a wide range of applicability, but the wastewater generated during its production is a major problem facing the world. The complex composition and high toxicity of acrylic wastewater pollutants, the presence of many oligomers and other difficult to biodegrade, as well as the presence of sulfate, sulfite and other bioinhibitory components, make the biochemical properties of acrylic wastewater very low, making it very difficult to treat. In this paper, we propose an acrylic wastewater treatment method based on deep oxidation technology to solve this problem, which can effectively carry out wastewater treatment and make it meet the national discharge standards.

1 Introduce

In recent years, the demand for acrylics has been increasing, and the mass production of acrylonitrile has led to increasing pollution problems in its production. Both the production process and the storage and transportation of acrylonitrile inadvertently lead to its effluent, which can cause pollution of water resources and pose a threat to human health if not treated. Various data show that due to the physical and chemical properties of acrylonitrile, trace amounts of acrylonitrile dissolved in water bodies can cause high toxicity and potential "tritoxicity" to water bodies. The high toxicity of acrylonitrile has led to its appearance on the list of priority toxic chemicals and the blacklist of priority pollutants in water in China. In recent years, many scholars from different countries have studied the degradation of this highly toxic wastewater. For the treatment of acrylonitrile wastewater, the research focuses on new processes that are efficient, resourceful and harmless. [1]

2 Common methods for biochemical treatment of acrylic wastewater

For the treatment of acrylic production wastewater, the current research is mostly focused on the development of end-of-pipe treatment technology. Pre-treatment mainly focuses on how to improve the biochemical properties of wastewater and remove biologically inhibitory components; subsequent biochemical treatment mainly focuses on reducing the discharge of COD and ammonia nitrogen in the effluent, and also includes the combination of these methods and the study of the adopted process route. The main purpose of acrylic wastewater treatment is to remove difficult biodegradable substances, and the main treatment technologies are coagulation, internal electrolysis, air floatation, advanced oxidation, ultrasonic wave, membrane filtration, catalytic ozone method, microwave method, bentonite and lime treatment of acrylic wastewater, photocatalytic oxidation technology, Fenton oxidation technology and catalytic wet oxidation technology.[2]The technologies studied in this paper are mainly those in coagulation and oxidation.

2.1. Coagulation technology

Coagulation is to add coagulant to the wastewater to make the colloid and suspended particles in the wastewater coalesce into larger flocs, so as to separate them from the water and achieve the purpose of purifying water quality. Coagulation includes two processes: coagulation and coagulation. Coagulation is the destabilization of colloidal particles coagulation into small flocs, coagulation is in the role of the catalyst will be small flocs coagulation into a large volume of coagulation body, so as to complete the entire coagulation process. Coagulation mechanism varies because of the coagulant and coagulation conditions. At present, the coagulation mechanism mainly includes: adsorption electro-neutralization, adsorption bridging, compression double layer, net trapping, etc. [3]The action forces mainly include hydrogen bonding, polarization, and the coagulation of the coagulant. The main forces are hydrogen bonding, polar groups, covalent bonds, electrostatic gravitational force and van der Waals force, etc. Which force plays a deterministic role is determined by the structure of the polymer itself and the characteristics of the colloid.

Different flocculants have different flocculation effect in treating acrylic process wastewater. For acrylic

* Corresponding author: liangwei1980@126.com

wastewater, inorganic polymer flocculants have better treatment effect than organic polymer flocculants, among which polymerized ferric sulfate (PFS) has the best flocculation effect; inorganic and organic compound flocculants have better coagulation effect than single inorganic and organic flocculants, and the best compound combination is PFS and cationic polyacrylamide, with a certain amount of After adding a certain amount of MgSO4, the modified polyacrylamide with MgSO4 and PFS can achieve a COD removal rate of 32.5%.

2.2 Fenton oxidation technology

Fenton advanced oxidation technology is a chemical oxidation water treatment technology invented by H.J.H. Fenton more than 100 years ago, which does not require high temperature and pressure and simple process equipment. When degrading organic matter with water, the hydroxide radicals can eventually oxidize the organic matter to the simplest H2O and CO2 molecules by initiating chain reactions. The reaction mechanism is shown in Figure 1.

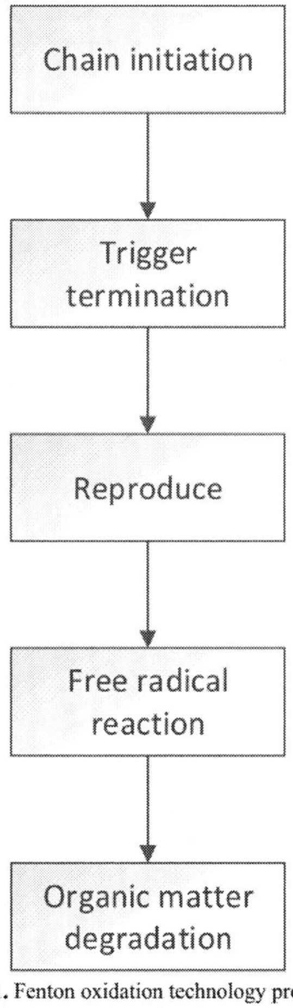

Fig. 1. Fenton oxidation technology process.

The organic degradation kinetic equation is:

$$-\frac{d[RH]}{dt} = \frac{K_1K_3}{K_2} \cdot \frac{[RH][H_2O_2]}{\frac{\Delta[H_2O_2]}{2\Delta[Fe^{2+}]}} \quad (1)$$

where [RH] is the concentration of organic matter. If K2>>K3, under acidic conditions, because Δ[H2O2] ≈ 2Δ[Fe2+], the organic degradation kinetic equation is:

$$-\frac{d[RH]}{dt} = \frac{K_1K_3}{K_2} \cdot [RH][H_2O_2] \quad (2)$$

The amount of H2O2 and Fe2+ salt input varies depending on the organic matter. fe2+ itself has flocculation effect and can remove COD from the effluent.

2.3 Wet catalytic oxidation technology

Catalytic wet oxidation technology is an advanced oxidation technology developed in the 1970s based on wet oxidation, using highly efficient and stable catalysts to reduce the activation energy of the reaction, thereby reducing the temperature and pressure required for the reaction and improving the treatment effect. Catalytic wet oxidation technology has been used industrially in Japan and other countries, but it has been studied later in China. According to the different states of catalysts, catalysts can be divided into two categories: homogeneous catalysts and non-homogeneous catalysts. The homogeneous catalyst, which was studied more in the early stage, is a soluble catalyst added to the reaction system to catalyze the reaction. It has good reaction performance, mild reaction and specific selectivity, and the current research focuses on non-homogeneous catalytic oxidation. The research of non-homogeneous catalytic wet oxidation focuses on efficient and stable multiphase catalysts.[4]

From the current research results, the more studied and most valued homogeneous catalytic oxidants are soluble transition metal salts, and the catalytic activity of copper is more obvious, and the addition of copper ions mainly triggers the free radical chain of oxidation reaction by forming intermediate complexation products after dehydrogenation. The reaction flow is shown in Figure 2.

There are many methods for the preparation of catalysts, and the following two methods were chosen for the preparation of copper-based catalysts in this design.

(1) Precipitation method: The configured Cu(NO3)2 solution was adjusted to pH 10 with ammonia as precipitant, aged in a water bath shaker at a constant temperature of 80°C for 2 h, then filtered and washed, and the filtered solid sample was dried in an oven and placed in a muffle furnace at 120°C for 1 h. Then the temperature was slowly and uniformly increased to 500-550°C (the set roasting temperature), and the time was started.

(2) Impregnation method: γ-A12O3 was chosen as the carrier of the catalyst. In order to remove impurities, the catalyst was impregnated with distilled water for 8 hours before use, and the impregnation process was continuously stirred to resolve as many magazines as possible. After the impregnation, the γ-A12O3 was cleaned with distilled water, and then placed in an oven at a temperature of 120°C and dried for 12 h, ready for use. Weigh a certain amount of dried γ-A12O3 particles into a triangular flask, add an excess of Cu(NO3)2 solution (impregnation solution), put it on an oscillator, and filter it after 8 hours of reaction. The filtered solid was dried and placed in a muffle furnace for 1 h at 120 °C, then the temperature was slowly and uniformly increased to 500-550 °C and the time was started. The roasted catalyst was crushed, sieved and placed in a desiccator.[5]

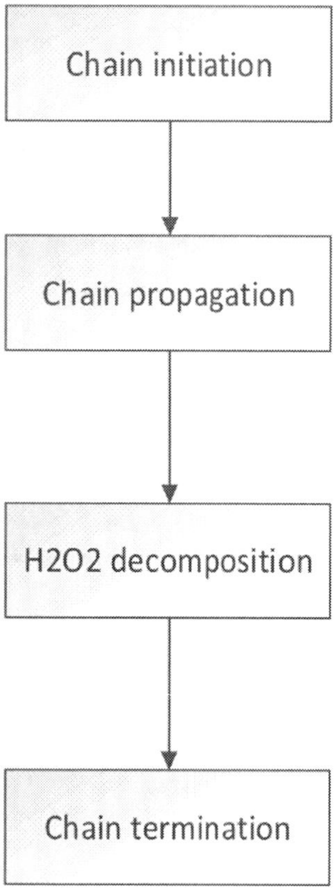

Fig. 2. Oxidation chain reaction flow.

3 Design of acrylic wastewater treatment method

3.1 Kinetic model of catalytic wet oxidation reaction process

The study of kinetics is an important part of the research on wet oxidation technology. By investigating the reaction kinetics, not only can we gain insight into the course of the reaction, understand the reaction mechanism and the control steps of the reaction, which can better promote the occurrence of the reaction, but also provide a basic theoretical basis for engineering design and process control through the establishment of mathematical models of kinetics. There are three main categories of kinetic models proposed so far: mechanistic, empirical and semi-empirical models. The most commonly used is the empirical model, whose reaction rate equation is:

$$-\frac{d[c]}{d[t]} = k_0 cxp\,(-E_a/RT)[c_1]^m[c_2]^n \quad (3)$$

Where:

[c] - molar concentration of reactants, mol/L or mg/L;

k0 - refers to the prefactor, its value is related to m, n;

Ea - reaction Ea - activation energy of the reaction, kJ/mol;

R - gas constant. 8.314J/(mol-K);

c1 - concentration of organic matter, which can be expressed in terms of specific organic matter concentration or comprehensive index of water quality (e.g. COD);

c2 --Oxidant concentration, can be expressed as liquid-phase DO or gas-phase oxygen partial pressure; m --reaction level of organic matter;

n --reaction index of oxidant;

t - reaction time, min or s;

T - reaction temperature, K.

The main investigation of this design is the effect of catalytic wet oxidation on the comprehensive index COD of water quality. Therefore, the kinetic model of wet catalytic oxidation of acrylic wastewater is as follows:

$$-\frac{d[c]}{dt} = k_0 \exp\,(-E_a/RT)[COD]^m[O_{2ads}]^n \quad (4)$$

3.2 Process design of treatment method

In view of the problems of the existing process and technology, the acrylic wastewater treatment method is formulated by combining the actual site and the characteristics of wastewater quality. Firstly, the chemical method is used to pretreat the hard-to-degrade acrylic wastewater, and then the deep oxidation method is used to treat it, so as to ensure the removal of COD fundamentally.

In the design process, the influencing factors of coagulation and Fenton oxidation were analyzed using the single-factor method to determine the best parameter conditions for treating acrylic wastewater. The influence of wet catalyst preparation on the catalytic efficiency was studied, and the single-factor method was used to analyze the influence of pressure and temperature on the treatment effect of wet catalytic oxidation and to obtain the optimal reaction conditions. The pretreatment and catalytic wet oxidation were combined to treat acrylic wastewater to determine the optimal process. The reaction kinetics of catalytic wet oxidation process was initially explored, the corresponding kinetic model was established, and the model parameters and activation energy were investigated.

The proposed design is shown in Figure 3.

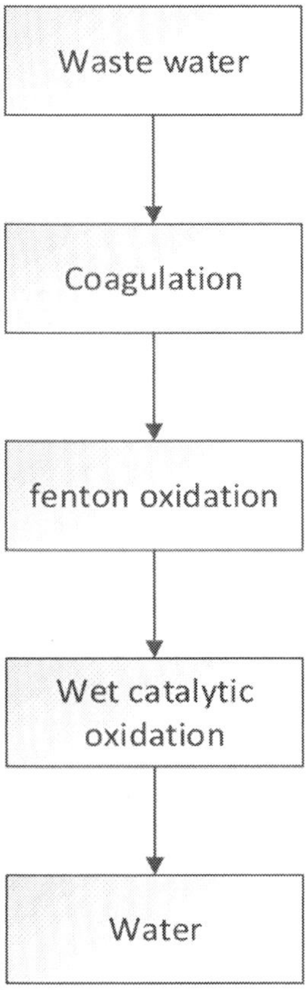

Fig.3. Wastewater treatment process

The experimental steps of each step treatment method are as follows.

(1) Firstly, the coagulation test was carried out by taking 200mL of wastewater, adding polyaluminium and polyacrylamide at concentrations of 600mg/L and 2mg/L respectively under the condition of room temperature and p H=8, stirring at 200r/min for 1min and then stirring at 60r/min for 15min to make the reaction adequate, and taking the supernatant after standing for 30min, and taking a small amount to measure the COD concentration.

(2) Take 50mL of supernatant for Fenton oxidation treatment. Add 30% hydrogen peroxide (H2O2) solution 0.6mL, then add FeSO4 solution 4.2mL, adjust the pH to 3.5, avoid light reaction for 120min, adjust the pH to 8 with sodium hydroxide respectively, terminate the oxidation reaction, stand for 30min, extract the supernatant, take a small amount to measure the COD concentration.

(3) The supernatant obtained from Fenton oxidation treatment was subjected to wet catalytic oxidation treatment. At a temperature of 250 ℃ and a partial pressure of oxygen of 2MPa, 200mg/L of catalyst prepared by impregnation was added, and the reaction time was 90min, and the COD concentration was measured by condensing a sample.

The test results show that the combined coagulation + Fenton oxidation + wet catalytic oxidation process has high stability, and the removal rate of COD in the wastewater by this combined process reaches more than 90%, and the COD value of the treated effluent is less than 60mg/L.

4 Conclusion

In this paper, we firstly studied the pretreatment process of acrylic wastewater, using a combination of coagulation and Fenton advanced oxidation to pretreat acrylic wastewater, and then focused on the preparation and influencing factors of catalysts for catalytic wet oxidation treatment of acrylic wastewater, and finally conducted a preliminary investigation on the reaction kinetics of copper-based catalysts for catalytic wet oxidation treatment of acrylic wastewater. Through the experimental study on the actual acrylic wastewater, the process technology scheme of this experiment was determined as: coagulation-Fenton oxidation-wet catalytic oxidation. After this process to treat acrylic wastewater with wet oxidation method alone, the catalyst was prepared by precipitation method and impregnation method, and the experiments showed that the catalyst prepared by impregnation method had better catalytic effect, so the catalyst was prepared by impregnation method. It can effectively degrade the oligomers in acrylic wastewater, and the COD concentration of treated wastewater can meet the discharge requirements. In summary, the deep oxidation wastewater treatment process proposed in this study is effective in treating acrylic wastewater, and the wastewater effluent can meet the national standard, which has explored a feasible path for the treatment of actual acrylic wastewater, and its economic and technical properties can meet the requirements of wastewater treatment, which provides a certain research basis for future industrialization research and has strong practicality.

References

1. L. W. Liang,L. Y. Liu,F. Yang. Research on pretreatment technology of acrylic industrial wastewater [J]. Advances in fine petrochemicals,2019,20(04):26-28.
2. Zhang Yulong,Shen Yan,Lv Zhengping,Xu Yejun. Study of A/O-MBR process treatment based on acrylic plant wastewater[J]. Contemporary Chemical Industry,2019,48(07):1471-1475.
3. Xu K. Analysis of acrylonitrile wastewater treatment technology[J]. Contemporary chemical research,2018(08):59-60.
4. Wang J.E.,Wu J.J.,Yu J.Q.,Lei T.S. Study of wet catalytic hydrogen peroxide oxidation for the treatment of acrylic wastewater[J]. Contemporary Chemical Industry,2017,46(05):838-840+845.
5. Bai Fan. Research on pretreatment of acrylic wastewater by sponge iron type "iron-carbon microelectrolysis coupled with Fenton reagent" method[D]. Lanzhou Jiaotong University,2017.

Analysis on the influence of water concentration on ADCP current measurement

Mingmin ZHANG[1,*], Yicheng LIU[1], Jinghan XU[2], and Hui ZHAO[1]

[1]Tianjin Research Institute for Water Transport Engineering, M.O.T., Tianjin, 300456, China
[2]Tianjin Dongfang Tairui Technology CO., LTD, Tianjin, 300456, China

Abstract: With the objective of improving the measure performance of Acoustic Doppler Current Profiler (ADCP) and promoting the level of marine environmental monitoring, the precision and accuracy of ADCP current measurement have been analyzed in detail based on the current flume with water concentration (0 - 20) kg/m^3. The experiments indicated that the precision of ADCP current measurement is enhanced with the uprising of water concentration. The precision of ADCP measurement was significantly influenced by the water concentration when the current was (0.7-0.9) m/s, and it was less affected by the water concentration when the current was 0.3m/s. In addition, with the increase of water concentration, the accuracy of ADCP measurement increased at first and then decreased. The accuracy of current measurement was higher when the concentration was (9-11) kg/m^3, and decreased in varying degrees outside this range. When the current was 1.1 m/s, the accuracy of ADCP was greatly affected by the concentration, while it was less affected by the concentration in the case that current is 0.3 m/s. The correlation analysis and conclusion can provide a reference not only for the research and development of ocean observation equipment, but also for marine scientific studies.

1 Introduction

Acoustic Doppler Current Profiler (ADCP) is a kind of acoustic instrument, which uses acoustic Doppler principle to measure the frequency shift information of scattered signal in layered water medium, and uses vector synthesis method to obtain the velocity of stratified water flow in vertical section[1]. With its advantages of high measuring accuracy, plentiful measuring points, strong adaptability and high automation, ADCP has been widely used in water transportation engineering, hydrological monitoring, Marine environment research and other aspects.More importantly, ADCP has been designated as one of the four most advanced ocean observation instruments by the IOC (International Oceanographic Commission) [2-6]. However, in practical application, the measurement results of ADCP are greatly affected by environmental factors such as depth, temperature, salinity, concentration, etc., often resulting in the absence or failure of measurement data, which brings great loss of human and financial resources to the marine survey work [7-10]. In order to ensure the precision and accuracy of ADCP measurement results, some scholars have conducted relevant studies. Among them, Yan Song et al. compared the ADCP's Current Measurements in the mixing of suspended particulate matter and clear water, founded that the current measurement data is more stable after mixing an appropriate amount of suspended particulate matter [11]. Wagner C R showed that the navigation ADCP could not be used under the condition of too shallow water [12].

The WHS600-I ADCP manufactured by Teledyne RD Instrument in the United States and the Aquadopp Profiler 2MHz ADCP manufactured by Nortek in Norway are selected herein as the prototype to comprehensively analyze the characteristics of the accuracy and precision of the ADCP current measurements in the water environment with a concentration of (0-20) kg/m^3. The analysis and conclusion of this study are not only of great value to the development and application of ADCP, but also of practical significance to the standardization of the detection methods of ADCP parameters, Marine environmental protection and scientific research.

2 Materials and methods

2.1 Experimental equipment and facilities

In this paper, the experimental equipment and facilities mainly include conductivity-temperature-depth (CTD), ADCP and verification device of ADCP.

(1) The Compact-CTD is a kind of self-contained CTD that equipped with high-precision pressure, temperature, salinity and concentration sensors. Compared with traditional concentration measuring instruments, it has the advantages of higher accuracy, higher efficiency and easy operation in field measurement.

(2)The WHS600-I ADCP is manufactured by Teledyne RD Instruments in the United States, which uses acoustic doppler principle to measure the velocity and flow

*Corresponding author:2570983389@qq.com

 https://doi.org/10.1051/e3sconf/202129001006

direction of ocean currents at different depths. TRDI's broadband signal processing technology enables the current measurement accuracy of WHS600-I ADCP to achieve 0.3%v ±0.3cm/s (v represents the current). In addition, the WHS600-I ADCP can be installed on buoys, ships or seabed, which makes it highly accurate for long term measurements.

(3) The AquAdopp Profiler 2MH ADCP consists of a 2MHz acoustic transducer and emits a short pulsed sound beam into the water. It has the characteristics of wide application range, small size and light weight, which is mainly used for nearshore Marine research, online monitoring and scientific research of rivers, lakes and rivers. The accuracy of current measurement is 1%v ±0.5cm/s (v represents the current).

(4) The verification device of ADCP (Fig. 1)mainly includes a current flume, a current trailer, a fixed support for ADCP and a suspended particle agitator, of which the size of the current flume is 75m (length) ×1.5m (width) ×1.5m (depth), and the maximum speed of the current trailer is 2m/s, also it is equipped with lifting and turning mechanism.

Fig. 1. Verification device of ADCP

2.2 Experimental principle

Two tracks are set up above the rectangular static flume with uniform cross-section, and a current trailer which can move back and forth is installed on the track. The ADCP is fixed on the current trailer by lifting measuring rod, which can rotate around the axis, so that the indicated direction of the ADCP is in the same direction with the traveling direction of the trailer. Under the condition that the water in the flume is stationary, when the ADCP is moving at different speeds on the track, it can be equivalent to the ADCP is stationary, while the water in the flume is flowing at a speed of trailer relative to the ADCP. On the premise that the speed of the current trailer is known, the current measurements of the ADCP in different water concentrations can be evaluated by analyzing the consistency between the measured velocity and the trailer speed as well as the dispersion between the measured values. The diagram of ADCP current measurement system is shown in Figure 2.

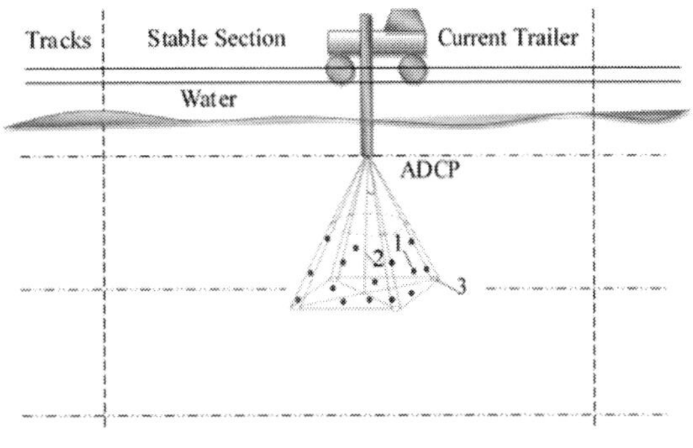

Fig. 2. The diagram of ADCP current measurement system
1—Scatterers in water; 2—Central axis; 3—Transmit beams.

2.3 Experimental procedure

The main purpose of this experiment is to obtain the current measurements of ADCP in the flume with different water concentration. The specific steps are as follows:

1) Start the current trailer, which drag the mixing device reciprocate several times until the suspended sediment is evenly dispersed;

2) Set up five concentration collection points (N_1, N_2...N_5) at equal intervals in the flume, and 30 concentration values at each collection point were measured with Compact-CTD;

3) Take the average of 30 measurements of Compact-CTD as the actual measured value of the collection point $\overline{N}_i$ (i=1、2、3、4、5);

4) Calculate the average value of five collection points and take it as the actual concentration of the water environment;

5) Fix the ADCP to the lifting rod of current trailer, and adjust the ADCP forward through the Angle rotary plate until it is in the same direction as the current trailer. Meanwhile, adjust the lifting mechanism so that the depth of the ADCP into water is 0.4 meters;

6) Within the trailer speed range, six speed points of 0.3 m/s, 0.5 m/s, 0.7 m/s, 0.9 m/s, 1.1 m/s and 1.3 m/s are selected to input into the current trailer control system. At the same time, the current trailer is controlled by wireless remote control equipment to pull the ADCP at the selected speed;

7) When the current trailer runs to the stable section, record 30 currents continuously measured by ADCP;

8) Add a certain amount of fine-grained sand to the flume, and repeat steps 1) -7).

The concentration point of flume were set at 1kg/m^3, 3 kg/m^3, 5 kg/m^3, 7 kg/m^3, 9 kg/m^3, 11 kg/m^3, 13 kg/m^3, 15 kg/m^3, 17 kg/m^3 and 19 kg/m^3. The actual measured values of each concentration collection point by the CTD are shown in Table 1.

Table 1. The actual measured values of each concentration collection point

Point 1 (kg/m^3)	Point 2 (kg/m^3)	Point 3 (kg/m^3)	Point 4 (kg/m^3)	Point 5 (kg/m^3)	Mean (kg/m^3)
1.009	1.002	1.008	1.010	1.005	1.0068
3.011	3.015	3.011	3.010	3.012	3.0118
5.014	5.019	5.019	5.016	5.021	5.0178
7.015	7.008	7.002	7.012	7.008	7.0090
9.005	9.002	9.003	9.007	8.999	9.0032

 https://doi.org/10.1051/e3sconf/202129001006

11.018	11.008	11.009	11.016	11.013	11.0128
13.014	13.010	13.014	13.020	13.017	13.0150
15.010	15.008	15.002	15.004	15.007	15.0062
17.000	17.003	17.000	16.996	17.006	17.0010
19.009	19.008	19.010	19.005	19.013	19.0090

2.4 Precision and Accuracy

2.4.1 Precision of current measurement

Precision refers to "the degree of consistency between the indicated or measured values that obtained from repeated measurements of the same or similar objects under specified conditions" [13]. Precision is usually expressed in numerical terms as the degree of imprecision, such as the standard deviation, variance, or coefficient of variation under specified measurement conditions. In this paper, the experimental standard deviation $s(x)$ was estimated by Bessel formula as

$$s(x)=\sqrt{\frac{\sum_{i=1}^{n}(x_i-\bar{x})^2}{n-1}} \tag{1}$$

Where x_i is the measured value of the ith current measurement, $\bar{x}$ is the arithmetic average of n measurements, which is used to evaluate the precision of ADCP current measurement finally.

2.4.2 Accuracy of current measurement

Accuracy refers to "the degree of agreement between the measured value and its true value" [13]. However, truth value is an ideal concept which is generally impossible to know exactly, but it can be replaced by conventional true value. Indication error is one of the most important metrology characteristics of a measuring instrument, and its essence reflects the accuracy of the measuring instrument. The larger the absolute value of the indication error, the lower the accuracy. In this paper, the average value of 30 ADCP current measurements is used as the measured value, and the speed of the current trailer in the stable section is used as the conventional true value. Finally, the absolute value of the indication error is used to evaluate the accuracy of the ADCP current measurement.

3 Results and Discussion

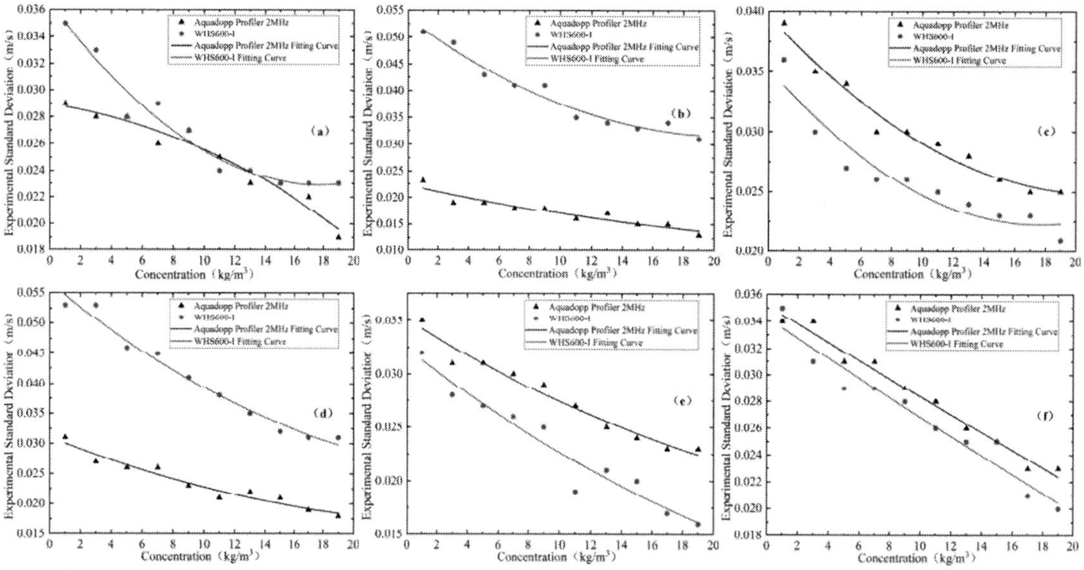

Fig. 3. The fitting curve between concentration value and experimental standard deviation of the current measurements (a) Current:0.3m/s; (b) Current:0.5m/s; (c) Current:0.7m/s; (d) Current:0.9m/s; (e) Current:1.1m/s; (f) Current:1.3m/s.

As schematised in Fig. 3, which shows the fitting curve between concentration value and experimental standard deviation of the current measurements, the experimental standard deviation of each current measurement is less than 0.055 m/s within the range of water concentration. The maximum experimental standard deviation of Aquadopp Profiler 2MHz ADCP is 0.039m/s (Fig. 3c) and the minimum is 0.013m/s (Fig. 3b), while the maximum experimental standard deviation of WHS600-I ADCP is 0.053 m/s (Fig. 3d) and the minimum is 0.016m/s (Fig. 3e).

In this experiment, under the condition that the current remains unchanged, the experimental standard deviation decreases as the concentration increases, that is, the higher the concentration, the higher the measurement precision. When the current is 1.3m/s (Fig. 3f), the precision of current measurement shows an obvious linear downward trend with the increase of concentration, while at 0.7m/s (Fig. 3c) and 0.9m/s (Fig. 3d), it shows a downward trend of quadratic curve. In addition, the precision of Aquadopp profiler 2MHz ADCP measurement is most affected by the water concentration when the current at 0.7m/s (Fig. 3c), and the experimental standard deviation decreased by 0.014m/s, from 0.039m/s at the water concentration of 1.007 kg/m³ to 0.025 m/s at the water concentration of 19.009 kg/m³. On the other hand, the precision of Aquadopp profiler 2MHz ADCP measurement is the least

https://doi.org/10.1051/e3sconf/202129001006

affected by the water concentration when the current is 0.3m/s (Fig. 3a) and 0.5m/s (Fig. 3b), and it only decreased by 0.010m/s. For the WHS600-I ADCP, the precision of current measurement is the most affected when current at 0.9m/s (Fig. 3d), which the experimental standard deviation reduced by 0.022 m/s, and it is the least affected when current is 0.3m/s (Fig. 3a), with a decrease of only 0.012 m/s.

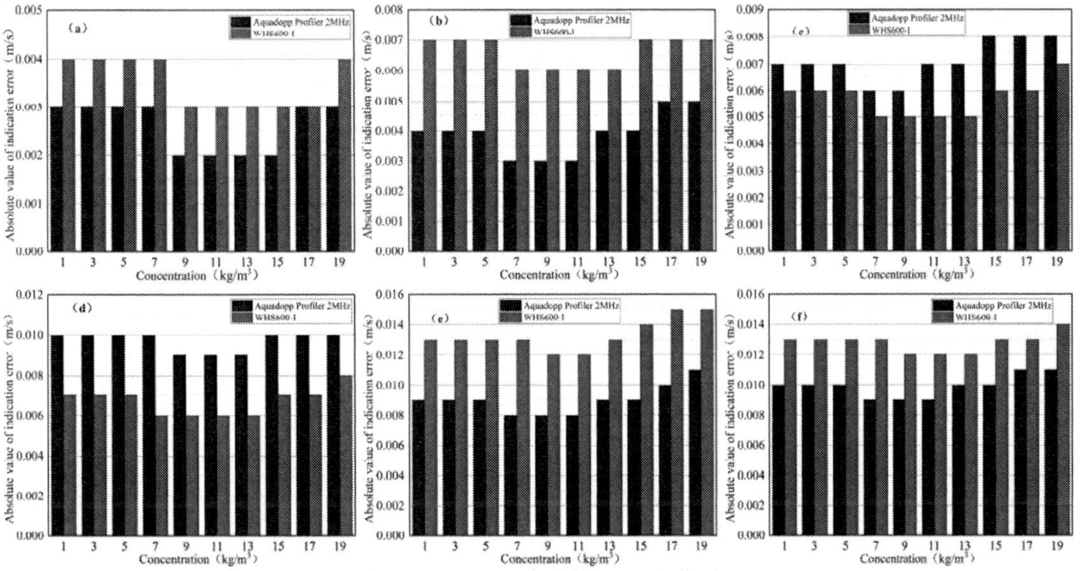

Fig. 4. The absolute value of ADCP current indication error
(a) Current:0.3m/s; (b) Current:0.5m/s; (c) Current:0.7m/s; (d) Current:0.9m/s; (e) Current:1.1m/s; (f) Current:1.3m/s.

As schematised in Fig. 4, which shows the absolute value of ADCP current indication error in each concentration, all the absolute values of the indication error are less than 0.016m/s within the range of water concentration. For the Aquadopp Profiler 2MHz ADCP, the maximum value of indication error is 0.011m/s (Fig. 4e,f) and the minimum is 0.002m/s (Fig. 4a), while the maximum value of indication error of the WHS600-I ADCP is 0.015 m/s (Fig. 4e) and the minimum is 0.003 m/s (Fig. 4a).

When the concentration remains unchanged, the absolute value of the indication error shows an increasing trend as the current increases.

Under the condition of constant current, the accuracy of current measurement shows a trend of first decreasing and then increasing with the increase of concentration. The Aquadopp Profiler 2MHz ADCP has the smallest absolute value of the indication error when the concentration is (9-11) kg/m^3, especially when the current is 0.3m/s (Fig.4a), the absolute value of the indication error is only 0.002m/s. When the concentration is outside (9-11) kg/m^3, the absolute value of indication error increases in varying degrees,such as when the current is 1.1m/s (Fig. 4e), the increase is most obvious, that is, it is most affected by the concentration, and the appreciation is 0.003m/s. In addition, when the current is 0.3m/s (Fig. 4a) and 0.9 m/s (Fig. 4d), the absolute value of indication error is less affected by concentration with the amount of change is 0.001m/s. For WHS600-I ADCP, as with Aquadopp Profiler 2MHz, the absolute value of the indication error is the smallest at the concentration (9-11) kg/m^3, that is, the accuracy of the current measurement reaches the highest. Especially when the current is 0.3m/s (Fig. 4a), the absolute value of the indication error is only 0.003 m/s. When the concentration is outside (9-11) kg/m^3, the absolute value of indication error increases in varying degrees, such as when the current is 1.1m/s (Fig. 4e), the accuracy of current measurement is most affected by the concentration, which increases by 0.003m/s, while the concentration has the least influence when the current is 0.3 m/s (Fig. 4a) and 0.5 m/s (Fig. 4b), the increment is only 0.001 m/s.

4 Conclusions

The WHS600-I ADCP and the Aquadopp Profiler 2MHz ADCP are selected herein as the prototype to comprehensively analyze the characteristics of the accuracy and precision of the ADCP current measurement in the water environment with a concentration of (0-20) kg/m^3. The following conclusions can be drawn from this study:

(1) Under the condition that the current remains unchanged, the precision of ADCP current measurement is enhanced with the uprising of water concentration.

(2) When the current is 1.3 m/s, the precision of current measurements shows an obvious linear downward trend with the increase of concentration, while at 0.7m/s and 0.9m/s, it shows a downward trend of quadratic curve.

(3) The precision of ADCP current measurement is greatly affected by concentration when the range of current is (0.7-0.9) m/s, while the current is 0.3 m/s, it is less affected by concentration.

(4) Under the condition of constant current, the accuracy of current measurement shows a trend of first decreasing and then increasing with the increase of concentration. The absolute value of the indication error is the smallest when the concentration is (9-11) kg/m^3, that is, the accuracy of the current measurement reaches the

highest. When the concentration is outside (9-11) kg/m^3, the accuracy of the current measurement decreases in varying degrees.

(5) When the current is 1.1 m/s, the accuracy of current measurement is most affected by the concentration, and it is least affected by the concentration when the current is 0.3 m/s.

Acknowledgments

This work was supported by National Key Research and Development Program(2018YFF0212202) and Research and Innovation Fund of Tianjin Research Institute for Water Transport Engineering, M.O.T.(TKS180104).

References

1. Y.X. Liu, Review on development of ADCP technology and its application. Hydrographic surveying and charting, 36(02): 45-49.(2016)
2. H.M. Zhang, Z.G. Chen, J.N. Zhao, ADCP integration meaurement based on external sensors. Gematics and infoemation science of wuhan university, 41(8):1131-1136.(2016)
3. D.A. Fong, S.G. Monismith,Evaluation of the Accuracy of a Ship-Mounted, Bottom-Tracking ADCP in a Near- Shore Coastal Flow.Journal of atmospheric and oceanic technology, 21(7):1121-1128.(2004)
4. R. Bozzano, E. Fanelli, S. Pensieri, Temporal variations of zooplankton biomass in the Ligurian Sea inferred from long time series of ADCP data. Ocean Science, 10(1):93-105.(2014)
5. F. Claude, Y. Wang, L. Leena, Empirical modeling of spatial 3D flow characteristics using a remote-controlled ADCP system:monitoring a spring flood. Water, 7(12):217-217.(2015)
6. H.L. Tang, H.T. Nie, J.G. Xiao, A quality contral procedure for shipboard ADCP data and its application. Advances in marine science, 38(1):153-161.(2020)
7. A. Sentchev, M. Yaremchuk, Monitoring tidal currents with a towed ADCP system. Ocean dynamics, 66(1):119-132.(2016)
8. Q.W. Zhou, S. Zhang, X.Y. Wang, Discussion on the calibration methods for acoustic doppler current profilers.Jounral of ocean technology, 35(4):31-35.(2016)
9. Y.M. Zhi, J. Shao, Y.P. Tang, Discussion on driving scheme of ADCP current meter verification trailer.China measurement and test, 42(11):79-83+93.(2016)
10. F. Jiang, S.H. Cheng, Y.C. Pang, Comparative analysis of in-suit ADCP key measuring performances.Hournal of ocean technology, (06)：59-64.(2020)
11. S. Yan, H. Wu, D.P. Sun, Application of acoustic doppler velocimetry in flume flow experiment. Research and exploration in laboratory, 36(5):9-13.(2017)
12. C.R. Wagner, D.S. Mueller, Comparison of bottom-track to global positioning system referenced discharges measured using an acoustic Doppler current profiler. Journal of Hydrology, 401(3):250-258.(2011)
13. China metrology and testing society group, Basic knowledge and professional practice of second-level registered metrology, China quality inspection publishing, The BeiJing.(2017)

Recognition and adsorption of uranyl ion by ion imprinted polymer based materials

Jin Bai[1,*],Yang Yi[1], Xilin Xiao[1],Yi Ge[1]

[1]School of chemistry and chemical engineering, Nanhua Universityt, 421000, PRC

Abstract. Ion imprinted polymer is a kind of three-dimensional material with special cavity structure. Its cavity site has the function of specifically recognizing and binding target ions. It is a kind of materials that can selectively combine, enrich and remove target ions in the environment. Uranium is an important radioactive element, which has been widely used in nuclear power, military industry, science and technology and other fields. However, the accumulation of uranyl ions in environmental water causes serious health hazards to human body. Therefore, the detection and enrichment of uranyl ions in water is of great significance. In this paper, the theory, technology and application status of ion imprinted polymer for micro detection and adsorption of uranyl particles in wastewater were reviewed, and the development of uranyl ion imprinted polymer was prospected.

1 Introduction

With the rapid development of economy, the demand for energy in the world is more and more intense. As a renewable and clean energy, nuclear energy has been widely interested and vigorously developed. Uranium is a kind of radioactive heavy metal element, which is widely used in nuclear power generation, nuclear weapon manufacturing, nuclear activation analysis and other fields. In the process of nuclear energy application, uranium containing dust, uranium containing sewage and other wastes will be produced. If the nuclear waste is not fully treated and discharged, it will cause serious pollution to the environment and endanger human health. When insoluble uranium dust is inhaled by human body, uranium compounds will enter the blood and cause damage to human body. Uranyl ion in uranium containing wastewater is easy to enter the human body along with the food chain. With the accumulation of uranyl ion, it can cause systemic diseases and multiple organ canceration. Therefore, it is very important to detect the content of uranyl ion in the environment and to enrich and remove it. At present, the removal methods of radioactive uranium in wastewater include adsorption, membrane separation, ion permeation, reverse osmosis, and so on

Chemical precipitation, etc. Among them, adsorption method is widely used because of its simple operation, high efficiency, high cost performance, and be not easy to cause secondary pollution. At present, the detection methods of uranyl particles in wastewater mainly include ion chromatography, spectrophotometry, fluorescence, seed activation analysis, etc. These methods have the advantages of high sensitivity and low detection limit, but their application requires high cost and expensive equipment. Electrochemical method has attracted wide attention because of its high sensitivity, simple operation and low price.

Ion imprinted polymers (IIPs) are a kind of materials with three position cavity structure, which can specifically combine with target ions, and have both specific recognition and selective adsorption. In recent years, IIPs have been widely used in electrochemical sensor systems because of their simple preparation, good stability and low price. Because of their high sensitivity and low detection limit, IIPs have been widely used in metal ion selective electrochemical sensors. At the same time, IIPs have strong adsorption capacity for target ions, so IIPs can be used as an excellent material for the detection and adsorption of uranyl ions.

2 Preparation of ion imprinted polymer

When the template ion and the functional monomer are physically self-assembled, the template ion will be surrounded by the monomer to form a chelate and fixed. After the cross-linking polymerization, the imprinted cavity with the same size and shape as the template ion is obtained by eluting the template ion in the polymer, and the obtained polymer can specifically bind with the template ion.

The preparation methods of IIPs can be mainly divided into: (1) template ions and functional monomers are mixed in suitable solvents to form covalent or noncovalent complexes; (2) polymers are formed in the presence of crosslinkers and initiators; (3) template ions are removed by appropriate elution methods to obtain cavity structures with specific recognition function (As shown in Figure 1)

* Corresponding author: 20184507234@stu.usc.edu.cn

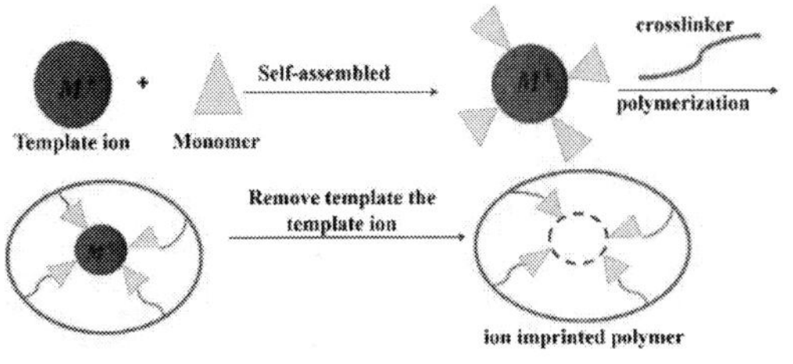

Fig. 1 Preparation of ion imprinted polymer

3 Application of Ion Imprinted Electrochemical sensor (iip-ce) to the determination of uranyl ion in Wastewater

The sensor will output signals after being stimulated by external physical or chemical stimuli. The electrochemical sensor can detect the concentration of uranyl ion in wastewater according to the changes of different electrochemical signals such as current, voltage and electrochemical impedance. Specific electrochemical sensors can be divided into voltammetry, impedance measurement and potential measurement. Wang et al. Constructed an ion imprinted electrochemical sensor with graphene modified carbon paste electrode, which can be used for the selective detection of trace uranyl ions. Using isophthalaldehyde-tetrapyrrole (IPTP) as the U ligand of uranyl ion, APTMS as the functional monomer and TEOS as initiator, the ion imprinted polymer was prepared by gel sol-gel method and modified on the surface of the graphene carbon paste electrode (GR-CPE) and then eluted to obtain uranyl ion sensor (IIP-GR-CPE). The linear range of uranyl ion detection is 1.0 × molto 1.0 × The recoveries were 97.3% - 102.9%. Majid kalate bojdi [2] and others developed an IIP carbon paste electrode (IIP CPE) sensor based on the high selectivity of IIP nanoparticle technology and the high sensitivity of differential pulse adsorption cathodic stripping voltammetry (dp-adcsv). Methacrylic and phenolic uranyl ligands were selected as functional monomers, EGDMA and AIBN as crosslinking agents and initiators, and the iip-cpe was constructed by precipitation polymerization. The detection limit was 0.03 μg and the linear range was 0.1 μg to 10 μg

Sevgi G ü Ney et al. [3] uranyl ion imprinted polymer (U-imp) and non uranyl ion imprinted polymer (N-imp) were synthesized by sol-gel copolymerization, and then modified on carbon paste electrode to obtain U-imp/CE.

Compared with the paste electrode containing n-imp, the paste electrode containing u-imp has higher reduction peak current. The selectivity of electrochemical sensor can ensure that it has a good application prospect in the detection of actual samples.

The application of ion imprinted polymers in the adsorption of uranyl ions in sewage: the imprinted polymer prepared by sol-gel method has the advantages of good mechanical effect, high repetition rate and good stability. The memory cavity structure formed during the preparation process can selectively adsorb specific ions in complex environment, and plays an important role in the adsorption of uranyl ions in sewage. Thayyath s. anirudhan [4] reported a new type of uranyl ion imprinted polymer. Using 4-vinylpyridine as functional monomer and 2,20-azobisisobutyronitrile as initiator, IIPs were prepared by bulk polymerization, and non imprinted polymer (NIP) was prepared under the same conditions. The results show that u (VI) - IIP has better selectivity and adsorption capacity. When pH = 7 and temperature is 30 ℃, the equilibrium adsorption capacity is 133.95 mg / g, u (VI) - IIP

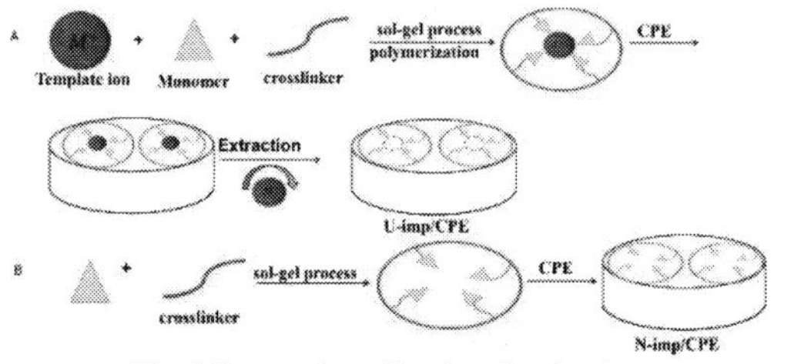

Fig. 2 Preparation of ion imprinted polymer

It has the potential to remove and recover U (VI) from wastewater. 5. E. pakade [5] et al. Prepared a novel acyl ion imprinted polymer by a simple and mild method, using 1 - (prop-2-en-1-yl) - 4 - (pyridin-2-ylmethyl) piperazine (ppmp) and MAA as functional monomers, edgma and accn as crosslinking agents and initiators, the optimal adsorption pH was 4.0 ~ 8.0, the amount of adsorbent was 20 mg, the contact time was 20 min, and 120 mg uranyl ion was retained per g IIP. Hu Meng [6] and others introduced a Si coated graphene oxide modified U (VI) imprinted polymer (go Si / IIP). Through intermittent experiments, compared with go Si / nip, it was found that the maximum adsorption capacity of go Si / IIP was 17.89 mgg-1, which was higher than 10.32 mgg-1 of go Si / nip, indicating that the ionic imprinted composite material had better adsorption performance.

After five adsorption desorption cycles, go / sio2-iip can be efficiently regenerated and reused, which shows that it has a good removal effect on template ions. The disadvantage of go Si / IIP is that the preparation process is cumbersome. The development and preparation of simple and efficient IIP composites is the focus of future research.

Table 1. Application of ion imprinted polymers in selective recognition and adsorption of other metal ions

Adsorbed ion	synthetic method	Decorative materials	Maximum adsorption capacity（mg/g）	reference
Cu^{2+}	Copolymerization	SiO_2	71. 400	【9】
Cu^{2+}	Surface Imprinting	-	39. 82	【10】
Cu^{2+}	Copolymerization	Magnetic nanoparticles Fe_3O_4	50.38	【11】
Pb^{2+}	Surface Imprinting	Sulfhydryl group	68. 03	【12】
Pb^{2+}	Copolymerization	Schiff Base	54. 9	【13】
Cd^{2+}	Surface Imprinting	Magnetized carbon nanotubes	-	【14】
Cd^{2+}	Copolymerization	Mesoporous silicon	32. 9	【15】
Cd^{2+}	Surface Imprinting	-	38.30	【16】
Hg^{2+}	Surface Imprinting	$Fe_3O_4@SiO_2$	78.3	【17】
Cr^{2+}	Surface Imprinting	MGO	311.95	【18】
Th^{4+}	Copolymerization	-	33.3	【19】
Au^{3+}	Copolymerization	Chitosan-MB	195±0.5	【20】

Magnetic nanomaterials (NMPs) are special materials with both magnetic properties and nanostructures. Magnetic nanomaterials have superparamagnetism, magnetic single domain structure and other special magnetic properties. They can be uniformly dispersed in aqueous solution and rapidly separated from wastewater solution with the help of external magnetic field. Therefore, magnetic nanomaterials are a kind of adsorbent with good development prospects, The adsorption of uranyl ion by several magnetic nanomaterials is listed in Table 2.

Table 2. Application of ion imprinted polymers in selective recognition and adsorption of other metal ions

Type of adsorbent	Maximum adsorption capacity（mg/g）	reference
CB/GO/Fe3O4	66.81	[21]
Fe3O4@TiO2	97.08	[22]
Novel magnetic graphene oxide	113.27	[21]
Fe3O4@SiO2@Ni-L	129.26	[22]
Amidoxime functionalization Fe3O4@SiO2 Core shell magnetic microspheres	160.00	[23]
Amino modified Fe3O4 / SiO2 Composites	170.00	[24]
Titanate nanotubes / cobalt ferrite / tetraethylenepentamine（TNTs/CoFe2O4/TEPA）	465.95	[25]

At present, the adsorption effect of some magnetic nano materials for uranyl ion is satisfactory, but the selectivity for uranyl ion is not high, so the development of magnetic nano composite modified ion imprinted polymer has a good application prospect. At present, there are few reports on the adsorption and removal of uranyl ions by magnetic nanomaterials modified ion imprinted polymers, mainly involving the selective and efficient adsorption of. Xiao [26] and others introduced a kind of magnetic graphene oxide composite metal ion imprinted polymer, which was used for selective adsorption and synthesis. Using dopamine (DA) as the functional monomer, the imprinted polymer was synthesized by surface imprinting technology. Compared with the adsorption effect of [7-9] and [10-11], MIPs have larger adsorption capacity, stronger selectivity and stability. FARZANEH abdollahi [27] et al. Reported a novel ion imprinted polymer based on core-shell structure, Fe3O4@SiO2@TiO2 The modified IIP was used to detect the serum concentration of patients with gallstones, and the dynamic linear range was 0.20 ~ 28.00 μ GL − 1, the detection limit was 0.05 μ g L−1。 The modification of magnetic nanoparticles enhances the affinity of imprinted polymers, which is a promising selective adsorbent.

4 Conclusion

Ion imprinted polymers (IIPs) can realize the specific recognition and adsorption of target ions. They have stable structure, strong regeneration and high reuse rate. They can be used for enrichment and recovery of heavy metal ions such as uranyl ions in environmental water, and widely used in electrochemical sensors due to their high sensitivity and low detection limit. However, IIPs have some functions in synthesis, and there are few kinds of monomers, which lead to their performance can not meet the ideal requirements; Moreover, IIPs composites have some problems, such as high cost, complicated preparation process . Generally speaking, the current ionic imprinting materials are still in the development stage. Using green new materials, simplifying the preparation process of IIPs, developing new functional monomers, improving the performance of ionic imprinted polymers, studying the preparation of magnetic ionic imprinted polymers and large-scale industrial application will be one of the future research directions.

Acknowledgments

Thank you for the financial support of the innovation training program of Nanhua University, project number: x202010555322

References

1. Wang , Z. Zhang,D. Xiao,X. Su,C.Li , Z. Xue ,J. Peng,P. Liao,L. Wang,H.(2020)A highly sensitive and selective sensor for trace uranyl (VI) ion based on a graphene-coated carbon paste electrode modified with ion imprinted polymer.Microchemical Journal,155:0026-265X.
2. Majid , K. Mohammad, B.Mostafa, N.Akbar, B.Fariborz ,O.Sara, S.(2015)Selective and Sensitive Determination of Uranyl Ions in Complex Matrices by Ion Imprinted Polymers-Based Electrochemical Sensor.Electroanalysis, 27: 2458 – 2467.

3. Sevgi ,G. Orhan ,G .(2016)A novel electrochemical sensor for selective determination of uranyl ion based on imprinted polymer sol-gel modifified carbon paste electrode.Sensors and Actuators B Chemical http://dx.doi.org/10.1016/j.snb.2016.02.119.
4. Thayyath ,A. Jayachandran ,N. Peethambaran ,L. Divya.(2014)Adsorption and separation behavior of uranium(VI) by 4-vinylpyridine-grafted-vinyltriethoxysilane-cellulose ion imprinted polymer.Journal of Environmental Chemical Engineering,465: 1–10.
5. Bojdi M K, Behbahani M, Sahragard A, et al. A palladium imprinted polymer for highly selective and sensitive electrochemical determination of ultra-trace of palladium ions[J]. Electrochim Acta, 2014,149:108-116.
6. Metilda P, Prasad K, Kala R, et al. Ion imprinted polymer based sensor for monitoring toxic uranium in environmental samples[J]. Anal Chim Acta, 2007, 582(1):147-153.
7. V. E. Pakade, E. M. Cukrowska, J. Darkwa, G. Darko, N. Torto and L. Chimuka. (2012) Simple and efficient ion imprinted polymer for recovery of uranium from environmental samples.Water Science & Technology, 65.4: 728-736.
8. Zhu,J. Liu,Q. Liu,J. Chen, R. Zhang,H. Yu,J. Zhang,M. Li,R. and Wang,J.(2018)A novel ion-imprinted carbon material induced by hyperaccumulation pathway for the selective capture of uranium.ACS Appl. Mater. Interfaces,DOI: 10.1021/acsami.8b09022.
9. Buhani, Narsito, Nuryono, et al . Desalination and Water Treatment, 2015, 55(5) , 1240 .
10. R en ZQ, Zhu X Y, Du J, et al . Applied Surface Science, 2018, 435, 574 .
11. Zhan,Y. Luo,X. Nie,S. et al . (2011)Selective Separation of Cu(II) from Aqueous Solution with a Novel Cu(II) Surface Magnetic Ion-Imprinted Polymer.Industrial & Engineering Chemistry Research,50: 6355–6361.
12. Zhu B, Zhou L Y, Zhang Q, et al . Desalination and Water Treatment, 2017, 86, 231 .
13. Fan H T, Sun X T, Li W X . Journal of Sol-Gel Science and Technology, 2014, 72(1) , 144 .
14. Zargar Behrooz , Khazaeifar Ali , Microchimica Acta, 2017, 184 (11) , 4521 .
15. Li W M, He R, Tan L, et al . Journal of Sol-Gel Science and Technology, 2016, 78(3) , 1 .
16. C. Feng, M. Li, M. Li, Q. Zeng, Q. Gan, H. Yang, Synthesis and characterization of a surface-grafted Cd (II) ion-imprinted polymer for selective separation of Cd (II) ion from aqueous solution, Applied Surface Science (2015), http://dx.doi.org/10.1016/j.apsusc.2015.01.201
17. Zhang Q G, Wu J Y, Luo X B . R SC Advances, 2016, 6(18) , 14916 .
18. Liang Q W, Geng J J, Luo H J, et al . Journal of Molecular Liquids, 2017, 248, 767 .
19. Liang,H.Chen, Q. Ma,J. Huang,Y.Shen,X.(2017)Synthesis and characterization of a new ionimprinted polymer for the selective separation of thorium(IV) ions at high acidity.RSC Adv., 2017, 7, 35394
20. M.Monier, DA Abdel-Latif, Fabrication of Au(III) ionimprinted polymer based on thiol-modifified chitosan, International Journal of Biological Macromoleculeshttp://dx.doi.org/10.1016/j.ijbiomac. 2017.07.098
21. SHAO L, WANG X, REN Y, et al. Facile fabrication of magnetic cucurbit[6]uril/graphene oxide composite and application for uranium removal[J]. Chemical Engineering Journal, 2016, 286: 311-319.
22. ZHANG Xiaofei. Synthesis of core/shell structural magnetic materials and their adsorption properties of uranium[D]. Harbin: Harbin Engineering University, 2014.
23. SHENG G, SHAO X, LI Y, et al. Enhanced removal of uranium(VI) by nanoscale zerovalent iron supported on Na-bentonite and an investigation of mechanism[J]. Journal of Physical Chemistry A, 2014, 118(16): 2952-2958.
24. HU Jianbang, YUAN Yali, TANG Qiong, et al. Preparation and adsorption of uranyl(VI) of amino-modified magnetic Fe3O4/SiO2 composite materials[J]. Applied Chemical Industry, 2012, 41(12): 2067-2070, 2074.
25. ZHU J, LIU Q, LI Z, et al. Efficient extraction of uranium from aqueous solution using an amino-functionalized magnetic titanate nanotubes[J]. Journal of Hazardous Materials, 2018, 353: 9-17.
26. Xiao, H.Cai,L.et al.(2020)Preparation of metal ion imprinted polymer on magnetic graphene oxide / mil-101 (CR) surface and its selective adsorption of Cu (II) and Pb (II).CHINESE JOU R NAL OF APPLIED CHEMIST R Y,DOI: 10 . 11944 /j . issn . 1000-0518 . 2020 . 09 . 200028 .
27. Farzaneh ,A.et al.(2019)Application of selective solid-phase extraction using a new core-shell-shellnmagnetic ion-imprinted polymer for the analysis of ultra-trace mercury in serum of gallstone patients.SEPARATION SCIENCE AND TECHNOLOGY,https://doi.org/10.1080/01496395.2019.1651337.

https://doi.org/10.1051/e3sconf/202129001008

Chiral phosphoric acid and its applications of asymmetric catalytic hydrogenation

Sun Zhuoxin[1] *

[1] School of Pharmaceutical Science and Technology, Tianjin University, Tianjin, 300072, China

Abstract. As an excellent representative of organic small molecular catalysts, chiral phosphoric acid played an important role in asymmetric catalytic hydrogenation. According to the skeleton of chiral phosphoric acid, the development and applications of chiral phosphoric acid were summarized. The applications of chiral phosphoric acid in asymmetric catalytic hydrogenation were introduced according to three kinds of hydrogen source systems used in chiral phosphoric acid catalytic reaction. It was found that chiral phosphoric acid has high activity and excellent selectivity. Chiral products with excellent yields and enantioselectivities were obtained in most reactions.

1 Introduction

In 2001, Professor Noyori, and Dr. Knowles won the Nobel Prize for their outstanding contributions in asymmetric catalytic hydrogenation. Since then, asymmetric catalytic hydrogenation of compounds has become a hot topic in the field of synthesis, and has been studied and reported by a large number of scientists and scholars. Asymmetric catalytic hydrogenation was first catalyzed by transition metals. Compared with small molecular catalysts, transition metal catalysts have the disadvantages of high toxicity, difficult to synthesize and recover.[1] In 1989, Professor Batra first used amino acids as catalyst to realize asymmetric transfer hydrogenation without metal participation, which opened a new door for the asymmetric reduction of small molecular catalysts.[2] Subsequently, the technology of small molecular catalysis has been developed rapidly, and the types of catalysts are becoming more and more abundant.

Chiral phosphoric acid is a kind of small molecular catalyst with excellent reaction activity and selectivity. It is divided into four types according to the difference of skeleton structures, which are: BINOL-type, H_8-BINOL-type, SPINOL-type, VANOL-type and VAPOL-type (**Figure 1**).[3]

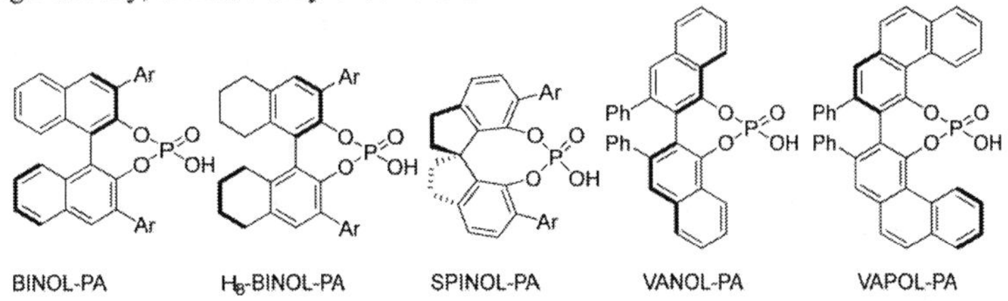

Figure 1 Skeletons of chiral phosphoric acids.

Based on these skeletons, chiral phosphoric acids were further expanded by changing the types of 3,3'-substituents. With the further research, the catalytic mechanism of chiral phosphoric acid has been further understood. Taking BINOL-type phosphoric acid as an example: (1) chiral phosphoric acid has strong acidity, and can activate the substrate by forming ion pairs. (2) The oxygen in the double bond of oxygen and phosphorus can be used as Lewis base, so the chiral phosphoric acid can be regarded as a bifunctional catalyst. (3) By changing the substituents of R-group, different space types were constructed, and the direction of the combination of catalyst and reactant was changed, so as to improve the selectivity (**Figure 2**).[4]

*Corresponding author's e-mail: SUNZX@tju.edu.cn

 https://doi.org/10.1051/e3sconf/202129001008

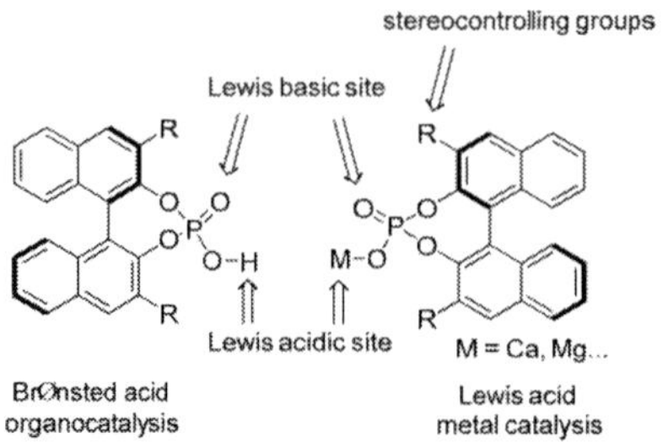

Figure 2 Potential of phosphoric acids and metal phosphates.

2 Development of chiral phosphoric acid

Although BINOL derived chiral phosphoric acids have been used in organic synthesis for nearly 50 years, their application as catalysts only started more than ten years ago. Akiyama and co-workers published a report on the Mannich reaction catalyzed by BINOL derived chiral phosphoric acid in 2004, which is widely considered as one of the pioneers in this field.[5] Subsequently, other types of chiral phosphoric acids were gradually applied to asymmetric synthesis. Until 2006, Gong and co-workers first used H_8-BINOL derived chiral phosphoric acid as catalyst to realize asymmetric Biginelli reaction, and obtained chiral products with excellent yields and enantioselectivities.[6] In 2002, Zhou and co-workers used SPINOL as ligand, which was the first time that SPINOL was used in asymmetric synthesis.[7] In 2010, Wang and co-workers successfully synthesized SPINOL derived chiral phosphoric acid for the first time, and used it as a catalyst to efficiently realize the Friedel-Crafts reaction of indole and imine.[8] The development of chiral phosphoric acid derived from VAPOL as catalyst in asymmetric synthesis was also very dramatic. In 2005, Antilla and co-workers successfully synthesized the chiral phosphoric acid with VAPOL for the first time, and used it as a catalyst for reductive amination to obtain chiral amines with excellent yields and enantioselectivities.[9] However, because the catalytic mechanism of chiral phosphoric acid was not fully understood at that time, we mistakenly thought that the chiral phosphoric acid derived from VAPOL was an effective catalyst in the reaction. Until 2013, Sala revealed that the real catalyst in the reaction should be chiral phosphate through a series of complex control experiments.[10] Subsequently, chiral phosphates derived from VAPOL have been well developed as catalysts for asymmetric reactions.

3 Asymmetric reduction catalyzed by chiral phosphoric acid

There are three kinds of hydrogen sources used in asymmetric transfer hydrogenation of chiral phosphoric acid: Hantzsch ester, benzothiazoline and borane. Hantzsch esters were widely used in asymmetric transfer hydrogenation catalyzed by chiral phosphoric acid.

In 2005, Reuping and co-workers first reported the asymmetric reduction of imines using chiral phosphoric acid as catalyst and Hantzsch ester as hydrogen source.[11] The chiral amines with excellent yields and enantioselectivities (**Figure 3**) were obtained by the reduction system. Since then, asymmetric reduction based on the system has been reported, including asymmetric reduction of nitrogen heterocyclic derivatives with high reduction difficulty.

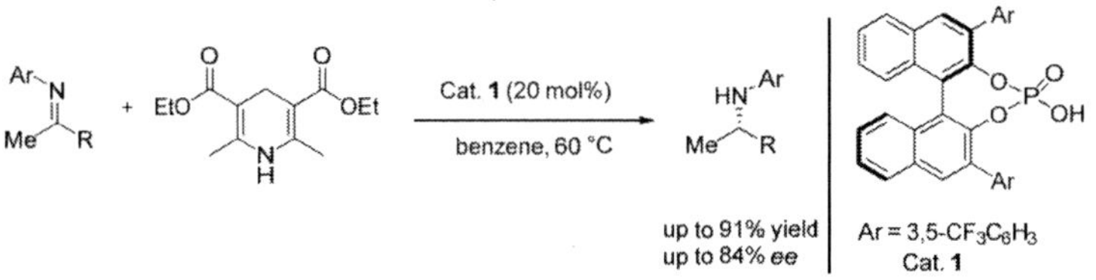

Figure 3 Asymmetric transfer hydrogenation of imines.

Benzothiazoline as a hydrogen source has also been shown in asymmetric reduction catalyzed by chiral phosphoric acid. In 2009, Akiyama and co-workers first used chiral phosphoric acid as catalyst and benzothiazoline as hydrogen source to asymmetric reduction of ketimides and obtained chiral products with excellent yields and enantioselectivities (**Figure 4**).[12] Then, asymmetric reduction reaction based on this reduction system has been widely used. They proposed that the reaction accorded with the bifunctional mechanism of coordination between benzothiazoline and phosphoric acid's Lewis basic site.

 https://doi.org/10.1051/e3sconf/202129001008

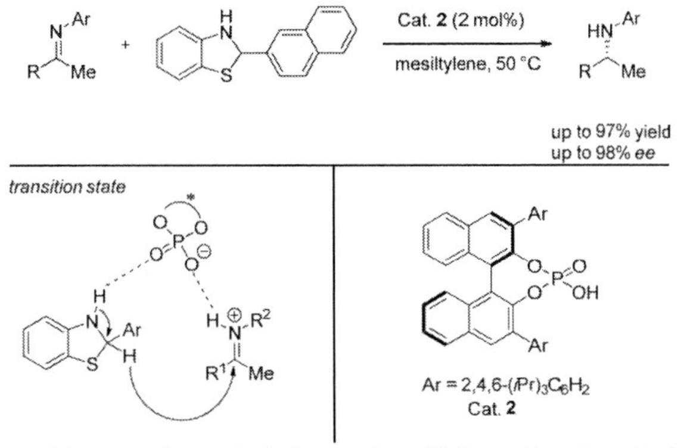

Figure 4 Asymmetric transfer hydrogenation with benzothiazoline of imines.

In the present reports, the asymmetric reduction reactions of Hantzsch esters or benzothiazolines as hydrogen sources mostly act on imines or imines, but there are few reports on the reduction of ketones. In 2011, Antilla and co-workers first used borane and chiral phosphoric acid to achieve asymmetric reduction of aromatic ketones, and obtained chiral products with excellent yields and enantioselectivities (**Figure 5**).[13] At the same time, a possible reaction mechanism was proposed. It was speculated that in this reaction, chiral phosphoric acid combined with borane to form a kind of chiral boro-phosphate catalyst in situ. The catalyst was a kind of bifunctional catalyst. Lewis acid in the boron center could activate ketones, and the oxygen in P=O could be used as Lewis base to improve the affinity of borane. At the same time, the hydrogen in unreacted borane attacked the activated ketones to form chiral products

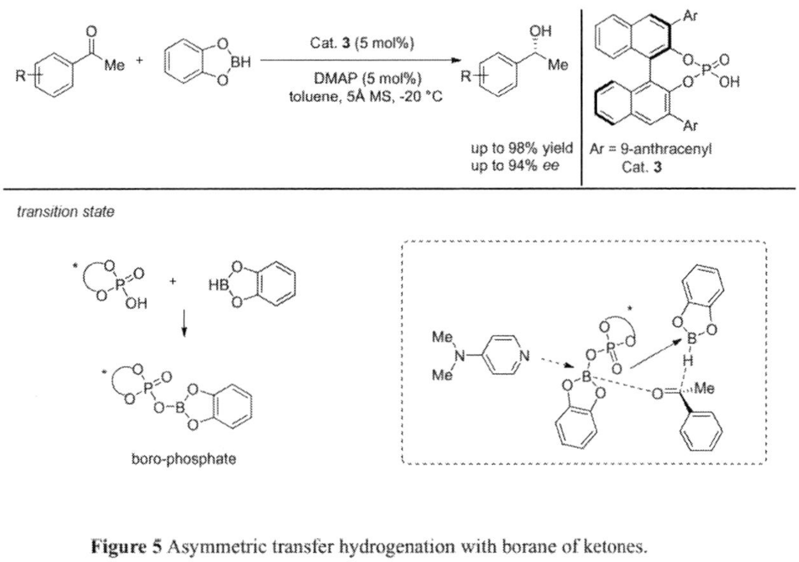

Figure 5 Asymmetric transfer hydrogenation with borane of ketones.

4 Conclusion

Chiral phosphoric acid is one of the outstanding representatives of organic small molecular catalysts. Compared with transition metals, chiral phosphoric acid has the advantages of low toxicity and easy synthesis and recovery, which provides a new idea for the synthesis of a variety of compounds with pharmacological activities. These catalysts performed well in asymmetric catalytic hydrogenation. Most of the chiral products obtained in asymmetric transfer hydrogenation have excellent yields and high *ee* values. However, it should also be noted that most of the reports have not reached the gram level reaction due to the limitations of the current catalytic reaction, and there are still many limitations in the realization of industrial production.

Acknowledgments

I would like to thank my supervisor for giving me a better understanding of the field of chiral phosphoric acid catalysis. I would also like to thank my team members for their help and my family for their support.

References

1. P. I. Dalko, L. Moisan, Angew. Chem., Int. Ed. 43, 5138 (2004)
2. S. Singh, U. K. Batra, Indian J. Chem., Sect. B: Org. Chem. Incl. Med. Chem. 28, 1 (1989)
3. D. Parmar, E. Sugiono, S. Raja, M. Rueping, Chem. Rev. 117, 10608 (2017)
4. M. Rueping, E. Sugiono, F. R. Schoepke, Synlett 6, 852 (2010)
5. T. Akiyama, J. Itoh, K. Yokota, K. Fuchibe, Angew. Chem., Int. Ed. 43, 1566 (2004)
6. X. Chen, X. Xu, L. Hua, L. Cun, L. Gong, J. Am. Chem. Soc. 128, 14802 (2006)
7. Y. Fu, J. H. Xie, A. L. Hu, H. Zhou, L.-X. Wang, Q.-L. Zhou, Chem. Commun. 5, 480 (2002)
8. F. Xu, D. Huang, C. Han, W. Shen, X. Lin, Y. Wang, J. Org. Chem. 75, 8677 (2010)
9. G. B. Rowland, H. Zhang, E. B. Rowland, S. Chennamadhavuni, Y. Wang, J. C. Antilla, J. Am. Chem. Soc. 127, 15696 (2005)
10. G. D. Sala, Tetrahedron 69, 50 (2013)
11. T. Akiyama, Chem. Rev. 107, 5744 (2007)
12. C. Zhu, T. Akiyama, Org. Lett. 11, 4180 (2009)
13. Z. Zhang, P. Jain, J. C. Antilla, Angew. Chem., Int. Ed. 123, 11153 (2011)

Study on Hydrolysis Reaction Rate of Several Chemical Warfare Agents

Dai Xuezhi[1,a]

[1] Institute of NBC Defense, Chinese PLA Army, Beijing, China

Abstract. According to the first-order kinetic reaction equation, the hydrolysis rate constants of homogeneous and heterogeneous phase of eight poisons under different conditions were calculated, and the relationship between hydrolysis rate constants and temperature was obtained according to the arrhenius formula, and the activation energy of hydrolysis reaction of some poisons was calculated. The evolution curve of homogeneous hydrolysis rate of 0.5 g/L HD with time was collected by T-135 method, and the hydrolysis rate was 0.1 min^{-1} and the half-life was 7 min. Poison concentration, pH of system, environmental temperature and structural composition of soil all affect the hydrolysis rate of Chemical Warfare Agents.

1 Introduction

Many disinfectants are formulated with water as one of their components. In these water-based systems, chemical warfare agents such as Tabun (GA), Sarin (GB), Soman (GD), VX, mustard gas (HD), Lewisite (L), phosgene (CG) and Bitz (BZ) will undergo hydrolysis reaction [1]. In the research and development of decontamination agents, though many scholars hope to find a faster, more efficient and mild disinfection method, instead of simply relying on hydrolysis to complete the decontamination of military chemical agents, it is very meaningful to understand the contribution of hydrolysis to decontamination compared with other decontamination agents with stronger activity. Therefore, this paper summarized some literatures about poison hydrolysis.

Chemical reaction kinetics, as an important component of chemical theory, can answer the question of the rate and mechanism of chemical reaction. Up to now, the first-order or second-order reaction kinetic model is often used in the literature to deal with the reaction process kinetic curve and find out the first-order or second-order reaction kinetic rate constant; The apparent activation energy can be calculated based on the rate constant at different temperatures and Arrhenius equation [2].

2 Materials and Methods

2.1 Materials

752 spectrophotometer, Shanghai Precision Scientific Instrument Co., Ltd.; Model 501A super thermostat, Shanghai Laboratory Instrument Works Co., Ld., micropipette, Eppendorf Co., Ltd., Germany; 25 mL test tube, Beijing glass instrument factory; HD (≥95%); Anhydrous ethanol (analytically pure), Dasen Chemical Products Sales Co., Ltd.; Thymol phthalide (analytically pure); Potassium hydroxide (analytically pure).

2.2 Methods

Take a 25mL (φ = 27 mm) test tube with stopper, inject about 5 ml of deionized water, put it in a thermostat at 25℃, add a certain amount of HD ethanol solution with micropipette, the concentration of HD in the reaction system is 0.5 g/L, ethanol is 4%, let it stand for a certain time, then immediately measure the residual concentration of HD in the system by T135 method, collect the residual concentration of HD at different times, and then calculate its hydrolysis.

3 Results and Discussion

3.1 Homogeneous hydrolysis of HD

The solubility of HD in water is low, only 0.8 g/L at 20℃. HD dissolved in water will hydrolyze quickly, and the main factor affecting the hydrolysis rate of HD is the dissolution and diffusion of HD. The final products of HD hydrolysis are thiodiglycol and two molecules of hydrogen chloride [3]. The reaction equation is:

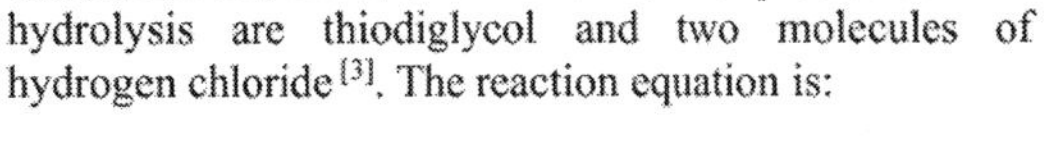

The hydrolysis of HD first produces half mustard gas, and then the half mustard gas continues to hydrolyze to produce thiodiglycol. In the process of generating semimustard gas, sulfonium ions are first generated, and the reaction between sulfonium ions and water molecules to generate semimustard gas is much faster than the

[a] Corresponding author: 297896810@qq.com

generation of sulfonium ions. However, semi-mustard gas is unstable and easily reacts with water molecules to form thiodiglycol. The decisive step of HD hydrolysis reaction is to generate sulfonium ions, which can be considered as bilateral hydrolysis, that is, complete hydrolysis to generate thiodiglycol [4].

Considering HD hydrolysis as a first-order reaction, there are:

$$-\frac{dC_A}{dt}=kt \Rightarrow \ln C_A=-kt+d \quad (1)$$

According to the linear relationship between $\ln C_A$ and t, the average rate constant k of HD hydrolysis reaction can be calculated. Homogeneous hydrolysis is dissolved HD. In order to ensure homogeneous hydrolysis of HD, HD can be dissolved in absolute ethanol first and then in water. At 25℃, the hydrolysis rate of 0.5 g/L HD increases exponentially with time, and the hydrolysis rate is fast at first and then slow. As shown in Figure 1, the hydrolysis rate of HD can be calculated as 0.1 min^{-1} and the half-life is 7 min by the above formula.

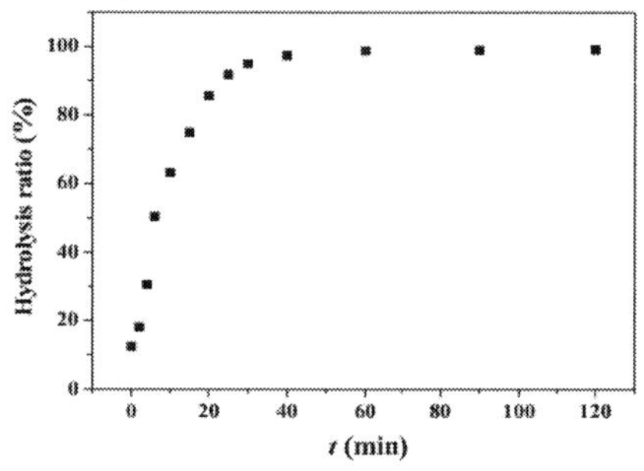

Figure 1. The change of hydrolysis rate of 0.5 g/L HD at 25℃ with time

3.2 Homogeneous hydrolysis of various poisons

According to the first-order kinetic reaction equation, the hydrolysis rate constants of homogeneous and heterogeneous poisons were calculated by using the experimental data of hydrolysis rates of various poisons given in the literature, and the relationship between hydrolysis rate constants and temperature was obtained according to arrhenius formula, and the activation energy of hydrolysis reaction was calculated. The rate constants and activation energies of homogeneous and heterogeneous hydrolysis of various poisons are listed in Table 1 and 2.

It can be seen from Table 1 that Tabun (GA) is attacked by parallel hydroxyl nucleophiles in the pH range of 3~9 to form pentacoordination intermediates, which are hydrolyzed to form cyanide ions and dimethyl amino phosphonic acid monoethyl ester. J.Epstein gave the hydrolysis rate constants of GA in seawater at different temperatures, and the activation energy was 71.1 kJ/mol, which was basically consistent with the hydrolysis rate constants given by Lennart Larsson, indicating that the effects of ionic strength and seawater on hydrolysis rate were limited. M.M.Demek determined that the hydrolysis rate constant of Sarin GB in seawater is 2.63 h^{-1}, which is nearly 40 times faster than that in pure water under the same experimental conditions. Ca^{2+} and Mg^{2+} from seawater can significantly catalyze the hydrolysis of GB. It was also found that the hydrolysis rate constant decreased with the increase of pH in the range of 1~4, but remained unchanged in the range of 4~6.5. When pH was above 6.5, the hydrolysis rate increased with the increase of pH. The hydrolysis product of GD is mainly pinanyl methylphosphonate, and both acidic and alkaline environments can accelerate the hydrolysis reaction of GD. In addition, it is found that GD can be completely hydrolyzed in 5 minutes in 5% NaOH aqueous solution and 45 minutes in 0.1 HCl aqueous solution at 20℃. VX mainly breaks P-S under acidic and alkaline conditions, and the hydrolysis product is β-diisopropylaminoethyl mercaptan. However, when P-O breaks under weak alkaline conditions, VX degreasing and hydrolysis reaction takes place to produce S-β-diisopropylaminoethyl thiomethylphosphonic acid. When VX concentration is in the range of 2~5.5 g/L, the hydrolysis rate constant is basically unchanged. Hydrolysis of CG is a part of nucleophilic substitution reaction or acylation reaction. Affinity substitution reaction occurs at acyl chloride site (-COCl), and the hydrolysis rate of dissolved CG is relatively fast. Adding alkali will accelerate the hydrolysis of CG. BZ is not easy to hydrolyze at room temperature, and heating or adding alkali can accelerate its hydrolysis reaction. The hydrolysis products are diphenylglycolic acid and quininol, but hydrolysis is slow under acidic conditions. Controlling pH value can accelerate the hydrolysis reaction rate.

Table 1. Homogeneous hydrolysis rate constant and activation energy of various poisons

Type	Experiment Condition			Speed Constant	Activation Energy (kJ/mol)	Remarks
	T(℃)	c (g/L)	pH	k_1 (h^{-1})	E_a	
GA	25	2.15	3	0.47		HCN-Lennart Larsson
			4	0.30		
			5	0.42		
	15			0.09	71.1	Seawater -J.Epstein.[5]
	20			0.16		
	25			0.24		

GB	0.2			2.63		Seawater - M.M.Demek.[6]
GD			3	0.03		T.V Healy.[7]
			7	8.66×10^{-3}		
			10	1.73		
VX	25	2.03		2.36×10^{-2}		Zhao Guohui [3] [3]
		2.51		2.30×10^{-2}		
		4.89		2.55×10^{-2}		
		5.25		2.43×10^{-2}		
	25			4.00×10^{-3}		Seawater - M.M.Demek.[6]
HD	20			5.16		Xue Jishen[8]
	10			1.61	85.5	
	15			3.0		
	20			5.57		
	25			11.12		
	30			16.7		
	5			0.24	84.2	Seawater -J.Epstein.[9]
	15			0.85		
	25			2.76		
CG	20	8.68×10^{-4}		0.18		Xue Jishen [8]
		2.44×10^{-3}		0.18		
		8.26×10^{-4}		0.21		
BZ	25	3.50×10^{-2}		3.35×10^{-3}		Xue Jishen [8]

Table 2. Rate constants and activation energies of heterogeneous hydrolysis of various poisons

Poison Type		Experiment Condition				Speed Constant	Activation Energy	Remarks
		T ℃	Viscosity s	Droplet mass mg (mg)	Type of Soil/Asphalt	k_1 h^{-1}	E_a kJ/mol	
GD	Gelatineous	0	1	10	Loam	0.01	52.3	Chen Haiping[10]
		20				0.05		
		40				0.25		
		0		20		0.01	53.1	
		20				0.05		
		40				0.26		
		0	6	10		0.01	53.0	
		20				0.05		
		40				0.26		
		0		20		0.01	52.5	
		20				0.05		
		40				0.26		
		0	5.2	20	Chernozem	0.01	54.8	
		30				0.11		

https://doi.org/10.1051/e3sconf/202129001009

		40				0.23		
	Industry	0				0.01	60.6	
	Industry	30				0.07		
	Industry	40				0.16		
VX		25		4	Humidified loam (enzymatic method)	0.14		Albert Verweij
					Humic sandy soil (enzymatic method)	0.02		
					Peat containing clay (enzymatic method)	0.04		
					Humic sandy soil (chromatography)	0.06		
					Peat and humic loam containing clay (chromatography)	0.15		
				10	Asphalt (outer)	3.92×10^{-3}		S.Gura
					Asphalt (inner and outer)	2.94×10^{-3}		
HD		20			Sand (air dried)	2.17×10^{-3}		Zhao Guohui[4]
					Chernozem (air dried)	4.42×10^{-3}		
					Loam (air-dried)	8.93×10^{-3}		
					Sandy soil (water content 25%)	8.91×10^{-3}		
					Chernozem (water content 25%)	2.45×10^{-3}		
					Loam (water content 25%)	2.82×10^{-3}		
		15				1.5	61.6	Xue Jishen[8]
		20				2.6		
		25				3.6		
L		25			Sand (air relative humidity 20%)	0.44		Chen Haiping[10]
					Sand (air relative humidity 80%)	1.23		
					Sandy soil (water content 9-68%)	2.25		

3.3 Heterogeneous hydrolysis of poison

The degradation of toxic agents in soil mainly occurs in heterogeneous hydrolysis of toxic agents, and the hydrolysis process is closely related to soil temperature, soil water content, soil structural composition and dispersion degree of venom drops. The hydrolysis reaction rate of colloidal GD accelerates with the increase of temperature, which accords with arrhenius formula relationship, and the activation energy is about 50 kJ/mol However, when the mass and viscosity of GD droplets change, the hydrolysis reaction rate basically changes; The hydrolysis rate of industrial GD is lower than that of colloidal GD because colloidal GD has stronger hygroscopicity. In loam and chernozem, the hydrolysis reaction of colloidal GD has the same apparent activation energy. The hydrolysis rate constant of HD in sandy soil, chernozem and loam is in the order of 10^{-3}, while the activation energy of heterogeneous hydrolysis in water is about 61.6 kJ/mol. The hydrolysate of Lewis gas is oxidized vinyl chloride arsine, which is easy to hydrolyze in pure water, air and soil. At the same time, it can be found that the hydrolysis is faster in soil with high water content.

4 Conclusions

For the eight different types of poisons discussed in this paper, the hydrolysis rate constants of homogeneous and heterogeneous phase can be calculated by first-order kinetic reaction equation, and the relationship between hydrolysis rate constants and temperature can be obtained according to arrhenius formula. Poison concentration, system pH, environmental temperature and soil structural composition will all affect the hydrolysis rate of poison. A full understanding of the hydrolysis behavior of poison is conducive to the development and use of disinfectant.

References

1. George O B. Handbook of Chemical and Biological Warfare Agent Decontamination.2012
2. Zhu B C. Chemical Reaction Engineering [M]. Beijing: Chemical Industry Press. 1991
3. Cheng Z X. Nuclear, biochemical decontamination agent and its application [M]. Beijing: Tsinghua University Press. 2015
4. J. Epstein, D. H. Rosenblatt. Summary Report on a Data Base for Predicting Consequences of Chemical

https://doi.org/10.1051/e3sconf/202129001009

Disposal Operations, EASP1200-12,1973.
5. M. M. Demek, G.T. Davis. Behavior of Chemical Agents in Seawater. Edgewood Arsenal, MD,1970.
6. T.V Healy. Kinetics of the Hydrolysis Pinacolyl Methane Fluorophonate (GD). Chemical Defence Research Establishment, Porton Down, UK, 1948.
7. J. Epstein, D. H. Rosenblatt. Summary Report on a Data Base for Predicting Consequences of Chemical Disposal Operations, Edgewood Arsenal, MD, 1970.
8. Chen H P, Wang X Y. Effect and Destruction of Chemical Weapons [M]. Beijing: Weapons Industry Press, 2002.

Synthesis and Structure of the Copper Complex $[Cu_2L_2^{ox}(SO_4)(H_2O)_4](H_2O)_4$

Zhaodong Wang* [1]

[1]College of Chemistry and Environmental Engineering, Chongqing University of Arts And Sciences, Yongchuan 402160, China.
Chongqing Key Laboratory of Environmental Materials & Remediation Technologies, Chongqing University of Arts and Sciences, Yongchuan 402160, China.

Abstract. The reaction of N-(pyridine-2-ylmethyl)pyrazine-2-carboxamide (HL) and copper(II) sulfate leads to the formation of a new copper complex, $[Cu_2L_2^{ox}(SO_4)(H_2O)_4](H_2O)_4$(1), in which the deprotonated amide L^- ligand was oxidized to a imide L^{ox-} exposed to the air in the presence of the Cu ion. The X-ray crystal analysis reveals that it consists of two mononuclear copper unit with different coordination modes, multiple O-H...O hydrogen bonding between water molecules or water and sulfate O atom to form 3D networks. CCDC: 2088935,**1**.

1 Introduction

Pyrazine-based amide ligands have played very important role in the copper coordination chemistry[1-5]. These ligands exhibit different coordination modes depending on a deprotonated base used or not in the synthesis of the title complexes. The neutral form of the ligands coordinate in either a monodentate fashion via the pyrazyl N donor or in a bridging mode through the N and carbonyl O atom. When a base was used to deprotonate the NH group, which renders the ligand as an anionic moiety to coordiante with the metal ions in a bidentate fashion through the pyrazyl and amidate N donors. Bis(terdentate)pyrazine based diamide ligands and their transition metal chemistry have been well studied by Dr. Brooker's and Dr. Stoeckli Evans's group[6]. As a homolog of the aforementioned diamide ligands, an alkylpyridine armed pyrazine-pyridine amide ligand HL^{pz} was prepared and its transition mental chemistry were also investigated by several different groups. Dr. Hubin firstly reported a trinuclear cobalt complex[7]. Khavasi's group also reported a novel binuclear manganese(II) complex and its catalytic property[8]. Later, they reported two temperature dependent complexes of 1D and 2D $HgCl_2$ coordination polymer with HL^{pz}[9]. Recently, my previous group reported the synthesis, characterization and structures of five copper complexes with HL and its homolog ligand[10]. As a complimentary work to my previous research, here, we report the synthesis, crystal structure of a new copper complex based on
the bispyrazyl amide ligand, HL, which was reported by Dr. Brooker's group. A a cobalt complex with abrupt, complete, hysterestic spin crossover properties by thermal, pressure, and electrochemical stimuli was well investigated[11]. The most interesting part of the work is that the HL ligand was oxidized to an imide due to the expose of the air during the reaction in the presence of the copper ions, which is consistent to the reported results[[12]. A mononuclear Cu complex with pyrazine-pyrazine imide ligand from the oxidation of an amide ligand has been reported by my group[13].Here, we synthesized and characterized a new mononuclear copper complex based on the same ligand.

2 Experimental

2.1 General procedures

All the chemicals commercially available were used as supplied without further purification. The HL was synthesized according to previously reported methods.[10] Infrared spectra were recorded with a Bruker VERTEX-70 spectrometer as KBr pellets in the frequency range 4000 to 400cm^{-1}. Elemental analyses were performed with a Perkin-Elmer 2400 Series II analyzer.

2.2 Synthesis of $[Cu_2L_2^{ox}(SO_4)(H_2O)_4](H_2O)_4$(1)

A solution of $CuSO_4{\cdot}5H_2O$ (25.8mg,0.103mmol) in methanol was added to a solution of HL(22.3mg,0.103mmol) in CH_3OH, affording a dark green solution. Keep stirring for 6h, after the filtration, the blue filtrate was obtained. Slow evaporation of the filtrate afforded green block-shpaed crystals of the title copper complex suitable for X-ray analysis. $C_{20}H_{28}Cu_2N_{10}O_{16}S$:Calcd.C 29.16, H 3.43, N 17.01%; found:C 29.42,H 3.28, N 16.91%. IR(KBr): 3 359(s), 2 407(w), 1 652(s), 1 356(s), 1 134(s), 1 067(s), 835(w), 673(w), 614(w), 467(w)(cm^{-1})

*Corresponding author's e-mail: 201200051@cqwu.edu.cn

2.3 X-ray crystal structure determination

Single crystal X-ray diffraction analyses of complex **1** was carried out with a Bruker SMART APEX IICCD diffractometer equipped with a graphite monochromated Mo-*Kα* radiation ($\lambda = 0.71073$ Å) by using a ω-scan mode. The structure of **1** was solved by direct methods and refined by full-matrix least-squares methods on *F*2 using the program SHEXL 97[14]. Using Olex2[15]. All non-hydrogen atoms were refined anisotropically. The hydrogen atoms were located by geometrically calculations, and their positions and thermal parameters were fixed during the structure refinement. The crystallographic data and experimental details of structural analyses for **1** are summarized in Table 1. Selected bond and angle parameters are listed in Table 2.

3 Results and discussion

3.1 Crystal structure of complex 1

Single crystal X-ray analysis reveals that complex **1** crystallizes in the tetragonal, I 41/a space group. In the title compound, it consists of two monocopper units and four free water molecules as shown in Fig1.The Cu1 is five-coordinated by three N atoms from the oxided imide ligand and two O atoms from water and sulfide anion with a square pyramidal geometry.The coordination mode of Cu2 is six-coordinated by three N atoms from the oxided imide ligand and three O atoms from water molecules with a octahedral geometry. The bond distance of Cu2-O10 is 0.2696nm, which is longer than other Cu-O bond distances assigned as a weaker bond interactions. There are multiple O-H.....O hydrogen bonding between water molecules or water and sulfide O atom to form 3D networks as shown in Fig2. The cabonyl groups in the title complex are available for the coordination with secondary metal ions to construct heteronuclear metal complexes.

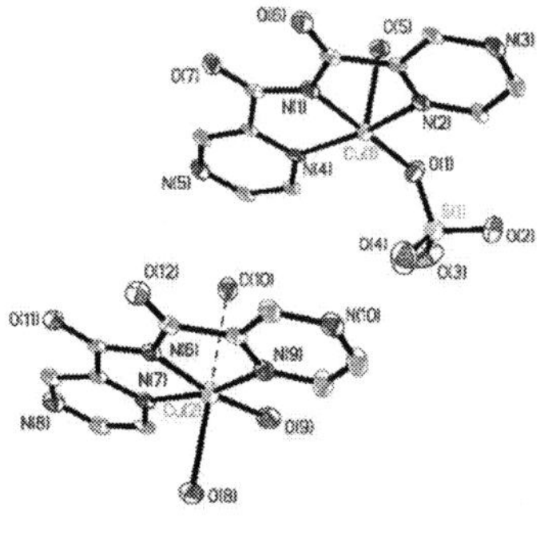

Fig.1 Molecular structure of the complex **1** shown with 30% probability displacement ellipsoids. Hydrogen atoms and the uncoordinated water molecule are deleted for clarity.

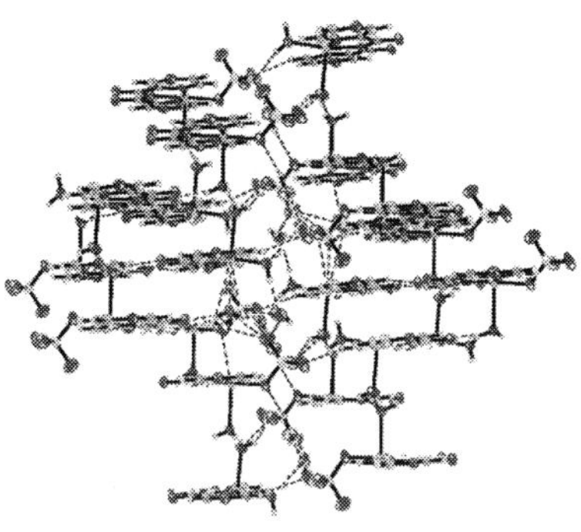

Fig. 2 Intermolecular hydrogen bonding diagram of complex **1**

Table 1 Crystal data and structure refinement for 1

Complex	1
Empirical formula	$C_{20}H_{28}Cu_2N_{10}O_{16}S$
Fourmula weight	586.44
T/K	296(15)
Crystal system	Tetragonal
Space group	*I 41/a*
a /nm	8.271 2(5)
b /nm	13.603 3(8)
c /nm	14.743(9)
α /(°)	110.958 (7)
β / (°)	102.332 (7)
γ / (°)	93.857 (7)
V /nm^3	1494.4(16)
Z	2
D_c/ (g. cm^{-3})	1.831
μ /mm^{-1}	1.587
F(000)	840
θ range/ (°)	2.553 ~ 26.021
Reflections collected/unique	5 814/442
Data / restraints / parameters	5 814/0/442
Final R indices (I >2δ(I))	R_1=0.0721, wR_2=0.1623
R indices (all data)	R_1=0.1308,wR_2= 0.1915

Table 2 Selected bond distances (nm) and angles (°) for 1

Cu1-N1	0.1937(4)	Cu2-N6	0.1944(5)
Cu1-N2	0.2009(5)	Cu2-N7	0.2005(5)
Cu1-N4	0.2006(4)	Cu2-N9	0.2009(5)
Cu1-O1	0.1936(4)	Cu2-O8	0.2374(5)
Cu1-O5	0.2438(4)	Cu2-O9	0.1953(4)

		Cu2-O10	0.2696(5)
O1-Cu1-O5	88.56(17)	O8-Cu2-O10	174.81(15)
O1-Cu1-N2	99.17(18)	O9-Cu2-O8	93.14(19)
O1-Cu1-N1	178.82(19)	O9-Cu2-O10	84.03(17)
O1-Cu1-N4	96.81(18)	O9-Cu2-N7	98.1(2)
N2-Cu1-O5	91.31(17)	O9-Cu2-N9	97.4(2)
N1-Cu1-O5	90.76(17)	N7-Cu2-O8	88.95(18)
N1-Cu1-N2	81.81(18)	N7-Cu2-O10	87.15(17)
N1-Cu1-N4	82.29(18)	N7-Cu2-N9	163.9(2)
N4-Cu1-O5	94.95(17)	N6-Cu2-O8	96.69(18)
N4-Cu1-N2	162.98(18)	N6-Cu2-O9	170.2(2)
N6-Cu2-O10	86.14(17)	N6-Cu2-N7	81.78(19)
N6-Cu2-N9	82.19(19)	N9-Cu2-O8	94.47(18)
N9-Cu2-O10	90.21(17)		

4 Conclusions

In summary, we successfully synthesized a new copper(II) complex derived from N-(pyridine-2-ylmethyl)pyrazine-2-carboxamide(HL), in which the deprotonated amide was oxidized to imide ligand. The X-ray analysis reveals that the coordination modes and geometries of the copper center change with the choice of the anion in the synthesis of the title complexes.

Acknowledgments

This research work is supported by the Chongqing University of Arts and Sciences major scientific research project Cultivation Fund (P2017CH10); Chongqing Yongchuan District Science and Technology Bureau Natural Science Foundation program funded project (2020nb0224)

References

1. Boonmak, J., Youngme, S., Chaichit,N., van Albada, G.A., Reedijk, J. (2009) Series of Copper(II) Coordination Polymers Containing Aminopyrazine and Different Carboxylato Bridges: Syntheses, Structures and Magnetic Properties. Cryst. Growth Des., 9: 3318-3326.
2. Cati, D.S., Ribas,J., Ribas-Arino, J., Stoeckli-Evans. H. (2004) Self-Assembly of CuII and NiII [2 × 2] Grid Complexes and a Binuclear CuII Complex with a New Semiflexible Substituted Pyrazine Ligand: Multiple Anion Encapsulation and Magnetic Properties Inorg. Chem., 43: 1021-1030.
3. Zhang, X.M., Fang.R.Q. (2005) Hydrothermal Syntheses and Structures of Two Mixed-Valence Copper(I,II) 2-Pyrazinecarboxylate Coordination Polymers. Inorg. Chem., 44: 3955-2959.
4. Nather,C., Wriedt, M., Jess.I. (2003) Dimorphism of a New CuI Coordination Polymer:Synthesis, Crystal Structures and Properties of Catena[CuI(2-Iodopyrazine-N)] and Poly[CuI(μ2-2-Iodopyrazine-N,N')].Inorg. Chem., 42: 2391-2397.
5. Hausmann, J., Jameson, G.B., Brooker. S.(2003). Control of molecular architecture by the degree of deprotonation: self-assembled di- and tetranuclear copper(II) complexes of N,N'-bis(2-pyridylmethyl)pyrazine-2,3-dicarboxamide. 24: Chem. Commun., 2992-2993.
6. (a) Hausmann J, Jameson G B, Brooker S. Control of molecular architecture by the degree of deprotonation : self-assembled di- and tetranuclear copper(II) complexes of N,N-bis(2-pyridylmethyl)pyrazine-2,3-dicarboxamide. (2003) *Chem. Commun.* 2992-2993. (b)Hausmann J, Brooker S.Control of molecular architecture by use of the appropriate ligand isomer: a mononuclear "corner-type" versus a tetranuclear[2x2] grid-type cobalt(III) complex.(2004) *Chem. Commun.*1530-1531. (c)Klingele J, Boas J F, Pilbrow J R, Moubaraki B, Murray K S, Berry K J, Hunter K A, Jameson G B, Boyd P D, Brooker S. A [2x2] nickel(II) grid and a copper(II) square result from differing binding modes of a pyrazine-based diamide ligand. (2007) *Dalton. Trans.* 633-645
7. Cockriel D L, McClain J M, Patel K C, Ullon R, Hubin T J. The design and synthesis of pyrazine amide ligands for the "tiles" approach to molecular weaving with octahedral metal ions (2008) *Inorg. Chem. Commun*,11:1-4.
8. Khavasi H R, Sasan K, Pirouzmand M, Ebrahimi S N. Highly Efficient Isobutyraldehyde-Mediated Exoxidation of Cyclic Alkenes with Dioxygen Catalyzed by a Novel Dimeric Maganese (II) Complex Containing an Easy-to-Prepare Flexible Carboxamide Ligand. (2009) *Inorg.Chem.* 48, 5593-5595.
9. Khavasi H R, Sadegh B M,Temperature-Dependent Supramolecular Motif in Coordination Compounds.(2010) *Inorg. Chem.* 49: 5356-5358.
10. Houser R P, Wang Z D, Powell D R,Hubin T J.Copper(I) and copper(II) complexes with pyrazine-containing pyridylalkylamide ligands N-(pyridin-2-ylmethyl)pyrazine-2-carboxamide and N-(2-pyridin-2-yl)ethyl)pyrazine-2-carboxamide *Journal of Coordination Chemistry,* **2013**,66:4080-4092.
11. Cowan, M. G.; Olguin, J.; Narayanaswamy, S.; Tallon, J. L.; Brooker. S.(2012). Reversible switching of a cobalt complex by thermal, pressure, and electrochemical stimuli: abrupt, complete, hysterestic spin crossover. J. Am. Chem. Soc.134: 2892-2894.
12. Hellyer, R.M.; Larsen, D.S.;Brooker. S. Cobalt and silver complexes of terdentate pyrazine-based amide ligands and assemble of monocobalt building blocks through a silver connector.(2009). Eur. J. Inorg. Chem.,9:1162-1171.

13. Wang Z., Xu Q., Ni H., Liu X. Crystal structure of bis((pyrazin-2-ylmethyl)(pyrazine carbonyl)amido-κ3N,N′,N″)copper(II), C20H16CuN10O2.(2017)Z.Kristallogr.N.Cryst.Struct.232:835-836.
14. Sheldrick G M, *SHELXTL NT*, version 5.1, Program for Solution and Refinement of Crystal Structures, University of Göttingen, Göttingen, Germany, **1997**.
15. Dolomanov O V, Bourhis L J, Gildea R J,Howard J A K, Puschmann H. A complete structure solution, refinement and analysis program. (2009) *J. Appl. Crystallogr.* *42*:339-341.

Effect of Fineness on Mechanical Properties of Cementitious Material Made from Magnesia Slag

Dan Cheng[1], Jianxun Ma[1,a], Yilei Zhang[1], Xuhui Chen[2] and Wenxiao Li[1]

[1] School of Human Settlements and Civil Engineering, Xi'an Jiaotong University, Xian, China

[2] Xi'an company, China Overseas Land & Investment Ltd, Xian, China

Abstract: Magnesium slag is a kind of solid waste discharged by magnesium alloy related industries in the process of producing magnesium products or their alloys. Its composition is similar to Portland cement clinker, so it has great potential to be used as a new cementing material. In this paper, magnesium slag cementitious material specimens with different fineness were made by imitating the actual production environment. The flexural strength and compressive strength of specimens with different curing ages were tested, and the influence of fineness on them was analyzed. The results show that the fineness has a significant effect on the mechanical properties of pure magnesium slag cementitious materials. Its effect on the early strength is greater than that on the later strength. But the flexural strength and compressive strength will not always increase with the increase of fineness. There is a reasonable range of fineness.

1 Introduction

With the development of industrial technology and the increase of energy demand, magnesium alloys have been widely used in automotive industry, aerospace industry and electrical communication industry. It is second only to the application of aluminum alloy. Magnesium slag is a kind of solid waste discharged by magnesium alloy related industries in the process of producing magnesium products or their alloys. Magnesium ore is reduced to coarse magnesium through relevant industrial chemical reaction. The remaining waste is magnesium slag [1]. Magnesium slag has strong hygroscopicity, easy to make soil salinization, resulting in soil hardening. At the same time, the powder magnesium slag particles are easy to mix in the air, causing dust pollution.

China is rich in available magnesium mineral resources, and the magnesium industry is developing rapidly. However, with the rapid development of magnesium industry, how to efficiently and reasonably recycle magnesium slag has become an urgent problem.

Fineness is an index to characterize the thickness of a substance.[2] For magnesium slag, the fineness of magnesium slag is the index of grinding degree or dispersion degree of magnesium slag. The chemical composition of magnesium slag is very similar to Portland cement, so for the cementitious materials made of these two materials, the influence factors on their mechanical properties should be the same[3]. Studies at home and abroad show that the fineness of cement has an important influence on the performance of concrete. Therefore, the effect of fineness on the mechanical properties of magnesium slag cementitious materials can not be ignored.

2 Test

2.1 Test raw materials

Magnesium slag, tap water, glass (grinding aid) and 100% pure magnesium slag were selected as test materials. Considering that the change of test materials and methods will cause the change of test results, the magnesium slag used in the test is the same batch of magnesium slag produced by the same company. The magnesium slag is white powder. Its main chemical composition is SiO_2, CaO and Fe_2O_3. The specific chemical composition is shown in Table 1. The test water was clean tap water. The water cement ratio is 0.275. There were only differences in particle size and age among different groups.

Table 1. Chemical composition of magnesium slag

Component	SiO_2	CaO	MgO	Al_2O_3	Fe_2O_3
Content (%)	21.89	60.07	3.86	3.56	8.14

2.2 Specimen size

The raw materials used in this experiment are magnesium slag, fly ash and water. The properties of specimens are similar to those of cementitious materials. The test piece size, mixing and pouring method refer to *"Test Method for Strength of Cement Mortar (ISO method)" GB/T*

[a] Corresponding author: majx@xjtu.edu.cn

https://doi.org/10.1051/e3sconf/202129001011

17671-1999. The specimen was a prism of 40mm×40mm×160mm.[4]

2.3 Test scheme

Three groups of pure magnesium slag cementitious materials with different fineness were prepared. Its fineness is 100 mesh, 150 mesh and 200 mesh respectively. The specific number of specimens is shown in Table 2. Under natural curing conditions, the specimens were cured for 3, 7, 14 and 28 days, respectively. Observe the curing results and measure the flexural and compressive strength. Record the failure of the specimen and analyze the mechanical test data.

Table 2. Summary of sample number of pure magnesium slag

Specimen number	Fineness /mesh	Curing period /day	Magnesium slag /g	Water /g
1-1-1		7		
1-1-2	100	14		
1-1-3		28		
1-2-1		7		
1-2-2	150	14	1350	371.3
1-2-3		28		
1-3-1		7		
1-3-2	200	14		
1-3-3		28		

2.4 Curing results

In this test, all specimens were cured and formed, and no expansion crack occurred. Through the maintenance results of this test, it can be seen that the magnesium slag produced by a company after magnesium smelting has certain hydration activity. After the activation is stimulated by grinding, the pure magnesium slag cementitious material can be cured under natural conditions. It can be completely demoulded after 2 days of pre curing, and has a certain hardness. The curing of specimens is shown in Figure 1.

Figure 1. Production of pure magnesium slag sample

2.5 mechanical property test

2.5.1 Flexural strength test

(1) Bending test instrument

The bending part of the electro-hydraulic servo synchronous testing machine was used in the test. The loading mode is three-point loading. Hinged bearings are adopted with spacing of 100 mm.

(2) Steps of bending test

When testing, wipe the specimen clean and measure the size. And check the appearance. The specimens that do not conform to the section characteristics are excluded. Then, the specimen is placed on the testing machine and loaded until the specimen breaks. Observe the cracks and record the test data.

(3) Bending test results

The bending test results are calculated according to Formula 1. The results were accurate to 0.1MPa.

$$R_b = \frac{3Pl}{2bh^2} \tag{1}$$

In the formula: R_b- flexural strength / MPa. P- failure load / N. l- bearing spacing / mm. b- section width / mm. h- section height / mm.

The calculation of flexural strength test results can refer to *"Cement Mortar Strength Test Method"*[4]. The average value of the flexural strength of each group of specimens is taken as the flexural strength of the group of specimens. If the flexural strength of one specimen in a group exceeds 10% of the average flexural strength of the group, it shall be excluded as invalid data. If there is still more than 10% of the average value after excluding the invalid data, the group of data is invalid. Repeat the test.

2.5.2 Compressive strength test

(1) Compression test instrument

The compressive part of the same testing machine is selected for the test. The loading rate is 2.4 KN/S.

(2) Compression test procedure

After cleaning the surface of the broken specimen in the bending test, place it on the compression machine. Load until the specimen is destroyed. Observe the section and record the test data.

(3) Compression test results

The compression test results are calculated according to formula 2.[4] The results were accurate to 0.1MPa.

$$f_C = \frac{F_C}{A} \tag{2}$$

In the formula: f_C- compressive strength /MPa. F_C- maximum load at failure /N. A- area of compression part / mm2, and its value is $40mm \times 40mm = 1600mm^2$.

The compressive strength test results can refer to *"Cement Mortar Strength Test Method"*[4]. The specific treatment method is the same as that of flexural strength.

3 Test result

The specimens are pure magnesium slag cementitious materials under natural curing conditions. It is made of pure magnesium slag after grinding treatment, and has strong compressive strength after 28 days, which has a certain application value. In order to further analyze the effect of fineness on the mechanical properties of magnesium slag cementitious materials, the mechanical test records of the above pure magnesium slag specimens were summarized. The test results are shown in Table 2, Figure 2 and Figure 3.[5]

Table 3. Summary of flexural strength and compressive strength of specimens

Specimen number	Curing period / day	Flexural strength / /MPa	Compressive strength / /MPa
1-1-1	7	2.3	14
1-1-2	14	3.1	17.4
1-1-3	28	3.4	17.2
1-2-1	7	3.2	16.4
1-2-2	14	3.5	20.6
1-2-3	28	3.6	21.3
1-3-1	7	2.7	15.7
1-3-2	14	3.3	18.4
1-3-3	28	3.5	19.1

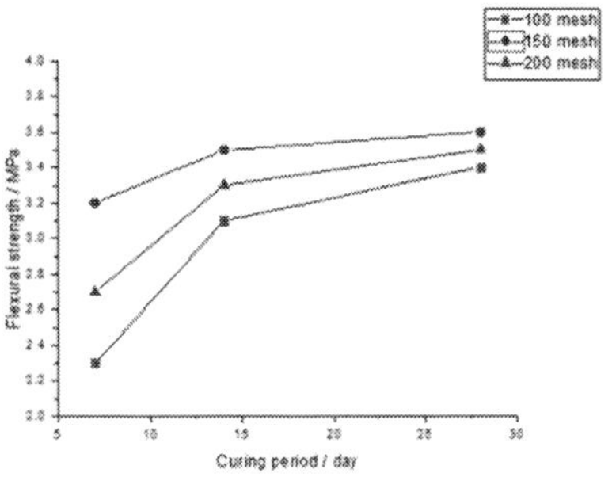

Figure 2. Summary of flexural strength

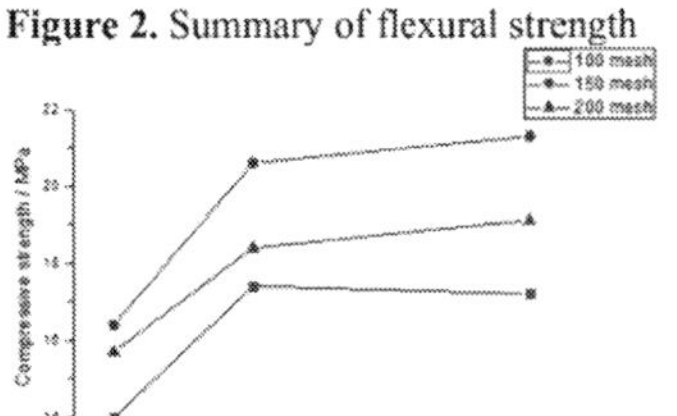

Figure 3. Summary of compressive strength

3.1 Analysis of flexural strength

It can be seen from Figure 2 that the flexural strength of specimens with different fineness groups is obviously different. The fineness has an effect on the flexural strength of pure magnesium slag.

It can be seen from the figure that the flexural strength of three kinds of fineness pure magnesium slag cementitious materials gradually increases with the increase of curing age. The trend of the curve is basically the same. With the increase of curing time, the rising trend slows down gradually. The flexural strength after 28 days is obviously higher than that after 3 days.

According to the three groups of data, among the three fineness of magnesium slag specimens, the strongest flexural strength is 150 mesh group, and the weakest is 100 mesh group. At the same time, it can be seen from the curve that the bending strength of the three kinds of fineness specimens is the biggest difference when they are cured for 7 days. At 28 days, the difference gradually decreased. Take the 150 mesh group with the largest intensity and the 100 mesh group with the smallest intensity as an example. At 7 days, the strength of 150 mesh is 39.1% higher than that of 100 mesh. 12.9% on 14 days. By 28 days, the intensity of 150 mesh was only 0.6% higher. [5]

3.2 Analysis of compressive strength

As shown in Figure 3, the broken line diagram of compressive strength of pure magnesium slag specimen is basically consistent with the bending curve. The compressive strength of 100 mesh, 150 mesh and 200 mesh specimens increased with the increase of curing age. The growth in the early stage is relatively large, and the change in the later stage is relatively stable. In the compressive strength, 150 mesh > 200 mesh > 100 mesh, consistent with the flexural strength.

4 Conclusion

By making pure magnesium slag specimens with different fineness, the influence of fineness on its mechanical properties under natural curing conditions was explored. Based on the analysis of the test results, the following conclusions are drawn.

(1) Under the condition of natural curing, the fineness has a significant effect on the mechanical properties of pure magnesium slag cementitious materials.

(2) There is an optimal fineness when exploring the influence of fineness. Through the test, it can be seen that the flexural strength and compressive strength of the specimen will not always increase with the increase of the fineness of the specimen, and there is a reasonable fineness range. For the magnesium slag selected in this test, the strength of 150 mesh magnesium slag cementitious material is better than that of 100 mesh and 200 mesh under the premise of keeping the water cement ratio unchanged.

(3) The influence of fineness on the early strength of magnesium slag cementitious material is greater than that on the later strength.[5]

https://doi.org/10.1051/e3sconf/202129001011

References

1. Li Shanshan. Research progress of magnesium slag concrete [J]. Sichuan cement, 2018.
2. Celik I B. The effects of particle size distribution and surface area upon cement strength development[J]. Powder Technology, 2009, 188(3): 272-276.
3. Ma Kunlin, he jionghuang, long Guangcheng, et al. Effect of steam curing temperature and its influence on thermal damage of cement-based materials [J]. Materials guide, 2017, 31 (23): 171-176.
4. Tang Zhongning. Test method for strength of cement mortar (ISO method) [J]. CNBM, 1999.
5. Xuhui Chen. The effect of fineness on cementitious material made from magnesia Slag[D]. xian: Xi'an Jiaotong University, 2019.

Polymer Degradation: Category, Mechanism and Development Prospect

Yuanxin He[1, a,*,†], Hongyu Li[1,b,*,†], Xiang Xiao[1,c,*,†] and Xinyu Zhao[1,d,*,†]

[1]School of Queen Mary University of London Engineering School, Northwestern Polytechnical University, 710100 Xi'an City Shaan Xi Province, China

[†]These authors contributed equally to this work.

Abstract. With the increasing demand for polymers, white pollution has become a serious concern all around the world. The admirable degradation methods of them are desirable for overcoming this problem. In the past several decades, numerous researches on polymer degradation have been reported. This review commits to different degradation strategies of polymers and four main degradation protocols firstly, including photodegradation, oxidative degradation, catalytic degradation, and biodegradation, are demonstrated in detail. Secondly, some specific samples are discussed for each kind of degradation. Finally, the outlook and future of polymer degradation are proposed. In particular, the comprehensive comparison of different degradation methods is covered to provide the best choice for dealing with different polymers wastes. These will be beneficial to the development of processing plastic and conversion of polymer wastes.

1 Introduction

With the rapid development of science and technology, polymer materials have been widely used in people's daily life and industrial production [1,2]. The birth of functional polymer materials also promotes the development of industrial production, agricultural production, medical and health fields in the direction of depth and refinement. According to IHS data, the global demand for polycarbonate in 2019 is about 4.4 million tons, and the total production capacity is about 5.6 million tons. Annual capacity is expected to increase by 42 percent over the next several years. It will increase from 5.6 million tons in 2019 to 8 million tons in 2024. Although polymer materials bring great convenience to human life, the hardly degradable polymer materials represented by "white pollution" also cause great biological harm and environmental pollution [3]. Therefore, how to degrade the polymer materials after use has become an important research topic for academic and industrial fields.

At present, polymer materials are mainly disposed of utilizing incineration, heap burial, and degradation. However, due to the toxic gases and greenhouse effect of burning, this method has been strictly regulated [4]. At the same time, with the serious white pollution and biological environment problems caused by heap burial, it is not an effective way of long-term disposal of waste polymer materials [5]. Therefore, the development of degradable polymer materials and the promotion of the development of polymer degradation technology have become important methods to dispose of waste polymer materials.

Many types of research have been conducted in the field of polymer degradation. Recent theoretical developments have shown that there are mainly four methods for polymer degradation, including thermal degradation, photodegradation, catalytic degradation, and biodegradation. Generally, energy from heat and light could be absorbed by the chains in polymer, which cause a series of chain-breaking reactions, leading to thermal cracking and light cracking, respectively [6,7]. The presence of oxygen would promote these two reactions because several types of radicals would be formed by oxidation, which can react with other chains [6,7]. Except for oxidation, another effective method for promoting cracking reaction is to use a catalyst, such as zeolite, which could further improve the reaction activity and produce value-added chemicals [8]. Furthermore, biodegradation is a very promising strategy, which is much more complex. Microorganisms such as bacterial could produce enzymes to convert polymer into low molecular weight fragments by hydrolysis and oxidative reactions, and then "eat" them and excrete small molecules like carbon oxide and water [9].

Understanding the mechanism of different cracking strategies is supposed to help people find better methods to dispose or recycle waste polymeric materials. Based on giving an overarching framework of polymer degradation approaches for beginners or providing systematic collection to support further study in this field, this work will review the mechanism and features of basic polymer cracking ways and make a comparison between them. The paper could provide important theoretical knowledge for readers to learn about or deepen their understanding of the degradation of the polymer.

[*] Corresponding author: [a]yuanxin.he@mail.nwpu.edu.cn [b]hongyu.li@mail.nwpu.edu.cn [c]jp2018303621@qmul.ac.uk [d]zhaoxinyu@mail.nwpu.edu.cn

 https://doi.org/10.1051/e3sconf/202129001012

2. Degradation strategy

2.1 Thermal degradation

Heat is one of the most common types of energy in nature. Thermal energy could promote the movement of the polymer chain, leading to the degradation and decompose of the polymer, as well as the decrease in a series of properties, such as mechanical property. At certain elevated temperatures, polymers' the physical and chemical properties of polymers will be changed spontaneously [6,10]. The chemical bonds absorb enough energy from heat in polymers, resulting in the breaking of these chemical bonds and then producing some free radicals. The formed free radicals are extremely active, and they would react with the molecules around. Hence, more chains can be further broken, and more free radicals are generated. Subsequently, many parameters of polymers would be changed, including molecular weight, molar weight distribution, degree of crystallinity, degree of branching, and so on. As the reaction process, the variation in mass, state, shape even color can be observed [6]. As a result, the working performance, such as mechanical or thermal performance of the polymers, would alter dramatically, being prone to cause the failure of the material [6].

Xu et al. have carefully investigated the thermal degradation mechanism of PVC material, indicating the products, include different hydrocarbons, are produced with HCl in the gas phase [11]. The team of Cleide Borsoi took Flynn-Wall-Ozawa method to explore the thermal degradation of cellulose nanofibers. The results showed that the degradation behaviors of cellulose are affected not only by morphology, crystallinity and degree of polymerization but also mechanical defibrillation and drying by lyophilization [12]. Wang et al. have verified that bonging temperature, dwell time, and thermal conductivity of mold material influence the cracking rate of nonwovens(PP/PET fiber) using FTIR and DSC methods [13].

2.1.1 Thermal degradation without oxygen

Even though it is almost impossible to ignore the presence and function of oxygen in terms of the chain-breaking reaction, the study in this field is still vital since plenty of heat treatments on polymeric materials are operated under non-oxygen condition, and the degree of thermal degradation determines the quality of the products directly [6,14,15]. There are mainly three types of methods for thermal degradation [16].

(1) Depolymerization

Depolymerization is also called unzippering reaction. The cracking starts at the end or the weak points of the main chains, and then the connecting monomers, which are the only products theoretically, break away from the backbones one by one [16]. For instance, the thermal cracking of polymethyl methacrylate (PMMA), as shown in Fig. 1a.

(2) Random chain-breaking

Random chain-breaking reaction or random scission reaction could start from any points in polymer chains. The main product is an oligomer, which refers to low molecular weight polymers [16]. For example, the thermal decomposes of polyethylene (PE).

In reality, for most polymers, the 2 methods of cracking above always happen simultaneously (Fig. 1b).

(3) Side group elimination

Side group elimination is sometimes called small molecule elimination. The reactions always begin as the elimination of side groups, producing small molecules instead of monomers. The formed small molecules are based on the groups attached to the polymer chains [16]. For example, hydrogen chloride (HCl) is formed at the early stage in the thermal cracking of polyvinyl chloride (Fig. 1c).

With the process of elimination, more and more weak points on the backbone would be exposed, and then the depolymerization and random chain-breaking reactions of polymers spread from these exposed weak points [16].

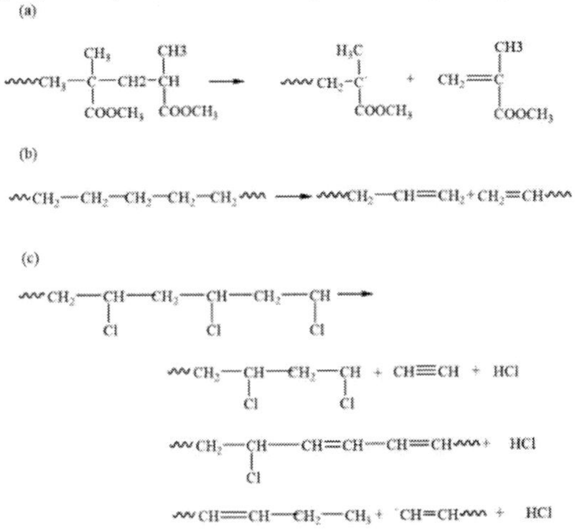

Fig. 1. Thermal cracking mechanism of (a) PMMA, (b) PE, and (c) PVC.

2.1.2 Thermal-oxidative degradation

In the processing, storage, and application of polymeric materials, contact with oxygen can hardly be avoided. In normal thermal degradation, some radicals formed from cracking could react and recover to chains again if the temperature is not high enough [8]. However, oxygen could react with these radicals to prevent them from forming polymer chains again, so the reaction rate can be increased with the function of oxygen. It is proved that high-density polyethylene (HDPE) cracking is faster at a considerably lower temperature in the air than that in a nitrogen atmosphere [16].

The mechanism of thermal-oxidative degradation is a sort of typical chain-step radical process, with initiation, propagation, and termination. The main feature is the"auto-oxidative" reaction [14].

(1) Mechanism-initiation (Fig. 2a)

With the function of heat and oxygen, the basic radical R· forms.

(2) Mechanism-propagation (Fig. 2b)

R· can react with oxygen then other chains, forming a peroxide, ROOH, which could decompose and produce new radicals to react with more other chains to obtain more basic radicals.

(3) Mechanism-termination (Fig. 2c)

Different radicals may react with each other, leading to coupling termination.

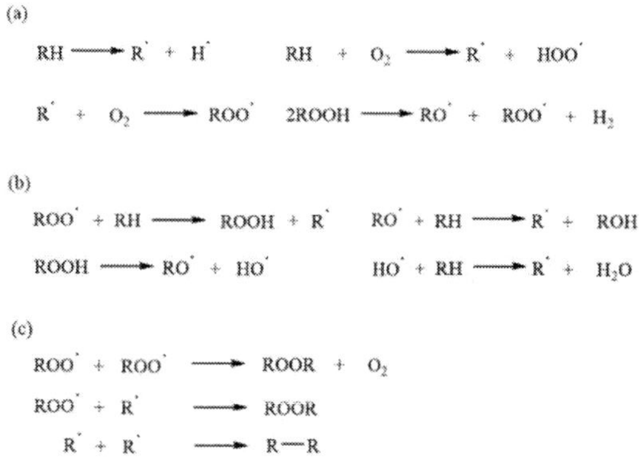

Fig. 2. Reaction mechanism of thermal-oxidative degradation [14].

2.2 Photodegradation

Photodegradation refers to the phenomenon of chain fracture caused by the existence of photosensitive groups in the polymer chains such as carboxyl, carbonyl, and unsaturated bonds due to the action of light and the presence of oxygen [7]. It is an important kind of method for polymer degradation in our daily life. Light is usually used as the source to provide energy for electrons in the polymer, and the polymer will be excited by light to promote the following chain-breaking reaction. Subsequently, the molecular weight of the polymer will decrease significantly. In the reaction process of polymer chain-breaking step by step, various properties of the polymer, such as mechanical properties and chemical properties, will also gradually change, which could achieve the degradation of different polymers to protect the environment.

2.2.1 Photodegradation without oxygen

Photodegradation without oxygen is the basic reaction of photodegradation. The photodegradation of polymers could be divided into two classes, including direct degradation and indirect degradation. For the direct degradation, the polymer could absorb photons directly, and polymer molecules will be directly transformed from the ground state to the excited state [16]. During this process, polymer molecular chains could be broken, and molecular weight gradually decreases. While for an indirect degradation process, the energy of photons is absorbed by a substance in the environment, firstly, which could promote the substance to an excited state. And then, the polymer molecules are induced to an excited state, which will result in the degradation of polymers by breaking down the chains.

There are two main types of chemical reactions during the degradation.

(1) Depolymerization

The photons with enough energy (especially for UV light) can break the chemical bonds, which can induce the generation of free radicals [17]. As shown in Fig.3a, polymer chains with free radicals are active. Some of them would release monomers from the end with free radicals under the attack of photons.

(2) Random chain-breaking

There are no free radicals for some polymers in the end, or chemical bonds in the polymer chains have similar energy. In this case, the photon with enough energy will randomly break the bonds of them, introducing free radicals and undergoing further reaction (Fig. 3b).

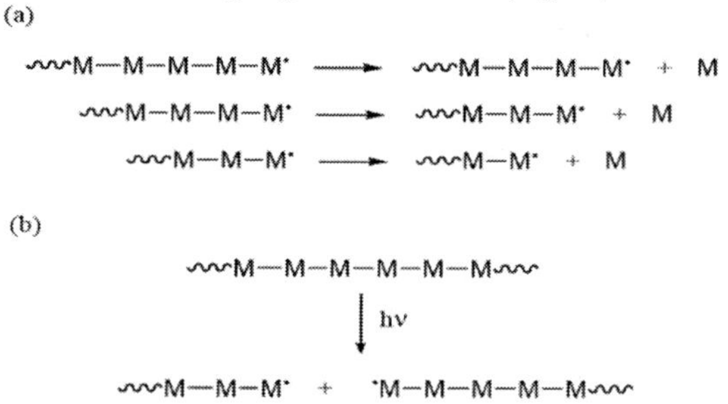

Fig. 3. The reaction mechanism of Depolymerization and Random chain-breaking [6,17].

2.2.2 Photo-oxidative degradation

The effect from oxygen is always hard to avoid [18]. Therefore, it is vital to focus on photo-oxidative degradation. Generally, photo-oxidative degradation goes through three steps as following (Fig. 4).

(1) Mechanism-initiation. Due to the function of light and oxygen, the basic radical R· is generated, which will be used in the next propagation process [10].

(2) Mechanism-propagation. R· react with oxygen and form a peroxide ROOH, which could decompose and produce new radicals. These new radicals decomposed by ROOH can go on next reactions. After that, there will be more new radicals produced and reacted with each other [10].

(3) Mechanism-termination. Different radicals may react with each other, leading to coupling termination [14].

$$RH \xrightarrow{h\nu} RH^{*} \longrightarrow R^{\cdot} + H^{\cdot}$$

$$R^{\cdot} + O_2 \longrightarrow ROO^{\cdot}$$

$$ROO^{\cdot} + RH \longrightarrow ROOH + R^{\cdot}$$

$$ROOH \longrightarrow RO^{\cdot} + HO^{\cdot}$$

$$ROOH \longrightarrow R^{\cdot} + {}^{\cdot}OOH$$

$$RO^{\cdot} + RH \longrightarrow R^{\cdot} + ROH$$

$$HO^{\cdot} + RH \longrightarrow R^{\cdot} + H_2O$$

$$ROO^{\cdot} + ROO^{\cdot} \longrightarrow ROOR + O_2$$

$$ROO^{\cdot} + R^{\cdot} \longrightarrow ROOR$$

$$R^{\cdot} + R^{\cdot} \longrightarrow R{-}R$$

Fig. 4. Reaction mechanism of photo-oxidative degradation.

2.2.3 Source of photo-oxidative degradation

Furthermore, some special structures in the polymer may affect reactions.

(1) ROOH — the very beginning of photo-oxidative degradation.

ROOH can be generated from photo-oxidative degradation. The radical ROOH can accumulate and decompose [10]. After the decomposition of ROOH, many radicals can be obtained (Fig. 5).

(2) Metal ions — sensitizer for the photo-oxidative degradation of polyolefin.

Metal ions, again, could work as catalysts, which are used to increase or decrease the rate of reactions. The photo-oxidative degradation of PP is related to the photosensitizer effect of the residual catalyst, which can take part in the reaction in the form of radical ions generated via redox of the excited state [10,19].

(3) Carbonyl groups —Norrish reaction

Moreover, if there are carbonyl groups in the polymer, a series of unique reaction called Norrish reaction may occur, including an α chain-breaking, α-β chain-breaking, and β hydrogen transfer reaction [8]. As shown in Fig. 6, these reactions could provide more position for the initiation for photo-degradation reaction.

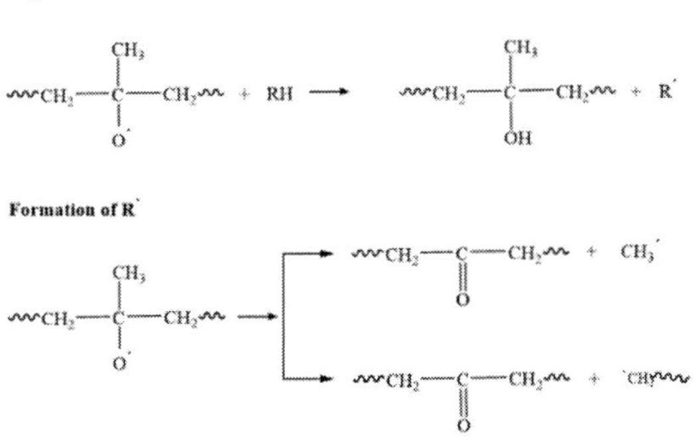

Fig. 5. The production and decomposition of ROOH in photo-degradation reactions [10].

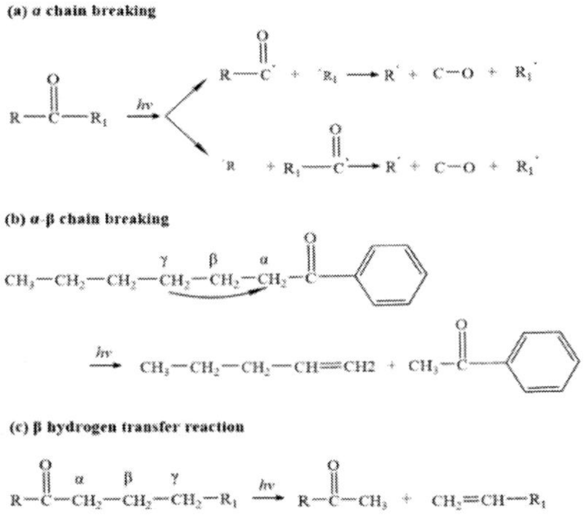

Fig. 6. Three different Norrish reactions for photo-degradation reaction [8].

2.3 Catalytic degradation

2.3.1 Oxidative degradation

With the development of population and technology, the demand for light alkenes such as propene and ethene grows with each passing day. The main method of producing light alkenes is steam cracking of plastic polymers, which wastes a lot of energy and emits much CO_2. Compared with steam cracking, oxidative catalytic cracking shows a promising future, which not only decreases the heat temperature but also produces some value-added chemicals.

The existence of oxygen will change the reaction thermodynamics, inhibit heavy productions, and accelerate reaction during oxidative alkene cracking. Besides, cracking is initiated by the single bond between two carbon or the formation of radicals. Oxygen presents in the initiation reaction during the oxidative cracking of the gas phase. It could oxidize part of the alkene and generate water and CO. Besides, it can also produce heat for the next stage [8].

Using catalyst during oxidative cracking needs lower temperature relative to not using a catalyst, which also adjusts the selectivity of forming light alkenes. Thus, a catalyst must promote the production rate. For example, with catalyst Li/MgO, the cracking rate of hexane could increase from 1wt% to 17wt% at the same conditions [8].

Besides, the mechanism of oxidative catalytic cracking could be divided into two steps. And the initiation mechanism largely depends on the active positions of the catalyst. The formation of alkyl radicals and extraction of hydrogen from alkane initiate the surface. This will be a β-scission reaction, and cracking productions will be produced. For example, the catalysts could be gold-containing, oxychlorides, and alkali. Under the same conditions, the gold-supported catalyst has higher propane conversion and selectivity to olefins than the platinum-supported catalyst [20]. The $Ce_{0.5}Zr_{0.5}O_2$ has shown a slight activity. However, when the catalyst was deposited with 0.2wt% gold, an increase in catalytic activity was observed. A further improvement in catalytic activity was achieved with gold-supported sulphated $Ce_{0.5}Zr_{0.5}O_2$ [20].

2.3.2 Zeolite degradation

Zeolite is a common acid solid catalysis, which is widely used for polymer cracking to get olefin. Light alkene has developed into the footstone of the petrochemical industry. Thus, there is a huge requirement for ethylene and propylene. The main method for producing them is the cracking of petroleum. The depletion of fossil resources has put tremendous stress on the production that relies on oil cracking. However, less than 9% of plastics have been recycled [21]. Under the circumstances, improving the recovery is necessary. Converting plastics selectively to light lubricants and fuels provides a new but reliable solution to plastics recycling. By zeolite cracking, olefins could be produced from polymer materials.

The architecture and morphology of zeolite influence the catalysis process. Lower temperature and less energy are needed with zeolite cracking. By employing zeolite catalyst, the reaction activity could be further improved. The activation energy of polymer cracking could be decreased, which has been carefully studied by Zhao and co-authors to crack LDPE with different morphology MFI zeolites [22]. Besides, the side branches MTW zeolites have also been used to crack LDPE, which could mainly produce the gasoline, compared with the conventional MTW zeolite [23,24]. Because of the selectivity of the microporous network, which makes it is possible for the selectivity of the product. For example, HY gives monomers a selectivity range, whose product has less diesel and aliphatic hydrocarbons, and naphthene [24].

2.4 Biodegradation

Biodegradable polymer refers to a kind of polymer, which can be completely degraded with the aid of microorganisms like bacteria or can be dissolved into water and degraded by CH_4(methane), CO_2, and metal-salt. Until now, the most common materials used are polylactic acid (PLA), polyvinyl alcohol (PVA) and polymethyl succinate (PBS) [25]. With the development of biodegradation technology, environmental problems can be gradually solved largely. At the same time, these wastes can be converted into different useful materials or chemicals, which could further promote the development of economics and life.

In recent years, the biodegradable polymer has been widely used as environmental-friendly protection materials (Mulch for agriculture, construction), packaging materials (Packaging film, garbage bags, food packaging bags), biologically functional materials, medical materials (bandages, surgical sutures), and many other fields. It also shows unique advantages compared with single-use plastic products.

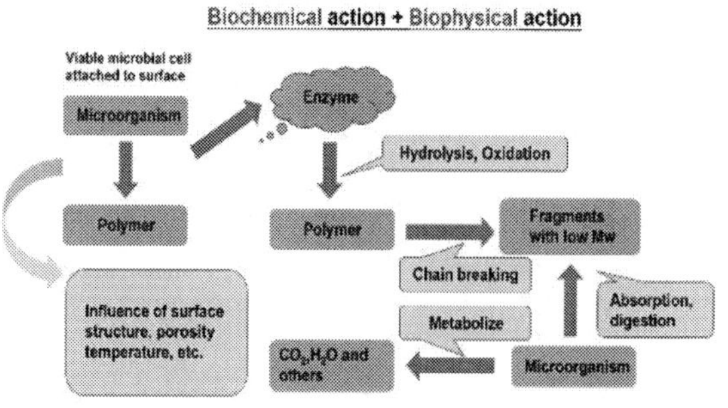

Fig. 7. Schematic diagram of biodegradation [9].

2.4.1 Mechanism of biodegradation

Biodegradation is a very complex process. It is a combination of biochemical and biophysical action, and still now, most of the mechanisms are not very well studied.

The process of biodegradation can be mainly divided into three stages (Fig.7). First, bacteria will attach to the surface of the polymer products. The surface structure and temperature will influence this process. For example, both superhydrophobic and super hydrophilic polymer surfaces increase bacterial absorbability. Secondly, the polymer chain will be broken by the enzyme secreted by bacteria with hydrolysis and oxidation reaction [9,26]. Then, Fragments with low MW (<500) will be absorbed and digested by bacteria. It goes through a series of metabolizing processes and eventually gets rid of it in carbon dioxide, water, and other substances [9].

2.4.2 Classification of biodegradable polymers

There are two ways to identify the types of biodegradable polymers. One is based on the degradation mechanisms and damage form of the polymer. The other one is based on the mode of production [27].

(1)According to the degradation mechanisms and damage form, biodegradable polymers are divided into completely biodegradable polymers and bio-destructive polymers, which could only be degrading to random fragments by microbial.

(2)According to the mode of production, there are three types of biodegradable polymers. It includes macromolecule production by microorganisms (PHBV, PHB), synthetic polymer material (PLA, PCL), and natural polymer materials (cellulose, starch, lignin, chitin). One of the more promising polymers is additive biodegradable plastics. This kind of plastic is made by adding biodegradable ingredients as additives to raw materials (PE, PP, PS) at a relatively low price [27].

2.4.3 Some specific biodegradable polymer materials

Polylactic acid

Polylactic acid (PLA) is one of the most biodegradable thermoplastic materials. It is made from starch, honey, and other raw materials through fermentation. The polymer was obtained by further chemical synthesis [28].

PLA degrades more easily than other plastics. Under some environmental conditions, PLA can hydrolyse after contacting water. The whole process does not require sunlight, and it is not sensitive to pH [28,29]. This is the big advantage of PLA as a biodegradable polymer material. At present, there are numbers of relevant PLA research being carried out, and some technologies have already entered the promotion stage. Some of the applications include extrusion sheets, injection moulding, blister products, spinning, bidirectional stretch film, and other processes. PLA is also widely used in medical treatment because of its good antimicrobial properties.

Polyhydroxy fatty acid esters

Polyhydroxy fatty acid esters (PHA) is a kind of biological polyester particles synthesized by microorganism. PHA has good biocompatibility, but its application has been greatly limited due to its poor mechanical properties and high cost [30].

PHB is another kind of PHA, which is widely used. It has a simple structure and can be completely decomposed within 12 months, only releasing pollution-free water and carbon dioxide. Therefore, PHB has become one of the preferred materials for green packaging.

3. Conclusion

In this paper, various problems and countermeasures encountered in the development of polymer degradation in recent years are summarized. Then, introducing the most important degradation types in the four fields of polymer degradation in detail makes a certain interpretation for the basic mechanisms and prospects of polymer degradation. In summary, the paper shows thermal degradation, photodegradation, catalytic degradation, and biodegradation. Thermal degradation (including thermal-oxidative degradation) is a common type of degradation. This paper expounds on the main mechanism and prospect of thermal degradation. It is difficult to control temperature and other conditions, such as pressure, required for degradation in real life. In the process of polymer thermal degradation, other substances such as additives can also participate in the response, causing the discharge of environmental pollutants. This is the problem thermal degradation is going to be solved. Photodegradation is relatively inexpensive, and the light sources needed for degradation are readily available. It is more environmental-friendly and has a very broad market. The research direction of photodegradation provides corresponding higher efficiency photodegradation reactions for different polymers. For example, different photosensitizer catalysts can be added to different polymers and improve the polymer structure to make it more photodegradable. Compared with steam cracking, catalytic oxidation degradation reduces the temperature required for the reaction, produces chemicals with added value, and meets the increasing demand for light olefin such as propylene and ethylene, which has a good application prospect. Biodegradation has shown its incomparable advantages in many fields, and it is a promising direction for the future development of polymer materials. But at the present stage, there still are some problems that need to be solved, such as high material cost, limited material uses, and lack of production technology. New degradation technology will reduce environmental pollution and make a breakthrough in navigation, aerospace, agriculture, medical instruments, biological applications, clean energy, and many other fields.

These summaries and explorations can provide a very good reference for people to understand the mechanism and development prospects of polymer degradation. At the same time, it can also provide a specific reference for the degradation of different types of polymer.

References

1. Shahinpoor, M.; Kim, K. J. Smart Mater. Struct. 2005, 14 (1), 197.
2. Alrifaiy, A.; Lindahl, O. A.; Ramser, K. Polymers 2012, 4 (3), 1349–1398.
3. Nicolas, J.; Mura, S.; Brambilla, D.; Mackiewicz, N.; Couvreur, P. Chem. Soc. Rev. 2013, 42 (3), 1147–1235.
4. E. Chiellini, R. Solaro Adv Mater, 1996, 8, 305-313
5. Y.M. Zhang, J.M. Jiang, Y.M. Chen Polymer, 1999, 40 (22), 6189-6198
6. Ray, Sudip. Handbook of Environmental Degradation of Materials || Thermal Degradation of Polymer and Polymer Composites. 2018, 185–206.
7. Kai Zhang, Amir Hossein Hamidian, Aleksandra Tubić, YuZhang, James K.H. Fang, Chenxi Wu, Paul K.S. Lame. Envriomental Pollution, 2021,274, 116554.
8. Boyadjian, C. , & Lefferts, L. . (2018). European Journal of Inorganic Chemistry.
9. A Moses, A. E. , EB Bassey, & Udoh, S. B. Risk Factors Associated With Oral Cavity Colonization By Candida Albicans, Among Immunosuppressed Persons In Uyo, South South Nigeria. Journal of Medical Laboratory Science, 2012, 21
10. Cuadri, A. A. , & JE Martín-Alfonso. Polymer Degradation and Stability, 2017, 141, 11-18.
11. Zhuo, X. A. , Ssk, A. , Sz, A. , Bz, A. , & MD B. Polymer Degradation and Stability, 176.
12. Borsoi, C. , Zimmernnam, M. , Zattera, A. J. , Santana, R. , & Ferreira, C. A. . (2016).. Journal of Thermal Analysis and Calorimetry.
13. Wang, X. Y. , & Gong, R. H. Journal of Applied Polymer Science, 2007, 104(1), 391-397.

https://doi.org/10.1051/e3sconf/202129001012

14. Rabek, Jan F. [Comprehensive Chemical Kinetics] Degradation of Polymers Volume 14 || Chapter 4 Oxidative Degradation of Polymers. 1975, 425–538.
15. Mcneill, I. C. Polymer Engineering and Science, 1980, 20(10), 668-674.
16. Pielichowski, K. , & Njuguna, J. Thermal Degradation Of Polymeric Materials. 2008.
17. S.M.A. Abel, K. Werner, Z. Christiane, H. Stefan, C. R M .Global Change Biol., 2018, 24 (4), 1405-1416.
18. M Galloway, T. S. , & Lewis, C. N. . (2016). Proceedings of the National Academy of Sciences of the United States of America, 2331.
19. R. Stephen Davidson, Dean Goodwin, Ph. Fornier de Violet. Tetrahedron Letters 1981, 22(26), 2485-2486
20. Narasimharao, K. , & Ali, T. T.. Catalysis Letters, 2013, 143(10), 1074-1084.
21. Liu, S. , Kots, P. A. , Vance, B. C. , Danielson, A. , & Vlachos, D. G. .Science Advances, 2021, 7(17), eabf8283.
22. Blay, V. , Louis, B. , Miravalles, R. , Yokoi, T. , Peccatiello, K. A. , & Clough, M. ACS Catal. 2017, 7, 6542-6566..
23. Zhao, Y. , Ye, Z. , Zhang, H. , Zhang, Y. , & Tang, Y. Industrial & Engineering Chemistry Research, 2019, 58(29), 13174-13181.
24. Zhao, Y. , Ye, Z. , Wang, L. , Zhang, H. , F Xue, & Xie, S. Crystal Growth & Design, 2018, 18, 2, 1101–1108.
25. Bassey, E. , Mo, B. C. , Aluyi, H. , & Anukam, K. . Journal of Medical Laboratory Science, 2004,13(1).
26. Mathew C. Celina, Jeffrey S. Wiggins, Norman C. Billingham., Polymer Degradation and Performance, ISBN: 978-0-8412-6978-1, Oxford University Press, 2009,67-75.
27. Taghavi Navid, Udugama Isuru Abeykoon, Zhuang Wei-Qin et al. Biotechnology Advances, 2021, 21-35.
28. Bordes, P. , Pollet, E. , & Averous, L.. PROGRESS IN POLYMER SCIENCE. 2009.
29. HAMAD K,KASEEM M,AYYOOB M, Progress in Polymer Science,2018,85, 85-127.
30. Chuah JA, Tomizawa S. Yamada M, Appl Environ Microbiol,2013,79(12), 813-3821.

Research on textile materials for X-ray shielding

Dong Liang[1 st, 1], Fu Shen[1 st, 1], Zizhen Bao[1], Yuchen Liu[1], Honghui Li*[1,a]

[1]China Institute for Radiation Protection, P.O. Box 120, Taiyuan 030006 , Shanxi, P.R. China

Abstract: X-ray radioactive rays are widely used with the continuous development of radioactive medicine and nuclear technology applications, as well as lead shielding material pollutions new no lead shielding material was needed. In this paper, the main properties of metal tungsten and bismuth as X-ray shielding materials were studied for the protection people avoid the 150 kV X-rays by the Monte Carlo method is used to study. According to simulation with 2 kg/m^2, results show that performance of single metal material tungsten iron is superior to that of bismuth material. Tungsten-bismuth better than bismuth-tungsten with the case of equal-quality double-layer metal. The protection performance is better when the metal-mixed tungsten-bismuth ratio is 0.5: 0.5 or the tungsten ratio is large. The research provides effective support for the development of textile radiation protection materials.

1. Introduction

In recent years, with the continuous development of national defense scientific research, radioactive medicine, and nuclear technology applications, various radioactive rays have been widely used. X-ray is a type of ionizing radiation, has been used in material X-ray inspection, human X-ray fluoroscopy, and X-ray analysis. Achieve successful application. While using radiation to make profits, the damage to the human body and the destruction of the environment by the rays are gradually recognized by people. Long-term exposure of staff to X-rays can cause damage to the gonads, breast, hematopoietic bone marrow, etc. Excessive doses can even cause cancer, posing a serious threat to the human body. Therefore, the research on anti-X-ray materials has become an important subject. However, traditional X-ray shielding materials are mainly lead-containing vulcanized rubber and metal film materials. It has the disadvantages of high lead toxicity, environmental pollution, weak X-ray absorption area, strong scattering of low-energy X-rays, and bulky materials. Rubber materials have poor air permeability, require vulcanization processing, are not easy to prepare thin and light products, vulcanized rubber cannot be recycled, and other shortcomings. Lead-free nuclear protection materials have become an inevitable trend.

The US RST company mainly uses metal tantalum wire composite fabrics to achieve lightweight and high-efficiency protection against X and gamma rays, but it faces problems such as the high cost of metal tantalum and complex tantalum wire processing technology. The American INFAN company and the German MI company add nano-heavy metal alloy powder to the fiber to prepare functional clothing for radiation protection to achieve the same protective effect. The weight is only about 30% of the traditional lead clothing, but the alloy still contains a certain proportion of metallic lead. Nano-alloy powders are complex to process and difficult to disperse in polymers, and fiber-forming and processing are difficult. The Japanese company adopts the skin-core structure melt spunbond spinning technology, taking into account the high addition of powder and the mechanical properties of non-woven materials, mainly by increasing the thickness of the fabric to improve the protective effect.

Domestic China Institute of Radiation Protection, Hubei Huaqiang Technology Co., Ltd., Beijing University of Chemical Technology, and other institutions have researched lead-free radiation protection materials[1,2,3]. The research focus of lead-free radiation protection materials is mainly on glass, Rubber, board, and other materials are added with heavy metal powder, which is mainly used for the protection and isolation of the radiation source body. And there is less development of individual radiation protection fibers and clothing fabrics. There is a lack of in-depth research on lightweight textile materials for lead-free radiation protection. Therefore, it is particularly necessary to conduct systematic research on the selection and modification of lead-free radiation protection materials, fiber molding technology, textile material structure design, and evaluation methods.

This article lacks in-depth research on lightweight textile materials for lead-free radiation protection. It is particularly necessary to conduct systematic research on the selection and modification of lead-free radiation protection materials, fiber molding technology, textile material structure design, and evaluation methods.

[a] Corresponding author: lihonghui@cirp.org.cn

https://doi.org/10.1051/e3sconf/202129001013

2. Principle and Method

Protective clothing has various structural forms. As shown in Figure 1, a representative type of protective clothing is mainly composed of an inner layer, an outer structural layer, and a protective layer with multiple protective functions. The shielding material of the protective layer has various structures. When X-rays with low energy represented by 150 kV interact with matter, the main use of high atomic number materials is to reduce the amount of energy of X-rays through the photoelectric effect of X-rays and Compton scattering.

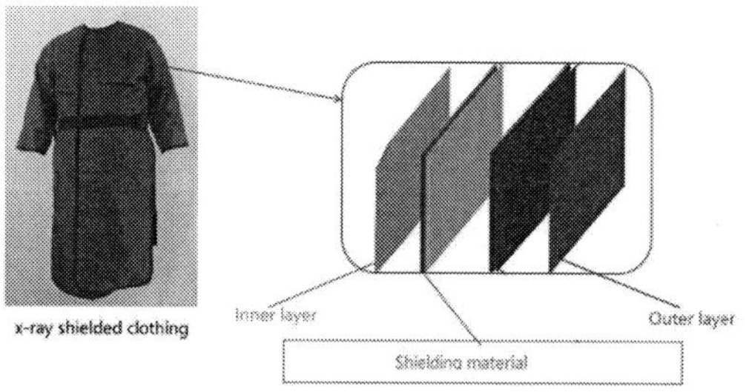

Figure 1 Schematic diagram of X-ray shielding protective clothing and its material structure

For personnel X-ray protection, it can be expressed as a reduction coefficient relationship to the dose, which is expressed as follows:

$$K = \frac{Dx_o}{Dx_i} \tag{1}$$

Where K is the protection weakening coefficient of the protective material against X-rays; Dxo represents the dose rate without X-ray protective materials (mSv/h); Dxi represents the dose rate with X-ray protective materials (mSv/h). The protective performance of the material can be determined by determining the dose rate relationship before and after the protection [4,5].

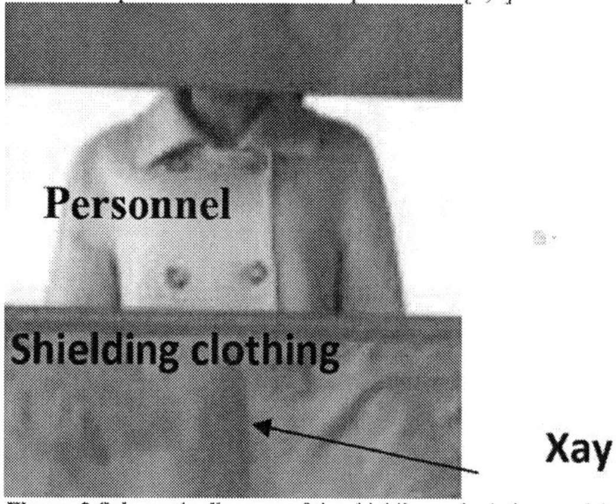

Figure 2 Schematic diagram of the shielding calculation model

The protective layer usually can use a single metal textile or additive materials, and can also use combined or mixed materials. To reduce the radiation hazards of material development, the Monte Carlo method is studied to simulate X-rays, the researched protective shielding materials, and the three-body physical model of the personnel, and the reasonable material plan is evaluated with formulas.

To reduce unnecessary analysis, studied and trial-produced the feasibility of metallic bismuth and tungsten materials as textile materials. Preliminary tests have shown that three structures of single metal (as shown in Figure 3), layered metal, and metal mixture can be realized. Therefore, the simulated protective shielding materials are used to carry out protective performance research based on these three determined ranges.

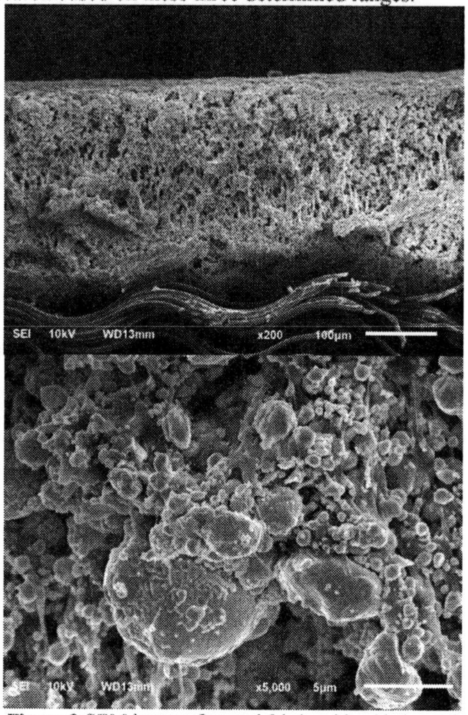

Figure 3 SEM image of coated fabric with added metal powder

3. Calculation and Analysis results

3.1. Single metal layer result

Considering that the application range of 150 kV X-ray is different from the filter device, the X-ray energy spectrum is different. Generally speaking, there are many kinds of X-ray equipment whose main energy spectrum is distributed at 150 keV and below. The 150 kV X-ray is investigated according to the target substance. There are obvious differences in the different energy spectra of the filter materials, but some investigations show that 80% or more of the energy spectra are mostly concentrated in 60~120 keV. The simulation carried out 150 keV and the following is divided into 60 keV, 70 keV, 80 keV, 90 keV, 100 keV, 110 keV, 120 keV, 150 keV. The dose rate of X-ray shielded by a textile protective material. Figure 4 shows the material model shown in Figure 3, the shielding

 https://doi.org/10.1051/e3sconf/202129001013

material is 2 kg/m^2 single metal tungsten, and the X-ray dose rate under 150 keV is used for shielding. Figure 5 shows the X-ray dose rate under 150 keV is shielded with a 2 kg/m^2 single metal bismuth material.

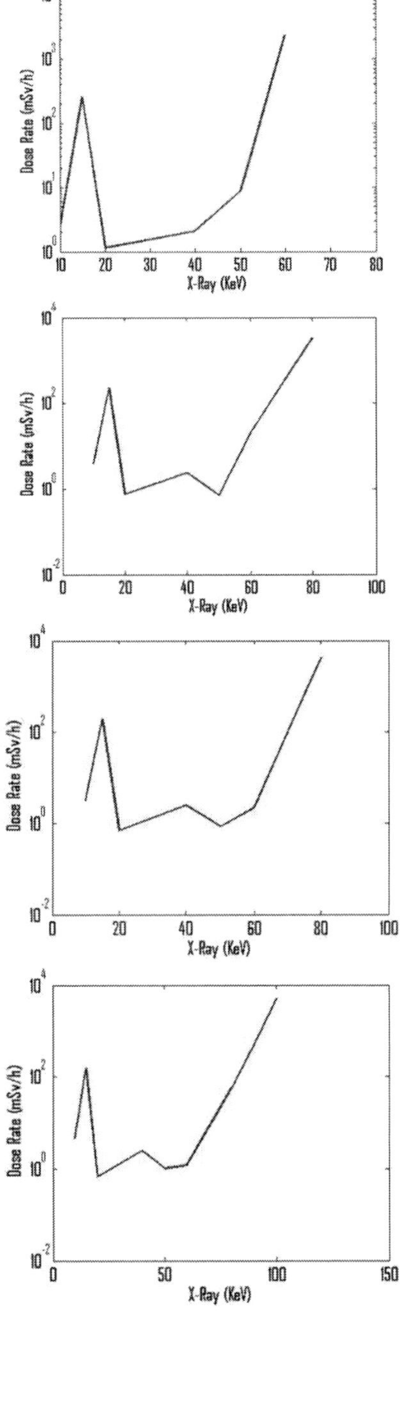

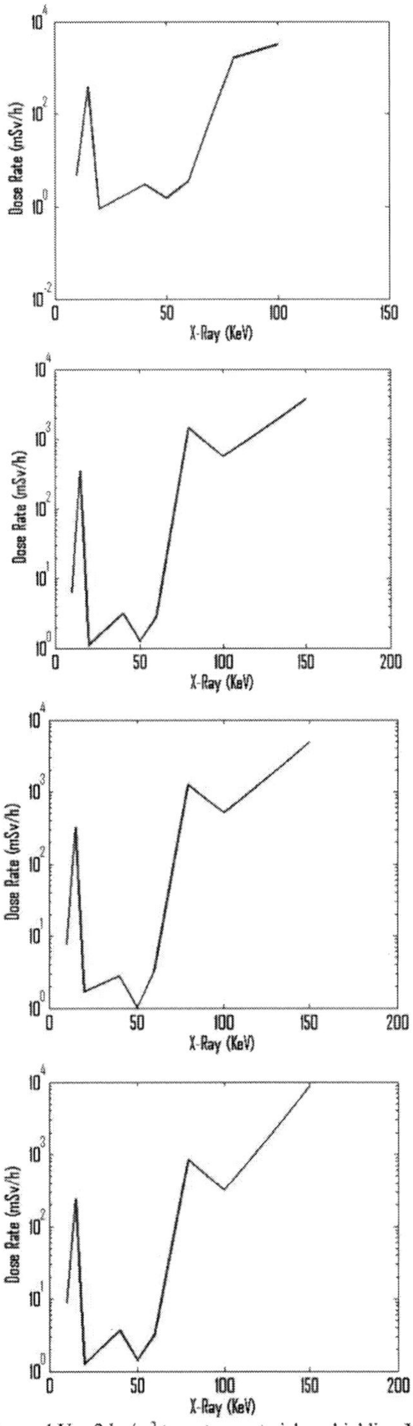

Figure 4 Use 2 kg/m^2 tungsten material as shielding X-ray dose rate under 150 keV

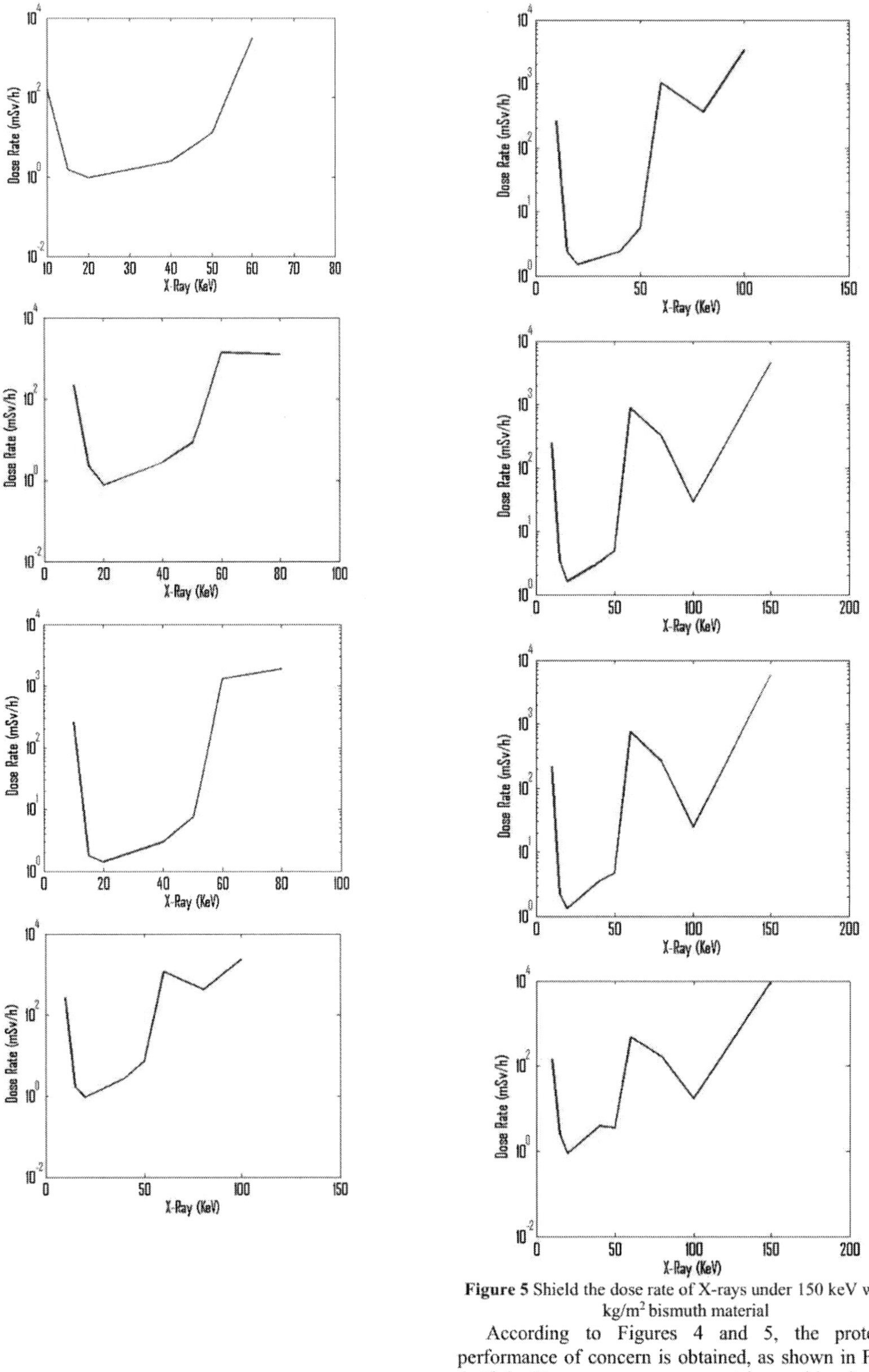

Figure 5 Shield the dose rate of X-rays under 150 keV with 2 kg/m^2 bismuth material

According to Figures 4 and 5, the protective performance of concern is obtained, as shown in Figure 6. If the X-ray energy component is considered at 150 keV, the protective weakening coefficient of all heavy metal materials against X-ray degradation can only reach between 1.22 and 1.26. The attenuation coefficient of 1.5 cannot be achieved, and the protection performance of X-ray reduction is not obvious. For the energy spectrum distribution, the main weight is concentrated below 100 keV, and the X-ray dose rate can be reduced by 1.5 by relying on Wu metal. If it is bismuth metal, it is possible to ensure that the X-ray energy is mainly concentrated at 70 keV and below to achieve the attenuation of X-rays to 1.5.

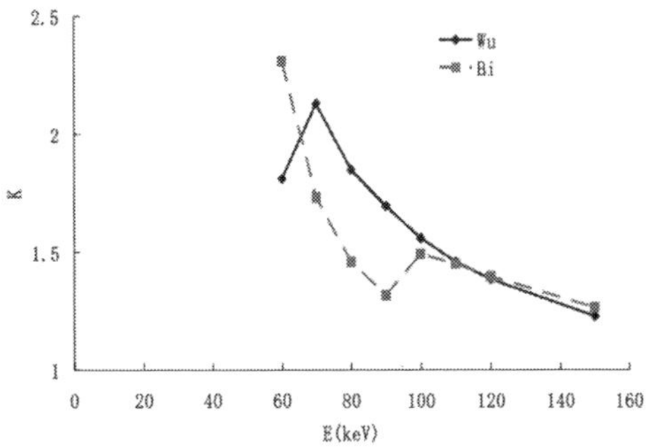

Figure 6 The ability to protect single-energy X-rays below 150 keV with 2 kg/m^2 tungsten or bismuth textile materials

3.2. Double-layer metal results

The protective fabric shielding material can also be an interlayer of two metal materials arranged in sequence. For this reason, the same model as shown in Figure 2 is adopted, and the shielding material adopts a double metal layer structure. The material uses a mass ratio of 1: 1 to simulate Bi/Wu double-layer shielding and Wu/Bi double-layer shielding. 150 keV and below, covering 60 keV, 70 keV, 80 keV, 90 keV, 100 keV, 110 keV, 120 keV, 150 keV The dose rate of X-ray shielded by textile protective materials, and then the same as a single layer of metal, the results are sorted out. The X-ray dose rate of the two layers of metal bismuth and tungsten arranged in different order is reduced. Figure 7 shows that the shielding material is 2 kg/m^2 metal tungsten and bismuth double-layer material for shielding. The X-ray dose is reduced to under 150 keV.

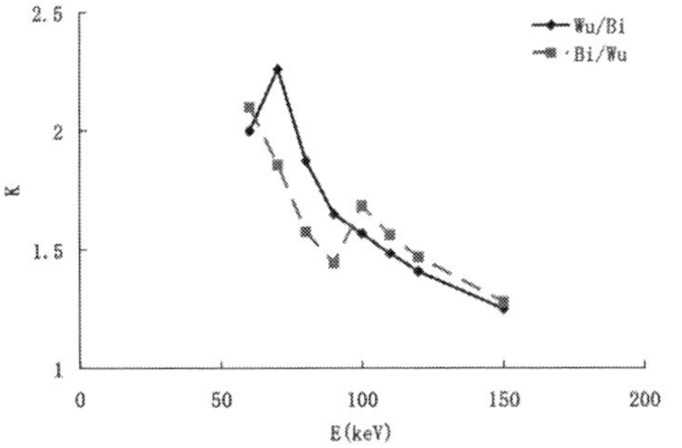

Figure 7 The attenuation coefficient of single-energy X-rays under different orders of 2 kg/m^2 tungsten-bismuth two-layer metal material

As shown in Figure 7, if the X-ray energy component is considered at 150 keV, it is also impossible for all heavy metal materials to achieve the ability to weaken X-rays by 1.5. If the energy of X-rays is mainly concentrated below 100 keV, it is possible to achieve the requirements that the dose rate of X-ray reduce by 1.5. Meanwhile, because of the difference in scattering and resonance absorption of different materials, even the same material, different combinations are different. Besides, considering that in clothing textile applications, separate metal or layered metal and its different inter-layer weaving are not ideal fabrics, so consider the mixed composite situation. To this end, the feasibility of metal mixing research has been carried out, and a simulation study has been done to calculate and analyze the difference between it and the layering, which can lay the foundation for the subsequent realization of the ideal fabric.

3.3. Metal hybrid structure results

The model is the same as in Figure 2. The shielding material adopts a bimetallic mixed structure. The mixed materials are determined by mass ratios, and the Wu: Bi ratios are respectively 0.2: 0.8, 0.5: 0.5, 0.8: 0.2 metal mixing. Simulate radiation 150 keV and below, covering 60~120 keV and 150 keV shielding performance dose rate calculation. Figure 7 shows the model of Figure 3, the shielding material is 2 kg/m^2, and the metal tungsten-bismuth mixing ratio is 0.2: 0.8, 0.5: 0.5, 0.8: 0.2. Three metal mixed structures (recorded as WuBiMix1, WuBiMix2, WuBiMix3 in the figure). The X-ray attenuation coefficient under the mixed material shielding 150 keV.

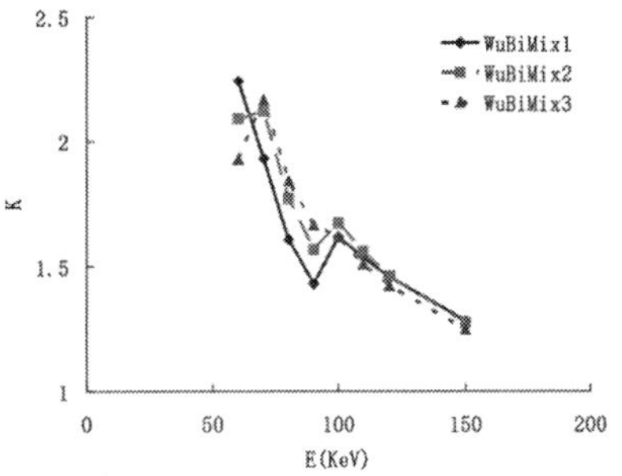

Figure 8 The areal density of 2 kg/m^2, the mixed material of Wu and Bi mass ratio 0.8: 0.2 shields the X-ray attenuation ability under 150 keV

According to the results of Fig. 8, if the X-ray energy component is considered at 150 keV, it is also impossible for mixed metal materials to achieve the possibility of a weakening factor of 1.5 for X-rays.

4. Analysis and Discussion

The performance of protective clothing of different materials is normalized to the relationship of Bi performance, and Figure 8 is obtained. The results show that the higher the proportion of bismuth when the radiation component is low in energy, the better the protective performance. As the energy increases beyond 70 keV, the more tungsten materials, the better the protective performance. Compared with a single metal, tungsten has poorer protection performance below 70 keV. Between 70~120 keV, the protection performance of tungsten is much better than that of Bi. When the X-ray energy is higher than 120 keV, the protection performance of bismuth begins to be better than that of metal tungsten. Except for the X-ray energy lower than 70 keV in the double-layer metal, interlayer material has better protection performance than a single metal bismuth material. When the energy is lower than 70 keV X-ray, the metal bismuth serves as the first layer of protection for the protection below 70 keV. The performance is better than tungsten metal, and the metal bismuth is used as the first layer when low energy is dominant. When X-rays are above 70 keV, tungsten as the first layer material has obvious advantages in protection performance, and the overall protection performance is better than single metal bismuth and bismuth as the first layer protection material. The performance of the mixed material is similar to that of the double-layer metal material, except that it is less than 70 keV, its protection performance is also better than that of a single metal bismuth material. If the ray energy is between 70~90 keV, the mass ratio of tungsten to bismuth is 0.8: 0.2. When the energy is high, the protection performance is best when the mass ratio of tungsten to bismuth is 0.5: 0.5, and when the X-ray energy is less than 70 keV, the mass ratio of tungsten to bismuth is 0.2: 0.8 as superior.

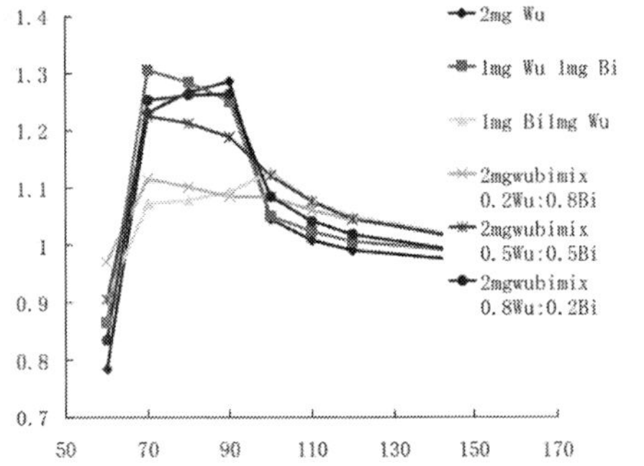

Figure 9 Comparison of protection effect under different energy

5. Conclusion

In this paper, combined with the feasibility of textile materials, the Monte Carlo method is used to analyze the 150 kV typical X-ray protection capacity of single-layer metal, double-layer metal, and different proportions of metal mixed textile materials. The results show that if the X-ray composition is mainly below 70 keV, then Regardless of the method used, metallic bismuth should account for the main proportion or directly use metallic bismuth. Above 70 keV, the protective performance of a single metal bismuth material is weaker than that of double-layer or mixed metal materials. For this reason, mixed materials should be used. Among them, the ratio of tungsten and bismuth is 0.5: 0.5 or 0.8: 0.2. For good protection performance and comprehensive low-energy protection effect, it is a better choice to use tungsten and bismuth ratio of 0.5: 0.5.

Acknowledgement

This research was supported by Foundation of The National Key Research and Developmeng program of China (2018YFC0810301).

References

1. S. Fu, Application of MCNP to Study the shielding Effect of Zinc Bromide Solution,[R], Progress in NUCLEAR and TECHNOLOGY, Vol. 1, p.98-101(2011)
2. S. Fu, E. Yao, Weibing Yang, etc. Study on shielding material (sands) for disused high activity radioactive sources conditioning [J]. Radiation protection. 2018(06).
3. S. Fu, etc. Shielding Calculation on Spent High Activity Radioactive Sources Conditioning [J]. Radiation protection newsletter. 2008(05).
4. General Administration of Quality Supervision, Inspection and Quarantine of the People's Republic of China, Basic Standards for Ionizing Radiation Protection and Radiation Source Safety [S], GB18871-2002.
5. S. Fu etc, A structurally optimized movable liquid outflow monitoring device[P]. ZL201420692457.9.

Polymeric Micelles for Cancer Drug Delivery and Targeting

Yuyan Wang[1,*]
[1]University of California Irvine, Physical Department, 92697, Irvine, United States

Abstract. Over the past years, the emergence of various therapeutic approaches has attracted so much consideration, but chemotherapy remains the most effective way in experimental practice. However, due to obstacles and difficulties, there is a limitation for the additional clinical presentation of the outdated chemical treatments such as paclitaxel and doxorubicin. Some of the barriers are severe side effects, profound cytotoxicity, and low therapeutic efficacy. Drug delivery systems (DDS) such as polymeric micelles have been technologically advanced over the recent decades to advance the treatment efficacy and lessen the side effects incurred. The polymeric micelles self-assembled from amphiphilic polymers serve to sum up hydrophobic treatments in the core and regulate the discharge profile of the drugs through response towards stimuli of the tumor microenvironments, which in this case are pH, temperature, light, and reduction.

1 Introduction

Recently, polymer micelles techniques have pulled in expanded consideration as a good vehicle for inadequately dissolvable medications in modern cancer treatment [1]. Polymer micelles are self-congregations of amphiphilic block copolymers in aqueous media. Meanwhile, many benefits of utilizing polymer micelles have been exhibited with their unique core-shell structure. The hydrophobic centres are isolated by hydrophilic shells from the aqueous exterior, so then hydrophobic medications can be solubilized into the hydrophobic centre designs of polymer micelles at focuses a lot higher than their characteristic water-solvency [2]. Polymer micelles have a high medication stacking limit, high water-dissolvability, and fitting size for a long course in blood. The hydrophilic shell encompassing the micellar centre can secure unfortunate marvels, for example, between micellar conglomeration or precipitation, protein adsorption, and cell bond. The substance organization of polymer micelles can be customized to have alluring strong compound properties for the drug solubilization. In most polymer micelles, hydrophobic medications are joined into the hydrophobic centre of micelles by hydrophobic collaboration just as other extra communications, for example, metal-ligand coordination holding an electrostatic connection. The drug solubility's ability depends on the compatibility of drug and the micelle core. This study focus on the structure of polymeric micelle, the procedure of polymeric micelle formation, and the methods used to stimulated the drug delivery, which included pH-sensitive, photosensitive, thermal-sensitive, and reduction.

1.1 Overview of self-assembly

1.1.1 Structure

Polymeric micelle has a small size, and usually their diameter ranges from 5 to 100 nm [3], which depends on the type of head and length of function groups. Polymeric micelles are based on block-copolymers with hydrophilic and hydrophobic units that self-assemble in an aqueous environment into structures These blocks can be arranged in different ways: A-B type, A-B-A type [4]. In the micelle structure formation process, the hydrophobic and hydrophilic are separated due to the interaction between the polar part of the head and surrounding water [5]. This formation process results in a flexible and porous structure of micelle. Now, polymer micelle is widely used as a high effectivity model of biological application and drug delivery systems because of the special function polymer micelle. For example, they can increase the drug solubility from the targeting ability, reduce the toxicity, and enhance the penetration on tissue. Polymeric micelles are composed of the amphiphilic molecules in an aqueous solution, then self-assemble into a core-shell structure that contains both hydrophobic and hydrophilic segments.

For the solution, the critical concentration required in forming polymeric micelle is called critical micelle concentration (CMC). The formation of micelle occurred when CMC value higher than the concentration needed [6]. The typical micelle formation is driven by dehydration of the hydrophobic tails. The final structure of polymeric micelles are spherical, supramolecular molecule, and micelle have a unique morphologies ability can be assembled in different sharp, for example, rods, spheres, and tubules, which depend on the temperature of the environment, chain's length, and quality of solvent. Polymeric micelle's function is not limited to cancer

* Corresponding author: yuyanw7@uci.edu

therapy, but also in different fields, such as material engineers and microelectronics.

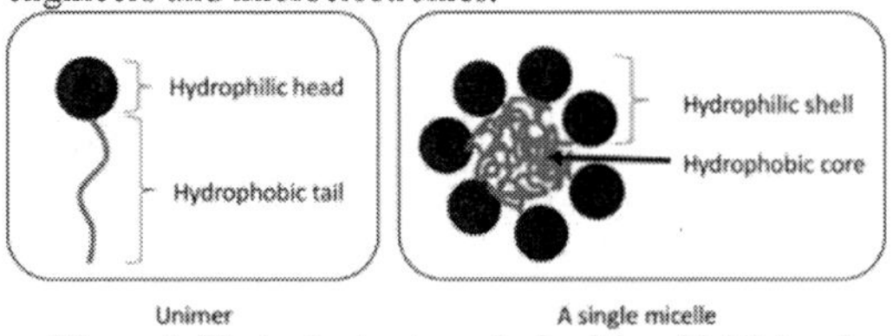

Figure 1. The basic structure of animal (amphiphile) and combine polymeric micelle [7].

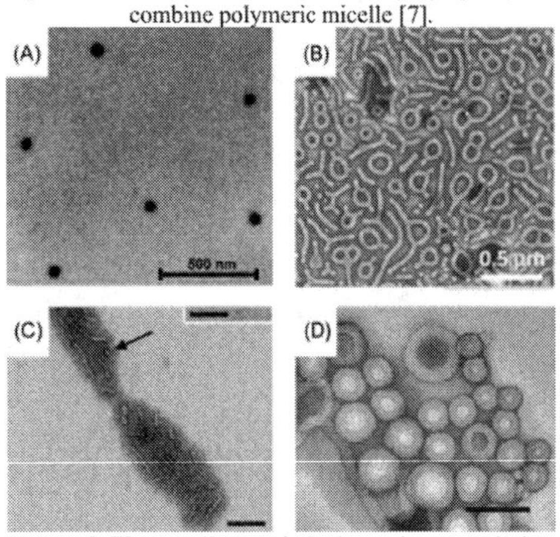

Figure 2. The micelle morphologies under transmission electron micrographs: (A) spheres, (B) toroids, (C) worm, (D) vesicles.

1.2 Synthesis of polymer

The synthesis of polymers has been a great interest and popular field in polymer chemistry. Among those polymer technologies, living polymerization techniques are widely used, as it can employ in the polymerization proceed under a lack of irreversible chain transfer and chain termination condition. Recently, cutting-edge technology called controlled/living radical polymerization (CRP) has been developed, as the radical process is more tolerant of functional groups and impurities. The rapid dynamic equilibration was the foundation of the new CRP methods, between a small amount of growing free radicals and many dormant species.

1.2.1 ATRP

One method focuses on transition metal-catalyzed atom transfer radical polymerization (ATRP). ATRP was based on the atom transfer step, which is a key element in the uniform growth of the polymeric chains [8]. ATRP was target to form 1:1 adducts of alkyl halides and alkenes, and its originated atom transfer radical addition (ATRA) [9,10]. ATRA is one kind of Kharasch additional reaction under the presence of light or conventional radical initiators. The ATRP mechanism is controlled by the equilibrium between propagation radicals and dormant functional groups, mainly in the form of initiating alkyl halides or macromolecular species. For the dormant species, they can react with the rate constant of activation (k_{eact}) with a transition metal, for example, copper, in the lower oxidation state. During the reaction, polymer chains grow in the constant rate of propagation k_p by adding the intermediate radicals to monomers in the solution, so the chains growth process is similar to conventional radical polymerization. In addition, termination reactions (k_t) are also shown in ATRP reaction, in the radical coupling and disproportionation process. However, only a meager percentage of polymeric chains pass the termination section. Less than 5% of the total growing polymer chains end the reaction in the lab during the initial stage of the polymerization. This process results in the growing radicals (Pn•) and formed transition metal complexes ($X\text{-}M_t^{n+1}$) in a higher oxidation state. The oxidized metal complexes can minimize the influence of termination by reducing the concertation of growing radicals. Under the reverse reaction (k_{deact}) condition, propagating radicals can reform the dormant species and the activator.

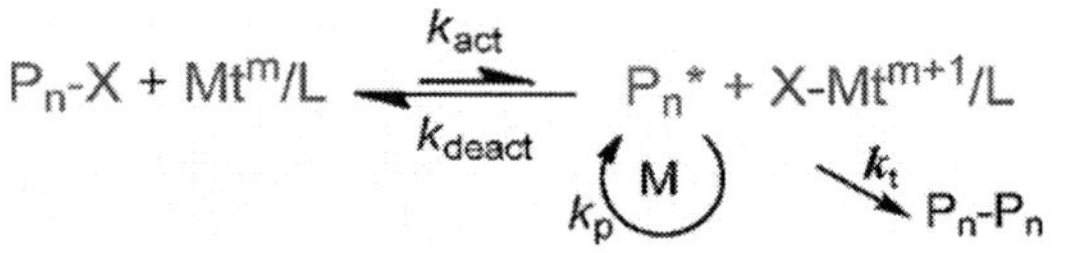

Figure 3. ATRP equilibrium mechanism.

ATRP is a catalysis reaction, and CuI/L and X-CuII/L are the most commonly used transition metals in response. Its rate is based on the rate constant of propagation, the concentrations of monomer, and growing radicals. The concentration of growing radicals is related to the attention of dormant species, activators, and deactivators. Equation 1 shows fast initiation and no chain termination under the ideal case. The dispersity of molecular weight (MW) (Mw/Mn) of polymers influenced by dormant groups' concentration (P_nX and X-CuII), the rate constants of propagation (k_p), and deactivation (k_{deact}), and monomer conversion (p).

$$R_p = k_p[M][P_n^*] = k_p K_{ATPR}\left(\frac{[P_nX]\left[\frac{Cu^{II}}{L}\right][M]}{\left[X\text{-}\frac{Cu^{II}}{L}\right]}\right) \quad (1)$$

1.2.2 RAFT polymerization

The reversible addition-fragmentation chain transfer (RAFT) process has been a powerful and popular polymerization technique in the modern field of complex polymeric synthesis processes [11,12]. RAFT belongs to reversible deactivation radical polymerization (RDRP), also called living radical polymerization, and it has the ability to predictable molecular weight and low molar mass dispersity. The RAFT mechanism (Figure 3 and Figure 4) is shown below. In step I, the radical source is added to the RAFT. It will enter to equilibrium between active and dormant species in later steps III and V. In the RAFT mechanism, which formed by the chain transfers steps, the reactions degenerate due to the involvement of eversible transfer of the functional chain end-group. For example, the thiocarbonylthio group, Z-C(=S)S-R. In the result part, all chains will result in a similar degree of polymerization (DP). And the whole process is based on the insert of monomers between the R- and Z-C(=S)S-groups of a RAFT agent and formed α and ω end-group at the resulting polymeric chains. However, using the ω-functionalization is not the best approach in some applications due to the instability of the C-S bond in the RAFT end-group.

The critical element in the RAFT mechanism is based on the correspondence between the number of chains that undergo bimolecular termination and the number of initial radicals in the whole system. Indeed, unlike other systems, the bimolecular termination event will not cause the loss of the ω-end thiocarbonylthio end-group, so the total amount of thiocarbonylthio end-group is constant in the whole process. As a result, the number of dead chains can be predicted and controlled by the number of radicals introduced in the system.

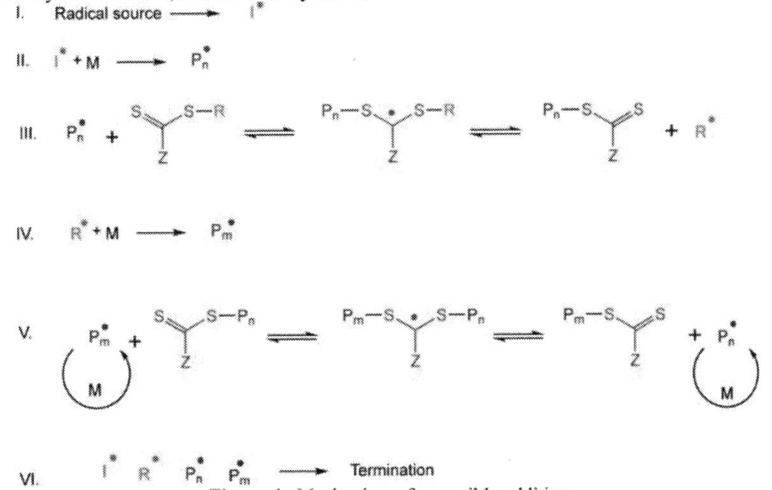

Figure 4. Mechanism of reversible addition.

2 Stimuli-responsive polymeric micelles

Nowadays, to enhance the efficient drug release rate in the tumor cells, the targeting drug release mechanism is developed in polymeric micelle system by controlling the methods of drug release when reach the target cells [13]. The stimuli-responsive of polymeric micelle has been used as an innovative technology for possession of drug release. The stimuli systems can be divided into two main parts: internal stimuli and external stimuli. For internal stimuli, it contains temperature, pH-value, enzymes, and so on. In this method, compound Thermo or pH-sensitive compound releases the inside drugs after reaching the targeting tumors. The external stimuli methods include light, ultrasound, and both types of stimuli-responsive methods show a superior ability on cancer therapy on accuracy on location and time of drug release.

2.1. pH-sensitive polymeric micelles

During the cancer therapy process, tumor microenvironment is one of the most important factors that can influence treatment. Tumor microenvironment (TME) demonstrates a tumor composition, and it has been focused as a highly effective technique to enhance human immunotherapy [14]. PH value has been considered an ideal trigger for anticancer drug release, based on the different pH values between primary tumors and normal tissue [15]. In a human vivo system, many intracellular components can maintain their pH value in normal conditions, such as cytoplasm, endosomes, lysosomes, and mitochondria. When the tumor overgrown, the expanding population of tumor cells caused the lack of oxygen. This reaction results in hypoxia, lactic acid production, and hydrolysis of ATP in the human Vivo system. Therefore, the acidic microenvironment can be direct to a pH-sensitive drug delivery system developed. The pH value in tumor cells is ~6.5, and the value of normal cells is ~7.4, and pH-value can decline further in the organelle such as endosomes and lysosomes [16]. In this way, planning the pH-delicate nanocarrier for focusing on drug conveyance to tumors that the nanocarrier is steady at physiological pH; however it should be distorted to work with the arrival of the medication under gentle acidic conditions at target tumors, may bring about essentially upgraded remedial adequacy and insignificant results [17].

2.1.1 Unimolecular polymeric micelles (UPM)

In this manuscript, the drug delivery systems of a pH-sensitive polymeric micelle are triggered by the protonation of carboxylic acid group. The pH value of polymeric micelle must be stabled at 7.4 to reduce the probability of the disassembly and destabilization in vivo. To prepare pH-sensitive polymeric micelle, positive-charged drugs are combined with negative-charged block copolymer to form a core. In the human vivo system with neutral pH value, the carboxylic acid units are shown as negative charge, allowing cooperative electrostatic interactions with the encapsulated compounds [18]. Once in a mildly acidic environment, protonation of the carboxylic acid gatherings brings about a net reduction in electrostatic interactions resulting in separation of the medication-containing centre. Another way to produce pH-sensitive PMs is to combine a drug in the uncharged hydrophobic centre introducing an ionized polyanion shell. For this situation, protonation of the body prompted an annoyance of the centre shell structure and eventually influenced the intracellular conveyance of the medication.

https://doi.org/10.1051/e3sconf/202129001014

There are mainly two pH-stimuli polymeric micelle types that are commonly used: unimolecular and multimolecular [18].

Amphiphilic star-shaped polymers were a novel method to the pH-sensitive of polymeric micelle by atom transfer radical polymerization. The new polymers show a traditional core-shell structure of PM, and have dilution stability as unimolecular polymeric micelles (UPM). UPM can be extracted from both dendrimers and star polymers (Figure 5A). A four-armed multifunctional initiator was used for polymerization, as the Star polymers can be produced in fewer steps than other technology. Atom transfer radical polymerization (ATRP) was used to synthesize UPM in the previous work (Figure 5B). A novel water-soluble UPM with an ionizable core was developed as potential carriers for the oral delivery of drugs. The compound was called Star-P(EMA-co-tBMA)-b-P(PEGMA). To investigate the function of UPM, the pH-value of different concentrations and percent of invitro drug release rate were tested. Titration was used to evaluate MAA content with hydrolyzed UPM and NaOH 0.05N. The value of α shown the degree of ionization, where α = Cb/CMAA and Cb is the concentration of NaOH, CMAA is the concentration of MAA(methacrylic acid) in solution. The result shows that the titration curve reaches the equivalent point when pH=9. Indeed, it also shows that UPM can be protonated in the stomach (pH 1 to 2), Ileum (pH 7). As the drug release rate depends on drug diffusion, the stability of carrier material, and polymer degradation, the introduction of pH-sensitive acidic functions were supposed to influence the polarity of pH response. In Figure 5D (a), the graph shows the percentage of drug release rate is between free drugs and packed drugs. After 60 min, more than 80% of non-packing drug release, and the diffusion rate reach 100% after 2h. Compared with the rate of drug release from UPM, it hardly reaches 20% after 60min, and the maximum rate is 45% in 8h. In figure 5D (b), the graph showed the pH-sensitive drug from UPM in pH 7 or 11 conditions. Based on the graph, the change of drug release amount from the micelles at higher pH was shown.

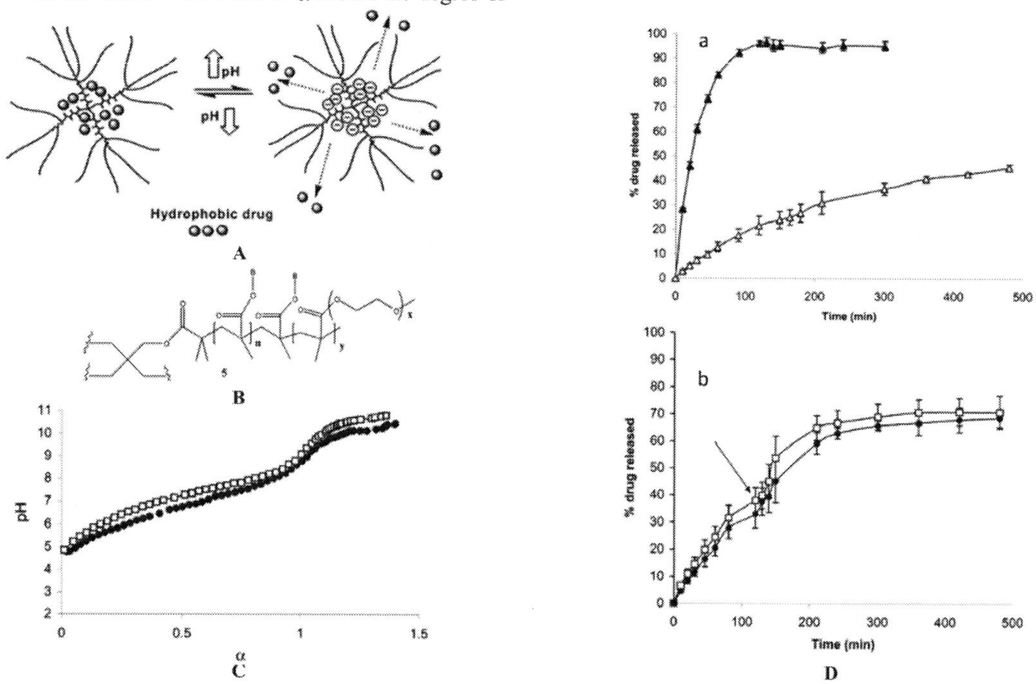

Figure 5. (A) The structure of pH-sensitive unimolecular polymeric micelles (UPM) release hydrophobic drug. (B) The chemical structure of pH-sensitive UPM. (C) The comparison of pH-value and UPM concentration of S_E18M17P442 (closed circles) and S_E24T23P229 (open squares). α was the ionization, and the polymer solution was titrated with NaOH solution. (D) (a) In pH 1.2 environment, vitro release of free (closed triangles) and S_E24M23P229-entrapped (5% w/w) (open triangles). (b)In pH 1.2, 7 (open squares) and 11 (closed circles) environment, S_E24M23P229-entrapped progesterone (5%w/w) released.

2.1.2 Unimolecular polymeric micelles (UPM)

pH-sensitive polymeric micelle also can be used as the pH probe for detecting acidic biological environment. pH imaging for tumor therapy was due to the unique acidic environment. To monitor tumor pH-value in the human vivo system, MR imaging/spectroscopy (MRI/MRS) was used as clinical applicability [19]. To reach a better pH-sensitive system, a mixed micelle pH probe composed of polyethylene glycol (PEG)-b-poly (l-histidine) (PHis) and PEG-b-poly (l-lactic acid) (PLLA) (Figure 6A).

To get a better result of tumor testing using MRs, the pKa value of the pH probe should range between weak acidic and neutral (pH 6.0-7.4). The PHis material can be changed from water-soluble to hydrophobic aggregates with increasing pH-value. As the pH-value and pass the pH 7.0 condition, the normalized integration percentage of the H2 peak decrease sharply, as H2 represented the chemical shift imidazole ring. With the molecular weight increase, the aqueous solubility of PHis is decreased (Figure 6B). The function of micelle probes was tested by using 1H-MRS to compare the pH-value and chemical shift of the H2 peaks in the graph. The pH 7.4 and 6.0 conditions were used to clarify the MRS data of the mixed micelles (Figure 6C). At pH=6.0, H2 and H5 peaks of the mixed micelle showed a visible rise compared with the pH 7.0 condition. The diverse micelle probe peaks coincided with the spectrum of blood itself. The results indicated the mixed micelle can provide pH-monitoring power in tumor pH-value range.

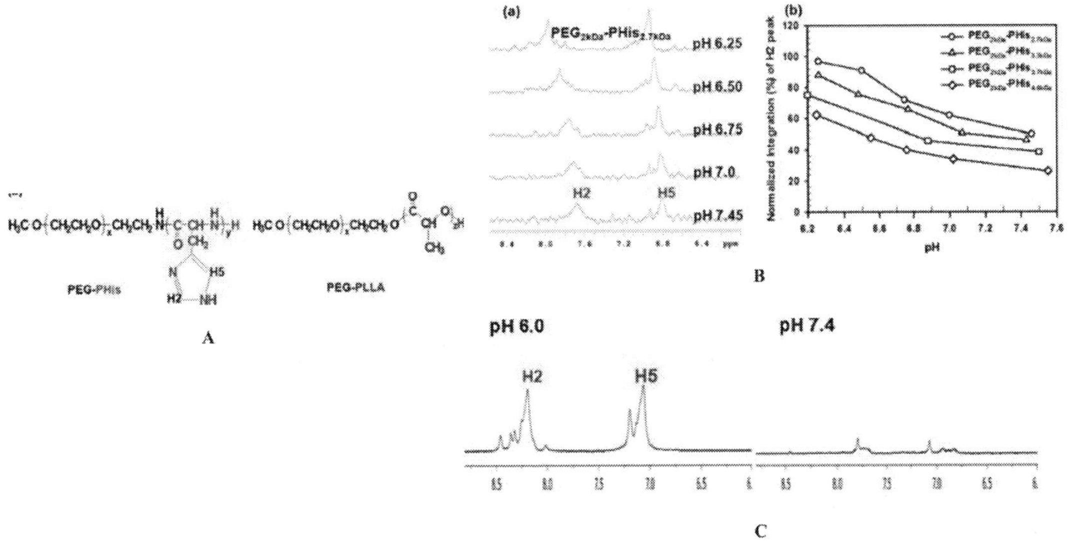

Figure 6. (A) The chemical structure of polyethylene glycol (PEG)-b-poly (l-histidine) (PHis) and PEG-b-poly(l-lactic acid) (PLLA). (B) (A) H2 and H5 HNMR peaks in PEG2kDa-PHis2.7kDa at different pH conditions. (B) pH-sensitive H2 and H5 peaks of PEG2kDa-PHis in different MWs of PHis. (C) 1H NMR spectra of H2 and H5 peaks in mixed micelles of PEG2kDa-PHis3.3kDa (90 wt %) having PEG2kDa-PLLA3kDa (10 wt %) in different pH condition (6.0 and 7.4).

2.2 Photo-sensitive polymeric micelles

For the photosensitive method to control the polymeric micelle remotely and accurately, which can focus into specific areas quickly. Based on modern technology, the use of light control factor for polymeric micelle was largely unexploited [20].

2.2.1 Poly(tert-butyl acrylate-co-acrylic acid) (tBA-AA)

Synthesis of an amphiphilic diblock polymer micelle was developed to stimulate by UV light and be reformed by later light exposure. In the synthesis of polymer micelles, it can be divided into two parts. For the hydrophobic block, its principal structure is a side-chain liquid crystalline polymethacrylate with azobenzene mesogens (PAzo). And the hydrophilic block is a random copolymer of poly(tert-butyl acrylate-co-acrylic acid) (tBA-AA) [21]. The hydrophobic, which is the core part, contains a photolabile chromophore as a pendent group to result in the polymeric micelle break under UV light. The bond-breaking reaction occurred between the chromophore and polymer under UV light and resulted in transformation of a hydrophobic block into a hydrophilic block.

To detect the bond-breaking and dissociation of polymeric micelle in UV light condition, scanning (SEM) and transmission electron microscope (TEM) technology were used to observe the reaction in the solution. The polymeric micelles were formed in the solution in non-UV light conditions, and the compared group was placed under UV light. In the result, no micelles remained after exposure to UV light condition, which indicated the polymeric micelle probe can monitor the dissociation of micelle (Figure 7A). Meanwhile, as the intensity of the irradiation increased, the transmittance increased. As the dissociated of micelles under UV light, the solution turbidity rate decline contributes to the rise of transmittance rate (Figure 7B). The result showed the detachment of dye pendant groups responded to photo solvolysis.

https://doi.org/10.1051/e3sconf/202129001014

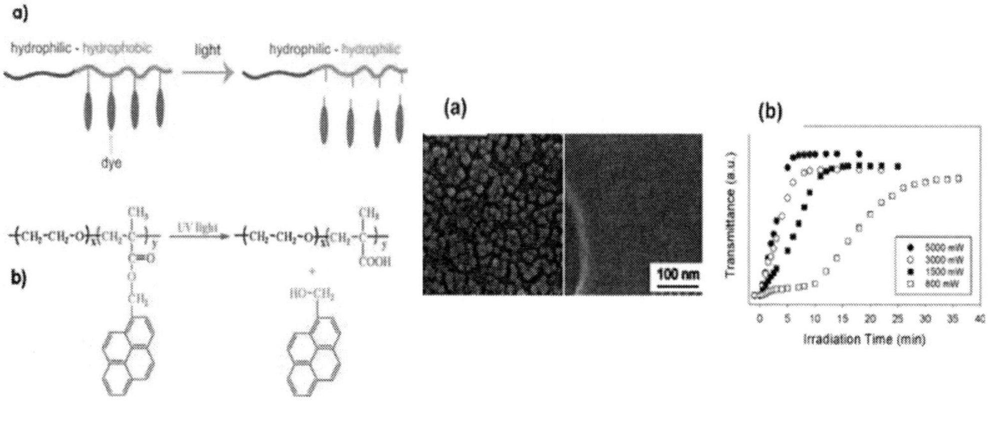

A B

Figure 7. (A) (a) The flow chart of the change of hydrophobic to hydrophilic due to detachment of dye pendant groups. (b) Chemical structure of the amphiphilic diblock copolymer and reaction under UV light. (B) (a) The scanning image of PEO-b-PPy (left) and the dissociation of micelles under UV light in the micellar solution(right). (b) The comparison of transmittance rate and time under different irradiation intensities.

2.2.2 UCNPs

Many photosensitive required a high-energy UV light during the reaction; however, the high intensity may induce negative effects on the human body. So longer-wavelength near-infrared (NIR) light, which has less penetration into cells, was a modern popular field. In the photoreaction, NaYF4:TmYb upconverting nanoparticles (UCNPs) was encapsulated in poly(ethylene oxide)-block-poly(4,5-dimethoxy-2-nitrobenzyl methacrylate) micelle under 980nm light [22]. By encapsulating NaYF4:TmYb upconverting nanoparticles (UCNPs) inside micelles of poly(ethylene oxide)-block-poly(4,5-dimethoxy-2-nitrobenzyl methacrylate) and exposing the micellar solution under 980 nm light, photons are emitted by the UCNPs. During the mechanism, photons are absorbed by o-nitrobenzyl groups on the micelle core-forming block, activating the photocleavage reaction and leading to the dissociation of BCP micelles core and release of co-loaded hydrophobic species (Figure 8A).

To determine NIR light whether can induce the release of hydrophobic species co-loaded with the UCNPs inside the micelles, micelles contained UCNPs and Nile Red (NR) were used to observe the result. The absorbance spectrum recorded the UCNP-loaded micellar solution before and after NIR light condition without NR add. No changes in absorbance without irradiation in 24h, so the result convinced the photocleaved when micelle break under NIR light (Figure 8B(a)). For the micelles loaded contain UCNPs and NR, the normalized fluorescence intensity of NR was measured under NIR light at 640nm, and the result showed a hydrophobic dye is much faster under NIR light of the micellar solution (Figure 8B(b)).

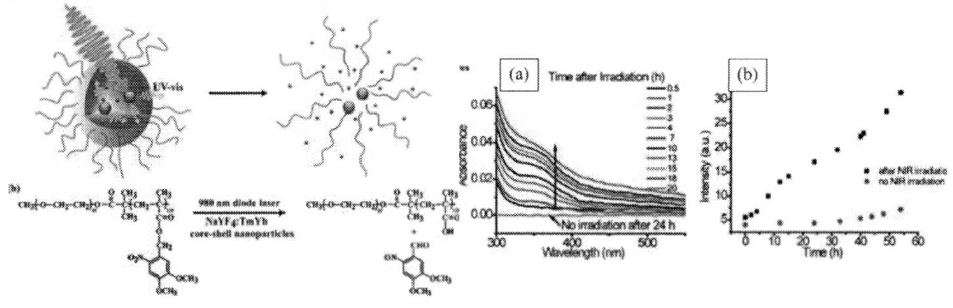

A B

Figure 8. (A) (a) The dissociation of BCP micelles in UCNPs under NIR light. (b) The photoreaction trigged by NIR light with BCP of PEO-b-PNBMA and UCNPs of NaYF4:TmYb. (B) (a) Absorption spectrum vs. time of dissociation of UCNP-loaded micelles under NIR light exposure (5W, 4h). (b) Intensity rate in the normalized fluorescence at 640nm vs. time for two micelle-loaded solutions contain NR and UCNPs. Black dot was subjected under NIR light condition, and the red dot was not.

2.2.3 Gemcitabine (GEM)

Light-cross-linked small-molecule micelles with enediyne units are designed to increase the efficiency of the drug delivery rate. Gemcitabine (GEM) was used as hydrophilic drug, and it can tethered with a maleimide-based enediyne (EDY), which can be used hydrophobic tail in the preparation of amphiphilic EDY–GEM under Bergman cyclization (Figure 9A). The process is an intramolecular cyclization of EDY under photo-triggering conditions and formed diradical type intermediates [23]. Confocal laser scanning microscopy (CLSM) technology had the ability to show the drug-loaded micelles and light-cross-linked micelles. Human cell lung cancer (A549) cells were used to react with light-cross-linked micelles (50 μM) in 6 h and shown fluorescence image (Figure 9B). In green and blue images (525nm and 450nm), the cellular uptake was established, which indicated the successful delivery of the drug (GEM) to the cancer cells and nucleus.

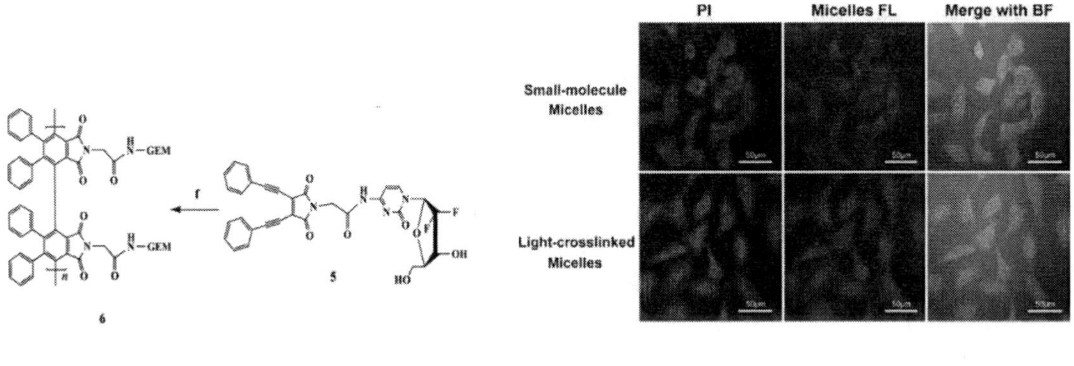

A B

Figure 9. (A) light-cross-linked small-molecule micelles chemical structure (6). (B) CLSM image of A549 with light-cross-linked micelles for 6 h. The cell dyed with propidium iodide (red), the micelles fluorescence in cells (green/blue), and overlapped two images into 3th column image. The concentration of [EDY–GEM] was 50 μM.

2.3 Temperature-sensitive polymeric micelles

Temperature plays a critical role in nearly everything in chemical reaction and regulation, for example, reaction requirement, cell division, and gene expression [24]. Compared with other external stimuli, such as pH-valued, light and so on, temperature control is one of the simplest drug release methods. The basic structure of thermo-sensitive polymeric micelle was formed by hydrophobic block and thermo-responsive hydrophilic block.

2.3.1 Poly(N-isopropylacrylamide) (PIPAAm)

Poly(N-isopropylacrylamide) (PIPAAm) is hydrophilic and known as a characteristic reversible temperature-responsive phase transition when it reaches the lower critical solution temperature (LCST) in aqueous [25]. By adding hydrophilic comonomers, such as N,N-dimethylacrylamide, the value of LCST of PIPAAm could be easily controlled to a near body temperature (Figure 10A).

To ensure the efficiently influenced efforts between time and temperature in the intracellular uptake, the micelles were incubated at two temperatures, 37 °C (below the LCST) and 42 °C (above the LCST) (Figure 10B). The result of fluorescence intensity showed the amount of intracellular increased sharply only when temperature above micelle LCST. This result is caused by the enhancment of the interaction between micelle and cells due to the thermo-sensitive phase of the micelle coronas.

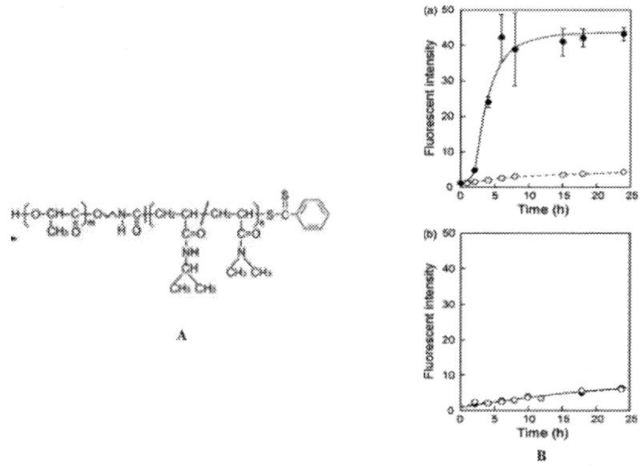

A B

Figure 10. (A) The chemical structure of polymeric micelle from thermo-sensitive polymer. (B) Comparison of time and fluorescent intensity of (a) OG-labelled thermo-responsive micelles and (b) the OG-labelled P(IPAAm-co-DMAAm) in different temperatures (37 °C, open circle) and above (42 °C, closed circle).

2.3.2 P(IPAAm-DMAAm)-b-PLA)

One kind of mechanism of thermo-responsive polymeric micelles is motivated by poly (N-isopropylacrylamide-co-N,N-dimethylacrylamide)-b-poly(d,l-lactide) (P(IPAAm-DMAAm)-b-PLA (Figure 11A), its LCST can be modulated for desirable temperatures, by the introduction of the hydrophilic or hydrophobic monomers into the PIPAAm leading chains [26]. Based on the corona-forming, the temperature-sensitive polymeric micelle can reversibly change structure and water solubility across the change on LCST.

The property of temperature-sensitive can influence the drug release rate of polymeric micelle. By controlling the variable of temperature in the range of 37–45 °C, the viable change in the fluorescent intensity occurs at 39.4 °C (Figure 11B). When the heat is provided above the LCST, the fluorescent intensity increases when the temperature increases to in the range of 39-43 °C. Also, a significant decrease in the percentage of cell viability was observed around 45 °C, and this result may cause by the extremely higher than limited temperature, so critical destruction of cellular functions of cellular proteins occurs. The sharply decrease cellular viability when the temperature approaches 45 °C also illustrated this point.

The repeated cycle of the intracochlear uptake of P(IPAAm-DMAAm)-b-PLA micelles in the temperature range of 37 to 42 °C (Figure 11C). The intracellular micelle uptake, which was indicated by the fluorescent intensity, was significantly increased by heating above the LCST. Moreover, when the LCST reaches 37 °C, the micellar internalization is limited due to P(IPAAm-DMAAm) reversibility, and these results achieved the temperature control intracellular micelle uptake.

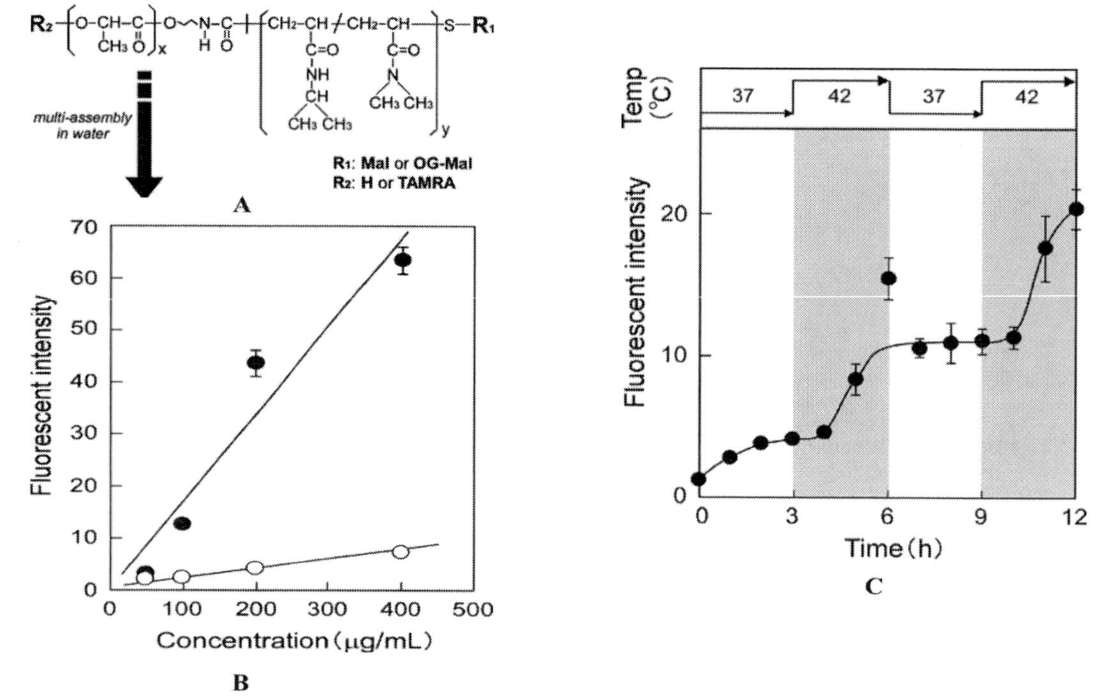

Figure 11. (A) Chemical structure of thermo-responsive block copolymers of P(IPAAm-DMAAm)-b-PLA. (B) The temperature effect the P(IPAAm-DMAAm)-b-PLA micelles (closed circles) and cellular viability (open circles). Cells were incubated with DMEM/FBS, including 200 μg/mL of OG-labelled P(IPAAm-DMAAm)-b-PLA micelles for 6 h at different temperatures. (C) The time effect the fluorescent intensity of P(IPAAm-DMAAm)-b-PLA micelles (closed circle) in the range of temperature between 37 and 42 °C. The micelles concentration is 200 μg/mL in DMEM/FBS.

2.4 Reduction control

Reduction-sensitive polymeric micelle is one of the modern class of biomedical material which can be applied for drug delivery. It can remain excellently stable in human circulation and in extracellular fluids, and in intracellular compartments, for example, cytoplasm and the cell nucleus, micelle can be easily degraded [27]. The common structure of reduction-sensitive polymers usually contains disulfide linkage(s) in the main chain, side chain, or cross-linker. In standard conditions, the disulfide bonds are stable in the circulation and in the extracellular milieu. However, under thiol-disulfide exchange reactions, the bond can be cleavaged from minutes to hours [8]. For example, compared with other polymers, aliphatic polyesters and polycarbonates take a long range of time to cleavage.

2.4.1 HECS-ss-OA

An O,N-hydroxyethyl chitosan–octylamine (HECS-ss-OA) was introduced as a conjugate to build disulfide linkages between the hydrophobic alkyl chains and hydrophilic chitosan backbone. And O,N-hydroxyethyl chitosan–octylamine conjugate (HECS-ss-OA) involving bioreducible 3,3'-dithiodipropionic acid (DPA), resulted in the formation of a redox-sensitive micellar system [28]. This system was used to trigger intracellular delivery of hydrophobic paclitaxel (PTX). In the experiment, a control group insensitive structurally analogous HECS-OA conjugates was synthesized, which is structurally

https://doi.org/10.1051/e3sconf/202129001014

similar to HECS-ss-OA, but lacked the redox-sensitive disulfide bond (Figure 12A and 12B). The disassembling process of the reduction-sensitive polymeric micelle was triggered by GHS, which is a natural reducing agent for disulfide bonds, based on the size change of the HECS-ss-OA and HECS-OA micelles. GHS is possibly connected with cancer, as the concertation of the GHS in tumor was found extremely higher than in normal cells. The mean of the size of the HECS-ss-OA micelles, which is the peak of the curve, increased when 10 mM or 20 mM GSH was added. And the rate of increase in mean particle size was controlled by the concentration of GHS. The result indicated the redox-sensitive micelles have the ability to remain stable in normal conditions, but in a high reducing agent concentration, micelles will be disassembled, which indicated the suitability of redox-sensitive micelles for specific redox-mediated intracellular delivery of anticancer drugs. The tendency of the curve indicated cumulative release percentage curves of PTX from the micelles at variable GSH concentrations. The PTX release from the HECS-ss-OA micelles was slow when the GSH level (10 μM) is low, and only 15.8% of PTX was released after 6 hours incubation. After 24h, 21.3% was released (Figure 12C). However, under high concentrations of GSH (20 μM simulating the tumor GSH concentration, the drug release rate from the redox-sensitive micelles increased sharply. For example, under 20 mM GSH condition, approximately 71.3% cumulative release was observed after 6 hours, and 95.4% release rate was observed after 1 day. The reduction-sensitive polymeric micelle can be rapidly released in the intracellular tumor environment, due to the explosion to high concentrations of GSH.

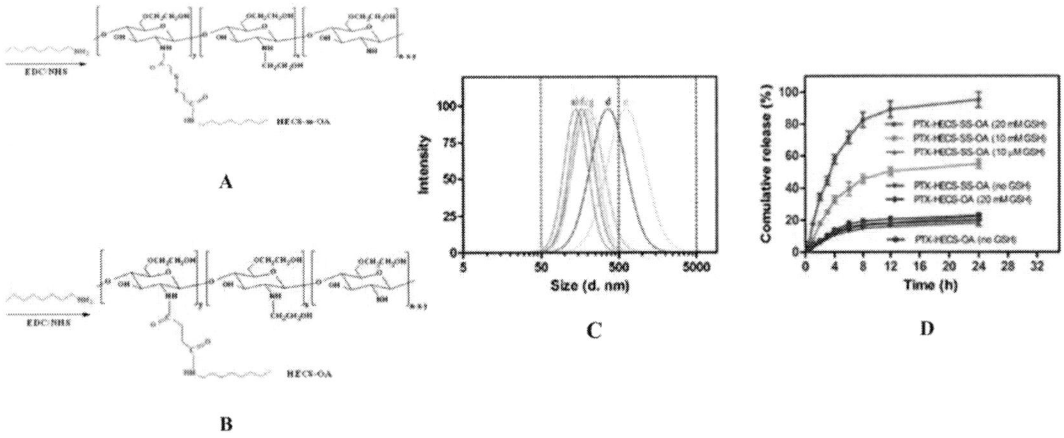

Figure 12. (A) The chemical structure of redox-sensitive HECS-ss-OA conjugate. (B) Non-redox-sensitive HECS-OA conjugate. (C) The comparison of size HECS-OA micelles and HECS-ss-OA micelles and intensity at different concentrations of GSH determined by DLS for 36h. (a) HECS-ss-OA micelles before GSH addition. (b) HECS-ss-OA micelles in the absence of GSH. (c) HECS-ss-OA micel fff le s were incubated alongside 10 μM GSH. (d) HECS-ss-OA micelles were incubated alongside 10 mM GSH. (e) HECS-ss-OA micelles were incubated alongside 20 mM GSH. (f) HECS-OA micelles before adding GSH. (g) HECS-OA micelles incubated alongside 20 mM GSH. (D) The plot of time (h) vs. the p ercentage of cumulative release of the release of HECS-OA micelles and HECS-ss-OA micelles trigged by GSH. The standard deviation represents in the error bar (n=3).

2.4.2 folic acid -decorated amphiphilic polymer–CPT

In the current study, folic acid (FA)-decorated amphiphilic polymer-CPT conjugates with disulfide linkages (FSC) were used to design novel reduction-sensitive polymeric micelles (Figure 13A). The active CPT was released by the linker between CPT and amphiphilic poly(ethylene glycol)-b-poly(ε-caprolactone) (PECL) copolymers to respond to GSH [29]. FA has been used widely, due to its property of accurate target to tumor cells, and the ability to avoid normal nonspecific attacks on normal tissues. The polymeric micelles were produced by FSC and FC conjugates. Moreover, FSC-16 micelles were synthesized from hexadecanol, and FSC (or FC) conjugates were dissolved in THF. To examine the responsiveness, the CPT release from FSC and FSC-16 micelles were investigated in the presence or absence of GSH and compared with those of FC and FC-16 micelles. GSH concentrations of 2 μM and 10 and 40 mM were used to represent the reductive microenvironment of extracellular space and cytoplasm of normal and tumour cells, respectively.

To determine the responsiveness of the reduction-sensitive polymeric micelles in tumor cells, the CPT release from FSC and FSC-16 micelles was investigated under contains GHS or no GHS condition. The concentrations of GHS included 2 μM and 10 and 40 mM were used to represent the environment of extracellular space and cytoplasm of normal and tumor cells. The rate of CPT release from FSC micelles is influenced by the concentration of GHS positively (Figure 13B). Meanwhile, in 10mM GHS condition, 38% of CPT release was detected for FSC micelles after incubation 100 h, shown the reductive sensitivity of CPT release. Advantages in premature release during circulation can influence cellular uptake efficiency in cancer cells and promote the intracellular release by designing amphiphilic

 https://doi.org/10.1051/e3sconf/202129001014

copolymers containing reduction-sensitive linkages between CPT and polymers. And FA is used as a targeting section.

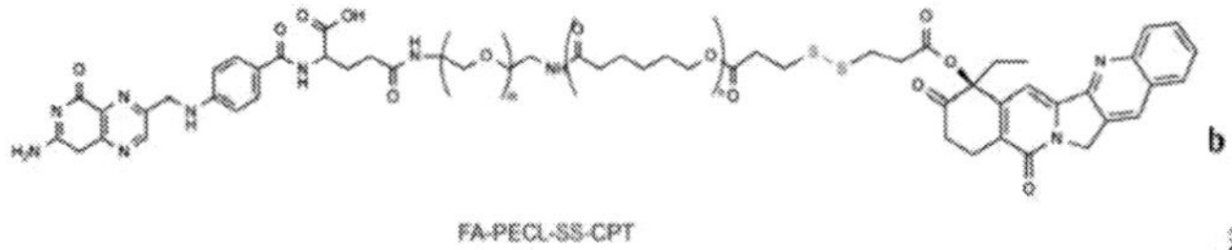

A

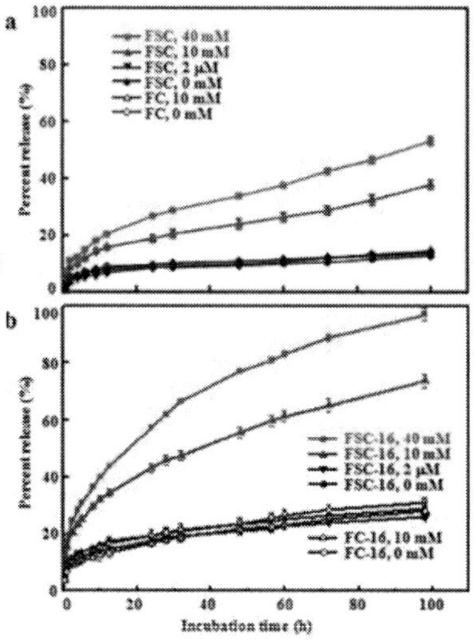

B

Figure 13. (A) The chemical structure of FA-decorated polymer–drug conjugates containing disulfide linkages (red line) between CPT and amphiphilic PECL copolymers. (B) The comparison of incubation time and percentage release of CPT from (a) FSC and (b) FSC-16 micelles incubation at 37 °C in standard PBS and PBS with 2 μM and 10 and 40 mM GSH, compared with FC and FC-16 micelles without disulfide linkage.

3 Modern technology

Recent technological advancements have fostered the use of polymeric micelles as drug delivery systems, particularly for lipophilic drugs with several benefits such as small size(10-100nm), prolonged time circulation, enhanced accumulation of drugs at tumor sites, and shell-core structure. Due to polymeric micelles being relatively more stable compared to surfactant micelles, it is preferred for its ability to solubilize hydrophobic molecules in their main core. They are also being used to transport antisense, plasmid DNA, oligonucleotides, and deliver diagnosis agents to particular organs in the body.

Polymeric micelles have emerged as potential carriers of poorly water-soluble drugs, and they provide attractive traits. For that reason, new technologies have led to the development of several methods for preparing polymeric micelles, such as the dialysis method, solid dispersion, oil in water emulsion, and microphase separation method. The dialysis methods involve adding water to the polymer solution and drugs in water-miscible organic solvents. The solid dispersion methods are conducted by dissolving the polymer and the drug in an organic solvent to obtain a solid polymer matrix after the evaporation of the solvent under reduced pressure. The oil in water emulsion solvent evaporation approach involved dissolving drugs and polymers in water-immiscible organic solvents and slowly adding distilled water under vigorous stirring to rearrange the polymer to create micelles. Ultimately, the microphase separation method is conducted by dissolving the polymer and the drug in an organic solvent and adding the solvent in water under magnetic stirring to spontaneously for polymeric micelles.

4 Conclusion

Polymeric micelles have been subject to various scientific studies over the last decade. They have been studies as versatile and efficient drug delivery systems. For that reason, new designs have been developed to create polymeric micelles to increase drug loading capacities. Various strategies have been formulated to enhance the interaction of polymeric micelles with carriers to maximize loading capacities, such as hydrotropic polymers. However, there is a need for more systematic studies on polymeric micelle structures and their activity relationships to enhance the understanding of the mechanisms of polymeric micelles and their interactions with carriers and the effect of their structure on the loading capacity of carriers.

References

1. T. Millera, S. Breyer, G.Colena, W. Mier, U. Haberkorn, S. Geissler, S.Voss, M. Weigandt, A. Goepferich, I.J.Pharm **445**, 118 (2013)

2. H. Cho, T. Lai, K. Tomoda, G. S. Kwon, AAPS PharmSciTech **16**, 13 (2015)
3. C. Oerlemans, W. Bult, M. Bos, G. Storm, J. W. Nijsen, W. E. Hennink, Pharm. R **27**, 2569 (2010)
4. V.P. Torchilin, Pharm. R **24**, 1 (2007)
5. Q. Zhou, L. Zhang, T. Yang, H. Wu, Int. M **13**, 2925 (2018)
6. M. Ghezzi, S. Pescina, C. Padula, P. Santi, E. Del Favero, L. Cantù, S. Nicoli, J. C. R **332**, 313 (2021)
7. M. Amin, A. M. Butt, M. W. Amjad, P. Kesharwani, *Nanotechnology-Based Approaches for Targeting and Delivery of Drugs and Genes*, 168 (2017)
8. K. Matyjaszewski, J. Xia, Chem. Rev. **101**, 2922 (2001)
9. J. Wang, K. Matyjaszewski, J. Am. Chem. Soc **117**, 5614 (1995)
10. K. Matyjaszewski, Macro. M **45**, 4016 (2012)
11. B. S. Lokitz, A. W. York, J. E. Stempka, N. D. Treat, Y. Li, W. L. Jarrett, C. L. McCormick, Macro.M **40**, 6473 (2007)
12. S. Perrier, Macro.M **50**, 7434 (2017)
13. N. Rapoport, Pro. Poly. Sci. **32**, 964 (2007)
14. H.Choudhury, P. Manisha, B. Gorain, B. Chatterjee, T. Madheswaran, S. Md, K.K. Mak, M. Tambuwala, M.K. Chourasis, P. Kesharwani, NETLC, 218(2019)
15. C.J.F. Rijcken, O. Soga, W.E. Hennink, C.F. van Nostrum, J. CR **3**, 136(2007)
16. I. F. Tannock, D. Rotin, PCR **49**, 4373 (1989)
17. A.E. Felber, M.H. Dufresne, J.C. Leroux, ADDR **64**, 985 (2012)
18. M.c. Jones, M. Ranger, J.C Leroux, Bio Chem **14**, 775 (2003)
19. Y.J. Lee, H.C. Kang, J. Hu, J.W. Nichols, Y. S. Jeon, Y.H. Bae, Biomacromolecule **13**, 2946 (2012)
20. H. Lee, J. Pietrasik, K. Matyjaszewski, Macro. M **39**, 3914 (2006)
21. J. Jiang, X. Tong, Y. Zhao, J. Am. Chem. Soc. **127**, 8290 (2005)
22. B. Yan, J. Boyer, N. R. Branda, Y. Zhao, J. Am. Chem. Soc. **133**, 19715 (2011)
23. B. Li, Y. Wu, Y. Wang, M. Zhang, H. Chen, J. Li, R. Liu, Y. Ding, A. Hu, ACS Appl. Mater. Interfaces **11**, 8897 (2019)
24. S. V. Aathimanikandan, E. N. Savariar, S. Thayumanavan, J. Am. Chem. Soc. **127**, 14922 (2005)
25. O. Jung, Soft Matter 7, 5096 (2011)
26. J. Akimoto, M. Nakayama, K. Sakai, T. Okano, Mol. Pharm. **7**, 928 (2010)
27. F. Meng, W. E. Hennink, Z. Zhong, Bio. **30**, 2180 (2009)
28. M. Huo, Y. Liu, L. Wang, T. Yin, C. Qin, Y. Xiao, L Yin, J Liu, J. Zhou, Mol. Pharm **13**, 1751 (2016)
29. C. Liu, J. Yuan, X. Luo, M. Chen, Z. Chen, Y. Zhao, X. Li, Mol. Pharm **11**, 4260 (2014)

Properties of heat induced blend gel of Peruvian squid (*Dosidicus gigas*) myofibril protein fortified with millet (*Panicum miliaceum L.*) flour

Tian Wang[1], Xuepeng Li[1], Jing Xie[2], Jianrong Li[1] and Geng Mi[1,2,a]
[1]College of Food Science and Technology, Bohai University, 121013 Jinzhou, Liaoning, China
[2]College of Food Science and Technology, Shanghai Ocean University, 201306 Shanghai, China

Abstract: Weak gelation ability and simple nutrient composition of squid myofibril protein limits its further development in surimi production. Millet flour is rich in nutrients and has a good viscoelasticity after heating. Physicochemical properties of Peruvian squid myofibril protein based blend gel fortified with millet four were studied. Rheological test results showed that all blend gels exhibit solid-like characters. With the addition of millet flour, the composite modulus of all the samples showed a trend of decreasing first and then increasing. The minimum value appeared around 48°C. Moreover, it was found that gel strength, water holding capacity, hardness, gumminess and chewiness showed a trend of increasing at first and then decreasing. The blend gel with 4% millet flour addition exhibit good performance in those testing. The subsequent analysis of water distribution further explained and proved the mechanism. In summary, the blend gel of squid myofibril protein fortified with 4% of millet flour addition was considered the most suitable amount for enhancing the quality of blend gel.

1 Introduction

Consumption of surimi food is one of the most distinguished cultures prevalent in Asia [1-2]. Peruvian Squid contains a large amount of salt-soluble protein and a low level of fat, making it an ideal source for surimi products. However, the weak gel strength and simple nutrient composition limits its further development. Millet flour (MF) is a popular health raw material for product cereal food, especially in the north of China. It has a favourable natural yellow colour and contains a lot of wholesome ingredients such as unsaturated fatty acid, dietary fiber, vitamin E, Iron and phosphorus [3]. Moreover, it was reported that the MF is significantly richer in essential amino acids than wheat flour [4]. There are two main types of MF according to the amylopectin content [5]. *Panicum miliaceum L* contains a higher proportion of amylopectin, which makes it exhibit higher viscoelasticity after cooking. Therefore, MF has potential application value as a functional additive. Adding MF into surimi can not only reduce the cost, enhance nutrition, but also has a potential ability to improve the quality of products. However, there is still a lack of relevant research.

In this study, the different levels of MF were added to Peruvian squid surimi and then formed gels after experiencing the heat process. The gel properties, chewing characteristics, moisture distribution and rheological properties of blend gels were investigated. The results of this research will provide useful information for the further development and utilization of MF, and provide reference for the development of additives with improved gel products.

[a] Corresponding author's e-mail: migeng1979@bhu.edu.cn

2 Materials and methods

2.1 Materials

Peruvian squid were purchased from a local Aquatic Products Market. The raw MF was purchased from Inner Mongolia Mengqing Agricultural Technology Development Co., Ltd. The starch content of the MF was 70-74 g/100 g, and the protein content was 12-14 g/100 g. All chemicals used were of analytical grade.

2.2 Preparation of MP/MF blends

The squid was thawed for 5 min and then cut into slices of 1 mm thickness. After soaking in 5% sodium bicarbonate at 4°C for 30 min, and then carrying out dehydration treatment. The prepared fish was put into a vacuum freezing chopper (UMC5, Stephan Machinery Co., Hameln, Germany). Then the slices were chopped at 1500 rpm for 2 min at 4°C and 6 mbar. About 3% NaCl and 4°C water was added to obtain 80% moisture of the myofibril protein (MP), and then continue chopped at 1500 rpm for 2 min. Finally, different levels of MF were added (0%, 1%, 2%, 4%, 6%, 8% and 15%), and chopped

https://doi.org/10.1051/e3sconf/202129001015

for 5 min at 3000 rpm. The blend samples were marked as PF_0, PF_1, PF_2, PF_4, PF_6, PF_8, PF_{15}, respectively.

2.3 Determination of rheological properties

The rheological properties of the raw blends were measured according to the method of Fan et al. [6]. The complex shear modulus (|G*|) is calculated according to the following equation:

$$|G^*| = \sqrt{(G')^2 + (G'')^2} \quad (1)$$

2.4 Gel strength and cooking properties

The prepared blend paste was extruded into a plastic casing with a diameter of 30 mm. After that, it was heated at 90°C for 20 min, put immediately in 0°C water for 10 min, then all the samples were stored in 4°C refrigerator for 12 h before testing [7]. The gel strength of the sample was measured by the method of Chen et al. [8]. The water holding capacity (WHC) of the sample was determined by a centrifugation method [9]. The whiteness was measured by the method of Petcharat et al. [10].

2.5 Chewing properties

The TPA test was measured according to the method of Alipour et al. [11].

2.6 Low-field NMR

The spin-spin relaxation time T_2 was measured by the method of Song et al. [12].

2.7 Statistical analysis

All the results were represented by mean ± std. (standard deviation). Statistical analysis was performed using IBM SPSS 22.0 (SPSS Inc, Chicago, USA). All experiments were repeated at least 3 times.

3 Results and discussion

3.1 Rheological properties

The results of temperature sweep of the gel are shown in Figure 1. The complex shear modulus (|G*|) is a combination parameter describing the intensity of the gel, a higher |G*| value indicates a high-intensity gel. Tanδ (G"/G') < 1 indicates that the elasticity of the gel is better than the viscosity [13]. In the whole gel heating period, the |G*| of all samples showed a trend of decreasing first and then increasing. It was found when all the samples were heated to about 48°C, the |G*| showed its lowest value. Previous studies have shown that there are large amounts of metallo-proteinase in squid, which will be activated around 40-50°C and damage the structure of protein network [14]. On the one hand, with the temperature increasing, the |G*| of the blends increased obviously. It is assumed that the protein network gradually formed. On the other hand, as the amount of MF added increased, it can be seen the |G*| shown a trend of decreasing first and then increasing. We speculated that there was a destructive effect of MF on the protein network. But at a high addition ratio, the starch in the MF formed its own continuous network, which enhanced the strength of the blend.

Tanδ represents the ratio of the elastic modulus to the viscous modulus of the blend. It can be seen from the result that the Tanδ of all samples showed the maximum value around 35°C, which may be due to the looseness of salt-soluble protein tail denaturation and unwinding at the temperature [15]. In addition, it can be seen that the Tanδ of the blends increased obviously with the increase of MF at the temperature of 50°C, and then began to decline near 80°C, finally tended to a similar value. This may be because the starch in the MF began to absorb water and swell around 50°C, and the viscosity of the system increased. When heated to around 80°C, starch gelatinization occurs, where the amylose in the system begins to crosslink, and the viscosity modulus of the system gradually decreases while the elastic modulus increases.

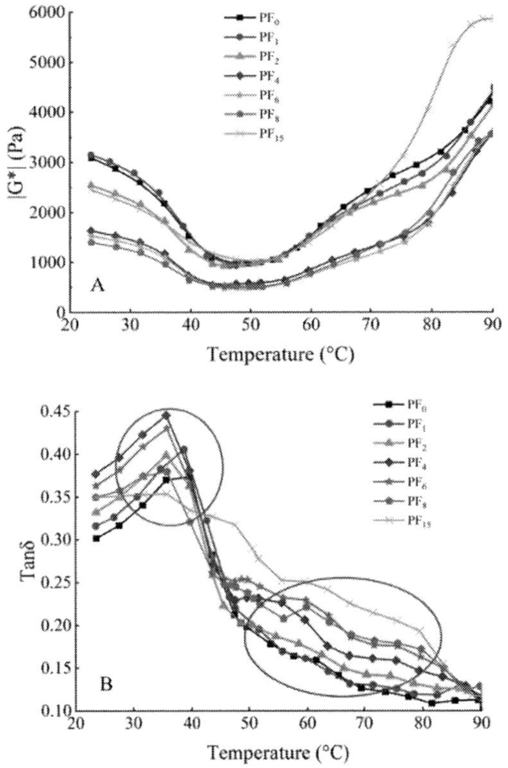

Figure 1. Effects of heating (5°C/min) on (A) complex shear modulus (|G*|) versus temperature (°C); and (B) loss factor (tanδ) versus temperature (°C) for Peruvian squid MP added with different concentrations of MF.
The MP added with different amounts of MF (0%, 1%, 2%, 4%, 6%, 8% and 15%), was marked as PF_0, PF_1, PF_2, PF_4, PF_6, PF_8, PF_{15}, respectively.

3.2 Cooking properties

The cooking properties of MP-MF blend gels are shown in Table 1. The results suggest that the gel strength increased first and then decreased with the addition of MF. At the addition level of MF was 4%, the gel strength increased its maximum value of 976.00 g·mm. At low MF levels of samples like PF_1 to PF_2, we speculate that MP formed a continuous network, and starch act as filler, which had been reported to have a significant effect in increasing gel strength [16]. However, too much of MF would reduce the gel strength. The millet flour used in our experiment was rich in amylopectin, which could cause the MF to exhibit higher viscidity after being cooked. This may account for the decrease in gel strength of samples like PF_6-PF_{15}.

It can be seen that the WHC of the gels also show a trend of first increasing and then decreasing. It was found when the additional amount of MF was 4%, the WHC of the gel reached the maximum value of 80.04%. This may be related to the structure of protein network of MP. Moreover, the addition of MF can significantly ($P < 0.05$) reduces the whiteness of MP gels (except for the 1% addition of MF). This may be due to the strong yellow colour of MF. Similar results also were described by Romero et al. [17], who used MF to make pasta and results indicated that the carotenoids and lutein contained in MF make the noodles attractive yellow.

Table 1. Gel strength and cooking properties of Peruvian squid MP added with different concentrations of MF.

Treatment	Gel strength (g·mm)	WHC (%)	Whiteness
PF_0	714.47 ± 16.00^{d}	69.71 ± 1.02^{c}	79.92 ± 0.35^{a}
PF_1	808.53 ± 29.07^{c}	73.78 ± 1.84^{bc}	79.83 ± 0.24^{a}
PF_2	904.32 ± 24.40^{b}	76.80 ± 1.34^{b}	77.46 ± 0.16^{b}
PF_4	976.00 ± 3.93^{a}	80.04 ± 0.15^{a}	76.62 ± 0.24^{bc}
PF_6	822.17 ± 2.82^{c}	79.78 ± 0.57^{a}	76.23 ± 0.41^{c}
PF_8	565.31 ± 16.85^{e}	78.98 ± 1.11^{ab}	73.40 ± 0.38^{d}
PF_{15}	408.56 ± 18.90^{f}	76.96 ± 0.52^{b}	68.44 ± 1.91^{e}

[a] Different letters within the same column mean significant difference ($p < 0.05$) between mean value.
[b] The MP added with different amounts of MF (0%, 1%, 2%, 4%, 6%, 8% and 15%), was marked as PF_0, PF_1, PF_2, PF_4, PF_6, PF_8, PF_{15}, respectively.

3.3 Chewing properties

The TPA test is also called a two-bite test. It simulates the bite movement of the human mouth, which is equipped with a probe to compress the sample twice to explore the changes of the sample when it is chewed. The chewing properties of the blend gels are shown in Table 2. The hardness, gumminess and chewiness of the gels first increased and then decreased with the increasing additional level of MF. When MF was 4%, the hardness, gumminess and chewiness reached the maximum values, which were 3276.53 g, 2530.28 g and 2244.44 g, respectively. When the additional amount exceeds 4%, the chewing properties of the gels became increasingly worse. The results are similar to the study of Yu et al. [18], their result showed that 15% acetylated distarch phosphate added into soy protein showed a similar trend. The reason for this reduced effect may be caused by destruction of MF on the network of MP.

Furthermore, the increasing level of MF of the blend samples leads to competition between protein and starch for water molecules, and it may hinder the MP to access enough water in order to form a three-dimensional gel network by heating processing [11,19].

Table 2. Chewing properties of Peruvian squid MP added with different concentrations of MF.

Treatment	Hardness(g)	Springiness	Cohesiveness	Gumminess(g)	Chewiness(g)	Resilience
PF_0	2403.29 ± 116.21^{cd}	0.89 ± 0.00^{bc}	0.79 ± 0.02^{ab}	1896.05 ± 114.71^{cd}	1691.00 ± 92.98^{c}	0.48 ± 0.01^{a}
PF_1	2928.63 ± 39.86^{b}	0.89 ± 0.02^{bc}	0.77 ± 0.00^{b}	2258.65 ± 38.08^{b}	2015.33 ± 50.02^{b}	0.45 ± 0.00^{b}
PF_2	2497.95 ± 30.52^{c}	0.87 ± 0.01^{c}	0.77 ± 0.01^{b}	1933.36 ± 35.97^{c}	1686.27 ± 52.65^{c}	0.46 ± 0.01^{b}
PF_4	3276.53 ± 83.40^{a}	0.89 ± 0.00^{bc}	0.77 ± 0.00^{b}	2530.28 ± 68.27^{a}	2244.44 ± 54.34^{a}	0.45 ± 0.00^{b}
PF_6	2303.66 ± 0.16^{d}	0.92 ± 0.01^{a}	0.78 ± 0.00^{ab}	1794.69 ± 9.90^{d}	1658.80 ± 35.09^{c}	0.45 ± 0.00^{b}
PF_8	1070.80 ± 19.83^{f}	0.91 ± 0.01^{ab}	0.80 ± 0.00^{a}	855.88 ± 13.49^{f}	779.35 ± 24.94^{e}	0.49 ± 0.01^{a}
PF_{15}	1655.62 ± 9.81^{e}	0.91 ± 0.00^{ab}	0.77 ± 0.01^{b}	1272.57 ± 3.57^{e}	1158.45 ± 2.62^{d}	0.46 ± 0.01^{b}

[a] Different letters within the same column mean significant difference ($p < 0.05$) between mean value.
[b] The MP added with different amounts of MF (0%, 1%, 2%, 4%, 6%, 8% and 15%), was marked as PF_0, PF_1, PF_2, PF_4, PF_6, PF_8, PF_{15}, respectively.

3.4 Water distribution analysis

The distribution of spin-spin relaxation time T_2 of the blend gels is shown in Figure 2, and their corresponding

https://doi.org/10.1051/e3sconf/202129001015

proportions are shown in Table 3. The results show that the proportion of P_{22} increased with the increase of MF. That may be due to water absorption and swelling of the starch in MF, which breaks the internal hydrogen bond and increases the monolayer of molecular water on the molecular surface. P_{23} showed an obvious trend of increasing at first and then decreasing. The maximum value appeared when 4% MF was added. We speculate that the starch in the MF has formed an interpenetrating gel network due to thermodynamic incompatibility with the salt-soluble proteins, which enhances the water retention capacity of the blend [6]. However, the starch in the MF may damage the gel network of the blend with the increase of the amount of MF, which reduces the water content in the network. The results are consistent with the previous studies results of Nishizawa et al. [20] and the WHC results in section 3.2.

Table 3. Peak ratio of populations of Peruvian squid MP added with different concentrations of MF.

Treat-ment	P_{22} (%)	P_{23} (%)	P_{24} (%)
PF_0	1.13 ± 0.08^{e}	92.61 ± 0.25^{d}	5.55 ± 0.40^{a}
PF_1	1.11 ± 0.12^{e}	93.75 ± 0.05^{b}	2.90 ± 0.10^{bc}
PF_2	1.38 ± 0.15^{de}	94.43 ± 1.02^{abc}	3.88 ± 1.16^{b}
PF_4	1.73 ± 0.11^{d}	94.47 ± 0.12^{a}	2.91 ± 0.20^{b}
PF_6	2.32 ± 0.31^{c}	93.95 ± 0.80^{abc}	3.27 ± 1.11^{b}
PF_8	2.94 ± 0.12^{b}	93.54 ± 1.71^{abc}	1.71 ± 0.00^{d}
PF_{15}	3.31 ± 0.45^{a}	93.30 ± 1.12^{c}	2.19 ± 0.75^{cd}

[a] Different letters within the same column mean significant difference ($p < 0.05$) between mean value.
[b] The MP added with different amounts of MF (0%, 1%, 2%, 4%, 6%, 8% and 15%), was marked as PF_0, PF_1, PF_2, PF_4, PF_6, PF_8, PF_{15}, respectively.

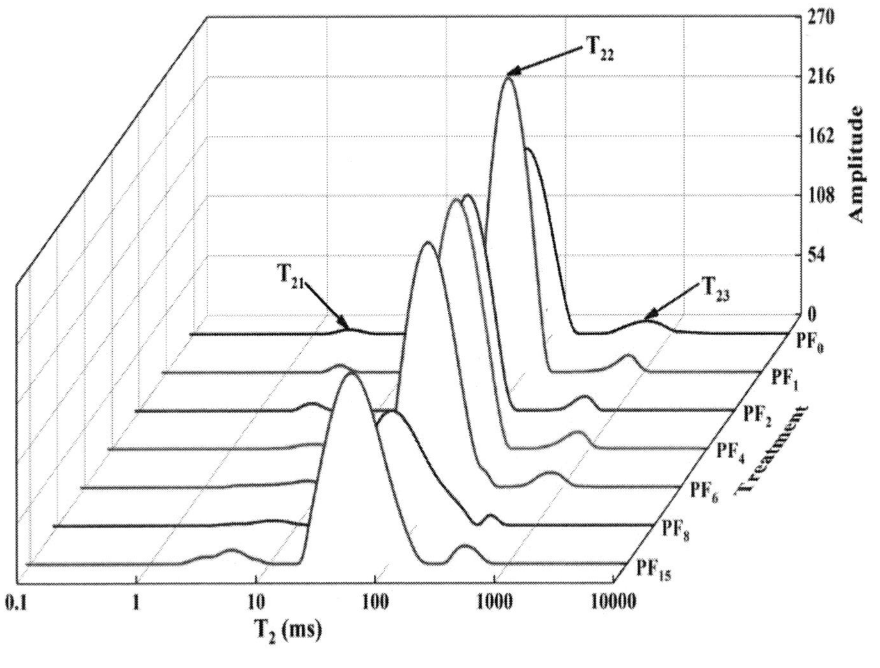

Figure 2. Water states of Peruvian squid MP added with different concentrations of MF.
The MP added with different amounts of MF (0%, 1%, 2%, 4%, 6%, 8% and 15%), was marked as PF_0, PF_1, PF_2, PF_4, PF_6, PF_8, PF_{15}, respectively.

4 Conclusions

This work mainly studied the effect of millet flour on the quality of squid myofibrillar protein blend gel, and demonstrated a feasible solution for preparing a new type of surimi product. The rheological analysis showed that the complex shear modulus of all samples showed a trend of first increase and then decrease and the lowest values appeared around 48°C. Moreover, all samples showed solid-like character after the blends experienced heat processing. With the increased addition of MF, the gel strength, WHC, and hardness, gumminess and chewiness of all samples also showed a trend of increasing at first and then decreasing. MF gave the blends a bright yellow colour. Furthermore, according to the results of moisture distribution analysis, 4% MF addition will enhance the water retention capacity of blend gel. In terms of application, 4% addition of MF was considered to be the most suitable for the production of blend gel products.

Acknowledgments

This work was supported by the National Natural Science Foundation of China (31972107, U20A2067).

References

1. J.S. Córdova-Ramos, U. Gonzales-Barron, L.M. Cerrón-Mallqui, LWT-Food. Sci. Technol., **93**, 506-510 (2018)
2. R.R. Olivas, O.R. Sández, N.F. Haard, R.P. Aguilar, J.M.E. Brauer, Eur. Food. Res. Technol., **219**, 312-315 (2004)
3. K. Lorenz, W. Dilsaver, L. Bates, Cereal. Chem., **57**, 16-20 (1980)
4. J. Kalinova, J. Moudry, Plant. Food. Hum. Nutr., **61**, 43-47 (2006)
5. Q.H. Yang, P.P. Zhang, Y. Qu, X.L. Gao, J.B. Liang, P. Yang, B.L. Feng, Food. Chem., **257**, 271-278 (2018)
6. M.C. Fan, T. Hu, S.M. Zhao, S.B. Xiong, J. Xie, Q.L. Huang, Food. Chem., **218**, 221-230 (2017)
7. F.Y. Sun, Q.L. Huang, T. Hu, S.B. Xiong, S.M. Zhao, Int. J. Biol. Macromol., **64**, 17-24 (2014)
8. Y.C. Chen, J. Jaczynski, J. Agr. Food. Chem., **55**, 9079-9088 (2007)
9. A.D. Westphalen, J.L. Briggs, S.M. Lonergan, Meat. Sci., **72**, 697-703 (2006)
10. T. Petcharat, S. Benjakul, Food. Hydrocolloid., **77**, 746-753 (2018)
11. H.J. Alipour, M. Rezaei, B. Shabanpour, M. Tabarsa, Food. Hydrocolloid., **74**, 87-96 (2018)
12. Y.K. Song, X. Zang, T. Kamal, J.R. Bi, S. Cong, B.W. Zhu, M.Q. Tan, Dry. Technol., **36**, 72-83 (2018)
13. D.Q. Zhang, T.H. Mu, H.N. Sun, Carbohyd. Polym., **176**, 345-355 (2017)
14. K. Konno, C. Fukazawa, J. Food. Sci., **58**, 1198-1202 (1993)
15. T.A. Gill, J.T. Conway, Agric. Biol. Chem., **53**, 2553-2562 (1989)
16. H.B. Mi, C. Wang, Q. Su, X.P. Li, S.M. Yi, J.R. Li, J. Texture. Stud., **50**, 571-581 (2019)
17. H.M. Romero, D. Santra, D. Rose, Y. Zhang, J. Cereal. Sci., **74**, 238-243 (2017)
18. B. Yu, F. Ren, H.B. Zhao, B. Cui, P.F. Liu, Int. J. Biol. Macromol., **142**, 237-243 (2020)
19. I. Sánchez-Alonso, M.T. Solas, A.J. Borderías, J. Food. Eng., **81**, 404-411 (2007).
20. N. Nishizawa, Y. Fudamoto, Biosci. Biotech. Bioch., **59**, 333-335 (1995)

Preparation and characterization of coal gangue/Al cermet materials

Guihua Wu[1,a], Lihua Tang[1]

[1]School of Mechanical Engineering, North University of China, Taiyuan, Shanxi

Abstract. In this study, the calcined coal gangue was mixed with Al powder at different mass fractions (5wt%～35wt%). After being pressed at 80MPa and sintered at 1200℃ for 2h, the coal gangue/Al cermet material was obtained. The phase composition, microstructure and mechanical properties of the material were studied. The results show that the main components of the prepared materials are Al, mullite, AlN, alumina and silicon. With the increase of coal gangue powder, the hardness of coal gangue/Al cermet gradually increases, while the static compression strength first increases and then decreases. When the mass fraction of gangue is 25wt%, the coal gangue/Al cermet material has the best performance. Density is 2.23g/cm^3, hardness is 752HV, static compression strength is 127.3MPa.

1 Introduction

These Coal gangue is a kind of solid waste which is separated in the process of coal production, and it increases and accumulates in large amount every year. It is estimated that China's coal gangue output will reach 729 million tons in 2020 [1]. At present, coal gangue is mainly used in filling materials and power generation, which have low utilization rate and low added value[2]. In order to solve the environmental problems caused by the accumulation of coal gangue, it is necessary to explore the efficient and high-value utilization of coal gangue. Through composition analysis, the main components of gangue removing organic impurities are SiO_2 and Al_2O_3 (>90%), which are the raw materials for preparing ceramics with excellent performance[3]. Therefore, the utilization of silica-alumina ceramic raw materials in coal gangue has become a research hotspot[4].

Cermet is a composite material consisting of a metal or alloy and one or more ceramic phases. Ceramics not only maintain the high hardness, wear resistance, corrosion resistance, high temperature resistance, oxidation resistance and chemical stability of ceramic materials, but also have high toughness and high thermal and electrical conductivity of metal materials[5,6]. The results show that adding ceramic particles into Al matrix can improve the strength, hardness, wear resistance and corrosion resistance[7,8]. At present, there are few studies on the preparation of coal gangue/Al cermet materials from coal gangue and Al.

In this study, the preparation of coal gangue/Al cermet materials was carried out by using coal gangue removed organic matter as raw material and adding it into Al powder according to different mass fraction. Then the effects of different formulations on phase composition, mechanical properties and microstructure of gangue/Al cermet materials were explored.

2 Materials and Characterization

2.1 Material preparation

Coal gangue (Pingshuo Opencast Coal Mine, Shanxi, China)and Al powder (45um, 99.9% purity, Zhongnuo New Material (Beijing) Technology Co., Ltd)as raw materials. The coal gangue is ground to micron level and calcined at 700℃ for 2 hours to remove the organic matter. The calcined powder was placed in a ball-milling tank with the mass ratio of ball: powder: ethanol = 3:1:1, and the powder was ball-milling for 3 h at 200 r·min-1. The milled powder was dried in a vacuum drying oven at 100 ℃ for 12 h, and then sieved through a 200 mesh steel screen. The ball-milled powders were added into Al powder (45um) at different mass fractions (5, 15, 25, 35wt%) and mixed in a mixer for 2h. The mixed powder was put into a mold with a diameter of 12 mm and pressed into shape at 80MPa for 30min. The blank was placed in an alumina crucible and put into a Maeboiling furnace (model SX2-8-16) at a heating rate of 10℃·min-1 to 1200℃ for 2 h. After cooling, it was taken out. The samples were polished on the metallographic polishing machine.

2.2 Characterization

BSA224S-CW electronic balance was used to measure the density. LEXT OLS4100 laser confocal microscope was used to observe the surface morphology. SmartLab-3KW X-ray diffractometer was used to test the image

[a] Corresponding author:guihuawu_nuc@163.com

composition of sintered materials. The microstructure of the samples was observed by SU5000 scanning electron microscope. The Vickers hardness of ceramics was measured by DHV-1000 digital microhardness tester. CMT5105 universal testing machine was used to test the static compression strength.

3 Results and discussion

3.1 Phase composition of coal gangue/Al cermet materials

XRD patterns of coal gangue/Al cermet materials are shown in Fig. 1. Phase composition of coal gangue /Al cermet materials,the phase composition of cermet material is Al (PDF#04-0787), mullite (PDF#15-0776), AlN (PDF#25-1133), alumina (PDF#10-0173) and silicon (PDF#37-1484). With the increase of coal gangue, the contents of AlN, alumina, mullite and silicon in the material increase with the increase of coal gangue, while the contents of Al decrease continuously. It is inferred that the Following chemical reaction occurred:

Mullite formation reaction:

$$Al_2O_3{\cdot}2SiO_2{\cdot}2H_2O \xrightarrow{450\sim550℃} Al_2O_3{\cdot}2SiO_2+2H_2O \quad (1)$$

$$2(Al_2O_3{\cdot}2SiO_2) \xrightarrow{>950℃} Si_3Al_4O_{12}+SiO_2 \quad (2)$$

$$3Si_3Al_4O_{12} \xrightarrow{>1050℃} 2(3Al_2O_3{\cdot}2SiO_2)+5SiO_2 \quad (3)$$

Silicon formation reaction:

$$4Al+3SiO_2 \xrightarrow{>860℃} 3Si+2Al_2O_3 \quad (4)$$

AlN formation reaction:

$$2Al+N_2 \xrightarrow{>800℃} 2AlN \quad (5)$$

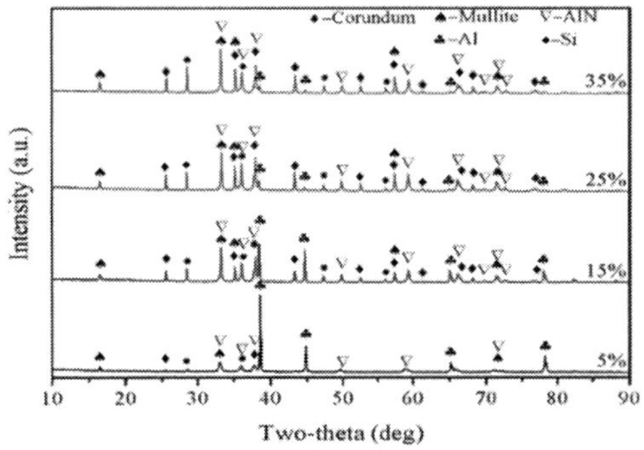

Figure 1. Phase composition of coal gangue /Al cermet materials.

It can be seen from XRD pattern that the sample with 5wt% coal gangue generated less AlN than other samples. The reason may be that less coal gangue is added, and sufficient Al forms a dense oxide film on the surface of the sample, which prevents Al from further reacting with air.

3.2 Density and surface morphology of coal gangue/Al cermet materials

Table 1 shows the density of samples obtained by different formulations. With the addition of coal gangue, the density of samples gets lower and lower, mainly because the density of coal gangue is lower than Al. At the same time, there are a lot of pores in the sample, which will reduce the density.

Table 1. Density of Coal Gangue/Al Cermet Materials

Addition of coal gangue wt%	Density g/cm^3
5	2.54
15	2.41
25	2.23
35	2.05

Fig. 2 shows the laser confocal micrograph of the sample surface with different formulas. It can be seen from the figure that there are black lumps in all samples, which is presumed to be ceramic strengthening phase. (a) There are a lot of scratches after grinding on the surface of the sample, and a small amount of black lumps exist on the surface. (b) There are still a few scratches after polishing on the surface of the sample, and there are a large number of black lumps. (c) scratches after grinding can hardly be observed on the sample surface, and black blobs are numerous and evenly distributed. (d) scratches after polishing were almost not observed on the surface of the sample. Compared with (c), there were more black masses and they were connected with each other.

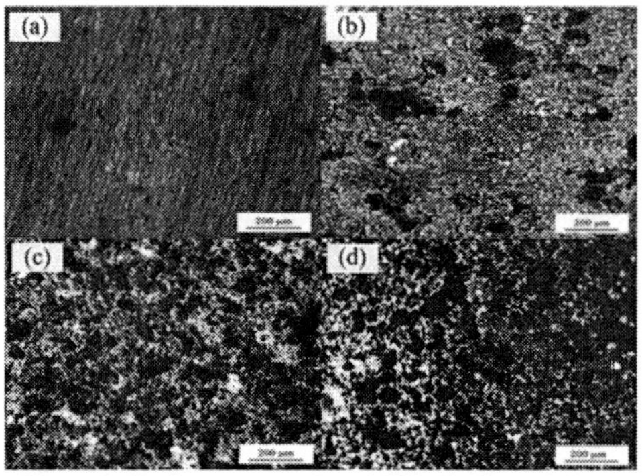

Figure 2. Surface topography of different formulations.

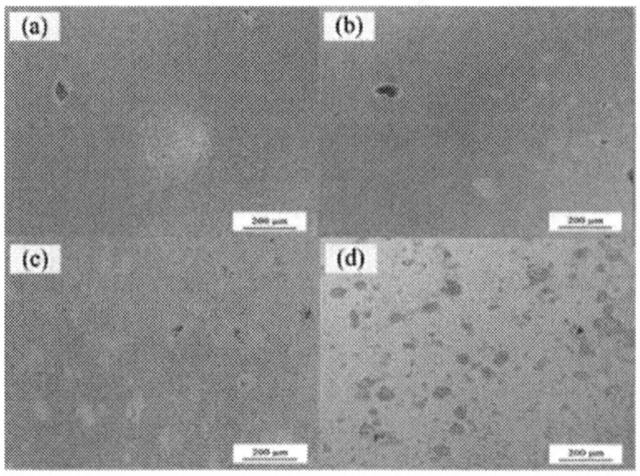

Figure 3. Surface heights of different formulations.

Fig. 3 shows the surface heights of samples with

different formulations. In the figure, each sample has surface depressing. Most of the samples do not show the deepest color in the pore's middle, but a combination of dark blue and light blue, suggesting that a network structure or other structures with different heights appear in the holes.

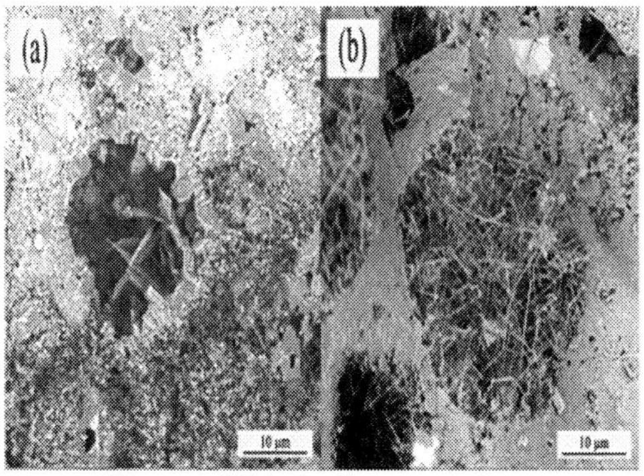

Figure 4. SEM of surface morphology.

By observing the SEM diagram of the sag area of each sample (Fig. 4), it can be found that most of the depressions have staggered columnar structures in (a) and interwoven reticular whiskers in (b).According to XRD analysis, the staggered columnar structure is mullite, and the interwoven network whiskers are AlN.

Table 2. Vickers hardness and static compression strength of Coal Gangue/Al Cermet Materials.

Coal gangue /wt%	Vickers hardness /Kg.mm-2	Static compression strength /MPa
5	154	61.8
15	513	89.6
25	752	127.3
35	723	109.1

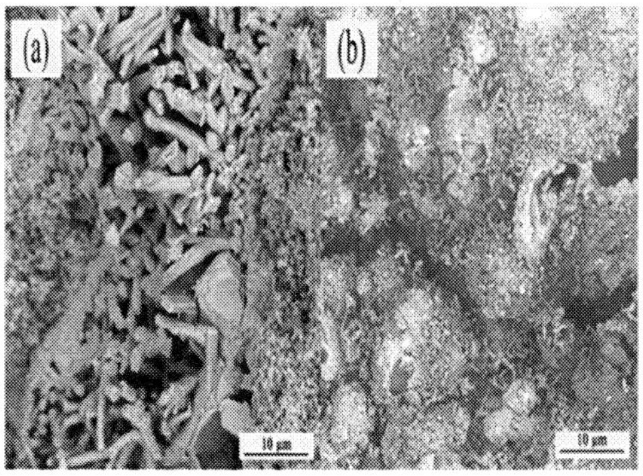

Figure 5. SEM of fracture surface.

4 Mechanical properties and structure of coal gangue/Al cermet

Table 2 shows the Vickers hardness and static compression strength of each sample. It can be seen that Vickers hardness increases with the increase of coal gangue, the reason is that the ceramic phase in the sample gradually increases, making the overall hardness of the material increase.

Among them, the Vickers hardness of the sample with 5wt% coal gangue is much lower than that of other samples, because its main constituent phase is Al. Static compression strength is also significantly related to the amount of coal gangue added, in which it can be seen that the sample with 5wt% coal gangue added is far less than the other three samples.

Through the SEM (Fig. 5) observation of the cross section of each formula sample, it was found that the columnar mullite and the AlN whisker were fractured, and the two structures increased with the increase of the addition amount of coal gangue. However, the sample with 25wt% coal gangue is higher than that with 35wt% coal gangue. It is inferred that the formation of a large amount of silicon will reduce the static compression strength of the sample.

5 Conclusion

1.When 25wt% coal gangue is added, the material performance is the best. The prepared cermet materials has a density of 2.23g/cm^3, a surface hardness of 752 HV and a Static compression strength of 127.3MPa.

2.Mullite can be formed from coal gangue at high temperature, and its interactive structure is beneficial to increase the compression resistance of cermet. At the same time, AlN whiskers formed by the reaction of Al and N_2 in the air also play a strengthening role.

3.Coal gangue can be used as the raw material source of strengthening phase of cermet, which can improve the wear resistance of metal and further broaden the utilization area of coal gangue, so as to reduce the environmental pollution caused by coal gangue accumulation.

References

1. Li J , Wang J . Comprehensive utilization and environmental risks of coal gangue: A review[J]. Journal of Cleaner Production, 2019, 239:117946.
2. Dong Z G, Li P, Jia Y X, et al.Research progress on comprehensive utilization and resource treatment of coal gangue [J]. Environmental Protection Frontiers, 2021, 11(2):9.
3. Lü, Q K, Dong X , Zhu Z , et al. Environment-oriented low-cost porous mullite ceramic membrane supports fabricated from coal gangue and bauxite[J]. Journal of Hazardous Materials, 2014, 273:136-145.
4. Lou G H, Jin B, Jiang W G, et al.Preparation of foamed ceramics from coal gangue [J]. Bulletin of Ceramics, 2020, v.39;No.283(04):276-280.

https://doi.org/10.1051/e3sconf/202129001016

5. Ru J J, He H.Preparation and research progress of ceramic particle reinforced metal matrix composites [J]. Science & Technology Innovation and Application, 2019, 275(19):122-123.
6. Cai X L,Xu Y H,Zhong L S.Fracture toughness of WC-Fe cermet in W-WC-Fe composite by nanoindentation[J]. J.Alloy.Compd.,2017,728:788-796.
7. Cong D L,Li Z S,He Q B.Wear behavior of corroded Al-Al_2O_3 composite coatings prepared by cold spray[J]. Surf.Coat.Tech.,2017,326:247-254.
8. Han Q,Geng Y,Setchi R.Macro and nanoscale wear behaviour of Al-Al_2O_3 nanocomposites fabricated by selective laser melting[J].Comp.Part B Eng.,2017,127: 26-35.

Mechanism of Oxygen Reduction Reaction for Sulfur-Doped Graphene Catalyst: First-principles research

Guifa Li[1,a], Hui Wang [1]

[1]Jiangxi University of Application Science, Institute of Intelligent Manufacturing Engineering,330100, Nanchang, Jiangxi, China

Abstract. The mechanism of oxygen reduction reaction for surfur-doped graphene catalyst was investigated by first-principles calculation. Several parameters, such geometry structure, reaction energy and electronic structure were used to analyze the reaction process. The results show that the OOH molecule absorbed on S-graphene with pyridine structure exhibits chemical adsorption. But the OOH molecule on S-graphene with pyrrole structure performs physical adsorption. The adsorbed energies of OOH molecule on S-graphene prove such phenomenon. The electronic structure shows that the excellent ORR of OOH molecule on S-graphene with pyridine structure comes from their adjacent energy level of valence electron no matter in highest occupied molecular orbital (HOMO) or lowest unoccupied molecular orbital (LUMO).

1 Introduction

Fuel cell, which can transform the chemical energy straightly into electronic energy, has high power density, quiet and non-pollution[1~3]. However, the oxygen reduction reaction (ORR) is the vital process in fuel cell, air-metal cell and other energy storages applications. Usually, the Pt/C is the main catalyst in ORR process. But it has some obvious defects, such as expensive, slow reaction process, low methanol resistance and unstability, which hinders the capability of cell and its commercialization. Then, the substitutes of gold metallic Pt attract many researchers attention. Exploring metalloid ORR catalyst, especially for doped carbon materials, has been found to a good substitute. Singla et al[4] investigated the sensing performance of grapheme by Cu/Pt dual-doping. He found that reactivity of graphene surface for hydrogen adsorption was found to be greatly enhanced with Pt@Cu co-doped graphene surface. Domga et al [5] studied the influence of N and P co-doped graphene in ORR under strong alkaline media in advanced chlor-alkali cell. He showed that the N and P co-doped graphene showed well exfoliated graphene oxide with few layers of graphene through the various active forms. Wu et al[6] produced the boron doped graphene for lithium-ion batteries anodes materials. And some elements have been doped in graphene to explore new low-cost catalyst in fuel cell[7~9]. In fact, all of the problems in doped graphene come from the electronic activity on their surface layer. Then in this paper, we research the mechanism of oxygen reduction reaction for sulfur-doped graphene catalyst by first-principles method.

2 Simulation models and method

Reported by Zhang et al[9], two sulfur hybridized graphene sheets ($C_{61}H_{19}S$ and $C_{60}H_{18}S$) were built with pyridine and pyrrole species as shown in Fig.1, respectively. No matter the carbon or sulfur atoms along the edge of graphene were terminated by hydrogen atoms. A molecular orbital package based on the density functional theory (DFT) is adopted to calculate the oxygen reduction reaction (ORR) process and electronic structures of graphene doping with sulfur. The oxygen reduction reaction processes were simulated firstly with the first electron transmission along with the intermediate molecule OOH formed. Following with the four-electron pathway presented by Zhang et al[9], the process of O2 reduced directly to water without involvement of hydrogen peroxide, $O_2+4H^++4e^-\rightarrow 2H_2O$, was only simulated in this paper. Then the ORR process of pure graphene and the other less efficient two-step two-electron pathway in which hydrogen peroxide was formed as an intermediate, $O_2+2H^++2e^-\rightarrow H_2O_2$, were not considered anymore. All of these impurity graphene models were relaxed to arrive at its global minimum energy. During optimization and total energy calculation, electronic exchange-correlation energy functions represented in reciprocal space with the Perdew-Burke-Emzerh (PBE) of functional type, which was based on a generalized gradient approximation (GGA), were used [10]. All electrons were included in the calculation.

[a] Corresponding author: L2877593958@163.com

 https://doi.org/10.1051/e3sconf/202129001017

Atomic wave functions were developed by double numerical plus d-functions (DND) basis sets [11]. Special K-point method in Monkhorst–Pack scheme [12] was used during the course of Brillouin zone integral. All atomic positions were relaxed according to total energy and force using the BFGS (Broyden, Fletcher, Goldfarb, and Shanno) scheme, based on the geometry optimization criterion ((root-mean-square) RMS force of 0.004 au/ Å and RMS displacement of 0.005 Å). The calculation of total energy and electronic structure was followed by the geometry optimization with Self-consistent field (SCF) tolerance of 2 x 10^{-5} au/atom.

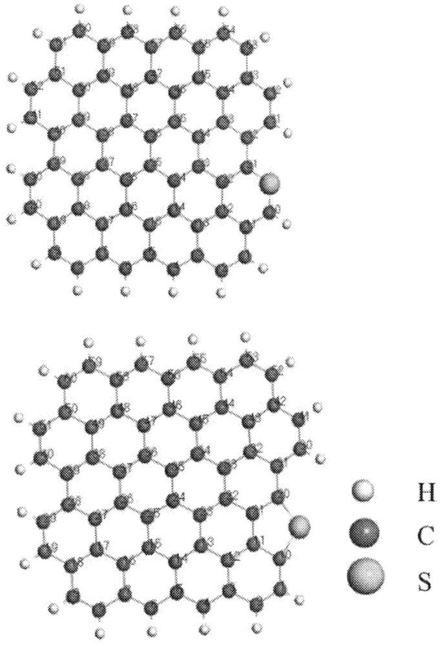

Fig.1. Sulfur-containing graphene sheets of (a) $C_{61}H_{19}S$ and (b) $C_{60}H_{18}S$

3 Results and discussion

3.1. Evolution of geometry structure

In this paper, we first investigate the ORR behavior of the S-graphene with pyridine structure (Fig. 2(a)~(h)). The variation of distance between different atoms is listed in Table 1.

In the first step of the reduction reaction, the OOH molecule moves from its original state on graphene and absorbs to a carbon atom near sulphur atom. Then the sulphur atom moves out of S-graphene plane to bond with oxygen atom (Fig.2(a)). The distance between the sulphur and oxygen atoms (O1) reduces from 3.30 Å to 1.85 Å (Table1), which the chemical bond is formed spontaneously. This is a vital step for S-graphene to obtain catalytical performance, because atomic absorption and formation of chemical bond are the precondition for the next ORR reaction.

After OOH absorbed on S-graphene, another H is introduced to such system. Randomly the H atom moves to an oxygen atom to form chemical bond. For two type of S-graphene with pyridine and pyrrole structures constructed in this paper, in the next part we will discuss both of their ORR. The ORR process is :

$$OOH \longrightarrow {}^*OOH \quad (1)$$

$$^*OOH + \mathrm{H}^+ + e^- \longrightarrow {}^*OH + {}^*OH \quad (2)$$

$$^*OH(\text{absorbed } 1) + \mathrm{H}^+ + e^- \longrightarrow H_2O \quad (3)$$

$$^*OH(\text{absorbed } 2) + \mathrm{H}^+ + e^- \longrightarrow H_2O \quad (4)$$

wherein aterisk * represents the graphene. This paper began with the reaction (1). Reactions (3) and (4) are the same reaction although the *OH is absorbed on different carbon atom near sulfur atom. After another H atom absorbs on OOH in reaction (2), the distance between two oxygen atoms and O1 –G don't change any more, but the distance between the other oxygen atom and graphene reduces to 1.80 Å from primary 2.65 Å, which means the *OH molecule prefers to bond with graphene. When the ORR process occurs at reaction (3), the first H2O molecule is produced after another H atom joining. Then the distance between two oxygen atoms increase from 1.39 Å to 2.96 Å, which means their chemical bond is broken. And we find the distance of O2-G is also increasing from 3.52 Å to 3.57 Å (Table 1). So the first H2O molecule is produced and escapes from S-graphene. Then more and more H atoms absorb on S-graphene, which produce much more H2O molecule.

About S-graphen with pyrrole structure(Fig. 2(a')~(d')), it can be found that its OOH molecular absorption ability is more hardly than that on the S-graphe with pyridine. From Table 1, the distance between the oxygen atom and S-graphene is more than 3.0 Å in every reaction steps, which means the OOH molecule don't absorb on graphene any more. Such result is different from the ORR in N- graphen[9].

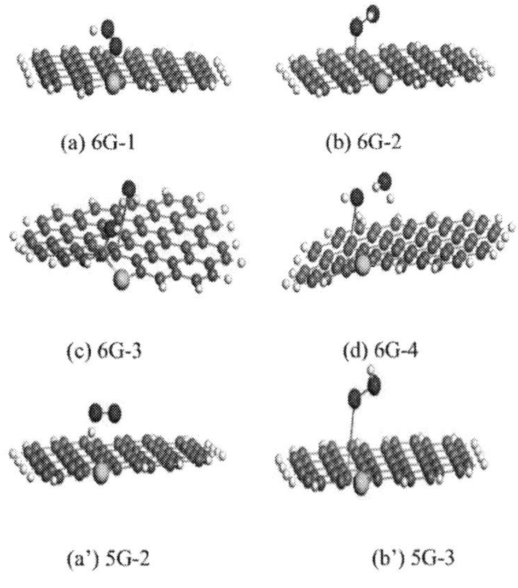

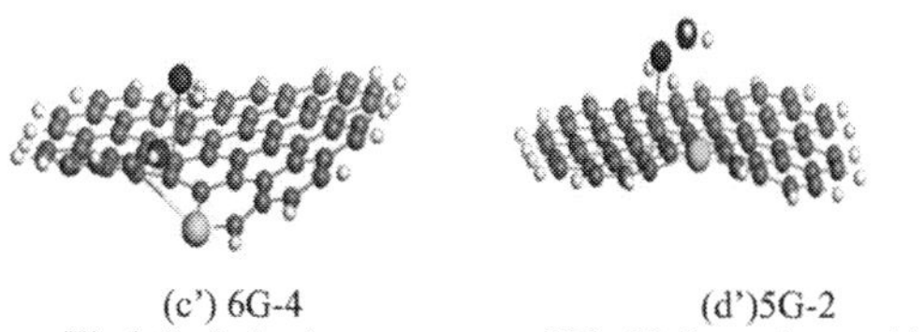

(c') 6G-4 (d')5G-2

Fig.2 Optimization structures of (a)~(d) S-graphene with pyridine and (a')~(d') S-graphene with pyrrole

Table 1 Atomic distance during each reaction steps (Å)

		Step1	Step2	Step3	Step4
S-Gr (pyridine)	*d*(O1-O2)	1.41	1.39	2.96	2.88
	d (O1-G)	1.85	2.76	1.45	3.72
	d (O2-G)	2.65	3.52	3.57	3.95
S-Gr (pyrrole)	*d* (O1-O2)	1.31	1.48	3.09	2.77
	d (O1-G)	3.10	3.50	3.37	3.43
	d (O2-G)	3.20	4.02	3.13	3.86

3.2 Reaction energy

Absorbed energies of molecules on S-graphene are calculated in each reaction process as shown in Fig.3. To be compared obviously, this paper takes the total energy of original S-graphene and OOH molecules as reference point to be 0. Then all of the other reaction energies subtract the reference energy to be their relative energy (Fig.3). About S-graphene with pyridine structure (as shown in Figure 3(a)), the relative energy of OOH molecule absorbed on S-graphene for the first reaction step is equal to -0.1326eV, which is larger than that (-0.0502 eV) on S-graphene with pyrrole structure. When another H atom was introduced into the system, the reaction energies of S-graphene with pyridine structure are equal to -0.7377eV, -1.4313eV, and -2.0575eV for each reaction steps respectively, wherein the corresponding reaction energies are equal to -0.6792eV, -1.9017 eV, and -2.4523eV for S-graphene with pyrrole (Fig.3(b)).

Conclusively, it can be found that all of the absorbed energies are negative, which means the S-graphene expresses ORR performance. Such results are consistent with the previous part analysis on S-graphene with pyridine structure. But it is opposite with the analysis of S-graphene with pyrrole structure. In Table 1, the oxygen atom doesn't absorb on S-graphene with pyrrole structure anymore because of their large distance. So the S-graphene with pyrrole structure doesn't have any influence on OOH molecule. Scrutinizing carefully, it isn't hard to find that the decrease of absorbed energy for S-graphene with pyrrole structure is originated from its deformed structure. So the S-graphene with pyrrole in ORR process can't keep its perfect geometry structure and performs unstable property.

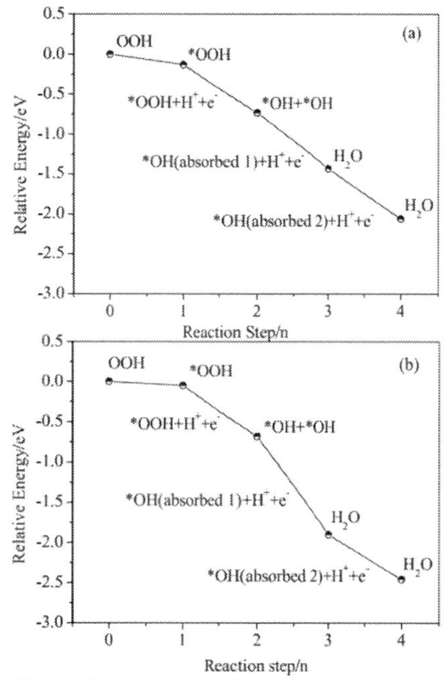

Fig.3 Relative energy of ORR reaction path way on two types of S-graphenes, wherein (a) pyridine and (b) pyrrole structure

3.3 Electronic density

In order to discover the influence of their electrons on ORR in different S-graphenes, this paper had calculated their highest occupied molecular orbital (HOMO) and lowest unoccupied molecular orbital (LUMO) along each reaction step. As well known, the OOH molecule has similar electron level with graphene. Following the reaction rule of quantum chemistry, if the reactants have adjacent energy level of valence electron, their will bond with each other very easily, and the chemical reaction will occur simultaneously. Compared with the HOMO and LUMO of OOH molecule absorbing on S-graphene with pyridine structure in the first reaction step, their HOMO and LUMO spatial distribution spread homogeneously after OOH →*OOH, which means such molecule state is stable (as shown in Figure 4(a) and (a')). After another H atom introduced in this system, the electrons of S atom moves to graphene and HOOH molecule. It is verified by the morphology of HOMO and LUMO spatial distribution around HOOH molecule (Figure 4(b) and (b')). Scrutinizing the third and fourth steps of producing H_2O, it is shown that their HOMO and LUMO spatial distribution are similar with each other(Figure 4(c), (c') and Figure 4(d), (d')). So the chemical reactivity of S-graphene with pyridine structure is obviously improved by sulfur doping because of their non-localized HOMO and LUMO distribution.

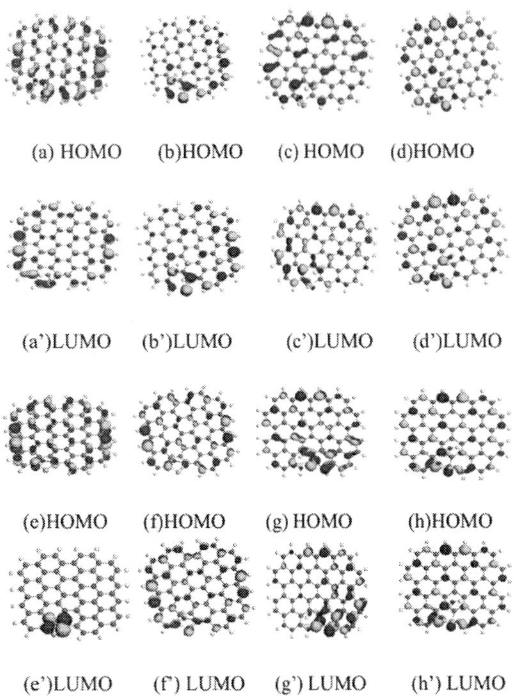

Fig.4 HOMO and LUMO spatial distribution of S-graphene with (a)~(d) or (a')~(d') pyridine and (e)~(h) or (e')~(h') pyrrole structure in ORR process of Step 1~4.

Focus on the S-graphene with pyrrole structure in ORR process, the LUMO spatial distribution is localized intensively in first step of ORR reaction (as shown in Figure 4(e) and (e')). So the OOH molecule is hardly absorbed on S-graphene, which is confirmed by previous analysis of geometry structure. In the next three reaction steps, they are similar with each other Figure 4(e~h) and (e'~h')). Their HOMO and LUMO spatial distribution don't have any change in HOOH→OH+OH and OH+H → H_2O reaction. So the chemical reactivity of S-graphene with pyrrole structure is not improved by sulfur doping (as shown in Figure 4(h) and (h'))

4 Conclusion

Based on the first-principles calculation, the mechanism of oxygen reduction reaction in S-graphene catalysis was researched deeply, the results show that:

1. In the first step of the reduction reaction, the OOH molecule moves from its original state on graphene and absorbs to a carbon atom near sulphur atom. After OOH absorbed on S-graphene, another H is introduced to such system. The distance between the sulphur and oxygen atoms (O1) reduces from 3.30 Å to 1.85 Å, which the chemical bond is formed spontaneously. About S-graphen with pyrrole structure, it can be found that its OOH molecular absorption ability is more hardly than that on the S-graphe with pyridine.

2. About S-graphene with pyridine structure, the relative energy of OOH molecule absorbed on S-graphene for the first reaction step is equal to -0.1326eV, which is larger than that (-0.0502eV) on S-graphene with pyrrole structure. When another H atom was introduced into the system, the reaction energies of S-graphene with pyridine structure are equal to -0.7377eV, -1.4313eV, and -2.0575eV for each reaction steps respectively, wherein the corresponding reaction energies are equal to -0.6792eV, -1.9017eV, and -2.4523eV for S-graphene with pyrrole.

3. Compared with the HOMO and LUMO of OOH molecule absorbing on S-graphene with pyridine structure in the first reaction step, their HOMO and LUMO spatial distribution spread homogeneously after OOH →*OOH. After another H atom introduced in this system, the electrons of S atom move to graphene and HOOH molecule. Focusing on the S-graphene with pyrrole structure in ORR process, the LUMO spatial distribution is localized intensively in first step of ORR reaction

Acknowledgment

This work was supported by the Innovation special fund project funding of Jiangxi graduate (Grant No.YC2018-S357) and Key Laboratory of Jiangxi Province for Persistant Pollutants Control and Resources Recycle (ST201522014).

References

1. P. Prabhu, V.Jose, J. M. Lee, Matter., **2**, 526 (2020).
2. X. Ji, Y. Mu, J. Liang, T. Jiang, J. Zeng, Z. Lin, Y. Lin, J. Yu, Carbon, **176**, 21 (2021).
3. J. Theerthagiri, G. Durai, T. Tatarchuk, M. Sumathi, P. Kuppusami, M. Y. Choi, Ionics, **26,** 2051 (2020).
4. M. Singla, D. Sharma, N. Jaggi, International J. Hydrogen Eng., **46**, 16188 (2021).
5. T. Domga, G. B. Noumi, M. J. Sieliechi, J. B. Tchatchueng, Carbon Trends., **4**, 100043 (2021).
6. Z. Y. Wu, , C. Y. Wu, J. G. Duh, Materials Letters **296**, 129875 (2021).
7. S. J. Lee, J. Theerthagiri, P. Nithyadharseni, P. Arunachalam, D. Balaji, A. M. Kumar, J. Madhavan, V. Mittal, M. Y. Choi, Renewable and Sustainable Energy Reviews, **143**, 110849 (2021).
8. H. Aghamohammadi, N. Hassanzadeh, R. E. Farsani, Ceramics International, 2021, (in press).
9. L. Zhang, Z. Xia, The J. Phys. Chem. C, **115**, 11170 (2011).
10. J.P.Perdew, K.Burke, M.Ernzerhof, Phys.Rev.Lett., **77**, 3865(1996).
11. Delley, B., J. Chem. Phys., **94**, 7245 (1991).
12. J. D. Pack, H. J. Mokhorst, Phys. Rev. B, **16**,1748 (1977).

Effect of Oil Content in Tea-seed Pancake on the Determination of Tea Saponin

Longfa Jiang[1,a], Lingyun Bai[2], Shumin Zhang[1], Heng zhang[1] and Xuming Tu[1]

[1] Technical Center of Nanchang Customs District, Nanchang 330013, China;
[2] Jiangxi Science and Technology Normal University, Nanchang 330013, China

Abstract. According to the national standard for the determination of tea saponin content in tea-seed pancake (GB/T 35131-2017), tea saponin was obtained by extracting oil with ethanol and removing solvent. The actual samples of tea-seed pancake did not completely remove the oil, and some of the residual oil content was as high as 9%. Due to the long period of national standard determination, people often use ethanol to extract tea saponin directly, instead of subsequent acid hydrolysis or alkali hydrolysis steps, so as to realize the rapid evaluation of tea-seed pancake quality. In this case, the residual oil has a greater impact on the results. If it was still tested according to the national standard method, the result will be much higher than the actual result. In this work, petroleum ether extraction process was selected before ethanol extraction to remove the residual oil. Experimental results showed that, the accuracy of the determination was effectively improved.

1 Introduction

Tea-seed pancake is a kind of cake left after extracting oil from camellia oil fruit, which is purple brown particles, as shown in Fig.1. Tea-seed pancake contains about 18% tea saponin, fresh cake can be as high as more than 20%, chemically refined cake only about 12%. The annual output of Camellia oleifera seeds in China has reached more than 800000 tons, and the annual output of pressed camellia oleifera seed meal can reach more than 500000 tons[1], which is the country with the largest variety, the widest distribution and the largest output of camellia oleifera seeds in the world. It is mostly produced in Jiangxi, Hunan and other places.

Fig.1 Tea-seed pancake (picture obtained from internet)

Tea saponin is a hemolytic toxin, which can dissolve the red blood cells of fish, so it can kill wild fish, loach, snails, mussels, frog eggs, tadpoles and some aquatic insects. Tea saponin is easy to dissolve in alkaline water, when used, adding a small amount of lime water, the effect is better. Due to the high protein content of tea meal, it is also a kind of high-efficiency organic fertilizer, which is widely used in crops and fruit trees with excellent effect. It can increase the fertility of the pond with less silt and poor bottom material. The use of tea-seed pancake as a pond clearing drug has more unique efficacy than other drugs.

Tea seed cake is widely used in fish pond, paddy field, high-grade lawn, earthworm, ground tiger and other pests. As a kind of green medicine, tea-seed cake can decompose by itself and remain non-toxic. It has no effect on human body and is safe to be used. The second is not to kill water plants, but also to promote the growth of water plants. Third, there are no side effects on shrimp and crab larvae, and the emergence rate of shrimp and crab larvae is higher.

Tea-seed pancake is also used in the production of feed and extraction of tea saponin from tea-seed pancake through deep processing. Tea seed powder is widely used in washing chemical industry, machine derusting, aquaculture, golf course, killing earthworms and so on.

Tea saponin is a kind of glycoside compounds extracted from tea seeds. The pure product is white columnar crystal, as shown in Fig.2. Tea saponin is a pentacyclic triterpenoid saponin [2-7], which consists of seven ligands, four glycosides and two organic carboxylic acids. It is a natural non-ionic surfactant, and has many physiological functions, such as antibacterial and anti-inflammatory, virus killing, antioxidant, anti hypertension, insecticidal and insect repellent, and promoting plant growth. Because of the above properties of tea saponin, it has a wide range of applications in

[a] Corresponding author: jianglongfa@126.com

agriculture, medicine, food, daily chemical and other fields[2-4]. Tea saponin can improve the wettability and suspension rate of wettable powder (≥ 75%) as a wetting agent of pesticide. As a natural non-ionic surfactant, tea saponin can significantly improve the physical and chemical properties of pesticide solution, increase the effective volume of pesticide on the target, and help to give full play to the efficacy of pesticide. Therefore, the application effect can be improved.

Fig.2 Tea saponin (picture obtained from internet)

China has established a number of manufacturers for the production of camellia oleifera seed oil and tea saponin. The production of tea saponin and its preparations has been greatly developed, but there are still a considerable number of resources that have not been fully utilized. Therefore, the development and extraction of tea saponin for daily chemical, building materials, agriculture, aquaculture, pharmaceutical industry, paper industry, leather industry and coating industry will produce good economic, social and ecological benefits[8].

Zhang Tuanjie [9] and others established phloroglucinol method, phenol sulfuric acid method, p-Dimethyl aminobenzaldehyde method, fluorescence spectrophotometry method, thin layer chromatography method, potassium bromate method, high performance liquid chromatography method and vanillin concentrated sulfuric acid method for the determination of tea saponins.

Tea saponin was extracted by a new technology, namely water extraction and alcohol precipitation. The optimum conditions of water extraction were as follows:

The extraction temperature was 60 °C. Liquid-solid ratio was of 10:1. The extraction time was 2 h.

Water extraction followed by alcohol precipitation. The results showed that the optimal conditions for this precipitation process were as follows: concentration of ethanol was 90%, volume ratio of ethanol to concentrated solution was 4:1, alcohol precipitation temperature was 75 °C and alcohol precipitation time was 2.5 h. Compared with water extraction and alcohol precipitation under these conditions, the extraction rate of water extraction and alcohol precipitation was 95.2%, and the purity was 69.9%, which improved the purity and was conducive to subsequent refining treatment.

With the wider use of tea-seed pancake, the price of tea-seed pancake is becoming higher and higher. Many manufacturers, after pressing the tea seed, no longer extract residual oil, but sale the tea-seed pancake directly. The residual camellia oil is mainly composed of palmitic acid, stearic acid, oleic acid, linoleic acid and linolenic acid [10], which has a great influence on the determination of crude tea saponin.

In the previous research of the extraction and determination of tea saponin from tea-seed pancake, the external factors, such as extraction solvent, solvent amount, extraction temperature and extraction time, are carefully discussed. However, oil content, as an internal factor, is relatively easy to be ignored. The influence of the oil content of the sample on the determination of tea saponin was studied in this work.

2 Experimental

2.1 Reagents and Instruments

Reagents and instruments used in the process of analysis and detection were listed in table 1.

Table1. Reagents and Instruments

Name	Specification/model	Manufacturer
anhydrous ethanol	AR	Sinopharm Group
petroleum ether	AR	Sinopharm Group
electronic balance	ME240E	Mettler Toledo
electric constant temperature drying oven	UFE500A0	Meimert company
constant temperature water bath	DK-S24	Shanghai Jing Hong company

2.2 Experimental Samples

The tea-seed pancake from different factories in Yichun area of Jiangxi Province was selected as the experimental sample, which was fully crushed and passed through 150 μm sieve, as shown in Fig.3.

Fig.3 Powders of tea-seed pancake

2.3 Experimental Method

Fig.4 shows the experimental device. Firstly, experimental sample with the weight of 5.0 g and 100 ml anhydrous ethanol was added into a 250 ml triangular ground flask sequentially. Then, a dry reflux condenser was installed, and refluxed for 2 h in a boiling water bath. After that process, the condenser was removed. The triangular grinding flask was placed on the hot water bath and let it stand for about 1 h (Fig.4 Step 1).

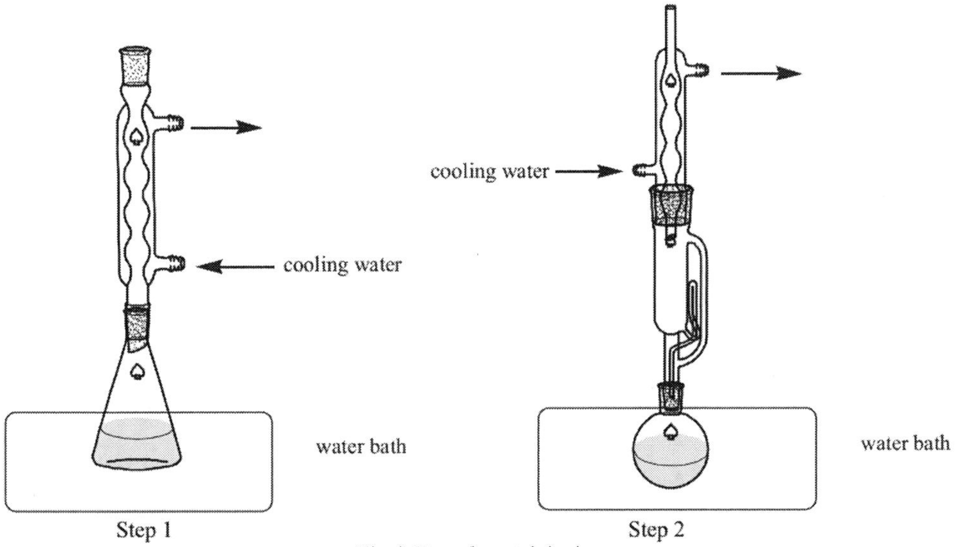

Fig.4 Experimental device

Secondly, the extract was filtered into a flat bottomed flask with constant weight at 105 °C through absorbent cotton when it was hot, and the insoluble matter was retained in the original triangular flask as far as possible. Then, 40 ml anhydrous ethanol was added and refluxed on the boiling water bath for 1 h. After that, the triangular grinded flask was tilted on the hot water bath pot and left standing for about 1 h, and the extract was filtered into the flat bottomed flask through the original funnel.

Thirdly, the flat bottomed flask containing the extraction solution was installed with Soxhlet extraction tube and reflux condenser (Fig.4 step 2). Anhydrous ethanol was evaporated into the extraction tube in a water bath for recovery until the anhydrous ethanol in the flask was nearly clear.

At last, the flask was moved into a constant temperature oven, dried at 105 °C for 2 h. Then, the flask was took out and cooled in a dryer. The product was weighed.

To examine the influence of residual oil, sample without oil removal with the weight of 5.0 g was put into the filter paper cartridge, put the filter paper cartridge into the extraction cartridge of Soxhlet extractor, and added petroleum ether from the upper end of the condenser tube of the extractor to two thirds of the volume of the bottle. Sample was extracted in water bath for 10 h. After extraction, the filter cartridge and the sample were placed in a 250 ml triangular flask to extract tea saponin according to the experimental method.

3. Results and discussion

The results of crude tea saponin determination with and without oil removal were shown in Table 2.

For the three experimental samples, the oil content was 5.4%, 7.6% and 8.4%, respectively. The ea saponin content was 18.0 %, 27.7 % and 24.9 %, respectively, when the detection method without residual oil removing was selected. On the other hand, when the residual oil was removed from the raw material, the corresponding results were 12.5 %, 18.7 % and 16.5%, respectively.

The relative error was 9.0 %, 9.7 % and 10.1 % for sample 1#, sample 2# and sample 3#.

It could be analyzed from table 2 that, the relative error increased with the increase of residual oil content. Due to the principle of similar compatibility, the residual camellia oil contains polar components and was easily soluble in hot ethanol. When tea saponin was extracted by hot ethanol, the residual oil was also extracted, which made the result higher.

The higher the oil content, the greater the relative error is. Considering the uncertainty of the residual oil content in tea-seed pancake samples, it is suggested that petroleum ether be used to extract and remove the residual oil first, and then hot ethanol be used to extract tea saponin for detection, so as to improve the accuracy of tea saponin content determination.

Table 2. Effect of different oil content on the determination results of tea saponin in tea-seed pancake

Tea saponin content (%) / Method	Sample 1#	Sample 2#	Sample 3#
Oil content	5.4	7.6	8.4
Without residual oil removing	18.0	27.7	24.9
With residual oil removing	12.5	18.7	16.5
Relative error	**9.0**	**9.7**	**10.1**

4. Conclusions

1) In this work, petroleum ether extraction process was added before ethanol extraction to remove the residual oil. Experimental results showed that, the accuracy of the determination was effectively improved.
2) The oil content of tea-seed pancake itself has a great influence on the extraction and determination of tea saponin. The higher the oil content, the greater the relative error is.
3) Considering the uncertainty of the residual oil content in tea-seed pancake samples, it is suggested that petroleum ether be used to extract and remove the residual oil first, and then hot ethanol be used to extract tea saponin for detection, so as to improve the accuracy of tea saponin content determination.

Acknowledgements

The research content was supported by the Key R & D projects in Jiangxi Province (20161BBE50103)

References

1. Tuanjie Zhang, Quantitative analysis of tea saponin from seed cake. Science and Technology of Food Industry, 2016:53-56
2. Tan Bo, The extraction and purifieation of tea saponin and the extraction of tea seed Polysaecharide in camellia cake, Central South University of forestry science and technology, 2009.
3. Chen Ying et al. Quantitative Analysis of Saponins in Camellia Seed Cake and Tea Saponins. Journal of the Chinese Cereals and Oils Association, 2012.27 (2): 105-111
4. Zhang Mingzhen and Liao Xiaoli, Extraction and Analysis of Tea Saponin in Camellia Seed Kernel. Analysis and Testing, 2021 (7): 176-178
5. Guo Li et al. Fatty Acid Compositions and Contents of Marketed Tea Oils. Food industry, 2020: 333-336
6. Zheng-Wei Ma, Liu Han-Qiao, Lü Qiu-Feng. Porous biochar derived from tea saponin for supercapacitor electrode: Effect of preparation technique[J]. Journal of Energy Storage, 2021, 40.
7. Yue Sun, Zou Minting, Li Changyi, et al. The solubilization of naphthalene using tea saponin as a biosurfactant: Effect of temperature[J]. Journal of Molecular Liquids, 2021, 335.
8. Mukai T, Horie H, Goto To. A simple method for determining saponins in tea seed[J].Chagyo Kenkyu Hokoku, 1992, 75: 29- 31.
9. Shimada K. Determination of tea- leaf saponins and water- soluble pectin in a green tea infusion [J].Nippon Kasei Gakkaishi, 2003, 54(11) : 957- 962.
10. Wang Hu, The Research on the extraction and Purification technology of Tea oil and High Purity Tea Saponin from the Camellia, Lanzhou University of Technology: Lanzhou, 2011.

Support effect on electrocatalytic performance of methanol oxidation over platinum catalysts

Ke Gu[1], JingWei Li[1,a], Jianguo Zhao, Yajuan Liu[1,a], Jun Qiao[1], Dongdong Chen[2], Siyu Hou[1], Ran Zhang[1]

[1]Institute of Carbon Materials Science, Shanxi Datong University, Datong, 037009, China
[2]Department of Chemistry & Chemical Engineering, Lvliang University, Lvliang, 033001, China

Abstract: Graphene oxide and carbon nanotubes supported Pt nanoparticles were successfully synthesized by wet chemical reduction method. The structures and methanol oxidation performance of the two catalysts were characterized by scanning electron microscope (SEM), transmission electron microscope (TEM), X-ray diffraction (XRD), Raman spectroscopy (Raman) and electrochemical workstation. The results show that graphene oxide supported Pt nanoparticles exhibits higher electrochemical active surface area, methanol oxidation activity and stability compared with carbon nanotubes. It is proposed that the well dispersed Pt on the graphene oxide plays an important role for the excellent performancefor methanol oxidation.

1. Introduction

Fuel cell is a kind of environment-friendly energy[1, 2], which is powered by renewable small molecular substances, such as hydrogen[3, 4], ethanol[2, 5], formic acid[6, 7], methanol. Methanol oxidation on the anodic surface of direct methanol fuel cell (DMFC) is a slow kinetic process, which involves multi-electron and multi-step elementary reactions[8-10]. It constitutes the rate control step of the total reaction of the fuel cell and determines the efficiency of the direct methanol fuel cell to a great extent. Nowadays, Pt supported catalyst is widely used as the commercialized catalysts in the fuel cell. At the same time, Pt catalysts are vulnerable to chemical and electrochemical corrosion. Therefore, a large number of precious metal platinum is needed to catalyze the occurrence of methanol oxidation reaction efficiently[11-13]. Based on the fact that Pt is fairly expensive and difficult to replace by other components, one of the biggest issues is how to improve the utilization and stability of Pt effectively. Carbon is the most widely used support in both anode and cathode of DMFC owing to its large specific surface area, good electrochemical stability and remarkable electrical conductivity. As a consequence, Pt/C catalysts have aroused growing interest. For Pt/C catalysts, the structure or properties of the catalyst carrier directly determine the dispersion, coordination, crystal structure, electronic properties and interaction of the active components, thus affecting the activity of the catalytic reaction. In this experiment, graphene oxide (GO) and carbon nanotubes (CNTs) were used as carriers to systematically study the catalytic performance of Pt catalysts supported on different kind of carbon supports for methanol oxidation.

In this paper, the preparation methods and catalytic performance for two kinds of Pt supported catalysts were compared. The platinum catalysts supported on graphene oxide were prepared by 1-ethyl-3-methylimidazolium tetrafluoroborate ([BMIM][BF_4]) assisted ethylene glycol reduction method[14, 15], denoted as Pt/GO; Pt supported on carbon nanotubes were prepared by wet chemical reduction method, denoted as Pt/CNTs. The physical property of the catalysts were characterized by TEM, SEM, Raman and XRD. The catalytic performance of the two catalysts for methanol oxidation were investigated by electrochemical methods.

2. Experimental

2.1. Preparation of Pt nanoparticles supported on graphene oxide

2.1.1 Pretreatment of GO

50 mg GO was added to a mixture solution of 25 μL [BMIM][BF_4], 50 mL deionized water and 100 mL ethylene glycol with vigorous stirring for 3 h. Then the GO suspension was transferred into a cell crusher. The setting parameters of the crusher are as follows: 60% power and stop for 2.5 s after ultrasonic 1 s for 2 h. After

[a] Corresponding author: bjyddx1988@163.com; dtdxliuyajuan@139.com

https://doi.org/10.1051/e3sconf/202129001019

crushing, GO was washed and isolated by centrifugation with a speed of 10000 rpm. The obtained sample was dried at 100 °C for 6 h and ground for use.

2.1.2 Preparation of Pt/GO

20 mg GO was added to a mixture solution of 20 mL deionized water and 40 mL ethylene glycol in a 100 mL round bottom flask. Then, the above GO suspension was placed in oil bath with vigorous stirring. When the temperature of the reaction solution rose to 125 °C, 3.62 mL $H_2PtCl_6{\cdot}6H_2O$ aqueous solution with a concentration of 15 mmol/L was dripped into it in three times and kept at 125 °C for 4 h. After cooling down to room temperature, the sample was washed three times with deionized water and dried overnight in vacuum at 60 °C.

2.2. Preparation of Pt nanoparticles supported on carbon nanotubes

2.2.1 Pretreatment of CNTs

The CNTs was pretreated by oxidation with the reflux of concentrated nitric acid. The slurry of CNTs and concentrated nitric acid (68 wt%) with the mass ratio of 1: 100 was put in oil bath and reflux at 100 °C for 5 h. Then, the mixture was filtered, washed and dried at80 °C for 6 h.

2.2.2 Preparation of Pt/CNTs

The Pt/CNTs catalyst was prepared by wet chemical reduction method. Typically, 0.1650 g CNTs was added to 10 mL ethylene glycol (EG). Then 10 mL H_2PtCl_6-EG solution of (8 mol/L) is added to the above suspension and the pH value was adjusted to 11 with NaOH. The resultant reaction mixture was transferred into a 100 mL Teflon-lined autoclave, which was then sealed and kept in the electric oven at 130 °C for 3 h. After cooling to room temperature, the precipitates were collected by centrifugation, repeatedly washed by deioned water, and dried in air at 100 °C for 12 h.

2.3 Materials Characterizations

The morphologies and microstructures were obtained by TEM (JEM2100) and SEM (TESCAN，MAIA3LMH). The XRD data were recorded on X-ray diffractometer (Rigaku, SmartLabSE). The crystal structure and degree of graphite were analyzed by Raman (Invia, RenishawInstruments, England).

2.4 Electrochemical measurements

The electrochemical test was carried out on the Shanghai Chenghua chi 760E electrochemical workstation. The whole test process was carried out using a three-electrode battery with glass carbon working electrode, and the counter electrode was composed of platinum wire and Ag/AgCl reference electrode. 2 mg catalyst was added to the solution of 450 mL ethanol solution, 450 mL distilled water and 50 mL Nafion (5 wt%) solution with stirring for 0.5 h, and then 5 uL above slurry was transferred to the surface of glassy carbon electrode to prepare working electrode. Before the measurement, high purity N_2 was injected into the electrolyte solution for 10 min to remove dissolved oxygen.

The cyclic voltammetric (CV) experiment was carried out in the electrolyte solution of 0.5 M H_2SO_4 with the scanning speed of 50 mV/s. Methanol oxidation activity was determined in 0.5 M H_2SO_4 and 1 M CH_3OH solutions. The chronoelectric current was measured in 0.5 M H_2SO_4 and 1 M CH_3OH solutions. All the experiments were carried out in high purity N_2 at room temperature.

3. Results and discussion

The morphologies and microstructures of Pt/GO and Pt/CNTs catalysts were observed by SEM and TEM as shown in Fig. 1 and 2. From Fig. 1, it can be seen the GO observed is of few-layer with varying lengths between 2 μm and 6 μm based on SEM analysis. In the case of CNTs, they exhibit average width of ca. 20 nm and length of 10 μm~100 μm. As indicated in Fig. 2, it can be clearly found that Pt nanoparticle were uniformly loaded on GO and CNTs carriers with the average size of 1.2 nm and 2.6 nm. The smaller particle size of Pt is more favorable on account of fully exposing the active sites in the catalytic process, thus improving the utilization of Pt.

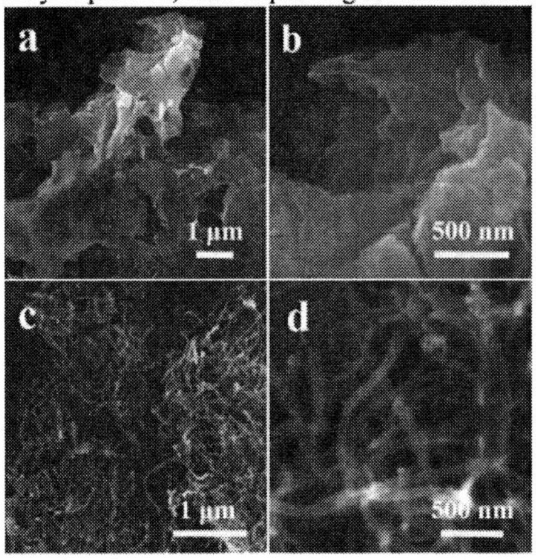

Figure 1. SEM images of Pt/GO (a, b) and Pt/CNTs (c, d) catalysts

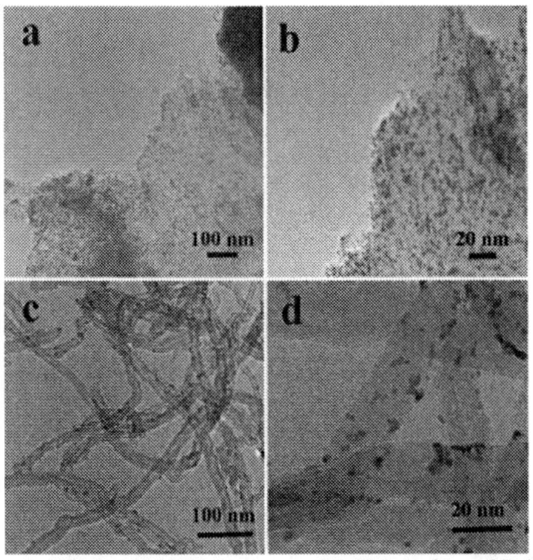

Figure 2. TEM or HRTEM images of Pt/GO (a, b) and Pt/CNTs (c, d) catalysts

Raman spectra are usually used to indicate the degree of graphitization of materials. Fig. 3 shows the Raman curves of the two catalysts. It can be seen that both catalysts have a D peak at 1300 cm^{-1} and a G peak at 1600 cm^{-1}. The D peak corresponds to the disorder degree caused by the defects in the material itself, while the G peak corresponds to the intrinsic peak of the graphite structure, which represents the stretching vibration of the sp^2 hybrid of the carbon atom. The disorder degree of graphite materials is usually characterized by the intensity ratio of D peak to G peak (I_D/I_G)[16]. The value of Pt/GO is 1.03, which is similar to that of Pt/CNTs (1.06), indicating that the two catalysts exhibit approximate degree of graphitization and electrical conductivity.

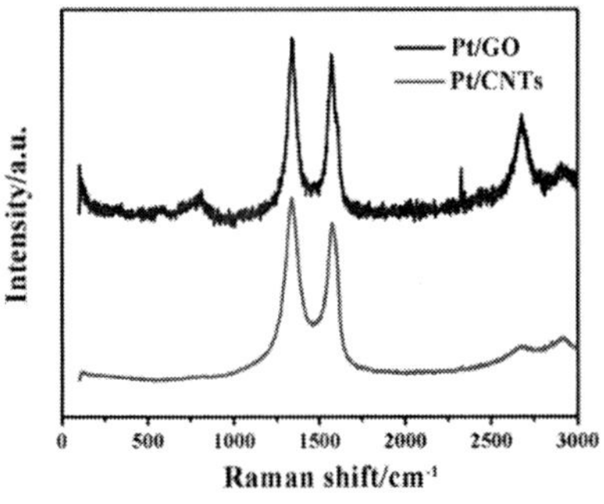

Figure 3. Raman spectroscopy characterization of Pt/GO and Pt/CNTs catalysts

Wide angle XRD patterns of Pt/CNTs and Pt/GO catalysts are showed in Fig. 4. The prominent peaks at 25°, 39.8°, 46.4° and 67.6° were observed. The peak near 25° belongs to the carbon (002) crystal plane, which usually indicates that the carbon material has good electrical conductivity. The peaks of 39.8°, 46.4° and 67.6° correspond to the crystal plane of Pt (111), (200) and (220) respectively[17], indicating that the precious metal Pt is completely deposited on the surface of the carrier.

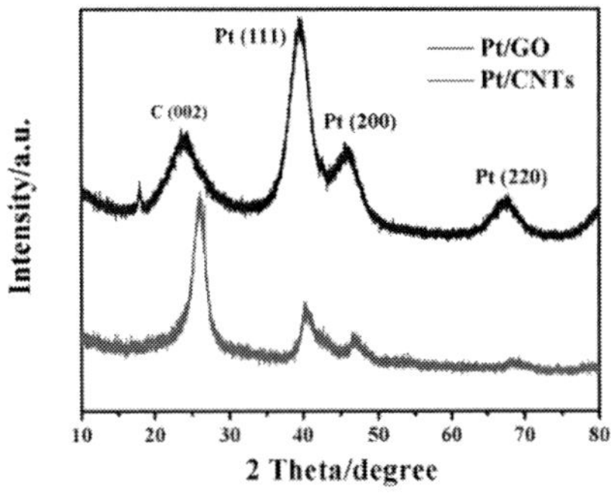

Figure 4. XRD patterns of Pt/GO and Pt/CNTs

The electrochemical active surface area (ECSA) of the catalysts was estimated by cyclic voltammetry (CV). The test environment was N_2 saturated 0.5 M H_2SO_4 solution with a scanning rate of 50 mV/s. The results are as shown in the Fig. 5. The ECSA of Pt/GO is 66.6 $m^2 g^{-1}$. As contrast, that of Pt/CNTs is only 3.7 $m^2 g^{-1}$. Such a ECSA of provided the Pt/GO electrochemical accessibility.

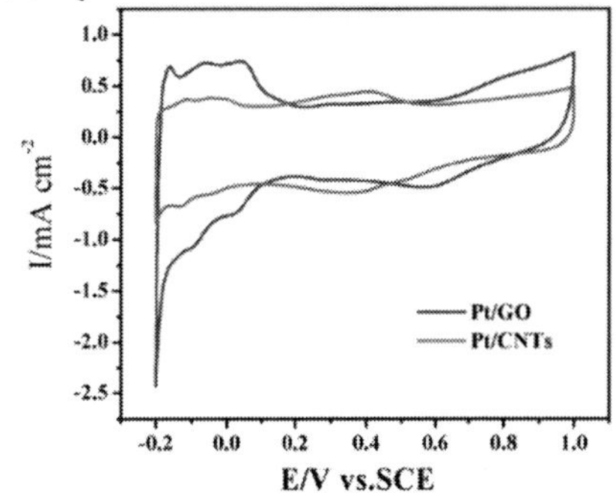

Figure 5. Cyclic voltammograms of the Pt/GO and Pt/CNTs

Fig. 6 exhibits the activities of the catalysts for methanol oxidation were tested in 0.5 M H_2SO_4 and 1 M CH_3OH solutions by means of CV test. The oxidation peak in the anodic scanning (forward scanning) process is caused by methanol oxidation, and that in the cathodic scanning (reverse scanning) process is caused by the further oxidation of the carbon-containing intermediates produced in the anodic scanning process[18]. It can be seen that the mass specific activities of Pt/GO and Pt/CNTs catalysts are 253.5 $mA \cdot mg^{-1}$ and 95.95 $mA \cdot mg^{-1}$, respectively. The electrocatalytic performance of Pt/GO catalyst for methanol oxidation is better than that of Pt/CNTs catalyst, which is mainly due to the better dispersion of Pt on GO carrier, smaller particle size of Pt nanoparticles and higher electrochemical active surface area.

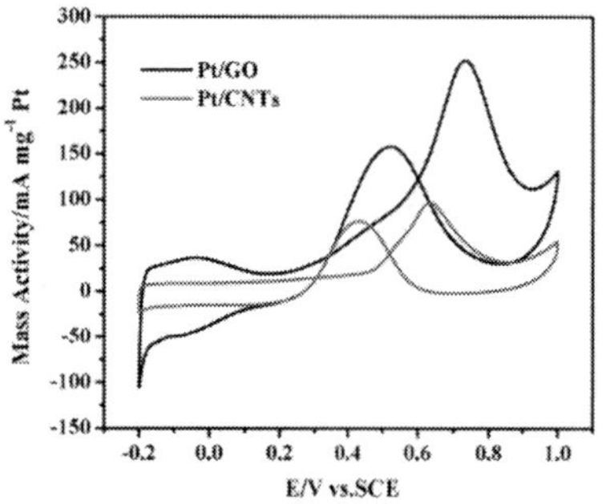

Figure 6. N_2 saturated 0.5 M H_2SO_4 in 1.0 M CH_3OH at a scan rate of 50 mV /s of the Pt/GO and Pt/CNTs

As shown in Fig.7, the stabilities of methanol oxidation catalyzed by the two catalysts were also tested by chronoamperometry at constant potential. With the extension of polarization time, the activity of both catalysts decreases, which may be due to the loss or accumulation of catalysts with the extension of reaction time, and the formation of insoluble carbonates on the surface of catalysts by possible carbonaceous residues[19]. The initial current density of Pt/GO is larger, and with the prolongation of polarization time, the decreasing trend of the curve is slow, and Pt/GO shows a higher steady-state current density. The above structure shows that Pt/GO has better catalytic stability for methanol oxidation. This is due to the better dispersion of Pt nanoparticles on GO carrier, which promotes the rapid transfer of reactants and products.

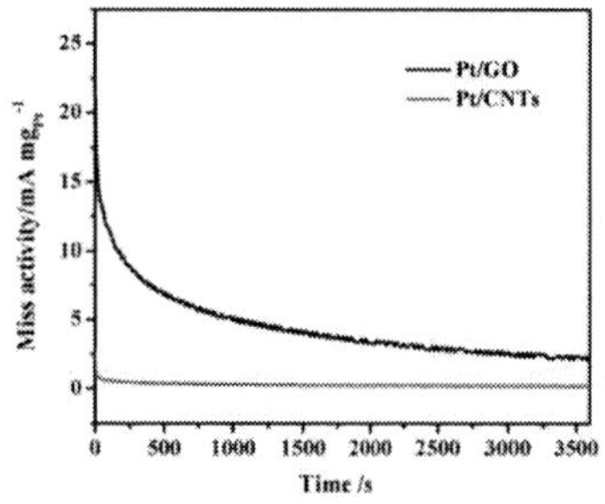

Figure 7. Chronoamperometric curves of Pt/GO and Pt/CNTs catalyst in at 0.6 V for 3600 s in the mix solution of 1 M CH_3OH+0.5 M H_2SO_4

4. Conclusions

In summary, the Pt nanoparticles supported on GO and CNTs were synthesized by wet chemical reduction method to catalyze methanol oxidation. The particle size of Pt supported on GO is much smaller than that on CNTs, which means that there are more active centres for methanol oxidation and the size effect of Pt is more significant.

Acknowledgments

The authors acknowledge Natural Science Foundation of Shanxi Province (201803D121122) and National Natural Science Foundation of China (51804191, 52071192).

References

1. K.M. Lin, F.C. Wang, Int. J. Hydrogen Energ., **46**: 21083-21097 (2021)
2. D. Alqdeimat, P.G. Pickup, ECS Transactions, **97**: 893-900 (2020)
3. C.F. Wang, Q.S. Li, C.M. Wang, Y.J. Zhang, W.L. Zhuge, Energy, **232**: 121038 (2021)
4. Z.F. Wang, Y. Yang, Z.F. Zhao, et al., Nat. Commun., **12**: 1982 (2021)
5. MTX Nguyen, MK Nguyen, PTT Pham, HKP Huynh, ST Nguyen, Electroanal. Chem., **888**: 115180 (2021)
6. N. Mardini, Y. Bicer, Int. J. Hydrogen Energ., **46**:13050-13060 (2021)
7. W. Yan. , Y. Xiang, J. Zhang, S. Lu, S.P. Jiang, Adv. Sustain. Syst., **4** (2020)
8. G.X. Xu, Z.G. Wu, Z.Y. Xie, Z.L. Wei, J. Li, K.G. Qu, Y. Li, W.W. Cai, Int. J. Hydrogen Energ., **46**: 9782-9789 (2020)
9. B.L. Garcia, VA Sethuraman, JW Weidner, R Dougal, RE White, J. Fuel Cell Sci. Tech., **1**: 43 (2020)
10. D. Lee, Sujin Gok, Youngkwang Kim, Yung-Eun Sung, Eunjik Lee, Ji-Hoon Jang, Jee Youn Hwang, Oh Joong Kwon, Taeho Lim, Acs Appl. Mater. Inter., **12** (2020)
11. Y.X. Bao, B.W. Hao, J. Am. Chem. Soc., **134**: 13934-13937 (2012)
12. H.C. Kim, R.K. Pramadewandaru, S. Lee, J.W. Hong, B. Korean Chem. Soc., **41**:2 (2020)
13. Zeynep Daşdelen, Ali Özcan, Int. J. Hydrogen Energ., **45**: 21881-21891 (2020)
14. W.L. Ding, X.L. Peng, Z.Z. Sun, K. Bi, Y. Zhang, Y.L. Wang, L. Ji, H.Y. He, Mater. Adv., **2**: 2009-2020 (2021)
15. D. Tomida, A. Kumagai, K. Qiao, C. Yokoyama, Int. J. Thermophys, **27**: 39 (2006)
16. X.L. Zhang, M.Y. Gang, J. Power Sources, **340**: 22-31 (2017)
17. M. Xiao, J. Zhu, J.J. Ge, C. Liu, W. Xing, J. Power Sources, **281**: 34-43 (2015)
18. X. Zhang, , J.X. Zhu, S.T. Chandra, Z.Y. Ma, H.J. Huang, J.F. Zhang, Z.Y. Lu, W. Huang, Y.P. Wu, ACS Appl. Mater. Interfaces, **17**: 10858-10865 (2016)
19. M.L. Xiao, L.G. Feng, Nanoscale, **21**: 9467-9471 (2015)

How Protein Labeling Applying

Bian Xinyu [1,*,†], Fan Sun [2,†], Zhixuan Xu [3,†]

[1]Reading College, Nanjing University of Information Science & Technology, 210044, Nanjing, China
[2]Shanghai Soong Ching Ling School, 201703, Shanghai, China
[3]Dulwich International High School-Suzhou, 215000, Suzhou, China
†These authors contributed equally to this paper

Abstract. In recent years, biosensing has played an irreplaceable position in scientific fields as well as our daily lives. A biosensor is an instrument that is sensitive to biological substances and converts its concentration into electrical signals for detection. It is an analysis tool or system composed of immobilized biologically sensitive materials as identification elements, appropriate physical and chemical transducers and signal amplification devices, and it has the functions of both a receiver and a converter. Pregnancy test sticks, blood glucose testing, etc. in life are all potential biosensors. The main point of this paper is to introduce several main techniques of biosensing for better understanding, like ELISA, fusion proteins, fluorescent probe and others. In the future, the bright future for them is also worth looking forward to if more researches and experiments can be processed in this area, for example, in the detection and treatment of cancer, medical technology can be further improved.

1 Introduction

Biological researches usually require the use of molecular labels covalently linked to the target protein to facilitate the detection or purification of the labeled protein, which is the general use of the protein labeling technique. Different labeling methods result in different molecules being covalently bound to the target protein. For some researches, specialized protein modifications are required to increase molecular mass, enlarge solubility for storage, or create a new functional group that can be targeted in the following stage. To study this labeling technique further, a review has been presented about protein labeling. The primary organization of this paper is divided into three parts, the comparison between enzyme linked immunosorbent assay (ELISA) and Western Blot, between fusion proteins and immune-fluorescence and between the fluorescent probe and quantum dots separately.

2 Protein labeling strategy: ELISA and Western Blot

The current mainstream biosensors include ELISA, fluorescent protein, western blot, etc., which are used for quantitative detection. The high efficiency and operability of ELISA detection make it one of the widely used detection techniques.

Western blot (WB) is a common method to detect and analyze proteins. It is based on a technique that transfers proteins from electrophoresis to membrane by electrophoretic separation and displays them specifically on the membrane. Agarose or polyacrylamide gel electrophoresis is one of the core technologies of gene manipulation. It can be used to isolate, identify and purify DNA fragments. This technique is simple and rapid and has become the technical basis of many common molecular biology research methods, for example, DNA recombination, DNA nucleotide sequence analysis, DNA restriction enzyme analysis, restriction enzyme mapping. When a molecule is placed in an electric field, it will move to the corresponding electrode at a certain speed, which is the migration speed of electrophoretic molecules under the action of the electric field. This mobility is called electrophoretic mobility. It is proportional to the electric field strength and the net charge carried by the electrophoretic molecule itself. That is to say, the higher the electric field intensity, the more net charges the electrophoretic molecules carry and the faster the migration speed, and vice versa. Due to the use of non-reactive stable supporting media, such as agarose gel and polyacrylamide gel, to reduce convective motion, the mobility of electrophoresis is inversely proportional to the coefficient of friction. It is well known that friction coefficient is a function of molecular size, polarity and medium viscosity. Therefore, according to the size, composition or shape of molecules and the amount of net charge, various components in protein or nucleic acid molecular mixture can be separated by electrophoresis. Under physiological conditions, the phosphate group in the phosphate backbone of nucleic acid sugar is in the ionic state. In this sense, DNA and RNA polynucleotide chains can be called polyanions. Therefore, when nucleic acid molecules are placed in an electric field, they migrate to the positive electrode. Due to the repetitive structure of

* Corresponding author: ma803008@student.reading.ac.uk

https://doi.org/10.1051/e3sconf/202129001020

the glycerophosphate backbone, the same amount of double-stranded DNA has almost the same net charge, so they can migrate to the positive electrode at the same speed. Under a certain electric field, the mobility of DNA molecule, that is, the mobility of electrophoresis depends on the size and configuration of the nucleic acid molecule. DNA molecules with smaller molecular weights migrate faster than those with larger molecular weights. This is the basic principle of DNA fragments separated by gel electrophoresis [1].

Samples were prepared and loaded into the gel. In the process of electrophoresis, the protein with a negative charge moves to the anode with a positive charge. To further analyze these proteins, the process of transferring them to the membrane is called imprinting. After transfer, the membrane is blocked to prevent unnecessary membrane protein interactions in the following steps. To visualize the proteins of interest, the first-order protein-specific antibody detection membrane is usually used, and then the labeled second-order antibody is used for detection. The film was imaged and the results were analyzed.[2]

By adding a separate marker solution to a hole in the gel, it can not only estimate the antibody interaction of a specific protein but also estimate the size of the protein. The separation of the gel depends not only on the size but also on the molecular charges, hydrophobic regions and denaturation. The experimental settings can be varied, which is most suitable for specific exploration. When analyzing the results, the load variation between lanes and the transfer rate between bolts must be considered. Besides, the nonlinear relationship of the generated signal in the range of sample concentration is also an aspect to be considered when interpreting the results.[3] The results of the leukocyte test depend on three important factors: the ability of the antibody to bind specific proteins, the intensity of interaction and the concentration of protein itself. Also, the specificity and low cross-reactivity of target binding are important characteristics. WB results are not always easy to explain because protein size may differ from theoretical weight due to post-translational modifications (such as glycosylation) or interactions with other proteins. However, WB is a quite common method, and almost all available commercial antibodies have been verified by this method.[4]

Western blotting is a technique for the analysis of antigens and antibodies. It has the following advantages.

1). The time of incubation and washing was significantly shortened.

2). To be used for multiple analysis and identification, multiple copies can be made at the same time.

3). The results can be displayed in the form of a chromatogram and can be stored for a long time.

4). Immunoprobe can be used to erase the probe by reducing the pH value and replace it with a second probe for analysis and detection.

5). The wet immobilized matrix membrane is flexible and easy to operate.

6). The immobilized biomacromolecules can approach all kinds of immune probes uniformly, and the pore size will not block them.

7). The amount of reagent needed for Western blot analysis was less. ELISA is a quantitative measurement method for specific substances. It connects antibodies to a specific enzyme for identification and measurement.

There are four major types of ELISA. Which are Direct ELISA (antigen-coated plate; screening antibody), Indirect ELISA (antigen-coated plate; screening antigen/antibody), Sandwich ELISA (antibody-coated plate; screening antigen) and Competitive ELISA (screening antibody). No matter it is direct or indirect ELISA, both of them have to fix the antigen on the plate.[5] When the antigen is added to the plate, it will be incubated at 37 degrees for one hour. The second step is very important, all unbound sites on the plate will be blocked with reagents such as animal protein. This step ensures that there will not be any non-specific antibodies bind to the plate later. After adding the buffer in the last step, wash the plate again and incubate for another hour after adding the antibody. Some enzymes that can change the color will be added to the direct ELISA. Compared with other detection methods, direct ELISA is simpler and more efficient. And it also contains the disadvantages of high reaction cost and low sensitivity. The steps of indirect ELISA are basically the same as those of direct ELISA, but compared with direct ELISA, the antibody is washed first and then added. Indirect ELISA uses two kinds of antibodies [6]. "A primary detection antibody that sticks to the protein of interest and a secondary enzyme-linked antibody complementary to the primary antibody." Compared to direct ELISA, indirect ELISA is cheaper and more sensitive. But indirect ELISA has a disadvantage, it can easily cause a chain reaction [7]. "Once the coating step is complete, the plates are washed with phosphate buffer saline (PBS), then buffered/blocked with bovine albumin (BSA). The buffer washes are carried out for at least 1-2 hours at room temperature. Finally, the plate is washed with PBS once again before the addition of the antigen." The disadvantage of Sandwich ELISA is that they require more time and cost. The last type of ELISA is competitive ELISA. Two types of antibodies are both added in the competitive ELISA, which leads to the competition of substrates. Its advantage is that this ELISA can more accurately measure smaller and more diverse antigens.

ELISA has many medical applications. For example, it is usually used for food allergies tests, pregnancy tests and HIV tests. The ELISA test usually has two test zones. One is for control, which makes sure that the test kit is working. The second test zone shows whether it is positive or negative. However, a second test is necessary for the HIV test, because the antibodies may haven't been secreted yet, which may lead to inaccuracy [7].

3 Protein labeling strategy: fluorescent labels

The attachment of fluorescing parts to proteins or other biomolecules is a common but useful strategy widely used for us to research protein structure and function in a better way. These fluorescent proteins have a broad range of

https://doi.org/10.1051/e3sconf/202129001020

applications, including immunochemistry, fluorescence in situ hybrids, cell tracing, receptor labeling, and fluorescence spectroscopy, etc. Two methods are introduced here, that are mainly used in the bioresearch area, one is the fusion protein method, and another is the immune-fluorescence method.

3.1 Fusion protein

The fusion protein is a kind of special protein which combined by mixing two or more genes that have been originally coded for separate proteins. The single or multiple polypeptides translated from this fusion gene have functional properties that are derived from each original protein. Among these proteins, the recombinant fusion protein is an important variable, which is created by recombinant DNA technology, and they are playing an irreplaceable role in the use of biological research or therapeutics area. And recombination technology, in which genes or gene fragments are spliced together for the formation of a gene fusion, has become an attractive tool for biotechnology, in particular in protein labeling [8].

At the mention of fusion proteins, an introduction about chimera mutant protein is necessary. Chimera mutant protein usually designates hybrid proteins made of polypeptides that have different functions. Chimeric mutant proteins can easily be prepared in vitro ——— Recombinant, which means by fusing the structural genes of the proteins in a suitable expression vector. There are essentially two types of fusion proteins (**Fig. 1**) [9].

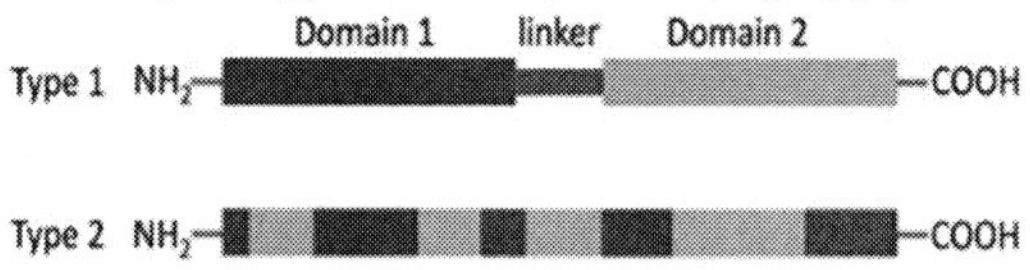

Figure 1. Basic types of fusion proteins.

Besides, chimeric mutant proteins can also be produced naturally when a complex mutation (a chromosomal translocation, tandem duplication, retro-transposition) creates a peculiar coding sequence containing coding sequences' parts from two different genes. Commonly occurring fusion proteins are found in cancer cells, where they may function [10].

Just because of such characteristics of the fusion proteins, there is a research team proposed a method to label antiglobulin test fusion proteins with different fluorophores in mammalian cells. This research shows that different specific fluorophores can be used to label AGT fusion proteins that have different localizations in cells, and the fluorophores can also be used for other applications like multicolor analysis of the dynamic process of labeling and fluorescence resonance energy transfer measurement. Due to the diversity of AGT substrates and the specificity of labeling reactions, this method is convenient in the field of studying protein functions in living cells.

The processes are simple, they synthesized AGT substrates firstly, which is prepared by reacting BG derivative with commercially available N-hydroxy-succinimide esters. And then is the construction of AGT fusion proteins, the AGT mutant here is based on human AGT. Fluorescence Labelling of AGT Fusion Proteins is the final step. All AGT fusion proteins were added to a special solution for transient transfection was performed.

What important are the results they get. AGT fusion proteins show the possibility to label individual fusion proteins through kinds of different fluorophores. In live cells, AGT fusion proteins can be labeled specifically to synthesize fluorophores sequentially. The localization of this kind of fusion protein they researched can be confirmed through mainly two methods, specific targeting sequences and the natural subcellular localization of the fusion position to which the AGT is linked. And the future of the stability of AGT fusion proteins labeled by fluorescence can be tracked in several hours. Besides, what important is that varieties of different fluorophores can be applied to label a single AGT fusion protein for the analysis of multicolor or FRET applications. And these characteristics also make this kind of fusion protein a vital object for fluorescence-based researches of how the protein functions in live cells [11].

3.2 Immune fluorescence

Then are a few things about the immune fluorescence method. Immunofluorescence is an important essay that is mainly used on biological samples and its official definition is a procedure that uses antibodies to detect antigens in cellular contexts. Immunofluorescence depends on the specificity of antibodies to the antigen. And it allows researchers to measure whether a particular sample cell expresses the antigen in research. What's more, researchers also can easily determine the target subcellular compartments expressing the antigen by the technology of immunofluorescence when a positive signal is found. Also, immunofluorescence can be applied to cultivating cell lines, tissue sections, or individual cells [12].

After some researches, it is widely known that hemolysis of the Sendai virus can be assayed in the erythrocytes system by poly-L-lysine. Multiple-step experiments can perform easily because of the immobilization of cells with the help of cooperation with lectins and antibodies. These ligands' binding had a remarkable inhibitory effect on viral hemolysis. These subtle molecular rearrangements are a prelude to cell fusion in which disintegration of the viral envelope structure is observed. This process could be characterized by the expression of virus and host-specific antigens with immune fluorescence techniques.

According to these molecular rearranging at the researched site, it is easier to make sense that the dispersal of viral compounds in the membrane. It could be observed by viral antigens' immune fluorescence directly or indirectly that the granular fluorescence shows diffuse staining in the researched position, indicating that viral parts conducted integration and dispersal in the membrane plane [13].

To conclude, both these two techniques have their value in different areas, with no better and worse differences. For fusion protein expression, it can be easily detected in protein labeling researches and separated for purifying, but it also may affect the structure of the original protein and destroy its natural function if wrong using. For the immunofluorescence method, like most fluorescence techniques, a significant problem that shouldn't be ignored is photobleaching, which can be controlled by increasing the density of fluorescence or limiting the intensity of light.

3.3 Fluorescent probe

The fluorescent probe, also called fluorophores or simply fluors, respond distinctly to light compared to other molecules. As shown below, a photon of excitation light is absorbed by an electron of a fluorescent particle. In excited via absorption of light energy to a higher energy state. Also called an excited state, the energy of the excited state, which cannot be sustained for long delays or decreases. Resulting in the emission of light energy. This process is called fluorescence. To flourish means to emit light via this process of law reform is a molecule that is capable of fluorescence and its ground state.[14] The fluorophore for a molecule is in relatively low energy, stable configuration, and it does not for us. When light from an external source hits a fluorophore, for a molecule, the molecule can absorb the light energy. If the energy absorbed a sufficient. The molecule reaches a higher energy state called an excited state. This process is known as excitation. There are multiple excited states or energy levels of the fluorophore before can attain, depending on the wavelength and energy of the external light source. Since the fluorophore of four is unstable and high energy configurations, it eventually adopts the lowest energy excited state, which is semi-stable.[15] The length of time that the fluorophore for his inexperience states is called the exciting lifetime, and it lasts for a very short time, ranging from ten to the minus fifteen to turn to the minus nine seconds. Next, the fluor for rearranges from the semi-stable excited state. Back to the ground state and the excess energy is released and admitted as light. The emitted light is of lower energy, and thus, the longer wavelengths than the absorbed light. This means that the color of the light that is emitted is different from the color of the light that has been absorbed.[16] Emission of light returns to the fluorophore, afford to its ground state, the fluorophore, can absorb light energy again and go to the entire process repeatedly. A cyclical fluorescents process can be summarized as one excitation of a fluorophore, or for it through the absorption of light.[17] To a transient, excited lifetime with some loss of energy. And three return of the floor, afford to its ground state, accompanied by the emission of light. The light energy emitted always has a longer wavelength, and the light energy is absorbed due to the energy lost during the transience, excited lifetime, as shown in step two. A fluorophore can repeatedly undergo the fluorescence process, in theory, indefinitely. This is extremely useful because it means that one fluorophore of four molecules can generate a signal multiple times. This property makes fluorescents a very sensitive technique for visualizing microscopic samples. Even a small amount of the stain can be detected. In reality, however, the fluorophore for structural instability during the exciting lifetime, makes it susceptible to degradation. Fluorescent probes with the best photophysical properties have the advantages of high sensitivity and fast response and are important research tools for biological imaging. [18]

Up to now, various fluorescent dyes have been used to construct various types of fluorescent probes. In particular, anthracane-based dyes, such as luciferin, rhodamine and Rodanil, have been developed over the past few years due to their excellent photophysical properties (high quantum yields and extinction coefficients, excellent water solubility and photostability). However, there are still some challenging problems to solve: most dyes containing heteranthracene emit green and red light (500-600 nm). In contrast, the near-infrared emission (650-900nm) has obvious advantages, such as less light damage to biological samples, deep penetration of tissues, and less interference from biomolecules to background autofluorescence. Therefore, it is necessary to prepare more near-infrared anthracene dyes. In recent years, heteranthracene nuclei have been modified by extending π conjugation or replacing epoxy atoms with other elements, such as C and Si. They produce some new red emission and near-infrared yellow raw dyes. However, like traditional anthracene dyes, they are still limited by the Stokes shift (usually less than 30 nm), which leads to severe fluorescence self-quenching and measurement errors in excitation backscattering. It is well known that the ideal near-infrared dyes for the construction of sensitive imaging probes should not only have high fluorescence quantum yield and large stokes shift, but also have optically tunable groups, which can be easily modified to obtain a variety of reaction-based fluorescent probes. Unfortunately, the number of anthracene dyes that meet the above criteria is still quite limited.

3.4 Quantum dots

Quantum dot is a kind of semiconductor material, which is generally spherical or quasi-spherical in diameter of 2-20nm. The more common quantum dots are composed of IV, II-VI, IV-VI or III-V elements. Specific examples include silicon quantum dots, germanium quantum dots, cadmium sulfide quantum dots, cadmium selenide quantum dots, cadmium telluride quantum dots, zinc selenide quantum dots, lead sulfide quantum dots, lead selenide quantum dots, indium phosphide quantum dots and indium arsenide quantum dots. As a kind of nanometer semiconductor material, quantum dots can emit light with a specific frequency by applying a certain frequency of electric field or light pressure. The frequency of the light will change with the size of quantum dots. By adjusting the size of quantum dots, the frequency of light emitted by them can be within the range of visible light, and the wavelength of light can be directly observed through the change of color. The quantum dots have large Stokes shifts. Another optical property of quantum dots different from organic dyes is the wide Stokes shift, which can avoid the overlap of the emission spectrum and excitation spectrum, and is conducive to the detection of fluorescence spectrum signal. Good biocompatibility. After a variety of chemical modifications, quantum dots can be specifically linked, which has low cytotoxicity and little harm to organisms, and can be used for in vivo labeling and detection. Among all kinds of quantum dots, silicon quantum dots have the best biocompatibility. For quantum dots containing cadmium or lead, it is necessary to package the surface of quantum dots before biological application. Quantum dots have a long fluorescence lifetime. The fluorescence lifetime of organic fluorescent dyes is only a few nanoseconds (which is equivalent to the decay time of autofluorescence in many biological samples). The fluorescence lifetime of quantum dots with a direct band-gap can last for tens of nanoseconds (20-50 NS), and that of quantum dots with a quasi-direct band-gap, such as silicon quantum dots, can last for more than 100 μs.[19] In this way, in the case of light excitation, most of the spontaneous fluorescence has decayed, while the fluorescence of quantum dots still exists. At this time, the fluorescence signal without background interference can be obtained. In a word, QDs have the advantages of wide and continuous excitation spectrum, narrow and symmetrical emission spectrum, adjustable color, high photochemical stability and long fluorescence lifetime. At present, water-soluble quantum dots and oil-soluble quantum dots can be coupled with antibodies, streptavidin and other molecules, which can be used in the fields of biomarker detection, high-throughput coding, in vivo imaging and dynamic tracing.[20] However, the size of water-soluble quantum dots combined with biological macromolecules is large (about 10 nm ~ 30 nm in diameter), which hinders its passage through the cell membrane structure. Therefore, it can only be used for the study of permeabilized cells or extracellular proteins or proteins that can be endocytosed by cells.

4 Conclusion

In general, as a relatively mature technology, ELISA has emerged in different branches. The advantages of direct and indirect ELISAs are their high efficiency, speed and low cost. Their operability has become the most extensive household biosensor on the market. For fusion protein expression, it can be easily detected in protein labelling researches and separated for purifying, but it also may affect the structure of the original protein and destroy its natural function if wrong using. For the immunofluorescence method, like most fluorescence techniques, a significant problem that shouldn't be ignored is photobleaching, which can be controlled by increasing the density of fluorescence or limiting the intensity of light. And QDs have the advantages of wide and continuous excitation spectrum, narrow and symmetrical emission spectrum, adjustable colour, high photochemical stability and long fluorescence lifetime.

References

1. Luo Yunbo. Introduction to food biotechnology, 3rd Edition: China Agricultural University Press, page 27 (2016.08)
2. Burnette WN., "Western blotting": electrophoretic transfer of proteins from sodium dodecyl sulfate--polyacrylamide gels to unmodified nitrocellulose and radiographic detection with antibody and radioiodinated protein A. Anal Biochem. PubMed: 6266278 (1981)
3. N Am J. Western Blotting technique, theory, and trouble shooting:Mahmood T et al., Western blot: technique, theory, and trouble shooting. PubMed: 23050259 DOI: 10.4103/1947-2714.100998 (2012)
4. Biotechnology. The first article describing electrophoretic transfer of proteins to membrane:Towbin H et al., Electrophoretic transfer of proteins from polyacrylamide gels to nitrocellulose sheets: procedure and some applications. (1992)
5. National Center for Biotechnology Information. PubChem Bioassay Record for AID 686974 (2001) Source: https://pubchem.ncbi.nlm.nih.gov/bioassay/686974.
6. Application Center for Biotechnology Information. PubChem Bioassay Record for AID 686974 (2019) Source: https://pubchem.ncbi.nlm.nih.gov/bioassay/686974.
7. Barnett, C. ELISA Applications. News-Medical Life Science. PubMed: 1422008. Source: https://www.news-medical.net/life-sciences/ELISA-Applications.aspx
8. Uhlén, M., Forsberg, G., Moks, T., Hartmanis, M., & Nilsson, B.. Fusion proteins in biotechnology. (2004, November 11). Source: http://www.sciencedirect.com/science/article/abs/pii/095816699290164E
9. Chimeric protein. (2011, May 3). Source: https://www.sciencedirect.com/topics/medicine-and-dentistry/chimeric-protein
10. Strohl, W.. Chimeric genes, proteins. (2006) Source: https://www.sciencedirect.com/science/article/pii/B9780128096338061999
11. Hub: Labeling of fusion proteins with Synthetic fluorophores in LIVE Cells: 10.1073/pnas.0401923101. (2004, May 24). Source: https://sci-hub.ren/https://www.pnas.org/content/101/27/9955.short
12. Principle of immunofluorescence. (2001, April 11) Source: https://www.sinobiological.com/category/principle-of-if
13. Bächi, T., Eichenberger, G., & Hauri, H. Sendai virus HEMOLYSIS: Influence of lectins and analysis by Immune fluorescence. (2015, Augest 11). Source: https://www.sciencedirect.com/science/article/abs/pii/0042682278904580

14. Fluorescent probe. (2002, July 23). Source: https://www.sciencedirect.com/topics/chemistry/fluorescent-probe
15. Immunofluorescences. (2001, February 21). Source: https://www.thermofisher.com/cn/zh/home/life-science/protein-biology/protein-biology-learning-center/protein-biology-resource-library/pierce-protein-methods/fluorescent-probes.html
16. Fluorescent probes. (2002, June 25) Source: https://www.thermofisher.com/cn/zh/home/life-science/protein-biology/protein-biology-learning-center/protein-biology-resource-library/pierce-protein-methods/fluorescent-probes.html
17. Fu, Y., & Finney, N. Small-molecule fluorescent probes and their design. (2018, August 14). Source: https://pubs.rsc.org/en/content/articlelanding/2018/ra/c8ra02297f
18. Quantum dot. (April 14, 2003), Source: https://www.sciencedirect.com/topics/materials-science/quantum-dot
19. Design resource molecule. (2006, December 11). Source: https://www.sigmaaldrich.com/technical-documents/articles/materials-science/nanomaterials/quantum-dots.html
20. Berger, M. What are quantum dots? (2019, October 15). Source: https://www.nanowerk.com/what_are_quantum_dots.php

https://doi.org/10.1051/e3sconf/202129001021

Universal Adhesives- Different Curing Methods and Applications

Wanying Dong[1, a, †, *], Xiaojing Gu[2, †], Jianghao Han[3, b, †, *], Liwen You[4, †]

[1]Chemical Engineering, Guangdong Technion-Israel Institute of Technology, 515063, Shantou, China
[2]Polymer material science and engineering, Beijing University of Chemical Technology, 100029, Beijing, China
[3]College of Chemistry, Nankai University, 300071, Tianjin, China
[4]School of Materials Science and Engineering, East China University of Science and Technology, 201424, Shanghai, China
[†]These authors contributed equally to this work.

Abstract. The adhesive is a kind of material with good adhesion properties which could join objects by adhesion of surfaces via forces of adhesion and cohesion. There are many kinds of adhesives. This paper mainly classifies them by curing method, from UV curing and heat curing to moisture curing and microwave curing, and describes the corresponding mechanisms. Furthermore, it covers the advantages and disadvantages, application ranges, and related chemical compounds of different curing ways for adhesives. This can broaden our understandings of different ways of adhesive curing. Besides, it could also help us choose the most appropriate curing method according to different requirements in daily life.

1 Introduction

Adhesives have played a leading role in structural materials due to their excellent properties, including high bonding reliability and simple operation process [1]. They are commonly used in anti-corrosive materials, biomedical materials. For instance, super-glue fuming is used in police crime investigations to detect potential fingerprints [2]. Moreover, they also have a place in the military field. Adhesives can be used in tank maintenance as they are faster and better than welding. However, because of its brittleness, how to cure the different adhesives has been a hot spot in this field nowadays [3].

The curing process is the most vital part of the adhesive characteristics, which stands as one of the most significant ways to judge the usage of one kind of adhesive. It depends on the circumstance where we want to apply the adhesive, how we store it, and whether we can get rid of it after using it. Different conditions in need of adhesive spawn diversified requirements of them. Therefore, clarifying and analysing the curing method is an important academic topic in this area, both in the past and the future. The theory developed for the systematization of curing has gone through for several years, trying to improve our understanding of the curing process. By focusing on the factors that lead to the transformation, the curing method can be mainly divided into light-initiated curing, thermosetting curing, moisture curing, etc. Mostly, light-initiating curing is activated by high-energy ultraviolet ray photons, while heating plays the role of activator in the thermosetting curing process. Moisture curing is commonly used in mild conditions, as the energy to initiate the reaction is supplied by the adhesive molecule itself instead of provided by extraneous factors.

Adhesive curing is a process in which the properties of adhesives are obtained and improved through chemical reactions (polymerization, crosslinking, etc.) [4]. According to the curing way, the types of curing agents can be divided into reaction adhesive, solvent sensitive adhesive, UV curing adhesives, and anaerobic adhesive. Firstly, reaction adhesive is mainly classified into thermosetting resin type, thermoplastic resin type, synthetic rubber type, and mixed type. It can react chemically under certain conditions. A solidified adhesive layer is further polymerized or cross-linked into a linear or body shape structure. Secondly, solvent-sensitive adhesives are adhesives containing volatile organic solvents. This adhesive dissolves the resin component in the solvent to give the adhesive fluidity and then becomes a kind of adhesive absorbed by the sticky place through the solvent volatilization. For example, sodium silicate, starch, vinyl resin adhesives, rubber adhesives [5]. Thirdly, UV curing adhesive is a one-component, solvent-free, UV, and visible curable adhesive and sealant that can be cured with a wide range of light sources, such as point, pulse, and high-density lamps. Finally, it comes to the anaerobic adhesive. It will not solidify when it is in contact with oxygen or air. Once it is isolated from air, coupled with the catalytic effect of the metal surface, it can quickly polymerize and solidify at room temperature, forming a firm bonding and good sealing. It can be stored for many years by the effect of oxygen. A liquid adhesive that remains liquid in the presence of air at room temperature and rapidly solidifies in isolation.

*Corresponding author. Email:[a]Wanying.dong@gtiit.edu.cn, [b]1812259@mail.nankai.edu.cn

This review summarizes the different curing methods of adhesives so far, including four different ways, UV curing, heat curing, moisture curing, and microwave curing. This review discusses the mechanisms, application ranges, advantages, and disadvantages of different curing methods. This summary can deepen the understandings of different ways of adhesive curing. Besides, we can choose the most appropriate curing method according to different requirements in industrial production and daily life.

2 Curing methods

2.1. UV curing

Ultraviolet curing (UV curing) is a processing technology that quickly cures the liquid materials employing ultraviolet light [6]. Under the irradiation of ultraviolet light, the polymerization of liquid adhesive material can quickly happen as the reactive substances are excited by UV light. Therefore, the curing of adhesive materials is very rapid within a short time. UV curing adhesives are widely used, such as epoxy acrylate adhesives, polyurethane acrylate adhesives, and polyester acrylate adhesives [7]. Compared with other traditional adhesives, UV curing adhesives have the advantages of convenient operation and use, fast curing speed, high production efficiency, no pollution, and energy-effective. They have been successfully applied in many industrial assembly fields, especially those that require a rapid assembly industry.

Generally, for a UV curing adhesive system, different components are included, such as photo-initiators, pre-polymers, reactive diluents, and other additives. The function of the photo-initiator is to absorb UV and initiate polymerization. The pre-polymer is the main component of the adhesives [8]. The molecular structure and molecular weight of the pre-polymer affect the final properties of the adhesive such as curing rate, flexibility, adhesion, hardness, and so on. Furthermore, the reactive diluent is used to adjust the system viscosity, and it also participates in the curing process. Other ingredients, including dyes, stabilizers, surfactants, etc., can be added in appropriate amounts according to the different applications of the adhesives.

The photo-initiator in the glue system is firstly excited by ultraviolet light to generate free radical active species. Then the polymerization process proceeds with the unsaturated pre-polymers and monomers through grafting or crosslinking processes to obtain a body-shaped network structured polymer. The active monomer acts as a bridge, which links the pre-polymers with different molecular weights together to form long molecular chains. For example, diethylene glycol diacrylate is triggered by ultraviolet light to produce a polymer with a network structure (Fig. 1).

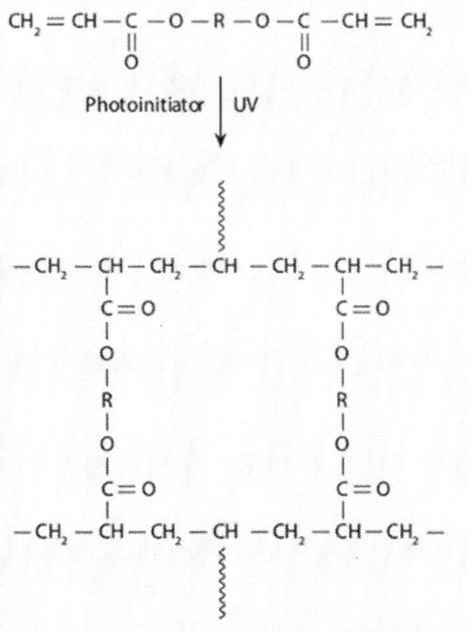

Fig. 1. Photocrosslinking of a diacrylate (simplified): formation of a network [8].

In general, UV curing adhesives have some advantages and disadvantages. Firstly, the process of UV curing can be carried out at room temperature, which indicates that UV curing is an energy-effective and cost-effective process, and the required equipment for the curing process is relatively simple. Furthermore, there is no organic solvent in the system in UV curing adhesives, which is an environmental-friendly process. In contrast, thermoplastic adhesives are processed through endothermic physical changes (such as solvent evaporation and melting). Moreover, some low-volatility monomers are added to the adhesive system to adjust its viscosity, and these monomers can fully participate in the curing. These monomers act as a bridge to connect the oligomers and finally produce a cross-linked substance. As a result, the solid content can be regarded as 100% [9]. Finally, UV-cured adhesives can cure in a very short time. For example, superfast curing (in less than 0.5 s) of acrylate adhesives for plastics has been achieved by a proper choice of pre-polymers, photo-initiators, and photo-sensitizers, which have been successfully used in blister packaging, as well as for bonding glass to plastics [6]. The reduced time of cure implies much higher productivity, which can increase profits and reduce costs. However, since UV light cannot pass through opaque materials, UV curing technology is not suitable for curing opaque systems. Furthermore, the light-curing rate is fast, the internal stress of the formed cured product is relatively large, and the strength and adhesion of the material will also be affected. Finally, the addition of monomers will also make the adhesive system have a greater shrinkage after curing.

2.2 Heating Curing

Heat curing adhesive is a kind of heat curing cross-linked polymers or heats curing plastics. Most heating curing adhesives are supplied in a two-component system, usually made up of a resin and a hardener. They are usually existent as liquid form or gel form, and the curing process will initiate when they are mixed. The reaction time usually takes in a range between 10 minutes to a few hours. Moreover, there are also some adhesives are supplied in single-component systems or one-component systems. In this system, the heat curing pastes must be cold-stored to delay the curing itself for certain applications, such as phenolic resin. One-component systems can offer advantages in some specific reaction processes. It is unnecessary to add any other additional substances into the system because all of the required components, say the main agent, curing agent and crosslinking agent, have been mixed within one component.

Epoxies are a class of the most common heat curing adhesives, and several different advantages are offered in distinct applications. For example, when used to bond the wood, better resistance to moisture and a higher ability in gap-filling are provided. And when it is employed to bond the metals and glass, epoxy resins are also unusual as they could create a strong bond. However, unmodified epoxy is a kind of hard and brittle solid, which is easily fractured from time to time. Consequently, it is vital to improving the toughness, flexibility, and fatigue resistance of adhesives. Ordinarily, different modifiers are utilized in most cases. For example, the addition of benzoic acid as a modifier could enhance the deoxidizing curing ability and compatibilization effect.

Additionally, some other polymers can be used as heat curing adhesives, such as polyester resin, which could be used as a lower-cost substitution to epoxy resins. However, the ability of it in gap filling and bond strength is not as outstanding as epoxy resins. Phenolic resins, one of the earliest forms of plastics, could also act as heat curing adhesive, typically used for paper and wood.

Different chemical reactions usually happen for the heat curing adhesives during the process of curing, which results in higher strength for obtained products. There are many other advantages as well. First of all, the gap filling ability is excellent. Moreover, the resistance to moisture, heat and other changes from the outside environment is also good. However, the impact strength and peel strength are sometimes low. In addition, the initial cohesiveness is not high. And the ability of water resistance is also a big problem that adhesives face nowadays. Hence, it is essential to add some modifications necessary in the process of curing to overcome these problems. For instance, Dai and co-authors have introduced a proper amount of dopamine and metal ion to improve the water resistance of the epoxy adhesives and endow them with excellent adhesion properties [10].

2.3 Moisture curing adhesive

In most common circumstances, the adhesive should be used in the open air, and the process of curing has to take place in a mild condition. An ideal situation is that the adhesive keeps liquid form in the bottle and turns hard as soon as being exposed to the air. H_2O molecule has a strong potential of nucleophilic attacking. Therefore, the major strategy for moisture curing is to provide a structure that can quickly react with water.

One of the most effective methods is introducing an isocyanate group, which can immediately react with the moisture from the atmosphere [11]. Isocyanate groups can be attacked by nucleophilic reagents, including those molecules with -OH or $-NH_2$ groups (Fig.2). And then, a decarboxylation process happens for carboxylic acid, and an anime is formed. It can then react with another isocyanate group. Eventually, the small molecules are linked together by an acylamide bond.

Like the alkoxy groups in alkoxysilanes, other kinds of groups are also able to realize such a similar process [12]. Firstly, silanol and alcohol are generated from the reaction. After that, the silanol group condenses together or catches other alkoxy groups to form the siloxane group, eliminating water or alcohol molecules (Fig.3). In this process, the H_2O can reach deeper under the surface to react with mole alkoxy groups.

A type of glue that is widely used in our daily life is called ethyl cyanoacrylate. In this molecule, two electron-withdrawing groups are linked to the same carbon atom in the monomer, creating a high tendency to be attacked nucleophilic reagents. Therefore, a very little amount of OH- from the moisture in the air can initiate an anionic polymerization, quickly transferring the system into a kind of mass that is hard and dry. The curing process is just the process of polymerization.

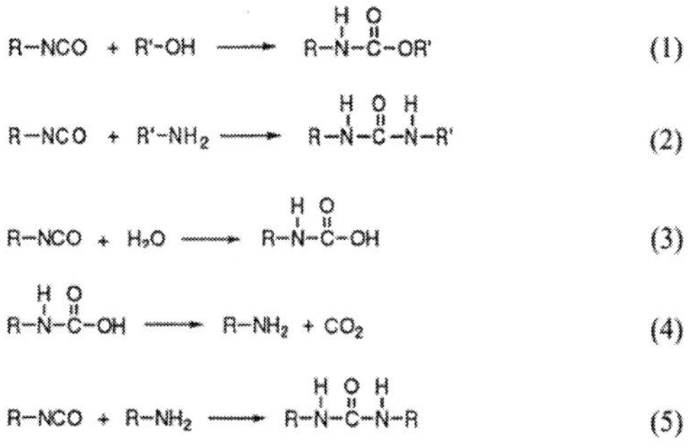

Fig.2. The whole process of the curing of compounds led by isocyanate groups. (3) take place as soon as facing the moisture, generating the substrate of (4). The anime that just formed can react with an unreacted NCO group (5), linking each other together (S[11].

$$\equiv Si\text{–}OR + H_2O \rightarrow \equiv Si\text{–}OH + ROH$$

$$\equiv Si\text{–}OH + \equiv Si\text{–}OH \rightarrow \equiv Si\text{–}O\text{–}Si\equiv + H_2O$$

$$\equiv Si\text{–}OR + \equiv Si\text{–}OH \rightarrow \equiv Si\text{–}O\text{–}Si\equiv + ROH$$

Fig.3. The mechanism of curing by alkoxysilane is shown in the graph. After reacting with the H2O molecule, the group transfer into a silanol group, which is ready to build a Si-O-Si bond to construct the polymer [12].

The most vital characteristic of moisture curing is effectiveness. The conversion of adhesive can take place immediately when it is exposed to the moisture air. Significant advantage as it is, the promptness of moisture curing also acts as the greatest limitation to the usage of this kind of adhesives because the efficacy of it will lose as soon as being exposed to the air. Therefore, it is crucial to avoid any possibility of touching the moisture during the manufacturing, transporting, and storing process.

2.4 Microwave curing

Microwave curing is also widely utilized for the curing of adhesive, such as E44/DDM, UPR, and so on. Generally, the microwave frequency is about 0.3~300 GHz, and it will produce heating, melting even chemical reaction, and other physical phenomena with the assistance of microwave. Therefore, microwave curing is more efficient than the conventional methods. It can make chemical reactions more than 3 times or even dozens of times faster at the same or even lower temperature. Due to such high efficiency, microwave curing can also be used for chemical synthesis [13]. Furthermore, microwave curing can easily convert the liquid resin into a solid state, which could benefit the curing of adhesive.

K.Johnston and co-authors have carefully investigated the microwave curing mechanism of epoxy. As shown in Fig 4, firstly, the epoxy reacts with amines through a ring-opening mechanism. Then during the curing process, an epoxy group reacts with a primary amine (RNH_2) to produce a hydroxyl group and a secondary amine (RNH). In the second step, the formed secondary amine reacts with another epoxy group to produce a hydroxyl group and a tertiary amine. Opening of the epoxy ring by an amine to form OH and CN groups [14]. It seems that the chemical mechanism of microwave curing is essentially similar to that of heat curing. However, the rate of microwave curing is much faster. Thus, microwave curing has a more complex effect on the curing process of adhesive, which is obviously different from the effect of heating curing just by increasing the reaction temperature [15]. Microwave curing can reduce the curing time and manufacturing costs. Moreover, the consumed energy of microwave curing is obviously reduced.

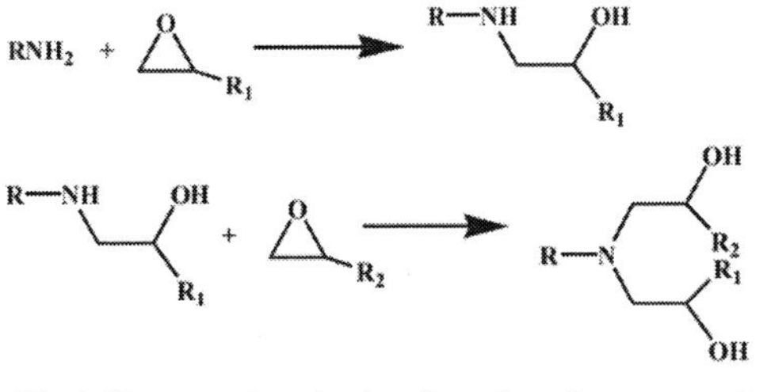

Fig.4. The general mechanism for curing of an epoxy resin with an amine curing agent [13]

3 Conclusion

This article has reviewed the research progress of adhesives through four curing methods: UV curing adhesives, heating curing adhesives, moisture curing adhesives, and microwave curing adhesives. These four curing methods have their own advantages and disadvantages. Moisture-curing adhesives are very easy to use. Compared with other curing methods, they do not require special equipment and can be cured in the air. However, moisture-curing adhesives are easily affected by humid environments and water, and they have poor resistance to polar solvents. In contrast, the curing process of thermal curing and UV curing adhesives requires specific equipment. Thermal curing adhesives have higher strength and excellent bonding performance. However, the curing speed is slow, and the curing process usually lasts from ten minutes to several hours. The curing speed of UV-curing adhesives is much faster than that of the thermal curing adhesives, and therefore it is more efficient, which is more conducive to industrial applications. Microwave curing is a new type of curing method that has been used in adhesives, and the curing speed is also extraordinarily fast. However, the high speed of the curing process also has a weakness. When the adhesive is cured, the volume shrinkage due to solvent volatilization or chemical reaction will generate internal stress, which will affect durability and bonding strength. As a result, though UV curing and microwave curing have higher efficiency, they will also cause greater internal stress in the material, thereby affecting the performance of adhesives. Most industrial applications have a great demand for adhesives. In future research, how to improve the bonding efficiency, bonding performance, and cost reduction of adhesives will be very important topics.

References

[1] S. Rimdusit, H. Ishida, Polymer 2000, 41(22), 7941-7949.

[2] R.C. Benson, D. Farrar, J.A. Miragliotta, Johns Hopkins APL Technical Digest (Applied Physics Laboratory) 2008, 28(1), 58-71.

[3] I. Fotiou, A. Baltopoulos, A. Vavouliotis, V. Kostopoulos, Journal of Applied Polymer Science, 2013, 129(5), 2754-2764.

[4] Eun Seon Kim, Jae Heung Lee, Dong Hack Suh, Woo Jin Choi, Macromolecular Research 2021, 29, 129-139

[5] Nasrin Moini,Maryam Khaghanipour,Kourosh Kabiri,Ali Salimi,Mohammad J. Zohuriaan-Mehr,Arash Jahandideh, 5. ACS Sustainable Chemistry & Engineering 2019, 7, 19, 16247–16256.

[6] DH Wang, L Jiang. Theory and Application of Ultraviolet Curing Materials [M]. Beijing: Science Press, 2001.

[7] YT Xue. Study on Preparation and Properties of UV-Curable Adhesive [D]. Fuzhou University ,2011.

https://doi.org/10.1051/e3sconf/202129001021

[8] V Alessandra, T Giuseppe, B Roberta. Reviews of Adhesion and Adhesives,2017,5(2).

[9] M Bajpai,V Shukla,A Kumar. J. Progress in Organic Coatings,2002,44(4), 271-278.

[10] F Dai, F. Chen, T. Wang, S. Feng, C. Hu, X. Wang, Z. Zheng Journal of materials Science, 2016, 51, 4320–4327

[11] Comyn and F Brady and R.A Dust and M Graham and A Haward. International Journal of Adhesion and Adhesives, 1998, 18(1), 51-60.

[12] Hai, N. , A. D. Skaja , and MD Soucek. J.Progress in Organic Coatings 2000,40(1), 175-184.

[13] Johnston K, Pavuluri S K, Leonard M T, Thermochimica Acta, 2015, 616(20), 100-109.

[14] S.K. Pavuluri, M.P.Y. Desmulliez, G. Goussetis, et al. Post cure behaviour of encapsulants for QFNpackages processed by an open-ended single mode resonant microwaveapplicator, 2011 IEEE 13th Electronics Packaging Technology Conference(EPTC), 2011, 573–578.

[15] R.C.Benson, D. Farrar, J.A. Miragliotta. Polymer adhesives and encapulants formicroelectronic application, Johns Hopkins Apl Tech. Dig. 2008, 28, 58-71.

Chemical behaviours of Arsenium, Chromium, Mercury, Lead, and Strontium in aqueous system

Qiuyu Zhao[1], Siyi Wang[1] and Jianjun Wang[1,a*]

[1]MOE Key Laboratory of Resources and Environmental Systems Optimization, College of Environmental Science and Engineering, North China Electric Power University, Beijing 102206, PR China

Abstract. Over the decades, the application of heavy metals and the expansion of civilization resulted in severe pollutions in aqueous solutions. The poor degradation of toxic heavy metal contaminants in hydrosphere undoubtedly posed a huge threat to environmental safety and biological health. In this review, as most common heavy metal pollutants, arsenium (As), chromium (Cr), mercury (Hg), lead (Pb), and strontium (Sr) were introduced in detail. The chemical behaviours, chemical status, biological toxicity, and migration of mentioned contaminants were summarized. This work highlighted and reviewed the basic information of five heavy metal pollutants, which provided a new direction of toxic heavy metal ion remediation.

1 Introduction

In the process of human development, the utilization of heavy metal greatly promoted the advancement of civilization. However, the heavy metal pollutions also arose several environmental problems because of the improper disposal, accidentally leaked and illegal discharge. Most of the heavy metal pollutants would cause varying degrees of damage to human beings and natural environment, thus, which aroused worldwide attention to address these issues [1]. The accumulating and enriching of heavy metal ions in aquatic organisms and microorganisms were widely existed. Adverse effects on the life activities of organisms will occur when the concentration and toxicity threshold of pollutants are exceeded [2, 3].

Arsenium (As), Chromium (Cr), Mercury (Hg), Lead (Pb), and Strontium (Sr) are most common heavy metal pollutants, which existed world widely. The environment problems caused by mentioned pollutants in aqueous system received great attention and numerous investigations had focused on them [4]. The worrisome aftereffects of drinking water pollution by these heavy metals have blowed up from the 19th century to the 20th century during the period of vigorous development of heavy industry [5, 6]. Thereinto, several typical cases attracting worldwide attention: itaiitai disease events (Cd), Japan minamata disease events (Hg) and the arsenic poisoning of Bangladesh groundwater.

Considering the hazards of heavy metal pollution, there is much talk about water crisis, of which the most worrying is that hundreds of millions of people lack access to clean and safe water just for their daily consume and the condition is becoming serious with the impetuous development of industry. This grim situation became an inevitable problem in the industrial society. This work reviewed chemical behaviours of common toxic metal ions in solution, and their transformation in aqueous system was briefly introduced. This review summarized the environmental behaviours of As, Cr, Hg, Sr, and Pb, which provided new direction of toxic heavy metal ion remediation.

2 Arsenium

Arsenic, an element widely existed in the environment, ranks 20th in nature and 14th in seawater [7]. As an essential trace element of human body, arsenic deficiency can lead to intrauterine growth retardation, bone mineralization, and significantly affect the activity of certain enzymes involved in neurological disorders and vascular diseases. Due to the low demand of arsenic (220 $\mu g \cdot day^{-1}$), human can receive adequate amounts from normal diets such as seafood and grains.

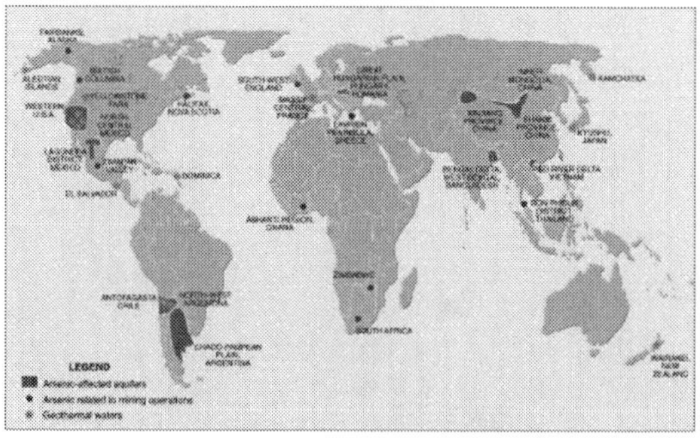

Figure 1. Distribution of Arsenic environmental problems due to mining and geothermal resources [8].

Arsenic is widely existed in groundwater (Figure. 1), rocks, and minerals [8, 9]. However, due to surging demand of human and rapid industrial development, large

[a] Corresponding author: wangjj2016@ncepu.edu.cn

https://doi.org/10.1051/e3sconf/202129001022

amounts of arsenic are released into nature. Major anthropogenic sources of As in environment are from industrial processes (mining, steel production, and fossil fuels), agriculture production (the use of pesticides, preservatives, and rodenticides), and infant industry (raw material for semiconductor and laser device) [6, 10]. According to WHO regulations, the concentration of arsenic in drinking water should be limited to 10 μg/L. However, in untreated groundwater, the dissolved arsenic is generally higher than the recommended concentration. Regardless of trivalent and pentavalent As, excessive As may cause arsenic poisoning. Among them, As(III) and its oxide (As_2O_3) are the most toxic because they can react with sulfur-containing compounds to generate reactive oxygen species. The clinical manifestations of acute and subacute arsenic poisoning are initially general discomfort, fatigue, headache, dizziness, etc., followed by nausea, abdominal pain, and diarrhoea.

Chronic arsenic poisoning increases the risk of lung, skin, bladder, and liver cancer [11]. Up to now, there is no effective medicine or treatment for arsenic poisoning. It is recommended to take a range of preventive measures, such as drinking qualified water, taking supplements such as vitamins and proteins, and exercising to improve body function [12]. The geochemistry of arsenic reveals that the mutual conversion and solubility of As(III) and As(V) in water depends on a number of factors, including redox potential (Eh), pH, organic content, and aquatic bioactivity. These environmental conditions undoubtedly affect the migration and transformation of many arsenic species in nature [13].

3 Chromium

Chromium widely exists in nature, and its natural source is mainly weathering of rocks, mostly in the form of Cr(III). Sources of man-made pollution include leather preparations, chrome plating of metal parts, industrial pigments, as well as the production of tanned leather, rubber and ceramic raw materials. As depicted in Figure. 2, different reaction mechanisms in atmosphere, water, and soil greatly affected the biogeochemical cycle of Chromium [14].

The chromium produced by industrial pollution is mainly a hexavalent compound, usually presents as chromate ion (CrO_4^{2-}). The migration and transformation behaviours in the environment of chromium ions in different valences are also different. For example, Cr(III) in water can be adsorbed on solid matter and also would exist in sediment. However, Cr(VI) is more soluble and stable than Cr(III) in water, and can be reduced to Cr(III) under anaerobic conditions. Cr(III) can undergo a hydrolysis reaction under neutral or alkaline conditions, producing insoluble $Cr(OH)_3$ then sinking into the bottom of the water. Therefore, Cr(III) and Cr(VI) can easily convert into each other in aqueous solution. Deserved to be mentioned, Cr(III) is a trace element necessary for human and animal, which is the subsequent component of digesting glucose, fat, and protein as well as can be used as a dietary enhancer. Studies have shown that the human body contains about 0.03 mg/L of Cr and needs to supplement 15-200 μg every day [15].

The lack of Cr may cause diseases such as arterial hypertension and cardiovascular disease. Meanwhile, Cr can also stimulate plant growth to increase yield. However, high doses of chromium, both Cr (III) and Cr (VI), are harmful to human health. Moreover, Cr(VI) is much more toxic than Cr(III), which is a strong mutagenic substance. Symptoms such as vomiting, diarrhoea and headache will occur when Cr(VI) is mildly toxic, but in severe cases, it can induce lung cancer and nasopharyngeal cancer. The dangerous dose of Cr(VI) is about 1 - 2 g, and the median lethal dose (LD50) of Cr(III) and Cr(VI) is 1900-3000 mg/kg and 50-100 mg/kg, respectively [15].

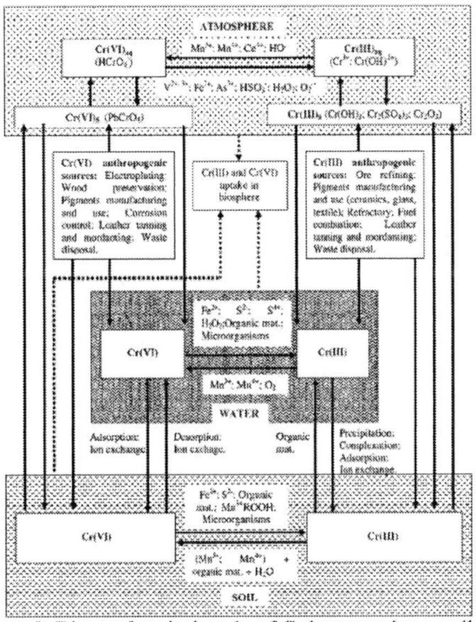

Figure 2. Biogeochemical cycle of Cr in atmosphere, soil, and water [14]

4 Mercury

Mercury is a natural element that widely found in biosphere with the form of metal elements, inorganic and organic matters. As an element with a variety of economic uses, metallic mercury and its compounds are widely used in chemistry, medicine, metallurgy, electronic instruments, military and other precision and technological fields [16]. Due to inappropriate mercury emissions, humans are exposed to various forms of mercury pollution in their daily lives. Mercury-containing wastewater poses a potential threat to human health caused by the accumulation of organisms. The most serious mercury poisoning accident was the Minamata Disease in 1956, which aroused continuous concern about mercury pollution worldwide.

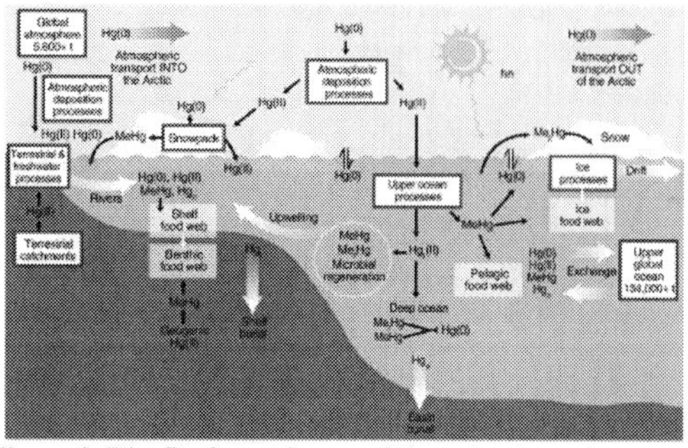

Figure 3. Distribution pathways of Hg to Arctic marine waters and their transformations [17]

There are many sources of mercury pollution in nature, mainly from natural sources and man-made sources. The major pathways and transformations of Mercury in Arctic was shown in Fig. 3, and the most existing form was inorganic Hg(II) [17]. Natural sources include volcanic activity, geothermal hot springs, natural weathering, soil emissions, and vegetation release. Typical man-made sources of mercury pollution include: (1) Medical waste: mercury alloys and compounds are used in dental filling materials and disinfectants; (2) Metallurgical industry: Mercury is used to dissolve other metals to form amalgam to extract gold, silver, thallium and other metals; (3) Electrical instrument industry: Mercury is used to make mercury salt dry batteries, mercury lamps, barometers, diffusion pumps, mercury switches and thermometers and other physical instruments; (4) Precision high-tech fields: Mercury is used for precision casting molds, plutonium nuclear reactor coolant, cadmium-based bearing alloys, etc.; (5) Fossil fuel combustion: The process of coal burning and garbage incineration in power plants is considered to be the main cause of global mercury emissions [18]. In view of the threats mercury poses to the environment and organisms, the United Nations Environment Programme adopted the Minamata Convention in 2013, an international convention aimed at global control and reduction of mercury emissions, and made detailed regulations on specific emission limits to reduce mercury exposure.

The prevention and control of mercury in water bodies should not only pay attention to the source of mercury pollution in the environment, but also focus on understanding its geochemical cycle in nature. Since the Minamata Convention came into effect, the source of mercury release has been effectively controlled. However, human activities may lead to the re-release of mercury and thus increase the efficiency of mercury methylation. In addition, the form of mercury gas makes it possible for long-distance transmission, even in remote areas like the Arctic without exhaust emissions. Moreover, changes in natural climate and land utilization have made it more difficult to determine the mercury cycle, methylmercury bioavailability, and mercury migration and transformation in the biological chain [19]. Therefore, more detailed and up-to-date data on the distribution and the health effects of mercury are needed to provide a reliable basis for assessing the hazard risk as well as predicting the migration and transformation of mercury, which will play an extremely beneficial role in preventing and controlling mercury pollution.

5 Lead

In recent years, with the rapid development of the steel, battery and gasoline industries, toxic heavy metal lead has been released into the environment. When the accumulation of lead in the water environment reaches the safety limit, it will cause great harm to the human body, aquatic organisms and crops. With the frequent occurrence of a series of lead pollution problems, scientists continue to conduct a more in-depth discussion on the harm of lead and environmental behaviour.

Generally speaking, in natural water bodies or deep seas, the lead content is not high. However, the lead concentration in coastal waters and lead mine drainage areas has generally increased, indicating that the contribution rate of lead pollution mainly comes from emissions from industries and households as well as the runoff from mineralized areas. With excellent physical and chemical properties, lead is widely used in smelting, painting, medicine and cosmetics [20]. For example, tetraethyl lead is added to gasoline to improve explosion-proof properties; metallic lead can let the paint brighter and moisture-resistant; and lead can also help cosmetics achieve brightening and whitening effects [21]. However, lead is relatively toxic in heavy metals, which can act on various systems and organs of the human body and is mainly neurotoxic. The toxicity of lead is closely related to its form and solubility in water. Lead nitrate and lead acetate are soluble in water and easily absorbed by human bodies; solid lead oxide and alkaline lead sulfate are easily soluble in acid and have high toxicity; lead sulfide and lead chromate have low toxicity because of the poor solubility; organic lead such as tetraethyl lead is more toxic than inorganic species. The accumulate of lead in the kidneys, brain, muscles, and bones, inevitably cause damage to the nervous, bone marrow hematopoietic, digestive, cardiovascular, kidney and reproductive as well as immune system. Both inorganic and organic lead at high levels in human body can cause deformity, cancer, genetic mutation and poor embryonic development [21-23]. After long-term exposure, lead may cause human diseases such as anaemia and hepatitis even at low level. According to the stipulation of WHO and USEPA, the concentrations of Pb in safe drinking water should be less than10 μg/L.

6 Strontium

Alkaline earth metal Strontium can be easily oxidized into a stable bivalent Sr(II). Strontium can be existed in natural environment as ^{84}Sr, ^{86}Sr, ^{87}Sr, ^{88}Sr, and 31 unstable isotopes [24]. Among them, ^{90}Sr has the longest life and it is the most common fission product of ^{235}U and ^{239}Pu, and also the most common radioactive pollutant in wastewater.

In addition to nuclear power plant operation, uranium and thorium mining, nuclear weapon production, strontium is also widely used in ceramics, sugar, medicine, chemical industry and other industrial fields [25]. Therefore, industrial production of strontium-containing radioactive wastewater should also be given high attention. Strontium is an indispensable trace element of human body, which is a normal part of human bones and can help enhance the excretion of sodium [26]. In addition, some radioisotopes of strontium also play an important role in pain treatment. However, since Sr is related to calcium (Ca), they have similar chemical properties. In biology, excessive Sr may interfere with calcium absorption and metabolism. After entering the human body, the radioactive ^{90}Sr will form $^{90}Sr_3(PO_4)_2$ through ion exchange with the Ca in the bone, which accumulates in the inorganic substance of the bone marrow cavity, thus seriously radiating the hematopoietic stem cells in the bone marrow cavity, and leading to a significant decrease in the number of white blood cells, red blood cells and platelets. Long-term ^{90}Sr radiation can cause leukaemia, osteosarcoma and other diseases. Although non-radioactive Sr is not radioactive and is also less toxic, it can seriously interfere with the calcification process of bone tissue, resulting in spontaneous fractures and systemic osteoporosis. Moreover, after radioactive strontium contaminate the mother, strontium can be transferred to the baby through placenta and milk, thus causing serious harm.

7 Conclusion

In this revies, we summarized the basic information of heavy metal ions (chromium, arsenic, lead, mercury, and strontium) in wastewater. The chemical status, biological toxicity, and migration were briefly introduced. Overall, the purification technology for practical applications to remediate these highly toxic and carcinogenic pollutants. In foreseeable future, advanced purification technology will be utilized for heavy metal removal, and this review could provide basic information and environmental mechanism of mentioned pollutants to protect public health and environmental, and fulfil this task.

Acknowledgments

The work was financially supported by National Natural Science Foundation of China (11875028).

References

1. J. O. Nriagum, Science **12,** 272 (1996)
2. H. Zhen, L. Jia, C. Huang, Y. Qiao, J. Li, H. Li, Q. Chen, Y. Wan, Environ. Pollut **263,** 114552 (2020)
3. E. G. Farmaki, N. S. Thomaidis, Global Nest J **10,** 366 (2008)
4. P. U. Singare, R. M. Mishra, M. P. Trivedi, Front. Sci **2,** 28 (2012)
5. C. Z. Ledezma, D. N. Bolagay, F. Figueroa, E. Z. Ledezma, M. Ni, F. Alexis, V. H. Guerrero, Environ. Technol. Inno **22,** 101504 (2021)
6. J. P. Vareda, A. J. M. Valente, L. Durães, J. Environ. Manage **246,** 101 (2019)
7. B. K. Mandal, K. T. Suzuki, Talanta **58**, 201 (2002)
8. P. L. Smedley, D. G. Kinniburgh, Appl. Geochem. **17**,517 (2002)
9. S. Aksentijević, J. Kiurski, M. V. Vasić, Environ. Monit. Assess. **184,** 335 (2012)
10. R. Eisler, Rev. Environ. Contam. T **180,** 133 (2004)
11. P. Tchounwou, J. Centeno, A. Patlolla, Mol Cell Biochem **255,** 47 (2004).
12. S. R. Mitra, A. Basu, D. N. Mazumder, Environ. Health Perspect. **112,** 1104 (2004)
13. H. Onishi, E. B. Sandell, Geochim. Cosmochim. Ac. **7,** 1 (1955)
14. M. Gheju, Water Air & Soil Poll **222,** 103 (2011)
15. K. G. Pavithra, V. Jaikumar, P. S. Kumar, P. Sundarrajan, J. Clean. Prod. **228,** 580 (2019)
16. J. Yu, B. Yue, X. Wu, Q. Liu, F. Jiao, X. Jiang, X. Chen, Environ. Sci. Pollut. Res **23,** 5056 (2016)
17. B. Braune, J. Chételat, M. Amyot, T. Brown, M. Clayden, M. Evans, A. Fisk, A. Gaden, C. Girard, A. Hare, J. Kirk, I. Lehnherr, R. Letcher, L. Loseto, R. Macdonald, E. Mann, B. McMeans, D. Muir, N. O'Driscoll, A. Poulain, K. Reimer, G. Stern, Sci. Total Environ. **509-510**, 67 (2015)
18. L. T. Budnik, L. Casteleyn, Sci. Total Environ. **654,** 720 (2019)
19. E. Ha, N. Basu, S. Bose-O'Reilly, J. G. Dorea, E. Mcsorley, M. Sakamoto, H. M. Chan, Environ. Res **152,** 419 (2016)
20. G. D. Corder, A. Golev, D. Giurco, Miner. Eng **76,** 2 (2015)
21. J. D. O. Brandão, O. J. Okonkwo, M. Sehkula, R. M. Raseleka, Toxicol. Environ. Chem. **94,** 70 (2012)
22. G. Lockitch, Clin. Biochem **26,** 371 (1993)
23. P. C. Hsu, Y. L. Guo, Toxicology **180,** 33 (2002)
24. S. Voerkelius, G. D. Lorenz, S. Rummel, C. R. Quétel, G. Heiss, M. Baxter, C. Brach-Papa, P. Deters-Itzelsberger, S. Hoelzl, J. Hoogewerff, E. Ponzevera, M. Van Bocxstaele, H. Ueckermann, Food Chem. **118,** 933 (2010)
25. A. Ahmadpour, M. Zabihi, M. Tahmasbi, T. R. Bastami, J. Hazard. Mater. **182**, 552 (2010)
26. V. Zaichick, Biol. Trace Elem. Res. **153,** 84 (2013)

Cold Model Study on Melting of Ice Made by KCl Solution in Gas-Water Two-Phase Plume Area

Tao Wang[1], Kanghua Pei[1], Jian Zhao[1], Zhao Li[1], Huan Wang[1], Rongwang Yang[1], and Chao Chen[1,a]

[1]College of Materials Science and Engineering, Taiyuan University of Technology, 030024 Yingze West Street No. 79, Taiyuan City, Shanxi Province, P.R. China

Abstract. With the increasing of scrap usage in steelmaking processes, the melting of scrap becomes a very important phenomenon that limits the productivity and tap-to-tap time. Ice-water systems have been widely used to study the melting of scrap and alloys. In this study, the melting rate of saturated KCl solution ice spheres in gas-water two-phase plume zone are studied as a function of height of location, gas flowrate, and melt temperature. The results show that the shape of the ice sphere gradually changes firstly from spherical to elliptical, and finally becomes an irregular state. 1) The decreasing of the distance between the ice sphere and the bottom plugs, 2) the increasing of the gas flowrate, 3) the slightly increasing of bath temperature will all benefits the melting rate of ice spheres.

1 Introduction

With the international trade trend becoming more unstable, international iron ore prices continue to rise. Iron ore prices rose more than 80 per cent from April 2020 to April 2021. As a result, steel scrap has become an important national strategic resource and the main raw material for steelmaking processes. Chinese scrap consumption has increased from 83.3 million tons in 2015 to 215.93 million tons in 2019[1]. Due to economic reasons, great efforts have been made to increase the scrap usage in basic oxygen furnace (BOF) and even in secondary refining ladles. Nowadays, to the authors' knowledge, the scrap ratio in the converter steelmaking increases to more than 20%. Consequently, the melting time of steel scrap in molten steel is greatly prolonged. Besides, the mixing of chemical compositions in BOF and ladle is also delayed.

Scrap melting has been widely studied. Penz and Schenk[2] summarizes the scrap melting studies by coupled heat and mass transfer, small-scale high temperature experiments, mathematical models, and numerical simulations. Based on the similarity theory, water model studies have been widely used to study the fluid flow, mixing, and mass transfer in metallurgical reactors. A few studies were focused on the ice and water systems.

R. Vanier et al.[3] examined the weight of ice particles in water with a cooling coil in a secondary compartment. Experimental results have been obtained for heat transfer to melting ice spheres by measuring the rate of change of apparent weight.

S. Taniguchi et al.[4] studied the effect of gas injection upon the melting rate of an ice sphere in a water bath. Heat transfer coefficient was obtained from the experiment in which the melting rate of an ice sphere was measured. The heat transfer coefficient increased with an increase in gas flow rate, while it was not influenced much by the orifice diameter and vessel shape.

J. Szekely et al. [5] concered the rate at which ice rods melt, when immersed in a pool of water, agitated by an ascending stream of gas bubbles. The melting rate at the quiescent zone and the surface were compared.

M. Iguchi et al. [6-7] measured the veloaty and turbulence around a sphere made of synthetlc resin submerged in water and bubbling jets in a cylindrical vessel with bottom blowing.The effect of velocity and turbulence intensity on Nusselt number were investigated, and empirical experession for mean Nusselt number was proposed as a function of Reynolds number and turbulence intensity.

Y. L. Hao et al. [8] examined the melting and heat transfer characteristics of a fixed ice particles in flowing water. The shape changes of melting ice spheres with time and temperatures of water and at different initial ice temperature are record. The time variations of local parameters, local heat transfer coefficient and local Nusselt number at various angular positions are obtained. In addition, an empirical correlation for average Nusselt in convective melting of a solid pariticle is obtained.

A. K. Shukla et al. [9] studied the melting of ice samples of different geometries and sizes in an argon stirred vessel containing water. The relationships between non-demensional groups ralated to heat transfer are derived for different conditions, and the heat transfer coefficient is estimated as a function of mixing power.

L. Cao et al.[10] used the ice samples with quartz sand to simulate scrap steel in BOF converter. The melting characteristics of scrap steel in the converter was studied through water model experiments.

However, in most of the water model experiments, ice

[a] Corresponding author: chenchao@tyut.edu.cn

spheres or columns are the main research objects[3-9], which can only simulate the light scrap steel with a density lighter than liquid steel. Besides, salt solutions for example KCl and NaCl solutions are used in water models to record the electro-conductivity and corresponding concentrations[11-12] . It was pointed out that the density of salt solutions is 16 to 18% heavier than water[11]. Therefore, the ice made by the denser KCl salt solution could be used to simulate the heavier scrap steel. The melting of ice made by KCl solution was rarely reported in literatures.

In this study, the melting of saturated KCl solution ice spheres in gas-water two-phase plume zone are studied for a better understanding of the melting of heavier scrap steel. The ice spheres are fixed at different locations above the plume. The shape evolution and melting rate are studied as a function of height of location, gas flowrate, and melt temperature. This study could provide theoretical basis and reference for the melting of scrap steel in the steel production process.

2 Experimental apparatus and procedure

Schematic diagram of the experimental device and model parameters are shown in Figure 1. and Table 1. respectively. The experiment is carried out in a cylindrical container made of plexiglass with a cubic outer vessel. The container is filled with water both inside the cylindrical vessel and outer cubic vessel. The outer cubic vessel serves to counteract the problem of light refraction when photographing the melting process of ice sphere. The air is blown by air compressor through the bottom porous plug into the vessel. The gas flow rate is controlled by the flowmeter. A gripper which is composed of iron wire is used to maintain the same height of the ice sphere. A camera is used to record the melting of the salt sphere.

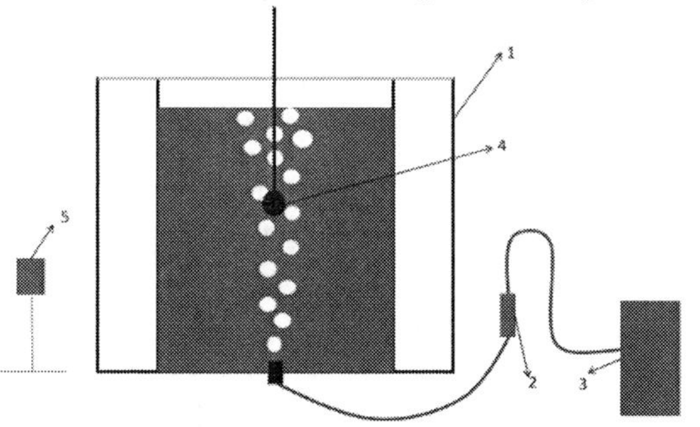

Figure 1. Schematic diagram of experiment devices. 1-Water model;2-flow meter;3-air compressor;4- KCl solution ice sphere;5-camera

During the experiment, the temperature of the bath was kept constant. The ice sphere was fixed at the same position. This position is precisely above the plug and the gas plume. Before the experiment, saturated KCl solution at room temperature were frozen to ice spheres in the refrigerator. The initial temperature of ice sphere was -24 ℃, and the bath temperature in this study were 20℃, 25℃ and 30℃ respectively. The ice spheres were placed 100mm, 200mm and 340mm above from the bottom plug location. The bottom gas flowrate increased from 0L/min to 10L/min. After the experiment, the transverse and vertical diameters of the ice spheres were obtained through the video for intervals of 5 or 10s. The dimensionless diameter is obtained and used to analysize the melting rate of ice sphere.

Table 1. Water model parameters.

Water level	440mm
Diameter of cylindrical vessel	445mm
Side length of cubic vessel	600mm
Diameter of saturated KCl solution ice sphere	50mm
Diameter of porous plug	20mm

3 Results and discussion

3.1 Shape change of saturated KCl solution ice sphere in the gas-water plume

The shape variation trend of ice spheres in the gas-water plume is shown in Figure 2. During the whole melting process, the shape of the ice sphere gradually changes firstly from spherical to elliptical, and finally becomes an irregular state. As shown in Figure 3(a), when the ice sphere is fixed above the porous plug, the ice sphere is washed by bubbles. The bubbles impinging the bottom of the ice sphere and rises around the ice sphere.

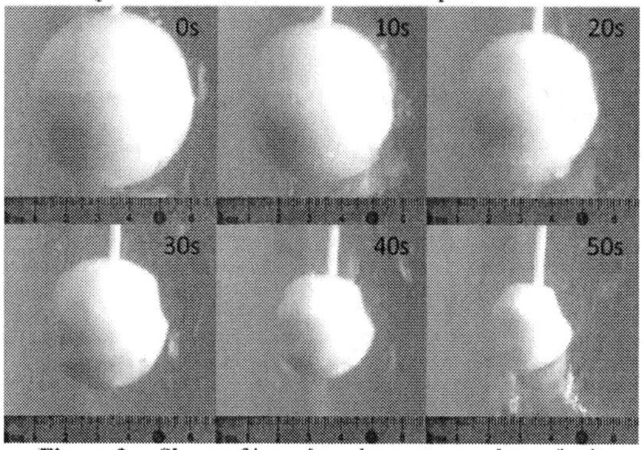

Figure 2. Shape of ice sphere in gas-water plume (bath temperature 25℃, location: 200mm from the bottom, gas flow rate 3L/min)

During the melting process, a layer of crystalline delamination appears at the bottom of the ice sphere, as shown in Figure 3(b). The forced convection, mass and heat transfer of the segregated ice sphere is a complex process which maybe the reason of this phenomenon. During the freezing of ice spheres, the heavier KCl solute deposits at the bottom of the sphere and results in the segregation of KCl solute in the sphere. The mass transfer of KCl solute might be dominant at the bottom of salt ice sphere and leading to the unmelted ice layer at the bottom.

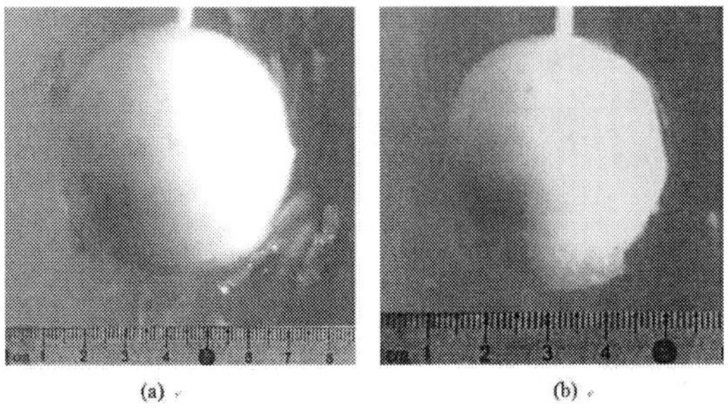

Figure 3. Typical shape of ice sphere melting: a-washed by air bubbles; b- stratification of ice sphere.

3.2 Effect of bottom blowing gas flow rate on ice sphere melting

As can be seen from Figure 4, both the transverse and vertical diameters of ice spheres as a function of time show a linear trend. The vertical diameter decreases faster than the transverse diameters for the same case. In other words, the melting rate of vertical direction is greater than that of the transverse direction. It's due to the gas washing effect for the vertical direction. Besides, the diameter change rate is gradually accelerated with the increase of bottom blowing gas flow rate.

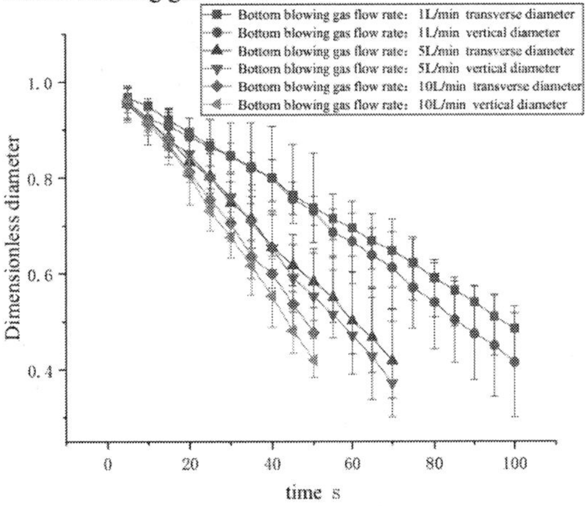

Figure 4. The variation of dimensionless transverse and vertical diameters of ice spheres with time for different bottom blowing flowrates (bath temperature 20℃, location: 200mm from the bottom).

After linear regression of the diameter-time curve, the rate of change of diamteres are calculated and summarized in table 2. Overall, both the rate of change of transverse and vertical diameters are close to each other. The difference between the increases with the increasing of gas flow rate. For better comparasion, the rate of change of the transverse diameter of the ice sphere will be used in the latter part.

Table 2. Comparison of rate of change of transverse and vertical diameter at different flow rates (bath temperature 20℃, location: 200mm from the bottom)

Rate of change	1L/min	5 L/min	10 L/min
Transverse diameter(mm/s)	0.253	0.415	0.5385
Standard error	0.00883	0.4252	0.02643
Vertical diameter(mm/s)	0.276	0.446	0.6055
Standard error	0.02442	0.04619	0.04603

3.3 Effect of ice sphere placement on ice sphere melting

The rate of change of the transverse diameter for three locations of salt ice spheres are compared in Figure 5.

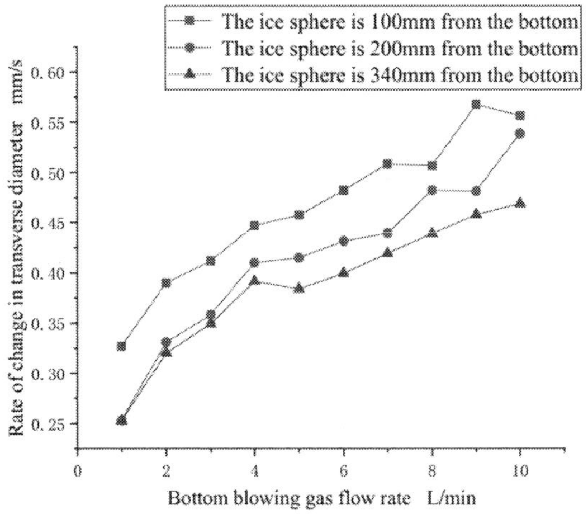

Figure 5. The rate of change of transverse diameters of ice spheres as a function of bottom blowing gas flow rate at different locations (bath temperature 20℃).

A general tendency could be found that the lower the locations of ice spheres the higher the melting rate. The results are not noteworthy for higher locations under small gas flowrates. As an example, the rate of change of diameter is very close for the cases that the ice sphere are 200mm and 340mm above the bottom under the gas flowrate of 1L/min to 3L/min. When the bottom-blowing gas flow rate is greater than 5L/min, the melting rates of ice spheres at these two locations are significantly different. However, the melting of ice spheres at lower position (100mm above the bottom) under the lower gas flowrate conditions is considerably higher than that at the height of 200mm and 340mm. This is due to the strong interaction between bubble plume and ice spheres. For higher gas flowrates, the bubble plume is well developed and the interaction between bubble plume and ice spheres is not significantly changed with the height of location of ice spheres. In all, the bubble plume rising, gas flowrates, and the location of ice spheres will all maters.

3.4 Effect of bath temperature on ice sphere melting

The effect of bath temperature on the melting rate of ice sphere is shown in Figure 6. Generally, the higher the bath temperature the faster the melting of ice sphere. That's easy to understand as the heat and mass transfer process is accelerated. However, the results are obtained under a smaller gas flowrate, i.e. 1L/min. As described before, the melting of ice spheres at lower position (100mm) is higher than the other cases. In Figure 6, the increased temperature can slightly improve the melting of ice sphere at higher positions.

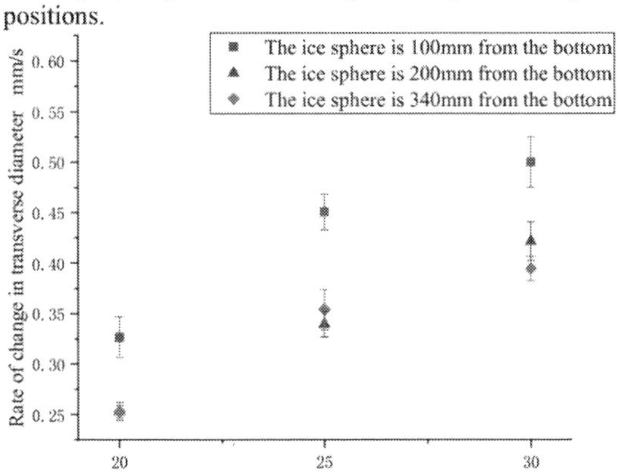

Figure 6. The melting rate of ice spheres at different bath temperatures (bottom blowing gas flow is 1L/min).

4 Conclusion

In this study, the melting rate of saturated KCl solution ice spheres in gas-water two-phase plume zone are studied as a function of height of location, gas flowrate, and melt temperature. During the whole melting process, the shape of the ice sphere gradually changes firstly from spherical to elliptical, and finally becomes an irregular state. The bubbles impinging at the bottom of the salt ice sphere will accelerate the melting process. This can be controlled by lower the height of ice sphere to the bottom gas blowing and by increasing the flowrate of gas plume. Besides, the bath temperature is playing an important role in the melting of ice sphere. Finally, the mass transfer and heat transfer mechanism of the melting of saturated KCl solution ice sphere should be paid more attention.

Acknowledgments

The authors are grateful to the financial support of the Na tional Natural Science Foundation of China project (5190 4204), Applied Fundamental Research Programs of Shan xi Province (201901D211013), Shanxi Provincial Scienc e and Technology Major Special Project (20191102004), and undergraduate innovation and entrepreneurship traini ng program of TYUT (20087).

References

1. Bureau of International Recycling Ferrous Division, World Steel Recycling in Figures 2015–2019: Steel Scrap—A Raw Material for Steelmaking.
2. F.M. Penz, J. Schenk, Steel Res. Int. **90**, 1900124 (2019)
3. C.R. Vanier, C. Tien, AIChE., **16**, 76–82 (1970)
4. S. Taniguchi, M. Ohmi, S. Ishiura, S. Yamauchi, Trans. ISIJ, **23**, 565-570 (1983)
5. J. Szekely, H. H. Grevet, N. EL-Kaddah, Int. J. Heat Mass Trans. **27**,1116-1121 (1984)
6. M. Iguchi, J.I. Tani, T. Uemur, Z. Morita, ISIJ Int. **29**, 658-665 (1989)
7. M. Iguchi, H. Takeuchi, H. Kawabata, T. Uemura, Z. Morita, ISIJ Int. **31**, 46-52 (1991)
8. Y.L. Hao, Y.X. Tao, J. Heat Trans. **124**, 891-903 (2002)
9. A.K. Shukla, R. Dmitry, O. Volkova, P.R. Scheller, B. Deo, Metall. Mater. Trans. B, **42**, 224-235 (2011)
10. L. Cao, Q. Liu, Y. Wang, W. Lin, J. Sun, L. Sun, W. Guo, Mater. Trans. **59**, 1829-1836 (2018)
11. C. Chen, G. Cheng, H. Sun, Z. Hou, X. Wang, J. Zhang, Steel Res. Int. **83**, 1141-1151 (2012)
12. C. Chen, Q. Rui, G. Cheng, Steel Res. Int. **84**, 900-907 (2013)

https://doi.org/10.1051/e3sconf/202129001024

Applications of transition state calculations in the key cyclization of small molecule natural product synthesis

Rongxiang Qiu[1,a]

[1]Tianjin University, School of pharmaceutical Science and Technology, 300072 Tianjin, China

Abstract. The diversification of the conformation and configuration of the carbocyclic skeleton of natural products is an important reason for the diversity and complexity of the structure of natural products. The corresponding cyclization synthesis has attracted much attention. Computational chemistry approaches have the advantages of non-toxic, harmless and relatively low cost, and they are increasingly used to model and understand molecular phenomena. The reaction mechanism and thermodynamic parameters determine the feasibility of the cyclization and the enantioselectivity of the cyclization products. The transition state calculations can provide these thermodynamic parameters, which helps to elucidate the cyclization mechanism, calculate the reaction rationality, predict the performance of the new synthesis method and provide a basis for the comprehensive synthesis design.

1 Introduction

Natural products are an important source of drugs and lead compounds as usually have unique structures and the ability to bind to specific targets. Avoiding the difficulty in extraction and separation, high cost and the ecosystem destruction, artificially synthesized natural products have gradually emerged and attracted much attention. In the early 20th century, computational chemistry approaches were introduced into the synthesis of natural products to model molecular structures and interpret experimental results. [1] With the development of computing technology, the calculation accuracy continues to improve, and the calculation model continues to diversify. Furthermore, the cost is constantly decreasing. A few decades ago, a typical transition state calculation would require months to run, in generally, while the same calculation now just takes a few seconds. [2] Appropriate calculation methods are employed to investigate various chemical and physical properties through molecular mechanics, ab initio calculations, semi-empirical and hybrid methods. [2-4] It is more and more helpful and meaningful for the comprehensive synthesis design.

Small molecules natural products with complex structures often have a carbocyclic skeleton structure, the corresponding cyclization is usually the key to synthesis. There are many conformational isomers and configurational isomers in these carbocyclic skeleton structures, [5] affected the feasibility and enantioselectivity of the cyclization, it is need to consider carefully in the synthesis design. Essentially, these isomers are a transition states or minimums on the potential energy surface, the transition state calculations can provide much information on them. Transition state energy and thermodynamic parameters can provide important kinetic information [6] to elucidate cyclization mechanism. Transition state structure can reveal the molecular interaction that controls the chemical results, and interpret the rationality, feasibility and enantioselectivity of the synthesis. The combination of synthesis and transition state calculations is leveraged for greater understanding and guiding cyclization synthesis.

2 Elucidating mechanism and predicting reactivity: 5-exo cyclization for Chromodorolide B

The diterpenoids are a large family of natural products, which are mainly from marine sources. [7] Chromodorolide is one of the diterpenoids, and it was reported that the chromodorolides have the in vitro nematocidal, antitumor, and antimicrobial activities. [8-10] In 2018, the Overman group reported a synthesis of (-) -chromodorolide B **4** through radical cyclization cascade. [11] Moreover, they elucidated the mechanism and optimized this reaction by analysing the transition state modelling.

The reaction before optimization is shown in **Scheme 1-A**. [11] The radical cleavage of redoxactive ester compound **1** forms intermediate **2** through the 1,4-addition to (R)-5-methoxybutenolide. The intermediate **2** under-went 5-exo-trig cyclization and C–Cl cleavage to form compound **3**, and then elaborated to obtain the (-) -chromodorolide B. However, the key cyclization 5-exo cyclization is nonselective and favoured the undesirable diastereomer, namely C8 epimer **5a** (**Scheme 1-B**), resulted in the yield for the overall reaction was low.

[a] Corresponding author: rx_qiu@163.com

https://doi.org/10.1051/e3sconf/202129001024

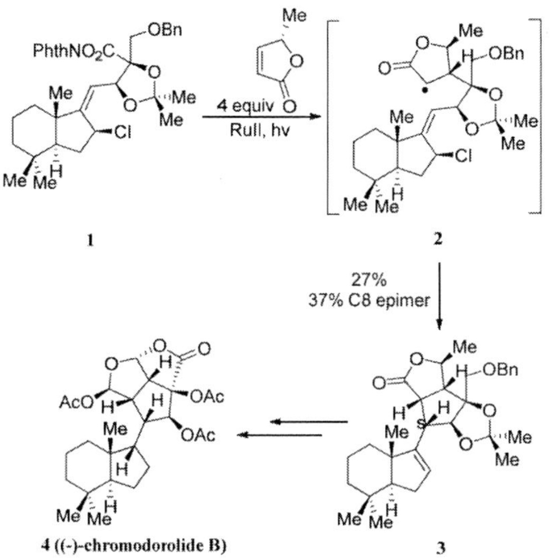

1-A. The route towards the synthesis of (-) - chromodorolide B before optimization.

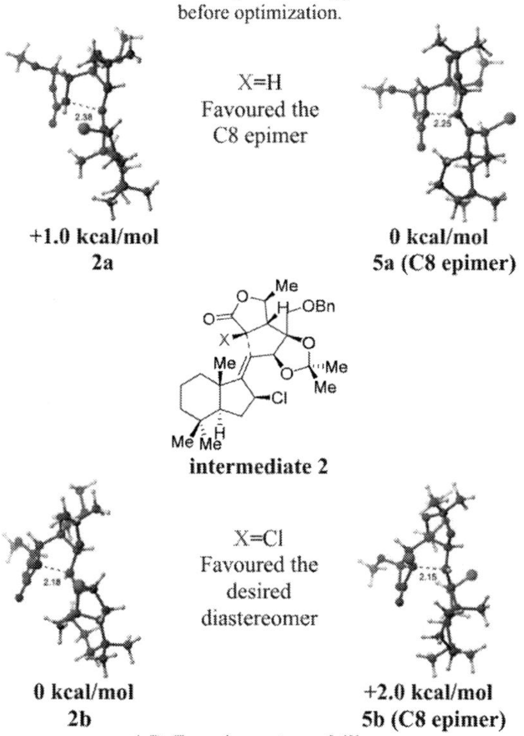

1-B. Transtion state modelling.

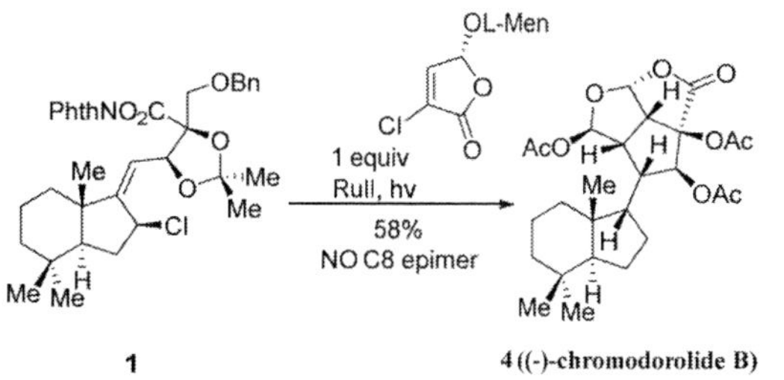

1-C. The route towards the synthesis of (-) -chromodorolide B after optimization.

Scheme 1. (A) The route towards the synthesis of (-) - chromodorolide B before optimization. (B) Transition state modelling. (C) The route towards the synthesis of (-) - chromodorolide B after optimization.

They came up with two possible mechanisms of transformation via 5-exo cyclization. One possible mechanism is radical mechanism, the lactone 1-radical could undergo direct 5-exo cyclization with the pendent alkene. The other one is anionic mechanism, the lactone 1-radica reduces to the lactone enolate and reacts with the allylic chloride through S_N2'. The transition state modelling was used to elucidate the cyclization mechanism, where the DFT methods are employed for the energy minimization (TPSS/def2-TZVP with BJ-damped D3-dispersion) [12,13] and single point calculation (TPPSh/def2-TZVP). The modelled result, shown in **Scheme 1-B**, indicates that it is possible for the two mechanisms thermodynamically and kinetically. Further analysis indicates that there are potential steric interactions sites that could change the reaction selectivity. If the X atom of intermediate **2** is hydrogen atom, corresponding to the radical mechanism, the 5-exo cyclization is favoured the C8 epimer. If the X atom change from hydrogen atom to the chlorine atom, the 5-exo cyclization is favoured the desired (-) - chromodorolide B, corresponding to the anionic mechanism. Furthermore, they predicted the ratio of two products. For the radical mechanism, the desired : undesired product ratio of 1 : 2.5. For the anionic mechanism, the desired : undesired product ratio of 1 : 1.3, consistent with the synthetic result. Therefore, they changed the (R)-5-methoxybutenolide to the chlorobutenolide to optimize the reaction (shown in Scheme 1-C), the optimized 5-exo-trig cyclization doesn't have the C8 epimer and the yield of chromodorolide B increases from 27% to 58%.

The Overman group successfully optimized the key cyclization in the synthesis of (-) - chromodorolide B through the transition state modelling. The potential transition state modelling approaches can elucidate mechanism and predict reactivity and guide synthetic chemists to propel their studies.

3 Understanding enantioselectivity: thioester oxy-Michael cyclization for 2,6-disubstituted THP

The 2,6-disubstituted tetrahydropyran (THP) rings are existed in a large number of structurally complex and biologically active natural products [14-18], including the lasonolide A, [19] phorboxazoles, [20] psymberin [21] and diospongins [22] and so on. One usual synthesis strategy is via oxy-Michael cyclization of a,b-unsaturated carbonyl group, the products have enantioselectivity, one is 2,6-cis-diastereomers and the other one is 2,6-trans-diastereomers. Clarke group [23] reported a synthesis of utilizing thioesters as electophiles in oxy-Michael reaction which has good enantioselectivity (**Scheme 2**). In addition, they elucidated the mechanism and enantioselectivity through the transition state calculations.

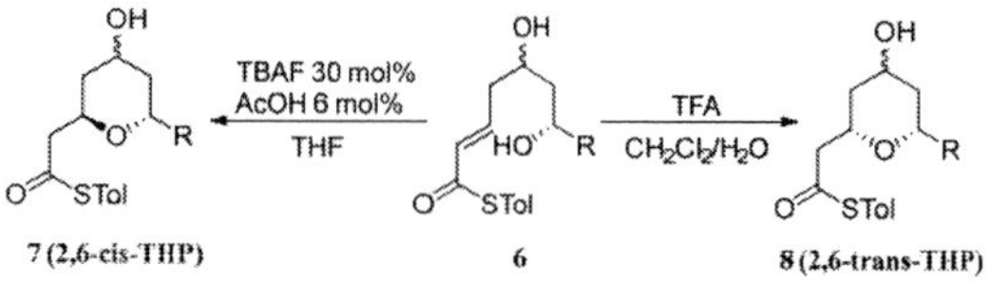

7 (2,6-cis-THP) 6 8 (2,6-trans-THP)

Scheme 2. The Route towards the synthesis of 2,6-cis- and 2,6-trans-THP.

TBAF and the Brønsted acid TFA are using to create a cyclization conditions respectively, the thioester substrates **6** with the 4-hydroxyl group form the 2,6-cis-THP **7** and 2,6-trans-THP **8**. Both TBAF-mediated and TFA-mediated have good selectivity. TBAF-mediated reactions mainly generated 2,6-trans-THP (2,6-cis- : 2,6-trans-THP is 20 : 1) , while TFA-mediated mainly generated the 2,6-cis-THP (2,6-cis- : 2,6-trans-THP is 20 : 1). The authors guessed the cyclization mechanism and employed the MacroModel to search the possible conformation via MMFF force field, and optimized the geometries by employing the DFT with B3LYP density functional [24,25], and split-valence polarized 6-31G*+ basis set with diffuse functions. [26,27] They searched the corresponding lowest energy isomerization pathways of TBAF-mediated reaction and TFA-mediated reactions.

For TBAF-mediated cyclization, transition state calculations indicated that the active molecule is the alkoxide **9**, which attacks the conjugate double bond to form cyclizing alkoxide with 4-hydroxyl group (cyclizing 4-alkoxide **10**) and then forms 2,6-trans-THP or of 2,6-cis-THP further (**Scheme 3-A**). The forming of cyclizing 4-alkoxide has two kinds of transition state pathway (**Scheme 3-B**), consistent with 2,6-trans-THP (**TS-a, TS-b, TS-c**) and 2,6-cis-THP (**TS-d, TS-e, TS-f**) (**Scheme 3-C**). **TS-a, TS-b, TS-d** and **TS-e** are boat-like transition state, and **TS-c** and **TS-f** are chair-like. For the pathway of 2,6-trans-THP, the lowest energy structure is **TS-a** (9.1 kcal/mol relative to the active alkoxide). For the pathway of 2,6-cis-THP, the lowest energy structure is **TS-c** (10.4 kcal/mol). Since the boat-like transition state energy is relatively lower than the chair-like, the active alkoxide tends to form a cyclizing 4-alkoxide through the boat-like transition state in the buffered TBAF mediated, which generates dominantly the 2,6-trans-THP.

For TFA-mediated cyclization, computational investigations indicated that the TFA protonates the thioester **6** to form the 1,3-TFA proton shuttle **11** and then form the 2,6-trans- and 2,6-cis-THP further (**Scheme 4-A**). For 2,6-cis-THP, the transition states have two types of conformations, namely chair-like transition state **TS-g** and boat-like transition state **TS-h** (**Scheme 4-B**). For 2,6-trans-THP, there are also two types of transition states, namely chair-like transition state **TS-i** and boat-like transition state **TS-j** (**Scheme 4-B**). Since the **TS-g** has lower energy than **TS-i** ((**Scheme 4-C**), the TFA tends to protonate thioester via the chair-like transition state in the buffered TFA mediated, which forms dominantly the 2,6-cis-THP.

The Clarke group successfully provided a thioester oxy-Michael cyclization in the synthesis of 2,6-cis- and 2,6-trans-THP, the reason for the enantioselectivity was elucidated via the transition state calculations. The combination of synthesis and transition state calculations

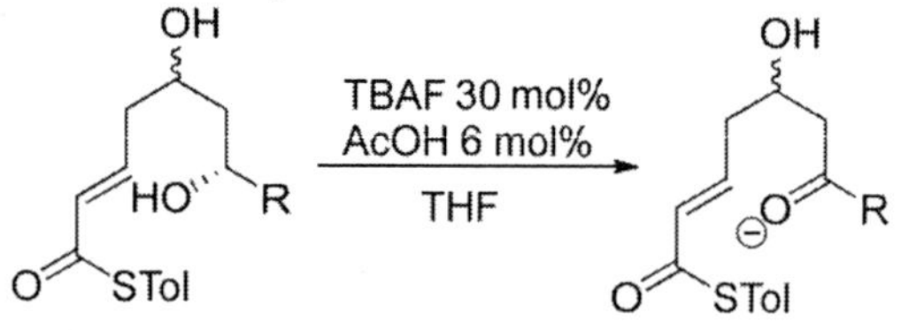

6 **9 (active alkoxide)**

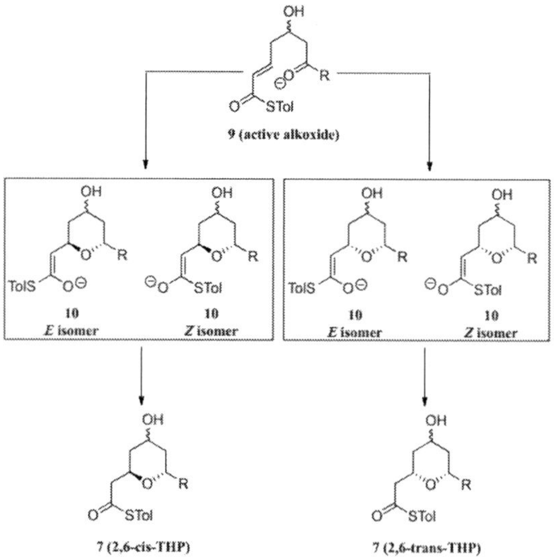

3-A. The mechanism of TBAF-mediated cyclization.

https://doi.org/10.1051/e3sconf/202129001024

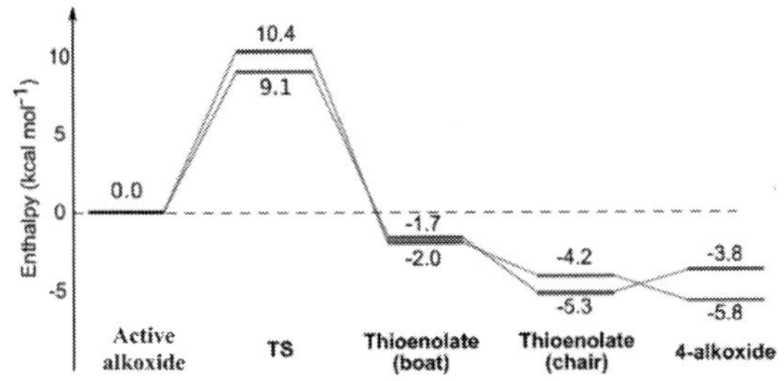

3-B. The two kinds of pathway.

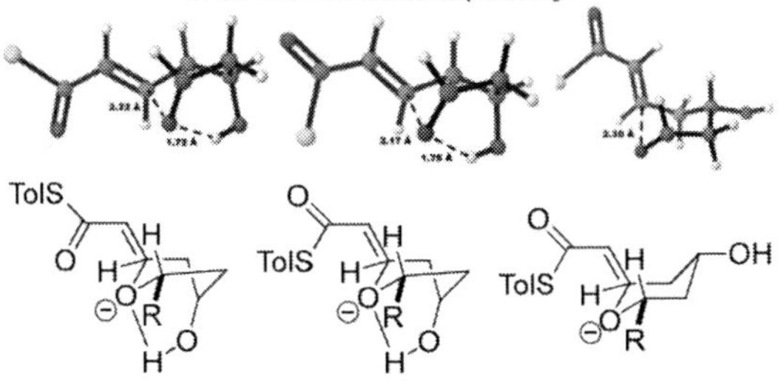

TS-a 9.1 kcal/mol TS-b 9.9 kcal/mol TS-c 19.4 kcal/mol

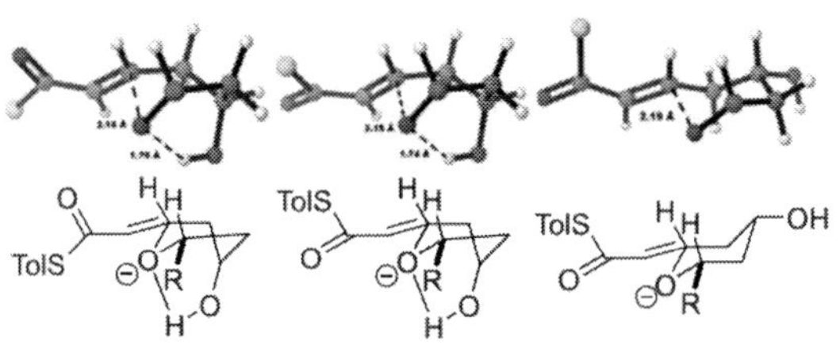

TS-d 10.4 kcal/mol TS-e 10.4 kcal/mol TS-f 18.7 kcal/mol

3-C. The structures of transtion states.

Scheme 3. (A) The mechanism of TBAF-mediated cyclization. (B) The lowest energy pathways to the 2,6-trans 7a (blue) and 2,6-cis 8a (red). (C) The structures of transition states.

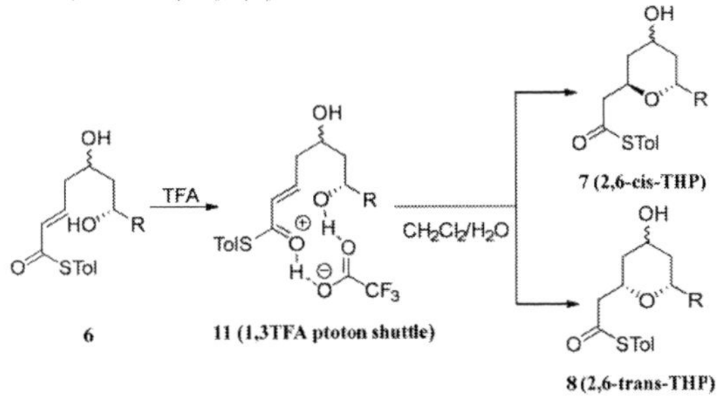

4-A. The mechanism of TFA-mediated cyclization.

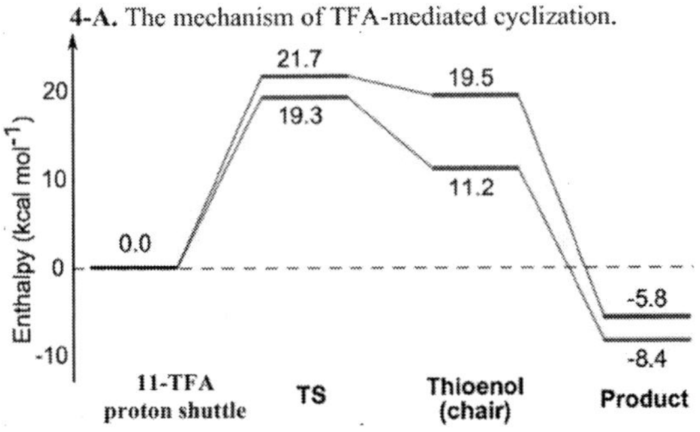

4-B. The two kinds of pathway.

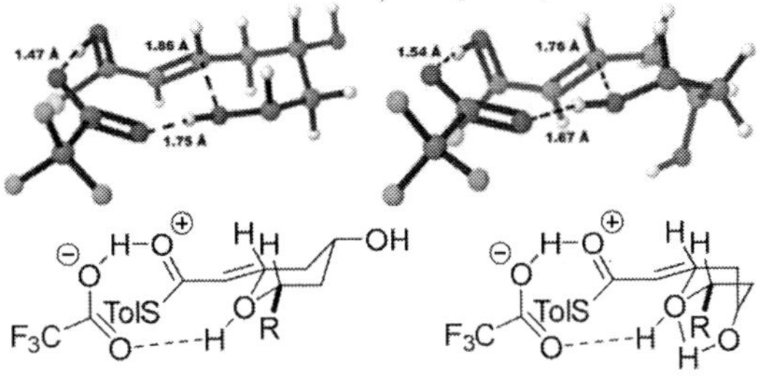

TS-g 19.3 kcal/mol TS-h 25.9 kcal/mol

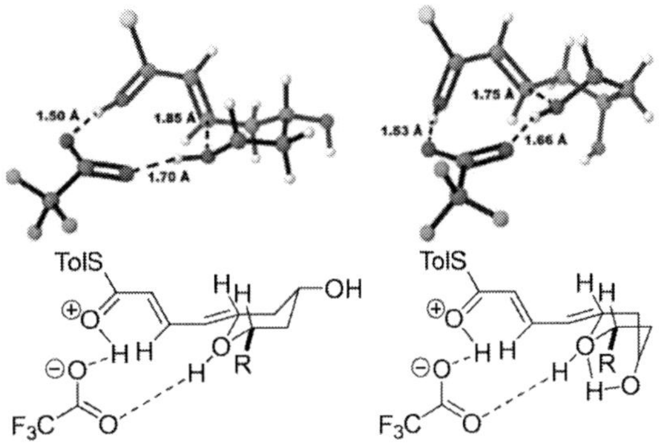

TS-i 21.7 kcal/mol TS-j 26.3 kcal/mol

4-C. The structures of transtion states.

Scheme 4. (A) The mechanism of TFA-mediated cyclization. (B) The lowest energy pathways to the 2,6-trans 7a (blue) and 2,6-cis 8a (red). (C) The structures of transition states.

will be extremely useful for understanding and synthesizing functionalized THP rings with high selectivity in the context of natural product synthesis.

4 Conclusion and prospects

Transition state calculations can provide useful and meaningful insight for the key cyclization in the small molecule natural product synthesis, drive the natural product synthesis to the level of high selectivity and efficiency, and deepen the basic understanding of cyclization feasibility, reactivity and rationality.

With the development of computational technique, especially the improvement of supercomputer computing power, the system size of transition state calculation applications increases continuously, which will provide better theoretical support for the synthesis of larger and more complex natural products with carbocyclic skeleton structures and further promote the industrial production of natural product synthesis.

Acknowledgements

I gratefully acknowledge my lab colleagues for their help in my research and my families for their supports.

https://doi.org/10.1051/e3sconf/202129001024

References

1. F.R. Ornellas, Theor. Chem. Acc. 131, 1086 (2012).
2. Q.N.N. Nguyen, D.J. Tantillo, Chem. – Asian J. 9,674 (2014)
3. T. Sperger, I.A. Sanhueza, F. Schoenebeck, Acc. Chem.Res. 49, 1311 (2016)
4. K. Jackson, S.K. Jaffar, R.S. Paton, Annu. Rep. Prog.Chem., Sect. B: Org. Chem. 109, 235 (2013)
5. E.J. Corey and X.-M. Cheng, The Logic of Chemical Synthesis (Wiley, New York, 1995)
6. G.-J. Cheng, X. Zhang, L.W. Chung, L. Xu, Y.-D. Wu, J. Am. Chem. Soc. 137, 1706 (2015)
7. R.A. Keyzers, P.T. Northcote, M.T. Davies-Coleman, Nat. Prod. Rep. 23, 321 (2006)
8. M.L. Houdhary, J. Clardy, J. Am. Chem. Soc. 111, 2712 (1898)
9. S.A. Morris, E. Dilip de Silva, R. Andersen, J. Can. J. Chem. 69, 768 (1991)
10. M.H. Uddin, M.K. Hossain, M. Nigar, M.C. Roy, Tanaka, J. Chem. Nat. Compd. 48, 412-415 (2012)
11. D.J. Tao, Y. Slutskyy, M. Muuronen, A. Le, P. Kohler, L.E. Overman, J. Am. Chem. Soc. 140, 3091 (2018)
12. H. Kruse, L. Goerigk, S. Grimme, J. Org. Chem. 77, 10824 (2012).
13. J.P. Wagner, P.R. Schreiner, Angew. Chem. Int. Ed. 54, 12274 (2015)
14. N.M. Nasir, K. Ermanis, P.A. Clarke, Org. Biomol. Chem. 12, 3323 (2014)
15. K.W. Armbrust, M.G. Beaver, T.F. Jamison, J. Am. Chem. Soc. 137, 6941 (2015)
16. B.A. Loy, A.B. Lesser, D. Staveness, K.L. Billingsley, L. Cegelski, P.A. Wender, J. Am. Chem. Soc. 137, 3678 (2015)
17. B.M. Trost, C.E. Stivala, K.L. Hull, A. Huang, D.R. Fandrick, J. Am. Chem. Soc. 136, 88 (2014)
18. H. Kraus, A. Français, M. O'Brien, J. Frost, A. Di´eguez-V´azquez, A. Polara, icordi, R. Horan, D.-S. Hau, T. Tsunoda, S.V. Ley, Chem. Sci., 4, 1989 (2013)
19. P.A. Searle, T.F. Molinski, J. Am. Chem. Soc. 117,8126 (1995)
20. P.A. Horton, F.E. Koehn, R.E. Longley, O.J. McConnell, J. Am. Chem. Soc. 116, 6015 (1994)
21. J. Yin, K. Kouda, Y. Tezuka, Q. Le Tran, T. Miyahara, Y. Chen, S. Kadota, Planta Med. 70, 54 (2004)
22. F.A. Cichewicz, P. Crews, Org. Lett. 6, 1951 (2004)
23. K. Ermanis, Y.T. Hsiao, U. Kaya, A. Jeuken, P.A. Clarke, Chem. Sci., 8, 482 (2017)
24. A.D. Becke, Phys. Rev. A: At., Mol., Opt. Phys. 38, 3098 (1988)
25. C. Lee, W. Yang, R.G. Parr, Phys. Rev. B. 37, 785 (1988)
26. R. Krishnan, J.S. Binkley, R. Seeger, J.A. Pople, J. Chem. Phys. 72, 650 (1980)
27. P.M.W. Gill, B.G. Johnson, J.A. Pople, M.J. Frisch, Chem. Phys. Lett. 197, 499 (1992)

Condensation polymers: synthesis, properties, and applications

Ziyi Cheng[1, a,†,*], Yang Liu[2, †], Zehao Li[3, †], Yixuan Song[4, b,†,*]

[1]Department of Materials science and Engineering, Beijing University of Chemical Technology, Beijing 100029, China

[2]Department of Physics and Mathematics, University College London, London WC1E6BT, United Kingdom

[3]Department of Chemistry and Chemical Engineering, Huazhong University of Science and Technology, Wuhan 430074, China

[4]Department of Chemical and Environmental Engineering, University of Nottingham Ningbo China, Ningbo 315100, China

[†]These authors contribute equally.

Abstract. With the widening of the application fields of polycondensation polymers, many types of modifiers have been used to modify the molecular structure of polycondensation polymers. In recent years, more and more researchers have focused on the study of the molecular structure modification of polycondensation polymers and gradually ignored the research on the synthesis methods and properties of polycondensation polymers. This review covers the synthesis strategies of four common polycondensation polymers in the market, including polyamide (PA), polycarbonate (PC), polyester (PET), and phenolic resin. Two general synthesis routes of PET starting from dimethyl terephthalate and terephthalic acid were suggested; the synthesis strategies of PA-66 and PA-6 were discussed; the synthesis methods of aromatic PC from bisphenol A polycarbonate by phosgene or transesterification were summarized; the general synthesis methods of thermoplastic phenolic resin and thermosetting phenolic resin were discussed. In addition, some new synthesis methods and modification methods of these polycondensation polymers were reviewed. Besides, the properties, application, and market value of these polycondensation polymers were summarized. It is hoped that it will be helpful for the new synthesis of polycondensation polymers.

1 Introduction

Condensation polymers are one of the largest polymer classes. They play an important role in the production of packaging, fibers, electronic materials, and insulating materials. While, it is still challenging to produce these kinds of polymers through the economic friendly method in a large scale. Therefore, it is essential to have a good understanding of the fundamental theory of these polymers, especially the synthetic mechanism, as more efforts should be devoted to developing the fabrication strategies for the industrial scale. Polyethylene terephthalate (PET), polyamide (PA), phenolic resin, and polycarbonate (PC) are the four representative polymers that have been widely studied in the past decades. Thus, it is necessary to look back to go through the developing history of the polymers and evaluate the recent progress. In this review, firstly, the preparation of raw materials, esterification/transesterification, condensation, and polycondensation process were elaborated. Secondly, some other advanced synthetic techniques developed recently would also be introduced in this review. Besides that, the properties of each condensation polymers and its practical applications were presented. Finally, current market volume and future prospects were talked.

2 Polyethylene Terephthalate (PET)

Polyester, one of the most important polymer categories in chemistry, plays an important role in our daily lives. One of the representatives in polyester is polyethylene terephthalate, which could be abbreviated PET. It could be formed by polymerizing the repeated monomer units of ethylene terephthalate. It is also worth noting that PET is of great commercial significance nowadays because of its wide range of outstanding features, such as high strength and light weight. That is why it is commonly used in the production of beverage bottles, fabrics and electronic devices. The production of PET has reached around 360 million tones and covered nearly 16 % of European consumption, and it was predicted to continue growing in the next few years [1,2].

*Corresponding author: [a]email: 2018020262@mail.buct.edu.cn [b]email: ssyys4@nottingham.edu.cn

https://doi.org/10.1051/e3sconf/202129001025

2.1 Synthetic routes of PET

2.1.1 Production of PET from Dimethyl Terephthalate

2.1.1.1 Preparation of the raw materials

The first way to produce PET relies on the application of dimethyl terephthalate (DMT). One is by the direct esterification of terephthalic acid, which is followed by further distillation. It is also worth mentioning that methanol can esterify the terephthalic acid instantaneously even without a catalyst. The second way to prepare DMT is through a multistep process between p-xylene (PX) and methanol. At first, PX and methyl toluate could be oxidized at 120-140°C with catalyst. Then the oxidized mixture would react with methanol, and the raw ester formed could be processed further by fractional distillation to extract DMT [3]. For the preparation of ethylene glycol (EG), EO needed to be prepared at first from the catalytic oxidation of ethylene with the function of silver. The hydrolysis of EO in the presence of water would directly lead to EG production [4].

2.1.1.2 Transesterification

DMT is pretreated with inert gas at first, and the treated mixture could react with EG at 150-200°C and 100 kPa [5]. The methanol produced during this process is continuously removed from the system. The reaction mechanism has been shown in Figure 1.

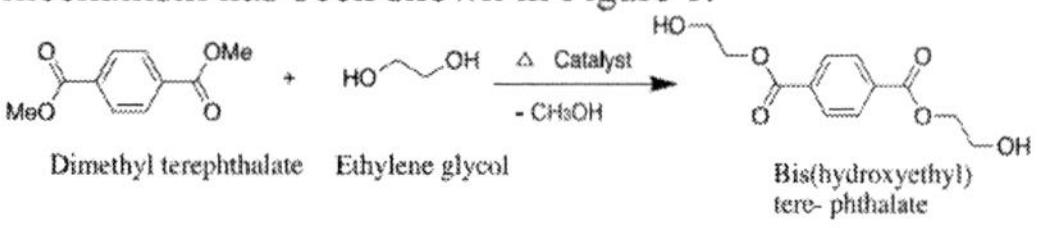

Fig. 1. A schematic illustration of the transesterification process. Ester linkage is formed, and methanol is eliminated at the end.

2.1.1.3 Pre-condensation and Polycondensation

In the pre-condensation process, the reaction conditions are at 250-280 °C and 2-3 kPa. At the polycondensation stage, the temperature would reach 290°C and 100 Pa [6]. The polycondensation must be accelerated by specific catalyst, such as antimony and germanium. The reason for that is the active catalyst would weaken the thermal stability of the final polymer, dampening the properties of PET [7]. When the defined melt viscosity is reached, the polycondensation would be terminated.

2.1.2 Production of PET from Terephthalic Acid

2.1.2.1 Preparation of the raw materials

The second way to synthesize PET relies on the use of EG and terephthalic acid (TA). TA is produced through the oxidation of para-xylene with cobalt-manganese-bromine catalyst in acetic acid. Compressed air and the mixture of p-xylene, acetic acid is constantly adding to the operator in order to maximize the high conversion rate. This process for exploiting terephthalic acid is widely used nowadays in the industrial scale because of the high yield (>95mol %) [8].

2.1.2.2 Esterification

Diacid terephthalic acid and the diol EG made from the esterification process are mixed under 3-6 bar and 250 Celsius [9]. One of the intriguing features of this reaction is the absence of need of catalyst, and the diacid functionalities are capable to self-catalyze the diol through esterification. See Figure 2.

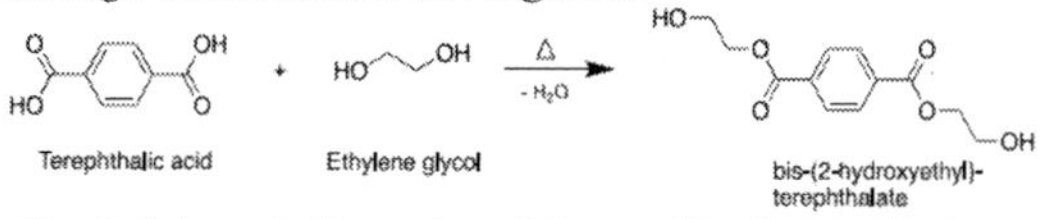

Fig. 2. Schematic illustration of the esterification mechanism. Ester linkage is formed, and water as the one of the products at the end esterification process.

2.1.2.3 Polycondensation

The polycondensation process is performed at a high temperature (~300°C) with the presence of antimony oxide (Sb_2O_3) [4]. In order to ensure the polymer growing linearly, it is vital to carefully control the stoichiometric ratio of the reactants. Simply adjusting the amounts of limiting reagents added to the system could effectively manipulate the molecular weight of the final polymer. However, the polycondensation process still faces some challenges by bringing about several side reactions when the formed DEG through the dimerization of EG reacted with the diacid functionalities of terephthalic acid [8], therefore leading to the difficulty of processing the final polymer.

Besides these two well-known processes to make PET, other advanced methods have been introduced nowadays. Gan et al. introduced the approach to produce nano-sized PET powder by using polyester fiber, and this new technology largely increased the crystallinity and reduced the need for milling [9]. Through the glycolysis method and electrospinning technique, the recycled PET (RPET) could be made from the wasted PET bottles [10]. RPET manufactured in this way showed high mechanical strength and low elastic modulus, which provided a promising future for applications in textile industries and filters.

Compared to DMT used for the production of PET, TA showed a faster reaction rate because of its self-catalysis nature in direct esterification process. From the economic angle, the smaller molecular weight of TA (about 30 g/mol lower than DMF) could bring about less storage cost. That is why TA process is the most common technique employed for the manufacture of PET.

2.2 Properties of PET

PET is unreactive and inert towards many reactants and solvents, such as water. But it is soluble in trifluoracetic acid and hexafluoroisopropanol [11]. PET possesses high mechanical strength and greater Young's Modulus (~3 GPa) [4]. Good-gas-barrier and thermal properties are present in PET as well. PET could form an excellent physical barrier with oxygen and carbon dioxide, which is the most important feature to preserve the taste for food. The large relative molecular mass and the stabilization of aromatic rings can retain the polymer's general stiffness, remaining unchanged at high temperatures.

2.3 Applications of PET

PET is widely used nowadays, especially in packaging, such as bottles. The good physical barrier formed between the liquid and gases, and inert nature made it capable of maintaining the flavor of the beverage. In addition, the light weight and transparency also play a role on the popularity of the manufacture of PET bottles. PET fabrics could be made through its monofilaments. The strong, flexible and low friction properties lead to the popularity of the production and consumption of curtains, bed sheets, and clothing. Because of the high dimensional stability and good electrical insulating features, PET is commonly employed in electrical fields, especially in the manufacture of instrument case, solar junction boxes and solenoids.

2.4 Market analysis and prospects

The demand for PET increased at an approximate rate (~ 1 million metric tons each year) in the past ten years. With that average annual growth, the global capacity would reach 42 million metric tons [12], and that huge consumption exerts more pressure on the recycling for PET. Nowadays, several advanced recycling techniques, such as the synthesis of PET based on the glycolysis of waste PET fiber, were introduced. Most of them showed a satisfactory result and could be utilized on an industrial scale. In the past few years, the average rate of recycling for PET in Europe had already been reached 26%, and it is aiming to achieve 60% for plastics by the end of 2030 [13]. The demand for recycled post-consumer products is the main driving force for that. Overall, the market of PET showed a promising future from the progress made on recycling technologies and growing environmental awareness from the citizens.

3 Polyamide(PA)

3.1 Polyamide-6(PA-6)

PA-6 is one of the most manufactured polyamide products in the world. Due to its high strength, good fatigue resistance, good water absorption, and chemical stability, PA-6 is now widely used in textiles and engineering plastics. The preparation of PA-6 was first described in 1937, which is by the polymerization of ε- caprolactam (CL), and first used as the material by DuPont in 1937 [14].

3.1.1 Synthesis route of PA-6

3.1.1.1 Production of linear PA-6 from CL

3.1.1.1.1 Synthesis of CL

The first generation of preparing CL was invented by Dutch company DSM and Germany company BASF, using benzene to make cyclohexanone oxime, then get CL through rearrangement reaction. The technology was not so ripe, atom utilization is low and pollution issue was severe. The main reaction is listed in Figure 3 [16].

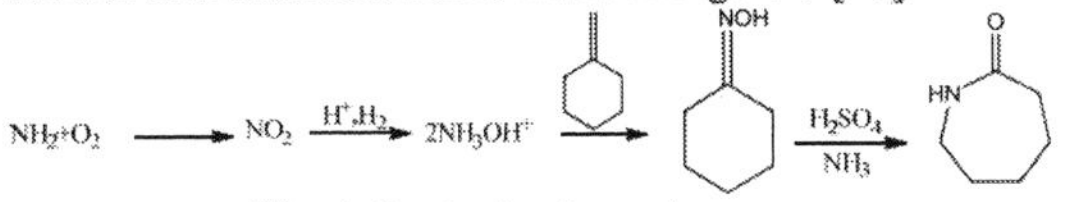

Fig. 3. Synthesis of caprolactam.

Second generation focused on improvement of oximation reaction. Using the molecular sieve to catalyze the reaction of cyclohexanone, ammonia and hydrogen peroxide, the second generation successfully improve the production rate and partly reduce pollution. The improved reaction can be checked in Figure 4 [17,18].

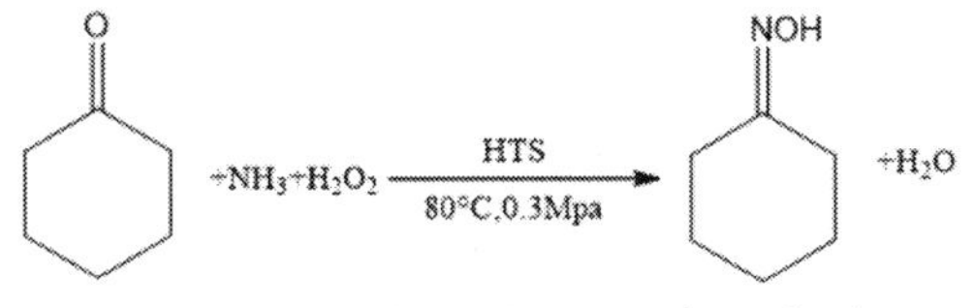

Fig. 4. Improvement of secondary generation technology.

Third generation, which is also the newest technology for now, is attempting to solve pollution and cost problems. By improving the way of forming Beckmann Rearrangement, it is possible to get CL in a green and environmentally friendly way. The technology was first invented by the Japanese company Sumitomo Chemical, and it was the only company that completely master this technology [18].

3.1.1.1.2 Hydrolysis and Polycondensation

The main reaction of making PA-6 via CL is shown in Figure 5.

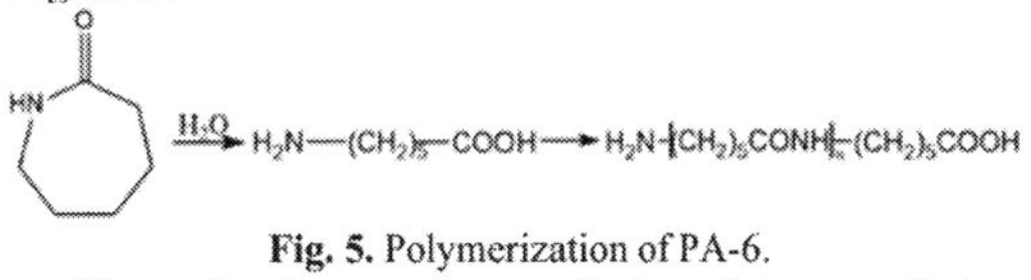

Fig. 5. Polymerization of PA-6.

The technology route was first used to manufacture PA-6 by Germany company called I.G. Farbon in 1943. There are several ways to prepare PA-6 by this technology route, and more common ones use VK tubes to form hydrolysis polymerization in atmospheric pressure. However, for better application, a product needs to have relative viscosity of 2.8-3.5, which the normal polymerization method cannot achieve. To meet different demands for product, different technologies were invented.

First technology is solid state polycondensation. It has two steps to form. First, polycondensation conventionally to get nylon chips in low viscosity, then put pure nitrogen to vacuum dryer. The nylon chip experience drying and crystallization, the chain structure will reform, after which viscosity of chip will raise. It is the most convenient method to increase viscosity of nylon-6 chip.

Other technology including two order polymerization. The first order is to mix the melted monomer and water in the first container, and raise the pressure to polymerization. viscosity can reach 1.7 in this step. Then contents flow to second container to polymerize in atmospheric pressure, for monomers will evaporate along with water, the polymerization degree will raise swiftly, finally output nylon-6 chips with high viscosity [19].

In spare of the industrial method to make nylon-6. Some laboratory described other possible ways to get nylon-6 with anion and cation as catalysts. Using amino hexanoyl and phosphoric acid will shorten the polymerization time, and metal ion like $SnCl_2$, $CuCl_2$ have the same effect to raise the reaction rate. Cation catalyzing polymerization is getting more and more common these days, for its outstanding ability to increase reaction speed and economize power.

3.1.2 Production of modified Polyamide 6

There are plenty of works to research different modified PA-6. Some representative works will be introduced briefly.

Nanocomposite PA-6. Polyamides are widely used as matrix material for composite. The first polyamide used as nanocomposite was PA-6, produced by Toyota company in 1987 [20]. It contains 4 wt% of nanoclay. Nowadays, various nanomaterials are able to be added to PA-6. By adding 2wt% of multi-walled carbon nanotube into PA-6, the modulus, strength, and hardness of PA6 was remarkably improve. This research led to invent electrically conducting nonwoven PA6 membranes. Another example is graphene-PA6 composites prepared by in situ ring opening polymerization, which has an excellent improvement in mechanical property for only 0.1wt% of graphene. Different properties of different nanocomposite PA-6 are listed in Table.1 [21].

Table.1 Different properties of various nanocomposites

Materials	Tensile strength)MPa)	Themal consuctivity (W/mk)	Electrial conductivity (S/m)
Graphene	130±10 GPa	(4.84±0.44)E3to (4.50±0.48)E3	7200
CNT	60-150 GPa	3.500E3	3000-4000
Nanosized steel	1769	5-6	1.35×E6
Plastic (HDPE)	18-20	0.46-0.52	Insulator
Rubber	20-30	0.13-0.142	Insulator
Fiber (Levlar)	3620	0.04	Insulator

Branched PA-6. Since 1978, Tomalia reported a method to make polyamidoamine (PAMAM). He used methyl acrylate to ammonia to give the triester. After starbranched, the product was called "starburst". The same method was used to polymerization of PA-6 and PAMAM, to make branched PA-6. It has higher gel point, and extra thermal stability, could be used in various areas [22].

3.1.3 Properties and Applications of PA-6

PA-6 has high strength, good fatigue resistance, good water absorbed ability and great and chemical stability. Thus, it is widely used as engineering thermoplastics. However, another application of Nylon-6 may be more usual to make synthetic fiber. The strength of nylon fiber is 1-2 times than cotton and 4-5 times than wool, which makes nylon a great material for textile. All these special properties have contributed to the importance of nylon-6 in plastic business ever since it was created.

3.2 PA-66

Nylon-66 is the earliest polyamide product to be manufactured in history. Since synthesis in laboratory in 1935 for the first time, PA-66 held the leadership of world polyamide market as materials of synthesis fabric and engineering plastics.PA-66 has various advantages, including high strength, high mechanical strength, easy to modified, etc. PA-66 and further product is now applied in various areas such as vehicle part, electronic appliances, packaging materials, playing an important role in everyday life.

3.2.1 Synthesis route of PA-66

3.2.1.1 Production of PA-66 via condensation

The production of PA-66 is not a direct polycondensation process, industrially, this research first synthesizes PA-66 salt, which is considered the monomer of PA-66, then polyamide-66 salt condensation to get PA-66 polymer.

3.2.1.1.1 Synthesis of Polyamide-66 salt

Polyamide-66 salt is actually hexamethylene diamine adipate. This route is mainly order to avoid blocking effect, making polymerization degree high enough. Main reaction of synthesize polyamide-66 salt is shown in Figure. 6 [23].

Fig. 6. Synthesis of PA-66 salt.

Using different solvent will leads to different PA-66 salt producing technology. Water solvent technology. Using water as the solvent to reaction, first solve hexanediamine into soft water, then add adipic acid slowly with stirring. Under nitrogen protection and process polymerization, after dehydration and concentration, product is finally made. Dupont Company,

Monsanto Company and some other uses this technology. It doesn't need organic solvent such as ethanol, so its advantages are convenience and safety. However, it has high demand for quality of intermediate, and is not easy for transport.

Solvent crystallization technology. Using methanol or ethanol as solvent, solve hexanediamine in organic solvent and slowly add adipic acid. Stir till the crystal comes out. This technology has significant requirement for the purity of monomers, and mainly used by British company ICI and BASF. The bright side of this technology is that it is easy to delivery and product quality is guaranteed. Short cone include sensitive to light, heat, oxygen, and humidity [24].

3.2.1.1.2 Polymerization of Polyamide 66

Polymerization of polyamide-66 is to polymerize polyamide-66 salt, using different polymerization method. This paper will briefly discuss them.

Continuous condensation technology. The polyamide-66 salt will be solved in water and react with slight heat, nitrogen, compressing to 1-2 times atmospheric pressure. Then polymerize for another two times or more to increase degree of polymerization, excepting steam, molding in some way to complete synthesis.

Melt batch polymerization technology. It is limited in one reactor; monomers are heat and polymerization in molten state. The technology is easy, but it has many weak points such as long technological process, require enough space, need heat resistance reactor. Other technology, take solid polymerization for example, pass the synthesis of PA-66 salt, process reaction directly with two monomers adipic acid and hexanediamine with low temperature and pressure. It is a simple and environment friendly way to manufacture nylon-66, so it raises many eyes of expectation. Recent study is focusing on metal catalysis, and industrial process is remaining unclear [25].

3.2.2 Production of modified Polyamide 66

PA-66 nanocomposites. Nanocomposites of polyamides are widely researched these years. Since work of Toyota laboratory came out, more and more group began to pay attention to this modification. PA-66 nanocomposites have significantly higher stiffness and strength than normal polyamide materials, and have extra stability, ionic conductivity, tunable biodegradability. What's more, depending on exact contents it contains, the nanocomposites have different useful properties, and can be applied in different areas. PA-66 nanocomposites with organoclay, with carbon nanotube, with cellulose, etc, are widely and deeply studied by research group around the world [26-28].

3.2.3 Properties and Application of Polyamide 66 products

Nylon-66 has better heat resistance, rigidity, water absorption than nylon-6, and has high crystal point, its strength is even higher than steel. Nowadays, nylon-66 industry has been a mature Industrial chain, nylon product is widely used for manufacturing film, filament, Staple fiber and mostly, engineering plastic. In 2007, nylon-66 yield has reached 2-2.4 million tons per year, makes nylon -66 one of the most important condensation polymer in the world.

4 Polycarbonate (PC)

4.1 Introduction

Polycarbonate (PC) is a typical kinds of condensation polymer containing carbonate ester group in the molecular chain. According to the structure of ester group, it can be divided into aliphatic group, aromatic group, aliphatic group - aromatic group and other types. The low mechanical properties of aliphatic and aliphatic-aromatic polycarbonate limit their application in engineering plastics. Only aromatic polycarbonate has been industrially produced.

Aromatic polycarbonate is a kind of carbonate polymer with aryl group in the molecular chain, and bisphenol A polycarbonate is the main type. It has good mechanical properties, dimensional stability, chemical stability and thermal stability, which means it has the practical value of polycarbonate. Among them, bisphenol A polycarbonate has become one of the important varieties of thermoplastic engineering plastics with excellent properties which are produced in large scale. It is prepared by the reaction of aromatic dihydroxy compounds with diphenyl carbonate or photo gasification. It can be used in machinery, electronics, instrument, instrument and military industry. Since it has particularities in structure, polycarbonate has become a common plastic used in industry. In this part of report, it would mainly focus on bisphenol A polycarbonate [29,30].

4.2 Synthetic mechanism of Bisphenol A Polycarbonate

Basically, there are two methods to produce bisphenol A polycarbonate: phosgene method and transesterification method.

4.2.1 Phosgene Method to produce Bisphenol A Polycarbonate

The phosgene method is also called the solvent method, which can be divided into the solution phosgene method and the interface phosgene method. The reaction is usually carried out in an organic solvent. The general reaction can be listed in Figure 7 [31].

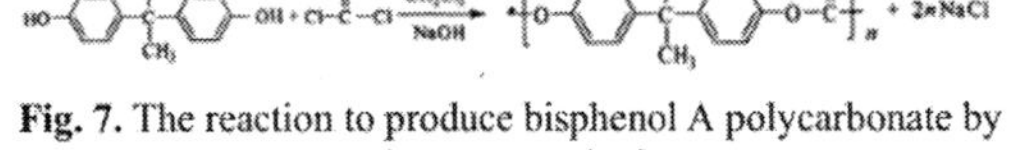

Fig. 7. The reaction to produce bisphenol A polycarbonate by phosgene method.

In the solution phosgene method, first add bisphenol A to dichloromethane to dissolve, and then pass phosgene into the dichloromethane solution of bisphenol A for absorption. At the same time, the hydrogen chloride gas produced by the reaction is neutralized with sodium hydroxide solution to achieve the purpose of treating harmful gases. Another method is called interface phosgene method, which is a method of interface condensation and the whole process can be divided into two stages. First, bisphenol A (BPA) forms a salt with an alkaline aqueous solution, and then reacts with phosgene to produce a low-molecular-weight polymer. After that, the low molecular weight polymer and phosgene undergo secondary polycondensation under the action of the catalyst.

4.2.2 Transesterification Method to produce Bisphenol A Polycarbonate

Currently, the most common use of transesterification method is the melt transesterification method. This reaction is also divided into two steps. First, under the catalysis of the quaternary ammonium salt, bisphenol A (BPA) and diphenyl carbonate (DPC) react to form methoxy carbonylated bisphenol A (MPC). Then, the disproportionation reaction further produces polycarbonate. The general reaction is shown in Figure 8.

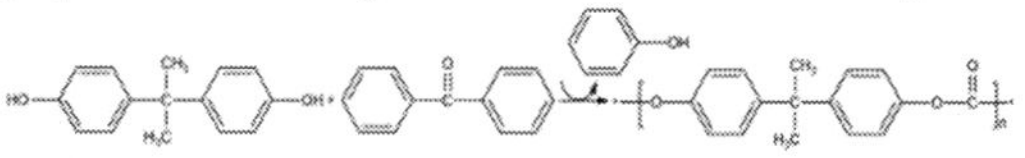

Fig. 8. Reaction to produce bisphenol A polycarbonate by transesterification method.

4.2.3 The Industrial Production

The main industrial production methods of bisphenol A type polycarbonate are transesterification and interfacial polycondensation. Due to the high purity, easy processing and relatively high molecular weight of the products produced by interfacial polycondensation, the interfacial polycondensation process is the main production process of polycarbonate in the world at present, so far about 90% of the polycarbonate production by interfacial polycondensation method [32].

4.3 Application and Prospects

Good mechanical properties (high impact resistance and creep resistance) over a wide range (15~130°C), high transparency, good dimensional stabilities, and electrical insulation. As engineering plastics, it is widely used in transparent materials, electrical parts, medical instruments and mechanical housings [33]. It can also be used for cooling water circulation and oil field water injection as scale inhibitor and dispersant [33].

Global polycarbonate supply is in an oligopolistic competition pattern. As of the end of 2019, there were approximately 38 major PC devices in the world, with a total production capacity of approximately 6.4 million tons. In terms of the regional distribution of production capacity, China continues to rank first in the world with a total production capacity of 1.66 million tons, accounting for 26% of the total global production capacity. The United States and South Korea rank second and third respectively, accounting for 14% and 13% of the world's total capacity respectively [6]. At present, the global PC production giants are still Covestro, Saudi SABIC and Japan's Mitsubishi. Their total production capacity accounts for 56.3% of the global production capacity, showing an oligopolistic competition pattern [29-30, 33].

5 Phenolic resin

5.1 Introduction

All phenolic compounds and aldehydes by condensation reaction resin is collectively referred to phenolic resin. Among them, the most commonly used phenolic resin is the condensation product of phenol and formaldehyde. This kind of phenolic resin is the first thermosetting resin to realize industrialization, which is also the phenolic resin mainly mentioned in this article. Phenolic resin is widely used in industry because of its easy availability of raw materials, convenient synthesis and its properties after curing. Two kinds of phenolic resins with different properties can be synthesized by controlling the molar ratio of phenol to formaldehyde and the pH value of the reaction system: thermoplastic phenolic resin without hydroxymethyl reaction functional group and self-curing phenolic resin with hydroxymethyl structure. In the following chapter or part, it is going to talk about general synthetic methods of thermoplastic phenolic resin and thermosetting phenolic resin, and two new synthetic methods of modified phenolic resin. Finally, the properties and market value of phenolic resin are analyzed briefly.

5.2 Synthesis of phenolic resin

5.2.1 Synthesis of thermoplastic phenolic resin

Thermoplastic phenolic resin is a kind of thermoplastic linear resin synthesized under the condition of pH < 7 and the molar ratio of formaldehyde to phenol is less than 1. This kind of resin does not contain hydroxymethyl groups in the molecule, As shown in Figure 9, the addition reaction between phenol and formaldehyde mainly generates ortho and para hydroxymethyl phenol. Hydroxymethyl phenol is unstable and will continue to react with phenol to form various isomers of bi-phenolic. The reaction rate of bi-phenolic with formaldehyde is approximately the same as that of phenol with formaldehyde [34]. Therefore, the isomer of bi-phenolic can continue to react with formaldehyde to further increase the molecular chain of the polycondensation product, and finally get linear phenolic resin through the para or ortho connection of the phenol ring. According to the different pH value, two kinds of phenolic resins can be obtained: normal phenolic resin and high-ortho phenolic resin, the possible idealized molecular structure is shown in Figure 10. Generally, under the condition of pH<3, the polycondensation reaction is mainly realized by the para position of phenolic hydroxyl group, and the content of 2,4'-linked products is 50-75%, which is

https://doi.org/10.1051/e3sconf/202129001025

normal phenolic resin. Under the condition of pH = 4-7 and under the action of special divalent metal catalysts (such as compounds containing manganese, cobalt, zinc, etc.), the condensation reaction position is mainly in the ortho position of phenolic hydroxyl group, and the final 2,2-linked products content can reach 75-100%, which is high-ortho phenolic resin [35].

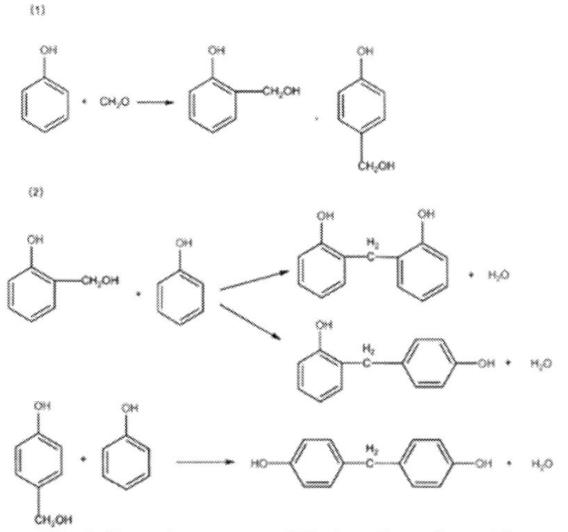

Fig. 9. Formation process of bi-phenolic methane: (1) Synthesis of ortho and para hydroxymethyl phenol (2) Synthesis of bi-phenolic methane.

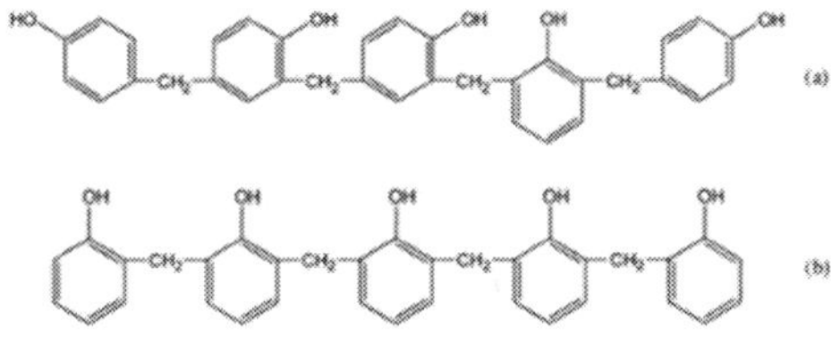

Fig. 10. Possible molecular structure of (a) normal phenolic resin and (b) high-ortho phenolic resin.

5.2.2 Synthesis of thermosetting phenolic resin

Thermosetting phenolic resin was synthesized by condensation of phenol and excess formaldehyde (molar ratio 1.1-1.5) with alkaline catalyst (pH = 8-11). The reaction process is shown in Figure 11. Firstly, a variety of hydroxymethyl phenols are generated by the addition reaction of phenol and formaldehyde. Then, the molecular weight of hydroxymethyl phenol was increased by further polycondensation. Usually when the reaction temperature is below 60°C, the polycondensation reaction rarely occurs, and the addition reaction is about 5 times the polycondensation reactions. When the temperature is higher than 60°C, the polycondensation reaction began to increase. The degree of reaction can be controlled by temperature and time [36]. Different degree of reaction will form different thermosetting phenolic resin. As the polycondensation proceeds, a linear resol resin is formed. The second-order phenolic resin with branched chain and partial cross-linking was obtained by further polycondensation. Because the second-order phenolic resin still has reactive hydroxymethyl group, the molecular chain will crosslink to form a three-dimensional network structure and form the third-order phenolic resin [37]. Thermosetting phenolic resin is a resin containing hydroxymethyl reactive groups, which can be cross-linked and cured under heat or acidic conditions. It is possible ideal molecular structure is shown in Figure 12.

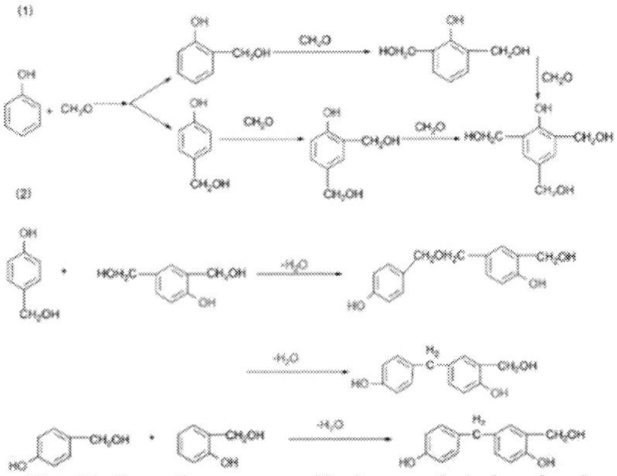

Fig. 11. Formation process of hydroxymethyl-phenol and polyhydric phenol alcohols: (1) Synthesis of hydroxymethyl-phenol (2) Synthesis of polyhydric phenol alcohols.

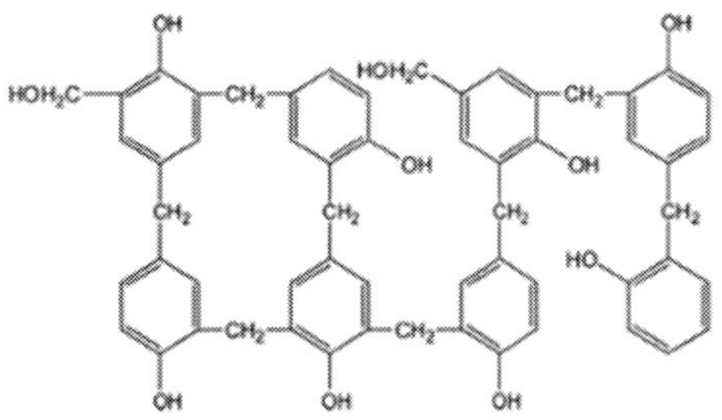

Fig. 12. Possible molecular structure of thermosetting phenolic resin.

5.2.3 Synthetic process of modified phenolic resin

In order to improve the poor heat resistance and flame resistance of phenolic resin, green production problems, researchers have selected a variety of modifiers to modify the molecular structure of phenolic resins. In this article, some modified synthesis methods of phenolic resin were reviewed, including the synthesis of nickel modified thermoplastic phenolic resin and the production of renewable phenolic resin with lignin-derived monomers instead of formaldehyde.

Thermoplastic phenolic resin is widely used in molding compounds, its heat resistance and properties at high temperature cannot meet the needs of some manufacturers. The heat resistance of phenolic resin can be improved by synthesizing Ni-modified thermoplastic phenolic resin. The main method is: under the acidic catalyst, formaldehyde and phenol are used as raw materials to produce thermoplastic phenolic resin at high temperature, then the temperature is lowered, and the Ni-modified thermoplastic phenolic resin is synthesized by adding modified nickel sulfate at low temperature. The Ni modified thermoplastic phenolic resin can also produce carbon nanotubes under the catalysis of Ni and carbonization at high temperature [38]. This is a new method to prepare carbon nanotubes from modified thermoplastic phenolic resin.

Most phenolic resins are synthesized from petroleum-based phenol and formaldehyde, which have adverse effects on the environment and human health. By using lignin-derived monomers to synthesize renewable phenolic resin, formaldehyde can be replaced. The main method is (Figure 13): hydroxymethyl is introduced into the aromatic benzene ring by reducing the ester and aldehyde groups in the para position of lignin-derived monomers. Then, vanillyl and syringic alcohols were directly polymerized into corresponding polymers without formaldehyde using oxalic acid as catalyst. This method of synthesizing phenolic resin is not only green and sustainable, but also has the potential to supplement or replace petroleum-based polymer materials [6].

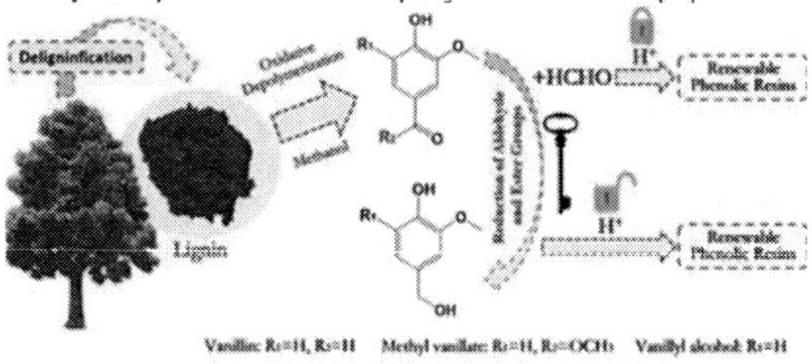

Fig. 13. Schematic illustration of the renewable phenolic resin synthesis based on lignin-derived monomers [39].

5.3 Properties and applications of phenolic resin

Because of its high strength and elastic modulus, corrosion resistance, electrical insulation, low curing temperature and good processability, phenolic resin is widely used in the manufacture of various plastics, coatings, adhesives, and synthetic fibers. By controlling the reaction degree, resin products suitable for different uses can be obtained. For example, if the reaction degree is low, the water-soluble phenolic resin with low average molecular weight can be obtained, which can be used as wood binder; When the controlling reaction is that the product dehydrates into semi-solid resin, this kind of product is called resol resin, which is soluble in alcohol solvent and suitable for coating and matrix material of composite materials; If the reaction is controlled to dehydrate into a solid resin, it can be used as phenolic molding compound or a binder for special purpose.

5.4 Market analysis and prospects of phenolic resin

As of 2016, there are about 200 phenolic resin manufacturers in China, with a production capacity of about 1.3 million t/a and a production of about 1.02 million tons. The main consumption fields of phenolic resin are phenolic molding plastics, wood processing, laminate, grinding, friction materials, fire resistance, insulation materials. Table 1 shows the specific consumption of phenolic resin in recent years in China [40].

Table 2. Consumption of phenolic resin in China in recent years.

Application area	2007	2008	2009	2010	2014
Phenolic molding compound/t	12.4	12.1	11.9	12.3	13.5
Wood processing, laminate/t	11.6	10.7	10.8	11.3	12.3
Grinding and friction materials/t	11	10	10	10.6	12
Fire resistant and insulating materials/t	10.6	8.8	9	9.8	11.3
Foundry materials/t	4.9	4.4	4.5	5	6
Coating/t	2.3	1.9	1.8	2	2.3
Other types/t	2.2	1.7	1.8	2	2.7
Total/t	55	49.6	49.8	53	60.1

With the application of phenolic resin wider, more and more researchers focus on modifying phenolic resin to obtain better performance of phenolic resin, which has a good promotion for the research and development of phenolic resin.

6 Conclusion

In summary, four important condensation polymers, including polyethylene terephthalate (PET), polyamide (PA), polycarbonate (PC), phenolic resin were described from synthetic process to application as representative examples in this review. Both synthetic laboratory strategies and industrial routes were discussed, respectively, with these four polymers. Two general synthesis routes of PET, starting from dimethyl terephthalate and terephthalic acid were suggested. TA synthesis path has priorities with higher molecule weight and faster reaction rate than DMT process. In the case of PA, the synthetic strategies for PA-66 and PA-6 were mainly discussed. It is foreseeable that cation catalyzed polymerization for PA-6 and PA-66 are of great industrialization interest. Some modified PA product like nanocomposites showed a greater improvement in strength and rigidity to go further to the edge. PC is also one of the most utilized condensation polymers and is majorly used in medical instruments and mechanical housings because of their high transparency and stable nature. Bisphenol, a polycarbonate, was chosen as a representative example of aromatic PC, which could be made either through phosgene or transesterification methods. In the case of phenolic resin , the general synthesis methods of thermoplastic phenolic resin and thermosetting phenolic resin were discussed. Besides that, properties (chemical, physical, and mechanical) and practical application of each condensation polymer were briefly introduced to facilitate the fundamental understanding of the polymers. Condensation polymers contribute to the consumption fields of packaging, insulation materials, and textile industries enormously.

Thus, it needs researchers to put more effort into the invention of new and greener synthetic processes for these four polymers. More importantly, making laboratory production be widely employed on an industrial scale.

Reference

1. *Plastics-the facts 2019-Plastic Europe*, Plasticseurope.org. Available on-line at: https://www.plasticseurope.org/application/files/9715/7129/9584/FINAL_web_version_Plastics_the facts2019_14102019.pdf. (Accessed 2 May 2021).
2. *Plastics packaging: European consumption 2007*, Plasteurope.com. Available on-line at:https://www.plasteurope.com/news/PLASTICS_PACKAGING_t209101/. (Accessed 2 May 2021).
3. Sheehan J, Richard. *Terephthalic Acid, Dimethyl Terephthalate, and Isophthalic Acid.* (2011)
4. R Nisticò. *Polyethylene terephthalate (PET) in the packaging industry.* Polym. Test. **90**:106707, (2020)
5. A F A, B D P. *Recycling of PET.* Eur. Polym. J. **41**, 1453-1477 (2005)
6. Gubbels E, Heitz T, Yamamoto M, et al. *Polyesters.* (2018)
7. Tomita K. *Studies on the formation of poly (ethylene terephthalate): 6. Catalytic activity of metal compounds in polycondensation of bis(2-hydroxyethyl) terephthalate.* Polymer **17**, 221-224 (1976)
8. RAF Tomás, Bordado J, Gomes J. *p-Xylene oxidation to terephthalic acid: a literature review oriented toward process optimization and development.* Chem. Rev. **113**, 7421-7469 (2013)
9. Gan L L, Xiao Z H, Pan H, et al. *Efficiently production of micron-sized polyethylene terephthalate (PET) powder from waste polyester fibre by physicochemical method.* Adv. Powder Technol. **32**, 630-636 (2021)
10. Chinchillas-Chinchillas M J, Orozco-Carmona V M, Alvarado-Beltran C G, et al. *Synthesis of Recycled Poly (ethylene terephthalate)/Polyacrylonitrile/Styrene Composite Nanofibers by Electrospinning and Their Mechanical Properties Evaluation.* J. Polym. Environ. **27**, 659-669 (2019)
11. Lepoittevin B, Roger P. *Poly (ethylene terephthalate)* **3**, 97 (2011)
12. *Demand for polyethylene terephthalate worldwide from 2010 to 2020, with a forecast for 2021 to 2030*, Statista, Available on-line at: https://www.statista.com/statistics/1128658/polyethylene-terephthalate-demand-worldwide/. (Accessed 2 May 2021).
13. Forrest M J. *Recycling of Polyethylene Terephthalate*, (2016)
14. Hanford, W.E. and R.M. Joyce, Polymeric amides from epsilon-caprolactam. Journal of Polymer Science, **3**, 167-172 (1948)
15. Fontaine, L., Handbook of Ring-Opening Polymerization. Edited by Philippe Dubois, Olivier Coulembier and Jean-Marie Raquez. Angewandte Chemie International Edition, **48**, 9780 (2009)
16. Russo, S.C., E., 4.14—Ring-Opening Polymerization of Cyclic Amides (Lactams). Polym. Sci. A Compr. Ref., **9**,331–396 (2012)
17. Mu XH, Z.B., Min EZ, Wang X, Wang Y, Zhang XX, Shu XT., Hydrogenation Catalyst and Its Preparation. US Patent, 6368996 (2002)
18. Lin M, S.X., Wang XQ, Zhu B., Titanium-Silicalite Molecular Sieve and the Method for Its Preparation. US Patent, 6475465 (2002)
19. Proceedings of the 5th Beijing international chemical fiber Conference,**29** (1994)
20. Y. Kojima, A.U., M. Kawasumi, A. Okada, Y. Fukushima,T. Kurauchi, and O. Kamigaito, J. Mater. Res.,**8**, 1185 (1993)
21. Faridirad, F., S. Ahmadi, and M. Barmar, Polyamide/Carbon Nanoparticles Nanocomposites: A Review. Polymer Engineering & Science, **57(5),** 475-494 (2017)
22. Matthews, O.A., A.N. Shipway, and J.F. Stoddart, Dendrimers—Branching out from curiosities into new technologies. Progress in Polymer Science, **23(1)**, 1-56 (1998)
23. Plant, S.L.C., Summary of nylon 6 salt production technology. Synthetic fiber industry (1979)
24. Pengzhou, B.R.L., Advances in the research on Nylon(polyamide) 66 polymerization technology. CHEMICAL INDUSTRY AND ENGINEERING PROGRESS, **33** (2014)
25. HUA Yang , L.Z.-m., LIU Quan-yi , ZHANG Li , ZHANG Wei, Present situation and development suggestion of nylon-66. CHINA ELASTOMERICS, **20(6)**, 78-82 (2010)
26. Guan, X., et al., Carbon Nanotubes-Adsorbed Electrospun PA66 Nanofiber Bundles with Improved Conductivity and Robust Flexibility. ACS Applied Materials & Interfaces,. **8(22),** 14150-14159 (2016)
27. Zhang, Q.-X., et al., Multiple melting and crystallization of nylon-66/montmorillonite nanocomposites. Journal of Polymer Science Part B: Polymer Physics, **41(22)**, 2861-2869 (2003)
28. Chavarria, F. and D.R. Paul, Comparison of nanocomposites based on nylon 6 and nylon 66. Polymer, **45(25),** 8501-8515 (2004)
29. J. Huygh , K. Clotman , G. Malarvannan , A. Covaci , T. Schepens , W. Verbrugghe , E. Dirinck , L. Van Gaal and P. G. Jorens , Environ. Int., 81, 61-74 (2015)
30. P. Gimeno , C. Spinau , N. Lassu , A. F. Maggio , C. Brenier and L. Lempereur , J. Sep. Sci., 38, 3729 — 3730 (2015)
31. S. Fukuoka , H. Hachiya , K. Matsuzaki and H. Miyaji , US ,0041712A1, (2008)
32. F. J. Feng , R.-X. Zhuo and X.-Z. Zhang , Prog. Polym. Sci., 37, 200 —229 (2012)

33. D. Kloukos and T. Eliades, Orthodontic Applications of Biomaterials, A Clinical Guide, 207-219, (2017)available at: https://www.sciencedirect.com/science/article/pii/B9780081003831000138 [Accessed on: 2021/5/02].
34. K. Tang, A. Zhang, T. Ge, X. Liu, X. Tang, Y. Li, Materials Today Communications **26** (2021) 101879
35. F. Huang, L. Wan, *Phenolic Resin and Its Application, 1st ed.*, Chemical Industry Press, Beijing, China, pp. 20–24 (2011)
36. A. Izumi, T. Takeuchi, T. Nakao, M. Shibayama, Polymer **52** (2011) 4355–4361
37. L. Huang, *Polymer materials, 2nd ed.*, Chemical Industry Press, Beijing, China, pp 65-69 (2010)
38. J. Liu, L. Wu, Y. Hu, *IOP Conf. Series: Materials Science and Engineering* **423** (2018) 012080
39. W. Yang, L. Jiao, X. Wang, Int. J. Biol. Macromol., **166,** pp 1312–1319 (2021)
40. H. Zhang, *Analysis of Production and Market Phenolic Resin in China*, Chemical Industry **36(02)**, pp 47-47+60 (2018)

 https://doi.org/10.1051/e3sconf/202129001026

Purification of tubulin and microtubules from mouse brain

Xiangxiao Li[1,*]

[1]School of Pharmaceutical Science and Technology, Tianjin University, Tianjin, 300072, China

Abstract. Purified tubulin or microtubules are widely used for in vitro assays related to their function, such as activities of their modifying enzymes and motor protein activity. This paper gave a review for purification tubulin and microtubules (MTs) using two different methods. One method of purification MTs depended on polymerization and depolymerization. MTs were polymerized at 37 °C with taxol in the presence. Another method is to purify tubulin by affinity column chromatography. Stu2p, a kind of MAPs, is able to bind with C-terminal of tubulin. Recombinant Stu2p was purified from expressing bacteria. Then Stu2p was coupled with NHS-activated resins, which were used to purify tubulin from mouse brain. Tubulins purified in this way can also be polymerized at 37 °C.

1 Introduction

Microtubule (MT) is composed by the heterodimers of α- and β-tubulin [1]. The tails of α- and β-tubulin from out surface of MTs support binding site for other protein, such as Microtubule Associate Proteins (MAPs) (Figure 1). They are an important part of the cytoskeleton and play an important role in maintaining cell morphology, cell division, signal transduction, and transport [2, 3]. Under physiological conditions, microtubules have the kinetic properties of polymerization and depolymerization [4]. The structure of cells, the positioning and function of organelles depend on the stability of the microtubule structure. In vitro, MTs can polymerize at high temperature (37 °C) and depolymerize at low temperature (0-4 °C). The polymerized MTs can be further irreversibly stabilized by chemicals, such as taxol.

Purified tubulin or microtubules are widely used for in vitro assays related to their function, such as activities of their modifying enzymes and motor protein activity. In this research, I gave a review related to two methods for tubulin purification. One method is to use the MT polymerization [5]. Another method is to use the affinity chromatography [6].

Members of the XMAP215/Dis1 family of microtubule-associated proteins (MAPs) are essential for promoting microtubule polymerization [7, 8]. MAPs in this family contained several 250 residue repeats, called TOG domain [9], which are thought to bind tubulin dimers and promote microtubule polymerization. Stu2p is a kind of lower eukaryotic homolog proteins of MAPs from Saccharomyces cerevisiae [10]. Widlund et al. designed a recombinant GST-stu2p to bind with tubulin [6]. GST-stu2p was firstly purified from Rosetta (DE3) using a GST affinity column. Then GST-stu2p protein was coupling with NHS-activated resins [11]. NHS-activated resin has NHS group, which can interact with free amino, such as the amino of N-terminus of GST-stu2p (Figure 2). The resins coupled with GST-stu2p are able to bind with tubulin.

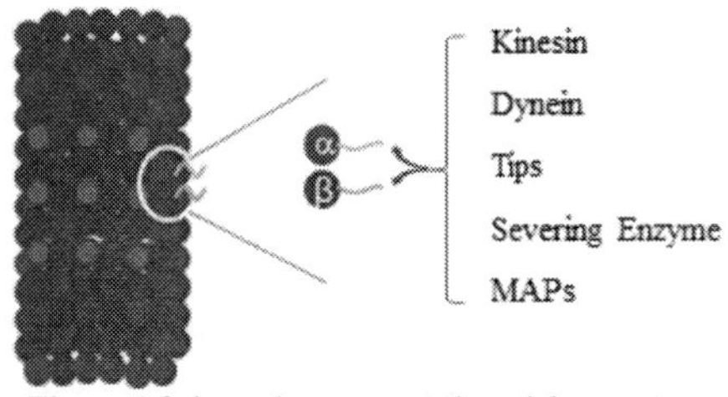

Figure 1 Schematic representation of the structure microtubules. Blue ball indicates α-tubulin, red ball represents β-tubulin.

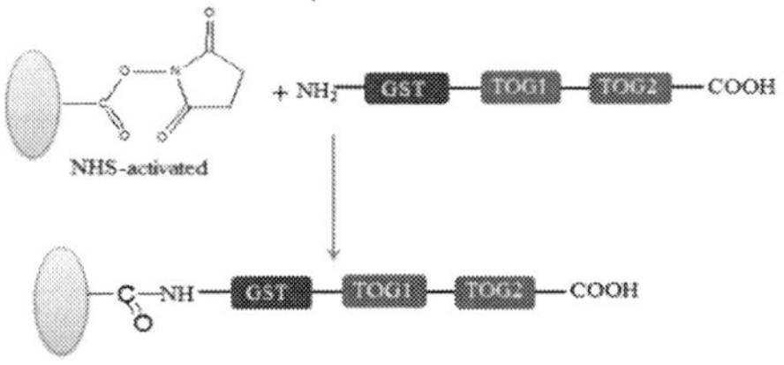

Figure 2 The principle of NHS-activated HP column coupling with stu2p.

2 Method

2.1 Expression and purification of stu2p protein.

pGEX6P1-stu2p was obtained from addgene (Catalog#: 38314). This vector was fusion with a GST tag at the N-terminus of the sequence. For protein expression, the plasmid was transformed into Rossetta (λDE3) and

* Corresponding author: 760101559@qq.com

induced with 0.2 mM isopropylthiogalactopyranoside (IPTG) at 16 °C for 18 h. The harvested cells were resuspended in 2x PBS containing 25 μg/mL Lysozyme, 25 μg/mL DNase and 1 mM PMSF. The cells were disrupted by sonication on ice using an ultrasonic processor (Ningbo scientz, China). The lysate was centrifuged at 40,000 xg for 40 min at 4 °C. The supernatant was loaded on 5mL equilibrated GST-resins (Genscript, China) and incubated with resins at 4 °C for 2 h. After washing with 2x PBS containing 1 mM dithiothreitol (DTT), 0.1% (v/v) tween 20 for 10 column volumes (CV) and 2x PBS, 10 mM $MgCl_2$, 5 mM ATP for 2 CV. The proteins were eluted with 5 mM glutathione in 2x PBS containing 1 mM DTT, pH 8.0. Monitored with SDS-PAGE followed by Coomassia Brilliant Blue (CBB) staining, the fractions containing the target protein were combined and dialyzed against 100 mM $NaHCO_3$, 100 mM NaCl, pH 8.2 with three changes. After concentrated using an ultrafiltration concentration tube (Millipore, USA), the protein concentration was determined using NanoDrop (Thermo Fisher Scientific, USA).

2.2 Preparation of TOG column.

Before coupling with NHS resin, the proteins were pooled to 100 mM $NaHCO_3$ (pH=8.2), 100 mM NaCl and 80 mM $MgCl_2$ (dialysis buffer). Firstly, NHS-activated fast-flow resins were packed in the column and then washed with 10 CV ice-cold 1 mM HCl. Next, GST-stu2p proteins were transferred to NHS-resin, and incubated at room temperature (RT) for 2 h. The flow through was transferred to other tube following binding. The resins were washed with 0.5 M ethanolamine and 0.5 M NaCl, pH=8.3 (block buffer) for 4 CV and blocked with 6 CV block buffer at RT for 1 h. The block buffer was discarded and washed with 9 CV 6x PBS, follow by 3 CV storage buffer (1x PBS containing 50% glycerol). At last, this column was stored at -20 °C.

2.3 Purification of tubulins from mouse brain using an affinity column.

Tubulins were prepared from mouse brains using an affinity column. Firstly, the brain tissue is homogenized with a glass homogenizer (20 stocks) in 1.5 volumes of BRB80 buffer. Next the homogenate was collected by centrifugation at 40,000 xg for 30 min at 4 °C. Then the supernatant was transferred to column, incubated it at 4 °C for 2 h. The flow through was transfer to other tube. The resins were washed with 1x BRB80 containing 10 μM GTP for (buffer A) 10CV, 1x BRB80 containing 100 μM GTP, 10 mM $MgCl_2$, and 5 mM ATP (buffer B) for 4 CV. And resins were incubated with buffer B at 4 °C for 20 min in order to remove MAPs. Then resins were washed with 1x BRB80 containing 0.1% Tween 20 and 10% glycerol (buffer C) for 5 CV. At last, tubulins were eluted by 1x BRB80 containing 10 μM GTP and 500 mM $(NH_4)_2SO_4$ and the concentration were monitored using Nanodrop. For desalting, target protein was desalted by PD10 desalting column (GE healthcare, USA) and eluted with BRB80.

2.4 Purification of MTs from mouse brain using taxol.

The brain tissue is homogenized with a glass homogenizer (20 stocks) in 1.5 volumes of BRB80 buffer. The homogenate is centrifuged at 30,000 xg in a TLA-150 rotor (Beckman) for 15 min at 2 °C. The supernatant (S1) is recovered and centrifuged at 180,000 xg for 90 min in a TLA-150 rotor at 2 °C. To the recovered supernatant (S2), taxol was added in a final concentration of 20 μM and GTP to 1 mM. The solution is warmed at 37 °C for 30 min, and transferred to a centrifuge tube. Equal volume of 10% sucrose containing 20 μM taxol and 1 mM GTP was cushioned in the bottom of mixture. The sample is centrifuged at 37 °C at 30,000 xg for 30 min. The pellet (MTs, designated as P3) was resuspended with BRB80.

3 Results

In order to prepare the affinity column for tubulin purification, recombinant GST-stu2p protein were purified from Rosetta (DE3) using GST resins according to Widlund [6]. The molecular weight of target protein is about 100 kDa. By comparing the supernatant (S) and flow through (FT), we can find that a large amount of protein was bound to GST-resin (Figure 3A). After monitored with SDS-PAGE, the eluate fractions majorly containing GST-Stu2p were combined and dialyzed against dialysis buffer at 4 °C overnight. These proteins were then coupled with NHS resin at RT for 2 h. The coupling efficiency was monitored using SDS-PAGE followed by CBB staining. The results showed that most of GST-Stu2p was bound to NHS resin (Figure 3B).

We purified tubulin or microtubules (MTs) using two methods, one method is through affinity column (Figure 4A). Using prepared affinity column, tubulins were prepared from mouse brains using an affinity column. The molecular weight of tubulin is 55 kDa. Assessed with CBB-stained SDS-PAGE, tubulin appears as the major band after purification (Figure 4A). Another is taxol-assisted polymerization (Figure 4B). After polymerization, we got microtubules (P3) from mouse brain (Figure 4B). The purity of tubulin from first method is higher than that from second method.

 https://doi.org/10.1051/e3sconf/202129001026

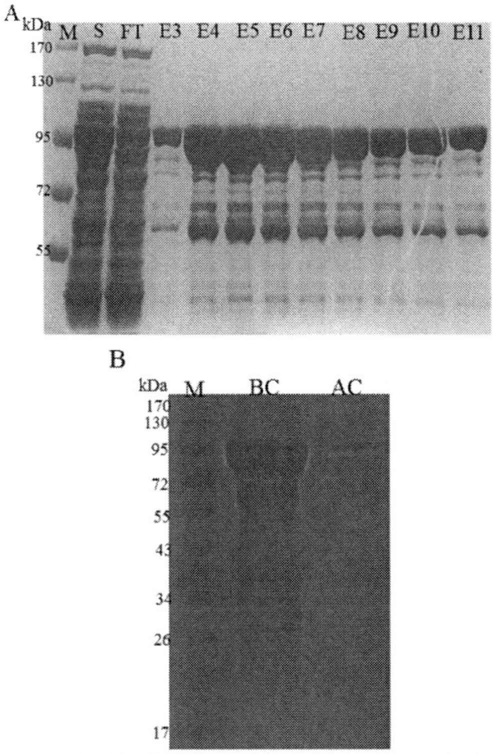

Figure 3 Purification of recombinant GST-stu2p protein for NHS-resin coupling. M means marker; (**A**) CBB-stained SDS/PAGE showed the proportion and purity of target proteins in soluble fraction (S), flow-through (FT), and elute (E) during purification of recombinant GST-Stu2p. (**B**) CBB-stained SDS/PAGE showed GST-Stup2 protein before coupling (BC) or after coupling (AC) with NHS-resins.

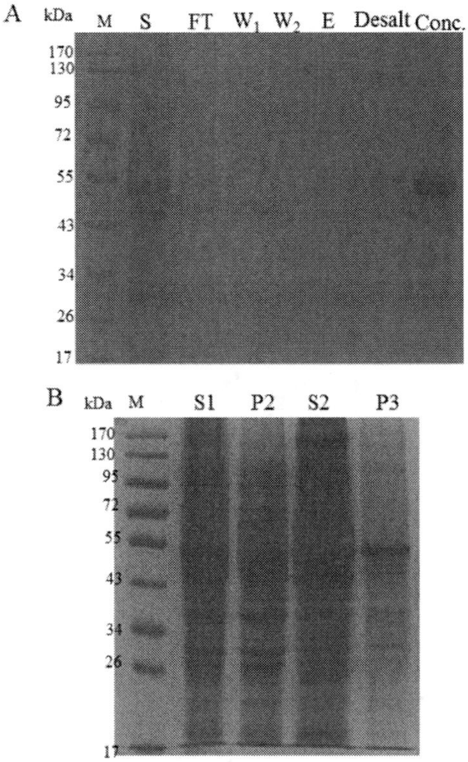

Figure 4 Purification of tubulin and MTs based on two different methods. M means marker; (A) CBB-stained SDS/PAGE showed the proportion and purity of tubulin in soluble fraction (S), flow-through (FT), wash (W) and elute (E) fraction. (B) Purification of MTs depended on polymerization and depolymerization. S1 is the brain lysate after homogenizer. S2 is the recovered supernatant after 1st run ultracentrifuge and P2 is the corresponding precipitation. P3 is the pellet obtained from S2 after addition of taxol.

4 Conclusion

A review related to purification of tubulin and MTs was given in this paper. First method is to purify tubulin using an affinity column according to Widlund [6]. Firstly, Stu2p, a kind of MAPs, was expressed and purified from Rosetta competent cells. Then, stu2p was coupled with NHS-activated resins. The resins coupled with stu2p were able to binding with tubulin. I purified tubulin from mouse brain based on this affinity column. Another method is taxol-assisted polymerization. MTs were polymerized at 37 °C with taxol presence. After polymerization, we got microtubules from mouse brain. The purity of tubulin from first method is higher than it from second method. 1.2 mg tubulin was prepared through the second method from 12 mouse brains.

References

1. Ku, S., Messaoudi, C., Guyomar, C., Kervrann, C., & D Chrétien. (2020). Determination of microtubule lattice parameters from cryo-electron microscope images using tubule J. Bio-Protocol, 10, 3814.
2. Vemu, A., Garnham, C. P., Lee, D. Y., & Rollmecak, A. (2015). Generation of differentially modified microtubules using in vitro enzymatic approaches. J. Methods Enzymol, 108(2), 451a-451a.
3. Souphron, J., Bodakuntla, S., Jijumon, A. S., Lakisic, G., Gautreau, A. M., & Janke, C., et al. (2019). Purification of tubulin with controlled post-translational modifications by polymerization–depolymerization cycles. J. Nature Protocols.
4. Mirco, Castoldi,, Andrei, V., & Popov. (2003). Purification of brain tubulin through two cycles of polymerization-depolymerization in a high-molarity buffer-sciencedirect. J. Protein Expression and Purification, 32(1), 83-88.

5. Vallee R. B. (1986). Purification of brain microtubules and microtubule-associated protein 1 using taxol. J. Methods Enzymol, 134(4):104-115.
6. Widlund, P. O., M. Podolski, S. Reber, J. Alper, M. Storch, A. A. Hyman, J. Howard & D. N. Drechsel (2012). One-step purification of assembly-competent tubulin from diverse eukaryotic sources. J. Molecular Biology of the Cell, 23(22), 4393-4401.
7. Al-Bassam, J. & F Chang. (2011). Regulation of microtubule dynamics by tog-domain proteins xmap215/dis1 and clasp. J. Trends in Cell Biology, 21(10), 604-614.
8. Greenlee, M., A. Alonso, M. Rahman, N. Meednu, K. Davis, V. Tabb, R. Cook and R. K. J. C. Miller. (2018). The tog protein stu2/xmap215 interacts covalently and non-covalently with sumo. J. Cytoskeleton.
9. Leano, J. B., & Slep, K. C. (2018). Structures of TOG1 and TOG2 From the Human Microtubule Dynamics Regulator CLASP1.
10. Al-Bassam, J., Larsen, N. A., Hyman, A. A., & Harrison, S. C. (2007). Crystal structure of a tog domain: conserved features of xmap215/dis1-family tog domains and implications for tubulin binding. J. Structure, 15(3), 355-362.
11. Healthcare G. E. (2006). HiTrap NHS-activated HP.

 https://doi.org/10.1051/e3sconf/202129001027

Effect of the Temperature on chromium precipitation in faucets

Huang Tingting[1,2,3]

[1]Nanan Testing Institute of Quality and Metrology, Quanzhou, Fujian, China
[2]National Quality Supervision and Inspection Center of Plumbing & Sanitary Ware (Fujian),Quanzhou, Fujian, China
[3]Nanan Market Supervision and Comprehensive Law Enforcement Brigade, Quanzhou, Fujian, China

Abstract. The research was studying the chromium precipitation in faucets which were extracted at ambient temperature of 20 ℃,40 ℃ and 60 ℃ by extraction solution, the results showed that in general, the chromium extraction in the faucets increased as the temperature rises. And the chromium extraction in different batches and different samples in the same batch varied greatly, mainly affected by materials and the processing technology. After further analysis, the detection showed that Cr VI in chromium precipitation was between 83% and 103% of the total chromium content. This indicated that Cr VI was the main form of chromium in extraction solution, which will be harmful to human health.

1 Introduction

Everyone's daily life is inseparable from water, water quality, especially drinking water quality is closely related to human health. As the endpoint devices of drinking water system, the precipitation of harmful contaminants from faucets directly affect the quality of water and the safety of residents' health. Most of the faucets in the electroplating process, often use chromate and dichromate as electroplating reagents, and most of the faucets material also have trace amounts of chromium elements, so the extraction of chromium in the faucets is inevitable. The precipitated chromium mainly exists in the form of of Cr III and Cr VI, of which Cr VI is a carcinogen, and is easily absorbed by many tissues and organs in the human body and accumulated , which will harm human health [1].

According to the research, in the daily use of room temperature, the small range of pH and electrical conductivity had little influence on the chromium precipitation, the influence factor is the extraction time [2]. With the development of science and technology and the improvement of living standards, now the general household faucets have a temperature control function. The precipitation of harmful contaminants from faucets is a complex electrochemical corrosion process, which is also affected by the temperature [3]. This research mainly studied the extraction of chromium in faucets in different temperature conditions and analysis the main form of chromium, aiming at providing scientific basis for the safety of drinking water for residents and industrial production.

2 Experiment

2.1 Instruments and Reagents

2.1.1 Instruments

Nexion[TM] 350X ICPMS (USA Perkin Elmer company), UV spectrophotometer (UV 2600, SHIMADZU) ,Mini-q Reference Ultra Pure Water system (USA Millipore company), PH meter (Mettler-toledo), Conductivity meter (Mettler -toledo).

2.1.2 Reagents

Standard solution:1 g/L chromium (GBW08614) was purchased from China Institute of Metrology, 1 g/L Cr VI (BW30030-100-N-50) was purchased from Beijing Tanmo.

Other reagents: waterless $NaHCO_3$ (AR), 5% sodium hypochlorite solution (AR), nitric acid (GR) were all purchased from Shanghai Guo pharmaceutical group chemical reagents Co. Ltd.; Ultra Pure Water (electrical resistivity $\geq$ 18.2 MΩ·cm).

2.2 Samples extraction

According to the GB18145-2014 Appendix B method [4], the extraction solution that simulates tap water is formulated. In order to carry on the test better, this experiment chose the inferior faucets as samples , a total of 5 batches, respectively, 1#, 2#, 3#, 4#, 5#. Each batch had 9 parallel samples. The samples were rinsed with tap water for 15 min,then 3 times with ultra pure water, the residue and dirt from the sample were washed out, and

Corresponding author: tthuang411@foxmail.com

then 3 times with extraction solution. Each batch of samples was taken in 3 parallel samples at an ambient temperature of 20 ℃, 40 ℃, 60 ℃ for extraction experiments. The samples were fully filled with extraction solution, and the sample handle switch was in a fully open state, and both ends of the sample were tightly plugged with a clean rubber plug covered with Teflon film. After the sample filled with the extraction solution had stabilized at its respective ambient temperature of no more than 72 h, the extraction solution in the sample was replaced in accordance with appendix B of GB18145-2014, and the held for 16 h extraction solution in the sample was collected for 5 days after stabilization. The 3# samples were repeated the above steps and collected the extraction solution for the next 5 days. The 3# samples originally incubated at 20 ℃ were then exchanged with the 60 ℃ samples, and the extraction solution is collected for the next 5 days. The collected extraction fluid was divided into two parts, one with thick nitric acid to make the solution pH <2, placed at room temperature of more than 24 h to ensure that the metal was completely dissolved prior to analysis, was used to measure the total chromium, and the other was for the detection of Cr VI.

2.3 Test method

2.3.1 ICP-MS

Following consideration of the factors of abundance, interference and sensitivity, selected ^{52}Cr isotope for testing, ^{45}Sc as the internal standard element, using the method of online marking.

Instrument working conditions: RF (1.6 KW), Carrier Gas (Argon): 0.90 L/min, Cooling gas (Argon): 15.0 L/min, Helium: 4.5 mL/min, Sweeps/reading (20), Readings/replicate (2), Number of Replicate (2), Detector Mode (Dual), KED mode.

2.3.2 UV spectrophotometer

Cr VI in extraction solution was detected according to GB/T 5750.6-2006: Diphenylcarbonyl dihydrazine spectrophotometric method [5], when the detected concentration exceeded the upper limit of the standard curve, the dilution and re-determination was made with fresh preparation extraction solution.

3 Results and discussion

3.1 Chromium precipitation in the faucets affected by temperature

The experiment selected a total of 5 batches, 9 parallel samples per batch, and 3 parallel samples at three degree ambient temperature. The concentration of total chromium of 5 days of extraction solution were detected by ICP-MS. Each batch of 3 parallel samples per day for the weighted average was defined as the derived value of the batch. The sample total chromium precipitation concentrations changed with temperature rises were shown in Figure 1.

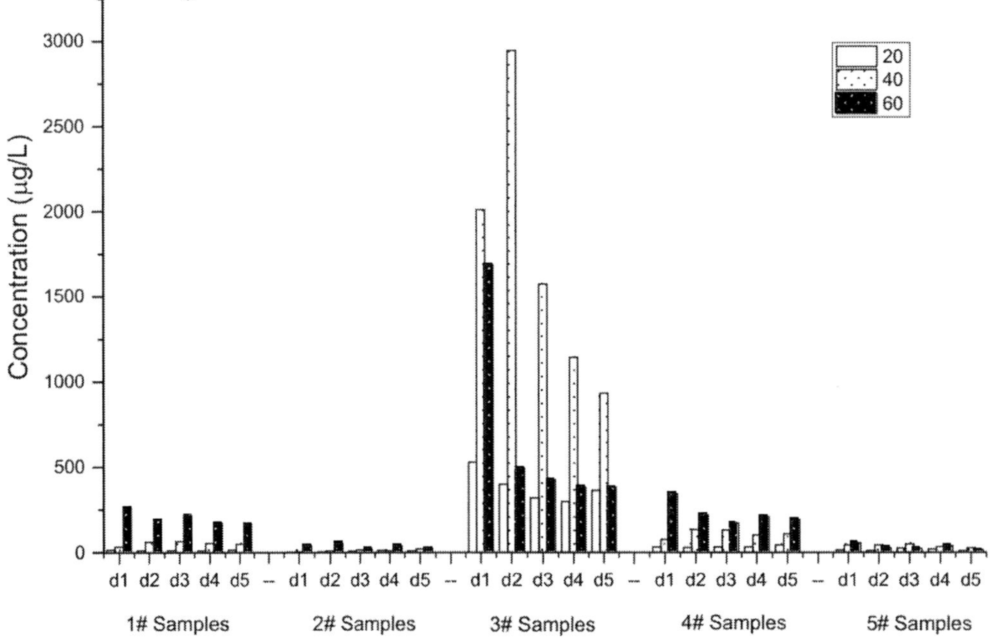

Figure 1. Total chromium precipitation in 5 batches samples for 5 days at 20 ℃, 40 ℃, 60 ℃

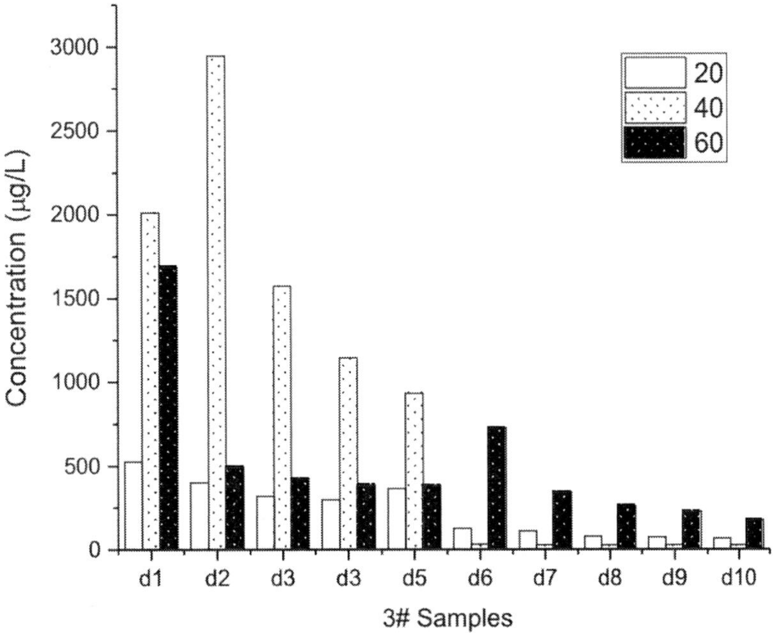

Figure 2. Total Chromium Precipitation in 3# samples extracted for 10 days

Figure 1 showed that the chromium precipitation in the 5 batches samples varied widely, especially the 3# samples were very high. Except for the 3# samples, the chromium precipitation of other samples generally increased with the rise of temperature, and the chromium precipitation did not show a regular and obvious reduction with the increase of extracting days. The concentration of chromium precipitation of the 3# samples showed a trend of 40 ℃ >60 ℃ >20 ℃ , and the precipitation had an obvious trend of decreasing with the increase of extracting days.

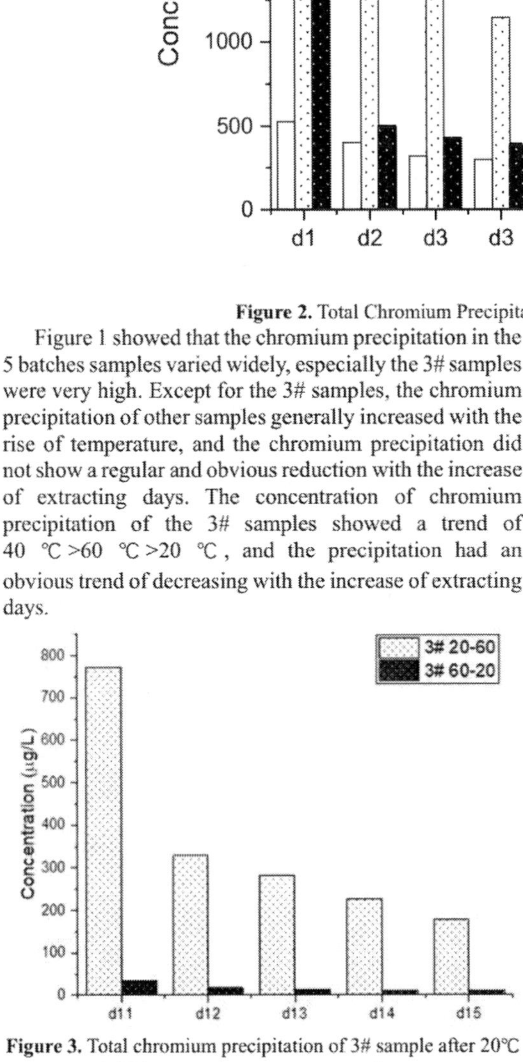

Figure 3. Total chromium precipitation of 3# sample after 20℃ and 60℃ extraction temperature replacement

In order to further study the total chromium precipitation of the sample 3#, the 5-day extraction solution of the next cycle was continuously collected and tested, as shown in Figure 2. The figure showed that in the second round of immersion extraction, the total chromium precipitation was 60 ℃>20 ℃>40 ℃, among which 40 ℃ showed a sharp decrease. For further investigated the effect of temperature on the precipitation of total chromium in faucets 3# samples which incubated at 20 ℃ for the first 10 days were selected and placed at 60 ℃. The samples extracted at 60 ℃ were replaced and incubated at 20 ℃ for the third round of extraction for 5 days. The results were shown in Figure 3. The samples (3#20-60) originally extracted at 20 ℃ showed a sharp increase of chromium precipitation at 60 ℃, and showed a decreasing trend with the increase of extracting days. However, the chromium precipitation of (3#60-20) which was originally incubated at 60℃ was very little at 20℃, which decreased from 35 μg/L to 10 μg/L as the days gone by.

The above experimental results indicated that the precipitation of chromium in the test samples was greatly affected by temperature, in general, the concentration of chromium precipitation increased at the higher temperature . The abnormal precipitation of 3# samples in the first 5 days should be affected by the process technology. The general temperature of decorative chromium in the electroplating process of ordinary faucets is 40-45 ℃ . If the faucets are not cleaned well after electroplating, the residual electroplating reagents should attached to the inner cavity of the faucets are more likely to extracted under suitable temperature conditions. Fig.4,Fig.5 and fig.6 respectively showed the 10-day total chromium precipitation of the three parallel samples of the 3# sample. The precipitation trends of the three parallel samples were similar, but the concentration of precipitation varied greatly, indicating that the chromium precipitation of individual faucet in the same batch varied greatly also.

3.2 Morphology analysis of chromium precipitation in samples

For further research of the morphology of the chromium which was extracted from faucets samples, the collected extraction solution was tested by Diphenylcarbonic dihydrazine spectrophotometric method. The result showed that Cr VI which was more toxic to human body accounted for 83%~103% of the total chromium precipitation. So it indicated that the extraction solution was harmful to human.

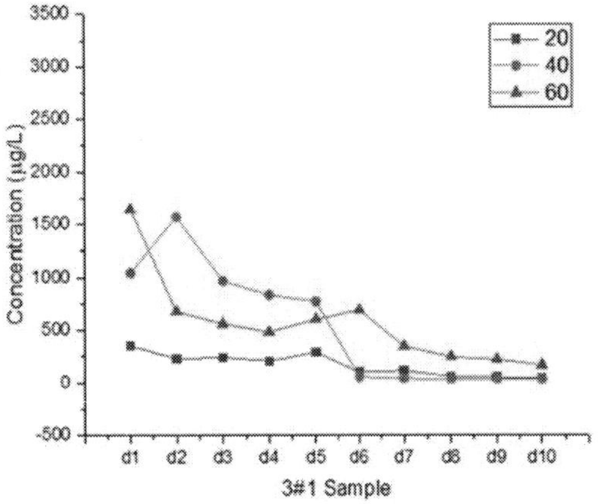

Figure 4. Total Chromium Precipitation of 3#1sample extracting for 10 Days

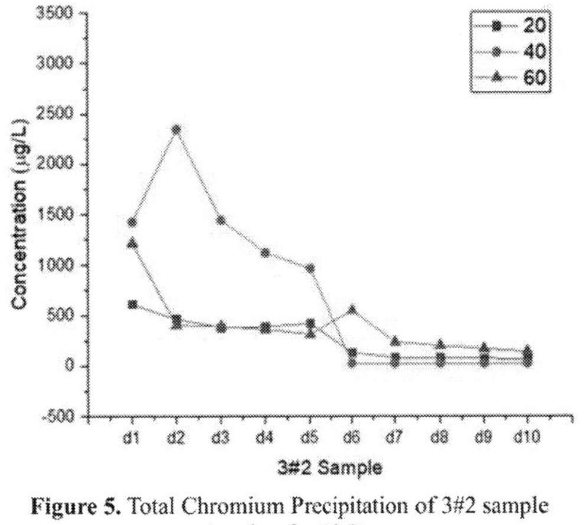

Figure 5. Total Chromium Precipitation of 3#2 sample extracting for 10 Days

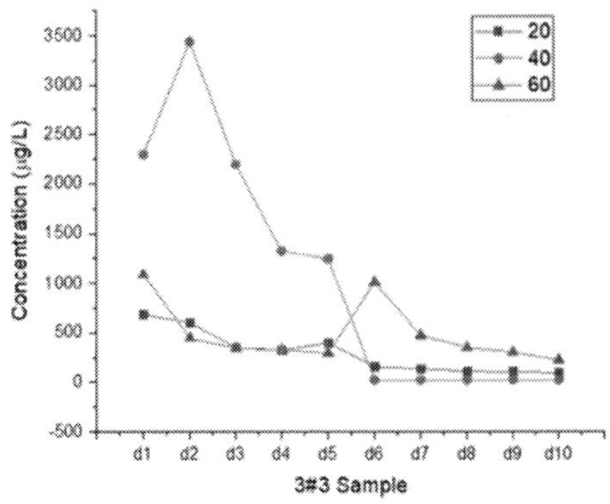

Figure 6. Total Chromium Precipitation of 3#2 sample extracting for 10 Days

4 Conclusions

In this study, through detecting the precipitation of chromium in faucets which were extracted at 20 ℃, 40 ℃, 60 ℃ environment temperature, the results showed that under normal circumstances, the chromium precipitation at 60 ℃ was the most, 20 ℃ was the least, indicating that the chromium precipitation in the faucets increases with the rise of temperature. And the precipitation between different batches and different samples from the same batch were greatly different, which was mainly affected by the faucets material and processing technology. Further detection showed that the precipitation of chromium mainly existed in the form of Cr VI, which will cause harm to human health. Therefore, in daily life, quality and qualified faucets should be selected, and in the process of use, the long-term retention of water in the faucets should be avoided as drinking water (especially under high temperature environment).

Acknowledgments

This research was supported by Quanzhou Science and Technology Project (Project No. 2018C093R).

References

1. X. Zeng, Y.G. Huang, J. H.B. Polytech. Univ., **36**, 2, 18-21(2020).
2. T.T. Huang, IOP Conf. Ser.: Earth Environ. Sci. **199** 032082 (2018).
3. L. Zlatanovic, J.P. Hoek, J.H.G. Vreeburg, Water Res., **123**: 761-72 (2017).
4. GB 18145-2014, Ceramic cartridge faucets[S].
5. GB 5750.6-2006, Standard examination methods for drinking water-Metal parameters[S].

 https://doi.org/10.1051/e3sconf/202129001028

Determination of Isoniazid by Fe(II)-2,2'-Bipyridine Spectrophotometry

Changqing Tu[1] and Xinrong Wen*[1]

[1]School of Chemistry and Environment, Jiaying University, Meizhou, Guangdong 514015, China
Guangdong Provincial Key Laboratory of Conservation and Precision Utilization of Characteristic Agricultural Resources in Mountainous Areas, Jiaying University, Meizhou, Guangdong 514015, China

Abstract. In acidic media, Fe(III) could be reduced to Fe(II) by isoniazid, Fe(II) reacts with 2,2'-bipyridine to form an orange complex with a maximum absorption wavelength of 522 nm. According to this principle, the content of isoniazid can be measured indirectly by measuring the amount of Fe(II). A new method for the determination of isoniazid by Fe(II)-2,2'-bipyridine spectrophotometry has been established. At the optimum conditions, the linear relationship is good between the mass concentration of isoniazid and absorbance of Fe(II)-2,2'-bipyridine complex in the range of 0.0004000-0.002800 mg/mL. The linear regression equation is A=0.0251+138.04C(mg/mL), and the linear correlation coefficient is 0.9995. This method is used to determine the content of isoniazid in isoniazid tablets, and the results are similar to those obtained by pharmacopoeia method.

1 Introduction

Isoniazid (the structure is shown in Figure 1) is mainly as one of the preferred drugs for effective treatment of various tuberculosis in clinical practice. But it has drug resistance, and can cause liver damage during the treatment process. Therefore, it is of great importance and significance for the determination of isoniazid content. So far, spectrophotometry[1-2], fluorescence spectrum method[3], flow-injection chemiluminescence analysis[4], capillary electrophoresis analysis[5-6], GC[7], HPLC[8-9], electrochemical method[10-11], and so on have been applied for the determination of isoniazid.

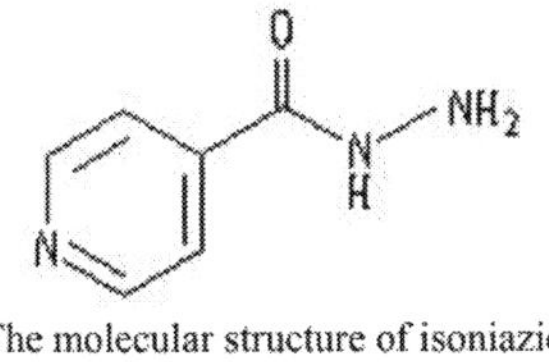

Fig. 1. The molecular structure of isoniazid

A new method for the determination of isoniazid by Fe(II)-2,2'-bipyridine spectrophotometry has been established. The various factors affecting the determination of isoniazid content are investigated. The results show that in acidic media, Fe(III) could be reduced to Fe(II) by isoniazid, Fe(II) reacts with 2,2'-bipyridine to form an orange complex with a maximum absorption wavelength of 522 nm. At the optimum conditions, the linear relationship is good between the mass concentration of isoniazid and absorbance of Fe(II)-2,2'-bipyridine complex, the content of isoniazid can be measured indirectly by measuring the amount of Fe(II). In the range of 0.0004000-0.002800 mg/mL, the linear regression equation is A=0.0251+138.04C (mg/mL), and the linear correlation coefficient is 0.9995. This method is used to determine the content of isoniazid in isoniazid tablets, and the results are satisfactory.

2 Experimental

2.1 Equipment and reagents

723S spectrophotometer (Shanghai Precision & Scientific Instrument Co,. Ltd) is used for photometric measurements. UV-2401 UV-visible spectrophotometer (The Shimadzu Corporation, japan) is used for scanning the absorption spectrum.

Isoniazid standard solution: 0.1000 mg/mL. Fe^{3+} solution: 0.1000 mg/mL, is prepared by $NH_4Fe(SO_4)_2 \cdot 12H_2O$ and appropriate amount of 3.0 mol/L H_2SO_4 solution. 2,2'-bipyriddine (BPY) solution:0.01998 moL/L. Buffer solutions of different pH was prepared as references 12.

All reagents are analytical reagent. Bidistilled water is used throughout.

2.2 Method

A given volume of 0.1000 mg/mL isoniazid standard solution or sample solution, 0.2000 mg/mL Fe^{3+} solution 1.40 mL are added into a 25 mL ground color comparison tube, added water to 10.00 mL and mixed well. Aftering this mixture reacted for 30 min at 80°C in water both and cooled back to room temperature, 0.01998 moL/L 2,2'-bipyridine solution 0.30 mL and

* Corresponding author: 198601012@jyu.edu.cn

pH=5.0 buffer solution 4.00 mL are added. The solution is diluted to 25.00 mL and mixed well, then, the absorbance is measured at 522 nm against the reagent blank after placing 10 min.

3 Results and discussion

3.1 Maximum absorption wavelength

According to the experimental method, the absorption spectrum(400-600 nm) of Fe(II)-2,2'-bipyridine orange complex is shown in Figure 2. The maximum absorption wavelength of Fe(II)-2,2'-bipyridine orange complex is at 522 nm. So, the absorption wavelength 522 nm is selected.

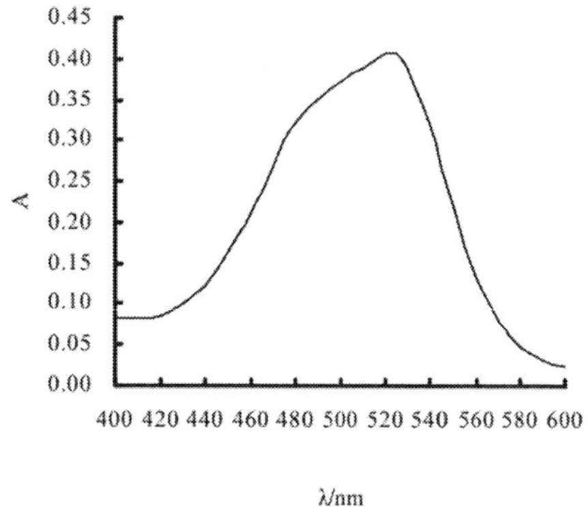

Figure 2. Absorption spectrum
Isoniazid solution:0.30 mL; Fe^{3+} solution:1.40 mL; BPY solution:0.30 mL; pH 5.0 buffer solution:5.00 mL; reaction temperature:80℃; reaction time:20 min; placing time:15 min.

3.2 Reaction temperature

The effect of reaction temperature on absorbance is seen in table 1. It can found that the absorbance of solution reaches greatest and almost constant when the reaction temperature is 70～100℃. Hence, 80℃ is used.

Table 1 Relationship between the reaction temperature and the absorbance

Temperature /℃	30	40	50	60
Absorbance	0.126	0.138	0.143	0.147
Temperature /℃	70	80	90	100
Absorbance	0.158	0.159	0.158	0.160

3.3 Reaction time and placement time

The relationship between the reaction time and the absorbance can be seen in table 2. It is found that the absorbance of solution reaches greatest and keeps constant when the reaction time is 30～60 min. Therefore, 30 min has been chosen.

Table 2 Relationship between the reaction time and the absorbance

Time/min	5	10	15	20
Absorbance	0.159	0.158	0.159	0.161
TimeT/min	30	40	50	60
Absorbance	0.168	0.166	0.166	0.165

When the placement time is 5～120 min, the experimental results of placement time on absorbance show that the absorption is not affected by the placement time. Thus, the placement time is selected for 10 min.

3.4 pH buffer solution and its dosage

The effects of pH buffer solution on absorbance are considered. The results show that the absorbance of solution are maximal and remain almost constant when the pH is 3.6～5.6. Consequently, pH 5.0 buffer solutions is used.

The experimental results of the dosage of pH 5.0 buffer solution on absorbance showed that the absorbance reaches its maximum value and remains constant when the dosage of pH5.0 buffer solution is 3.00～8.00 mL. Therefore, the optimum dosage of pH buffer solution is 4.00 mL.

3.5 The dosage of 2,2′-bipyridine

The effect of the dosage of 2,2′-bipyridine can be seen in Figure 3. From Figure 3, we can see that the absorbance reaches its maximum value and keep basically unchanged when the dosage of 2,2′-bipyridyl is 0.20～0.60 mL. So, 0.30 mL is identified as the optimum dosage of 2,2'-bipyridine.

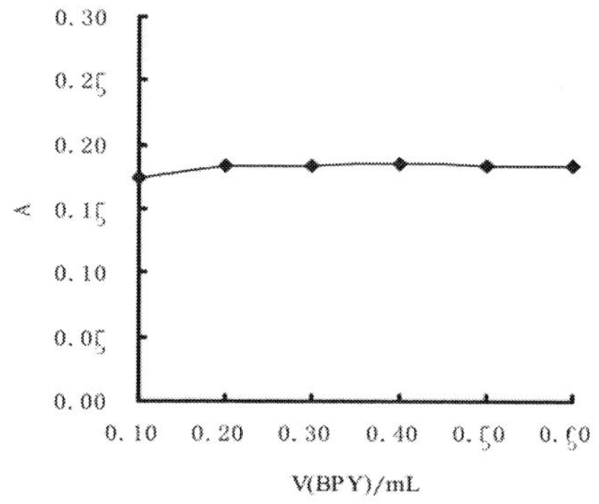

Figure 3. Effect of the dosage of 2,2′-bipyridine
Isoniazid solution:0.30 mL; Fe^{3+} solution:1.40 mL; pH 5.0 buffer solution:4.00 mL; reaction temperature:80℃; reaction time:30 min; placing time:10 min.

3.6 The dosage of Fe(Ⅲ)

The relationship between the the dosage of Fe(III) and the absorbance is showed in Figure 4. It is found that the absorbance reaches greater and remains constant when the dosage of Fe(III) is 1.20 mL~1.60 mL. So, 1.40 mL of Fe(III) is employed.

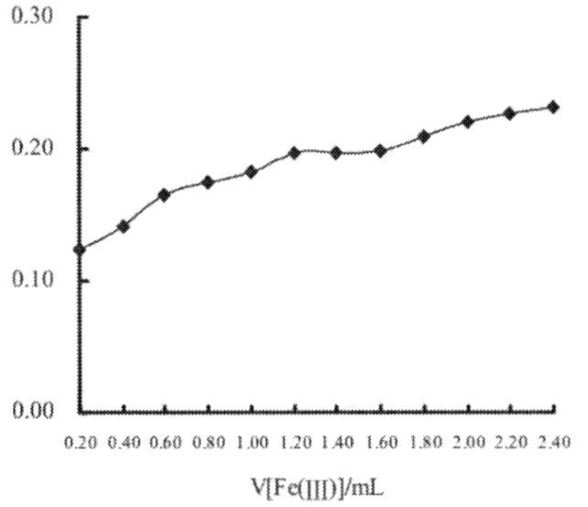

Figure 4. Effect of the dosage of Fe(III)
Isoniazid solution:0.30 mL; BPY solution:0.30 mL; pH 5.0 buffer solution:4.00 mL; reaction temperature:80℃; reaction time:30 min; placing time:10 min.

3.7 Standard curve

According to the experimental method, under the selected conditions (0.2000mg/mL Fe^{3+} is 1.40 mL, 0.01998mol/L 2,2′- bipyridine is 0.30 mL, pH 5.0 buffer solutions is 4.00mL, reaction temperature is 80℃, reaction time is 30min, placement time is 10 min), the different concentration of isoniazid standard solutions are prepared and the absorbances of these solutions are measured at 522 nm against the reagent blank. Then draw with concentration as horizontal coordinate and absorbance as vertical coordinate, the standard curve is showed in Figure 5. In the range of 0.0004000-0.002800 mg/mL, the Beer's law is obeyed between the concentration of isoniazid and the absorbance, the linear regression equation is A= 0.0251+138.04C (mg/mL) and the correlation coefficient is 0.9995.

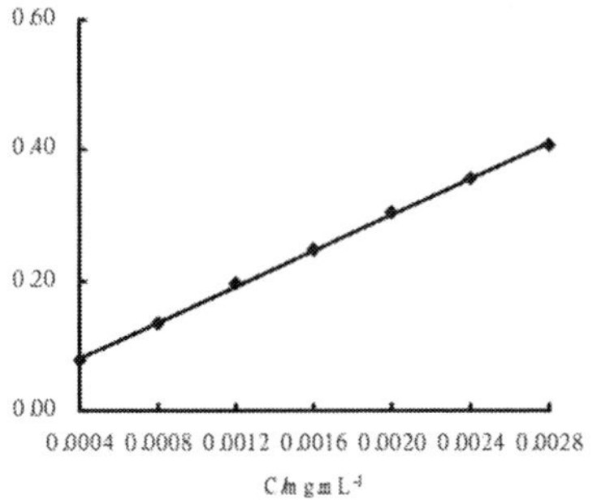

Figure 5. Standard curve
Fe^{3+} solution:1.40 mL; BPY solution:0.30 mL; pH 5.0 buffer solution:4.00 mL; reaction temperature:80℃; reaction time:30 min; placing time:10 min.

3.8 Sample analysis

20 tablets of isoniazid tablet are weighed 2.4039 g, 0.5098 g powder of isoniazid tablet is weighed precisely after grinding and biending. The powder of isoniazid tablet is dissolved in bidistilled water and is transferred into a 250 mL volumetric flask, the solution is diluted to 250.0 mL, mixed well and filtered. 10.00mL filter liquid is transferred into a 200 mL volumetric flask, then this filter liquid is diluted to 200.0 mL with bidistilled water, mixed well. This is the isoniazid sample solution, and the solution is preserved at 4℃, shielding from light.

When isoniazid sample solution(0.30 mL), 0.2000 mg/mL Fe^{3+} solution(1.40 mL), 0.01998 moL/L 2,2′-bipyridine solution(0.30 mL) and pH=5.0 buffer solution (4.00 mL) are added. Based on the experimental method, the content of the isoniazid in isoniazid tablet is determined by controlling the reaction temperature is 80°C, reaction time is 30 min and placing time is 10 min. Meanwhile, the recovery tests of standard addition are performed and the content of isoniazid in isoniazid tablet is determined by pharmacopoeia method. The results as show in Table 3.

Table 3 The content of isoniazid in isoniazid tablets n=5

Sample	Isoniazid tablet
Proposed method(mg·tablet^{-1})	100.7
RSD（%）	1.3
Pharmacopoeia method[13] (mg·tablet^{-1})	96.1
Added (μg·mL^{-1})	0.4000 0.8000
Recovered(μg·mL^{-1})	0.3997 0.7724
Recovery(%)	99.9 96.6

It is seen from table 3 that the content of isoniazid in isoniazid tablet is 100.7 mg·tablet^{-1} by this proposed method, and the content of isoniazid in isoniazid tablet is 96.1 mg·tablet^{-1} by pharmacopoeia method. It shows that the two results are similar. It is indicates that the content of isoniazid in isoniazid tablet can be accurately determined by Fe(II)-2,2'-bipyridine spectrophotometry.

4 Conclusion

In this paper, a new method for the determination of isoniazid by Fe(II)-2,2'-bipyridine spectrophotometry is reported. This method has been successfully used for the determination of isoniazid in isoniazid tablets, and the results are similar to the results of pharmacopoeia method. It is clearly obvious that the determination of isoniazid by Fe(II)-2,2'-bipyridine spectrophotometry has certain practical significance and application prospect.

References

[1] ZARGAR B, HATAMIE A . (2013) Localized Surface Plasmon Resonance of Gold Nanoparticles as Colorimetric Probes for Determination of Isoniazid in Pharmacological Formulation[J]. Spectrochim. Acta. A., 106:185-189.

[2] ZHANG H, WU L L, LI Q M, et al . (2008) Determination of isoniazid among pharmaceutical samples and the patients' saliva samples by using potassium ferricyanide as spectroscopic probe reagent[J]. Anal. Chim. Acta., 628: 67-72.

[3] RUI A S L, JOSE L F C L, JOAO L M S. (2000) Fluorimetric Determination of Isoniazid by Oxidation with Cerium(IV) in a Multicommutated Flow System[J]. Anal. Chim. Acta., 419: 17-23.

[4] HAGHIGHI B, BOZORGZADEH S. (2010) Flow Injection Chemiluminescence Determination of Isoniazid using Luminol and Silver Nanoparticles [J]. Microchem. J., 95: 192-197.

[5] LIU Y, FU Z F, WANG L. (2012) Capillary Electrophoresis Analysis of Isoniazid using Luminol-peridoate Potassium Chemiluminescence system[J]. Luminescence., 26(6):397-402.

[6] ZHANG X F, XUAN Y L, SUN A M, et al. (2009) Simultaneous Determination of Isoniazid and p-Aminosalicylic Acid by Capillary Electrophoresis using Chemiluminescence Detection[J]. Luminescence., 24: 243–249.

[7] KHUHAWAR M Y, ZARDARI L A. (2008) Ethyl Chloroformate as a Derivatizing Reagent for the Gas Chromatographic Determination of Isoniazid and Hydrazine in Pharmaceutical Preparations[J]. Anal. Sci., 24: 1493-1496.

[8] CHELLINI P R， LAGES E B， FRANCO P H C， et al. (2015) Development and Validation of an HPLC Method for Simultaneous Determination of Rifampicin， Isoniazid， Pyrazinamide， and Ethambutol Hydrochloride in Pharmaceutical Formulations[J]. J. Aoac. Int. 98(5):1234-1239.

[9] AYYAPPAN J， UMAPATHI P， QUINE S D． (2011) Development and Validation of a Stability Indicating High-performance Liquid Chromatography (HPLC) Method for the Estimation of Isoniazid and its Related Substances in Fixed Dose Combination of Isoniazid and Ethambutol Hydrochloride Tablets[J] . Afr. J. Pharm. Pharmacol., 5(12):1513-1521.

[10] ABSALAN G， AKHOND M， SOLEIMANI M， et al． (2016) Efficient Electrocatalytic Oxidation and Determination of Isoniazid on Carbon Ionic Liquid Electrode Modified with Electrodeposited Palladium Nanoparticles[J]. J. Electroanal. Chem., 761:1-7.

[11] RASTOGI P K， GANESAN V， AZAD U P. (2016) Electrochemical Determination of Nanomolar Levels of Isoniazid in Pharmaceutical Formulation using Silver Nanoparticles Decorated Copolymer[J]. Electrochimica. Acta., 188:818-824.

[12] CHANG W B, LI K A. (1981) Brief Handbook of Analytical Chemistry[M]. Beijing:Beijing University Press, 262. (In Chinese)

[13] Chinese Pharmacopoeia Commission. (2015) Pharmacopoeia of People's Republic of China (2015 edition, two) [M]. Beijing:China Medical Science Press, 413. (In Chinese)

Polymeric Micelles: Morphology, Synthesis, and Pharmaceutical Application

Kun Bai[1,†], Anda Wang[2,*,†]

[1] Queen Mary University of London Engineering School, Northwestern Polytechnical University, Xi'an, Shaanxi Province, China.

[2] 1010 West University Ave, Urbana, IL, 61801, USA.

[†]These authors contributed equally.

Abstract. Polymeric micelles (PMs) are nanosized core-shell molecules formed by the self-assembly of amphiphilic block copolymers. Their unique morphologies, biocompatibility have allowed them to serve as drug carriers and have found wide applications in the pharmaceutical industry. This review summarizes the recent progress of PM development and its applications in the pharmaceutical field. Firstly, we discussed the fundamentals of polymeric micelles, the formation mechanism, and their relative morphologies. Then we listed various common synthetic methods, including the solvent-free method, solvent-switch method, microfluid method, etc. Secondly, we illustrated the application of polymeric micelles in pharmaceutical applications on targeting, imaging, and stimuli-responsive drug release. In the end, we summarize the fundamental aspect of how to prepare polymeric micelles, their resulting morphology, and their applications in the medical field. Finally, we provide insights into polymeric micelles' future developments on their longer shelf life and better environmental adaptability.

1 Introduction

Polymeric micelles (PMs) are nanosized polymer capsules with membranes generally considered to have a hydrophobic bilayer structure similar to phospholipids [1]. In the last two decades, PMs have experienced rapid development [2-8]. Various polymeric micelles have been synthesized, and they have found potential applications in target positioning and as drug carriers for cancer, gene therapy, and even catalysts and building materials.

In this review, we summarized the recent progress of synthesis and applications of these PMs. We first discuss the basic definition of polymeric micelles, followed by their formation mechanism and relative morphology. Testing methods of CMC (critical micelles concentration) were introduced, which significantly impacted the synthesis of polymeric micelles. The morphology of PMs was determined by several factors, including the ratio of hydrophobic to hydrophilic segments, temperature, and concentration of the solution.

We further introduced synthetic methods, which are an important subject in the study of polymer vesicles. There are many alternative methods to prepare PMs. Among them, solvent-switching techniques and solvent-free techniques are the most widely used technologies [5], but they still have many drawbacks. Therefore, this review will also introduce some other new technologies in recent years, including PISA (Polymerization-induced self-assembly) [3-6]. In addition, complex or large vesicles are also a hot topic in recent years. The synthesis mechanism and technology of some such vesicles will also be introduced [7,8].

Finally, several pharmaceutical applications of PMs would be discussed in this review. The most important contribution is its applications as drug carriers due to its high similarity to cell membranes. However, after the drug enters the blood vessel, the cells that need to be released must be accurately located. Otherwise, serious side effects may occur. Therefore, some tumor-targeting technologies and applications, including passive targeting and active targeting, will be introduced in this article. Additionally, PM imaging methods include nuclear imaging, CT, and MRI, which are also briefly mentioned. Finally, the review will discuss several stimulus-responsive PMs used as drug carriers in detail.

2 Fundamentals of PMs

2.1 Micelles

According to IUPAC, micelles are "particles of colloidal dimensions that exist in equilibrium with the molecules or ions in solutions from which it is formed." Micelles have great application value in the pharmaceutical field because they can carry drugs into the liquid to make some slightly soluble or insoluble substances more soluble. The hydrophilic head is generally combined with the solution

*Corresponding author: anda2@illinois.edu

https://doi.org/10.1051/e3sconf/202129001029

and self-assembled into a spherical shape in the aqueous solution. Aggregation number is an index used to describe the average number of monomers required to form a micelle, usually between 50-200. The formation of micelles is determined by the number and morphology of amphiphilic molecules and the temperature and properties of the solvent. Continuously adding amphiphilic molecules to the solvent, when the concentration of amphiphilic molecules exceeds a certain value, they will self-assemble to form micelles. The driving force for the formation of micelles is the difference in the free energy of the system. This specific concentration is called the critical micelle concentration (CMC). Below CMC, amphiphilic molecules will preferentially gather at the air-water interface. At CMC, the amphiphilic molecules adsorbed at the air-water interface are saturated, and micelle monomers begin to form [9].

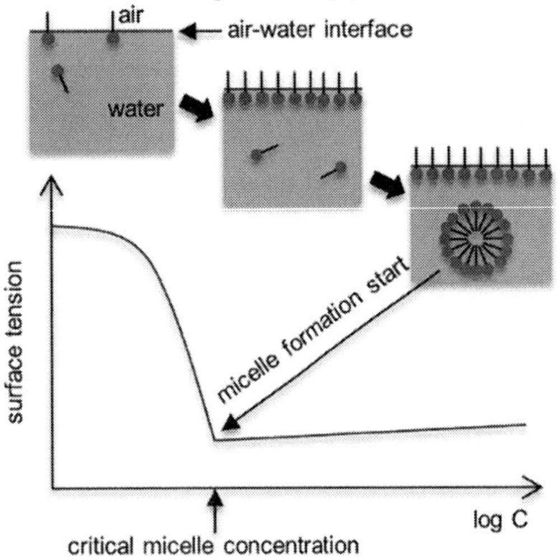

Fig. 1. Surface tension and surfactant concentration relationship. After CMC, the surface tension no longer changes, which indicates that the amphiphilic molecules on the surface of the solution are saturated [10].

2.2 Polymeric micelles

Polymeric micelles are a type of micelles, but the amphiphilic molecules formed in solution are macromolecules. According to the definition of IUPAC, polymer micelles are "organized self-assembled bodies formed in liquid and composed of amphiphilic macromolecules, usually amphiphilic and amphiphilic blocks made of hydrophilic and hydrophobic blocks, diblock or triblock copolymer." PMs are usually spherical. Compared with traditional micelles, PMs have the following characteristics. (1) Larger volume, usually 10-100nm in diameter. (2) Lower CMC, but still in equilibrium with isolated macromolecules. (3) PMs show better thermodynamic and kinetic stability. [11] In addition, due to the biological stability and diversity of polymers, the hydrophilic and hydrophobic blocks of PMs are usually easier to design. For instance, the hydrophilic shells originate from poly (ethylene glycol) (PEG), poly(vinyl alcohol) (PVA), poly(N-vinyl-2-pyrrolidone) (PVP), poly(-acrylic acid), poly(acrylamide) (PAAm), poly(glycerol) (PG), poly(amino acid) (PAA), and polysaccharides. The hydrophobic blocks can usually be designed by polyesters (i.e., poly (glycolic acid) (PGA), poly(ecaprolactone) (PCL), and poly (D, L-lactic acid) (PDLLA)) or polyethers (i.e., poly (ethylene oxide) (PEO) and poly(propylene oxide) (PPO)). [12]

2.3 Formation mechanism

Similar to the formation mechanism of micelles, the formation of PMs is also based on the self-assembly of amphiphilic compounds. The resulting micelles consist of a hydrophilic block shell and a hydrophobic block core. After reaching the CMC, PMs can also self-assemble generation. Compared with general micelles, the advantage of PMs is that their hydrophobic blocks are larger and consequently have a lower CMC value. Therefore PMs are more stable in dilute solutions. During the formation process below the CMC, the hydrophobic block will avoid contact with the solution. Because of that, the amphiphilic macromolecules will be concentrated at the air-solution interface. When CMC is reached, the hydrophobic heads of amphiphilic macromolecules will gather into nuclei due to the saturation of the air-solution interface. This can be explained by the "like dissolve like" theory. In addition, to keep the hydrophilic block in contact with the solution at all times, it will generate an inward force, ensuring that the amphiphilic macromolecules will not combine indefinitely into a single phase. Therefore, PM is formed by the balance of two opposite forces. Besides CMC, there is another important indicator called the critical micelle temperature (CMT). Only when the temperature is greater than this critical temperature, and the concentration is greater than CMC, polymeric micelles can self-assemble and form.

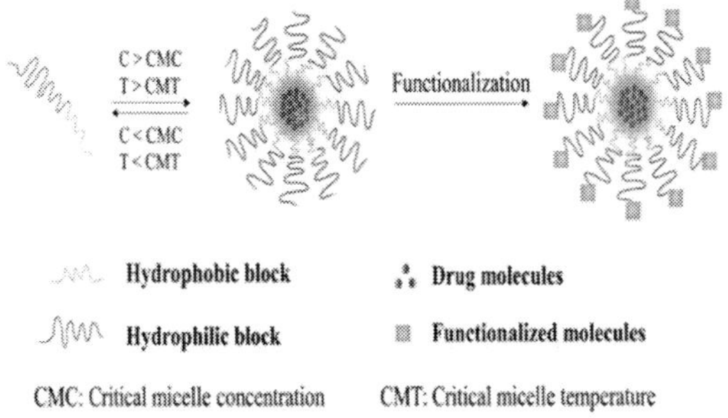

Fig. 2. Mechanism of polymeric micelles formation [12].

2.4 Morphology

2.4.1 Typical morphology of polymeric micelles

The shape of polymeric cells is mostly determined by the size ratio of the hydrophobic segment to the hydrophilic segment and the concentration of surfactants. When block copolymer in dilute solution, they will likely form the morphology of spherical shape [13]. In most block copolymers, the hydrophilic group will have a larger size than the hydrophobic group. This feature results in a large hydrophilic "head group" and forms a cone shape as a hydrophobic group is located at the cone's tip. All the

https://doi.org/10.1051/e3sconf/202129001029

hydrophobic tips gathered to form a spherical shape. Shapes of micelle other than spherical (rods, vesicles, etc.) can be produced by increasing temperature or concentration [14].

2.4.2 Morphology of polymeric micelles in drug-delivery application

The effect of different shapes of polymeric micelles on drug delivery remained mainly unexplored. Spheric polymeric micelles are the most explored ones for drug delivery [15]. However, polymeric micelles other than spherical shape in drug delivery are also worth further research. Since polymeric micelles are required to release drugs under specific conditions and specific time lengths, the morphology of polymeric micelles can be used as one of the physical parameters to improve their effectivity. For example, rod shapes have different loading capacities compare to spherical shapes. Moreover, rod shapes are more suitable in the aerosol formulation in the lung compare to spherical shapes because their slim tubular structures are more easily expand in the lung

3 Synthesis

To prepare polymersomes, six synthesis methods could be used. The most commonly used method is Solvent-switch technique and organic solvent-free technique. The former needs to dissolve polymers into the organic solvent, while the latter only needs to dissolve polymers into the water. This paper will also introduce other preparation methods such as polymerization-induced self-assembly (PISA), centrifugation-induced self-assembly, and the microfluid method.

3.1 Solvent-switch methods

The solvent-switch method is also known as the "phase inversion" technique. Using this method, firstly, polar organic solvents such as N,N-dimethylformamide (DMF), and tetrahydrofuran (THF) are needed for the solution of polymers [1,2,17,18]. Then slowly add the second solvent (usually water) to the polymer solution or inject the polymer solution into the second solvent to hydrate the solution. Due to the insolubilities of the hydrophobic segment of the amphiphilic copolymer water, it will increase the interfacial tension between the hydrophobic segment and water, thereby making the copolymer self-assemble to form polymer vesicles. To change the particle size and distribution of polymer vesicles, different solvents can be chosen. However, the polymersomes prepared by this method is usually contained considerable residual organic solvent, which required further purification. This step is highly time-consuming and will increase the cost. In addition, as mentioned above, although changing the solvent can change the particle size and distribution of polymer vesicles, the change like the solvent during the dialysis process can also cause unexpected morphological changes.

3.2 Solvent-free methods

3.2.1 Rehydration method

The rehydration method [19] usually contains film swelling and bulk swelling, sonication, and extrusion. For film swelling, the first step is similar to the solvent-switch method. Polymers need to dissolve into an organic solvent such as chloroform, followed by the evaporation of organic solvent to form a thin film. During the hydration, the water is used to penetrate the polymer di-layers by the hydration interaction force made the copolymer swell and extrude, and self-assemble to form polymer vesicles with larger particle size and wider distribution. Some specific methods can be used to improve the properties of polymer vesicles. Generally, the polymer vesicles are squeezed through filters of different pore sizes under high pressure to produce the vesicles with smaller particle size and distribution [20]. Additionally, Battaglia and Ryant increase the speed of water diffusion into the polymer film through an alternating electric field action, thereby preparing polymer vesicles with a narrow particle size distribution [21].

For bulk swelling, with the assistance of some water-soluble polymers, there is no need to use organic solvents when preparing polymer vesicles.[22] Jianzhong and Steven P. Armes reported a method for preparing block copolymer vesicles in pure water by preparing and using poly(ε-caprolactone)-block-poly[2-(methacryloyloxy) ethyl phosphorylcholine], PCL-b-PMPC, and then through the process of sol-gel chemical stabilization in the capsule membrane [23].

Ani T. Nikova et al. also reported a method of "stuffing" the aliphatic region of self-assembled aggregates with hydrophobic homopolymers to more easily prepare self-assembled macromolecular structures with controllable properties. The filled membranes made of hydrates and hydrophobic homopolymers are hydrated, and "filled" vesicles and various emulsions can be formed. The film thickness of micelles can be adjusted by changing the content of hydrophobic homopolymer in the polymer mixture [23].

3.2.2 pH tuning in pure water

For pH-sensitive polymer, polymersomes can be formed by increase or decrease the pH value of solution due to the ratio of hydrophobic section and hydrophilic section will change with pH value, and polymer could self-assemble polymersomes. For instance, Jianzhong Du et al. reported a method for successfully preparing polymer vesicles using A Biocompatible Zwitterionic Copolymer, PMPC25-b-PDPA120 diblock copolymer. By increasing the pH of the solution, the self-assembly formation of polymer vesicles can be observed in TEM [24,25].

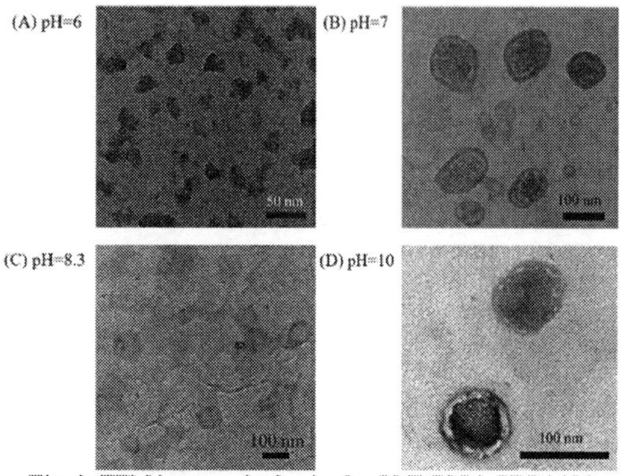

Fig. 3. TEM images obtained using PMPC25-b-PDPA120 diblock copolymer at different solution pH. (A) At pH=6.0, no obvious vesicles are observed in the microstructure. (B) At pH=7.0, a few small vesicles appear. (C) At pH=8.3, more vesicles appear. (D) At pH=9.0, a completely large vesicle structure appears.

3.2.3 PICsomes

PICsomes (Polyion Complex Vesicles), first developed by Kishimura et al. [26-30,33-35]. It is formed by the self-assembly of a complex of oppositely charged polyelectrolytes in an aqueous solution. PICsomes are simple to prepare, have semi-permeability and high biocompatibility, and provide technical support for the future use of polymer vesicles as drug carriers. Akihiro Kishimura introduced a method for preparing nano-PICsomes. Compared with traditional PICsomes, nano-PICsomes have a uniform structure and can control their size. Using homopolymer with fPEG value ⩽ 10% to form PIC, mixing homo-P (Asp-AP) and PEG-b-PAsp (fPEG ~8%), the desired hollow vesicles and vesicle wall structure was obtained. It is a PEG-PIC-PEG three-layer structure with a size range of up to 300nm. This achievement is crucial for future applications in nanomedicine [3]. Takahashi et al. also reported that by adding anionic neutral di-hydrophilic block copolymer (AP) and cationic neutral di-hydrophilic block copolymer (MP) to the solution to achieve classically unstable reversibility, vesicles and spherical gels could be produced. The morphology between the bundle's changes and forms PICsomes.[30]

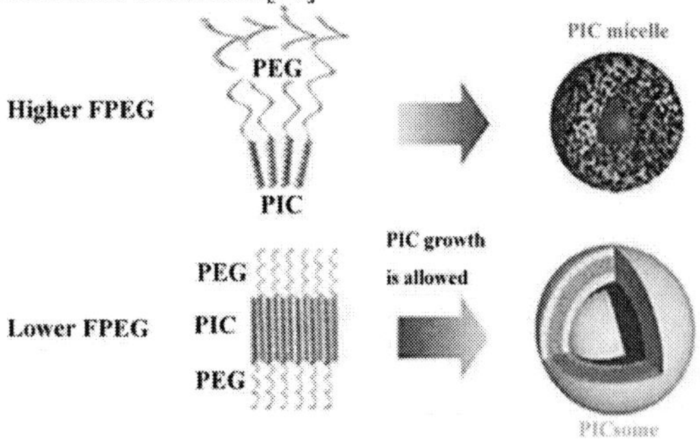

Fig. 4. Schematic representations of the structural control of polyion complex (PIC) assemblies.[3]

3.3 Other synthesis methods

3.3.1 Polymerization-induced self-assembly (PISA):

General polymer self-assembly needs to be carried out in a dilute solution with a volume fraction of less than 1%, reducing the yield and limiting its use in practical applications [3-6]. Efficient synthesis of block copolymer vesicles with fixed morphology in concentrated solution is recognized as a solvable technical problem. Recently, amphiphilic diblock copolymers can be formed by radical polymerization of water-immiscible monomers (such as styrene, methyl methacrylate, or n-butyl acrylate) [31]. The general PISA reaction mechanism is shown in Fig. 5.

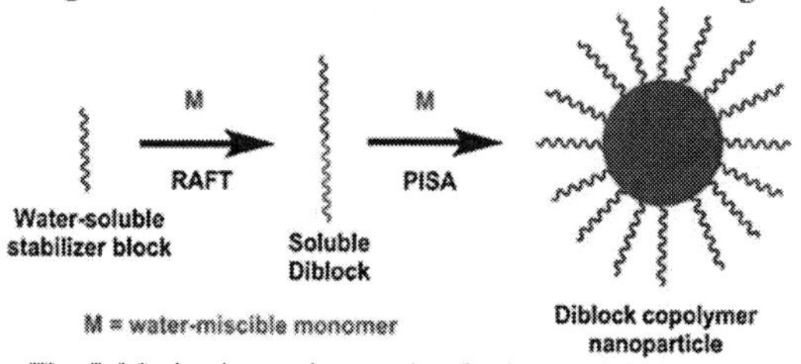

Fig. 5. Mechanism and example of polymerization-induced self-assembly in aqueous solution [6].

For example, A Blanazs reported a method of self-assembled vesicles using poly(glycerol monomethacrylate)-blockpoly(2-hydroxypropylmethacrylate), PGMA-b-PHPMA, through reversible addition-fragmentation chain transfer (RAFT). In this case, self-assembly and polymerization reactions occur simultaneously. Since the PGMA block is hydrophobic, as the polymerization reaction progresses, there are more and more PHPMA blocks, making the ratio of hydrophilic and hydrophobic ends larger and larger, and promotes the progress of self-assembly [4]. However, only hydrophilic monomers that form a hydrophobic polymer after the polymerization reaction can use this method.

PISA is an important method to form polymer vesicles, which solves the low efficiency of traditional methods. It can be self-assembled in a concentrated solution. It has great application potential for high-efficiency micro-encapsulation carriers and as drug carriers [36].

3.3.2 Centrifugation-induced self-assembly

Centrifugation is an important way to form vesicles. In the simplest case, applying agitation or centrifugation in a water/oil or oil/water system can produce complex vesicles. Inspired by this, stable polymer vesosomes (polymersomes in polymersomes) can be prepared by simple centrifugation. Using oil phase dispersion PB-b-PEO diblock copolymer and water phase as aqueous sucrose (380 mOsm) solution to produce W/O emulsion followed by pouring it into another water phase as aqueous sucrose (380 mOsm) solution. With centrifugation and gravity force assistance, W/O droplets would come out and penetrate the emulsion to form polymer vesosome macromolecules [7].

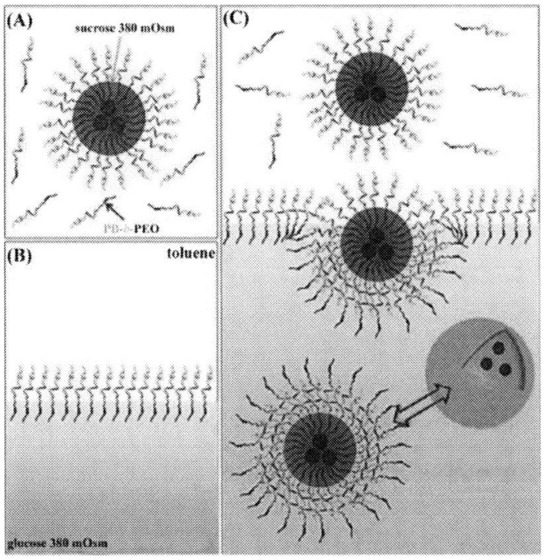

Fig. 6. Schematic representation of the centrifugation-induced process generating giant polymersomes or polymer vesosomes [8].

3.3.3 Microfluid method

The Microfluid method is another significant method to prepare giant polymer vesosomes. The mechanism of producing macromolecular vesicles is similar to that of centrifugation [32]. Both force the water-in-oil system to pass through the oil-in-water system to form a W/O/W system or a more complex system. This method uses a typical device, as shown in Fig. 7. The left arm is an O/W emulsion. The ring structure allows the water-in-oil system to pass through the aqueous solution and form stable droplets. At the "three forks", the water-in-oil colloid containing the droplets and the aqueous solution meet and form a stable fluid. After the fluid is stabilized, the triangular column guides the water at the position indicated by the green circle to make the aqueous droplets enter the water from the oil phase and form large vesicles [31].

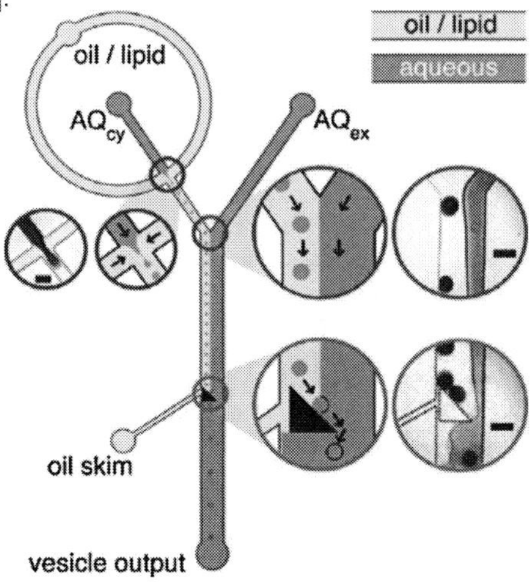

Fig. 7. Microfluid device schematic diagram

4 Pharmaceutical application

Nowadays, around 90 percent of drugs generated by the pharmaceutical industry have poor water solubility [37]. While anti-cancer drug potency will be greatly reduced after oral administration, intravenous administration of these drugs is also challenging due to the specific formulation of organic solvent and surfactants [38]. This challenge result in the failure of reaching the market due to their poor solubility even though they have desired pharmaceutical properties. Recent research of polymeric micelles, however, provides an alternate novel way to deliver these medicines. Drugs can be loaded by trap the target in the hydrophobic core of polymeric micelles.

4.1 Tumor targeting

Polymeric micelles are generally more stable in physiological solutions than surfactant micelles because Polymeric micelles are formed from amphiphilic block copolymers [37]. The small size (<100 nm) of polymeric micelle gives a surface-smoothing effect and allows longer retention time circulating in the blood and results from a higher accumulation of drug at the target site. Passive and active targeting has been used as delivery strategies to target anti-cancer drugs to interested physiological sites. Those strategies can improve their therapeutic index and reduce the harm to other non-target organs.

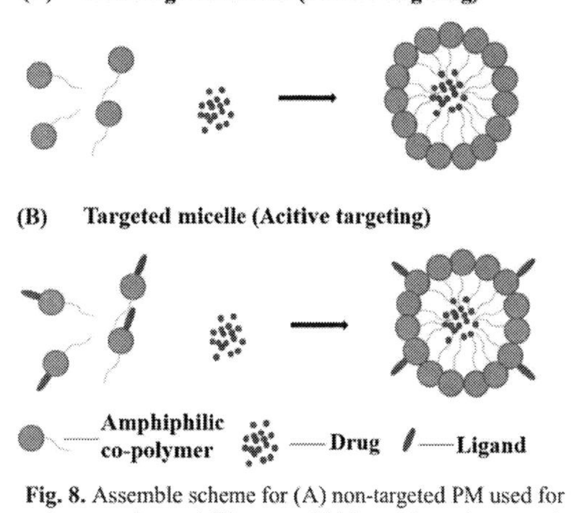

Fig. 8. Assemble scheme for (A) non-targeted PM used for passive targeting and (B) targeted PM used for active targeting

4.1.1 Passive targeting

It is now a well-known fact that under certain conditions such as inflammation or hypoxia, which is typical for tumors, the endothelium of blood vessels becomes more permeable compared to normal blood vessels [39]. Those blood vessels in tumor sites will leakier than normal ones due to their incomplete tumor vasculature culture. Therefore, those cultures allow selectively enhanced permeation of macromolecules, but small molecule drugs will not be affected due to their short circulation time. This phenomenon is called the EPR effect. Thus,

https://doi.org/10.1051/e3sconf/202129001029

polymeric micelles can be used to encapsulate those small-molecule drugs and enhance their prolonged systemic circulation to provide tumor selectivity and decrease side effects. This targeting method is called passive targeting because it relies on carrier characteristics and tumor biology, but it does not contain any specific ligands to bind target tissue or organ.

4.1.2 Active targeting

Active targeting can greatly increase the number of delivered drugs to the target than passive targeting [39]. Active targeting is achieved by modifying the surface of polymeric micelles with specific ligands that bind to receptors on the target tumor cells. Active targeting is especially effective in treating cancer due to the over-expression of cancer cell receptors. The attachment of specific ligands to the hydrophilic surface of block polymers can be used to produce active targeting micelles [37]. Such modification will improve the affinity of the polymeric micelles for the target tumor, enhance the drug efficiency and reduce side effects to normal tissues. Another approach of active targeting is to manipulate polymeric micelles' response to specific stimuli unique to disease conditions. Drugs can be released by internal stimuli such as pH, temperature, by external stimuli such as light, heat, or by a combination of both stimuli.

4.2 Imaging

Nuclear imaging, magnetic resonance imaging (MRI), and X-ray computed tomography (CT) are used to diagnose cancer and disease evaluation.[38] Contrast agents are introduced to the body to greatly enhances the area of interest in the body. Polymeric micelles have the application of used as delivering contrast agent. For nuclear imaging, contrast agents are made by polymeric micelles loaded with gamma emitters. As a contrast agent of MRI, iron oxide particles are loaded into micelles. CT is less suitable for molecular imaging due to its high concentration of contrast agent requirement. Micelles modified for tumor imaging using CT have not been reported.

4.3 Stimulus Responsive Drug Release

Even though Polymeric micelles are suitable for drug carriers, some problems remain. Polymer micelles are highly associated with premature and insufficient drug release [40]. Ideally, polymer micelles should hold the drug until they are present in the target area, but in reality, they will start to release after accumulating in the tumor interstitium. The way to solve this problem is to increase the interaction between the drug and the hydrophobic core. To achieve adequate drug release, stimuli-responsive PMs are used in tumor-targeted drug delivery

4.3.1 pH-responsive

pH-responsive polymeric micelles have become important nanocarriers for drug delivery. Due to increased metabolic rates and aerobic glycolysis, most tumor tissues perform acidic (pH 6.5-7.2) environments [41]. While blood is relatively basic (pH 7.4), such difference can be exploited to drug release using pH-responsive polymeric micelles. One possible approach to provide pH sensitivity to polymeric micelles is to incorporate the ionizable groups into polymeric micelles. Once the groups are ionized in acidic conditions, polymeric micelles will decompose and cause the drug to release into tumor tissues.

4.3.2 Thermo-responsive

Temperature is one of the most investigated stimuli in drug delivery [40]. The change of temperature on tissue is internal (e.g., inflammation, infection) or external (apply heating device). Since tumor cells can also be damaged by increasing temperature, thermo-responsive polymeric micelle are the most investigated drug-release approach in cancer therapy. Polymers made from thermo-responsive polymeric micelles can undergo a sharp change of physical properties with temperature change, so polymeric micelles will deform and release drugs when the temperature is changed. Those polymers typically have low critical solution temperatures (LCST). The polymer becomes water-soluble below the LCST, becomes insoluble above LCST due to the disruption of hydrogen bonds between water and polymer chains. LCST can be modulated by adding hydrophobic or hydrophilic comonomers to obtain the desired LCST range (37°C–42°C), which higher than normal body temperature but lower than tumors.

4.3.3 Ultrasound-responsive

Ultrasound has been used in therapeutic and diagnostic applications for many years due to its low cost, ease of administration, and deep penetration [42]. Both hyperthermia and non-thermal effects take place when the tissue is exposed with ultrasound. Studies have shown that ultrasound can be used to target micelles and trigger drug release [43].

5 Conclusion

In this review, we introduced the morphology, synthesis method of PMs, and their applications in the medical field. From the initial hollow vesicles to the now complex and large smart vesicles, the technology of polymer vesicles is changing with each passing day. In addition to solvent switching technology and Solvent-free technology, more synthesis techniques are developed. Synthetic methods like PISA are likely to become the main way to synthesize polymer vesicles in the future. The double-layer membrane structure of polymer vesicles has good compatibility with biological membranes, and it is an ideal drug carrier in vivo. PMs have good histocompatibility and cell penetration like liposomes when used as drug carriers, but they have higher stability and chemical versatility than liposomes. After the polymer vesicle encapsulates the drug, it can play the role of synergism, attenuation, slow-release, and increase drug stability. Its hydrophilic surface shell has a steric stabilizing effect, which can prolong the circulation time in the body and enhance passive targeting Ability: After being attached to the surface of biological molecules, it can also give vesicles active targeting function or other biological effects. However, it should be noticed that the difference between lab experiments and practical applications is still considerably large. The polymer vesicles themselves need to be biodegradable and biocompatible, but their degradation products also need biocompatibility. The degradation products need to be non-toxic and not have harmful side effects on other cells. Moreover, the current PMs are independently developed by various laboratories. They do not have interoperability, which means they only have a single function, resulting in high costs when introducing practical applications. The transportation and storage of polymer vesicles also require further research. Finally, only limited companies can manufacture PMs for practical applications. Therefore, further research needs to emphasize producing PMs with a longer shelf life and better environmental adaptability.

References

1. P.L. Soo, A. Eisenberg, J. Polym. Sci. B. Polym. Phys. **42**, 923 (2004).
2. Q. Liu, S. Chen, J. Chen, J. Du, Macromol. **48**, 739 (2015).
3. A. Kishimura, Polym. J. **45**, 892(2013).
4. A. Blanazs, J. Madsen, G. Battaglia, A.J. Ryan, S.P. Armes, J. Am. Chem. Soc. **133**, 16581 (2011).
5. N.J. Warren, O.O. Mykhaylyk, D. Mahmood, A.J. Ryan, S.P. Armes, J. Am. Chem. Soc. **136**, 1023 (2014).
6. N.J. Warren, S.P. Armes, J. Am. Chem. Soc. **136**, 10174 (2014).
7. M. Marguet, O. Sandre, S. Lecommandoux, Langmuir **28**, 2035 (2012).
8. M. Marguet, L. Edembe, S. Lecommandoux Angew. Chem. Int. Ed. **51**, 1173-6 (2012).
9. Y. Moroi Micelles: Springer international ed. New York: Springer **41-50**, 2005.
10. M. Ueno, N. Isokawa, K. Fueda, S. Nakahara, H. Teshima, N. Yamamoto, H. Yokoyama, Y. Noritsugu, K. Shibata, K. Miyagawa, S. Tanaka, T. Hirano, A. Fujito, A. Takashima, K. Kanno. World J. Chem. Edu. **32-44** (2016).
11. Vert M. et al. IUPAC Recommen **84**, 377 (2012).
12. D.T. Pham, A. Chokamonsirikun, V. Phattaravorakarn, et al. J. Mater. Sci. **56**, 2016 (2021).
13. P.C. Heimenz, R. Rajagopalan, Principles of Colloid and Surface Chemistry, Third Edition, (1997).
14. G. Wanka, H. Hoffman, W. Ulbricht, Macromol. **27**, 4145 (1994).
15. C. Allen, D. Maysinger, A. Eisenberg, Colloids. Surf. B **16**, 3 (1999)
16. D.L. Johnson, M. Polikandritou-Lambros, T.B. Martonen, The Knee **3**, 9 (1996)
17. E. Themistou, G. Battaglia, S.P. Armes, Polym. Chem. **5**, 1405 (2014).
18. M.T. Popescu, M. Korogiannaki, K. Marikou, C. Tsitsilianis, Polym. **55**, 2943 (2014).
19. J. Du, R.K. O'Reilly, Soft Matter **5**, 3544 (2009).
20. J. Lee, H. Bermudez, B.M. Discher, M.A. Sheehan, Y.Y. Won, F.S. Bates., D.E. Discher,. Biotechnol. Bioeng. **73**, 135 (2001).
21. J. Du, S.P. Armes, Langmuir **25**, 9564 (2009).
22. J. Du, S.P. Armes, Langmuir **24**, 13710 (2008).
23. A.T. Nikova, V.D.Gordon, G. Cristobal, M.R.Talingting, D.C. Bell, C.Evans, M.Joanicot, J.A.Zasadzinski, D.A Weitz, Macromol. **37**, (2004).
24. J. Du, Y. Tang, A.L. Lewis, S.P. Armes., J. Am. Chem. Soc. **127**, 17982 (2005).
25. J. Chen, Q. Liu, J. Xiao, J. Du. Biomacromol. **16**, 1695 (2015).
26. A. Koide, A. Kishimura, K. Osada, W Jang, Y. Yamasaki, K. Kataoka, J. Am. Chem. Soc. **128**, 5988 (2006).
27. Y. Anraku, A. Kishimura, M. Oba, Y. Yamasaki, K. Kataoka, J. Am. Chem. Soc. **132**, 1631-6 (2010).
28. S. Schrage, R. Sigel, H Schlaad, Macromol. **36**, 1417 (2003).
29. S. Ohno, K. Ishihara, S. Yusa, Langmuir, **32**, 3945 (2016).
30. R. Takahashi, T. Sato, K. Terao, S. Yusa,. Macromol., **49**, 3091 (2016).
31. S. Hauschild, U. Lipprandt, A. Rumplecker, U. Borchert, A. Rank, R. Schu-bert, S. Förster, Small **1**, 1177 (2005).
32. H.C. Shum, J.W. Kim, D.A. Weitz, J. Am. Chem. Soc. **130**, 9543 (2008).
33. Q. Yi, G.B. Sukhorokov, J. Ma, X. Yang, Z. Gu, Int. J. Polym. Sci. **2015**, 756237 (2015).
34. 18H. Gao, O.A. Goriacheva, N.V. Tarakina, G.B. Sukhorukov, ACS Appl. Mater. Interfaces **8**, 9651 (2016).
35. S.M. Hartig, R.R. Greene, J. DasGupta, G. Carlesso, M.M. Dikov, A. Prokop, Davidson JM, Pharm. Res. **24**, 2353 (2007).
36. C. Gonzato, M. Semsarilar, E.R. Jones, F. Li, G.J.P. Krooshof, WymanP, O.O. Mykhaylyk, R. Tuinier, S.P. Armes, J. Am. Chem. Soc. **136**, 11100 (2014).
37. Y. Zhu, Bo Yang, C. Shuai, et al., Prog. In Polym. Sci. **64**, (2016).
38. S. Movassaghian, O. M. Merkel, V.P. Torchilin, WIREs Nanomed Nanobiotech. **7**, 691 (2015).
39. Q. Zhou, L. Zhang, T. Yang, H. Wu, Int. J.of Nanomedi. **13**, 2921 (2018).

40. C. Oerlemans, W. Bult., M. Bos., G. Storm., J Frank W Nijsen, W.E. Hennink, Pharm. Res. **27**, 2569 (2010).
41. J.F. Nijsen, G.C. Krijger. A.D.van het Schip, Anticancer Agents Med. Chem. **7**, 271 (2007).
42. G.A Husseini, W.G Pitt, J. of Pharm. Sci. **98**, 795-811 (2009).
43. M. Kanamala, W. R. Wilson, M. Yang, B.D. Palmer, Z. Wu, Biomater. **85**, 152 (2016).

Influence of graphene-based materials on invertebrate and vertebrate

Zhihui Huang[1], Ning Gao[1], Jianing Xing[1], Siyi Zhang[1] and Jing Hou[1a]

[1]MOE Key Laboratory of Resources and Environmental Systems Optimization, College of Environmental Science and Engineering, North China Electric Power University, Beijing, 102206, PR China

Abstract. With the rapid increased application of graphene and related materials (GRMs) in the industrial community, its release in the environment is gradually increased. The toxic effects of GRMs has drawn our attention. We characterised the toxic effects of GRMs on invertebrates and vertebrates and the common toxic mechanism, which can be classified as physical and chemical damage. This work gives a brief understanding of the toxic effects and mechanisms of GRMs.

1 Introduction

The application of Graphene and related materials (GRMs) including graphene, reduced graphene, graphene oxide and some other related hybrids have been increased rapidly in recent decades. In medicine, GRMs have been proposed for many potential applications like antibacterial agents, drug carriers, and cancer treatment, etc.[1-3]. Thus, the chances of direct contact with human body are greatly increased. At the same time, the negative effects of GRMs have been constantly reported. Therefore, we summarized the characteristics and general patterns of toxic effects of GRMs on invertebrates and vertebrates. Meanwhile, we also gave a classifiable exposition of its toxicity mechanisms. This work contributes to the environmental risk assessment of GRMs.

2 Toxic effects of GRMs on invertebrates

D. magna is commonly used invertebrate model species for evaluating the aquatic toxicity. Some of the toxic effects of GRMs on invertebrates and vertebrates are shown in table 1. Oxidative stress, growth inhibition, reproductive toxicity, neurotoxicity, the regenerative capacity damage and some other physiological impacts were the common negative effects of GRMs on invertebrates[4-7]. Generally speaking, the effects on invertebrates are time-dependent. For example, the ROS levels of *D. magna* increase apparently with exposure time as reported by Lv et al. When exposed to 5 mg/L and 20 mg/L GO, the fluorescence intensity of *D. magna* at 24 h was significantly lower than that at 48 h[8]. The concentration of GRMs in the actual environmental is usually lower than its acute lethal concentration of GRMs on organisms, so toxicities caused by low concentration cannot be ignored. The reproductive toxicity of GO on *C. elegans* reported by Chatterjee et al. addressed that GO exposure at 5 mg/L activated *POP-1* and then suppressed the expression of *ELG-5* (gene) through wingless type-MAPK crosstalk pathway, consequently leading to reproductive failure[4]. The reproduction of organisms is an important factor for the stability of ecosystems, so the reproductive toxicity of GRMs should not be neglect. In addition, there is a study on invertebrates found that GRMs with lower C/O atoms ration are more toxic, which may be due to the decrease of O-functional groups leading to the decreased activity[9]. This is of reference significance for the selection of GRMs with low toxicity.

3 Toxic effects of GRMs on vertebrates

D. rerio is a fluently used model organism in toxicological studies. The impact of GRMs on vertebrate can better reflect the potential impact on human to some extent than other organisms because vertebrates are higher grade animals[10, 11]. The increase of ROS levels, cell apoptotic, necrotic, pericardial edema, malformation and tissue damage were all negative effects of GRMs that have been reported on zebrafish or its larva[12-14]. Besides, zebrafish larvae exposed to ultra-trace GO (0.01 mg/L) induced a loss of more than 90% of dopamine neurons, a 69% to 522% increase in lewy bodies and other parkinson's disease-like symptoms[15]. These are all symptoms that may occur in human beings. Meanwhile, there was report said brain might be one of the primary targets of graphene[16]. There were also concentration-

[a] Corresponding author: houjing@ncepu.edu.cn

https://doi.org/10.1051/e3sconf/202129001030

dependent effects of GRMs was reported. Immunoreactivity studies found that interleukin-1 and interleukin-6 (proinflammatory cytokines) increased with GO concentration after exposure of zebrafish to 1, 5, 10 and 50 mg/L GO for 14 days[17]. These two kind of cytok-

Table 1. Toxic effects of GRMs on different invertebrates and vertebrates.

GRMs	Species	Exposure condition	Efftcts	Refe-rence
GO	*D. magna*	Exposed to 0, 0.01, 0.1, 1.0, 10, 50 mg/L. The mediums were renewed every 48 h	21 d LC_{50}: 3.3 mg/L; ROS: 136% of the control; SOD and CAT activities were increased	[18]
GO-HA	*D. magna*	Three concentrations of HA (5, 10, 25 mg/L)	21 d LC_{50}: 9.7 mg/L; Reduced the time to produce the first brood; Production of neonates increased; SOD and CAT activities were decreased with increasing concentrations of HA.	[18]
GO	*A. salina*	10 instar nauplii (6–24 h old) were placed per well containing 2 mL (24-well plates) of GO solutions at 24°C	LC_{50}: 489.30 ± 19.41 mg/L	[19]
GO	*D. magna*	Exposed to 0, 0.1, 1.0, 10 and 100 mg/L GO at 20 ± 2°C	EC_{50}: 0.58 (0.35–2.07) mg/L	[20]
GO	*A. salina*	Exposed to GO at 20 ± 2°C for 96 h with daily renewal of the test solutions.	EC_{50}: > 100 mg/L	[20]
GO	*C. sancticaroli.*	Exposed to test solution at 23 to 26°C in a 16/8 light/dark cycle and mild aeration	EC50: >100 mg/L	[20]
GO	*D. magna*	10 individuals exposed to 0, 1, 5, 10, 20, and 50 mg/L GO at 25 ± 2°C with natural light/dark cycle	Immobility and mortality displayed dose- and time-dependent; 72-h LC_{50}: 45.4 mg/L; GO filled most of the gut after 12 h of exposure	[8]
GO	*A. salina*	The cysts exposed to GO at 28 °C using 24-well plates with shaking	Dose-dependent decreases in hatching rates and swimming speed; Dose-dependent increases in mortality rates; MDA content, ROS and T-AOC increased	[21]
GO	*C. dubia*	Chronic assay: exposed to sub-lethal concentrations of 0, 0.05, 0.1, 0.2, 0.4 and 0.8 mg/L GO with renewal of exposure medium every 48 h	48 h-EC_{50}: 1.25 mg/L; ROS increased in all concentrations; Significantly decreased in the filtration and ingestion rates; GO was accumulated mainly on C. dubia digestive tract.	[22]
GO	*D. rerio embryos*	Exposed to GO (1, 10 and 100 μg/L) from 2.5 hpf to 7 dpf	Increased malformation rates; Changed the levels of proteins associated with the cytoskeleton/muscle system and mitochondrial respiratory chain; 100 μg/L GO decreased the GSH content	[23]
GO	*D. rerio embryos*	Exposed to 0, 5, 25, 50 mg/mL GO at 28 ± 5°C	Hatching rates slightly decreased at 5 mg/mL, and slightly increased at higher concentration (25, 50 mg/mL)	[24]
GO	*D. rerio embryos*	Exposed to 0, 1, 10, and 100 mg/L GO for 96 h	Larvae exposed to 100 mg/L GO were significantly shorter; The exposure to GO did not change the CAT or GST activity	[25]
bw-GO	*D. rerio embryos*	Exposed to 100 mg/L bw-GO for 96 h	Exposed to 100 mg/L bw-GO were approximately 3% bigger than those exposed to 100 mg/L GO; A 32% decrease in the GST activity	[25]
GO	*D. rerio*	Exposed to GO for 24 or 72 h in the short-term assays and for 14 days in the chronic assay	Did not reveal DNA damage; Induced significant ROS production; Caused some histological disturbances	[12]

Caption: Graphene oxide (GO); GO-humic acid (GO-HA); Base washed GO (bw-GO)

ines also play a role of immune regulation in human body, so zebrafish is a good choice for the speculation of the effects of GRMs on human being because of its high homology with human genes.

4 Toxic mechanisms of GRMs

The toxic mechanism of GRMs can be classified as physical and chemical influences. Common physical damage involves mechanical cutting of membrane, extraction of phospholipids, adsorption of extracellular nutrients and shadowing effect on photosynthetic organisms[26-29]. The chemical mechanisms are mainly chemical interaction between materials and organism subsequently causing oxidative stress, metabolic disorders, DNA damage, inflammatory reactions, apoptosis, and necrosis, etc.[12, 19, 30]. Generally, GRMs adsorb on living organisms due to electrostatic attraction, other intermolecular forces or random contact. They enter cells through physical interaction and cause various physiological reactions due to chemical influences.

https://doi.org/10.1051/e3sconf/202129001030

Most GRMs have sharp or rough edges, which allow them penetrate cell membrane or cell wall easily, resulting in cellular contents flow out or organelle destruction[31, 32]. Both experimental and characterization results confirmed that GO with vertically aligned has stronger antibacterial properties than the random array, which indirectly illustrates that physical film damage is one of the toxicity mechanisms[33]. Another important physical mechanism is phospholipid extraction, which had been testified though both experiment and computer simulation. Phospholipid extraction is the result of the realignment of phospholipid molecules on the nanosheets caused by the special two-dimensional structure of graphene nanosheets and the hydrophobicity of phospholipid molecules[27]. The other two common physical damage mechanisms are the shadowing effect on photosynthetic organisms and the adsorption of extracellular nutrients result from intermolecular forces. Many studies have shown that GRMs could cause a shadowing effect on algae and further reduce its photosynthesis and chlorophyll level resulting the death of cells[29, 32, 34]. Lack of essential elements will have adverse effect on cell growth. Research of yu et al. showed that GO could adsorb extracellular iron of unicellular organisms and destroy metabolism where iron is necessary causing the cell growth and metabolism inhibition due to iron deficiency[28].

GRMs as foreign substances can react with cellular biomolecules, trigger the immune response and cause oxidative stress. To be specific, graphene, rGO, hydrogenated graphene and many other GRMs have been confirmed to replace the role of metal ions such as Fe^{2+} to catalyze a Fenton-like reactions, schematic of which can be seen in figure1A and 1B[35-37]. GRMs act as Fenton-actives to catalyze the H_2O_2 to form ·OH or other radials, which increase the ROS level. Early study has shown that several graphene-based nanomaterials cause glutathione deactivation under extracellular aerobic conditions[38]. Glutathione is a vital intracellular antioxidant, and if it is inactivated, the ROS level will increase significantly. In addition, Li et al. found that GRMs can extract respiratory chain electrons, causing ATP generation to be blocked and thus dying from energy depletion[39]. Whereas, these are merely the biomolecules we have found that interact with GRMs. Because of the redox activity and a variety of functional groups of GRMs and adsorption effect resulted from electrostatic or intermolecular forces, it may also affect the function of other substances in the cells. In short, the chemical interaction between GRMs and biomolecules was the main chemical mechanism.

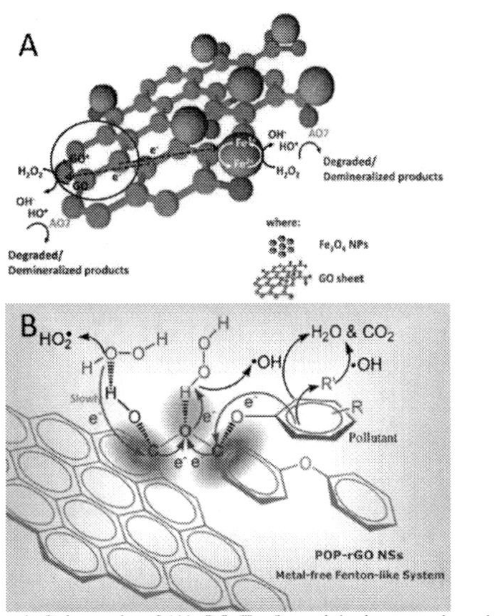

Figure 1. Schematic of (A) GO-Fe_3O_4 and 4-phenoxyphenol-functionalized (B) rGO catalyzed the metal-free Fenton-like reaction. Reuse with permission from reference[40]and[41].

5 Conclusions

Oxidative stress, growth inhibition, membrane damage, cell apoptotic, necrotic and tissue damage were the most fluently reported toxic effects of GRMs. The harmful effects of graphene-based products should be assessed before they are widely used. At present, researches about the effects of GRMs at low-concentration and long-time exposure is not adequate. The main toxic mechanisms of GRMs are the physical and chemical interaction between materials and organisms. There are usually both physical damage and chemical damage in every negative effect. This work is of reference for the research of the GRMs toxicology.

References

1. M. Yousefi, M. Dadashpour, M. Hejazi, M. Hasanzadeh, B. Behnam, M. De La Guardia, N. Shadjou, A. Mokhtarzadeh, Materials Science & Engineering C-Materials for Biological Applications **74**, 568 (2017)
2. X. Yang, X. Zhang, Y. Ma, Y. Huang, Y. Wang, Y. Chen, J Mater Chem **19**, 2710 (2009)
3. Z. Liu, J.T. Robinson, X. Sun, H. Dai, JACS **130**, 10876 (2008)
4. N. Chatterjee, Y. Kim, J. Yang, C.P. Roca, S.W. Joo, J. Choi, Nanotoxicology **11**, 76 (2017)
5. M. Qu, Y. Li, Q. Wu, Y. Xia, D. Wang, Nanotoxicology **11**, 520 (2017)
6. L. De Marchi, V. Neto, C. Pretti, E. Figueira, L. Brambilla, M.J. Rodriguez-Douton, F. Rossella, M. Tommasini, C. Furtado, A.M.V.M. Soares, et al.,

Comparative Biochemistry and Physiology C-Toxicology & Pharmacology **193**, 50 (2017)
7. A.L. Fernandes, M.E. Josende, J.P. Nascimento, A.P. Santos, S.K. Sahoo, F.M. Rodrigues Da Silva, Jr., L.A. Romano, C.A. Furtado, W. Wasielesky, J.M. Monserrat, et al., Toxicology Research **6**, 205 (2017)
8. X. Lv, Y. Yang, Y. Tao, Y. Jiang, B. Chen, X. Zhu, Z. Cai, B. Li, Environ Pollut **234**, 953 (2018)
9. Y. Liu, W. Han, Z. Xu, W. Fan, W. Peng, S. Luo, Environ Pollut **237**, 218 (2018)
10. X. Xie, J.L. Ross, J.K. Cowell, Y. Teng, Future Med Chem **7**, 1395 (2015)
11. B.C. Das, L. Mccormick, P. Thapa, R. Karki, T. Evans, Future Med Chem **5**, 2103 (2013)
12. J.P. Souza, J.F. Baretta, F. Santos, I.M.M. Paino, V. Zucolotto, Aquat Toxicol **186**, 11 (2017)
13. L. Mu, Y. Gao, X. Hu, Environ Sci Technol **50**, 8530 (2016)
14. Y. Chen, C. Ren, S. Ouyang, X. Hu, Q. Zhou, Environ Sci Technol **49**, 10147 (2015)
15. C. Ren, X. Hu, X. Li, Q. Zhou, Biomaterials **93**, 83 (2016)
16. A.L. Fernandes, J.P. Nascimento, A.P. Santos, C.A. Furtado, L.A. Romano, C.E. Da Rosa, J.M. Monserrat, J. Ventura-Lima, Chemosphere **210**, 458 (2018)
17. M. Chen, J. Yin, Y. Liang, S. Yuan, F. Wang, M. Song, H. Wang, Aquat Toxicol **174**, 54 (2016)
18. Y. Zhang, T. Meng, L. Shi, X. Guo, X. Si, R. Yang, X. Quan, Sci Total Environ **649**, 163 (2019)
19. J. Lu, X. Zhu, S. Tian, X. Lv, Z. Chen, Y. Jiang, X. Liao, Z. Cai, B. Chen, Chemosphere **211**, 390 (2018)
20. V.L. Castro, Z. Clemente, C. Jonsson, M. Silva, J.H. Vallim, A.M. Zigiotto De Medeiros, D.S.T. Martinez, Environ Toxicol Chem **37**, 1998 (2018)
21. S. Zhu, F. Luo, W. Chen, B. Zhu, G. Wang, Sci Total Environ **595**, 101 (2017)
22. J.P. Souza, F.P. Venturini, F. Santos, V. Zucolotto, Chemosphere **190**, 218 (2018)
23. W. Zou, Q. Zhou, X. Zhang, L. Mu, X. Hu, Ecotoxicol Environ Saf **159**, 221 (2018)
24. C.-J. Lu, X.-F. Jiang, M. Junaid, Y.-B. Ma, P.-P. Jia, H.-B. Wang, D.-S. Pei, Chemosphere **184**, 795 (2017)
25. Z. Clemente, V.L.S.S. Castro, L.S. Franqui, C.A. Silva, D.S.T. Martinez, Environ Pollut **225**, 118 (2017)
26. C.H. Deng, J.L. Gong, G.M. Zeng, P. Zhang, B. Song, X.G. Zhang, H.Y. Liu, S.Y. Huan, Chemosphere **184**, 347 (2017)
27. Y. Tu, M. Lv, P. Xiu, T. Huynh, M. Zhang, M. Castelli, Z. Liu, Q. Huang, C. Fan, H. Fang, et al., Nature Nanotechnology **8**, 594 (2013)
28. Q. Yu, B. Zhang, J. Li, T. Du, X. Yi, M. Li, W. Chen, P.J.J. Alvarez, Nanotoxicology **11**, 1102 (2017)
29. M. Garacci, M. Barret, F. Mouchet, C. Sarrieu, P. Lonchambon, E. Flahaut, L. Gauthier, J. Silvestre, E. Pinelli, Carbon **113**, 139 (2017)
30. J. Chen, S. Li, J. Luo, Y. Zhang, W. Ding, Journal of Nanoscience and Nanotechnology **17**, 186 (2017)
31. X. Jia, M. Hofmann, V. Meunier, B.G. Sumpter, J. Delgado, J.M. Romo-Herrera, H. Son, Y.P. Hsieh, A. Reina, J. Kong, et al., Science **323**, 1701 (2009)
32. P.F.M. Nogueira, D. Nakabayashi, V. Zucolotto, Aquat Toxicol **166**, 29 (2015)
33. X. Lu, X. Feng, J.R. Werber, C. Chu, I. Zucker, J.H. Kim, C.O. Osuji, M. Elimelech, Proc Natl Acad Sci U S A **114**, E9793 (2017)
34. S. Du, P. Zhang, R. Zhang, Q. Lu, L. Liu, X. Bao, H. Liu, Chemosphere **164**, 499 (2016)
35. J.C. Espinosa, S. Navalon, A. Primo, M. Moral, J. Fernandez Sanz, M. Alvaro, H. Garcia, Chemistry-a European Journal **21**, 11966 (2015)
36. J.C. Espinosa, S. Navalon, M. Alvaro, H. Garcia, Chemcatchem **8**, 2642 (2016)
37. Y. Zhao, W.F. Chen, C.F. Yuan, Z.Y. Zhu, L.F. Yan, Chinese Journal of Chemical Physics **25**, 335 (2012)
38. X. Liu, S. Sen, J. Liu, I. Kulaots, D. Geohegan, A. Kane, A.A. Puretzky, C.M. Rouleau, K.L. More, G.T.R. Palmore, et al., Small **7**, 2775 (2011)
39. J. Li, G. Wang, H. Zhu, M. Zhang, X. Zheng, Z. Di, X. Liu, X. Wang, Sci Rep **4**, (2014)
40. N.A. Zubir, C. Yacou, J. Motuzas, X. Zhang, X.S. Zhao, J.C.D. Da Costa, Chem Commun **51**, 9291 (2015)
41. L. Lyu, G. Yu, L. Zhang, C. Hu, Y. Sun, Environ Sci Technol **52**, 747 (2018)

Three basic types of fluorescence microscopy and recent improvement

Xinyi Wang[1, *, †], Yunyan Lai[2, †]

[1]Research School of Chemistry, Australia National University, Canberra, ACT, 2601, Australia
[2]Guangzhou Foreign Language School, Guang Zhou, Guang Dong, 511455, China
[†]These authors contributed equally.

Abstract. Fluorescence microscopy is one of the most used imaging modalities in molecular biology and living specimens. To increase image contrast and spatial resolution, different type of fluorescence microscopy has been developed. This review introduces three main types of fluorescence microscopy: wild-field microscopy, confocal microscopy, and total internal reflection fluorescence microscopy. The basic principles are similar but with different modifications, which also indicates their attributes and limitation. The recent improvement on these microscopies is also discussed. Some most recent techniques show advance in overcoming common fluorescence microscopy's weakness, and future perspectives are also discussed.

1 Introduction

Sir George Gabriel Stokes first described fluorescence in 1852. It is a phenomenon that organic and inorganic specimens absorb light and subsequently re-radiate light. Because some energy is lost during fluorescence, the emitted light has less energy than absorbed. Light with a longer wavelength shows less energy than light with a shorter wavelength, so the emitted light usually has a longer wavelength than absorbed, called Stokes shift. The emission and absorption are simultaneous and usually about 10^{-9} to 10^{-8} seconds. Molecules that show fluorescence are called fluorophores or fluorochromes, which have a characteristic absorbance and emission spectrum. Fluorochromes were introduced in biological investigations in the 1930s and stimulated the development of fluorescence microscopy (FM). They are considered stains that attach to visible or sub-visible structures with high specificity in their attachment targeting, such as proteins, lipids, or ions [1, 2].

Due to the different energy between emitted light and excitation light, the goal of the FM is to separate these 2 lights and only allow emitted light to reach the detector (Fig. 1). The basic function of FM is to irradiate the specimen with its characteristic band of wavelengths and then separate its characteristic emitted light from the excitation light. Because the excitation light has higher energy, it is necessary to filter out this very bright excitation light to view the relatively weak fluorescence. Optical filters generally achieve this separation. Dichroic mirror serves as the primary optical element to reflect and separate the light, but it is not perfect. To enhance the selection, excitation and emission filters are commonly used. The excitation filter ensures only the characteristic wavelength of excitation light is transmitted. The emission filter is used to block excitation light, pass emitted light, and make sure only the emission light can reach the detector. However, to create a useful image, FMy needs to obtain sufficient light; filters always cause loss of light intensity. Using a long-pass filter, it is possible to collect all the light from the emission spectrum [3]. The emitted fluorescence can be viewed by the eyes or captured electronically.

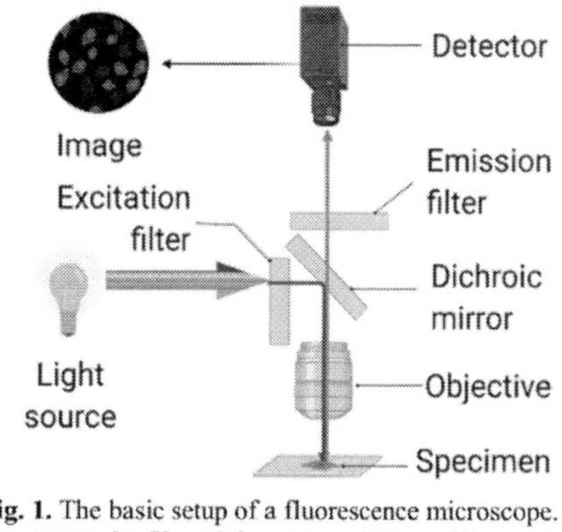

Fig. 1. The basic setup of a fluorescence microscope.
The source is filtered by an excitation filter and then reflected off the dichroic mirror toward the specimen. A dichroic mirror and emission filter fluorescence. Only the emitted light reaches the detector.

Due to not being available in traditional optical microscopy, fluorescence microscopy has become an essential tool in biology, biomedical sciences, and material science. In the fields of botany, petrology, and the semiconductor industry, autofluorescence has been thoroughly exploited. However, in biology and biomedical sciences, the situation is often complicated

Corresponding author's e-mail:[1]u6581146@anu.edu.au

https://doi.org/10.1051/e3sconf/202129001031

due to the extremely faint or bright, nonspecific autofluorescence in animal tissues and pathogens. Fluorochromes are often considered in the latter studies. By introducing fluorochromes in non-fluorescing material, identify cells and sub-microscopic cellular components have been achieved. The specificity of fluorochromes makes it possible to identify several target molecules simultaneously within the same sample by using multiple fluorescence labeling simultaneously. Some fluorescence microscopes can show the 3-dimensional interior of cells and organisms [4].

The major weakness is the limitation of the resolution. FM cannot provide spatial resolution below the specimen's diffraction limit [5]. The diffraction limit restricts optical resolution in about half the wavelength of the emitted light [6]. Therefore, the resolution of FM is fundamentally limited by the wavelength of light. In addition, the wavelength of the excitation light and the microscope optics also limits spatial resolution [3]. Its dependence on probes also limits FM. For molecular structures that are incapable of autofluorescence, suitable fluorochromes are required. Some fluorochromes show cytotoxicity and damage DNA [7]. It spends time and money to design and select suitable fluorochromes. In addition, the simultaneous use of multiple fluorescences may lead to a poorer signal-to-noise ratio (SNR). This is because the induced interactions between proximate probes can detract from the effectiveness [8]. In some cases, sample cells can be damaged by exposure to light called phototoxicity [9]. The lower wavelengths of light used for excitation are particularly damaging to cells due to their higher energy. In other cases, fluorochromes may lose their capacity to fluoresce, called photobleaching [10]. This is due to the accumulated chemical damage from the electrons excited during fluorescence. Despite this, FM is still a popular technique in live-cell observation and structure elucidation of biomolecules in tissues and cells in situ.

The basic FM is wide-field (WF) fluorescence microscopy, which is excellent for 2D images of specimens, and the entire field can be captured at once. Confocal microscopy is an advanced FM that can show 3D images with higher SNR than WF fluorescence microscopy. Total internal reflection fluorescence microscopy (TIRFM). This report will introduce these 3 major types of fluorescence microscopies and discuss their attributes and shortages how scientists try to overcome these shortages and improve their properties. The report will also discuss the most recent technique improvement in FM.

2 Wile-field fluorescence microscopy

2.1 Principle of WF fluorescence microscopy

Wide-field (WF) Fluorescence Microscopy is a basic fluorescence microscope that cell biologists commonly use. For this microscope, a parallel beam of light illuminates the whole specimen at once to excite the fluorophore. All the resulting fluorescence of specimens can be vied simultaneously, allowing simple and fast imaging. For multiple-prob specimen, all the Fluorescence can be viewed at once. Due to all parts of the specimen can be viewed at once, it allows a quick selection of fluorescent cells to image.

2.2 Limitation

The wide-field illumination and view not only focus information from the corresponding section of the specimen but also allows out-of-focus light to arrive camera, leading to low contrast and spatial resolution [4]. For thick samples, light from sample planes above and below the focal plane will also be detected. The proximity of fluorochrome also contributes to low SNR. Due to optical diffraction, each fluorochrome shows a broad fluorescence zone that can interact with nearby fluorescence and cannot be resolved. Therefore, if fluorochromes are closely associated, it will result in a bright blur rather than discrete objects. Another shortage is its limitation in 3D imaging. The WF microscope doesn't gather sufficiently complete information to allow 3D imaging due to its non-specific data collection, including out-of-focus blur [11]. The resolution of the z-axis is about 1um.

2.3 Recent improvement

However, Wide-field Fluorescence Microscopy is well suited in thin specimen imaging, like single-cell layer imaging and macroscopic structure. These specimens closely adhered to a glad surface can limit out-of-focus light arriving. By using spatially structured illumination, the resolution of WF microscopy can be increase. Gustafsson et al. [4] applied structured illumination microscopy in 3D and doubled the 3D resolution in WF microscope. Xue et al. applied Computational Miniature Mesoscope (CM2) to the wide-field microscope and successfully presented 3D imaging [12]. By using computational algorithms, they augmented the optics and expanded imaging capability. This combination enlarges miniscope systems imaging areas from lower than $1mm^2$ to $56mm^2$ and wide-field measurements depth of field to 2.5-mm and resolution to ~7-μm lateral and better than 200-μm axial.

3 Confocal Fluorescence Microscopy

3.1 Principle of confocal microscopy

Thin specimens limit WF microscope. As specimen thickness adds, out-of-focus light increases and hinders imaging. Confocal microscopy can eliminate out-of-focus light from images and allow thick specimens to be viewed with high resolution [13]. The best resolution that a confocal microscope can attain is about 200nm [14]. The major difference of confocal microscopy in the region of illumination and detection. A confocal microscope only illuminated a diffraction-limited region of the specimen at one time and only accepted a signal from that region [15]. The confocal microscope has been widely used in immunofluorescence and greatly benefits in overcoming background glare [15].

3.2 Laser scanning confocal microscopy

Laser scanning confocal microscopy (LSCM) place a pinhole conjugated to the focal plane (Fig. 2). By closing the pinhole to various extents, the out-of-focus light is blocked, and only the light from the focused point in the specimen can reach the detector. By a sequential scanning method of the excitation and detection point, fluorescence intensity across the specimen can be collected, and images can be sequentially generated [3]. The excitation light and pinhole are stationary, but the region they focus on is optically moved by an orthogonal pair of oscillating mirrors, x-axis mirror, and y-axis mirror. Therefore, out-of-focus elements cannot affect the background because they receive little illumination, and a pinhole rejects any out-of-focus light.

In addition, confocal microscopy also provides more resolution in depth. Closing down the pinhole increases the resolution in Z-axis. By assembling a series of thin specimens along the z-axis, it is possible to build a 3D image [16]. Comparing with a WF microscope, the confocal microscope can provide a 3D structure with a higher resolution, and the z-axis resolution can attain 700nm.

Moreover, by using a laser scanning system, dwell time decreases, which reduces the damage of photobleaching and photodamage. Dwell time is the time that excitation light remains in one location in the specimen. For LSCM, the laser beam continually scans across the sample, and the image is generated from discrete pixels that follow the laser scan proceeds. The dwell time is related to the image's recorded speed and pixel resolution. The example indicated by Sanderson, Smith, Parker & Bootman explains this relationship clearly: if the detector records 30 images per second, and each image has 512 horizontal pixels and 521 vertical lines, the dwell time would be 33ms/512*512=127ns. In this case, each location of the sample only needs to be illuminated for a very short time for LSCM, while, for WF microscopy, the whole sample needs to be exposed for 33ms [3]. Therefore, confocal microscopy decreases the exposure time of specimens to light and reduces the damage of photobleaching and phototoxicity.

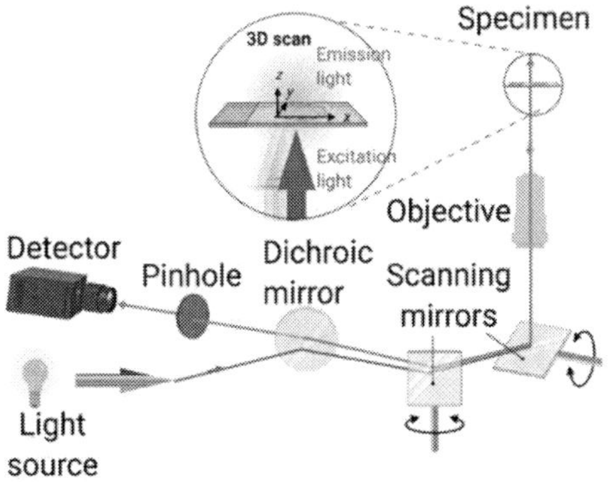

Fig. 2. The basic setup of laser scanning confocal microscopy.

Light from the laser is scanned across the specimen by the scanning mirrors. The emitted light passes through a pinhole to the detector. Size of the pinhole limits out-of-focus light to the detector.

3.3 Spinning disk confocal microscope

LSCM uses a single-point scanning system which needs much time to scan the whole specimen. To capture image at high speed, spinning disk confocal microscopy are developed [17]. Spinning disk confocal microscopy allows a multiple-point scanning system. In this instrument, the sample is both illuminated and viewed through a spinning disk with rows of pinholes arranged. As the disk spins, each pinhole on the spinning disk serves as a point source of light to scan across specimen; similarly, the emitted light also passes through a corresponding pinhole before separated by DM. Spinning disk microscopy indicates a faster image acquisition rate and lowers light requirement than LSCM. Drawbacks include the loss of resolution due to cross-talk between multiple fluorescence points and fixed pinhole size, which is optimized for a specific objective. One recent technique is ribbon scanning confocal microscope [18]. By employing resonant scanners and high-precision x, y-axis scan mirrors continuously capture data across samples, reducing time to data collection.

3.4 Limitation

However, the attributes of confocal microscope also connote shortages. Confocal microscope blocks out-of-focus light by keep pinhole small, but much in-focus light is also discarded. There is a tradeoff between light-collection efficiency and resolution. Considering the weak fluorescence of biological samples, the loss of in-focus light usually outweighs resolution benefits. The best resolution is 0.2μm on the x-axis and 0.6um on the y-axis. The x-axis resolution in confocal microscopy is worse than the WF microscope. In practice, confocal microscopy often uses wider pinholes to generate a similar x-axis resolution as the WF microscope [4]. Similarly, the high resolution of the Z-axis is achieved by closing the pinhole, which also sacrifices the level of light reaching the detector. In addition, similar to WF microscopy, confocal microscopy is also limited diffraction effects. Another consideration is the image speed. In the WF microscope, image capture speed is mainly determined by the technology of image acquisition and brightness of the image. However, with confocal microscopy, an additional factor can be the speed of the laser raster scan of the specimen. Due to inertia, the speed of scan mirrors is limited; thus, the most confocal system can only capture 1 to 10 images per second, and dwell time also increases substantially [19]. In fact, LSCM performs worse in photobleaching and phototoxicity than WF microscopy. The higher price is another shortage. Generally, confocal microscopy costs 2-7 times more than WF microscopy.

3.5 Recent improvement

To increase the resolution of confocal microscopy, Airyscan technology is applied [20]. Airyscan has a 32-channel detector array that can reassign pixels and summate images from all detectors. This system provides 1.7 times higher resolution in the x-, y- and z-axis. A re-scan confocal microscope includes an additional pair of scanning mirrors (re-scanner) which improves x-axis resolution by 1.4, but y-axis resolution shows no improvement [20].

4 Total Internal Refraction Microscopy

4.1 Principles of TIRFM

Total Internal Refraction Microscopy (TIRFM) is based on the excitation of fluorophores by evanescent wave or field when the laser beam internally refracts. Total internal refraction (TIRF) is an optical phenomenon that the light will reflect instead of refracting when the light travels from a medium of high refractive index to that of low refractive index, and the incident angle should be greater than the critical angle. This phenomenon follows Snell's Law.

In the case of TIRF, reflected light turns into an electromagnetic field at the interface, and an evanescent field forms through the media of low refractive index, which amplitude decays exponentially (Fig. 3). Therefore, the penetration depth of the evanescent field can only be around 100 nm. In a typical experimental setup, the fluorophores near the interface having the capacity of electronic transition with or close to the wavelength of the laser beam can be excited by the evanescent field.

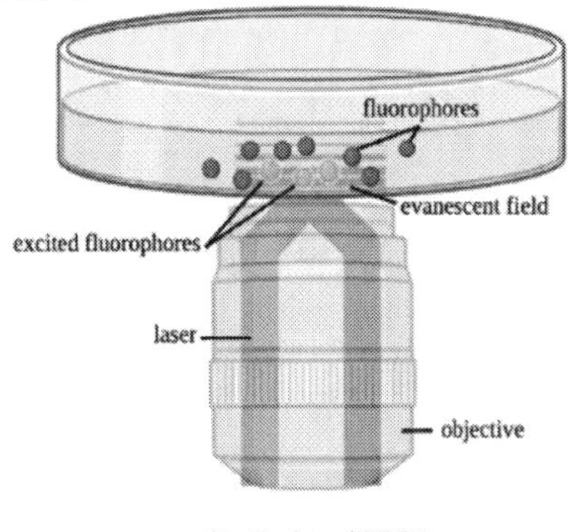

Fig. 3. Objective-based TIRFM.

The incident laser produces an evanescent field, which excites the fluorophores close to the interface and generates high SNR images.

4.2 Prism-based TIRFM and objective-type TIRFM

The prism-based method employs a prism to generate evanescent waves and collects fluorescence images with an objective lens [21]. But because of its structure, this method has a deficiency in complex operations of specimens, such as adding drugs and medium. On the other hand, the objective-type approach depends on a large numerical aperture (NA) objective to produce an evanescent field and to image [22]. Compared with a prism-based microscope, objective-type TIRFM is more prevalent in the market, for it is much easier to operate and has higher precision.

4.3 Advantages

Because of the low penetration depth of evanescent wave (usually less than 100nm), there is almost no background fluorescence, causing super z-axial resolution and high SNR. In addition, since only a small portion of the specimen is exposed to evanescent waves, the effect of toxic gas and phototoxicity drastically weakened, extending experiment time. Meanwhile, TIRF is generally accepted and only restricted by available laser beams, so it is relatively easier to perform live-cell imaging techniques like FRET. TIRF microscopy was widely adapted to the situation where specimen contains many fluorophores, and the research needs imaging, such as the investigation of cell membrane and endocytosis.

The research about protein to single sites of endocytosis was always hindered due to the lack of direct images of this process. In a study about Plant Science on clathrinid, TIRF microscopy spatially and temporally probes the process of endocytosis and helps to find out the protein resulting in endocytosis in plants [23]. In another research to inhibit the infection of HIV, combined with cryo-electron tomography (cryoET), TIRFM can image the fusion of blebs from native cell membranes as targets and pseudo virion of HIV envelope glycoprotein with or without Serinc host restriction factors. According to the images, researchers can investigate the fusion behavior and thus recognize which Serinc can inhibit HIV fusion [24].

4.4 Limitation

Because the energy of the evanescent wave decreases exponentially from the interface, the fluorescence signal is not only strictly limited near the total reflection interface (generally the interface between the glass slide and the sample, in the range around 100 nm) but also too weak to detect.

Also, the low penetration depth of the evanescent field contributes to the limitation of specimens. Specimens can only be applied to specific cell types and lie flat on the coverslip, such as epidermal cells [23]. At the same time, TIRFM relies on the specimen's operation, such as fluorescent labeling of proteins, which may change the structure and function of specimens.

4.5 Recent improvement

TIRF is used in combination with many other microscopes to compensate for its shortcomings and achieve better results.

In recent research, fast two-dimensional total internal reflection fluorescence, high-resolved structured illumination microscopy (SIM), and traction force microscopy (TFM) are combined for higher spatial and temporal imaging. 2D TIRF-SIM-TFM microscopy supplies information of cellular structure and magnitude and the direction of force produced by cells. Thus, the microscopy has >2-fold spatially and >10-fold temporally resolution of planar cellular force probing [24].

Also, the combination of fast high-resolution spinning disk and total internal reflection fluorescence microscopy using a new light path can fast produce higher temporal resolution images about the dynamic cellular processes between membrane and cytoplasm [25].

5 Conclusion

WF, confocal microscopy, and TIRFM are 3 basic types of fluorescence microscopy. Each of them shows different attributes and limits. WF is a quick and simple technique with an advantage in single-cell layer imaging, but it has relatively low SNR and a shortage in 3D imaging. By using spatially structured illumination, recent research has shown great success in WF 3D imaging. Confocal microscopy shows an advantage in 3D imaging but with long image speed. Spinning disk confocal microscopy shows reduction of time of data collection. TIRFM is an advanced FM with high SNR and relatively low photobleaching and phototoxicity but shows specimen limits. New techniques have been combined with the basic type of fluorescence microscopies and show great improvement in their drawbacks. However, some shortages still limit the imaging in all three basic FM.

The main shortage of all these 3 fluorescence microscopies is the diffraction limit which limits the resolution. New fluorescence microscopy has been developed to overcome this weakness. Based on WF, TIRF, or confocal microscope setups, super-resolution microscopy (SRM) greatly improves optical resolution. By changing the way of excitation and detection, SRM bypasses the diffraction limit to nanometre scales and allows imaging of subcellular organization. SRM makes it possible to view cellular structure in detail of electron microscopy level and keep sample activity. SRM can also view 3D structure details and show single-molecule localization, for example, how proteins are distributed and organized in cells.

Although extensively adopted for a long time in cellular biology, FM still has challenges. High cost, type of fluorochrome, and high noise level, and other limitations are worth exploring. Because different specimens need specific fluorochrome, we propose investigating how to get or design proper fluorochrome to specimen efficiently. Meanwhile, noise level always deteriorates results of algorithms, which is shown in the combination of CMOS and SMLM. In past years, the most adopted way to overcome limitations is to combine different microscopes or cameras to achieve better results. In this way, prospective research can try different combinations for greater images. For example, researchers can combine low-cost with other microscopes, producing low-cost and high-resolution images.

Reference

[1] Giepmans, B. N., Adams, S. R., Ellisman, M. H., & Tsien, R. Y. (2006). The fluorescent toolbox for assessing protein location and function. science, 312(5771), 217-224.

[2] Palmer, A. E., & Tsien, R. Y. (2006). Measuring calcium signaling using genetically targetable fluorescent indicators. Nature protocols, 1(3), 1057.

[3] Sanderson, M. J., Smith, I., Parker, I., & Bootman, M. D. (2014). Fluorescence microscopy. Cold Spring Harbor Protocols, 2014(10), pdb-top071795.

[4] Gustafsson, M. G., Shao, L., Carlton, P. M., Wang, C. R., Golubovskaya, I. N., Cande, W. Z., ... & Sedat, J. W. (2008). Three-dimensional resolution doubling in wide-field fluorescence microscopy by structured illumination. Biophysical journal, 94(12), 4957-4970.

[5] Hell, S. W. (2003). Toward fluorescence nanoscopy. Nature biotechnology, 21(11), 1347-1355.

[6] Pawley, J. (Ed.). (2006). Handbook of biological confocal microscopy (Vol. 236). Springer Science & Business Media.

[7] Zhao H, Traganos F, Dobrucki J, Wlodkowic D, Darzynkiewicz Z. Induction of DNA damage response by the supravital probes of nucleic acids. Cytom Part A. 2009;75A:510–9. https://doi.org/10.1002/cyto.a.20727.

[8] Simeonov, A., & Davis, M. I. (2018). Interference with fluorescence and absorbance. Assay Guidance Manual [Internet].

[9] Wäldchen, S., Lehmann, J., Klein, T., Van De Linde, S., & Sauer, M. (2015). Light-induced cell damage in live-cell super-resolution microscopy. Scientific reports, 5(1), 1-12.

[10] Becker, P.L. Quantitative fluorescence measurements. In Fluorescence Imaging Spectroscopy and Microscopy; Wang, X.F., Herman, B., Eds.; John Wiley & Sons, Inc.: New York, 1996; 1–29.

[11] Agard, D. A., Hiraoka, Y., Shaw, P., & Sedat, J. W. (1989). Fluorescence microscopy in three dimensions. Methods in cell biology, 30, 353-377.

[12] Xue, Y., Davison, I. G., Boas, D. A., & Tian, L. (2020). Single-shot 3D wide-field fluorescence imaging with a Computational Miniature Mesoscope. Science advances, 6(43), eabb7508.

[13] Inoué, S. (2006). Foundations of confocal scanned imaging in light microscopy. In Handbook of biological confocal microscopy (pp. 1-19). Springer, Boston, MA.

[14] Semwogerere, D., & Weeks, E. R. (2005). Confocal microscopy. Encyclopedia of biomaterials and biomedical engineering, 23, 1-10.

[15] Amos, W. B., White, J. G., & Fordham, M. (1987). Use of confocal imaging in the study of biological structures. Applied optics, 26(16), 3239-3243.

[16] Brakenhoff, G. J., Voort, H. V. D., Spronsen, E. V., & Nanninga, N. (1986). Three-Dimensional Imaging by Confocal Scanning Fluorescence Microscopy a. Annals of the New York Academy of Sciences, 483(1), 405-415.

[17] Petráň, M., Hadravský, M., Egger, M. D., & Galambos, R. O. B. E. R. T. (1968). Tandem-scanning reflected-light microscope. JOSA, 58(5), 661-664.

[18] Watson, A. M., Rose, A. H., Gibson, G. A., Gardner, C. L., Sun, C., Reed, D. S., ... & Watkins, S. C. (2017). Ribbon scanning confocal for high-speed high-resolution volume imaging of brain. PloS one, 12(7), e0180486.

[19] Tsien, R.Y.; Bacskai, B.J. Video-rate confocal microscopy. In Handbook of Biological Confocal Microscopy, 2nd Ed.; Pawley, J.B., Ed.; Plenum Press: New York, 1995; 459–478.

[20] Korobchevskaya, K., Lagerholm, B. C., Colin-York, H., & Fritzsche, M. (2017, September). Exploring the potential of airyscan microscopy for live cell imaging. In Photonics (Vol. 4, No. 3, p. 41). Multidisciplinary Digital Publishing Institute.

[21] Mandracchia, B., Hua, X., Guo, C., Son, J., Urner, T., & Jia, S. (2020). Fast and accurate sCMOS noise correction for fluorescence microscopy. Nature Communications, 11(1), 1–12. https://doi.org/10.1038/s41467-019-13841-8

[22] Moreno, I., Araiza, J. J., & Avendano-Alejo, M. (2005). Thin-film spatial filters. Optics Letters, 30(8), 914–916.

[23] Johnson, A., & Vert, G. (2017). Single event resolution of plant plasma membrane protein endocytosis by TIRF microscopy. Frontiers in Plant Science, 8, 612. https://doi.org/10.3389/fpls.2017.00612

[24] Barbieri, L., Colin-York, H., Korobchevskaya, K., Li, D., Wolfson, D. L., Karedla, N., … Fritzsche, M. (2021). Two-dimensional TIRF-SIM–traction force microscopy (2D TIRF-SIM-TFM). Nature Communications, 12(1), 1–14. https://doi.org/10.1038/s41467-021-22377-9

[25] Zobiak, B., & Failla, A. V. (2018). Advanced spinning disk-TIRF microscopy for faster imaging of the cell interior and the plasma membrane. Journal of Microscopy, 269(3), 282–290. https://doi.org/10.1111/jmi.12626

Experimental Study on the Catalytic Reduction of NO_X by Fe and Ce During the Coke Combustion Process

Qi Wang[1,2], SongYang[2,3], Shoujun Liu[2,3], Lili Chen[1,2], Tingting Jiao[1,2] and Ju Shangguan[1,2,*]

[1] State Key Laboratory of Clean and Efficient Coal Utilization, Taiyuan University of Technology, Taiyuan 030024, Shanxi, China
[2] Shanxi Engineering Center of Civil Clean Fuel, Taiyuan University of Technology, Taiyuan 030024, Shanxi, China
[3] College of Chemistry and Chemical Engineering, Taiyuan University of Technology, Taiyuan 030024, Shanxi, China

Abstract. NO_X emissions during the combustion of coal would cause serious pollution to the atmosphere. It is particularly important to reduce NO_X pollution during the combustion process by replacing loose coal with clean coke. The use of clean coke for civil is one of the main ways to reduce NO_X emissions, which is mainly formed by the mixed dry distillation of metal additives and coal. The coke samples with different proportions of metal additives were prepared by physical mixing method, and the NO_X emission law of clean coke at different temperatures was studied in a high-temperature tube furnace reactor. The results show that the addition of Fe alone has a significant effect on the control of NO_X emissions, and the co-doping with Ce can further promote the reduction of NO_X, so as to achieve the final low emission of NO_X. While the reaction temperature is 1000 ℃ and the addition ratio is 2:1, the NO_X reduction rate is 73%. It has important practical value and scientific significance for the clean utilization of coal and the treatment of NO_X in the atmosphere.

1 Introduction

As a primary energy source, direct combustion of coal is the main way of utilization. However, the contribution of coal combustion to the emission of NO_X pollutants is huge[1]. The emitted NO_X can form acid rain, photochemical smog, also cause the greenhouse effect[2, 3], and ultimately endanger human health. Therefore, it is imperative to control NO_X emissions during coal combustion. But the combustion of pulverized coal is very complicated, and the combustion of coke is the most influential link, therefore. Many studies have focused on the coke combustion stage[4]. Transition metal oxides also show excellent catalytic performance and can effectively promote the migration of fuel nitrogen to the gas phase[5]. Therefore, it can be boldly guessed that the metal additives also have a catalytic effect on the control of NO_X emissions during the combustion process. Zhao et al.[6] found that both Na and Fe loading reduced NO_X emissions during coke combustion. Zhong et al.[7] believed that iron can effectively catalyse the reduction of NO_X during the process of high temperature combustion of coal char and significantly reduce NO_X emissions. Literature[8] believed that different metals have different catalytic activity factors, and the catalytic activity of iron depends on temperature. Ce can change the reducibility of the catalyst surface to improve the catalytic activity. Current research is still focused on the effect of single-metal additives on combustion denitrification, and there is a lack of research on the combined effect of composite metal additives on the combustion process. Therefore, this work has studied the effect of transition metals and rare earth metals in the combustion of co-catalysis NO_X reduction, and obtained the corresponding laws.

2 Materials and Methods

2.1 Coal sample modification

Select coking coal as the experimental sample. Firstly, because of the complex mineral composition of coal, in order to simplify the research, the demineralized coal was selected as the research object. In this paper, HCl-HF acid pickling was used to remove minerals, soak in a water bath at 60 °C for 4 hours, and rinse with deionized water until Cl^- is not detected, and demineralized coal is prepared. Next the coal was baked in an oven at 105 °C for about 6 hours. Then it is crushed and sieved to select a coal sample with a particle size of about 0.2mm. Afterwards, 1%, 2%, 3% and 4% Fe additives were loaded on the demineralized coal using a physical mixing method. Make the Fe additives and demineralized coal as homogeneous as possible.

2.2 Coke preparation

The pyrolysis was completed in a high-temperature tube furnace, the sample was heated at a rate of 10 ℃/min to 1000 ℃, and was subjected to high-temperature pyrolysis in an Ar atmosphere of 50 mL/min for 1 h, and then naturally cooled to room temperature. Finally, the coke is sieved to 40~80 mesh. The industrial analysis and

*Corresponding author: shanggj62@163.com

elemental analysis of raw coal char (RC) and demineralized coal char (DC) are shown in Table 1. The nitrogen content of demineralized coke is higher than that of coke. This may because the minerals in coal promote the migration and emission of nitrogen-containing substances to gas during coal pyrolysis.

Table 1. The proximate analysis and ultimate analysis of raw char (RC) and demineralized char (DC)

	Proximate analysis (*wt* %)				Ultimate analysis (*wt* %)				
	M_{ad}	A_d	V_{daf}	FC_{ad}	C_{daf}	O_{daf}	H_{daf}	N_{daf}	S_{daf}
YC	0.15	9.99	1.78	88.08	79.09	19.12	0.23	1.16	0.40
DC	0.20	0.49	1.69	95.72	89.45	8.77	0.20	1.26	0.32

2.3 Experimental process

The combustion experiment was carried out in a tube furnace. Before each reaction, weigh about 0.5g of the coke, and place it in the middle constant temperature zone of the reaction tube, The Ar gas is flowed at a flow rate of 50 mL/min, and purge for 5 minutes to purge the air in the pipe, heat up to the combustion temperature required for the experiment, after the constant temperature is stable for 10 minutes, switch the O_2/Ar mixture, adjust the flow rate to 200mL/min, and the tail gas end is connected to a flue gas analyzer (KANE9506) for online detection of NO_X. After the experiment is completed, coal ash is collected, and XRD is used to characterize the crystal form of additives in the ash sample. Each experiment was repeated twice, and when the error of parallel experiment data was less than 2%, the average value was taken.

The amount of NO_X produced and the decrease ratio are shown in formulas (1) and (2).

$$m_{NO_X} = Q \times 10^{-6} \int_0^t C_{NO_X}(t)dt \qquad (1)$$

m_{NO_X} — the amount of NO_X produced by coke combustion, mg

Q — the volume flow of air, mL/min

C_{NO_X} — the mass concentration of NO_X in flue gas at time t, mg/m^3

$$\varphi = \frac{W_{NO_X(\text{coke-0})} - W_{NO_X(\text{coke-1})}}{W_{NO_X(\text{coke-0})}} \qquad (2)$$

φ — the reduction ratio of NO_X of clean coke, %

$W_{NO_X(\text{coke-0})}$ — the amount of NO_X produced during the combustion of coke, mg

$W_{NO_X(\text{coke-1})}$ — the amount of NO_X produced during the combustion of clean coke, mg

3 Results & Discussion

3.1 Effect of iron loading ratio on NO_X emissions from demineralized coke combustion

When the demineralized coke with different iron loading ratios is burned separately, the NOx emission and decrease ratio are shown in Figure 1 and Figure 2, respectively. It can be seen from the Figure 1 that NOX emissions are negatively correlated with the temperature range, that is, as the temperature increases, the emissions decrease. It shows that increasing the temperature is beneficial to reduce NOX emissions. However, the higher the load, the better NOX reduction. When the load ratio exceeds 4%, the NOX emissions will be higher than 3% at the same temperature. And it can be seen from that the best temperature is 900 ℃, the emission of NOx of was calculated to be 1.18 mg.

In order to express the decrease of NOX by loading additives more vividly, compare the release of NOX in various proportions with the demineralized coke without additives. The decrease ratio at each temperature is shown in Figure 2. The calculation can be obtained at 900 ℃, 3% with the addition of Fe additives, the NO_X emission is 1.18mg, and the NO_X decrease rate is 60.5%. Under these conditions, the NO_X emission is the least. It can be inferred that a 3% Fe load has the best effect on catalytic reduction of NO_X release. Therefore, in this experiment, clean coke with a load of 3% Fe_2O_3 co-pyrolysis releases the least amount of NOx, which proves that this load of additives is most conducive to reducing the release of NO_X. Meanwhile, the catalytic effect of Fe in the NO-coke reaction is proved[9], that is, the addition of Fe_2O_3 is beneficial to promote the reduction of NO_X, thereby reducing the release of NO_X.

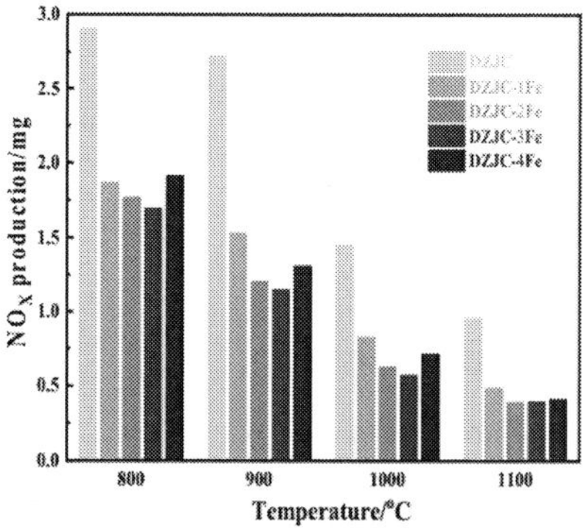

Figure 1. NOx production with different iron loading ratio

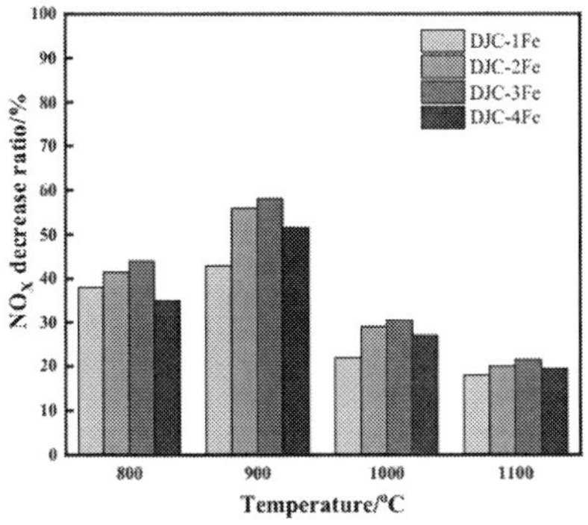

Figure 2. NOx decrease ratio with different iron loading ratios

3.2 Effect of Fe-Ce composite ratio on NOx emissions from demineralized coke combustion

The NOx emissions and decrease ratio during the combustion process with different Fe-Ce composite ratios are shown in Figure 3 and Figure 4, respectively.

It can be seen from the figure that the effect of single additive CeO_2 loading is far less than that of single Fe_2O_3 loading. Compared with DC-3Fe and DC-3Ce, the NOx emission of coke combustion under the composite auxiliary is reduced However, when the two metals are combined loading, the release of NOx at (800-1100 ℃) is effectively reduced. When the recombination ratio is less than 2, the NOx reduction rate increases with the increase of CeO_2 load, but a downward trend appears when the temperature rises to 1100 ℃. When the CeO_2 to Fe_2O_3 load ratio exceeds 2, the reduction of NOx is directly related to the temperature, and this trend is proportional to temperature. The reduction of NOx emissions from coke combustion after the composite addition of Fe and Ce oxide additives is more obvious, and the NOx reduction rate of DJC-2FeCe is the highest, and the effect is the most significant. Therefore, the experimental results show that the doping of cerium is not always beneficial, and a small amount of doping of cerium is beneficial to further reduce combustion NOx release on the basis of supporting iron-based additives.

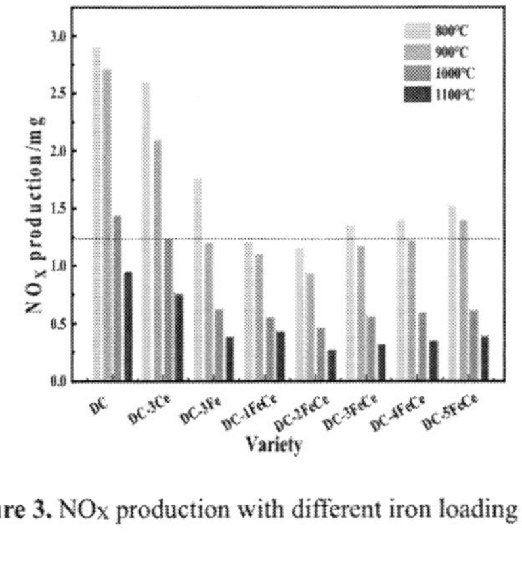

Figure 3. NOx production with different iron loading ratios

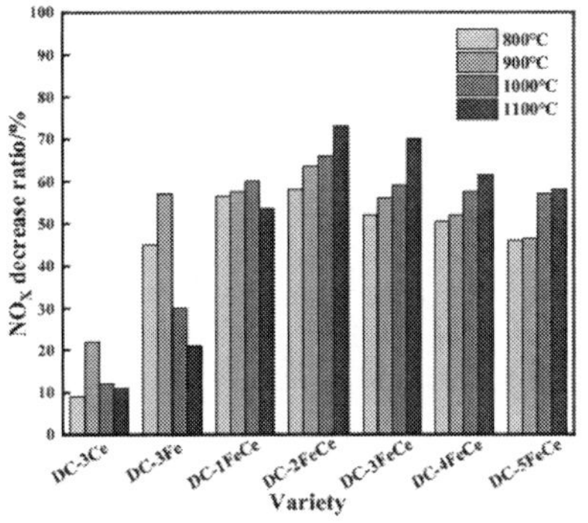

Figure 4. NOx decrease ratio with different iron loading ratios

3.3 Effect of temperature on NOx emissions during the combustion of demineralized coke with the optimum ratio of Fe-Ce

The NOx emissions and decrease ratio during the combustion process of demineralized coke with Fe-Ce composite ratio of 2:1 are shown in Figure 5.

Obviously, during the combustion process, the NOx emissions of demineralized coke with the optimal load ratio gradually decreases, which corresponds to a linear increase in the NOx reduction rate. This shows that temperature is a key factor affecting combustion, and high temperature can significantly reduce NOx emissions during combustion. When the combustion temperature increases to 1100 ℃, the amount of NOx emitted by the coke is 0.28mg, and NOx reaches 73%. This indicates that high temperature is beneficial to reduce the release of NOx in coke. The literature[10] believes that as the high temperature time passes, the coke enters the burn-out stage, and the O concentration decreases. At this time, although the coke NOx is continuously formed, it decomposes quickly at the same time, and part of the NOx that has been generated is reduced by metal decomposed into N or N_2, Ce doping improves the reducibility of mixed oxides, resulting in a reduction in overall NOx emissions.

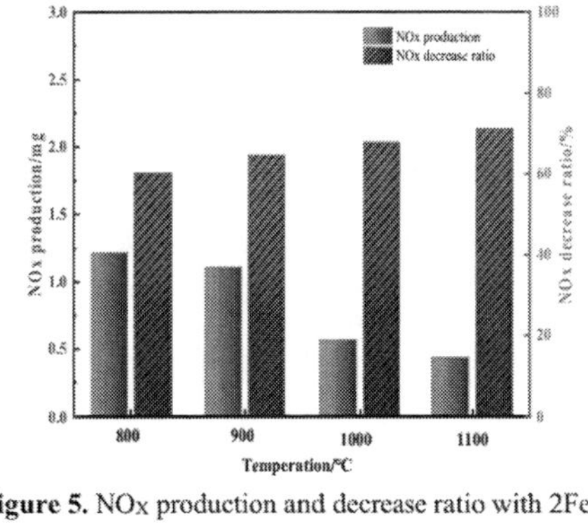

Figure 5. NOx production and decrease ratio with 2FeCe

3.4 Phase morphology analysis of loading additives before and after coke combustion

The XRD analysis of the auxiliary agent in the demineralized coke and in the ash after combustion are shown in Figure 6 and Figure 7, respectively.

As shown in the Figure 6, XRD characterization analysis was carried out on the coke to determine the phase morphology of the auxiliary agent, the coke doped with Fe has obvious Fe_2O_3 and Fe diffraction peaks, and the intensity of the diffraction peaks is strong, which may be the reason for the low catalytic efficiency of crystalline iron oxide. After the co-doping of Ce, the peak strength of iron diffraction peak of each load auxiliary sample is still gradually weakening, which shows that the addition of Ce can effectively change the crystallinity of iron. Researches have shown[11] that when metal oxides interact, cerium oxide can promote the storage and release of oxygen improves the mobility of oxygen and forms surface and structural oxygen vacancies, thereby improving the redox properties of the catalyst[12].

In order to clarify the phase morphology of the composite additives after combustion, the burned ash was subjected to XRD characterization, the result is shown in Figure 7. Through the analysis of the ash content, it is found that the burned ash contains almost no other components, except for the added compound additives. Only the phase structures of iron and cerium are obtained by the characterization, and they are in their respective oxide forms after combustion, namely Fe_2O_3, CeO_2, and there are almost no elemental Fe compounds in the phase. This is related to the combustion reaction is a strong oxidizing system, and the composite assistant as a catalyst is finally oxidized under this oxidizing condition.

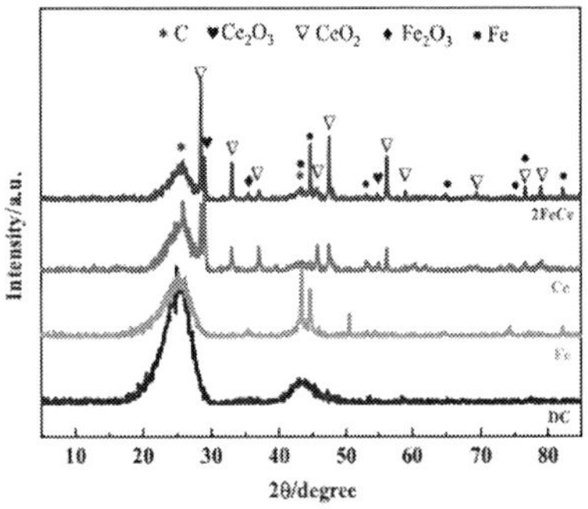

Figure 6. Fe-Ce phase transition law before combustion

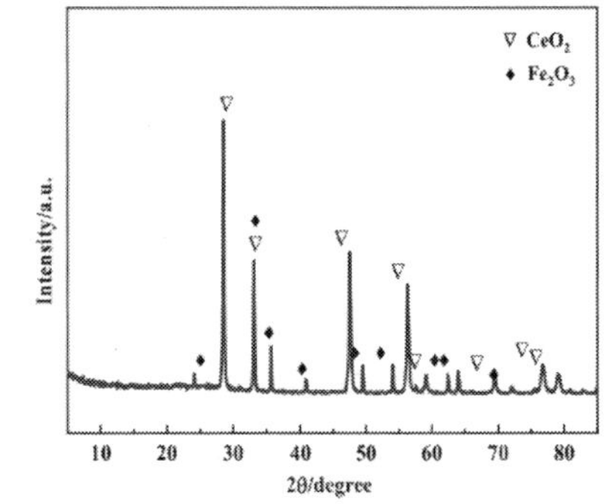

Figure 7. Fe-Ce phase transition law after combustion

4 Conclusions

The effect of composite metal promoters in the catalytic reduction of NO_X was studied in a high-temperature tubular furnace reactor. The results are as follows: In the combustion process of demineralized coke, the addition of metal additives can not only inhibit the production of NO_X, but also reduce the release of NO_X, and ultimately achieve lower NO_X emissions. During the combustion of coke, the catalytic effect of Fe-Ce composite is greater than that of iron or cerium alone. While the reaction temperature is 1000 ℃ and the addition ratio is 2:1, the catalytic effect on NO_X decrease ratio is optimal, the NO_X emissions is 0.28 mg, and the NO_X decrease ratio is 73%, which indicate that the modification by a proper amount of Ce doping is beneficial to further improve the effect of iron additives. The composite addition method can effectively reduce NO_X emissions and control air pollutants, which has important practical significance for achieving clean and efficient utilization of coal.

Acknowledgments

This work was supported by National Natural Science Foundation of China (21878210), Shanxi "1331" Civil Clean Fuel Engineering Research Center, Scientific and Technological Innovation Programs of Higher Education Institutions in Shanxi (2019L0313), Patent Promotion and implementation in Shanxi Province (20200719) and sponsored by Taiyuan Green Coke Energy Co., Ltd. (China).

References

1. H. Ma, L. Zhou, S. Ma, Z. Wang, H. Du, J. Li, W. Zhang and P. Guo. J. Energy, **32**, 1464-1473(2018)
2. W. Fan, Y. Li, Q Guo, C. Chen and Y. Wang. J. Energy, **125**, 417-426(2017)
3. X. Liu, Z. Luo and C. Yu, Fuel, **242**, 389-397(2019)
4. J. Xu, R. Sun, T.M. Ismail, S. Sun and Z. Wang. Energy Fuel, **31**, 7502-7509(2017)
5. J. Cheng, F. Zhou, X. Xuan, J. Liu, J. Zhou and K. Cen. Fuel, **181**, 820-826(2016)

6. Z. Zhao, W. Lia, J. Qiu and B. Li. Fuel, **81**, 2343-2348(2002)
7. B. Zhong and H. Tang. Int. J. Chem. React. Eng, **6**, (2011)
8. P. Li, Y. Xin, Q. Li, Z. Wang, Z. Zhang and L. Zheng. Environ. Sci. Technol, **46**, 9600-9605(2012)
9. Z. Lei, J. Yan, J. Fang, H. Shui, S. Ren, Z. Wang, Z. Li, Y. Kong and S. Kang. Energy, **216**, (2021)
10. R. Brackmann, C. Alves, B. Woyames, F.S. Toniolo and M Schmal. Appl. Catal. A: Gen, **600**, (2020)
11. B. Gradoń and J. Lasek. Fuel, **89**, 3505-3509(2010)
12. H. Chen, A. Sayari and A. Adnot. Appl. Catal. B, **32**, 195-204(2001)

Interesterification for biofuel synthesis over HJ-2# caly-supported SO_4^{2-}/ZrO_2 solid acid catalyst

Dong Lixin [1], Zhang Xueqiong [1], Chen Jing [1], Hao Yinan [1], Pang Liwen[2] and Ding Lijun [1,a]

[1]College of Science, Inner Mongolia Agricultural University
[2]Inner Mongolia Vocational College of Chemical Engineering

Abstract. This study makes use of soybean oil to produce biodiesel using SO_4^{2-}/ZrO_2-HJ-2# caly solid acid catalyst (SZ-HJ-2#). It was through coprecipitation and impregnation method that the catalyst was prepared which was then characterized by means of FE-SEM, XRD, EDS, BET, FT-IR, ICP-MS, NH_3-TPD and XPS. The catalytic property of the synthesized catalyst was determined by using it to produce biodiesel from soybean oil. A study was carried out to find the effect of the different preparation condition of catalyst affecting the process. For SZ-HJ-2#, Optimized condition of 0.5 mol/L(zirconium salt solution), 1.5 mol/L (the concentration of sulfuric acid impregnating solution) and 450°C(calcination temperature). Optimized conditions of 8.32:1 methanol to soybean oil molar ratio and catalytic loading of 1 wt% at 55°C with a stirring rate of 500 rpm for a reaction duration of 10 h gave a maximum yield of 89.6 wt%. Moreover, the further investigation indicated the catalytic activities were closely related to the ratio of Brönsted acid sites and intensity on catalysts. Besides, the excellent performance of the catalyst during recycling was shown by conducting reusability study.

1 Introduction

In recent years, the increasing demand for fossil fuel has led to energy shortage and environmental degradation[1]. According to the BP's statistics show that the oil reserve and production ratio in 2019 is only 49.9 years[2]. The development of cheap and eco-friendly energy source has become one of the main challenges of the current century[3,4]. Biodiesel as a clean renewable fuel is considered a good alternative for nonrenewable petroleum-based fuel, which has outstanding advantages of high energy density, biodegradable, sustainable supply and non-toxicity[5].

Biodiesel is generally produced by the transesterification of vegetable oil[6], animal fat, recycled oil[7], etc., with methyl alcohol in the presence of a catalyst including basic, acidic or enzymatic catalysts. Therefore, the choice of catalyst is crucial. Conventionally homogeneous basic catalysts such as NaOH and KOH are used. However, there are drawbacks associated with homogeneously catalyzed processes, including corroding equipment, not being reusable and not being easy to separate. Heterogeneous catalysts, on the other hand, makes product separation easier and the reusable[8]. So far, solid acid catalysts, one kind of heterogeneous catalysts, are widely applied in transesterification reactions[9].

SO_4^{2-}/ZrO_2 (SZ) is one of the SO_4^{2-}/M_xO_y type solid super acids, which has the advantages of strong acidity, high activity, easy separation, good thermal stability and regeneration stability[10]. Nevertheless, SO_4^{2-}/M_xO_y solid acid catalysts still suffer from rapid deactivation due to the sulfate leaching and coke deposition, which limit their application in industries.[11, 12] To solve these problems, some researchers supported SZ on a molecular sieve with a high specifc surface area, such as a MFI framework of ZSM-5[13], ordered mesoporous molecules of MCM-41[14] and SBA-15[15]. However, the stability of catalyst will be affected. Consequently, it remains a big challenge to develop high active and stable heterogeneous catalysts for transesterification reactions. Hangjin 2# clay, which is composed of two layers of silica-oxygen tetrahedra and a layer of aluminum (magnesium)-oxygen octahedra, is a 2:1 type of clay minerals, with the advantages of small particle size and large specific surface area[16]. It has a larger specific surface area, average pore size and total pore volume than montmorillonite and kaolin[17]. In addition, Hangjin 2# clay is inherently acidic, Therefore, it will be an excellent support of solid acid catalysts[18].

Herein, we realized the improved catalytic activity and stability of solid acid catalyst by synthesizing SO_4^{2-}/ZrO_2-Hangjin 2# Clay (SZ-HJ-2#) with co-precipitation-impregnation method. The influence of zirconium solution concentration, sulfuric acid impregnation solution concentration and roasting temperature during the preparation of the catalyst on the structures and acidic properties of SZ-HJ-2# catalyst were studied in detail. The relationship between the structures, properties and the activity results was systematically discussed by combining the characterizations and catalytic performances. It is well known that the reusability is of

[a] Correspondence: dlj78@163.com

vital importance to heterogeneous acid for industrial application. Thus, the recycling test was taken in this work. Furthermore, characterization of used catalyst were employed to study the stability of sulfate species and explain the slight deactivation. This work may provide referential values in the application of raw HJ-2# clay in preparation of eco-friendly, low-cost and highly efficient catalysts and explore their application in the conversion from biomass to green bio-fuels.

2 Materials and Methods

2.1 Materials

HJ-2# clay was supplied from Hangjin Banner, Ordos City, Inner Mongolia Autonomous Region, China. It was screened for 200 mesh and dried. The average molecular weight of Soybean oil was obtained at 883.46 g/mol based on the acid value (0.16 mg KOH/g) and saponification value (190.66 mg KOH/g)(see the method of calculation in the Supporting Information). $ZrOCl_2 \cdot 8H_2O$(A.R) was purchased from Shanghai Macklin Biochemical Co., Ltd. Analytically pure methanol, petroleum ether and 25%~28% ammonia were obtained from Tianjin Dengke Chemical Reagent Co. Ltd. H_2SO_4 (95-98.0%) was purchased from Sinopharm Chemical Reagent Co. Ltd.

2.2 Synthetic procedures

The In this study, the SO_4^{2-}/ZrO_2-HJ-2# caly solid acid (SZ-HJ-2#) catalyst was prepared by coprecipitation and impregnation. Firstly, 10 g HJ-2# clay was mixed with 500 mL 3 mol/L H_2SO_4, and stirred at 90°C for 4 h. After stirring, the product was washed with water to pH= 5~6 of the filtrate and dried. The product obtained at this time was activated white clay. Secondly, after mixing the activated white clay with different concentrations of $ZrOCl_2 \cdot 8H_2O$ solution (0.1 mol/L, 0.3 mol/L, 0.5 mol/L, 0.7 mol/L, 0.9 mol/L) at 5:1 (g/mL) for 1 hour, the pH of the mixture was adjusted to 9~10 with 6mol/L $NH_3 \cdot H_2O$ and aged 24 h, the Cl^- was removed by water washing (assay with $AgNO_3$ solution) and dried, followed by precalcined at 400°C for 3 h. The composite carrier obtained was called Zr/HJ-2# and then impregnated with different concentrations H_2SO_4 solution (0.5 mol/L, 1.0 mol/L, 1.5 mol/L, 2.0 mol/L, 2.5 mol/L) for 4 h and aged 24 h. Finally, the samples were calcined at different temperatures (250°C, 350°C, 450°C, 550°C, 650°C) for 3 h to obtain SZ-HJ-2# catalysts.

2.3 Catalytic Activity

Catalytic activities were tested for biodiesel synthesis. 50g of soybean oil, 18.83 mL of methanol and 2.5 g of SZ-HJ-2# were put into a reactor and reacted at 60°C for 6 h. After extraction, washing and drying, the mass of biodiesel obtained was recorded, The catalyst activity was expressed in terms of biodiesel yield and the yield Y(%) was calculated as follows[19].

$$Y(\%) = \frac{m_2}{m_1} \times 100$$

where m_1 is the initial mass of soybean oil and m_2 is the mass of biodiesel in the product.

3 Results and Discussion

3.1. Catalyst characterization

Figure 1. SEM images of HJ-2# (a), activated clay(b), Zr/HJ-2# before precalcined(c), Zr/HJ-2#(d), SZ-HJ-2#(e) and SZ-HJ-2# after 5 times use(f)

The general morphologies and microstructures of sample were investigated by SEM analysis as shown in Figure 1. The raw Hangjin 2# clay(HJ-2#)(Figure 1a) exhibited typically plate-like morphology which was formed by many parallel nanoflakes. After the sulk morphology of the activated clay as shown in Figure 1b. The smaller bulk were stuck together with some relate activation, there was an obvious fragmentation in semble needles and flakes. After the introducing of Zr (Figure 1c), the number of fine particles decreased significantly, and the particles showed irregular morphology, relatively rough surface, and a large number of free growing small grains. After the precalcination (Figure 1d), the grains grew larger result from recrystallization. Impregnated by sulfuric acid and calcined, SZ-HJ-2# (Figure 1e) were prepared. The catalysis were block and the surface was loose and uneven. In addition, the aggregation of catalyst is serious.

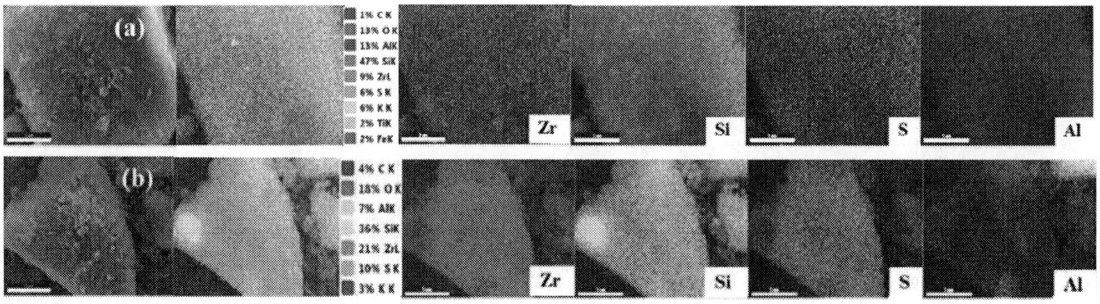

Figure 2. EDS mapping images of SZ-HJ-2#(a) and SZ-HJ-2# after 5 times use(g)(The percentages in the legend are weight percentages)

Table 1. Font styles for a reference to a journal article.

Element	Fresh	Used
C	2.08	15.86
O	54.74	57.92
Al	7.37	2.21
Si	25.48	13.40
Zr	3.05	5.38
S	3.29	4.57
K	2.70	0.66
Ti	0.48	--
Fe	0.81	--

According to mapping images in Figure 2(a), the SZ-HJ-2# catalysts well maintained the block morphologies and the roughly surface. The distinction of four elements(Si, S, Zr and Al) was given by the mapping image, respectively (Figure 2). The four elements(Si, S, Zr and Al) in SZ-HJ-2# catalysts were highly dispersing. It suggested that the experiment condition would facilitate the high dispersion of active components. EDS analysis results were carried out on SZ-HJ-2# in Table 1. Combined with the elemental analysis for the region, the ratio of S and Zr atomic percentages in the fresh catalyst and the used catalyst were 1.08 and 0.85, respectively. It suggested that the content of S decreases on the surface of used SZ-HJ-2#. SO_4^{2-} was an important condition for the formation of acid sites to sustain the activity in SZ-HJ-2#. Therefore, the reduction of S content will seriously affect the catalytic activity of catalysts. Meanwhile, SZ-HJ-2# catalysts still contain some Fe and Ti.

XRD measurement was used to further explore the structural insight by analyzing the crystallographic structure, as shown in Figure 3. The activated clay samples showed the characteristic peaks of raw HJ-2# (JCPDS No.14-0164), moreover, some impurity peaks including $CaCO_3$ at 29.38° disappeared after the treatment of sulfuric acid.

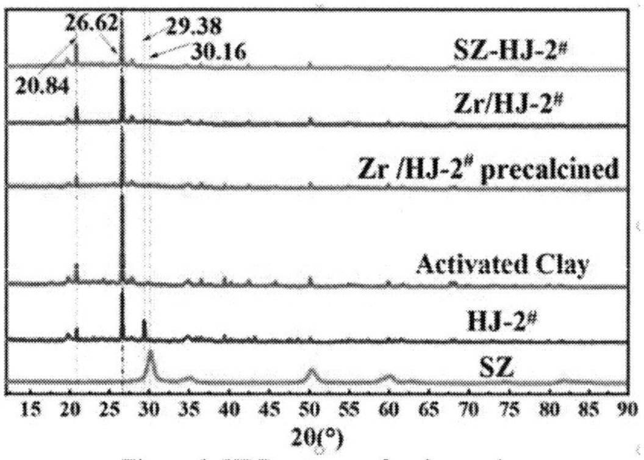

Figure 3. XRD patterns of each sample

The strongest diffraction peaks at 20.84° and 26.62° of SiO_2 (PDF#89-8934) were clearly displayed in all samples associated with the activated clay, indicating the structures of the activated clay were well maintained in supported catalysts. Comparing with SZ (Zirconia impregnated with sulfuric acid), the peak 2θ=30.16° of t-ZrO_2 (PDF#88-1007) did not occur in SZ-HJ-2# catalysts. It was implied that the Zr in the catalyst may form Zr-O-Si bond on the surface and facilitate to form the efficient catalysts. Introducing ZrO_2 and H_2SO_4 might result in the aggregation of the block structure as shown in FE-SEM morphology in Figure 1. Compared with the peak height of the curve, the impurity was obviously decreased by the SZ loading and almost disappeared in SZ-HJ-2# catalyst. The intensity of SiO_2 peaks sharply decreased when introducing the SZ components. As determined by XRD, it also proved that the structure of SZ-HJ-2# was different from HJ-2# and SZ.

The N_2 adsorption-desorption isotherms of the SZ-HJ-2# catalyst was shown in Figure 4 to study the texture. According to the IUPAC classification, the isotherm was ascribed to the representative type IV adsorption-desorption with the hysteresis loop in the relative pressure (P/P_0) ranging from 0.7 to 0.99, reflecting the presence of the macro-porous structure. The hysteresis loop was assigned to the type H3, which represented the formation of slit-shape pores due to the flakes [20,21]. Thus, it suggested that the macro-porous structure in SZ-HJ-2# catalyst were generated from the flake-like HJ-2# supports, which had been also observed in other HJ-2#-based materials[22]. The formed macro-porous structure had been reported to significantly promote the mass transfer during the reaction and consequently improved the catalytic activities[23]. Besides, it also performed the surface areas was 32.07 m^2/g, average pore size was 19.14 nm and pore volume was 0.0817 cm^3/g.

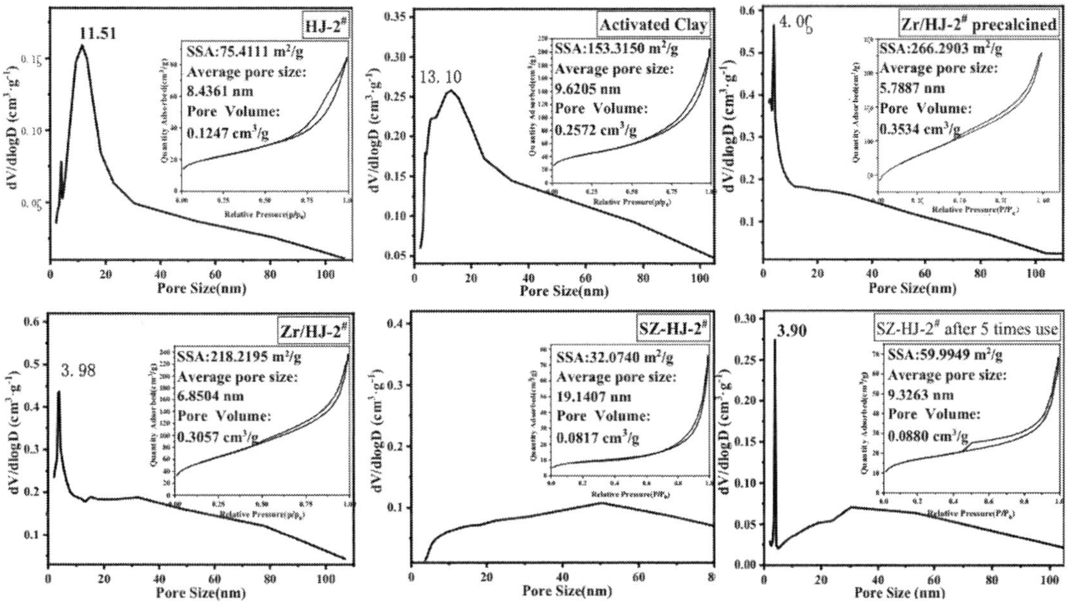

Figure 4. BET patterns of each sample

FT-IR studies were carried out for the structure of samples (Figure 5a). By compared to the raw HJ-2#, the activated clay, Zr/HJ-2# and SZ-HJ-2# catalysts exhibited some intense bands in the range of 1000-1200 cm^{-1} due to the strong interaction between sulfate species and metal ions. Among them, the bands at 1088 and 1053 cm^{-1} were assigned to the stretching of S=O bond, indicating that a chelating bidentate structure had been formed between surface sulfate species and Zr ions and contained amount of Brönsted acid[24]. It was noted that the band of S=O stretching was shifted to high wavenumber gradually by introducing of Zr ions, implying that there might be some synergistic interaction between Zr and HJ-2# components influencing the molecular structure of species. These bands were correlated to surface active sites and explained the superior catalytic activities. The bands at 1026 and 987 cm^{-1} were assigned to Si-O vibrations[25]. The widen absorption peak in Zr/HJ-2# and SZ/HJ-2# at 1026 cm^{-1} appeared obviously, and the location of the peak occurred red shift it to a certain extent compared to the HJ-2# (987 cm^{-1}). It was shown that the three-dimensional structure of the silicon oxide was broken and the content of amorphous SiO_2 increases. It suggested that Zr has doped into the structure of HJ-2#. The bands at 1637 and 3350 cm^{-1} were due to OH bending vibrations and stretching vibrations, respectively of water that was adsorbed from the air on the surface of the catalyst[26]. Additionally, the spectra of HJ-2# displayed strong band around at 798 cm^{-1} was characteristic of Si-O structure. It further confirmed the HJ-2# structure was maintain in catalysts well, which was consistent with the XRD results (see Figure 3). Besides, the weak bands at 3654 and 3568 cm^{-1} were due to the stretching vibration of -OH on surface metal ion which were easily covered by the broad band from water molecules.

To understand the amount and strength of acid sites, the SZ and SZ-HJ-2# catalysts were investigated by NH_3-TPD, as shown in Figure 5(b). It is well acknowledged that the area of desorption peak represents the amount of acid sites, while desorption temperature indicates the acid strength. Generally, the strength of acid sites is classified into weak (< 250°C), medium (250-400°C) and strong (> 400°C) strength[27]. The broad peak of SZ-HJ-2# catalyst indicated the existence of weak and medium strength acid sites on the surface. As for the SZ and SZ-HJ-2#, it implied that the supports could generate a few strong acid sites after sulfate activation. Among them, the SZ possessed more the strong acid sites than SZ-HJ-2# catalyst. By comparing the peak area of SZ-HJ-2# (1789.5) with SZ (2977.0), it was suggested that the number of acid sites in SZ-HJ-2# is significantly less than that in SZ, but the intensity of acid sites in SZ-HJ-2# is higher than that in SZ.

https://doi.org/10.1051/e3sconf/202129001033

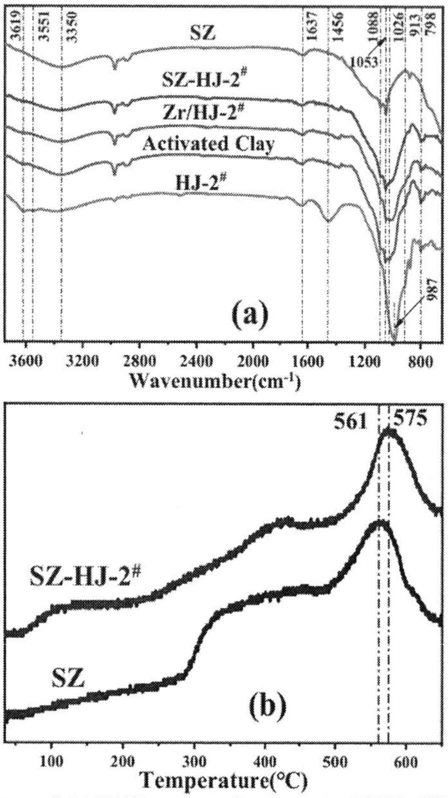

Figure 5. (a) FT-IR patterns of each sample. (b) NH_3-TPD patterns of SZ-HJ-2# and SZ

Chemical valence states were detected by XPS analysis, as shown in Figure 6. The binding energy spectrum of sulfur elements can be divided into 169.0eV and 170.0eV, both of which assigned to the sulfur oxidation state of +6.[28] The S 2p peak has closely spaced spin-orbit components (Δ=1.16eV)。Effect of spin-orbit splitting, either splitting between components or asymmetric peak shape can often be observed. The S $2p_{3/2}$ peak were at 168.9 and 169.3eV in SZ-HJ-2# and SZ, respectively. The high oxidation state was necessary to prove the formation of bidentate sulfate species, confirming that catalyst possessed acid sites, which may be the main source of the formation of different acid centers in the catalysts. The peaks at 183.2eV and 185.6eV were corresponded to Zr $3d_{3/2}$ and Zr $3d_{5/2}$ in the sample before calcined due to the formation of Zr-O-Si. The binding energy of Zr decreased slightly after calcined due to part of the Zr-O-Si bond formed ZrO_2. Furthermore, the peak of Zr $3d_{3/2}$ at 183.55eV, which shifted to high bonding energy of 0.55eV(the bond energy of Zr-O-Si was 183eV). This was because Zr in SZ-HJ-2# was affected by the strong electron-absorbing action of SO_4^{2-}, which reduced the density of the outer electron cloud of Zr, weakened the shielding effect, and increased the electron binding energy, indicating that strong interaction is formed between Zr and SO_4^{2-}. While, the peak of Zr $3d_{3/2}$ (183.55eV) in SZ was lower than that of SZ-HJ-2# because it was only affected by sulfate radical but not affected by electron absorption of Si. By comparison, it indicated that the density of electron cloud in the outer layer of Zr atom in SZ-HJ-2# was lower than that in SZ, thus the SZ-HJ-2# had stronger acidity. The results of NH_3-TPD test also showed that the acid strength of SZ-HJ-2# was higher.

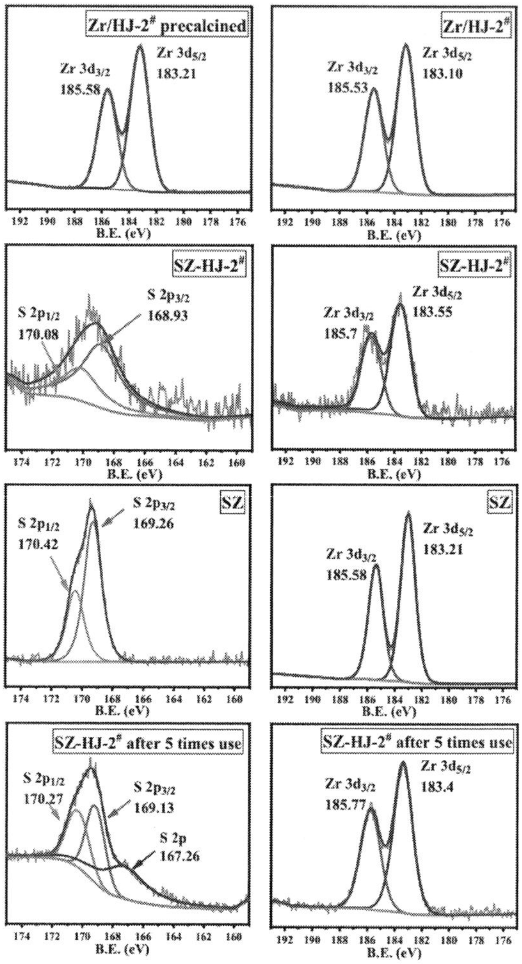

Figure 6. High-resolution XPS spectra of Zr 3d and S 2p in each sample

3.2 Catalytic performances

The optimization of preparation of SZ-HJ-2# process was carried out by the controlling variable method. Several offers were presented to optimize the conditions that the parameters are chosen such as the concentration of zirconium salt solution (cZr), the concentration of sulfuric acid impregnating solution (cH_2SO_4) and the calcination temperature of catalyst. In the present study(see Figure 7(a), (b) and (c)), optimum conditions were chosen as 0.5 mol/L, 1.5 mol/L and 450°C, respectively. The experimental biodiesel yield of 89.6% was achieved with adopting above SZ-HJ-2#.

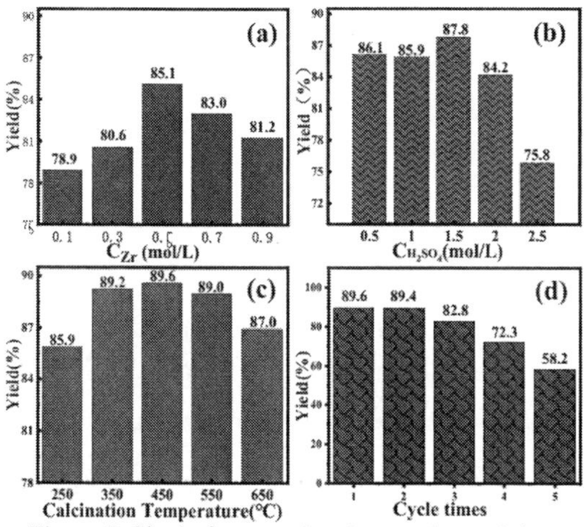

Figure 7. Single-factor and cycle experimental data

3.3 Reusability

The reusability of SZ-HJ-2# in optimum conditions was assessed in Figure 7(d). After each reaction, the catalyst was washed two times with methanol and petroleum ether and then dried at 90°C for 3 h. According to Figure 7(d), the catalyst keeps its activity for the fourth use. The yield decreased to 72.3% (NO. 4) which could be caused by either the solubility of some sulfate ions on the catalyst surface in mixture products, or the fillings of the catalysts pore spaces. However, the yield sharply reduces after the fifth used (NO. 5). The cause of the decrease has been attributed to the leaching of active phases and poisoning of catalyst surfaces with the reactants and products produced in the interesterification reaction[29–31]. In the present study, the effects of reactant and product, probably, were meaningfully eliminated by washing methanol and petroleum ether of used catalyst after each reaction. Therefore, the leaching of active phases may be the main reason for deactivation of SZ-HJ-2#.

Table 2. Font styles for a reference to a journal article

Element		Zr	Al	C
Weight %	Fresh	9.94	4.19	0.12
	Used	9.35	4.02	0.77

In order to explore the cause of deactivation of SZ-HJ-2#, a series of structural characterizations of the SZ-HJ-2# which has used for 5 times was carried out such as FE-SEM, EDS, BET, FT-IR, XPS and ICP-MS. According to the SEM (see Figure.1f) and mapping image, the morphology and element distribution of SZ-HJ-2# catalyst did not change significantly before and after catalysis reaction. The N_2 adsorption-desorption isotherms of the used SZ-HJ-2# catalyst were shown in Figure 4. The isotherm was ascribed to the representative type IV adsorption-desorption, reflecting the presence of the macro-porous structure. The hysteresis loop was still assigned to the type H3, which represented the formation and structure of used SZ-HJ-2# was consistent with that before catalysis.

EDS analysis results were carried out on used SZ-HJ-2# in Table 1. It can be found that the composition of surface element changes greatly. By analyzing the relative content, it can be seen that the relative content of Zr increases, while the content of Si, Al and S decreases in varying degrees on the surface of used SZ-HJ-2#. According to the ICP-MS date (see Table 2), the overall elemental content of the catalyst does not change significantly, so this elemental loss may be more on the surface of the catalyst. SO_4^{2-} was an important condition for the formation of acid sites to sustain the activity in SZ-HJ-2#, therefore, the reduction of S content will seriously affect the catalytic activity of catalysts. As evidenced from infrared (IR) spectra(see Figure 5), the stretching vibration peaks of S=O bond near 1088 cm^{-1} and 1053 cm^{-1} were obviously wider, indicating that the S=O bond is indeed damaged. Besides, the XPS analysis of used SZ-HJ-2# is shown that the shift of S 2p peak to high electron binding energy obviously. It indicated that the chemical environment has changed though S was still +6 and Si-O-Zr bond still existed. In addition, EDS had shown that the content of C increased rapidly. Combined with XPS analysis, it can be considered that there was carbon deposition on the surface of the catalyst, which was also one of the reasons for catalyst deactivation.

4 Conclusions

SZ-HJ-2# solid acid catalysts by using the raw HJ-2# clay as supports were synthesized and applied in esterification of soybean oil with methanol for the biofuels synthesis. The optimal SZ-HJ-2# solid acid well maintained the effective structures of both HJ-2# clay and SO_4^{2-}/ZrO_2 components. These structures brought about the most intensity acid sites and the easier mass transfer in the reaction. As a result, the SZ-HJ-2# solid acid performed the best catalytic activities of 89.6% conversion of soybean oil. Moreover, the relationship between acidic properties and activities clearly indicated the intensity of Brönsted acid on catalysts were favorable for the esterification reactions. Besides, the SZ-HJ-2# solid acid showed an excellent performance in reusability conversion of soybean oil in the fifth cycle. The excellent reusability was due to the stable surface sulfate and Brönsted acid sites in the recycling process. The present work indicated that the raw HJ-2# clay was constructive in synthesizing a low-cost and eco-friendly, efficient solid acid, which is of vital significance for production of green bio-fuels.

Acknowledgments

This work was supported by the Project of Inner Mongolia Science and Technology Department (2019CG018), the Scientific Research Project of Colleges and Universities in Inner Mongolia Autonomous Region (NJZY21346).

https://doi.org/10.1051/e3sconf/202129001033

References

1. L. Wang and F. S. Xiao, Green Chem 17, 24-39 (2015)
2. P.l.c BP. Bp statistical review of world energy 2020, (P.l.c BP, 2020)
3. Q. Shu, G. Q. Tang, F. S. Liu, W. Q. Zou, J. F. He, C.X. Zhang and L. X. Zou, Fuel 209, 290-298 (2017)
4. G. Vicente, M. Mercedes and J. Aracil, Bioresource Technology 92, 297-305 (2004)
5. A. Tiwari, V. M. Rajesh and S Yadav, Energy Sustain Dev 43, 143-161 (2018)
6. W. T. Wang, P. Lu, H. Tang, Y. M. Ma and X. F. Yang, New J. Chem 41, 9256-9261 (2017)
7. J. M. Chang, X. Y. Guan, S. Y. Pan, M. L. Jia, Y. Chen. and H. J. Fan, New J. Chem 42, 13074-13080 (2018)
8. M. M. Aboelhassan, A. F. Peixoto and C. Freire, New J. Chem 41, 3595-3605 (2017)
9. Y. M. Sani, W. M. A. W. Daud and A. R. Abdul Aziz, Catal. A-Gen 470, 140-161 (2014)
10. R. Z. Raia, L. S. Silva and S. M. P. Marcucci, Catal. Today 289, 105-114 (2017)
11. J. K. Satyarthi, S. Darbha and P. Ratnasamy, Energy Fuels 24, 2154-2161 (2010)
12. C. C. Huang, C. J. Yang and P. J. Gao, Green Chem 17, 3609-3620 (2015)
13. Y. M. Sani, P. A. Alaba, A. O. Raji-Yahya, A. R. A. Aziz and W. M. A. W. Daud, Taiwan Inst. Chem. E 60, 247-257 (2016)
14. F. Ye, Z. W. Dong and H. J. Zhang, Catal. Commun 10, 2056-2059 (2009)
15. W. M. Hua, Y. Yue and Z. Gao, J Mol Catal A-Chem 170, 195-202 (2001)
16. S. Bai, M. Tegshi, X. Wang and Z. Bao, ChemistrySelect 2, 4847-4851 (2017)
17. S. Uchiyama, T. Isobe, S. Matsushita, K. Nakajima, M. Hara and A. Nakajima, Journal of Materials Science 47, 341-349 (2012)
18. X. G. Li, Y, Zhao, B. D. Xi, X. G. Meng, B. Gong, R. Li, X. Peng and H. L. Liu, J. Environ. Sci 52, 8-17 (2017)
19. J. Medina-Valtierra and J. Ramirez-OrtizFront, Front. Chem. Sci. Eng 7, 401-407 (2013)
20. L. Zhang, L. Yang, B. P. Jelle, Y. Wang and A. Gustaysenf, Constr Build Mater 162, 576–83 (2018)
21. G. E. Christidis, S. Kosiari and E. Petavratzi, Acid Activation and Bleaching Capacity of Bentonites from the Troodos Ophiolite, Cyprus.(Amsterdam, Elsevier Science Bv, 2003)
22. M. F. González, C. S. Stull, F. Lópezlinares and P. Pereiraalmao, Energy Fuels 21(1), 234–41 (2007)
23. Z. W. Wu, C. Chen, Q. R. Guo, B. X. Li, Y. G. Que, L. Wang, H. Wan and G. F. Guan, Fuel 184, 128–35 (2016)
24. K. Saravanan, B. Tyagi, R. S. Shukla and H. C. Bajaj, Appl Catal B 172, 108–115 (2015)
25. M. Khairy, H. A. Ayoub, F. A. Rashwan and H. F. Abdel-Hafez, Appl Clay Sci 153, 124–33 (2018)
26. H. Ghafuri, A. Rashidizadeh, B. Ghorbani and M. Talebi, New J. Chem 39, 4821-4829 (2015)
27. V. G. Deshmane and Y. G. Adewuyi, Appl. Catal. A-Gen 462, 196-206 (2013)
28. N. W. Liu, X. Wang, L. Shi and X. Meng, New J. Chem 43, 3625-3632 (2019)
29. J. Kansedo and K. T. Lee, Biomass Bioenerg 40, 96-104 (2012)
30. K. Ramachandran, P. Sivakumar, T. Suganya, S. Renganathan, Bioresour. Technol 102, 7289-7293 (2011)
31. Y. Zhang, W. T. Wong and K. F. Yung, Bioresour Technol 147, 59-64 (2013)

https://doi.org/10.1051/e3sconf/202129001034

Toxic effects of graphene and related materials on bacteria

Zhihui Huang[1], Ning Gao[1], Siyi Zhang[1], Jianing Xing[1] and Jing Hou[1a]

[1]MOE Key Laboratory of Resources and Environmental Systems Optimization, College of Environmental Science and Engineering, North China Electric Power University, Beijing, 102206, PR China

Abstract. Graphene and related materials (GRMs) have been reported to have extensive applications in many areas. The widespread use of GRMs makes them inevitably enter the environment through various links causing adverse effects on organisms. Bacteria were representatively used to review the toxicological effects of GRMs on biological organisms in this paper. We comprehensively summarize the recent researches about negative effects of GRMs on bacteria. This is conductive to the evaluation of the ecological risk assessment of GRMs.

1 Introduction

Graphene is a two-dimensional material composed of carbon atoms arranged in a net of fused hexagons[1], including single-layer graphene, few-layer graphene and multi-layer graphene[2]. Graphene, graphene oxide (GO), reduced graphene oxide (rGO) and graphene-based materials formed by combining with various functional groups, organic compounds, metal ions, metal particles, etc. are collectively referred to as graphene and related materials(GRMs)[3-5]. GRMs have a wide range of potential applications in the fields of medicine, electronics, chemicals, water treatment, etc. because of its exceptional optical, mechanical and electrical properties[6-8]. The widespread use of GRMs makes them inevitably enter the water environment through various links resulting in exposure of organisms to the GRMs.

There have been many studies reported the negative effects of GRMs on different types of living organisms and a great possibility of bioaccumulation[9, 10]. Studies using the body burden factor (BBF) to evaluate the accumulation of 14C-labeled graphene in *E. coli*, T. thermophila, *D. magna*, and *D. rerio* have showed that organisms exposed to graphene-containing culture medium have high potential of graphene accumulating[11]. Researches on the ecological safety of GRMs have increased rapidly and received extensive attention. However, these studies are chaotic and lacking in sorting and induction. Therefore, we summarized the characteristics and general patterns of toxic effects of different kinds of GRMs on bacteria, which were selected as representatives. The review of this article is worthy of reference for the future exploration of the environmental safety of GRMs.

2 Toxicity characteristics of GRMs on bacteria

2.1 Common toxic effects of GRMs on bacteria

Table 1 summarized the effects of GRMs on bacteria. On the whole, the antibacterial properties of GRMs has been recognized. They can reduce the cell viability, destroy cell wall integrity, inhibit the formation of bacterial biofilm, alter energy metabolism and cause oxidative damage on bacteria[12-15]. What's more, there are studies reported that its antibacterial effect is comparable to certain antibiotics[16]. In any case, more researches are needed about the antibacterial properties of different types GRMs to take advantage of this excellent feature. A disk diffusion test conducted by Kadiyala et al. showed that the diameter of inhibition zone (DIZ) of GO on gram-negative (*E. coli*) and gram-positive bacterial strain (*S. aureus*) are 19 mm and 21 mm, respectively, which is bigger than gentamicin (9.3 mm and 12.17 mm respectively)[17]. But it is unsure whether GRMs exert greater effects on all the gram-negative bacteria than gram-positive ones. GRMs could be the promising materials for inhibiting resistant bacteria, because one of its antibacterial mechanisms is based on the physical damage. GO with concentrations higher than 10 mg/L could damage the resistant plasmids of *E. coli HB101* consequently reduce its resistance to antibiotics[18]. Once GRMs were successfully used to inactivate various resistant bacteria, there will be a leap forward in the treatment of resistant bacteria. Among the influencing factors, the shape, size and surface coating etc. are factors that are relatively easy to regulate.

[a] Corresponding author: houjing@ncepu.edu.cn

https://doi.org/10.1051/e3sconf/202129001034

2.2 Effect of surface modification on the toxicities of GRMs

Surface modification of GRMs can both increase and decrease the toxicities of GRMs. Understanding the changes of toxic effects of surface modification is useful

Table 1. Toxic effects of different GRMs on different bacteria.

GRMs	Species	Exposure condition	Efftcts	Refe-rence
Ag-GO	Natural aquatic microorganisms	Exposed to Ag-GO concentration of 10, 50, or 100 mg/L at mimic natural aquatic environment	Reduced enzymes activities; Decreased nitrification rate; ROS was generated continuously	[19]
Cu_2O-Graphene	*V. cholerae*	The bacteria culture medium was incubated with 100 or 300 μg contained Cu2O-Graphene disc loaded at 4°C for 24 h	300 μg-DIZ: 20 mm; 200 μg-DIZ: 14 mm; 300 μg-DIZ: 20 mm	[20]
Cu_2O-Graphene	*S. pyogenes*	Incubated with 100 or 300 μg contained Cu2O-Graphene disc loaded at 4°C for 24 h	100 μg-DIZ: 0 mm; 200 μg-DIZ: 14 mm; 300 μg-DIZ: 22 mm	[20]
AuNPs-rGO-NC	*E. coli*	Incubated with 100 ± 15 μg AuNPs-rGO-NC contained disk placed on at 32°C for 24 h and incubated with 10 to 100 μg/L AuNPs-rGO-NC in an orbital shaker maintained at 180 rpm at 30°C	DIZ: 33 mm; AI: 3.548; MIC: 30 μg/mL; MBC: 40 μg/mL	[17]
AuNPs-rGO-NC	*B. subtilis*	Incubated with 100 ± 15 μg AuNPs-rGO-NC contained disk placed on at 32°C for 24 h and incubated with 10 to 100 μg/L AuNPs-rGO-NC in an orbital shaker maintained at 180 rpm at 30°C	DIZ: 30 mm; AI: 2.323; MIC: 40 μg/mL; MBC: 50 μg/mL	[17]
AuNPs-rGO-NC	*S. aureus*	Incubated with 100 ± 15 μg AuNPs-rGO-NC contained disk placed on at 32°C for 24 h and incubated with 10 to 100 μg/L AuNPs-rGO-NC in an orbital shaker maintained at 180 rpm at 30°C	DIZ: 29 mm; AI: 3.382; MIC: 50 μg/mL; MBC: 60 μg/mL	[17]
GO	*B. subtilis*	Incubated with 100 ± 15 μg GO contained disk placed on at 32°C for 24h and incubated with 10 to 100 μg/L GO in an orbital shaker maintained at 180 rpm at 30°C	DIZ: 21 mm; AI: 1.626; MIC: 60 μg/mL; MBC: 80 μg/mL	[17]
rGO	*E. coli*	50 μL rGO solution were placed into a well with diameter of 1 cm	DIZ: 18 ± 0.5 mm; Cell wall integrity reduced with increasing time	[15]
ZnO-rGO	*E. coli*	50 μL ZnO-rGO solution were placed into a well with diameter of 1 cm	DIZ: 28 ± 0.7 mm; A sudden decrease of integrity after 20 min	[15]
GO	*E. coli*	Dispersed GO was inoculated with bacteria of 70μl 108 CFU/ml, incubated at 37°C; Bacteria of 108 CFU/ml were cultured on nutrient agar plates supplemented with GO of 250 ppm	MIC: 125 ppm; Removal efficiency: about 70%	[21]
GO	*S. aureus*	Dispersed GO NPs was inoculated with 70 μl 108 CFU/ml of the bacteria, incubated at 37°C	MIC: 225 ppm	[21]
G-Fe_3O_4	*E. coli*	Dispersed G-Fe3O4 NPs was inoculated with 70 μl 108 CFU/ml of the bacteria, incubated at 37°C; Bacteria of 108 CFU/ml were cultured on nutrient agar plates supplemented with G-Fe3O4 of 250 ppm	MIC: 100 ppm; Removal efficiency: about 92%	[21]
G-Fe_3O_4	*S. aureus*	Dispersed G-Fe3O4 NPs was inoculated with 70 μl 108 CFU/ml of the bacteria, incubated at 37°C	MIC: 200ppm	[21]
rGO-ZnO	*K. pneumonia*	RGO-ZnO nanocomposite was loaded (20, 30, 50 μL) in 4 mm diameter of agar wells, incubated at 38°C for 24 h	DIZ: 14 ± 0.44 mm	[22]
rGO-ZnO	*B. thuringiensis*	rGO-ZnO nanocomposite was loaded (20, 30, 50 μL) in 4 mm diameter of agar wells, incubated at 38°C for 24 h	DIZ: 14 ± 0.73 mm	[22]

rGO-ZnO	*B. cereus*	rGO-ZnO nanocomposite was loaded (20, 30, 50 μL) in 4 mm diameter of agar wells, incubated at 38°C for 24 h	DIZ: 15 ± 0.55 mm	[22]
rGO-ZnO	*P. aerunginosa*	rGO-ZnO nanocomposite was loaded (20, 30, 50 μL) in 4 mm diameter of agar wells, incubated at 38°C for 24 h	DIZ: 12 ± 0.65 mm	[22]
GO-Ag	*E.coli*	Bacterial suspensions with different concentrations of GO-Ag (from 1.0 to 100.0 μg/mL) added were incubated at 37°C for 18 h	MIC: 0.01 mg/mL	[23]
GO-Ag	*S. aureus*	Bacterial suspensions with different concentrations of GO-Ag (from 1.0 to 100.0 μg/mL) added were incubated at 37°C for 18 h	MIC: 0.17 mg/mL	[23]
GO-Ag	*E. coli*	Bacteria were cultured in connect with foils coated with GO-Ag at 37°C	Antibacterial activity: 88.6%; Membrane Integrity: 66.3%; Enhanced ROS	[14]
GO-Ag	*S. aureus*	Bacteria were cultured in connect with foils coated with GO-Ag at 37°C	Antibacterial activity: 79.6%; Membrane Integrity: 59.4%; Enhanced ROS	[14]
Spindle-shaped GO	*E. coli*	Cell solutions added spindle-shaped GO inoculated at 37°C for 12 h; Put disks on the bacteria contained agar at 37°C for 12 h	MIC: 125 μg/mL; R: 15 ± 0.2%	[24]
Spindle-shaped GO	*S. typimurium*	Cell solutions added spindle-shaped GO inoculated at 37°C for 12 h; Put disks on the bacteria contained agar at 37°C for 12 h	MIC: 125 μg/mL; R: 15 ± 0.2%	[24]

Caption: Diameter of inhibition zone (DIZ); Activity index (AI); Minimum inhibition concentration (MIC); Minimum bactericidal concentration (MBC); The normalized viability ratio (R)
Graphene oxide (GO); Graphene (G); AuNPs-rGO nanocomposite (AuNPs-rGO-NC); Reduced graphene oxide (rGO); Zinc oxide decorated rGO (ZnO-rGO)

for choosing appropriate types of modifier to achieve the specific purpose of toxicity regulation. The factors determining the modification effect include the kinds of surface coating and the ratio of GRMs to modifiers. Common types of modifying include metal or metal oxide nanoparticles, metal ion, inorganic salts and other organics. Generally speaking, the compound is more toxic when it is combined with a substance that itself is antibacterial[25, 26]. For instance, GRMs decorated with Au-NPs, Cu-NPs, Cu_2O, curcumin or polyindole have stronger inhibitory effects than the bare GRMs[17, 20, 27-29]. This may be the synergy effect of GO and modifiers. But there are also exceptions. Unlike this enhanced toxicity, Gao et al. reported the decreased joint toxicity while investigating the combined toxicity of GO and heavy metal cation (Me(II)) (Cd^{2+}, Co^{2+} and Zn^{2+})[30]. The authors gave some reasons for this phenomenon like the adsorption of GO reduced the content of free Me(II), Me(II) weakened the sharp edges of GO and bacteria are difficult to assimilate the complexes. In other word, it is because the combination of graphene and Me(II) reduces the probability of Me(II) entering cells. At the same time, the deeper reason may be that Me(II) need to enter the inside of bacteria firstly and then bind to the biomacromolecules, while substances which have a synergistic effect with GRMs can engender toxic effects through contact with the surface of bacteria. The other factor is the ratio of GRMs to modifiers. The average growth inhibitory rate of *E. coli* for Graphene-Fe_3O_4 nanohybrid with Graphene/Fe_3O_4 weight ratios of 1-1, 1-3 and 1-5 at concentration of 250 mg/L are 75.6%, 82.8% and 92.1%, respectively[21]. But there is no research that has gone into the influence of the ratio of other coatings to GRMs on the antibacterial properties of composites, which deserves further research to get more information about the effect of composition ratio on antimicrobial properties.

2.3 Effect of inherent properties of GRMs on its toxicities.

Some inherent properties of GRMs such as size and morphology will affect its antibacterial characteristics. Generally speaking, materials with smaller size exert greater effects on organisms. Smaller-sized GO sheet showed stronger antibacterial property due to larger defect density, while measuring the antibacterial activity on GO-coated surface[31]. This could also be because the smaller-sized has a larger specific surface area with a stronger adsorption effect, and is easier to enter the organism accordingly. In term of morphology, wrinkled GO nanosheets with different roughness grade have different antibacterial capabilities, which is related to the relative dimension between the surface roughness grade and the bacterial size[32]. But, there is a lack of comparison of antibacterial properties between GRMs with the same size but wrinkled and smooth surfaces, respectively.

3 Conclusion

In general, there are many factors that contribute to the toxicities of GRMs on bacteria. Toxicities of GRMs with different purity, morphology, thickness, lateral dimension, surface charge and coating may vary greatly.

So, we can select appropriate types of GRMs according to our needs in practical. GRMs with higher toxicity can be chosen in the process of sterilization and elimination of harmful algae, but for materials that are frequently or directly in contact with human beings, the GRMs with lower toxicity should be selected through optimization. The researches on the toxic effects of GRMs make great contributions to the utilization and environmental risk assessment of GRMs.

References

1. A.K. Geim, K.S. Novoselov, Nature Materials **6**, 183 (2007)
2. A. Bianco, H.M. Cheng, T. Enoki, Y. Gogotsi, R.H. Hurt, N. Koratkar, T. Kyotani, M. Monthioux, C.R. Park, J.M.D. Tascon, et al., Carbon **65**, 1 (2013)
3. S. Stankovich, D.A. Dikin, G.H.B. Dommett, K.M. Kohlhaas, E.J. Zimney, E.A. Stach, R.D. Piner, S.T. Nguyen, R.S. Ruoff, Nature **442**, 282 (2006)
4. P. Wick, A.E. Gaume, M. Kucki, H.F. Krug, K. Kostarelos, B. Fadeel, K.A. Dawson, A. Salvati, E. Vazquez, L. Ballerini, et al., Angewandte Chemie-International Edition **53**, 7714 (2014)
5. V.C. Sanchez, A. Jachak, R.H. Hurt, A.B. Kane, Chem Res Toxicol **25**, 15 (2012)
6. R.K. Shervedani, A. Amini, Electrochim Acta **173**, 354 (2015)
7. K. Liu, J.J. Zhang, F.F. Cheng, T.T. Zheng, C. Wang, J.J. Zhu, J Mater Chem **21**, 12034 (2011)
8. B. Nam, H.J. Lee, H. Goh, Y.B. Lee, W.S. Choi, J Mater Chem **22**, 3148 (2012)
9. C.J. Lu, X.F. Jiang, M. Junaid, Y.B. Ma, P.P. Jia, H.B. Wang, D.S. Pei, Chemosphere **184**, 795 (2017)
10. C. Wang, X.L. Chang, Q. Shi, X. Zhang, Environ Sci Technol **52**, 12133 (2018)
11. S. Dong, T. Xia, Y. Yang, S. Lin, L. Mao, Environ Sci Technol **52**, 541 (2018)
12. J. Chen, S. Li, J. Luo, Y. Zhang, W. Ding, Journal of Nanoscience and Nanotechnology **17**, 186 (2017)
13. N. Yadav, A. Dubey, S. Shukla, C.P. Saini, G. Gupta, R. Priyadarshini, B. Lochab, Acs Omega **2**, 3070 (2017)
14. S. Jaworski, M. Wierzbicki, E. Sawosz, A. Jung, G. Gielerak, J. Biernat, H. Jaremek, W. Lojkowski, B. Wozniak, J. Wojnarowicz, et al., Nanoscale Research Letters **13**, (2018)
15. P.K. Sandhya, J. Jose, M.S. Sreekala, M. Padmanabhan, N. Kalarikkal, S. Thomas, Ceram Int **44**, 15092 (2018)
16. K. Krishnamoorthy, M. Veerapandian, L.H. Zhang, K. Yun, S.J. Kim, The journal of physical chemistry C **116**, 17280 (2012)
17. N.K. Kadiyala, B.K. Mandal, S. Ranjan, N. Dasgupta, Materials Science & Engineering C-Materials for Biological Applications **93**, 191 (2018)
18. M.T. Guo, G.S. Zhang, Chemosphere **183**, 197 (2017)
19. K. Ko, M.J. Kim, J.Y. Lee, W. Kim, H. Chung, Sci Total Environ **651**, 1087 (2019)
20. J. Athinarayanan, V.S. Periasamy, R. Krishnamoorthy, A.A. Alshatwi, Materials Science & Engineering C-Materials for Biological Applications **93**, 242 (2018)
21. A.N. Mahmoodabadi, A. Kompany, M. Mashreghi, Mater Chem Phys **213**, 285 (2018)
22. R. Rajeswari, H.G. Prabu, Journal of Inorganic and Organometallic Polymers and Materials **28**, 679 (2018)
23. J. Mohammadnejad, F. Yazdian, M. Omidi, A.D. Rostami, B. Rasekh, A. Fathinia, Eng Life Sci **18**, 298 (2018)
24. L. Zhong, H. Liu, M. Samal, K. Yun, Journal of Photochemistry and Photobiology B-Biology **183**, 293 (2018)
25. K. Prasad, G.S. Lekshmi, K. Ostrikov, V. Lussini, J. Blinco, M. Mohandas, K. Vasilev, S. Bottle, K. Bazaka, K. Ostrikov, Sci Rep **7**, (2017)
26. L. Ouyang, Y. Deng, L. Yang, X. Shi, T. Dong, Y. Tai, W. Yang, Z.-G. Chen, Macromol Biosci **18**, (2018)
27. C.H. Deng, J.L. Gong, G.M. Zeng, P. Zhang, B. Song, X.G. Zhang, H.Y. Liu, S.Y. Huan, Chemosphere **184**, 347 (2017)
28. F. Bugli, M. Cacaci, V. Palmieri, R. Di Santo, R. Torelli, G. Ciasca, M. Di Vito, A. Vitali, C. Conti, M. Sanguinetti, et al., Interface Focus **8**, (2018)
29. M. Shoeb, M. Mobin, M.A. Rauf, M. Owais, A.H. Naqvi, Acs Omega **3**, 9431 (2018)
30. Y. Gao, X. Ren, J. Wu, T. Hayat, A. Alsaedi, C. Cheng, C. Chen, Environmental Science-Nano **5**, 362 (2018)
31. S. Liu, M. Hu, T.H. Zeng, R. Wu, R. Jiang, J. Wei, L. Wang, J. Kong, Y. Chen, Langmuir **28**, 12364 (2012)
32. F. Zou, H. Zhou, D.Y. Jeong, J. Kwon, S.U. Eom, T.J. Park, S.W. Hong, J. Lee, Acs Applied Materials & Interfaces **9**, 1343 (2017)

The optimization of standard test method for chemical composition of ceramic materials by spectrophotometry

Jinzhang Cai[1,a]

[1]Nanan Testing Institute of Quality and Metrology, Quanzhou. Fujian, China

Abstract. Chemical composition of ceramic materials is one of the key properties and its determination is complex and time-consuming. The melting of silicon dioxide is the most important and difficult step for its determination. This paper is about optimizing the pretreatment method from the test standard, QB/T 2578-2002 <Standard test method for chemical composition of ceramic materials by spectrophotometry>. It is not convenient to use mixed flux (lithium hydroxide and boric acid) for melting ceramic materials. We replaced it by lithium metaborate and inspected the influence. After optimization, ceramic materials can be melted more completely. The test steps don't change much except CaOandthe test result is the same as before.

1 Introduction

Chemical composition of ceramic materials has important influence on physical and chemical properties of ceramics. Soit is important to test chemical composition of ceramic materials. Ceramic materials usually contain SiO_2, Al_2O_3, Fe_2O_3, TiO_2, CaO, MgO, Na_2O, K_2O and so on. There are two common test methods at present, conventional chemical analytical methods and advance instrumental analytical method[1-3].It takes more chemical reagents, more test steps and more times to complete the test by chemical analytical method. And more toxic wastes are produced. In contrast, instrumental analytical method is sample, quick and friendly to environment because of simultaneous determination of multi-element and less reagents [4-5].The common step of both method is sample dissolution, and it is difficult to dissolve SiO_2. There are two usual methods for SiO_2 dissolution, alkali fusion and acidolysis. Alkali fusion is more safer and friendly to environment because acidolysis needs HF. The common agents of alkali fusion are lithium metaborate [6], sodium carbonate[7] and mixed flux [7-8]. The mixed flux on QB/T 2578-2002 is lithium hydroxide and boric acid. The defect of mixed flux is heterogeneous. This paper discusses the replace of mixed flux by lithium metaborate.

2 Experiment

2.1 Instruments and reagents

Ultraviolet-visible spectrophotometer: Shimadzu UV-2600. Muffle furnace: SX3-10-14, Xiangtan Instrument Co., Ltd.

Reagents: lithium metaborate, mixed flux(lithium hydroxide and boric acid, 4:1), chromogenic reagents, hydrochloric acid, sulfuric acid and so on. Standard substances: standard substance of kaolin(GBW03121a), SiO_2 standard solution, Al_2O_3 standard solution, Fe_2O_3 standard solution, TiO_2 standard solution, CaO standard solution, MgO standard solution, and the concentration of all standard solution is 1000mg/L.

2.2 Methods

Pretreatment: Add 0.33g standard solution (or 1g mixed flux) and 0.05g samples into platinum crucible, mix up them, and put the platinum crucible into muffle furnace at 950 degree. After 15 min's melting, take out the crucible, cool it to room temperature and put it into 1L dry beaker. Add 500mL soak solution into the beaker and solve samples by ultrasonic wave until samples are completely solved.

Ultraviolet-visible spectrophotometer: SiO_2 by silicon molybdenum blue spectrophotometry (the wavelength is 530nm); Al_2O_3 by chlorophosphonazo-I spectrophotometry (the wavelength is 580nm); Fe_2O_3 by 1,10- Phenanthroline spectrophotometry (the wavelength is 510nm); TiO_2 by diantipyrylmethane spectrophotometry (the wavelength is 420nm); MgO by chlorophosphonazo-Ispectrophotometry (the wavelength is 580nm);CaO, GBHA spectrophotometry(the wavelength is 510nm).

3 Result and discussion

3.1 Sample fusion

Fusion by lithium metaborate: samples melted

[a]515604900@qq.com

 https://doi.org/10.1051/e3sconf/202129001035

completely and became colorless and transparent.

Fusion by mixed flux: samples melted completely melted and became colorless and transparent. But there were some white spots.

The mixed flux is consist of lithium hydroxide and boric acid, and lithium hydroxide excesses. So themixed flux is more alkaline than lithium metaborate. According to the melt results, lithium metaborate is alkaline enough to dissolve samples and strong alkalinity may corrode platinum crucible. In addition, the heterogeneity of mixed flux may lead to insufficient reaction in local areas.

To sum up, it is better to dissolve samples by lithium metaborate.

3.2 Determination of oxides in ceramic materials

The alkalinity of samples dissolved by lithium metaborate is weaker than the alkalinity by mixed flux. The solution pH value is key to the color reactions of oxides. SiO_2, Al_2O_3, Fe_2O_3 and TiO_2 need acidic condition to take color reactions. The replace will not affect the color reactions because of weaker alkalinity of lithium metaborate. So the test process doesn't change. Their color reactions were normal and the linearity of the standard curves is quite well. See Figure 1-8.

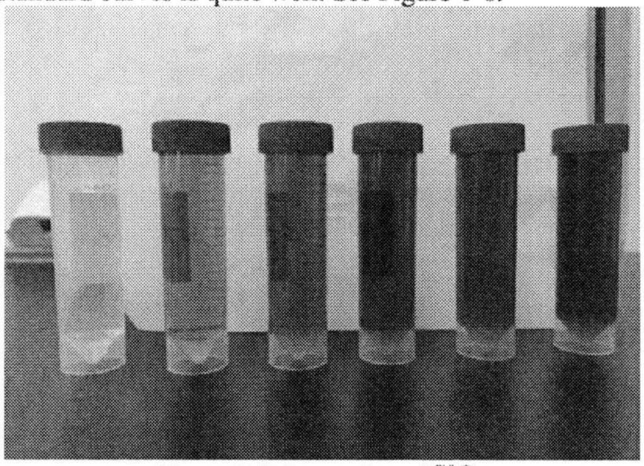

Figure 1. Color reaction of SiO_2

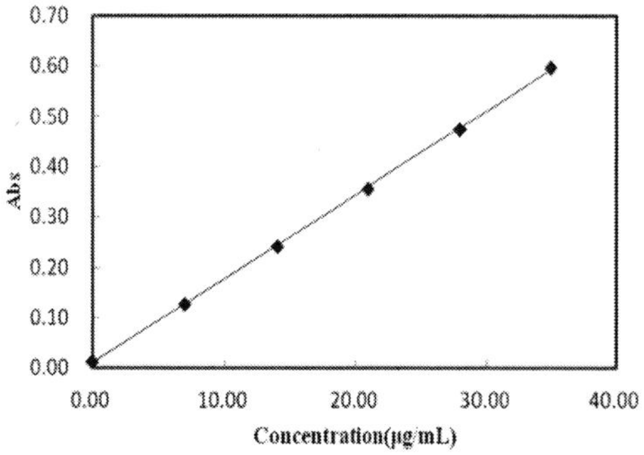

Figure 2. The standard curve of SiO_2

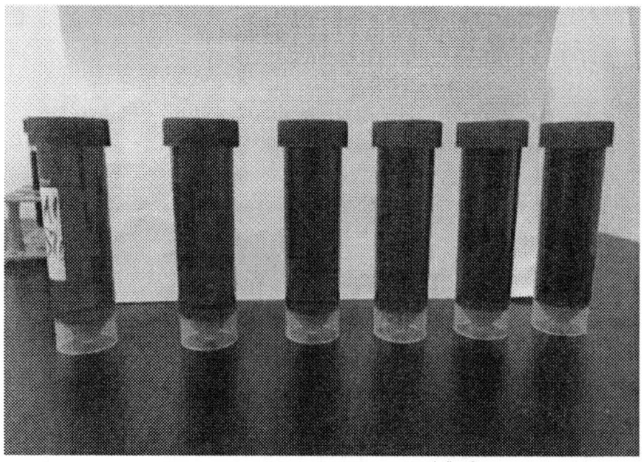

Figure 3. Color reaction of Al_2O_3

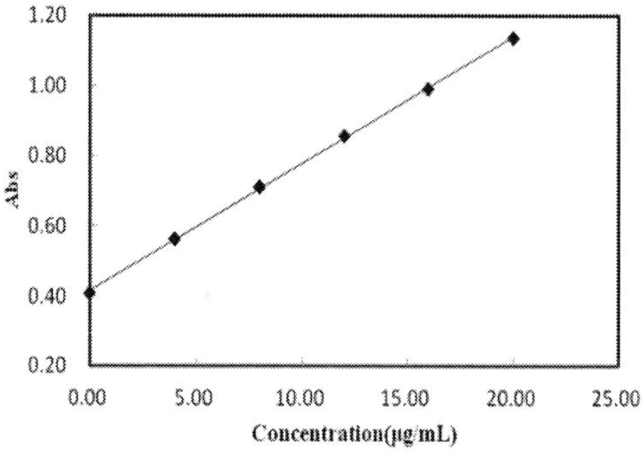

Figure 4. The standard curve of Al_2O_3

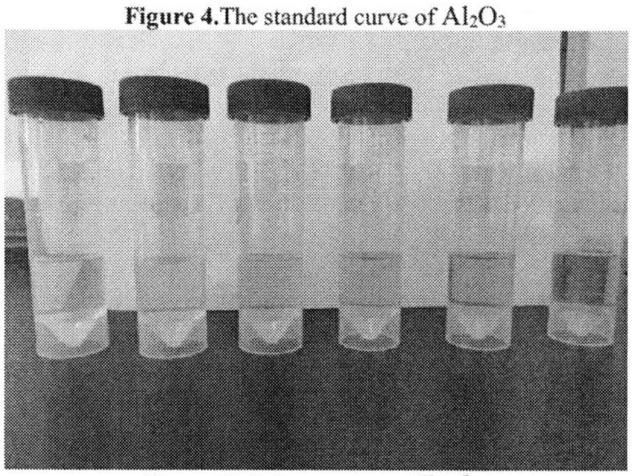

Figure 5. Color reaction of Fe_2O_3

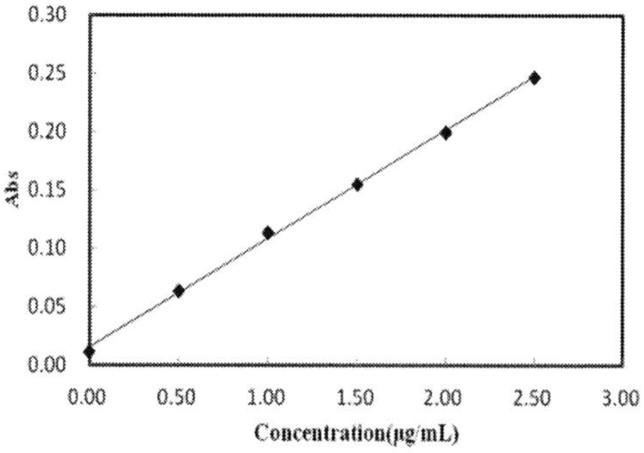

Figure 6. The standard curve of Fe_2O_3

https://doi.org/10.1051/e3sconf/202129001035

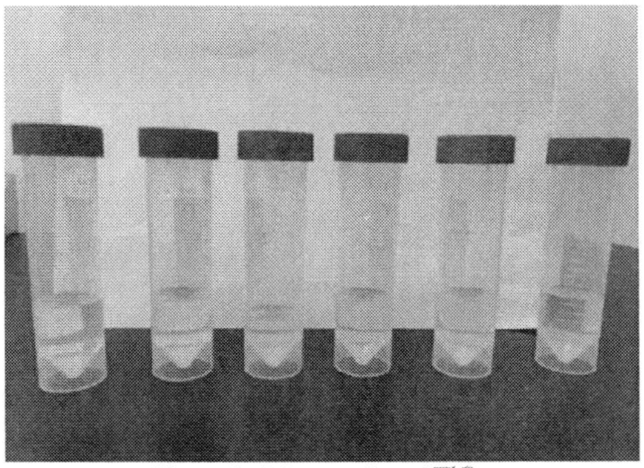

Figure 7. Color reaction of TiO_2

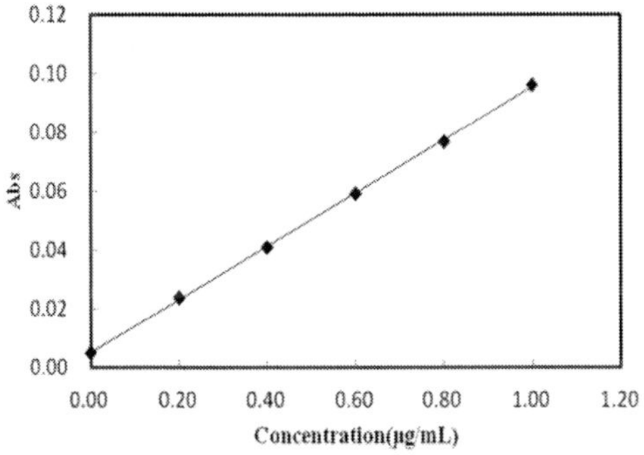

Figure 8. The standard curve of TiO_2

The standard doesn't set a pH range for MgO's color reaction. We just did the test as the standard required. We found the color reaction is normal and the linearity of the standard curve is good. See Figure 9-10.

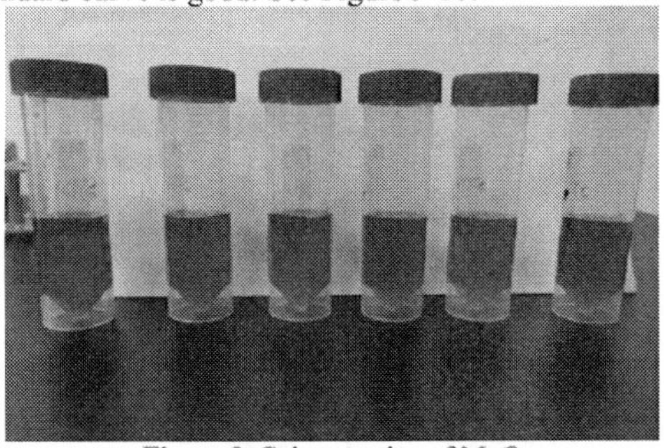

Figure 9. Color reaction of MgO

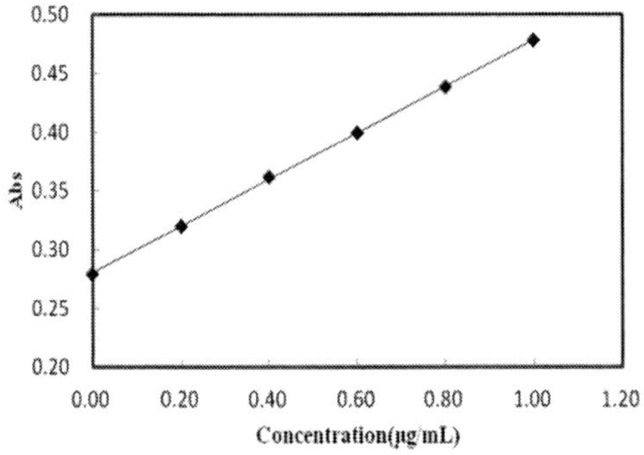

Figure 10. The standard curve of MgO

The color reaction of CaO occurs at NaOH solution of above 0.1mol/L. If we don't adjust pH of sample solutions and standard solutions, the tangerine color will occur after adding Ca color developing agen. This is caused by the amount of NaOH is not enough to neutralize HCl. As a result, we must add proper amount of LiOH to adjust the pH of solutions. After adding LiOH, the tangerinecolor occurs and the linearity of the standard curve is good. See Figure 2.

Figure 11. Color reaction of CaO

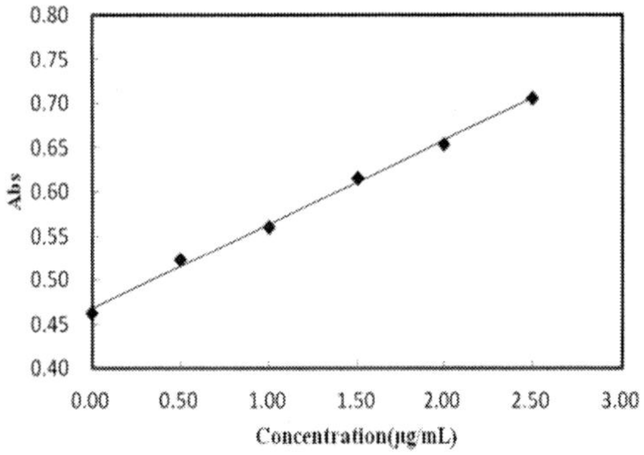

Figure12. The standard curve of CaO

The test results of the oxides in the kaolin(GBW03121a) are as follows.

Table 1. The parallel tolerance

	Test values		**Parallel tolerance**	**Parallel tolerance limit**
SiO_2	43.18%	43.06%	0.12%	0.30%
Fe_2O_3	1.55%	1.41%	0.14%	0.15%
TiO_2	0.32%	0.30%	0.02%	0.05%
Al_2O_3	34.20%	34.23%	0.03%	0.40%
MgO	0.061%	0.071%	0.01%	0.05%
CaO	N.D.	N.D.	/	/

 https://doi.org/10.1051/e3sconf/202129001035

Table 2. The test errors compared to standard values

	Test Values	Standard Values
SiO_2	43.12%	43.41%
Fe_2O_3	1.48%	1.50%
TiO_2	0.31%	0.25%
Al_2O_3	34.22%	34.70%
MgO	0.066%	0.069%
CaO	N.D.	0.038%

As the tables show, the Parallel tolerances of all oxides are within the limits and test values are all close to standard values. Detection limit for CaO is 0.1%, so the result is N.D. So it is OK to replace mixed flux by lithium metaborate.

4 Conclusion

In conclusion, it is better to use lithium metaborate as flux than mixed flux and the replacing can bring several benefits.

Firstly, it becomes more easier by omitting the mixing step. The heterogeneity of mixed flux can be avoided, then the inadequacy of melting reaction and corrosion to crucible will not happen. Secondly, the replacing does not bring many changes. The testing steps almost keep the same except CaO. The only change is adding pH adjust for determination of CaO in ceramic materials. Finally, the accuracy and stability of testing do not change. The Parallel tolerances of all oxides are within the limits and test values are all close to standard values.

References

1. Pan, Y.L. (2002) Contrast and selection of analysis methods of SiO_2. China Ceramics, 38: 33-35
2. Zheng, P.F. (2013) Contrast of analysis methods of SiO_2 in ceramic materials. Ceramics, 10: 11-13
3. Shi, X.T., Yi H.Y., Wang, H.J., Huang, Z. G., Lei, L.W., Xiao, H., Determination of titanium dioxide in raw materials of ceramics by microwave digestion spectrophotometry. Chemical Analysis and Meterage, 26:47-50
4. Feng, L., Chen, Z.W. (2008) Research progress of methods to the analysis of ceramic raw materials. China Ceramics, 44: 14-16
5. Lin, S. (2012) The analysis on the applicability of several method for chemical composition of ceramic materials. Foshan Ceramics, 22: 28-31
6. GB/T 14506.31-2019 Methods for chemical analysis of silicate rocks-Part 31: Determination of 12 components including silicon dioxide etc.-Lithium metaborate fusion-inductively coupled plasma atomic emission spectrometry
7. GB/T 4734-1996 Methods of chemical analysis for ceramic materials and articles
8. QB/T 2578-2002 Standard test method for chemical composition of ceramic materials by spectrophotometry

Nickel-Iron Layered Double Hydroxide for Highly Efficient Oxygen Evolution Reaction

Qiuyu Zhao[1] Wenke Xi[1] and Jianjun Wang[1*]

[1]MOE Key Laboratory of Resources and Environmental Systems Optimization, College of Environmental Science and Engineering, North China Electric Power University, Beijing 102206, PR China

Abstract. It is extensively accepted that electrolysis of water producing regenerable energy is a crucial substitution of traditional fuel strategy. Herein, we report a wet-chemical route to Ni-Fe layered double hydroxide (LDH) interconnected nanosheets with large surface area with the support of vertically aligned ZnO microrods arrays on the nickel foam (Ni-Fe LDH@ZnO/NF) for oxygen evolution reaction. Owing to the 2D Ni-Fe LDH nanosheets distributed in 3D space, the Ni-Fe LDH@ZnO/NF demonstrates excellent OER performances with a reasonably low overpotential of 271 mV at a current density of 10 mA·cm^{-2}, and a long-term stability up to 29 hours. This work proposes a new strategy to prepare Ni-Fe LDH/NF as efficient OER catalyst.

1 Introduction

With the development of industry and increase of population, it is of great urgency to product clean regenerable energy to replace fossil fuels. Hydrogen is considered as one of the most promising candidates because of its environmental friendliness and recyclability [1, 2]. Among the various methods of hydrogen production, electrolysis of water can be coupled with the wind energy or photovoltaic system which has been considered as the main way of hydrogen production in the future [3]. However, the important half-reactions in the water electrolysis process—oxygen evolution reaction (OER)—involving a four-electron reaction is constrained a sluggish kinetics, which generally requires a high overpotential to drive the reaction [4, 5]. Currently, RuO_2 and IrO_2 have been demonstrated to be effective electrocatalysts for OER [6, 7], but high cost and resource scarcity hinder their practical applications [8, 9]. Consequently, it is of elementary significance to engineer competent, low-cost and stable OER catalysts [10, 11].

Up to now, the improvement of the performance of electrocatalytic catalysts contains three main approaches: (1) increasing the catalytic specific surface area of material to expose more catalytic active sites; (2) enhancing the stability through material structure design; (3) improving the catalyst's electrical conductivity by integrating other components [12]. However, the mentioned enhancement strategies towards electro-catalytic materials still require numerous efforts to decrease the produced and maintained cost, prevent the secondary pollution, and promote the practical applications.

Among of Ni-Fe-based catalysts [13-17], particularly Ni-Fe layered double hydroxides (LDHs) have been extensively investigated, with a significant improvement of OER activity by structural design [18-22]. For example, Lou's group synthesized Ni–Fe LDH double-shelled nanocages exhibited an overpotential of 246 mV in 1.0 M KOH, which is lower than 261 mV by conventional Ni–Fe LDH single-shelled nanocages at 20 mA·cm^{-2} [18]. Shi et al. reported the immobilization of Ni-Fe LDH nanoplates reduced graphene oxide. Interestingly, the Ni-Fe LDH/3D-ErGO showed a higher OER activity with an overpotential of 259 mV than that of Ni-Fe LDH/2D-ErGO (278 mV) at 10 mA·cm^{-2} because of the large electrochemically active surface area of Ni-Fe/3D-ErGO [19]. However, improvement of poor stability and low conductivity of Ni-Fe LDH [20, 21] is still challenging towards its practical application.

Herein, we reported a wet-chemical route to synthesis Ni-Fe LDH@ZnO on nickel foam (denoted as Ni-Fe LDH@ZnO/NF), and then we investigated their electrochemical properties via serial characterizations and experiments. The result illustrated that Ni-Fe LDH@ZnO/NF exhibited excellent oxygen evolution reaction (OER) activity and long-term stability, which was ascribed to the effective enhancement catalytic surface area after the introduction of Ni-Fe LDH onto ZnO microrods arrays.

2 Experimental section

Growth of ZnO microrods arrays on nickel foam (ZnO/NF): Commercial nickel foam (NF) was ultrasonically washed with concentrated HCl (~37 wt %), ethanol and ultrapure water for 10 min, successively. Treated NF was then dropped into Teflon-lined stainless reactors (50 mL) containing 40 mL ultrapure water, 0.030 M $Zn(NO_3)_2·6H_2O$, and 0.025 M $C_6H_{12}N_4$. After the

[a] Corresponding author: wangjj2016@ncepu.edu.cn

https://doi.org/10.1051/e3sconf/202129001036

hydrothermal reaction conducting at 100 ℃ for 6 h, the uniform ZnO microrods were generated on NF.

Synthesis of Ni-Fe LDH@ZnO/NF: $NiFeS_x$ was grown on the surface of ZnO microrods by electrochemical cathode deposition at 70 °C in the 0.010 M $NiCl_2$ and 0.001 M $FeCl_3$ + 0.100 M $Na_2S_2O_3$ (10 mL) aqueous solution at 1.0 mA·cm^{-2} for 600 seconds. After drying, the $NiFeS_x$ was soaked in a 2-methylimidazole aqueous solution (0.10 M, 20 mL) at 90 ℃ for 2 h to obtain Ni-Fe LDH@ZnO/NF. For comparison, Ni-Fe LDH on NF (Ni-Fe LDH/NF) was synthesized by using the same method without ZnO microrods arrays.

3 Results and discussion

The Ni-Fe LDH@ZnO/NF mesostructured microrods were synthesized *via* wet-chemical route. Firstly, Hexagonal prism ZnO microrods arrays with an average lateral length of 1.2 μm were *in-situ* grown on NF (ZnO/NF) by hydrothermal method. Secondly, a NiFeSx layer was electrodeposited on the surface of ZnO/NF (Figure. 1(a, b)). Finally, Ni-Fe LDH grew on the outside of ZnO *via* the reduction of NiFeSx by using 2-methylimidazole (Figure. 1(c, d)). As we can see, Ni-Fe LDH was uniformly distributed with a thickness of 2 nm with a distinct boundary between ZnO microrods arrays.

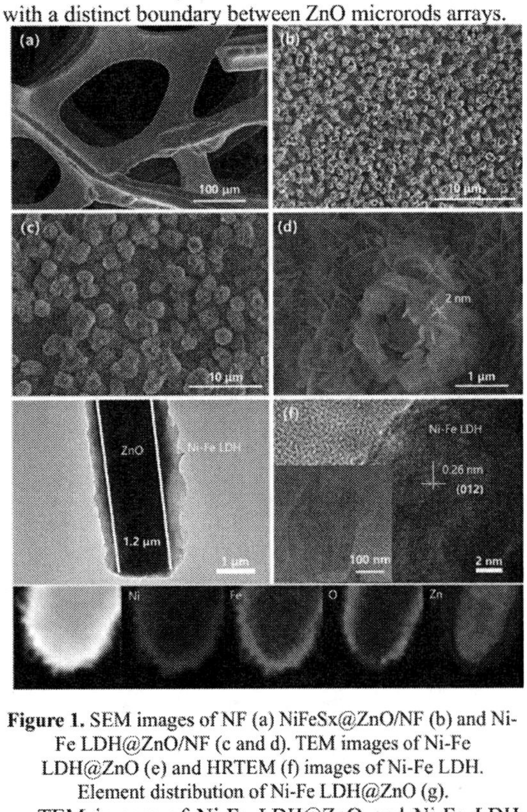

Figure 1. SEM images of NF (a) NiFeSx@ZnO/NF (b) and Ni-Fe LDH@ZnO/NF (c and d). TEM images of Ni-Fe LDH@ZnO (e) and HRTEM (f) images of Ni-Fe LDH. Element distribution of Ni-Fe LDH@ZnO (g).

TEM images of Ni-Fe LDH@ZnO and Ni-Fe LDH were shown as Figure. 1(e, f). The layered structure of Ni-Fe LDH was well maintained after decorated on the surface of ZnO microrods arrays. High-resolution image of TEM (HRTEM) revealed a lattice fringe distance of 0.26 nm, which corresponded to the spacing of (012) lattice plane of Ni-Fe LDH [22]. The image of TEM element mapping (Figure. 1(g)) indicated a homogeneous elemental distribution of Ni, Fe, Zn, and O, which further verified that the successful synthetization of Ni-Fe LDH/ZnO.

The X-ray diffraction (XRD) patterns of Ni-Fe LDH@ZnO/NF, Ni-Fe LDH/NF, ZnO/NF, and NF were displayed in Figure. 2(a). The peaks at 2θ =13.1 °, 19.5 ° and 33.7 ° can be attributed to (003), (006) and (009) planes of Ni-Fe LDH nanosheets (Figure. 2(b)) [23], indicating that LDH was successfully loaded on the ZnO (2θ = 32 °, 34.6 ° and 36.4 °) [24]. To investigate the electrocatalytic OER active sites, quasi real-time Raman spectra were obtained at different potential as shown in Figure. 2(c). The Raman bands at 484 cm^{-1} and 564 cm^{-1} correspond to γ-NiOOH (Ni^{3+}) [25]. The band intensities were changed with the increase of potential, indicating that the content of oxyhydroxides increased at the high potential. This revealed that oxyhydroxide was the active catalytic substance of OER. Thus, in line with the previous reports [23, 30], the oxyhydroxide was active for OER in our study. Besides, the peak at 1060 cm^{-1} was assigned to intercalated carbonate in Ni-Fe LDH.

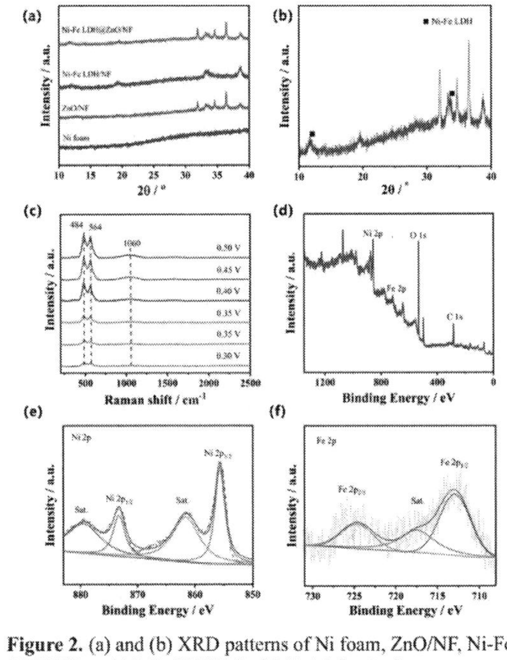

Figure 2. (a) and (b) XRD patterns of Ni foam, ZnO/NF, Ni-Fe LDH/NF, and Ni-Fe LDH@ZnO/NF. (c) *In-situ* Raman spectra of Ni-Fe LDH@ZnO/NF recorded at varied potentials during OER process. (d) XPS survey spectrum of Ni-Fe LDH@ZnO/NF. (e) and (f) High-resolution Ni 2p spectrum and Fe 2p of Ni-Fe LDH@ZnO/NF, respectively.

The X-ray photoelectron spectroscopy (XPS) spectra recorded on the surface of Ni-Fe LDH@ZnO/NF revealed the presence of Ni 2p, Fe 2p, and O 1s (Figure. 2(d)). The high-resolution XPS spectrum of Ni 2p in Figure. 2(e) showed a binding energy difference between two spin-orbit peaks, which was related to Ni-O (Ni $2p_{1/2}$ at 873.43 eV) and Ni-OH (Ni $2p_{3/2}$ at 855.73 eV) [26]. Figure. 2(f) showed the divided Fe $2p_{3/2}$ and Fe $2p_{1/2}$ were affiliated with iron oxides or hydroxides such as Fe_2O_3, Fe_3O_4, and

FeOOH. Though oxygen evolution reaction mechanisms demonstrated that the active variety was Ni^{2+} in the vicinity of Fe ions. With the Fe incorporation, overpotential of OER was significantly reduced, which was attributed to Ni^{3+} reduced to Ni^{2+} by H transfer around Fe^{2+}. Both Ni 2p and Fe 2p exhibited same ingredient of Ni-Fe LDH, verifying the existence of LDH. Zn 2p was not detected by XPS and we believe that it is due to the dense coverage of Ni-Fe LDH nanosheets (~200 nm) on the surface of ZnO which can be seen from the TEM image.

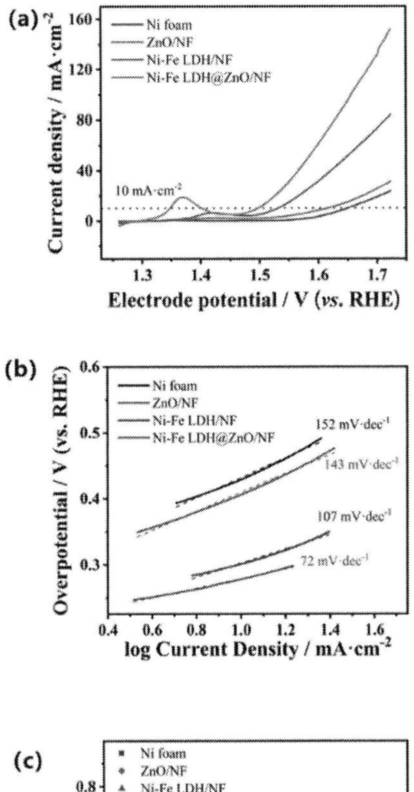

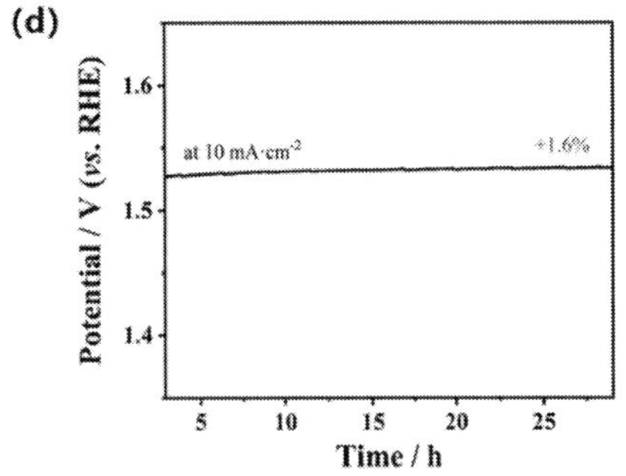

Figure 3. (a)-(c) LSV curves, Tafel plots and Double-layer capacitance (Cdl) plots of NF, ZnO/NF, Ni-Fe LDH/NF, and Ni-Fe LDH@ZnO/NF. (d) Stability measurement of Ni-Fe LDH@ZnO/NF at the current density of 10 mA·cm^{-2}.

The OER performances of Ni-Fe LDH@ZnO/NF was evaluated in 1.0 M KOH at the ambient condition. Figure. 3(a) showed its linear sweep voltammetry (LSV) curves. Ni-Fe LDH@ZnO/NF exhibited the lowest overpotential of 271 mV at 10 mA·cm^{-2} (NF 419 mV, ZnO/NF 386 mV and Ni-Fe LDH/NF 303 mV). It was worth noting that, the overpotential of Ni-Fe LDH@ZnO/NF was reduced by 32 mV in comparison with Ni-Fe LDH/NF which resulted from the introduction of ZnO. The Tafel slope of Ni-Fe LDH@ZnO/NF was calculated to be 72 mV·dec^{-1} (Figure. 3(b)), which was lower than that of Ni-Fe LDH/NF (107 mV·dec^{-1}), ZnO/NF (143 mV·dec^{-1}), and NF (152 mV·dec^{-1}). This indicated that Ni-Fe LDH@ZnO/NF possessed a faster kinetics towards OER due to the increase of γ-NiOOH and 3D structure. The ECSA was studied through the double-layer capacitance (Cdl) (Figure. 3(c) and Figure. 4(a-d)). ECSA with values of 2.8 μF·cm^{-2}, 1.2 μF·cm^{-2}, 0.68 μF·cm^{-2}, 0.35 μF·cm^{-2} were obtained for Ni-Fe LDH@ZnO/NF, Ni-Fe LDH/NF, ZnO/NF and NF, respectively. Ni-Fe LDH@ZnO (3D) had a larger ECSA than Ni-Fe LDH (2D) with the existence of ZnO. The best OER performance of Ni-Fe LDH@ZnO/NF results from the largest surface area of the materials. We also evaluated the stability of Ni-Fe LDH@ZnO/NF at 10 mA·cm^{-2} and it exhibited a long-term stability up to 29 h (with a potential increase of 1.6%) (Figure. 3(d)).

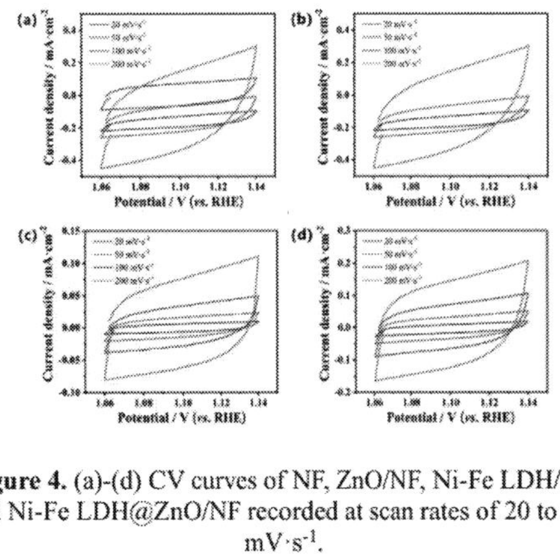

Figure 4. (a)-(d) CV curves of NF, ZnO/NF, Ni-Fe LDH/NF, and Ni-Fe LDH@ZnO/NF recorded at scan rates of 20 to 200 mV·s^{-1}.

4 Conclusion

A novel wet-chemical method has been successfully developed for Ni-Fe LDH nanosheets on ZnO microrods. Multiple characteristic technologies such as SEM, TEM, XRD, and XPS were conducted to conformed the accessibility of this method. The Ni-Fe LDH@ZnO/NF exhibited excellent OER catalytic ability with an overpotential of 271 mV at 10 $mA \cdot cm^{-2}$ as well as long-term stability up to 29 h with potential increase of 1.6%. The excellent OER activities of LDH@ZnO/NF were owing to a larger catalytic surface area and more catalytically active sites.

Acknowledgments

The work was financially supported by National Natural Science Foundation of China (11875028).

References

1. R. Zhang, X. Wang, S. Yu, T. Wen, X. Zhu, F. Yang, X. Sun, X. Wang, W. Hu, Adv. Mater **29**, 1605502 (2017)
2. M. S. Dresselhaus, I. L. Thomas, Nature 414, 332 (2001)
3. I. Roger, M. A. Shipman, M. D. Symes, Nat. Rev. Chem **1,** 0003 (2017)
4. C. C. L. McCrory, S. Jung, J. C. Peters, T. F. Jaramillo, J. Am. Chem. Soc **135**, 16977 (2013)
5. M. G. Walter, E. L. Warren, J. R. McKone, S. W. Boettcher, Q. Mi, E. A. Santori, N. S. Lewis, Chem. Rev **110**, 6446 (2010)
6. J. Kibsgaard, I. Chorkendorff, Nat. Energy **4**, 430 (2019)
7. J. W. D. Ng, M. Garcia-Melchor, M. Bajdich, P. Chakthranont, C. Kirk, A. Vojvodic, T. F. Jaramillo, Nat. Energy **1**, 16053 (2016)
8. E. A. Paoli, F. Masini, R. Frydendal, D. Deiana, C. Schlaup, M. Malizia, T. W. Hansen, S. Horch, I. E. L. Stephens, I. Chorkendorff. Chem. Sci **6**, 190 (2015)
9. D. Y. Kuo, J. K. Kawasaki, J. N. Nelson, J. Kloppenburg, G. Hautier, K. M. Shen, D. G. Schlom, J. Suntivich, J. Am. Chem. Soc **139,** 3473 (2017)
10. F M. S. Faber, S. Jin, Energ. Environ. Sci **7**, 3519 (2014)
11. X. Zou, Y. Zhang, Chem. Soc. Rev **44**, 5148 (2015)
12. J. Zhu, L. Hu, P. Zhao, L. Y. S. Lee, K. Wong, Chem. Rev **120**, 851 (2020)
13. F. Dionigi, P. Strasser, Adv. Energy Mater **6**, 1600621 (2016)
14. Y. Li, G. Zhang, W. Lu, F. Cao, Adv. Sci **7**, 1902034 (2020)
15. Z. Liu, C. Dong, Y. Huang, J. Cen, H. Yang, X. Chen, X. Tong, D. Su, Y. Wang, S. Wang, J. Mater. Chem. A **7**, 14483 (2019)
16. C. Huang, J. Nie, Z. Xu, X. Zhang, J. Tang, B. Wang, J. Huang, C. Du, J. Chen, Int. J. Hydrog. Energy **46**, 12992 (2021)
17. X. Feng, Y. Shi, J. Shi, L. Hao, Z. Hu, Int. J. Hydrog. Energy **46**, 5169. (2021)
18. J. Zhang, L. Yu, Y. Chen, X. F. Lu, S. Gao, X. Lou, Adv. Mater **32**, 1906432 (2020)
19. X. Yu, M. Zhang, W. Yuan, G. Shi, J. Mater. Chem. A **3**, 6921 (2015)
20. L. Wu, L. Yu, F. Zhang, D. Wang, D. Luo, S. Song, C. Yuan, A. Karim, S. Chen, Z. Ren , J. Mater. Chem. A **8**, 8096 (2020)
21. N. Todoroki, T. Wadayama, Acs. Appl. Mater. Inter **11**, 44161 (2019)
22. G. Zhang, J. Yuan, Y. Liu, W. Lu, N. Fu, W. Li, H. Huang, J. Mater. Chem. A **6**, 10253 (2018)
23. Trotochaud, L.; Young, S. L.; Ranney, J. K.; Boettcher, S. W. J. Am. Chem. Soc **136**, 6744 (2014)
24. L. Han, S. Dong, E. Wang, Adv. Mater **28**, 9266 (2016)
25. J. Feng, J.Wu, Y. Tong, G. Li, J. Am. Chem. Soc **140**, 610 (2018)
26. Y. Han, Z. Liu, Z. Yang, Z. Wang, X. Tang, T. Wang, L. Fan, K. Ooi, Chem. Mater **20**, 360 (2008)
27. X. Wang, S. Zhou, W. Xing, B. Yu, X. Feng, L. Song, Y. Hu, J. Mater. Chem. A **1**, 4383 (2013)
28. X. Fan, M. Zhang, I. Shafiq, W. Zhang, C. Lee, S. Lee, Adv. Mater **21**, 2393 (2009)
29. Y. Luo, Y. Wu, D. Wu, C. Huang, D. Xiao, H. Chen, S. Zheng, P. Chu, ACS. Appl. Mater. Interfaces **12**, 42850 (2020)
30. S. Klaus, Y. Cai, M. Louie, L. Trotochaud, A. Bell, J. Phys. Chem. C **119**, 7243 (2015)

Recent developments: self-healing polymers based on quadruple hydrogen bonds

Guangpu Zhang[1,a], Zhe Sun[1], Miaomiao Li[2]

[1]School of Chemistry and Chemical Engineering, Nanjing University of Science and Technology, Nanjing 210094, P. R. China
[2]Shanghai Space Propulsion Technology Research Institute, Huzhou 313000, P. R. China

Abstract. The microcrack of materials was inevitable in the process of transportation, storage and utilization, which may cause functional failure and resources waste. Inspired by nature, self-healing polymers have attracted significant attention owing to widespread applications in wearable electronics, cartilage replacement, coatings and elastomer. Compared with extrinsic healing, intrinsically healable polymers offer multiple self-healing by supramolecular reversible interactions, such as host-guest interactions, π-π stacking, ionic interactions and hydrogen-bonding. Self-healing polymers based on quadruple hydrogen bonds have been extensively investigated due to its high thermodynamic stability and rapid kinetic reversibility, and have been well developed for the past two decades. In this paper, the strategies and designs of self-repairing polymers based on quadruple hydrogen bond were classified and summarized, including main-chain self-healing polymers, side-chain self-healing polymers and supramolecular self-healing polymers. It is expected that quadruple hydrogen bonding can be construct more robust, highly tough, multi-stimuli-responsive, and fast self-healing supramolecular polymer, and is potential to be applied to numerous civilian and military fields in the future.

1 Introduction

Microcrack and damage are inevitable in the process of material storage and utilization, which cause functional failure, resources waste, and even security issues[1,2]. Therefore, the ability to spontaneously heal defect is essential to engineering materials, and can improve their survivability, lifetime, and economic effect. Inspired by nature, like cucumbers[3], mussel[4,5], and titin[6], scientists and engineers have developed numerous self-healing materials over the past decades, which can be divided into extrinsic healing and intrinsic healing materials[7]. The latter, which includes host-guest interactions, π-π stacking, ionic interactions and hydrogen-bonding, attracted public and scientific attention owing to its multiple self-healing and mild repairing condition.

Self-healing polymers based on hydrogen-bonding has been extensively investigated because of high affinity and directionality[8]. Ureidopyrimidone (UPy)(Figure1), which is a classic moiety of multiple H-bonding, famous for its high thermodynamic stability ($\Delta G \approx 10$ kcal mol^{-1}), rapid kinetic reversibility ($k_{off} \approx 8$ s^{-1}) and high dimerization ($K_{dim}=6\times10^{7}$ M^{-1}, chloroform), have been well developed these years[9-11]. Ureidopyrimidone (UPy) moiety can not only effectively enhance mechanical strength, but also endow engineering materials fascinating self-healing capacity, dramatically influence the safety, service life, intelligent manufacturing of the materials[12,13].

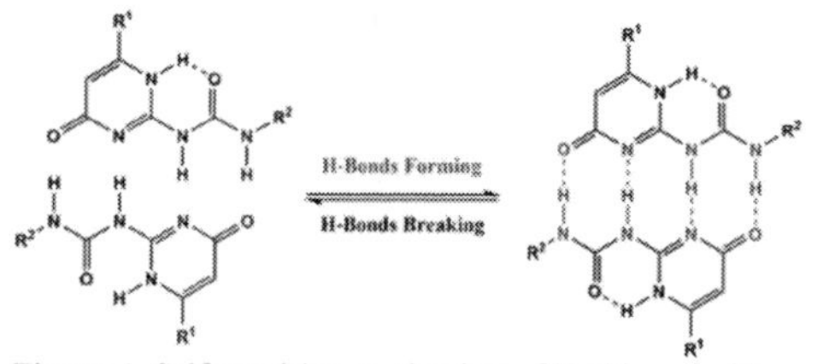

Figure 1. Self-repairing mechanism of Ureidopyrimidone (UPy) groups

Therefore, this paper summarized the research progress in self-healing polymers based on quadruple hydrogen bond, mainly including main-chain self-healing polymers, side-chain self-healing polymers and supramolecular self-healing polymers, and introduced the application prospect of self-healing polymers based on quadruple hydrogen bond. The future development of self-repairing materials involved smart material, composites material, and energetic materials is proposed. The mechanical properties, self-healing ability, and energy properties should be improved together with the introduction of quadruple hydrogen bond, indicating its potential for future technology development and industrial progress.

2 Research progress of self-healing polymers based on quadruple hydrogen bonds

The high dimerization and rapid kinetic reversibility of

[a] Corresponding author: gpzhang@njust.edu.cn

quadruple hydrogen bonds endow supramolecular polymers excellent mechanical strength and fascinating self-healing capability. The self-healing polymers based on quadruple hydrogen bonds can be spontaneously recovered original mechanical properties, electrical conductivity, and functional performance, improving material reliability, durability, and lifetime. The strategies and designs of self-healing polymers based on quadruple hydrogen bonds have been well pioneered and developed for the past two decades, which can be divided into main chain self-healing polymers, side chain self-healing polymers and supramolecular self-healing polymers according to the molecular structure and polymer designs.

2.1 Main-chain self-healing polymer

When main-chain self-healing polymer is damaged by external forces, the quadruple hydrogen bond between the polymer chains breaks first to form non-associative hydrogen bond groups. Under mild conditions, the quadruple hydrogen bond can be reformed to realize the self-healing of the material. Fan et al. introduced 2-ureido-4[1H]-pyrimidinone (UPy) group into the PPG chain segment and prepared a robust self-healing polyurethane elastomer PPG-mUPy[14]. By adjusting the PPG segment length and UPy content, the tensile strength of optimized sample PPG_{1000}-$mUPy^{50\%}$ can reach 20.62 MPa, and the self-healing efficiency can reach over 93%. The repair process is shown in the figure2. Yang et al. prepared waterborne polyurethane dispersion which used poly (1,4-buthylene-neopentylene adipate glycol) as soft segment, 2-ureido-4[1H]-pyrimidinone (UPy) functionalized monomer (5-(2-hydroxyethyl)-6-methyl-2-aminouracil [HMA]), isophorone diisocyanate, blocking agent, and hydrophilic monomer as the hard segment[15]. The self-healing time of the scratched film is inversely proportional to the content of HMA, and the mechanical properties can be restored to more than 90% in 2.5h at 80°C. Wang et al. synthesized high-performance PUs with self-healing ability and reprocessability combining 2-ureido-4-[1H]-pyrimidinone (UPy) and iron−catechol (DOPA−Fe^{3+}) cross-links[16]. These two dynamic forces, as sacrifice bonds, can effectively dissipate energy under external stress load, which makes the double-physical cross-linked network higher toughness and elongation at break. However, the self-healing temperature of PU−UPy−DOPA−Fe is 115°C and the self-repair efficiency is less than 80%, which attributes to additional physical cross-linking structure formed by metal coordination. Fan et al. achieved visible light assisting supramolecular polymers combining dynamic ditelluride bonds and 2-ureido-4[1H]-pyrimidinone (UPy) moieties in main chain. Improved mechanical properties and fast recoverability attributes to the physical cross-links formed by quadruple hydrogen bonded UPy moieties[17].

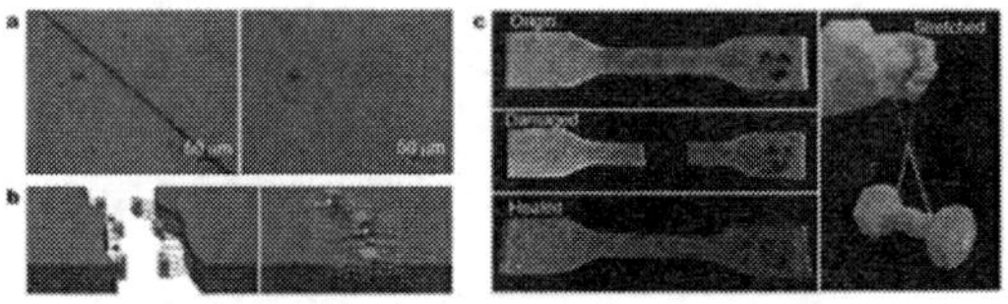

Figure 2. Self-healing process of PPG-mUPy

2.2 Side-chain self-healing polymers

Side-chain self-healing polymers can form a physical cross-linking network through the interaction of quadruple hydrogen bonds side groups, which can often be prepared by controllable free radical copolymerization or polyurethane condensation methods[18-21]. Zhu et al prepared self-healing polymers by free radical copolymerization of 2- (3- (6-methyl-4-oxo-1,4-dihydropyrimidin-2-yl) ureido) ethyl acrylate (MAUPy), hydroxyethyl acrylate (HEA) , PEG360MA and PEG500MA[22]. The chain length of flexible chain determines the self-healing ability of the supramolecular polymer (Figure3). Specifically, a long flexible chain permits a lower glass transition temperature and sufficient chain segment, and thus ensure the self-healing properties at room temperature. Yan et al. developed a tough and multi-stimuli-responsive supramolecular polymer networks based on RAFT polymerization, which integrated the functions of the shape memory and self-healing[23]. The polymer network consisted of two cross-linked networks, one is a chemical cross-linked network formed by the polymerization of PEG and PCL, and another is dynamic physical cross-linking network formed by the quadruple hydrogen bond of MAUPy. Chemical cross-linking and dynamic double cross-linking network, not only make it have dynamic reversible and self-repairing properties, but also avoid permanent irreversible deformation under stress.

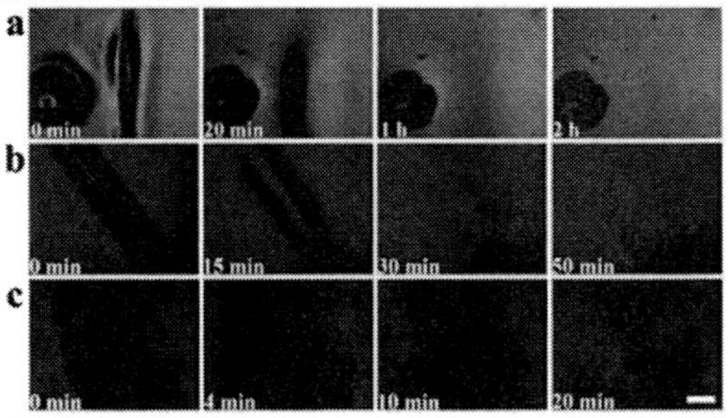

Figure 3. Self-healing processes of PHEA-UPy at ambient temperature with a relative humidity of 50 %: a) PPEG360-UPy; b) PPEG500-UPy films; and c) without special humidity treatment

Side-chain self-healing polymers can also be achieved by polycondensation of oligomer, isocyanate, and dihydroxyl chain extender with quadruple hydrogen bond[24-26]. Hu et al. reported a strategy of combining dynamic main chain and dangling UPy side group to fabricate a robust self-healing polyurethane[27] (Figure4). When the content of quadruple hydrogen bonds is low, the effective binding of quadruple hydrogen bond will decrease and the self-repair property will decrease. If the content of quadruple hydrogen bond is high, the irregularity of structure and the blocking effect of hydrogen bond will be enhanced, and then the mechanical properties of quadruple hydrogen bond will be reduced. As the dangling side group, the quadruple hydrogen bond group intensified the chain diffusion ability at high temperature, and accelerated the recovery of tensile strength. Moreover, Side chain polymers can be obtained by the reactions of epoxide functionalized-PEG and 2(6-isocyanatohexylaminocarbonylamino)-6-methyl-4[1H]-pyrimidinone (UPy-NCO) with the hydroxyl of PVA reported by Ye Hyang Jo et al [25]. The optimal PVA-UPy-PEG750 electrolyte material with low glass transition temperature (T_g), good self-healing ability and high ionic conductivity with shape memory. The PVA-UPy-PEG750 battery has good compatibility with lithium metal electrode and Li/PVA-UPy-PEG750/LFP can achieve coulomb efficiency of 99% with 0.1C at 60 °C and a discharge capacity of 145 mA h g^{-1}.

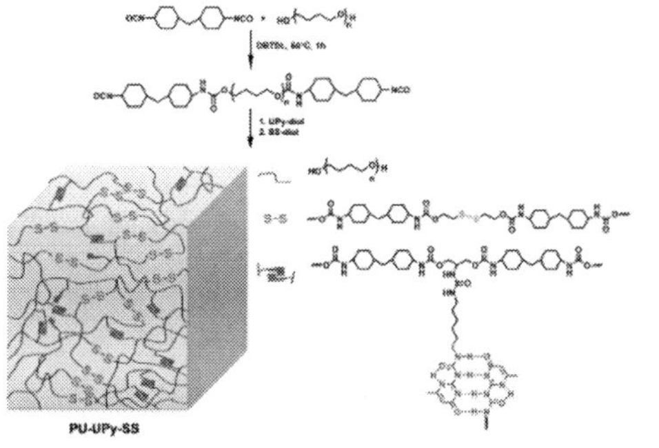

Figure 4. The synthetic routes of PU-UPy-SS samples with dynamic dangling UPy side chains

2.3 Supramolecular Self-healing polymers

The chain propagation of supramolecular self-healing polymer was achieved by quadruple hydrogen bond moiety, and the self-healing capability were accomplished by reversible UPy groups in the ends of the oligomer at both ends of the oligomer[28-36]. Yang et al. synthesized Ureidopyrimidinone functionalized polyethylene glycol (UPy-PEG-UPy) by mimicking the self-healing ability of living organisms and verified its application in anode materials for Li-ion batteries[37]. The self-healing electrode can be automatically repaired at room temperature after 1.5 h and exhibited excellent electrochemical performance, with an initial Coulomb efficiency of 81% and average capacity decay of only 0.04% after 400 cycles. Ren et al. prepared a linear polyurethane from UPy-Capped Poly(dl-Lactic Acid) Macrodiol, which exhibited attractive properties than covalent polyurethane[35]. The UPy-Capped polyurethane can be basically healed within 80 s and completely healed within 270 s at body temperature, and is expected to be applicated in the field of medical materials (Figure5). Dai et al. introduced the reversible imine bonds and quadruple hydrogen bonding (UPy) moieties into polymer networks to construct self-healable triboelectric nanogenerators based on infrared radiation through human skin[32]. Electrification layer almost completely recovered to the initial performance after 24 h, which both imine bond and quadruple hydrogen bond contributes to remarkable self-healing performance. KatelynR. Houston et al. investigated the effect of the UPy group, oligomer molecular weight on the structure-property relationships of supramolecular self-healing polymers by end-group functionalization of glycol modified PET with Ureidopyrimidinone [31]. The enhancement of materials properties of lower molecular weight polymer were remarkable, and the quadruple hydrogen bond with alkyl substituent can significantly improves the thermomechanical and viscoelastic properties due to the large substituent hindrance and competitive effect of aromatic ring substituent.

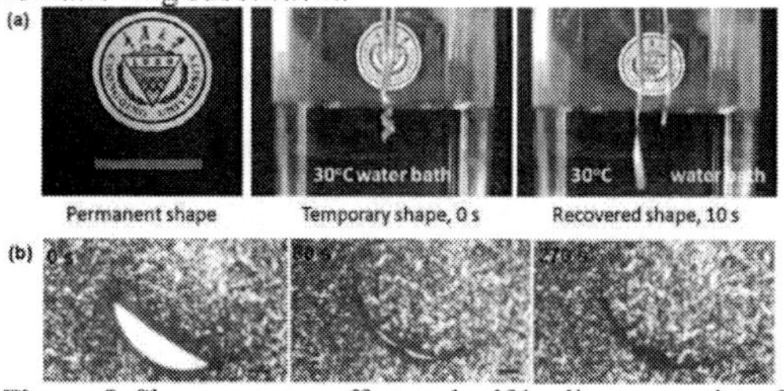

Figure 5. Shape memory effect and self-healing properties of UPy-PDLLA-PEG400-PDLLA4K-UPy

Bo Zhu et al. synthesized thermo-reversible supramolecular polyurethanes by condensation polymerization of isocyanate-terminated oligomer PTMG and 2-ureido-4-[1H]-pyrimidinone (UPy) with amine group[38]. The UPy groups consequently serve as thermo-reversible associations or the hard blocks and endow supramolecular polyurethane self-assemble capability and excellent mechanical properties. Bo Qin et al. proposed a new strategy of incorporating low-content quadruple hydrogen bonding to construct tough and multi-recyclable cross-linked supramolecular polyureas[39]. Through the incorporation of quadruple hydrogen bonded diamine monomer, the transparent sample are endowed with superior toughness and excellent multiple recyclability which can recover more than 95% properties even after five generations of recycling processes (Figure6), providing a novel and general strategy for the preparation of reprocessible and recyclable cross-linked materials.

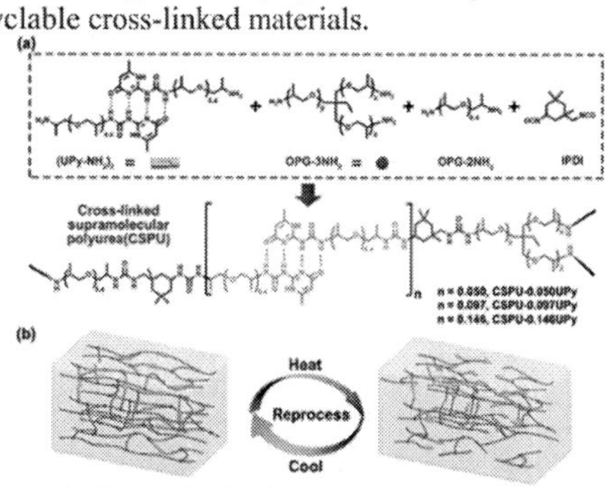

Figure 6. The synthesis, chemical structures, and schematic diagram of the recycling process of CSPU incorporating UPy groups

The above synthesis method required the precursor polymer with hydroxyl, or amine, or carboxylic groups, which also limits its application due to the difficult synthesize of UPy synthon. Sachin Bobade et al.

combined ATRP and CuAAC click chemistry for the synthesis of telechelic oligomers functionalized with Propargyl-UPy and Azido-UPy[40]. The quadruple hydrogen bond provides the self-healing ability, and the triazole group can strengthen the UPy dimerization to enhance the mechanical properties of the polymer. However, complex copper used in CuAAC reaction were necessary to be removed by electrochemical avenue from the polymer (>98%), for which standard polymer purification procedures fail[41]. Xu et al. realized the fabrication of cross-linked supramolecular polymers through photoinitiated thiol-ene click reaction of quadruple hydrogen bonded supramonomers[42]. It was found that the cross-linked supramolecular polymer has a higher critical polymerization concentration than the linear supramolecular polymer and the supramolecular interfacial polymerization can effectively control thermodynamic properties of the prepared quadruple hydrogen bond self-healing polymer[43].

3 Application

The self-healing polymer based on quadruple hydrogen bonds, through the reformation of hydrogen bonds, endows the material self-repairing capacities which can be repaired at room temperature. It has significant influence in practical application, including self-healing hydrogels, self-healing elastomers, shape memory polymers based on 4D printing with longer service life, lower maintenance cost, and higher safety.

Hydrogel, an outstanding hydrophilic polymer, has been widely used in biomass materials, electronic sensors, and artificial organs, which attracted scientists in self-healing field. Lin et al. successfully prepared a polyurethane hydrogel with UPy side groups by free radical copolymerization reaction[44]. The crack in hydrogel can be automatically repaired within 10 minutes at room temperature, and exhibited enhanced mechanical properties than traditional hydrogels (such as SR, NC, DN, and MMC hydrogels). Aránzazu del Campo et al. synthesized a self-healing based on quadruple hydrogen bonds with cracks almost can be disappeared within 2 minutes, representing an alternative to physical thermo-gelling hydrogels for injectable materials for biomedical applications[20]. The sufficient chain movement provides hydrogels excellent self-healing, and the emergence of self-healing hydrogels show enormous potential in drug carriers, agricultural materials, food field.

Elastomer consists of soft matrix and hard phase, widely used in industrial products including rubber shoes, adhesives and electronic components. The emergence of self-healing technology endows elastomer fascinating character. Jian et al. synthesize a self-healing polyurethane elastomer which used PTMEG as the soft segment, IPDI as the hard segment, and HEDS as the chain extender. Disulfide exchange increase the mobility of the polymer chain, and hydrogen bond strengthen both mechanical properties and repairing capacity[45]. The repaired elastomer can reach to 5.01 MPa, almost recovering original tensile strength. Zhang et al. conclude self-healing capacity of polyurethane elastomer weaken when hard segment redundancy owing to insufficient hydrogen bonds. Song et al. prepared a self-healing elastomer based on quadruple hydrogen bonds in side chain. The toughness of the elastomer can reach 345 MJ/m^{-3} after 48 hours of healing at 80°C, showing its excellent self-healing effectiveness[46]. The application of self-healing elastomer undoubtedly exerts profound and lasting influence in future.

4D printing technology provides material capacity with the shape and structure can be dynamically changed over time, attracting intense attention all over the world[47]. However, the weak connection between the interfaces and terrible mechanical properties of printed products results in unsatisfactory printing structure and even printing failure. The introduction of self-healing technology into 4D printing can effectively resolve these defects[48,49]. Marta Invernizzi et al. explored digital light processing (DLP) printing with self-healing polymers based on quadruple hydrogen bonds[50,51]. The printed sample was shown in Figure7. The mechanical properties of printed samples are comparable to traditional processing products, and the self-healing efficiency is higher than that of cast samples. The self-healing of cracks can be achieved by placing it at 80°C for 1 hour, because the more effective interpenetration between PCLDMA and UPyMA results more quadruple hydrogen bonds in the damaged defect. Although the research on 4D printing based on quadruple hydrogen bonds is still in infancy, and the printing effect is not even ideal, however, the prospect is brightness due to the following advantages: first, it has greatly improved the mechanical properties, rheological properties, processability and other aspects of polymer materials; second, due to the dynamic reversibility of hydrogen bonds and the characteristics of macroscopic thermoplastics, polymers that could not be melt-processed (such as cross-linked resins) can now be processed by melt molding, which greatly expands the types of melt-processed materials; Third, endowing the material with reproducible, self-repairing ability, and recycling characteristics.

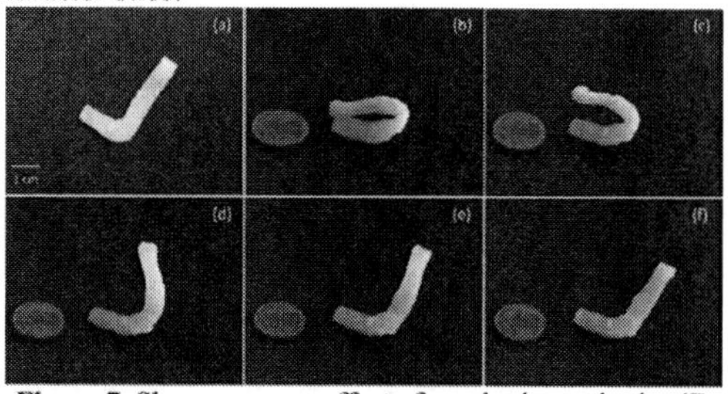

Figure 7. Shape memory effect of repaired samples by 4D printing

4 Conclusion

Self-healing polymers based on quadruple hydrogen bonds attracted public and scientific attention thanks to its excellent mechanical properties, charming self-healing capability, which can automatically repair damage by recombination of dynamic non-covalent quadruple hydrogen bonds under mild condition and improve the service life, stability and safety of engineering materials.

Numerous endeavors have been dedicated to the research and exploration in self-healing polymers based on UPy groups, which can be classified into main chain, side chain, and supramolecular self-healing polymer according to the effect of UPy moieties in polymer chain. It was found that low prepolymer molecular weight, appropriate UPy content, and advisable polymer type could endow materials excellent self-healing efficiency. The combination of self-healing covalent bonds, crosslinking network structure could strengthen mechanical strength and broaden the application, however, these strategies may confine polymer chain movement and require higher temperature to achieve self-healing. The quadruple hydrogen bonding is expected to be construct more robust, highly tough, multi-stimuli-responsive, and fast self-healing supramolecular polymer by sophisticated molecular designing, and is potential to be applied to numerous civilian and military fields in the future.

Acknowledgements

This work was supported by the Natural Science Foundation of Jiangsu Province (BK20200471) and Industry-University-Research Cooperation Foundation of Shanghai Academy of Spaceflight Technology (SAST2020-093).

References

1. G. Chen, Y. Yu, X. Wu, G. Wang, J. Ren, Y. Zhao, Adv. Funct. Mater. 28, 1801386. (2018)
2. Z. Liu, Z. Zhang, R. O. Ritchie, Adv. Mater. 30, 1705220. (2018)
3. D. Yuan, S. Delpierre, K. Ke, J. Raquez, P. Dubois, I. Manas-Zloczower, ACS Appl. Mater. Inter. 11, 17853-17862. (2019)
4. S. Song, H. Yang, Y. Cui, Y. Tang, Y. Chen, B. Yang, J. Yuan, J. Huang, Polymer 198, 122528. (2020)
5. P. Zhao, C. Yin, Y. Zhang, X. Chen, B. Yang, J. Xia, L. Bian, J. Mater. Chem. A 8, 12463-12471. (2020)
6. J. Schuetz, W. Peng, P. Vana, Polym. Chem.-UK 6, 2150. (2015)
7. C. Heinzmann, C. Weder, L. M. de Espinosa, Chem. Soc. Rev. 45, 342-358. (2016)
8. Q. Pei, A. Ding, Prog. Chem. 31, 257-274 (2019)
9. G. Armstrong, M. Buggy, Materials Science & Engineering C-Biomimetic and Supramolecular Systems 18, 45-49. (2001)
10. R. Sijbesma, F. Beijer, L. Brunsveld, B. Folmer, J. H. K. Hirschberg, R. Lange, J. Lowe, E. Meijer, Science (New York, N.Y.) 278, 1601-1604. (1997)
11. E. Berda, E. J. Foster, E. Meijer, Macromolecules 43. (2010)
12. S. C. Cummings, O. J. Dodo, A. C. Hull, B. Zhang, C. P. Myers, J. L. Sparks, D. Konkolewicz, ACS Applied Polymer Materials 2, 1108-1113. (2020)
13. L. Wang, Z. Gong, S. Li, W. Hong, Y. Zhong, D. Wang, L. Wan, Angewandte Chemie International Edition 55, 12393-12397. (2016)
14. C. Fan, Z. Huang, B. Li, W. Xiao, E. Zheng, K. Yang, Y. Wang, Science China Materials 62, 1188-1198. (2019)
15. Z. Yang, D. Sun, J. Appl. Polym. Sci. 137, 49413. (2020)
16. F. Wang, Z. Yang, J. Li, C. Zhang, P. Sun, ACS Macro Lett. 10, 510-517. (2021)
17. W. Fan, Y. Jin, L. Shi, W. Du, R. Zhou, S. Lai, Y. Shen, Y. Li, ACS Appl. Mater. Inter. 12, 6383-6395. (2020)
18. C. Heinzmann, U. Salz, N. Moszner, G. L. Fiore, C. Weder, ACS Appl. Mater. Inter. 7, 13395-13404. (2015)
19. Y. C. Gang Wanga, Angew. Chem. Int. Edit. 59, 2055-2060. (2019)
20. J. Cui, A. D. Campo, Chem. Commun. 48, 9302. (2012)
21. I. Jeon, J. Cui, W. R. K. Illeperuma, J. Aizenberg, J. J. Vlassak, Adv. Mater. 28, 4678-4683. (2016)
22. Q. Y. Dandan Zhu, Polym. Chem. 5086-5092. (2015)
23. M. L. Z. W. Jing Yan, Chem. Eng. J. 389, 123468. (2019)
24. Y. S. GuoLiang Li, Angew. Chem. Int. Edit. 57, 13838-13842. (2018)
25. B. Z. Ye Hyang Jo, Polym. Chem. 6561-6569. (2019)
26. Y. Chen, Y. Meng, Y. Luo, Y. Wang, Macromol. Mater. Eng. 305, 2000042. (2020)
27. J. Hu, R. Mo, X. Jiang, X. Sheng, X. Zhang, Polymer 183, 121912. (2019)
28. B. J. B. Folmer, R. P. Sijbesma, R. M. Versteegen, J. A. J. V. Rijt, E. W. Meijer, Adv. Mater. 12, 874-878. (2000)
29. R. Liu, X. Yang, Y. Yuan, J. Liu, X. Liu, Prog. Org. Coat. 101, 122-129. (2016)
30. B. Qin, S. Zhang, Z. Huang, J. F. Xu, X. Zhang, Macromolecules 8b-289b. (2018)
31. A. A. S. J. Katelyn R. Houston, Polym. Chem. 6744-6751. (2016)
32. X. Dai, L. B. Huang, Y. Du, J. Han, Q. Zheng, J. Kong, J. Hao, Adv. Funct. Mater. 1910723. (2020)
33. H. M. Keizer, R. V. Kessel, R. P. Sijbesma, E. W. Meijer, Polymer 44, 5505-5511. (2003)
34. F. Gao, J. Cao, Q. Wang, R. Liu, S. Zhang, J. Liu, X. Liu, Prog. Org. Coat. 113, 160-167. (2017)
35. W. Ren, Z. Li, Y. Chen, H. Gao, W. Yang, Y. Wang, Y. Luo, Macromol. Mater. Eng. 304, 1800491. (2019)
36. A. Watts, M. A. Hillmyer, Biomacromolecules 20, 2598-2609. (2019)
37. J. Yang, L. Zhang, T. Zhang, X. Wang, Y. Gao, Q. Fang, Electrochem. Commun. 87, 22-26. (2018)
38. B. Zhu, Z. Feng, Z. Zheng, X. Wang, J. Appl. Polym. Sci. 123, 1755-1763. (2012)

39. B. Qin, S. Zhang, P. Sun, B. Tang, Z. Yin, X. Cao, Q. Chen, J. Xu, X. Zhang, Adv. Mater. 2000096. (2020)
40. S. Bobade, Y. Wang, J. Mays, D. Baskaran, Macromolecules **47**, 5040-5050. (2014)
41. N. Jasinski, A. Lauer, P. J. M. Stals, S. Behrens, S. Essig, A. Walther, A. S. Goldmann, C. Barner-Kowollik, ACS Macro Lett. **4**, 298-301. (2015)
42. B. Q. J. X. Xiaoguang Liu, X. Z. Wang, Journal of Photochemistry and Photobiology A: Chemistry **355**, 414-418. (2018)
43. S. Z. Bo Qin, Angew. Chem. Int. Edit. **56**, 7639-7643. (2017)
44. Y. Lin, G. Li, J. Mater. Chem. B 2014, 2, 6878-6885.
45. X. Jian, Y. Hu, W. Zhou, L. Xiao, Polym. Advan. Technol. **29**, 463-469. (2018)
46. Y. Song, Y. Liu, T. Qi, G. L. Li, Angewandte Chemie International Edition **57**, 13838-13842. (2018)
47. H. WEI, X. WAN, Y. LIU, J. LENG, SCIENTIA SINICA Technologica **48**, 2-16. (2018)
48. D. P. Street, W. K. Ledford, A. A. Allison, S. Patterson, D. L. Pickel, B. S. Lokitz, J. M. Messman, S. M. Kilbey, Macromolecules **52**, 5574-5582. (2019)
49. Y. Zhu, J. Hu, Y. Liu, The European Physical Journal E **28**, 3-10. (2009)
50. M. Invernizzi, S. Turri, M. Levi, R. Suriano, Eur. Polym. J. **101**, 169-176. (2018)
51. R. Suriano, R. Bernasconi, L. Ma gagnin, M. Levi, J. Electrochem. Soc. **166**, B3274- B3281. (2019)

Advantage analysis of aluminum honeycomb composite used in motorhomes

Xinghai Shao[1,a], Qianqian Yang[1,2], Yuan Liu[3] and Lei Cao[1,2]

[1]Luoyang Advanced Manufacturing Industrial R&D Center, Tianjin Research Institute for Advanced Equipment, Tsinghua University, Luoyang 471004, China
[2]Qingyan Tecai technology (Luoyang) Co., Ltd, Luoyang 471004, China
[3]Tsinghua University, Beijing 100083, China

Abstract. In the light weight, high weather resistance and environmental protection requirements of the panel used for the motorhome, aluminum honeycomb composite, as a new material with mature industrialization and excellent performance, has a good application prospect in the quality improvement and material upgrading of the panel used for the motorhome. In this paper, the existing problems of the panel used for the motorhome and the properties of aluminum honeycomb composite are discussed, the advantages of applying aluminum honeycomb composite to the board of the motorhome are analyzed. The application of aluminum honeycomb composite materials to motorhomes will meet the requirements for improving the quality of panels , such as reduced in weight, environmentally friendly, non-combustible, recyclable, improved weather resistance and increased service life.

1 Introduction

The sample survey of China tourist vehicle and Vessel Association shows that, in 2020, China's self-driving travel group still reached 2 billion 240 million passengers under the influence of the new crown pneumonia epidemic, and the scale of self driving travel industry market was 125 billion 300 million RMB, up 10.4% from the same period last year[1]. With the normalization of epidemic prevention and control, and the growing demand for self driving travel, China's motorhome industry has ushered in the spring of rapid development, and the prospect is worth looking forward to.

Along with the development comes product upgrading and market competition. To a certain extent, the competition of the current motorhome industry is mainly distributed among product design, product material quality and reliability. For example, the lightweight, cost-effective, weatherability, environmental protection and fire resistance of the panel used in the passenger compartment are the necessary conditions for the passenger compartment to remain invincible in the market competition. In the demand of high-quality motorhome, aluminum honeycomb composite materials and panels with outstanding advantages, have excellent lightweight, cost-effective and environmental protection characteristics, and have good application prospects in the upgrading of compartment materials. Especially in the process of vigorously promoting the key parts technology autonomy, it is urgent to develop and master the high-quality panel manufacturing technology of light weight, weather resistance and environmental protection for residential vehicles. The lightweight and durable aluminum honeycomb composite materials and technology are widely used in the material upgrading of the panel of the motorhomes, which meets the development and application needs of the lightweight composite materials in the high-tech field [2-4].

2 Page layout

In vehicles, there are many kinds and quantities of panels used in motorhome, mainly including exterior compartment panel and interior furniture panel. At present, the main materials for the panel used in the motorhome are metal panels, polymer foam composite panels and wood panels. Based on the actual use environment of vehicles stored and walking in the open air, there are some requirements for the panels used in the motorhome, such as weight reduction, environmental protection, incombustibility, recyclability, weatherability improvement and service life improvement.

The main components and key components of the motorhome are compartment and interior furniture. By taking the materials and their preparation process used by a Chinese motorhome joint venture company as an example, the problems and demands for quality improvement of existing panels are analyzed. The company used a German production line, mainly produces European models of motorhome. According to the quality requirements of the vehicle, all carriage panels were imported. The following is a comparative analysis of the existing technology and service life of the carriage panel.

[a] Xinghai Shao: shaoxh@lamic.com.cn

2.1 Existing technology and problems of carriage panel

The panel for the motorhome mainly includes the panel for the outer wall of the carriage and the panel for the interior furniture, including the panel for the outer wall covering six sides of the carriage and the door panel, bed panel and cabinet panel for the interior furniture. Environmental protection, light weight, weathering resistance, aging resistance and UV resistance are important indicators of the panel. The general requirements of the composite panel are to meet the requirements of 60-70 ℃ and comply with the environmental protection standard. The company's panels are all imported materials. In the domestic market, more than 5% of the imported materials and 95% of the domestic panels are used in a comprehensive way. The advantages of the imported panels in appearance quality and service life are much higher than those of the domestic panels.

The following is an example to illustrate the problems existing in the current vehicle panel and the gap with foreign technology. Most of the carriage panels are composite structures, which are divided into panel core and panel veneer.

(1) Comparison from the material: ①The imported carriage panel adopts 3mm veneer + insulation layer + 2-3mm PVC material. The insulation layer material of the core is composed of two kinds of polymer composite materials with different hardness, which are used for structural strengthening and filling respectively. The insulation material can be recycled. The veneer of imported panel can be made of different materials, such as metal, wood and PVC, and the veneer is generally not recyclable. ② The core filling material of domestic panel is ordinary thermal insulation material with small thickness. The inner veneer uses rolled metal veneer instead of PVC layer, so as to reduce the production cost, but the quality of panel decreases.

(2) In terms of service life, the domestic panels are generally 30-50mm thick. At present, the main problems are short service life and poor weather resistance, and even easy to blister and seep after one year of use. The total thickness of the imported panel is about 50 mm, and its service life is generally 8 years.

(3) In terms of technology, there are two kinds of composite methods: positive spinning and vacuum negative pressure rolling. The former has poor surface quality but high bonding strength, while the latter has good surface quality but easy peeling. In addition to rolling veneer, color aluminum veneer is also used in a small amount of composite panel. At present, due to less application and immature technology, rolled veneer and color aluminum panel are difficult to repair. In comparison, PVC is easier to repair. But in the future, with the popularization and application of metal panel in motorhome and the development of repair technology, this problem will be gradually improved.

Generally speaking, the main problem of the current domestic panel used in the motorhome is poor weather resistance, and the advantage is low cost, which is mainly used as low-end panel; The imported panel has good weather resistance, but the cost is very high, so it is often used as high-end panel. The furniture panel of the motorhome is mainly composite panel. Like the composite panel used in the carriage, there is also a big difference in quality between home and abroad. The high-end furniture panel is generally imported.

2.2 Demand for quality improvement

In addition to the quality gap between domestic and foreign panels, there is still an urgent need for lightweight, weatherability and cost-effective improvement of the currently used car panels and furniture panels.

At present, the service life of the panel used in the passenger compartment is less than 8 years, and the service life of the furniture panel is 2-3 years. The polymer material and PVC veneer used have poor fire resistance rating, low strength and general density. In comparison, the aluminum honeycomb composite panel has lower density, higher specific strength, higher weatherability, more than 10 years or even decades of service life, and significantly improved fire performance. At the same time, the aluminum alloy preparation process is environmentally friendly, the product is non-toxic, tasteless, and the material can be recycled. The high strength, high hardness and low deformation characteristics of the metal material itself can significantly reduce the later maintenance cost of the material. Therefore, the material performance advantages and cost performance are significantly improved.

It will be an effective way to upgrade the materials and improve the quality of the panel for the motorhome by using the aluminum honeycomb structure composite panel with lighter weight, better weather resistance and non-flammability.

3 Aluminum honeycomb composite

Aluminum honeycomb composite material mainly refers to the sandwich composite structure, formed by making aluminum foil into honeycomb like structural material and bonding wood, stone and metal veneer on the upper and lower surfaces. This kind of aluminum honeycomb composite panel is equivalent to the I-beam structure, with outstanding compression and bending resistance, and the weight is only 10% ~ 20% of the same volume of metal panel, and has the characteristics of high specific strength, high weather resistance, fire resistance, high temperature resistance and environmental protection[5,6]. At present, hexagonal honeycomb is commonly used in aluminum honeycomb composite panel[6].

Due to the perfect structural strength of honeycomb, people have been committed to the bionic application research of honeycomb structure for many years. Since the 1950s, honeycomb materials have been gradually applied to the industrial field. The invention of aluminum honeycomb panel has brought revolution to the development of aviation industry. Due to the difficulties in manufacturing and processing, aluminum honeycomb panel is mainly used in high-tech and high value-added products such as aircraft and high-speed trains. In the field of special vehicles, it is mainly used in special parts and special high-end vehicles, and less used in ordinary civil vehicles.[7-11]

Aluminum honeycomb structure composite panel has the advantages of fire prevention, light composite, energy absorption and shock absorption, long service life, environmental protection and recyclability, which can fully meet the needs of improving the quality of the panel of the motorhome, which provides an effective reference for the application research of the motorhome[7-13]. Table 1 lists the performance comparison between the commonly used panel and aluminum honeycomb panel in the current motorhome. Aluminum honeycomb composite panel has outstanding advantages in density, service life, weather resistance and environmental protection.

Table 1. Pperformance Comparison of Common Panel and Aluminum Honeycomb Panel for Motorhome.

performance	High polymer composite panel for carriage	Interior furniture panel	Aluminum honeycomb composite panel
density	0.2-0.6 g/cm3	0.45-0.6g/cm3	0.1-0.4 g/cm3
life	5-8 years, aging easily	Domestic 2-3 years, imported 3-8 years	More than 10 years
combustion performance	combustible material	combustible material	Incombustible material
Environmental protection	strong smell, recyclable	Odorous, not recyclable	Non-toxic, tasteless and recyclable
Wet deformation	yielding	yielding	No deformation, washable
UV resistant	yielding	yielding	Not affected
Corrosion Resistance	Certain corrosion resistanc	yielding	Anti corrosion, no aging

In terms of density, strength, weather resistance, fire resistance and environmental protection, the quality of aluminum honeycomb composite panel is obviously better than the panel material currently used in the motorhome. The application of aluminum honeycomb composite to the motorhome can further reduce the weight of the motorhome, and improve the environmental protection, light weight, fire resistance and weather resistance of the compartment panel and the interior furniture panel of the motorhome. Compared with the imported products of the same kind, the aluminum honeycomb composite has a certain cost advantage, increase service life and significantly improve cost performance.

The following are some application cases of aluminum honeycomb composite.

(1) aviation and aerospace equipment materials

In the aviation sector, aluminum honeycomb material is widely used in helicopter power cabin fairing, bomber rudder, transport aircraft floor, assault aircraft flap and civil aircraft spoiler. In the aerospace field, aluminum honeycomb materials are widely used in the internal and external components of spacecraft such as satellites.

(2) Rail vehicle

Aluminum honeycomb composite panel with adhesive structure is widely used in the door, floor and roof of rail vehicles.

(3) Civil field

At present, aluminum honeycomb composite technology has been applied in building wallpanel, floor and roof. It has excellent properties such as high wind pressure resistance, shock absorption, sound insulation, heat preservation, flame retardant and high strength. It can be used for high-end civil furniture panel and engineering decoration panel after compounding the pattern plate and aluminum honeycomb. The pattern panel can imitate a variety of marble, stone and wood patterns, and has a wide range of applications. For example, aluminum honeycomb curtain wall panel is an advanced decorative material used on the external wall of high-rise buildings in recent years. It belongs to a light and thick hollow sandwich structure, with high strength and high specific strength, and good sound and heat insulation performance.

4 Advantages of aluminum honeycomb composite used in motorhome

4.1 Processing and appearance advantages

The panel of aluminum honeycomb composite panel is generally metal panel, which is pressed by mold. The processed material has smooth and regular surface, minimal deformation, high dimensional accuracy, good texture and good painting effect. Aluminum honeycomb structure material is suitable for the forming requirements of simple shape panel material on vehicles, and its molding technology is excellent. However, most of the traditional metal panels and profiles are welded in the process of pressing or bending, which is slightly inferior to the aluminum honeycomb parts in the aspects of smoothness, deformation and dimensional accuracy. The application of aluminum honeycomb composite technology in the motorhome can improve the overall appearance quality of the panel.

4.2 Decorative stability and weatherability

The stability of surface decoration is mainly analyzed from the anti-aging property of the material surface. In the production process of aluminum honeycomb panel, by ensuring the thickness of the coating and using high-grade and high-quality paint or powder spraying, it can meet the requirements of surface decoration stability to keep the color unchanged. In the process of compounding the panel and the aluminum honeycomb core, the outward facing surface decoration of the panel will not be damaged.

The aluminum alloy panel of aluminum honeycomb composite panel has good antistatic adsorption property, cleanliness and cleanliness, cleanliness and abrasion resistance, and excellent UV resistance, moisture resistance and anti-aging performance. The application of aluminum honeycomb composite technology in the motorhome can improve the decorative stability and weatherability of the panel.

4.3 Environmental protection and safety

Aluminum honeycomb and aluminum panel of aluminum honeycomb composite panel are all aluminum. There is no pollution and harmful gas volatilization in the production, processing, composite and final recovery of raw materials, so the environmental performance is excellent. By using aluminum honeycomb composite panel to replace polymer materials and wood, the fireproof performance of the panel can be improved to a level, and the vehicle safety performance can be significantly improved.

4.4 Energy saving and weight reduction

Low density and high-strength aluminum honeycomb composite panel is used as a structural part in the motorhome. It can completely replace the external compartment panel and internal furniture panel to reduce the weight of the whole vehicle and save energy consumption.

5 Conclusions

When the aluminum honeycomb composite panel is applied to the motorhome, the quality of the compartment and interior furniture panel will be significantly improved, which is mainly reflected in the improvement of strength, fire performance, weatherability, environmental protection and recyclability. Therefore, the motorhome is more high-end, reliable and durable. In view of the light weight, high weather resistance and environmental protection requirements of motorhomes, aluminum honeycomb composite material, as a new type of material with mature industrialization and excellent performance, has a very good application prospect.

The use of aluminum honeycomb composite panel can reduce the use of wood and polymer materials used in the original RV, reduce the felling of trees, and the environmental pollution caused by the production and waste of polymer materials. Furthermore, the aluminum honeycomb composite panel material can be recycled with high utilization efficiency, greatly reducing the probability of fire, which is a kind of green material. In other aspects, based on the advantages of low density and high specific strength of aluminum honeycomb composite, the motorhome can achieve weight reduction, further reduce the energy consumption in the process of traveling, and reduce the air pollution of the vehicle.

References

1. China Industry Research Institute . In-depth analysis of the self-driving tourism industry and investment strategy research consulting report in 2021-2026 [R] . China : Annual research and consultation report of panorama survey and investment strategy on china industry, 2021.2.
2. Senuma T. Physical Metallurgy of Modern High Strength Steel Sheets [J]. ISIJ International. 2001, 41(6): 520-532.
3. Banhart J, Seeliger H W. Aluminium foam sandwich panels: Manufacture, metallurgy and applications[J]. Advanced Engineering Materials, 2008, 10(9): 793-802.
4. Ning H, Janowski G M, Vaidya U K, et al. Thermoplastic sandwich structure design and manufacturing for the body panel of mass transit vehicle[J]. Composite Structures, 2007, 80(1): 82-91.
5. Herrmann A S, Zahlen P C, Zuardy I. Sandwich structures technology in commercial aviation[M]. Sandwich structures 7: Advancing with sandwich structures and materials. Springer, Dordrecht, 2005: 13-26.
6. Prof, Dehankar. Implementation Of Aircraft Wardrobe Using Honeycomb Composite Over Aluminium [J] . International Journal of Innovative Science, Engineering & Technology, Vol. 2 Issue 10, October 2015.
7. Aminanda Y, Castanié B, Barrau J J, et al. Experimental Analysis and Modeling of the Crushing of Honeycomb Cores. Applied Composite Materials. 2005, 12(3): 213-227.
8. Ashab A S M, Ruan D, Lu G, et al. Experimental investigation of the mechanical behavior of aluminum honeycombs under quasi-static and dynamic indentation. Materials & Design.2015, 74: 138-149.
9. Newton J F, Martin T D, Westre W N, et al. Lightweight honeycomb panel structure: U.S. Patent 5,445,861[P]. 1995-8-29.
10. Charles E H, Starnes Jr J H, Mark J S. An assessment of the state-of-the-art in the design and manufacturing of large composite structures for aerospace vehicles[J]. 2001.
11. Metschkow B. Sandwich panels in shipbuilding[J]. Polish Maritime Research, 2006: 5-8.
12. Wang D. Impact behavior and energy absorption of paper honeycomb sandwich panels[J]. International Journal of Impact Engineering, 2009, 36(1): 110-114.
13. Chen W, Hao H. Experimental and numerical study of composite lightweight structural insulated panel with expanded polystyrene core against windborne debris impacts[J]. Materials & Design, 2014, 60: 409-423.

Spatial Association characteristics of facilities around scenic spots considering distance and orientation: A case study of 3A and above scenic spots in Beijing

Hansong Cao[1,2,3,4] , Jiahui Qin[1,2,3,4] , Disheng Yi[1,2,3,4], Yusi Liu[1,2,3,4] and Jing Zhang[1,2,3,4,*]

[1]College of Resources Environment and Tourism, Capital Normal University, Beijing, 100048, China
[2]Beijing Laboratory of Water Resources Security, Capital Normal University, Beijing, 100048, China
[3]3D Information Collection and Application Key Lab of Education Ministry, Capital Normal University, Beijing, 100048, China
[4]Beijing State Key Laboratory Incubation Base of Urban Environmental Processes and Digital Simulation, Capital Normal University, Beijing, 100048, China

Abstract. In the urban tourism and service industry, the POI data with coordinate and attribute information of the major map platforms constitute one of the important data sources of the urban tourism and service industry. In this paper, the spatial data transaction database under four distances was established based on the gate buffer of 3A and above scenic spots in Beijing. The Apriori algorithm was used to calculate the lifting degree to obtain the distance for mining the best association features of 3A, 4A and 5A scenic spots, and then the association features of the three scenic spots in different directions were analysed.

1. Introduction

With the acceleration of modernization process and the development of mobile portable devices, LBS acquired by mobile devices provides massive data, which provides a strong theoretical basis and data support for the mining of spatial knowledge and spatial relations hidden in spatial data. Spatial association rules (SAR, Spatial Association Rules is an important part of Spatial data mining. Many researchers have carried out a lot of urban calculation work based on big data, using POI data, floating car track data and microblog check-in data to conduct research on the division of urban functional areas [1], travel characteristics of residents [2], urban vitality [3] and other aspects.

After the concept and algorithm of association rules [4] were proposed, scholars conducted a large number of researches, which mainly focused on the application field and the mining algorithm. From the perspective of application fields, studies are mainly focused on emergency events [5], environment [6], and urban functions [7]. In terms of mining algorithm, researches mainly focus on optimizing algorithm efficiency by using vertical data format [8], binary constraint [9], removable window [10] and other conditions. However, there are few research examples of spatial association analysis of urban facilities, and the current research is mainly based on the fuzzy perspective to study the association characteristics of urban services. In this paper, using POI point data of urban public facilities and considering the influence of distance and orientation, the spatial association characteristics of facilities in urban scenic spots are analysed. The main research contents are as follows :(1) Calculate the average promotion degree and analyse the optimal distance for mining the optimal association characteristics in different levels of scenic spots. (2) Calculate the similarities and differences of the association features of different levels of scenic spots in different directions and corresponding distances.

2. Materials and Methods

2.1. Overview of the study area and data acquisition

In this paper, the spatial association characteristics of 3A and above scenic spots in Beijing were analysed. As shown in the figure1, as of June 26, 2020, Beijing has a total of 149 scenic spots, including 72 3A-level scenic spots, 56 4A-level scenic spots and 8 5A-level scenic spots. In addition, the POI data of different operators offer abundance, there is a difference, to the POI classification system is not the same, Scott map as one of the most popular maps usage, this article from Beijing university open data platform for the scope of gold map of Beijing in 2018, a total of about 850000 POI data, selected from 29 class POI data contains the public service facilities, And about 420,000 taxi OD data on a rest day for experimental analysis.

*Corresponding author's e-mail: zhangjing@mail.cnu.edu.cn

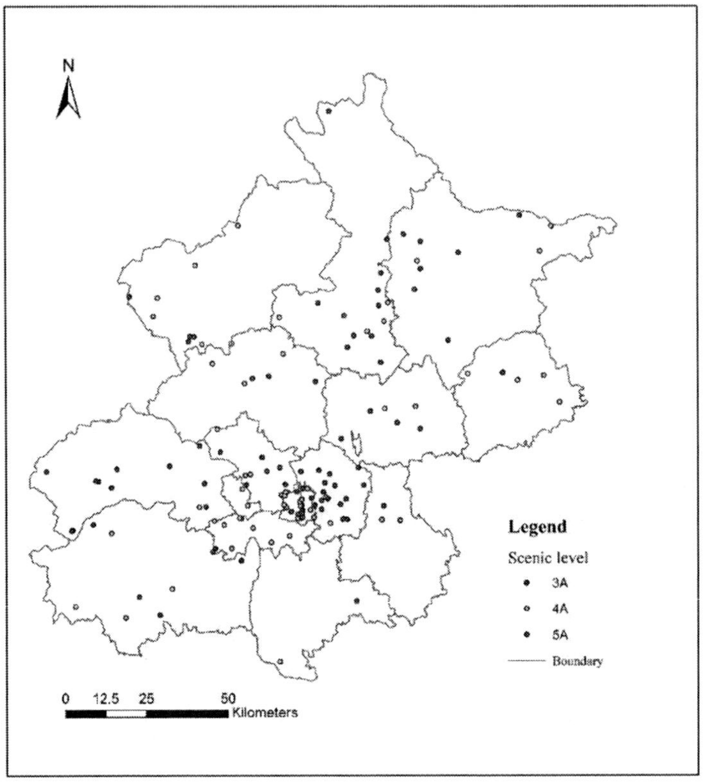

Figure 1. Study area

2.2. Overview of the study area and data acquisition

Association analysis is one of the core technologies of data mining. Association rule model and data mining algorithm were first proposed by Agrawal in 1993 [11]. The goal is to find interesting associations or interrelationships between projects in large amounts of data. The most classic case of association rule data mining is the story of "Beer and Diaper" in Wal-Mart. By mining the association of various goods in customers' "shopping basket", customers' shopping habits can be analysed to help merchants make better sales strategies. In basket mining, an item is an item, a collection of multiple items is called an item set, and each purchase record is called a transaction. Association rules are generally written in the form of X => Y, where the left item set X is the prerequisite and the right item set Y is the corresponding association result, which is used to represent the implied association in the data. For example, {scenic spot} => {parking lot}, indicating that the two spatial entities, scenic spot and parking lot, have a certain association within the distance threshold. For the intensity of association rules, whether scenic spots and parking lots are more concentrated or social security agencies and bus stations are more concentrated. The evaluation is mainly carried out by the concepts of support, confidence and lift [12].

Support refers to the possibility of appearing in all item sets {X, Y}, namely the probability of containing both X and Y, as shown in Equation (1).

$$Support(X \rightarrow Y) = P(X,Y) \tag{1}$$

Where, Support(X→Y) is the degree of Support, and the probability that P(X, Y) item set contains both X and Y. Support is the first requirement for mining strong association rules. Its significance lies in eliminating meaningless rules with low probability by setting minimum threshold, and reserving items with relatively frequent occurrence as spatial frequent item sets. The above example data is taken as an example. Assuming that the minimum support is 0.05, {parking lot, Chinese restaurant} =80/1000=0.08, {parking lot, convenience store} =10/1000=0.01. Since {parking lot, Chinese restaurant} meets the minimum support requirements, it is retained as the frequent item set. At the same time, the spatial association rule parking lot => convenience store is retained, while the rule corresponding to {parking lot, convenience store} is eliminated.

Confidence refers to the probability of occurrence of association result Y under the condition of occurrence of prerequisite X of association rule, that is, the item set containing X contains the possibility of both containing Y, as shown in Equation (2).

$$Confidence(X \rightarrow Y) = P(Y \mid X) / P(X) \quad (2)$$

Where, Confidence(X→Y) is Confidence; P(Y | X) is the probability of occurrence of the corresponding association result Y under the condition that the precondition X of the association rule appears; P(X) is the probability of Y. Confidence is the second requirement for generating strong association rules, which measures the reliability of association rules. It is also necessary to set the lowest confidence level to continue filtering. In the above example, when the minimum confidence is set as 0.6, the confidence of parking lot => Chinese restaurant is 80/100=0.8, and this rule is retained; the confidence of parking lot => convenience store is 10/100=0.1, and the rule is eliminated.

The promotion degree refers to the ratio of the possibility of both containing Y under the condition of containing X to the possibility of the item set containing Y without such a condition, namely, on the basis of the occurrence probability P(Y) of Y itself, the promotion degree of the occurrence probability P(Y | X) of Y by the appearance of X, as shown in Equation (3).

$$Lift(X \rightarrow Y) = P(Y \mid X) / P(Y) \quad (3)$$

In the formula, Lift(X→Y) is the lifting degree; Lifting degree can make up for the deficiency of confidence and can be regarded as a complementary index of confidence. When the degree of promotion value is 1, it means that instance X and Y are independent, and the occurrence of X has no effect on the possibility of occurrence of Y. The larger the degree of promotion value (confidence >1) is, the greater the degree of promotion of X to Y is, that is, the stronger the spatial association is.

The association analysis of tourist attractions and public service facilities can be similar to the analysis of shopping basket. Each facility type is equivalent to the goods in the shopping basket, and the facility type in each unit is equivalent to a shopping record.

3. Results & Discussion

In this paper, the Apriori algorithm is used to analyse the tourist attractions in Beijing at four distances of 100, 150, 250 and 500 meters. The minimum support is set as 0.2 to calculate the average promotion degree of association rules of the three types of scenic spots at different distances, as shown in Figure 1. It can be found that the 3A and 4A scenic spots have the highest average lift degree at a distance of 150 meters, and the 5A scenic spots have the highest average lift degree at a distance of 250 meters. Therefore, 150 meters is the best distance to excavate the association characteristics of the 3A and 4A scenic spots, and 250 meters is the best distance to excavate the 5A scenic spots.

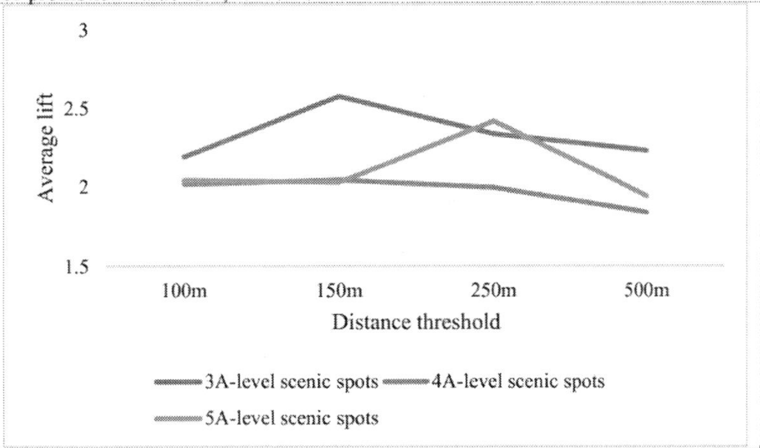

Figure2. Trend of average promotion degree of association rules in scenic spots at all levels with distance

In the distance, on the basis of in this paper, three kinds of scenic area according to the north and south, east and west four different bearing characteristics of the spatial association analysis, through many experiments, the 3 a level scenic spot in the 150 - meter distance, the select 0.1 as the minimum support degree of the east and the south, 0.2 as a scenic spot in west and north minimum support degree, minimum confidence level of 0.6. Due to space constraints, only the first five items in each direction are listed in the table.

Table 1. The top five rules of 3A-level scenic spots are in descending order of support.

Orientation	Association rules	Support	Confidence	Lift
East	{Social security agency} => {parking lot}	0.22	1	3.85
	{parking lot} => {social security agencies}	0.22	0.85	3.85
	{social security agencies, 3A-level scenic spots} => {parking lot}	0.22	1	3.85

https://doi.org/10.1051/e3sconf/202129002001

	{parking lot, 3A-level scenic spots} => {social security agencies}	0.22	0.85	3.85
	{Medical facilities} => {parking lot}	0.18	1	3.85
West	{Medical facilities} => {Chinese restaurants}	0.35	1	2.12
	{Chinese restaurants} => {Medical facilities}	0.35	0.75	2.12
	{auto service} => {Chinese restaurants}	0.35	1	2.12
	{Chinese restaurants} => {auto service}	0.35	0.75	2.12
	{illegal parking sign} => {parking lot}	0.35	1	2.83
South	{Chinese restaurants} => {parking lot}	0.36	1	2.27
	{parking lot} => {Chinese restaurants}	0.36	0.83	2.27
	{Chinese restaurants, 3A-level scenic spots} => {parking lot}	0.36	1	2.27
	{parking lot, 3A-level scenic spots} => {Chinese restaurants}	0.36	0.83	2.27
	{Medical facilities} => {parking lot}	0.31	0.86	1.97
North	{pick-up points} => {drop-off points}	0.4	1	1.52
	{drop-off points} => {pick-up points}	0.48	0.73	1.52
	{3A-level scenic spots, pick-up points} => {drop-off points}	0.48	1	1.52
	{3A-level scenic spots, drop-off points} => {pick-up points}	0.48	0.73	1.52
	{parking lot} => {drop-off points}	0.42	0.93	1.42

In the distance of 150 meters, the overall support degree of association rules is at a low level. Strong association rules related to parking lots are found in 3A-Level scenic spots in all four directions, indicating that their confidence degree is above 0.9. In the east direction, {social security agencies, 3A-level scenic spots} => {parking lot} is taken as an example. The probability of parking lots within 150 meters of 3A-level scenic spots and social security agencies in Beijing is about 22%. It can be seen from Table 1 that in the eastern position of the 3A-level scenic spot, the association features are mainly social security institutions, parking lots and medical facilities, while in the eastern position, the social security institutions are relatively complete. In the west, the related features are mainly medical facilities, parking lots, Chinese restaurants, and car services. The appearance of car rental and maintenance services indicates to some extent that car rental and maintenance services will be provided for self-driving tourists at the west gate of the 3A-level scenic spots. In the south direction, the association features are mainly Chinese restaurants, parking lots and medical facilities. Different from the other locations, the northern location attracts the characteristics associated with the pick-up and drop-off points of taxis, and the northern location is more likely to attract tourists to take taxis in the four locations of the 3A-level scenic spots.

Table 2. The top five rules of 4A-level scenic spots are in descending order of support.

Orientation	Association rules	Support	Confidence	Lift
East	{public toilets} => {parking lot}	0.25	0.69	1.42
	{social security agencies} => {parking lot}	0.25	1	2.05
	{Medical facilities} => {parking lot}	0.25	1	2.05
	{public toilets, 4A-level scenic spots} => {parking lot}	0.25	0.69	1.42
	{social security agencies, 4A-level scenic spots} => {parking lot}	0.25	1	2.05
West	{hostel} => {Chinese restaurants}	0.41	1	2

	{Chinese restaurants} => {hostel}	0.41	0.81	2
	{Medical facilities} => {parking lot}	0.41	1	2
	{parking lot} => {Medical facilities}	0.41	0.81	2
	{public toilets} => {Chinese restaurants}	0.41	0.9	1.8
South	{bus stations} => {parking lot}	0.36	0.9	1.83
	{parking lot} => {bus stations}	0.36	0.75	1.83
	{bus stations,4A-level scenic spots} => {parking lot}	0.36	0.9	1.83
	{parking lot,4A-level scenic spots} => {bus stations}	0.36	0.75	1.83
	{public toilets} => {parking lot}	0.34	0.85	1.73
North	{drop-off points} => {pick-up points}	0.73	1	1.36
	{pick-up points} => {drop-off points}	0.73	1	1.36
	{4A-level scenic spots, drop-off points} => {pick-up points}	0.73	1	1.36
	{4A-level scenic spots, pick-up points} => {drop-off points}	0.73	1	1.36
	{parking lot} => {drop-off points}	0.65	1	1.3

As shown in Table 2, in the four directions at a distance of 150 meters, the northern position has the highest overall support, followed by the south, the west, and the east. According to the analysis of sub-direction, under the eastern position of the 4A level scenic spot, the associated features are mainly public toilets, social security organs, parking lots, scenic spots and other related features, while under the eastern position, the scenic spot itself attracts more facilities. In the western bit, {hostel}=> {Chinese restaurant} is the highest association rule, which shows strong association characteristics. In the south, bus stations, parking lots, public toilets and scenic spots are the main associated features, which also show that the scenic spots attract other facilities. Similar to the 3A spots, the 4A spots are also more attractive to taxi pickup points in the north.

Table 3. The top five rules of 5A-level scenic spots are in descending order of support.

Orientation	Association rules	Support	Confidence	Lift
East	{Chinese restaurants} => {bus stations}	0.42	1	2.33
	{bus stations} => {Chinese restaurants}	0.42	1	2.33
	{Chinese restaurants} => {parking lot}	0.42	1	2.33
	{parking lot} => {Chinese restaurants}	0.42	1	2.33
	{Chinese restaurants} => {the convenience store}	0.42	1	1.75
West	{Chinese restaurants} => {bus stations}	0.6	1	1.6
	{bus stations} => {Chinese restaurants}	0.6	1	1.67
	{Chinese restaurants} => {parking lot}	0.6	1	1.25
	{parking lot} => {Chinese restaurants}	0.6	0.75	1.25
	{bus stations} => {parking lot}	0.6	1	1.25
South	{Chinese restaurants} => {parking lot}	0.76	1	1.18
	{parking lot} => {Chinese restaurants}	0.76	0.91	1.18
	{Chinese restaurants,5A-level scenic spots} => {parking lot}	0.76	1	1.18
	{parking lot,5A-level scenic spots} => {Chinese restaurants}	0.76	0.91	1.18

	{bus stations} => {parking lot}	0.69	1	1.18
North	{Chinese restaurants} => {public toilets}	0.57	1	1.75
	{public toilets} => {Chinese restaurants}	0.57	1	1.75
	{Chinese restaurants} => {social security agencies}	0.57	1	1.75
	{social security agencies} => {Chinese restaurants}	0.57	1	1.75
	{Chinese restaurants} => {Ordinary hotel}	0.57	1	1.75

As shown in Table 3, the support degree of association rules generated in the four directions is at a high level and the gap is small. The support degree of the south direction is the highest, followed by the west and north and south directions, and finally the east direction. The 5A scenic spots show strong association characteristics of Chinese restaurants, bus stations and parking lots in the east, west and south directions. Different from other level scenic spots, the characteristics of the highest association rules are less attracted by 5A level scenic spots and more attracted by each other's public facilities, resulting in the phenomenon of aggregation.

The 3A-level scenic spots will be taken as an example to illustrate the difference of spatial association characteristics in different directions. This paper used arulesViz visualization package drawing association rules as a result, the horizontal axis LHS (left) the rules of the first two facilities said in the group of the most frequent two types, the vertical axis RHS (right) and the left side of the associated facilities. The size of the circle indicates the degree of support, and the shade of color indicates the degree of improvement.

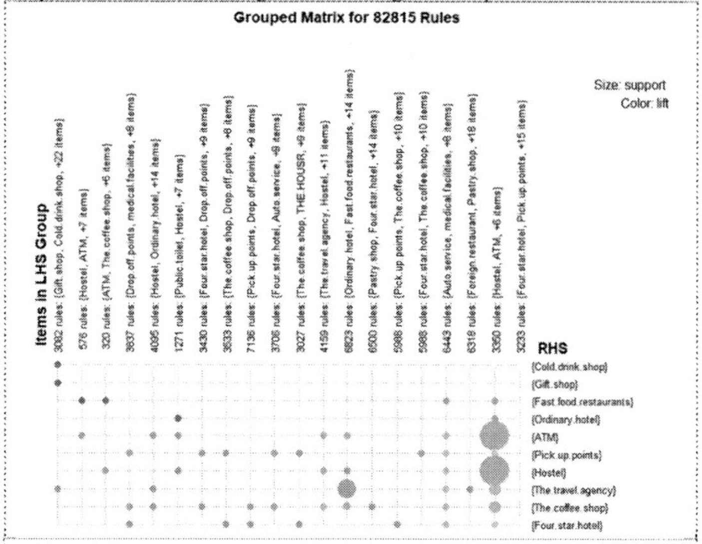

Figure3. East association rules of 3A-level scenic spots in different directions.

As can be seen from Figure3, the association characteristics of {gift shop, cold drink shop, etc.} and {3A-level scenic spots, hostel, ATM, etc.} are obvious. In the east, scenic spots will attract hostel, fast food restaurants, ATM and other facilities. In addition, the 3A-level scenic spots show the aggregation phenomenon between different catering types and different accommodation types in the east.

https://doi.org/10.1051/e3sconf/202129002001

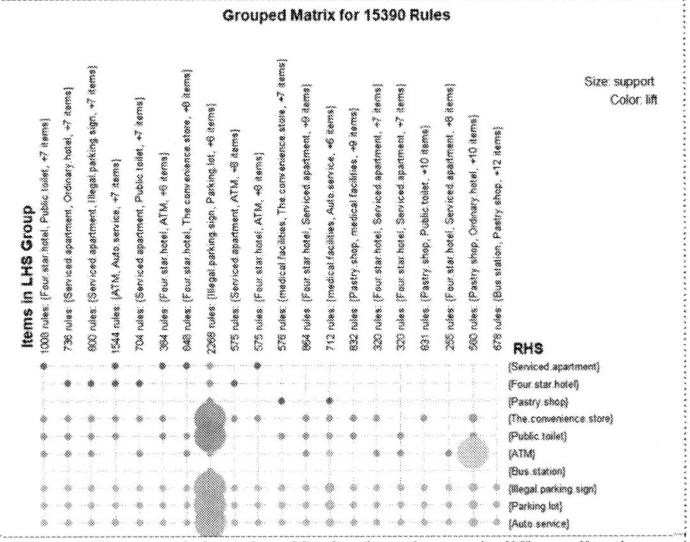

Figure4. West association rules of 3A-level scenic spots in different directions.

Under the western level, Figure4, {four-star hotel, ordinary hotel, serviced apartment, etc.}, {illegal parking signs, parking lots, public toilets, etc.} have obvious characteristics, and there is an obvious clustering phenomenon of accommodation types under the western level, which indicates that the accommodation industry layout is compact under the western 3A-level scenic spot 150 meters away. Practitioners tend to locate with competitors.

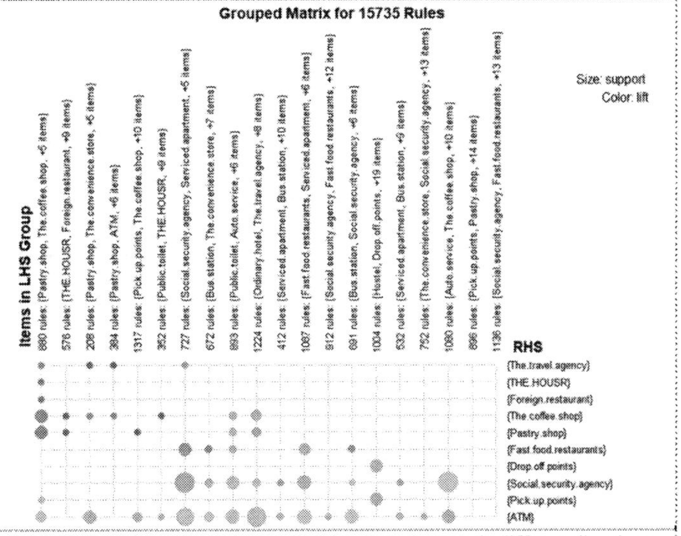

Figure5. South association rules of 3A-level scenic spots in different directions.

In the south direction, Figure5, {cafe, bakery, tea house, etc.}, {public toilet, tea house, fast food restaurant, etc.} and other types of leisure catering facilities have obvious association characteristics. In this direction, a variety of catering types of facilities show spatial aggregation characteristics.

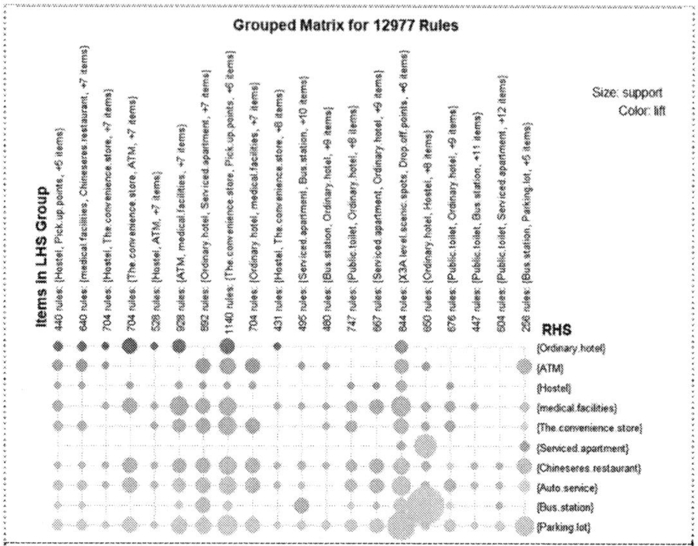

Figure6. North association rules of 3A-level scenic spots in different directions.

In the north, Figure6, {convenience stores, ATM machines, ordinary hotels, etc.}, {convenience stores, pick-up points, ordinary hotels, etc.}, {3A-level scenic spots, car services, ordinary hotels, etc.} have obvious features, and there are many types of facilities associated with the scenic spots. In the north, the scenic spots also attract other types of facilities. Different from the other three locations, the northern location does not have strong correlation characteristics with the catering types such as pastry shop, cold drink shop, fast food restaurant and tea house, indicating that the catering types in the northern direction are relatively single.

4. Conclusions

In the face of a large number of data based on spatial location produced in the era of big data, mining of spatial association rules is an important part of urban research. This paper uses POI data and taxi data to calculate the association characteristics among facilities and study the spatial distribution among industries. In each city, facilities entities generally have symbiosis, connection and other topological relations. Apriori algorithm, as a classic spatial association rule mining method, can well mine the spatial association characteristics of urban facilities, indicating the simultaneous occurrence of facilities and the frequent simultaneous occurrence of facilities. The practical significance of the results in this paper is as follows :(1) For tourists, the association features of taxi trips in the results can provide some reference for tourists when taking a taxi trip, and the features of illegal parking signs can make self-driving tourists carefully consider when choosing appropriate places to park their cars. (2) For scenic spots, the experimental results also reflect the problem of unbalanced distribution of public service resources in scenic spots. From downtown to suburban areas, the abundant allocation of public service resources in scenic spots gradually decreases. The experimental results provide reference for urban management departments to reasonably arrange scenic resources and meet the needs of tourists.

Acknowledgments

This research was funded by the National Nature Science Foundation of China (Grant No. 41771477).

References

1. Hu S, He Z, Wu L, et al. A framework for extracting urban functional regions based on multiprototype word embeddings using points-of-interest data[J]. Computers, Environment and Urban Systems, 2020,80:101442.
2. Luan X, Cheng L, Song Y, et al. Better understanding the choice of travel mode by urban residents: New insights from the catchment areas of rail transit stations[J]. Sustainable Cities and Society, 2020,53:101968.
3. Yue W, Chen Y, Thy P T M, et al. Identifying urban vitality in metropolitan areas of developing countries from a comparative perspective: Ho Chi Minh City versus Shanghai[J]. Sustainable Cities and Society, 2021,65:102609.
4. Koperski K, Han J. Discovery of Spatial Association Rules in Geographic Information Databases[J]. 1995.
5. Bo F. Hybrid spatial data mining methods for site selection of emergency response centers[J]. Natural Hazards, 2014,70(1):643-656.
6. Ma W, Xue C, Zhou J. Mining time-series association rules from Western Pacific spatial-temporal data[J]. IOP Conference Series: Earth and Environmental

Science, 2014,17(1):12224-12226.

7. Chi J, Jiao L, Dong T, et al. Quantitative identification and visualization of urban functional area based on POI data[J]. Journal of Geomatics.
8. Sheikh A S, Patidar M R. A NOVEL ALGORITHM DUFP for MINING ASSOCIATION RULES in DYNAMIC DATABASES[J].
9. Gang F, Wei Z, Qian Y. An Algorithm of Constrained Spatial Association Rules Based on Binary: International Symposium on Neural Networks: Advances in Neural Networks, 2008[C].
10. Li X. Mining spatial association rules in spatially heterogeneous environment[J]. International Society for Optics and Photonics, 2008.
11. Agrawal R. Mining Association Rules Between Sets of Items in Large Databases, SIGMOD Conference[J]. proc sigmod, 1993.
12. Merdun H, Ozturk Y. arules - A Computational Environment for Mining Association Rules and Frequent Item Sets[J]. Journal of Statistical Software, 2005,014.

https://doi.org/10.1051/e3sconf/202129002002

Evaluation and Optimization of Urban Basic Education Facilities Layout based on GIS——A Case Study in Xuzhou, China

Ma Yue[1], Shen Shan[1*]

[1]School of Geography, Geomatics, and Planning, Jiangsu Normal University, Xuzhou , 221116, China

Abstract. The layout of educational facilities has become a hot topic in urban geography research. Guides by rational demand consensus, typical human settlements extraction and matching of physical property planning, this paper constructs an evaluation system for the spatial layout of urban basic education facilities from the perspective of " Homo-Urbanicus " theory. The content includes the evaluation of accessibility, evaluation of bearing pressure, classification of layout types, selection and analysis of excellent communities. This paper conducts an empirical research on the basic education facilities of kindergartens, elementary schools and middle schools in the main urban area of Xuzhou City. And then, we delineated excellent community, standard communities, under-covered communities, overstressed communities, undeveloped communities on the basis of evaluation results. Finally, we selected sample communities for analysis, and put forward relevant optimization strategies.

1. Introduction

The equity of basic education is important to social equity as educational distribution is to man educational foundation [1,2]. Chinese Education Modernization 2035 [3] proposes to promote the high-quality and balanced development of compulsory education as well as promote the universal and beneficial development of basic education. At the same time, scientific layout of basic education facilities plays an important role in promoting high-quality urban development and improving urban quality.

Existing studies mainly focus on the accessibility of educational facilities, ethnic differences, and the spatial effect of education on communities [4,5,6], etc. O'Brien [7] advocated the establishment of an educational layout system that is integrated with efficiency and equity to improve the accessibility of educational facilities. Lange [8] explored the correlation between educational layout and residential differentiation from the perspective of economics and sociology. McDonald [9] discussed the influence of school site selection and service on surrounding communities by sorting out the standards of school construction. In addition, some studies have shown that differences in the allocation of educational resources exist among different social groups. Peiyang [10] believed that differences in the allocation of educational resources lead to residential migration and class agglomeration, thus aggravating the spatial differentiation of social classes.

Based on the informed research and the actual issues of the basic educational facilities in China, we will explore several issues

(1) How to seek consensus between residents' needs and facility layout? How to consensus between residents' needs and facility layout can be coordinated, and the optimal balance interval can be found.

(2) How to accurately identify the mismatch between supply so as to demand and develop the corresponding optimal layout strategy?

(3) How to coordinate ideal choice and realistic constraints to build a stable supply-demand matching relationship of basic education facilities?

2. The theoretical framework and methods

2.1. The theory of "Homo-Urbanicus"

By referring the theory of "economic man" in the western classical economics and the theory of "human settlement science" by Doxiadis, Hok-Lin Leung put forward the theory of "Homo-Urbanicus" which is people-oriented.[11] " Homo-Urbanicus " are who rationally choose to live in clusters to pursue spatial contact, and the core of urban planning is to match typical "Homo-Urbanicus" with typical "human settlement environment".

2.2. Research method

Due to basic education facilities as a typical spatial contact, its layout should meet the demand of people's property and rational, this can be explained by " Homo-Urbanicus ".

2.2.1. Rational needs consensus. According to the theory of "Homo-Urbanicus", the rationality of the

*Corresponding author's e-mail: 2020190053@jsnu.edu.cn

residents is embodied in "self-preservation" and "coexistence", Therefore, a rational balance between "self-existence" and "coexistence" should be sought to improve the spatial contact opportunities within the consensus range.

2.2.2. Identification of typical layout types. The identification of typical human settlements is an important method to analyse "Homo-Urbanicus". [11] It is more helpful to propose an optimal strategy by classifying different types of community layout through multiple variables and analyzing the causes of different types.

2.2.3. Planning and matching of physical properties. The match between residents and basic educational facilities is the optimization of "Accessibility" and "Bearing capacity ". The significance of planning is to enable more residents to have equal access to basic educational facilities through spatial planning and layout.

This paper firstly took the community as the evaluation unit, secondly divided the layout types of basic education facilities, again selected sample communities for layout reference and identified the mismatched communities, finally put forward optimization strategies from the evaluation of accessibility and bearing pressure.

3. A case study of Xuzhou city

3.1. Research area and data

Xuzhou is a typical resource-transformation city in China. The scope of data acquisition in this study is defined in the main urban area of Xuzhou. The data adopted in this study mainly include the spatial location data of basic education facilities POI (point of internet) in Xuzhou, the information of residents' population and spatial distribution, the current urban road network, and the policies and norms on the configuration standards of kindergartens and primary and secondary schools. POI spatial data are derived from the open platform of Amap, including the locations of kindergartens, primary schools and middle schools and their related attributes. The population data comes from the regional version of well-known real estate websites in China, and Python crawler program is used to obtain the location and number of households of each community in the main urban area of Xuzhou. The data of community boundary and urban road network were obtained from the Laboratory of Complex and Sustainable Urban Networks (CSUN) of the University of Illinois at Chicago, USA.

3.2. Layout evaluation of basic educational facilities

3.2.1. The evaluation unit. In reality, due to the basic education facilities and services are often divided according to community boundaries. Therefore, this paper takes community boundaries as a unit to evaluate the layout of basic education facilities. With reference to the Standard for Planning and Design of Urban Residential Areas, the community evaluation is divided into different level, the natural discontinuity method was used to visualize each evaluation result.

3.2.2. Layout Type Partitioning. In order to effectively identify the layout characteristics of basic education facilities in different communities and put forward corresponding optimization strategies, community space of basic education facilities in Xuzhou City was summarized into 5 types based on community space boundary and overlay the results of accessibility evaluation and stress-bearing evaluation (Fig. 1). (1) Excellent community, at least one evaluation of accessibility and Bearing pressure higher than ordinary level; (2) Standard community, accessibility evaluation and facility bearing pressure evaluation are average; (3) Under-covered community, lower accessibility evaluation and above average stress evaluation of facilities; (4) Overstressed community, with an average or above level of accessibility and a lower than average level of facility bearing pressure; (5) Undeveloped community, where the evaluation of accessibility and stress of facilities are lower than the average level.

3.2.3. Results of evaluation. The superposition results showed that: (1) The comprehensive evaluation of three types of basic educational facilities in kindergarten and primary school was higher, while that in middle school is worse; (2) The comprehensive the evaluation of kindergarten is relatively high, and also showed that the current situation of urban center is under too much pressure and the marginal area is not covered enough, this meant the needs of the periphery of the city to study nearby; (3) Primary schools have the highest comprehensive evaluation among the three types of basic education facilities, but some communities located at the edge of the main urban area have under-covered and overstressed, indicating that the overall layout of kindergartens tends to the center of the urban area, and can not meet the needs of the periphery of the urban area to go to school nearby; (4) Middle school comprehensive evaluation results as a whole is low, insufficient coverage and pressure overload community is more, but, in contrast to the primary school, under-covered communities focused on the city center, overstressed communities in urban fringe, illustrate the middle school lack of overall layout, and there are more accessibility is insufficiency, the large bearing pressure and the undeveloped community.

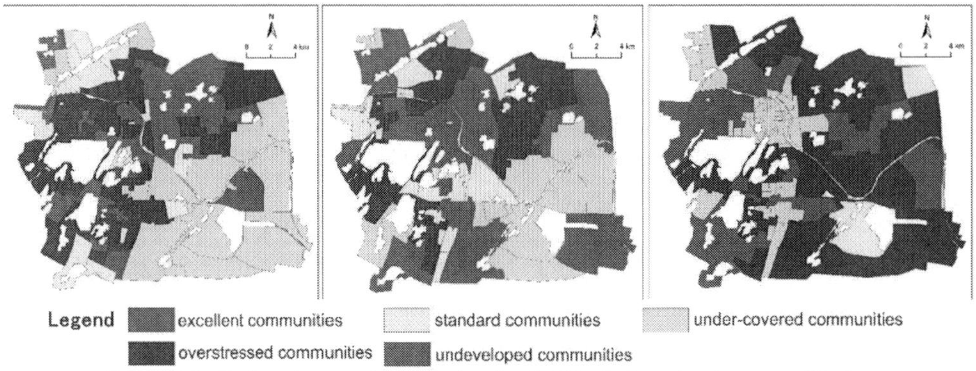

Figure 1 Distribution of community basic education facilities in the main urban area of Xuzhou

3.3. Identification and analysis of the layout of educational communities

This paper have selected sample communities of education layout as practical reference and normative layout demonstration. Based on the review results in the above section, this paper selects demonstrative communities and standard communities with strong representation for the three types of basic educational facilities as typical samples.

Table 1 Community evaluation results of educational layout samples

Case Type	Community Name	Education Facilities Type	Accessibility	Services Population (people)
Excellent community	Wenwo Community	Kindergarten	15min	17722
	Xiyuan Community	Primary School	15min	40984
Standard Community	Fenghuangshan Community	Kindergarten	30min	25806
	Jiulonghu Community	Primary School	30min	6541
	Pantao Community	Middle School	30min	25522

In the two excellent communities and the three standard communities, the average service access standards of residents for 15min and 30min have been reached respectively, and the facilities bear less pressure (Table1). Most of the demonstration and standard community is located in Xuzhou city central zone of the population density is bigger, development speed is relatively mature, still have to face the uncertainty of future problems, such as the aging of educational facilities and the opening of two-child policy, the issues that led to the increase in the number of the education demand should therefore dynamic monitoring population change, improve, improve education facilities configuration. Overstressed communities are partly caused by geographical conditions, but the main problem is still that the accessibility is not up to standard. Therefore, increasing the layout of educational facilities and optimizing road connectivity are the next urgent work to be completed.

3.4. Optimization Suggestions

3.4.1. Combined with the needs of residents, adopt measures to upgrade the standard community.

We need to monitor changes in population data dynamically to ensure the fairness of the spatial distribution of basic education facilities. Then we will make full use of the existing basic education facilities to optimize the structure of the existing facilities and improve the level of educational facilities and services.

3.4.2. Accurately identify under-covered communities and improve access to educational space. For the under-covered communities, the density of the road network can be improved, and the travel environment can be improved to increase the opportunities for spatial access to education. Improving the urban pedestrian transportation system and increasing the density of travel network can properly break the closure of communities as well as improve the accessibility of basic education facilities.

3.4.3. Remitting the overstressed communities and establishing the dynamic flow and sharing mechanism of educational resources. We can increase the capacity allocation of educational facilities in densely populated urban central communities to enhance the service capacity of schools through sporadic update capacity expansion and quality improvement. Then we can optimize the stock, revitalize the existing educational resources to change the situation of idle and waste of high-quality educational resources.

3.4.4. Focusing on optimizing the undeveloped communities to fill the gaps in the layout of basic education. We will increase basic educational facilities and give priority to meeting the basic educational needs of the school-age population to achieve full coverage of education.

4. Conclusion and discussion

This paper took the main urban area of Xuzhou City as an example to evaluate the layout of three types of basic education facilities by using the theory of "Homo-Urbanicus", then classify the layout types of basic education facilities in different communities, last put forward targeted optimization strategies. The analysis results are as follows.

The accessibility of primary school is better than kindergarten and middle school, and the layout of primary school and kindergarten is significantly better than middle school. The excellent communities and the standard communities should continue to prompt the service quality as the reference samples for the education layout and development of other communities. Under-covered communities should increase the distribution of educational facilities and improve the space access opportunities. Overstressed communities should increase the capacity of education and relieve the overcrowded population. Undeveloped communities as the focus of the next education facility layout, they should be implemented as soon as possible to fill the shortage of education allocation.

This study evaluated quantitatively the layout of basic educational facilities in community space and extracted the excellent communities and standard communities with relatively perfect services. On this basis, we identified the imbalance in the development of community education layout, include under-covered communities, overstressed communities and undeveloped communities, to provide the empirical basis for decision-making and ease the education fairness. However, there are still some limitations in this study. First of all, due to the limitation of data acquisition, the scale of basic education facilities, service capacity, travel mode and other factors of the educated did not have to be considered, so there are still errors in the calculation results. Secondly, this paper lacked cause and mechanism discrimination why the distribution of basic educational facilities is unfair, and we will take community residents as the object and adopt the methods of questionnaire survey and community interview to obtain more detailed micro data, in order to obtain more scientific and reasonable research results in the future.

Funding

This research was financially supported by the Priority Academic Program Development of Jiangsu Higher Education Institutions.

References

1. Malczewski J, Jackson M. (2000) Multicriteria spatial allocation of educational resources: An overview. Socio-Economic Planning Sciences, 34 (3) :219-235.
2. Hongqi C. (2020) What Kind of Educational Equity We Need in the New Era——Consummating the Domain of Research Questions and the Policy Toolbox of Educational Equity. Educational Research, 41(02):4-16.
3. CPC Central Committee, (2019) State Council. China's education modernization. http://www.gov.cn/zhengce/2019-02/23/content_5367987.htm.
4. Talen E. (2001) An empirical investigation of access to elementary schools in West Virginia.
5. Lenntorp B. (1978) A time-geographic simulation model of individual activity programs. Human activity and time geography vol.2.
6. Agarwal S, Rengarajan S, Sing T F, et al. (2016) School allocation rules and housing prices: A quasi-experiment with school relocation events in Singapore. Regional Science and Urban Economics, 58.
7. O'Brien. R J. (1969). Model for planning the location and size of urban schools. Socio-Economic Planning Sciences, 2(2-4), 141-153.
8. Lange W J, Koslinski M C, Alves F. (2013) Educational inequalities in urban contexts: a study of the geography of educational opportunities in the city of Rio de Janeiro. Educao & Sociedade. 35 (125): 200-220.
9. Mcdonald, C N. (2010) School Siting: Contested Visions of the Community School. Journal of the American Planning Association, 2(76):184-198.
10. Peiyang C. (2015) School district and Its Socio-Spatial Consequences in Urban China. Urban Studies, 22 (8): 55-60.
11. Hok-Lin L. (2020) Homo-Urbanicus: Basic logic and operating procedure. City Planning, 44(02):68-76.

 https://doi.org/10.1051/e3sconf/202129002003

Application of random walk to model temperature in 60 African cities for the 20th century

Shaomin Yan[1], and Guang Wu* [1]

[1]National Engineering Research Center for Non-Food Biorefinery, State Key Laboratory of Non-Food Biomass and Enzyme Technology, Guangxi Biomass Engineering Technology Research Center, Guangxi Key Laboratory of Biorefinery, Guangxi Academy of Sciences, 98 Daling Road, Nanning, 530007, Guangxi, China

Abstract. The 20th century is marked with climate change led by global warming. So far, many models have been applied to analyze the temperature change. However, a simple but interesting model, a random walk is hardly used in his regard. In this study, we use the random walk to model the temperature in the format of random walk that is the conversion of recorded temperature and the real recorded temperature in 60 African cities including 53 capitals for the 20th century. The results show that the random walk can satisfyingly model either temperature in the format of random walk or real recorded temperature although the fitted results from other climate models are unavailable for comparison in these 60 cities. As nothing else besides random numbers is involved in this modelling, the results seem somewhat counter-intuitively.

1 Introduction

The climate change is a topic too important to use just a few models to reach a solid conclusion, thus every effort should be made to use all the possible models to address this issue carefully and throughout until we exhaust all our already-known models. Highly likely, we have not yet exhausted all our already-known models to analyze and explain the temperature changes.

Although random walk model was proposed to deal with the global warming in the 90s in the 20th century [1], the real application of random walk model was impeded by the lack of computation power.

In the 20th century, the computation power is no longer the limiting factor for such work, so we have the opportunity to use the random walk to model the real temperature.

The currently very complicated climate models seem to fall behind the public expectation that public can verify the temperature change using models because the social network, advocated-democracy and Wiki phenomenon give strong impulses to public to verify what they are told by either scientists or politicians. In this context, one cannot verify the theory of global warming without supercomputer and related expertise. Therefore, it is important and pressing to apply some simple model to treat the temperature changes.

In this context, we apply the random walk to model the temperature in 60 African cities including 53 capitals for the 20th century in this study. In the random walk model, the random factors will no longer be considered as minor players as in other models [2], but as the only mechanism for temperature change.

2 Materials and methods

2.1. Data

Sixty cities of Africa are chosen according to Wikipedia [3], which are listed in Table 2. Based on 0.5° by 0.5° latitude and longitude grid-box basis cross globe, the temperatures in these 60 African cities recorded for the 20th century are obtained from the website of Oak Ridge National Laboratory [4]. The latitudes and longitudes of these 60 cities are determined using Find Latitude and Longitude [5].

2.2 Temperature in the format of random walk

Although temperature recorded along the time course is decimal number, it can be simplified as higher or lower or equal to that in previous recording, which results in 1, –1 and 0 in numerical term. The addition of ±1 along the time course is the one-dimensional random walk starting from zero [6]. In plain words, the random walk can be viewed as a sequential result of tossing a fair coin, by which we record the head as 1 and the tail as -1, and then we add the results together along the time course.

For this purpose, we convert the temperature into a format similar to random walk as shown in Table 1. For example, the average temperature in 1901 was 15.825 C (cell 2, column 2, Table 1), the temperature in 1902 is higher than that in 1901 (16.275 C vs. 15.825 C), so we assign 1 (cell 3, column 3, Table 1). Then, the temperature in 1903 is lower than that in 1902 (15.675 C

* Corresponding author: hongguanglishibahao@gxas.cn

https://doi.org/10.1051/e3sconf/202129002003

vs. 16.275 C), so we assign –1 (cell 4, column 3, Table 1), and then we add them together to construct temperature walk (column 4, Table 1)

Table 1. Procedure of converting the temperature in Algiers for the 20th century into its format of random walk and generation of random walk.

Year	Temperature C	Converted Temperature	Temperature Walk	Generated Random Number	Random Walk Step	Random Walk
1901	15.825		0	-0.47192		0
1902	16.275	1	1	-0.88720	-1	-1
1903	15.675	–1	0	0.57078	1	0
1904	16.108	1	1	0.20363	–1	–1
1905	15.667	–1	0	–0.46987	–1	–2
1906	15.433	–1	–1	–0.36344	1	–1
1907	15.417	–1	–2	0.64925	1	0
1908	16.033	1	–1	0.09431	–1	–1
1909	15.075	–1	–2	0.86238	1	0
1910	15.175	1	–1	0.42282	–1	–1
...	...	...	...	...	...	...
1989	18.892	1	4	–0.21946	–1	2
1990	18.658	–1	3	0.73550	1	3
1991	17.567	–1	2	0.55552	–1	2
1992	17.500	–1	1	–0.74005	–1	1
1993	17.642	1	2	–0.30770	1	2
1994	18.833	1	3	–0.93889	–1	1
1995	18.475	–1	2	–0.36484	1	2
1996	17.883	–1	1	0.27173	1	3
1997	18.642	1	2	–0.41865	–1	2
1998	17.925	–1	1	0.72083	1	3

2.3 Generation of random walk

So far, difference methods have been used to generate the random numbers, for example, the Monte-Carlo methods in SigmaPlot [7]. The second part of Table 1 shows how to generate a random walk. A sequence of random numbers was generated using SigmaPlot (column 5, Table 1). The second random number is –0.8872 (cell 3, column 5, Table 1), which is smaller than the first random number (cell 2, column 5, Table 1), so we assign –1 (cell 3, column 6, Table 1). Following the similar procedure in the last section, we construct the random walk (the last column in Table 1).

2.4 Searching for seed

For any random number generator, the command usually as follows: seed, number of generated random numbers, up and low ranges. If a seed that can generate random numbers whose random walk is most similar to temperature walk, then this seed would be best. However, this is not an easy task, therefore we used a brutal force method to try each sequence of random numbers with 1 million seeds.

2.5. Extension of random walk

In reality, temperature is a decimal number rather than 1, 0, –1 as those in random walk. Similarly, the random number generator generates decimal numbers rather than 1, 0, –1. These provide the possibility to use a sequence of random numbers to construct a random walk, and use the recorded temperature to construct a temperature walk.

2.6. Comparison

The least squares method is used for comparison between the temperature in the format of random walk and random walk, and between temperature and decimal random walk.

3 Results and discussion

As the random mechanism is the only player in the random walk model, thus the use of random walk to model either temperature in the format of random walk or temperature is not difficult, and every one can in principle do this job easily and conveniently. This is in clear contrast to the current climate models, where random factors are generally considered as noises [2].

Figure 1 shows the application of random walk model to the temperature in the format of random walk (left panel) and the temperature (right panel) in 5 representative cities at different locations in Africa.

On the left panel of Figure 1, the temperature in the format of random walk can take several steps either up or down, which would be arguably the trend for underlined temperature trend as we have already seen the temperature increased over some years. Although the temperature in the format of random walk is the conversation from recorded temperature, it can indeed be modelled by a random walk whether it goes up or down.

In fact, we would not expect a perfect fitting for 100 years because the chance is too small, $(1/2)^{100}$, to achieve as tossing a fair coin has 1/2 probability to be head up or down. As these fittings were done in all 60 cities without any exception (columns 3 and 4 in Table 2), it would suggest that the temperature in the format of random walk could be attributed to a random mechanism.

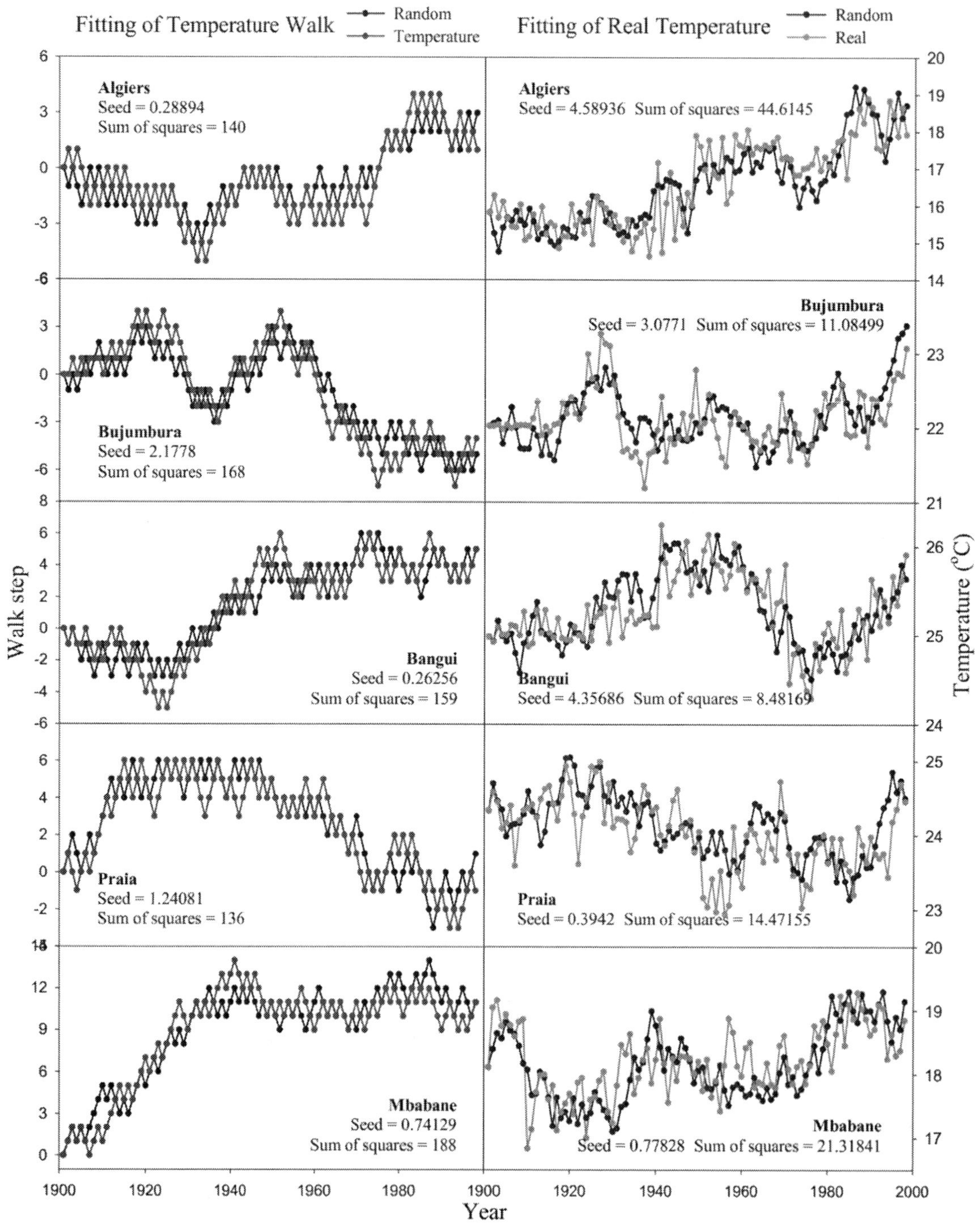

Fig. 1. Comparison between the temperature in the format of random walk an random walk (left panel) and between temperature and decimal random walk (right panel).

https://doi.org/10.1051/e3sconf/202129002003

Table 2. Seeds for generation of random numbers for modelling of temperature in the format of random walk (column 3) and recorded temperature (column 4).

State/Island/City	Capital/Main City	Seeds	Seeds
Algeria	Algiers	0.28894	4.58936
Angola	Luanda	0.06963	1.02646
Benin	Porto-Novo	4.92860	0.45102
Botswana	Gaborone	3.97033	2.24124
Burkina Faso	Ouagadougou	1.67443	2.73101
Burundi	Bujumbura	2.17780	3.07710
Cameroon	Yaoundé	1.12887	0.84994
Canary Islands (Spain)	Las Palmas de Gran Canaria	5.23634	2.23310
Cape Verde	Praia	1.24081	0.39420
Central African Republic	Bangui	0.26256	4.35686
Ceuta (Spain)	—	5.81828	3.49371
Chad	N'Djamena	1.06187	0.52598
Comoros	Moroni	0.76109	0.01219
Côte d'Ivoire	Abidjan	6.60339	7.34362
Democratic Republic of the Congo	Kinshasa	5.33755	0.29761
Djibouti	Djibouti	1.52317	3.94442
Egypt	Cairo	8.52921	2.41488
Equatorial Guinea	Malabo	2.32397	3.54964
Eritrea	Asmara	1.21252	5.96245
Ethiopia	Addis Ababa	1.78082	4.26952
Gabon	Libreville	0.52118	3.84746
Ghana	Accra	0.50938	4.52341
Guinea	Conakry	0.48103	2.80122
Guinea-Bissau	Bissau	0.45082	7.20704
Kenya	Nairobi	1.38852	1.00904
Lesotho	Maseru	2.02340	8.35252
Liberia	Monrovia	1.29483	0.80564
Libya	Tripoli	0.63052	0.07157
Madagascar	Antananarivo	7.39082	3.79777
Madeira Islands (Portugal)	Funchal	0.71346	0.53171
Malawi	Lilongwe	3.31921	2.19555
Mali	Bamako	3.58343	5.96289
Mauritania	Nouakchott	0.95814	3.39094
Mauritius	Port Louis	4.90808	0.01219
Mayotte (France)	Mamoudzou	0.48937	4.26447
Melilla (Spain)	—	2.06516	2.23310
Morocco	Rabat	2.58505	0.98484
Mozambique	Maputo	0.78381	0.52342
Namibia	Windhoek	1.44089	0.00980
Niger	Niamey	3.10076	0.26236
Nigeria	Abuja	2.57323	1.63191
Republic of the Congo	Brazzaville	5.33755	0.29761
Réunion (France)	Saint-Denis	0.08955	1.04547
Rwanda	Kigali	0.29356	2.61825
Santa Cruz de Tenerife	—	0.16051	2.23310
São Tomé and Príncipe	São Tomé	8.77678	3.84746
Senegal	Dakar	1.39995	0.06963
Seychelles	Victoria	1.39836	1.68698
Sierra Leone	Freetown	0.88954	0.97418
Somalia	Mogadishu	0.70028	0.43083
South Africa	Pretoria	4.39973	1.53916
Sudan	Khartoum	6.17171	2.13654
Swaziland	Mbabane	0.74129	0.77828
Tanzania	Dodoma	2.04458	1.30204
The Gambia	Banjul	2.37399	2.10617
Togo	Lomé	0.69901	0.43821
Tunisia	Tunis	7.96571	3.99012
Uganda	Kampala	2.55225	2.51977
Zambia	Lusaka	0.71026	1.46179
Zimbabwe	Harare	1.27065	3.88437

On the right panel of Figure 1, we used the random walk model to fit the real recorded temperature. As can be seen, the fittings by random walk go along the general trend of recorded temperature. The fitted results are listed in columns 5 and 6 in Table 2.

Unfortunately, we have not the results in these 60 cities using other climate models for comparison, however, our results can be considered satisfying because the fluctuated fitting curve cannot be produced by any deterministic model operating on differential equations.

If the temperature in the format of random walk is a deliberate and artificial conversation from the recorded temperature, then the fittings in the right panel of Figure 1 can exclude such a possibility. This not only simply means that the random walk model can fit the real recorded temperature, but also decisively means that the random mechanism can be considered as the driving force for the temperature change. If so, it is very natural that the temperature can continue to increase or decrease for sequential years before a turning point.

Also, the random walk model is very simple, which may dramatically reduce the uncertainty in comparison with the current climate models [8].

Although the 20th century is marked with climate change led by global warming, African continent should be an exemption because Africa did not experience the industrial revolution, which polluted Europe. Thereafter, the rise of Japan, China, South Korea as well as East Asian countries increased the carbon emission in Asia. On the other hand, Africa could suffer the climate change severely, especially for sub-Saharan, where droughts occur frequently.

Table 2 shows the seeds used to generate random numbers for random walk modelling. As can be seen in Table 2, no repetition in seeds can be found, thus the temperature in each city has its own profile with a different seed.

Because there is no prior knowledge on the best seed, we have to search it from one million seeds from 0 to 1 although the real best seed might well be off this range. However, a compromise should be made in order to balance the computational time with the time required to finish a study.

As seen in Table 2, the seeds are usually smaller than 5, which could provide a safeguard to our searching range from 0 to 1.

As a matter of fact, this study finished our endeavour to use the random walk model to fit/simulate the temperature in different parts of the world [9-15]. The results seem to be similar across different continents.

4 Conclusion

Although the 20th century is marked with climate change led by global warming, the 21st century should be marked with its consequence and resolution. This study demonstrates the possibility to model the temperature changes using random walk, which could suggest a simpler mechanism.

Acknowledgments

This study was supported by Scientific Development Fund of Guangxi Academy of Sciences (2021YFJ1203).

References

1. A. H. Gordon, J. Clim., 4, 589–597 (1991).
2. P. D. Williams, Phil. Trans. R. Soc. A 363, 2931–2946 (2005)
3. Wikimedia Foundation, Inc. Wikipedia, the free encyclopedia, http://en.wikipedia.org/wiki/Countries_of_Africa. (2020)
4. M. New, M. Hulme, P. Jones, J. Clim. 13, 2217–2238 (2000)
5. https://www.findlatitudeandlongitude.com/. (2020)
6. Wikimedia Foundation, Inc. Wikipedia, the free encyclopedia, http://en.wikipedia.org/wiki/Random_walk (2020)
7. SPSS Inc. SigmaPlot 2002 for Windows Version 8.02, (1986-2001)
8. G.C. Cawley, G.J. Janacek, M.R. Haylock, S.R. Dorling, Neural Netw. 20, 537–549 (2007)
9. S. Yan, G. Wu, Guangxi Sci. 17, 148–150 (2010)
10. S. Yan, G. Wu, Natural Sci. 2, 1425–1431 (2010).
11. S. Yan, G. Wu, Guangxi Sci. 20, 115–120 (2013)
12. S. Yan, G. Wu, New Developments in Global Warming Research. ed. C.B. Keyes, O.C. Lucero, Nova Science Pub Inc., New York, Chapter 7, 143–156 (2013)
13. S. Yan, G. Wu, 2020 IEEE 3rd International Conference on Automation, Electronics and Electrical Engineering (AUTEEE 2020), AE345:. IEEE Xplore. 121–124
14. S. Yan, G. Wu, Proceedings of the 3rd International Forum on Geoscience and Geodesy (IFGG 2020) IOP Conference Series: Earth and Environmental Science (IOP Conf. Ser.: Earth Environ. Sci). 658: 012037 (2021)
15. S. Yan, G. Wu, Proceedings of the 3nd International Conference on Environmental Prevention and Pollution Control Technologies （EPPCT 2021） IOP Conference Series: Earth and Environmental Science (in press)

A Retinex-based Enhancement method for Ocean Remote Sensing Image

Tongyu Li[1], Jie Chen[2,*], Pu Cheng[2], and Lu Yu[2]

[1]CSSC Ocean Exploration Technology Institute Co.,Ltd, 214135 Wuxi, China
[2]PLA Naval Submarine Academy, 266199 Qingdao, China

Abstract. In order to effectively monitor important sea areas, one of the key issues is the detection of small dynamic targets such as ships. Besides monitoring the hull itself, the monitoring of small targets can also be achieved by detecting ship wakes, which means the sea image with clear texture feature is required. This paper firstly reviews the multiscale retinex (MSR) method, commonly used to enhance the image contrast. Then, it proposes a novel contrast enhancement algorithm base on the subband-decomposed multiscale retinex (SDMSR) method. Experimental results show that our proposed method can make a remarkable enhancement effect for ocean remote sensing images with clouds and whitecap, etc.

1 Introduction

Since there are abundant material resources in the ocean, how to develop, utilize and protect ocean resources has been a hot issue in many countries. China is a country with vast ocean territory, however, our understanding of the ocean is still very insufficient. On the other hand, we are faced with extremely complex regional disputes, especially in the South China Sea. Therefore, it is crucial to solve the problems of how to effectively monitor important sea areas such as disputed sea areas and route sea areas, ensure the safety of sea traffic and transportation, and maintain the integrity of ocean territory. One of the key issues is the detection of small dynamic targets such as ships.

In addition to monitoring the hull itself, the monitoring of small targets can also be achieved by detecting ship wakes, which means the sea image with clear texture feature is required. Because of high spatial resolution, visible light remote sensing method is an important way to monitor small targets in the sea. The image processing is necessary since visible light remote sensing image can be easily influenced by the factors such as clouds, rain, fog, and light.

At present, image enhancement method can be divided into two types: the frequency domain enhancement and the spatial domain enhancement. In recent years, researchers pay much attention to the image enhancement method based on retinex, which is named for an abbreviation of retina and cortex. As an image enhancement algorithm based on the human visual system, the retinex method performs well in edge enhancement, color fidelity, and dynamic range compression, and can achieve adaptive enhancement for multiple types of images.

The retinex method was originally proposed by Land [1, 2]. Then Jobson et al. evolved Land's method to single-scale retinex (SSR) method [3] and multiscale retinex (MSR) method [4, 5]. Especially, the MSR scheme is effective in dynamic range compression and local contrast enhancement. Then there have been several trials to improve the MSR method in the aspects of edge sharpness [6], halo effects [7], and color correction [8, 9]. Taking the sub-band characteristics into consideration, Jang et al. proposed subband-decomposed multiscale retinex (SDMSR) method [10, 11]. Experimental results showed that the proposed algorithm can effectively distinguish the details in both highlights and shadows and greatly enhance the overall contrast. We applied the SDMSR method to ocean remote sensing image enhancement, and the image enhancement effect is remarkable in most cases. However, the image enhancement effect is not ideal while the ocean remote sensing image contains clouds, whitecap, etc.

In this paper, we propose a modified SDMSR method which can effectively enhance the ocean remote sensing image with clouds and whitecap. This paper is organized as follows. In section 2, we briefly review the MSR method, which is the basis of the SDMSR method, and our improved algorithm based on the SDMSR method. In section 3, experimental results are presented for several ocean remote sensing images. Finally, the conclusion of this paper is in section 4.

2 Materials and Methods

2.1 Multiscale Retinex

As an extended SSR method with multiple kernel windows of different sizes, the output of MSR method is a weighted sum of several different outputs of SSR method [4]. The MSR output is represented as

* Corresponding author: grass2009@163.com

$$\log(R(x,y)) = \sum_{n=1}^{N} \omega_n (\log I(x,y) - \log(I(x,y) * G_n(x,y))) \quad ,(1)$$

where N denotes the scale number, $I(x, y)$ represents the input image, and $G_n(x, y)$ denotes a surround function. The symbol "*" denotes the convolution operation, and ω_n denotes the weight. If N=3, then $\omega_n = 1/3 \quad (n = 1,2,3)$. The surround function is denoted as following

$$G_n(x,y) = ke^{\frac{-(x^2+y^2)}{\sigma_n^2}} \quad , (2)$$

where σ_n is a Gaussian surround space constant and k is determined, so that $G_n(x, y)$ satisfies the following condition

$$\iint G_n(x,y)dxdy = 1 \quad . (3)$$

2.2 Proposed Algorithm

Based on the MSR method, the subband-decomposed multiscale retinex (SDMSR) method is proposed. The SDMSR method utilizes the subband characteristics of retinex outputs in the aspects of a hybrid intensity transfer function (HITF), subband decomposition, space-varying subband gain, and automatic gain control [12, 13]. Based on the SDMSR method, we introduce a modified SDMSR method, the processing steps are as follows

The first step is dynamic range compression. To enhance details in both highlights and shadows, the SDMSR method applied a combination of two complementary logarithmic functions, which is called hybrid intensity transfer function (HITF), to an input image. The HITF is represented as following

$$h\_\log(I(x,y)) = \omega \cdot \bar{I}(x,y) + (1-\omega) \cdot \tilde{I}(x,y)$$
$$\bar{I}(x,y) = \log(I(x,y)+1)$$
$$\tilde{I}(x,y) = \log L - \log(L - I(x,y)) \quad , (4)$$

where L denotes the image dynamic range (e.g., L is 65536 for a 16-bit image), $I(x, y)$ represents the input image, and ω denotes weighting parameter of contrast enhancement, which is produced as

$$\omega = \frac{1}{M} \sum_{(x,y)\in\Omega} \frac{\bar{\zeta}(x,y)}{\bar{\zeta}(x,y) + \tilde{\zeta}(x,y)} \quad , (5)$$

where Ω denotes a set of all pixel positions, M is the total number of image pixels. $\bar{\zeta}(x,y)$ and $\tilde{\zeta}(x,y)$ represent the local standard deviations (LSDs) at the position (x, y) of $\bar{I}(x,y)$ and $\tilde{I}(x,y)$, respectively. A 5×5 window is applied to obtain the local statistics.

In the second step, the retinex output $R_n(x, y)$ is decomposed into nearly no overlapping spectral bands. Based on the assumption that $R_{n-1} < R_n$, the subband decomposed retinex output $\bar{R}_n(x,y)$ is calculated as following

$$\bar{R}_n = \begin{cases} R_1, & n = 1 \\ R_n - R_{n-1}, & 2 \le n \le N \end{cases} \quad , (6)$$

where N denotes the scale number, and the retinex output $R_n(x, y)$, n=1, 2, ..., N, is represented as

$$R_n = h\_\log(I(x,y)) \text{-} h\_\log(G_n(x,y) * I(x,y)) \quad , (7)$$

where "*" represents the convolution operator, and $G_n(x, y)$ is the nth scale surround function, which is defined as following

$$G_n(x,y) = ke^{-(x^2+y^2)/\sigma_n^2} \quad ,(8)$$

where σ_n is a Gaussian surround space constant and $\sigma_{n+1} > \sigma_n$, n=1, 2, ..., N-1. k is determined so that $G_n(x, y)$ satisfies

$$\iint G_n(x,y)dxdy = 1 \quad . (9)$$

The third step is space varying subband gain. The enhanced subband decomposed retinex output $\tilde{R}_n$ in the original SDMSR method is defined as

$$\tilde{R}_n = P_n(x,y) \cdot \bar{R}_n(x,y) \quad , (10)$$

where $P_n(x, y)$, which denotes the subband gain function, is defined as

$$P_n(x,y) = \left(\frac{1}{\hat{R}_n(x,y) + b}\right)^{1-\hat{\sigma}_n} \quad , (11)$$

where

$$\hat{R}_n(x,y) = \frac{|\bar{R}_n(x,y)|}{|\bar{R}_n|_{\max}} \quad , (12)$$

$$\hat{\sigma}_n = \frac{\sigma_n}{\sigma_{\max}} \quad , (13)$$

where

$$|\bar{R}_n|_{\max} = \max_{(x,y)\in\Omega}\{|\bar{R}_n(x,y)|\} \quad , (14)$$

$$\sigma_{\max} = \max_{n\in\{1,2,\ldots,N\}}\{\sigma_n\} \quad . (15)$$

In equation (12), $\hat{R}_n(x,y)$ represents the normalized subband decomposed retinex output, $\hat{\sigma}_n$ represents the normalized Gaussian surround space constant. Constant b in equation (11) is set to 0.2 in the experiment.

In our modified SDMSR method, the normalized subband decomposed retinex output $\hat{R}_n(x,y)$ in equation (12) is revised as following

$$\hat{R}_n(x,y) = \frac{|\bar{R}_n(x,y)|}{F_{\bar{R}_n}^{-1}(c)} \quad , (16)$$

where constant c denotes the cumulative probability of $\bar{R}_n(x,y)$, and is set to 0.97 in our experiment. $F_{\bar{R}_n}^{-1}(c)$ represents the inverse function of the cumulative probability function of $\bar{R}_n(x,y)$.

Then enhance the details of high-frequency subband by using the detail adjustment function. The enhanced output $\hat{I}(x,y)$ of the SDMSR method is given by

$$\hat{I}(x,y) = DAF(x,y) \cdot \tilde{R}_1(x,y) + \sum_{n=2}^{N} \tilde{R}_n(x,y) \quad , (17)$$

where $DAF(x,y)$ is the detail adjustment function and is defined as

$$DAF(x,y) = 1 - \lambda \cdot (\hat{\varsigma}(x,y) - 0.5)^2, \quad 0 \le \lambda \le 4 \quad , (18)$$

where $\hat{\varsigma}(x,y)$ represents the normalized LSD, which is calculated as following

$$\hat{\varsigma}(x,y)=\frac{\varsigma(x,y)}{\varsigma_{\max}},\quad where\,\varsigma_{\max}=\max_{(x,y)\in\Omega}\{\varsigma(x,y)\} \quad ,(19)$$

where $\varsigma(x,y)$ represents the LSD at pixel position (x, y) of the input image $I(x, y)$, and is obtained by using a 5×5 window.

In the last step, for the final enhancement to display, the automatic gain control is needed to transform the output $\hat{I}(x,y)$ of the SDMSR method from the logarithmic domain to the display domain. The automatic gain control in the original SDMSR method is defined as

$$I_{out}(x,y)=\frac{L-1}{\hat{I}_{max}-\hat{I}_{\min}}\cdot\left[\hat{I}(x,y)-\hat{I}_{\min}\right] \quad , (20)$$

where $I_{out}(x,y)$ represents the final enhanced image, $\hat{I}_{\min}=\min_{(x,y)\in\Omega}\{\hat{I}(x,y)\}$, and $\hat{I}_{\max}=\max_{(x,y)\in\Omega}\{\hat{I}(x,y)\}$.

In our modified SDMSR method, the automatic gain control in equation (20) is revised as

$$I_{out}(x,y)=\frac{L-1}{F_{\hat{I}}^{-1}(c)-\hat{I}_{\min}}\cdot\left[\hat{I}(x,y)-\hat{I}_{\min}\right] \quad , (21)$$

where $\hat{I}_{\min}=\min_{(x,y)\in\Omega}\{\hat{I}(x,y)\}$, constant c denotes the cumulative probability of $\hat{I}(x,y)$, and is set to 0.97 in our experiment. $F_{\hat{I}}^{-1}(c)$ represents the inverse function of the cumulative probability function of $\hat{I}(x,y)$.

3 Results

To evaluate the proposed method, we adopted two 16-bit ocean remote sensing image with clouds and whitecap, and compared our algorithm with SSR method, MSR method, and original SDMSR method. The experimental results are shown in fig. 2. In this experiments, the scale factor of Gaussian surround function in the SSR method is σ=100, and in other methods is σ=30,100,300. The convolution kernel size of Gaussian surround function in all methods is 200×200.

As shown in fig. 1, the contrast in the original image is weak. By applying the SSR method, the MSR method, and the original SDMSR method to the remote sensing image, respectively, the contrast of the image has been improved to a certain extent. However, the texture of the enhanced images is still not clear enough. Compared with the other three methods, our method highlights the details of the ocean remote sensing image better.

(a) original image

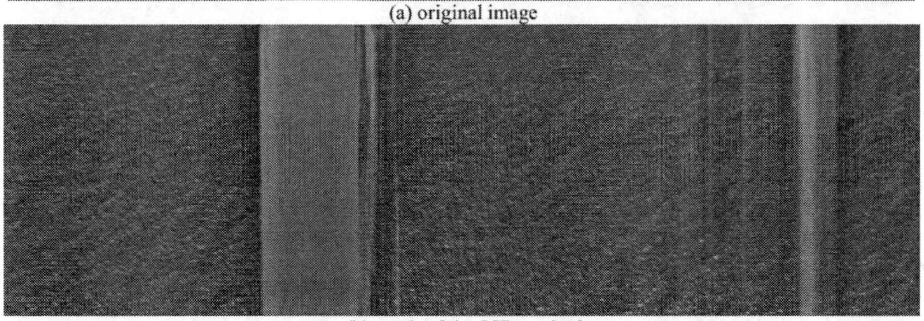

(b) result of the SSR method

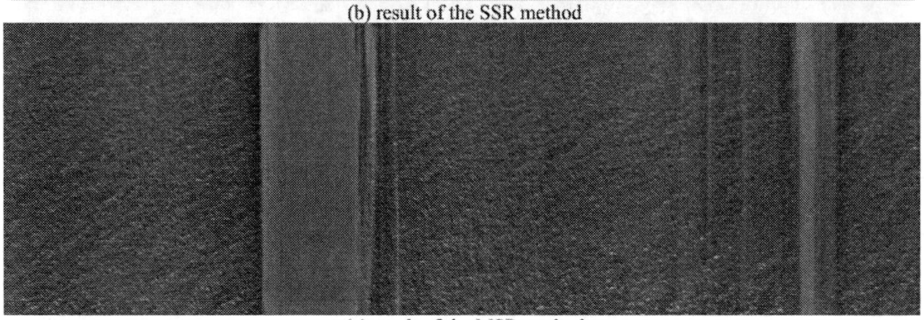

(c) result of the MSR method

(d) result of the original SDMSR method

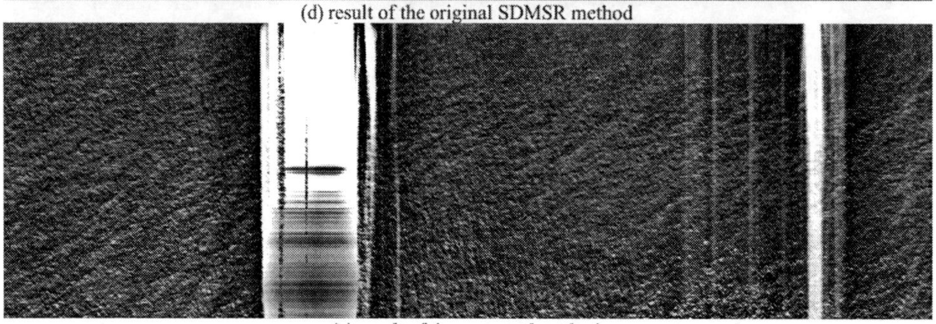

(e) result of the proposed method

Fig. 1. The results of ocean remote sensing image enhanced by different methods

4 Conclusions

In order to effectively monitor important sea areas, the sea image with clear texture feature is required. However, the image enhancement effect based on the SDMSR method is not ideal while the ocean remote sensing image contains clouds, whitecap, etc. Therefore, in this paper, we modified the original SDMSR method by revising the subband gain function and the automatic gain control function. Experiment results show that our proposed method can effectively enhance the clarity of the ocean remote sensing image with clouds and whitecap.

References

1. E. H. Land, and J. J. McCann. Lightness and retinex theory. *J. Opt. Soc. Am*, **61**: 1–11 (1971).
2. E. H. Land. An alternative technique for the computation of the designator in the retinex theory of color vision. *Proc. Nat. Acad. Sci.*, **83**: 3078–3080 (1986).
3. D. J. Jobson, Z. Rahman, and G. A. Woodell. Properties and performance of a center/surround retinex. *IEEE Trans. Image Processing*, **6**: 451–462 (1997).
4. D. J. Jobson, Z. Rahman, and G. A. Woodell. A multiscale retinex for bridging the gap between color images and the human observation of scenes. *IEEE Trans. Image Processing*, **6**: 965–976 (1997).
5. Z. Rahman, D. J. Jobson, and G. A. Woodell. Retinex processing for automatic image enhancement. *J. Electronic Imaging*, **13**: 100–110 (2004).
6. G. Orsini, G. Ramponi, P. Carrai, and R. Di Federico. A modified retinex for image contrast enhancement and dynamics control. In: *Int. Conf. Image Process., Barcelona*. pp. 393-396 (2003).
7. B. Sun, W. Chen, H. Li, W. Tao, and J. Li. Modified luminance based adaptive MSR. In: *IEEE ICIG, Chengdu*. pp. 116-120 (2007).
8. K. Barnard and B. Funt. Analysis and improvement of multiscale retinex. In: *Fifth Color Imaging Conference: Color Science, Systems and Applications, Scottsdale*. pp. 221- 226 (1997).
9. T. Watanabe, Y. Kuwahara, A. Kojima, and T. Kurosawa. Improvement of color quality with modified linear multi-scale retinex. In: *15th SPIE Symposium on Electronic Imaging, Santa Clara*. pp. 59-69 (2003).
10. J. H. Jang, Y. Bae, and J. B. Ra. Multi-sensor image fusion using subband decomposed multiscale retinex. In: *16th IEEE Int. Conf. Image Process., Cairo*. pp. 2177-2180 (2009).
11. J. H. Jang, Y. Bae, and J. B. Ra. Contrast-Enhanced Fusion of Multisensor Images Using Subband-Decomposed Multiscale Retinex. *IEEE Transactions on Image Processing*, **21**: 3479-3490 (2012).
12. J. H. Jang, S. D. Kim, and J. B. Ra. Enhancement of optical remote sensing images by subband-decomposed multiscale retinex with hybrid intensity transfer function. *IEEE Geosci. Remote Sensing Lett.*, **8**: 983–987 (2011).
13. J. H. JANG, B. CHOI, S. D. Kim, et al. Sub-band decomposed multiscale retinex with space varying gain. In: *IEEE Int. Conf. Image Process., San Diego*. pp. 3168-3171 (2008).

Research on Characteristics of Foreshocks of the Dengta M5.1 Earthquake Sequence in Liaoning

Xiaoshan Wang[1, a], Yimei Sun[2], Liang Wang[2], and Li Ma[2,*]

[1] Hebei Earthquake Agency, 050021 Shijiazhuang Hebei, China
[2]Liaoning Earthquake Agency, 110031 Shenyang Liaoning, China

Abstract. With the observed earthquake sequence data and the waveform data in the hypocenter region of the M5.1 earthquake that hit Dengta County in Liaoning on 23rd January, 2013, this study calculated the amplitude spectrum correlation coefficient before and after the M5.1 earthquake, and performed data clustering. Meanwhile, through comparing the focal mechanism of the mainshock and the maximum foreshock, this study explored whether the Dengta M5.1 Earthquake had foreshocks, and found that the foreshocks of Dengta earthquake had a consistent focal mechanism.

1 Introduction

On the evening of 21st May, 2021, Yangbi County of Dali Prefecture of Yunnan was hit by several above-M4.0 earthquakes, with the maximum earthquake reaching a magnitude of 6.4. The earthquake left three dead and 32 injured. Earthquakes often lead to immeasurable damages to human life and properties, and hence earthquake prediction has become an eternal topic among seismologists. Foreshocks are considered the most effective indicator among all short-imminent seismic precursors for earthquake prediction[1]. Early researchers, from the angle of seismic source spectra, maintained that there were abnormal high-frequency parts in the amplitude spectra of foreshocks[2], and this argument was confirmed by the experimental studies later[3]. The earthquake focal mechanism consistency, which is considered the most important measure for foreshock detection, is a salient feature of foreshock sequences[4]. Recent studies on foreshock sequences based on observed data also revealed highly similar seismic waves of foreshocks[5-6]. Among other methods that followed the same logic is the amplitude spectrum correlation coefficient[7-8]: a high amplitude spectrum correlation coefficient in fact indicates a consistent earthquake focal mechanism. The occurrence of foreshocks is related to the formation environment. For instance, the Yingkou – Haicheng – Xiuyan region in Liaoning has a high possibility to see foreshock sequences, and except for the Magnitude 7.3 earthquake in Haicheng in 1975, the Xiuyan – Haicheng M5.6 Earthquake that took place in 1999 showed typical foreshocks[9]. On 23rd January, 2013, a M5.1 earthquake hit Dengta County in Liaoning. Likewise, this earthquake also took place in Liaoning. In this study, several seismologic methods were employed to explore whether there were foreshock sequences in the Dengta Earthquake.

2 The earthquakes in Dengta area

On 23rd January, 2013, a M5.1 earthquake hit the junction between Dengta County of Liaoyang City and Sujiatun District of Shenyang City in Liaoning. This region had seen few earhtquake before this M5.1 Earthquake. Though in 2001, several earthquakes with magnitude 1–2 hit this region, no earthquake with a magnitude over 2.5 was recorded. Between 2002 and 2008, this region showed no obvious earthquake activity, and since 2009 it has seen increasing earthquakes with increasing magnitudes (Figure 1).

3 The focal mechanism consistency

The Dengta M5.1 earthquake sequence were sequenced by the time: the earthquake n and the preceding (n-1) earthquakes were grouped into one sequence, which was moved with 1 as the step size to calculate the arithmetic mean of the amplitude spectrum correlation coefficient and hence the amplitude spectrum correlation coefficient that changed with the time[8] as well as the clustering tree. The difference of earthquake focal mechanism between earthquakes was therefore obtained to explore the spatio-temporal change characteristics of the earthquake focal mechanism of the Dengta earthquake sequence.

Specific research steps are as follows. First, four monitoring stations with an interspace angle with the stations around the area of principal shocks smaller than 180°were selected, the data of earthquakes with a magnitude smaller than 2.3 was removed because the phase of earthquakes with a small magnitude is unclear and will result in errors in the calculation result. An amplitude spectrum correlation coefficient analysis

* Corresponding author: rongcuoxue@163.com.
a pragueboy@163.com

https://doi.org/10.1051/e3sconf/202129002005

program was employed to analyze the data, and the seismic waves of 31 earthquakes were extracted. Different lengths of time windows were selected to ensure that the main energy of the P waves and S waves was within the time window. A higher recognition level of the waves was preferred, and the length was usually set at 2000, and the length would be extended in the case of strong earthquakes. After horizontal correction of the wave shape, the P wave sequence and S wave sequence of different weights in the seismic wave were computed to obtain the Fourier spectra of the observed waves: the velocity spectra and the displacement spectra. The amplitude spectrum correlation coefficients of 31 earthquakes were obtained (Figure 2).

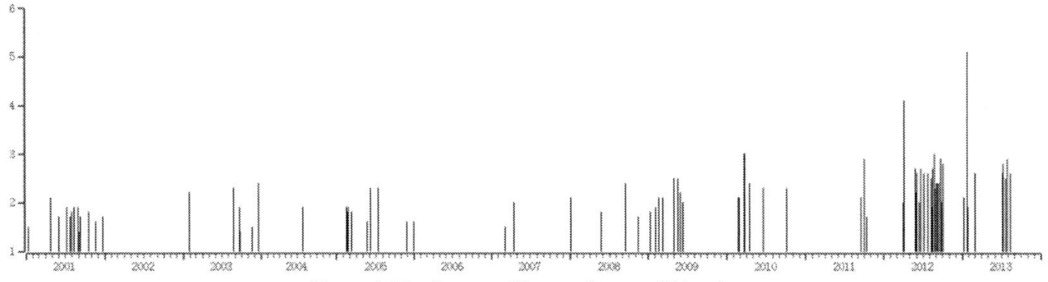

Figure 1. M-t diagram of Dengta County of Liaoning

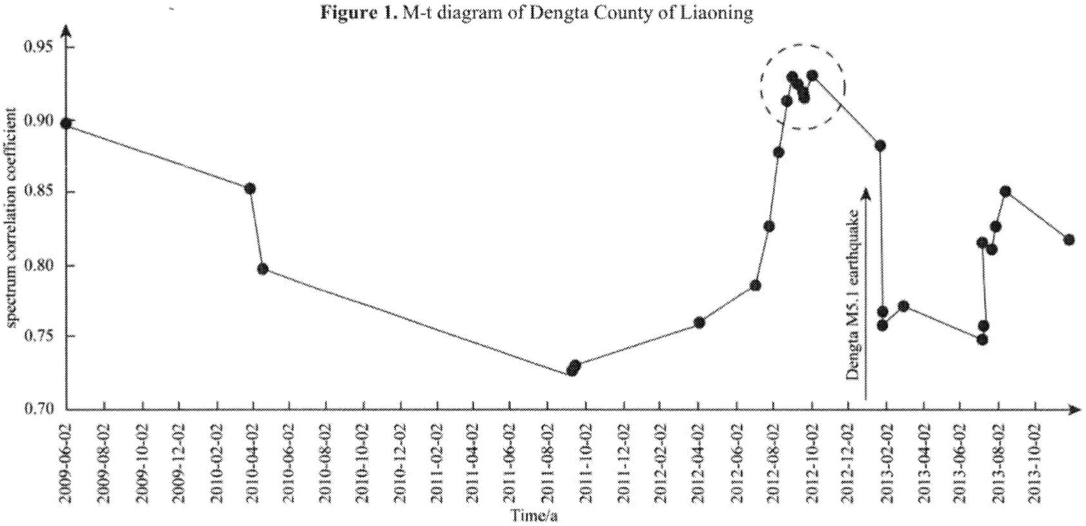

Figure 2. Time series of the multi-spectrum correlation coefficient

As Figure 2 shows, there is tangible earthquake focal mechanism consistency in the earthquakes before the Magnitude-5.1 Earthquake that hit Dengta on 23rd January, 2013. The coefficients of the several earthquakes circled in Figure 4 were above 0.9; the arrow marked the time point where the principal shock turned up; t he amplitude spectrum correlation coefficient post the earthquakes were all below 0.85, indicating that there was no longer obvious earthquake focal mechanism consistency. This presents the time series changing characteristics of the earthquake focal mechanism consistency of the seismic sequence of the Dengta Earthquake.

The multi-spectrum correlation coefficients of 31 earthquakes underwent clustering analysis, and the maximum-length method was used as the clustering method. This method was a successive clustering method, and the binary-tree cluster diagram for the last cluster was obtained. Based on the 31 measured values, the distance between clusters was calculated, and the adjacent two observed values were grouped into one cluster, while single observed values that had not been grouped were set as one cluster. Figure 3 shows the clustering result.

As Figure 3 shows, there were four sets of events in the clustering trees. In the first set, the maximum length is within 0.1, which is considered to have similar earthquake focal mechanism within the set; in the second set, the maximum distance is within 0.15, which is considered to have relatively similar earthquake focal mechanism within the set; in the third and fourth sets, the maximum distance is above 0.2, which is considered to have dissimilar earthquake focal mechanism in the set. Overall, most seismic events had similar earthquake focal mechanisms, and the events that showed inconsistency of earthquake focal mechanisms were 5, 25, 24, 7, 8, and 21.

In sum, during the foreshocks, the earthquake focal mechanism presented changes from a scattered pattern to a consistency pattern. From 1st January, 2008 to 23rd January, 2013 when the Magnitude-5.1 earthquake that hit Dengta, there were a total of 19 earthquakes, among which only three had dissimilar earthquake focal mechanisms. There were 12 earthquakes that happened after the Magnitude 5.1 Earthquake, among which an increasing proportion of earthquakes showed inconsistent earthquake focal mechanisms. Thus, it could be concluded that there was advantageous earthquake focal mechanism distribution in the Dengta earthquake sequence, which is correlated to the geometric features of the causative faults. The small earthquakes that took place before the principal shock showed similar earthquake focal mechanisms, which is the features of foreshocks; at the start of the sequence, the earthquake focal mechanism was disordered and became divergent.

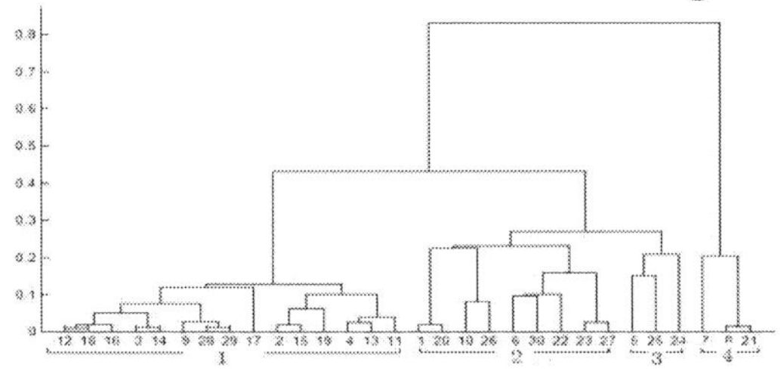

Figure 3. Results of clustering trees

4 The focal mechanism solution of the mainshock and the maximum foreshock

The recorded seismic waves observed from 26 monitoring stations with a high signal-to-noise ratio and within 200 km from the hypocenter were selected, the P-wave first motion inversion method was employed to invert the focal mechanism solution of the Dengta M5.1 Earthquake. This earthquake is a strike-slip earthquake. For nodal plane #1, the strike is 284°, the dip is 84°, and the slip angle is -176°; for nodal plane #2, the strike is 15°, the dip is 87°, and the slip is -5°. On the P axis, the azimuth angle is 240°, and the plunge is 6°; on the T axis, the azimuth angle is 149°, and the plunge is 3°; on the B axis, the azimuth angle is 36°, and the plunge is 83°. The Dengta M5.1 earthquake is a strike-slip earthquake. The calculation result shows that the stations were evenly distributed, and the results concentrated on the P and T axes, with the contradiction ratio 0 (Figure 4).

Before the Dengta M5.1 Earthquake, the maximum earthquake that hit the region was a M4.1 earthquake that took place on 4th April, 2012. By the same method, the P-wave first motion data were used to calculate the focal mechanism solution of this earthquake, and the earthquake was identified as a strike-slip earthquake. The specifics of this earthquake are as follows: on the nodal plane #1, the orientation is 204°, the dip is 85°, and the slip is 3°; on the nodal plane #2, the strike is 294°, the dip is 87°, and the slip is 5°; on the P axis, the azimuth angle is 69°and the plunge is 1°; on the T axis, the azimuth angle is 159°and the plunge is 6°; on the B axis, the azimuth angle is 326°and the plunge is 84°. Figure 5 shows the calculation result.

Comparison of the focal mechanism between the two earthquakes shows that both are strike-slip earthquakes, with an plunge of P and T axes larger than 25°, and horizontal compression as the major cause. As regional stress field reflected by the small earthquake composite fault-plane solution in the Shenyang-Liaoyang area since 1974 shows, the composite earthquake focal mechanism solution of the many small earthquakes in history on the P-axis direction is between 52° and 86°, averaged at 72°; on the T-axis direction, the solution ranges from 149° to 176°, with a mean of 161°. The plunge of the P axis is between 2° and 46°, with a mean of 16°; the plunge of the T axis is between 2° and 12°, with a mean of 7°. In all, the M5.1 earthquake that hit Dengta in 2013 and the M4.4 earthquake that hit Dengta in 2012 showed that the Shenyang-Liaoyang stress field and the regional stress field were consistent, the principal pressure axis that constructed the stress field was the NEE direction, the principal tensile stress axis is along the NNW direction, and the principal stress mainly works along the horizontal direction. Under this stress field, the northeasterly faults are prone to dextral strike-slips, and the northwesterly faults are prone to sinistral strike-slips.

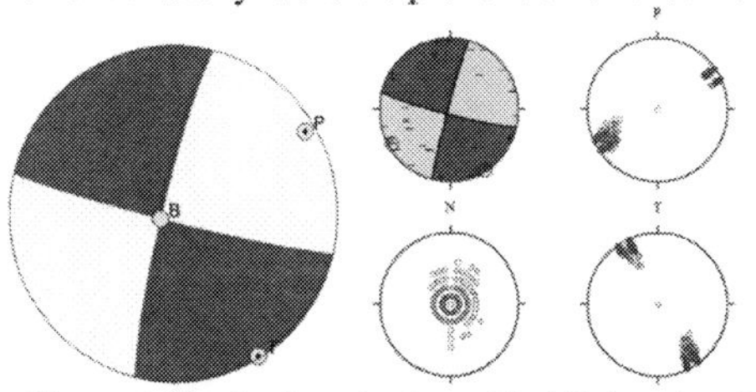

Figure 4. The focal mechanism of the M5.1 earthquake in Dengta on 23rd January, 2013

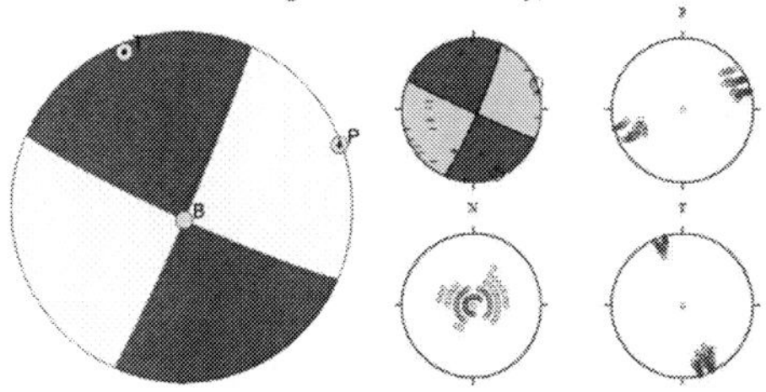

Figure 5. The focal mechanism solution of the M4.1 Earthquake on 4th April, 2012

5 Conclusion and discussions

In this study, the body-wave amplitude spectrum correlation coefficient method was calculated and it was found that the amplitude spectrum correlation coefficients of Dengta Earthquake on 23rd January, 2013 and several foreshocks had an amplitude spectrum correlation coefficient above 0.9, while the coefficient after this earthquake was below 0.85, which indicates that the earthquakes after the Dengta M5.1 Earthquake had no obvious earthquake focal mechanism consistency. The foreshocks of the M5.1 earthquake that hit Dengta on 23rd January, 2013 showed obvious earthquake focal mechanism consistency; as the earthquake focal mechanism solution of the mainshock and the maximum foreshocks was similar, the focal mechanism of foreshocks within the hypocenter area changed from a scattered pattern to a consistent pattern, and showed advantageous focal mechanism distribution, which is considered to be related to the geometric features of the causative fault. Therefore, the sequence had foreshocks, and the small shocks before the mainshock might have similar focal mechanisms. Meanwhile, the focal mechanisms of aftershocks presented scattered.

It is, however, illogical to rely on this single feature to identify whether there are foreshocks, and more in-depth analysis of the earthquake sequence should be performed. As the nucleation of an earthquake starts, it will result in instability[10-11]. Foreshocks are an important part of the nucleation process of an earthquake, i.e., the direct evidence of accelerated deformation before instability[12], and the expected typical micro-seismic events before the sub-instability stage in the sub-instability theory[13]. Therefore, effective detection of foreshocks is the direct evidence to evaluate whether there will be subsequent macro-instability. Despite the lots of studies having been done in the past, there is still no effective foreshock recognition method available [14]. The rupture growth and strike-slip displacement of the nucleation of the foreshocks expand with the time, and may incur migration of small shocks towards the principal shock hypocenter. In contrast, there are no significant expansion or migration of earthquakes in common earthquake swarms[15].

Acknowledgements This research is funded by the seismic research project of Changning, Sichuan (IGCEA1919).

References

1. Jones, L. M., Molnar, P. (1979) Some characteristics of foreshocks and their possible relationship to earthquake prediction and premonitory slip on fault. J. Geophys. Res., 84(B7): 3596-3608.
2. Zheng, J. C., Wang, P., Xu, C. T., Xu, F. F. (2015) Analysis of seismic spectra of the 2014 Yutian Ms7.3 Earthquake sequence and foreshock recognition. Earthquake Research in China, 31(02): 253-261.
3. Jiang, H. K. (2000) Typical fault complexes and sound emission characteristics of rock deformation under different temperatures and pressures. Institute of Geology, China Earthquake Administration, Beijing, 28-56.
4. Chen, Y. (1978) Earthquake focal mechanism consistency as a new parameter for seismic activity. Chinese Journal of Geophysics, 21(2):46-159.
5. Ni, S. D., Wang, W. T., Li, L. (2010) Yushu Earthquake on 14th April, 2010: A destructive earthquake with foreshocks, Scientia Sinica(Terrae), 40(5): 535-537.
6. Bouchon, M., Karabulut, H., Aktar, M., Özalaybey, S., Schmittbuhl J., Bouin, M-P. (2011) Extended nucleation of the 1999 Mw 7.6 Izmit earthquake. Science, 331(5): 877-880.
7. Lund, B., Boevarsson, R. (2002) Correlation of microearthquake body wave spectral amplitudes. Bull. Seismol. Soc. Am., 92(6): 2419-2433.
8. Cui, Z. J., Li, Z. X., Chen, Z. L. Zhao, C. P., Zheng, S. H., Zhou, L. Q. (2012) Research on new methods for classification of small earthquake swarm sequences - amplitude spectrum correlation analysis, Chinese Journal of Geophysics, 55(5): 1718-1724.
9. Yu, L. W., Li, F., Lu, L. Y., Chi, G. C., Wang, H. Y. (2000) Scientific summary of successful forecast of the Magnitude-5.6 Xiuyan-Haicheng Earthquake on 29th November, 1999. Seismological Research of Northeast China, 16(2): 14-26.
10. Das, S., Scholz, C. H. (1981) Theory of time dependent rupture in the earth, J Geophys Res, 86(B7): 6039- 6051.
11. Kanamori, H. (1981) The nature of seismicity patterns before large earthquake. In: Simpson, D. W., Richards, P. G. (Eds.), Earthquake Prediction: An International Review, American Geophysical Union, (4): 1-19.
12. Scholz, C. H. (2002) The mechanics of earthquakes and faulting. Cambridge University Press, New York.
13. Ma, J., Sherman, S. I., Guo, Y. S. (2012) Recognition of stress sub-instability before earthquakes - an experimental case study on the temperature evolution of a 5° bend fault. Scientia Sinica(Terrae), 42(5): 633-645.
14. Jiang, H. K.,Zhou, S. H. (2020) Foreshocks: the value of forecast and recognition methods. Seismological and Geomagnetic Observation and Research, 41(05): 222-225.
15. Qu, J. H., Jiang, H. K., Li, J., Zhang, Z. H. Zheng, J. C. Zhang, Q. (2015) A brief analysis of the seismogenic structure of the earthquake sequence that hit Rushan of Shandong from 2013 to 2014. Chinese Journal of Geophysics, 58(6):1954-1962.

PS-InSAR based surface subsidence analysis in Changchun metropolitan area

He WANG[1], Jiuchang MAO[2], Shijun ZHAO[2], Xinguo NING[3], Qiong WU[1,*]

[1]College of Geo-exploration Science and Technology, Jilin University, Changchun 130026, China
[2]China Water Northeastern Investigation, Design&Research Co. Ltd, Changchun 130062, China
[3]Changchun Institute of Surveying and Mapping, Changchun 130021, China

Abstract The monitoring of urban land surface subsidence is the main task of urban disaster prevention and mitigation. In this paper, the land surface subsidence process in Changchun metropolitan area is calculated based on PS-InSAR technology by using the SLC data of Sentinel-1B with 50 scenes within the metropolitan area of Changchun from 2016 to 2020. The results show that the metropolitan surface of Changchun is basically in a stable state, the main settlement is located in the southeast of the metropolitan around the Century Square, the maximum settlement rate is 20.45mm/a, the average settlement rate is 2.98mm/a, the correlation coefficient between the settlement results calculated by PS-InSAR and the second-class leveling verification results reaches 0.67, which indicates that the regional settlement results calculated by PS-InSAR have high reliability. The subsidence area is dominated by the geological structure of the Mesozoic Cretaceous Yaojia Formation, with many intercalations of fine silty sand rocks and locally containing medium-coarse gravel sandstone. The overall structural stability is poor, which may be the main reason for subsidence in this area.

1 Introduce

Urban surface subsidence has the characteristics of slow formation, long duration, and irreversibility [1]. There are many reasons for the settlement, including unstable urban geological structure, urban construction, and over-exploitation of groundwater [2-5]. Large-scale urban ground subsidence may cause foundation collapse, underground pipeline damage and road damage [6]. Therefore, monitoring urban surface settlement is of great significance to urban disaster forecasting and urban maintenance.

Leveling measurement and global navigation satellite system (GNSS) measurement are the main ground surface monitoring methods for land subsidence, both of which can achieve millimeter-level measurement accuracy in urban areas. However, these two methods have high operating costs and long operating times, and can only be obtained. The settlement value of point-like and linear areas has low spatial resolution [7]. Synthetic Aperture Radar Interferometry (InSAR) has the characteristics of large monitoring range, low cost, high spatial resolution, and unaffected by climatic conditions [8-10]. Conventional synthetic aperture radar differential interferometry is susceptible to spatial decoherence, time decoherence, and atmospheric delay phase, and has certain limitations in application [11]. PS-InSAR recognizes permanent scatterers based on the time series of SAR images covering the same area and the amplitude dispersion index threshold method, which can maintain the scattering characteristics on long-term series and is almost free from the influence of time-space miscorrelation, which overcomes the differential synthetic aperture radar Insufficiency [12], it can monitor the long-term series of regional settlement processes with millimeter-level accuracy [13-14]. Sentinel, as free radar data, is widely used in urban land surface settlement monitoring. It has a good monitoring effect in the vertical direction of the area, and can reach a monitoring accuracy of 8mm. This article uses Sentinel-1B SLC data, based on PS-InSAR technology to monitor the surface settlement process in Changchun City from 2016 to 2020, identifies the main settlement areas and analyzes the main factors that occur.

2 Data and methods

2.1. Data sources

This article uses SLC data of 50 scenes of Sentinel-1B in the urban area of Changchun from October 2016 to December 2020, IW mode, VV polarization method, one scene is taken every month, and the DEM is SRTM-1 data with a resolution of 30m. , The accurate orbit ephemeris data comes from the Alaska radar data download network, and the ground verification data is the second-class level monitoring data in the urban area of Changchun City in November 2019 and September 2020.

* Corresponding author: Wuqiong@jlu.edu.cn

2.2. Calculation method

PS-InSAR extracts the phase of the PS point after differential interference processing for modeling and calculation. After subtracting the error components in the differential interferogram, the residual phase diagram is obtained, and the time-space filtering method is used to recover the residual phase time series, and various error phases are separated. Finally, the overall deformation is superimposed to obtain the deformation time series at each PS point.

When two adjacent PS points are modeled by neighborhood difference, the phase difference between them in the i-th interferogram is:

$$\Delta\phi_i(x_l,y_l,;x_p,y_p;T_i)=\frac{4\pi}{\lambda Rsin\theta}B_i^{\perp}\Delta\varepsilon(x_l,y_l,;x_p,y_p)$$
$$+\Delta\varphi_i^{res}(x_l,y_l,;x_p,y_p;T_i)+\frac{4\pi}{\lambda}T_i\Delta_v(x_l,y_l,;x_p,y_p) \quad (1)$$

Where $B_i^{\perp}$ and T_i are the spatial vertical baseline and time baseline of the interference pair, respectively; λ, R and θ are the wavelength, the distance from the sensor to the target and the radar incident angle, respectively; $\Delta\varepsilon(x_l,y_l,;x_p,y_p)$, $\Delta\varphi_i^{res}(x_l,y_l,;x_p,y_p;T_i)$, and $\Delta_v(x_l,y_l,;x_p,y_p)$ are the elevation error and residual phase increment, respectively And LOS direction deformation rate increment.

Under the condition of $|\Delta\varphi_i^{res}|<\pi$, maximize the following objective function to find the solutions of $\Delta\varepsilon$ and Δ_v.

$$\gamma=|\frac{1}{M}\sum_{(i=1)}^{M}(cos\Delta\omega_i+i\,sin\Delta\omega_i)|=max \quad (2)$$

In the formula, γ is the model coherence coefficient between adjacent PS points; $i=\sqrt{-1}$; $\Delta\omega_i$ is the difference between the observed value and the fitted value, namely

$$\Delta\omega_i=\Delta\phi_i-\frac{4\pi}{\lambda Rsin\theta}B_i^{\perp}\Delta\varepsilon-\frac{4\pi}{\lambda}T_i\,\Delta_v \quad (3)$$

After separating the linear deformation rate and the elevation error, the residual phase is expressed as:

$$\varphi_i^{res}=\varphi_i^{nl}+\varphi_i^{a}+\varphi_i^{n} \quad (4)$$

In the formula, φ_i^{res} is the residual phase; φ_i^{nl} is the nonlinear deformation phase; φ_i^{a} is the atmospheric delay phase; φ_i^{n} is the noise. Perform spatial low-pass filtering and time domain filtering on the residual map to identify the nonlinear deformation phase. The true deformation phase is the sum of the linear deformation and the nonlinear deformation phase:

$$\varphi_{def}=\varphi_l+\varphi_{nl} \quad (5)$$

In the formula, φ_{def} is the total deformation; φ_l is the linear deformation; φ_{nl} is the non-linear variable.

After the Sentinel-1B SLC data format is converted and cropped, the main image is the image on November 2, 2018, and the 5 times critical baseline threshold is used to make other data and the main image establish a master-slave data pair, and then perform an interference workflow to combine all images It is registered to the main image, and 4 times sampling is selected in the distance to avoid the generation of fast-changing interference fringes. Use DEM data to flatten, and then perform radiometric calibration to determine the amplitude dispersion index. The first step is inversion to obtain the displacement rate and residual terrain to flatten the synthesized interferogram. The second step is inversion to estimate and remove the atmospheric phase component to obtain the final deformation rate. After geocoding, the coherence threshold is greater than 0.45 on the surface The deformation result is projected to the vertical direction to make a vertical deformation rate grid graph.

3 Results

In the settlement monitoring calculation results, 87.94% of the PS points have a correlation greater than 0.7. The settlement change value of the SAR image from December 2019 to October 2020 is calculated, and 46 level monitoring points that overlap with the image calculation result patches are selected and used The monitoring value is compared with the patch value for accuracy verification (Figure 1). The verification results show that the maximum difference between the monitoring values obtained by the two monitoring methods is 8.98mm/a, the minimum is 0.04mm/a, the average error is 1.13mm/a, and the root mean square error is 3.09mm/a. Correlation The coefficient is 0.67, and the settlement trends calculated by the two are basically the same, indicating that the settlement results obtained from the SAR images in the area calculated by using the PS-InSAR technology are highly reliable.

Most of the urban surface of Changchun City is stable without changes in subsidence. Most of the urban geological structure is located on the first and second terraces of the Cenozoic Quaternary, and the terraces are stable and not prone to subsidence. Areas with small-scale settlements are located in Jilin Overpass, Century Plaza, Guoxinmeiyi and Ecological Plaza (Figure 2). The maximum settlement rate is 20.45mm/a, and the average settlement rate is 2.98mm/a.

The average settlement rate in the Jilin overpass area is 4.23mm/a, and the cumulative settlement is 16.92mm (Figure 3). Since 2012, three new buildings have been built in this area. Although all of them were completed before 2016, they may also have a certain impact on the regional surface settlement.

The subsidence area of Century Plaza is relatively large and the velocity distribution is uneven. The average settlement rate in the northern region is 11.60mm/a and the

https://doi.org/10.1051/e3sconf/202129002006

cumulative settlement is 46.41mm. The average settlement rate in the southern region is 8.89mm/a and the cumulative settlement is 35.56mm. The average settlement rate in the eastern region is 6.72mm/a. The cumulative settlement is It is 26.89mm, the average settlement rate in the western region is 1.82mm/a, the cumulative settlement is 7.29mm, the average settlement rate in Zhaojiadian area is 5.83mm/a, and the cumulative settlement is 23.31mm. There are many new buildings in the Century Plaza area, there are many factories, and the settlement is relatively concentrated.

The average settlement rate in Guoxinmeiyi area is 9.37mm/a, and the cumulative settlement is 37.48mm. The geological structure in this area is the structure of the Mesozoic Cretaceous Yaojia Formation. The top of this structure is composed of gray siltstone interbedded, with gravel-bearing medium-coarse sandstone in some parts; brown-red mainly, thick argillaceous rock layer, fine siltstone There are few interlayers and a small amount of calcareous nodules; thin argillaceous siltstone layers are occasionally seen at the bottom. Relative to other structures, the stability is poor. Two earthquakes of magnitude ML1.4 and ML1.3 occurred in the southern part of this area in 2001 and 2006 respectively [15], and geological activities may be relatively frequent.

The average settlement rate in the ecological square area is 4.65mm/a, and the cumulative settlement is 18.59mm. Three new buildings and an elementary school have been built in this area, and expressway construction projects have also been carried out, which may have a certain impact on the surface settlement of the area.

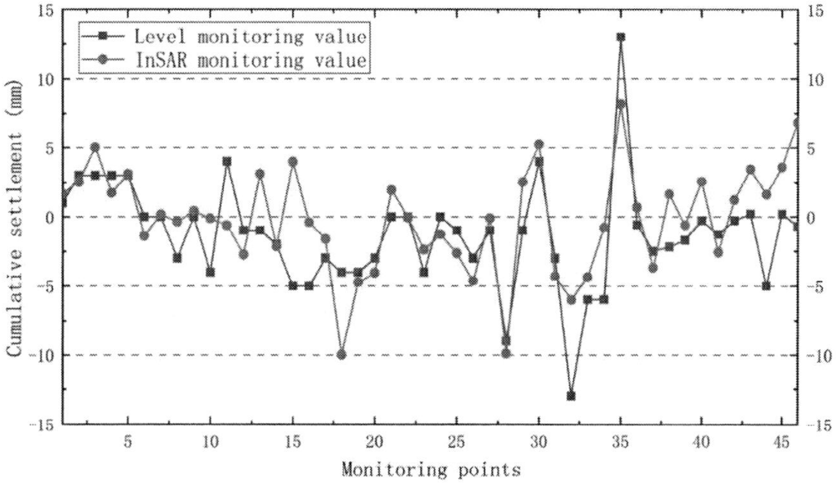

Fig.1. Comparison of InSAR monitoring values and level monitoring values

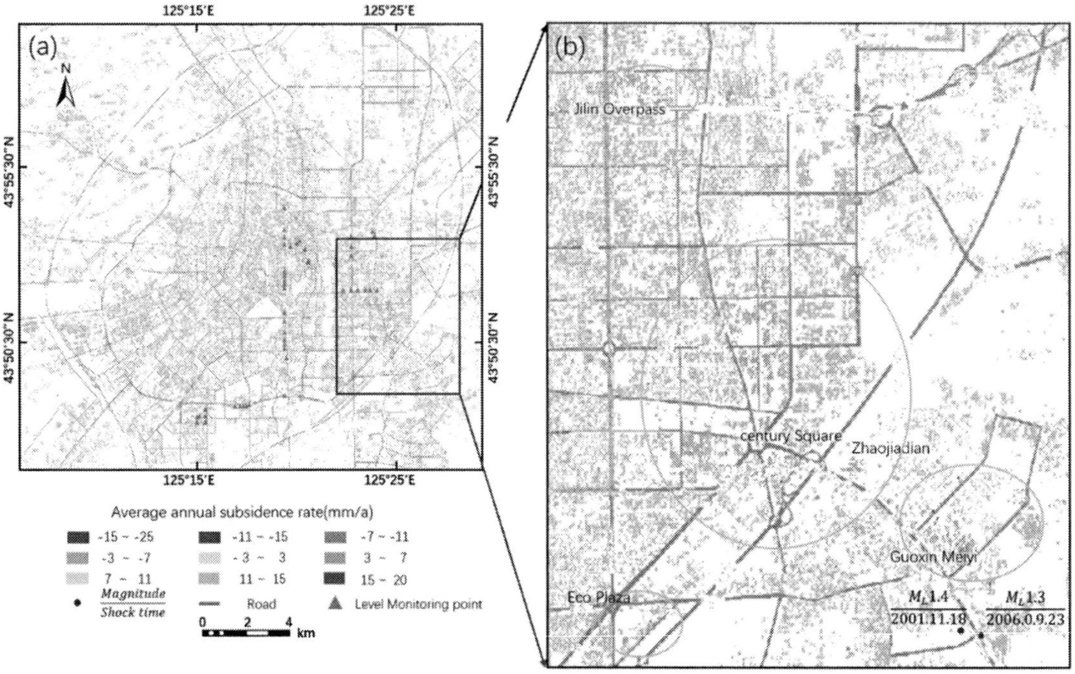

Fig.2. (a) 2016-2020 settlement rate in urban areas of Changchun,(b) Settlement rate in small scale settlement area

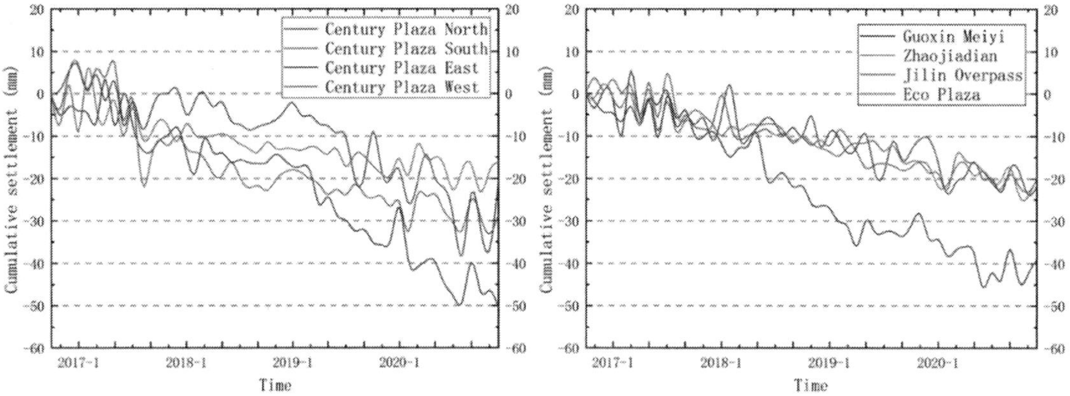

Fig.3. Accumulated settlement in each area

4 Conclusion

(1) Most of the surface of Changchun City is stable without changes in subsidence;

(2) There are four small-scale settlement areas in the southeast of Changchun City: Jilin Overpass, Century Plaza, Guoxinmeiyi and Ecological Plaza;

(3) Engineering construction and unstable geological structure in some areas may be the main cause of settlement in Changchun City.

(4) The key settlement monitoring area of Changchun in the future is the southeast area with Century Square as the centre, and the influence of construction on surface deformation should be minimized.

References

1. Wu Xueyu, Zhao Zhisheng, Bao Ziyu. Research on inversion method of urban land subsidence using SBAS technology [J]. Modern Surveying and Mapping,2019,42(06):9-12.
2. WU Q , JIA C , CHEN S , et al. SBAS-InSAR Based Deformation Detection of Urban Land, Created from Mega-Scale Mountain Excavating and Valley Filling in the Loess Plateau: The Case Study of Yan'an City[J]. Remote Sensing, 2019, 11(14):1673-1683.
3. Ao Meng, Zhang Lu, Liao Mingsheng, Zhang Li, Adaptive fusion deformation measurement of multi-source InSAR data based on variance component estimation [J]. Chinese Journal of Geophysics, 2020, 63(08): 2901-2911.
4. Shi W , Chen G , Meng X , et al. Spatial-Temporal

https://doi.org/10.1051/e3sconf/202129002006

Evolution of Land Subsidence and Rebound over Xi'an in Western China Revealed by SBAS-InSAR Analysis[J]. Remote Sensing, 2020, 12(22):3756-3776.

5. Lu Yanyan. Land subsidence monitoring and influencing factors analysis based on multi-source SAR data in Su-Xi-Chang area [J]. Acta Geodaetica et Cartographica Sinica, 2019,48(07): Backinsert 1.
6. Chen Qinghua, Bai Guifeng, Chen et al. Study on land subsidence monitoring in Changchun city based on PS-InSAR technology [J]. Geomatics & Spatial Information Technology, 2017, 40(09):185-190.
7. CIAMPOLI L B , GAGLIARDI V, FERRANTE C , et al. Displacement Monitoring in Airport Runways by Persistent Scatterers SAR Interferometry[J]. Remote Sensing, 2020, 12(21):3564-3577.
8. Mo Ying, Zhu Yufeng, Jiang Liming, et al. Land subsidence monitoring based on Sentinel-1A time series InSAR in Nanchang City [J]. Journal of Geodesy and Geodynamics, 2020, 40(03):54-59.
9. CHEN D,CHEN H,WEN Z, et al. Characteristics of the Residual Surface Deformation of Multiple Abandoned Mined-Out Areas Based on a Field Investigation and SBAS-InSAR: A Case Study in Jilin, China[J]. Remote Sensing, 2020, 12(22) : 3752-3752.
10. Dou Chao, Zhao Lijiang, Zhang Shengpeng, et al. Surface Subsidence Monitoring Based on InSAR Data in Golmud City [J]. Bulletin of Surveying and Mapping,2020,(10):123-126.
11. Wang Shengyan, PS_INSAR based monitoring of urban land surface subsidence [J], Geoscience and Spatial Information, 2020,18(9):113-121.
12. Zhou Yuchen, LI Jinping, LU Haojian, et al. Study on land surface subsidence and its mechanism in Changchun city based on SBAS-InSAR [J]. Journal of Heilongjiang Institute of Technology, 2019, 33(03):20-25.
13. You Hong, Mi Hongyan, Zuo Xiaoqing, et al. InSAR based on PS and SBAS for subsidence monitoring in the main urban area of Guiyang [J]. Urban Surveying, 2020,179(04):79-83.
14. Li Lu, Hong Youtang. Study on land subsidence monitoring in Taiyuan based on PS-InSAR technology [J]. Mine Surveying, 2020,208(04):54-59.
15. Zhang Jianwei, Wang Xikui, Sui Weiguo, et al. Seismological and geological tectonic activity analysis in Changchun area [J]. Inland Seismology, 2010, 024(003): 269-274.

The Temporal and Spatial Variations of Stable Isotopes in Precipitation on Different Scales in Yangtze River Basin

Ruiheng Meng*[1]

[1]WLSA Shanghai Academy, 200243 Yangpu District, Shanghai, China

Abstract. Analysis is conducted on the temporal and spatial variations of stable isotopic composition in precipitation, using the data of IAEA/WMO sampling stations, in Yangtze river basin. The correlations between $\delta^{18}O$ in precipitation and air temperature/amount of precipitation are analyzed at individual stations and in the whole valley on monthly or annual scales. There is a notable amount effect, the distinct inverse relationship between $\delta^{18}O$ and precipitation, on monthly scale in the Yangtze valley and at the selected stations except Chongqing. Compared with monthly scale, the positive correlation between $\delta^{18}O$ and air temperature is more marked on annual scale. The interannual variations of $\delta^{18}O$ reflect the change of large-scale climate and environment, and are mainly controlled by large-scale weather conditions.

1. Introduction

Two stable isotopes deuterium (D) and oxygen-18 (^{18}O) occer naturally in waters. Stable isotopes of hydrogen and oxygen are an integral part of the water molecule itself, and are commonly seen as ideal tracers. To apply the stable isotope tool to the study of water cycle mechanisms, such as the mutual transformation of surface water and groundwater, groundwater replenishment, etc., the most important prerequisite is to deeply understand the composition of atmospheric precipitation isotopes, because the input item of the water cycle-atmospheric precipitation is the fundamental source of inland water resources. In the 1950s, the International Atomic Energy Agency (IAEA) and the World Meteorological Organization (WMO) jointly established the Global Network of Isotopes in Precipitation (GNIP), which was officially launched in 1961 to continuously monitor the stable isotopic composition of precipitation worldwide and it is still in progress. According to the GNIP database, China has joined the GNIP in a total of 31 sites, of which 8 belong to the Yangtze River Basin, namely Chengdu, Chongqing, Zunyi, Guiyang, Kunming, Nanjing, Wuhan, and Changsha. Past related studies in China [2] were mostly based on fund projects and conducted local (or single-site) studies with a short period of time. It is still not clear about the relationship between long-term atmospheric precipitation isotopes, temperature and precipitation in China's inland areas. This paper uses the monitoring data in the GNIP database to analyze the long-term trend of stable isotopic changes of atmospheric precipitation in the Yangtze River Basin in China. The composition and changes of atmospheric precipitation isotope in the Yangtze River Basin at different scales were shown, which will be helpful to establish a water cycle model in the Yangtze River Basin, clarify the characteristics of the water cycle, identify the temporal and spatial distribution of water resources, and determine the sustainable development model of water resources.

2. Data Collection and Processing

The sampling time, geographic location and altitude of the atmospheric precipitation at the 8 IAEA/WMO precipitation stations in the Yangtze River Basin (in Table 1) shows that the stations are unevenly distributed. Especially in the upper reaches of the basin, there is no GNIP observation station. In order to improve the accuracy of the analysis conclusions, this paper makes full use of the atmospheric precipitation isotope data of 31 stations that China has participated in the GNIP, when analyzing the spatial distribution characteristics of precipitation isotope. Environmental isotopes are expressed in thousandths relative to the Vienna-standard mean ocean water V-SMOW (Vienna-standard mean ocean water):

$$\delta^{18}O(‰)=\left[\left(R_{sample}-R_{VSMOW}\right)/R_{VSMOW}\right]\times10^{3} \quad (1)$$

Where in formula (1): R is the isotope ratio, R=D/1H or $^{18}O/^{16}O$; δ is the thousandth difference between the stable isotope ratio and the standard isotope ratio.

Table 1. List of basic data of 8 IAEA/WMO precipitation stations selected.

Serial number	Sampling point	sampling time	latitude	longitude	Altitude/m
1	Chengdu	1986-1999	30° 40' 12"	104° 1' 12"	506
2	Kunming	1986-2003	25° 1' 00"	102° 40' 59"	1892
3	Wuhan	1986-1998	30° 37' 12"	114° 7' 48"	23

* Corresponding author: 1740829420@qq.com

4	Chongqing	1992-1993	29° 37' 12"	106° 35' 60"	192
5	Changsha	1988-1992	28° 11' 60"	113° 4' 01"	37
6	Zunyi	1986-1992	27° 41' 60"	106° 52' 48"	844
7	Guiyang	1988-1992	26° 34' 60"	106° 43' 01"	1071
8	Nanjing	1987-1992	32° 10' 48"	118° 10' 48"	26

3 The Relationship Between δD and $\delta^{18}O$ in Atmospheric Precipitation in the Yangtze River Basin

In 1961, Craig [3] discovered the linear law of global atmospheric precipitation δD and $\delta^{18}O$, expressed by mathematical formula as $\delta D=8\delta^{18}O+10$, which is the Global Meteoric Water Line (GMWL). However, due to the different sources of water vapor and meteorological conditions in the local circulation system in some areas, there will be significant deviations between the Local Meteoric Water Line (LMWL) and the global atmospheric water line GMWL.

$$\delta D=7.323\delta^{18}O+5.336 \quad (2)$$

Equation (2) is relatively close to the global atmospheric precipitation line GMWL. Figure 1 shows that the change of stable isotopic abundance of precipitation in the Yangtze River Basin is closely related to the cause of precipitation, and is mainly affected by the type of monsoon and the source and nature of precipitation cloud clusters.

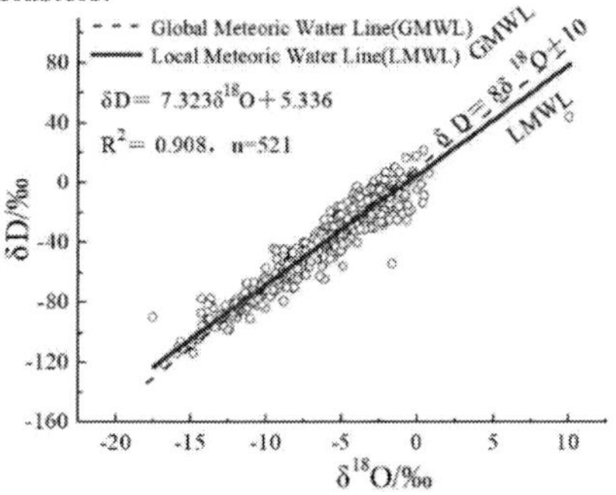

Fig. 1. the Local Meteoric Water Line in the Yangtze River.

4. The relationship between atmospheric precipitation $\delta^{18}O$ and temperature and precipitation under different time scales

4.1 The relationship between precipitation $\delta^{18}O$ and temperature and precipitation on a monthly scale

The change of atmospheric precipitation $\delta^{18}O$ value is a physical process of evaporation and condensation, which is related to geographical factors of temperature and precipitation, and has the characteristics of regular changes. There are temperature effects (showing a positive correlation between temperature and precipitation $\delta^{18}O$) and precipitation effects (showing a negative correlation between precipitation and precipitation $\delta^{18}O$), etc. [4, 5].

According to the multi-year isotope data of 8 IAEA/WMO precipitation stations in the Yangtze River Basin, the monthly precipitation $\delta^{18}O$, monthly average temperature and monthly precipitation of the 8 GNIP stations and the Yangtze River Basin after calculation are shown in Table 2 and Figure 2. The change of $\delta^{18}O$ is negatively correlated with the temperature on the whole. On the general trend, $\delta^{18}O$ decreases with the increase of precipitation P, indicating that atmospheric precipitation $\delta^{18}O$ in the Yangtze River Basin presents a more significant precipitation effect, with greater precipitation effect than the temperature effect. The rainfall in the Yangtze River Basin is relatively high, and the air humidity is relatively high. The effect of atmospheric precipitation δ^{18} at the GNIP site can't be reflected on temperature changes, but on the effect of precipitation. The $\delta^{18}O$ value of the Chongqing site increases with the increase of precipitation, with the precipitation effect masked in Chongqing, due to its extremely high temperature amid the rainy season.

Table 2. Correlation between monthly precipitation $\delta^{18}O$ and monthly average temperature and monthly precipitation.

	Site	The temperature effect of atmospheric precipitation $\delta^{18}O$	The precipitation effect of atmospheric precipitation $\delta^{18}O$
1	Chengdu	$\delta^{18}O=-0.037T-5.028$	$\delta^{18}O=-0.02P-3.107$
2	Kunming	$\delta^{18}O=-0.488T+0.251$	$\delta^{18}O=-0.03P-4.98$
3	Wuhan	$\delta^{18}O=-0.085T-4.421$	$\delta^{18}O=-0.01P-4.737$
4	Chongqing	*/(Temperature data missing)*	$\delta^{18}O=0.012P-12.56$
5	Changsha	$\delta^{18}O=-0.182T-2.478$	$\delta^{18}O=-0.001P-5.453$
6	Zunyi	$\delta^{18}O=-0.354T-0.987$	$\delta^{18}O=-0.03P-4.01$
7	Guiyang	$\delta^{18}O=-0.287T-2.33$	$\delta^{18}O=-0.025P-4.716$
8	Nanjing	$\delta^{18}O=-0.065T-6.264$	$\delta^{18}O=-0.01P-6.132$
Yangtze River Basin		$\delta^{18}O=-0.16T-3.878$	$\delta^{18}O=-0.02P-4.726$

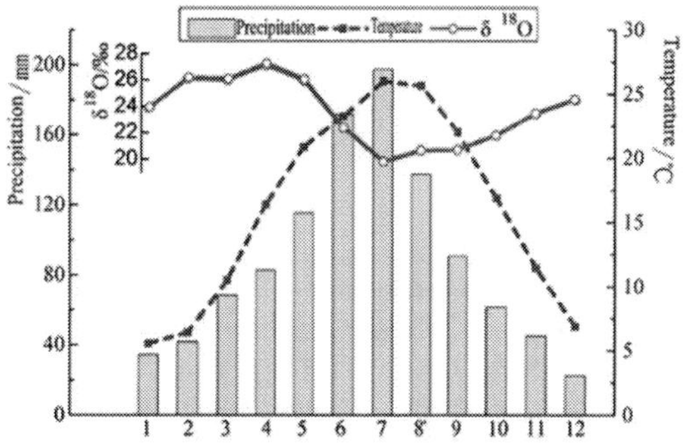

Fig. 2. The seasonal variation of monthly precipitation $\delta^{18}O$ and monthly average temperature and monthly precipitation.

4.2 The Relationship between Annual Atmospheric Precipitation $\delta^{18}O$ and Temperature and Precipitation

4.2.1 The Relationship between Annual Declining Water $\delta^{18}O$ and Temperature and Precipitation

From the above analysis, it can be seen that on the monthly scale, except for the Chongqing site, the precipitation effect of atmospheric precipitation $\delta^{18}O$ at the remaining seven GNIP monitoring sites is greater than the temperature effect, with the temperature effect masked. The annual weighted average atmospheric precipitation $\delta^{18}O$, annual average temperature and annual precipitation, and multi-year weighted average atmospheric precipitation $\delta^{18}O$, multi-year average temperature and annual average precipitation of the 7 stations with significant precipitation effects in the Yangtze River Basin are all plotted in Figure 3.

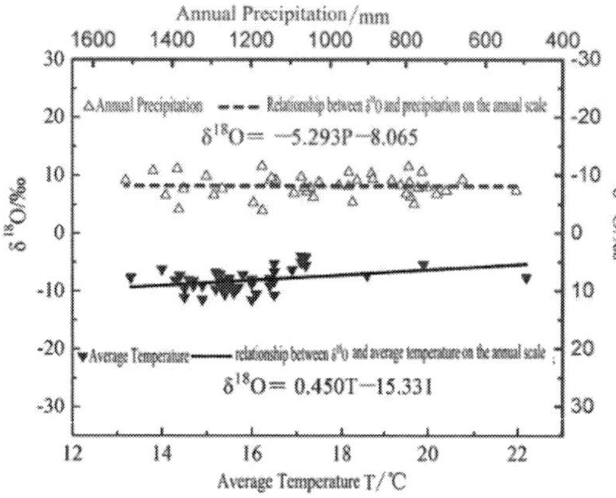

Fig. 3. The correlation between annual weighted average $\delta^{18}O$ of sampling stations with precipitation effect and annual average temperature and annual precipitation.

Compared with a single station, the isotope of atmospheric precipitation in the Yangtze River Basin not only has a precipitation effect (the correlation is $\delta^{18}O$ = -5.293-8.065), but also shows an interannual temperature effect (the correlation is $\delta^{18}O$ =0.450T-15.331), which shows that the atmospheric precipitation $\delta^{18}O$ is mainly subject to large-scale meteorological conditions on the interannual scale. It also reflects that environmental and meteorological conditions are large-scale spatial changes, rather than local changes. The above analysis results show that the stable isotopes of atmospheric precipitation in the Yangtze River Basin have a precipitation effect on the monthly scale. They also have a temperature effect trend, in addition to the precipitation effect on the interannual scale. Taking into account the complexity of the $\delta^{18}O$ impact factors of atmospheric precipitation in the Yangtze River Basin, the temperature and precipitation changes at a single site are an important part of the changes in atmospheric precipitation $\delta^{18}O$ in the basin.

4.2.2 Changes of Annual Atmospheric Precipitation $\delta^{18}O$ and Temperature and Precipitation

In order to analyze the variation trends of the stable isotope $\delta^{18}O$ of atmospheric precipitation in the Yangtze River Basin and the annual average temperature and annual precipitation over time, three GNIP sites (Zunyi, Guiyang, Kunming) with a long monitoring time were selected to analyze the trend of $\delta^{18}O$ with climate change. Figure 4 shows the annual weighted average $\delta^{18}O$ and annual average temperature and annual precipitation changes of 3 typical stations. The annual weighted average $\delta^{18}O$ and the annual precipitation generally show a reversed change, indicating that the atmospheric precipitation isotope in the Yangtze River Basin has a precipitation effect on the annual scale. The annual weighted average $\delta^{18}O$ and the annual average temperature generally show obvious simultaneous changes, indicating that the temperature effect of atmospheric precipitation $\delta^{18}O$ in the Yangtze River Basin is not obvious on the monthly time scale. There is also a correlation on the annual time scale, showing the temperature effect on the annual scale. Although the monitoring time of Kunming Station is relatively long, it shows local asynchronous and anti-phase changes, mainly due to the lack of monitoring data from 1993 to 1995. Moreover, the precipitation effect of atmospheric precipitation $\delta^{18}O$ at Kunming site is more obvious than that of Zunyi and Guiyang sites.

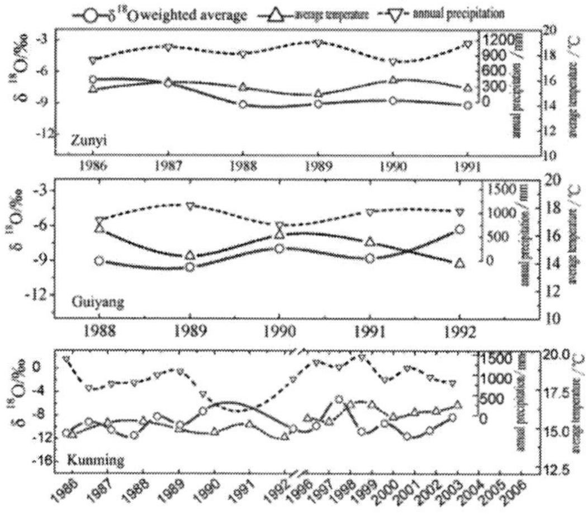

Fig. 4. The inter-annual variation of the average annual atmospheric precipitation $\delta^{18}O$ and the annual average temperature at typical sites.

5 Analysis

Due to climatic and geographic factors, in terms of the spatial scale of the entire Yangtze River basin, the atmospheric precipitation isotope $\delta^{18}O$ can reflect the geographical and regional characteristics of atmospheric precipitation and temperature to a certain extent. Due to the difference in topography of the Yangtze River Basin, the annual average temperature is higher in the east and lower in the west, higher in the south and lower in the north, and higher in the middle and lower reaches than in the upper reaches. The source area of the Yangtze River is the lowest temperature area, while the Sichuan Basin, Yunnan-Guizhou Plateau and Jinsha River Valley are closed high and low temperature central areas. Due to the vast area, complex topography and typical monsoon climate, the annual precipitation is unevenly distributed in time and space. The source area of the Yangtze River is a dry zone, while most of the watershed is a humid zone. It can be seen that the climatic diversity has caused the difference in the spatial distribution of $\delta^{18}O$ in the atmospheric precipitation, coupled with the multi-step topography of the Yangtze River Basin, the upper reaches of the west have high terrain and large drops, and the middle and lower reaches of the central and eastern parts are low and flat, which is conducive to the deepening of the ocean airflow in the southeast direction. The effect of precipitation in China exceeds the effect of temperature to some extent.

6 Conclusion

The atmospheric precipitation line equation in the Yangtze River Basin is $\delta D=7.323\delta^{18}O+5.336$, which is basically consistent with the global atmospheric precipitation line $\delta D=8\delta^{18}O+10$.

On the spatial scale, the Yangtze River Basin shows a clear monsoon effect on the whole, affected by the topography and monsoon climate, and the temperature effect is concealed by the monsoon precipitation effect. However, the effect of atmospheric precipitation is more obvious in the middle and lower reaches than in the upper reaches and in the south of the Yangtze River than in the north of the Yangtze River. Among the 8 GNIP stations, there are 7 stations with a precipitation effect. There is a negative correlation between precipitation and precipitation $\delta^{18}O$.

On the monthly time scale, the change of atmospheric precipitation $\delta^{18}O$ and temperature show a negative correlation, with the precipitation effect greater than the temperature effect. As the amount of precipitation increases, the value of precipitation $\delta^{18}O$ decreases. However, on an annual scale, the entire watershed not only has a precipitation effect, but also exhibits a temperature effect. Stations where the $\delta^{18}O$ temperature effect is not significant on the monthly time scale exhibit temperature effects on the annual time scale. This correlation will provide an important scientific reference for further understanding of climate change and hydrological cycle laws at different time scales in the Yangtze River Basin.

The problem with the study of stable isotopes in the Yangtze River Basin is that there are few data at present, and the established perennial observation stations are relatively short in time and few in the number of stations, so comprehensive quantitative analysis is subject to certain restrictions. It is recommended to set up more stations in the entire watershed, especially in the upstream area, so as to obtain larger spatial and temporal scale observation data, to better understand the hydrological cycle process, and to study the evapotranspiration characteristics of the watershed.

References

1. Dansgard W. Stable isotopes in precipitation. Tellus, 16(4): 436-468 (1964)
2. Wang Yongsen, Chen Jiansheng, Wang Jiyang, et al. Study on the Theoretical Relationship between Hydrogen and Oxygen Stable Isotopes during Precipitation process. Advances in Water Science, 20(2): 204-208 (2009)
3. Craig H. Isotopic variations in meteoric waters. Science, 133: 1702-1703 (1961).
4. Vallet-Coulomb C, Gasse F Sonzogni C, et al. Seasonal evolution of the isotopic composition of atmospheric water vapor above a tropical lake: Deuterium excess and implication for water recycling. Geochim Cosmochim Acta, 72: 4661-4674 (2008).
5. Jouzel J, Merlivat L. Deuterium and oxygen-18 in precipitation: modeling of the isotope effects during snow formation. Journal of Geophysical Research, 89(D7): 11749-11757 (1984).

The Application of Radon Measurement In the Active Fault Detection of Xiadian Fault

Lu Lina[1,2*], Li Jing[1,2], and Yang Ming[1,2]

[1]Institue of Disaster Prevention, 065201 Sanhe, China
[2]Hebei Key Laboratory of Earthquake Dynamics, 065201 Sanhe, China

Abstract. The Xiadian active fault is one of the most important seismological faults in the vicinity of Beijing-Tianjin-Hebei region. This paper mainly focuses on the geochemical characteristics of the fault to unveil and evaluate its activity. Via the application of soil gaseous radon measurement, a detailed field study at Qixinzhuang segment of Xiadian fault is proposed. Based on all the radon impulse value we acquired of 3 profiles in the study area. We plotted the radon impulse trend chart of 3 profiles, analyzed anomaly distribution of soil gaseous radon and discussed the scale and activity of the Xiadian fault. The radon multiple-peaks anomaly of Qixinzhaung segemnt demonstrates the hidden faults underground exist.

1 Introduction

The Xiadain fault is an important hidden active fault of Beijing plain area [1-2], located in the north of the North China Plate, general trend is NE45°, extends 45 km. Its formation and evolution play an important role in the tectonic evolution of Beijing plain areas. Among all the active fault exploration methods, measuring fault gas has the merits of easy, shortcut, economical and effective. Active fault as the natural channel of connecting the Earth's deep interior and deep gas release, fault-gas permeates in the vicinity of fault zone. A large number of data shows that the more active the fault is, the more fault-gas it has [3-6]. Based on this idea, In this paper, we used the method of soil gaseous radon measurement in Qixinzhuang section and carried on a detailed field survey on site. According to the radon impulse trend charts, we analyzed the characteristics of soil gaseous radon anomaly distribution and the spacial location of the fault in Qixinzhuang section and also discussed the scale and activity of Xiadian fault.

2 Geological setting

The Xiadain active fault, which is located in the northern North China Plate (Fig. 1), is one regional deep fault belt with a NNE trend of the lithosphere scale and located at the east gate of Beijing. Its formation and evolution plays an important role on Beijing plain region tectonic evolution. The fault zone is located north from Yuegezhuang area of Pinggu District, southwest by Ma Fang and Pangezhuang area, and south to the Yongle village. The whole fault is Northeast fortyfive degree and fortyfive kilometeres long. The biggest Cenozoic fault distance can be up to 3 or 4 kilometers during Cenozoic geological time. It is a basin boundary fault that controls the formation and evolution of the Dachang depression and constitutes the boundary between the Dachang depression and the Daxing uplift [7-8]. The 1679 Sanhe Earthquake caused the surface rupture of the Xiadian active fault zone, and the new Xiadian fault was formed on the fault zone. For the sake of discussion, the new Xiadian fault (NE trending) and the old Xiadian fault (NNE trending) are collectively referred to as the Xiadian active fault in this paper. In 1679, Sanhe earthquake formed a fault scarp of about 10 kilometers long on the earth's surface. The scarp is located west from Liuhetun area, through Pangezhuang and Qixinzhuang, and east to Dongxingzhuang area. This fault displays normal fault nature. The main lithology strata in the study area are shown in table 1. During the field work, the typical fault plane and significant fault scarp can be observed on the Qixinzhuang artificial exploratory trench profiles can be observed. According to the field measurement, the Qixinzhuang pofile of the fault has a strike of 45, a dip of 135 and a dip angle of 74. Both sides of the fault are Quaternary accumulation. The horizontal bedding is developed at the bottom of the upper wall of the fault, and there is a traction in the thin clay silt layer.

3 Methods and Results

3.1 Measuring line layout

After the field survey, we can observe the typical fault plane and abvious fault escarpment in Qixinzhuang artificial exploratory trench profile. So we chose Qixinzhuang section (N39°58′54.32″, E116°55′54.32″) as the study area of the gaseous radon measurement. We set 3 isometric measuring profiles perpendicular to the

* Corresponding author: lulina@cidp.edu.cn

fault strike. They are QX-1, QX-2 and QX-3 (Fig.1) and include 181 measuring points. And QX-2 profile lies in the south of QX-1 profile, instead, QX-2 profile is located in the north of QX-1 profile. The adjacent two measurement point interval is generally 10m in every measuring line which extends 300m at each side of the fault. The adjacent two parallel measuring line interval is 20m.

3.2 Instrument introduction

The most appropriate way to measure soil gaseous radon measuring method is conventional instantaneous method. The FD3017RaA gaseous radon measuring machine satisfies the requirement. It's very easy to operate and portable to carry. In addition, it has a high accuracy.

Table 1. Study area stratigraphic classification [9-10]

System	Series	Formation	Codename
Q	Qh	Tianjin	QhT
	Qp	Ouzhuang	QpO
		Yangliuqing	QpY
		Gu'an	QpG
N	N2	Minghuazhen	N_2M
	N1	Guantao	N_1G
E	E3	Dongying	E_3D
		Shahejie	E_3S
	E2	Kongdian	E_2K

Operating principle: The first generation of radon decay-RaA is electropositive. A negative pressure resulting from the high voltage electric field will absorb RaA on the metal wafer. RaA concentration on matal wafer is positively correlated with soil radon concentration. We can acquire the concentration of soil gaseous radon by calculation. The apparatus' limit of detection sensitivity is less than 0.1 Eman.

The operating condition for gas extraction is 60cm~80cm, depending on subsoil depth. The surveyor should take 1.5L gas within 2 min.

Before taking radon values using this instrument, we should choose the detection profiles first. Usually, three profiles are appropriate and caculated for one section of a fault. A profile should crosscut one fault, and 90 degree is the best crosscut angle. The surveying site between another surveying site is 10 meters as usual, and the perpendicular diatance between two profiles is 20 meters. Therefore obtaining these statistics data using this instument legitimately is the precondition for this fault activity study. Additionally, when you use this apparatus, you should don't use your tellphone or keep it far away; when the sampling apparatus is inserted into the soil, the surface of the soil should be sealed for the gas gathering; during the radon testing, the metal sheet can be touched with the sheetmetal margin, and the surface can't be touched, otherwise you will obtain invalid radon value of this testing.

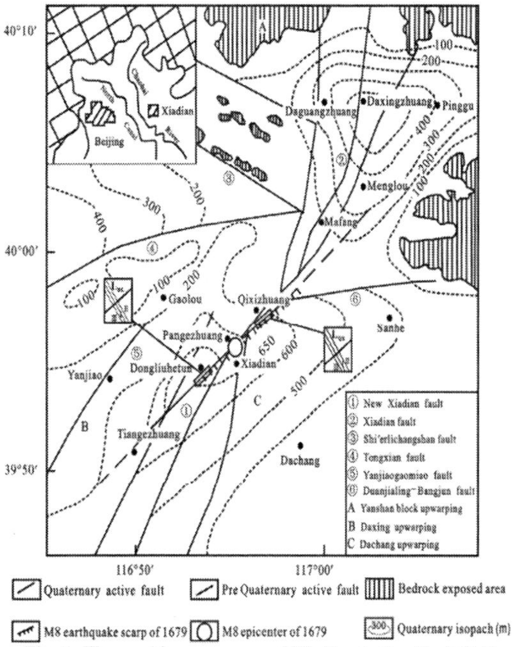

Fig. 1. The working area map of Xiadian Active Fault [11]

3.3 Analysis of the result

This study attained all 181 soil radon geochemical sample data in Xiadian fault, which is the measurement result of the 3 profiles. Each profile owns 60 sample data approximately. In accordance with measurement result, we ploted the impulse graph of each profile (Fig. 2, 3, 4).

We can see from the anomalous curve (Fig. 2, 3, 4) in Qixinzhuang section that the profile QX-1, QX-2 and QX-3 all have conspicuous peak value in each radon impulse graph which is the direct reflection of radon concentration anomaly in each profile. The higher the peak value is, the radon anomaly is more evident. The figures 2, 3 and 4 show that, every one of profile has a higher radon impulse anomaly at the first half part.

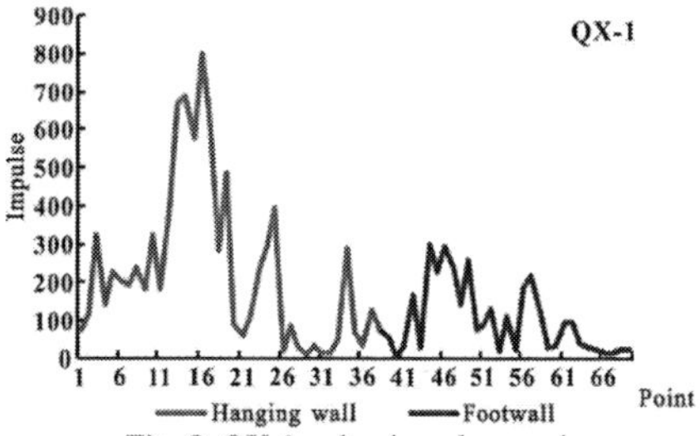

Fig. 2. QX-1 radon impulse graph

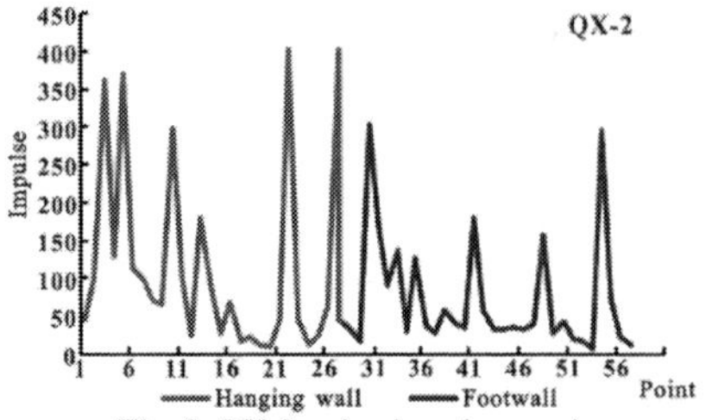

Fig. 3. QX-2 radon impulse graph

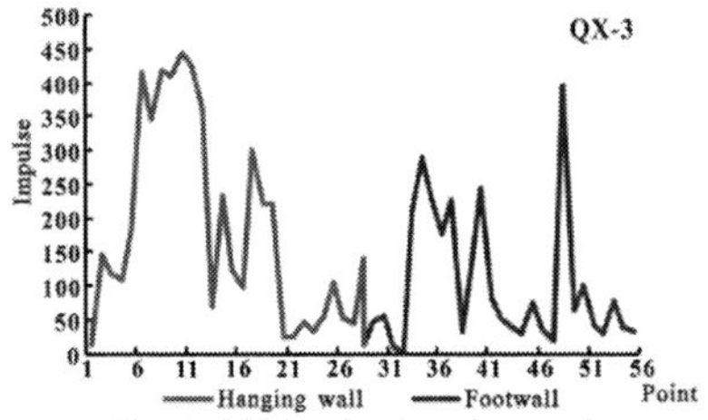

Fig. 4. QX-3 radon impulse graph

4 Discussions

4.1 Fault spacial location analysis

Our work is based on predecessors' work [12] and proceed to carry on exploratory trench excavation. We detected the multiple peaks anomaly of Rn impulse in 3 profiles (QX-1, QX-2 and QX-3) at Qixinzhaung section. And then it unveiled the multiple peaks anomaly of Rn concentration. Multiple peaks anomaly suggests that there are more faults lying underground. Multitudes of abnormal secondary peaks (Fig. 2, 3, 4) imply the location of the branch hidden faults of the major fault. Among the 3 parallel profiles, the abonomal secondary peaks of each profile exists in different dot mark, which demonstrates that the relavent hidden fault exists in different location. On account of that the hanging wall is more active than the footwall and has more fissure than the footwall. Combined with Rn geochemical characteristics, we can acquire that No.1 to 37 stands for the hanging wall of QX-1 profile. In the same way, the hangng wall of QX-2 and QX-2 are No.1 to 26 and No.1 to 27. And the rest of the dot marks of each profile represents the footwall. The analysis fits the practical observation. The peak anomaly in the hanging wall is much higher than the footwall. According to the exploratory trench location and fault strike, we infer that No.6 to 51 of Qixinzhuang profile is the location of the main fault.

4.2 Fault scale and activity analysis

The accurate location and activity natures of this blind active fault zone are detected by kinds of comprehensive detection and trench digging. Therefore, uncertainties can be excluded in the progress of the soil gas radon characteristics for the Xiadian buried fault zone, and data interpretation is also excluded. Practice indicates that the more big dimension of the fault is, the more radon anomaly plots are and the higher anomaly peak is. On the contrary, the more small dimension of the fault is, the even less radon anomaly plots are and the lowerer anomaly peak is, and usually the single and double peak [13]. According to the radon geochemistry characteristics of the Xiadian fault and the collected data [14] , the Qixinzhuang section is larger than the Liuhetun section on the scale for the blind fault under the earthground.

The practice shows that for the large scale fault, the more radon anomalous values are, then the higher points of the peak value is. On the contrary, the smaller fault scale has less anomalous points, and the peak is lower usually with mono-peaks or double-peaks [15] anomaly. Multiple peaks anomaly generally exists in deep fault zone. The fault zone may have multiple branch hidden faults [16]. According to the main fault location and the multiple peaks anomaly of radon impulse, it indicates that Qixinzhuang section of Xiadian fault's activity is intense. Reaserch show that fault activity nature is closed bound up with radon gas concentration. If the regional geological setting is generally consisitent, the high radon concetration, the large gradient and many anomalies show powerful fault active nature, conversely the low radon concetration, the small gradient and lack of anomalies show weak activity [17]. The radon impulse values are higher on the Xiadian profile than that on the Liuhetun of the Xiadian buried fault. The radon contration ratio conditon is just like that of the radon impulse value. The anomalies plots of the Qixinzhuang profile are more than that of the Liuhetun of the Xiadian hidden active fault. Also, the Qixinzhuang profile has more remarkable radon change of gradient than the Liuhetun profile. These data and phenomena abundantly indicate that the Qixinzhuang profile is more active than the Liuhetun of the Xiadian blind fault zone.

5 Summary

The soil gaseous radon impulse-value distribution plots (Fig.2, 3, 4) of Xiadain fault clearly reflect the location and activity of the fault. The soil gaseous radon impulse anomaly of Xiadian fault indicates that the multiple peaks anomaly of Radon impulse value at Qixinzhaung section where multiple hidden faults exist. The dot mark No.1 to 37, No.1 to 26 and No.1 to 27 of each profile respectively reprensents the hanging wall of QX-1,QX-2 and QX-3 profile. The dot mark No.6 to 51 is the range of soil gaseous radon which implys the location of the main fault.

This research and collected data show that there is a blind active fault which is named the Xiadian fault zone from the soil gas randon impulse graphs. The fact that the soil gas radon concentration show many anomalies with lots of peaks indicates that the Xiadian fault zone is composed of some blind faults under the Qixinzhuang and Liuhetun profile. According to radon concentration isoline graph, the whole strike at the Qixinzhuang and Liuhetun of the Xiadian buried fault zone is forty-five degree. The characteristics of many anomalies and high peak value infer that the active intensity of the

Qixinzhuang profile is more strong than the Liuhetun profile of the Xiadian hidden fault zone. Therefore, the Qixinzhuang pofile of the Xiadian fault zone should be the key locality for seismological monitoring in future.

Acknowledgements

This reserach was jointly funded by the National Natural Science Foundation of China [No. 41802084] and the State Key Laboratory of Ore Deposit Geochemistry Foundation of Institute of Geochemistry, Chinese Academy of Sciences [No. 201705].

References

1. Jiang W.L, Hou Z.H., Xiao Z.M, Xie X.S.. (2000) Study on Paleoearthquakes of Qixinzhuang Trench at the Xiadian Fault, Bejing Plain. Seismology & Geology, 22(4):413-422.
2. Han X.K, Li Y., Du J.G., Zhou X.C., Li X.Q.. (2013) Geochemical Characteristics of Soil Gas in the Central South Segment of Xiadian Fault. Geophysical & Geochemical Exploration, 37(6):976-982.
3. Wei B.L, Huang H.S, Chen P.L., Li F.G., Xu H.L.. (1998) A Study on Fault Activity and Statutory by Geochemical Prospecting. South China Journal of Seismology, (4):50-58.
4. Jia G.X., Zhao Y.F., Yao J.Q., Yan Z.J., Li S.J.. (2005) Research and Applicaton of Radon Gas Geochemical Exploration:Interference Factors and Its Eliminating Method[J]. Mineral Resources & Geology, 19(6):653-659.
5. Liu J.H., Wang Z.W., Liu S.T., Wang X.L.. (2006) The Evaluation Method of Soil Radon and Mercury Gas Measurement about Urban Active Fault Zones. Journal of Jilin University, 36(2):295-297.
6. Wang X.L., Liu J.H., Wang Z.W, Zhang Y.L.. (2007) Numerical Simulation of Radon Movement in Quatenary Overburden over the Fault. Geophysical & Geochemical Exploration, 27(4):186-91.
7. He F.B., Bai L.Y., Wang J.M, Liu Y..(2013) Deep structure and quaternary activities of the Xiadian fault zone. Seismology and Geology, (03): 490-505.
8. Ran Z.J., Li H., Lv G.J., Wen C., Yang L., Yang Q.Y., Meng L.P., Peng Y.Q., Luo Y.X.. (2013) Seismic detecting for the shallow tectonic features of the XiaDian fault. China Earthquake Engineering Journal, (03): 656-663.
9. Ju Z.Y., Li Q., Zhang Y.H.. (1989) Regional Geology of Beijing, Tianjin and Hebei Province. Geology Press, Beijing.
10. Zhang L.L. (2012) Quaternary stratigraphic division and Contradistinction study of Langfang city. A master degree paper, 1-70.
11. Gao Z.W, Xu J., Song C.Q., Sun J.B.. (2001) The Segmental Character of Zhangjiakou-Penglai Fault. North China Earthguake Sciences, (01): 35-42+54.
12. Jiang W.L., Hou Z.H., Su Y.Z., Zhang Y.L., Xie X.S., Xiao Z.M.. (2000) Crustal tectonics and crustal press collection. Institute of Crustal Dynamics, China Earthquake Administration, Beijing: 1-15.
13. Li L., Zhang S.H., Cai X.M., Yang T.S., Wang J.M., Wu H.C.. (2011) Magnetic investigation on activity of the Xiadian fault in the Beijing plain. In: The 27th Symposium of Chinese Geophysical Society. Anhui. 161.
14. Lu L.N., Yang M., Li J., Li H.L., Duan M.F., Li B.J.. (2018) The application research of soil-gas mercury detection in the Xiadian fault zone. Geology and Exploration, 54(1): 112-120.
15. Li H.Y., Zheng C.B., Liang S.R., Zhang Y.D.. (2011) Application and Influencing Factors of Variational Curve of Soil Radon Concentration in Fracture Zone. Shanghai Land & Resources, 32(1):78-83.
16. Lu L.N., Li J., Shen J., Sheng S.Z., Yang M., Li B.J., Gong L.W.. (2016) Geochemical Characteristics of Soil Radon in the Xiadian Active Fault. Tchnology for Earthquake Disaster Prevention, 11(4) : 736-746.
17. Liu J.H., Wang Z.W., Liu S.T., Wang X.L.. (2006) The evaluation method of soil radon and mercury gas measurement about urban active fault zones. Journal of Jilin University(Earth Science Edition), (02): 295-297.

Evaluation of fusion methods for GF-6 aiming to water body observation

YiZhe Wang[1], Guo Liu[2,3,*], Bai Xue[1], Li Guo[1] and XueLi Zhang[4]

[1] Land Satellite Remote Sensing Application Center, Ministry of Natural Resources, Beijing, China
[2] National Engineering Research Center for Geographic Information System, China University of Geosciences (Wuhan), Wuhan, China
[3] Geoscience Documentation Center of China Geological Survey, Beijing, China
[4] East China Mineral Exploration and Development Bureau, Nanjing, China

Abstract. Gaofen-6 (GF-6) has the advantages of wide coverage, multiple resolutions, and multiple bands, and can provide richer information for remote sensing interpretation. Image fusion method is the key process in high resolution remote sensing application. Guangdong-Hong Kong-Macao Greater Bay Area was selected as experiment area and four image fusion methods of HPF (High-Pass Fliter), NND (Nearest Neighbor Diffusion), GS (Gram-Schmidt) and Pansharp were employed to process panchromatic and multispectral imaging. In order to evaluate the result performances, firstly, four kinds of fusion results were evaluated by visual, and then established mean, standard deviation, entropy, average gradient, correlation coefficient and spectral distortion for quantitative evaluation of fusion results. The results demonstrated that The Pansharp algorithm and the GS algorithm have the best comprehensive evaluation of the GF-6 satellite fusion effect, taking into account the color effect of the image and the enhancement of spatial details, which can meet most fusion requirements. For land-based fusion areas, the Pansharp fusion method can be selected, and the GS fusion method can better reflect the water body information and is the most suitable fusion method for GF-6 satellite water body information enhancement among the four algorithms.

1 Introduction

In recent years, China's satellite observation application system has been gradually improved, and satellite data acquisition capabilities have been greatly improved. In 2010, a major special project of high-resolution earth observation system was formally implemented, and a perfect ending was achieved in 2020. China has formed a space-based earth observation capability with high spatial resolution, high time resolution, and high spectral resolution. It greatly meets the needs of my country's high-resolution remote sensing data application. With the rapid development of high-resolution remote sensing satellites, how to make full use of the spatial and spectral information of remote sensing images has become an important issue in the processing and application of high-resolution remote sensing. Fusion of high spatial resolution panchromatic images and low spatial resolution multispectral images can effectively superimpose spatial detail information and color spectrum information, which is a feasible way to solve this problem.

To evaluate the fusion methods of different feature types on the remote sensing images, researchers have done more algorithm comparisons one after another. Zhou Weifeng et al. used WorldView-2 images as the data source, and the Xiangshan Port in Zhejiang Province as the research area, compared Principal Component Analysis (PCA), Gram-schmidt (GS), NND, and Brovey transformation, Wavelet Transform 5 fusion methods for subjective qualitative and objective quantitative evaluation and analysis and found the method of PCA is the most suitable fusion method for coastal oyster farming remote sensing applications [1]. Using GF-1 and GF-2 images as data sources, Chen Jiwei et al. evaluated Brovey transformation, GS, NND and PCA four fusion methods. Results show that the NND method is best for the GF-1 images and the GS method is best for the GF-2 images on waveform fidelity [2] . Du Yong et al. used the three images of WorldView-2 as the data source to evaluate the fusion method. The results indicate that GS is the ideal fusing method in this study. The detailed information of suspended sediment transport is constructed using the fused images [3]. Guan Yuqing et al. chooses domestic GF-1 satellite 2 m full-color/8 m multi/spectral imagery to study the remote sensing image fusion method for enhancing cultivated land information. The results showed that when facing cultivated land information, the MHIS method was the best fusion image [4]. Lin Xue et al. based on the Pleiades-1 image, and the study found that the results of Pansharp and GS fusion have good visual effects, strong ability of spectral preservation with the correlation

* Corresponding author: liuguo@mail.cgs.gov.cn

coefficient up to 0.8. The two fusion algorithm images have their own advantages in the information extraction ability of different forest land levels. The Pansharp fusion result has a total classification accuracy of 86.55% at the forest land level, and the GS fusion result has the highest total classification accuracy of 78.66% at the forest type level. The selection of the specific fusion algorithm needs to be determined according to the information extraction level of its application [5]. Chen Chao et al. verified the applicability and effectiveness of four commonly used pixels level fusion algorithms, and found that the IHS transformation method distorted the spectral characteristics of the source image, so the feature spectra were easily to become vestigial; Wavelet transformation had superiority on maintaining spectral characteristics, but the results easily became block and blurred; PCA transformation maintained most detail texture and architectural feature of the source image, but it will lose some of the detail spectral information; the tasseled cap transformation method fusion results in clear edges of ground objects, but it is not for the subtle spectrum Weak information retention[6]. At present, a variety of remote sensing image fusion algorithms have been implemented in the mainstream commercial remote sensing software. Although they can improve the image spatial resolution better, for different sensor types, remote sensing algorithms have their own advantages and disadvantages, and there is no unified recognition fusing algorithm.

With the specific goal of facing urban water bodies, this paper analyzes the image fusion results of the Gaofen-6 satellite (GF-6) to provide a basis for water remote sensing applications. The GF-6 satellite has the advantages of high resolution, wide coverage, high quality imaging, high performance imaging, and high localization rate. It has been in operation for three years, and a large number of high-quality remote sensing images of Earth observation have been obtained, but there are few reports on the applicability of the GF-6 satellite panchromatic and multispectral fusion algorithm. This paper selects GF-6 satellite images of the Guangdong-Hong Kong-Macao Greater Bay Area and uses four fusion methods of HPF, NND, GS and Pansharp to conduct fusion experiments, compare and evaluate visual and quantitative evaluation indicators, and analyze the applicability of the GF-6 satellite fusion method. The results can provide a reference for GF-6 satellites in the application of resource and environmental observation.

2 Data and research area overview

2.1 Introduction of GF-6 satellite

GF-6 is the sixth satellite of China's high-resolution earth observation system, developed by China Aerospace Science and Technology Corporation. It was successfully launched by Jiuquan Satellite Launch Center on June 2, 2018 by the Long March 2D rocket. On March 21, 2019, GF-5 and GF-6 were officially put into use, marking that the high-spectrum special-purpose hyperspectral capability has been formed. GF-6 satellite carries a 2m panchromatic / 8m multispectral resolution camera with a width of 90km and a 16m multispectral resolution wide coverage camera with a width of 800km. It is a sun-synchronous orbit with a design life of 8 years. It is the first multi-spectral remote sensing satellite of china with red-edge spectral bands and domestic CMOS imaging devices.

GF-6 is a high-resolution optical satellite with high maneuverability in China's high-resolution ground-observing system that takes into account both census and detailed investigation capabilities. The satellite and GF-1 satellite network has realized the revisited observation of China's land area for 2 days, which greatly improves the scale and timeliness of the acquisition of remote sensing data, and effectively makes up for the existing domestic and foreign high-resolution multispectral satellite resources. And the self-sufficiency rate and application range of domestic remote sensing satellite data have been improved.

2.2 Research area overview

In this paper, domestic multi-spectrum PMS sensor of GF-6 satellite (multi-spectrum 1 ~ 4 band (resolution 8m) and panchromatic band (resolution 2m) are selected for fusion experiments. The experimental image date is September 18, 2019. The multi-spectral data and panchromatic data were spatially registered before the experiment.

The Guangdong-Hong Kong-Macao Greater Bay Area is a world-class urban agglomeration established by relying on the unique geographical location advantage of the "Bay", with outstanding advantages in marine resources. The study area is located in the waters of the Guangdong-Hong Kong-Macao Greater Bay Area, mainly consisting of three different types of waters. Area 1 is the waters near the artificial island of the Zhuhai-Macao Port, which is close to the city and contains more suspended sediment; Area 2 is the waters on both sides of the east entrance of the submarine tunnel, with strong water bodies; Area 3 is the fishery coastline waters, with aquaculture as a the Lord.

The study area has unique geomorphic conditions and complex water and sand conditions. It is also one of the waters with the densest coastal routes and the highest density of ships in my country and the world. In recent years, the marine ecological and environmental protection of the Guangdong-Hong Kong-Macao Greater Bay Area has faced tremendous challenges. Therefore, the comprehensive use of satellite remote sensing and marine monitoring and other technical means to build a coastline dynamic monitoring and monitoring network will help the Guangdong-Hong Kong-Macao Greater Bay Area to implement a comprehensive, Real-time, dynamic and three-dimensional monitoring is of great significance for promoting the sustainable development of the Guangdong-Hong Kong-Macao Greater Bay Area.

3 Fusion algorithm and effect evaluation method

3.1 Fusion algorithm

(1) HPF (high-pass filtering fusion)

The HPF method is to perform a positive FFT transform on a multispectral image and replace the high-frequency part of the multispectral image after the FFT with a high-resolution panchromatic band image. Spectral fusion image [7]. The fusion effect is affected by the size of the filter. This paper chooses a high-pass filter window of 3 * 3.

(2) NND (Nearest Neighbor Diffusion)

The HPF method down samples the panchromatic band to low resolution, performs linear regression on pixels in each band to obtain the contribution vector T, and calculates the neighboring pixel difference factor using the original panchromatic band to generate the initial high-resolution multi- Spectral images, and finally use T as the weight to normalize the spectrum [8].

(3) GS (Gram-Schmidt)

The GS method first uses spectral resampling to simulate the first component, transforms the multispectral image to orthogonal space, and replaces the first component with a high spatial resolution panchromatic image. Then obtain the fused image by inverse transform. The GS fusion algorithm has no band limitation on the image, and can fuse all multi-spectral bands in one fusion process [9].

(4) Pansharp

The Pansharp method uses the least squares method to best match the gray value relationship between the fusion of the original multispectral image, the panchromatic image and the fused multispectral image and the panchromatic image. And the use of mathematical statistics to improve the integration process automation. When applying this fusion method, it is best to designate all the multispectral bands as reference bands to reduce the deviation between the fused image and the original multispectral image [10].

3.2 Fusion algorithm and effect evaluation method

3.2.1 Qualitative evaluation

Qualitative evaluation mainly performs subjective evaluation from visual effects, that is, visual method. Through visual comparison of image spectrum, edges, texture and overall brightness before and after fusion, it can intuitively and conveniently reflect the spatial resolution and clarity of different fusion images. However, this evaluation method is relatively subjective, and is greatly affected by the observer's experience and technology, image types, application occasions, and environmental conditions, lacks stability, and does not guarantee the repeatability of the evaluation method.

3.2.2 Quantitative evaluation

Quantitative evaluation method is an objective method to quantitatively analyze images based on the statistical characteristics of the images. It can overcome the influence of human subjective factors and make scientific and objective evaluations of the performance of various image fusions for further research. Quantitative evaluation indicators are mainly divided into two categories, one is single-factor evaluation, including image mean, standard deviation, and information entropy; the other is comprehensive evaluation indicators, including average gradient, correlation coefficient, and relative deviation. According to the characteristics of GF-6 image, this experiment selects the mean value, standard deviation, information entropy and average gradient as the quality evaluation index of the fusion image, and quantitatively describes the spectral fidelity, information amount and image clarity of the fusion image.

(1) The mean is the average of all the pixel values in the image and reflects the average reflection intensity of the features.

$$u = \frac{\sum_{i=1}^{M}\sum_{j=1}^{N} F(i,\ j)}{MN} \qquad (1)$$

u is the average value of the image, $F_{(i,j)}$ is the gray value at the position of the pixel (i, j), and M and N are the number of image rows and columns, respectively.

(2) The standard deviation reflects the discrete situation of the gray level of the image relative to the average gray level average value. It is generally believed that the larger the standard deviation, the more dispersed the gray level distribution, the better the separability between features, and the more image information.

$$std = \sqrt{\frac{\sum_{i=1}^{M}\sum_{j=1}^{N}(F(i,\ j)-u)^2}{MN}} \qquad (2)$$

std is the standard deviation of a certain band of the image, M and N are the number of rows and columns of the image; $F_{(i,j)}$ is the gray value at the position of the pixel (i, j); u is the average gray of the image after fusion.

(3) Information entropy reflects the richness of image information. The larger the information entropy, the more information the image contains.

$$H = -\sum_i p(i)log_2[p(i)] \qquad (3)$$

H is the information entropy of a certain band of the image, p (i) is the probability of the pixel whose gray value is i; it is expressed by the ratio of the number of pixels with gray value i and the total number of pixels in the image.

(4) The average gradient is the average value of the grayscale change rate of the image, which can reflect the detail expression ability of the image.

$$G = \frac{1}{(M-1)(N-1)} \cdot \sum_{i=1}^{M-1}\sum_{j=1}^{N-1}\sqrt{\frac{\left[\frac{\partial F(i,j)}{\partial x}\right]^2+\left[\frac{\partial F(i,j)}{\partial y}\right]^2}{2}} \qquad (4)$$

G is the average gradient of a certain band of the image, $F_{(i,j)}$ is the gray value of the image at the point (i, j), ∂F (i, j) / ∂x, ∂F (i, j) / ∂y are the partial derivatives of the image at the point (i, j) in the row and column directions.

(5) The correlation coefficient reflects the degree of preservation of the spectral information of the fused image. The larger the correlation coefficient, it means that the spectral features of the image before and after the fusion

are approximately close and the information is maintained approximately well.

$$C=\frac{\sum_{i=1}^{M}\sum_{j=1}^{N}\left(Mul_{(i,j)}-\overline{Mul}_{(i,j)}\right)\left(F_{(i,j)}-\bar{F}_{(i,j)}\right)}{\sqrt{\sum_{i=1}^{M}\sum_{j=1}^{N}\left(Mul_{(i,j)}-\overline{Mul}_{(i,j)}\right)^{2}\sum_{i=1}^{M}\sum_{j=1}^{N}\left(F_{(i,j)}-\bar{F}_{(i,j)}\right)^{2}}} \quad (5)$$

$Mul_{(i,j)}$ is the gray value of the original multispectral image, $F_{(i,j)}$ is the gray value of the image after fusion.

(6) The degree of spectral distortion directly reflects the degree of spectral distortion of the fusion image and the original multi-spectral image quality inspection. The greater the spectral distortion, the smaller the fidelity, the smaller the distortion, the better the guarantee.

$$D=\frac{1}{M\times N}\sum_{i=1}^{M}\sum_{j=1}^{N}\left|F_{(i,j)}-M\acute{u}l_{(i,j)}\right| \quad (6)$$

$M\acute{u}l_{(i,j)}$ is the gray value of the original multispectral image, $F_{(i,j)}$ is the gray value of the image after fusion.

4 Results and analysis

This paper selects the panchromatic image with a resolution of 2m and the multispectral image with a resolution of 8m from the Gaofen-6 satellite as experimental data. The study area selects three different types of waters near the Hong Kong-Zhuhai-Macao Bridge. Area 1 is the water area near the artificial island of the Zhuhai-Macao Port, which is close to the city, and has a large amount of suspended sediment; Area 2 is the water area on both sides of the east entrance and exit of the submarine tunnel, and the water body is highly fluid; Area 3 is the fishery shoreline waters, mainly aquaculture.

The geographic imaging accelerator GXL software is used to perform orthorectification and geometric registration of the image, and the correction error is within 1 pixel. Then HPF fusion, NND fusion, GS fusion, Pansharp fusion were used to carry out fusion experiments on the registered panchromatic and multispectral images. The fusion effect of each method is shown in Figure 1~3.

(a) Gram-Schmidt

(b) HPF

(c) NND

(d) Pansharp

Fig. 1. Comparison of 4 kinds of fusion method of study area 1

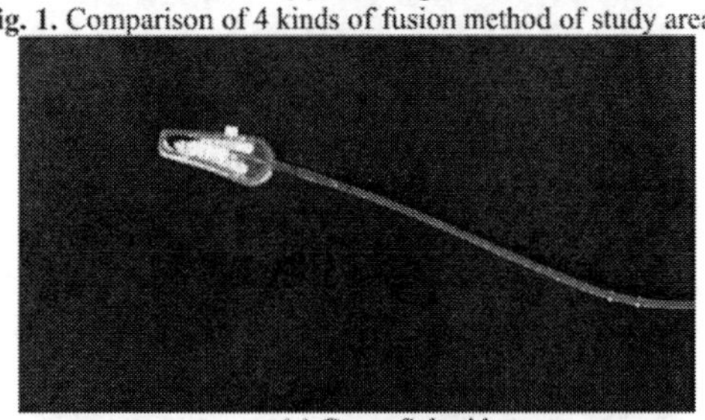

(a) Gram-Schmidt

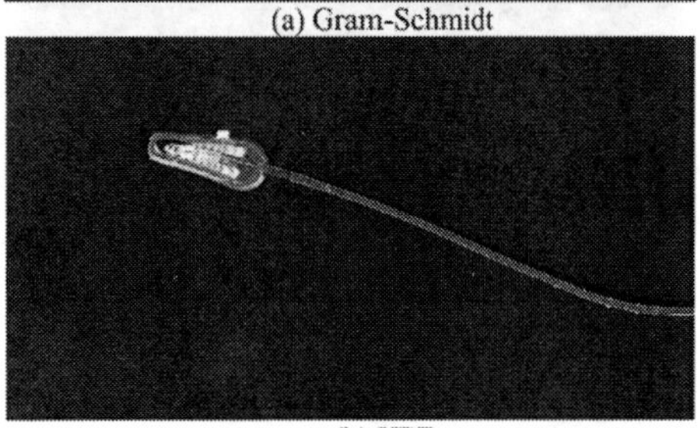

(b) HPF

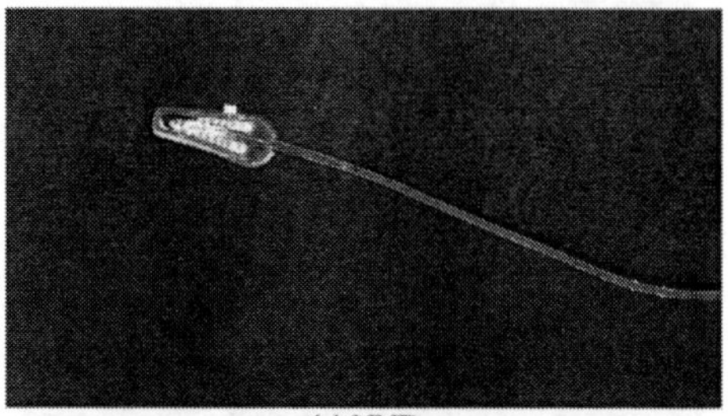

(c) NND

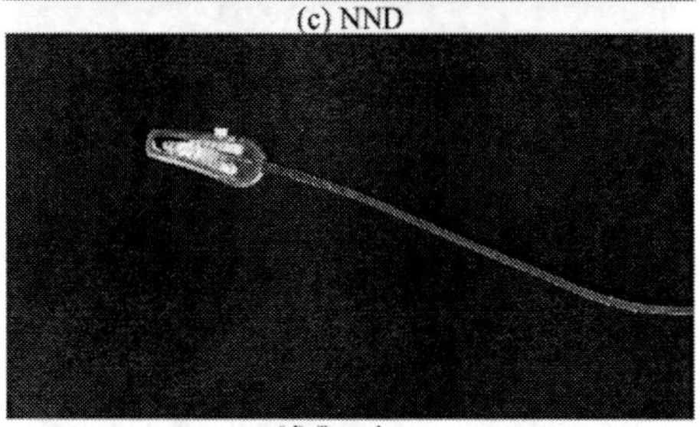

(d) Pansharp

Fig. 2. Comparison of 4 kinds of fusion method of study area 2

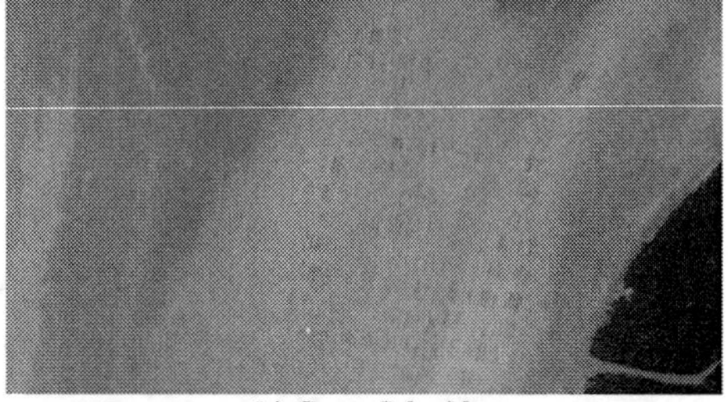

(a) Gram-Schmidt

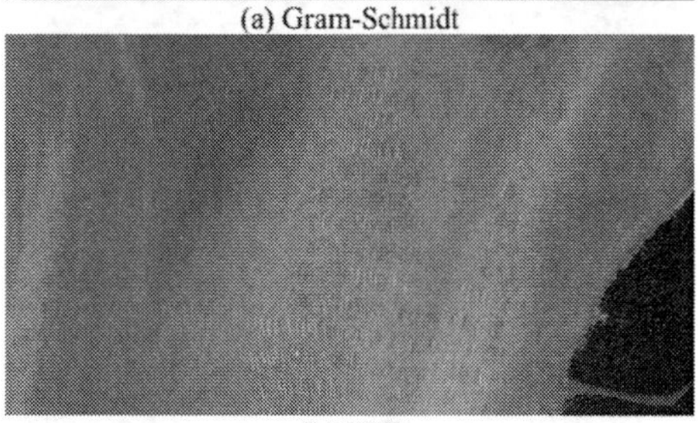

(b) HPF

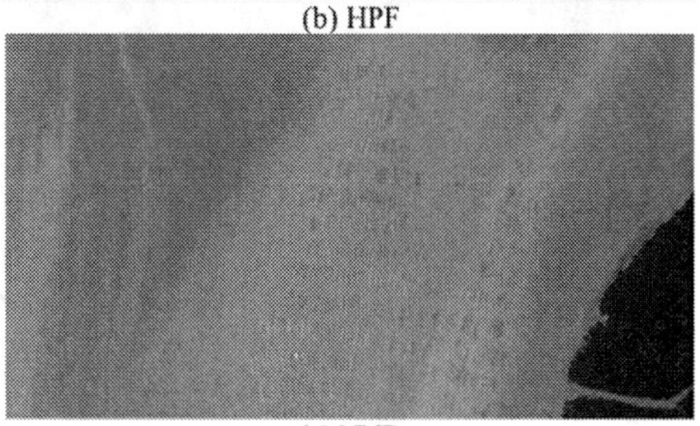

(c) NND

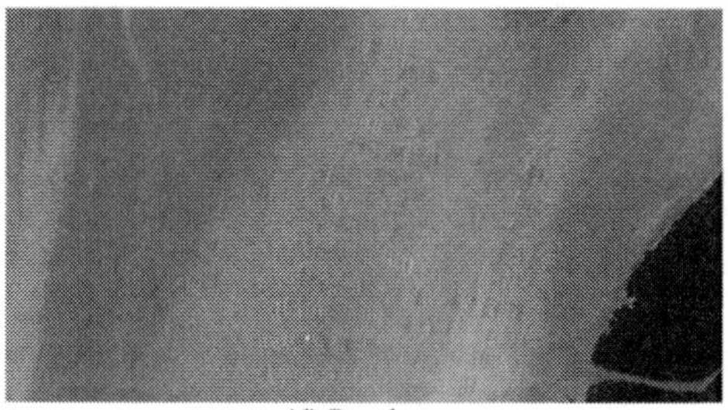

(d) Pansharp

Fig. 3. Comparison of 4 kinds of fusion method of study area 3

4.1 Qualitative evaluation

The visual interpretation of the fusion image shows that the four fusion algorithms have enhanced the resolution of the multispectral image by comparing Figures 1-3. Area 2 uses the NND algorithm to merge the construction land to be yellowish, and the color of the water body has a large deviation from the original MS image. The fusion result of GS algorithm is the closest to the multi-spectral MS image. However, from the perspective of details, the architectural part is overexposed and the detail texture is lost more. It is preliminarily judged that the NND algorithm is not suitable for the study of water bodies in area 2. Pansharp and HPF algorithm fusion image water body brightness is moderate, water body flowing ripples, road lanes are clear, while the spectral information of the original MS image is well preserved, which is suitable for region 2 water body fusion.

The colors of the results of the four fusion algorithms in area 1 are roughly the same as the original MS images. The suspended sediment texture information in the water bodies of Pansharp and GS algorithms is rich, and the boundaries of the buildings are distinct. The fusion result of the HPF algorithm has obvious graininess and a feeling of over-sharpening. After the NND algorithm is fused, the overall image is slightly blue. HPF and NND algorithms are obviously not suitable for area 1 fusion.

The overall effect of the four fusion algorithms in area 3 is not much different. From the enlarged view of the cultured floating raft, the fusion result of the GS algorithm and the original MS image have the highest retention in terms of image color. The second is the fusion result of Pansharp algorithm. The result of the fusion with the NND、HPF algorithm is the biggest difference compared with the original MS image color. In terms of spatial resolution, Pansharp and GS algorithms have clear edges, moderate brightness, and distinct vegetation layers.

Therefore, the overall visual results show that Pansharp and GS algorithms have better fusion effects on the three regions in terms of spectrum preservation and spatial information enhancement.

4.2 Quantitative evaluation

In order to further compare the differences of several fusion algorithms, the mean, standard deviation, entropy, average gradient, correlation coefficient and spectral distortion of the four fusion results of the whole image, water image and excluded water image are respectively evaluated. The results are shown in Table 1.

Table 1. Objective evaluation results of the entire image

		MS	PSH	GS	NND	HPF
Mean	Image 1	730.73	721.69	734.52	784.14	716.73
	Image 2	653.04	651.43	654.34	736.82	650.34
	Image 3	676.90	674.88	682.24	741.45	670.72
Standard eviation	Image 1	127.26	128.08	126.95	142.33	125.39
	Image 2	43.03	42.18	42.45	45.95	41.88
	Image 3	39.77	41.09	45.55	56.91	35.79
Entropy	Image 1	7.53	7.49	7.56	7.69	7.53
	Image 2	4.76	5.16	4.76	6.28	5.19
	Image 3	6.40	6.50	6.48	6.91	6.46
Average gradient	Image 1	5.46	10.64	5.79	15.88	11.88
	Image 2	1.31	3.41	1.45	7.48	3.38
	Image 3	0.82	2.73	1.08	4.62	2.94
Correlation coefficient	Image 1		0.955	0.988	0.876	0.910
	Image 2		0.946	0.990	0.784	0.913
	Image 3		0.982	0.996	0.930	0.952
Spectral distort-ion	Image 1		20.50	10.84	69.10	28.08
	Image 2		5.81	2.60	85.62	7.04
	Image 3		5.07	7.86	66.48	9.91

Table 2. Objective evaluation results of land part

		MS	PSH	GS	NND	HPF
Mean	Image 1	238.08	234.37	238.32	248.34	230.57
	Image 2	27.51	27.10	27.55	28.78	26.87
	Image 3	63.42	62.59	61.46	63.75	64.69
Standard deviation	Image 1	386.75	381.95	387.29	408.39	376.06
	Image 2	149.88	147.81	149.91	157.86	146.66
	Image 3	175.23	173.17	169.93	176.79	179.48
Entropy	Image 1	3.60	3.60	3.61	3.68	3.63
	Image 2	0.51	0.51	0.50	0.53	0.52
	Image 3	1.10	2.11	1.18	1.39	1.36
Average gradient	Image 1	5.95	9.68	6.07	13.11	11.16
	Image 2	1.47	1.62	1.34	1.90	1.65
	Image 3	1.14	2.11	1.38	2.67	2.49
Correlation coefficient	Image 1		0.994	0.997	0.986	0.988
	Image 2		0.987	0.990	0.981	0.983
	Image 3		0.994	0.995	0.992	0.989
Spectral distortion	Image 1		14.94	7.72	26.30	21.61
	Image 2		2.22	1.25	3.42	2.87
	Image 3		2.70	2.57	4.12	4.50

Table 3. Objective evaluation results of water part

		MS	PSH	GS	NND	HPF
Mean	Image 1	492.59	486.90	495.76	535.36	485.75
	Image 2	623.14	622.05	624.50	705.45	621.29

	Image 3	637.19	635.46	643.46	701.11	630.14
Standard deviation	Image 1	318.24	314.76	320.56	346.06	314.00
	Image 2	123.46	123.03	123.44	140.37	122.69
	Image 3	178.15	177.96	180.33	196.44	176.04
Entropy	Image 1	5.15	5.12	5.19	5.22	5.10
	Image 2	4.49	4.89	4.49	6.06	4.93
	Image 3	6.06	6.04	6.09	6.15	5.97
Average gradient	Image 1	1.86	3.27	5.15	2.08	3.21
	Image 2	3.04	4.67	8.51	3.01	4.52
	Image 3	2.07	3.38	4.96	2.15	3.30
Correlation coefficient	Image 1		0.996	0.996	0.995	0.997
	Image 2		0.955	0.956	0.950	0.951
	Image 3		0.993	0.993	0.990	0.990
Spectral distortion	Image 1		7.02	4.57	44.28	8.03
	Image 2		6.12	3.96	84.50	6.96
	Image 3		4.41	7.29	64.77	8.65

By comparing the statistical data of the original multispectral image and the four fusion images in the table (1-3), the following conclusions are obtained.

(1) The four algorithms have little difference in the mean, standard deviation, and information entropy after the fusion of the entire image. When comparing the average gradients, the average gradient of the NND is the largest, followed by the PSH and the HPF, indicating that the three algorithms have better spatial details of the entire fused image. However, when comparing the correlation coefficient and the spectral distortion, it is found that the NND has the worst spectral fidelity effect. Taken together, the PSH algorithm has the best overall fusion effect among the five fusion methods.

(2) In the fusion results of land part, the mean standard deviation of the NND and GS is higher than that of the MS image, and the fusion result is brighter; the mean standard deviation of the PSH is slightly lower than that of the MS image, and the fusion result is darker. Information entropy and average gradient are important indicators for evaluating image information and spatial details. It can be seen from the table that the information entropy of the NND is the highest, indicating that the information amount of the fusion image has increased; the information entropy of the other four algorithms is close to that of the original MS image, and the information amount of the original image is well maintained. The average gradient of NND, PSH and HPF is significantly higher than the original MS image; the average gradient of NND is the largest, and the spatial detail enhancement effect is the best. The correlation coefficient and spectral distortion directly reflect the effect of maintaining the spectrum. The correlation coefficient between the NND and the original MS image is the smallest, the spectral distortion is the largest, and the fidelity of the spectrum is the worst.

(3) In the fusion results of water part, the NND algorithm has the largest mean standard deviation, the image brightness is overexposed, which is not conducive to observing the water information, and the spectral distortion is the largest. The NND algorithm is the least ideal water fusion algorithm. The GS algorithm has the smallest spectral distortion and the largest average gradient. It is the most suitable method for water fusion among the four methods.

5 Conclusion

Remote sensing image fusion is an effective method to improve image resolution, enhance image utilization, and fully mine image information. This paper takes water as the research area and uses GF-6 satellite data to perform data fusion using HPF, NND, GS, and Pansharp four fusion algorithms. On the basis of subjective evaluation, from the perspective of spectral fidelity, image clarity, and image information, six objective evaluation indexes were selected to evaluate the fusion result. The experimental results show that, Pansharp algorithm and GS algorithm perform well in enhancing spatial details; GS algorithm is better in maintaining spectral information; HPF algorithm fusion results are over-sharpened, the visual effect is not good, and it is easy to cause texture loss. The NND algorithm has a small correlation coefficient and the largest spectral distortion. Compared with the other three algorithms, the effect of the spectral consistency with the source image is poor.

Therefore, according to the three regional evaluation results of this study, the Pansharp algorithm and the GS algorithm have the best comprehensive evaluation of the GF-6 satellite fusion effect, taking into account the color effect of the image and the enhancement of spatial details, which can meet most fusion requirements. The Pansharp algorithm can be selected for the land-based fusion area, and the water-based fusion area GS algorithm can better reflect the water body information. However, in actual work, integration needs are diversified, and specific problems should be analyzed in detail, and appropriate integration methods should be selected. This research selected four classic and practical methods to conduct experiments and analysis based on GF-6 satellite data,

https://doi.org/10.1051/e3sconf/202129002009

which has certain guidance and reference significance for the future production, application and scientific research of GF-6 satellite data.

References

1. Zhou Weifeng, Cao Li, Li Xiaoshu. Assessments of fusion methods using WorldView-2 satellite images for coastal oyster culture observation [J]. Remote Sensing Technology and Application, **103-109**, 33(1) (2018)
2. Chen Jiwei, Shen Qian, Zhu Li. Influence of different fusion methods based on GF Images on the spectral fidelity of urban water [J]. Journal of Beijing Polytechnic University, **677-682**, 43(5)(2017)
3. Du Yong, Zhang Xiaoyu, Huang Dasong. Evaluation of fusion methods for Worldview-2 aiming to water body observation [J]. Journal of Zhejiang University (Engineering Science), **993-1000**, 5 (2015)
4. Guan Yuqing, Mi Wentian. Research on GF-1 satellite remote sensing image fusion method for cultivated land [J]. Hubei Agricultural Sciences, **104-107,113**, 5(2019)
5. Lin Xue, Peng Daoli, Huang Guosheng. Evaluation analysis of fusion algorithms for Pleiades-1 data oriented to extraction of forestland information [J]. Engineering of Surveying and Mapping, **18-24**, 25(11)(2016)
6. Chen Chao, Qin Qiming, Wang Jinliang. Comparison of quality evaluation methods for image fusion of farmland remote sensing [J]. Transactions of the Chinese Society of Agricultural Engineering, **95-100**, 10(2011)
7. Wei Jun, Li Bi-cheng. Remote-sensing image fusion based on ihs transform、 wavelet transform and high pass filtering [J]. Journal of Information Engineering University, **46-50**, 4(2)(2003)
8. Sun W, Messinger D. Nearest-neighbor diffusion-based pan-sharpening algorithm for spectral images[J].Optical Engineering, **013107**, 53(1)(2014)
9. Laben C A, Brower B V. Process for enhancing the spatial resolution of multispectral imagery using pan - sharpening: US, 6011875[P], (2000-01-04)
10. Pang Jinjiao, Sun Rui, Wang Wen. Influence of registration error on land cover classification and change detection in high-resolution remote sensing image[J] . Remote Sensing Technology and Application, **498-505**, 29(03) (2014)

A New Method for Estimating Elastic Parameters of Clay Rock Host Rock in Geological Disposal of High-Level Radioactive Waste in China

GENG WEI RAO[1,2,a*] , XIAO DONG LIU[1,2,3,b], HAI AN LIANG[1,2] & PING HUI LIU[1,2]

[1]School of Geosciences, East China University of Technology Nanchang 330013, China
[2]State Key Laboratory of Nuclear resources and environment East China University of technology, Nanchang 330013, China
[3]Jiujiang University, Jiujiang,332000,China

Abstract. Taking the clay rock from the Tamusu pre-selected area of China's high-level radioactive waste geological repository as the research background. Using the methods of triaxial compression testing, X-ray diffraction and microscopic analysis, the main mineral composition, structure and mechanical properties of clay rock under different confining pressures were systematically obtained. Combined with the classical elastic parameter estimation methods at home and abroad, the elastic modulus of the clay rock in the Tamusu pre-selected area is estimated. It finds that the classical elastic parameter estimation method has a large error with the actual test value when calculating the elastic modulus of Tamusu clay rock. The maximum error can reach 788%, and the error decreases with the increase of confining pressure, however, the minimum error is still also 46%. In order to establish the deformation evaluation index suitable for the clay rock in the Tamusu pre-selected area of high-level radioactive waste, considering the influence of the mineral properties and structure of the clay rock on the macro-mechanics properties, a new method for estimating the elastic parameters of the clay rock was established. Compared with the classic elastic parameter estimation method, the maximum error is reduced to 33% and the minimum error is only 2%. Therefore, we suggest that the new elastic parameter estimation method proposed here may be used to evaluate and predict the elastic modulus parameters of the clay rock in the Tamusu pre-selected area for geological disposal of high-level radioactive waste in China.

1 Introduction

The deep geological disposal of high-level radioactive waste (HLW) is a systematic and complex project. At present, the most in-depth research on the host rocks of the HLW geological repository in Finland, Sweden, China and other nuclear countries is granite and clay rocks (Pan et al., 2009; Laurence, 1997; Ma et al., 2020). The research on granite as the host rock of the HLW geological repository is much earlier than that of clay rock. On May 6, 2019, State Administration of Science. Technology and Industry for National Defense, China, officially approved the construction proposal of the underground laboratory for the geological disposal of high-level radioactive waste in Beishan, Gansu, marking the geological disposal of HLW granite in China has officially entered the underground laboratory stage (Wang, 2019). Clay rock has been recognized internationally as one of the alternative wall rock types for geological repositories of HLW due to its significant advantages in terms of low permeability, strong adsorption, and good self-healing ability. The research on geological repository of HLW clay rock host rock in China is in its infancy. Since 2014, State Administration of Science. Technology and Industry for National Defense, China, approved the "Clay Rock Project" and made the Tamusu area of the Bayin Gobi Basin in Inner Mongolia as the pre-selected area for the HLW geological repository, the research mainly focuses on the site selection and suitability evaluation of the pre-selected area (Yuan et al., 2018; Xiang et al., 2018; Wang et al., 2018; Gong, 2017). But the research on the mechanics of the clay rock in the area is relatively basic (Wan, 2018; Hu, 2014; Hu et al., 2014; Hu et al., 2020; Xue et al., 2016; Rao, 2018).

At present, the research on the correlation between the clay rock mineral composition and the elastic modulus in the Tamusu pre-selected area are based on the classic elastic parameter estimation method of Voigt (1910) and Reuss (1929) et al. (see Table 1 for specific methods) (Wan et al., 2017; Wang et al., 2018). However, the above-mentioned elastic parameters of clay rock estimated on the basis of volume average of mineral properties have large errors when compared with actual

* Corresponding author: email address:[a] 79313666@qq.com
[b]liuof99@163.com

test values, which makes it impossible to correctly evaluate the deformation characteristics of clay rock in the Tamusu pre-selected area for geological disposal of HLW.

Rock is a complex aggregate of various minerals, and its mineral composition and structure simultaneously affect the mechanical properties of the rock. Voigt and Reuss et al.'s classical elastic parameter estimation method test samples are both hard rocks such as igneous rocks, while clay rock in the Tamusu pre-selected area is a kind of sedimentary rock, which is structurally different from hard rocks.

In this paper, TAW-2000 microcomputer-controlled rock servo triaxial compression testing machine, X-ray diffraction and microscopic analysis are used to study the main mineral components and mechanical properties of clay rock in the Tamusu pre-selected area and their mechanical properties under different confining pressures. On the basis of comparing clay rock as the structural difference between sedimentary rock and igneous rock, considering the volume average of mineral properties in clay rock and the influence of interstitial materials on the rock during cementation, a new method for estimating the elastic parameters of Tamusu clay rock is established. It is of positive significance to the research and development process of the actual related engineering of the clay rock host rock of the HLW geological repository.

2 Materials and Methods

2.1. Classic elastic parameter estimation methods at home and abroad

Different scholars have different understandings on the estimation of rock elastic modulus. The classical estimation methods are based on the estimation of rock elastic parameters, and then calculate the elastic modulus of rock according to the elastic modulus conversion formula (Formula 1). In the formula, K is the bulk modulus of the rock, G is the shear modulus of the rock, and E is the elastic modulus of the rock. The specific estimation method is shown in Table 1.

$$E = \frac{9KG}{3K + G} \tag{1}$$

Table1 Summary of classical elastic parameters estimation methods at home and abroad

Elastic parameter estimation method	Meaning and explanation of formula
Estimation method of elastic parameters proposed by Voigt in 1910	$K_V = \sum_{i=1}^{N} K_i \bullet V_i$, $\mu_V = \sum_{i=1}^{N} \mu_i \bullet V_i$, N is the kind of mineral in the rock, the percentage of the ith mineral in the rock volume is $V_i (i=1,...,N)$. Rock bulk modulus K_V, shear modulus μ_V.
Estimation method of elastic parameters proposed by Reuss in 1929	$K_R^{-1} = \sum_{i=1}^{N} K_i^{-1} \bullet V_i$, $\mu_R^{-1} = \sum_{i=1}^{N} \mu_i^{-1} \bullet V_i$, N is the kind of mineral in the rock, the percentage of the ith mineral in the rock volume is $V_i (i=1,...,N)$. Rock bulk modulus K_R, shear modulus μ_R.
Estimation method of elastic parameters proposed by Hill in 1952	$K_{VRH} = \frac{1}{2}(K_R + K_V)$, $\mu_{VRH} = \frac{1}{2}(\mu_R + \mu_V)$, K_V and K_R are bulk modulus parameters obtained by Voigt and Reuss estimation methods respectively, μ_V and μ_R are shear modulus parameters obtained by Voigt and Reuss estimation methods respectively. K_{VRH} and μ_{VRH} are bulk modulus and shear modulus respectively.
Estimation method of elastic parameters proposed by Kumazawa in 1969	$K_{goom} = (K_R \times K_V)^{1/2}$, $\mu_{goom} = (\mu_R \times \mu_V)^{1/2}$, K_V and K_R are bulk modulus parameters obtained by Voigt and Reuss estimation methods respectively, μ_V and μ_R are shear modulus parameters obtained by Voigt and Reuss estimation methods respectively. K_{goom} and μ_{goom} are bulk modulus and shear modulus respectively.
Estimation method of elastic parameters proposed by Jianping ZUO in 2015	$K_{Zuo} = \alpha K_V + \beta K_R$, $\mu_{Zuo} = \alpha \mu_V + \beta \mu_R$, α, β is the percentage of hard minerals and soft minerals in the whole rock. K_V and K_R are bulk modulus parameters obtained by Voigt and Reuss estimation methods respectively, μ_V and μ_R are shear modulus parameters obtained by Voigt and Reuss estimation methods respectively. K_{Zuo} and μ_{Zuo} are bulk modulus and shear modulus respectively.

2.2. Geological background and sample collection

The Tamusu pre-selected area is located at the southern end of the Bayin Gobi Basin in Inner Mongolia, China. The basement in the area is composed of Archean, Proterozoic and Paleozoic metamorphic rock series, with simple structure and stable sedimentation (Zhang er al., 2015; Guan et al., 2014). The upper member of the Cretaceous Bayin Gobi formation (K_1b^2) is the key target formation in the Tamusu pre-selected area for geological disposal of HLW, mainly composed of gray white dark gray clay rock. The 15 clay rock samples studied in this paper are all collected from the designed borehole (TZK-2) of "Clay Rock Project" in Tamusu pre-selected area. The borehole location is shown in Fig 1. The collected samples were vacuum packaged on site to ensure their freshness and integrity.

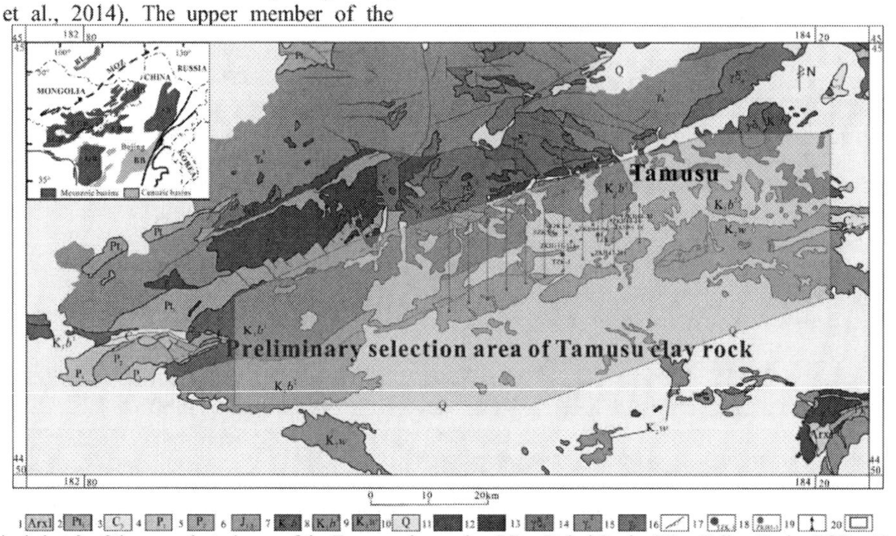

Fig1 Geological sketch of the pre-selected area of the Tamusu clay rock BB = Bohai Basin, EB = Erlian Basin, EGB = East Gobi Basin, HB = Hailar Basin, OB = Ordos Basin, SB = Songliao Basin,BGB = Bayingobi Basin, BL = Baikal Lake, MOZ = Mongol-Okhotsk Zone.
1. Proterozoic; 2. Archean; 3. Carboniferous System; 4. Lower Permian; 5. Upper Permian; 6. Lower middle Jurassic system; 7. Lower Cretaceous Bayingobi formation; 8. Upper section of the Lower Cretaceous Bayingobi formation; 9. Upper Cretaceous Wulansuhai formation; 10. Quaternary; 11. Late Caledonian granite; 12. Middle Variscan gabbro; 13. Late Variscan granodiorite; 14. Late Variscan granite; 15. Indosinian granite; 16. Crack; 17. Project team drilling; 18. Drilling by Nuclear 208 in China; 19. Seismic line-metry; 20. pre-selected area

2.3. Rock sample preparation and test scheme

The rock mechanics test were completed in the State Key Laboratory of Nuclear resources and environment (East China University of technology). The drilled core retrieved on site was processed into a standard sample of φ50×100 mm. Using TAW-2000 microcomputer control rock servo triaxial compression testing machine. According to the Chinese GB/T 50266-99 and SL 264-2001 test standards, the rock sample is loaded with a radial deformation of 0.005 mm/min, and the indoor conventional uniaxial and triaxial compression test at 5 MPa and 10 MPa on triaxial.

The analysis of rock mineral composition were completed in the State Key Laboratory of Nuclear resources and environment. Selecting broken samples from mechanical testing and grind them to below 200 mesh. Using German Brewer D8 ADVANCE polycrystalline ray diffractometer for X-ray powder crystal diffraction (XRD) method analysis, the accuracy of the goniometer is 0.0001°, the accuracy is ≤0.02°, the test peak record 2θ angle is 5-70° and the Bruker-Diffrac EVA software was used for semi-quantitative mineral composition estimation.

The observation experiment of the rock under the microscope were completed in the State Key Laboratory of Nuclear resources and environment. Selecting broken samples from mechanical testing and send them to Langfang Shangyi Rock and Mine Testing Technology Service Co Ltd to preparation of microscopic thin sections of rock. The rock and mineral analysis of the sample were carried out under single polarized light and orthogonal polarized light by Leitz polarized light microscope. The microscopic morphology of the rock was carried out under a scanning electron microscope. The experimental instrument was a JSM-35CF scanning electron microscope, the instrument model was Nova Nano SEM 450. The natural section of the broken sample was selected from mechanical testing, sprayed with platinum, and glued to the sample holder with conductive glue, scanning electron microscopy (SEM) observation was performed at an acceleration voltage of 10 kV.

3 Results

3.1. Analysis of mechanical characteristics

The photos of the clay rock standard samples in the Tamusu pre-selected area and their damaged state are

 https://doi.org/10.1051/e3sconf/202129002010

shown in Fig 2. The results of indoor conventional uniaxial, triaxial 5 MPa and triaxial 10 MPa compression tests are shown in Table 2. The stress-strain curves of some rock samples are shown in Fig 3.

(a)Standard sample

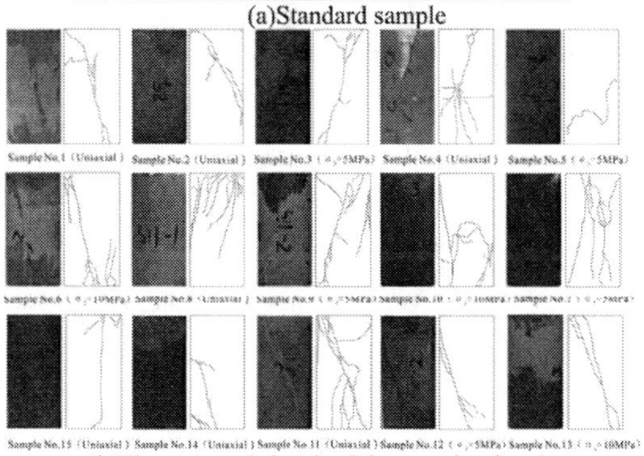

(b)The state and sketch of the sample after destroy

Fig2 Failure diagram of Tamusu clay rock standard sample and uniaxial (triaxial) compression test

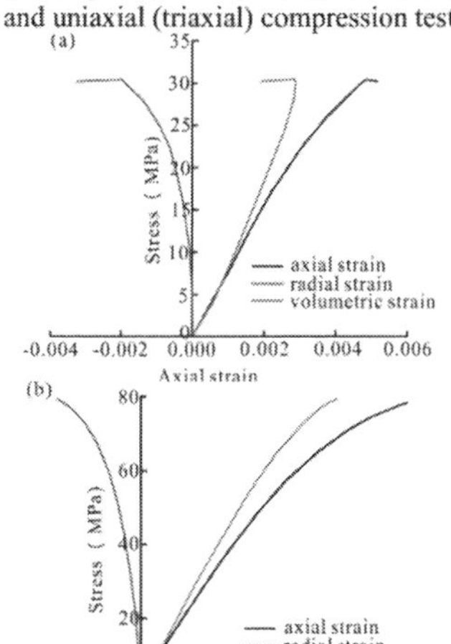

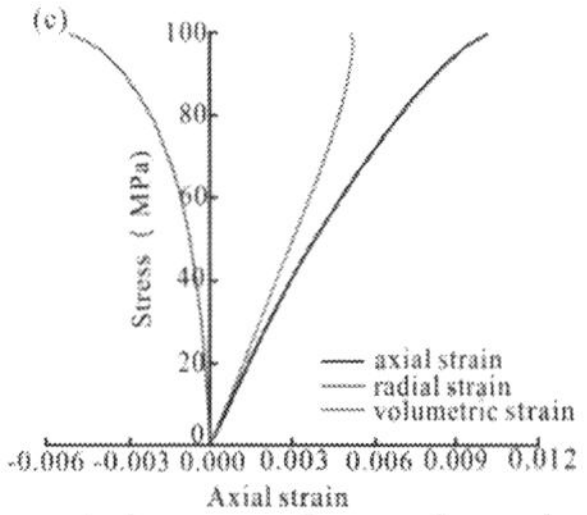

Fig3 Stress-strain curves of some clay rock samples in Tamusu: (a) The uniaxial stress-strain curve of sample No.1; (b) triaxial 5MPa stress-strain curve of sample No.12 at room temperature; (c) triaxial 10MPa stress-strain curve of sample No.13 at room temperature

The fracture characteristics and joint fractures of clay rock determine the strength of the clay rock mass in the entire Tamusu pre-selected area (Guo et al., 2019). Combining the damaged state of the sample (Fig. 2b) and its stress-strain curve (Fig. 3), it can be seen that under the conventional uniaxial compression test conditions, the deformation of the rock sample is mainly elastic deformation. After the axial stress reaches the elastic limit state. The rock sample begins to produce small local damage, and the stress-strain curve fluctuates. The axial stress reaches the peak stress. The rock sample is compacted under the axial stress, and the bearing capacity continues to rise until brittle failure.

Under the experimental conditions of 5 MPa at room temperature, the deformation of the rock stress-strain curve is mainly plastic deformation. When the peak strength is reached, the stress-strain curve of the rock sample drops sharply. The end face of the rock sample is crushed under pressure, and there are potential structural planes inside the rock sample, which will break along the structural plane when stressed. The rock sample increases with the confining pressure, but the brittle failure characteristics are not significant. The bearing capacity decreases slowly after reaching the peak, and gradually develops towards ductile deformation failure.

Under the experimental conditions of 10 MPa at room temperature, the deformation of the rock stress-strain curve is mainly elastic. The rock is brittle failure, reaching the peak stress, and the rock sample is broken. Under the action of confining pressure, and the damaged rock sample still maintains a high bearing capacity. The rock sample has an overall penetrating tensile fracture. The end face of the rock sample is seriously damaged, and a large number of joint cracks appear.

According to the mechanical test data of the rock samples in Table 2, the compressive strength and elastic modulus of the pre-selected clay rock have no significant correlation with depth changes. The compressive strength under various confining pressure conditions is relatively good, with the increasing trend of confining pressure. But the elastic modulus is relatively small, it shows that the macro-mechanics properties of rock are mainly affected by rock structure and mineral composition.

Table2 Results of uniaxial (triaxial) compression test of clay rock samples in the Tamusu pre-selected area

Sample number	Depth (m)	Compressive strength(Mpa)	Poisson's ratio(μ)	Elastic modulus (Gpa)	Experimental conditions
1	160.3	30.41	0.13	6.98	Uniaxial

 https://doi.org/10.1051/e3sconf/202129002010

2	255.6	34.46	0.18	6.08	Uniaxial
3	263.5	56.48	0.17	5.65	Triaxial 5 MPa
4	303.2	35.51	0.19	7.79	Uniaxial
5	303.5	66.05	0.22	8.50	Triaxial 5 MPa
6	304.3	71.38	0.18	9.12	Triaxial 10 MPa
7	314.5	60.99	0.18	9.62	Triaxial 5 MPa
8	401.2	37.94	0.25	6.93	Uniaxial
9	401.5	56.98	0.17	8.23	Triaxial 5 MPa
10	401.9	60.94	0.18	9.64	Triaxial 10 MPa
11	403.4	61.56	0.19	10.27	Uniaxial
12	403.6	79.27	0.16	6.51	Triaxial 5 MPa
13	404.8	99.67	0.20	11.33	Triaxial 10 MPa
14	515.4	68.59	0.20	8.44	Uniaxial
15	685.6	81.91	0.23	8.45	Uniaxial

3.2. Correlation analysis of mineral composition and mechanics

The XRD test results of the rock samples are shown in Table 3. According to the analysis of mineral composition, the main minerals of rock samples are dolomite, analcime, clay minerals, albite, quartz and so on. Clay rock at different depths have basically the same mineral composition, but there are differences in mineral content. The elastic modulus of the rock sample does not increase with the increase of the depth, and is less affected by the depth of burial and compaction, but it is different due to the different mineral components. The compressive strength of the rock sample is significantly affected by the depth of burial and compaction, and it has an upward trend as the confining pressure and depth of burial increase.

Table 3. Main mineral composition and content of clay rock samples in the Tamusu pre-selected area(%)

Sample number	Quartz	Analcime	Montmoril -lonite	Kaolinite	Chlorite	Illite	Calcite	Albite	Dolomite
1	0.4	2	—	1.8	9.7	15	0.8	8.7	61.6
2	4.3	8.3	—	3.6	2.9	10.6	49.1	9.3	11.9
3	2.6	24.1	—	0.7	2.9	5.1	—	7.6	57
4	14.1	28.4	0.3	2.6	5.2	9.7	2.7	11.2	25.8
5	4.5	37.5	—	0.9	2.3	5.5	1.8	8.1	39.4
6	3.4	59.5	—	0.8	1.8	5.6	1	8.3	19.6
7	8	20.3	0.5	1.9	6	6.9	3.4	6.9	46.1
8	6.9	20.6	0.6	1.7	4.9	6.3	—	10.1	48.9
9	4.1	42.8	—	1.2	2.9	5.7	1.8	8.8	32.7
10	2.7	48.1	—	1.3	2.3	5.6	1.9	8.7	29.4
11	2.8	52.2	—	1.3	2.3	5.1	1.7	7.6	27
12	3.6	56.2	—	1.7	2.1	8.4	1.6	7.3	19.1
13	2.9	53.2	—	1.6	2.2	5.3	1.7	7.5	25.6
14	2	39.5	—	0.9	2.7	4.8	—	9.4	40.7
15	1.8	30.4	—	1.4	2.8	4.2	—	12.1	47.3

Comparing the changes of various minerals and mechanical parameters, it is found that the changes in the content of quartz, calcite, albite and other minerals in the clay rock in the Tamusu pre-selected area have little effect on the macro-mechanics characteristics. It can be seen, it will have a greater impact on the overall mechanical properties of the rock, when the mineral content reaches a certain value. Therefore, for dolomite, analcime and clay minerals that have significant effects on mechanics in rock samples, the correlation between mineral components and mechanics is explored under the conditions of conventional uniaxial, triaxial 5 MPa and triaxial 10 MPa (Fig. 4).

The analysis shows that the overall mechanical parameters of each mineral increase with the increase of confining pressure. The elastic modulus of rock samples increases first and then decreases with the increase of dolomite content. When the dolomite content exceeds 40%, it begins to make a negative contribution to the elastic modulus. The relationship between rock compressive strength and dolomite content also shows the same trend. The analcime content of the rock sample is positively correlated with the elastic modulus as a whole, but when the confining pressure increases, the elastic modulus of the rock sample decreases with the increase of the analcime content. The compressive strength varies slightly with the content of analcime, but increases significantly with the increase of confining pressure. The content of clay minerals in rock samples has a significant influence on mechanical parameters. When the clay mineral content is less than 15%, there is no obvious linear relationship with the elastic modulus and compressive strength, and the range of change will fluctuate greatly due to the different types of clay minerals. However, when the content of clay minerals is greater than 15%, the compressive strength and elastic modulus of rocks have a significant downward trend, indicating that clay minerals have a negative contribution to the elastic modulus and compressive strength of rocks. The clay minerals in the rock sample the increase will weaken the mechanical properties of the rock mass.

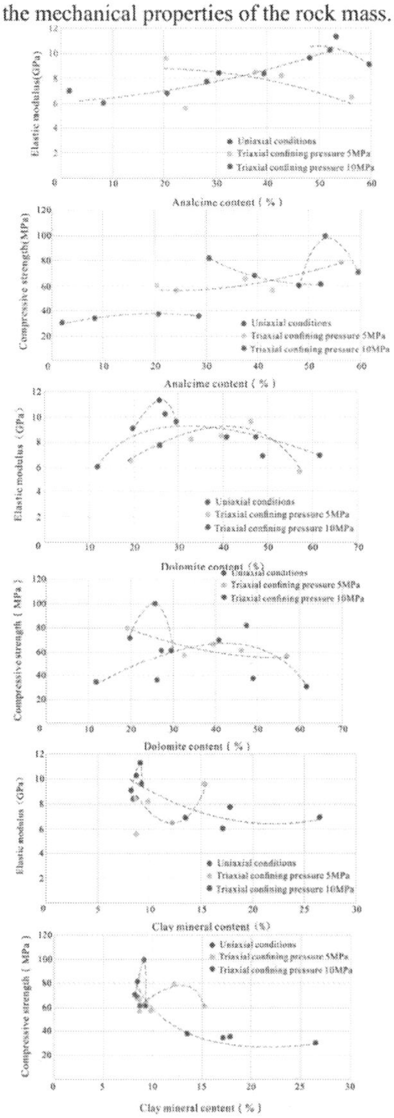

Fig4 Relationship between main minerals and macro-mechanics behavior under different conditions

3.3. Characteristics and structure analysis of rock under microscope

The macro photos of the clay rock in the Tamusu pre-selected area and the photos under the indoor microscope can be seen in Fig 5. The macroscopic characteristics of the rock in the field are mainly shown as continuous and uniform gray-white clay rock, with horizontal bedding structures (Fig. 5a). The analysis of rock minerals on the clay rock in the Tamusu pre-selected area by Leitz polarized light microscope shows that the sample has a higher mud content, and the whole is relatively clean and uniformly distributed under the microscope. The cement is mainly composed of carbonate minerals such as dolomite and calciteand argillaceous minerals. The rock structure mainly includes laminar structure, embedded crystal structure, comb-like structure and secondary large edge structure (Fig. 5c-f). As the argillaceous cement in the clay rock cannot be clearly distinguished under the polarized light microscope, the microscopic morphological characteristics of the rock sample are observed by SEM. It can be seen that the argillaceous cement is mainly clay minerals, and the surface is clean and the self-shape is good, it is an authigenic clay mineral (Fig. 5b).

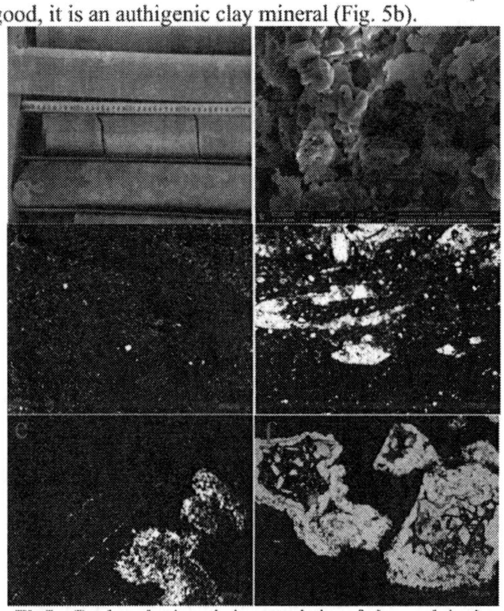

Fig5 Rock and mineral characteristics of clay rock in the Tamusu pre-selected area: (a) Clay rock drill cores in the Tamusu pre-selected area, showing horizontal bedding structures (signs of weak hydrodynamics); (b) Authigenic clay minerals with good self-shape under SEM; (c) Uniform and single argillaceous cement; (d) The secondary large edge structure of carbonate cement (calcite); (e) The cement grows vertically on the surface of the clastic particles in the form of fibers or short columns, showing a comb-like structure; (f) The coarse crystals formed by the crystallization of the cement enclose several dolomites to form an embedded crystal structure

4 Discussion

4.1. Establishment of estimation method of clay rock elastic parameters

As a kind of sedimentary rock, clay rock is different from hard rock such as igneous rock in structure. The texture and structure of igneous rocks are mainly affected by the degree of crystallinity, crystal grain size and aggregate morphology of minerals in the rock. The texture of sedimentary rock is affected by the nature, size, shape and relationship of particles, while structure is the spatial distribution and arrangement of various components in sedimentary rock (Shu, 2010). Therefore, in order to establish a method for estimating the elastic parameters of the clay rock in the Tamusu pre-selected area for geological disposal of HLW, based on the volume average analysis of the mineral properties of the sedimentary rock, the influence of sedimentation on the

https://doi.org/10.1051/e3sconf/202129002010

rock structure should also be considered.

Sedimentary rocks are mainly affected by compaction and cementation during consolidation and diagenesis. The compaction reduces the number of pores in the sediments, and the pores are reduced, causing the water to be squeezed out, which makes the thickness smaller and the sediments harder. This effect is especially obvious in muddy sediments. Cementation is the main factor that fills the interstitial materials between the particles, cement and solidifies the deposits and makes them hard. Different from the matrix in igneous rocks (small mineral particles in the rock) (as shown in Fig. 6), the fine clastic interstitial material filled in the grain voids of the sediments are the matrix (heterobase), while the cemented and consolidated sediments the chemical interstitial material is cement. Therefore, distinguishing the main forms of cementation interstitial materials is the key to exploring the structure of clay rock.

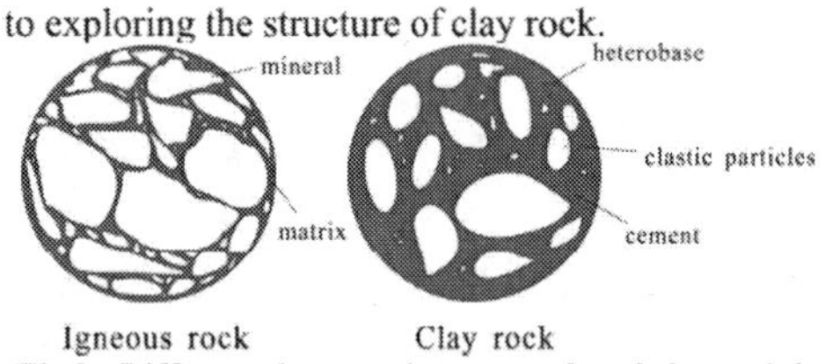

Fig6 Difference between igneous rock and clay rock in cement and matrix

Observed by Leitz polarized light microscope, it can be seen that the interstitial materials of the rock sample are mainly composed of embedded crystal structure, comb-like structure and secondary large-edge structure, which are the structural forms of cement. The argillaceous minerals under SEM are mainly authigenic clay minerals, combining with the paleo-sedimentary environment and sedimentary facies analysis of the clay rock in the Tamusu pre-selected area (Gong, 2017; Rao, 2018; Huang, 2018; Li, 2010; Li, 2018). The interstitial materials in the clay rock in the Tamusu pre-selected area are mainly cement, and the cement is mainly argillaceous cement (authigenic clay minerals) and carbonate cement (calcite, dolomite, etc.) of the syngenetic and diagenetic period.

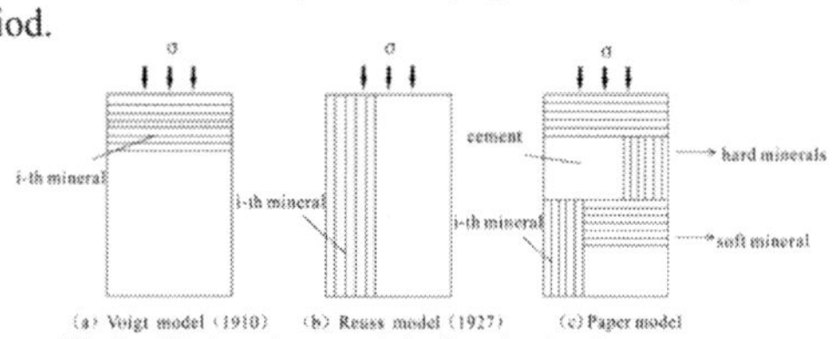

Fig7 Estimation model of rock elastic parameters

According to the classic elastic parameter estimation method of Voigt and Reuss et al. (Fig. 7a, b), a new elastic parameter estimation method was established for the clay rock in the Tamusu pre-selected area for geological disposal of HLW. Based on the volume average analysis of rock mineral properties, the main minerals are divided into soft and hard minerals based on physical and mechanical properties, and it is assumed that the macroscopic mechanical parameters of the rock sample will be affected by the cement in the clay rock. Therefore, it is assumed that the stress and strain of the soft and hard minerals in the clay rock have almost equivalent effects on the macro-mechanics parameters of the rock, and the elastic parameters in the clay rock are mainly affected by the cementation during the sedimentation and diagenesis process (Fig. 7c). The new method of elastic parameter correction estimation is shown in formula 2, in the formula, K_N and μ_N are the bulk modulus and shear modulus estimated by the clay rock evaluation method, α and β are the percentages of hard minerals and soft minerals in the entire rock, and γ is the proportion of cement. Table 4 shows the proportion of soft and hard minerals and the percentage of cement in each rock sample.

$$\begin{cases} K_{\mathrm{N}} = \dfrac{(\alpha+\beta)\times(K_V+K_R)}{2}\bullet\gamma \\ \mu_{\mathrm{N}} = \dfrac{(\alpha+\beta)\times(\mu_V+\mu_R)}{2}\bullet\gamma \end{cases} \quad (2)$$

Table4 The proportion of cement and the division ratio of soft and hard minerals in clay rock samples(%)

Sample number	Hard mineral content	Soft mineral content	Proportion of cement	Experimental conditions
1	74	27	27	Uniaxial
2	83	17	17	Uniaxial
4	82	18	28	Uniaxial
8	87	14	27	Uniaxial
11	91	9	52	Uniaxial
14	92	8	40	Uniaxial
15	92	8	30	Uniaxial
3	91	9	24	Triaxial 5 MPa
5	91	9	38	Triaxial 5 MPa
7	85	15	35	Triaxial 5 MPa
9	90	10	43	Triaxial 5 MPa
12	88	12	36	Triaxial 5 MPa
6	92	8	60	Triaxial 10 MPa
10	91	9	48	Triaxial 10 MPa
13	91	9	53	Triaxial 10 MPa

4.2. Trial calculation of various elastic parameter estimation methods

The main mineral crystals that make up the clay rock in the Tamusu pre-selected area for geological disposal of HLW have been studied in detail in (Dortmann, 1985; Tiziana et al., 2003; Deer et al., 2009) and the average

 https://doi.org/10.1051/e3sconf/202129002010

elastic parameters of each mineral are shown in Table 5. Although chlorite is a clay mineral, it is listed separately because of its relatively high bulk modulus and shear modulus. Zeolite is a clay mineral, so analcime (a kind of natural zeolite) with relatively small bulk modulus and shear modulus is classified as clay mineral for analysis.

Table5 Basic elastic parameters of main minerals

Types of minerals	Elastic Modulus *E*/GPa	Poisson's ratio μ	Bulk modulus *K*/GPa	Shear modulus *G*/GPa
Clay minerals	12.66	0.27	9	5
Chlorite	72.90	0.35	81	27
Quartz	114.57	0.30	96.4	44.0
Dolomite	23.5	0.26	16.3	9.3
Calcite	68.8	0.31	74.4	26.3
Albite	69.0	0.28	57.0	27

The classical elastic parameter estimation methods of predecessors were used to calculate the theoretical values of bulk modulus and shear modulus of Tamusu clay rock. According to the elastic modulus conversion formula, the theoretical elastic modulus estimates of rock samples under different confining pressures of conventional uniaxial, triaxial 5 MPa, and triaxial 10 Mpa are obtained. The specific data and related comparison can be seen in Table 6 and Fig 8.

The test value under normal uniaxial conditions is between 6.08 and 10.27 GPa. The test value under normal temperature triaxial conditions under 5Mpa is between 5.65 and 9.62 GPa. The normal temperature triaxial conditions under 10 Mpa conditions is between 9.12 and 11.33 GPa. The estimation method proposed by Reuss, which is closest to the actual test value. The estimated value of elastic modulus is between 16.75 and 42.74 GPa under conventional uniaxial conditions. The estimated value is between 16.02 and 21.72 GPa under normal temperature triaxial conditions of 5 MPa. The estimated value is between 16.05 and 17.22 GPa under normal temperature triaxial conditions of 10 MPa.

Table6 Comparison of the theoretically estimated elastic modulus (E) and the test result value (unit: GPa)

Sample number	Test value	Voigt method	Reuss method	Hill method	Kumazawa method	Jianping Zuo method	Experimental conditions
1	6.98	31.18	22.80	27.01	26.67	28.97	Uniaxial
2	6.08	54.00	31.34	42.74	41.17	50.15	Uniaxial
4	7.79	41.04	21.13	31.11	29.46	37.51	Uniaxial
8	6.93	33.81	21.44	27.64	26.93	32.26	Uniaxial
11	10.27	25.23	16.75	21.00	20.56	24.49	Uniaxial
14	8.44	26.17	18.19	22.19	21.82	25.50	Uniaxial
15	8.45	28.29	19.65	23.98	23.58	28.68	Uniaxial
3	5.65	27.64	20.14	23.89	23.60	26.99	Triaxial 5 MPa
5	8.5	28.65	18.65	23.66	23.12	27.78	Triaxial 5 MPa
7	9.62	35.46	21.72	28.61	27.77	33.37	Triaxial 5 MPa
9	8.23	28.29	18.02	23.16	22.58	27.28	Triaxial 5 MPa
12	6.51	24.83	16.02	20.44	19.95	23.76	Triaxial 5 MPa
6	9.12	24.72	16.05	20.39	19.92	24.01	Triaxial 10 MPa
10	9.64	26.13	17.22	21.69	21.22	25.32	Triaxial 10 MPa
13	11.33	25.06	16.59	20.83	20.39	24.29	Triaxial 10 MPa

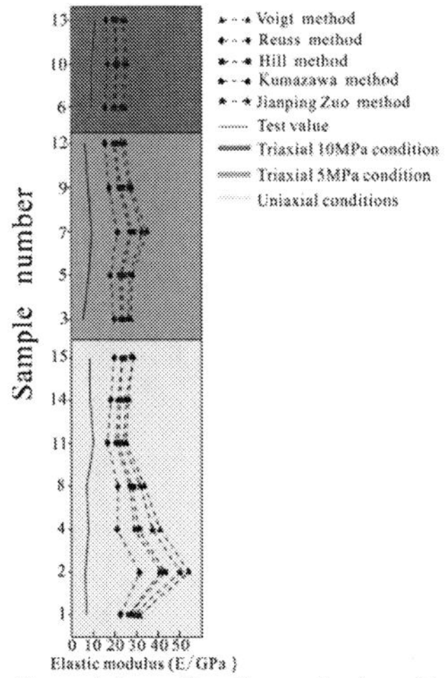

Fig8 Comparison of classical elastic modulus estimation methods and test value under different conditions

The theoretical value calculated by the elastic parameter estimation method proposed in this paper is compared with the actual test value. See Table 7 and Fig 9 for details. The estimated values of elastic modulus of the proposed method are 7.16 and 10.96 GPa under normal uniaxial conditions, 8.76 and 10.04 GPa at normal temperature triaxial 5 MPa, and 10.43 and 12.13 GPa under normal triaxial 10 MPa conditions.

Table7 Comparison of theoretical estimation value and actual test value based on the proportion of clay rock cement(GPa)

Sample number	1	2	4	8	11	14	15	3	5	7	9	12	6	10	13
Experimental conditions	Uniaxial							Triaxial 5 MPa					Triaxial 10 MPa		
Actual test value	6.98	6.08	7.79	6.93	10.27	8.44	8.45	5.65	8.5	9.62	8.23	6.51	9.12	9.64	11.33
Theoretical estimate	7.16	7.31	8.83	7.43	10.96	8.76	7.29	5.76	8.87	10.04	9.91	7.4	12.13	10.43	11.08

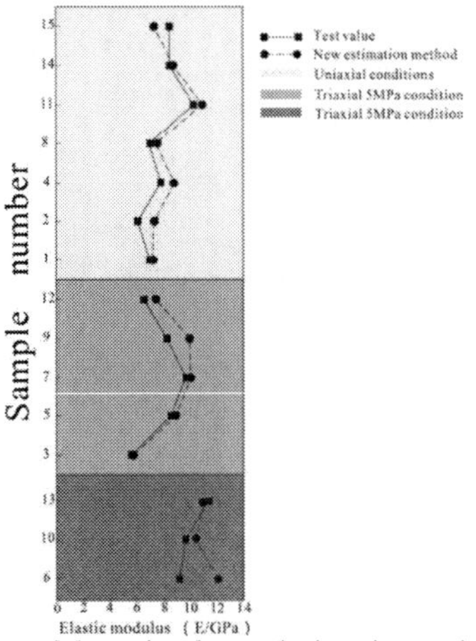

Fig9 Comparison of elastic modulus estimation methods and test value under different conditions

Comparing the actual elastic modulus test value with the theoretical value calculated by each elastic parameter estimation method, it can be seen that with the increase of confining pressure, the error of the theoretical value obtained by the classical elastic parameter estimation method is reduced compared with the test value, but the overall error is still large. Under the same test conditions, the estimated value of the rock sample elastic modulus obtained by the elastic parameter estimation method proposed in this paper is more consistent with the actual test value, which can more accurately predict and evaluate the elastic modulus of the clay rock host rock in the geological disposal for HLW.

4.3. Comparison and analysis of errors between theoretical and test values

Table 8 shows the error comparison data between the actual test value of the elastic modulus of the clay rock sample in the Tamusu pre-selected area and the theoretical value obtained by each elastic parameter estimation method. The classical elastic parameter estimation method has a large overall error when estimating the elastic modulus of clay rock samples, the maximum error can reach 788%, the minimum is 46%, and the error range varies greatly. As the confining pressure increases, the sedimentary rock is obviously compacted, which reduces the overall error. The maximum error is reduced to 389% under the condition of normal temperature triaxial 5 MPa, and the maximum error is reduced to 171% under normal temperature triaxial condition of 10 MPa, but there is still a large error compared to the actual test value. Compared with the actual test value of the rock sample, the elastic parameter estimation method established in this paper has the minimum error of only 2% and the maximum error of only 33%, which is more consistent with the actual test value. It is superior to the classical elastic parameter estimation method in the evaluation.

Table8 The relative error between the estimation method of elastic parameters and the actual test value

	Under different test conditions		
Estimation methods of various elastic parameters	Uniaxial	Triaxial 5 MPa	Triaxial 10 MPa

Voigt method	146%~788%	237%~389%	121%~171
Reuss method	63%~415%	119%~256%	46%~79%
Hill method	104%~603%	178%~323%	84%~125%
Kumazawa method	100%~577%	172%~318%	80%~120%
Jianping Zuo method	138%~725%	227%~378%	114%~163%
The method of this paper	3%~20%	2%~20%	2%~33%

5 Conclusion

Through the analysis of rock minerals and mechanical characteristics of the rock samples, it can be known that the main mineral components of the Tamusu clay rock are dolomite, analcime, albite, calcite, quartz, and clay minerals. The rock structure is mainly affected by cementation, and the interstitial material are mainly argillaceous cement and carbonate cement. The conventional uniaxial mechanical strength is relatively low, mainly manifested as elastic deformation, and the failure form is brittle failure. As the confining pressure increases, the rock deformation becomes more obvious plastic deformation, and the failure form is plastic and ductile. When the confining pressure increases to 10 MPa, the rock deformation tends to be more elastoplastic instead.

By distinguishing the structural differences between sedimentary rocks and igneous rocks and other hard rock structures, the soft and hard minerals are classified for the Tamusu clay rock. According to the analysis under the microscope, combined with the material source, sedimentary form and formation environment of the Tamusu clay rock, the main form of cement is determined. Based on the volume average analysis of the mineral properties of sedimentary rocks, and considering the impact of sedimentation on the structure of the rock, a new method for estimating the elastic parameters of Tamusu clay rock is established.

Through the trial calculation of the elastic modulus of each elastic parameter estimation method and the comparison analysis with the actual test value error, it can be seen that the classical elastic parameter estimation method has a large error in estimating the elastic modulus of Tamusu clay rock, with a maximum error of 788%, the minimum is 46%, and the new elastic parameter estimation method established in this paper has a maximum error of only 33% and a minimum error of only 2% when estimating the elastic modulus of rock samples. Compared with the classical elastic parameter estimation method proposed by the predecessors, the error is smaller, and it is more consistent with the actual measured value, which can more accurately estimate and evaluate the elastic modulus parameters of the clay rock in the Tamusu pre-selected area for geological disposal of HLW.

Acknowledgements

This article is one of the phased research results of National Defense Science and Technology Industry Financial Support Project "Research on the Screening and Evaluation of Clay Rock Sections in the Northwestern Region of the Geological Repository for High-level Radioactive Waste" (Second Division of Science and Engineering [2014] No. 1587) and the Jiangxi Province Graduate Student Innovation Fund Project "Research on the Characteristics of Clay Rock/Mudstone Host Rock in the Northwest Preselected Area of the Geological Repository for High-level Radioactive Waste" (YC2019-B107).

References

1. Deer, W.A., Howie, R.A. & Zussman, J. 2009. Rock-forming minerals: Layered silicates excluding micas and clay minerals. London: Geological Society.
2. Dortmann. 1985. Physical properties of rocks and minerals.1th ed. Trans: Jiang Hongyao. Beijing: Science Press.
3. Gong, Y. 2017. Study on regional geological conditions of Tamusu candidate area, inner mogolia-siting of clay rock high-level radioactive waste disposal repositoty. Dissertation. Nanchang: East China University of technology.
4. Guan, W.C., Liu, X.D. & Liu, P.H. 2014. Geological characteristics of clay rocks in Tamusu area of Bayingobi Basin. World Nuclear Geoscience 31 (02): 95-102.
5. Guo, Q.F., Wu, X., Cai, M.F., Xi, X., Ren, F.H. & Miao, S.J. 2019. Experiment on the strength characteristics and failure modes of granite with pre-existing cracks. Chinese Journal of Engineering 41 (01): 43-52.

6. Hill R. 1952. The elastic behavior of a crystalline aggregate. Proceedings of the Royal Society of London A A65:349-354.
7. Hu, H.Y. 2014. Mechancial properties of clay rocks Tamusu area. Dissertation. Nanchang: East China University of technology.
8. Hu, H.Y., Liu, X.D., Yang, T., Wang, G.B. & Huo, L.2014. Experimental study and mechanism analysis of mechanical properties of clay rock in Tamusu area. Geotechnical Investigation & Surveying 42 (12): 9-13.
9. Hu, Q.B., Liang,H.A., Yang, T., Liu, X.D, Wang, Y. & Zhang, L.P. 2020. Permeability and Strength Characteristics of Tamusu Deep Clay under Confining Pressure and Osmotic Pressure. Journal of Yangtze River Scientific Research Institute. 25(11):1-12.
10. Huang, H.E. 2018. Study on Hydrogeological Conditions of Clay rock Pre-selected Section of High-level Radioactive Waste Repository-Taking Tamusu and Suhongtu as Examples. Dissertation. Nanchang: East China University of technology.
11. Kumazawa M. 1969. The elastic constant of polycrystalline rocks and nonelastic behavior inherent to them. Journal of Geophysical Research 74: 5311.
12. Laurence S C.1997. Site selection and characterization processes for deep geological disposal of high-level nuclear waste. Open-file report. Albuquerque Sandia National Laboratories.
13. Li, S.Z. 2018. Geological Characteristics and Comprehensive Evaluation of Clay rock Pre-selected Section of Bayingobi Formation in Ingejing Depression Inner Mongolia. Dissertation. Nanchang: East China University of technology.
14. Li, X.D. 2010. Sedimentary facies analysis of upper member of Bayingobi formation in Tamusu area of Bayingobi Basin. Journal of Henan Polytechnic University(Natural Science) 29 (S1): 177-180.
15. Ma, J.Y., Zhang, H.Y. & Chen, L. 2020 Research status of public participation in high level radioactive waste disposal abroad: a case study of Finland and Sweden. World Nuclear Geoscience 37 (03): 250-256.
16. Pan, Z.Q. & Qian, Q.H. 2009. Strategic Study on geological disposal of high-level radioactive waste. 1th ed. Beijing: Atomic Energy Press.
17. Rao, G.W. 2018. Study on Characteristics of Surrounding Rock of High-level Radioactive Waste Geological Disposal Reservoir: A Case Study of Mudstone in the Upper Part of the BayinGobi Formation in the Preselected Area of Tamusu in Inner Mongolia. Dissertation. Nanchang: East China University of technology.
18. Reuss A. 1929. Berechnung der fliegrenze vom misch kristallen. Journal of Applied Mathematics and Mechanics 9:49-58.
19. Shu, L.S. 2010. Physical Geology. 3th ed. Beijing: Geological Publishing House.
20. Tiziana, V. & Manika, P. 2003 Amos Nur. Elastic properties of dry clay mineral aggregates, suspensions andsandstones. Geophysical Journal International. 155(1) : 319-326.
21. Voight W. 1910. Lehrbuch der Kristallphysik. 1th ed Leipzig: Teubner-Verlag.
22. Wan, C. 2018. A Dissertation Submitted to East China University of Technology for Master's Degree. Dissertation. Nanchang: East China University of technology.
23. Wan, C., Liang, H.A., Wang Y., Chen, Q. & Zhang,J. 2017. Study on the Influence of Mineral Composition and Microstructure on the Mechanical Properties of Clay. Construction Quality 35 (06): 78-82 + 86.
24. Wang, J. 2019 Progress of geological disposal of high-level radioactive waste in China in the 21st century. Atomic Energy Science and Technology 53 (10): 2072-2082.
25. Wang, X.L., Han, Z.H., Zhang, L.Q & Zou, J. REGIONAL ENGINEERING GEOLOGY SUITABILITY ASSESSMENT FOR HIGH-LEVEL RADIOACTIVE WASTE DISPOSAL OF PRE-SELECTED ALXA AREA. Journal of Engineering Geology 26 (06): 1715-1723.
26. Wang, Y., Liang, H.A., Hu, Q.B., Cheng, H.K.,Wan, C. & Xie, Z.2018. Study on Correlation Between Mineral Composition and Elastic Modulus of Clay Rock in Tamusu. Chongqing Architecture 17 (03): 18-21.
27. Xiang, L., Liu, P.H., Liu, X.D., Liu, S. & Liang, H.A.2018. Study on screening of favorable areas for high level radioactive waste claystone/mudstone disposal repository in Yingejing sag. China Sciencepaper 13 (21): 2499-2504.
28. Xue, K.X., Liu, S., Liu, X.D., Hu,X.Y., Tiwari, B., Wei, Y.Q. & Liang, H.A. 2016. Review of the Physical,Mechanical and Engineering Properties of Clay Rocks. Science Technology and Engineering 16 (16): 110-121.
29. Yuan, G.X., Zhang, L.Q., Zheng, Q.L., Huang, Z.Q., Li, J.Y., Wang, H.J & Deng, X.B. 2018. PREDICTION OF ROCK MASS QUALITY IN TARGET DEPTH FOR TAMUSU AREA OF ALXA PRE-SELECTED REGION FOR PREDICTION OF ROCK MASS QUALITY IN TARGET DEPTH FOR TAMUSU AREA OF ALXA PRE-SELECTED REGION FOR GEOLOGICAL DISPOSAL OF HIGH-LEVEL NUCLEAR WASTE. Journal of Engineering Geology 26 (06): 1690-1700.
30. Zhang, C.Y., Nie, F.J., Hou, S.R., Wang, J.L., Zhang, L. & Deng, W. 2015. Ore-Controlling Factors and Metallogenic Model of Tamusu Sandstone Type Uranium Deposit in Bayingobi Basin, Inner Mongolia. Bulletin of Geological Science and Technology 34 (01): 140-147.

31. Zuo, J.P., Chai, N.B., Zhao, C. & Liu, M. 2015. Investigation on the Relationship Between of Micro/meso Mineral Composition and Macro Mechanical Behavior of Moutougou Basalt. Journal of Basic Science and Engineering 23 (05): 97-106.

The Groundwater Vulnerability Researches of Horqin District in Tongliao City Based on the AHP Method

Li Xuguang[1], Dai Yajian[1,*], He Haiyang[1]

[1]Shenyang Center of Geological Survey, China Geological Survey, Shenyang,Liao Ning, China

Abstract. On the basis of DRASTIC model index system, combined with the actual situation in Horqin district in Tongliao City, this paper removed the influence of the topography, added land use effects, modified the new model DRASLIC, used analytic Hierarchy Process (AHP), and calculated the new evaluation results. Groundwater management and protection decisions in Tongliao Horqin area provide a reliable basis for the application to design the groundwater monitoring network, site selection of waste landfill, evaluation and control of groundwater pollution, and provide a basis for other work. such as groundwater contamination and protection.

1 INTRODUCTION

As China releases relevant policies about revitalization of the old northeast industrial base, the economy in the Northeast Region is booming in recent years, with an increase of population and an expansion of cities. However, the contradiction between supply and demand of water resources is becoming more and more prominent. In particular, the decrease of precipitation in recent years has caused a serious shortage of surface water in the region, and groundwater resources seem particularly important in the Northeast Region. However, the groundwater level in the region gradually decreases with the exploitation of groundwater in quantity. Centralized water supply in urban areas results in the heavy groundwater exploitation. In addition, groundwater is polluted to varied degrees under the influence of industrial waste water, domestic sewage and agricultural irrigation. Tongliao City is one of the most important prefecture-level cities on the West Liaohe Plain in the Northeast Region. Particularly, Horqin Region, Tongliao has concentrated urbanization, dense population, developed industry and high pollution. Groundwater is the main water supply for residents in the region. The pollution of groundwater environment will seriously affect the production and life of local residents. The phreatic aquifer in the region is very thick, and shallow groundwater is distributed continuously, which is closely related to the circle of human activities on the surface and is most vulnerable to pollution. It is also closely associated with the hydraulic power of deep groundwater and easily contaminates other aquifers through interlayer hydraulic connection after pollution. Hence, it is crucial to evaluate the antifouling performance of shallow water.

2 OVERVIEW OF STUDY AREA

2.1 Natural conditions

The study area is located in the central part of the West Liaohe Alluvial Plain and the hinterland of Horqin Grassland. With regard to the terrain, Horqin Region is high in the south and north and low in the central part, with a height above sea level of 144.9-260 m. The south part has high topographic relief and is dominated by sand hump land. East-west closed strip-shape depressions are common among humps. In the north, the topography is undulating and humps and recesses are connected; there are local closed depressions. The central part is a valley plain with wide and flat terrain inclined towards the northeast from the southwest. The study area is located in the central part of the West Liaohe Plain and belongs to Horqin Region, Tongliao, Inner Mongolia. It is the political, cultural, economic and transportation center of the east part of Inner Mongolia, with excellent traffic locations and a three-dimensional transportation network consisting of civil aviation, highways and railways. It is also a main transportation hub in the Northeast Region.

2.2 Hydrogeology

The study area is located in the hinterland of the West Liaohe Plain, in which there are extremely thick Quaternary loose sediments and simple exposed strata that are all Upper Pleistocene and Holocene strata of the Quaternary system. The aquifer structure and lithology in

Author: Li Xuguang, male, born in 1982, Master Degree Candidate, Senior Engineer, mainly engaged in groundwater pollution, endemic disease and urban geology. Email: john2011@163.com.
* Corresponding author:Dai Yajian, male, born in 1980, Master Degree Candidate, Senior Engineer, mainly engaged in groundwater pollution and pedogeochemistry. Email: 337056992@qq.com.

https://doi.org/10.1051/e3sconf/202129002011

the area have zoning rules from the west and south to the northeast consistent with the stratigraphic structure and lithology. In terms of the lithology, the aquifer in the west and south regions is featured with thick-layer fine sand, medium sand, coarse sand and sand pebbles and gravels with coarser grains and a single structure. The aquifer in the northeast is featured by silt and fine sand, increased and thickened interlayers of cohesive soil, finer grains and increased silt composition. The thickness of the aquifer group gradually decreases from west to east. From the west to the northeast and the southeast, the aquifer thickness shows a zoning rule of 130- <60 m and 130- <120 m respectively. The buried depth of groundwater varies greatly and is 3-10 m in most areas. That in the sand hump land in the south of Honghe River is extremely variable due to high topographic relief and frequent changes. It is generally 5-10 m and is greater than 10 m in some areas in the southeast. The buried depth of groundwater in the valley plain is generally small, 1-3 m in most areas.

3. RESEARCH METHOD AND CONTENT

3.1 Evaluation method

The antifouling performance of groundwater is a process with dynamic changes. Different evaluation periods will lead to different results. At present, it is common to evaluate the antifouling performance of groundwater with the DRASTIC method jointly developed by the US Environmental Protection Agency (USEPA) and the US National Wells Association (NWWA) in 1987, which combines the experience of more than 40 credible experts in hydrogeology and is an empirical model for macroscopic evaluation of the antifouling performance of regional groundwater on a large scale. The DRASTIC model was first used in 40 counties in the United States and in areas with different hydrogeological conditions. It was then used to evaluate the antifouling performance of groundwater in a wider range in different countries and regions and was improved to be applicable to the evaluation of the antifouling performance of groundwater under various hydrogeological conditions.

The traditional DRASTIC model adopts seven factor systems related to groundwater vulnerability evaluation:

(1) Depth to groundwater table D;

(2) Net recharge R;

(3) Aquifer media A;

(4) Soil media S;

(5) Terrain T;

(6) Impact of unsaturated zone I;

(7) Hydraulic conductivity C;

The expression (Formula 1) in the traditional DRASTIC model is:

$$DI=D_W D_r + R_W R_r + A_W A_r + S_W S_r + T_W T_r + I_W I_r + C_W C_r \quad ......$$

......Formula 1

Where,

DI is the antifouling performance index in DRASTIC (groundwater vulnerability index);

Dr, Rr, ..., Cr are the scores of evaluation index factors, respectively;

Dw, Rw, ..., Cw are the weighted values of evaluation index factors, respectively;

The higher the DI result (groundwater vulnerability index) calculated with the model, the worse the antifouling performance of groundwater (higher groundwater vulnerability) and the higher the risk of groundwater pollution; vice versa.

3.2 Selection of evaluation factors

All the seven parameters involved in the evaluation system are generally used as evaluation indexes in the evaluation of antifouling performance. However, it is necessary to make appropriate adjustments objectively and increase or decrease evaluation indexes based on the actual situation and the actual data collected and collated for different regions and evaluation work.

1. Soil lithology: the part of the evaluation system that is connected with the surface and most affected by human activities. Different soil lithology is different in structure, composition, thickness and natural humidity which determine its self-purification capacity, that is, the ability to adsorb and degrade pollution components. The stronger the self-purification capacity of soil, the higher the antifouling performance of groundwater. Therefore, soil lithology is one of the main indexes determining the antifouling performance.

2. Vadose zone: the medium between the soil medium and the medium of the water-bearing rock formation and the main path for pollution components to enter the groundwater. The lithology in the vadose zone determines the dilution function, biodegradation function, neutralization function and chemical reaction strength of pollution components before they enter the groundwater. Hence, it determines the antifouling performance of groundwater to a large extent.

3. Aquifer: a complex system, the vulnerability of which varies spatially. During analysis of the lithology of an aquifer, its hydraulic conductivity coefficient should also be regarded as an important index, which restricts the migration of groundwater and controls the migration of pollution components in groundwater. In addition, the buried depth of groundwater also has great impacts on the evaluation results in the process of phreatic aquifer vulnerability evaluation.

4. Terrain: the change of surface slope or land use type in the study area.

5. Recharge: the amount of water per unit area seeping into the aquifer from groundwater recharge sources. Net recharge is also an index reflecting aquifer vulnerability.

Based on the collected actual data of Horqin Region, Tongliao, we chose 7 index factors to build a model for evaluating the antifouling performance (vulnerability) of groundwater in the region, including the depth to groundwater table (D), net recharge (R), aquifer media (A), soil media (S), land use (L), impact of the vadose zone (I) and hydraulic conductivity (C).

Six factors reflecting the groundwater properties in the working area were selected for the groundwater vulnerability evaluation based on DRASTIC model - the traditional groundwater vulnerability assessment method. The factors land use reflecting the pollution load was added. Therefore, 7 factors in total were used for the evaluation of regional groundwater pollution risks. The modified model is DRASLIC, the abbreviation of the following.

(1) D (Depth to groundwater table):

(2) R (Net recharge):

(3) A (Aquifer media):

(4) S (Soil media):

(5) L (Land use):

(6) I (Impact of the vadose zone):

(7) C (Hydraulic conductivity):

Depth to groundwater table (D). It was directly obtained from groundwater statistical and survey data in the evaluation, m.

Net recharge (R). The main recharge of groundwater in the area is the infiltration of atmospheric precipitation, accounting for more than 80% of the total recharge. For the convenience of calculation, the net recharge was approximately replaced by the infiltration of atmospheric precipitation, mm.

Aquifer media (A) and impact of the vadose zone (I). They were modified based on loose strata since they are classified by rock (bed rock) in the Technical Requirements for Groundwater Vulnerability Assessment (GWI-D3) and the strata in the study area are all Quaternary loose strata. The level and characteristic value of the modified aquifer media (A) and impact of the vadose zone (I) were used.

Soil media (S). A value was assigned to it according to the level and characteristic value corresponding to the surface lithology and the surface lithology distribution diagram.

Land use (L). Its category and score were determined according to the land use status subject to remote sensing interpretation, the classification and scoring of land use in USGS and SI methods and the actual situation of the working area .

Hydraulic conductivity (C). A value was directly assigned according to the hydraulic conductivity distribution diagram of the aquifer, m/d.

3.3 Weight calculation

The weight of each evaluation factor can usually be obtained from the table when the DRASTIC evaluation method is used.

However, the table is based on the information about hydrogeological characteristics of the United States, so the weight of each factor is not always appropriate for Horqin Region, Tongliao. In order to make the evaluation result more accurate, the idea of AHP and the method of factor weight calculation in it are used to obtain the weight of each factor of DRASLIC.

Weight calculation method

Analytic Hierarchy Process (AHP) is a systematic and hierarchical analysis method combining qualitative and quantitative indexes, which was put forward by T.L. Satyr, a famous operations research expert and professor in the Department of Mathematics, the University of Pittsburgh, in the middle 1970s. It is used to calculate the combined weight of components of each hierarchy to the target layer in four steps under the premise of clear and correct understanding of problems: setup of a structural model , construction of a judgment matrix, single hierarchical sorting and total hierarchical sorting, thus obtaining the comprehensive evaluation value of different operational schemes. AHP is a simple, flexible and practical multi-criteria decision-making method, which has been applied to decision-making issues in all walks of life since it was proposed.

Index weight calculation by APH

(1) Construct a hierarchical structure model according to the evaluation index system;

(2) Compare the factors (a1、a2、⋯⋯an) in each target layer (C) pairwise and construct a fuzzy consistent matrix (Formula 2) R according to the membership:

$$\begin{matrix} C & a_1 & a_2 & \cdots & a_n \\ a_1 & r_{11} & r_{12} & \cdots & r_{1n} \\ a_2 & r_{21} & r_{22} & \cdots & r_{2n} \\ \cdots & \cdots & \cdots & \cdots & \cdots \\ a_n & R_{n1} & R_{n2} & \cdots & r_{nn} \end{matrix} \quad \text{.........Formula 2}$$

Where,

r_{ij} means that elements ai and aj have a fuzzy relation "… much more important than…" when they are compared with target C.

To quantitatively describe the relative importance of any two schemes with respect to a criterion, corresponding values can be given based on the scale of 0.1-0.9 in Table 1 below.

Table 1 0.1-0.9 number scale

Scale	Definition	Description
0.5	Equally important	The two elements are equally important according to comparison.
0.6	Slightly more important	One element is slightly more important than the other according to comparison.
0.7	Significantly more important	One element is significantly more important than the other according to comparison.
0.8	Much more important	One element is much more important than the other through comparison of the two elements.
0.9	Extremely important	One is extremely more important than the other according to comparison.
0.1,0.2,0.3, 0.4,0.5	Negative contrast	If the element ai is compared with the element aj to judge rij, the judgment obtained from the comparison is rj i = 1 - r_{ij}.

(3) According to the fuzzy consistent matrix, the consistency of the matrix shall be checked first. After mathematical manipulation, the following formula is used to solve the weight value wi of elements in each layer (Formula 3).

$$w_i = \frac{1}{n} - \frac{1}{2\alpha} + \frac{1}{n\alpha} \times \sum_{k=1}^{n} r_{ik}, \ \ i \in \Omega \quad \text{......Formula 3}$$

Where, n is the order of matrix R, $\alpha = (n-1)/2$.

If rij is scaled according to Table 2.1 and rij = 1 - rji, R is a fuzzy consistent matrix. That is, it is unnecessary to check the consistency of the matrix.

(4) If the fuzzy consistent matrix is satisfied, the weight value wi of elements in each layer can be solved.

Determination of the weight of antifouling performance evaluation factors

The evaluation system includes two levels according to the evaluation of antifouling performance of groundwater in the study area, with reference to the steps above:

Level 1 is the target layer of evaluation, namely the antifouling performance of groundwater;

Level 2 is the factor layer of evaluation, namely D, R, A, S, L, I and C in the model above.

Hence, the evaluation factors in level 2 were pairwise compared. The importance scale of factors (membership) obtained from pairwise comparison was scored by several experts in hydrogeology. The average value of the scores was obtained after removal of the highest and lowest values. The fuzzy consistent matrix R of antifouling performance of groundwater in Tongliao was constructed based on AHP, as shown in Table 2.

Table 2 Fuzzy consistent matrix R of groundwater vulnerability in Horqin Region, Tongliao

Index	D	R	A	S	L	I	C
D	0.5	0.6	0.7	0.7	0.9	0.5	0.7
R	0.4	0.5	0.6	0.7	0.8	0.4	0.6
A	0.3	0.4	0.5	0.6	0.8	0.4	0.5
S	0.3	0.3	0.4	0.5	0.7	0.3	0.4
E	0.1	0.2	0.2	0.3	0.5	0.2	0.2
I	0.5	0.6	0.6	0.7	0.8	0.5	0.4
C	0.3	0.4	0.5	0.6	0.8	0.6	0.5

The weight of each index for the evaluation of antifouling performance of groundwater in Horqin Region, Tongliao can be obtained according to the fuzzy consistency test and calculation, as shown in Table 3.

Table 3 Weights of evaluation indexes for the evaluation of antifouling performance of groundwater in Horqin Region, Tongliao

Evaluation index	D	R	A	S
Weight	0.195	0.167	0.143	0.114
Evaluation index	L	I	C	
Weight	0.057	0.171	0.152	

3.4 Scores of index factors

Different index factors should be quantified uniformly into scores from 1 to 10 in the model calculation due to their different properties and dimensions during the evaluation of antifouling performance of groundwater. Their scores reflect the influence of index factors on the vulnerability of groundwater. Generally, 1 indicates the smallest influence, and 10 the biggest influence. Table 4 shows the specific scores.

Table 4 Category and score of each index factor in DRASTIC model

D		R		A		S		L		I		C	
Scope (m)	Score	Scope (m)	Score	Scope (m)	Score	Scope (m)	Score	Scope	Score	Scope	Score	Scope	Score
> 30	1	0	1	Clay	1	Non-swellable and non-condensing clay	1	Wasteland	2	Mainly clay	1	0	1
30—22.5	2	51	2	Silty clay	2	Rubbish	2	Lake and reservoir	3	Silty clay	2	4.1	2
22.5—15	3	71.4	3	Silt	3	Clayey subclay	3	Residential site	6, 7	Silt	3	12.2	3
15—9	5	91.8	4	Silt	4	Silty gravel subclay	4	Farmland (excluding (paddy field)	7	Silt	4	20.3	4
9—4.5	7	117.2	5	Silty fine sand	5	Subclay	5	Farmland (excluding (paddy field)	8	Silty fine sand	5	28.5	5
4.5—1.5	9	147.6	6	Fine sand	6	Gravelly subclay	6	Land for mining and industry	9	Fine sand	6	34.6	6
1.5—0	10	178	7	Medium sand	7	Swellable or condensing clay	7	Paddy field	10	Medium sand	7	40.7	7
		216	8	Coarse sand	8	Peat	8			Coarse sand	8	61.1	8
		235	9	Sand gravel	9	Sand	9			Sand gravel	9	71.55	9
		254	10	Gravel cobble	10	Gravel	10			Gravel cobble	10	81.5	10

Note: In addition to the pH, the units of the concentration of the rest chemical mass parameters are mg/L.

4 SUMMARY

MAPGIS software was used to process all evaluation factors in the groundwater vulnerability assessment. The attribute library management function and spatial data analysis function were mainly used as analysis tools for the evaluation of groundwater vulnerability in Horqin Region, Tongliao, which simplified the analysis process and improved the calculation accuracy .

Attribute fields were added to the grading chart of evaluation factors. On this basis, the region-to-region combined analysis function in the MAPGIS spatial analysis module was used for regional spatial superposition operation of evaluation factors. Then, the result was substituted into the new evaluation model for evaluation.

Evaluation results

Due to the change of factors in the groundwater vulnerability evaluation, the DRASTIC model was modified to build a new evaluation model (Formula 4):

DI=DrDw+RrRw+ArAw+SrSw+LrLw+IrIw+CrCw= DRASLIC······ Formula 4

The evaluation results were obtained according to the new model. A result chart of groundwater vulnerability evaluation was drawn as below according to the levels in Table 5.

Table 5 Correspondence of level, vulnerability and antifouling performance conclusions

Level	1	2	3	4	5	6	7	8	9	10
Vulnerability description	Extremely low	Very low	Relatively low	A little low	Slightly low	Slightly high	A little high	Relatively high	Very high	Extremely high
Antifouling performance description	Extremely high	Very high	Relatively high	A little high	Slightly high	Slightly low	A little low	Relatively low	Very low	Extremely high
Degree of pollution	Extremely difficult to pollute	Very difficult to pollute	Relatively difficult to pollute	A little difficult to pollute	Slightly difficult to pollute	Relatively easy to pollute	A little easy to pollute	Relatively easy to pollute	Very easy to pollute	Extremely easy to pollute

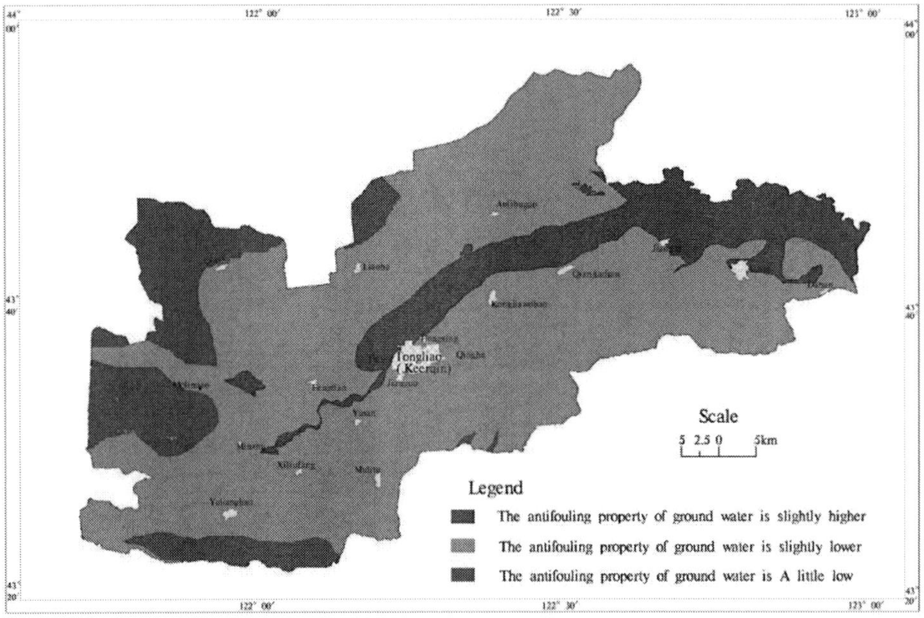

Figure 1 Map of groundwater vulnerability assessment results

As shown in Figure 1, the study area can be divided into 3 levels according to the vulnerability assessment results of groundwater in the study area and the correspondence shown in Table 5: areas with slightly low vulnerability (slightly difficult to pollute; the evaluation result is 4-5), areas with slightly high vulnerability (slightly easy to pollute; the evaluation result is 5-6) and areas with a little high vulnerability (a little easy to pollute; the evaluation result is 6-7).

Areas with a little low antifouling performance of groundwater (i.e. a little high groundwater vulnerability) are mainly distributed in Qinghe Town, Horqin Region, the west area of Molimiao Sumu, Taipingzhuang, Yuliangbao Town - Tangjia Wopu near Kezuohou Banner, East Huadeng Village and West Huadeng Village, Mulitu Town as well as the north of Shuangpaozi Village, Liaohe Town near Kezuozhong Banner. The buried depth of groundwater in these areas is basically less than 4m. That in the Molimiao Reservoir area is less than 3 m. Such areas account for 18.25% of the study area. The depth to

https://doi.org/10.1051/e3sconf/202129002011

groundwater table of the areas is the minimum. In terms of the lithology, the vadose zone is featured with medium sand, medium fine sand or fine sand and small thickness. The surface soil is mainly sandy with poor natural protection capability. The land is mainly dry farmland. Surface water and some pollutants have strong vertical permeability. The phreatic aquifer has high permeability and rich water. Contaminants are easily diffused and groundwater is relatively easy to be contaminated in the working area.

Areas with slightly high groundwater antifouling performance (i.e. slightly low groundwater vulnerability) are mainly distributed in the main stream valley plain area of West Liaohe River in Horqin Region, Dahan Town-Dalin Town-Jiaogan Town and the east area of Molimiao Town. The buried depth of groundwater in these areas is basically greater than 7 m. Such areas account for 16.53% of the study area. The depth to groundwater table of the areas is the maximum. In terms of the lithology, the surface soil is featured with silty silt, and the vadose zone is featured by shaly sand or silty fine sand, with large thickness. The phreatic aquifer has the minimum permeability and good natural protection capacity. The land is mainly rivers, wastelands and some underdeveloped urban land. The surface water and some pollutants have weak vertical permeability, and it is not easy for contaminants to diffuse. The groundwater is relatively difficult to be contaminated in the working area.

Areas with slightly low groundwater antifouling performance (i.e. slightly high groundwater vulnerability) are other areas except the above two types of areas, which are widely distributed in Horqin Region, Tongliao. Such areas account for 65.22% of the study area. The depth to groundwater table of the areas is large, 5-7 m. In terms of the lithology, the vadose zone is featured with fine sand or shaly medium and fine sand. There has been a small quantity of shaly particles in the vadose zone, and the rainfall infiltration coefficient also begins to decrease. The surface soil is mainly shaly silt, and some areas are of sandy lithology. The possibility for groundwater contamination in the working area is between that of the two areas above.

5 CONCLUSION

The impact of terrain was removed and land use status was added to modify the DRASTIC model into a new model DRASLIC based on Chinese and overseas literature and actual conditions in the study area. AHP was applied to figure out new weights. The evaluation result is that the groundwater vulnerability in the study area can be divided into three grades. The subdivision of groundwater antifouling performance obtained is consistent with the regional environmental hydrogeological conditions.

The results obtained by the combination of the antifouling performance index system with GIS technology provide a basis for the decision-making on groundwater management and protection in Horqin Region, Tongliao and can also be extensively applied to the planning and design of a groundwater monitoring network, the site selection of waste landfills, the evaluation and control of groundwater pollution, the selection and protection of water sources, etc.

References

1. Zhang Xuegang and Mao Yuanyuan, et al. (2009) Assessment of Groundwater Vulnerability and Pollution Risk in Zhangji . Hydrogeology & Engineering Geology, 1: 51~55.
2. Zhao Shubai. (1988) Practical Decision-making Method - Analytic Hierarchy Process. Tianjin: Tianjin University Press.
3. Li Yong and Feng Yong. (2003) Determination of Objective Weight in Multi-Objective Decision . Journal of Gansu University of Technology, 29(3):118-119.
4. Song Guangxing and Zou Ping. (2001)Method of Determining the Weight of the Decision-maker in Multi-Attribute Group Decision-making . Systems Engineering, 19(4):84-89.
5. Zhang Haitao, Liu Chaoying and Tian Shui. (2004)Subjective and Objective Combination Method of Weight Determination . Journal of Jianghan University (Natural Science), 32(4):63-65.
6. Zhu Yin, Meng Zhiyong and Shan Shuyu. (1999)Weight Calculation by AHP . Journal of Beijing Jiaotong University, 23(5):119-122.
7. Wang Jing and Zhang Jinsuo. (2001)Comparison of Several Methods for Determining Weight Vector in Comprehensive Evaluation. Journal of Hebei University of Technology, 30(2): 52-57.
8. Wang Wei and Zhang Yuzhi. (2006,)Study on Weight Vector Solving Method of Fuzzy AHP . Control and Decision, 21(2):184-188.
9. Zhu Jianjun. (2005)Research on Some Problems of Analytic Hierarchy Process and Its Application . Doctoral thesis of Northeastern University.
10. Li Song, Qiu Wei, Zhao Qingliang and Liu Zhengmao. (2006)Application of Analytical Hierarchy Process to Assessment of Eco-Environment Quality in Heilongjiang Province . Environmental Science, 27(5):1031-1034.
11. Zhu Xiaohua and Yang Xiuchun. (2001)Study on Application of Analytic Hierarchy Process in Regional Ecological Quality Assessment . Scientific and Technological Management of Land and Resources, (5): 45-45.
12. Hu Anyan, Guo Sheng Lian, Liu Yan and Lin Kairong. (2006)Fuzzy Comprehensive Evaluation of Ecological Environment Quality in Inland River Basin in Arid Areas . Journal of Geological Hazards and Environment Preservation, 17(2): 69-73.
13. Yang Shu, Chu Yu, Yang Xiangkui and Lou Benjun. (2005)Application of Analytic Hierarchy Process (AHP) in the Evaluation of Geo-Environmental Quality in the Sanjiang Plain . Geological Bulletin of China, 24(5):485-490.
14. Cui Jian, Li Xiao and Du Jizhong, et al. (2011)GIS-based Assessment of Antifouling Performance of Shallow Groundwater in Alluvial Fan of Hunhe River . Geology and Resources, 20(2): 137-140.

Sedimentary Facies of the Chang 2 Member of Yanchang Formation in Area A of Northern Shaanxi

Yingzhi Liu[1], Yanyan Liu[2], Ying Chen[1], Shasha Yang[3,*] and Bowang Wang[3]

[1]Zhidan Oil Production Plant, Shaanxi Yanchang Petroleum (Group) Co. Ltd., Yan'an, Shaanxi, 716000, China
[2]Changqing Downhole Technology Operation Company, CNPC Chuanqing Drilling Engineering Co.Ltd., Xi'an, Shaanxi , 710018, China
[3]Shaanxi Key Laboratory of Safety and Durability of Concrete Structures, XiJing University, Xi'an, Shaanxi, 710123, China

Abstract. In order to promote the rolling exploration and development of oil reservoirs in northern Shaanxi, a fine study of sedimentary facies in area A has been carried out. Using core and logging data, on the basis of identifying different levels of sequence interfaces and lake flooding, the Chang 2 Member in Area A was divided into two base level cycles, and a high-resolution sequence stratigraphic correlation was established. On the basis of core description, the analysis of single-well sedimentary facies, continuous-well sedimentary facies and planar sedimentary facies are carried out. It is believed that the Chang 2 member of the study area is a delta plain deposit, dominated by distributary channels, inter-distributary depressions, and natural dike microfacies. There are two to three main rivers in the study area. Both the vertical and the plane show that the length of Chang 2_2 sand bodies is more developed than the length of Chang 2_1 sand bodies. Among the three small layers of Chang 2_2, Chang 2_2^1 sand body is relatively well developed, and the channel width and thickness are large, making it the most favorable reservoir.

1 Introduction

The Ordos Basin is the second largest basin in China. The Triassic Yanchang Formation in the basin made a major breakthrough in the early stage of exploitation, and has been the main target layer for oil and gas exploration ever since. According to sedimentary cycles, paleontology, lithology and other characteristics, the Yanchang Formation is divided into 10 oil-bearing groups from top to bottom, of which Chang 2 is the main oil-bearing series of the basin[2-4]. By the Chang 2 stage, the plain facies or plain distributary channels developed, the lake basin had shrunk, and the sedimentation was strengthened to form a set of sand-rich sediments[5,6]. Area A is located in the middle-eastern part of the northern Shaanxi slope of the Ordos Basin. The Chang 2 oil layer group is the main oil-bearing series of the northern Shaanxi slope, but the development effect in recent years has not been satisfactory. Therefore, it is necessary to study in detail the sedimentary characteristics of the Chang 2 reservoir and its sand body distribution characteristics [7].

Area A is located in the northern Shaanxi area of the Ordos Basin. The landform of this area is classified as the landform of the northern Shaanxi Loess Plateau. The flat top is wide, and the surrounding valleys are cut. The ground elevation is about 1.1~1.35 km, and the height difference is about 150~250 m. The regional tectonic unit in the study area belongs to the middle-eastern part of the northern Shaanxi slope of the Ordos sedimentary basin. The general structural pattern of the Chang 2 oil layer group in the eastern part of the northern Shaanxi slope is a large, gentle monoclinic with gentle tectonic activity and a small dip angle. As a result of compaction, a series of low-amplitude nose-like uplifts that dip from east to west develop on the slope belt [8].

2 The distribution of sedimentary microfacies and sand bodies

2.1 Types of sedimentary facies in the study area

Sedimentary facies is a combination of the sedimentary environment and the characteristics of the sedimentary rocks formed in the environment. Since the depositional conditions are the material basis for the formation of reservoirs, the formation of a good reservoir must have favorable sedimentary facies belts. Sedimentary facies not only determine the production status of the reservoir, including the horizon, reservoir cap conditions, reservoir scale and distribution form, but also determine the internal structure and rock type of the reservoir, thereby controlling the distribution of oil and gas.

Area A is located in the middle-south section of the Yishan Slope, which is the broadest in the Ordos Basin, and is one of the main areas for oil exploration and development in the basin. According to previous research results, with reference to the regional sedimentary background, combined with actual data, a comprehensive analysis of the Chang 2 oil layer group of the Yanchang Formation in the A area is a delta plain sedimentary system.

* Corresponding author: 20190100@xijing.edu.cn

The delta plain sedimentary environment and sedimentary characteristics have many similarities with fluvial facies. Due to the flat terrain of the delta plain area, a large number of bifurcated river channels, small river energy, shallow water depth, and relatively large width of the distributing channel, the grain size of the sediments is small and the separation is good. The lithology is mainly sandstone and silt and mudstone. At the same time, due to the flat and humid terrain in the plain area, the vegetation is relatively developed. There are many plant rhizomes and plant fragments in the sediments, and sometimes thin coal seams can be seen. The delta plain is composed of distributary channels, depressions between distributaries, and natural dikes. The distributary channels and channels constitute the main body of the study area. The division of the microfacies is shown in Table 1.

Table 1. Sedimentary facies division table of Chang 2 in area A

Sedimentary facies	Sedimentary subfacies	Sedimentary microfacies
Delta	Delta plain	Diversion channel
		Depressions between diversions
		Natural embankment

2.2 Sedimentary microfacies in the study area

2.2.1 Diversion channel

The distributary channel deposition is the skeleton part of the delta plain facies, which has the characteristics of general channel deposition. It is mainly fine-grained feldspar sandstone, with a small amount of siltstone and argillaceous siltstone, sorted, etc. On the plane, it is pointed and belt , strip-like distribution, lenticular in the section, with many scour surfaces at the bottom, containing mud and gravel. Sedimentary structures can be seen in various cross bedding such as groove cross bedding, plate cross bedding, parallel bedding, etc., On the vertical the thinning sedimentary sequence shows a positive rhythm.

2.2.2 Depression between diversions

Depression between diversions refer to the storage depressions behind the banks of distributary channels and between adjacent distributary channels. In low-lying water-rich places, lush vegetation can form marsh deposits, which are the limited environmental deposits between distributary channels. The depressions and swamps between the diversions are the "body" of the delta plain subfacies. The microfacies of the inter-distributary depressions are mainly formed by the cross-bank deposition of fine suspended matter carried by the river. It is dominated by dark gray and brown gray mudstone with a small amount of fine sand and silt deposits. Horizontal bedding is seen in the mudstone, and plant fossil fragments are often preserved. The natural gamma curve has a high-value micro-tooth shape.

2.2.3 Natural embankment

Natural embankments are developed on both sides of the distributary channel. When the flood overflows out of the riverbed during the flood, the suspended silt and fine sand are deposited on both banks of the channel. The main lithology is fine siltstone interbedded with mudstone or interbedded with thin mudstone. The grain size becomes finer as the distance away from the river channel, and the mudstone increases. Wave-like bedding, sand-line bedding and climbing laminae, horizontal bedding are common, and plant roots, stems and plant fragments can be seen. The natural gamma curve is generally a low-amplitude toothed clock Shape, box shape.

2.3 Single-well facies analysis in the study area

Based on the classification of sedimentary facies and the characteristics of sedimentary microfacies, in order to reveal the law of vertical evolution and planar distribution of sedimentary facies in the Chang 2 oil layer group in the study area, a single-well sedimentary facies analysis was carried out on two wells in the study area.

(1) B95 single well facies analysis

B95 is located in the northwest of the study area. The distributary channels, interdistributary depressions and natural dike microfacies frequently overlap each other in the bottom-up delta plain. Chang 2_2^3 mainly develops distributary channel microfacies, Chang 2_2^2 develops interchannel microfacies, and Chang 2_2^1 sand The thickest body, the curves are mostly box-shaped, and distributive channel microfacies are also developed. In comparison, the long 2_1^3 and long 2_1^2 curves are mostly positive anomalies, and the microfacies between the channels are relatively developed, and the lower part of the long 2_1^1 develops a distributive channel , The microfacies between the river channels are developed in the upper part.

(2) D116 single well facies analysis

D116 is located in the middle of the study area. The distributary channels, interdistributary depressions and natural dikes in the delta plain from the bottom to the upper are micro-overlapping and inter-channel microfacies. The lower part of Chang 2_2^3 develops distributary channel microfacies, and the upper part of Chang 2_2^3 to the middle part of Chang 2_2^2 develops interchannel microfacies. From the middle of Chang 2_2^2 to the top of Chang 2_2^1, the curves are mostly box-shaped, and distributary channel microfacies are also developed. In comparison, the long 2_1^3 curves are mostly positive anomalies, and the microfacies between the channels are relatively developed, ranging from 2_1^2 to long 2_1^1. The development of distributary channel microfacies.

2.4 Analysis of the connected profile of the sand body in the study area

Profile facies analysis is based on the analysis of single well sedimentary facies, making full use of electrical logging data for comparison, establishing the phase sequence relationship between adjacent wells, and determining the distribution characteristics of sedimentary facies in two-dimensional space.

Combining the structural location and geological evolution background of the study area, based on a large number of exploration and development practices, two continuous well profiles that can reflect the sedimentary facies of the area have been established. The two sections show that the Chang 2 formation is not undulating. The sand bodies of Chang 2_1^1 and Chang 2_2^1 are the best developed, with a thickness of more than 10 m, and the sand bodies have good connectivity. Chang 2_1^1 develops a set of thicker sand bodies, while Chang 2_1^1 and Chang 2_2^1 are several sets of sand bodies superimposed on each other. The Chang 2_2^3 sand body is well developed, with an average thickness of about 6 m, and the sand body has good connectivity. Only small lenticular sand bodies develop in other small layers.

2.5 Sedimentary microfacies and distribution characteristics of sand bodies

The study of the planar characteristics of sedimentary microfacies is an important geological basis for reservoir development and analysis. It is a key step in the study of reservoir heterogeneity and remaining oil distribution. It is also the main purpose of microfacies research. It is restricted by a series of factors, including Factors such as paleo-climatic conditions, source area direction and paleo-topographic features, as well as the development status and energy of river systems. On the basis of determining sedimentary facies, subfacies, and microfacies, following the point-line-plane research method, starting from the single-well facies diagram analysis, the well facies diagram analysis is connected again, and finally the sedimentary microfacies of the target layer in the study area is carried out. Accurate analysis and summary of phase plane characteristics.

During the Chang 2 deposition period, the sediments in the study area mainly came from the northeast, and the delta plain distributary channel was dominant, and two large sedimentary cycles developed, namely Chang 2_2 and Chang 2_1, of which the distributary channel of Chang 2_2 was more developed, and the sand bodies were distributed close to each other. Northeast-southwest direction, and the distributary channel of Chang 2_1 is not as well developed as that of Chang 2_2. Three secondary cycles developed in the Chang 2_2 period, namely Chang 2_2^3, Chang 2_2^2, and Chang 2_2^1. The role of the distributing channel from the lower sedimentation is strong, and it gradually changes upwards. It is mainly composed of inter-channel sedimentation, and then develops into distributive channel upwards. During the Chang 2_1 period, three secondary cycles were also developed, namely Chang 2_1^3, Chang 2_1^2 and Chang 2_1^1, which were deposited between the lower channels and gradually became dominated by distributary channel deposits upward. By drawing the sedimentary microfacies plane distribution map and sand body thickness contour map of each small layer of the Chang 2 oil layer group in this area, some understanding of the sedimentary microfacies and sand body distribution characteristics of each small layer of the Chang 2 oil layer group in this area can be obtained. Two representative sub-levels are selected below.

(1)Distribution characteristics of Chang 2_2^1 sedimentary microfacies and sand bodies

The Chang 2_2^1 period also continued the development of the lower strata. Three main channels developed in the study area, extending from northeast to southwest. In plane, these three diversion channels flow in from the northeast direction, and flow out from the southwest direction and the south direction. Among them, the channel in the middle of the study area is the best developed, with a channel width of up to 550m, the channel in the northwest of the study area is well developed, with a width of about 510m, and the channel in the southeast of the study area is the worst developed, with a width of about 240m, at K36 There are two distributary channels developed at the well, and the thickness of the sand bodies of the three main channels in this small layer are all above 15m .

(2)Distribution characteristics of Chang 2_1^1 sedimentary microfacies and sand bodies

In the Chang 2_1^1 period, delta plain deposits were still developed, and three main river channels were developed in the study area, extending from northeast to southwest. In plane, these three diversion channels flow in from the northeast direction, and flow out from the southwest direction and the south direction. Among them, the channel in the middle of the study area is the best developed, with a channel width of up to 740m, the channel in the southeast of the study area is well developed, with a width of about 700m, and the channel in the northwest of the study area is the worst developed, with a width of about 510m. The thickness of the sand bodies of the three main channels of the layer is more than 10m.

3 summary

1. The Chang 2 stratum in the study area is mainly composed of delta plain deposits controlled by provenance in the Northeast, with distributary channels, inter-distributary depressions and natural dike microfacies. Each sub-layer develops two to three main channels in the study area.

2. Vertically, the two sedimentary cycles of Chang 2_2 and Chang 2_1 all started with the development of distributary channels. Among them, the sand body of Chang 2_2 is more developed than that of Chang 2_1. Among the three small layers of Chang 2_2, the Chang 2_2^1 sand body is the best developed. The sand body of the channel is more than 15 m thick, and the Chang 2_2^3 sand body is well developed. The sand body of the channel is more than 10 m thick.Chang 2_2^2 sand body is the worst developed. Among the three small layers of Chang 2_1, the Chang 2_1^1 sand body is the best developed, and the sand body thickness of the channel is more than 10 m. In comparison, the Chang 2_1^3 and Chang 2_1^2 sand bodies are poorly developed, and the sand body thickness of the channel is greater than 6m.

3. On the plane, the sand body of Chang 2_2 is more developed than Chang 2_1, and the channel width is larger.

In the three small layers of Chang 2_2, Chang 2_2^1 sand bodies are relatively well developed, with a channel width of 550 m, and Chang 2_2^3 and Chang 2_2^2 sand bodies are relatively poorly developed, with a channel width of about 500 m. The Chang 2_1^1 sand body is relatively best developed among the three small layers of Chang 2_1, with a channel width of 700 m, the sand body of Chang 2_1^2 is well developed, and the channel width is about 500 m. The sand body of Chang 2_1^3 is the worst developed, and the channel width is about 450m.

References

1. Yang Junjie. Tectonic evolution and oil and gas distribution in the Ordos Basin[M]. Beijing: Petroleum Industry Press. 2000.
2. Wang Guicheng, Ma Weimin, Zhao Hong, Yang Renchao. Sedimentary characteristics of the Triassic Yanchang Formation in Fuxian Exploration Area, Ordos Basin[J]. Journal of Northwest University (Natural Science Edition), 2003. (5): 608-612.
3. Fu Jinhua, Guo Zhengquan, Deng Xiuqin. Upper Triassic Yanchang Formation sedimentary facies and petroleum geological significance in southwestern Ordos Basin[J]. Journal of Palaeogeography, 2005. (1): 34-44.
4. Yang Hua, Liu Xianyang, Zhang Caili, Han Tianyou, Hui Xiao. The main controlling factors and distribution of low-permeability lithologic reservoirs in the Triassic Yanchang Formation in the Ordos Basin[J]. Lithologic Reservoirs, 2007. (3) :1-6.
5. Wang Jianmin, Wu Changrong. The sedimentary characteristics of remote sandy braided rivers in Chang 2+3 oil-bearing group in the eastern part of northern Shaanxi [J]. Minerals and Rocks, 2007. (4): 92-97.
6. Li Wenhou, Pang Jungang, Cao Hongxia, Xiao Li, Wang Ruogu. The Late Triassic Sedimentary System and Lithofacies Paleogeography Evolution in the Late Triassic of Ordos Basin[J]. Journal of Northwest University (Natural Science Edition), 2009.39(3):501 -506.
7. Guo Chunen, Zuo Yi, Yu Cundong. Application of Sedimentary Microfacies Research Technology in Oilfield Development[J]. Tuha Oil & Gas, 2006. (3): 230-232, 243
8. Zhao Jingzhou, Yang Xianchao, Wu Fuli, Shi Baohong. On the control of uplift background on the formation and distribution of Triassic reservoirs in the northern Shaanxi slope area of the Ordos Basin[J]. Acta Geology, 2006. (5): 648 -655.

Numerical Analysis of Wave Characteristic In the Freak Wave-- "New Year Wave"Formation

Yu Xiang-jun[1,*], Li Qing-hong[1], and Li Mao-lin[1]

[1]Department of Military Oceanography and Hydrography, Dalian Navy Academy, Dalian, Liaoning, 116018, China

Abstract. Freak waves are both extremely large waves and highly transient time. Such a wave may lead to damage of ships to deaths. In this paper ,to describe the connection between freak wave and wave essential factor, we use WAVEWATCH III model simulating "New Year Wave" in the North Sea to explore freak wave, with the importing of ECMWF re-analysis wind field. By this way, we successfully simulate the formation of freak wave in the random wave.Analysis shows large wave steepness and small directional spread angle are necessary conditions for freak waves to easily occur.By analyzing the wave spectrum, it is found that the wave energy is distributed in a small range, and the propagation direction is relatively concentrated.

1 Introduction

Freak wave is considered to be wave with unusually large wave heights that suddenly appear on the sea, accompanied by deep troughs in front of or behind huge wave crests. For centuries, freak waves have existed in folklore. Seafarers called them "walls of water" or "deep sea caves", but the marine community generally doubted whether they really existed until the 1995 "New Year Wave" on the Daupner platform in the North Sea was observed. Only then did people begin to generally accept and pay attention to freak wave. The European Union initiated a freak wave observation project called "MaxWave" in 2000. The project analyzed the 10-meter-accuracy global satellite data for three consecutive weeks in 2001, and found more than a dozen freak waves, all of which have wave heights greater than 25 meters. Since then, people's understanding of deformed waves has reached a substantial stage. With the enrichment of observational data, the research on the mechanism of freak waves has also made substantial progress.

This paper selects a typical freak wave event--the "New Year Wave" event, and uses the WAVEWATCH III model to numerically simulate the wave field when the event occurs. By analyzing the wave height, wave steepness, directional spread, spectral peak frequency, average wavelength and time change law of the freak wave occurrence point, the relationship between the freak wave and these statistical factors is explored to verify the theoretical results. Analyze the one-dimensional wave spectrum and the two-dimensional wave spectrum that reflect the inherent characteristics of the waves, and study the relationship between them and the freak waves.

2 "New Year Wave"

On January 1, 1995,Freak wave hit the Draupner oil platform in the North Sea with a depth of 70m, causing some damage to the equipment. The position of the oil platform is 2o28'E, 58o11'N (see Figure 1), and the occurrence time is approximately 15:20 UTC. The wave meter on the platform recorded a complete sequence of freak waves. This wave is known as the "New Year Wave" and is currently the most complete recorded deformed wave. At that time, the effective wave height of the sea surface was about 11.9m, the observed abnormal wave height was 25.6m, the height of the wave crest was 18.4m, and the height of the wave trough was 7.2m. The weather situation at that time was that southern Sweden was mainly controlled by a low pressure, forming a strong northerly wind covering the entire North Sea and Norwegian waters; in addition, there was a relatively weak low pressure moving south in the North Sea, which strengthened in the western part of the North Sea. Wind field. As the weak low pressure moves to the south, the strongest wind area (see Figure 2) moves to the south. At 1200 UTC, the wind reaches its extreme value and reaches the wind level of a hurricane.

* Corresponding author: dl_yuxj@163.com

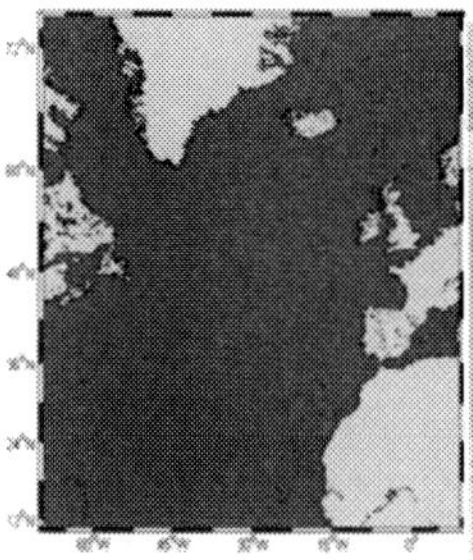

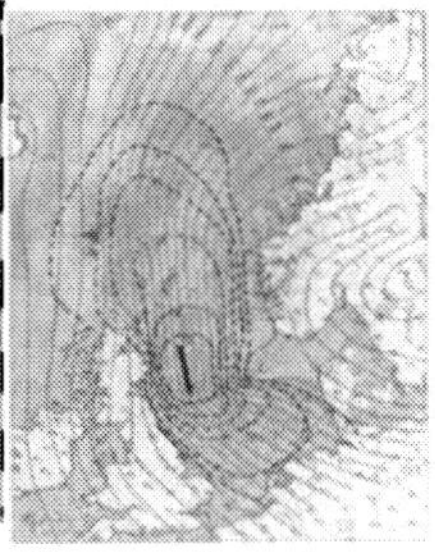

Pic1. the sea area where " New Year wave" event occurred
Pic2. Weather chart at 12:00 on January 1, 1995

3 WAVEWATCH III wave model

3.1 Governing equation

In the WAVEWATCH III wave model[6], the evolution of the action density is governed by the action balance equation. The wave action spectrum N(ω,θ) is the ratio of the energy spectral density E(ω,θ) and the relative frequency ω.In the Cartesian coordinates, the conservation equation for wave action can be expressed as

$$\frac{\partial N}{\partial t} + \nabla_x \cdot \dot{\vec{x}} N + \frac{\partial}{\partial k} \dot{k} N + \frac{\partial}{\partial \theta} \dot{\theta} N = \frac{S}{\sigma} \quad (1)$$

$$\dot{\vec{x}} = \vec{c}_g + \vec{U} \quad (2)$$

$$\dot{k} = -\frac{\partial \sigma}{\partial d}\frac{\partial d}{\partial s} - \vec{k} \cdot \frac{\partial \vec{U}}{\partial s} \quad (3)$$

$$\dot{\theta} = -\frac{1}{k}\left(\frac{\partial \sigma}{\partial d}\frac{\partial d}{\partial m} - \vec{k} \cdot \frac{\partial \vec{U}}{\partial m}\right) \quad (4)$$

where cg is given by cg and θ, s is a coordinate in the direction θand m is a coordinate erpendicular to s.

3.2 Source terms

In deep water, the net source term S is generally considered to consist of three parts, a wind-wave interaction term Sin, a nonlinear wave-wave interactions term Snl and a dissipation (`whitecapping') term Sds. The input term Sin is dominated by the exponential growth term, and this source term generally describes this dominant process only. For model initialization, and to provide more realistic initial wave growth, and linear input term Sln can also be considered in WAVEWATCH III.

In shallow water additional processes have to be considered, most notably wave-bottom interactions Sbot (e.g., Shemdin et al., 1978). In extremely shallow water, depth-induced breaking (Sdb) and triad wave-wave interactions (Str) become important. Also available in WAVEWATCH III TM are source terms for scattering of waves by bottom features (Ssc) and a general purpose slot for additional, user defined source terms

This defines the general source terms used in WAVEWATCH III as

$$S = S_{\ln} + S_{in} + S_{nl} + S_{ds} + S_{bot} + S_{db} + S_{sc} + S_{xx} \quad (5)$$

Other source terms are easily added. These source terms are defined for the energy spectra. In the model, however, most source terms are directly calculated for the action spectrum. The latter source terms are denoted as $S \equiv \frac{E}{\sigma}$.

3.3 parameter settings

In this paper, the WAVEWATCH III wave model is used for numerical simulation of this freak wave event. The calculation area is shown in Figure 3. The driving wind field is the ECMWF mid-term reanalysis wind field, the resolution of the wind field is, the terrain uses etop1 terrain data, and the source function parameter settings use WW3 (v6.07) default settings. The calculation period is from December 29, 1994 to January 4, 1995. The calculation area, time step and grid resolution are set as follows: 5°W-5°E, 55°-65°N; 900s; 900s; 900s; 300s; resolution 1/6 degree.

4 Result analysis

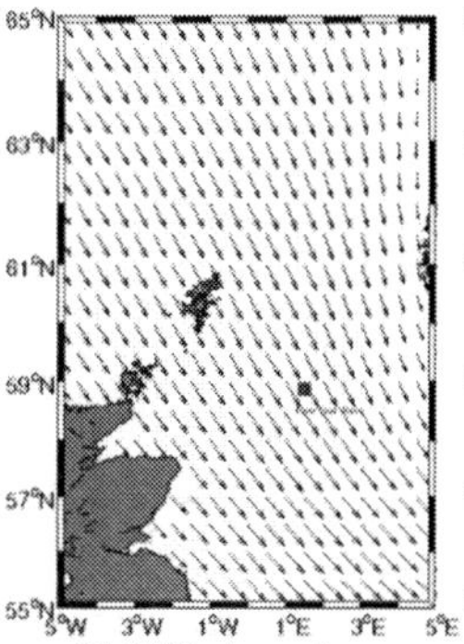

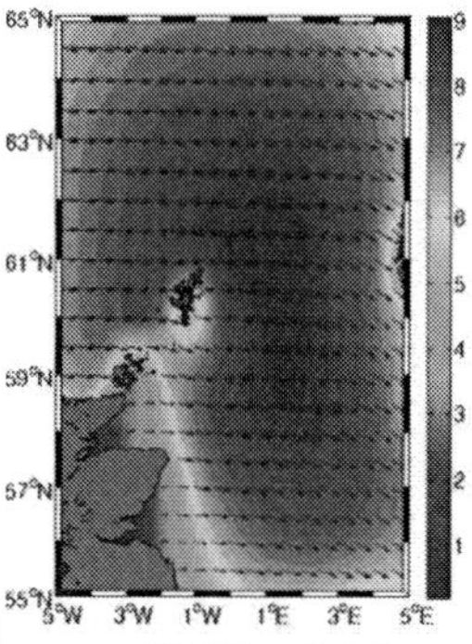

Pic3. Wind speed **Pic4.**Wave height

Figures 3-4 show the results of this simulation. Based on the simulation results, this paper conducts a statistical analysis of the element characteristics of the New Year wave and its occurrence background.

Figure 3 is a numerical simulation of the wind speed field diagram at the time of the New Year wave event. The wind speed at the event point was about 20m/s, and a large area around it was in a strong northerly gale state, which was in line with the meteorological conditions under the control of a low pressure to form a strong northerly gale covering the entire North Sea and the Norwegian Sea. Figure 4 shows the distribution of the effective wave height and wave direction at the moment when the freak wave occurs. The figure shows that the wave height of the New Year wave event point is about 8.5m, which is somewhat different from the measured freak wave height of 11.9m. The reason for this gap may be The input reanalysis wind field data is smaller than the actual measured data, but it has little influence on the general trend of the wave field. It can be seen from the figure that the wave heights in most of the North Sea are above 8m.

https://doi.org/10.1051/e3sconf/202129002013

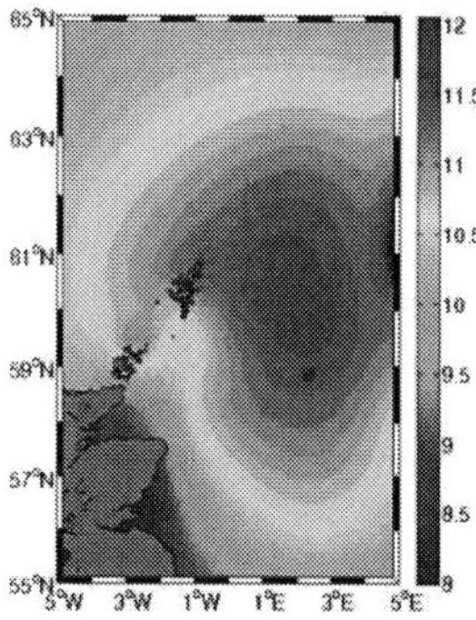

Pic5. Wave period

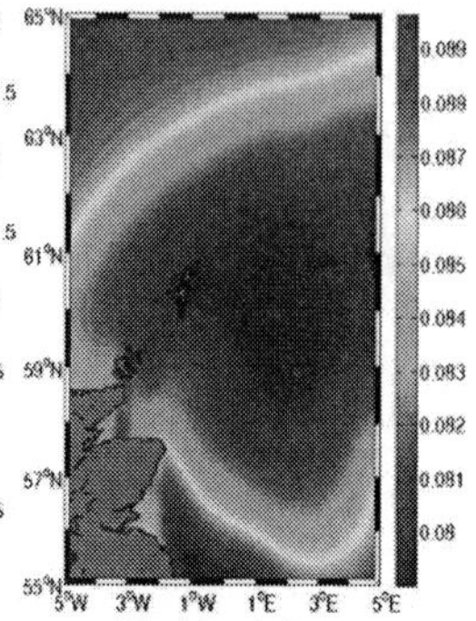

Pic6. Peak frequency

Figure 5 shows the wave period diagram when a freak wave occurs, with a longer wave period. Figure 6 shows the peak frequency diagram when the freak wave occurs, near the minimum value of the peak frequency.

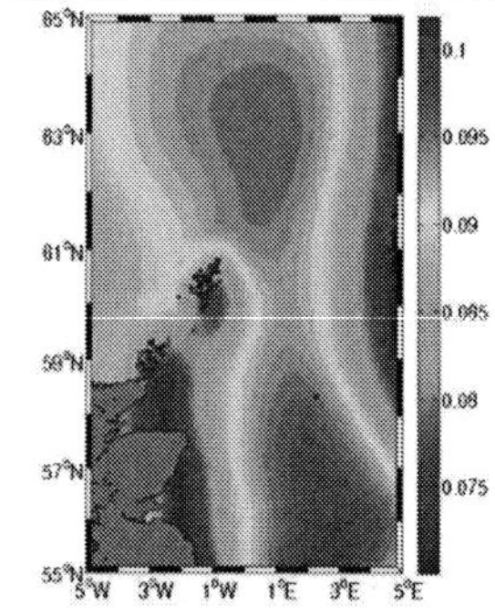

Pic7. Average wave steepness

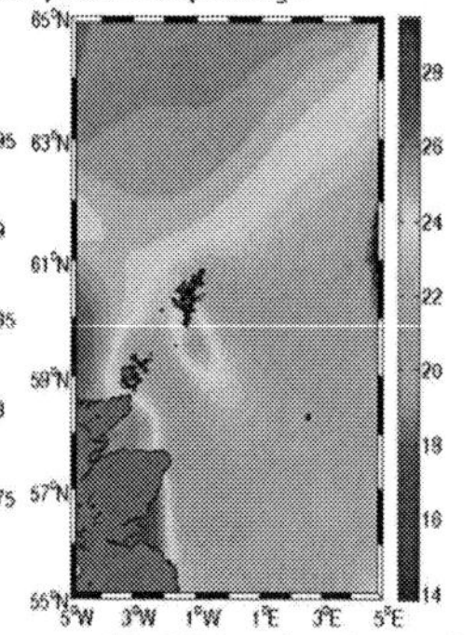

Pic8. Directional spread

Figure 7 shows the distribution of average wave steepness. The figure shows that the wave steepness at the event occurrence point is about 0.095, and the wave steepness is relatively large. Figure 8 is the distribution of the average wave direction spread angle, the distribution of the direction spread angle is in a small area in the southern sea area. The directional spread of the incident point is very small, about 20 degrees. From the spatial distribution map of wind speed, wave height, wave steepness and directional spread angle, we can know the moment when the freak wave occurs. The observation platform is in a relatively poor sea state, the effective wave height is about 8.5m, the wave steepness is very large, and the directional spread angle Very small. Both large wave steepness and small directional spread angle are necessary conditions for the occurrence of freak waves.

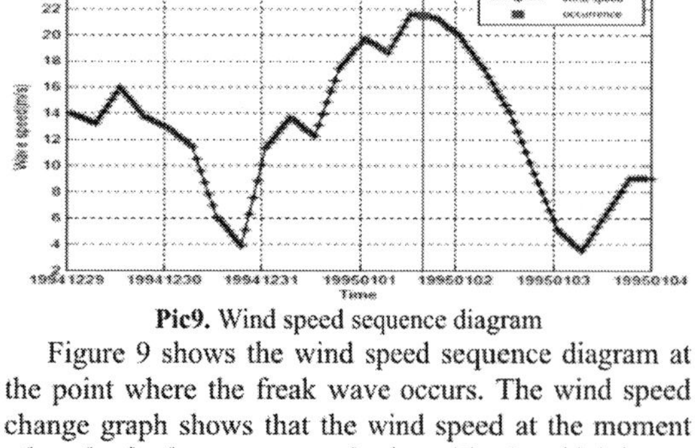

Pic9. Wind speed sequence diagram

Figure 9 shows the wind speed sequence diagram at the point where the freak wave occurs. The wind speed change graph shows that the wind speed at the moment when the freak wave occurs is about 20m/s, which is at a maximum value.

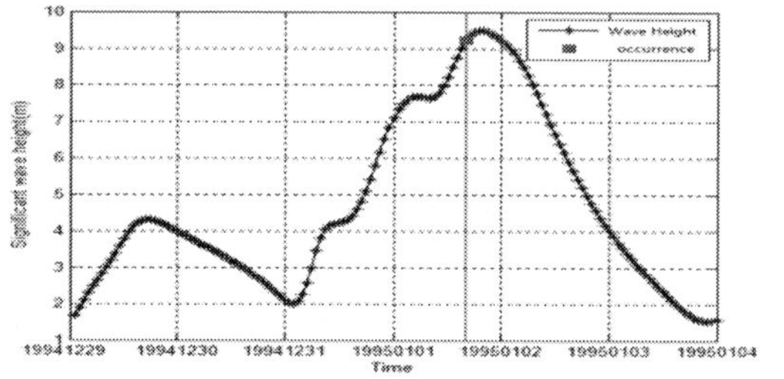

Pic10. Wave height sequence diagram

The wave height change graph in Figure 10 shows that when the event occurred, the wave height at the event point was about 8.5m. The wave height was in the rapid rise stage some time ago, and it was close to the maximum value of the wave height.

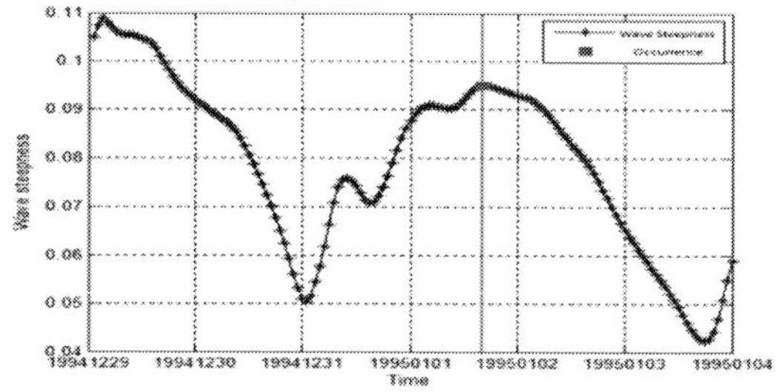

Pic11.wave steepness sequence diagram

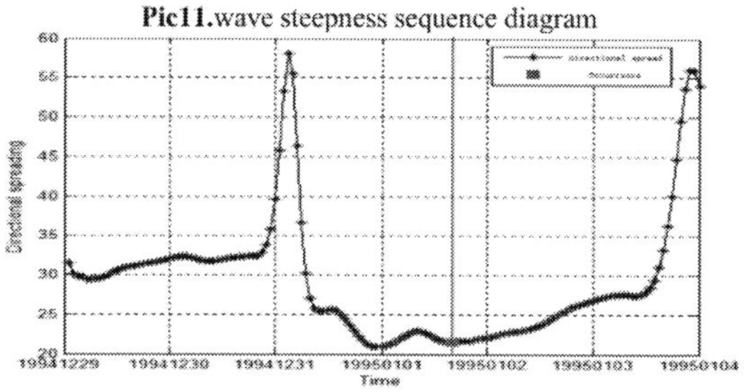

Pic12. Directional spread sequence diagram

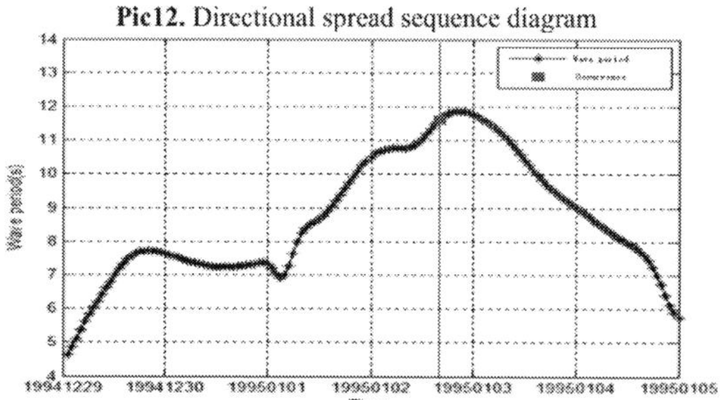

Pic13. Wave period sequence diagram

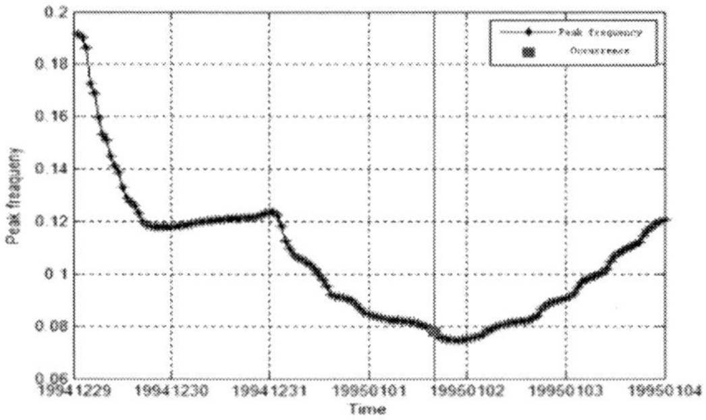

Pic14. Peak frequency sequence diagram

Figures 11-13 show that the change of wave steepness is similar to the change of wind speed and wave height, and it is also at a maximum value, which is about 0.095 at the time of the event. Figure 14 shows that the directional spread angle at the moment of the event is at a minimum, about 20 degrees.

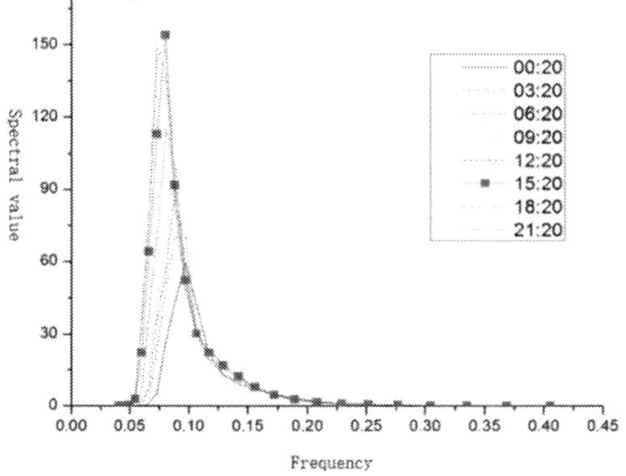

Pic15. one-dimensional ocean wave spectrum

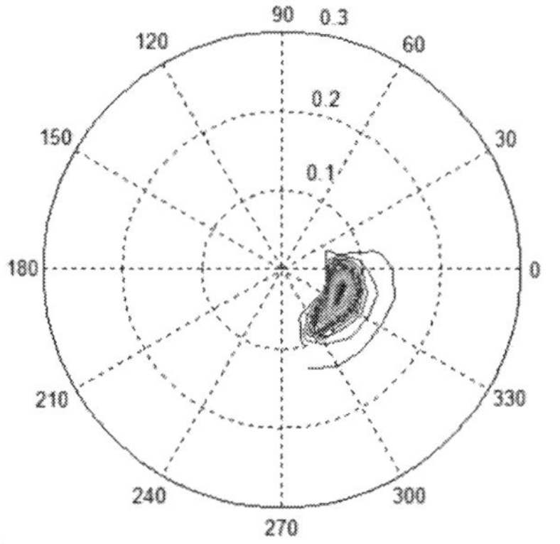

Pic16. two-dimensional ocean wave spectrum

Figure 15 is a time variation diagram of the one-dimensional wave spectrum. It can be seen that the one-dimensional wave spectrum has a sharp and narrow spectrum at the moment of the event. The wave spectrum is a single-peak spectrum, and the peak frequency is about 0.075. Most of the wave energy is concentrated in a small frequency range. Figure 16 is a spectrum pattern diagram of a two-dimensional ocean wave spectrum, showing that the energy of the ocean wave is concentrated in a small frequency band, and the propagation direction is relatively concentrated.

5 Conclusion

Through analysis of various statistical elements of sea waves, it is found that the observation platform is in a relatively bad sea condition at the moment when the freak wave occurs. The effective wave height is 8.918m, the wavelength is 205.9m, the wave period is 11.33s, the wave steepness is 0.095, and the directional spread angle is 21.97o. Through analysis, we can get the following conclusions:

1. When a freak wave appears, the wave height is very large. Under severe sea conditions, the wave steepness is large and the directional spread angle is small. Large wave steepness and small directional spread angle are necessary conditions for freak waves to easily occur.

2. By analyzing the one-dimensional wave spectrum, it is found that the spectrum is sharp and narrow. The wave spectrum is a single-peak spectrum, and most of the wave energy is concentrated in a small frequency range. By analyzing the two-dimensional wave spectrum, it is found that the wave energy is distributed in a small range, and the propagation direction is relatively concentrated.

References

1. Sverre Haver. A Possible Freak Wave Event Measured at the Draupner Jacket January 1 1995 [C]. Rogue waves 2004, Brest France, 2004:1-8
2. W. Rosenthal and S. Lehner. Rogue Waves: Results of the MaxWave Project [J]. Journal of Offshore Mechanics and Arctic Engineering.2008,130:67-74.
3. N. Mori, P.C. Liu, T. Yasuda. Analysis of freak wave measurements in the Sea of Japan[J]. Ocean Engrg. 2002,29 :1399–1414.
4. Didenkulova I., A. V. Slunyaev, E. N. Pelinovsky, and C. Kharif. Freak waves in 2005.Nat. Hazards Earth Syst. Sci. 2006,6: 1007-1015.
5. Lavrenov, I. V. The wave energy concentration at the Agulhas current of South Africa[J], Natural Hazards .1998,17: 117–127.
6. Tolman, H. L. User manual and system documentation of WAVEWATCH III version 3.14. NOAA / NWS / NCEP / MMAB Technical Note 276, 2009:194
7. Miles, J. W., 1957: On the generation of surface waves by shear ows. J.Fluid Mech., 3.
8. Janssen, P. A. E. M., 1982: Quasilinear approximation for the spectrum of wind-generated water waves. J. Fluid Mech., 117, 493-506.
9. Janssen, P. A. E. M., 1991: Quasi-linear theory of wind wave generation applied to wave forecasting. J. Phys. Oceanogr., 21, 1631-1642.
10. Ardhuin, F., B. Chapron and F. Collard, 2008: Ocean swell evolution from distant storms. Nature Geoscience
11. Abdalla, S. and J. R. Bidlot, 2002: Wind gustiness and air density effects and other key changes to wave model in CY25R1. Tech. Rep. Memomrandum R60.9/SA/0273, Research Department, ECMWF,

https://doi.org/10.1051/e3sconf/202129002013

Reading, U. K.

12. Liu, P. C., U. F. Pinho, 2004: Freak waves-More frequent than rare! Ann. Geophys. 22,1839-1842.

Hydrochemical Characteristics and Evolution Laws of Shallow Groundwater in Shuangliao City

Li Xuguang[1], Zhao YanDai[1,*], He Haiyang[1]

[1]Shenyang Center of Geological Survey, China Geological Survey, Shenyang,Liao Ning, China

Abstract. Shuangliao City is an important part of the Xiliaohe Plain, and one of the most important bases of grain production in the north of China. Therefore, it is important to ascertain the hydrochemical characteristics of groundwater and their causes and evolution laws in the Xiliaohe Plain to provide guidance to agriculture development and ecological improvement. After collection of detailed data and identification of the groundwater flow field, we studied the causes and evolution of the identified hydrochemical types by zone with mathematical statistics, correlation analysis, ion proportional coefficient and other methods. The results show that the concentrations of HCO3-, Cl-, and Na+ are relatively high, and these of Ca2+, Mg2+, SO42-, and NO3- are relatively low. The concentration of TDS increases gradually along the flow direction of groundwater, and TDS is positively correlated to the variation in concentration of Cl-, Na+, Mg2+, and SO42-. Along the flow direction of groundwater, the hydrochemistry of shallow groundwater show the evolution law from HCO3-Ca·Mg to HCO3·Cl-Na·Ca and HCO3·Cl-Na·Mg, and then to Cl·HCO3-Na·Mg. The hydrochemical types are formed mainly due to the mineral dissolution and deposition, and reaction of cation exchange and adsorption in the aquifer, and the hydrogeochemical processes include leaching, evaporation and concentration, and mixing.

1 Introduction

Shuangliao City is an important part of the Xiliaohe Plain, and one of the most important bases of grain production in the north of China. Groundwater is the main source of water for the residents' life and industrial and agricultural uses. Especially as the Xiliao River has dried up in recent years, Shuangliao developed and utilized groundwater for agricultural uses to a greater extent. The importance of groundwater resources to the region is reflected not only in the water demand, but also the requirement for high quality groundwater. The hydrochemical characteristics and evolution of groundwater have been an important field of research in evaluating the quality of groundwater resources. Related research is of great significance to the utilization and management of groundwater resources and the improvement and protection of ecosystem and environment in the region[1]. The chemical characteristics of groundwater are determined by the sedimentary environment and characteristics of sedimentary combinations of sources, and are affected by natural and human factors such as flood, ecological water transfer, agricultural irrigation[2-5]. Therefore, it is helpful to identify the laws of distribution of groundwater hydrochemical types in understanding the regional hydrogeochemical characteristics, quality and distribution of groundwater in the region, and in guiding the sustainable development and utilization and comprehensive management of groundwater resources in the region[6-8].

By now, Chinese and foreign scholars have carried out a lot of research of hydrochemical characteristics of groundwater, mainly through mathematical statistics[9], Piper or Durov diagrams[10-11], ion proportional coefficient[12], mineral saturation index[13], isotope tracing[14], hydrogeochemical simulation and other techniques to study the chemical characteristics of groundwater and their causes and evolution[15-17]. Generally, many methods are used in related studies, but there is a lack of comprehensive research, especially that of zoning and evolution of regional hydrochemical types. For this reason, after collection of detailed data and identification of the groundwater flow field, we studied the hydrochemical characteristics by zone with mathematical statistics, correlation analysis, ion proportional coefficient and other methods to identify the current hydrochemical characteristics of shallow groundwater in the region, formation of hydrochemical composition and its evolution laws, uncover the main actions forming the hydrochemical composition of groundwater, and provide a scientific basis and management ideas for agricultural production and ecological protection in Shuangliao City and even the whole Xiliaohe Plain.

Author: Li Xuguang, male, born in 1982, master, senior engineer, mainly engaged in research of groundwater pollution, endemic diseases, urban geology, etc. Email: john2011@163.com.
* Corresponding author:Zhao Yan, male, born in 1982, master, senior engineer,mainly engaged in research of groundwater pollution, water resources evaluation, etc. Email: 337056992@qq.com.

2 Overview of study area

Shuangliao is a city of Jilin Province located in the transition between semi-humid climate and semi-arid climate, with frequency windy and sandy weather. Its annual average temperature is 6.6 °C, annual precipitation is 450-550 mm, and annual evaporation is as high as 1,600 mm. Its location is subject to serious land salinization, land desertification and grassland degradation, and is one of the key areas of ecological restoration in Jilin Province.

The aquifers in the study area are mainly of a Quaternary loose rock pore aquifer system, which is composed of three aquifers, namely the Quaternary Holocene, Upper Pleistocene, Middle Pleistocene phreatic aquifers.

(a) Holocene phreatic aquifer: This aquifer is distributed under the floodplains and terraces of the Xinkai River and Xiliao River. It is mainly composed of medium and fine sand with thickness of 10–25 m. The groundwater type is HCO_3-Ca•Na with pH value of 7.34–8.26.

(b) Upper Pleistocene phreatic aquifer: This aquifer is distributed under the low plains and the Songliao watershed. It is composed of sandy loam, silty sand and fine sand with thickness of 3–17m. The type of groundwater is Cl•HCO_3 and SO_4•HCO_3 with pH value of 7.47–8.47.

(c) Middle Pleistocene confined aquifer: This aquifer covers the whole study area, and its lithology is composed of fine sand, silty fine sand and sandy loam with thickness of 5.57–23.5m. The hydrochemical type of groundwater is mainly HCO3 with pH value of 7.37–8.49.

3 Water sample collection and testing

The shallow groundwater samples were collected from June to September 2012. According to the principles of plane control and local densification, the samples were collected with a density of 1.1 group / 100 km^2. 32 groups of shallow groundwater (phreatic water and micro-confined water) samples were collected. Certain protective agents were added and test and analysis were conducted in a timely basis for different test indicators. The samples were tested for 42 indicators, including 7 physical and chemical indicators tested on the field, and 35 inorganic indicators, including "three nitrogen indicators", heavy metals and conventional ions, tested in laboratory.

4 Results and analysis

After the scientific and systematic sampling, testing and analysis of the shallow groundwater across the study area, the methods of descriptive statistics, correlation analysis and ion proportional coefficient in the hydrochemical software (AquaChem4.0) were used to comprehensively and systematically study the hydrochemical characteristics of the shallow groundwater, explore the formation of hydrochemical composition and the evolution laws of regional hydrochemical types, and uncover the main reactions controlling evolution of the groundwater quality and the hydrogeochemical processes.

4.1 Statistical characteristics of hydrochemical parameters

Statistical analysis can reflect the overall picture of groundwater chemical composition in a certain area or a certain period of time[18]. To understand the hydrochemical characteristics of groundwater in this area, the 32 groups of samples collected from June to September 2012 were tested and analyzed, and the statistics of hydrochemical parameters were obtained (Table 1).

Table 1 Statistics of hydrochemical parameters of shallow groundwater

Component	Maximum	Minimum	Average Value	Standard Deviation	Coefficient of variation (%)
Na^+	997.60	10.58	107.65	177.06	164.48
Ca^{2+}	203.50	10.83	78.52	56.60	72.09
Mg^{2+}	176.30	4.31	31.95	32.70	102.34
HCO_3^-	927.86	132.13	347.94	174.97	50.29
Cl^-	1114.39	2.54	102.21	205.14	200.71
SO_4^{2-}	642.37	0.39	89.07	128.68	144.48
NO_3^-	379.63	0.00	50.83	99.48	195.71
TDS	1857.09	184.66	670.27	693.03	103.40
pH	8.3	6.56	7.21	0.35	4.91

Note: In addition to the pH, the units of the concentration of the rest chemical mass parameters are mg/L.

As shown in the above table, that the average concentration of HCO_3^- is the largest, followed by that of Na+ and Cl^-, indicating that these ions are dominant ions in the groundwater. The standard deviations of Cl^-, Na^+ and HCO_3^- are large, indicating that their absolute content in groundwater is high. This is consistent with the larger average concentrations of the three types of irons. The coefficients of variation of Ca^{2+}, Mg^{2+} and HCO_3^- are small, indicating that their content is relatively stable in the groundwater of this region, and the content of HCO_3^- is high and stable. The coefficients of variation of Na^+, Cl- and NO3- are large, indicating that their concentrations vary significantly in different areas and easily affected by the aquifer media, topography, hydrometeorological conditions and human activities. The standard deviation and coefficient of variation of TDS are high, indicating that the content of TDS is high in this region, and the concentrations vary considerably in different areas.

4.2 Correlation characteristics of hydrochemical parameters

Correlation analysis is intended to uncover the similarity and difference in groundwater hydrochemical parameters and the consistency and difference of their sources [1]. With the test results of the 32 groups of groundwater hydrochemistry, statistical software was used to calculate the correlation of typical components (Table 2).

Table 2 Correlation matrices of hydrochemical parameters of shallow groundwater

Component	Na^+	Ca^{2+}	Mg^{2+}	HCO_3^-	Cl^-	SO_4^{2-}	NO_3^-	TDS	pH
Na^+	1.000	0.338	0.918	0.774	0.950	0.898	0.350	0.947	0.179
Ca^{2+}		1.000	0.464	0.362	0.505	0.557	0.764	0.603	-0.014
Mg^{2+}			1.000	0.744	0.922	0.903	0.533	0.952	0.160
HCO_3^-				1.000	0.636	0.637	0.179	0.740	0.207
Cl^-					1.000	0.915	0.504	0.968	0.101
SO_4^{2-}						1.000	0.537	0.952	0.204
NO_3^-							1.000	0.594	-0.097
TDS								1.000	0.141
pH									1.000

From the correlation coefficient matrices of hydrochemical parameters, one can see that the correlation coefficients of Cl-, Na+, Mg2+ and SO42- with TDS are high, and the regression equations are shown in formulas (1)–(4).

$$\rho_{TDS}=3.2713\rho Cl^-+335.92 \quad R^2=0.968 \tag{1}$$

$$\rho_{TDS}=3.7067\rho Na^++271.22 \quad R^2=0.947 \tag{2}$$

$$\rho_{TDS}=20.175\rho Mg^{2+}+25.652 \quad R^2=0.952 \tag{3}$$

$$\rho_{TDS}=5.1275\rho SO_4^{2-}+213.57 \quad R^2=0.952 \tag{4}$$

Where, ρ_{TDS}, ρ_{Cl^-}, ρ_{Na^+}, $\rho_{Mg^{2+}}$ and $\rho_{SO4^{2-}}$ stand for the concentrations of TDS, Cl^-, Na^+, Mg^{2+} and SO_4^{2-} respectively, in mg/L; R is the correlation coefficient, dimensionless.

From formulas (1)–(4), one can see that the correlation coefficients between TDS and Cl^-, Na^+, $Mg^{2+}$and SO_4^{2-} are above 0.947, while the correlation coefficients between Cl^-, Na^+, Mg^{2+} and SO_4^{2-} in Table 2 are all higher than 0.9. This indicates that the change trend of TDS is consistent with the change trends in concentration of these ions. The correlation coefficients between HCO_3^-, Ca^{2+} and NO_3^-, and the correlation coefficients between them and other ions are poor. Especially, the content of HCO_3^- in groundwater is the highest and stable, and the correlation coefficient between it and TDS is only 0.74, which indicates that the change of TDS in the process of hydrochemical evolution in this region mainly depends on the change of content of Cl-, Na+, Mg^{2+} and SO_4^{2-}. Therefore, it is important to understand the change patterns of these ions for understanding of the evolution process of groundwater in this region.

4.3 Proportion coefficient characteristics of ions

In the chemical composition of groundwater, the content proportional coefficient between various components is often used to study some hydrogeochemical issues[10]. Proportional coefficients can be applied to determine the sources and formation process of groundwater hydrochemical composition. Compared with traditional hydrochemical type analysis along, they enable more deep description and depiction of the evolution process of water quality in spatial dimensions as well as typical analysis of hydrogeochemical evolution.

The γNa/γCl coefficient is called the casual coefficient of groundwater, which is a hydrogeochemical parameter to characterize the enrichment of sodium ions in groundwater. The average value of the γNa/γCl coefficient of standard seawater is 0.85. The γNa/γCl coefficients of low salinity water are high [γNa/γCl > 0.85], while those of high salinity water are low [γNa/γCl<0.85][19]. As shown in Fig. A, almost all the analysis points of water samples are distributed above the straight line 1:1, which indicates that the mg equivalent concentration of [Na^+] is generally greater than that of [Cl^-]. That is, the γNa/γCl coefficient is greater than 1. In the process of runoff, groundwater continuously weathers and dissolves rocks and minerals through hydrolysis and acidification, and Na^+ is thus released from feldspar. As a result, the [Na^+] mg equivalent concentration is greater than that of [Cl^-].

As shown in Fig. B and C, most of the analysis points are located above the straight line 1:1, which indicates that the mg equivalent concentrations of [Ca^{2+}] and [Mg^{2+}] are generally greater than that of [SO42-] concentration, and the mg equivalent concentration of [Ca^{2+}] is higher than that of [Mg^{2+}]. Along the flow direction of groundwater, TDS gradually increases, which facilitates the dissolution of calcite and dolomite. In Fig. D, most of the analysis points are located above the straight line 1:1, which indicates that the dissolved amount of calcite is higher than that of dolomite. Fig. E shows that the analysis points are below the straight line 1:1, which indicates that the mg equivalent concentrations of [Ca^{2+}+Mg^{2+}] are lower than that of [HCO_3^-+SO_4^{2-}]. This suggests the weathering-dissolution of carbonate rocks occurring in the process of transport by groundwater.

Cl^-, SO_4^{2-} and HCO_3^- characterize the dissolution of rock salt, gypsum and dolomite respectively, so $\gamma(Na^+-Cl^-)$ and $\gamma(HCO_3^-+SO_4^{2-}-Ca^{2+}-Mg^{2+})$ represent the dissolution of the three minerals. If there is cation exchange and adsorption, the ratio of $\gamma(Na^+-Cl^-)$ to $\gamma(HCO_3^-+SO_4^{2-}-Ca^{2+}-Mg^{2+})$ should be close to 1[20]. Most analysis points in Fig.

https://doi.org/10.1051/e3sconf/202129002014

F are located near the straight line 1:1, this indicates that cation exchange and adsorption occurs in the process of transport by groundwater.

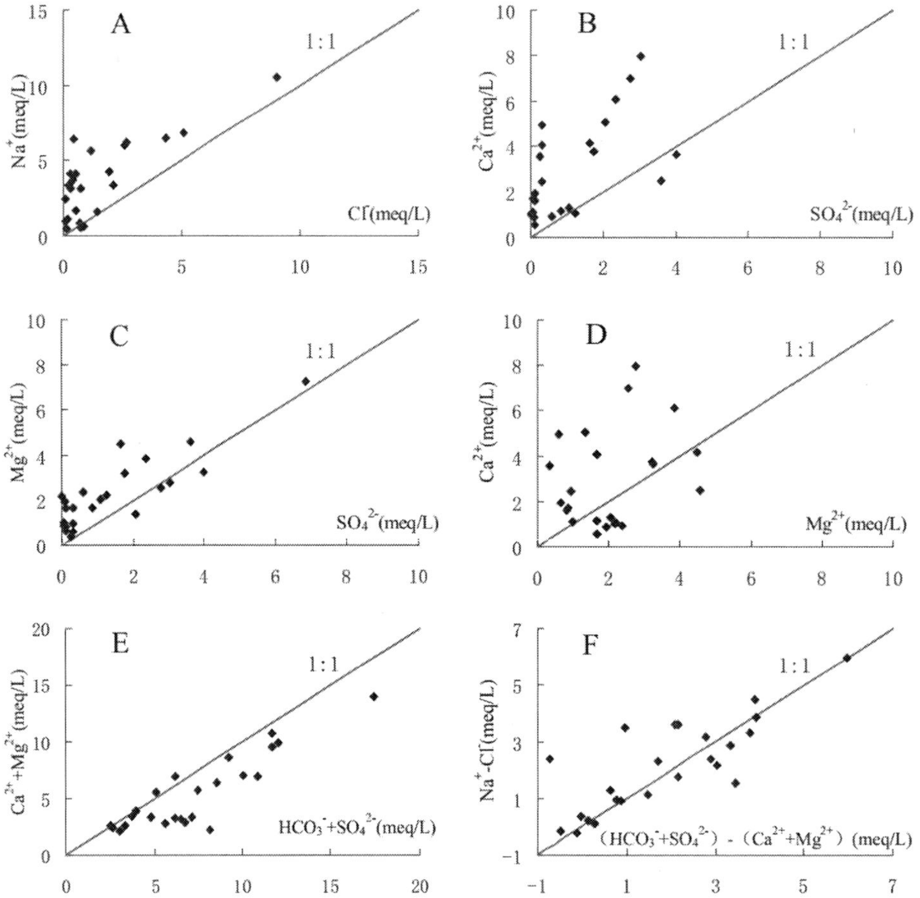

Fig. 1 Hydrochemical relationships between the selected ions of shallow groundwater

As shown in the left diagram of Fig. 2, when TDS<500mg/L, the γNa/γCl coefficient trends to increase along with the increase of TDS. This indicates that in low salinity water bodies, with the increase of groundwater flowing distance, the concentration of Na^+ from mineral weathering and dissolution gradually increases, resulting in the increase of Na^+ content in groundwater, and gradually increases downstream. When TDS>500mg/L, the γNa/γCl coefficient generally increases with the increase of TDS. This indicates that when TDS in groundwater reaches a certain concentration, Na+ starts iron exchange with Ca2+ and Mg2+ adsorbed by clay minerals in aquifers, resulting in the decrease of Na+ concentration and increase of Cl^- in groundwater, so the γNa/γ coefficient decreases.

As shown in the right diagram of Fig. 2, when TDS<500mg/L, the ratio of γ(Na-Cl)/γ(Ca+Mg-HCO_3-SO_4) changes remarkably, and only a few points have ratio values close to 1. When TDS>500mg/L, the ratio of γ(Na-Cl)/γ(Ca+Mg-HCO_3-SO_4) falls on a straight line, which indicates that cation exchange and adsorption occurs. In this case, the concentration of Na^+ ions decreases and the concentrations of Ca^{2+} and Mg^{2+} increase, which is consistent with the change law of the γNa/γCl coefficient along with TDS.

With the increase of TDS concentration, not only the dissolution and precipitation of rock salt, gypsum and dolomite, but also cation exchange and adsorption occur between groundwater and aquifer media. Therefore, the dissolution and precipitation of minerals and the cation exchange and adsorption are of great significance to the formation of groundwater hydrochemical characteristics.

https://doi.org/10.1051/e3sconf/202129002014

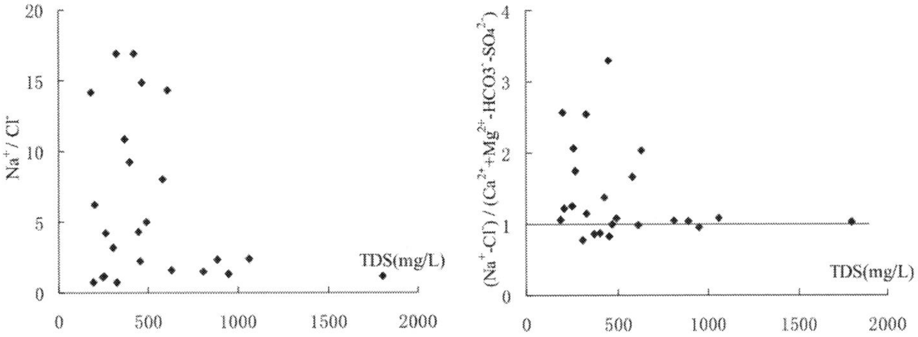

Fig. 2 Chart of γNa/γCl, γ(Na-Cl)/γ(Ca+Mg-HCO₃-SO₄) with the changing concentration of TDS

4.4 Regional characteristics of groundwater hydrochemical evolution

From the analysis above, we know the hydrochemical characteristics of shallow groundwater and explore its causes, but its spatial distribution and evolution laws are not presented in a sufficiently intuitive manner. Therefore, according to the results of test data with water samples, we applied the Sukharev classification method to classify the hydrochemical types of shallow groundwater, and the results of classification are shown in Fig. 3.

As shown in Fig. 3, shallow groundwater flows from northeast to southwest. By comparing the test results of TDS against the direction of groundwater flow, the concentration of TDS is found to gradually increases from northeast to southwest. Generally, the concentration of TDS is less than 500 mg/L in the northeast, 500–1000 mg/L in the middle and higher than 1,000 mg/L in southwest. In the northeast of the region, the runoff conditions of groundwater is favorable, and the water exchange is relatively rapid. The soluble components Cl^-, Na+ and SO_4^{2-} in the aquifer are continuously leached and transported with the water flow. The ions are mainly Ca^{2+} and HCO_3^-, forming low salinity HCO_3-Ca·Mg water. This is consistent with the conclusion in 3.3 that when TDS < 500mg/L, the concentration of Na^+ in the groundwater flowing downstream increases due to weathering and dissolution of minerals . The groundwater flow slows down in the middle of the region. With the increase of groundwater flowing distance and residence time, there is sufficient time of reaction for the dissolution of calcite, dolomite and other minerals, which makes the concentration of Ca^{2+} and Mg^{2+} ions in groundwater increase. Since the TDS this area is 500–1,000, it is possible to infer from 3.3 that cation exchange and adsorption occurs in slow runoff area, which further increases the content of Ca^{2+} and Mg^{2+} ions in the groundwater, and forms medium salinity HCO_3·Cl-Na·Ca and HCO_3·Cl-Na·Mg water. In the southwest of the region, TDS is higher than 1,000 mg/L, in which case Ca^{2+} and Mg^{2+} ions are easy to form carbonate precipitation. Moreover, the amount of calcite precipitation is higher than that of dolomite, resulting in decrease of concentrations of Ca^{2+}, Mg^{2+} and HCO_3^- ions in the groundwater, forming high salinity Cl·HCO_3^-Na·Mg water.

Along the flow direction of groundwater, the hydrochemistry of shallow groundwater show the evolution law from HCO_3-Ca·Mg to HCO_3·Cl-Na·Ca and HCO_3·Cl-Na·Mg, and then to Cl·HCO_3-Na·Mg. The hydrochemical types are formed mainly due to the mineral dissolution and deposition, and reaction of cation exchange and adsorption in the aquifer, and the hydrogeochemical processes include leaching, evaporation and concentration, and mixing.

5 Conclusions

(1) The concentrations of HCO3-, Cl- and Na+ in the shallow groundwater of Shuangliao City are relatively high, while the concentrations of Ca^{2+}, Mg^{2+}, SO_4^{2-} and NO_3^- - are relatively low. The concentration of TDS increases gradually along the flow direction of groundwater, and TDS is positively correlated to the variation in concentration of Cl-, Na+, Mg2+ and SO_4^{2}-.

(2) Along the flow direction of groundwater, the hydrochemistry of shallow groundwater show the evolution law from HCO3-Ca·Mg to HCO3·Cl-Na·Ca and HCO3·Cl-Na·Mg, and then to Cl·HCO3-Na·Mg.

(3) The hydrochemical types are formed mainly due to the mineral dissolution and deposition, and reaction of cation exchange and adsorption in the aquifer, and the hydrogeochemical processes include leaching, evaporation and concentration, and mixing.

 https://doi.org/10.1051/e3sconf/202129002014

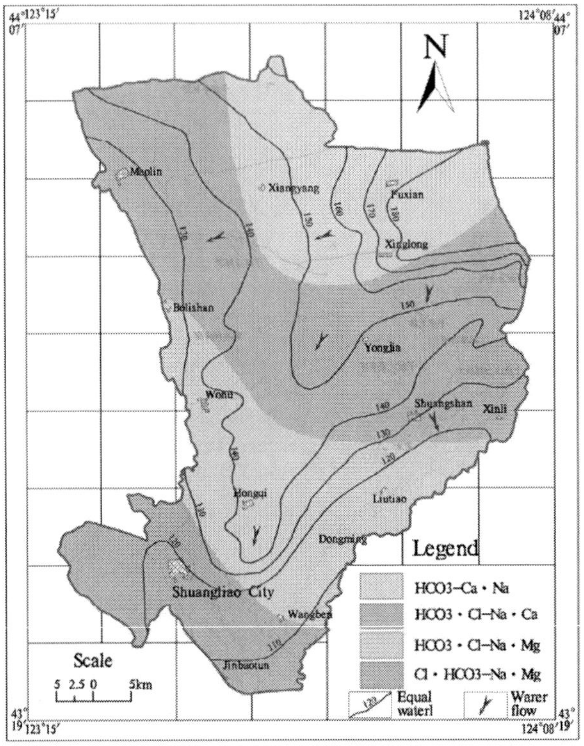

Fig. 3 Hydrochemical type diagram in shallow groundwater

References

1. Zhang, G, Deng, W, He, Yan, et al. (2006) Hydrochemical characteristics and evolution laws of groundwater in Songnen Plain, Northeast China . Advances In Water Science, 17(1) : 20-28.
2. Xie, Z, Liu, K, Li, Z, et al. (2010)Analysis of groundwater chemical characteristics based on sediment provenance analysis: A case study of Beijing Plain . Earth Science Frontiers, 17(6):81-87.
3. Xu, H, Zhao, T, Meng, H, et al. (2011)Relationship between groundwater quality index of physics and chemistry in riparian zone and water quality in river . Environment Science, 32(3) :632-640.
4. Chen, Y, Chen, Y, Li, W, et al. (2006)Three stages of the groundwater chemical properties reacting on the intermittent water deliveries in lower Tarim River, China. Environment Science, 27(7):1299-1304.
5. Ma F, Yang Y S, Yuan R, et al. (2007)Study of shallow groundwater quality evolution under saline intrusion with environmental isotopes and geochemistry. Environmental Geology, ,51(6): 1009-1017.
6. Ding, H, Zhang, J. (2005)Goechemical properties and evolution of groundwater beneath the Hexi corridor, Gansu province. Arid Zone Research, 22(1):24-28.
7. Liao, Z, Lin, X. (2004)Chemical characteristics and variations of groundwater quality in Songnen Basin . Journal of China University of Geosciences, 29(1): 96-102.
8. Li, C, Gao, X, Wang, Y. (2015)Hydrogeochemistry of high-fluoride groundwater at Yuncheng Basin, northern China. Science of the Total Environment, (508): 155-65.
9. Pu, T, He, Y, Zhang, T, et al. (2013)Isotopic and geochemical evolution of ground and river waters in a karst dominated geological setting: A case study from Lijiang basin, South-Asia monsoon region. Applied Geochemistry, 6(33): 199-12.
10. Martin B Goldhabera, Christopher T Millsb, Jean M Morrisonb, et al. (2014)Hydrogeochemistry of prairie pothole region wetlands: Role of long-term critical zone processes. Chemical Geology, 10(387): 170-83.
11. Han, D, Song, X, Matthew J Currell, et al. (2011)A survey of groundwater levels and hydrogeochemistry in irrigated fields in the Karamay Agricultural Development Area, northwest China: Implications for soil and groundwater salinity resulting from surface water transfer for irrigation . Journal of

Hydrology, 8(405):217-234.

12. Charles N Alpers, Donald O Whittemore. (1990)Hydrogeochemistry and stable isotopes of ground and surface waters from two adjacent closed basins, Atacama Desert, northern Chile. Applied Geochemistry, 10 (5):719-734.
13. Guo, Q. (2012)Hydrogeochemistry of high-temperature geothermal systems in China: A review . Applied Geochemistry, 10(27):1887-1898.
14. Maurizio Barbieri , Marco Morotti. (2003)Hydrogeochemistry and strontium isotopes of spring and mineral waters from Monte Vulture volcano, Italy. Applied Geochemistry, 1 (18):117-125.
15. Su, Y, Feng Q, Zhu G, et al. (2007)Identification and evolution of groundwater chemistry in the Ejin Sub-Basin of the Heihe River, northwest China [J]. Pedosphere, 17(3) : 331-342.
16. Belkhiri L, Boudoukha A, Mouni L, et al. (2010)Application of multivariate statistical methods and inverse geochemical modeling for characterization of groundwater-A case study: Ain Azel plain (Algeria). Geoderma, 159(3-4): 390-398.
17. Bennetts D A, Webb J A, Stone D J M, et al. (2006)Understanding the salinisation process for groundwater in an area of south-eastern Australia, using hydrochemical and isotopic evidence [J]. Journal of Hydrology, 323(1-4): 178-192.
18. Jiang, L. (2009)Study on the formation and evolution mechanism of groundwater chemical composition in arid oases -- a case study of Yaoba Oasis in Alashan Chang'an: Chang'an University.
19. Yang, T, Wang, M, Chen, L. (1998)Hydrogeochemical characteristics and causal analysis of Dongying Basin. Proceedings of the 30th International Geological Congress Vol. 22. Beijing: Geological Publishing House.
20. An, L, Zhao, Q, Ye, S, et al. (2012)Hydrochemical characteristics and formation mechanism of shallow groundwater in the Yellow River Delta. Environment Science, 33(2):370-378.

Study on the impact of calculation radius on IDW gravity modelling

Yangtao Meng[1], Ruihang Yu[1,*], Juliang Cao[1], Shaokun Cai[1], Guowei Pan[1]

[1]College of Intelligence Science and Technology, National University of Defense Technology, Changsha, Hunan, 410073, China

Abstract. Gravity modelling is a necessary process of applying gravity data to practice, and a vital research area in resource exploring and geodesy. Inverse distance weighted (IDW) interpolation is a widely used algorithm in gravity modelling and many other griding procedures. Based on the gravity anomaly data acquired from the airborne gravimetry over a region in the Northwest of China, IDW method is analysed and implemented with varying calculation radius in gravity modelling. The results show that over the study area, with a proper calculation radius, the IDW can give a more precise prediction than quadric surfaces fitting (QSF) method. The distances between sampled points and unsampled points are the only information taken into consideration in the IDW interpolation, so the analysis of the influence by changing the calculation radius of the unsampled points in this test provides a new research direction of error eliminating when using IDW.

1 Introduction

Last decades have seen large progresses made in the area of gravimetry, with different means of gravimetry developed rapidly, which makes it possible to acquire a huge amount of gravity data like gravity anomalies and disturbances. As a result, there is an urgent need to comprehensively use those gravity data to build gravity models both regionally and globally. Data used in the recovery of gravity field is usually discrete and randomly distributed and sometimes with different resolutions due to various of ways of measurements[1]. As a consequence, algorithms of interpolation are various, including inverse distance weighted, Kriging interpolation, quadric surfaces fitting and least squares collocation[2].

Different ways of modelling have different features. The advantage of IDW is that it only takes distance information into calculation when predicting the unknown points, which largely simplifies the calculation process and easy to implement. While the main weakness of this algorithm is the absence of relevant physical information over the study region which results in a low accuracy of the prediction[3]. Least squares collocation is a theory of function approximation, which has been developed gradually since the 1960s[4]. Least squares collocation is a complex and precise method in gravity modelling, especially when dealing with two or more kinds of gravity data, but the amount of calculation is usually large because of the computation of inversing large matrixes. The applicable scope of Kriging method is that the regional variables have spatial correlation, that is, if the results of variogram and structural analysis show that the regional variables have spatial correlation, then Kriging method can be used for interpolation, otherwise, it is not feasible. Therefore, considering the complexity and convenience of the calculation, IDW and QSF are selected.

The main objective of this study is to figure out the relation between the calculation radius and the prediction's accuracy, which is of vital importance to the further research on the determination of calculation radius. IDW and QSF are implemented to generate a regional gravity model based on a set of discrete gravity anomaly data over the survey region[5]. Datasets are extracted from the original data collected during the flight tests of airborne gravimetry conducted by National University of Defense Technology (NUDT), using a strapdown airborne gravimeter, which has an internal accuracy within 1.6mGal[6]. To see the impact of different calculation radiuses of IDW, the radius is set to grow step by step from 0.02° to 0.55° (considered in geodetic coordinate system, so the angular unit is used). This range of radius shows the comprehensive outcome according to the varying radius. With a certain properly selected radius in IDW, the error can be largely eliminated, performing better than QSF method.

2 Datasets and theories

2.1 Data source

In August, 2016, flight tests were conducted over a region in the Northwest of China, which is about 1600km^2. In the datasets from the task of airborne gravimetry, the flying height data is averagely 1500m, and the range of the latitude covers about 0.35° and the range of the longitude covers about 0.40°.

There are 1075 points of gravity anomaly in the dataset extracted from the original dataset. For convenience, we

* Corresponding author: yuruihang@nudt.edu.cn

assign a name for the dataset with 1075 points of gravity data, called DS, then we divided DS into two parts, DS1 and DS2. DS1 is employed as the experimental set, DS2 is used as the test set, which means the gravity model obtained from DS1, can be validated by the DS2. There are similar numbers of points in both DS1 (538 points) and DS2 (537 points). These two datasets are uniformly distributed in the study area.

Table 1. Standard deviation, Mean, Minimum and Maximum of DS, DS1 and DS2 (mGal).

	Std	Mean	Min	Max
DS	10.85	-103.94	-129.08	-72.76
DS1	10.80	-103.98	-129.09	-72.76
DS2	10.92	-103.90	-125.51	-73.30

Table 1 shows both DS1 and DS2 are similar to DS statistically, indicating three datasets have nearly the same statistical features, which means the idea of grouping is feasible.

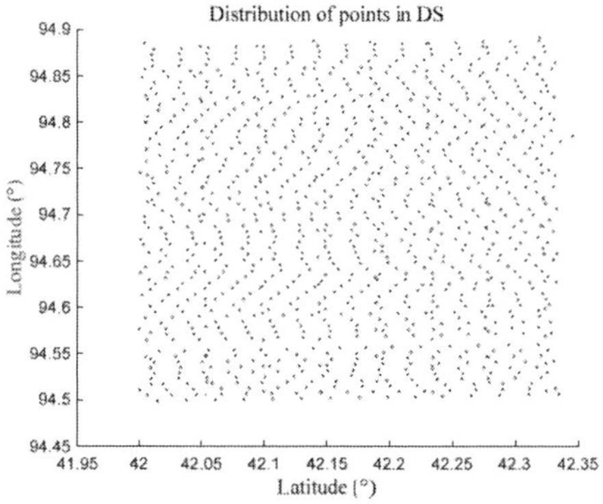

Fig. 1. The number of points in DS is 1075, which covers the whole survey region evenly.

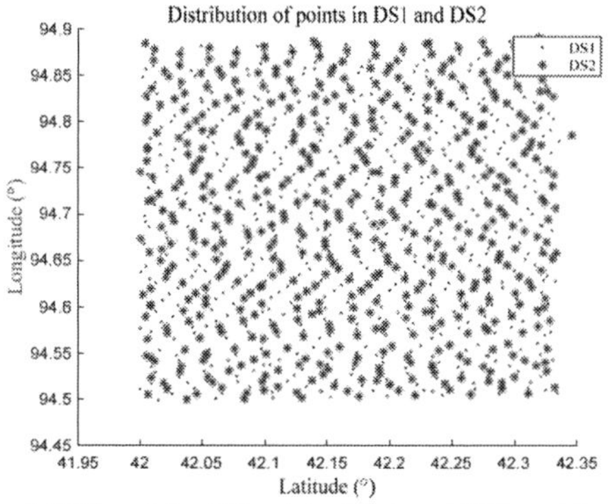

Fig. 2. DS1 contains 538 points of data and DS2 contains 537 points of data. Both datasets are distributed uniformly over the study area.

2.2 Theories of methods

2.2.1 Inverse distance weighted (IDW)

IDW is a widely used interpolation algorithm in the research of gravity field modelling and the computation of GPS height anomaly. IDW is applied to calculate the unsampled points according to other known points over the survey region. The basic principle is that the nearer a known point is, the bigger influence it has on the unsampled point, while the larger the distance is, the less impact it has on the unknown point.

First, distances between the unsampled point and the rest of the points need to be calculated within the two-dimensional plane.

$$V_{(X,Y)} = \frac{\sum_{i=1}^{n}\left(\frac{V_i}{Dis_i}\right)}{\sum_{i=1}^{n}\left(\frac{1}{Dis_i}\right)} \tag{1}$$

In equation (1), $V_{(X,Y)}$ is the value of a point needs to be estimated and the position of this point is (X,Y); Dis_i is the distance between the ith point and the unsampled point. When applying IDW in gravity modelling, all distances between the sampled data and the unknown points should be calculated. When there is not a large number of points of gravity data, it would be difficult to figure out the drawbacks of the IDW method. When the amount of data is large, the computation amount of the algorithm grows rapidly, which makes it a low efficiency algorithm. Mostly, the amount of gravity data is large enough that it takes a long time to accomplish the computation work. As a result, selecting a limitation for the calculation radius can help improve the efficiency.

2.2.2 Quadric surfaces fitting (QSF)

The mathematical model of quadric surfaces fitting can be described as the formula below:

$$\xi(x,y) = a_0 + a_1x + a_2y + a_3x^2 + a_4y^2 + a_5xy \tag{2}$$

In formula (2), (x,y) is the position of points in DS1, and a_i stands for the 6 coefficients in the model, so there should be no less than 6 known points in the DS1. In fact, the quantity of the data in DS1 is much larger than 6, so when the number of known points is n, which is larger than 6, the error equation is as follows:

$$\mathbf{V} = \mathbf{AX} - \mathbf{L} \tag{3}$$

In equation (3)

$$\mathbf{A} = \begin{bmatrix} 1 & x_1 & y_1 & x_1^2 & y_1^2 & x_1y_1 \\ 1 & x_2 & y_2 & x_2^2 & y_2^2 & x_2y_2 \\ \vdots & \vdots & \vdots & \vdots & \vdots & \vdots \\ 1 & x_n & y_n & x_n^2 & y_n^2 & x_ny_n \end{bmatrix}, \quad \mathbf{X} = \begin{bmatrix} a_0 \\ a_1 \\ a_2 \\ a_3 \\ a_4 \\ a_5 \end{bmatrix}$$

$$\mathbf{L} = \begin{pmatrix} \varphi_1 & \varphi_2 & \cdots & \varphi_n \end{pmatrix}^T, \mathbf{V} = \begin{bmatrix} V_1 \\ V_2 \\ \vdots \\ V_n \end{bmatrix}$$

According to the principle of least square algorithm, we can solve the equation and give the solution of the coefficients vector $\mathbf{X}$.

$$\mathbf{X} = \left(\mathbf{A}^T\mathbf{A}\right)^{-1}\mathbf{A}^T\mathbf{L} \tag{4}$$

 https://doi.org/10.1051/e3sconf/202129002015

In formula (4), **L** is the vector of the magnitude of gravity anomaly in DS1.

3 Methods validation and results

There are several steps in the whole process of gravity modelling and precision analysis.

1. Implementing the traditional IDW method without setting a radius limitation. Calculating of the gravity anomaly model by using quadric method.

2. Analysing the precision of the results.

3. Conducting the IDW with different limitations on the calculating radius. Comparing the results with the results of quadric method acquired in the second step.

3.1 Implementation of IDW and QSF

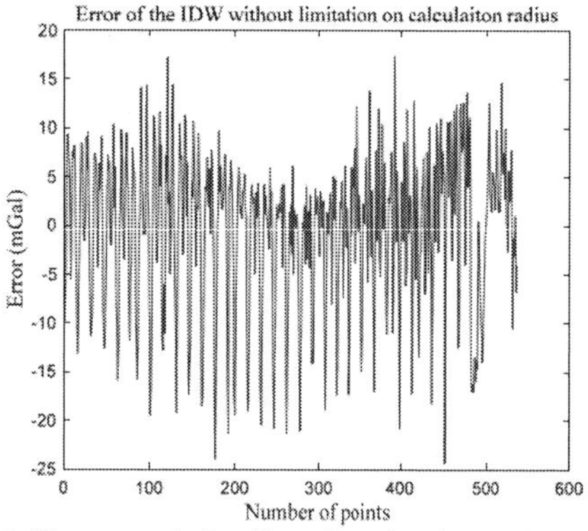

Fig. 3. The error calculated by subtracting the gravity anomaly in DS2 from the prediction of IDW based on DS1.

Table 2. Statistical properties of the error using IDW (mGal)

Std	Mean	RMSE	Min	Max
7.98	-0.38	7.98	-24.31	17.43

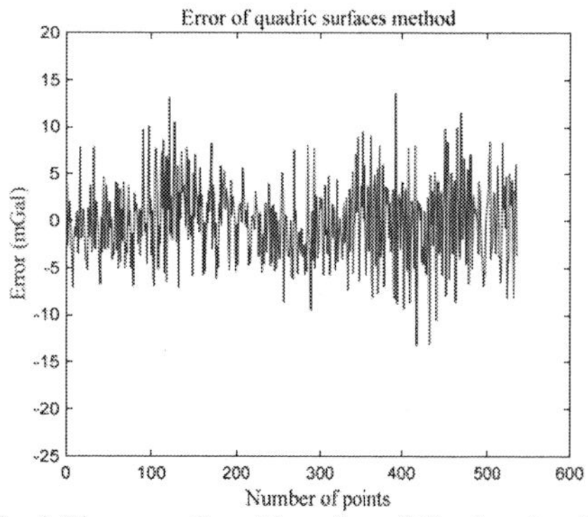

Fig. 4. The error of quadric surfaces fitting based on DS1

Table 3. Statistical properties of the error using QSF (mGal)

Std	Mean	RMSE	Min	Max
4.20	-0.03	4.20	-13.16	13.61

The root mean square error (RMSE) of IDW without limitation of calculation radius is 7.98mGal. As is shown in table 2, standard deviation, mean value, minimum and maximum are 7.98, -0.38, -24.31 and 17.43 respectively.

The RMSE of QSF is 4.20mGal, which is smaller than the result of IDW without radius limitation. Table 3 shows the standard deviation, mean value, minimum and maximum are 4.20, -0.034, -13.16 and 13.61 respectively.

By comparing the results from the tests of IDW and quadric method, it can be figured out that IDW algorithm results in a greater RMSE than the latter does. Besides, QSF method performs better in terms of the mean value and max-min value of modelling error.

3.2 IDW with limitation on calculation radius

In the algorithm of IDW, the distance between the sampled points and the unsampled points is the only information taken into calculation. Results show the accuracy is improved by controlling the calculating radius at a certain low level.

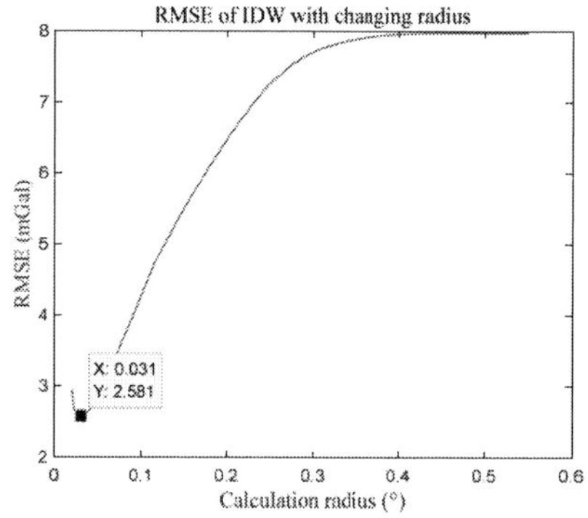

Fig. 5. The RMSE of IDW with changing calculation radius

Table 4. Statistical characteristics of the error using IDW when the calculation radius equals to 0.031°. (mGal)

Std	Mean	RMSE	Min	Max
2.58	-0.05	2.58	-11.19	9.26

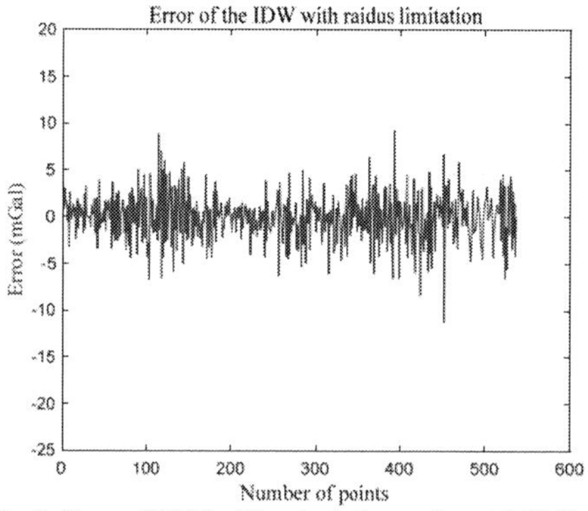

Fig. 6. Error of IDW with calculation radius at 0.031°, the RMSE equals to 2.58mGal.

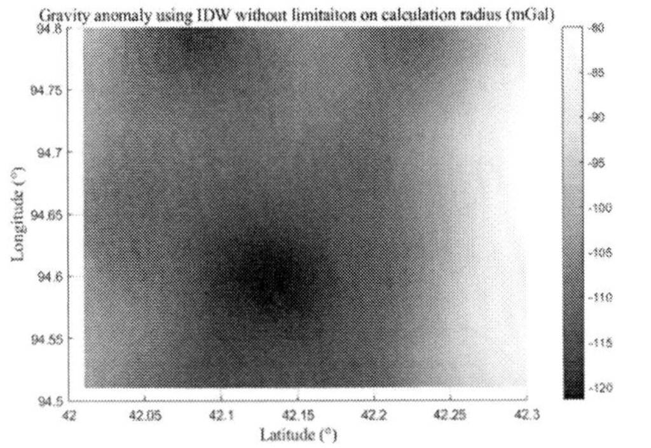

Fig. 7. Gravity anomaly over the study region after griding by IDW without limitation on calculation radius.

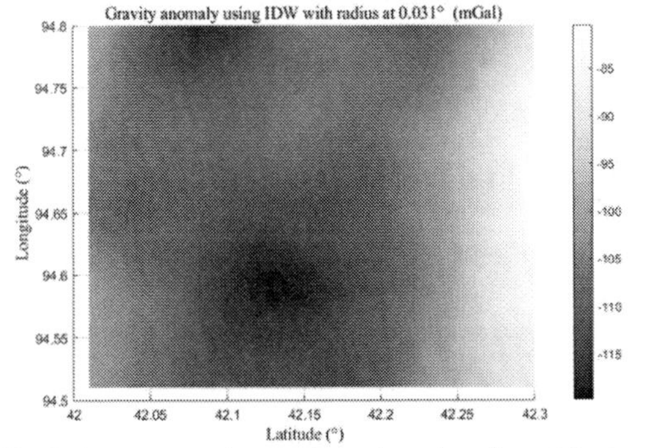

Fig. 8. Gravity anomaly over the study region after griding by radius limited IDW.

Figure 5 shows a minimum RMSE of the IDW method is attainable when the calculation radius is selected properly, in this context, the radius should be set at 0.031° and the minimum RMSE is 2.58mGal, which is much lower than that of QSF. When the limitation of the radius grows more than 0.031°, error of this method is increased gradually. Figure 7 and 8 show that the whole picture by IDW with and without limitation on calculation radius. The comparison between these two figures indicates that the improved method gives more details than the normal IDW over the survey region.

Error of the result from radius constrained IDW performs ideal statistical properties that mean value is -0.05mGal, standard deviation is 2.58mGal, min-max value is 20.45mGal, indicating that with a constrained calculating radius, results of IDW can be stable and ideal.

4 Conclusion

The results from the calculation based on the airborne gravity anomaly data show that the calculating distance has a significant impact on the precision of the unsampled points' prediction. According to this study, when the radius rises to a level higher than 0.031 °, the error increases with the distance growing. On the contrary, with a distance less than 0.031°, IDW performs worse or even unable to complete the calculation due to the lack of sampled points. With a shorter radius computed in the IDW, amount of the whole calculation in the algorithm is decreased due to fewer points of data are taken into calculation.

As a result, with a properly chosen calculation radius when applying IDW, the prediction's accuracy can be improved to 2.58mGal, which is better than the accuracy of QSF method with 4.20mGal of RMSE. Selecting a best radius for the algorithm is a nearly impossible task when constructing a model for the gravity data in a real scenario without knowing the data to be estimated, while in this study, these two datasets from the same original dataset perform similar statistical properties, so it is possible to figure out a proper radius by assigning one of the datasets a predicting dataset and another a validating dataset to give a similar result of precision as in a real predictive calculation. The quantitative relation between the radius constraint and the RMSE of the IDW will be the focus in future research.

Acknowledgement

This paper is supported by the National Key Research and Development Program of China (Grant No. 2017YFC0601703 and Grant No. 2017YFC0601701), the Natural Science Foundation of Hunan Province, China (Grant No. 2020JJ5658).

References

1. M. Kern, K.P. Schwarz, N. Sneeuw, A study on the combination of satellite, airborne, and terrestrial gravity data [J], Journal of Geodesy, 77, 3-4: 217-225(2003).
2. M.M. Ruan, S. Chen, J.C. Han. Regional gravity field model constructed by the least squares collocation [J]. Acta Seismologica Sinica, 42, 1: 53-65(2020). (in Chinese)
3. X.D. Yu, Y. Wu, L.M. He. Improvement and comparison of inverse distance weighted grid interpolation algorithm [J], Chinese Journal of Engineering Geophysics, 10, 6: 900-904(2013). (in Chinese)
4. S. Ramouz, Y. Afrasteh, M. Reguzzoni, A. Safari, A. Saadat. IRG2018: A regional geoid model in Iran using Least Squares Collocation [J]. Studia Geophysica & Geodaetica, 63, 2: 191-214(2019).
5. J.W. Lü, Z.H. Zhang, X.X. Zhang, Z.Y. Liu. Application of combination model of quadric surfaces and least squares collocation in GPS height anomaly fitting [J], Bulletin of Surveying and Mapping, 5: 127-129, 133(2020). (in Chinese)
6. S.K. Cai, M.P. Wu, K.D. Zhang, et al. The first airborne scalar gravimetry system based on SINS/DGPS in China [J]. SCIENCE CHINA Earth Sciences, 56, 12: 2198-2208(2013).

https://doi.org/10.1051/e3sconf/202129002016

A Statistical Study of lyrics of "Huar" from Northwest China

Yan Xu[1], Qianqian Tang[2*] and Ying Yuan[3]

[123]School of information science, Beijing Language and Culture University, Beijing, Beijing, 100083, China

Abstract. As an intangible cultural heritage, "Huar" from Northwest China is a folk song created and shared by many ethnic groups in Gansu, Qinghai and Ningxia provinces. It is a precious card of Chinese national culture. However, current research on "Huar" is mainly based on qualitative methods. This paper uses statistical methods to study the lyrics of "Huar". First, the word frequency of the lyrics of "Huar" is analysed statistically. Then, the lyrics of Hezhou Huar and Taomin Huar are compared and analysed from the perspective of quantitative linguistics ("Huar" mainly includes Hezhou Huar and Taomin Huar). By comparing three quantitative indicators, it is concluded that the lexical richness of Taomin Huar is higher than that of Hezhou Huar. Based on the frequency distribution of parts of speech, the similarities and differences of the use of parts of speech between Hezhou Huar and Taomin Huar are found. This paper uses statistical methods to analyse "Huar", which has certain research value and social value.

1 Introduction

"Huar" is a regional folk song widely popular in Gansu, Qinghai, Ningxia, Xinjiang and other provinces and regions in China. It is a regional folk song performed in Chinese by nine echic groups, including Han, Hui, Tibetan, Tu, Salar, Dongxiang, Bao'an, Yugur and Mongolia. It is a landmark oral art in folk songs of western China and even Chinese folk songs [1]. "Huar" has a history of hundreds of years. In 2009, it was rated as one of the world intangible cultural heritage projects by UNESCO. Originated from ancient river state and spread in the Ming Dynasty, it was widely spread through the silk road. "Huar" is the best way for people to express joys and sorrows. It comes from life; Its melody is high and generous, melodious and graceful; Its lyrics are improvised with emotion. It is a wonderful flower in Chinese and world folk songs.

But as a world-class cultural card in China, "Huar", a folk culture brand with local color and flavor, has faced an awkward situation in recent years. In the 1980s, a large number of folk singers once emerged in the northwest "Huar", which had a glorious history and a broad mass base. However, since the 1990s, the number of "Huar" singers in Northwest China has gradually decreased, and there are many problems in the inheritance and development of "Huar". There are few high-level "Huar" singing talents. At present, China has evaluated 11 national representatives and 50 provincial representatives of "Huar". Because most of them are dead and old, in view of the current grim situation, the digitization and networking of "Huar" performance art is particularly important. As a school with rich cultural heritage, Beijing Language and Culture University shoulders the important mission of spreading and vigorously carrying forward the excellent traditional culture and promoting the development of Chinese excellent traditional culture.

The study of "Huar" has experienced the period of germination, formation, development and prosperity, and its research is becoming more and more comprehensive. But it is mainly qualitative. At present, there is little research on the text features of "Huar" from the perspective of statistics and metrology. This paper studies the lyrics of "Huar" by statistical methods. Firstly, the frequency of the lyrics of "Huar" is analyzed statistically. Then, the lyrics of Hezhou and Taomin Huar are compared and analyzed from the perspective of quantitative linguistics ("Huar" mainly includes Hezhou Huar and Taomin Huar). Through the comparison of three measurement indexes, it is concluded that the lexical richness of Taomin Huar is higher than that of Hezhou Huar; Through the statistics of the usage frequency of parts of speech, we find the similarities and differences in the use of parts of speech between Hezhou Huar and Taomin Huar, and further verify that there are obvious differences in the use of auxiliary words, location words, numerals and adverbs between the two schools through clustering experiments. It is of great significance to study “Huar” in Northwest China by statistical methods.

2 Related work

From the beginning of the 21st century to now, the research on "Huar" has entered a prosperous period. Some folklore scholars, anthropologists, language scholars and social scholars have joined the research team of "Huar", and the scope of the research has been expanded accordingly. Some high-level research has also been set up. For example, Caoqiang was established in 2009 by the

*201921198622@stu.blcu.edu.cn

https://doi.org/10.1051/e3sconf/202129002016

National Social Science Fund Project "Huar" language folk custom research. Scholars began to study the flowers from many angles. Based on the previous studies on "Huar", the author studies it from the perspectives of anthropology, musicology, folklore, sociology and communication. The representative monographs and papers are: the general theory of Chinese "Huar" by wuyulin [2], Taomin Huar research by qixiaoping: the "Huar" survey of Tiancun in the perspective of living space, and "a brief analysis of the embodiment of Hehuang agricultural folk custom in flowers" by wangyoufu. Zhuxiaofeng has "Taomin 'Huar' research in the sociological perspective" and Cao Qiang 'the word use of the words of the Huar' lyrics from the perspective of communication science.

In the research on "Huar", school research is a very important topic. In his essay on the characteristics, schools and rules of "Huar", Mr. Liu Kai said: "the 'Huar' are generally recognized as two schools: Qinghai, Linxia (formerly known as Hezhou) and Ningxia "Huar". This group of "Huar" is commonly known as "Hehuang Huar" because it is popular in the Yellow River areas of Gansu, Ningxia and Qinghai and the coastal areas of Huangshui River in Qinghai. "Huar" in Taomin area of Gansu Province is another school. After many studies, Professor wuyulin agreed with the division, and called these two schools "Hezhou type Huar" and "Taomin type Huar".

Zhang Xiaojin studies "Huar" from the perspective of Quantitative Linguistics [3]. The study uses the method of quantitative linguistics to investigate the word frequency distribution and the word character of the rhyme, thus verifying the internal law of the "Huar" text; And the author analyzes "Huar" and the May 4th New Poetry Movement, and finds that there is a close relationship between them. Therefore, the measurement method can not only provide objective verification for the traditional qualitative method, but also find some text features and rules that are difficult to find by traditional methods. "Huar", as a special northwest folk song, is not only rich in content but also in various forms. Its characteristics and internal rules are worth studying and exploring. The method of quantitative linguistics will open the door to the new world for the study of flowers. But the research on "Huar" is mainly qualitative. At present, the research on the text features of "Huar" from the perspective of measurement is still rare.

3 Research based on statistics

Quantitative linguistics is a branch of linguistics that studies the structure and development of language through quantitative methods based on real corpus. With the help of the method of quantitative linguistics, we can extract the data of various language units from the text, and make a quantitative analysis of their characteristics, so as to get an objective and empirical conclusion.

3.1 Corpus

3.1.1. Data Acquisition. Due to the scarcity of "Huar" text resources, data acquisition is very difficult. First of all, we found five professional books of "Huar" from the Superstar Digital Library, and then generated Huar corpus from the books. The process includes three steps: OCR recognition of the text in the book, python extraction of "Huar" text, and manual proofreading.

3.1.2. Data cleaning. Remove unnecessary markers in the text, such as "male:" and "female:" etc., and de-duplicate the texts.

3.1.3. Word segmentation and part of speech tagging. Use a professional tool Jieba to carry out word segmentation and part of speech tagging on the texts. At the same time, we extract 3324 "Huar" vocabulary and from books and papers, and constructed a dictionary of "Huar", to improve the accuracy of word segmentation and part of speech tagging.

Based on the above steps, we got a small Huar corpus of the following size below:

Table 1. The size of Huar corpus.

	Hehuang Huar	Taomin Huar	Total
Texts	1909	570	2479
Words	65385	41387	106772
Characters	109263	73036	182299

3.2 word frequency analysis

According to the statistics of the word frequency of the lyrics of "Huar", we arrange them according to the frequency from large to small(See Figure 1):

Figure 1. The frequency of words in Huar's Lyrics.

According to the statistics of usage frequency, nouns account for the highest proportion of "Huar", especially those with gender characteristics, such as "a ge", "ga mei", "mei mei", "daughter-in-law", etc.

3.3 lexical richness

Vocabulary richness is an index to measure the vocabulary size of a text. In this section, three parameters are selected to measure the vocabulary richness: the type token ratio, the proportion of notional words, and the relative repetition rate. The type token ratio is the ratio of type to example in text; The proportion of notional words is based on the H-point estimation of the ratio of notional words in the text; The repetition rate is the estimation of text repetition based on word frequency. The relative repetition rate is the standardized form of repetition rate, which makes the calculation results of different texts comparable; The three parameters are proportional to lexical richness. The measurement results of the two types of texts are as follows:

Table 2. Measurement of vocabulary richness of Hezhou Huar and Taomin Huar.

school	Type case ratio	Proportion of notional words	Relative repetition rate
Hezhou Huar	0.1534	0.5966	0.9074
Taomin Huar	0.1897	0.6687	0.9368

From this table, the three indexes show that the vocabulary richness of Taomin Huar is slightly higher than that of Hezhou Huar.

3.4 usage of word

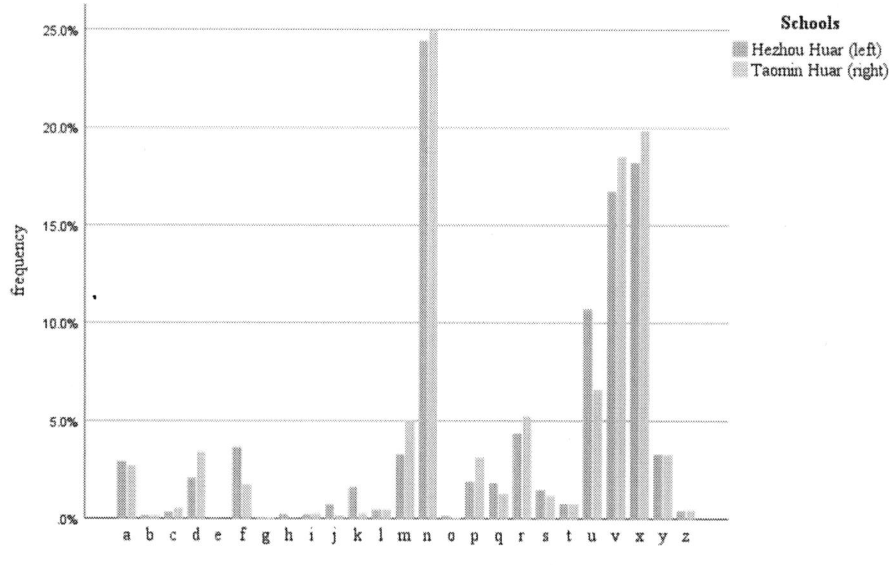

Figure 2. Frequency distribution of parts of speech.

Research shows that the frequency of specific words in text can distinguish different text styles. For example, Liu Ying and Xiao Tianjiu [4] show that the frequency of using nouns, verbs, pronouns, adjectives, adverbs, prepositions, auxiliary words, numerals, directional words and conjunctions can effectively distinguish the novels texts of Jin Yong and Gu Long. Is there any difference in the usage frequency of specific parts of speech between Hezhou and Taomin Huar? Can these differences distinguish the text styles of the two schools? In order to answer these questions, this section first investigates the distribution of parts of speech of Hezhou and Taomin Huar, calculates the usage frequency of each part of speech, and draws the frequency distribution of parts of speech, as follows:

From this figure, the usage of the words of Hezhou and Taomin Huar is both the same and different. The similarities are as follows:

- Firstly, nouns appear most frequently. This is because the rhetoric of "Fu Bi Xing" is widely used in "Huar", which makes people use a lot of "images" when they create "Huar" Image is mainly expressed as noun in language, so there are abundant nouns in "Huar".
- Secondly, there are many verbs and few adjectives in Huar texts. By calculating the busman coefficient of two kinds of text, we find that both texts show strong initiative.
- Thirdly, there exist a large number of functional words.

The difference is that there are obvious differences in the usage frequency of the seven parts of speech, namely, auxiliary words, verbs, orientation words, numeral words, adverbs, prepositions and pronouns between Hezhou Huar and Taomin Huar.

4 Conclusion

The research on "Huar" in northwest China generally adopts qualitative method, and seldom uses statistical or quantitative linguistic methods. This paper uses statistical methods to study the lyrics of "Huar". Firstly, the frequency of the lyrics of "Huar" is statistically analyzed. It is found that nouns account for the highest proportion of "Huar", especially those with gender characteristics, such as "a ge", "ga mei", "mei mei", "daughter-in-law", etc; Then from the perspective of quantitative linguistics, this paper makes a comparative analysis of the lyrics of Hezhou Huar and Taomin Huar; Based on the statistics of the usage frequency of parts of speech, the similarities and differences of the use of parts of speech between Hezhou Huar and Taomin Huar are found. This paper uses statistical methods to analyze "Huar", which has research value and social value.

Acknowledgments

This research is supported by the Fundamental Research Funds for the Central Universities, and the Research Funds of Beijing Language and Culture University (Approval number: 21YCX175), by "Construction of Ningxia Huar digital promotion platform" project of China National Arts Fund (No. 2018-a-03-(177)-0521).

References

1. Zhao Z.F.(1989), General theory of Huar. Qinghai People's publishing house, Xining.
2. Wu Y.l.(2008), General theory of Chinese Huar. Ningxia people's publishing house, Yinchuan.
3. Zhang X.J., Liu H.T. (2017) Quantitative characteristics of Chinese folk song "Huar". Journal of Ningxia University (HUMANITIES AND SOCIAL SCIENCES),39 (05): 76-80
4. Liu Y., Xiao T.J.(2014). Research on Metrological style of dream of Red Mansions. Journal of dream of Red Mansions, 2014 (04): 260-281

https://doi.org/10.1051/e3sconf/202129002017

Forecast China's monthly urban fixed asset investment based on the ARIMA model

Jianglei Yuan[1*], Yaning Li[2] and Yayu Li[3]

[1] School of Economics and Management, Beijing Jiaotong University, Beijing, 100000, China
[2] School of Law and Public Administration, Hebei Normal University, Shijiazhuang City, Hebei Province, 050000, China
[3] Quzhou County Shangshu Middle School, Handan City, Hebei Province, 057250, China

Abstract. In recent years, as China's urbanization level has risen, China's urban fixed asset investment has also been rising. Judging from monthly data, China's urban fixed asset investment has shown a volatile upward trend, with an obvious 11-month cycle. And in each cycle, the fluctuation range of the investment amount is getting larger. This paper uses an ARIMA model with an additive seasonal effect to fit the monthly urban fixed asset investment sequence and predict the future investment. In the end, this paper established a fitting model for China's urban fixed asset investment, and obtained a good forecasting effect.

1 Introduction

As the level of China's economic development continues to rise, the level of urbanization in China is also constantly being provided. The increase in the level of urbanization includes not only the increase in the urban housing population, but also the increase in the area of urban housing and the expansion of infrastructure. In the process of accelerating the pace of urbanization, China's annual investment in urban fixed assets is also increasing. Investment in fixed assets mainly includes investment in road construction and park expansion. As shown in Figure 1, in recent years, China's urban fixed asset investment level has continued to increase as a whole, from 1301.403 billion yuan in February 2010 to 49792 billion yuan in July 2019. This total of 104 months of investment data is a typical time series data. Moreover, this process of improvement shows the characteristics of rising volatility, showing the characteristics of cyclical fluctuations with an 11-month cycle. This is because the data obtained in this article contains only 11 months per year, and the investment in urbanization construction will change with the changes in the months of the year. For example, in 2016, the highest investment amount was 7,068,898 million yuan in June, and the lowest was 3,800.78 billion yuan in February. The goal of this article is to predict the future data (August 2019-July 2020) on the basis of these existing data (February 2010-July 2019).

The ARIMA model is a model specially used to predict this kind of time series data. It mainly captures the characteristics of past historical data, such as growth trends and volatility characteristics, to predict the future data of the model. As shown in Figure 1, the annual fluctuation range of the investment amount is becoming wider and wider, showing that the fluctuation range has a strong increasing trend, so this article uses the ARIMA model with additive seasonal effect to fit the model.

2 Sample data processing

2.1 Draw a line chart

This paper selects China's monthly urban fixed asset investment (100 million yuan) data as sample data, and selects the time interval from February 2010 to July 2019. As shown in Figure 1, the line chart shows that the monthly urban fixed asset investment increases month by month in the form of an approximate linear function curve, and has a seasonal effect with an 11-month cycle. The intra-season volatility increases with the increase in investment, but this variation does not show a drastic change. Therefore, this paper sets the model as an additive seasonal model.

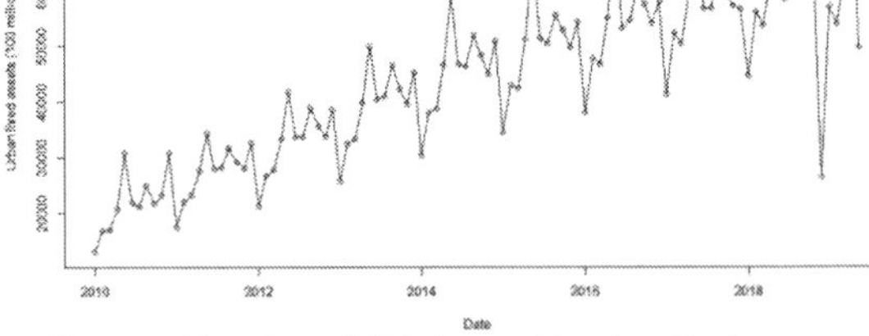

Figure 1. Line chart of China's monthly urban fixed asset investment

*Corresponding author's e-mail: 19120500@bjtu.edu.cn

2.2 Sequence factor decomposition

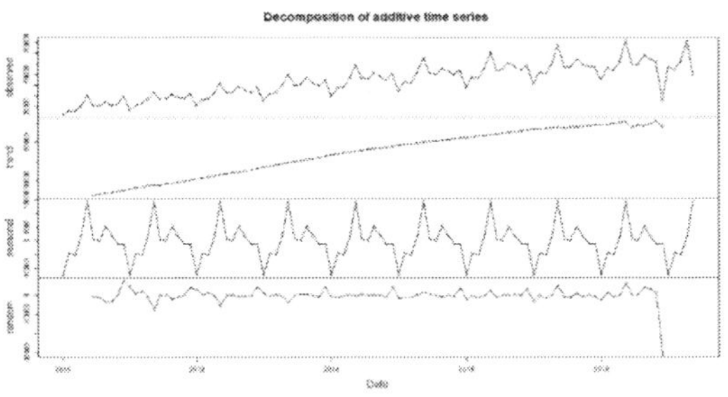

Figure 2. China's monthly urban fixed asset investment sequence factor breakdown diagram

As shown in Figure 2, this article decomposes China's monthly urban fixed asset investment into three parts: trend factors, seasonal factors, and random factors. It can be seen from the trend factors that China's monthly urban fixed asset investment has an obvious linear long-term increasing trend; from the seasonal factors, it can be seen that the time series has obvious seasonal factors with an 11-month cycle, so an 11-step difference process will be considered for this time series.

3 design model

3.1 Stationarity test: ADF test

Considering that China's monthly urban fixed asset investment has an obvious linear growth trend, this paper makes a first-order difference on the original series, and then makes a lagging 11-period first-order difference on the difference series. After extracting the trend effect and seasonal effect of the original sequence, the sequence after the difference is similarly stable. The ADF stationarity test is performed on the sequence after the difference. At the 5% confidence level, the sequence is stable after the difference.

3.2 Model ordering

Draw the autocorrelation graph and partial autocorrelation graph of the sequence after the difference, as shown in Figure 3. The autocorrelation graph shows the autocorrelation coefficient 1st order and 4th order truncation, the partial autocorrelation graph shows 2nd order truncation or 3rd order truncation, so try to use ARMA((1,4),(1,11),2), ARMA((1,4),(1,11),3) extracts the short-term autocorrelation information of the sequence after the difference.

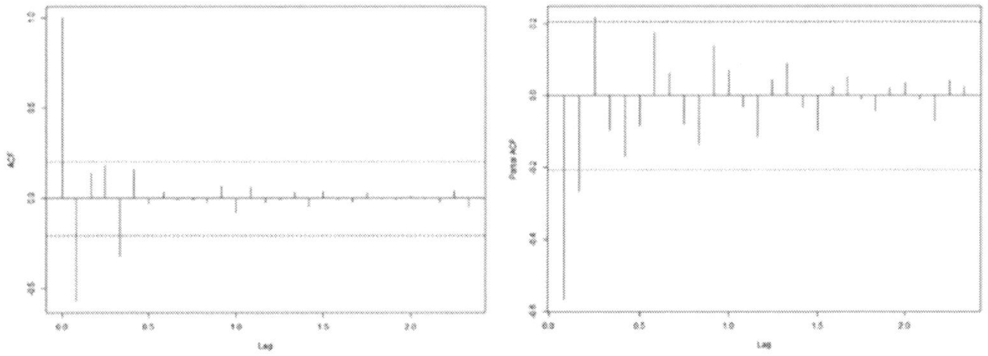

Figure 3. Sequence autocorrelation graph and partial autocorrelation graph

Based on the above information, a total of two addition models are established, as shown in Table 2.

4 predictions and conclusions

The model fitting is generally more accurate. Regardless of the time series stabilization in the early stage or the white noise test of the model residuals in the later stage, the model passed perfectly, so the final prediction results given by the model are convincing and credible. At the same time, it must be noted that there are some areas for improvement in this model fitting: sharp fluctuations at the end may reduce the confidence of the prediction results.

As shown in Figure 4, there are sharp and sudden violent fluctuations at the end of the sequence after the difference, so the predictions made also fluctuate greatly with the violent fluctuations at the end of the data. But from the overall view of the sequence, the predicted volatility is obviously too large to be suspicious. The reason why the model paper does not show the significance of each coefficient of the model is because the

Table 1. Model fitting expression.

Model type	model	Model expression
Additive model	M1	$(1-B)(1-B^{11})x_t = \frac{(1+1.2147B-0.3643B^4)\varepsilon_t}{(1+0.4854B-0.0099B^2)}$
	M2	$(1-B)(1-B^{11})x_t = \frac{(1+1.2356B-0.3731B^4)\varepsilon_t}{(1+0.4896B-0.1074B^2-0.3991B^3)}$

 https://doi.org/10.1051/e3sconf/202129002017

model coefficients fitted in this paper are less significant. Although the quality of the overall model is relatively good, the deficiencies in this part also bring some flaws to the overall model construction.

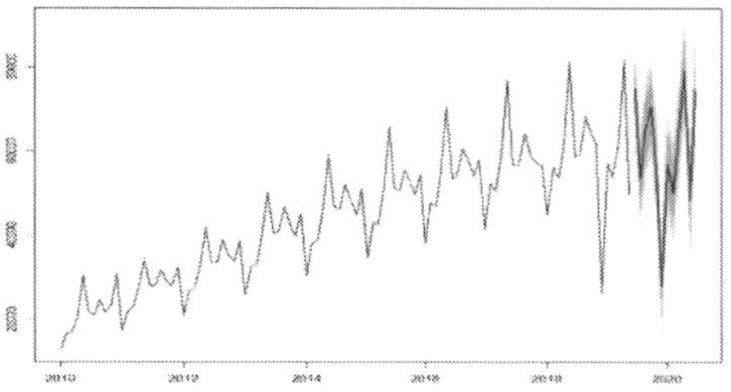

Figure 4. Predicted value map of Model 2

References:

1. Zou Shenghua."Research on GDP Forecast of Ji'an City Based on ARIMA Model." Open Journal of Social Sciences 08.12(2020):. doi:10.4236/JSS.2020.812029.
2. Yan Jingjing."Evaluating the healthcare practice by defining healthcare principles: An autoregressive integrated moving average model based on time series." The International journal of health planning and management.(2020):. doi:10.1002/HPM.3101.
3. Yan Jingjing."Evaluating the healthcare practice by defining healthcare principles: An autoregressive integrated moving average model based on time series." The International Journal of Health Planning and Management 36.2(2020):. doi:10.1002/HPM.3101.
4. Md. Salauddin Khan,and Umama Khan."Comparison of Forecasting Performance with VAR vs. ARIMA Models Using Economic Variables of Bangladesh." Asian Journal of Probability and Statistics .(2020):. doi:10.9734/AJPAS/2020/V10I230243.

Risk Assessment of Flood Disaster in Linyi City Based on GIS

Zheng Huang[1,*]

[1]School of Geomatics and Urban Spatial Informatics, Beijing University of Civil Engineering and Architecture, Beijing, China

Abstract. With the continuous acceleration of the urbanization process, the flood problem seriously affects the economic and social development of the city and the daily life of citizens. Therefore, taking Linyi City as an example, the flood disaster risk assessment is carried out, the Linyi City flood disaster assessment index system is constructed, and the methods of mathematical statistics, analytic hierarchy process and spatial overlay analysis are used to calculate the environmental sensitivity of Linyi City flood disasters with the help of GIS technology. , Risk of hazard factors, vulnerability of hazard-bearing bodies, disaster prevention and mitigation capabilities. The research results show that the hazard levels of flood disasters are increasing from west to east, and high-risk areas are distributed in Feixian and Junan counties; sensitivity is affected by topography and gradually increases from north to south; Lanshan District, Luozhuang District, etc. The city center is highly vulnerable due to its significant population and economic status, and has strong disaster prevention and mitigation capabilities.

1 Introduction

Flood disaster is a kind of severe natural disaster with strong suddenness, high frequency and wide distribution area. In our country, about 50% of the population and 70% of the property are distributed in the flood threat area [1]. In order to effectively achieve the purpose of flood fighting and disaster relief, to accurately analyze and evaluate the possibility of flood disasters and possible losses, and to better formulate disaster prevention and mitigation countermeasures and measures, flood disaster risk research is particularly important [2-4]. In recent years, a large number of scholars at home and abroad have carried out a lot of research work on flood disaster risk assessment and zoning [5-8] and established a set of more mature flood disaster risk assessment methods. Huang Guoru et al. [9] summarized the application status, scope of application, advantages and disadvantages of the four flood risk assessment methods; Wu Shuqi et al. [10-11] established a flood disaster risk assessment model in Zhejiang Province at the county scale for comprehensive evaluation and analysis. Comprehensive management provides a certain reference;

This paper takes Linyi City as the research object, uses existing data, and comprehensively considers the four evaluation indicators of flood hazard hazard factors, hazard environment sensitivity, hazard-bearing body vulnerability, and disaster prevention and mitigation capabilities, and uses GIS spatial analysis The technology spatializes each index factor, and combines the analytic hierarchy process to determine the weight of each influencing factor, and conducts the flood disaster risk assessment of Linyi City. This research can provide methodological basis for the flood disaster early warning and disaster prevention and reduction decision-making in Linyi City.

2 Overview of the study area and data sources and processing

2.1. Survey of the study area

Linyi City is located between 34°22′ ~ 36°13′ north latitude and 117°24′ ~ 119°11′ east longitude. It is located in the southeast of Shandong Province, close to the Yellow Sea, with a total area of 17,191.2 square kilometers and a permanent population of 10.667 million. The urbanization rate is 52.75%. It has jurisdiction over 3 districts, 9 counties, and 3 development zones. It is the largest and most populous city in Shandong Province. The administrative division map is shown in Figure 1. The topography is centered on the Yi and Shu Rivers, surrounded by mountains in the west, north and east, forming a fan-shaped alluvial plain to the south. The area ratio of mountains, hills, and plains is 2:4:4; Linyi has a temperate monsoon climate, with obvious seasonal precipitation. The precipitation is concentrated in summer, with the most precipitation in July. In recent years, Linyi City has been severely affected by flooding disasters. In the summer of 2019, many areas in Linyi City were affected. Hedong District and Yishui County were flooded. The city's affected population was 358,600, and the affected area of crops was 5,9257.3 hectares.

* Corresponding author: 2108521519023@stu.bucea.edu.cn

https://doi.org/10.1051/e3sconf/202129002018

Figure 1. Administrative District Map of Linyi City

2.2 Data source and processing

The data used in this paper mainly include the daily precipitation records of 9 meteorological stations in Linyi City from 2010 to 2019, which come from the China Meteorological Data Network. The DEM data of Linyi City comes from the geospatial data cloud, and the GDP per capita and GDP per area come from the Linyi City Statistical Yearbook.

Flood disaster risk assessment indicators have different dimensions. Because of the need to superimpose the raster data of each indicator, in order to eliminate the differences caused by the different data dimensions and maintain reasonableness, it is necessary to standardize the evaluation index data, that is, data quantification processing. In this paper, formula (1) is used to normalize each index into a calculable non-vector index between 0 and 1:

$$X'_i = \frac{X_i - X_{imin}}{X_{imax} - X_{imin}} \quad (1)$$

Where X'_i is the normalized value of the i-th index, X_i is the i-th index value, and X_{imin} and X_{imax} are the minimum and maximum values of the i-th index, respectively.

3 Research method

3.1. Evaluation index system construction

In the selection of indicators for flood disaster risk assessment in Linyi City, not only the objectivity and availability of each indicator data must be considered, but also the indicators selected must be able to reflect the actual situation of flood disaster risk in the region. The risk of hazard factors mainly depends on meteorological factors and underlying surface factors. In this paper, two indicators of rainfall intensity and rainfall frequency are selected as hazard factors. The sensitivity of hazard-pregnant environment mainly depends on the external environment. This study uses ground elevation and slope to determine the hazard-pregnant environment sensitivity index. The vulnerability of disaster-bearing bodies mainly depends on social and economic factors such as population and economy. This paper selects the average population and GDP per area to evaluate the vulnerability. Disaster prevention and mitigation capabilities mainly depend on pre-disaster protection and post-disaster response. This study uses per capita GDP to evaluate the disaster prevention and mitigation capabilities of the study area.

3.2. Index weight determination

The Analytic Hierarchy Process (AHP) is often used to assign values to various indicators and factors, and then to make decisions on fuzzy and complex problems. The main steps are: ①construct the judgment matrix; ② calculate the index weight; ③check the consistency. This article uses analytic hierarchy process and expert scoring method to analyze the relationship between the factors in the system and establish the hierarchical structure of the system. Experts are asked to score the factors in pairs and compare the elements of the same level with respect to a certain criterion in the previous level. The importance is compared in pairs, and a pairwise comparison matrix (judgment matrix) is constructed. Calculate the relative weight of the compared element to the criterion from the judgment matrix, and conduct a consistency test (CR<0.1), and finally get the comprehensive weight, and calculate the risk index (Table 1).

Table 1. Flood risk assessment indicators and their weights in Linyi.

Target layer	Criterion layer	Criterion-level weight	Index layer	Index layer weight
Flood disaster risk assessment	Hazard risk	0.489	First-degree flood frequency	0.06
			Secondary flood frequency	0.08
			Three-level flood frequency	0.15
			Four-level flood frequency	0.25
			Five-level flood frequency	0.46
	Disaster environmental sensitivity	0.305	Ground elevation	0.67
			slope	0.33
	Vulnerability of disaster-bearing body	0.127	Average population	0.83
			GDP per land	0.17
	Disaster prevention and mitigation capabilities	0.079	GDP per capita	1

4 Result analysis

4.1. Hazard risk assessment

The hazards of hazard factors mainly consider the rainfall intensity and rainfall frequency in Linyi City. The precipitation process takes at least 1 day of continuous precipitation and precipitation greater than 30mm as a process, and then the precipitation in the whole process is accumulated as the total precipitation of a heavy rain process. Calculate the daily precipitation data of all stations in Linyi City from 2010 to 2019 during the rainstorm process of 1, 2...7 days (including 7 days or more), and establish 7 precipitation process sequences of different time lengths, and use percentages The digit method calculates the precipitation corresponding to 98%, 95%, 90%, 80%, and 60% of the precipitation of the 7 time series, and divides the intensity of the rainstorm into five levels: the precipitation is between 60% and 80% The floods are level 1 floods, those between 80% and 90% are level 2 floods, those between 90% and 95% are level 3 floods, those between 95% and 98% are level 4 floods, and those between 98% and 98% are level 4 floods. It is a level 5 flood. Taking the Yishui County site as an example, the specific values are shown in Table 2 (other sites are omitted). The higher the level of rainfall intensity, the greater the effect on the formation of floods. Five levels of weight are given to rainfall intensity, namely 1/15, 2/15, 3/15, 4/15, and 5/15. The weighted comprehensive evaluation method is used to calculate the risk indicators of disaster-causing factors at each site, as shown in Table 3. Spatial interpolation is performed in ArcGIS to obtain a distribution map of the risk levels of flood-causing factors in Linyi City, as shown in Figure 2.

Table 2. Rainfall range of different grades of rainstorm intensity (R) in Yishui County/mm

Heaven	First level	Second Level	Third Level	Fourth Level	Fifth Level
1	89≤R≤156	156 < R≤240	240 < R≤247	247 < R≤507	R > 507
2	122≤R≤213	213 < R≤333	333 < R≤394	394 < R≤604	R > 604
3	208≤R≤328	328 < R≤540	540 < R≤835	835 < R≤1093	R > 1093
4	239≤R≤328	328 < R≤432	432 < R≤739	739 < R≤923	R > 923
5	372≤R≤754	754 < R≤819	819 < R≤854	854 < R≤977	R > 977
6	256≤R≤663	663 < R≤697	697 < R≤901	901 < R≤1228	R > 1228
7	761≤R≤1353	1353 < R≤1751	1751 < R≤1880	1880 < R≤2621	R > 2621

Table 3. Risk indicators of hazard factors at each site in Linyi City

Site	index
Yishui	13.9
Linyi	13.4
Cangshan	13.3
Feixian	14
Junan	14.1
Linshu	13.8
Mengyin	13.2
Yinan	13
Pingyi	12.3

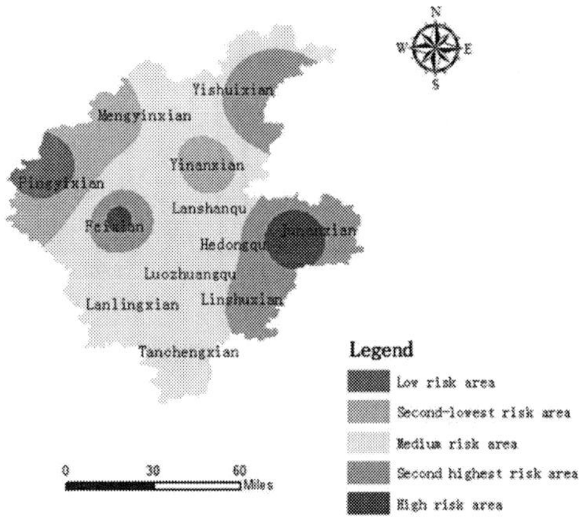

Figure 2. Distribution map of hazard levels of flood-causing factors in Linyi City

4.2. Environmental Sensitivity Assessment of Disaster Pregnancy

Disaster-generating environment refers to the natural environment that breeds flood disasters, such as topography, geology, soil, vegetation, water system, etc. The terrain of Linyi City is high in the northwest and low in the southeast. It is mountainous in the north and flat in the middle and south. The topography makes the rainfall in this area greatly affected by the topography, and it is easy to converge from the high terrain to the low terrain. In the case of large areas, long-term rainfall or short-term heavy rainfall is likely to cause water accumulation in mountainous depressions, which is conducive to the formation of floods. In this paper, ground elevation and slope are selected as disaster-pregnant environmental factors. A 30m resolution digital elevation model and a 3×3 rectangular window selected from the grid calculation module of ArcGIS are used to express the relative elevation standard deviation. The lower the elevation, the smaller the standard deviation of elevation, which means that flood disasters are more likely to occur, and the greater the impact value, and the comprehensive impact degree assignment of topographic factors is obtained (Table 4). Analyze the slope and reclassify it. According to the weights given in Table 1, the weighted comprehensive evaluation method is used to obtain the disaster-pregnancy environmental sensitivity level distribution map, as shown in Figure 3. The results show that the disaster-pregnancy environmental sensitivity of Linyi City is low in the north and high in the south, and the areas with the strongest sensitivity are mainly Distributed in the south of Tancheng County, Lanshan District, Hedong District, etc., while the north of Yishui County, Mengyin County, etc. are less sensitive. The reason is that the central and southeast parts of Linyi City are low in terrain and the comprehensive influence index of topographic factors is the largest Therefore, it is the most sensitive area for flood disasters.

Table 4. Linyi City Elevation and Elevation Standard Deviation Combination Assignment

Terrain elevation/m	Elevation standard deviation		
	First level (≤1)	Second Level (1 ~ 10)	Third Level (≥10)
First level (≤100)	0.9	0.8	0.7
Second Level (100 ~ 300)	0.8	0.7	0.6
Third Level (300 ~ 700)	0.7	0.6	0.5
Fourth Level (≥700)	0.6	0.5	0.4

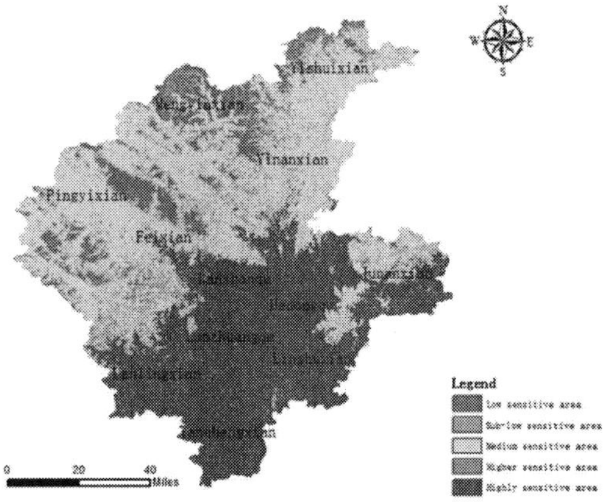

Figure 3. Distribution Map of Environmental Sensitivity Level of Disaster Pregnancy in Linyi City

4.3. Vulnerability assessment of the carrier

The occurrence of flood disasters has a certain relationship with the carriers of disasters. The losses caused by disasters generally depend on the economic conditions and population density of the disaster-affected areas. The economy is developed, and the densely populated places suffer from flood disasters. This article selects the average GDP and the average population in Linyi City in 2019 as the risk assessment indicators. After the data is standardized, the index values are assigned weights according to Table 1, and the weighted comprehensive evaluation method is used to obtain the vulnerability distribution of the carrier. Figure, see Figure 4. It can be seen from the figure that Lanshan District has the densest population and developed economic level and is prone to disasters, followed by Yishui County and Lanling County. Mengyin County and Linshu County have lower population density and have the lowest vulnerability in the city.

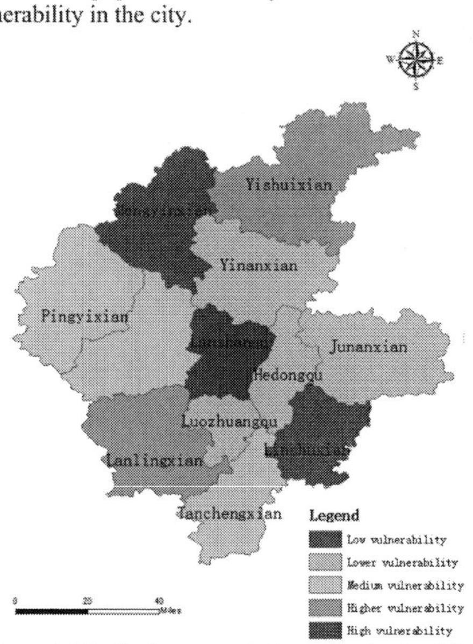

Figure 4. Distribution map of vulnerability levels of flood bearing bodies in Linyi City

4.4. Disaster prevention and mitigation capability assessment

According to the "Linyi City Statistical Illustrations in 2019", per capita GDP is selected as the evaluation index of disaster prevention and mitigation ability and standardized processing. It is divided into 5 grades on the ArcGIS platform to obtain the distribution map of Linyi City's flood prevention and disaster mitigation ability level, as shown in Figure 5. Comprehensive analysis shows that Linyi's central urban area has strong disaster prevention and mitigation capabilities, while Junan and Mengyin counties have weaker disaster prevention capabilities.

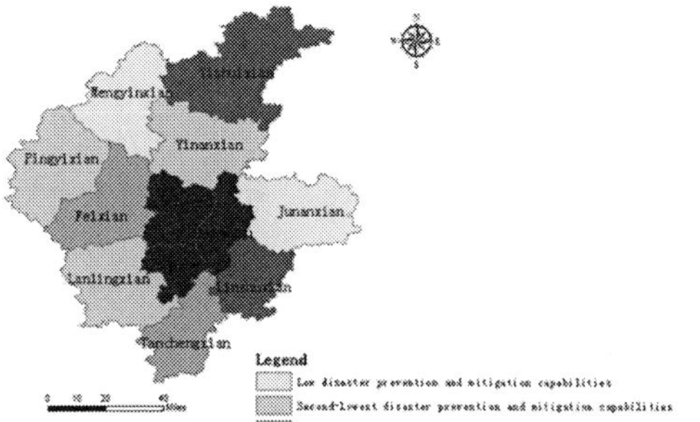

Figure 5. Distribution map of Linyi City's flood and waterlogging disaster prevention and mitigation capabilities

5 Conclusion

This paper builds a flood disaster risk assessment index system, comprehensively considers precipitation, topography, population density, GDP and other information, and uses GIS technology to assess the flood disaster risk in Linyi City, which can more objectively reflect the flood disaster risk in different regions. reality. However, there are still a series of problems such as incomplete basic information, disaster mechanism analysis and assessment techniques need to be strengthened. Therefore, in the future, the breadth and depth of risk assessment will be expanded to provide better support for risk assessment.

References

1. Chen, H., Chen, G., Ding, G. (2003) Risk assessment of regional flood disaster based on GIS. Yangtze River., 06: 49.
2. Zhao, W., Huang, J., Zhang, Z. (2017) Research on Comprehensive Risk Assessment of Flood Disaster in Chongqing City Based on GIS. Journal of Southwest China Normal University(Natural Science Edition)., 42(02): 34–42.
3. Li, C., Tian, J., Shen, R. (2020) Research progress of flood disaster risk assessment. Journal of Catastrophology., 35(03): 131–136.
4. Cheng, L., Fu, M., Wang, L. (2019) Risk Assessment of Flood Disaster at County Level Based on RS and GIS. South-to-North Water Transfers and Water Science & Technology., 17(06): 37–44+68.
5. Chen, W., Wang, X., Deng, S. (2019) Integrated urban flood vulnerability assessment using local spatial dependence-based probabilistic approach. Journal of Hydrology., 575: 454–469.
6. Li, S. (2019) Risk Assessment and Zoning of Rainstorm and Flood Disasters in Southeastern Inner Mongolia: A Case Study of Tongliao City. Meteorology Journal of Inner Mongolia., 01: 23–28.
7. Xue, X., Ma, J., Li, H. (2012) Gis-based risk assessment and regionalization of township flood disaster: a case study in Linzi District, Zibo, Shandong province. Journal of Catastrophology., 27(04): 71–74+79.
8. Zhu, Z., Zhang, Y. (2021) Flood disaster risk assessment based on random forest algorithm. Neural Computing and Applications., : 1–13.
9. Huang, G., Luo, H., Lu, X. (2020) Summary of urban flood disaster risk analysis and zoning methods. Water Resources Protection., 36(06): 1–6+17.
10. Wu, S., Zhao, W., Wang, Z. (2020) GIS-based flood disaster risk assessment and zoning: Taking Zhejiang Province as an example. China Rural Water and Hydropower., (06): 51–57.
11. Camarasa-Belmonte A. M., Soriano-García J. (2012) Flood risk assessment and mapping in peri-urban Mediterranean environments using hydrogeomorphology. Application to ephemeral streams in the Valencia region (eastern Spain). Landscape and Urban Planning., 104(2): 189-200.

Numerical simulation of wave spectrum during a typhoon

Congying KONG [1,2,3,a] Hao LIU[4]

[1]Tianjin Port Engineering Institute Co., Ltd. of CCCC First Harbor Engineering Co., Ltd., Tianjin, 300222,China
[2]CCCC First Harbor Engineering Company Ltd., Tianjin, China
[3]Key Laboratory of Coastal Engineering Hydrodynamic, CCCC, Tianjin, 300222,China
[4]Tianjin Nangang Industrial Zone Water Supply Co. Ltd, Tianjin, 300000,China

Abstract: The third-generation wave model WAVEWATCH-III was used to numerically simulate the wave under the influence of a typhoon in the coastal area of China. The wave spectrum at the buoy point was output, and the characteristics of the wave spectrum were analyzed. The change of the wave spectrum during the typhoon process reflected the growth process of typhoon formation, development and extinction. The relationship between the wave spectrum and the wind direction was intuitively shown by the directional spectrum, indicating the coexistence of wind waves and swells in the sea area during the typhoon process.

1 Introduction

Wave spectrum is an important statistical property of random waves. It not only contains the second-order information of waves, but also gives the distribution of wave energy relative to frequency and direction. It is the most effective way to describe waves. At present, the wave spectrum has been widely used in the strength and stability calculation of various offshore platforms. Compared with the characteristic parameters such as wave height and period, the wave spectrum can more effectively reflect the internal structure of the wave field. At the same time, each wave element can be obtained by each order spectral moment. The study of wave spectrum is of great significance for the construction and maintenance of marine engineering, and it is also an important basis for the study of wave prediction, coastal sediment movement and offshore hydraulic structure design[1].

Modern wave observation mainly relies on the wave meter, but large-scale long-term sequence of sea observation, the price is expensive, and in extreme weather (cold wave, typhoon) operability is not strong, so the use of wave numerical simulation method to study the time and spatial distribution of waves, to make up for the lack of measured wave data has become an effective way. The wave spectrum obtained by numerical simulation can be used for the prediction of large-area waves, the diffraction and refraction of waves, the force and vibration of hydraulic structures, the response of hull, buoy and other floating bodies to waves, and the study of sediment movement[2].

2 Model Setting and Verification

The WAVEWATCH- III model for large-scale wave simulation in the full spectral space jointly developed by the U.S Atmospheric and Oceanic Administration and the U.S Environmental Prediction Center is based on the solution of the spectral action balance equation[3].

Within WAVEWATCH-III, the basic spectrum is the wavenumber-direction spectrum $F(k, \theta)$, which has been selected because of its invariance characteristics with respect to physics of wave growth and decay for variable water depths. The output, however, consists of the more traditional frequency-direction spectrum $F(fr, \theta)$. The different spectra can be calculated from $F(k, \theta)$ using straightforward Jacobian transformations.

In a general sense, wave action $A = E/\sigma$ is conserved(Whitham, 1965; Bretherthon and Garrett, 1968). This makes the wave action density spectrum $N(k, \theta) \equiv F(k, \theta)/\sigma$ the spectrum of choice within the model. Wave propagation then is described by $DN/Dt = S/\sigma$ (1)

Where D/Dt represents the total derivative (moving with a wave component) and S represents the net effect of sources and sinks for the spectrum F.

The study area is 110°E~139°E, 13°N~40°N, and the terrain data are derived from the ETOPO1; The wind field adopts CCMP wind field with spatial resolution of 0.25°×0.25° and time interval of 6h. The model was calculated from June 21 to 27, and the results were output at a 1 hour interval. The measured data come from buoy1 and buoy2. Typhoon path map and buoy position are shown in Figure 1.

Figure 2 is the comparison between the calculated

* Corresponding author: [a]kongcongying@tpei.com.cn

https://doi.org/10.1051/e3sconf/202129002019

and measured significant wave height at the buoy site. From the effective wave height process curve, the measured wave height is in good agreement with the calculated wave height, which can accurately reflect the trend of wave size. The correlation coefficient R between the calculated values and the measured values of buoy1 and buoy2 were 0.956 and 0.971, respectively, and the root mean square error RMSE were 0.331 and 0.368, respectively. From the data, the model calculation results are reasonable and credible.

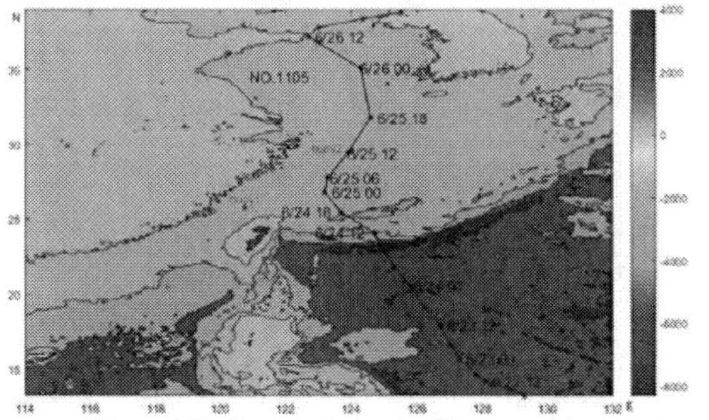

Figure1 Typhoon path map and buoy position

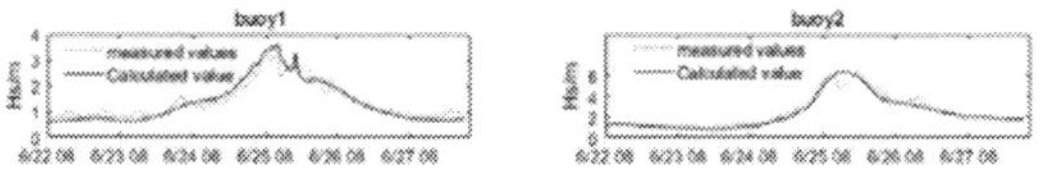

Figure2 the comparison between the calculated and measured significant wave height at the buoy site

3 results and analysis

3.1 One-dimensional spectrum characteristics analysis

Figs. 3 and 4 represent the one-dimensional wave spectrum, and Fig. 5 shows the spectral peak corresponding to the period with large typhoon influence. From figures 3 and 4, the wave spectrum is a single peak spectrum at 12 :00 on 23 th, which indicates that the wave field is a single wind wave or surge, and the peak frequency at each buoy point is above 0.15 Hz. Typhoon 1105 on 24th and 25th is in the growth period. Affected by the sea surface wind, the energy of wind waves and swells increases, and the energy of waves gradually moves to the low frequency part, and the frequency of spectral peaks is less than 0.1 Hz. On 26 and 27 days, the typhoon intensity gradually weakened, the spectral peak significantly decreased, the spectral shape gradually expanded, and the spectral peak frequency shifted to the right.

In Fig5, the maximum spectral peak of buoy1 was 45m^2·s at 4:00 on 25th, and the peak frequency was 0.0453 Hz. The buoy2 also reached the maximum spectral peak of 52.8 m^2·s at 4:00 on 25th, and the peak frequency was 0.073 Hz; This is because 24 and 25 days are the strongest periods of typhoon development. From the comparison of typhoon path in Figure 1, the buoy is close to the typhoon center during this period, and the buoys are in an open ocean with sufficient energy growth. Fig5 shows that the spectral peak first increases and then decreases, which exactly reflects the growth process of typhoon formation, development and extinction.

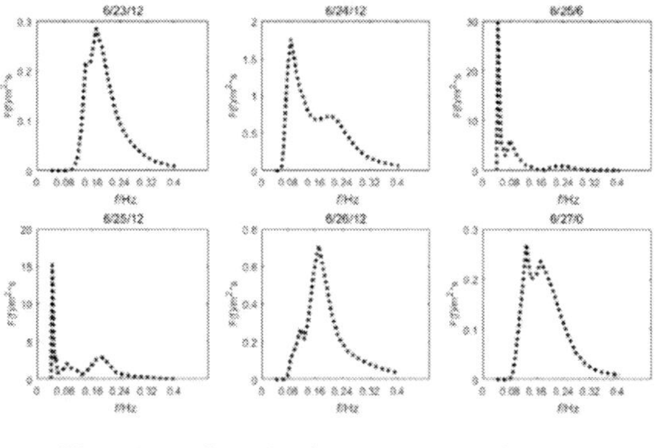

Figure3 one-dimensional wave spectrum at buoy1

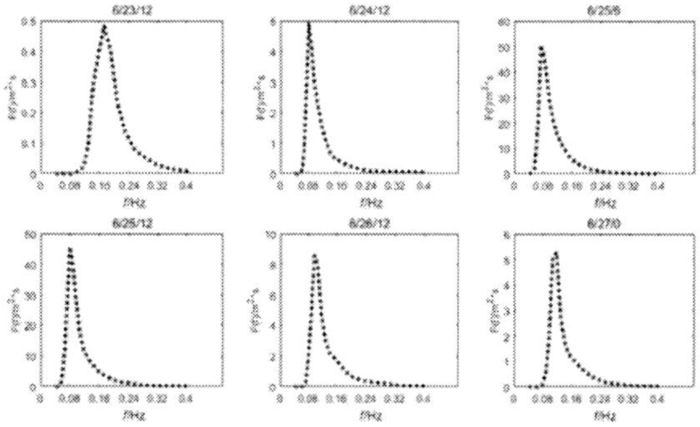

Figure4 one-dimensional wave spectrum at buoy2

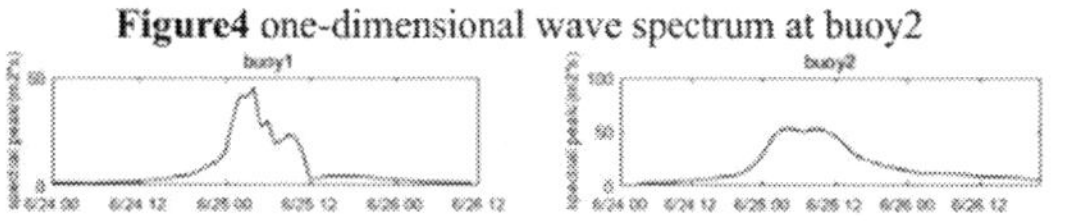

Figure5 the spectral peak corresponding to the period with large typhoon influence

3.2 Analysis of directional spectrum characteristics

The spectrum that describes the composition of waves along different directions is called directional spectrum. Directional spectrum is a function of frequency and direction, which can describe the distribution of wave energy relative to frequency and direction, and some statistical characteristics of wave space.

$$S(\omega,\theta) = S(\omega)D(\omega,\theta)$$

$S(\omega)$ is frequency spectrum; $D(\omega,\theta)$ reflects a function of energy relative to the direction distribution,θ is the angle between the constituent wave and the main wave direction[4].

Figs 6~11 are directional spectra at six different times. The radius of the circle in the graph represents the frequency (Hz), and the scale on the arc is the direction (degree). Directional spectrum value is expressed in color stripes. And arrows represent wind direction.

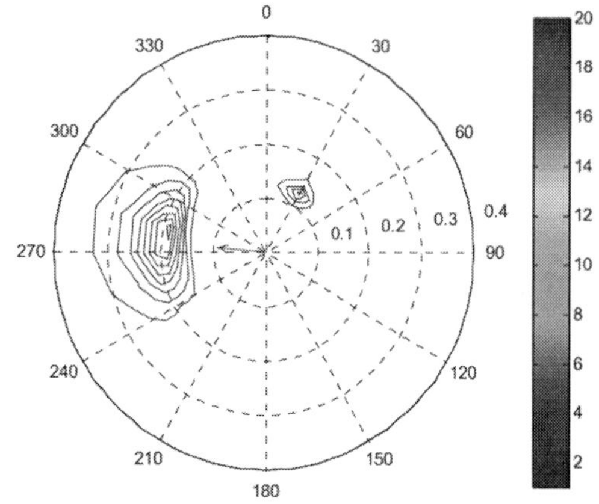

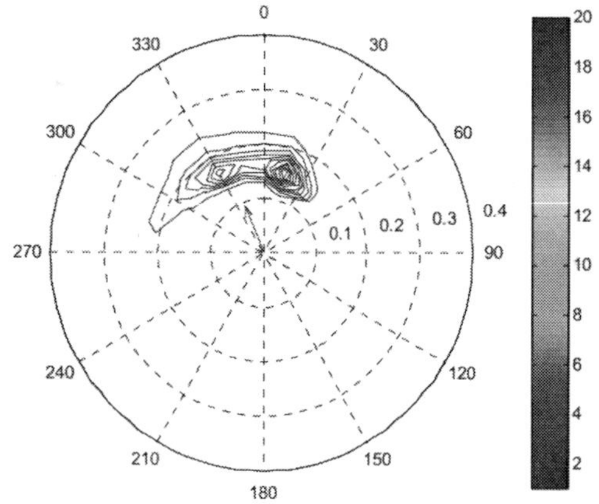

Figure6 Directional spectrum of buoys 1 and buoy2 varies with frequency and direction at 12:00 on 23rd

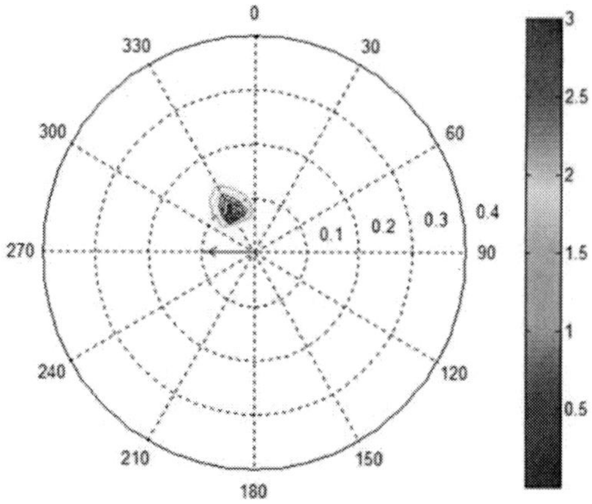

Figure7 Directional spectrum of buoys 1 and buoy2 varies with frequency and direction at 12:00 on 24rd

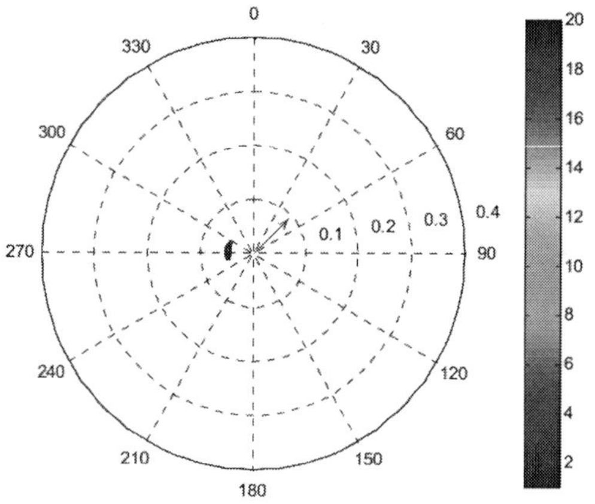

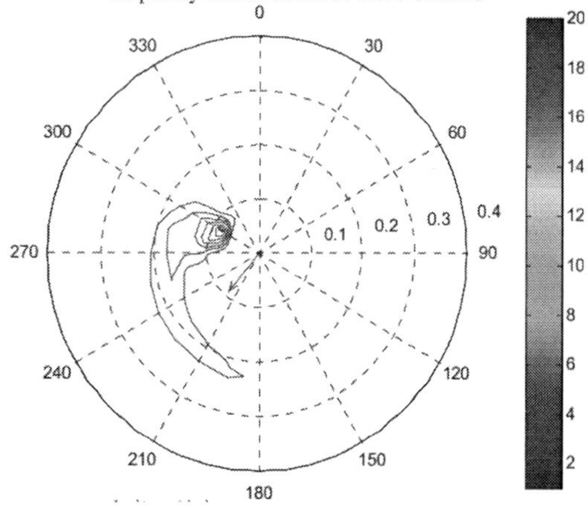

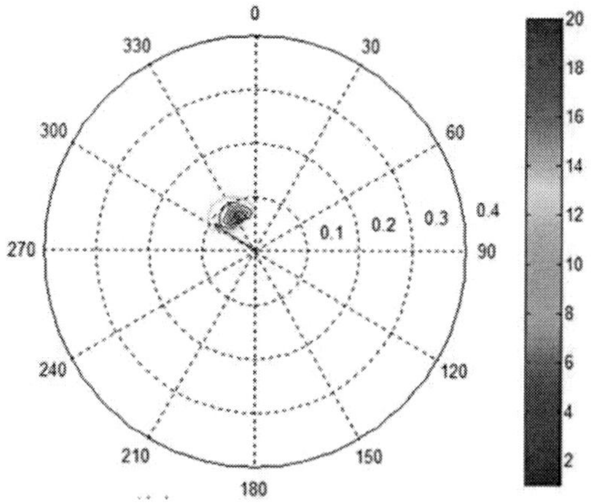

Figure8 Directional spectrum of buoys 1 and buoy2 varies with frequency and direction at 06:00 on 25rd

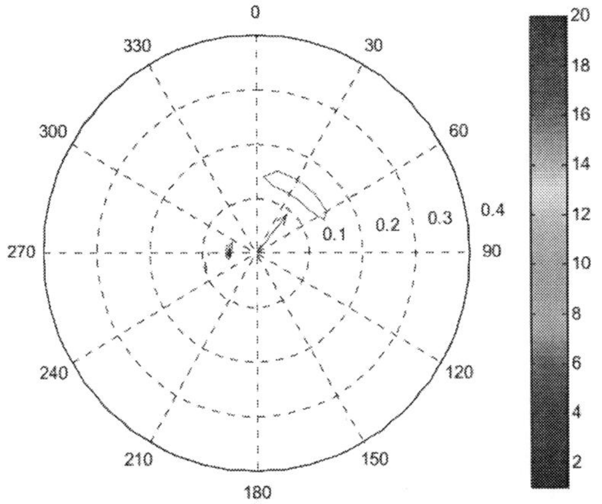

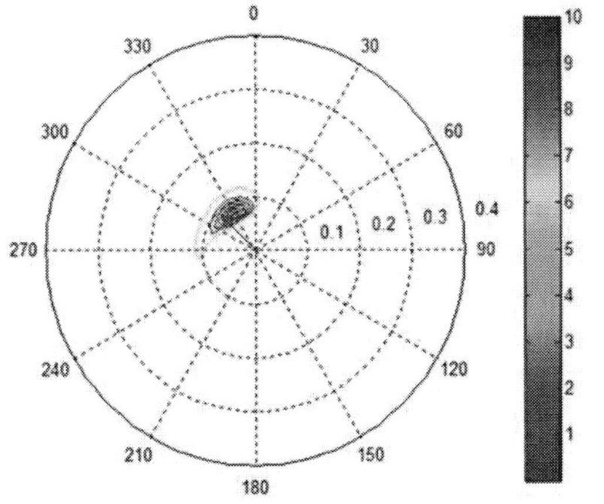

Figure9 Directional spectrum of buoys 1 and buoy2 varies with frequency and direction at 12:00 on 25rd

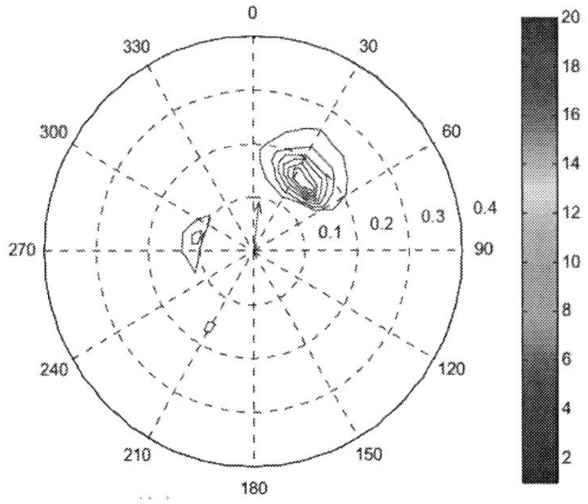

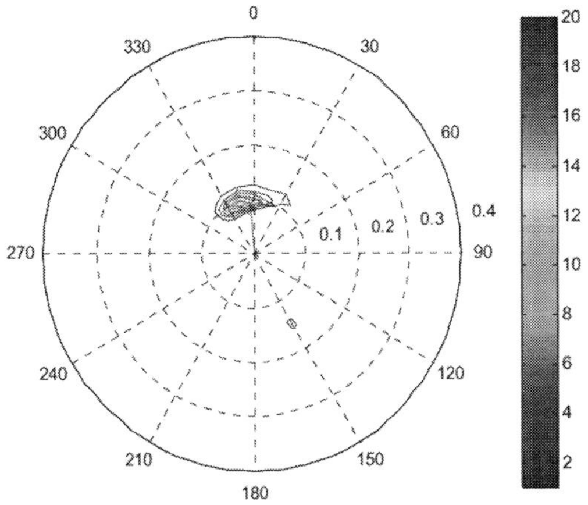

Figure10 Directional spectrum of buoys 1 and buoy2 varies with frequency and direction at 12:00 on 26rd

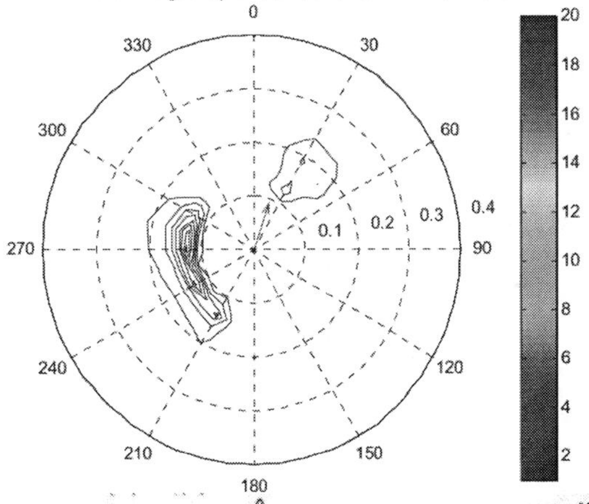

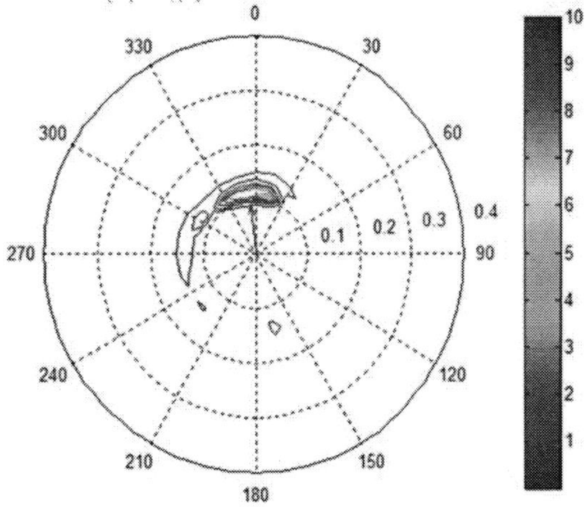

Figure11 Directional spectrum of buoys 1 and buoy2 varies with frequency and direction at 00:00 on 27rd

At 12:00 (Figure 6) on 23rd, it can be found that the frequency range of the spectrum is between 0.15Hz and 0.3Hz, the frequency range is wide, and the direction span is also large. The above characteristics show that the wave energy is more dispersed at this time, which is corresponding to the one-dimensional spectral characteristics of this time. With the increase of typhoon wind speed in the sea area from 24rd to 25rd (Fig7 ~Fig9), the energy of buoy points increased, and the energy of wave spectrum reached the maximum (the color strip represents the directional spectrum value).

In Figs. 6~11, the direction of the wave spectrum energy concentration of buoy 1 is inconsistent with the wind direction, while the place of the wave spectrum energy concentration of buoy 2 is gradually consistent with the wind direction, which reflects the relative position relationship between the typhoon center path and the two buoys : the direction of the typhoon movement is far from buoy 1, and towards buoy 2.

Taking the two-dimensional spectrum of buoy 1 at each time as an example, part of the wave spectrum energy in the figure is concentrated between 250° and 300°, and another part of the energy changes with the wind direction. The propagation direction of swell in the wave is not affected by the wind, and the propagation direction of wind wave changes with the wind direction. Therefore, it indicates that there are wind waves and swells in the sea area at the same time. When the angle between the two is large, the direction separation of wind waves and swells occurs (the part pointing to the arrow and the part far from the arrow in the figure), which exactly reflects the bimodal spectrum of one-dimensional spectrum at each time in Figs.3~4[5].

4 Conclusions

WAVEWATCH-III model can well respond to the change process of typhoon waves in the sea area, and the model calculation results are reasonable and reliable. One-dimensional frequency spectrum reflects the typhoon from growth to extinction, with the increase of the severity of the sea, the peak frequency of the wave spectrum moves to the left, and the energy is concentrated in the low frequency part ; In the decay process, the spectral peak gradually decreases, and the peak frequency tends to move to high frequency.

The directional spectrum shows the distribution of waves in various directions and frequency ranges ; The relationship between the direction of energy propagation and the wind direction is more intuitive to show the coexistence of wind waves and swells in the typhoon process and the characteristics of wind waves and swells. The obtained wave spectrum plays an important guiding role in air-sea interaction, upper ocean dynamics, wave prediction, ocean remote sensing and ocean engineering.

References

[1] Kong Congying.2013. Numerical simulation of typhoon waves around the waters of China's offsea J. Marine Environmental Science, Vol. 32 No. 3, P. 419-423.

[2] YANG Shengqiang. 2015.Observed typhoon wave spectrum in northern South China Sea. J. Chinese Journal of Oceanology and Limnology,Vol. 33 No. 5, P. 1286-1294.

[3] Hendrik L. Tolman. 2009.User manual and system documentation of WAVEWATCH III. Environmental Modeling Center Marine Modeling and Analysis Branch,the U.S.

[4] ZHANG teng. 2016. A Comparative Study of Stationary Waves Spectrum Analysi. D. Dalian University of Technology.

[5] YANG zhong-liang.2017.The wave characteristics and the analysis of wave spectrum off the southeastern coast of Zhaitang Island. D. Journal of Marine Sciences.

Side Scan Sonar Shipwreck Image Recognition Algorithm based on Transfer Learning and Deep Learning

Boyu Zhang [1], Xiao Wang [1, 2,*], Shudong Li[1], Jinghua Yang[1]

[1]School of Marine Technology and Geomatics, Jiangsu Ocean University, Lian yungang 222000 China;
[2]Jiangsu Institute of Marine Resources, Lian yungang 222000 China;

Abstract. Current underwater shipwreck side scan sonar samples are few and difficult to label. With small sample sizes, their image recognition accuracy with a convolutional neural network model is low. In this study, we proposed an image recognition method for shipwreck side scan sonar that combines transfer learning with deep learning. In the non-transfer learning, shipwreck sonar sample data were used to train the network, and the results were saved as the control group. The weakly correlated data were applied to train the network, then the network parameters were transferred to the new network, and then the shipwreck sonar data was used for training. These steps were repeated using strongly correlated data. Experiments were carried out on Lenet-5, AlexNet, GoogLeNet, ResNet and VGG networks. Without transfer learning, the highest accuracy was obtained on the ResNet network (86.27%). Using weakly correlated data for transfer training, the highest accuracy was on the VGG network (92.16%). Using strongly correlated data for transfer training, the highest accuracy was also on the VGG network (98.04%). In all network architectures, transfer learning improved the correct recognition rate of convolutional neural network models. Experiments show that transfer learning combined with deep learning improves the accuracy and generalization of the convolutional neural network in the case of small sample sizes.

1. Introduction

Marine surveying and mapping is the basis of all underwater activities and the premise of seabed surveys, seabed mineral explorations, and marine economic development. Side Scan Sonar (SSS), a high resolution, multi-purpose and low-cost marine survey equipment, is widely used for surveying oceans, ports, rivers, lakes, and other waters. The main applications of SSS are hydrographic surveys, underwater engineering site selection, detection of underwater targets such as aircraft wreckage, shipwrecks, torpedoes, mines, containers, chemical barrels, anchor chains, sunken rocks, buoys, anchorage obstructions, and seabed bottom material classification, marine survey and marine scientific research [1]. Currently, commercial SSS data processing software for underwater target detection and recognition relies on manual interpretation, which is inefficient and does not guarantee accuracy. Traditional target recognition and detection methods of SSS images mainly include recognition methods based on Markov Random Fields (MRF) [2], gray histogram and geometric features [3], extracting features in the frequency domain [4], fuzzy clustering [5], detection methods based on template matching [6], threshold [7], sparse representation [8], and classification models based on Support Vector Machine (SVM) [9], among others. The general workflow of such methods includes: feature extraction, model construction, and model application. Subject to the complex marine environment and the presence of a large number of noises in SSS images, the model constructed in one sea area may not be applicable to another. Moreover, due to the difficulty in obtaining typical features and distinguishing them, the recognition efficiency is low.

Convolutional Neural Network (CNN)[10] is a widely used artificial intelligence algorithm for image classification, image recognition, image segmentation and other fields [11-14] because of its advantages such as not requiring manually selected features and having excellent feature extraction skill. There are many architectures of convolutional neural networks [15-20]. However, an excellent convolutional neural network can only be obtained with a large amount of appropriate training data. For common study objects, such as cars, plants and animals, it is easy to collect a large number of images [21]. However, in special cases, such as in the case of underwater shipwrecks, obtaining a large number of samples is very difficult.

To overcome the problem of insufficient samples in deep learning, Matias et al. [22] proposed a method of training convolutional neural networks with small samples, which can achieve an accuracy of 80% when the

Correspondence author: *wxsdau2005@163.com;

Tel: +86-137-7548-9830

samples are less than 20. Liang et al. [23] proposed a small sample underwater target recognition method based on a convolutional neural network, and validated it with fish image data. The experiment achieved an accuracy recognition rate of 85.08%. Mehrotra et al. [24] combined transfer learning with deep learning methods to train the network with a small sample data of brain tumors, successfully distinguishing malignant from benign tumors. Zhen et al. [25] studied how to identify power towers in remote sensing images based on photos of natural power towers. In the experiment, the photos of natural power towers were used to train the network architecture, and then the training parameters were extracted and imported into the new network, and then the power towers in remote sensing images were used for network training. The research results show that transfer learning is helpful for network training. Zhu et al. [26] built a submarine model at a scale of 100:1 and performed simulation experiments on underwater submarine recognition. During the research, the parameters of the sonar were adjusted to the corresponding intensity, and the simulation data were collected in an underwater acoustic pool. Multiple pre-training networks can be used to perform transfer learning training on the data. Transfer learning can achieve higher recognition rates and faster convergence rates in the case of little data. These manuscripts provide a theoretical solution on how to apply CNN model well for SSS shipwreck that has small sample image.

The marine environment is complex and noisy, and sample image characteristics are not obvious. With very few marked samples, deep learning applications are limited. Transfer learning provides a good solution, but what kinds of data can be used for transferring the SSS shipwreck image and which CNN models are suitable for underwater shipwreck target detection and recognition have not yet been identified. We propose a method of combining transfer learning with a convolutional neural network. Through design experiments, the SSS data were transferred to a network trained with public data from different sources and different characteristics, and to a network trained with ship data from different sources but the same characteristics. The aim of this study was to obtain a convolutional neural network model suitable for underwater shipwreck recognition, and solve the problems related to the difficulty in obtaining large SSS image samples under complex marine environment, and in typical feature extraction.

The remainder of this paper is organized as follows: In Section 2, the source of the data set used in the experiment is introduced, the classification of the data, and the relevance of the data are described. In Section 3, briefly describes the experimental principle, experimental method, experimental procedure and accuracy evaluation method. Section 4 describes the experimental results. Discussion and the conclusion of the paper is given in Section 5.

2. Data

2.1 SeaShips-3

Shao et al. [21] created SeaShips, a ship dataset, by placing fixed cameras in river courses and estuaries to record a large number of samples through video screenshots. This dataset, with a total of 31455 pictures and 6 classes, has been used for research on target recognition algorithms for ships on water. Since the number of samples in each class is inconsistent in this dataset, three classes were randomly selected from SeaShips (Figure 1), each with 800 pictures, and the selected dataset was named SeaShips-3 for this experiment.

(a) (b) (c)

Figure 1. Examples of (a) Bulk cargo carrier samples; (b) General cargo ship samples; (c) Fishing boat samples.

2.2 CIFAR-3

Cifar-10 is a commonly used public dataset, which is composed of 10 classes and 60,000 color images with a size of 32 x 32 pixels. Three classes were randomly selected from CIFAR-10, and 1000 images were randomly selected for each class as the experimental data and named CIfar-3. Example images are shown in Figure 2.

(a) (b) (c)

Figure 2. Examples of (a) Cat samples; (b) Dog samples; (c) Frog samples.

2.3 SSS-3

The SSS-3 is a small dataset of SSS images (see examples in Figure 3). It consists of three classes: Underwater shipwreck, pure seabed background, and marine fish. Each class contains 55 sample pictures.

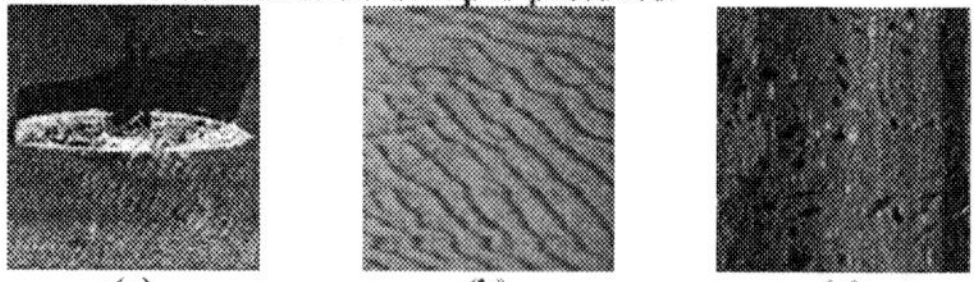

(a) (b) (c)

Figure 3. Examples of (a) Underwater shipwreck samples; (b) Pure seabed background samples; (c) Marine fish samples.

The above three datasets were divided into training data and testing data according to a ratio of 7:3. The training data was used for training the convolutional neural network, and the testing data was used for testing the trained network.

2.4 Data Correlation

Data correlation refers to the relationship between samples. In general, as long as the data are of the same source and type they are considered to be correlated. For example, when cats, leopards, and lions in the cat family appear in the data at the same time, they are considered to be strongly correlated data. Bicycles, electric bicycles, motorcycles, etc. are also considered to be strongly correlated data because they all have common characteristics. In contrast, cars, dogs, and apples are considered to be weakly correlated or irrelevant data.

In addition to subjective correlation, the degree of similarity between two images can be calculated using a hash algorithm. A number of sample images are randomly selected from the data, hash values are calculated between different data, and the final result is processed into a heat map that visualizes the correlation between the data. Three datasets were used in the experiment, namely Seaship-3, CIFAR-3 and SSS-3. We used the following methods to identify strong and weak correlations between the datasets:

(1) Randomly select 100 pictures from SeaShip-3 to form subset A. A $\subset$ SeaShip-3, $a_i \in A$. Similarly, 100 pictures were randomly selected from SSS-3 to form subset B. B $\subset$ SSS-3, $b_j \in B$.

(2) Use the Hash algorithm to calculate the similarity between a_i and b_j, and record the data in the matrix $m[i,j]$:

$$m[i,j] = \sum_{i,j}^{100} Hash(a_i, b_j) \quad (1)$$

(3) Calculate the average value of the matrix elements and convert the matrix $m[i,j]$ into a grayscale image.

(4) Repeat steps (1) to (3) for datasets CIFAR-3 and SSS-3.

Through the above steps, the correlation between SeaShip-3 and CIFAR-3 with SSS-3 can be calculated (Figure 4). The more similar the two pictures are, the greater the hash value; a greater hash value is indicated with a lighter shade of grayscale in the image. The average hash value of SeaShip-3 and SSS-3 is 129.1188 while the average hash value of CIFAR-3 and SSS-3 is 127.7184. This indicates that SeaShip-3 and SSS-3 are more similar than CIFAR-3. From Figure 4, it can also be seen that the color of the correlation heat map of SeaShip-3 and SSS-3 is brighter than that of CIFAR-3 and SSS-3. indicating that SeaShip-3 is strong correlate to SSS-3, and CIFAR-3 is a weak correlate SSS-3.

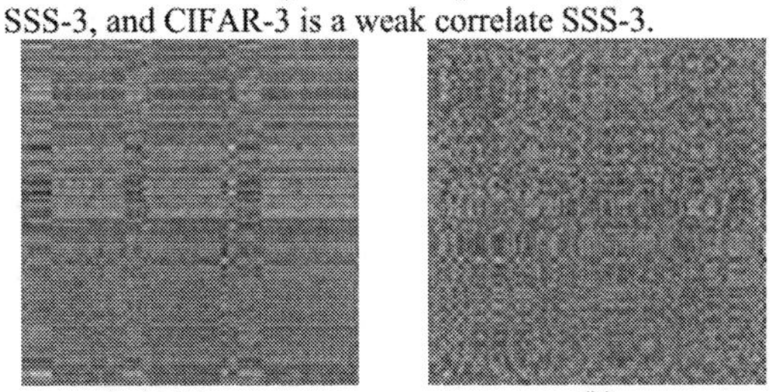

(a) (b)

Figure 4. (a) The result of the hash algorithm between SeaShip-3 and SSS-3; (b) The result of the hash algorithm between CIFAR-3 and SSS-3.

3. Model and Experiment Method

3.1 Transfer Learning

Applications of transfer learning mainly include computer vision, text classification, behavior recognition, natural language processing, indoor positioning, video surveillance, public opinion analysis, and human-computer interaction [27-32]. The principle of transfer learning is to start with the model developed by task A and reuse it in the process of developing a model for task B. A complete learning task is composed of domains and tasks. A domain D consists of two components: a feature space x and a marginal probability distribution $P(x)$, so that $D = \{x, p(x)\}$. Defining the tag space y and the model prediction function $f(\cdot)$, Task $T = \{y, f(\cdot)\}$. The definition of transfer learning is: Given a source domain D_s and a learning task T_s, a target domain D_T and a learning task T_T, transfer learning is dedicated to using the knowledge of D_s and T_s to help improve the learning of the target prediction function $f(\cdot)$ in T_T. D_s and D_T can be equal or unequal, and T_s and T_T can also be equal or unequal. The specific classification is shown in Table 1.

Unspervise transfer learning was used in this paper.

Table 1. Classification of types of transfer learning.

Domain	Task	Learning setting	
$D_s = D_T$	$T_s = T_T$	Traditional transfer learning	
$D_s = D_T$	$T_s \neq T_T$	Inductive transfer learning	Transfer learning
$D_s \neq D_T$	$T_s \neq T_T$	Unsupervised transfer learning	
$D_s \neq D_T$	$T_s = T_T$	Transductive transfer learning	

3.2 Convolutional Neural Network

3.2.1 Principle

A convolutional neural network consists of an input layer, a hidden layer, and a fully connected layer. The main task of the input layer is to input data. The input layer of the one-dimensional network can process one-dimensional or two-dimensional data, while the input layer of the two-dimensional network can process two-dimensional or three-dimensional data. In addition, the size of the input layer is fixed and should be consistent with the network. The fully connected layer is the output layer of the network. In the hidden layer of the convolutional neural network, a large number of parameters are accumulated through the calculation of the convolutional layer and the pooling layer. A large number of features are extracted from the fully connected layer for nonlinear combination, and the results are finally output. The hidden layer of a convolutional neural network is the main computing part

 https://doi.org/10.1051/e3sconf/202129002020

of the network and is composed of multiple parts. Its main architecture can be divided into the following types:

(1) Convolutional Layer

The convolutional layer is composed of a convolution kernel and an activation function (Figure 5). The convolution kernel is mainly used to extract features. The convolution kernel has no size limit and is mostly a regular rectangle. A convolution layer contains multiple convolution kernels, and each convolution kernel provides a set of parameters for passing to the next layer. The function of the activation function is to assist the convolution kernel in expressing complex features. Commonly used activation functions include ReLU, sigmoid, dropout, among others. A large number of experiments have shown that the existence of the activation function can improve the classification accuracy [16].

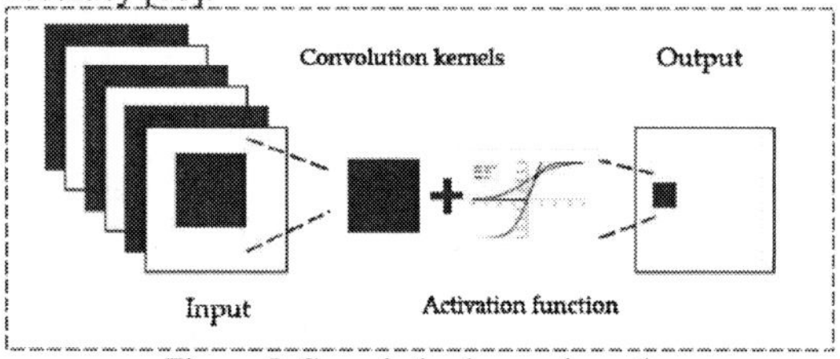

Figure 5. Convolution layer schematic.

(2) Pooling Layer

The function of the pooling layer (Figure 6) is to reduce the dimensionality of the data, remove redundant information, and, at the same time, expand the receptive field and achieve invariance, including translation invariance, rotation invariance, and scale invariance. Pooling also has the effect of simplifying the complexity of the network, reducing the amount of calculation, and reducing memory consumption [16].

The convolutional neural network finally realizes the classification of the image through the calculation of the above layers.

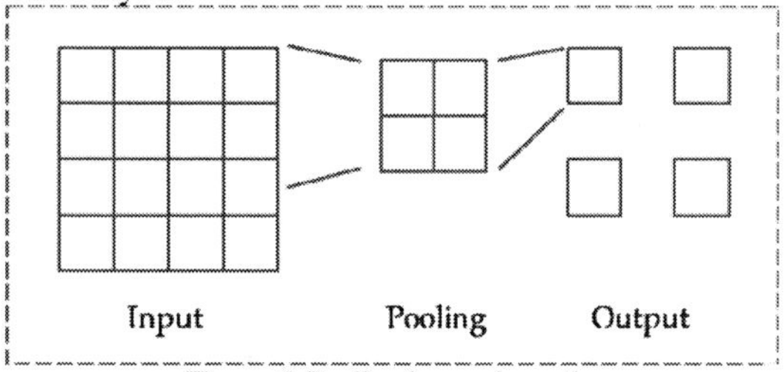

Figure 6. Pooling layer schematic

3.2.2 Architecture

There are many architectures of convolutional neural networks. LeNet-5 [15] is the first CNN, with excellent performance in handwriting recognition but, limited by the hardware conditions at the time it was devised, the network cannot be applied well to all types of images. AlexNet [16] introduced the ReLU function in the network architecture, which, to a certain extent, solved the problem of gradient disappearance. GoogLeNet [17] integrates multi-scale convolutional transformation through the idea of splitting, transforming, and merging. The traditional convolutional layer is replaced with small blocks. The concept of auxiliary learner is also introduced to accelerate the network convergence speed. ResNet [18] introduced the concept of residual learning in the network architecture, and performs well in image recognition and localization tasks. VGG [20], with a depth of 19 layers, is , at present, the network architecture with the most parameters. The small-scale filter used in the network replaces the large-scale filter, so that it achieves good image classification results and spatial positioning.

3.3 Method

The data were used for training the combined transfer learning and convolutional neural network. The goal was to train a convolutional neural network with strong generalization and high accuracy, which can be used for unfamiliar sonar band data, and overcome the limitations of training a convolutional neural network with small samples of sonar images. We first used the SSS-3 dataset for network training, and saved the results as a control group for subsequent experiments. We then used the weakly correlated data to train the network, extracted and saved the network parameters, used them on the new network, and then used the SSS-3 data to train the network. Finally, we used the strongly correlated data to train the network, and saved the network parameters, put them on the new network, and then trained with the SSS-3 data. Under normal circumstances, the training process of the convolutional neural network is shown in Figure 7. The training process of combined transfer learning and convolutional neural network used in the experiment is shown in Figure 8.

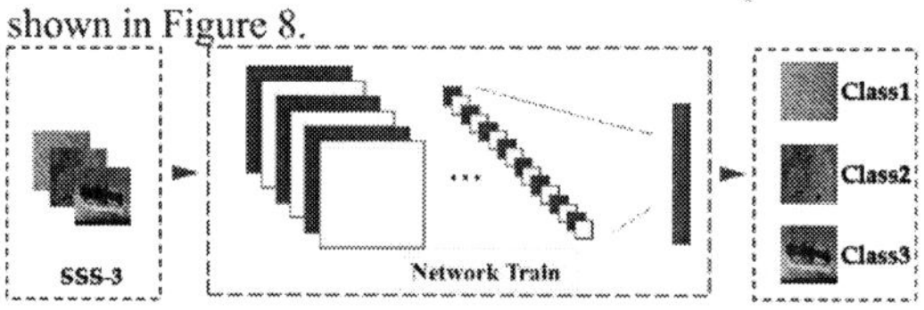

Figure 7. Convolutional neural network training process.

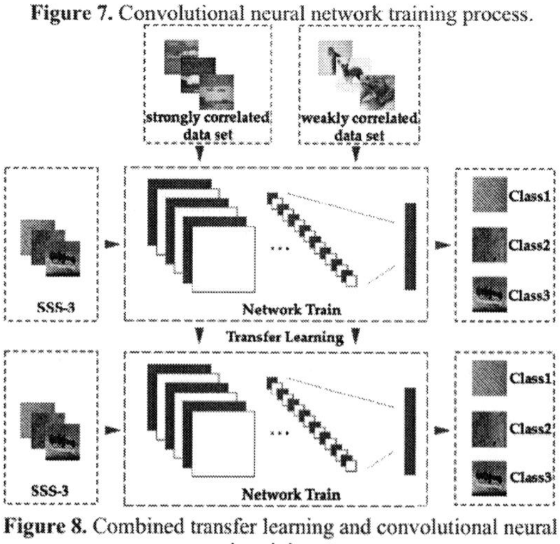

Figure 8. Combined transfer learning and convolutional neural network training process.

3.4 Experiment

The experiments were all run on OMEN on an HP Laptop 15-dc1061TX equipment. The hardware configuration includes an Intel i7-9750H CPU and NVIDIA GTX 1650 GPU. The software configuration includes Pytorch 1.4.0 and Python 3.7.6. The following three experiments were performed on five network models: LeNet-5, AlexNet, GoogLeNet, ResNet, and VGG:

(1) Non-transfer: Train the network with SSS-3.

(2) Weakly correlated data transfer: Train the network with weakly correlated data, extract the parameters into the new network, and then train the network with SSS-3.

(3) Strongly correlated data transfer: Train the network with strongly correlated data, extract the parameters into the new network, and then train the network with SSS-3.

There were fifteen groups in all. The network model was trained with training data. After the training was completed, it was tested with the test data. The accuracy of the network on the test data was used as the final evaluation index.

3.5 Precision Evaluation

A confusion matrix is an important tool for model accuracy evaluation. It is represented by a matrix with N rows and N columns. The number of correctly classified examples, that is, the number of instances that are actually positive and classified as positive by the classifier is TP (True Positives). The number of incorrectly classified as positive examples, that is, the number of instances that are actually negative but classified as positive by the classifier is FP (False Positives). The number of incorrectly classified as negative examples, that is, the number of instances that are actually positive but classified as negative by the classifier is FN (False Negatives). The number of negative examples that are correctly classified, that is, the number of examples that are actually negative and classified as negative by the classifier is TN (True Negatives). The correct rate is a commonly used evaluation index. A higher correct rate corresponds to a better model effect. The calculation formula is as follows:

$$\text{Accuracy}=\frac{(TP+TN)}{(TP+FN+FP+TN)} \quad (2)$$

4 Results

4.1 LeNet-5

The accuracy of the experiments on LeNet-5 was 78.43% without transfer (Figure 9a). The accuracy was 80.39% when the weakly correlated data was transferred (Figure 9b). The accuracy was 84.31% when the strongly correlated data was transferred (Figure 9c). The performance of the three data of experiments on the test data is summarized in Figure 9d and shows that the transfer of the strongly correlated data is higher than the transfer of the weakly correlated data, and it is higher than for the non-transferred.

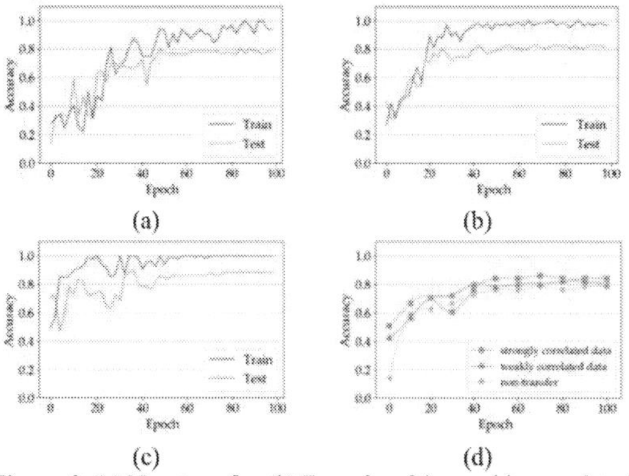

Figure 9. (a) Non-transfer; (b) Transfer of the weakly correlated data; (c) Transferred of the strongly correlated data; (d) The performance of the three data of experiments on the test data.

4.2 AlexNet

The accuracy of the experiments on AlexNet was 76.47% without transfer (Figure 10a). The accuracy was 82.35% when the weakly correlated data was transferred (Figure 10b). The accuracy was 88.24% when the strongly correlated data was transferred (Figure 10c). The performance of the three data experiments on the test data is summarized in Figure 10d, showing that the transfer of the strongly correlated data is higher than the transfer of the weakly correlated data, and it is higher than the for the non-transferred.

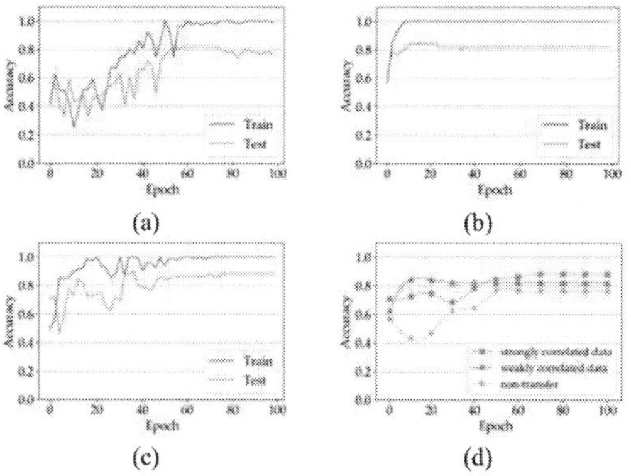

Figure 10. (a) Non-transfer; (b) Transfer of the weakly correlated data; (c) Transferred of the strongly correlated data; (d) The performance of the three data of experiments on the test data.4.3 GoogLeNet

4.3 GoogLeNet

The accuracy of the experiments on GoogLeNet was 82.35% without transfer (Figure 11a) while it was 86.27% when the weakly correlated data was transferred (Figure 11b). The accuracy was 88.24% when the strongly correlated data was transferred (Figure 11c). The performance of the three data of experiments on the test data is summarized in Figure 11d which shows that the transfer of the strongly correlated data is higher than the transfer of the weakly correlated data, and it is higher than the non-transferred.

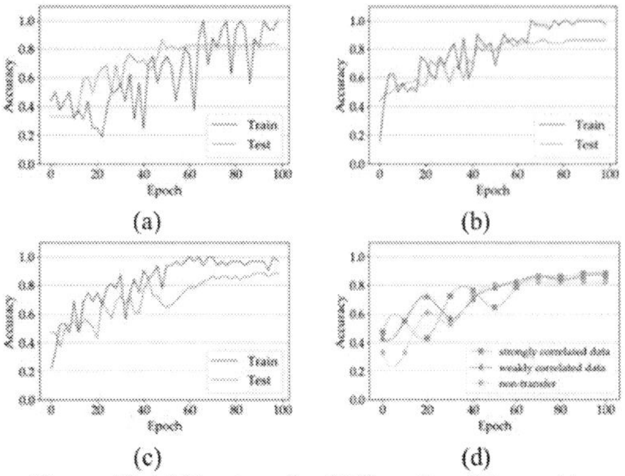

Figure 11. (a) Non-transfer; (b) Transfer of the weakly correlated data; (c) Transferred of the strongly correlated data; (d) The performance of the three data of experiments on the test data.

4.4 ResNet

The accuracy of the experiments on ResNet was 86.27% without transfer (Figure 12a). The accuracy was 86.27% when the weakly correlated data is transferred (Figure 12b). The accuracy was 90.20% when the strongly correlated data is transferred (Figure 12c). The performance of the three data of experiments on the test data is summarized in Figure 12d, which shows that the transfer of the strongly correlated data is higher than the transfer of the weakly correlated data, and it is higher than the non-transferred.

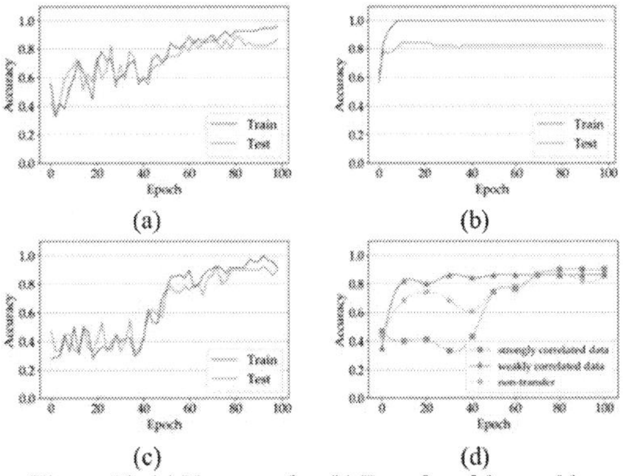

Figure 12. (a) Non-transfer; (b) Transfer of the weakly correlated data; (c) Transferred of the strongly correlated data; (d) The performance of the three data of experiments on the test data.

4.5 VGG

The accuracy of the experiments on VGG was 80.39% without transfer (Figure 13a). The accuracy was 92.16% when the weakly correlated data was transferred (Figure 13b). The accuracy was 98.04% when the strongly correlated data was transferred (Figure 13c). The performance of the three data experiments on the test data is summarized in Figure 13d, which shows that the transfer of the strongly correlated data is higher than the transfer of the weakly correlated data, and it is higher than the non-transferred.

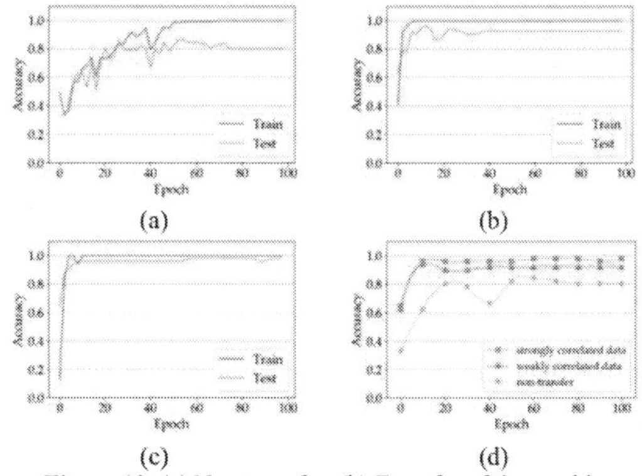

Figure 13. (a) Non-transfer; (b) Transfer of the weakly correlated data; (c) Transferred of the strongly correlated data; (d) The performance of the three data of experiments on the test data.

The accuracy of the above-mentioned experiments on the test set is summarized in Table 2.

Table 2 Test data accuracy

	Non-transfer	Weakly correlated data set	Strongly correlated data set
LeNet-5	78.43%	80.39%	84.31%
AlexNet	76.47%	82.35%	88.24%
GoogLeNet	82.35%	86.27%	88.24%
ResNet	86.27%	86.27%	90.20%
VGG	80.93%	92.16%	98.04%

In LeNet-5, the classification accuracy of weakly correlated data increased by 1.96%; the classification accuracy of strongly correlated data is increased by 5.88%; At the same time, the classification accuracy of strongly correlated data is 3.92% higher than that of weakly correlated data.

In the AlexNet architecture, compared with the non-transferred experimental group, the classification accuracy of weakly correlated data was improved by 5.88%. The classification accuracy of strongly correlated data increased by 11.77%. At the same time, the classification accuracy of strongly correlated data is 5.89% higher than that of weakly correlated data.

In the GoogLeNet architecture, compared with the non-transferred experimental group, the classification accuracy of weakly correlated data was improved by 3.92%. The classification accuracy of strongly correlated data increased by 5.89%. At the same time, the classification accuracy of strongly correlated data is 1.97% higher than that of weakly correlated data.

In the ResNet architecture, compared with the non-transferred experimental group, the classification accuracy of weakly correlated data was improved by 0%. The classification accuracy of strongly correlated data increased by 3.93%. At the same time, the classification accuracy of strongly correlated data is 3.93% higher than that of weakly correlated data.

In the VGG architecture, compared with the non-transferred experimental group, the classification accuracy of weakly correlated data was improved by 11.77%. The classification accuracy of strongly correlated data increased by 17.65%. At the same time, the classification accuracy of strongly correlated data is 5.88% higher than that of weakly correlated data.

The comparisons of difference value between various data sets were shown in Table 3.

Table 3 difference value between various data sets

	W-N	S-N	S-W
LeNet-5	1.96%	5.88%	3.92%
AlexNet	5.88%	11.77%	5.89%
GoogLeNet	3.92%	5.89%	1.97%
ResNet	0%	3.93%	3.93%
VGG	11.77%	5.88%	5.88%

Remark: N=Non-transfer, W=Weakly correlated data, S=Strongly correlated data.

5 Discussion

5.1 Performance Comparison of Transfer Learning Methods and Non-transfer Learning Methods

Comparing the three data experiments under different architectures, the accuracy after using transfer learning is higher than that without using transfer learnjng (Figure 14).

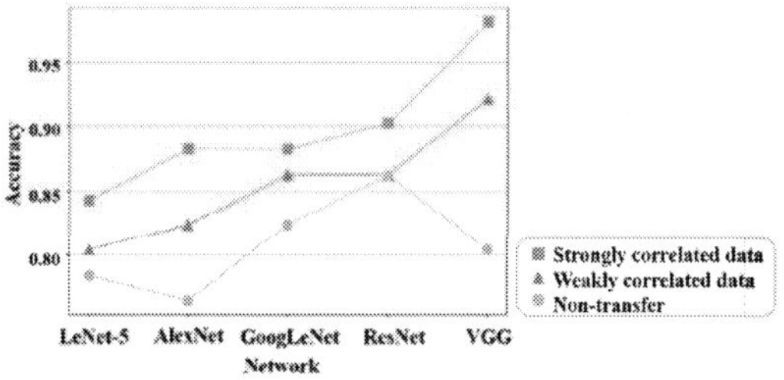

Figure 14. Classification accuracy of LeNet-5, AlexNet, GoogLeNet, ResNet, and VGG with strongly correlated and weekly correlated data.

5.2 Performance comparison between strongly correlated data transfer and weakly correlated data transfer

It is not difficult to conclude that, in all experiments, the accuracy of the strongly correlation data is better than that of the weakly correlation data (Figure 15). This shows that the method of using strongly correlation data transfer is greatly beneficial to training the network with small samples.

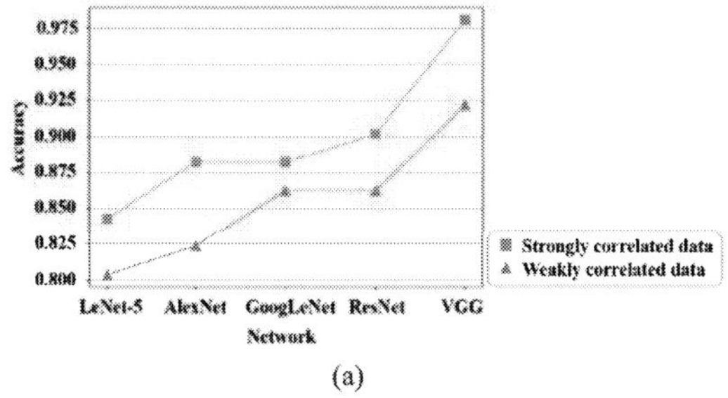

(a)

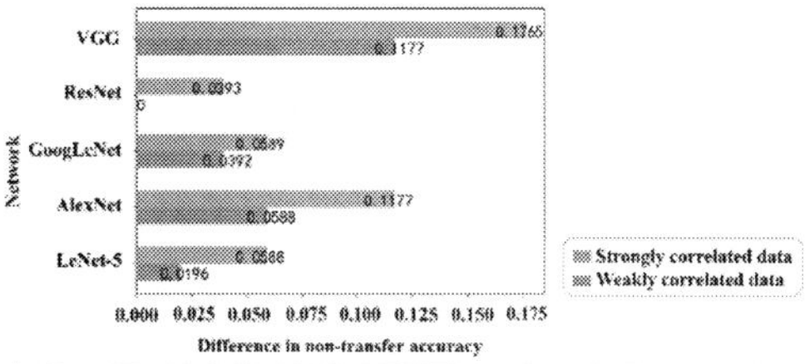

(b)

Figure 15. (a) Transfer learning accuracy; (b) Difference between the accuracy of strongly and weakly correlated data and the accuracy of non-transfer.

5.3 Initial Accuracy Analysis

At the core of transfer learning in deep learning is the transfer of network parameters. Generally, the initial parameters of the network are random. As the network training algorithm continues to modify the network parameters, the goal of accurate image classification and recognition is finally achieved. The initial parameters of each network are shown in Figure 16. In most cases, the initial accuracy of transfer learning is higher than that of the non-transferred experiment. In all cases, the initial accuracy of the strongly correlated data after transfer was higher than that of the experiment without transfer. This shows that transfer learning has a positive effect on network training.

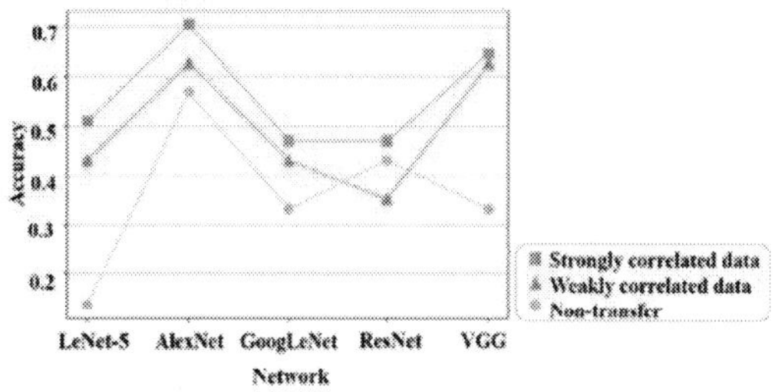

Figure 16. Initial accuracy of different networks with strongly and weakly correlated data and with non-learning transfer.

6 Conclusions

It is very difficult to obtain and label shipwreck side scan sonar images, and the small sample of data used for convolutional neural network training has additional limitations. Aiming at overcoming the classification problems of shipwreck sonar images, we proposed a solution combining transfer learning with a convolutional neural network under small samples. Comparing the experimental results, the following conclusions can be drawn:

(1) Compared with the case of non-transfer learning, the transfer of strongly correlated data and the transfer of weakly correlated data can effectively improve the accuracy of training convolutional neural networks with small sample data.

(2) In all experiments, the transfer accuracy of strongly correlated data is higher than that of weakly correlated data. In the VGG network, the classification

accuracy is as high as 98.20% after the transfer of the strongly correlated data.

(3) The initial accuracy of the strongly correlated data transfer is obviously better than that of the non-transferred case, indicating that the strongly correlated data transfer effectively transfers the features from the data to the network.

(4) The VGG network after the transfer of the strongly correlated data performs better than other methods in terms of accuracy and generalization. It is the most suitable convolutional neural network for small-sample shipwreck sonar image data.

The method of combining transfer learning with a convolutional neural network not only overcomes the difficulty of training for shipwreck sonar image recognition, but also provides an effective solution for other small sample classification problems. In the future, the method of combining transfer learning and convolutional neural network will be used in side scan sonar image mapping and real-time image recognition of unfamiliar shipwreck strips.

Acknowledgement:

This research was supported by the National Natural Science Foundation of China (41806117), and A Project Funded by the Priority Academic Program Development of Jiangsu Higher Education Institutions (PAPD).

References

1. X. Wang, A. Wang, T. Jiang, Y. Yang, and B. Zhang.(2019) Review of application areas for side scan sonar image. Bulletin of Surveying and Mapping, 01: 1-4.
2. S. Reed, Y. Petillot, and J. Bell.(2003) An automatic approach to the detection and extraction of mine features in sidescan sonar. IEEE Journal of Oceanic Engineering, 28: 90-105.
3. F. Yang, Z. DU, Z. Wu, J. Li, and F. Chu.(2006) Object Recognizing on Sonar Image Based on Histogram and Geometric Feature. Marine Science Bulletin, 05: 64-69.
4. S. Reed, Y. Petillot, and J. Bell.(2004) Model-based approach to the detection and classification of mines in sidescan sonar. Applied Optics, 43:237-246.
5. H. Guo, T. Tian, G. Liu, and F. Lu.(2002) A Recognizing Method Based on Fuzzy Clustering on the Sonar Image of a Small Target on the Sea Bed. Journal of Chinese Computer Systems, 02: 139-141.
6. G. J. Dobeck, J. C. Hyland, and L. Smedley.(1997) Automated detection and classification of sea mines in sonar imagery. in: Spie - Int Soc Optical Engineering. Bellingham. pp. 90-110.
7. L. Zheng and K. Tian.(2018) Detection of small objects in sidescan sonar images based on POHMT and Tsallis entropy. Signal Processing, 142: 168-177.
8. E. Dura, Z. Yan, L. Xuejun, G. J. Dobeck, and L. Carin.(2005) Active learning for detection of mine-like objects in side-scan sonar imagery. IEEE Journal of Oceanic Engineering, 30: 360-371.
9. F. Liu, S. Wang, and Y. Zhang.(2018) Overview on Models and Applications of Support Vector Machine. Computer Systems & Applications,27: 1-9.
10. A. Khan, A. Sohail, U. Zahoora, and A. S. Qureshi.(2020) A survey of the recent architectures of deep convolutional neural networks. Artificial Intelligence Review, 53: 5455-5516.
11. S. Peng, W. Jiang, H. Pi, X. Li, H. Bao, and X. Zhou.(2020) Deep Snake for Real-Time Instance Segmentation. in: 2020 IEEE/CVF Conference on Computer Vision and Pattern Recognition (CVPR). pp. 8530-8539.
12. M. Ding.(2020) Learning Depth-Guided Convolutions for Monocular 3D Object Detection. in: 2020 IEEE/CVF Conference on Computer Vision and Pattern Recognition (CVPR). pp. 11669-11678.
13. E. Shelhamer, J. Long, and T. Darrell.(2017) Fully Convolutional Networks for Semantic Segmentation. Ieee Transactions on Pattern Analysis And Machine Intelligence,39: 640-651.
14. Y. Taigman, M. Yang, M. Ranzato, and L. Wolf.(2014) DeepFace: Closing the Gap to Human-Level Performance in Face Verification. in 2014 IEEE Conference on Computer Vision and Pattern Recognition. Columbus. pp. 1701-1708.
15. Y. Lecun, L. Bottou, Y. Bengio, and P. Haffner.(1998) Gradient-based learning applied to document recognition. Proceedings of the IEEE, 86: 2278-2324.
16. A. Krizhevsky, I. Sutskever, and G. E. Hinton.(2017) ImageNet Classification with Deep Convolutional Neural Networks. Communications Of the Acm, 60: 84-90.
17. C. Szegedy. Going deeper with convolutions. 2015 IEEE Conference on Computer Vision and Pattern Recognition (CVPR). Boston. pp. 1-9.
18. K. He, X. Zhang, S. Ren, and J. Sun.(2016) Deep Residual Learning for Image Recognition. 2016 IEEE Conference on Computer Vision and Pattern Recognition (CVPR). Las Vegas. pp. 770-778.
19. V. Badrinarayanan, A. Kendall, and R. Cipolla.(2017) SegNet: A Deep Convolutional Encoder-Decoder Architecture for Image Segmentation. Ieee Transactions on Pattern Analysis And Machine Intelligence, 39: 2481-2495.
20. K. Simonyan and A. Zisserman.(2015) Very Deep Convolutional Networks for Large-Scale Image Recognition. Computer Science.
21. Z. Shao, W. Wu, Z. Wang, W. Du, and C. Li.(2018) SeaShips: A Large-Scale Precisely Annotated Dataset for Ship Detection. IEEE Transactions on Multimedia, 20: 2593-2604.
22. M. Valdenegro-Toro.(2017) Best practices in convolutional networks for forward-looking sonar image recognition. in: OCEANS 2017. Aberdeen. pp. 1-9.
23. H. Liang, L. Jin, and C. Yang.(2019) Research on Underwater Target Recognition Based on Depth Learning with Small Sample. journal of Wuhan University of Technology (Transportation Science & Engineering), 43: 6-10.
24. R. Mehrotra, M. A. Ansari, R. Agrawal, and R. S. Anand.(2020) A Transfer Learning approach for AI-based classification of brain tumors. Machine Learning with Applications, 2.
25. X. Zhen, B. Pan, and J. Zhang.(2020) Power tower detection in remote sensing imagery based on deformable network and transfer learning. Acta Geodaetica et Cartographica Sinica, 49: 1042-1050.
26. Z. Zhu, X. Fu, and Y. Hu.(2020) A sonar Image Recognition Method Based on Convolutional Neural

Network Trained through Transfer Learning," Journal of Unmanned Undersea Systems, 28: 89-96.

27. S. J. Pan and Q. A. Yang.(2010) A Survey on Transfer Learning. Ieee Transactions on Knowledge And Data Engineering,22: 1345-1359.
28. J. Yosinski, J. Clune, Y. Bengio, and H. Lipson.(2014) How transferable are features in deep neural networks ?. Advances In Neural Information Processing Systems, vol. 27.
29. A. S. Razavian, H. Azizpour, J. Sullivan, and S. Carlsson.(2014) CNN Features Off-the-Shelf: An Astounding Baseline for Recognition. in: 2014 IEEE Conference on Computer Vision and Pattern Recognition Workshops. Columbus. pp. 512-519.
30. J. Howard and S. Ruder.(2018) Universal Language Model Fine-tuning for Text Classification. in: Proceedings Of the 56th Annual Meeting Of the Association for Computational Linguistics. Melbourne. pp. 328-339.
31. Z. Wang, X. XU, and L. Zhang.(2020) Study of deep transfer learning for SAR ATR based on simulated SAR images. ournal of University of Chinese Academy of Sciences, 37: 516-524.
32. J. Liu, N. Liu, Z. Pan, and X. You.(2019) Domain adaptive methods for device diversity in indoor localization. Journal of Southeast University(English Edition), 35: 424-430.

Study on the Fatigue Failure Law of Gas Storage Caprock Under Periodic Injection and Production

Liu Lingyan[1], MuWenman[1], ZhaiYujia[1]
[1]Huabei Oilfield Company Gas Storage Management Office, Langfang 065000

Abstract. Whether qualitative or quantitative, whether macroscopic or microscopic, Predecessors' research on the sealing ability of caprocks is concentrated on the static model of the caprock. A few scholars have also proposed the sealability evaluations of the caprock dynamic evolution process, they are all aimed at the dynamic evolution of geological historical periods, and there are few studies on the sealability of gas storage caprocks under periodic injection and production. In this paper, the characteristics of stress and strain under alternating loads of rock and non-rock materials are investigated and experimentally studied to clarify the law of fatigue failure of gas storage caprocks under periodic injection-production. The results show that it is possible that the microcracks may develop and propagate in the rock after the gas storage has experienced many periodic injections and productions, when the stress is far less than the rock strength limit, And then a macroscopic deformation is formed. The accumulated plastic deformation of the rock will increase, and the plastic deformation rate will gradually decrease, as the number of injection and production increases.

1 Introduction

For gas storage, the caprock's plugging capacity refers to the ability of the gas storage to prevent natural gas from escaping, and it controls the vertical distribution, abundance and working pressure of natural gas in the gas storage. Taking the capillary sealing mechanism and hydraulic sealing mechanism of caprocks proposed by Watts in the United States in 1987 as a symbol [1], the sealing mechanism of caprocks and faults was systematically studied and qualitatively evaluated. Downey pointed out that the study of caprocks should be carried out from both macroscopic and microscopic aspects [2]; Grunau pointed out for the first time that evaporites such as gypsum and dolomite and shale are all effective caprocks [3], depositional environment, diffusion Speed, fracture formation and evolution are important parameters for evaluating caprocks. In 1996, Lu Yanfang and others proposed the overpressure sealing mechanism and the hydrocarbon concentration sealing mechanism [4], and established corresponding displacement pressure, effective stress and tensile strength, abnormal pore fluid pressure and diffusion coefficient according to different sealing mechanisms. The method of evaluating the sealing capacity, on this basis, establishes the industry standard for the evaluation of the sealing capacity of the caprock based on the macroscopic development characteristics and the microscopic sealing capacity of the caprock. The book "Research on Oil and Gas Reservoir Sealing" published by Lv Yanfang established a quantitative evaluation method for the sealing capacity of the caprock[4], and proposed the idea of evaluating the integrity of the caprock for the first time, and pointed out that the integrity of the caprock was damaged. The factors are structural fracture and hydraulic fracture[5-8]. Gradually realized that the petrological characteristics of the cap rock itself, especially the brittle-ductile characteristics, are the key factors that determine the integrity of the cap rock. Based on the density-strain, BRI and OCR [9], Byerlee friction law and Goetze's criterion, a quantitative judgment is proposed. Evaluation criteria for brittle-ductile transition of shale and gypsum [10]. Regardless of qualitative or quantitative caprock evaluation, predecessors paid more attention to the static evaluation method of caprock. Jin Zhijun proposed its sealing ability evaluation method for the dynamic evolution process of shale caprock [10].

Scholars generally believe that the study of the dynamic sealability of gas reservoir caprocks needs to be comprehensively evaluated from the macroscopic rock mechanics integrity and the microscopic capillary force sealability [11]. (1) The macro sealing ability depends on the lithology, thickness and the continuity of the lateral distribution of the cap layer; (2) the micro sealing ability depends on the permeability, porosity, permeability, specific surface and displacement pressure of the cap layer.

You Xiuling established the relationship between rock porosity, particle size median radius, burial depth and breakthrough pressure, and proposed a quantitative evaluation method that combines the macro and micro

*Liyi:654549059@qq.com

levels [12]. Fu Guang et al. divided the levels according to the size of influence through the micro and macro aspects, and discussed the method of comprehensively evaluating the sealing ability of the cap layer itself [13]. Scholars such as Sun Mingliang calculated the lithology, thickness and displacement pressure of the caprocks of more than 40 gas reservoirs in my country, and established the evaluation criteria for the caprocks of medium and high-efficiency gas reservoirs [14]. Zhang Lihan and Zhou Guangsheng calculated the actual data of 46 large and medium-sized gas fields in my country, and clarified the controlling factors that affect the sealing capacity of the gas reservoir caprock: the thickness of the caprock, fracture, displacement pressure and other influencing factors, and proposed the caprock Evaluation method of sealing ability [15]. Fan Ming et al. combed through the relationship between specific surface area, displacement pressure and caprock sealability, and proposed a comprehensive evaluation plan for the specific surface area and displacement pressure to judge the sealability of mudstone caprock [16].

However, for gas storage, previous studies on the sealing capacity of the cap rock, whether qualitatively or quantitatively, whether macroscopic or microscopic, are concentrated on the static model of the cap rock, and a few scholars have also proposed it. The evaluation of the sealing ability of the caprock dynamic evolution process is all based on the dynamic evolution of the geological history period. There are few studies on the sealing capability of the gas storage caprock under the action of periodic injection and production. Different from the development of conventional gas reservoirs, gas storages built from depleted oil and gas reservoirs require strong injection and production. The cap rock pressure is subject to periodic changes and changes frequently. The formation of microscopic rock fissures is the cause of the gas reservoir cap rock. The root cause of the failure [17].

2 Mechanical mechanism of rock failure under static action

The caprock lithology includes shale, dolomite and gypsum rock. Its brittleness and failure mechanism can be described by the method of full stress-strain characteristics. The peak strain can indicate the difficulty of rock failure. For mudstone caprocks, The brittleness of the rock is related to the slope and residual stress in the second half of the curve peak [17].

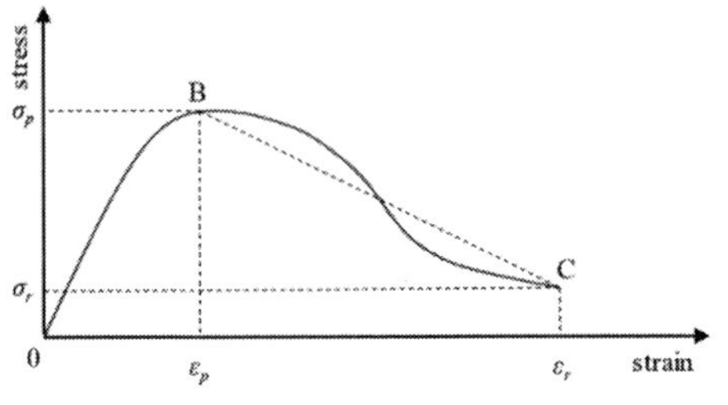

Fig.1 Model diagram of stress-strain curve of mudstone caprock

In the formula, point B corresponds to the peak stress, point C corresponds to the residual stress, and, are the peak stress and the residual stress, respectively;, are the peak strain and the residual strain, respectively.

The calculation formula of using rock brittleness index to describe the curve shape of cap rock at the back of the peak is:

$$I_1 = \frac{\sigma_p - \sigma_r}{\sigma_p} \tag{1}$$

$$I_2 = \frac{\varepsilon_r - \varepsilon_p - \varepsilon_{\max r-p}}{\varepsilon_{\min r-p} - \varepsilon_{\max r-p}} \tag{2}$$

$$BRIT = I_1\omega + I_2(1-\omega) \tag{3}$$

In the formula, are the peak stress and the residual stress, respectively;, are the peak strain and the residual strain, respectively; is the minimum peak strain minus the residual strain, is the maximum peak strain minus the residual strain, and ω is the relative weight. Through experiments, we can initially understand the mechanical mechanism of rock failure under static action of the caprock.

3 Mechanical mechanism of rock failure under dynamic action

By referring to the strain characteristics of non-rock materials under the action of periodic alternating stress, we can understand the stress-strain curve mode of the rock under the action of cyclic stress. In the first stage, after starting to load the stress, the curve from the initial residual stress point A should have been extended along the elastic loading line (the dotted line in the figure), but the actual situation has shifted; in the second stage, after the unloading starts, The unloading line should be parallel to the loading line, but the actual unloading line did not deviate along the elastic unloading line (the dotted line in the figure), thus forming a hysteresis loop; in the third stage, repeated loading and unloading is sufficient After more times, the hysteresis loop gradually stabilizes [18].

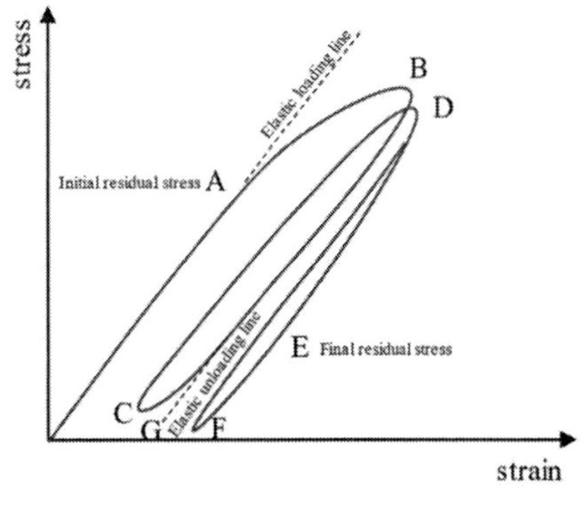

Fig.2. Stress-strain curve model diagram under cyclic stress

Applying periodic alternating stress to the sandstone, the hysteresis loop gradually shifts toward the direction of increasing strain, and the shape closure of the hysteresis loop is getting higher and higher, revealing that the plastic deformation of the rock becomes larger, but the plastic deformation rate is gradually reduced [19]. For sandstone gas storages, reservoirs and faults will increase the risk of local plastic deformation after a period of intense injection and production.

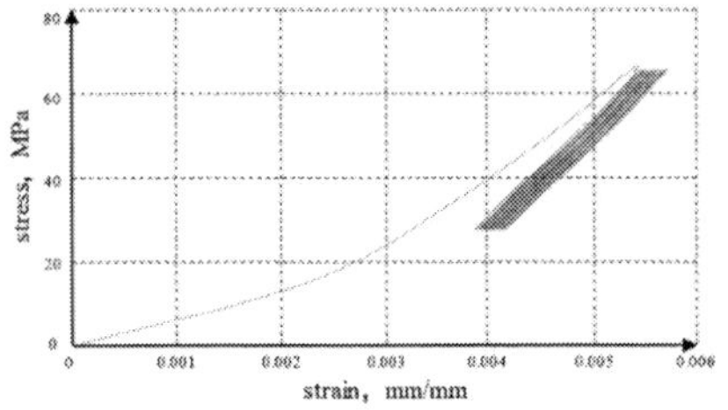

Fig.3. Typical stress-strain hysteresis curve of sandstone

Similar to the change of the stress-strain curve of sandstone, the hysteretic loop of mudstone gradually shifts toward the direction of increasing strain, and the shape closure of the hysteretic loop is getting higher and higher, revealing two points: 1. With the number of cycles Increasing, the cumulative plastic deformation increases, but the rate of plastic deformation decreases; 2. During the cycle of loading and unloading, the damping ratio of mudstone finally converges to a fixed value. In addition, as the number of cycles increases, the dynamic elastic modulus of mudstone gradually increases and eventually tends to converge[20].

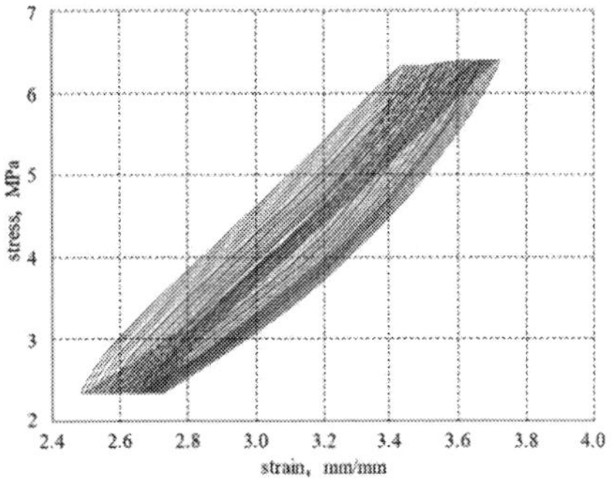

Fig.4. Typical stress-strain hysteresis curve of mudstone

4 Fatigue failure of rock

The in-situ stress field of gas storage also changes periodically with the change of periodic injection and production. In addition to different degrees of elasto-plastic deformation, it may also cause local stress concentration. This stress concentration will cause the rock to accumulate and form fatigue damage[21].

Therefore, under the action of cyclic loading, from a microscopic point of view, we should not only consider the creep of the rock, but also the possibility of fatigue damage to the rock [22]. "Fatigue failure" refers to the phenomenon that structural failure occurs after a certain number of cycles, far from reaching the strength limit.

Periodic disturbance stress is the external cause of fatigue and a necessary condition for fatigue. Fatigue failure occurs at a certain point in the rock. Fatigue failure starts from somewhere with higher stress. Once the microscopic deformation of the structure begins to accumulate, it will eventually lead to fatigue failure. There are many differences between static damage and fatigue damage. Periodic injection and mining are easy to cause fatigue damage: 1. Static damage is the instantaneous damage produced when the stress reaches the strength limit. Fatigue damage is one In the long-term process, under the action of cycles, after the accumulation of long-term stress, micro-cracks are generated locally from the fatigue source and gradually expand outward; 2. The condition for static failure is to reach the strength limit, but fatigue failure does not need to reach The strength limit is likely to occur, because the stress can gradually accumulate under the action of cycles [24]; 3. Due to the different triggering mechanism, the fractures of the two are also different. The fracture of fatigue damage can be seen in a flat picture, mainly It is caused by repeated squeezing and relaxation, and repeated friction, which is also an important sign to distinguish static failure from fatigue failure; 4. From the perspective of determinants, fatigue failure is not only related to the nature of the rock itself, but also It is related to external factors such as rock burial depth, fluid charging, fluid properties, and fracture development degree.

5 Conclusion

(1) Under static action, the difficulty of damage to the mudstone caprock of a gas storage can be expressed by the magnitude of the peak strain, and the magnitude of its brittleness is related to the slope and residual stress in the second half of the curve peak.

(2) After multiple rounds of injection and production in the gas storage, it is possible that microcracks may develop and propagate in the rock when the stress is far less than the rock strength limit. After the occurrence of microcracks, continuous periodic injection and mining will gradually accumulate the damage amount, and after accumulation to a certain extent, it will eventually show up as a macroscopic plastic deformation.

(3) When the injection-production intensity remains unchanged, the increase in the number of cycles will increase the accumulated rock plastic deformation, but the plastic deformation rate will gradually decrease.

References

1. Watts N L. Theoretical aspects of cap-rock and fault seals for singleand two-phase hydrocarbon columns[J].Marine and Petroleum Geology, 1987, 4 (4) :274-307.
2. Downey M W. Evaluating seals for hydrocarbon accumulations[J].AAPG Bulletin, 1984, 68:1752-1763.
3. Grunau H R. A worldwide look at the cap-rock problem[J].Journal of Petroleum Geology, 1987, 10

(3) :245-266.
4. Lv Yanfang, Fu Guang, Gao Daling, et al. Study on the cap rock of reservoir[M].Beijing:Petroleum Industry Press, 1996:118-120.
5. Skerlec G M. Evaluation top and fault seal[C]||Liu Delai, Xue Liangqing. Hydrocarbon traps exploration. Beijing: Petroleum Industry Press, 2002:395-444.
6. Jin Zhijun, Yuan Yusong, Liu Quanyou, et al. Controls of Late Jurassic-Early Cretaceous tectonic event on source rocks and seals in marine sequences, South China[J].Science China Earth Sciences, 2013, 56 (2) :228-239.
7. Ingram G M, Urai J L, Naylor M A. Sealing processes and top seal assessment[J].Norwegian Petroleum Society Special Publications, 1997, 7:165-174.
8. Ingram G M, Urai J L. Top-seal leakage through faults and fractures:the role of mudrock properties[J]. Geological Society, 1999, 158 (1) :125-135.
9. Corcoran D V, Dore A G. Top seal assessment in exhumed basin settings-some insights from Atlantic Margin and borderland basins[C]||Koestler A G, Hunsdale R.Hydrocarbon Seal Quantification.NPF Special Publication, 2002:89-107.
10. Fu Xiaofei, Jia Ru, Wang Haixue, et al. Quantitative evaluation on the fault-caprock sealing capacity of Dabei-kelasu structural belt in Kuqa depression, Tarim basin, NW China[J]. Petroleum Exploration and Deveopment, 2015, 42 (3) :300-309.
11. Ma Xinhua, Zheng Dewen, Shen Ruichen, et al. Key technologies and practice for gas field storage facility construction of complex geological conditions in China [J]. Petroleum Exploration and Development, 2018, (03): 489-499.
12. You Xiulin. Study on assessment method of caprocks in natural gas pools[J]. Oil & Gas Geology, 1991,(03):261-275.
13. Fu Guang, Chen Zhang Ming, Jiang Zhen Xue. Evaluation of the sealing ability of caprock and its application. [J]. Petroleum Exploration and Development, 1995,(03):46-50+127.
14. Sun Ming Liang, Liu Guang Di, Li Jian. Features of cap rocks of gas pools and criteria of identification [J]. Natural Gas Industry, 2008, (08): 36-38+137.
15. Zhang Lihan, Zhou Guangsheng. Improvement and Application of the Methods of Gas Reservoir Cap Sealing Ability [J]. Acta Sedimentologica Sinica, 2010,28(02):388-394.
16. Fan Ming, Chen Hongyu, Yu Lingjie, et al. Evaluation standard of mudstone caprock combining specific surface area and breakthrough pressure [J]. Petroleum Geology & Experiment, 2011, 33(01): 87-90.
17. Lin Jianpin, Jia shanpo, Liu Tuanhui, et al. Comprehensive evaluation of sealing ability of mudstone cap rock for xing 9 depleted gas reservoir in reconstructing underground gas storage [J]. Chinese Journal of Rock Mechanics and Engineering, 2015,(S2):4099-4107.
18. LuoBailu. Relaxation Performance of Welding Residual Stress under Alternate Loadings[D]. Wuhan University of Technology, 2010.
19. Wang Hong. Experimental Research on Deformation Characteristics of Rock under Cyclic Loading [D]. Chongqing University, 2007.
20. Chen Fan, pan danguang, Wang Ke, et al. Tests on Nonlinear Characteristic Parameters of Mudstone under Cyclic Loading with Step Amplitude [J]. Journal of Xinjiang University(Natural Science Edition), 2020, (01): 86-93.
21. Ren song, Wang Xiao Shu, Gao Si Xian, et al. Experimental Study on Fatigue Damage of Sandstone Based on NMR and Acoustic Emission [J]. Transactions of Beijing Institute of Technology, 2019, 39(08): 792-799.
22. Research on the Usability of Underground Salt Rock Gas Storage in the Process of Gas Injection and Production[D]. Graduate University of Chinese Academy of Sciences, 2009.
23. Wang Guojun, Hu Renxi. Guidance course of nsoft fatigue analysis theory and application examples [M]. China Machine Press, 2007
24. Sui Yiyong, Lin Tangmao, Liu Xiang, et al. The influence of alternating load on sand production law of injection/production wells of underground gas storages[J]. Oil Gas Storage and Transportation, 2019, 38(03): 303-307.

Study on sedimentary facies characteristics of Xujiahe Formation in Sichuan Basin

WANG Zhiguo*1

1Exploration and Development Research Institute of Daqing Oil Field Co. Ltd, Daqing City, Heilongjiang Province, 163712,China

Abstract. Sedimentary facies is the summation of all primary sedimentary features in a sedimentary unit, which is the product of sedimentation in a specific environment. It emphasizes the material performance of sediment formation conditions, including mineralogy, petrology, paleontology, geochemistry and geophysics. Therefore, it has unique sedimentary features in this environment. In this paper, the types and units of sedimentary facies which can reflect the basic characteristics and geometry of sedimentary system are established, and based on the comprehensive analysis of rock types, lithological assemblages, sedimentary structures, palaeontological assemblages and section structures, the sedimentary facies types of Xujiahe formation of Upper Triassic in the study area mainly include semi-arid flooded basin facies, delta facies, beach bar-lagoon system and lake facies, among which delta facies and beach bar-lagoon system are the main sedimentary facies types. Because sedimentary facies is the historical product of basin formation, evolution and extinction, it can not only reflect the formation conditions, temporal and spatial distribution and evolution law of related strata and rock types, but also be the basis for analyzing the nature and structural background of sedimentary basin. At the same time, the analysis of sedimentary facies is of great significance for understanding and analyzing the evolution history of basin and guiding the exploration and development of related oil and gas resources.

1 Introduction

Sichuan Basin is a multicycle sedimentary basin on the Yangtze block. From Sinian to Middle Triassic, it has been a stable large depression area. The Indosinian tectonic compression at the end of Middle Triassic made the surrounding island chain strongly thrust to form a mountain system. In the middle and late late Late Triassic, Xichang, Panzhihua and Central Yunnan areas transformed into foreland basins, Thus, a large coal bearing sedimentary basin, the upper Yangtze sedimentary basin in the Late Triassic, is formed by connecting it with the "Sichuan Basin", covering an area of 10 × 10^4km^2. The present "Sichuan Basin" is only a part of the upper Yangtze sedimentary basin in the Late Triassic. During the sedimentary period of Xujiahe formation, Sichuan Basin was a large bay connected with the ocean only in the southwest. During the sedimentary period of Xujiahe formation, the basin experienced three ups and downs, forming a sandwich structure between large open flow delta deposits and deep lake deposits. The first, third and fifth members of the Xujiahe formation are developed with coal shale deposits. The second, fourth and sixth members of the Xujiahe formation are sandstone and conglomerate[1][2][3].

2 Review of sedimentary facies markers

In the process of studying the depositional system of the study area, first of all, a detailed analysis of the surface profile of the study area, combined with the actual field observation of the adjacent areas studied by predecessors, and a comprehensive study of the drilling data (core data, logging data, core analysis data, etc.) were carried out to master the sedimentological information reflected by various data, Based on the principles of sedimentology and well logging geology, sedimentary facies types and facies units reflecting the basic characteristics and geometry of sedimentary system are established[4][5].

The main identification marks of sedimentary facies and subfacies in the study area are: color characteristics, structural characteristics, sedimentary structure, logging curve characteristics, paleontological fossils, etc.

2.1 Color mark

The color of rock, especially the primary color of mudstone, can be used as a direct indicator to judge the climate condition, water medium oxidation-reduction conditions and the quality of source rock. When the primary color is involved in the naming of rocks, the paleoclimate and the physical and chemical conditions of fluid medium of related rocks have been given

* Corresponding author: 47310372@qq.com

information. From the primary color change characteristics of mud and shale in Xujiahe formation of Upper Triassic, lower Jurassic, middle and Upper Jurassic, it reflects that northeast Sichuan has the characteristics of warm and humid (T_3x), The paleoclimate changes from hot and semi-arid (j_1z-$j_{1-2}l$) to hot and arid (j_2xs-j_3p), and the corresponding fluid medium conditions and physicochemical properties of the sedimentary basin also change from reduction (T_3x), weak reduction (j_1z-$j_{1-2}l$) to oxidation (j_2xs-j_3p).

2.2 Rock structure mark

The grain size of rock can reflect the hydrodynamic conditions of sedimentary environment and the distance of debris transportation. The grain size distribution can be used as one of the means to judge the sedimentary environment. The structural characteristics of rocks are the direct reflection of the medium conditions during deposition. The sediments formed under different medium conditions have different structural characteristics. Even if they are formed under the same medium conditions, the sediment particles will change from coarse to fine as the hydrodynamic conditions change from strong to weak. In addition, the speed of deposition and the time of transformation are also reflected in the sediment structure. The fourth to sixth members of Upper Triassic Xujiahe Formation in shiguansi section of Wanyuan around the study area are characterized by a large number of conglomerate accumulated in the fan, with poor sorting and roundness.

2.3 Sedimentary structure mark

Sedimentary structure is one of the important characteristics of sedimentary rocks. They are macroscopic characteristics caused by the heterogeneity of composition, structure and color of sediments. Its scale is generally large, and it can be directly observed and described in field outcrops and cores. Among them, primary sedimentary structures refer to those structures formed during or shortly after sediment deposition and before consolidation. They can provide information about the properties of sedimentary media and energy conditions during the depositional period. As the development of sedimentary structure is directly related to sedimentary velocity, water flow action mode and medium condition, the primary sedimentary structure and its combination or sequence have become the most important marks for distinguishing sedimentary environment and dividing sedimentary facies, subfacies and microfacies. The primary sedimentary structures in the study area mainly include bedding structure and bedding structure.

The biological fossils in the sediments or strata can not only be used to identify the geological age of the strata, divide and correlate the strata, but also be one of the important marks for the analysis of sedimentary environment. At present, the application of paleontological markers mainly plays an important role in the division of large environment, such as the emergence of a large number of plant fossils, especially plant rhizomes, carbonaceous mudstone and coal seams, reflecting the continental environment.

The development of the bottom scour structure is related to the sudden change of hydrodynamic conditions from weak to strong. Generally, the hydrodynamic conditions of the sedimentary environment above the scour surface are significantly enhanced compared with those below. Therefore, the sediments accumulated before the development of the scour surface are subject to different degrees of undercutting erosion during the process of bottom scour, The bottom scour surface is characterized by an uneven scour surface and a lithologic mutation surface. The rock grain size in the upper part of the scour surface is obviously coarser than that in the lower part, or contains mud and gravel from the underlying layer. Therefore, the erosion surface can represent a different degree of erosion discontinuity, which is usually developed in the sedimentary environment with strong and weak hydrodynamic conditions and frequent changes, such as river and delta.

Bedding is one of the most important appearance features of sediments or sedimentary rocks, and it is also the main sign that sedimentary rocks are different from magmatic rocks and some metamorphic rocks. Bedding is a layered structure formed by the change of rock properties along the direction of sediment accumulation (material composition, color, structure, etc.). It can be further divided into cross bedding, wavy bedding, grain order bedding, parallel bedding, horizontal bedding and composite bedding (wavy bedding, lenticular bedding, etc.). All kinds of bedding are very developed and colorful in the study area.

2.4 Paleontological mark

The biological fossils in the sediments or strata can not only be used to identify the geological age of the strata, divide and correlate the strata, but also be one of the important marks for the analysis of sedimentary environment. At present, the application of paleontological markers mainly plays an important role in the division of large environment, such as the emergence of a large number of plant fossils, especially plant rhizomes, carbonaceous mudstone and coal seams, reflecting the continental environment. There are plant stem fossils and minable coal seams in fluvial strata of Xujiahe formation of Upper Triassic in the study area.

3 Sedimentary facies type

According to the comprehensive analysis of rock type, lithologic association, sedimentary structure, paleontological Association and section structure, the following sedimentary facies types are mainly developed in Xujiahe Formation Jurassic system of Upper Triassic in the study area.

3.1 Semi-arid flooded basin facies

The flood basin facies in the semi-arid area mentioned in this study mainly refers to fluvial facies, which is widely

 https://doi.org/10.1051/e3sconf/202129002022

distributed in the upper Triassic to Jurassic strata in the study area, mainly in fluvial basin and fluvial basin. It has obvious vertical zonation and plane zonation distribution characteristics, and is obviously controlled by paleotopography, mainly as follows: in the upper reaches or the lower part of the profile, the fluvial basin is mainly developed, while in the lower reaches or the upper part of the profile, the fluvial basin is mainly developed; In plane, the fluvial basin is mainly distributed near the source area at the edge of the basin, while the fluvial basin is mainly developed at the transitional position from the edge of the basin to the center of the basin. Among them, the microfacies sandbodies of the central beach in the fluvial basin and the marginal beach in the fluvial basin are one of the important reservoir sandbodies in the study area, while the microfacies of the inter River depression, especially the mire microfacies of the marsh inter River depression, mud and shale are the important source rocks and effective regional cap rocks in the study area.

3.2 Delta facies

Delta is a large cone-shaped sedimentary system formed at the confluence of rivers and lakes (or basins), which can be divided into the top (also called delta plain subfacies), progradars (also known as delta front slope subfacies), and bottom (also known as pre delta subfacies). The continental lake delta can be divided into Braided River Delta, meandering river delta and net River Delta according to the river type controlling the sedimentation of the delta. Braided river delta facies in the study area are mainly distributed in Xujiahe formation, in which the sand bodies of Xuhe section 2 and xusi are thick, and the mud content is small, mainly composed of sand, and the mud interlayer in the channel sand body is rarely seen. Generally, a large set of sandstone appears, which is formed by multi-stage River superimposition. Therefore, we think it is braided river delta, The main reason for this feature is that the shallow flow of braided river is fast, the flow is unstable, and the lateral migration of the channel is frequent, which results in the fine matter deposited on the top of the river channel being washed away by the new stage of river channel before consolidation, and the accumulation of coarse-grained material. The continuous vertical stacking of the river leads to the characteristics of large deposition thickness and less fine-grained material content in braided river.

The origin of braided river delta is related to the fact that the river from the mountain pass still flows into the lake as braided river after passing through the narrow fan delta plain of alluvial fan. Its grain size is obviously coarser than that of Meandering River Delta. It is a kind of near source coarse clastic delta developed adjacent to alluvial fan, The braided river delta is further divided into two subfacies: plain and front. At the end of the braided river delta front, with the disappearance of underwater braided distributary channel and the formation of estuary bar distal bar, it gradually enters into the pre delta subfacies. The corresponding subfacies can be further subdivided into several microfacies. The braided river delta plain in this study area is mainly developed in the fourth member of Xujiahe formation, and also in the middle and upper part of the second member of Xujiahe Formation in some areas. In addition, braided river delta plain braided channel sedimentary environment is also developed at the bottom of the first member of Xujiahe Formation in lixiya section, with coarse lithology, mainly medium coarse sandstone, intercalated with breccia and gravelly sandstone. The main reason for this difference is that the lixiya section is closer to the provenance, closer to the edge of the basin, with higher terrain and no sea water coming from it. The braided river delta front is mainly developed in the second member of Xujiahe formation. The former Braided River Delta is not developed in this study area, and the lithology is mainly silty mudstone.

3.3 Beach bar-lagoon system

This kind of system is mainly marine terrestrial interaction. Marine terrestrial interaction has its own characteristics. In the transition area between the ocean and the land, on the one hand, the sea water continuously scours the coastline to make the sediment become finer and purer. On the other hand, the land direction continuously provides a large number of material sources. After crossing the coastline, it is blocked by the sea water and then deposited to accept the seawater transformation. When the wave action is stronger than the river action, the land river channel is constantly affected by the tide, and then widens; When the fluvial action is strong, the channel extends to the sea continuously, and the sedimentary body has the characteristics of progradation. The transitional facies between sea and land in this area only occurs in the lower Jurassic (Fig. 1).

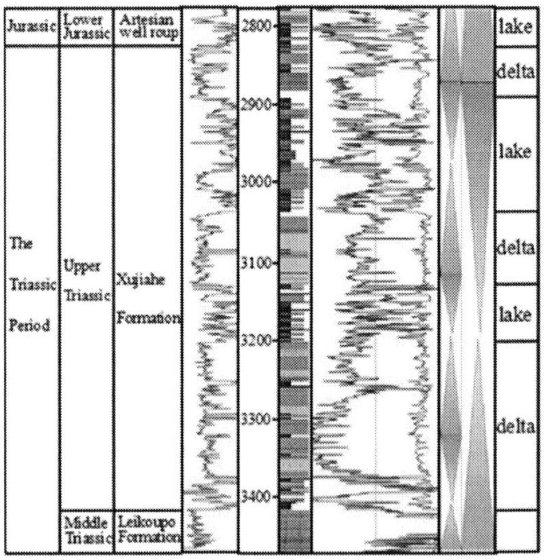

Fig. 1. Sedimentary characteristics of Upper Triassic beach bar lagoon system.

The main lithologic characteristics are fine siltstone and argillaceous sediments with good roundness and sorting. It includes lenticular bedding, vein bedding, wavy bedding and pinnate cross bedding, as well as plate and trough cross bedding formed by river action.

3.4 Lake facies

Lakes are relatively low-lying areas on the continent where water flows. The lake facies of Xujiahe Formation in the foreland basin of Northeast Sichuan Province is developed, mainly seen in xusan, Xuwu and Xu2 "belt Zi" of Lower Jurassic and Xujiahe formation. According to the water depth change determined by various facies signs and combined with the actual situation of the study area, it is divided into a sub phase of shore shallow lake.

The waterfront subfacies are located between the flood shoreline and the dry water shoreline, that is, the water depth decreases to the area where the waves break and form a strong wave, and the scour and backflow wash the lake bank repeatedly. Therefore, the hydrodynamic force of the lakeside environment can be strong (in flood level), weak (in the dry water level), and it is often oxidation environment, which is not suitable for biological survival. The sub facies of shore shallow lake is characterized by frequent interbedding of sand and mud, with good sand mud differentiation and obvious stratification. However, the lateral change of lithology and thickness is fast and the continuity is poor. The development of sand bodies in each part is different, so the characteristics of logging curve and seismic facies also change greatly.

4 Main sedimentary assemblage facies model

On the discussion of sedimentary system of Xujiahe formation, this paper considers that in addition to delta facies sand (Fig. 2), beach bar lagoon facies is also the main sedimentary facies type of Xujiahe Formation[6] (Fig. 3). The formation process of the delta system is that the river transported materials from the source to the sink, unloaded near the estuary and accumulated continuously; The construction process of beach bar lagoon system is a process in which the detrital materials from various sources, such as river unloading materials, are repeatedly washed and transformed by wind waves and tides, and redistributed under the action of wind waves, tides and coastal currents. The former is a vertical geological process, while the latter is a horizontal geological process. In the geological process chain of "initial unloading transformation redistribution", the delta system and the beach bar system are at two ends, and the initial unloading material is the main body of the formation, namely the delta system; There are also a series of transition types between the two different systems.

5 Conclusion and understanding

Through research and analysis, sedimentary facies types and facies units are established, which can reflect the basic characteristics and geometric morphology of sedimentary system. The main identification marks are: color characteristics, structural characteristics, sedimentary structure, logging curve characteristics, paleontological fossils, etc.

According to the comprehensive analysis of rock type, lithologic association, sedimentary structure, paleontological Association and section structure, the sedimentary facies types of Xujiahe Formation in the study area mainly include flood basin facies, delta facies, beach bar lagoon system and lake facies in the semi-arid area, in which delta facies and beach bar lagoon system are the main sedimentary facies types.

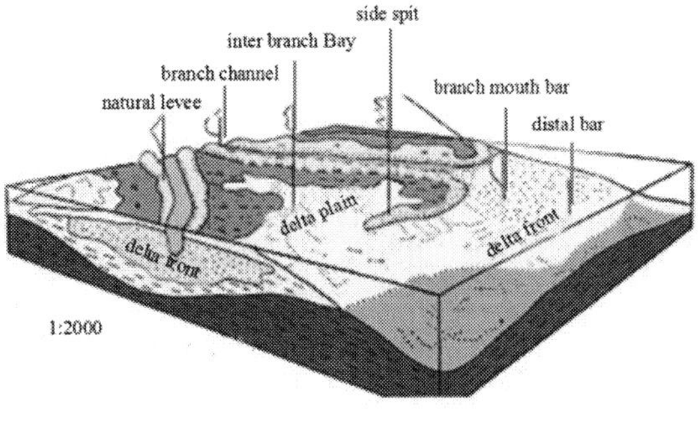

Fig. 2. Schematic diagram of delta system.

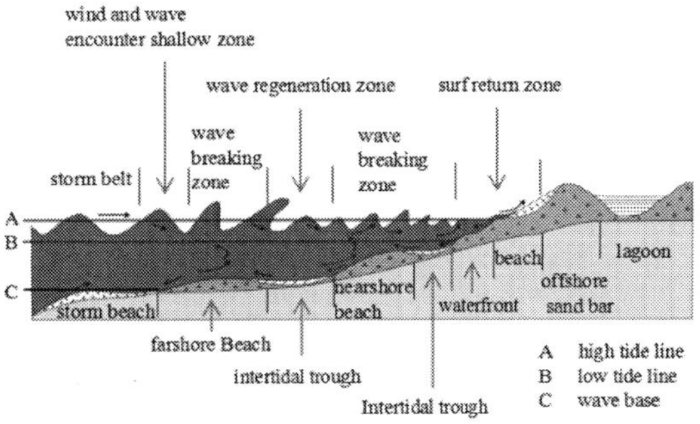

Fig. 3. Section diagram of beach bar—lagoon system.

References

1. Zhao Xiafei, Zhang Wenlin. A re-discussion on the origins of tidal deposits in the Xujiahe Formation of the Sichuan Basin:Further evidences and sequence analysis (Chinese:Natural Gas Industry) 2011, 31 (9) : p 25-30
2. LI Guoyu, CHEN Qilin, BAI Yunlai, etc. Marine sediment: The future expectation of China ' s petroleum industry (Chinese:Northwest Oil & Gas Exploration) 2014, 26 (6) : p 1-7
3. Dong Dazhong, Gao Shikui, Huang Jinliang, etc. A discussion on the shale gas exploration & development prospect in the Sichuan Basin (Chinese:Natural Gas Industry) 2014, 34 (12) : p 1-15
4. Yang Yuanping. A Study of Shale Gas Resources Potential of the Upper Triassic Xujiahe Formation in the Dazhou Region (Chinese:Acta Geologica Sichuan) 2019, 39 (1) : p 72-74
5. Chen Guo, Liu Gehang, Li Hongxi, etc. Resource

https://doi.org/10.1051/e3sconf/202129002022

potential of continental shale gas in Upper Triassic Xujiahe Formation,Sichuan Basin (Chinese:Natural Gas Technology and Economy) 2019, 13（5）： p 21-29

6. CJi Youliang, Liu Junlong, Wang Tianyun, etc. Distributing pattern and mapping method of delta and beach-bar composite sand-bodies in continental lacustrine basin (Chinese:Journal of Palaeogeography) 2016, 18（4）： p 615-630

Advances in Remote Sensing Extraction of Urban Roads

Liu Shujun[1*]

[1]Jiangxi University of Science and Technology, Ganzhou, Jiangxi, China

Abstract: As early as 1970s, the United States has begun the research of remote sensing image processing technology. In recent ten years, the research of road remote sensing extraction in China has also advanced by leaps and bounds. High resolution remote sensing images have been widely used in many fields, such as urban development planning, environmental monitoring and evaluation, and public announcement information services. The main application goal of remote sensing image is to extract the information of the object of interest, then identify it and complete the image understanding. Road is the most important and basic transportation mode of human beings, which provides a lot of support for the development of human civilization. road extraction is important for traffic management, including urban planning, road monitoring, GPS navigation, map updating, image registration, etc. extracting roads from high-resolution remote sensing satellite images is not only a challenging research direction, but also of great practical value.

1 Introduction

With the rapid development of remote sensing technology, we can obtain more image information more quickly, and how to deal with and interpret these massive image information quickly and accurately is an important problem. As the skeleton of the city, the urban road is also a very important basic geographic information. It plays an important role in the economic activities of the city. The rapid development of image processing technology makes it possible to extract the road based on remote sensing image automatically [1].

2 Basic Features of High Resolution Remote Sensing Image Roads

In high-resolution remote sensing images, roads have four basic features [2]

2.1. Geometric characteristics

The road is long and arranged in strips. Roads are usually long and curved. The width of the same road is almost constant, and the edges are parallel to each other. Generally speaking, the contrast between road and non-road features is very obvious, but in high-resolution remote sensing images, there may be green belts, cars, tall buildings on the road.

2.2. Topological features

Roads are not isolated, roads cross with other roads to form a road network, especially in cities, the road network is generally more dense.

2.3. Radiation characteristics

The same section is usually built with the same material, and the gray value of the road is the same. But roads in China, especially urban roads, are mostly obscured by vegetation and buildings.

2.4. Contextual features

Also known as background features. Roads in different places have different backgrounds. For example, rural roads are dominated by farmland and sparse small houses; urban roads have many vehicles, many green belts around the roads, and dense distribution of tall buildings, which also leads to a large number of shadows on urban roads.

3 Current Situation of Remote Sensing Image Road Extraction Based on Traditional Methods

The process of road extraction is to use a certain method to obtain road points, eliminate other interference, and connect road sections to the road network [3]. There are many road extraction algorithms. Different remote sensing image resolution and different imaging methods lead to the diversification of road extraction methods. Scholars at home and abroad put forward many algorithms for this problem [4].

*Corresponding author's e-mail: liushujun@jxlgdhccdsq999.onexmail.com

3.1. Watershed Algorithm

Watershed algorithm is a segmentation method of mathematical morphology based on topology theory. In the process of image segmentation, the similarity between adjacent pixels is regarded as a very important reference basis, and the pixel points with similar position and gray value are connected to form a set with boundaries [5].

In 2008, Pan Tingting proposed a watershed algorithm based on the basic characteristics of remote sensing image road and the regional characteristics of watershed algorithm[6]. In this method, the remote sensing image is first denoised by filter, then the road information is extracted by watershed algorithm, and finally some small spots are removed from the above operation. During 2011, Yu Changhui and Yi Yaohua first used watershed algorithm to segment remote sensing images into small areas, then marked each small area, and finally extracted the road information on the image by object-based Markov random field (MRF) method[7]. In 2013, Cai Hongyue and Yao Guoqing proposed an improved watershed road extraction algorithm based on object-oriented and mathematical morphology theory. This method can use mathematical morphology to remove or correct bad results[8].As shown in figure 1.

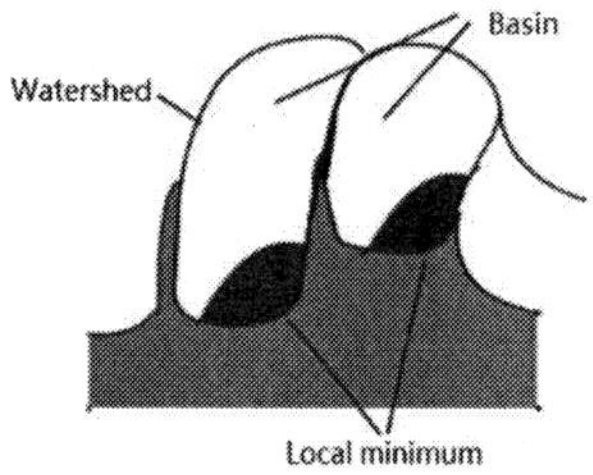

Fig.1 modle of watershed

3.2. Edge-based road extraction

Because different geographical locations and ground objects have different gray values in remote sensing images, this paper proposes an edge detection method based on different gray values. It can greatly reduce the amount of data, remove the information independent of the target, and only retain the structure of the image. The operators commonly used by researchers in edge extraction generally include: Sobel operator, Prewitt operator, Laplacian operator and Canny operator [9].

R.Gaetano, J.Zernbia and G.Scarpa were first proposed to use edge detection operators in remote sensing image segmentation Canny 2011[10]. In this method, firstly, the edge detection operator is used to detect the edge information of remote sensing image.Secondly, the edge center line information is calculated by using Euclidean distance function.Finally, the road information of the processed remote sensing image is extracted by using the graph cutting theory. Zeng Faming, Yang Bo and Wu Dewen also used the Canny edge detection operator in 2013 to extract the road in the mining area from the remote sensing image[11]. In this method, the edge information of the image is first detected by the edge detection operator. Then match the image edge information to find the edge of the road, and finally locate and vectorize the road in the mining area. In 2014, Xu Nan and Zhou Shaoguang used the direction template to detect the edge point information, then divided the image into blocks, and then used the search method to extract the sub-lines of the image after the block[12]. By 2016, Tan Yuan, Huang Huixian and Xu Jianmin improved the Sobe operator according to the continuous and complete characteristics of the road edge and the bad detection effect of the traditional edge detection operator. The continuity of edge information detected by this improved operator is better than that without improvement[13].As shown in figure 2.

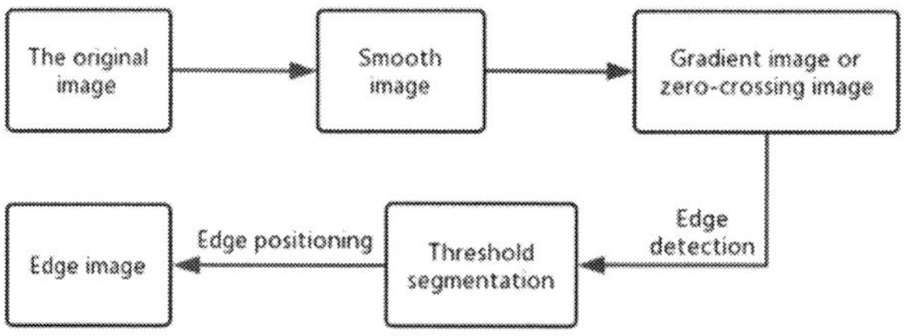

Fig.2 Image edge detection flow chart

3.3. Texture analysis

Unlike pixel-based color features, texture features represent the distribution of pixels in the region. In addition, texture feature is a global feature, which not only has good noise resistance, but also has rotation invariance [14].

In 2011, Wu Xuewen and Xu Hanqiu proposed a method to extract urban roads with high resolution remote sensing images by combining the speed function in the fast moving method with the gradient feature and texture feature of the road in the remote sensing image [15]. Zhou Shaoguang, Chen Chao, He Chunxiao proposed a road extraction method based on two shape priors using Graph Cuts computation to fuse them in 2013 [16]. In this method, the texture feature and input image are first extracted as the segmentation eigenvalue, then the road point map is extracted by Graph Cuts, and then the initial segment graph of the road is obtained by the method of straight line matching. Finally, the original segment of the road is extrapolated to obtain the complete remote sensing image road.As shown in figure 3.

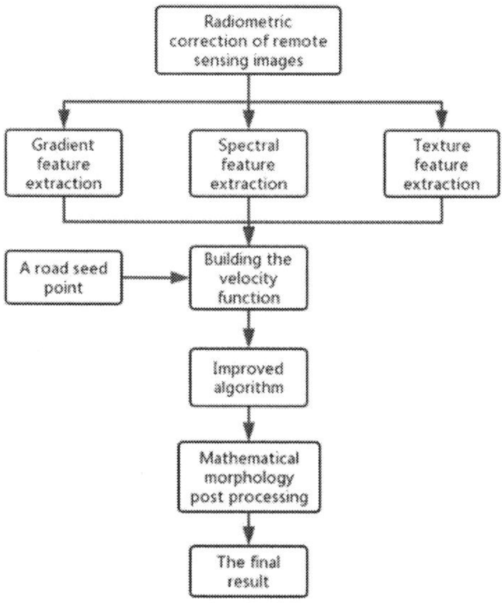

Fig.3 Technology Roadmap

3.4. Line segment matching method

W.Z.Shi, C.Q.ZhuW proposed line segment matching method for road extraction on high resolution remote sensing satellite images in 2002, which is a quasi-automated method[17]. This method is based on the characteristics of high resolution remote sensing images, the basic knowledge of roads and mathematical techniques. does not need to pre-select seed points as most semi-automatic method extraction methods do, although it needs to set some parameters. In 2004, Zhu Changqing, Wang Yaoge and Ma Qiuhe proposed a method to extract gray morphology based on mathematical morphology to facilitate the extraction of roads from high-resolution remote sensing images [18]. In this method, the image segmentation is carried out on the gray image through the morphological features to obtain the road network contour, and then the road network of the remote sensing image is obtained by using the line segment feature matching method to extract the road on the segmented image.

4 Current Situation of Remote Sensing Image Road Extraction Based on Deep Learning

Deep learning combines abstract high-level features by extracting simple low-level features of multiple layers, so that information such as categories and features of images can be obtained. It is developed by artificial neural network.The simplest artificial neural network consists of a data input layer, an output layer, and a hidden layer[19]. Deep neural network is a neural network containing multiple hidden layers. For deep learning, the number of layers of neural network is from dozens to hundreds, and some ResNet even reach more than 1000 layers[20]. In depth learning, the data set is trained, a loss function is set up, and the weights and deviations of the neural network are continuously calculated and updated to optimize the loss function and improve the prediction accuracy of the network. After continuous development, deep learning has been applied to face recognition, license plate recognition and other aspects. It is not superior, it has changed people's daily life and improved the efficiency of social production. Because of more and more research on depth learning, researchers in remote sensing field also apply depth learning to remote sensing image processing.

In 2015, Wang Gang, Chen Jinyong and Gao Feng et al .designed a deep neural network to solve the problem of target detection of remote sensing images with high score No .1[21] . They accurately detected playgrounds, airports and other infrastructure on remote sensing images of High Score One. It is proved that the application of neural network in remote sensing image target detection is feasible.

In 2017, Pei Lei, Li Guoqing and Zeng Yi proposed a hierarchical convolution feature in image registration by studying the main processes of image registration and point transformation[22]. First, the convolution neural network is used to adaptively extract the hierarchical convolution features of feature points, and then the filter is used to analyze their correlation to see their correlation and calculate the location of feature points. Finally, the registration feature points are used to calculate the changed parameters for remote sensing image stitching. Compared with the traditional scale invariant feature transformation method, when convolution neural network is introduced into image registration, the accuracy and robustness of remote sensing image stitching are higher.

In 2017, Han Jie, Li Shengyang and Zhang Tao applied to remote sensing image classification and urban development analysis because of the high accuracy of deep learning and the lower misjudgment rate than traditional methods[23]. Compared with other methods, the correct rate is much higher, which provides a technical reference for urban planning.

In 2017, solemnly, Zhang Jingdong and du Jianhua proposed to apply the deep neural network method to the study of ground collapse in order to solve the problem of poor efficiency of traditional ground collapse identification method[24]. The depth belief network is used to continuously adjust the parameters of the network model, and then compared with the traditional method, it is proved that the depth learning network is more efficient and the recognition accuracy is better.

The application of depth learning to remote sensing image processing has been very successful, but the road extraction of high resolution remote sensing images is still in the initial stage.

Asemi-automatic method was proposed to extract

remote sensing images based on deep learning in G.Cheng, C.Yang, X.Yao , 2015 [25]. They use deep convolutional neural networks to train a large number of road remote sensing images to obtain the directional features of each image, then use artificial methods to obtain the starting point of the road, and use trained neural networks to judge the direction of the road. Finally, the road is extracted from the image by finite state machine.

P.K.Li, Y.Zang,C.Wang, etc . designed a neural network with only 3 layers of convolution layer and 1 layer of full connection layer to determine whether each pixel of a remote sensing image slice with a size of 32 x32 is a pixel of the road, and then use line integral convolution to post-process the results obtained by neural network judgment in 2017[26].

G.L.Cheng, Y.Wang,S.B.Xu proposed a cascade end-to-end convolutional neural network in 2017, which solves the problem that some roads will be blocked by trees and cars when extracting the road centerline on high-resolution remote sensing images[27] . The network is divided into two parts, one is to carry out road detection, the other is cascaded with it to process the results of road detection and extract the center line of the road.

In 2018, Sun Zhenhui, Meng Qingyan, Sun Yunxiao and others proposed a neural network based on pulse coupled neural network to solve the influence of shadow on road extraction when high resolution remote sensing images are proposed[28]. First, it is necessary to eliminate the interference of shadow and water body in the near infrared band, then use neural network to segment the processed image, then use morphological method to eliminate the influence of buildings and vegetation on road extraction. Finally, the method of road extraction is extracted. This method has good effect on dealing with shadow interference, and the extracted urban road information is continuous and complete.

In 2019, Hong M, Guo J, Dai Y and others propose a new FMH model, main road information is extracted from high-resolution remote sensing images in urban areas[29]. Huang F, Yu Y, Feng T et al. considered that the potential of multi-source remote sensing data was not fully exploited, Low - level features are not effectively organized, IDE accuracy is low, Poor automation. In this paper, a new method of automatic extraction of impervious surface based on deep learning and multi-source remote sensing data is proposed, using the minimum noise fraction (MNF) method to de-noise multi-source data, extracting representative features of neighborhood windows using convolutional neural networks (CNN), trained ELM classifiers and MOR-speech operations, Distinguish the city's impervious surface from other surfaces, Three methods are compared and selected [30].

References

1. Liu Dong. A Study on Parallel Extraction of Road Information from High Resolution Remote Sensing Images Based on Mathematical Morphology [D].Jilin University. 2011
2. George Vosselman,Jurrien de Knecht.Road tracing by profilematching and Kalman filtering[J].] Proc. Automatic Extraction of Man-Made Objects, 1995,1995: 265-274.
3. Zhou. A Study on the Method of Road Information Extraction Based on Knowledge [D]; and Sichuan normal University.
4. Li Jianfei. A Study on the Method of Road Information Extraction and Expression in High Resolution Remote Sensing Image [D].
5. Cai Huifen. A Study on Image Segmentation Algorithm Based on Watershed Algorithm and Isochronous Theory [D].
6. Pan Tingting. Application of mathematical morphology and watershed algorithm in remote sensing image target recognition [D].]; and Jiangnan University ,2008.
7. Yu C ,Yi Y .Object-based Road Extraction in Remote Sensing Image Using Markov Random Field[J].] Geomatics &Information ence of Wuhan University, 2011,36(5): 544-547.
8. Cai Hongyue, Yao Guoqing. Method of Regional Division of Dynamic Remote Sensing Monitoring of Land Use Based on Development Pressure State —— Taking Jinnan District of Tianjin as an example [J].]; and Remote Sensing of Land and Resources 25(3):25-29.
9. Yang Xiaoyan. Research and Application of Image Edge Detection Algorithm Based on Wavelet and Morphology [D]. Chengdu University of Technology.
10. Gaetano R ,Zerubia J ,Scarpa G ,et al.Morphological road segmentation in urban areas from high resolution satellite images[C]//2011] 17 th International Conference on Digital Signal Processing (DSP). IEEE,2011.
11. Zeng Invention, Yang Bo, Wu Dewen, et al. Mining road extraction [J].] based on Canny edge detection operator Remote Sensing of Land and Resources ,2013(04):77-83.
12. Bao Yansong, Mao Fei, Min Jinzhong, et al. Extraction [J].] of Remote Sensing Road Edge Lines Based on Image Partition and Line Segment Voting Remote Sensing of Land and Resources , 2015,27(1):55-61.
13. Tan Yuan, Huang Huixian, Xu Jianmin, et al. Method for Road Edge Detection of Remote Sensing Image Based on Improved Sobel Operator [J].]; and Remote Sensing of Land and Resources ,2016,028(003):7-11.
14. Zhou Chunyan. A Study on Local Matching Algorithm for Stereo Image Based on Color Information and Pixel Feature [D].].(A Shanghai Jiaotong University.
15. Wu Xuewen, Xu Hanqiu. An Algorithm for Extracting Main Road Information from High

Resolution Remote Sensing Images [J].] Based on Level Set Method Journal of Astronautics, 2010,31(005):1495-1502.

16. Zhou Shaoguang, Chen Chao, He Chunxiao. New Methods of Road Segmentation Based on Shape Apriori and Graph Cuts Principle [J]. Mapping Bulletin ,2013,000(012):55-57.
17. Wenzhong,Shi,Changqing,, etc. The Line Segment Match Method for Extracting Road Network From High-Resolution Satellite Images.[J]. IEEE Transactions on Geoscience &Remote Sensing, 2002.
18. Zhu Changqing, Wang Yaoge, Ma Qiuhe, et al. High Resolution Remote Sensing Image Road Extraction [J].] Based on Morphology Segmentation Journal of Surveying and Mapping ,2004,33(4):347-351.
19. Li Xinyi. Image Feature Extraction and Application [D].] Based on Depth Network
20. He K ,Zhang X ,Ren S ,et al.Deep Residual Learning for Image Recognition[C]//IEEE Conference on Computer Vision &Pattern Recognition.IEEE Computer Society,2016.]
21. Wang Gang, Chen Jinyong, Gao Feng, et al. A Study on Target Detection of Remote Sensing Image Infrastructure Based on Depth Learning [J]. Radio Engineering ,2018,048(003):219-224.
22. Li Guoqing, Zeng Yi. An Improved Remote Sensing Image Splicing Method [J].] Based on Deep Learning Computer Engineering and Applications ,2017,53(20):180-186.
23. Han Jie, Li Shengyang, Zhang Tao. A Study on Urban Extension Method of Remote Sensing Image Based on Deep Learning [J].]; and manned spaceflight ,2017(3).
24. Solemn, Zhang Jingdong, du Jianhua. A Study on Ground Collapse Recognition Method in Remote Sensing Image Based on Depth Learning [J]. Modern Commerce and Industry ,2017(35):189-192.
25. Cheng G ,Yang C ,Yao X ,et al.When Deep Learning Meets Metric Learning: Remote Sensing Image Scene Classification via Learning Discriminative CNNs[J]. IEEE Transactions on Geoence and Remote Sensing,2018: 2811-2821.
26. K., Y.Zang,C.Wang,et al.Road network extraction via deep learning and line integral convolution[C]//2016IEEE International Geoscience and Remote Sensing Symposium (IGARSS). IEEE,2016.
27. Cheng G ,Wang Y,Xu S ,et al.Automatic Road Detection and Centerline Extraction via Cascaded End-to-End Convolutional Neural Network[J]. IEEE Transactions on Geoscience &Remote Sensing, 2017,55(6): 3322-3337.
28. Sun Zhenhui, Meng Qingyan,Sun Yunxiao, et al. Extraction of Urban Road from Remote Sensing Image of Pulse Coupled Neural Network [J]. Surveying and Mapping Science, 2018, 043(001):145-152.
29. Hong M ,Guo J ,Dai Y ,et al.A novel FMH model for road extraction from high-resolution remote sensing images in urban areas[J]. Procedia Computer Science,2019,147: 49-55.
30. Huang F ,Yu Y ,Feng T .Automatic extraction of urban impervious surfaces based on deep learning and multi-source remote sensing data[J]. Journal of Visual Communication and Image Representation, 2019.

The practical application of information technology in marketing

Yanyang Ma [1]

[1]Xichang University, 615013

Abstract. Marketing is a discipline closely related to social and economic development. In the teaching practice of marketing, we should scientifically and comprehensively reform the teaching methods of education and actively apply information technology to effectively and effectively improve the overall quality of marketing. And level, to better improve the overall literacy and cognitive abilities of students.

1.Introduction

At this stage, the continuous progress of the information age makes information technology bring new challenges to classroom teaching reform, and put forward brand new requirements for teachers' teaching. When carrying out teaching work, teaching concepts, teaching models and teaching methods must adapt to the development of the times. Marketing itself is a core characteristic course that closely fits the practice. In the education and teaching practice of marketing, it should be fully and comprehensively combined with information technology to form a complete education program, and better Improve the quality and effectiveness of education and teaching, and comprehensively improve students' information literacy and overall cognitive ability. Information-based teaching methods are mainly based on modern information technology, applying multimedia technology, network technology, virtual reality technology, artificial intelligence technology, etc. to the teaching process. [1]

2. Relevant overview of information teaching

With the continuous and rapid development of information technology, information-based teaching has become the main method of teaching at this stage. Informatization teaching is developed and evolved in the process of continuous development of information technology, and it is an efficient teaching organization mode. In the teaching practice of marketing, information technology should be used scientifically and comprehensively to innovate and reform the education and teaching mode. In information teaching, the current teaching thought is regarded as the guiding ideology, information technology is the support, and modern teaching methods are adopted. Through information-based teaching, teachers act as the disseminators of knowledge and convey knowledge to students. Students are changed from external stimulators and knowledge transmitters to information processors and active knowledge builders. The media has also become a cognitive tool that transforms teachers' explanations and presentations into students' self-directed learning and exploration and solving practical problems. Students can use this to search for information and search for information. The focus of marketing teaching reform is to change the previous form of indoctrination in the classroom. Informatization teaching is a very effective mode in the exploration of marketing teaching reform.

3. The necessity of the reform of marketing informatization teaching

Marketing is a characteristic subject closely integrated with practice. In the education and teaching practice of marketing, it should actively and effectively conform to the development trend of the times, effectively promote the in-depth information teaching, and comprehensively carry out the reform of information teaching . It can be said that in the educational practice of marketing, it is of very important practical significance to thoroughly and comprehensively do a good job in the reform of informatization teaching. On the one hand, actively and effectively promote the reform of marketing informatization teaching, which can enrich the course content to a large extent, and at the same time can improve students' practical literacy and cognitive vision in an all-round way, and better consolidate their learning effectiveness. It can be said that the requirements of the future society for marketing students are very high, not only require students to have solid practical literacy, but also require them to have a solid theoretical understanding, strong practical ability, market insight ability and research Ability and so on. Through the in-depth advancement of information-based teaching, it can not only provide students with broad learning resources, but also optimize students' cognitive literacy to a large extent. It can also enhance students' autonomous learning ability in an all-round way, and better cultivate social needs. High-quality talent. In the practice of informatization teaching, the

5213457744jqzy@dlvtc.edu.cn

subjective status of students has been fully embodied. Under the guidance of teachers, they can consciously and actively use the information platform to carry out autonomous learning. This efficient teaching organization model can further consolidate students' learning effectiveness, improve their learning quality in an all-round way, and further promote their growth and development. On the other hand, in the teaching practice of marketing, the active use of information technology can also effectively change the teaching organization mode and better stimulate students' interest in learning. It is undeniable that in the teaching process of traditional marketing, teachers and students are confronted with each other. Teachers subjectively teach courses, students passively accept and learn knowledge, which often affects to a large extent. And it restricts students' learning effectiveness, and it is difficult to truly optimize their cognitive literacy. Through informatization teaching reform, it can promote the effective conversion of the roles between teachers and students, can better play the guiding role of teachers, and can also enhance students' subjective initiative in an all-round way.[2]

4. Requirements for the integration of marketing and information technology

In the education and teaching practice of marketing, the traditional teaching organization model has big drawbacks and shortcomings, which affects the quality of education and teaching to a large extent, but also is not conducive to improving the overall quality of students. In the process of continuous development of information technology, a thorough and comprehensive grasp of the internal relevance between marketing and information technology, and actively and effectively promote the reform of information teaching, not only can improve the overall educational effectiveness of marketing, but also To a large extent, it consolidates students' practical literacy and optimizes students' cognitive level in an all-round way. In the course of practice, in order to effectively integrate marketing and information technology, it is necessary to accurately and comprehensively grasp the internal requirements and consolidate the quality of education and teaching to the greatest extent.

4.1. Update of teachers' teaching philosophy

In the process of informatization teaching of marketing, teachers are very important participants. Their own thinking and concepts, their own teaching concepts, etc., are often related to the education quality of marketing and directly related to the level of information teaching. In order to better promote the in-depth informationization of teaching, we must focus on the comprehensive update and transformation of teachers' teaching concepts, and carry out efficient adjustments and changes from the ideological and cognitive level, so as to consolidate teachers' teaching quality to the greatest extent and further improve them. Professional ability. In order to ensure the reform and promotion of marketing information teaching, the school should focus on creating an efficient talent training system, effectively cultivating a group of high-quality teachers with strong professionalism, and effectively improving the teaching skills of teachers and the development and design of teaching media. ability. At the same time, the school should also set up a sound teaching and research team to carry out a full range of innovations and discussions on the integration of marketing and information technology. In addition, the school must use systematic vocational skills training and competition systems to effectively stimulate the enthusiasm of teachers to participate, and to consolidate their practical literacy and teaching organization and management capabilities to the greatest extent.

4.2. Make necessary investment in teaching software and hardware

In the process of continuous development of information technology, actively and effectively integrate marketing and information technology to better provide the necessary conditions for information teaching, and the investment and management of software and hardware should be effectively promoted. Hardware investment is the fundamental and prerequisite for the development of informatization teaching. The effective replacement of equipment and the purchase of hardware provide a solid foundation for the school to create modern educational technology features. According to current technology, the fundamental characteristics of education informatization are digitization, network intelligence and multimedia.[3] Schools with a certain level of strength need to build multimedia classrooms such as electronic reading rooms and digital libraries. According to calculations in the China Education Statistics Yearbook 2019, nationwide, the Internet access rates of elementary schools, junior high schools, and ordinary high schools have reached 68.06%, 92.43%, and 96.26%, basically reaching the level of popularization. It can be said that in the process of advancing informatization teaching, in-depth and comprehensive strengthening of the investment and construction of software and hardware facilities is a very necessary investment and a relatively large investment in education. In order to better improve the overall quality and effectiveness of informatization teaching of marketing, and to consolidate the overall strength of informatization teaching in an all-round way, it is necessary to make efficient and scientific software and hardware investment.

5. Practical application of information technology in marketing

Marketing itself is a characteristic subject with quite practical attributes. In the education and teaching practice of marketing, information technology methods should be thoroughly and comprehensively applied to further improve the overall teaching quality of marketing and better consolidate students. Practical literacy and cognitive ability. It can be said that informatization teaching has become an important organization method of marketing teaching, and it has become an important teaching form of marketing. In order to give full play to the overall application of information technology, it is necessary to fully grasp the characteristics of marketing and apply information technology in a scientific and comprehensive manner.

5.1. Efficient use of marketing information platform

In the education reform practice of marketing, the efficient application of information technology plays a very important practical role. In order to better improve the effectiveness of education and teaching, it is necessary to accurately and comprehensively grasp the information technology means, and build a scientific and comprehensive marketing information platform. It can be said that the construction of an information platform for marketing can lay a solid foundation for the advancement of information teaching, and it can also help deepen the reform of marketing education, as shown in Figure 1. At the current stage, the construction of the marketing informatization curriculum system has achieved outstanding results. In the network platform, the type, quantity, content, etc. of marketing information education resources are relatively complete and extensive, which has laid a solid foundation for education and teaching. In order to make better use of the information teaching platform, an efficient course organization should be carried out to effectively improve the effectiveness and level of marketing information teaching.

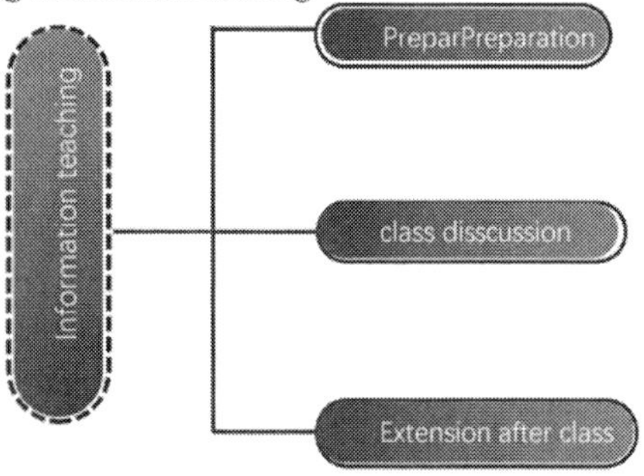

Figure 1: The architecture system of informatization teaching

5.1.1. Realize pre-class preparation with micro-classes, etc.

The deepening of the reform of marketing informatization teaching provides good conditions for students to learn independently, and at the same time, it also greatly enhances students' subjective initiative. In order to make better use of the overall application effectiveness of the marketing information platform, teachers should fully and comprehensively play the role of teaching guide, scientifically and comprehensively make fine and comprehensive pre-class preparations. It can be said that efficient pre-class preparation can clarify the learning direction and goals for students, and can also guide students to use the information platform to familiarize themselves with and recognize courseware resources in advance. In the process of practice, with the teaching characteristics and requirements of marketing as the core, students are required to use the information platform to learn or download related teaching content before teaching, familiarize themselves with the teaching outline in advance, and then set up relevant questions to guide students to think according to the outline requirements. According to the case in the network-assisted course, students are required to collect materials that they are interested in when collecting materials, and prepare for class discussion. Through the independent study of students, they can prepare for study in advance according to the relevant content in the information platform.[4]Relying on this approach, it can not only fully mobilize students' interest in learning, but also help students recognize the problems and obstacles in autonomous learning in advance, so as to better consolidate the effectiveness of students' learning.

5.1.2. Conduct classroom discussions with multimedia technology

In the education and teaching practice of marketing, the in-depth and comprehensive use of information technology means, and the active development of extensive and in-depth classroom discussions and interactions can consolidate students' learning effectiveness to a large extent, and at the same time create good results for students. Classroom discussion situation. In the specific process of classroom discussion, teachers should combine the needs of the curriculum and make effective use of multimedia technology. It can be said that, compared with traditional classroom lectures and flat board newspapers, the application of multimedia teaching situations can fully display the relevant knowledge and content of marketing to students, and guide them to conduct in-depth and effective learning and exploration. . Relying on pre-class learning, in the process of organizing students to discuss in class, teachers can use multimedia to display relevant cases or courseware resources, etc., to guide and encourage students to conduct in-depth and systematic discussions and learning about these cases or courseware resources. In classroom discussions, teachers can divide students in the class into groups and guide them to conduct in-depth and extensive interactions and exchanges in groups to better consolidate the effectiveness of classroom learning. For example, teachers can use multimedia teaching technology to effectively adopt different methods such as videos, animations, pictures, audios, etc., to effectively and effectively enhance students' cognitive interest, and better enhance students' subjective initiative. In addition, in information teaching, teachers should also combine the professional qualities and learning cognitive level of different students to effectively create differentiated teaching situations and truly realize personalized teaching.

5.1.3. Use an information platform to organize after-school discussions

In the education and teaching practice of marketing, the scientific and efficient application of information technology can create a scientific and good learning platform for students, and at the same time, it can better guide students to discuss and interact. For this reason, in the after-school stage, teachers should actively create tasks for after-school discussions based on the information platform, encourage and guide students to use the after-school platform for in-depth communication and interaction. In the process of after-class extension and discussion, students can use a wide range of network platforms to communicate and interact with teachers and students at a wide range of levels. At the same time, in the extension and expansion stage after class, teachers should also set up an online test section based on the information platform to guide students to actively participate in online answering questions based on the learning results of marketing before and during class, and quickly test their learning ability and Level of learning. Teachers can use the online platform to grasp the learning progress and learning situation of students, so as to give students necessary teaching guidance in time.[5]

5.2. Application of Social Network Analysis in Marketing

The analysis of social network is a kind of social relationship structure mainly produced by the interconnection between individuals. It is a scientific, reasonable and standardized social relationship structure, which can effectively analyze the social relationship structure and emphasize the relationship between social subjects. The relationship between, reveals the subject matter in the society, and then conducts a quantitative analysis on it. In recent years, my country's computer technology and network technology have been rapidly developed, and the level of social network analysis has gradually improved, and it has been widely used in marketing. With the continuous and rapid development of information technology, in the education and teaching practice of marketing, marketing and information technology should be efficiently integrated to form a complete and systematic teaching organization model to better improve students' Learning effectiveness. It can be said that the application of information technology provides perfect ideas for the development of marketing and also creates a broad platform. In the practice of informatization teaching, social network analysis is a very important foothold. If there is no comprehensive and systematic social network analysis, or no scientific curriculum content as a support, it will inevitably affect and restrict the deepening of the teaching of marketing to a large extent. To this end, teachers should combine practical needs to scientifically and comprehensively create teaching content. In practical applications, opinion leaders should be discovered scientifically and comprehensively. At the same time, big data technology should be used to re-evaluate customer value, and big data technology should be fully relied on to consolidate the teaching effectiveness of the course.

5.3. Use mature websites to use the "content + service" model

Marketing is a highly practical subject. In the course of course teaching, whether it is to adopt traditional teaching mode or construct an information-based teaching platform, we should pay attention to in-depth and comprehensive implementation of good practical teaching. In order to effectively enhance students' practical ability and to further consolidate students' curriculum literacy, teachers should base themselves on the curriculum characteristics of marketing and make effective use of various information-based teaching techniques. On the one hand, mature websites can be fully utilized in information teaching. Use professional marketing websites, such as: National Bureau of Statistics, China Marketing Network, China Marketing Network, etc. Among them, the National Bureau of Statistics can provide a large amount of economic data for marketing environment analysis and market research, so that students can grasp the latest and most comprehensive marketing macro environment, and refer to these big data in the process of product selection to help make correct decisions.

For example, teachers create corresponding practical tasks for students of different literacy, guide and encourage them to use the information platform to search and develop related resources, complete their respective exploration tasks effectively, and better improve students' practical literacy. On the other hand, in the educational reform of marketing, teachers should also use the "content + service" model to guide students to actively use a broad platform to optimize their learning literacy.

6. Conclusion

In the education and teaching practice of marketing, the application of information technology has a very important practical effect. In order to better improve the teaching effectiveness of marketing, and to better consolidate students' practical literacy, we should scientifically and comprehensively promote the in-depth information teaching work, and better improve the overall teaching quality of marketing.

References:

1. Mao Linglin, He Xin. Research on the Development and Application of SPOC Curriculum of "Marketing" in Local Colleges and Universities under the Background of Information Technology[J]. Teaching and Education (Higher Education Forum), 2019(4): 63-65.
2. Zhu Hong. Research on the Strategy of Improving the Teaching Quality of "Marketing" Courses in Secondary Vocational Schools under the Visual Threshold of Digital Aggregation[J]. Seeking Knowledge Guide, 2018(15):103-105.
3. Zhang Hongjun. Reform of "Marketing" teaching system in applied colleges and universities[J]. Teaching and Education (Higher Education Forum), 2019(7): 80-81.

4. Wang Jingjing. Discussion on the ideas and methods of the teaching reform of the "Marketing" course[J]. Science Education Journal-Electronic Edition (mid-day), 2019(4):127.
5. Li Yunqin, Zhang Li. Research on marketing teaching reform under the background of "Internet +"[J]. Rural Economy and Technology, 2018, 29(16): 300.

Research on the Application of Computer in the Design of Arts and Crafts

Zhengjuan Hu[1]

[1]Jingdezhen Ceramic University, Jingdezhen, Jiangxi, ZIP code, China

Abstract. Firstly, this paper expounds the design methods of traditional arts and crafts, including the traditional design methods of sculpture crafts, glass crafts and ceramic crafts. Then it analyzes the advantages of computer technology in the design of arts and crafts, including improving the product performance of arts and crafts, improving the design interest of arts and crafts designers and improving the modern flavor of arts and crafts; then it expounds the application of computer technology in different stages of arts and crafts design in detail, including the early preparation, concrete implementation, design scheme optimization and in-depth design stage; Finally, it analyzes the specific application of computer technology in arts and crafts design, including the application of computer technology in sculpture crafts design, glass crafts design and ceramic tea set modeling design.

1 Introduction

Arts and crafts retain the aesthetic feeling of art and have the practicality of handicrafts. It has high aesthetics and practicality, and is more easily accepted by the audience. It is the product of the integration of spirituality and aesthetics under the needs of the audience, and organically combines art with manual crafts. Traditional forms of arts and crafts can no longer meet the needs of modern people for arts and crafts. It is necessary to combine the characteristics of the times, use modern technical means and integrate modern elements, so as to produce brand-new modern arts and crafts that are needed and accepted by the society. The design of modern arts and crafts can not be separated from the assistance of computer technology. Arts and crafts are both practical and ornamental works of art. There are many kinds of arts and crafts, including arts and crafts, woodenware, lacquerware, metal crafts, glass crafts, sculpture crafts and ceramic crafts, which can be seen everywhere in people's daily life and work.

2 Design Method of Traditional Arts and Crafts

2.1 Traditional design methods of sculpture crafts

Traditional design methods of sculpture crafts are particularly troublesome and prone to many defects. In the lending stage of design and production, small samples of sculpture crafts need to be made by hand. In this process, it is necessary to select clay, paper pulp, gypsum or sculpture clay which are easy to be molded, which is complicated and time-consuming, or the designer's ideas are presented on paper to draw a draft drawing of arts and crafts to be manufactured, which will also produce more time cost. Then, the designer needs to make the sculpture clay sculpture according to the finished draft or handicraft sample, and replace the clay sculpture with materials such as resin or gypsum. We also call this process the turnover in the process of making the sculpture handicraft. The traditional sculpture creation cycle is very long, the design and manufacturing process is very complicated and the design and manufacturing process is extremely cumbersome. It takes a long time and more money to support the design and manufacturing of sculpture crafts, and the professional requirements of designers are also high. In the initial design stage of sculpture handicraft, it is necessary to modify the handicraft sample many times to achieve the best effect. Compared with other crafts, the design and production process of sculpture crafts takes longer time, and the workload of design and production is relatively large. Generally, sculptors will participate more in the design process, paying attention to the production of small samples and drafts of sculpture crafts, but sculptors will not participate too much in the production process. In this way, even if the design is no better and the samples of sculpture crafts are no better, the finished sculpture crafts may not achieve the expected results. The disconnection between design and the production process of sculpture crafts has a negative impact on the molding of sculpture crafts[1].

2.2 Traditional design method of glass crafts

In the process of designing and manufacturing traditional glass crafts, the real needs of the market and audience

1910012008@stu.jci.edu.cn

https://doi.org/10.1051/e3sconf/202129002025

customers will be investigated, and then the overall design scheme of glass crafts will be determined according to the survey results. It is necessary to select suitable processing facilities and techniques for making materials before drawing creative plane sketches, and then use professional techniques to make glass handicrafts. There are two ways to make traditional handmade glass, including "grafting" and "changing while doing". In the traditional form, when dewaxing and casting colored glaze, the corresponding plane design drawing should be drawn first, and then the three-dimensional model should be made according to the specific design drawing. It needs to go through a variety of procedures, including turning over the impression, pouring wax solution, trimming the corresponding wax mold, wrapping it with gypsum, dewaxing at high temperature, injecting glass frit into the molded gypsum mold, then firing finally, and stripping the gypsum mold after firing and molding. A series of treatments, including grinding, sandblasting, polishing, etc., are carried out on the fired glass, and the whole manufacturing process takes a long time, which also prolongs the design cycle.

2.3 Traditional design method of ceramic crafts

In the design process of traditional ceramic handicrafts, firstly, designers need to design ceramic decoration and ceramic modeling, and then show the concrete design in the form of hand-drawn drawings, and make ceramic physical models according to the drawn design drawings. In the process of designing and manufacturing this kind of ceramic handicraft, the designer's professional skills and experience are required to be high[2]. The designer needs to hand-painted and hand-made in the whole process, and the design and production process is complicated and difficult. At the same time, the selected raw materials will also affect the designer's design and production. Once there are any problems in the design process, it is necessary to modify them in time, so as to finally determine the design draft, ensure the feasibility of the design scheme and ensure that the design effect can reach the expectations. The molding process limits and affects the design and manufacture of ceramic handicrafts. Because of too many restrictions, it is difficult for designers to break through innovation in the design of ceramic handicrafts. The design of traditional ceramic crafts is limited by the molding process. When manufacturing ceramic crafts, it is necessary to make a middle mold first, and select the manufacturing methods of clay molding and turning. At this time, the shrinkage after firing and drying should be considered, and then the ceramic mold should be designed and manufactured. The design and production of traditional ceramic crafts pay more attention to the decoration part, and pay more attention to the treatment of three-dimensional sculpture of three-dimensional ceramics. Generally speaking, the relief decoration on ceramic sculpture crafts is treated by means of marking, engraving and printing, which can make ceramic crafts more hierarchica.

3 Advantages of Computer Technology in the Design of Arts and Crafts

3.1 Improve the product performance of arts and crafts

The practical value and artistry of arts and crafts including glass, ceramics and sculptures should be considered in the design process, which requires designers to have sufficient knowledge of the processing regularity of arts and crafts and the materials of processing tools and crafts. Using computer technology as an aid, better design of arts and crafts is conducive to improving the product performance of arts and crafts. In the design of arts and crafts, the use of computer aided design can incorporate modern elements, which improves the quality, practicality and appreciation of arts and crafts. Three-dimensional digital technology is gradually increasing in assisting the design of arts and crafts. Computer can be used as a medium to simulate the actual effect of arts and crafts design. Designers are no longer restricted by the physical layer, and can use computers to split, reorganize and adjust the building models of handicrafts. The traditional form has been broken because of the inconvenience of observing the arts and crafts due to their large size or small size, which greatly improves the product performance of arts and crafts[3].

3.2 Enhance the design interest of arts and crafts designers

The design process of traditional arts and crafts is cumbersome, and the designer's design ideas are limited. In the design process, most of them are repetitive work, boring and boring, and the designer's enthusiasm gradually dissipates. Using computer technology to assist the design of arts and crafts can simplify the design process, embody the designer's unconstrained ideas, and the design process is simple and interesting, which greatly increases the enthusiasm of design. In the design process of arts and crafts, the computer technology can make the drawing work simpler, at the same time, the drawing data is not prone to errors[4]. Once any problems appear in the design process, they can be solved in time. The computer can be used to search the design records, adjust and modify the completed drawings, and the divergent creative thinking of the designer becomes a reality. Once the designer generate gives creative inspiration, the computer can be used for drawing, which greatly increases the designer's design enthusiasm. With the help of computer technology, the process of creating and designing arts and crafts becomes simpler. Designers can use computers to design different styles and styles, adjust the angles, sizes, positions and components of works in time, and create other design alternatives by using computer technology such as splicing, cutting and reorganizing. It makes the designer's technique and style not limited by physical space, restores the designer's inspiration to the greatest extent, and effectively mobilizes the enthusiasm of design.

https://doi.org/10.1051/e3sconf/202129002025

3.3 Enhance the modern flavor of arts and crafts

Computer-aided design of arts and crafts can enhance the modern flavor of works. The design of arts and crafts can fully grasp the regularity of materials, tools and materials, and innovatively produce arts and crafts with rich patterns, shapes and structures. In the design of arts and crafts, the auxiliary function of computer technology is gradually strengthened, which can be used to design and manufacture ceramics, handicrafts, sculpture handicrafts and glass handicrafts, and at the same time integrate modern elements into the design process of arts and crafts. Glass crafts, ceramic crafts and sculpture crafts have a stronger modern fashion flavor. In the design process, various decorative techniques such as painting, relief, decal and printing can be combined, and then integrated into the designer's subjective thoughts to create modern arts and crafts with rich shapes[5].

4 Application of Computer Technology in Different Stages of Arts and Crafts Design

4.1 Pre-preparation of arts and crafts design

Computer technology runs through the whole process of arts and crafts design. In the early stage of design, computers can be used to sort out all kinds of materials and materials needed for hand design, and then assist designers to determine specific design ideas. Finally, designers use computer-related software to draw three views of design.

4.2 Implementation of arts and crafts design

The specific implementation stage of arts and crafts design can not be separated from the assistance of computer, and 3D design software is used frequently in this stage. In the process of using 3D software tools to model 3D models, it is necessary to combine the modeling features of arts and crafts with the three views drawn in the initial preparation stage[6]. In this process, designers need to master 3D software design tools skillfully. In the process of designing arts and crafts, if surface modeling is needed, 3D Rhino software is needed. Generally speaking, in the process of designing arts and crafts, when designing complex shapes, most of them will use computer Boolean operations and 3D scanning tools to generate shapes.

4.3 Optimization of design scheme for arts and crafts

The design scheme of arts and crafts can not be optimized without the help of computer. The designer needs to evaluate the original design scheme, find out the defects, and modify the design scheme based on the relevant requirements of the design to ensure the best design scheme after the change.

4.4 In-depth design stage of arts and crafts

In the final stage and in-depth design stage of arts and crafts design, computer assistance plays a decisive role in this stage. Designers need to make final analysis on the schemes determined in the first three stages, and then further improve, check and confirm the data of the three-dimensional model[7]. At the same time, they need to use computers to analyze the model data. And finally determine which molding process to choose, the shrinkage ratio of finished product production and the quantity and volume of required raw materials, so as to ensure the success rate of arts and crafts production to the maximum extent and reduce the consumption of raw materials.

5 Application of Computer Technology in Design of Arts and Crafts

5.1 Application of computer technology in sculpture crafts design

With the continuous development of computer technology, computer technology occupies an important position in all sectors of society at present. Three-dimensional digital technology is gradually integrated into the design process of ceramic crafts. In the design of sculpture crafts, it is particularly important to use three-dimensional digital technology to assist design, which can enhance the creativity of sculpture crafts. In the design of sculpture crafts, 3D digital technology can assist the construction of digital sculpture model, and then render the sculpture model of components, so as to obtain the rendering of 3D sculpture model. Designers can also use 3dsmax and other 3D software to build sculpture models and complete complex modeling design in sculpture design. In sculpture design, designers can use ZBrush software to build the skeleton of the sculpture model, and need to use ZSphere to build the skeleton of the model, and then use different commands including "Rotato", "Scale" and "Move" to add and adjust the angle size and position of the Z ball. After the final editing, you can use "Rota to" and "Move" to adjust the overall dynamics of the sculpture and complete the model construction[8].

5.2 Application of computer technology in glass crafts design

In the design process of glass crafts, computer technology also runs through the whole process, among which 3D printing technology is widely used. The combination of arts and crafts and 3D technology can create unique coloured glaze crafts. Designers of glass crafts need to create virtual digital models by using 3D modeling software, and directly print paraffin models by using 3D printing technology. After the paraffin model is printed, the next demoulding manufacturing and other subsequent processes are carried out, so that the process of paraffin making seed film in traditional glass manufacturing is replaced. Complex processing links are simplified, which shortens the design and manufacturing cycle, and enables glass designers to try to design and manufacture glass arts and crafts with different shapes by using computer 3D printing technology and other technologies, thus broadening the design and manufacturing channels of glass arts and crafts[9].

5.3 Application of computer technology in ceramic tea set modeling design

Ceramic tea set is a kind of ceramic handicraft with relatively complex shape. Computer technology is widely used in the modeling design of ceramic tea set, which can assist the design of the curved surface of ceramic tea set. The application of Rhino software can simplify the curved surface design, and at the same time greatly reduce the error of the teapot curved surface rate. Computer three-dimensional model can assist the drawing of ceramic tea set, and draw in sequence by using the subordinate relationship between different parts of arc distance. When drawing the curve of silent section, the tools of arc and curve will be used. In order to get the three-dimensional solid modeling of kettle body, the rotational moulding tool can be used. Then, the curve drawing and solid modeling of the spout are carried out. Finally, other details of the ceramic teapot, including the lid and button, are created[10]. After that, the created Chaohu Lake model needs to be carefully checked. Once the damaged surface is found, it should be repaired in time. Using computer software to render the ceramic teapot model, the rendering can simulate the visual effect of the real ceramic tea set, display the animation effect of the ceramic tea set model in an all-round way, and construct the overall coordinated ceramic tea set model, as shown in the following figure:

Figure 1. 3D Print Ceramic Tea Set Model

6 Conclusion

Looking at the long history of the development of arts and crafts, we find that the expression forms of arts and crafts are different in different times, and the artistic creation means and expression forms are restricted or promoted by different factors. At present, the development and inheritance of art fields are affected by the development of science and technology. With the wide application of computer technology in all walks of life, people's ways of living, working and studying have gradually changed, and the application of computer technology in arts and crafts design has become more and more extensive. The field of arts and crafts design has gradually developed into a bottleneck according to the traditional design and production. Computer technology has brought new opportunities to the development of arts and crafts design, which can enhance the practicality and artistic charm of arts and crafts and be liked by more people.

References

1. Zhang Mian.Dissemination of Art Works in the Internet Age[J].Beauty & Times,2021(02): 115-116.
2. Dong Yezi.Literature Review of Traditional Arts and Crafts Inheritance under the Background of New Media[J].Art Education Research,2021(02):50-51.
3. Wang Xingyang, Zhu Yutong.The Importance of Integrating Arts and Crafts in Landscape Design[J].China Place Name,2020(03):37.
4. Yu Zhu, Yan Ying.Analysis on the Application of Computer-Aided Design in the Innovation and Development of Arts and Crafts[J].Hubei Agricultural Mechanization, 2019(22):65.
5. Mu Jinming, Wang Chunhua.Research on the Application of Computer Technology in the Design of Arts and Crafts[J].Information Recording Materials,2019,20(10):131-133.
6. Tian Yuguo.Application of Computer in Design of Arts and Crafts[J].China Computer & Communication,2018(08):4-5+8.
7. Wang Bin.Application of New Media in Art and Crafts Painting[J].Youth Magazine,2018(01):49.
8. Zhu Liye.Analysis of Creativity in Contemporary Arts and Crafts Design[J].MING (Attitude),2017(18):63.
9. Peng Chenglu. Research on Internet Communication of Traditional Arts and Crafts Value[D].Chinese National Academy of Arts,2018.
10. Ma Jia.Computer-Aided Teaching of Arts and Crafts[J].Youth Diary: Research on Education and Teaching,2015(10):11.

The Application of Big Data Technology in the Practical Research Work of Ideological and Political Courses

Yuhong Bai[1]

[1]Hulunbuir University, Hulunbuir 021008, Inner Mongolia

Abstract. This article analyzes the application advantages, application core and technical nature of big data technology. The author studies the specific application of big data technology in the allocation of topics, pre-class research, data collection and processing, data statistical analysis, and data rational application. This article studies how to do well in survey method training, how to improve the data management system, how to strengthen the curriculum construction of colleges and universities, and how to actively learn from successful cases. The author's purpose is to enhance the application value of big data technology and improve the reliability of the practical survey results of ideological and political courses.

1.Introduction

With the continuous improvement of the quality of social life, people's requirements for the quality of social research are also increasing. If traditional research methods are still used for data collection, the integrity of the data information obtained will also be greatly affected. Big data technology has good promotion value in the practical research work of ideological and political courses because of its advantages of fast data collection and high accuracy. The application of big data technology to the practical research work of ideological and political courses can not only improve the integrity of the data collation results, but also has a positive significance for improving the scientificity of the research results.

2.Big Data Technology Related Content Analysis

2.1. Application Advantage

Big data has the characteristics of large volume, fast speed, diversification and high value. Firstly, large volume is the most significant feature that distinguishes big data from traditional data. According to statistics and calculations, by 2022, the global data will reach 40ZB (1ZB=1 trillion GB). Such a huge data resource will become the most solid foundation for big data technology to play a role. Secondly, fast speed refers to the fast frequency of data generation, replacement, and transmission. In the information age, the amount of data is continuously increasing at a rate of 1%-10%, and the frequency of data update is also controlled at the ns level, and the total amount of data is very rich. Thirdly, diversification means that the sources and types of data are very rich, and they are constantly increasing. From the number of rail transit operations to the mileage of passenger planes, from doctors' prescriptions to patients' medicines, the types of data are all-encompassing and will continue to increase over time. Fourth, high value means that big data technology can play a greater role than the original data after integrating and utilizing massive amounts of data. Although the value density of a single data is low, the relationship between the data can be found through the processing of scientific systems and algorithms, which breaks through the value limitations of the original data.

2.2. Application Core

The key to the realization of the value of massive data lies in its forecasting function. Through the sorting and summarization of data, the mastery of dynamic laws can realize the formal forecast of a certain stage in the future. For example, the famous American Netflix film and television company judges the British original "House of Cards" and the United States based on the user's viewing habits such as which node to pause at, which node to fast forward, or which actor searched for after a certain film, etc. The famous actor Kevin Spacey was the hot spot pursued by the market at that time. That's why the hit American drama "House of Cards" succeeded. The massive amount of user data that Netflix had at the time was a very valuable resource. Each individual user data may be meaningless, but when thousands of users are searching for a certain actor's work, it can explain the problem. In the context of the big data era, the convenience of data acquisition is also improving. The most critical content is how to use big data rationally and sort out the correlation between data. In this way, the data can be optimized and sorted, and the regularity between the data can be found to meet the data forecasting requirements.

2003511151@dlvtc.edu.cn

2.3. Technical Essence

Analyzing the essence of technology, the essence of big data technology is "hidden communication." More than 95% of the information sources of big data are the products of social life and activities, but the ultimate purpose of the services generated is to provide people with high-quality services. This also means that big data is essentially the information communication between people in the application, and this type of communication is not direct information exchange in the application. It is a kind of "hidden communication" based on the data information mastered under the background of complex data application. Moreover, people's communication with each other has also changed with the change of the times, and there have been changes at different levels. In the initial period, more than 60% of communication was accomplished through physical movements. In the follow-up, when the initial language appeared, many words were not used as an aid. For example, many ethnic minorities in modern China rely on oral transmission to pass on their culture. Later, when writing and language were formed, relying on the Internet and telephone, people's communication methods also showed rich characteristics. Big data, as a necessary product during the current stage of social development, not only provides a more convenient way of communication, but also satisfies people's basic requirements for information content.

3.The Application of Big Data Technology in the Practical Research work of Ideological and Political Courses

3.1. Assign Topic

In order to successfully complete the practical research work of ideological and political courses, the first task in practical application is to complete the related work of assigning topics and clarify the specific rules of the work. In the specific practice process, teachers can use the sorting function of big data technology to filter out topics that meet the needs of students according to the professional characteristics of students. For example, for students majoring in computer science, teachers can assign topics such as "Network Changes Life" and "Science and Technology Development Value". In this way, while satisfying the purpose of ideological and political teaching, it can also improve the professional ability of personnel. For students majoring in agriculture, teachers can assign topics such as "ecological planting development" and "biological control". Based on big data technology, teachers can initially grade the difficulty of the selected topic, and provide the topic selection in the order from simple to difficult, and gradually cultivate the practical research ability of college students. For each topic, the teacher will arrange 1-2 groups of people to complete it. The group members will be limited to 3-5 according to the difficulty of the topic, so as to avoid too many members who are not actively participating. In addition, according to statistics in 2020, more than 95% of students will use WeChat as their main social tool, and almost everyone has QQ. Therefore, teachers can use WeChat or QQ to form groups for problem communication and problem solving, which can ensure that problems can be solved smoothly.

3.2. Pre-class Research

After completing the topic selection, enter the pre-class research stage. That is to analyze the content of the pre-class research and sort out the detailed content of the research work, so as to improve the orderliness of the research process. Based on big data technology, it can provide necessary reference for the advancement of pre-class research work. Teachers can organize students to conduct group training after organizing the content of important test sites. Moreover, teachers will also conduct group tutoring to ensure the smooth progress of practical activities. In the specific evaluation process, the content flow is as follows. First, use some actual cases to express the value of social research content to students, and analyze the commonly used research methods. This allows students to choose the most appropriate homework method based on the actual situation. Second, introduce the format and content of the research report to the students to develop the application design. This can also further improve the reliability of the analysis results. Third, select 30%-40% of the excellent works from the previous works for display, to provide students with some new ideas. In addition, teachers also need to analyze various theories including Marxism, Engelsism, Leninism and so on. Teachers can use this to complete the analysis of real problems, leaving enough time for the group to discuss, so as to meet the group's independent research task.

3.3. Data Collection and Processing

During the traditional investigation method, the total amount of data collected in order to improve the reliability of the evaluation results is relatively high. In the process of applying big data technology, the ability to obtain valuable data is relatively strong. Especially in the process of sorting non-relational database data, relevant staff can use the data platform established by big data technology to conduct detailed analysis on the data types. Under normal circumstances, ETL tools are used to complete the distribution and sorting of content in the process of data information collection, and it also has a good application in the processing of heterogeneous data, plane control data, and related data. All collected data information will be directly put into the middle layer for cleaning, conversion, integration and other processing, and will also be directly loaded into the data warehouse after finishing the data. In this way, the analysis and processing of data information can be completed online smoothly, and the basic value of the data information mined can be improved. Compared with the previous data collation model, the shared platform established by big data technology can also use redundant design mechanisms and cloud computing mechanisms. In this way, information can be sorted out in an ultra-short time and the use value of data analysis results can be improved [1].

3.4. Statistical Analysis of Data

3.4.1. Regression Model Variable Selection

In the variable selection of the regression model, teachers need to determine the corresponding content of different indicators. Under normal circumstances, teachers will use students' research ability as the dependent variable in the entire regression model, denoted as y. This indicator can be divided into two groups of applied indicators, which are scoring for teachers (denoted as y1) and students' self-evaluation (denoted as y2). Take a case as an example, as shown in Figure 1. The abscissa represents the interval between the fights, and the ordinate is the effective number of people who have obtained the score. After calculating according to the content of Figure 1, it can be concluded that 15.31% of the students in the class get a "pass (60-69 points)" score. The number of people who got "good (70-84 points)" scores accounted for 53.31%, and the number of people who got "excellent (85-8100 points)" scores accounted for 32.41%. The content of the students' self-evaluation is based on the ability improvement evaluation form. The quantitative indicators generally need to be considered include classroom learning effect (denoted as a1), students' understanding of the textbook after reporting (denoted as a2), practical link impact (denoted as a3), and practical result impact (denoted as a4). From this, two sets of variable calculation formulas can be obtained, which are divided into y=y1+ky2 (k is a constant value) and y2=a1+a2+a3+a4[2].

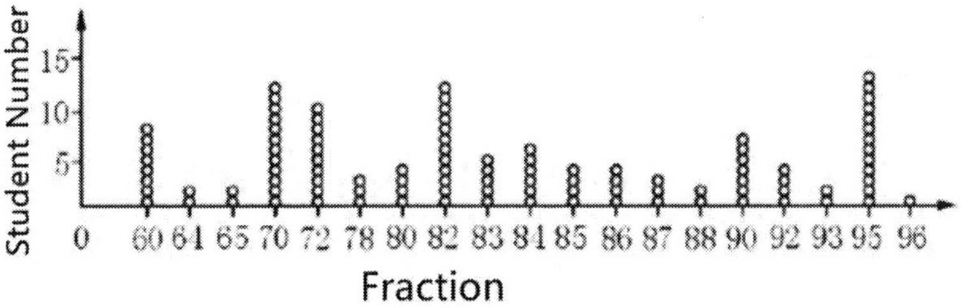

Fig.1 Teacher Evaluation Score Distribution

3.4.2. Linear Regression Model Design

Teachers need to design the corresponding management formula in advance when designing the linear regression model. The specific formula is as follows: y=b1x1+b2x2+b3x3+b4x4+a. Among them, b1-b4 are commonly used coefficients, x1-x4 represent independent variables, and a represents the error correction term after calculation. Enter the prepared variable information into the SPSS software for analysis, and obtain the required linear regression equation. This process also requires big data technology to be able to continuously provide calculated data, and at the same time to analyze the value of the dependent variable to prove the application value of the collected data variable. In the process of data analysis and application, statistics related knowledge is also involved. Teachers can use the calculation of P value to discuss the correlation between different variables. This is also the basic requirement for improving students' comprehensive ability [3].

3.5. Reasonable Application of Data

Based on the application advantages of big data technology, when dealing with problems, it can have a greater impact on the knowledge framework, thinking logic, and behavior content of the course content. Big data can enrich network data sources, microblog processing information, WeChat official accounts, video images, and photo information. The processing capabilities of big data technology can accurately process this information, and in the process, big data can also quickly filter out value data. This is also an important measure to enhance the attractiveness of research content. For example, in the "Fudan University Poisoning Case", a Baidu account began to conduct "science popularization" on the effects of some drugs and how to use daily necessities to configure poisons. This has also brought a very bad influence, and it is also an intuitive manifestation of personality distortion. This also requires that the practical research work of ideological and political courses will also be fully applied to Internet information resources in the development process, extract key information from the information, and spread positive energy topics. This also provides reliable conditions for the smooth reform of theoretical and practical courses [4].

4.Suggestions on Optimizing the Practical Investigation of Ideological and Political Course

4.1. Do A Good Job in Survey Method Training

Training on investigation methods is conducive to the smooth development of investigation work, thereby improving the reliability of investigation results. When students use big data technology to carry out practical research work, the primary task is to use appropriate survey methods to complete social practical activities, so as to improve the rationality of the research activities. For example, a university conducted group training for students before the start of work. Moreover, the school also printed the "Social Research Work Guide" in the curriculum to assist students in learning related content. The teacher will organize the group leader to conduct a second training after the student group training, so that they can systematically recognize the work content and help the group better complete the research task. In addition, the school uses the overall planning function of big data technology to make overall arrangements for the work content. This not only improves the use value of the research results, meets the corresponding management needs, but also improves the practicability and reliability of the management results [5].

4.2. Improve the Data Management System

By improving the data management system, the guiding value of the system content itself can be improved, and the rationality of extracting data content can be improved. The primary task of improving the reliability of course research results is to organize big data information with the help of a complete data management system, so as to give full play to the application value of the data itself. Moreover, the formulation of the system can also ensure the authenticity and rationality of the collected data, prevent others from maliciously modifying the data, and improve the application efficiency of the secondary use of data. At present, Internet technology has provided a lot of convenience to people's lives, and at the same time, it has also caused some undesirable problems such as privacy leakage and information theft. Ideological and political education is a course that correctly guides the development of personal thoughts. Its primary task when using big data technology to collect social survey data is to strictly abide by the content of the system and ensure the rationality of data sources. In this way, the data information can be sorted out better, and the timeliness of the data application results can be improved [6].

4.3. Strengthen the Construction of College Courses

By strengthening the curriculum construction of colleges and universities, the pertinence of the practical research work can be improved, and the use value of the curriculum content itself can be improved. In the specific practice process, we need to focus on the optimal design of the theoretical course content, so as to enhance ideology and meet the needs of the smooth development of follow-up research work. The ideological and political theory courses based on the application advantages provided by big data technology need to fit the current social development trends and trends. We need to incorporate digital content into the curriculum design, and at the same time we need to optimize the teaching methods. For example, micro-classes and short videos are used to implement theoretical knowledge into practical activities. In this way, the students' attention can be better attracted during the teaching process, and the students' knowledge learning ability can be gradually improved. Otherwise, schools can use the assessment system established by big data technology to adjust the teaching progress of college courses based on the assessment results. This not only enables it to have sufficient theoretical knowledge, but also enables it to be better applied in practice to meet the application requirements of college courses [7].

4.4. Actively Learn from Successful Cases

By actively learning from successful cases, it can provide guidance for students to carry out research work and reduce the appearance of unreasonable problems. At present, many universities have used big data technology to assist the research work of ideological and political courses. Moreover, there are also many successful cases uploaded in the Internet field. At this point, we can use big data technology to organize such content. Teachers will also conduct a detailed analysis of the content of these cases when intensively training students, and pick out the shortcomings and strengths of such cases. Besides, teachers will also simulate the research environment to allow students to draw up research plans and exercise their comprehensive abilities through virtual practice. This can not only lay the foundation for the smooth development of follow-up work, but also ensure the orderly and practicality of the process of carrying out such activities [8].

5.Conclusion

In summary, good investigation method training is conducive to the smooth development of investigation work. Improving the data management system can enhance the guiding value of the system content itself. Strengthening the curriculum construction of colleges and universities can improve the pertinence of the process of practical research work. Actively learning from successful cases can provide guidance for students to carry out research work. The application of big data technology to the research work of ideological and political courses has positive significance for accelerating the speed of research work.

Acknowledgments

Fund project: The 2020 Hulunbuir College Scientific Research Project "Research on the Cultivation Mechanism of University Network Cultural Works in Major Epidemic Prevention and Control-Taking Hulunbuir College as an Example", Item Number: 2020SZC10

References

1. Wei Chonghui, Xu Mengping. Evaluation of Ideological and Political Course Teachers in Colleges and Universities: Theoretical Basis, Realistic Dilemma and Construction Path [J]. Heilongjiang Higher Education Research, 2021, 39(04): 113-116.
2. Liu Chunqing. The practice of ideological and political reform in colleges and universities in the new era of heritage resource innovation [J]. Cultural Industry, 2021(09): 126-127.

3. Cai Xiaokui, Chu Shuijiang. Research on the current situation of the unification of theory and practice of ideological and political courses[J]. Journal of Guangdong Institute of Light Industry and Technology, 2021, 20(01): 69-73+80.
4. Gou Rong, Liu Wei, Li Chen. The construction and implementation of practical courses in ideological and political theory courses in colleges and universities[J]. Journal of Taiyuan Urban Vocational and Technical College, 2021(03): 122-124.
5. Shen Feiwei, Chen Xiaoling. Practical logic and optimization strategies for improving the teaching quality of ideological and political courses in colleges and universities on the platform of "learning to strengthen the country"[J]. Shenzhen Social Sciences, 2021, 4(02): 147-158.
6. Zhang Nan. Exploration of the Informationization of Ideological and Political Education in Colleges and Universities in the Era of Big Data——Comment on "Theory and Practice of Flipped Classroom in Ideological and Political Courses in Colleges and Universities in the Internet Era"[J]. Science and Technology Management Research, 2021, 41(03):218.
7. Huang Weihong. Exploring the innovation of the five-year higher vocational ideological and political course teaching in the era of big data[J]. Journal of Jiamusi Vocational College, 2021, 37(01): 111-113.
8. Zhang Haidong, Li Qianghua. Research on the construction of a sharing platform for practical teaching resources of ideological and political courses in colleges and universities[J]. Journal of Luliang University, 2020, 10(06): 60-62.

A Study on the Formation and Evolution Model of Social Psychology in Major Public Health Emergencies

Chenhe Yi[1], Shan Li[1*]

[1]School of Public Administration, Xiangtan University Xiangtan, Hunan P.R. China

Abstract. The occurrence of local or regional major public health emergencies has seriously damaged human health and life safety, and has an impact on the psychological state of the public. The study introduces"social amplification of risk"effect and the mechanism of emotional infection, and constructs a process model of psychological prediction, psychological cognition and psychological behavior in the generation of social psychology of major public health emergencies, and from the level of individual emotions, group emotions and social emotions analyze the evolutionary model of social psychology; Research has found that information disclosure, government trust and social confidence are the main factors influencing the formation of social psychology of major public health emergencies.

1 Introduction and Literature Review

Major public health emergencies refer to major infectious disease outbreaks, mass diseases of unknown causes, major food and occupational poisonings, and other events that seriously affect public health that occur suddenly and cause or may cause serious damage to public health [1]. It usually occurs suddenly without human beings' knowledge, making it difficult to prevent, and has the characteristics of crisis events such as complex diversity, widespread transmission, and severity of harm. It generally affects and changing the mental structure and mental activities of social members, causing the public to produce mental states such as panic, anxiety, and questioning, and even forming lingering psychological shadows and psychological behaviors, which may spread and continue to sudden public health emergencies daily life and behavior after the event. Based on this, the social psychological impact and intervention work in public emergencies has gradually attracted the attention of the Chinese government. For example, government departments have clearly stipulated the level of social mental health services from the perspective of laws and regulations[2-4]. However, in fact, there are many difficulties in the intervention and regulation of social psychology in major public health emergencies, especially in the modern complex social environment, the speed of information dissemination has accelerated the production and change of social psychology in public health emergencies. Therefore, grasping the general laws of the generation and evolution of social psychology in public health emergencies will help the government to correctly understand and scientifically guide the nature and trend of the generation and evolution of social psychology.

Compared with the influence of social politics, economic industry and public security, the impact of major public health emergencies on social value and psychological state is relatively unique and hidden, and it has attracted more and more researchers' attention. First of all, the existing research on the evolution of social psychology is mostly analyzed from the cognitive level or the perceived value dimension. For example, Some scholars have found that social psychology is the general public perception and cognition of social reality in a specific environment or a specific event.[5-7]Some scholars think in public health emergencies, social psychology usually resonates at the same frequency as online public opinion and group behaviors, and has an amplifying effect under the influence of risk perception bias and emotional infection mechanisms [8-9].Related research shows that there is a universal logical relationship between major public health emergencies and social psychology. Many behavioral factors such as information disclosure, response measures, prevention and control methods will affect the occurrence and dissemination of major public health emergencies. The social psychological state, emotional expression and behavior patterns constitute the influence. This mechanism of action and influence includes not only short-term effects such as physiological emotions and emotional tone, but also long-term effects such as value orientation, behavior habits, and social customs. Therefore, it is extremely necessary to study the social psychology and mentality adjustment in response to major public health emergencies, especially the analysis of the social psychology generation and evolution mechanism of major public health emergencies. It will have important guiding significance for the adjustment of social mentality and the reshaping of social values in response to public health emergencies. Therefore, one of the goals of this research is to analyze the main

Corresponding Author: 996253673@qq.com

influencing factors of social psychology in major public health emergencies and obtain the relationship between public risk perception and coping behaviors and psychological changes in emergencies, in order to help the public and the government to form an accurate understanding of social psychology. another goal of the research is to build a model of the generation process and evolutionary situation of social psychology in major public health emergencies through the understanding and grasp of social psychology.

2 Research design

2.1 Research method

The specific research method can be summarized as the five processes of "determining the research problem determining the appropriate text overall-text content analysis and classification-selecting representative texts-obtaining qualitative results "[10-11].First of all, because this research is an interdisciplinary basic research, it is necessary to use the basis of text content study to scientifically screen the overall text, and the original text is generally based on "public health emergencies", "social psychology in public emergencies", and "crisis scenarios" “Social Psychology”,etc.,conduct preliminary grouping, eliminate invalid and duplicate data, and extract and refine representative texts that are highly consistent or highly related to research content and research purposes; secondly, the extracted representative texts are classified according to the logical relationship between the categories, and the text content can be formed for comparative research, which can be used to sort out and analyze the social psychology generation process and evolutionary context of major public health emergencies.

2.2 Research steps

This research focuses on exploring the general laws of the generation and evolution of social psychology in major public health emergencies, and constructs a theoretical model of the generation and evolution of social psychology in major public health emergencies, in an attempt to provide the government with a correct understanding of social psychology. And then provide reference suggestions for determining social psychological early warning and intervention indicators, establishing and continuously improving the early warning system of social psychological behavior indicators. Specifically, the first step is to determine the main influencing factors of the generation of social psychology based on the logical relationship and change characteristics between the subject and the object of social psychology in major public health emergencies; the second step is to determine the main influencing factors of social psychology. Importance, combined with risk management theory and "cognition-evaluation" theory, etc., to construct a model of the generation and evolution of social psychology in major public health emergencies.

3 Theoretical model construction

3.1 Socio-psychological influence factors

Generally speaking, social psychology is the general public's perception and understanding of social phenomena and social activities. It is the core demand and intrinsic motivation for stimulating and regulating public behaviors for the opportunities, challenges and threats faced in social life [12]. In response to major public health emergencies, social psychology is different from the usual characteristics of social psychological changes, and is more susceptible to developments. It mainly stems from the public's information release, strategic decision-making and society in the prevention and control of major public health emergencies. First, information disclosure. American scholar Shannon pointed out that "information is used to eliminate random uncertainties" [13-14] Especially in major public health emergencies, information determines the public's cognitive level and psychological judgment of public health emergencies. Practice has proved that people are always full of distance with unfamiliar emergencies, and they are used to self-digesting and understanding with guessing, doubting, and even negative psychology. In this case, if the official government information is not released in the first time, it will be eliminated. The subject's one-sided cognition of the object's information will flood the society with relevant false information and rumors, leading to social suspicion, panic, confrontation and other cognitive biases and behaviors, which will seriously affect the correct social psychology in response to major public health emergencies. form. Second, government trust. From a psychological point of view, trust is an individual's perception, judgment and subjective attitude towards outsiders and things, and its essence is interpersonal trust. As an extended type of interpersonal trust, government trust is the core content of the concept of social influence, especially when the public interacts with the government, when facing external crises or risks, they still remain positive about the reliance and reliability of government intentions or administrative actions look forward [15]. In fact, the public's trust in the government largely affects social psychology and behavior. Different groups only have differences in the degree of trust in the government, and there is no difference between them. During the special period of major public health emergencies, in the face of sudden social risks, the public will show different levels of trust in the government's administrative capabilities and expected effects in the prevention and control of public health emergencies. Third, social confidence. Confidence can neither be produced independently nor can it exist independently. It is the result of human beings' positive expectations for a specific behavior or the development and evolution of a certain thing. It is a subjective judgment of future expectations based on realistic cognition. In response to major public health emergencies, public confidence refers to the optimistic evaluation of the public based on the social system's cognition and grasp of the state and evolution of the event, as well as the trust in the social investment power of event governance, including individuals. The confidence index also includes the group confidence index.On the contrary, it will form negative collective consciousness and cognitive savings, and lead to social structural resentment [16].

3.2 Social psychology generation model

The social psychology of major public health emergencies mainly comes from the public's psychological judgment on the risk or potential risk of public health emergencies and the existing risk prevention and control technology. Constructing a theoretical model of the generation of social psychology in major public health emergencies can link major public health emergencies with risk assessment, risk awareness, risk perception, etc., and analyze the logical relationship between the public's psychological judgments and decision-making behaviors. variable. The subjective risk of major public health emergencies is the public ' s subjective judgment and prediction of the potential harm of public health emergencies. It will follow the disclosure of prevention and control information and technical upgrades, as well as the public's psychological prediction and psychological recognition of the risk Knowledge and psychological behaviors show a partial normal distribution curve trend. According to the trend of risk perception curve and the analysis framework of "risk social amplification" [17], can represent the process model of social psychology in major public health emergencies as shown in Figure 1.

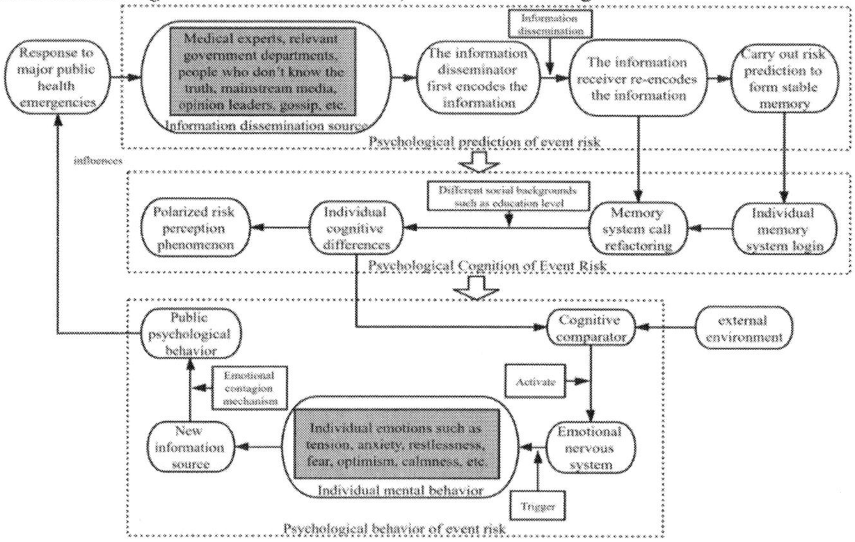

Figure 1: Schematic diagram of the process model of social psychology in major public health emergencies

The model includes three generation stages of psychological change process: First, the psychological prediction of event risk. In the early stage of major public health emergencies, especially when a new or rare public health event is involved, due to the uncertainty of the source of information dissemination, information transmitter will receive different information according to their own knowledge level and society. Encode information on experience, emotional experience, etc., and publish or publicize the processed information through the Internet, self-media, etc. As the receivers of information and the direct associates of event risks, the general public will, in the absence of relevant professional knowledge and reliable truth information, form psychological predictions and directional thinking inertia for event risks based on the relevant information they have obtained. Second, the psychological perception of event risk. The psychological cognition stage of the risk of public health emergencies mainly includes cognitive steps such as individual memory system login, memory system recall and reconstruction, and risk assessment.[18] During the occurrence of major public health emergencies, people's awareness of the risk of the event will be based on the individual's knowledge level, values, and professional differences, as well as the development of different states of the event, mainstream media information reporting, and incident prevention strategies. Third, the psychological behavior of event risk. Psychological cognition reflected in the result is the implementation of social psychological behavior. As far as the individual is concerned, after the second stage of psychological cognition, an optimistic or pessimistic psychological cognition of the risk of the event will be formed, and then through the cognitive comparator in the human brain system, the emotional nervous system will be activated to trigger the implementation of individual psychological behaviors.

Take the new crown pneumonia outbreak that broke out in early 2020 as an example. After the outbreak, the Social Psychology Research Center of the Institute of Sociology of the Chinese Academy of Social Sciences conducted two rounds of investigations on social psychology during the epidemic (the first round: 16: January 24, 2020 16: 40—January 25, 2020 12:50; Second round: January 25, 2020 12:51 — January 29, 2020 12:50),[19] the survey comparison results are shown in Figure 2:

https://doi.org/10.1051/e3sconf/202129002027

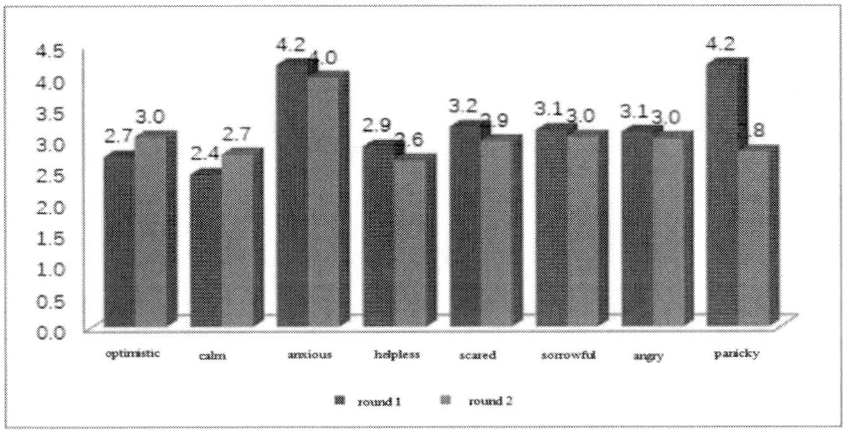

Figure 2: Comparison of social and psychological changes during the new crown epidemic

Data source: Social Psychology Research Center, Institute of Sociology, Chinese Academy of Social Sciences

As shown in Figure 2, the public forms different risk memories after psychologically predicting event risks. At this stage, social psychology is dominated by negative emotions such as worry, helplessness, and fear; however, with the development of events and government risk prevention. With the upgrading of control technology, the public has developed new psychological perceptions of events, and the negative memory storage in the brain has also transformed into positive emotions such as "government trust"and"social confidence". Accordingly, the public's positive social psychological behavior began to rise. As shown in the figure above, although the difference between the two rounds of surveys is only one day apart, the average scores of the public's "optimism" and "calm" mentality have both increased by 0.3 points, while negative mentalities such as "helpless" and "panic" have varying degrees decline, especially the average score of "panic" mood dropped by 1.4 points.

3.3 Social psychological evolution model

Through the above analysis, it is found that the generation and change of social psychology and its external performance characteristics in major public health emergencies reflect the role of emotion in the formation and evolution of social psychology. Therefore, when studying and analyzing the generation and evolution mechanism of social psychology, it can be assumed that the basic condition of social psychology is the concentration of social public emotions, and the activation of public emotions is the basic path and way of social psychology evolution, individual psychological emotions form a more common collective psychological cognitive state and behavior through other individual emotional effects (infection, transmission, co-op, imitating, etc.), which evolve into the public's social psychological state. The specific evolution mechanism model is shown in Figure 3.

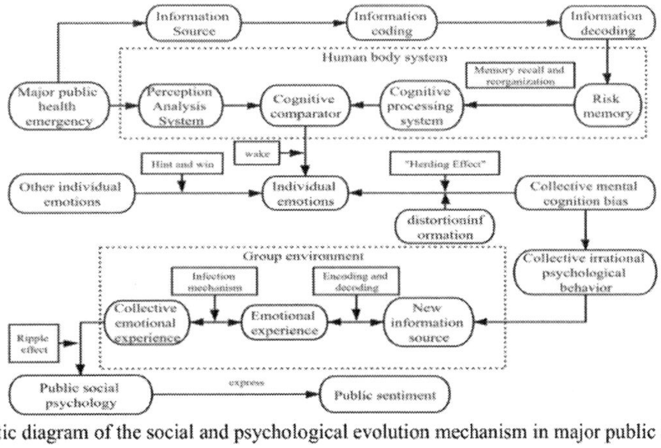

Figure 3: Schematic diagram of the social and psychological evolution mechanism in major public health emergencies

First, Individual emotions. According to Lindsay and Norman, an emotional arousal model was established to explain the generation of emotions under the cognitive processing system [20]. Individuals first encode and decode information about public health emergencies and related phenomena to complete cognitive judgments. If the perception analysis system feels the external environment and the conclusions drawn from the analysis of the cognitive processing system are sufficiently consistent, the event will develop steadily and will not generate individual emotions. But due to the contingency and complexity of major public health emergencies, it is different from individual risk memories, usually, the occurrence of events will exceed expectations, predictions, or even violate individual experience, memory and expectations, thereby stimulating the neural activation system to change the state of brain activity and awaken individual emotions. Second, Group emotions. The "herd effect" emphasizes that a single individual can easily lead to a herd mentality when it is guided or pressured by an external group, and therefore changes its original views, judgments and behaviors to maintain an attitude consistent with most other individual members. Third, Social emotions. In the case of a major public health emergency, the public cannot obtain accurate information about the event, resulting in information cascade and information distortion, and collective psychological cognitive bias, this cognitive bias is the main cause of the irrational psychological behavior of collectivity.As some researchers have pointed out, in the process of emotional agglomeration, the original emotions will be further presented in the public opinion field through new stimulus factors and deepen the collective emotional agglomeration of the group [21] and eventually, social psychology is generated and externalized into social emotions. For example, during the new crown pneumonia epidemic (January 24, 2020 to February 10, 2020), with the changes in the degree of publicity of the epidemic and related information, the public's social sentiment has also shown a trend of constant fluctuations, [22]as shown in Figure 4 show.

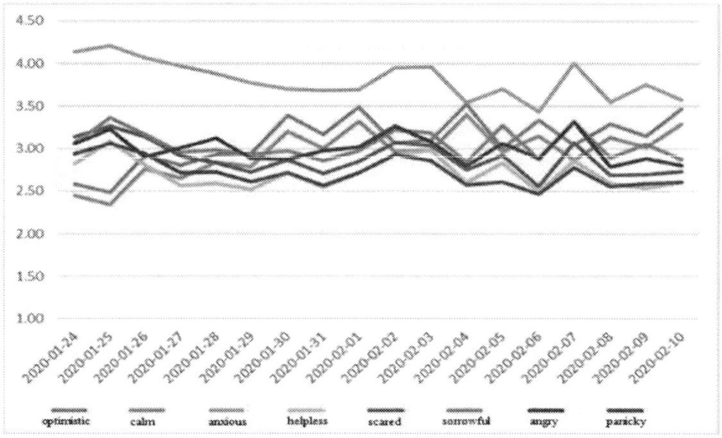

Figure 4: Trends in public social sentiment during the COVID-19 outbreak (2020.1.24-2020.2.10)

Data source: Social Psychology Research Center, Institute of Sociology, Chinese Academy of Social Sciences

4 Conclusion

By applying the theory of emotional contagion to group scenarios, combining the characteristics of major public health emergencies and social psychological changes, the study found that information disclosure, social trust and public confidence are the core influencing factors in the formation of social psychology; and on this basis, the theory of "risk social amplification" and "emotional arousal model" are used as intermediary variables to explore the general laws of the generation and change of social psychology, and construct a theoretical model of the generation and evolution of social psychology.The research conclusions drawn mainly include the following two aspects: 1) The generation process of social psychology in major public health emergencies goes through three stages of "psychological prediction of event risk-psychological cognition of event risk-psychological behavior of event risk". In this process, relevant information sources, information dissemination mechanisms, emotional infection mechanisms, and individual subjective factors of major public health emergencies all play an important role. 2) The evolutionary mechanism of social psychology in major public health emergencies is a continuous process of "infection" and transmission, which will respectively present three general social psychological states of individual emotion, group emotion and social emotion. Specifically, a single individual generates individual emotions through the cognitive system under the stimulation of external information sources; individual emotions in the public opinion field through the mobilization mechanism and the infection mechanism, produce collective psychological cognitive bias; finally, through the ripple effect and The linkage mechanism realizes collective irrational behavior in the group environment, which evolves into the public's social psychological state.

Acknowledgment

This research is supported by the National Social Science Foundation of China (No.16BGL177)

References

1. Feng Zhang. Research on risk management of major public health emergencies based on big data. Theoretical Vision, (09):67-73 (2020).
2. The General Office of the State Council of the People's Republic of China. Notice on Further Strengthening Guidance Opinions on Mental Health Work[EB/OL].http://www.gov.cn/xxgk/pub/govpublic/mrlm/200803/t20080328_32404.html.
3. The website of the Central People's Government of the People's Republic of China. The Mental Health Law of the People's Republic of China[EB/OL].http://www.gov.cn/jrzg/2012-10/26/content_2252122.htm.
4. The website of the Central People's Government of the People's Republic of China 22 departments issued the "Guiding Opinions on Strengthening Mental Health Services"[EB/OL].http://www.gov.cn/xinwen/2017-01/24/content_5162861. htm#1.
5. Tingqian Zhang. The ethical form and generation logic of the qualitative differentiation of social mentality. Jianghan Forum,(02): 41-49 (2021).
6. Xiaomi Zhou.A review of social psychological analysis and research in mass incidents. Reform and Opening, (02):84-85+89 (2018).
7. Binyi Zhang.Social and psychological guidance inpublic health emergencies.Cities and Disaster Reduction,2021,(01):5-8.
8. Sohail, M. T., Mahfooz, Y., Azam, K., Yen, Y., Genfu, L., & Fahad, S. Impacts of urbanization and land cover dynamics on underground water in Islamabad, Pakistan. Des, and water treat, 159, 402-411(2019).
9. Sohail, M. T., Mahfooz, Y., Aftab, R., Yen, Y., Talib, M. A., & Rasool, A. Water quality and health risk of public drinking water sources: a study of filtration plants installed in Rawalpindi and Islamabad, Pakistan. Des. and water treat., 181, 239-250 (2020).
10. Xiaoping Jiang,Ning Guo.The policy goals and evolutionary laws of the equalization of basic publicservices in my country——Based on the textual analysis of the central policy since the 18th National Congress of the Communist Party of China. Public Management and Policy Review,40(06):33-42 (2020).
11. Sohail, M.T., Ullah, S., Majeed, M.T. Pakistan management of green transportation and environmental pollution: a nonlinear ARDL analysis. Environl Sci and Poll Res, 1-10, (2021).
12. Xuefeng Chen,et al.Basic social motivation and social psychological service system construction[J].Advances in Psychological Science,2020,(01):13-21
13. Liangrong Li,Yuqiao Zhou.Constructing the academic discourse system of journalism in continuous exploration[J].Journalism University,2019,(10):46-54.
14. General Office of the State Council of the People's Republic of China. Regulations of the People's Republic of China on Disclosure of Government Information[EB/OL].http://www.gov.cn/zhengce/content/2019-04/15/content_5382991.htm.
15. Sohail, M. T., Delin, H., Talib, M. A., Xiaoqing, X., & Akhtar, M. M. An analysis of environmental law in Pakistan-policy and conditions of implementation. Research Journal of Applied Sciences, Engineering and Technology, 8(5), 644-653 (2014).
16. Li Zhu.A Social Psychological Interpretation of the Gathering and Scattering of Public Confidence. People'sForum, (05):11 (2013).
17. Yen, Y., Wang, Z., Shi, Y., Xu, F., Soeung, B., Sohail, M. T., ... & Juma, S. A. The predictors of the behavioral intention to the use of urban green spaces: The perspectives of young residents in Phnom Penh, Cambodia. Habitat International, 64, 98-108 (2017).
18. Yan Zhang,Jiuchang Wei,Wei Qi.Research on the Social and Psychological Impact Model and Governance Mechanism of Emergencies——Based on the Integrated Analysis of Virtual Risk Experience and Risk Society Amplification Theory. China Emergency Management, (06):34-38 (2011).
19. Junxiu Wang,etc.Epidemic prevention and control andchanges in social mentality. Beijing: Institute of Sociology, Chinese Academy of Social Sciences.2020:1-30.
20. Peter H Lindsay,Donald A Norman.Human information Processing.An introduction to psychology[M].New York.Academic Press,1977.
21. Min Zhang,Xuerui Liu,Yan Zhang.Analysis on the formation mechanism of audience emotion reversal of social public opinion in major disease online rescue events[J].Journal of Information,37(11):122-130 (2018).
22. Junxiu Wang,etc.The 18-day evolution of social mentality under the epidemic[R].Beijing: Institute of Sociology,Chinese Academy of Social Sciences :2-18 (2020).

Taking AI Neural Network to Analyze Labor Rights and Labor Disputes

Jia Liu[1], Ying Yang[1], Bao-Yao Xiao[1], Zhi-Tao Huang[1], Xiao-Hui Nie[1], Wen-Jie Liu[1] and Shih-Feng Chang[*1]

[1]Nanfang College of Sun Yat-Sen University, Guangzhou, 510970, China

Abstract. Exploring labor rights on the cognition of relationship with labor dispute is not only from the direct role of workers and HR, but also from the fight for labor rights of college students. However, many labor disputes in labor process are mainly related to labor rights. Therefore, we explore the relationship between labor rights and labor disputes, use neural network analysis and questionnaire survey method, and execute AI combining with big data analysis tools to collect first-hand data from two aspects of college students, workers in the industry, HR and labor dispatch personnel. Finally, the results show that: (1) there is a negative correlation between social security and labor disputes; (2) there is a negative correlation between labor safety and labor disputes; (3) there is a negative correlation between wages and labor disputes; (4) there is a negative correlation between labor contract signing and labor disputes; (5) there is a positive correlation between rights awareness and labor disputes.

1 Introduction

In recent years, with China's social and economic development entering a new era, great changes have taken place in labor relations, workers' awareness of safeguarding their rights is gradually enhanced, the pursuit of their own interests is more diversified, and the number of labor disputes has remained high for a long time, and the handling of labor disputes is increasingly complicated. According to the data of intelligent research consulting, from 2010 to 2019, the agency business scale of accepting workers' appeal cases in China has increased from 558,900 to 934,000, which is enough to verify that the labor disputes in enterprises are still growing.

In addition, according to the "GuoShuang 2019 Big Data Analysis Report on Labor Dispute Cases" (the total number of judgment documents is 190,832), the cases of labor disputes are mainly concentrated in Guangdong (accounting for 9.69%), Jiangsu, Beijing (accounting for 7.88%), Sichuan and Shandong. It can be seen that labor disputes mostly occur in provinces with more developed economy and more floating population.

Besides, since the outbreak of cowid-19 in 2020, the intensity, breadth, depth and sustainability of the epidemic have had a strong impact on social and economic development, destroyed the normal life and production order, and also made the problem of labor relations more prominent. However, small and medium-sized enterprises are the main body of employment in China, providing more than 80% of urban labor employment, but they are facing serious problems during the epidemic period. The rights of workers are also affected to varying degrees. There are more and more disputes between employers and employees. However, affected by the epidemic situation, only a small number of labor dispute cases are tried online, and most of them are stagnant. However, after the epidemic situation tends to stabilize, the number of cases may increase sharply.

Under the above background, we attempt to analyze which labor rights are the main causes of labor disputes through the investigation of college students and in-service workers who are going to work in Guangdong Province, and explore the prevention and solutions, so as to play a positive role in building a harmonious labor relationship in society. At the same time, we also provide reference for other provinces and enterprises in China under the new economic form. It can be used for reference to solve labor disputes effectively.

2 Literature Review

Domestic research methods of labor dispute are mainly based on literature analysis, comparative analysis and model test. Whether workers are infected COVID-19 pneumonia in their jobs or whether they are "work-related injuries" (Wu Xu, 2020) or whether social security should be compensated will result in labor disputes [4]. After all, it is related to human nature. Due to the lack of social security, labor disputes are often caused by the high attention of high-risk professionals, especially after accidents. At the same time, some first tier cities need to pay enough years for social security to settle down or buy a house locally, but enterprises evade their responsibilities or set unreasonable insurance conditions (Ai-Yu Liu, 2011) [1]. Due to the influence of urbanization and the idea of "entering the city", people's demand for social security has a strong conflict with the practice of enterprises to reduce

*Corresponding author's e-mail: zhangshf@nfu.edu.cn

https://doi.org/10.1051/e3sconf/202129002028

costs.

According to the compensatory wage theory, under the same conditions, the wage level of workers engaged in dangerous work is often higher. However, the measurement results show that the two are negatively correlated and not significant, indicating that even if migrant workers engage in dangerous work, they do not get a relatively high income level (Wen-Jiao Xue, 2017; Yang Wang, 2013; Yong Xie, 2008) [3] [5] [6], but the labor disputes caused by them account for a large proportion. Therefore, the more frequent the personal and health violations, the more likely they are to fight for rights (Ai-Yu Liu, 2011) [1].

As an important internal variable in the operation of labor relations, wages have a direct impact on the occurrence of labor disputes. As a form of conflict in the operation of labor relations, labor disputes are bound to be closely related to the legal system of labor relations. Therefore, it is their pursuit to get as much wages as possible, resulting in contradictions and conflicts between the two sides (Shao-Ping Zhang, 2012) [2]. From 1991 to 2009, labor disputes caused by labor remuneration accounted for 29.64% of all labor disputes. It can be seen that the higher the wage, the less the number of labor disputes. Most of them control the existing resources, so as to reduce the demand for wages. On the contrary, the lower the wage, the more difficult it is to meet the wage, which directly or indirectly leads to a large number of labor disputes. They always feel that their own "labor" is not worth the price. Under the same thing, they feel lost to the wage comparison, so they are not satisfied, and then the possibility of labor disputes is higher.

The law allows employers and workers to agree on the conditions for termination of labor contract. In this case, employers can rely on their strong position in the labor market to agree on harsh conditions for termination of labor contract (Shao-Ping Zhang, 2012) [2]. Therefore, the pre-work game and post-work deduction between workers and capitalists in the labor contract form the main manifestation of labor disputes. Although the signing of a more clear and reasonable labor contract will reduce the number of labor disputes and lasting consumption costs, in fact, the more clear the position and enterprise of the labor contract, the less labor disputes. Moreover, the labor contract has the function of legal basis, which can be directly decided in arbitration and appeal quickly, resulting in the reduction of a large number of similar cases of labor disputes in a short time, or that the duration of labor dispute is not long.

In recent years, the collective cases of labor disputes accepted by Huizhou Intermediate People's Court have increased significantly. Group or series of labor disputes are in the trend of frequent occurrence, especially in the public service industry such as transportation or construction industry. It can be seen that the stronger people's awareness of rights protection, that is, the more they know about labor rights, the more exploitation traps they will find in capital, and the more labor disputes they will cause.

3 Research Method

We collect and sort out the domestic and foreign labor disputes in social security, labor safety, wages, labor contract signing and rights awareness as the five influencing factors of labor rights, deeply explore the influence of college students, HR and working workers in the process of fighting for labor rights on labor disputes, and on the basis of the research, construct the theoretical model of this research. In addition, we mainly use use neural network analysis, Python 3 and SPSS 25.0 to explore the collected sample data to verify the theoretical hypothesis.

A total of 227 questionnaires were collected, 4 of which were invalid, and 223 were available. The effective rate was 98.2%. Among them, 200 questionnaires in the student edition were checked, 3 invalid questionnaires were eliminated (all the same options and screening options were randomly filled in, for example, under 18 years old or over 65 years old), 197 valid questionnaires were obtained, and the effective rate was 98.5%. In addition, 23 questionnaires were collected from the industry version, 0 invalid questionnaires were eliminated, and 23 valid questionnaires were obtained, with an effective rate of 100%.

4 Research Results

According to the results of the questionnaire data analysis, the respondents were school students, on-the-job workers, labor dispatch personnel and on-the-job HR. A total of 223 valid questionnaires were collected.

Among them, 72.65% of the respondents' income is below 2,500 yuan, 21.08% of the respondents' income is between 2,500 yuan and 4,499 yuan, and a few people's salary is higher than 4,500 yuan. Moreover, almost half of the people are not satisfied with their current salary, but 44.84% think that their current salary is average and acceptable, and less than 8% are satisfied with their current salary.

However, in the case that half of the people are not very satisfied with their wages, most of the labor disputes caused by their wages are zero, about 90.13%; only a very small number of people have labor disputes caused by their wages, and once labor disputes caused by their wages, accounting for 8.07%.

Otherwise, 30.94% of them have signed labor contracts with enterprises, about 29.6% of them have not signed labor contracts with enterprises due to personal reasons, only 8.07% of them have not signed labor contracts due to enterprises, and 14.35% of them have not signed labor contracts with enterprises due to other reasons, but 17.04% of them are still unclear whether they have signed labor contracts with enterprises.

Besides, 77.58% of the people think it is very important to sign a labor contract, and 18.39% think it is more important to sign a labor contract. From this viewpoint, we can see that most of the people think that signing a labor contract is important, while only a few people think that signing a labor contract is not important.

However, almost half of the people have a general

understanding of the labor contract. More than 26% of the people do not understand the labor contract, while more than 24% understand the labor contract. Moreover, 35.87% of the factors in the item "which part of the labor contract do you pay most attention to" thought that labor remuneration was the most important factor, followed by 19.28% who paid attention to work content and working hours, and finally 17.04% pointed out that labor protection and working conditions were more important (see Table 1).

5 Conclusions and Suggestions

Labor relations management is one of the six major modules of human resources, and it is also the difficulty of HR work in many enterprises. How to properly handle labor disputes and reduce disputes between enterprises and employees has been troubling many enterprises. At the same time, it also greatly affects the relationship between employees and the company, and even some employees leave because of labor disputes. Therefore, it is still an important step to explore the causes of labor disputes and try to reduce the impact.

Based on this situation, we mainly stand in the perspective of workers and the college students who will be employed, starting from the degree of attention of employees to labor rights, discuss the relationship between labor rights and labor disputes, and select five influencing factors, such as social security, labor safety, wages, labor contract signing and rights awareness, through questionnaire survey and statistical analysis, so as to find out the relationship between labor rights and labor disputes. It is confirmed that labor rights are indeed related to labor disputes, that is, the damage of labor rights will increase the occurrence of labor disputes.

Table 1. Which part of labor contract is the most concerned.

Item	Percentage
Term of Labor Contract	7.17 %
Work Content and Working Hours	19.28 %
Labor Protection and Working Conditions	17.04 %
Labor Remuneration	35.87 %
Social Insurance Benefits	7.17 %
Enterprise Discipline and Rules and Regulations	2.24 %
Termination Conditions of Labor Contract	5.83 %
Liability for Breach of Labor Contract	4.93 %
Other	0.45 %

References

1. Ai-Yu Liu (2011) Research on the damage of labor rights and the choice of action: A comparison between two generations of migrant workers. Journal of Jiangsu Administration College, 1: 68-75.
2. Shao-Ping Zhang (2012) Economic analysis on the influencing factors of labor disputes. Journal of China Institute of Labor Relations, 26(3): 37-41.
3. Wen-Jiao Xue (2017) Research on the influencing factors of labor disputes in Gansu Province. Modern Commerce, 22: 182-183.
4. Wu Xu (2020) COVID-19 pneumonia: Legal issues on employment. Journal of the Open University of Fujian, 4: 75-78.
5. Yang Wang (2013) Analysis of the influencing factors of labor disputes in China during the transition period. Economic and Management Research, 4: 71-80.
6. Yong Xie (2008) An empirical study on the influencing factors of migrant workers' labor rights: A case study in Nanjing. China Population Science, 4: 70-75+98.

Reconstruction of Hanqing Stadium Scenery Based on Virtual Reality Technology

ShiYu Zhang [1], YingHao Dong [2], Tingting Xie [1], Xinying Si[1] and JinZe Li[3*]

[1]College of Economics and Management University, Shenyang Aerospace University, Shen Yang, liao Ning, 110135,China
[2]College of Information and Computer Engineering, Northeast Forestry University, Harbin, Heilongjiang, 150040, China
[3]College of Mechanical and Electrical Engineering, Shenyang Aerospace University, Shen Yang, liao Ning, 110135,China

Abstract. In order to reproduce the grand look of Hanqing Stadium and restore the most realistic Hanqing Stadium internal layout and structure, a simulation system based on the virtual reality technology for the internal structure layout design of the stadium was designed and developed. Based on the current development trend of virtual reality technology, a three-dimensional dynamic space model was established to restore the real conditions of Hanqing Stadium. By integrating emerging technologies such as panoramic images, panoramic views, and stereo vision into virtual reality technology, the sense of reality was enhanced. 3Dmaxs was used to establish the Hanqing Stadium scene design model, the surrounding and internal scene models were reconstructed, and the experiencers' feelings about the real scene were restored 1:1. The use of technologies such as panoramic images and views enhances the realism and three-dimensionality of the picture.

1 Introduction

Located on Taishan Road in Huanggu District, Shenyang, Hanqing Stadium was built in 1929 and has a long history. In 2001, Hanqing Stadium was listed as a national key cultural relics protection unit. As the first modern stadium in the country at that time, it could hold more than 30,000 people. Its rostrum is of the Romanesque architectural style, and the horseshoe-shaped stands open to the south, so it is also known as the "Roman-style horseshoe stadium", which fully reflects the design style of the integration of Chinese and Western cultures [1]. However, due to the relocation of Shenyang Sport University in 2007, the fate of Hanqing Stadium became rugged. Because it is a cultural relic site and developers cannot develop it, Hanqing Stadium has begun to become dilapidated. The plastic field suffered severe damage, the house near the stand was on fire, and the text on the stone tablet had become blurred. However, as a national key cultural relics protection unit, Hanqing Stadium still has its historical research value.

Using virtual reality technology, through the simulating construction of Hanqing Stadium's style, appearance, structure composition, and spatial details, the historical grand scene of Hanqing Stadium was restored. Scientific and technological means were applied to display the building through visualization technology [2]. The application of this technology has made a great contribution to the historical research and educational research value of Hanqing Stadium.

Virtual reality technology is generated using computer graphics systems and various control system equipment. It relies on an interactive environment [3] to present users with visual, auditory, smell, touch and other immersive sensory simulations. The rise and development of virtual reality technology is a leap forward for the development of science and technology in China, and it has contributed a new force to the development of computers. Therefore, it has been applied to many current fields including medical, engineering, military, aviation and navigation, etc. The development of this technology is becoming more and more popular.

The emergence of VR technology [4] has also made important contributions to the development of new products. The advantages of VR technology, such as low cost, high efficiency, and ultra-high transmission speed, have made scholars in many countries notice this technology. China has also increased its research efforts on this technology. With the rapid development of computer science and technology in China, virtual reality technology has been widely used in the realization of venues, architecture, and design concepts. [5] On the basis of informatization design methods, the characteristics and advantages of VR technology are used to analyze the design of Hanqing Stadium, so as to construct realistic scenes, which enables users to form a brand-new visual experience, [6] effectively improves visual effects and quality, [7] thus offering immersive experiences.

The existing papers related to Hanqing Stadium [8] mainly focus on the analysis of the environmental structure, research significance and value of the stadium, [9] and have not systematically proposed research methods and specific technologies. This paper proposes a specific

*Corresponding author's e-mail: 183406040107@email.sau.edu.cn

technical plan for the design of the stadium's scene reproduction, so as to implement the overall display of the stadium and make up for the shortcomings of current related academic papers.

2 System design plan

2.1 Overview of the system

The recurring stadium solution is based on the VR somatosensory interactive virtual reality device HTC VIVE. Based on HTC VIVE's somatosensory interaction technology[10], it has the ability to track gesture information to complete operations, and can provide users with a better experience. [11] Virtual reality technology (VR) and environment design provide an experience of immersion, interaction and simulation. The SQL structured query language was used to establish contacts and communicate with multiple databases. In order to visualize the image scene and realize the visual management of the assembly relationship, a three-dimensional model was established to truly restore the stadium picture. The addition of panoramic images and panoramic videos made the stadium look stronger. The technology system was finally packaged in desktop-level programs through C# for people to use and conduct research with.

2.2 The key technology of the system

2.2.1 HTC VIVE interactive device, Unity3D engine

To fully display the various venues and sports facilities in the stadium and determine the location of objects, Unity3D was used as the main development platform, and interactive equipment HTC VIVE was used to complete the development of the Hanqing Stadium prototype system. [12] The HTC VIVE head-mounted interactive device consists of a head-mounted display, a laser sensor locator, and a handle controller. The device can display with different effects through the binoculars, so as to present immersive 3D virtual pictures after different scenes are superimposed when the user is watching with binoculars. [13] It combines cloud computing, Internet of Things, big data, and other technologies to realize the intelligent experience and viewing process of Hanqing Stadium, and it takes AI intelligent space navigation, intelligent dispatch control, real-time detection of space environment, emergency handling of emergencies and other technologies as the reference basis for space location.

2.2.2 Acquisition of 3D data and stadium modeling

As one of the technologies used in the 3D scene reproduction of the stadium, the acquisition of 3D data is a key part that restricts the location information. The visual presentation of Hanqing Stadium is an experience process from the overall scene to the partial scene, from the large-scale surface to the small-scale orientation. The establishment of a three-dimensional stadium model is an important part of three-dimensional visualization.

Firstly,3DMAX software was used to build a three-dimensional model, and the basic simple geometry was used to construct the complex geometry required for the visual presentation of Hanqing Stadium to meet the realistic visual experience. Secondly,In the use of complex geometry modeling, by using multiple two-dimensional geometry and three-dimensional geometry through a series of superimposition and combination, the three-dimensional complex geometry required for the construction of the stadium was obtained. Finally, a three-dimensional model was generated from the geometry and the three-dimensional image through Boolean operations. [14] Through the data collection and analysis of various facilities, buildings, runways, and sports equipment in the stadium, three-dimensional models of different areas such as outdoor sports ring, natural ecological ring, national fitness ring, and garden tour ring were established to provide users with an intuitive and visual stadium environment.

2.2.3 Acquisition of panoramic images and panoramic videos

To solve the issue of the monotonousness of traditional 3D modeling data, based on panoramic image and panoramic video acquisition technology, [15] a dynamic interactive virtual space was built. For the panoramic image, it was planned to adopt two methods to obtain the panoramic image. The first is the method of collecting multiple sets of pictures to form a panoramic image, using panoramic image splicing technology based on image rendering; the second is to use a professional panoramic camera, set appropriate camera calibration parameters in advance, and automatically obtain panoramic images. For the panoramic video, the 3D camera was used to obtain the 360-degree all-round shooting of the required field and scenery. Panoramic images and panoramic videos are new types of three-dimensional data acquisition methods, which have the characteristics of full viewing angle, strong sense of reality, and high current status, and can display various topography and landforms in more detail. It has a high resolution and three-dimensional effect combined with virtual reality to enhance the immersive experience for users.

2.2.4 Demonstration of part of the operating system

According to the content involved in the realization of virtual reality technology in the stadium, the operating system consists of five parts for demonstration, including demonstration of the external scene of the stadium, demonstration of the outline scene of the stadium, demonstration of the layout of the stadium, demonstration of the internal facilities of the stadium, and dynamic demonstration of the functions of the stadium.

https://doi.org/10.1051/e3sconf/202129002029

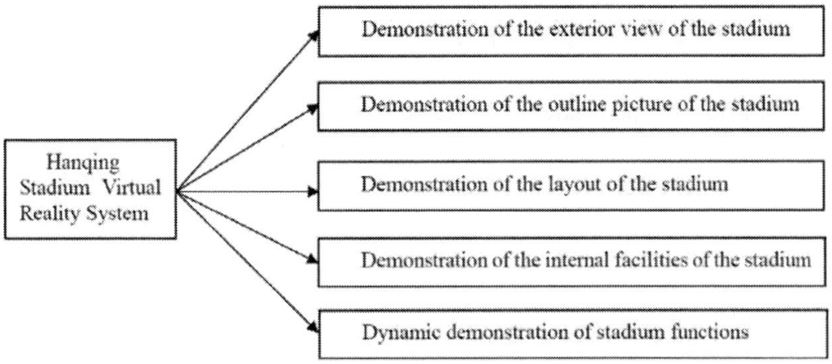

Fig1. Demonstration diagram of Hanqing Stadium virtual reality system

2.2.5 Model building process analysis [16]

The overall visualization model of Hanqing Stadium was established by 3DMaxs. 3DMaxs is suitable for a series of modeling productions such as architectural design, 3D animation, and dynamic scenes, which have strong flexibility and are professional tools for making architectural models. 3DMaxs has good guidance nature, maneuverability and practicability in the finished renderings, and has detailed modeling parameters. Compared with 3DMaxs, Maya is more suitable for model establishment of film and television scenes and animation environments, and 3DMaxs has high work efficiency and strong operability. On this basis, for Hanqing Stadium's in-field model design, 3DMaxs modeling software was used, and PixyzStudio was imported to specifically subdivide different venues or facilities in the stadium, and the model was revised, optimized, and exported. After further processing the Hanqing Stadium model with Cinema4D, it was imported into Unity for use, which effectively realized the lightweight conversion and format conversion of the model data.

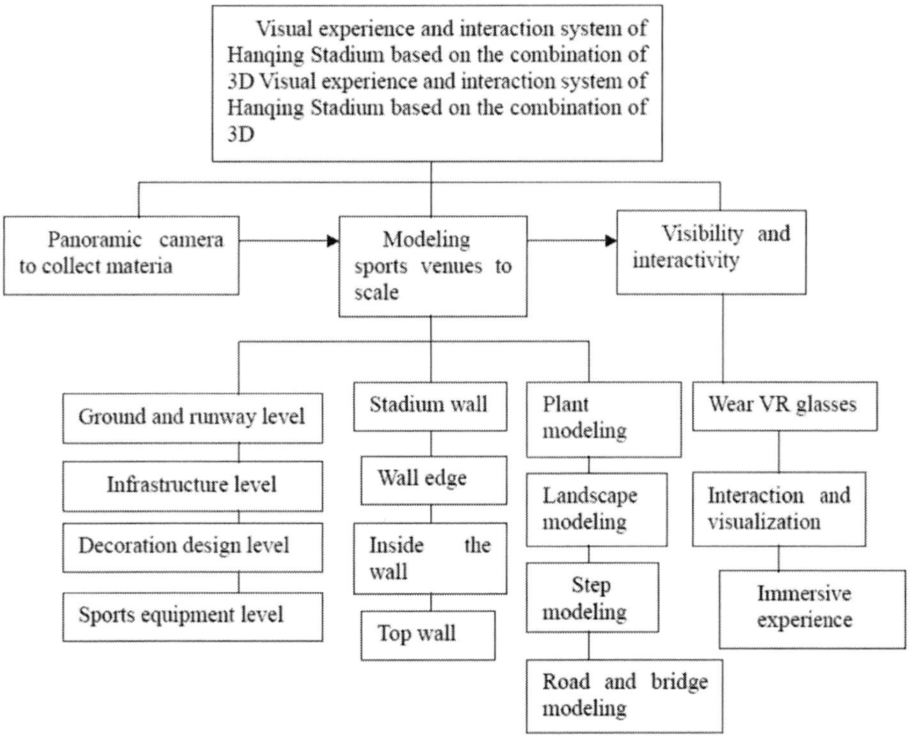

Fig2. Hanqing Stadium interactive experience system block diagram

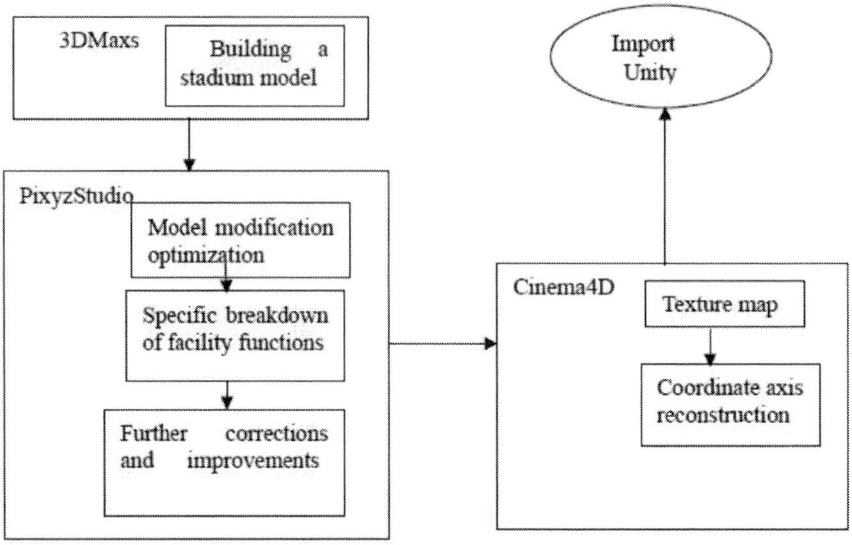

Fig3.Hanqing Stadium model export process diagram

3 Running results [17]

The abandoned Hanqing Stadium still has values in many senses. For example, the overall design of Hanqing Stadium perfectly embodies the integration of the central culture. It covers a large area and is a national key cultural relics protection unit, which makes the reconstruction of Hanqing Stadium possible. However, Hanqing Stadium is still in an abandoned state. It is difficult for the public to experience the grandeur of the stadium. Therefore, it is feasible to reshape the whole picture of Hanqing Stadium through virtual reality, which is an effective measure for the public to experience the stadium. Using VR virtual reality simulation, the appearance of the stadium was reproduced, and the large stadium of more than 3,000 square meters was reproduced, so that users can experience and feel the stadium. Light perception was applied to achieve such a user's experience.

As the system is running, the user enters the simulated real world, enters the main page, and sees the view outside Hanqing Stadium. In virtual reality, on the left-handed handle, the user can see the floor plan of Hanqing Stadium. During the experience, the user holds the two handles with both hands, puts the handle on the text on the floor plan, and releases it to the scene corresponding to the text. In the scene, the user can manipulate the handle to walk and watch the simulated real world in 360°.

During the simulation of Hanqing Stadium in virtual reality, with multiple realistic designs such as sound effects and light perception, Hanqing Stadium can be reproduced truthfully. Sound effects and light perception can simulate the real effects of theaters, bars, and circular racetracks in real stadiums. In order to enhance the experience, the control of the light perception was made, and the simulation of the cloudy sky was strengthened, so as to bring the user into the stadium. The aperture and animation were strengthened, so that when users walk in the stadium, there are apertures and arcs. The interactive design of HTC Vive interactive device was used to fully demonstrate the real effect of simulation. During of scene transition, the connection between the animation effect and the scene is strengthened to reflect the authenticity of the scene and make the experience smoother. All of the measures are to enhance the details, strictly control each part, improve the environment atmosphere, emphasize comfort, interactivity, and authenticity, and enhance the sense of immersion.

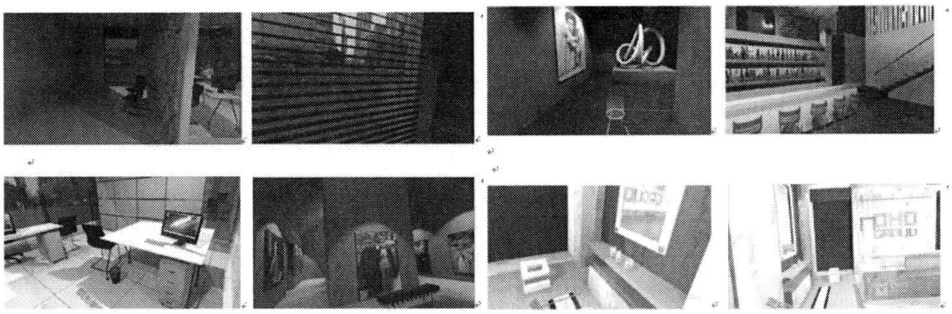

Fig4. Part of the scene

4 Conclusion

Based on virtual reality technology (VR) this system reproduces all-round scenes of Hanqing Stadium's internal venues and facilities. Based on the development of the existing 3D visualization technology, this study has realized the human-computer interactive upgrade, which brings the users natural and comfortable experiences, and enhances the sense of reality and realism in the projection. The acquisition technology of panoramic images and panoramic videos solves the issue of the monotonousness of traditional 3D modeling, and meticulously depicts the different topography of Hanqing Stadium, giving users an immersive experience, letting the users know about the truthful look of Hanqing Stadium without leaving home.

Based on the data analysis obtained from the experiment, the use of virtual reality technology (VR) to analyze the structure of Hanqing Stadium has the following research value. (1) Experiencers can cross geographical restrictions and have a comprehensive experience of the stadium's internal environment without leaving their homes. (2) This system technology can be used to pre-design the renderings of the relevant building products, which is convenient, true and fast. (3) In reality, it can avoid damage to the stadium facilities when the experiencer visits the stadium. (4) The sports culture and related sports knowledge of the stadium can be popularized through the simulated physical objects, models, various experience facilities and demonstration facilities of the stadium. (5) Breaking the visit limitation that may occur when traditional users visit the building and the limitation of incomplete information from the narration of human staff, this system can present the full picture of the stadium in all directions and multiple levels for detailed introduction. (6) The autonomy of the user is fully manifested, and the user can experience, watch and learn at any time and area suitable by actual needs. (7) The system integrates the fun and interactivity in the experience process, and brings the experience with visual shock and surprise.

Reference cited

1. Sui Hao, Chen Guanru. Research on the Overall Display System of Fengtian Hanqing Stadium [J]. Popular Art,2018(22):104-105.
2. Han Zhiren, Liu Xiaonong, Wu Meng, Liu Baoming. Research on 3D Visualization Technology of Aircraft Component Assembly Schedule Based on CATIA Composer [J/OL]. Aviation manufacturing technology: 1-9 [2021-05-10]. http://kns.cnki.net/kcms/detail/11.4387.V.20210510.0900.004.html.
3. Wang Peng. Contextual framework based on interactive machine translation environment [J]. Journal of University of Science and Technology Beijing (Social Science Edition), 201,37(02):138-146.
4. Liu Ping. Application of Virtual Reality Technology in 3D Product Detail Display [J]. Modern Electronic Technology, 201,44(09):162-165.
5. Yu Zhiqiang. Modern Building Environment Design Supported by Virtual Reality Technology [J]. Sichuan Cement,2020(09):324-325.
6. Hu Xiaoling, Tan Yongping. Application of virtual reality technology in architectural decoration design practice teaching [J]. Guangxi Education,2020(43):187-189+192.
7. Shen F X. Research on image compression sensing reconstruction algorithm based on visual communication effect [J]. Microcomputer Applications, 201,37(04):61-65.
8. Gu Aihua, Tong Xi, Zhou Huan. Study on the Realization Path of Material Cultural Heritage Value in Shenyang -- Discussion on the Feasibility of Establishing Northeastern University Site Park [A]. Shenyang Municipal Committee of the Communist Party of China, Shenyang Municipal People's Government, International Academy of Production Engineering, China Mechanical Engineering Society. Shenyang Municipal Committee of the Communist Party of China, Shenyang Municipal People's Government, International Production Engineering Academy, Chinese Mechanical Engineering Society: Shenyang Association of Science and Technology,2019:4. Design and Implementation of 3D Virtual Tour System Based on Panoramic Technology [J]. Technology and Market,2021,28(03):92-93.
9. Li Yao. Study on the Interaction between Architectural Structure and Interior Space under Parametric Design -- Taking Alien Space Design as an Example [J]. Tomorrow Style,2021(05):118-119.
10. Lei Zhang, Xiaobin Xu, Jia He, Kaiyuan Zhu, Minzhou Luo, Zhiying Tan. Two-dimensional LiDAR

and Camera Calibration Method Based on Indoor Structure Characteristics [J]. Acta Photonica Sinica,2020,49(12):67-78.

11. Yuan Hao. Virtual Reality Building Teaching Method Based on HTC VIVE [J]. Shanxi Architecture,2017,43(24):256-257.
12. SHU Jun.Implementation of Campus Display and Interactive System Based on Unity3D and HTC Vive [D]. Nanchang Hangkong University,2019.
13. Mi Dan, Wang Qin, Wang Chunjie. Virtual reality roaming of model house based on Unity3D [J]. Modern Computer,2019(24):97-100.
14. LUO Xu. Method Analysis and Technique Discussion on 3DMAX Modeling [J]. Computer Knowledge and Technology,2020,16(35):207-208+213.
15. Zheng Xiaoli. Design and Implementation of 3D Virtual Tour System Based on Panoramic Technology [J]. Technology and Market,2021,28(03):92-93.
16. Ma Leilei. Modeling and Hierarchical Rendering System for Large-scale Scenes [D]. Shandong University,2020.
17. ZHAO Bing. Analysis of VR natural disaster scene experience simulation system [J]. Electronic World,2020(08):23-24.

Research on the Professional anxiety of News Editors in the era of Intelligent Media

Yuanke Jin[1], Yan Liu[1*]

[1] School of Humanities, Jiangxi University of Traditional Chinese Medicine, Nan Chang, Jiang Xi, China

Abstract. Intelligence-intensive media are in the ascendant driven by algorithms and data. New technology not only leads news editors into the intelligent era, but also has an impact on them. Artificial intelligence carves up the work of news editors, while virtual reality forces the news editing industry to change, and the new concussion of business format causes the professional anxiety of news editors. Based on the background of media intelligence, this paper analyzed the psychological mechanism of professional anxiety caused by intelligent media to news editors from three aspects: cognition, emotion and motivation, it also put forward the scientific cognitive model, the psychological model of the combination of defense and confrontation, and the efficiency improvement model of multi-agent participation.

1 Introduction

Intelligent media is a self-strengthening ecosystem based on new technologies such as mobile Internet, big data, virtual reality and human-computer interaction, which realizes the intelligent matching of information and user needs[1]. Intelligent media is widely used in business, games, medical, education and other fields, showing certain advantages in massive information retrieval, complex data collation, automatic compilation and visual information presentation. Intelligent media can complete a series of manual editing work, which means that the professional value of news editors will be weakened, and the resulting sense of urgency makes news editors have professional anxiety.

The publishing industry is an ancient industry, which promotes the spread of human knowledge, ideas and culture for a long time. In recent decades, with the rapid development of Internet technology, the development of artificial intelligence is particularly eye-catching, and a large amount of capital has poured into the industry. Scholars continue to explore the influence of smart media on news editors, but most of them focused on the logical framework of macro industry analysis, such as Peng Lan's macro summary of the connotation, denotation, categories and characteristics of smart media[2]. Yu Guoming analyzed its market prospect motivation from the perspective of industrialization[3]. Few scholars discussed the micro-research of intellectual media on the psychology of news editors, which gives this paper important research significance. This paper analyzed the psychological mechanism of news editors' professional anxiety in the context of intellectual media, expounded the news editors' professional anxiety according to the three levels of industry environment, psychological environment and solution path, and explored the solution that can reduce professional anxiety.

2 Industry environment analysis

The change of industry environment directly affects the psychological state of news editors.

2.1 Artificial Intelligence

The definition of artificial intelligence is controversial. Some scholars have proposed that artificial intelligence belongs to the computer discipline, trying to explore the essence of "intelligence" and produce an intelligent machine that can respond in a way similar to human intelligence[4]. AI is applied to the field of news editing, which shows that intelligent machines simulate manual editing to complete some tasks. Intelligent machines enter the news editing industry, reshaping the internal operation paradigm of the industry. The development of the Internet has profoundly changed people's way of life, communication and reading habits. We can make two judgments: one is that the information explosion will not stop, and time fragmentation and reading fragmentation will become the norm. In addition, big data allows people's character and preferences to be clearly depicted through clicking, reading, buying and other behaviors, personalized communication and push will only become more and more accurate.

2.2 Virtual reality

Virtual reality (VR) is a new technology that constructs a simulation environment based on high simulation system, fully dispatches human senses and realizes users' high-

*Corresponding author's e-mail: 349620816@qq.com

quality information experience. On the one hand, it is a backlash to the macro news editing industry. Under the condition of VR new technology, news editors need to bring users into the virtual scene through "Internet + sensor" to make them get a sense of on-the-spot experience, but this also causes users to choose excess space and deviate from the news center of gravity track. In the smart media environment, news editors are liberated from the solidified information and have to devote themselves to the creation of visual content. At the same time, editors should also absorb the opinions of the audience and comprehensively adjust the production direction in the process of interaction. On the other hand, it is the pressure on the individual news editors. The production of VR news requires that in addition to the existing planning, arrangement, proofreading and design skills, news editors should also use VR technology to reasonably control the news content.In recent years, with the development of virtual reality technology, augmented reality technology and mixed reality technology under the background of artificial intelligence, the media began to try to use VR and AR technology to produce "immersive news" to create an "immersive" sense of presence for the audience. The production of "immersive news" is through the use of VR and AR technology in news reports, through collection and recording of news facts, and later technical editing presented on the client side of VR and AR, the user is "directly exposed" to the news scene, realizing the integration of the real scene and the virtual scene, and allowing the user to temporarily "peel off" the environment and place himself in the "reality" of the news reporting environment. So as to satisfy the audience's grasp and objective cognition of the news reporting scenes and facts.

3 Explore the psychological environment

Zhang Genjia defines professional anxiety as editors' psychological state of nervousness and fear because they are unable to achieve their work goals or overcome difficulties. From the perspective of mental health, occupational anxiety is essentially a psychological imbalance among editors in the process of value comparison, which is closely related to cognition, psychological expectation, sense of efficacy and other factors.

3.1 Cognitive block imbalance: cognitive bias leads to occupational anxiety

The utility of intellectual media depends on the psychological cognition of news editors, and cognition is affected by complex factors, so it is inevitable that there will be cognitive bias. Cognitive bias refers to the error caused by the cognitive subject to the cognitive object in a specific cognitive situation.

The cognitive barrier between news editors and intellectual media comes from cognitive resources and cognitive styles. First, the cognitive resource theory put forward by Fidel and Garcia expounds the important influence of pressure and cognitive resources on the effectiveness of leadership decision-making. Due to the limitation and influence of inherent cognitive resources, news editors often deviate in the choice of cognitive path of intellectual media, especially in the source of information acquisition. In addition, the accumulation of a large number of wrong information leads to the cognitive deviation of news editors from the reality, coupled with the sudden breaking of career stability by intellectual media, which leads to the disorder of their work rhythm and employment direction. Second, the wrong cognitive style misleads the news editor. Smart media brings a brand-new and unknown media environment, and news editors are prone to mistakes in judgment or subjective and objective cognitive dislocation in such a vague cognitive situation, such as magnifying the ability of machine deep learning or worry about the ethical and moral problems of the machine, and so on. Just as the miscognition emphasized by "halo effect" and "halo effect" in psychology is generalized, cognitive bias is also a kind of wrong cognitive logic to infer the whole from the part.

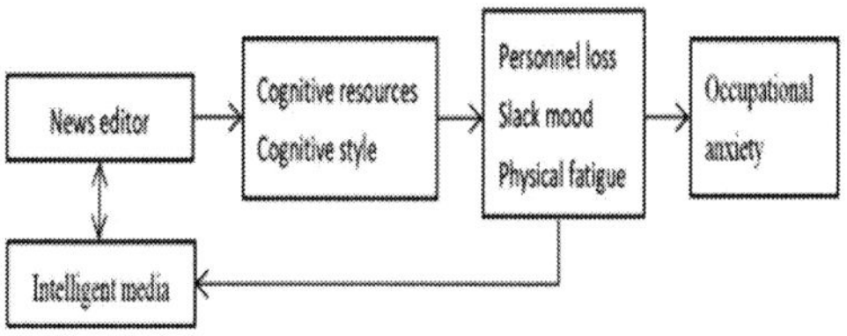

Fig.1. Cognitive bias Model of News Editors.

3.2 Imbalance of emotional blocks: negative expectations lead to occupational anxiety

Emotion is the subjective psychological feeling produced by individuals when they encounter stressors, which is affected by psychological expectations. When considering the impact of intellectual media on their careers, editors have a pre-evaluation, and they will make negative predictions according to the current situation of the development of intellectual media, mainly in the following two aspects. New technology leads to the expansion of stressors. Fierce competition among peers and laymen, as well as the invention and application of intelligent devices make it easy for machines to complete editing activities, which extends competitors from people to things. The work intensity is deepened and the psychology is overloaded.

3.3 Imbalance of motivation blocks: occupational anxiety induced by lack of efficacy

Motivation is affected by both inducement and internal drive. Self-efficacy belongs to the level of internal drive, which is the intermediary factor to stimulate motivation. Research shows that there is a strong negative correlation between self-efficacy and professional anxiety[6]. When applied to the field of news editors, the higher the sense of self-efficacy, the more confident editors are about their own work ability. News editors subconsciously lose confidence in solving new problems, and have a strong anxiety and panic, and then take an evasive way to respond.

4 Suggestions

The solution to the professional anxiety of news editors should return to psychology itself and get out of the predicament of anxiety with the help of scientific psychological methods. therefore, it is necessary to put forward targeted suggestions to news editors from three aspects: cognitive model, psychological model and efficiency improvement mode. to help them better adapt to the new environment of the new era.

4.1 Scientific cognitive model

The information obtained by news editors based on perception is shallow and extensive irrational judgment, which will inevitably be disturbed by the first cause effect, social stereotype, thinking stereotype and other factors, and fail to understand things in an all-round way. This requires news editors to cultivate a scientific cognitive model and overcome cognitive bias[7]. News editors should establish a correct view of technology and professionalism. The concept of technology needs to be correctly mastered from three aspects. Firstly, news editors should broaden the cognitive channels of intellectual media, establish online group communication, use modern communication tools to collect authoritative intellectual media information interpretation, and accumulate correct cognitive resources. Secondly, news editors should realize that smart media, are still in the "low age" stage, and the lack of abstract abilities such as imagination and innovation is still its fatal weakness. Finally, news editors realize that smart media can help them deal with complicated and time-consuming data, and news editors can devote themselves to the part of the machine blind zone to show their self-worth.

4.2 Psychological model of the combination of defense and confrontation

"Defensive" refers to a positive psychological model to relieve anxiety. A survey on editors' defense style and mental health shows that healthy groups are more likely to use mature defense styles to deal with psychological problems than symptom groups[8]. This will be divided into two types: one is communication, that is, news editors interact with professional psychologists to deal with problems with a normal mind by mobilizing various positive psychological elements, and they should also take the initiative to communicate with non-professional personnel to broaden their thinking; second, in the face of the negative impact of intellectual media, news editors can separate their focus from negative thinking and devote themselves to the cultivation of interest in work. "adversarial" means to stimulate the defensive psychology of news editors to fight anxiety. Intelligent technology is a potential threat to news editors, and news editors can regard the pressure caused by this threat as a reverse stimulus to stimulate their potential strength.

4.3 The efficiency improvement mode of multi-agent participation

In the actual operation, the difficulty faced by the publishing house is that the publishing house is unable to obtain a large amount of data.

From the terminal point of view, whether online bookstores or physical bookstores, it is difficult to share reader purchase data with publishers; from the supply side, manufacturers do not share data with publishers. As a result, it is difficult for publishers to have enough data to use artificial intelligence to improve decision-making and operations, both on the production side and on the sales side. Multiple subjects include individual, peer and internal management. First of all, to establish an individual effectiveness model with technical training as the core, the company introduces advanced equipment, news editors do hands-on VR and intelligent machines, and gradually accumulates intellectual media experience by setting stage goals for intelligent media learning, so as to enhance the sense of efficacy and control of technology. Secondly, based on the sharing mode of peer-to-peer learning interaction, under the impact of new technology, news editors need to learn successful cases of dealing with technology anxiety among peers or colleagues.We can see that artificial intelligence has made great progress in language through deep learning and other technical means. Although it can not reach the level of artificial practitioners, with the progress of technology, the error rate is getting smaller and smaller. The user experience is getting better and better. This is very valuable for publishers. Finally, news editors and managers need to build the exclusive culture of the enterprise and set up a management model with professional atmosphere as the core.

5 Conclusion

In the new era of intellectual media, the news editing industry has experienced a period of transition, which requires news editors to actively transform to fit the new environment, and psychological adjustment is the top priority of the transformation. Because the mental health of news editors is not only related to their own career development, but also to the construction of cultural industry and the effective transmission of spiritual civilization. Therefore, to help news editors get out of the psychological predicament of professional anxiety, it is more necessary to use long-term vision, focus on long-term interests, and promote reform.

References

1. Quanzhong, G.(2016) The characteristics and construction of intelligent media [J]. News and Writing, (3): 59-62.
2. Lan, P.(2016) Smart Media: future Media Wave: new Media Development trend report [J].

https://doi.org/10.1051/e3sconf/202129002030

International Press, 2016 (11): 7-23.

3. Guoming, Y., Meina, L., Wei, L.(2017) Intelligence: the core logic of future communication mode innovation: also on the basic operation paradigm of artificial intelligence+media [J]. News and Writing,(3): 41-45.
4. Qinying, C.(2017) A brief Analysis of the influence of artificial Intelligence on the Publishing Industry and its Countermeasures [J]. Technology and Publishing, (11): 7-10.
5. Spectrum.(2015) The person in charge of the New York Times headline is a chat robot [EB/OL]. [2015-08-17]. Http://tech.ifeng.com/a/.0817/414197380.shtml.
6. Yan, B., Cao, J.(2015) A Review of the Research on Cognitive bias at Home and abroad [J]. Zhengzhou normal Education, (5): 30-Qing.
7. Kai, Yao.(2008) A review of research on self-efficacy: a new trend in the development of organizational behavior [J]. Journal of Management, (5): 463-467.
8. Yanhua, G., Zengxue, Z., Shiguang, S.(2005) The relationship between defense style and mental health of editors of university journals [J]. Hebei Medical Journal of staff and Workers, (9): 28-30.

Business-oriented data management in conglomerates: insights from government data sharing

Zhepeng Lv[1,2,*]

[1] Guanghua School of Management, Peking University, Beijing, 100871, China

[2] Postdoctoral Research Workstation, China Everbright Group, Beijing, 100033, China

Abstract. Nowadays, the management of big data has become an emerging question for companies. Managers in conglomerates, or corporations made up of many businesses, concern more about data management as data integration and sharing inside the corporate can bring about product innovation, risk control and operation efficiency. To solve the technical and managerial problems that hinder data sharing, this study drew reference from federal enterprise architecture framework implemented constructed by the United States. By exploring this framework, this study suggests a methodology for data sharing in conglomerates, and stresses on the business-driven principle and a top-down architecture design with supporting from high-level decision-making.

1 Introduction

Nowadays, as more and more data are produced and consumed in the Internet, how to manage these big data has become an emerging question. Especially, managers concern about how to share data between different departments or companies, as data sharing could promote the interconnection of information, strengthen relationship between corporations, promote knowledge diffusion, optimize the allocation of resources, and contribute to the creation of innovation.

Conglomerates, in other words diversified business groups, pay special attention to big data sharing. A conglomerate is a corporation made up of a number of different and sometimes unrelated businesses. In the process of business development, each subsidiary corporation of a conglomerate has accumulated different dimensions and different types of user data. These data accumulate naturally with the business development of subsidiary corporation, and they are separated from each other, forming "information isolated island" within a conglomerate. Sharing these data between different subsidiary corporations can strengthen cross selling, optimize operation and management, and promote product innovation and risk control [1]. Therefore, conglomerates have strong motivation to integrate and share these big data.

However, the influence of sharing of big data remains unknown in practice. Many conglomerates, such as PingAn financial group in China, began to lay out data sharing and system connectivity in 2009, built the infrastructure of data sharing to integrate data resources, and finally gained rapid development for more than 10 years. However, the industrial also see failed cases in many conglomerates who constructed data sharing platform without ideal results.

Thus, here comes to the question that what is the challenges of data sharing in a conglomerate between subsidiary corporates. Furthermore, how those conglomerates could solve these problems and facilitate data sharing. This study draws reference from big and open linked data sharing in US government, and concludes on insights of federal enterprise architecture (FEA) framework on conglomerates in finance industry. This study provides a basis for data management and integration in conglomerates, and the suggested FEA framework would facilitate data sharing and information exchange across subsidiary corporation boundaries.

2 Literature review

From the perspective of social technology theory, researchers pointed out that social process and technological development both affect the integration and sharing of information resources [2]. The process of information resources integration and sharing will be accompanied by workflow reorganization and organization structure adjustment, which would be expensive and time-consuming [3]. By studying e-government, some researchers demonstrate that the integration and sharing of government data are hindered by resistance to change, IT infrastructure incompatibility (hardware, software, etc.), organization goal diversity, and environment complexity [4]. In a subsequent study, they found that the integration and sharing of data between government departments are closely related to the degree of e-government construction in both developed and

*Corresponding author's e-mail: lvzp@bj.ebchina.com

https://doi.org/10.1051/e3sconf/202129002031

developing countries.

Researchers have investigated the antecedents of data sharing among different government departments, suggesting the crucial role of technology development and management factors. In terms of technology, the main problem is IT infrastructure incompatibility and data specifications inconsistency. In terms of management, switching costs at individual level, conflicts of interest among departments at organizational level, lack of reasonable incentive mechanism at motivation mechanism level are main factors [5].

3 FEA introduction

3.1 The background

In late 1980s, with the tide of administrative reform and the application of enterprise architecture in enterprise information digitization, the United States began to explore e-government. The world-shaking terrorist attack in 2001 highlighted the urgency of data sharing within government departments, and then application of information technology to promote inter-department communication was accelerated. In 2002, Federal Enterprise Architecture (FEA), the top-level design of e-government of the United States, came into being [6,7]. The FEA framework contains not only architecture framework methodology (i.e., FEA reference models), but also detailed implementation guidelines and management tools, which transform the conceptual design into concrete operational practice.

3.2 Five reference models

The FEA reference model consists of five reference models: performance, business, service component, data and technology reference models. Among them, the crucial models are business reference model, performance reference model and data reference model.

Business reference Model (BRM) is a framework that puts forward from the business perspective rather than the department perspective, aiming to clarify the business lines for subsequent data sharing. It describes around the business scope and promotes the cooperation between departments and agencies. The model consists of three levels, that is, business domain level, business line level and subordinate function level. When applied to the United States e-government, business domain level specifically refers to four categories of government business operation, namely, citizen service, service providing mode, service supporting condition and government resource management. Each business domain is composed of business lines that provide similar activities and services with similar information definition and data structure (see figure 1). Each business line is composed of a sub-function set containing similar business processes and units.

Performance reference model (PRM) is a set of performance evaluation standards used to understand how and to what extent technology can output and results. It consists of evaluation field, evaluation category and general evaluation index. In the U.S. e-government project, the evaluation field includes tasks and business achievements, customer achievements, processes and activities, human capital, technology and other fixed assets. The evaluation categories are grouped according to the attributes and characteristics described in each evaluation field. General evaluation index are customized indicators designed for specific departments, which need to be defined under the guidance of FEA team.

Data reference Model (DRM) connects data taxonomy and management with BRM, and uses the concept of business association to identify the business environment in which data exists. The three levels of DRM are data context, data description and data sharing. Data context refers to the information to the purposes for which the data was created and used. Data description standardizes the data representation structure and semantics, with the purpose of data representation, discovery, sharing and utilization. Data sharing standardizes data exchange specifications and data access interface in order to exchange data among different departments and agencies. By standardizing data in the above fields, it effectively promotes the cross-department identifying, using and sharing data by the federal government (see figure 2).

https://doi.org/10.1051/e3sconf/202129002031

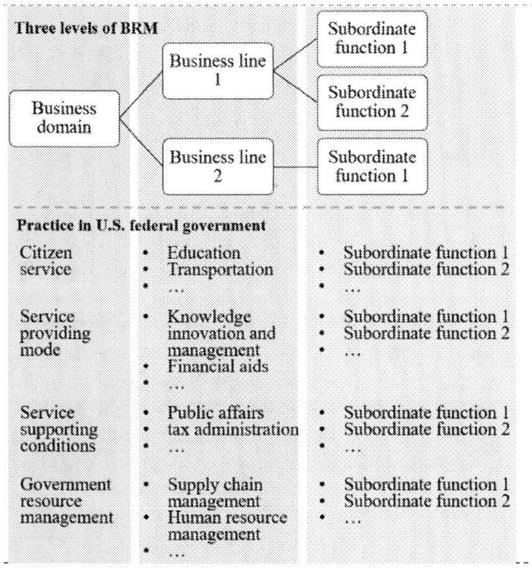

Figure 1. FEA business reference model (BRM) structure

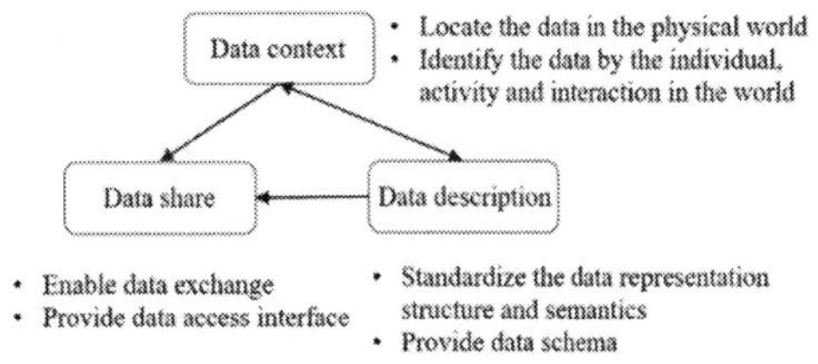

Figure 2. FEA data reference model (DRM) structure.

Service component reference model (SRM) refers to the IT system structure framework to support the functions of business. Technical reference models (TSM) describes technologies and standards that may be used in building the integrated e-government system.

4 FEA Solution for Data Sharing Problem

4.1.FEA Framework Characteristics First, FEA architecture is a top-level design driven by business. Business-oriented classification of government activities can make the government cover citizens' needs more completely, and thus ensure better government performance [8]. Second, FEA framework strengthens the concept of government as a whole and weakens the attention for a single department performance by facilitating business collaboration. The business classification in the FEA framework does not depend on departments classification, meaning that two or more departments may complete some business together. Third, top-down design is combined with bottom-up management. The government accomplishes the goal of data sharing by analyzing business structure and constructing suggested IT infrastructure using a top-down approach, and fills the framework with detailed metrics and data formats by working with specific departments.

4.2. The Operation and Management of FEA Framework

The implementation of the FEA framework mainly depends on the office of management and budget (OMB) of the United States with IT project approval mechanism. After the announcement of FEA architecture, OMB cooperated with the federal chief information officers council (CIO) and the general affairs administration (GSA) on developing a series of guidelines and management tools, including federal segment architecture methodology and enterprise architecture assessment framework.

Specifically, OMB manages FEA framework implementation by the following ways. First, it regularly carries out reporting and evaluation work, including determining the annual evaluation content, metrics and scoring standards. Then departments at all levels can submit segment reports and self-evaluation reports quarterly. Second, OMB announces some key business lines that will be included in the president's management list. These business lines show a higher possibility to be approved for IT budge. Third, the office examines whether the e-government investment and construction of departments at all levels conform to the FEA framework, in order to reduce the redundant and repeated IT investment.

4.3. Solutions for Technical and Management Problems in Data Sharing

FEA architecture solves the technical problem of data sharing among departments by coordinating technical architecture and standardizing data formats. The FEA architecture specifies common technical standards through technical reference models, so that different departments can identify and reuse IT assets. The data reference model also provides standardized architecture for data description, sharing behavior and taxonomy methods, which advance information and data interaction between different departments.

To solve the management problem, the business-oriented FEA framework can promote value redistribution across departments. The business classification in FEA framework is driven by business in practice, so is performance evaluation, which establishes the connection between business components and results. Therefore, at the motivation mechanism level, compared to the performance evaluation of departments respectively, the FEA framework guides benefit distribution by business results, which can effectively lead the motivation mechanism to encourage cross-institutional activities. At the organizational level, under the guidance of business reference model (BRM), the government can better decompose the business domain and business line from the top-level design, and subdivide them into specific sub-functions. Departments that need to work together are clarified responsibilities in order to complete those sub-functions. Under this circumstance, the clear responsibilities help reduce the problem of vague duties and help quantify the benefit and cost of each role those departments play.

5 FEA framework applying in conglomerate data management

5.1.A proposed FEA framework for conglomerates in financial industry

Data sharing in conglomerates faces similar problems as that in a government, including technical problems such as incompatible technical structures of subsidiaries, as well as management problems such as conflict of interests. The methodology and framework of data sharing among government departments can be used to improve conglomerates data management.

To practice the FEA framework, this study focuses on conglomerates in financial industry, or financial holding company in other words, and suggests a data management and data sharing architecture for the financial holding company with reference to FEA framework.

In the business reference model (BRM), this study explores wealth management field. According to FEA framework, BRM has three levels, that is, the business domain, business lines and subordinate functions. The business domains in this situation include wealth management, service delivery mode, service supporting conditions and resource management (see figure 3). Each business domain is composed of multiple business lines. The principle of creating a business line is that a business line is not necessarily conducted by a subsidiary corporation, but the information definition and data structure should be quite similar. This ensures the facilitation of information exchange and data management between subsidiaries.

In the performance reference model (PRM), this study designs performance metrics for credit loan business in retail finance services. This study develops performance metrics of credit loan business in the financial dimension, customer dimension, internal operation dimension and innovation growth dimension by combining FEA framework guidance and the bank literature [9].

In the data reference model (DRM), this study designs the data labels in the credit loan business for data context identification, and suggests the data quality standards for data description and data formats for data sharing and transmission (see figure 4).

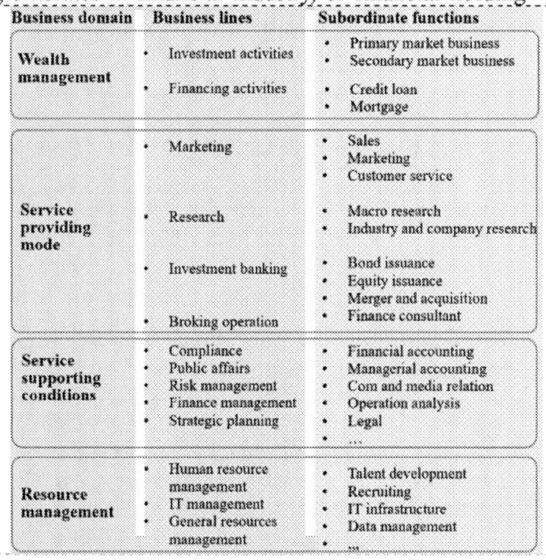

Figure 3. Business reference model for wealth management business

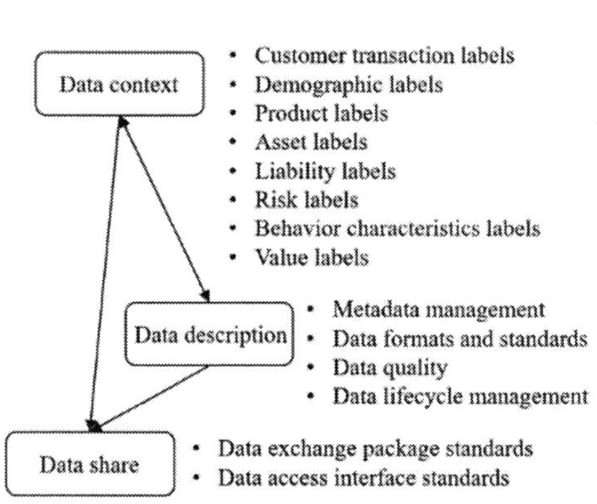

Figure 4. Data reference model in credit loan business

Considering that service component reference model and technical reference model are supportive models in financial holding companies, which are not distinguished from technical models in other companies, so this study does not elaborate on these two models.

5.2. Suggestions for implementing FEA framework in conglomerates

First, in order to implement big data sharing in conglomerates, the project should follow the business-driven principle. The fundamental purpose of data sharing in conglomerates is to improve the overall service capability. In fact, customers pay little attention to which subsidiary corporations provide products or services, but only remember the output of conglomerates as a whole. Therefore, when sharing and integrating data, managers should keep consumers' need in mind and underline the demand of the business instead of subsidiary boundaries

Second, FEA framework provides a complete and comprehensive information technology resource sharing framework. Data management and sharing are not only tasks to IT departments, but also closely associated with business departments. If data taxonomy and data description are ignored by business departments, data sharing construction is unable to adapt to business processes and market dynamics. In this situation, the data sharing system may turn out to be additional burden on employees in business apartment. Hence, related performance evaluation of business and relevant data sharing is an important means to collaborate IT and business departments. In addition, overall assessment and management of IT resources in the FEA framework can effectively reduce repeated IT project construction and incompatible structures. The five reference models of FEA framework focus on establishing a common business language to encourage cross-department communication and collaboration [10].

Third, data sharing in conglomerates requires a high-level decision-making mechanism. Data sharing is a systematic project which needs the coordination of various subsidiaries, otherwise it is difficult to be promoted and successfully implemented. The reconstruction of business segments, redesign of business process and development of data sharing cannot be completed by only IT department. Therefore, data sharing construction should be encouraged by high-level decision-making, and its implementation team is empowered to manage relevant issues across subsidiary corporations.

References

1. Dawes, S. S. (1996) Interagency information sharing: Expected benefits, manageable risks. Journal of Policy Analysis and Management, 15(3): 377–394.
2. Chua, A. (2002) The influence of social interaction on knowledge creation. Journal of Intellectual Capital, 3(4), 375–392.
3. Davenport, T. H., Short, J. E. (1998) New industrial engineering: Information technology and business process redesign. Sloan Management Review, 31(4): 11–27.
4. Gil-García, J. R., Schneider, C. A., Pardo, T. A. (2004) Effective Strategies in Justice Information Integration: A Brief Current Practices Review. In Center for Technology in Government.
5. Zhu, X. H. (2008) Main barriers in sharing the e-government information resources and countermeasures. Journal of Dongguan University of Technology, 15(6): 45–48.
6. Li, G. Q. (2021) Government data and government data integration. China Policy Review, (1): 101–104.
7. Wang, J. X., Yu, S. Y., Yang, D. L. (2011) Top level design of e-government: research on FEA methodology. E-Government, (8): 19–29.
8. Tupper, C. D. (2011) Enterprise Architecture Frameworks and Methodologies. In: J. Niles (Ed.), Data Architecture. Morgan Kaufmann, Burlington, MA. pp. 23–55.
9. Chen, Y. H. (2019) Current situation, bottleneck and breakthrough path of retail business of commercial banks. South China Finance, (7): 89–97.
10. OMB. (2006) FEA consolidated reference model document version 2. https://www.reginfo.gov/public/jsp/Utilities/FEA_CRM_v23_Final_Oct_2007_Revised.pdf

Research Hotspots and Evolution of Fusion of Blockchain and Supply Chain in China Based on Visualization Analysis

Xuedong Liang[1, 2], and Jinghong Sun[1*]

[1]Business School, Sichuan University, Chengdu, Sichuan,610000, China

[2]Sichuan Haina Rendong Science and Technology Co. Ltd.

Abstract. Based on 837 literatures of CNKI from 2015 to 2020, this paper conducts visualization analysis about domestic fusion of blockchain and supply chain with the utilization of Citespace, and reveal the research hotspots in the field of fusion of blockchain and supply chain in the past five years. The study finds that, firstly, China Finance, Southwest Finance, and Business & Economy are the journals with higher quantity of publication in this field; secondly, Song Hua and Xu Peng are the leading scholars in this research area; thirdly, the research institutions in this field are mainly concentrated in the School of Economics and Management, Law School and School of Information Management in major universities; fourthly, application scenarios, technology development, applied agent and fund are the research hotspots in this area. This study reveals the basic dynamics of domestic fusion of blockchain and supply chain research in the past five years, which not only can promote the theoretical development of this field, but also point out the research hotspots and future research direction for scholars in this area, and provide reference and reference for all walks of life who plans to enter this field.

1. Introduction

In recent years, blockchain and supply chain have been developing rapidly, but the blocked information flow, difficult transaction disputes, and the lack of trust between the parties often occur. A large number of scholars have conducted in-depth research on how to solve these problems and found that the blockchain and supply chain have a natural "dual-chain integration" mechanism that can effectively solve a series of problems. In 2016, Liu Dacheng of Tsinghua University put forward the concept of "dual-chain integration" for the first time, that is, "blockchain + supply chain". We can solve the pain points of the supply chain through the advantages of block chain and use the strengths of the supply chain to make up for the shortcomings of block chain to achieve effective fusion.

Domestic scholars have conducted in-depth research on issues related to blockchain and supply chain. In the supply chain field, Zhao Kan et al. (2017) used word frequency analysis method and Note-Express to summarize the research focus of domestic scholars in supply chain management in the past ten year[1],Xue Yanxiao (2020) used CiteSpace to analyse the domestic supply chain management research literature from 2010 to 2020, and obtained the research hotspots of the supply chain and the future trend of the supply chain research [2]. Aiming at the field of blockchain, Zhai Tong and Li Wenlan (2019) made analysis on the Chinese patent literature of blockchain technology, and obtained the research hotspots and current status of the domestic blockchain industry[3]. Regarding the dual-chain fusion field, Wang Junyu (2019) analysed the current research status of the domestic blockchain field from five aspects including finance, medical care, energy Internet, supply chain, and education[4]; Li Xudong (2019) summarized the research direction of application of blockchain in the field of integrated logistics, partial logistics and supply chain logistics [5]; Yi Qiang et al. (2020) analysed the application status of block chain technology in fields of cross-border e-commerce, supply chain, and intelligent logistics[6].

It can be seen from the above studies that domestic scholars have conducted a lot of exploration in the field of supply chain and blockchain, and have obtained rich research results. However, it can be seen from the existing literature that the current research has certain shortcomings: firstly, systematic research on the "hot spots of this field and its evolution" is lacking. This type of research can reflect the focus of dual-chain fusion research and promote the development of related theories of dual-chain fusion. Secondly, the analysis of literature data is not detailed enough. In view of this, this paper studies and analyses the domestic research literature on blockchain and supply chain from 2015 to 2020 in the "CNKI" by using the co-occurrence network analysis method.

*Corresponding author's e-mail: 2020225020090@stu.scu.edu.cn

2. Research design

2.1. Research method

Based on the mixed research method of bibliometrics, this study summarizes the past research hotspots and evolution process of domestic dual-chain fusion, and predicts the future development trend. The specific research process is as follows: first, this article analyses the core author, journal and institution in the domestic research field of dual-chain fusion based on the Citespace software, presenting the basic situation of research of this field; secondly, it analyses the domestic research literature of the domestic research based on the co-occurrence analysis method; then it further demonstrates the current domestic research subdivision areas of the dual-chain fusion, which will help us to further deepen our understanding of the structure, law and distribution of the scientific knowledge, and explores the development of the discipline knowledge field and its research hotspots, frontiers and trend.

2.2. Data collection

This study focuses on the structure and evolution of domestic dual-chain fusion research, and the sample literature comes from the "CNKI". First, we clarified that domestic research on this field started in 2015 through a preliminary search; then, we set the retrieval time as 2015~2020, "blockchain + supply chain" and "dual-chain fusion" as keywords, and conducted multiple retrieval with "topic"; additionally, we set the sources of journal as "CSSCI Journal Sources", "CSCD Journal Sources" and "Peking University Core Journal Sources", excluding irrelevant literatures. After statistics, we obtain 837 valid literatures.

3. Basic statistical analysis of domestic dual-chain fusion

3.1. Journal

Through the statistic, we find that the journals with higher publication volume in this field are: *China Finance*, *Southwest Finance*, *Journal of Commercial Economics*, *China Business and Market*, *Construction Economy*, *Computer Engineering and Applications*, and *Computer Engineering*. Among them, *China Finance*, *Southwest Finance,* and *Journal of Commercial Economics* have the highest number of publications with 10 articles each, accounting for 1.19% respectively. And we come to a conclusion that there is no centralized journal cluster in the research of this field, and the research on it is currently in a situation of interdisciplinary and diverse research topics. In addition, these journals also show different theme representations and content biases. For example, the main content of the literature on this field in *China Finance*, and *Southwest Finance* involves the application and research of blockchain in supply chain finance; *Journal of Commercial Economics* focuses more on the application mode of supply chain based on blockchain technology; *China Business and Market* is biased towards the optimization of supply chain management under the blockchain technology; the main content of *Construction Economy* is the research on the construction supply chain information management system under blockchain technology; *Computer Engineering and Applications*, and *Computer Engineering* tends to study the development of blockchain and supply chain technology as well as the supply chain traceability system based on blockchain

3.2. Author

Through the statistic of prolific authors on domestic dual-chain fusion research, we find that among the high-yield authors of domestic dual-chain research, there are 21 authors with more than 3 papers published. And it should be pointed out that Song Hua and Xu Peng are at the center of the high-yield authors, who are the leading scholars in the study of dual-chain fusion with 19 and 11 papers respectively. And we conclude that scholars in this field lack communication.

3.3. Institution

Through the statistical analysis, we find that the top 15 institutions published 125 papers, accounting for 14.93%. Among them, Renmin Business School issued 22 articles, accounting for 2.63%; Renmin University China Law School issued 14 articles, accounting for 1.67%; School of Economics and Management of Taiyuan University of Science and Technology published 11 articles, accounting for 1.31%; Antai College of Economics and Management of Shanghai Jiao Tong University and School of Management Tianjin University both issued 10 articles, accounting for 1.19%. At the same time, we draw a conclusion that the research institutions in this field are mainly concentrated in the School of Economics and Management, Law school, and Information Management School in major universities.

4. Visualization analysis of domestic research on dual-chain fusion

4.1. Visualization analysis of research hotspot identification based on keyword co-occurrence

Through statistical analysis, the results show that there are 36 keywords with frequency greater than or equal to 5. The top 20 keywords, in order, are: block chain, supply chain finance, block chain technology, funding constraints, supply chain, digital currency, intelligent contracts, financial technology, supply chain financing, online finance, decentration, artificial intelligence, the precision of poverty alleviation, big data, credit risk, the small and medium-sized enterprises, supply chain coordination, legal digital currency, regulatory sandbox, financial regulation, the corresponding frequency are:340, 185, 79, 53, 41, 39, 39, 36, 28, 25, 22, 19, 19, 17, 17, 17, 17, 16, 15, 13. Based on the above co-occurrence keywords, this paper further summarized the hot topics in the field:

supply chain finance, supply chain collaborative management, and food safety traceability. The internal logic of the above themes is as follows: supply chain finance solves the financing problems of enterprises, supply chain collaborative management highlights the comparative advantages of dual-chain fusion, food safety traceability is the main application scenario of dual-chain fusion.

4.1.1. Supply Chain Finance. Finance has always been an important application field of dual-chain fusion applications. At present, some scholars have explored the research of dual-chain fusion in financial scenarios, which can be divided into the following two aspects:

(1) Research on the application model of supply chain finance

At present, the rapid development of Internet technology provides powerful technical support and application scenarios for supply chain financial services. It is a development trend of the financial industry for Internet companies and traditional financial companies to cooperate and develop together.

Regarding the supply chain financial model, scholars have conducted research in recent years from the application model of the Internet supply chain finance era. Shi Jinzhao and Guo Jue (2015) investigated whether the banking business includes e-commerce and divided the supply chain finance into three patterns based on the bilateral interactive perspective of bank and e-commerce platform.[7] Zhuang Lei et al. (2019) analyzed the model of blockchain technology embedded in the financial industry and proposed the development path of applying block chain technology to drive financial business innovation[8]. Deng Aimin et al. (2019) conducted a coupling analysis of blockchain technology and supply chain factoring.

(2) Research on the financing model of supply chain finance[9].

There are generally four participants in supply chain financial activities, so there are four sources of funds in the financing process of supply chain finance, and these four financing modes of fund supply also have different characteristics. Specifically: financial institutions dominated by banks generally have sufficient capital reserves and more professional risk management technology; e-commerce platforms can implement precise financing for companies and monitor subsequent risks by virtue of their absolute data and information advantages; supplier participants have an absolute advantage in improving the efficiency of the supply chain; finally, because logistics provider is embedded in all links of the supply chain, logistics providers can carry out accurate value evaluation of goods.

First, financial institutions dominated by mainly banks act as capital providers. Chen Yonghui et al. (2018) designed a trade credit financing model based on the differences among retailers in the supply chain and studied the role of the differences of retailers in determining the supply chain financial model of the core enterprises[10]. Second, e-commerce platforms act as the provider, Tu Jianping et al. (2013) studied and compared four models of supply chain finance under the background of e-commerce platforms and traditional financing models, then designed an e-commerce platform performance evaluation index system [11]. Furthermore, the direct participants in the supply chain are the providers. Wang Mingzheng et al. (2017) discussed the supply chain coordination problem when retailers may default and they are in a competitive state on the background of trade credit between several retailers and one supplier[12]. Finally, logistics service providers are used as suppliers. So far, the relevant domestic research has not been retrieved.

4.1.2. Collaborative management of supply chain. In supply chain management, blockchain technology can effectively reduce the transaction cost and complete the information sharing of the whole chain members, then ensure the performance of the supply chain. At present, scholars have carried out researches in the field of collaborative management of supply chain under the background of block chain.

In terms of supply chain management, Zhu Jianming et al. (2016) studied the application of blockchain technology and designed a simplified transaction structure for each transaction subject in the supply chain[13]. In the supply chain information resources, Wang Chuanlei et al. (2017) conducted research on the coupling relationship between blockchain and supply chain logistics information resources, and designed the supply chain logistics information ecosystem model based on block chain to promote the sustainable development of supply chain logistics information resource management[14]. In the supply chain trust management, Zhao Gongmin et al. (2019) proposed a supply chain trust management mechanism that includes the blockchain contract layer, consensus layer and data layer by applying block chain technology in the supply chain trust management system[15].

4.1.3. Food safety traceability. The emergence of blockchain provides a solution for food supply chain management, some scholars have launched relevant research on this topic. Blockchain is conducive to the traceability of food safety. Wang Puqing (2019) explored the root causes of food safety problems and discussed the role of blockchain technology in food safety management[16]. In terms of the design of food traceability system, Li Mingjia et al. (2018) studied the food safety traceability system using block chain technology and designed the scheme of using block chain technology in the food traceability system by analyzing the technical framework[17]. Gao Shengqiao et al. (2020) built a dual-chain storage optimization model of food supply chain data based on blockchain and verify that the program can effectively adapt to the supply chain and optimize storage resources through simulation experiments[18].

5. Research conclusions and prospects

5.1. Main research conclusions

This paper uses Citespace software to conduct visualization analysis and data mining on 837 domestic literatures in the field of dual-chain fusion from 2015 to 2020, and analyses the research journals, authors, institutions, co-occurrence of key words and research hotspots. The main conclusions of this paper are as follows:

(1) Domestic journals in the field of dual-chain fusion have their own emphasis on theoretical basis, theme and content, empirical methods, and practical application. Among them, *China Finance*, *Southwest Finance*, and *Journal of Commercial Economics* focus on the practical application of blockchain in supply chain related fields. *China Business and Market*, and *Construction Economy* focus on empirical research. *Computer Engineering and Applications*, and *Computer Engineering* focus on the development of related theories and technologies.

(2) Scholars such as Song Hua and Xu Peng are leading scholars in the field. And domestic research in the dual-chain fusion field is in a state of high diffusion degree and low concentration degree, and there are obvious differences in disciplines between research teams, and scholars lack close contact and sufficient communication, so they are in a relatively independent research state.

(3) Research institutions in field of dual-chain fusion are mainly concentrated in the School of Economics and Management, Law School, and School of Information Management of universities. And these research institutions are relatively concentrated, but the cooperation among institutions is lacking.

(4) The research hotspots of Chinese scholars in the field in the past five years mainly focused on application scenarios, technological development, application entities, and funding. And this paper further summarizes the hot research topics in the field of dual-chain fusion: supply chain finance, supply chain collaborative management and food traceability.

5.2. Shortage of research

This paper only selects literatures from "CSSCI Journals", "CSCD Journals" and "Peking University Core Journals" as the research object, which may omit some research result. Second, due to the huge amount of data, this paper can only study and analyse from the macro and overall perspective, and it is difficult to analyse more detailed fields. Therefore, in the following research, we will continue to optimize data sources and data screening standards to further improve the quality of research.

5.3. Future research outlook

Double-chain fusion has been extended to finance, security, product traceability and other aspects, and has a huge impact on the supply chain. Therefore, hot issues in this field still need to be further explored and analysed.

(1) The composition of the supply chain contains different subjects. In the continuous exchange and cooperation between these subjects, the trust problem is inevitable. However, the current domestic research on the trust problem of the supply chain is relatively lacking. In the future, we can explore the integration of blockchain technology with other technology to improve the transparency of companies in the supply chain through the immutability of blockchain, and deal with the lack of trust in the supply chain at the technical level.

(2) We can analyse the solution to the problem of low efficiency of core enterprises in the field of supply chain finance based on block chain technology, so as to provide a strong guarantee for the interests of enterprises in the supply chain.

(3) The current application scenarios of blockchain technology are mainly food, agricultural products, and fresh supply chain fields. In the future, blockchain can be applied to more scenarios, such as supply chain traceability systems in other scenarios, including drug traceability, steel traceability, wood Traceability, traceability of automobile parts and so on.

(4) Double-chain integration can effectively solve the problem of information opacity in the supply chain, but at the same time it will also increase the risk of some fraud and crime. In the future, based on the characteristics of blockchain, a regulatory system can be established in the supply chain companies to supervise the supply chain systematically and comprehensively, and effectively avoid the risk of fraud and crime.

Funding Resources

This paper was supported by the Department of Science and Technology of Sichuan Province Fund (2021JDRC0002).

References

1. Zhao K et al. (2017) A Review of Supply Chain Management in China in Recent Ten Years-Based on Word Frequency Analysis (in Chinese). Modern Business Trade Industry, 30: 37-39.
2. Xue Y X. (2020) Hotspots and Trend of Supply Chain Management Based on CitesSpace (in Chinese). Logistics Engineering and Management, 42: 102-105.
3. Zhai T, Li W L. (2019) Research Focus and Current Situation of Blockchain Industry in China- Based on Analysis of Chinese Patent Documents (in Chinese). China Invention & Patent, 16: 49-54.
4. Wang J Y, Wu Q L, Cao H Y. (2019) Overview of Research on Typical Application of Domestic Block Chain (in Chinese). Science & Technology and Economy, 32: 1-6.
5. Li X D. (2019) The Research Review of "Block-chain + Logistics" (in Chinese). Journal of Guangdong Communication Polytechnic, 18: 42-47.
6. Yi Q, Yang H M, Wang K X. (2020) Review of Researches on Blockchain Technology Application in Logistics Field (in Chinese). Logistics Technology, 39: 129-133.
7. Shi J Z, Guo J E. (2015) Study on the Development and Domestic Practice of Supply Chain Finance from the Perspective of the Internet (in Chinese). Journal of Xi'an Jiaotong University (Social Sciences), 35: 10-16.

https://doi.org/10.1051/e3sconf/202129002032

8. Zhuang L, Zhao Y F. (2019) Research on application mode and development path of block chain technology (in Chinese). Journal of Finance and Economics, 40: 33-38.
9. Deng A M, Li Y F. (2019) "Intelligent Factoring" Business Model and Game Analysis in the Supply Chain Based on Block Chain (in Chinese). Management Review, 31: 231-240.
10. Chen Y H, Meng Z L, Zeng Y. (2018) Trade Credit Loan Pricing and Supply Chain Finance Pattern Choice Based on Retailers' Heterogeneity (in Chinese). Systems Engineering-Theory & Practice, 38: 2479-2490.
11. Tu J P, Yang X. (2013) Research on Performance Evaluation of Supply Chain Financing Model Based on E-commerce Platform (in Chinese). Management World, 29: 182-183.
12. Wang M Z, Zhou L, Liu W W. (2017) Revenue-sharing Trade Credit Contract for Coordinating Multiple Competing Retailers When Incorporating Default Risk (in Chinese). Operations Research and Management Science, 26: 1-11.
13. Zhu J M, Fu Y G. (2016) Supply Chain Dynamic Multi-center Coordination Authentication Model Based on Block Chain (in Chinese). Chinese Journal of Network and Information Security, 2: 00019-1-00019-7.
14. Wang C L et al. (2017) A Model of Logistics Information Ecosphere of Supply Chain Based on Block Chain (in Chinese). Information Studies: Theory & Application, 40: 115-121.
15. Zhao G M et al. (2019) Study on the Trust Management Mechanism of Supply Chain Based on Blockchain (in Chinese). Credit Reference, 37: 25-31.
16. Wang P Q et al. (2019) Application of Blockchain Technology in Food Safety Management (in Chinese). Journal of Agrotechnical Economics 38: 82-90.
17. Li M J. (2019) Food Safety Tracing Technology Based on Block Chain (in Chinese). Food Science, 40: 279-285.
18. Gao S Q, Liu X L, Gao Y P. (2020) Optimized Model of Dual-chain Storage of Food Supply Chain Data Based on Blockchain (in Chinese). Food & Machinery, 36: 63-74.

Research on the Application of Virtual Reality Technology in Psychotherapy

Lulu Liu [1], Yan Liu,*

1 School of Humanities, Jiangxi University of Traditional Chinese Medicine, Nanchang, Jiangxi, China

Abstract. The popularization and application of virtual reality technology provides a new way for psychotherapy. This paper expounds the advantages of virtual reality technology and its specific application in the treatment of psychological diseases such as social anxiety, phobia and eating disorder. At the same time, aiming at the psychological pressure problem faced by modern people, a set of psychological adjustment system for improving emotional experience and virtual scene is designed based on behaviourism-oriented psychotherapy theory and virtual reality technology.

1 Introduction

Psychotherapy, in a broad sense, refers to any technique and measure that can slow down or treat psychological, emotional or behavioural disorders. Traditional psychotherapy techniques, such as psychoanalytic therapy, behavioural therapy, humanistic psychotherapy, cognitive therapy, etc., play an important role in psychological counseling and treatment and have achieved fruitful results. However, the traditional psychotherapy technology inevitably has some limitations, such as: the treatment environment is single and it is difficult to restore the scene, relying too much on visitors to dictate their memories and experiences, which may lack authenticity. The development and application of virtual reality can make up for some shortcomings of traditional psychotherapy [1].

Burdea, a pioneer in the field of virtual reality, believes that "virtual reality is a synthetic computer user interface that simulates reality through a variety of perceptual channels, such as sight, hearing, touch, smell, and so on." others think that virtual reality is a virtual three-dimensional space generated by computer simulation, which enables users to interact with the virtual world immersively and triggers a series of events [2]. This technology has gradually developed from the initial stage of the lack of human-computer interaction to the flourishing period of commercialization and systematization. More and more people are aware of the broad prospects of this technology and apply it to different industries and fields, one of them includes the medical profession. Many researchers at home and abroad have applied this technology to the field of psychotherapy and evaluation, which undoubtedly brings convenience and good news to psychotherapy.

2 Application Advantages of Virtual Reality Technology in Psychotherapy

Virtual Reality technology is a new technology developed in recent 20 years, which mainly uses high-tech equipment such as computer, helmet display, data glove and so on to build a real-time interactive 3D virtual world for users. It is immersive, interactive and imaginative. When it is applied to the field of psychotherapy, it shows some advantages [3] as follows:

2.1. Security

Virtual reality technology can simulate dangerous situations and significantly improve treatment conditions, and effectively ensure the safety of patients. Different from traditional therapy, virtual reality provides a safe environment for visitors to experience and explore boldly. In the course of treatment, visitors can explore and improve their skills in various mistakes, difficulties and even unpredictable events without any harm.

2.2 Controllability

Virtual reality is also controllable. It can not only design the treatment process and create a personalized treatment environment according to the actual situation of visitors and the degree of psychological barriers, but also better understand the characteristics of patients in memory, attention, imagination, emotional response, etc., so as to correctly diagnose its etiology and formulate corresponding treatment plans.

2.3 Realistic Sense of Presence

Virtual reality technology can provide visitors with

* Corresponding author: 349620816@qq.com

https://doi.org/10.1051/e3sconf/202129002033

various intuitive sensory stimuli. In the virtual environment, they can see various scenes, hear different sounds, touch different objects and smell different tastes, thus intuitively feeling the environment and events. Realistic presence is conducive to the presentation of psychological problems of visitors and the improvement of treatment effect.

2.4 Experiential Nature

In the virtual scene, visitors can not only experience various situations in the real society, but also cross the boundaries of time and space, and experience scenes that can't appear in the real society, such as experiencing war or disaster, so as to gain wonderful insights and change personal cognition

3 Application of Virtual Reality Technology in Psychotherapy

3.1 Treatment of Social Anxiety Disorder

Virtual reality is used to make specific virtual scenarios and expose patients to them, thus causing anxiety and achieving the purpose of treating anxiety disorder. Slater et al. first applied VR technology to the study of social anxiety. They designed a virtual social scene, established a virtual character to engage in social activities with patients, and the patients were placed in the virtual social scene for treatment. In the virtual environment, the anxiety situation of patients is gradually exposed, and the anxiety threshold of patients is gradually raised, thus improving the symptoms of patients.

3.2 Treatment of Phobia

In the treatment of phobia, VR technology is mainly used to create a virtual situation that makes patients feel fear, and put patients in this virtual situation to implement exposure therapy [4]. In addition to using exposure therapy to treat acrophobia in virtual scenes, virtual reality technology can also be used to treat mental diseases such as flight phobia, claustrophobia and agoraphobia. Compared with the real situation, the use of virtual situation in the treatment process can prevent patients from appearing too strong fear reaction and ensure the personal safety of patients at the same time.

3.3 Treatment of Emotional Distress

In the form of combining VR technology with art therapy, the virtual reality helmet and the brush matched with the helmet are used to create, so that people can freely enter the "Painting world" to create art, and let patients release their anxiety, ease emotional conflicts and clarify past experiences through creation. At the same time, in the process of transforming ideas into concrete images, personal needs and emotions are conveyed, so that their personality can be adjusted and treated.

3.4 Treatment of Eating Disorders

Eating disorder is a mental disorder with abnormal eating behaviour. Most eating disorders are due to the misunderstanding of one's own body image. Riva [5] and others first combined VR technology with cognitive behavioural therapy to form experiential cognitive therapy, which was applied to the treatment of eating disorders. Their treatment mainly uses VR technology to create five virtual and distorted realistic environments and combines Socrates' questioning method in cognitive therapy to help patients have a more realistic cognition of their own body image. Using VR technology, researchers can not only correct patients' wrong body image cognition, but also deal with patients' evasive behaviour and negative emotions caused by food.

3.5 Treatment of Schizophrenia

The treatment of schizophrenia with VR technology is the research direction only in recent years, and the research results are still few, and it is still in the experimental stage of development. However, foreign researchers have used VR technology to treat schizophrenia patients with positive symptoms, and found that treatment can effectively alleviate the hallucinations and delusions of schizophrenia patients. In this way, VR technology can be used to make schizophrenic patients simulate the hallucinations in their minds and create an illusion avatar. Then let the patients face this illusion incarnation, and make them realize that their illusion is untrue and unreasonable, which is caused by morbid psychology. For those who can't get a radical cure, this method can help them adapt to their own hallucinations and improve their social function. The following is a sketch of the application of VR technology in specific psychological problems.

In the application of VR technology, in the face of different parties, we can flexibly use the theory and technology of psychotherapy schools. Behavioral therapy can use virtual situations to activate the psychological disorders of the parties, cognitive therapy can use VR technology to evaluate situational memory or disintegrate habitual selective attention patterns, psychoanalytic school therapy can use it as a complex symbol system to evoke or release emotions, expressive art therapy can use VR technology to build an artistic expression platform to ease emotional conflicts. Although VR technology can be widely used in the treatment of mental illness, it still needs the correct guidance of therapists.

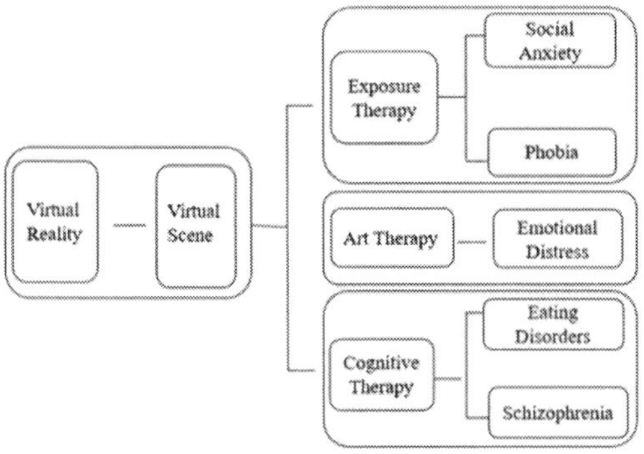

Fig 1. Application of VR technology in psychotherapy

4. Design of Virtual Visual System for Psychotherapy Process

4.1 Design Ideas

With the development of society, people's pace of life is accelerating, the competition for survival is becoming increasingly fierce, and interpersonal relationships are becoming more complex. More and more people feel strong psychological pressure, frequently appear negative emotions such as depression, anxiety, depression, sadness, and even produce psychological distress and psychological obstacles. This design is based on behaviorist relaxation therapy, combined with meditation technology, using virtual reality technology to build a three-dimensional virtual scene system suitable for meditation and relaxation. The system can effectively attract the attention of visitors, make them focus their attention on their feelings, experience a sense of relaxation, completely immerse themselves in meditation, relieve their inner troubles, release their emotions, and achieve the role of psychological intervention and adjustment[6]. Although the scene is virtual, the experience it gives people is real. The system can effectively attract the attention of visitors, make them pay attention to listen to their own physical feelings, experience a sense of relaxation, completely immerse themselves in meditation, relieve their inner troubles, release their emotions, and achieve the role of psychological intervention and adjustment.

4.2 Frame Composition

The system consists of two parts: software and hardware. the hardware part includes display stereo vision equipment helmet display, digital equipment, data gloves and input equipment keyboard and mouse. The hardware is connected with the software system through the computer interface. The software package consists of five parts, which is composed of environment, special effects, sound, prompts and system detection. It mainly provides three types of services, namely, mindfulness meditation, natural scenery and music relaxation[7]. The system framework is shown in figure 2. In a word, visual environment, auditory environment, active treatment experience and psychological adjustment hints work together to form a psychological regulation system suitable for relaxation and meditation therapy. Hardware data gloves, mouse keyboard and virtual reality helmet are used to realize the experience of virtual reality psychotherapy system, using sensing devices to achieve continuous man-machine information updates, see stereoscopic pictures, and hear simulated sounds. Achieve an immersive relaxation experience and get a complete three-dimensional relaxation meditation therapy experience.

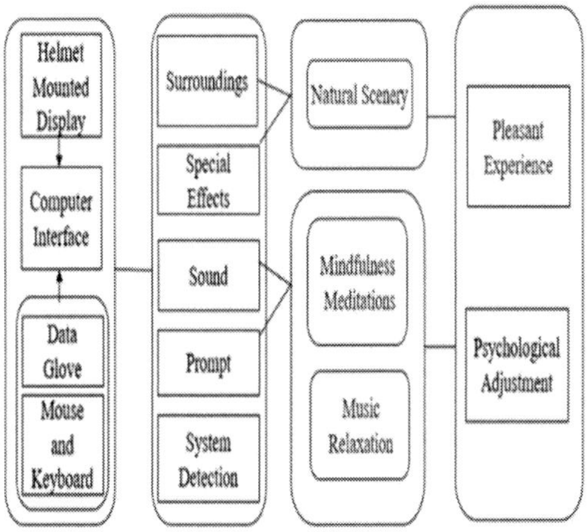

Fig 2. Design of VR psychological adjustment system

4.3Treatment Process

The visitor sits on a very comfortable chair, sees the virtual scene directly through the virtual vision device, interacts with the scene through the interactive device data glove, mouse and keyboard, roams the virtual venue freely from his own point of view, browses, experiences and listens, and fully enjoys the beauty and interest. Moreover, the emotional responses and cognitive changes that occur in this process can be transferred to real life. If you have virtual reality equipment, you can experience extraordinary things at home, even if you don't leave the house. Virtual life, the use of computers and immersive three-dimensional helmet equipment, get immersive relaxation experience and meditation experience, release and relieve the psychological pressure from work, study, life and other aspects, make the mind and body happy and relaxed.

5. Conclusion

With the free ride of Internet technology, the development of mental health service industry is also changing with each passing day. Virtual reality technology has the advantages of immersion, interaction and imagination. The application of virtual reality technology in psychotherapy can not only improve the treatment conditions, ensure the safety and privacy of patients, but also facilitate therapists to manipulate the treatment process, and greatly improve the efficiency of treatment. However, the virtual reality system is only a standardized program, and its flexibility and pertinence are poor, so it can not be used for every visitor. And psychological counseling or therapy emphasizes the relationship between counselors and visitors. VR technology is only an auxiliary means but can not replace the role of counselors. In addition, it takes a long time to build a new virtual environment, and requires a high cost of research and development, so it can not be used for special phobia or some rare mental diseases. VR equipment is also expensive and needs regular maintenance, which will increase the cost of treatment for patients.

This paper mainly introduces the application and development trend of virtual reality technology abroad from the point of view of clinical practice, and on the basis of the full application of VR technology and the advantages and disadvantages in the field of psychology, this paper designs a virtual scene system for healthy people, which can be used in the process of psychotherapy, which can help visitors release negative emotions and relieve all kinds of stress, so as to realize the beauty of life with a healthy state of mind.

References

1. Wu Hao. (2016). VR/AR technology will break through human psychological perception and bring the ultimate image realistic experience. News Research Guide (19),335.
2. Ni Bin. (2018). Application of Virtual Reality Technology and VR Equipment in Psychological Intervention Therapy. Computer Products and Circulation (02),168-169.
3. Li Jian. (2017). Research on the Application of VR Virtual Reality Technology in the Field of Psychological Rehabilitation.
4. Xiao Haichuan,Liao Ying,Che Jingshang. (2019). New progress in VR virtual reality technology and treatment of acrophobia. Electronic Technology and Software Engineering (02),119-121
5. Zhang Xiaoyu. (2017). Application of Virtual Reality Technology in Medicine. Technology and Innovation (03),161.
6. Fan Dongmei. (2019). Design and Implementation of Psychological Relaxation System Based on VR Technology. Overseas Chinese Garden (07),124.
7. Gu Chuhan. (2019). Application of Virtual Reality Technology in Medicine. Science and Technology Communication (06),129-130.

A Study on the Emotional Analysis of Abandoned Surrogacy Events Based on Text Mining

Guanlan Liang[1], Xunbing Shen[1,*]

[1]Jiangxi University of Chinese Medicine, Nanchang, Jiangxi, China

Abstract. In late January 2021, news that actress Zheng Shuang had surrogacy abroad and had wanted to give up her children sparked a public outcry. This paper takes Zheng Shuang's comments on the topic of surrogacy and her abandonment as the research object. Firstly, the web crawler technology is used to grab and mine the comment text, and then the ROSTCM software is used to analyze the text data to explore the comment topics of Weibo network users after the abandonment event and the analysis of their emotional tendencies to the event.

1 Introduction

According to the 47th Statistical Report on Internet Development in China released by China Internet Network Information Center (CNNIC) in February 2021, as of December 2020, the number of Internet users in China has reached 989 million, with the Internet penetration rate reaching 70.4%. More and more users choose to access news or hot events through the Internet in the first place. On January 18, 2021, news surfaced that actress Zheng Shuang had given birth to two children by surrogacy abroad and then wanted to give them up to be raised abroad. In a very short period of time, Weibo trending topics such as "Zheng Shuang abandoned her surrogate mother" and "China bans surrogacy in any form" topped the charts, garnered more than 5 billion views.

As the main body of this incident is a public figure, and involves surrogacy laws and regulations and special social ethical issues, it has aroused heated discussion among netizens. Therefore, this paper analyzes netizens' Weibo comments on this event, explores the topics of Weibo users' comments, and further understands the changes of netizens' emotional positioning.

2 Literature review

At present, there are two main methods of text sentiment analysis, one is based on sentiment dictionary, and the other is based on machine learning.

With the help of machine learning, Pang et al. used Naive Bayes (NB), Maximum Entropy (ME) and Support Vector Machine (SVM) to conduct sentiment analysis[1]. Yang et al. divided comments on Twitter into three categories: positive, neutral and negative[2]. Lin et al. uses a Bi LSTM and Attention model to reduce the dependence of sentiment analysis methods on humans[3]. Li Lei et al. also used this model and compared it with the analysis effect based on deep learning[4]. Luo et al. propose to merging naive Bayes and causal rule for small sample data sentiment analysis scenarios[5]. S. Kumano and K. Nomura proposed a multitask RS removal method, which was an extension of the method in the field of emotional computing[6]. K. Chawla et al. proposed techniques to incorporate affect lexica, which captured fine-grained information about a word's psycholinguistic and emotion orientation, into the training for Word2Vec SkipGram, Word2Vec CBOW, and GloVe using a Joint Learning approach[7]. There is also a wealth of research based on sentiment dictionaries. Liu et al. used K-means clustering algorithm to construct sentiment dictionary and further carried out sentiment analysis on online product reviews[8]. Mohamad and Moattar proposed a new approach that combines neural networks with emotion dictionaries[9]. Felipe et al. added the variable of time evolution[10].

Due to the particularity and diversity of Chinese grammar, the research on the emotional tendency of text appears relatively late in China, and the research mainly relies on the Web platform. In the early stage, Zhou Shengchen et al. summarized the research on Chinese microblog sentiment analysis and put forward the need to focus on the development of relevant dictionaries or corpora for the filtering and emotion mining of network language[11]. At present, through the literature review of the past ten years, Zhong Jiawa et al. found that there are mainly three text sentiment analysis methods, which are based on sentiment dictionary and rules, based on traditional machine learning and based on deep learning, and all of them have advantages and disadvantages. At the same time, the method based on multi-strategy mixing gradually becomes an important direction for improvement[12]. For example, Song Yan took Weibo topics as an empirical case to study and verify the fine-grained viewpoint mining process and system functional modules, which further extended the traditional fine-

* Corresponding author: 20060923@jxutcm.edu.cn

grained viewpoint mining model[13]. Fan Hao and Li Pengfei proposed a model based on word vector and two-way GRU cyclic neural network, which improved the accuracy of sentiment classification in networked short texts[14]. Du Yixian et al. used HowNet sentiment dictionary and sentiment vocabulary ontology to conduct sentiment analysis of public opinion in the COVID-19 epidemic[15]. Zhou Ziyu and Liu Fang'ai proposed a novel neural network, named Filter Gate Network based on Multi-head attention (FGNMH) to solve the problem that the attention mechanism in ASC brings noise and captures contextual words that are irrelevant to the current aspect[16].

To sum up, the research at home and abroad mainly focuses on the theoretical and technical exploration of data mining technology, and there are few empirical research results based on typical cases. At the same time, through sorting out the research process of the emotional tendency of text, it can be found that there is no most suitable method for the identification of the emotional tendency of Chinese microblog comment text. The current mainstream method is to use the emotion dictionary combined with Chinese grammar rules to score. Therefore, this paper uses text data mining technology to capture Weibo comments related to Zheng Shuang's surrogacy abandonment, and then uses ROSTCM to conduct emotional orientation analysis, so as to perceive the impact of this event on the public.

3 Sentiment analysis of microblog comments

The basic research idea of the article is to first use a web crawler to collect user comment information on topics related to "Zheng Shuang surrogacy and abandonment". After pre-processing such as cleaning the collected data, word frequency is counted and word cloud map is drawn to facilitate intuitive analysis of hot and concerned words in the later stage. Then perform sentiment analysis on the topic comment information, judge the positive and negative sentiment ratio of the topic, and conduct a more specific analysis based on this.

3.1 Text Mining and Pre-processing

Text mining refers to the discovery of hidden patterns in large text collections. Web text mining technology includes text capture, text collection, text preprocessing, text analysis and other processes. This article uses Gooseeker web crawler software to capture microblog comments on related microblog topics. The software can capture effective related data such as blog content, comments, user ID, and save the captured data as Excel, which is conducive to later data analysis. Fig. 1 shows the process of mining the comment text of Zheng Shuang's surrogacy and abandonment.

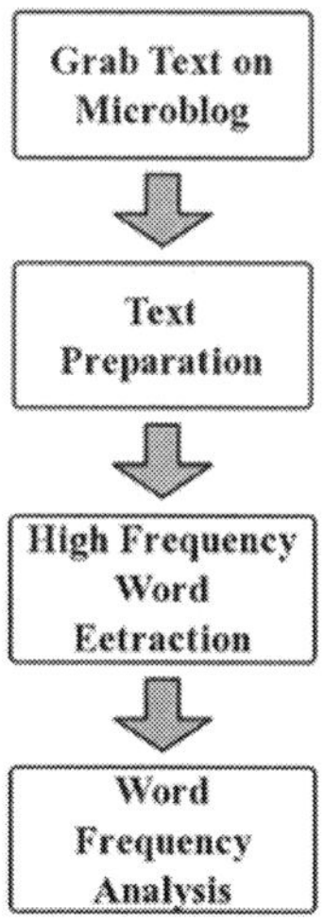

Fig. 1. Flow chart of microblog comment text mining.

3.2 Word Frequency Statistics

This paper searched the keyword "Zheng Shuang's Surrogacy Abandoned", and obtained a total of 814 comments on blog posts from January 18 to 22, 2021.

The ROST Content Mining software is used for word segmentation of text. In the process of word segmentation, we need to make custom word segmentation for the obtained text. The words involved in this study, such as "Zheng Shuang", "Zhang Heng" and "surrogacy", need to be added to the "custom word segmentation filter word list", and then carry out word frequency statistical analysis. The first 27 high-frequency words are sorted out as shown in Table 1 below.

Table 1. Some words and word frequency statistical table.

Words	Num.	Words	Num.	Words	Num.
Ban	104	Sentencing	31	Recording	23
Moral	103	Speak up	30	Law	23
Children	96	Artist	29	Cheating	22
Bottom line	81	Woman	29	Parents	22
China Television Artists Association	79	China	28	Man	22
Beyond	77	Entertain-ment	27	Oppose	21
Female	67	Country	26	Strongly	21
Surrogacy Agency	59	Star	25	Prohibit	17
Fans	31	Disgusting	23	Public Opinion	15

According to the word frequency statistics, in the topic of "Zheng Shuang's surrogacy and abandonment", Weibo users are more concerned about whether the actress will be banned, the moral issues of surrogacy and

https://doi.org/10.1051/e3sconf/202129002034

abandonment, the future of the children, and the rights of women and men. The two words with the highest frequency are "Ban" and "Moral", which shows the attitude of Internet users. In addition, from the words "Disgusting", "Oppose" and "Prohibit", we can initially perceive that most netizens maintain a negative attitude towards this matter, accompanied by anger and boredom.

3.3 Text Sentiment Analysis

ROSTCM has a custom word list, emotion calculation, positive emotion, negative emotion, neutral emotion classification calculation and other functions.

This software was used to conduct emotional analysis on the text data of "Zheng Shuang's surrogacy and abandonment", which was divided into three categories of positive emotions, negative emotions and neutral emotions. Statistical results of emotional distribution were obtained (Table 2), and the results were plotted into a pie chart (Fig. 2).

Table 2. A table of sentiment analysis on Weibo texts of surrogacy and abandonment incidents.

	Affective Tendency Value	Number	Percentage
Positive Emotions	(5 , +∞)	366	44.96%
Neutral Emotions	[5 , 5]	19	2.34%
Negative Emotions	(-∞ , 5)	429	52.70%
Segmental Statistical Results of Positive Emotions			
Low	(5 , 15]	220	60.11%
Medium	(15 , 25]	98	26.78%
High	(25 , +∞)	48	13.11%
Segmental Statistical Results of Negative Emotions			
Low	[-15 , 5)	383	89.28%
Medium	[-25 , -15)	32	7.46%
High	(-∞ , -25)	14	3.26%

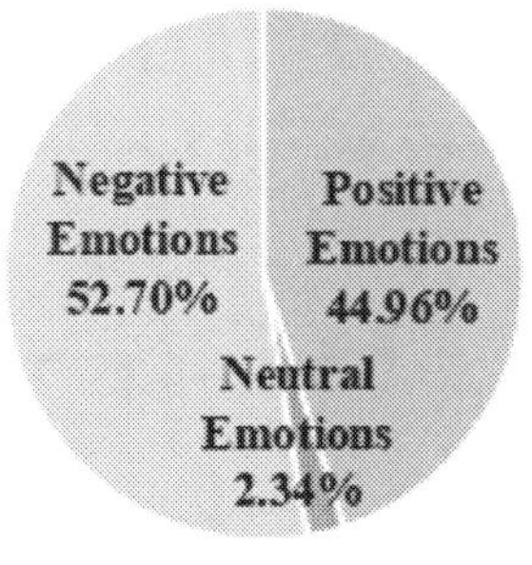

Fig. 2. A figure of Sentiment analysis on surrogate pregnancy abandonment event.

According to the rules of text emotion calculation, 814 microblog texts were calculated and counted, and then their polarity was judged. In this event, there were 429 microblogs with negative emotions, accounting for 52.70% of the total, which is more than the sum of the microblogs with positive emotions (366) and intermediate emotions (19). It can be seen that netizens' emotional tendency towards this event is more negative, and this event also has a certain negative impact on the society. By observing the data of negative emotions and combining with the analysis of Weibo users' historical information, it is found that most Weibo users with negative comments are female. On the one hand, they are emotional, on the other hand, they are familiar with laws and secular common sense. Therefore, it is only natural for people to express their condemnation of abandoning their children and their disdain for surrogacy. However, it should also be noted that there are a lot of comments with positive emotions. In addition to fans' support for their idols, most of them are comments like "strongly support the punishment of both parties in surrogacy transaction". Especially after the official media made their comments in the later stage, positive comments on microblogs dominated. It shows the netizens' approval of seriously dealing with the behavior of surrogacy and hope that the stars can play a positive guidance.

4 Summary

Starting from the event of "Zheng Shuang's surrogacy abandoning", this paper, based on web crawler technology, conducts word frequency statistics and judgment of emotional polarity intensity on relevant real-time microblog comment texts.

According to the statistics of high-frequency words, netizens have launched heated discussions on this topic, including Zheng Shuang was banned from broadcasting and television, whether surrogacy should be punished, the bottom line of morality, surrogacy agencies and so on. It can be seen that this surrogacy abandonment event is a social event involving a series of issues such as class, ethics and gender. Most netizens stand in opposition to surrogacy, condemn the abandonment of care, denounce the standpoint of surrogacy agencies. A small number of netizens believed that it was not all the woman's fault, and that the man was also to blame, and felt distressed and understanding. In addition, from the China Television Artists Association said Zheng Shuang beyond the moral bottom line and other official media can be seen in their opposition to surrogacy abandoned attitude.

Through the analysis of emotional tendency and the extraction of polarity intensity, it can be seen that the majority of netizens' attitudes are negative, which is far more than positive emotions. This is understandable, after all, surrogacy is banned in China. Abandoning is against social morality and public order good customs. Some positive emotions, including fans' affection for their idols and netizens' approval of the official ban, are not always in favor of surrogacy. Most netizen masses are not willing, do not accept, also do not allow this kind of surrogacy to abandon the behavior.

This incident, on the one hand, exposed the huge risks and ethical crisis of surrogacy, on the other hand, it also pushed the dialectical needs of surrogacy to the stage again. There is an objective market demand for surrogacy, such as infertility and sexual minorities. If surrogacy were to be banned altogether, it would likely drift further underground and into darker corners, even abroad. But both official and public opinion are firmly opposed to commercial surrogacy. In a word, the management system of surrogacy needs to be further improved.

https://doi.org/10.1051/e3sconf/202129002034

Acknowledgments

This study was partially supported by the grants from the Planed Project of Social Sciences in Jiangxi Province (No. 18JY24) and the project of "1050 Young top-notch talent" of Jiangxi University of Traditional Chinese Medicine.

References

1. Pang B, Lee L, Vaithyanathan S, et al. (2002) Thumbs up: Sentiment Classification using Machine Learning Techniques. Empirical Methods in Natural Language Processing, 79-86.
2. Yang M, Tu W, Wang J, et al. (2017) Attention Based LSTM for Target Dependent Sentiment Classification. National Conference on Artificial Intelligence, 5013-5014.
3. Lin Xingmin, Ho Chunheng, Xia Luting, Zhao Ruoyi. (2021) Sentiment analysis of low-carbon travel APP user comments based on deep learning. Sustainable Energy Technologies and Assessments, 44.
4. Li Lei, Wu Xuhui, Liu Ji. (2021) Sentiment Analysis Model of Bi-LSTM with Key Opinion Target Recognition and Deeper Selfattention. Journal of Chinese Computer Systems, 42(03): 504-509.
5. Luo Yongjian, Yang Xiaohua, Ouyang Chunping, et al. (2021) Merging Naive Bayes and Causal Rules for Text Sentiment Analysis. Journal of Physics: Conference Series, 1757(1).
6. S. Kumano, K. Nomura. (2019) Multitask Item Response Models for Response Bias Removal from Affective Ratings. 2019 8th International Conference on Affective Computing and Intelligent Interaction (ACII), Cambridge, UK, 1-7.
7. K. Chawla, S. Khosla, N. Chhaya, K. Jaidka. (2019) Pre-trained Affective Word Representations. 2019 8th International Conference on Affective Computing and Intelligent Interaction (ACII), Cambridge, UK, 1-7.
8. Liu F, Wei F, Yu K, et al. (2017) Sentiment classification of reviews on automobile website by combining word2vec and dependency parsing. Proceedings of the International Conference on Smart Computing and Communication, 206-221.
9. Mohamad Beigi Omid, Moattar Mohammad H.. (2020) Automatic construction of domain-specific sentiment lexicon for unsupervised domain adaptation and sentiment classification. Knowledge-Based Systems, 213.
10. Felipe Bravo-Marquez, Arun Khanchandani, Bernhard Pfahringer. (2021) Incremental Word Vectors for Time-Evolving Sentiment Lexicon Induction. Cognitive Computation (prepublish).
11. Zhou Shengchen, Qu Wenting, Shi Yingzi, et al. (2013) Summary of Chinese Microblog Sentiment Analysis. Computer Applications and Software, 30(03): 161-164+181.
12. Zhong Jiawa, Liu Wei, Wang Sili, et al. (2021) text sentiment analysis methods and application were reviewed. The data analysis and knowledge discovery,1-15.
13. Song Yan. (2020) Research on Multi-level Fine-grained Attribute Mining of Social Media Text Information. Information Science, 38(11): 98-103.
14. Fan Hao, Li Pengfei. (2021) Short Text Sentiment Analysis Based on FastText Word Vector and Bidirectional GRU Cyclic Neural Network -- A Case Study of Microblog Comment Text. Information Science, 39(04): 15-22.
15. Du Yixian, Xu Jiapeng, Zhong Linying, et al. (2021) Multi-dimensional Feature Analysis and Visualization of Online Public Opinion Situation and Sentiment: A Case Study of COVID-19 Epidemics. Journal of Geo-information Science, 23(02): 318-330.
16. Zhou Ziyu, Liu Fang'ai. (2021) Filter gate network based on multi-head attention for aspect-level sentiment classification. Neurocomputing, 441: 214-225.

Innovation of Traditional Creation Wisdom in Intelligent Age

Zonghua Zhu[1]

[1]Jingchu University of Technology, Jingmen, Hubei, 448001, China

Abstract. In recent years, with the idea of carrying forward traditional culture gradually gaining popularity, people pay more attention to the traditional creation wisdom contained in traditional artifacts, and on this basis, actively explore more innovative design methods, so as to realize the organic combination of traditional creation wisdom and intelligent era, so that traditional creation wisdom can be better inherited and carried forward. Based on this, this paper discusses the traditional creation wisdom, combines the development and improvement of modern times, takes the design of artistic furniture as a case, the paper analyzes how to combine and innovate the traditional creation wisdom in the product design under the current intelligent era by the combination of traditional creation wisdom and the emerging "Unity 3D" technology, in order to better inherit and carry forward the traditional creation wisdom.

1 Introduction

The traditional creation wisdom is an important part of China's excellent traditional culture, which contains profound ideological essence and value essence. These ideological values still have a certain degree of advancement in modern society, and they must also have important application value in modern society, so we still need to make further in-depth research on it. Of course, because the background of these creative wisdom has changed significantly, in order to better promote its application in contemporary life and make traditional creative wisdom deeply integrate with modern society, we need to think about how to innovate traditional creative wisdom based on the background of intelligent era, and truly realize "making the past serve the present", so that the wisdom contained in traditional creative wisdom can really inspire modern innovative design, promote China's leading role in the design field, and then make our traditional creative wisdom..

2 Overview of the Traditional Creations Wisdom

2.1 Connotation of Traditional Creation Wisdom

The connotation of traditional creation wisdom is divided into two levels: surface and deep. From the surface dimension, the generation of traditional creation wisdom can be said to be a historical necessity, which is based on people's basic living needs[1]. In the development of history, the wisdom of the working people has created a large number of traditional instruments, which generally have high practical value and play a great role in promoting productive labor. Behind the scenes, each instrument has its own unique background story, which records the birth and development of civilization[2].

With the gradual development of traditional culture, traditional creation wisdom has gradually broken through the category of original function and gradually developed towards the dimension of deep connotation. It not only focuses on basic practical value, but also usually explains the layout structure of objects with unique creativity and pursues natural design goals, which contains more historical culture behind it. Therefore, it is necessary to carefully study the traditional creation wisdom, with the emphasis on tracing back to the source, making an in-depth study on it from the level of history and culture, understanding the value of these traditional creation wisdom in essence, and comparing it with modern design to realize the inheritance and innovation of traditional wisdom of creation [3].

2.2 Characteristics of the Traditional Creation Wisdom

2.2.1 Make the Best Use of Everything. Making the best use of everything can be said to be the primary feature of the traditional creation wisdom. Since the origin of traditional creation, this feature has run through the whole process of traditional creation. It is mainly based on the following two reasons: first, the low level of productivity in ancient times; second, people summed up a lot of natural laws in their daily production and life. The characteristics of making the best use of everything were reflected in ancient China. In some agricultural civilization sites in China, archaeologists discovered the accumulation layer and carbonization layer of rice husk, which proved that rice husk, a by-product of rice threshing, had been fully utilized at that time, and there was no waste

1910012008@stu.jci.edu.cn

of materials. It can be seen that the ancients have made full use of all kinds of materials donated by nature in the unfavorable environment with extremely limited conditions, and minimized the waste of materials, which is quite similar to the concept of "zero waste" in modern society. Obviously, this traditional creation wisdom is also of great value to modern society [4].

2.2.2 Function First. In the traditional concept of creation, there are not too many requirements for form and structure, which are usually more concise, but more emphasis on the actual function of objects. For a long time, the traditional way of creation in our country has been influenced by this concept, and has continued to this day, becoming a design concept which is obvious and excellent in the wisdom of traditional creation and worth inheriting[5].

2.2.3 Tibetan Ritual. Since ancient times, China has enjoyed the reputation of "an ancient civilization" and "a country of etiquette", which is also related to many etiquette systems in ancient China. China's traditional etiquette system is also based on various traditional etiquette objects. Such as bronzes, sacrificial utensils and court display utensils in Shang and Zhou Dynasties. Generally speaking, ceremonial objects are generally characterized by majesty and solemnity. These ceremonial objects have been given more meaning of "ceremony" in the development of etiquette system. With the development of the times, the cultural value behind these ceremonial objects has been further sublimated, reflecting people's morality and ideals, as well as the rules and order of the real world.

3 The Innovative Strategy of Chinese Traditional Creation Wisdom in the Intelligent Era——Taking Art Furniture Design as an Example

In recent years, with the further development of the Internet, people have entered an intelligent era, and artificial intelligence technology has been able to assist human decision-making. Obviously, the future era will develop from the traditional ternary space to the quaternary space composed of human, physical world, intelligent machine and virtual information world. Under this development background, the future development of the times will pay more attention to meeting people's individual needs. Although most of the traditional artifacts have entered the museum, the traditional wisdom of creation should never disappear, but should be combined with various new technologies, so that our traditional wisdom of creation can exert greater influence in this intelligent era and promote the innovation of design and creation in our country. It is predicted that with the further development of science and technology, the design of objects in the future will be carried out under the framework of the Internet of Things system[6]. During the process of using products, users will have more frequent data and information interaction with products, and then through the integration and analysis of big data technology, more effective suggestions will be provided for users, which will improve the product use efficiency and further improve the user experience. At present, in daily life, there are still some shortcomings in many product designs, which need to be solved with more creative ideas. For this reason, traditional wisdom of creation can be used to solve them [7-8].

3.1 Overall design position

Based on the current market research results and project demand analysis, the design has the following positioning: First, positioning the product category, which mainly aims at the living room furniture in art furniture, such as sofa and coffee table, because it is frequently used furniture. Second, the overall positioning, which mainly uses the virtual design method to determine the material, shape and texture individually. Thirdly, we should make a functional orientation to ensure that the traditional wisdom of creation contained in it can adapt to the changes of modern lifestyle and meet the requirements of ergonomics, so as to get final recognition. Fourth, the structural orientation, in terms of structure, the traditional mortise and tenon structure in the traditional wisdom of creation is still used to realize the connection of furniture.

3.2 Overall development process

This design mainly uses the emerging "Unity 3D" virtual reality technology in computer technology to design and display the target products, and will use the rendering technology of Unity 3D platform to improve the simulation effect of art furniture virtual design. This is helpful to meet the individual needs of different groups for furniture design, and it also provides new ideas and methods for subsequent design work[7].

3.2.1 Establish engineering and import resources. First, start Unity 3D, create a new engineering project, and then click the Asset packages... button, and the Unity resource package required by the new project will be automatically introduced. Finally, click the Create Project button to create a new Unity 3D project, and import a specific Unity 3D resource package to improve work efficiency. Of course, it should be noted that the path and name of the project must use English instead of Chinese, otherwise mistakes will easily occur.

3.2.2 Scene construction and optimization. After the Unity 3D resources are imported, you can directly see the imported resources such as 3D models, mapping materials and texture pictures in the Assets directory of Project view, and then click the File-New Scene option to create a new scene. In the process of scene construction and optimization, in order to ensure a more realistic simulation effect, the maps and materials of art furniture should be baked first. After baking, the model can be directly dragged into the newly created scene view.

In the view, three different scenes need to be established, each scene corresponds to a different style of art furniture, and the three styles can be switched between each other. After adjusting the positions of the three different models, you can add lights and adjust the cameras in the scene. Because each scene corresponds to different types of content, it is necessary to flexibly adjust the light position, size and camera angle before setting up the scene.When the three styles can be switched freely, the material can be selected and converted. This function is mainly realized by C# script, and the key segment code is as follows.

```
//Color wheels
    m_ColorWheels=new ColorWheelsSettings
    {
    mode=FindSetting((Settings x)=>x.colorWheels.mode)
    log=FindSetting((Settings x)=>x.colorWheels.log)
    linear=FindSetting((Settings x)=>x.colorWheels.linear)
    }
    //Curves
    m_Curves = new CurvesSettings
    {
    master = FindSetting((Settings x) => x.curves.master.curve),
    red = FindSetting((Settings x) => x.curves.redxurve),
    green = FindSetting((Settings x) => x.curves.green.curve),
    blue = FindSetting((Settings x) => x.curves.blue.curve),
    hueVShue = FindSetting((Settings x) => x.curves.hueVShue.curve),
    hueVSsat = FindSetting((Settings x) => x.curves.hueVSsat.curve),
    satVSsat = FindSetting((Settings x) => x.curves.satVSsat.curve),
    lumVSsat = FindSetting((Settings x) => x.curves.lumVSsat.curve),
      currentEditingCurve = FindSetting((Settings x) => x.curves.e
      CurrentEditin   gCurve),
    curveY = FindSetting((Settings x) => x.curves.e_CurveY)?
    curveR = FindSetting((Settings x) => x.curves.e — CurveR),
    curveG = FindSetting((Settings x) => x.cnrves.e__CurveG),
    curveB = FindSetting((Settings x) => x.curves.e_CurveB)
    };
```

Finally, the "first person character controller" is directly imported into the environment, and the user can adjust the controller parameters according to the needs of human body size. After all the above steps are completed, the game view can be run to roam the virtual environment, and users can intuitively see the form and material of furniture virtual design and its matching degree with the environment in the virtual environment[8].

3.2.3 Visual design and implementation of the interface. In Unity 3D, most interactive design processes will be carried out. Obviously, in terms of visual design and implementation of interface, it mainly includes several modules: the choice of artistic furniture style, the choice of artistic furniture material, the display of furniture details and indoor roaming.

In the interface design, it is mainly embodied in the form of icons and information pop-ups. The entry interface of the system should contain some summary information about furniture virtual design; and in the main interface of the system (that is, the first interface after entering the system), the detailed display contents should be included.

3.2.4 Publishing and display of virtual design. After the virtual design of art furniture is completed, it needs to be released and displayed on the PC platform. The work in this link is relatively simple. First, click the "PC" icon in "Build Setting" under the menu bar, then set "icon" in "Player Setting", click "Build", export the relevant files in the format of "exe", rename them, save them in a specific folder, and then run them directly on the PC and realize the corresponding functions.

4. Conclusion

On the whole, Chinese traditional creation wisdom is profound, and there are still many outstanding points which have not been discovered by us. These outstanding points undoubtedly still have important practical significance for the innovation of modern product design. Looking at the current product design, it is clear that it has further developed in the direction of humanistic connotation. Therefore, it is still a problem worth studying how to make the traditional wisdom of creation with more humanistic connotation manifest on more artifacts from the technical level. Of course, there are more than one standard answer to things. The current excellent design cases should play the role of attracting valuable suggestions. In the future, further research and innovation in related fields are needed to better promote the development and inheritance of traditional creation wisdom.

Acknowledgments

Fund-The Second Batch of Scientific Research Teams of Jingchu University of Technology-Research on Traditional Creation Culture in Central Hubei (No. TD202005).

References

1. Wu Xinlin. The Form of Chinese Traditional Creation Wisdom and Its Modern Translation[A]. Tongji University, Hunan University of Science and Technology, National Major Social Science Projects Research on Chinese Artisan Cultural System and Its Inheritance and Innovation Research Group.Symposium on Chinese Design Theory and Technological Innovation——Proceedings of the 4th Chinese Design Theory and the 4th National High-end Forum on Cultivating Chinese Artisans[C].Tongji University, Hunan University of Science and Technology, National Major Social Science Projects Research on Chinese Artisan Cultural System and Its Inheritance and Innovation Research Group: College of Design and Innovation,Tongji University, 2020:8.
2. He Sheng.Study on Enlightenment and Innovative Design of Chinese Traditional Creation Wisdom in the Intelligent Age[J].Packaging Engineering,2020,41(16):353-359.
3. Li Bing, Yan Siyuan.The Path Study of Traditional Creation Wisdom Enlighten Modern Product Innovation Design[J].Packaging Engineering,2020,41(22):95-99.
4. Peng Wei. Research on the Application of Traditional Creation Wisdom in Daily Home Product Design[D].Sichuan Fine Arts Institute,2019.

https://doi.org/10.1051/e3sconf/202129002035

5. Liang Bing.Exploration on Creation Wisdom and Design Thought of Traditional Handicrafts[J].Art Panorama,2017(11):70-71.
6. Wu Xinlin.From The Artificers' Record to T'ien-kung k'ai-wu:The Concession of Chinese Traditional Creation Thought from the Perspective of Art Comparison[J].Art & Design Research,2018(03):79-83.
7. Pan Tianbo.Chinese Examination of Technology: History, Logic and Form[J].Ethnic Art Studies,2019,32(04):91-98.
8. Yu Ding.Chinese School of Creation of Art Management in 5G Times[N].Chinese Culture News,2019-08-18(006).

Characteristics of Atmospheric Dustfall and Soil Magnetic Susceptibility and Environmental Significance in Xining City

Qiang Peng[1,2], Yongjuan Sun[1,2,3*]

[1]Qinghai Provincial Key Laboratory of Physical Geography and Environmental Processes, School of Geographical Science, Qinghai Normal University, Xining, P.R. China
[2]MOE Key Laboratory of Tibetan Plateau Land Surface Processes and Ecological Conservation, Qinghai Normal University, Xining, P.R. China
[3]Academy of Plateau Science and Sustainability, People's Government of Qinghai Province & Beijing Normal University, Xining, P.R. China

Abstract. Magnetic susceptibility is an important means to quickly and economically monitor the enrichment of surrounding environmental elements. As a regional central city on the Qinghai-Tibet Plateau, the relationship between atmospheric dust magnetic susceptibility and heavy metal content in Xining area is relatively lacking. In this paper, the magnetic susceptibility and geochemical element content of the collected dust samples, loess samples, surface soil samples and soil profile samples obtained through long-term observations are analyzed. The results show that the magnetic susceptibility of the dust-fall samples is the highest, the soil samples are the second, and the loess samples are the lowest; the magnetic susceptibility of the dust-fall samples in Xining City has a significant correlation with the elements Cd, Cr, Ni, Cu, Pb, Zn and Hg, and the magnetic susceptibility can be used as an indicator of heavy metal in the soil; the elements Cr, Ni, Cu, Zn and As in the soil profile are enriched at 20 cm or 30 cm, which is related to the leaching of surface elements.

1 Introduction

Environmental magnetism, developed in the 1980s, was a borderline science between environmental science, earth science and magnetism [1-2]. It is widely used in soil science and environmental research, and has the characteristics of simple, fast, low-cost, and non-destructive [3]. Magnetic susceptibility is a physical concept that refers to the ability of a substance to magnetize under the action of an external magnetic field [4]. One of the current research directions of environmental magnetism is to combine magnetic susceptibility with geochemical elements to explore the changes of the environment. Chen et al. [5] found that the magnetic susceptibility characteristics of roadside soil in Xi'an are significantly positively correlated with geochemical elements, Xiao [6] found that the magnetic susceptibility of soil in Fuzhou is correlated with the content of heavy metal elements. Jiang et al. [4] found that the magnetic susceptibility value of the soil profile and the content of heavy metal elements are generally decreasing downward from the surface. The current research on urban atmospheric dust fall focuses on the physical and chemical properties and heavy metal pollution. Yang et al. [7] used factor analysis to study the pollution sources of atmospheric dust in Lanzhou, and the results showed that the main source of atmospheric dust in Lanzhou was coal. Feng et al. [8] found that the method of using magnetic susceptibility can effectively monitor urban pollution in Lanzhou. However, there are relatively few studies that combine urban atmospheric dust with loess and soil to analyze the relationship between magnetic susceptibility and geochemical elements. Cui et al. [9] studied the levels of heavy metals in atmospheric dustfall in Shijiazhuang City and found that the main sources of heavy metal elements in dustfall are coal burning activities, road traffic, industrial and mining enterprise exhaust emissions and soil particulate matter.The heavy metal content in high-value areas of atmospheric dustfall is consistent with the distribution of industrial areas. Zhao et al. [10] used correlation analysis and principal component analysis methods to analyze the atmospheric dustfall samples collected in Jinan city in order to find out the content characteristics and pollution sources of the elements in the near-surface atmospheric dustfall in Jinan City. It is found that the elements Cd, Pb, Zn, and Se in the atmospheric dust may come from burning coal, the elements Cr and Ni come from soil dust, the elements CaO and As come from building dust, the element F mainly comes from automobile exhaust emissions, and Hg is affected by the factors such as combustion, coal, smelting and transportation.

Xining is a regional center city on the Qinghai-Tibet Plateau, with high population density, fragile ecological environment, and strong aeolian activities in the surrounding area. The average annual atmospheric dust flux is up to ~400 mg/m^2 [11]. There is abundant

*Corresponding author: yongjuansun@163.com

environmental information in atmospheric dust, but the interpretation of atmospheric dustfall environmental information in Xining is relatively lack, especially the relationship between magnetic susceptibility and geochemical element content. This article compares and analyzes the magnetic susceptibility and element characteristics of atmospheric dust, loess and soil in Xining City, and discusses the environmental significance indicated by the magnetic susceptibility, so as to provide a valuable reference for the interpretation of the atmospheric environment.

2 Materials and Methods

2.1 Sample collection

The atmospheric dust-full samples were collected from November 2013 to November 2015. A sampling site was set up on the top of the Science and Technology Building on the Chengxi Campus of Qinghai Normal University, Xining City. A total of 8 atmospheric dust samples were collected, divided into winter and dustfull (KJL-D) and summer dustfull (KJL-X). A sampling site was set up in Duoba Town, Huangzhong County, Xining City, and 7 samples were collected, which were divided into winter dustfull (DB-D) and summer dustfull (DB-X). The sampler is a high-neck glass cylinder with a diameter of 20 cm and a height of 40 cm.

Four loess samples (MFJ-1, MFJ-2, MFJ-3, MFJ-4) were collected at intervals of 5 cm from a section with a height of 20 cm in Beishan, Xining City. Three soil samples (XQDJ-1, XQDJ-2, XQDJ-3) were collected at the surface 5 cm on the roadside of Xiaoqiao Street, Xining City. Two soil samples (QHSD-1, QHSD-2, respectively) were collected at the surface 5 cm and two soil profiles with a total of 11 soil samples (KJLT1-1, KJLT1-2, KJLT1-3, KJLT1-4, KJLT1-5, KJLT1-6, KJLT2-1, KJLT2-2, KJLT2-4, KJLT2-5, KJLT2-6) were collected in the Chengxi campus of Qinghai Normal University. A total of 24 samples were collected.

2.2 Sample analysis

The collected samples were placed in a ventilated and dark place to dry naturally.

The magnetic susceptibility index was measured with the magnetic susceptibility meter (MS2, UK) in the particle size analysis room of Qinghai Province Key Laboratory of Natural Geography and Environmental Processes. The instrument can measure the low-frequency magnetic susceptibility (470 Hz) and the high-frequency magnetic susceptibility (4700 Hz) of the sediment, using a 2×2×2 8 cm^3 magnetic special plastic box. The specific processes are as follows:

(1) The natural air-dried samples, filtered through a 20-mesh sample sieve in order to remove the gravel, plant roots and other biological residues, were mixed evenly, and then closely put it in the plastic box of known weight. The weight of the sample can be obtained by weighing the box;

(2) In order to minimize the interference of the external environment, it is necessary to stay away from magnetic materials when using the magnetic susceptibility instrument. We adjust the instrument to the low-frequency gear and start the measurement until the instrument is stable. We measure the background value before testing the samples;

(3) The low-frequency mass magnetic susceptibility and high-frequency mass magnetic susceptibility were calculated.

The element indicators were determined in the Xi'an Geological Survey Center, and the pre-processing work is carried out in the pre-processing branch of the Qinghai Provincial Key Laboratory of Natural Geography and Environmental Processes, Qinghai Normal University. The specific steps are as follows:

The natural air-dried samples were passed through a 20-mesh sample sieve, and then mixed thoroughly and grind to below 200 mesh with a ball mill, and 10-20 g of the mixed sample was sealed in a airtight bag and made a mark, and then sent it to the Xi'an Geological Survey Center for testing. Among them, the instrument used to measure the elements As and Hg is the AFS-2202E atomic fluorescence spectrometer of Beijing Haiguang Instrument Co., Ltd., and the standard is "DZ/T 0279-2016"; the instrument used to measure the elements Cd、Cr、Ni、Cu、Pb and Zn is the iCAP RQ inductively coupled plasma mass spectrometer of American Thermoelectric Company, and the standard is "GB/T14506.30-2010".

3 Results and Discussion

3.1 Magnetic susceptibility analysis

The magnetic susceptibility is an index that reflects the degree of difficulty of magnetization of a material, and it is numerically expressed as the ratio of magnetization to magnetic field strength [4]. From Table 1, we find that the magnetic susceptibility of dust-fall samples is the highest, followed by soil, and loess is the lowest, and the low-frequency magnetic susceptibility (χlf) of the three are slightly higher than the high-frequency magnetic susceptibility (χhf) (Table 1). Generally, the value of χhf is higher than that of χlf, but as the frequency of the magnetic field increases, the relaxation time of ultrafine magnetic minerals will be greater than the magnetic susceptibility test time, and this part of the particles will be locked and no longer contributes to the magnetic susceptibility, so its low-frequency magnetic susceptibility is greater than the high-frequency magnetic susceptibility [12].

Because of the same variation trend of low-frequency magnetic susceptibility and high-frequency magnetic susceptibility, the case of low-frequency magnetic susceptibility, referred to as magnetic susceptibility (χ), was discussed. The value of magnetic susceptibility varies greatly among all the samples. The magnetic susceptibility values of the dustfall samples collected in Xining in winter and summer are the highest, which are $616.27\times10^{-8}m^3\cdot kg^{-1}$ and $504.45\times10^{-8}m^3\cdot kg^{-1}$, respectively. The magnetic susceptibility values of the loess samples collected in

Beishan, Xining are the lowest, which are $32.67\times10^{-8}m^3\cdot kg^{-1}$ and $34.73\times10^{-8}m^3\cdot kg^{-1}$, respectively. There are many factors that affect the magnetic susceptibility of the soil. Current studies have found that the increase in the magnetic susceptibility of the soil is related to the accumulation of organic matter, excessive use of pesticides and fertilizers, industrial pollution, and the discharge of domestic garbage [12-14].

Table1 Average value of high-frequency magnetic susceptibility and low-frequency magnetic susceptibility of dust, loess and soil

sample name	$\chi hf(10^{-8}m^3\cdot kg^{-1})$	$\chi lf(10^{-8}m^3\cdot kg^{-1})$
dust	330.66	336.10
loess	37.71	41.00
soil	115.97	121.99

3.2 Correlation analysis between magnetic susceptibility and geochemical elements

In order to understand the correlation between magnetic susceptibility and geochemical elements, the correlation analysis between magnetic susceptibility and elements Cd, Cr, Ni, Cu, Pb, Zn, As, Hg on 24 samples was carried out (Table 2). The results show that the magnetic susceptibility is positive correlation with the elements Cd, Cr, Ni, Cu, Pb, and Zn, and the correlation coefficients are 0.821, 0.860, 0.872, 0.905, 0.870, 0.778 respectively, which are significantly correlated within the 99% confidence interval; the magnetic susceptibility is positive for the element Hg Correlation, and the correlation coefficient is 0.426, which is significantly correlated within the 95% confidence interval; the correlation between magnetic susceptibility and As is not obvious. Among them, the dust collection point of Xining City is located in the city, and the intensity of human activities nearby is high. The burning of residential fossil fuels, the emission of waste gas from industrial areas, and the exhaust gas produced by urban vehicles have increased the heavy metal elements, thereby increasing the magnetic susceptibility of dust.

During the analysis of samples of atmospheric dust fall within the urban area of Jinan City, it was found that the heavy metal elements in atmospheric dust may come from fuel combustion, soil dust, construction dust, and automobile exhaust emissions [10]. This is consistent with the research in this article.

Table2 Correlation between magnetic susceptibility value and element content

Magnetic susceptibility and elements	Correlation coefficient	Significance (99%)	Significance (95%)
Cd	0.821	√	√
Cr	0.860	√	√
Ni	0.872	√	√
Cu	0.905	√	√
Pb	0.870	√	√
Zn	0.778	√	√
As	0.390	No correlation	
Hg	0.426	×	√

3.3 Changes of elements in the soil profile

According to the analysis of profile 1 and profile 2 (Figure 1 and Figure 2), the mass fraction of element Cd in 11 soil samples is 0.14-0.76 ug/g, of which in profile 1 from the surface to the 60 cm below the ground showed an overall downward trend. Only the mass fraction of Cd at 20 cm in profile 2 is 0.76 ug/g, which is slightly higher than the national soil environmental quality standard value (0.6 ug/g). It may be because the element Cd is leached after being enriched on the surface of the soil, and is enriched at a depth of 20 cm. The mass fraction of element Hg is 0.08-0.35 ug/g, which does not exceed the human health standard of soil Hg of 1 ug/g, indicating that the pollution of element Hg in soil in Xining City is less. But the content is higher at 30 cm in section 1, and there is an enrichment effect. The mass fraction of element Pb in the two sections is 22.9-37 ug/g, which is lower than the national soil environmental quality standard (170 ug/g). The mass fraction of element Pb at 20 cm in section 1 is 28 ug/g, which is enriched. Profile 2 shows an overall downward trend from the surface to 60 cm underground, which is consistent with the change trend of magnetic susceptibility. The mass fractions of elements Cr, Ni, Cu and Zn are 56.3-105.7 ug/g, 31.2-63.6 ug/g, 22.5-66.4 ug/g and 64.8-113.8 ug/g, respectively. The maximum limit values of the national standard agricultural land soil for these four elements are 250 ug/g, 190 ug/g, 100 ug/g, and 300 ug/g. The four elements in both profiles are enriched at 20 cm or 30 cm, but they are all lower than the national standards for agricultural land soils. The mass fraction of element As is 11.9-67.0 ug/g, the national standard (GB15618-2018) maximum limit value of agricultural land soil is 25 ug/g. The mass fraction of profile 1 is highest at 20 cm and 30 cm, and the mass fraction of As at 20 cm in profile 2 is 67.0 ug/g, which is slightly higher than the national soil environment quality standard value [15-17].

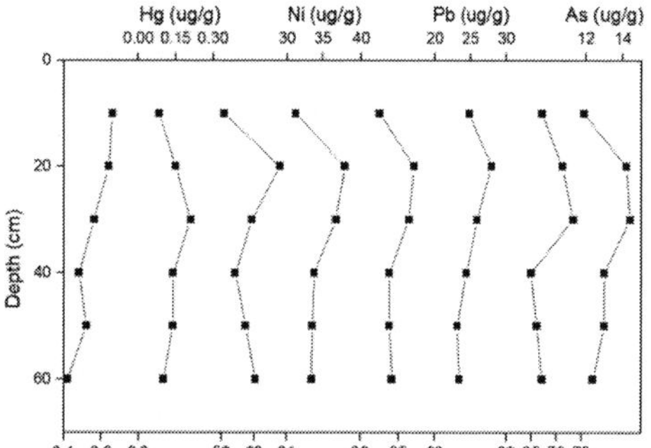

Figure1 Distribution of heavy metal elements in profile 1

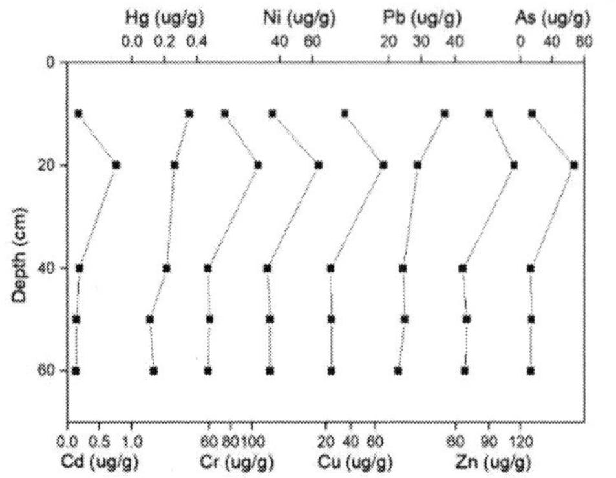

Figure2 Distribution of heavy metal elements profile 2

4 Conclusions

The distribution and influencing factors of the magnetic susceptibility and the content of heavy metal elements in atmospheric dustfall, loess and soil in Xining City, and the correlation analysis between the magnetic susceptibility and the content of heavy metal elements in soil profile was discussed in this study, the following conclusions are obtained:

The magnetic susceptibility of the samples in the study area has a significant correlation with the elements Cd, Cr, Ni, Cu, Pb, Zn, and Hg. The magnetic susceptibility is a convenient indicator of the content of heavy metals.

The heavy metal elements Cr, Ni, Cu, Zn and As in the soil are enriched at 20 cm or 30 cm, which is the result of leaching of surface elements.

Acknowledgments

This article is one of the phased achievements of the National Natural Science Foundation of China " Reconstruction of flood intensity frequency in Hehuang Valley in 10,000, 1,000, and 100 years and early warning under climate warming " (2021-ZJ-749)

References

1. Oldfield Frank.(1991) Environmental magnetism-A personal perspective[J]. Oldfield Frank,10(1).
2. Pan Y.X., Zhu R.X. (1996) Status and progress of environmental magnetism research[J]. Progress in Geophysics, (04): 87-99.
3. Jiang Y.H., Yin H.F., Wang R.H. (2004) Theories, methods and research progress of environmental magnetism [J]. Acta Geosciences, (03): 357-362.
4. Jiang Y.H., Yin H.F., Wang R.H., Kang X.J. (2005) Study on the distribution of soil magnetic susceptibility and heavy metal elements and their correlation in Huzhou City[J]. Journal of Jilin University (Earth Science Edition), (05): 653-660.
5. Chen J.H., Lu X.W. (2011) The characteristics of roadside soil magnetic susceptibility in Xi'an and its environmental significance[J]. Journal of Shaanxi Normal University (Natural Science Edition), 39(05): 76-82.
6. Xiao L. (2018) The characteristics of surface soil magnetic susceptibility and the magnetic response of heavy metal pollution in Fuzhou City[D]. Fujian Normal University.
7. Yang L.P., Chen F.H. (2002) Research on the source of atmospheric dustfall pollutants in Lanzhou City[J]. Acta Scientiae Circumstantiae, (04):499-502.
8. Feng H. (2012) Research on the magnetic characteristics of atmospheric dustfall in Lanzhou City[D]. Lanzhou University.
9. Cui X.T., Luan W.L., Li S.M., Song Z.F. (2012) Source analysis of heavy metal elements in atmospheric dustfall in Shijiazhuang City[J]. Geology in China, 39(04): 1108-1115.
10. Zhao X.Q., Wang Z.H., Wang C.L., Dai J.R., Liu H.F., Ji S.L. (2016) Geochemical characteristics and pollution evaluation of dustfall elements in the near-surface atmosphere in Jinan City[J]. Geophysical and Geochemical Exploration, 40(01): 154-159
11. E C.Y., Xi Y.S., Sun Y.J., Zhao Y.J., Yang L., Lv S.C. (2016) Analysis of dust flux and particle size characteristics in the northeastern part of the Qinghai-Tibet Plateau[J]. Salt Lake Research, 24(02): 62-67.
12. Deng C.L., Yuan B.Y., Hu S.Y., Liu D.S., Kenneth L.V., Han J.M., Guo Z.T., Lu H.Y. (2000) A review of some research advances in environmental magnetism[J]. Marine Geology and Quaternary Geology, (02):93- 101.
13. Yi Y.L., Gu W.W., Zhang D.G., Li J. (2008) Study on the distribution of soil magnetic susceptibility and heavy metal elements and their correlation in Huludao City[J]. Soil, (05):806-811.
14. Li X.Q., Hu X.F., Sun W.M., Zhang G.L. (2006) Research on Magnetic Monitoring of Urban Soil Pollution[J]. Soil, (01): 66-74.
15. Soil environmental quality agricultural land soil pollution risk management and control standards (excerpts) [J]. Humic Acid, 2018(04): 58-61.
16. Liu Z. (1998) Soil trace elements in China[J]. Advances in Earth Science.
17. Zhou A.G., Cai H.S. (1998) The theory and application of geological environmental quality evaluation[M]. China University of Geosciences Press.

Current Status and Development Trends of Sludge Disposal Technology

Siyan Huang[1], Liu Yang[1*], and Lianguo Ji[1]

[1]College of Geoscience and Surveying Engineering, China University of Mining and Technology (Beijing), 100083 Beijing, China

Abstract. Sludge is a typical solid waste of sewage treatment plants, and its disposal methods have attracted much attention. This paper summarized the current technical routes for sludge disposal in China, including anaerobic digestion, anaerobic fermentation, incineration, landfill, land utilization, and building materials utilization. In the context of energy saving and carbon reduction and the promotion of resource recovery, enhanced pretreatment technology of sludge anaerobic digestion and anaerobic fermentation technology for fuel gas production are the first choices for sludge disposal and resource utilization. In addition, the development trend of sludge disposal technology was analyzed.

1 Introduction

The daily sewage treatment capacity in China reached 128 million tons [1]. The huge sludge volume and high disposal cost have brought difficulties to the economic operation of sewage treatment plants [2]. The Chinese government has successively issued the Action Plan for Water Pollution Prevention and Control, the Construction Plan of the National Urban Sewage Treatment and Recyclation Facilities during the 13th Five-Year Plan Period, and Implementation Plan for Strengthening and Weaknesses of Urban Domestic Sewage Treatment Facilities, which put forward requirements for the environmentally friendly treatment of sludge. This paper systematically summarized the current technical route of sludge disposal in China, and analyzed the development trend of sludge disposal technology.

2 Typical sludge disposal technology

2.1 Anaerobic digestion technology of sludge

The anaerobic digestion technology of sludge is a biodegradation method, which can successfully treat more than 1/3 of the sludge. General sludge contains high sediment content and low organic matter content, and organisms can only decompose the organic matter in the sediment, but cannot decompose the sediment. In addition, the biodegradability of sludge is poor, and the operation of anaerobic sludge digestion equipment is unstable. Therefore, traditional sludge anaerobic digestion technology cannot be directly applied to sludge treatment and disposal, and some pretreatment technologies are needed, including ultrasonic, electrochemistry, high temperature and high pressure thermal hydrolysis and thermo-alkali pretreatment, etc.

2.1.1 Ultrasonic pretreatment

Ultrasonic pretreatment can improve the anaerobic digestibility of sludge by disturbing the original physical and chemical properties of sludge. It can effectively improve the sludge dewatering performance, promote the release of nitrogen and phosphorus, and reduce the amount of sludge. Liu et al. found that the gas and methane production of the heat-alkali ultrasonic pretreatment of straw were higher than that of the non-pretreatment [3]. Gao et al. verified that ultrasonic alkali hydrolysis could dissolve more sludge proteins by destroying sludge floc and microbial cell structure, which provided a potential application prospect for sludge protein extraction [4].

2.1.2 Electrochemistry pretreatment

Electrochemical pretreatment method is to make the nitrogen, phosphorus and cytoplasm in cells enter the supernatant by changing the permeability of microbial cell membrane in the sludge, which can be used by microorganisms in the process of aerobic digestion. Microbial metabolic speed is increased, and the aerobic digestion time is shortened. This method effectively improved the destruction of extracellular polymer matrix and anaerobic digestion. In this way, the effects of poor digestibility and slow degradation rate of waste activated sludge on the performance of sludge anaerobic digestion can be reduced [5].

* Corresponding author: yang_l@126.com

https://doi.org/10.1051/e3sconf/202129003002

2.1.3 High temperature and high pressure thermal hydrolysis pretreatment

The high temperature and high pressure thermal hydrolysis pretreatment breaks the microbial cell wall in the sludge, dissolves the colloidal structure, releases organic matter, and reduces the viscosity. The output and energy efficiency of anaerobic digestion are improved, with a strong operability. Experiments have confirmed by Hong et al. that due to thermal hydrolysis, the polymer can be decomposed into low molecular weight compounds, the pH value is reduced, and the operation is more stable [6]. Deng et al. applied the high-temperature and high-pressure thermal hydrolysis pretreatment method to the synergistic anaerobic synthesis of banana straws with high organic content, which significantly increased the methane yield of low-organic sludge [7].

2.1.4 Thermo-alkali pretreatment

After thermo-alkali pretreatment, the organic matter in the sludge was basically released into the liquid phase. The concentrations of Solluted Chemical Oxygen Demand (SCOD), soluble protein and polysaccharide in the filtrate increased significantly. The yield of anaerobic sludge fermentation Short Chain Fatty Acids (SCFAs) under alkaline conditions was much higher than that under acidic or neutral pH conditions, which could enhance the hydrolysis and acidification process of sludge and inhibit the generation of methane [8]. Hu et al. designed the process of thermal-alkali pretreatment of sludge and semi-continuous flow anaerobic fermentation sludge treatment process, and found that the Volatile Fatty Acids produced can meet the carbon source requirements of sewage treatment plants for nitrogen and phosphorus removal upgrades [9].

2.2 Landfill

The main landfill forms include separate landfill, mixed landfill and special landfill. Mixed land landfill technology is mostly used in China, which has low cost, high efficiency and strong universality. However, it has the disadvantages of high transportation cost and pollution of the land. This method goes against China's increasing emphasis on environmental protection and will gradually be eliminated in the future.

2.3 Incineration and land utilization

Incineration is the preferred method for rapid sludge treatment. Under sufficient oxygen, the sludge is fully burned to generate H_2O, CO_2 and ash, and harmful viruses and bacteria in the sludge are eliminated. According to the survey, approximately 20% and 25% of the sewage sludge produced in the United States and the European Union are incinerated, respectively [10]. In addition, The sludge after incineration contains a lot of beneficial elements (such as P, N and K), which can be recycled for land use [11].

2.4 Building materials utilization

After sludge drying and dewatering, porous energy saving brick can be prepared by high temperature calcination. It is generally used to make ceramsite, cement, brick, biochemical fiberboard and other infrastructure materials. Using sludge in cement construction materials can effectively reduce the negative impact on the environment [12].
The main process routes of sludge disposal market in China are shown in Fig. 1.

3 Development trend of sludge disposal technology

3.1 Anaerobic fermentation technology

Anaerobic fermentation is an environmentally friendly waste activated sludge disposal technology that can realize resource recycling (CH_4, H_2). However, the traditional sludge anaerobic fermentation efficiency is low. Freezing and potassium ferrate can be combined for sludge pretreatment to promote the production of hydrogen by anaerobic fermentation. Zhao et al. proposed the strategy of adding lye to the AD system driven by biogas recirculation for sludge management (Fig. 2), which can not only achieve biogas upgrade, but also achieve good settling/dewatering capacity of sludge and improve its stability [14].

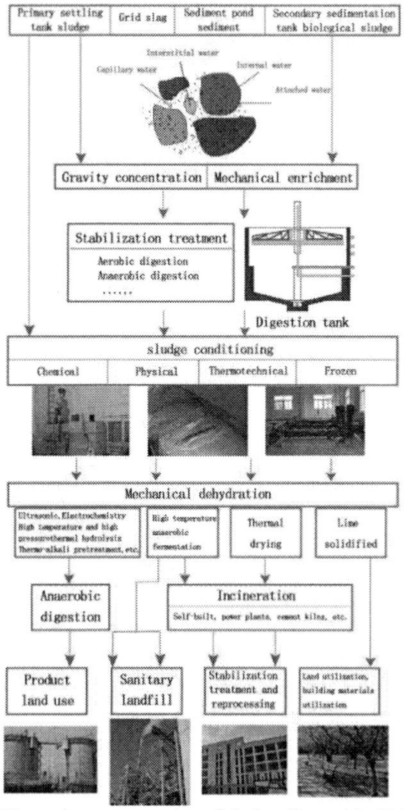

Fig. 1. The main process routes of sludge disposal in China [13]

3.2 Preparation of degradable plastics

The interaction of certain types of microorganisms and sludge can promote the reproduction of microorganisms. By adding carbon sources and inorganic salts to the sludge to accumulate polyhydroxy fatty acids, it can be used to prepare degradable plastics. Chen obtained nutrients such as Poly-3-hydroxybutyrate and Acetic Acid, which can be used for microbial culture, as raw materials of biodegradable plastics through pyrolysis of sludge at high temperature [15]. Chua et al. designed anaerobic-aerobic activated sludge process model to successfully produce polyhydroxyalkanate materials (Fig. 3) [16].

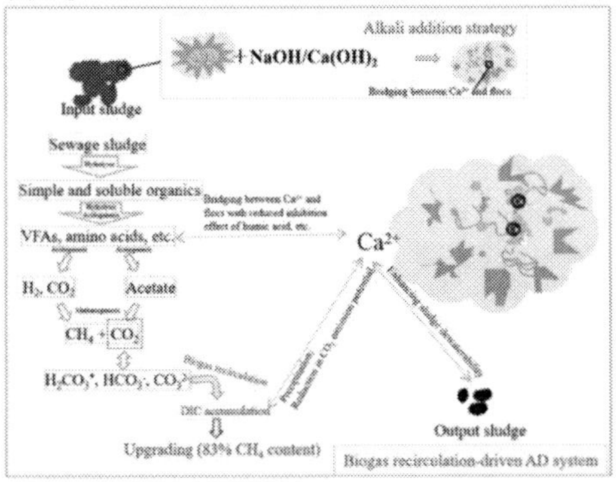

Fig. 2. The strategy of adding lye to the AD system driven by biogas recirculation for sludge management [14]

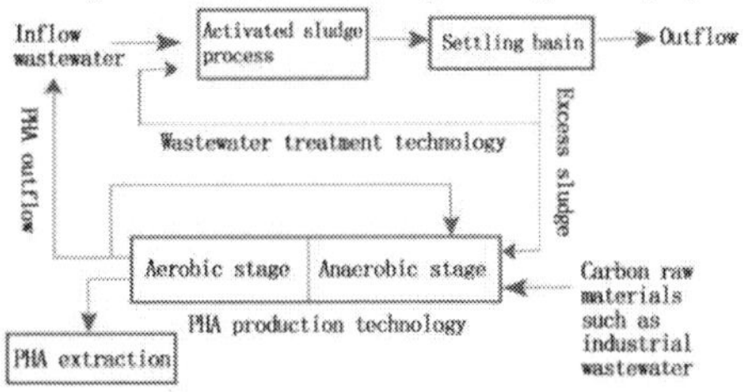

Fig. 3. Process flow model of PHA production by anaerobic-aerobic activated sludge [16]

3.3 Preparation of activated carbon

Activated carbon prepared from sludge has good application prospects in the fields of wastewater, waste gas and soil improvement. Zeng et al. used sewage sludge and corn stalks as raw materials to prepare sludge-based activated carbons (SACs), modified with ferric nitrate, and treated landfill leachate with unmodified and modified SAC as adsorbents to remove COD and organic matter in the leachate [17]. Du et al. discussed the effects of different sludge sources such as oil sludge and municipal sludge, as well as different pyrolysis processes such as microwave pyrolysis and hydrothermal carbonization, as well as the types of activators and additives on the performance of activated carbon from sludge [18]. Wu et al. designed an experiment using sludge-poplar sawdust co-pyrolysis as raw material, and successfully prepared activated carbon by KOH chemical activation method [19].

4 Summary

At present, most of the sludge treatment plants in China still use traditional methods such as aerobic composting, incineration and landfill, land utilization and other technical means. However, in the long run, the traditional methods do not conform to China's sustainable development strategy. Anaerobic fermentation technology is the most potential sludge disposal method, which has the advantages of sludge reduction and resource utilization. The comprehensive utilization of sludge based on multiple and multi-stage treatment is worthy of further exploration.

References

1. MOHURD, *China Urban Construction Statistical Yearbook 2017* (Beijing: China Statistics Press, 2018)
2. W. Guo, S. Yang, W. Xing, et al, Minimization of excess sludge production by in-situ activated sludge treatment processes — A comprehensive review. Biotechnol. Adv. 31, 1386-1396 (2013)
3. P. Liu, H. Li, Z. Chen, Effect of thermoalkali and ultrasonic pretreatment on anaerobic digestion of straw and sludge. Water & Wastewater Engineering. 46, 120-124 (2020)
4. J. Gao, Y. Wang, Y. Yang, Z. Li, Ultrasonic-alkali method for synergistic breakdown of excess sludge for protein extraction. J. Cleaner Prod. 295, 126288 (2021)
5. H. Huang, Q. Zeng, P. Heynderickx, G. Chen, D. Wu, Electrochemical pretreatment (EPT) of waste activated sludge: Extracellular polymeric substances matrix destruction, sludge solubilisation and overall digestibility. Bioresour. Technol. 330, 125000 (2021)
6. E. Hong, J, Park, B. Lee, W. Shi, H. Jun, Improvement of Waste Dehydrated Sludge for Anaerobic Digestion through High-Temperature and High-Pressure Solubilization. J. Cleaner Prod. 13, 88 (2020)
7. H. Deng, J. Zhang, P. Xian, Q. Fang, Z. Meng, Improving Anaerobic Digestibility of Sludge Pretreated by Thermal Hydrolysis and Banana Straw Added. Environmental Engineering. 38, 144-149 (2020)
8. H. Yuan, Y. Chen, H. Zhang, S. Jing, Q. Zhou, G. Gu, Improved Bioproduction of Short-Chain Fatty Acids (SCFAs) from Excess Sludge under Alkaline Conditions. Environ. Sci. Technol. 40, 2025-2029 (2006)
9. J. Hu. B. Guo, Z. Li, Z. Wu, W. Tao, Freezing pretreatment assists potassium ferrate to promote hydrogen production from anaerobic fermentation of waste activated sludge. Sci. Total Environ. 781, 146685 (2021)
10. S. Ducoli, A. Zacco, E. Bontempi, Incineration of sewage sludge and recovery of residue ash as

https://doi.org/10.1051/e3sconf/202129003002

building material: A valuable option as a consequence of the COVID-19 pandemic. J. Environ. Manage. 282, 111966 (2021)

11. P. Ma, C. Rosen, Land application of sewage sludge incinerator ash for phosphorus recovery: A review. Chemosphere. 278, 129609 (2021)
12. L. Swierczek, B. M. Cieslik, P. Konieczka, Challenges and opportunities related to the use of sewage sludge ash in cement-based building materials-A review. J. Cleaner Prod. 287, 125054 (2021)
13. J. Sun, Y. Ye, X. Zheng, X. Zhang, K. Qu, S. Li, Y. Jiang, Y. Liu, Research and Discussion on the Sludge Disposal Technology of Kunming WWTPs. China Water & Wastewater. 36, 108-112 (2020)
14. J. Zhao, T. Hou, Z. Lei, K. Shimizu, Z. Zhang, Performance and stability of biogas recirculation-driven anaerobic digestion system coupling with alkali addition strategy for sewage sludge treatment. Sci. Total Environ. 783, 146966 (2021)
15. J. Chen, From waste to treasure: turning activated sludge into bioplastic poly-3-hydroxybutyrate. Chinese Journal of Biotechnology. 33, 1934-1944 (2017)
16. A. Chua, H. Takabatake, H. Satoh, T. Mino, Production of polyhydroxyalkanoates (PHA) by activated sludge treating municipal wastewater: effect of pH, sludge retention time (SRT), and acetate concentration in influent. Water Research. 37, 3602-3611 (2003)
17. F. Zeng, X. Liao, J. Lu, D. Pan, Q. Qiu, K. Ding, M. Luo, Effect of iron nitrate modification on elimination of organic matter from landfill leachate by sludge-based activated carbon. Waste Manage. Res. 0734242X211009966 (2021)
18. M. Du, C. Lu, F. Wang, C. Qu, J. Li, Preparation of activated carbon from sludge and its application in environmental treatment. Applied Chemical Industry. 47, 2777-2780+2785 (2018)
19. S. Wu, K. Jiang, Y. Ye, Y. Liu, D. Wang, S. Li, Preparation of Activated Carbon from Co-pyrolysis Char of Sewage Sluge and Poplar Sawdust and Its Application in Wastewater Treatment. Chemistry and Industry of Forest Products. 39, 56-64 (2019)

Concentration, distribution and occurrence of mercury in Chinese coals

Wenying Ren[1], Liu Yang[1,*], Qingyi Cao[1], and Chaoming Liang[1]

[1] College of Geoscience and Surveying Engineering, China University of Mining and Technology (Beijing), 100083 Beijing, China

Abstract. Mercury in coals is one of the important sources of atmospheric mercury, which is potentially harmful to the ecological environment. Based on the data of 970 coal samples, the concentration, spatial distribution and occurrence of mercury in Chinese coals were analyzed. The main conclusions are as follows: The distribution of mercury concentration in Chinese coalfields is uneven; medium and high mercury coals are mainly distributed in southwest China and eastern Inner Mongolia. The mercury concentrations in various coal-forming periods are as follows: K (0.320 mg/kg) > P_2 (0.220 mg/kg) > C_3 (0.179 mg/kg) > J (0.177 mg/kg) > D (0.165 mg/kg) > P_1 (0.136 mg/kg) > C_1 (0.090 mg/kg) > E (0.086 mg/kg) > T_3 (0.066 mg/kg). The mercury concentrations in different coal ranks are as follows: Lignite (0.164 mg/kg), long flame coal (0.078 mg/kg), non-caking coal (0.256 mg/kg), weakly caking coal (0.086 mg/kg), gas coal (0.151 mg/kg), fat coal (0.122 mg/kg), coking coal (0.171 mg/kg), lean coal (0.393 mg/kg), meagre coal (0.161 mg/kg), anthracite (0.160 mg/kg). Sulfide bound state is the main form of mercury in coals, and pyrite is the main occurrence medium.

1 Introduction

Mercury is a highly toxic and volatile heavy metal, which is widely present in terrestrial, atmospheric, freshwater and marine ecosystems [1-2]. Anthropogenic mercury emissions account for 60-80% of the total global emissions, of which more than 90% of mercury comes from fuel combustion [3-4]. China is the largest mercury emitter, accounted for more than 20% of global anthropogenic mercury emissions [5]. Coal combustion is one of the important anthropogenic mercury sources. In China, approximately 400 tons of mercury were emitted annually due to coal burning, accounting for nearly half of anthropogenic atmospheric mercury emissions [6-9].

Due to the massive consumption of coal and the high toxicity of mercury, it has become one of the most concerned harmful trace elements in coals. Mercury can stay in the atmosphere for 0.5 to 2 years, and participates in the global cycle [10-11]. The concentration, distribution and occurrence of mercury in coals have attracted the attention of many scholars. Zheng [12], Dai [13], Wu [14], Huang [15], Song [16], Gao [17], and Wang [18] et al. have studied mercury content in coals from different regions successively. However, there are relatively few studies on the distribution of mercury in coals based on the national scale. In this paper, the concentration and spatial distribution of mercury in Chinese coals were analyzed, and the occurrence mode was discussed.

2 Method

The data of 970 coal samples were used in this paper, which obtained from the Trace Elements in Coal of China (TECC) database [19]. ArcGIS software was used to carry out inverse distance weight interpolation analysis on the data, and create a spatial distribution map of mercury in Chinese coalfields. The specific calculation formula is shown in Formula 1:

$$z = [\sum_{i=1}^{n} \frac{z_i}{d_i^j}] / [\sum_{i=1}^{n} \frac{1}{d_i^j}] \quad (1)$$

Where Z is the estimated value of the predicted point; Z_i is the i-th (i=1, 2, 3,..., n) measured value; d_i is the distance between the interpolation point and the sample point; j is the power value of the distance, which is generally 1 or 2. This study used the default value of 2; n is the number of sample points involved in interpolation, n=970.

3 Distribution of mercury in Chinese coals

3.1 Spatial distribution of mercury in Chinese coals

Based on 970 coal samples, the mercury concentration of most Chinese coals ranges from 0.002 to 56.20 mg/kg. The arithmetic average value is 0.14 mg/kg, which is

* Corresponding author: yang_l@126.com

lower than the content value calculated by Yang et al. [20] (0.20 mg/kg, n=1793) and Ren et al. [21] (0.195 mg/kg, n=1413). It is close to the result calculated by Zhao et al. [22] (0.20 mg/kg, n=1466). And it is between the world coal (0.1 mg/kg) [23] and the USA coal (0.17 mg/kg) [24], which is 2.3 times of the Australia coal (0.06 mg/kg) [22].

Table 1. The classification of mercury content in coal in Chinese standards [25].

Content classification	Range (mg/kg)
Ultra-low mercury coal	≤ 0.150
Low mercury coal	> 0.150~0.250
Middle mercury coal	> 0.250~0.600
High mercury coal	> 0.600

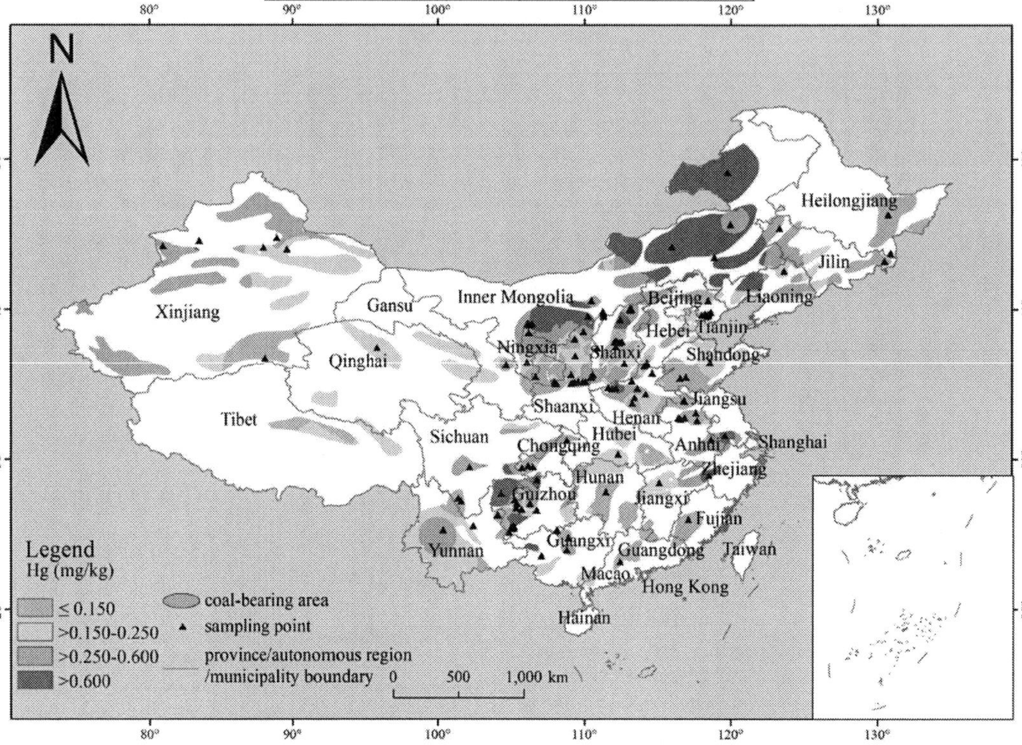

Fig.1. The distribution of mercury content in Chinese coalfields.

To facilitate the discussion of the mercury content distribution in Chinese coals, mercury was divided into four categories according to China national standard *"Classification for content of harmful elements in coal-Part 4: Mercury"* (GB/T 20475.4-2012) [25] (Tab. 1). The distribution of mercury content in Chinese coals is uneven (Fig. 1). Medium and high mercury coals are mainly concentrated in the southwest and northeast regions, including Guizhou, Yunnan and eastern Inner Mongolia. In addition, some medium-high mercury coals also exist in northern Ningxia, southern Anhui, northern Shanxi and western Henan. Low mercury and ultra-low mercury coals are widely distributed, mainly in Hubei, Hunan, Guangxi and Jiangxi in southern China, as well as northern China, including Shaanxi Province, Henan province, central Shanxi, northern Anhui, and western Shandong.

The mercury enrichment in coals in most areas is caused by magma intrusion. In the process of hydrothermal intrusion, the mercury activated will enter the coal seam along with the volatile matter, resulting in the increase of mercury content in coals. Dai et al. [13] analyzed the coal sample content of Xingren county, Guizhou province, which reached 12.1 mg/kg, and believed that the reason for high concentration of mercury in coals was the effect of low-temperature hydrothermal fluids. Huang et al. [15] studied the distribution of mercury in Wolong Lake coal mine and found that the closer to the magma intrusion, the greater the mercury content in coals. Moreover, hydrothermal activity may also be an important factor causing mercury enrichment in coals [26].

3.2 Mercury content in coals of various coal-forming periods

The 970 coal samples analyzed cover all coal-forming periods in China, including Devonian (D), Early Carboniferous (C_1), Late Carboniferous (C_3), Early Permian (P_1), Late Permian (P_2), Late Triassic (T_3), Jurassic (J), Cretaceous (K), Paleogene (E). The distribution of mercury in C_3, P_1, and P_2 periods is relatively concentrated. The mercury concentration in coals varies greatly in different coal-forming periods (Fig. 2). The decreasing order of mercury concentration in coals: K (0.320 mg/kg) > P_2 (0.220 mg/kg) > C_3 (0.179 mg/kg) > J (0.177 mg/kg) > D (0.165 mg/kg) > P_1 (0.136 mg/kg) > C_1 (0.090 mg/kg) > E (0.086 mg/kg) > T_3 (0.066 mg/kg).

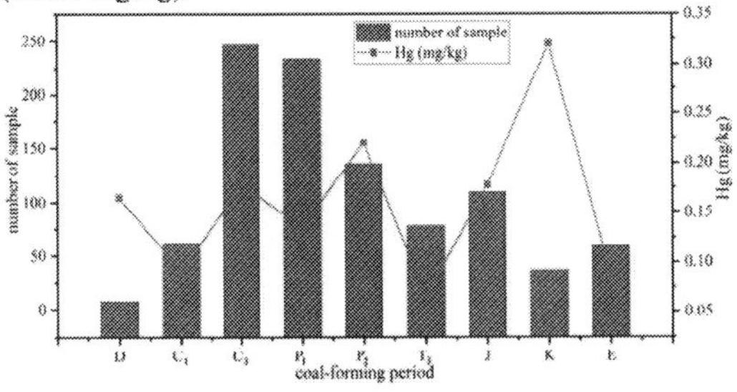

Fig. 2. Concentration of mercury in coals of different coal-forming periods.

3.3 Mercury content in coals of different ranks

Fig. 3 shows the mercury concentration in different coal ranks. Mercury is more widely distributed in gas coal, coking coal, anthracite, long-flame coal and lignite. The concentration of lean coal is the highest, while in long flame coal is the lowest. High rank meagre coal and anthracite have similar mercury concentrations. The mercury concentrations of different coal ranks are as follows: Lignite (0.164 mg/kg), long flame coal (0.078 mg/kg), non-caking coal (0.256 mg/kg), weakly caking coal (0.086 mg/kg), gas coal (0.151 mg/kg), fat coal (0.122 mg/kg), coking coal (0.171 mg/kg), lean coal (0.393 mg/kg), meagre coal (0.161 mg/kg), anthracite (0.160 mg/kg).

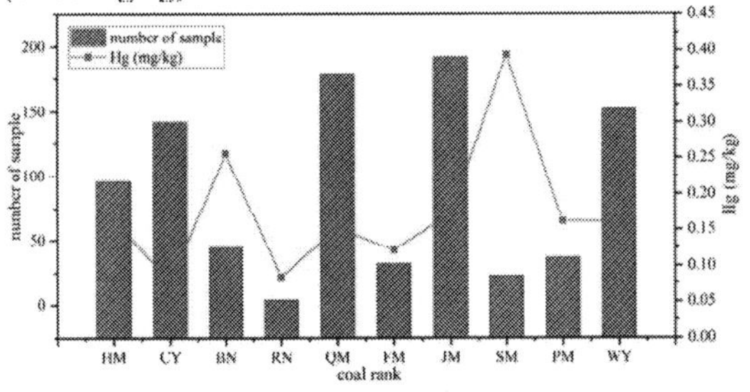

Fig. 3. Mercury content in coals of different ranks. (HM, CY, BN, RN, QM, FM, JM, SM, PM, WY represent lignite, long flame coal, non-caking coal, weakly caking coal, gas coal, fat coal, coking coal, lean coal, meagre coal, anthracite, respectively.)

4 Occurrence of mercury in coals

4.1 Correlation analysis of mercury and ash content in coals

Based on the data of 385 coal samples, the relationship between mercury and ash content in coals was analyzed (Fig. 4). There is no obvious correlation between mercury and ash in coals (R=0.193, $p<0.01$, n=385). Similar to the results of this analysis, Ge et al. [27] pointed out that the correlation between mercury and ash in Huainan coal mines is poor, and it does not exist in natural minerals, but exists in the form of organic matter. However, Cheng [4], Zhao [28], Pan [29], and Dai [30] et al. showed a positive correlation between mercury and ash content. Overview, the occurrence of mercury in coals has not yet produced a unified conclusion.

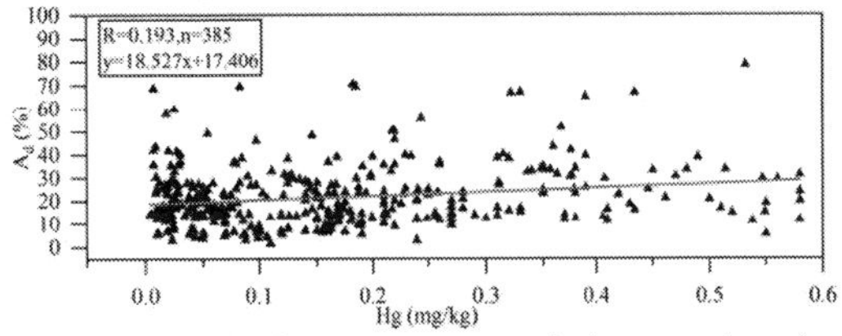

Fig. 4. Correlation between mercury and ash content in coals.

4.2 Correlation analysis of mercury and sulfur content in coals

Fig. 5 shows the relationship between different forms of sulfur and mercury content in coals. Mercury concentration is closely related to total sulfur content, with the correlation coefficient (R) of 0.328 ($p<0.01$, n=425) (Fig. 5a). Specifically, the correlation coefficient between mercury and pyrite sulfur is significant (R=0.451, $p<0.01$, n=198), showing a strong positive correlation (Fig. 5b). However, there is no significant correlation between mercury content in coals and organic sulfur and sulfate sulfur (Fig. 5c, Fig. 5d). These indicate that pyrite may be the most important occurrence carrier of mercury. Dai [30], Kolker [31], Dziok [32], and Bai [33] et al. have also reached the same conclusion.

In fact, scholars have a variety of views on the occurrence of mercury. Most scholars recognize the close relationship between mercury and pyrite. Liu et al. [34] found that compared with organic sulfur, mercury in Yanzhou coalfield was more correlated with pyrite sulfur. It was reported that mercury was mainly combined with pyrite sulfur, and the presence of organic mercury in the form of secondary. Song et al. [16] analyzed the relationship between mercury and inorganic elements in Xinmi coal, and inferred that mercury was mainly related to pyrite or kaolinite. Kostova I et al. [35] found a strong relationship between mercury and pyrite in high-sulfur Maritsa-West lignite, especially epigenetic pyrite. In addition, some scholars believe that mercury can be bound by organic matter or dispersed and tightly mixed in the form of submicron particle size inside organic matter, but it is not organic mercury in the true sense [18, 27, 28].

In summary, the sulfide bound state is considered to be the main form of mercury, especially pyrite as the main occurrence medium.

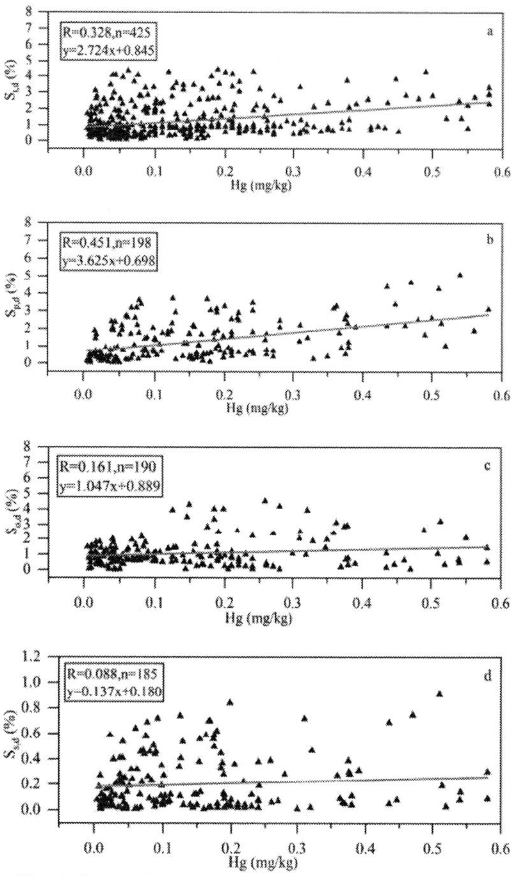

Fig. 5. Correlation between mercury and different forms of sulfur content in coals.

5 Conclusion

Based on 970 coal samples, the average content of mercury in Chinese coals is 0.14 mg/kg. The distribution of mercury content is unevenly distributed. The middle and high mercury coals are mainly distributed in southwest and northeast China. Low mercury coals and ultra-low mercury coals are mainly concentrated in north China. The mercury concentration decreases in the following order in different coal-forming periods: K (0.320 mg/kg) > P_2 (0.220 mg/kg) > C_3 (0.179 mg/kg) > J (0.177 mg/kg) > D (0.165 mg/kg) > P_1 (0.136 mg/kg) > C_1 (0.090 mg/kg) > E (0.086 mg/kg) > T_3 (0.066 mg/kg). The mercury concentrations of various coal ranks are as follows: Lignite (0.164 mg/kg), long flame coal (0.078 mg/kg), non-caking coal (0.256 mg/kg), weakly caking coal (0.086 mg/kg), gas coal (0.151 mg/kg), fat coal (0.122 mg/kg), coking coal (0.171 mg/kg), lean coal (0.393 mg/kg), meagre coal (0.161 mg/kg), anthracite (0.160 mg/kg). Mercury in coals has no obvious correlation with ash, organic sulfur and sulfate sulfur content, but has a strong affinity with total sulfur and pyrite sulfur. Pyrite is the main carrier of mercury in coals.

References

1. X.B. Feng, J.B. Chen, X.W. Fu, H.Y. Hu, P. Li, G.L. Qiu, H.Y. Yan, R.S. Yin, H. Zhang, W. Zhu, Progresses on environmental geochemistry of mercury. Bull. Mineral., Petrol. Geochem. 32, 503 (2013) (in Chinese)
2. T.K. Tsui, J.D. Blum, S.Y. Kwon, Review of stable mercury isotopes in ecology and biogeochemistry. Sci. Total. Environ. 716, 135386.1 (2020)
3. J.T. Dvonch, J.R. Graney, G.J. Keeler, R.K. Stevens, Use of elemental tracers to source apportion Hg in South Florida precipitation. EnViron. Sci. Technol. 33, 4522 (2014)
4. Z. Cheng, C. Gang, T. Yang, G. Lu, C. Mak, D. Kelly, Z. Xu, An investigation on mercury association in an Alberta subbituminous coal. Energ. Fuel. 21, 485 (2007)
5. Q.R. Wu, G.L. Li, S.X. Wang, K.Y. Liu, J.M. Hao, Mitigation options of atmospheric Hg emissions in China. EnViron. Sci. Technol. 52, 12368 (2018)
6. W.J. Huang, Z. Qu, W.M. Chen, H.M. Xu, N.Q. Yan, An enhancement method for the elemental mercury removal from coal-fired flue gas based on novel discharge activation reactor. Fuel. 171, 59 (2016)
7. G. Wang, F. Cao, B. Shan, M. Meng, W. Wang, R.Y. Sun, Sources and assessment of mercury and other heavy metal contamination in soils surrounding the Wuda underground coal fire area, Inner Mongolia, China. Bull. Environ. Contam. Toxicol. 103, 828 (2019)
8. J. Chen, G.J. Liu, Y. Kang, B. Wu, R.Y. Sun, C.C. Zhou, D. Wu, Atmospheric emissions of F, As, Se, Hg, and Sb from coal-fired power and heat generation in China. Chemosphere. 90, 1925 (2013)
9. Q.R. Wu, S.X. Wang, G.L. Li, S. Liang, C.J. Lin, Y.F. Wang, S.Y. Cai, K.Y. Liu, J.M. Hao, Temporal trend and spatial distribution of speciated atmospheric mercury emissions in China during 1978-2014. Environ. Sci. Technol. 50, 13428 (2016)
10. P.A. Ariya, M. Amyot, A. Dastoor, D. Deeds, A. Feinberg, G. Kos, A. Poulain, A. Ryjkov, K. Semeniuk, M. Subir, Mercury physicochemical and biogeochemical transformation in the atmosphere and at atmospheric interfaces: A review and future directions. Chem. Rev. 115, 3760 (2015)
11. S.F. Dai, D.Y. Ren, C.L. Chou, R.B. Finkelman, V.V. Seredin, Y.P. Zhou, Geochemistry of trace elements in Chinese coals: A review of abundances, genetic types, impacts on human health, and industrial utilization. Int. J. Coal Geol. 94, 3 (2012)
12. L.G. Zheng, G.J. Liu, C.C. Qi, Y. Z, M.H. Wong, The use of sequential extraction to determine the distribution and modes of occurrence of mercury in Permian Huaibei coal, Anhui province, China. Int. J. Coal Geol. 73, 139 (2008)
13. S.F. Dai, R.S. Zeng, Y.Z. Sun, Enrichment of arsenic, antimony, mercury, and thallium in a Late Permian

anthracite from Xingren, Guizhou, Southwest China. Int. J. Coal Geol. 66, 217 (2006)

14. J. Wu, J.H. Zhang, S. Zheng, Y.Y. Zhang, W.G. Pan, J.X. Ren, S. Chen, K. Ding, T. Zhang, Experimental study on the relationship between mercury enrichment coal and the depth of coalseam. Adv. Mater. Res. 268, 351 (2011)
15. X.Y. Huang, L.G. Zheng, Q.W. Zhang, H. Chu, H. Chu, X.X. Yan, Y.Y. Han, Distribution and modes of occurrence of mercury in coal seams Altered by magmatic hydrothermal from Wolonghu coal mine. Geol. J. China Univ. 21, 280 (2015) (in Chinese)
16. D.Y. Song, C.H. Song, B.Y. Song, C.B. Yang, Y.B. Li, Geochemistry of mercury in the Permian tectonically deformed coals from Peigou mine, Xinmi coalfield, China. Acta Geol. Sin.-Engl. Ed. 91, 2243 (2017)
17. L.B. Gao, Y.B. Wang, Q.W. Huang, S.Q. Guo, Modes of occurrence and thermal stability of mercury in different samples from Guandi coal preparation plant. Fuel. 200, 22 (2017)
18. X.D. Wang, D. Wu, G.J. Liu, R.Y. Sun, Variation of Hg content in low sulfur coals in relation to the coal-forming environment: A case study from Zhuji coal mine, Huainan coalfield, North China. Environ. Earth Sci. 77, 703 (2018)
19. Q.Y. Cao, L. Yang, W.Y. Ren, Y. L. Song, S.Y. Huang, Y.T. Wang, Z.Y. Wang, Spatial distribution of harmful trace elements in Chinese coalfields: An application of WebGIS technology. Sci. Total. Environ. 755, 142527 (2021)
20. A.Y. Yang, Z.C. Yan, R.T. Hui, Z.Y. Shen, K. Zhuang, The Abundance, distribution, and modes of occurrence of Hg in Chinese coals. Sci. Tech. Engrg. 15, 93 (2015) (in Chinese)
21. D.Y. Ren, F.H. Zhao, S.F. Dai, J.Y. Zhang, K.L. Luo, Trace element geochemistry of coal (Science press, Beijng, 2006) (in Chinese)
22. J.Y. Zhao, X.Y. Tang, W.H. Huang, Abundance of trace elements in coal of China. Coal. Geol China. 14, 6 (2002) (in Chinese)
23. Y.E. Yudovich, M.P. Ketris, Mercury in coal: A review. Part 1. Geochemistry. Int. J. Coal Geol. 62, 107 (2005)
24. R.B. Finkelman, Trace and minor elements in coal. Org. Geochem. (1993)
25. State Administration for Market Regulation. Classification for content of harmful elements in coal-Part 4: Mercury. (GB/T 20475.4-2012) (in Chinese)
26. S.F. Dai, W.G. Zhang, C.R. Ward, V.V. Seredin, J.C. Hower, X. Li, W.J. Song, X.B. Wang, H. Kang, L.C. Zheng, P.P. Wang, D. Zhou, Mineralogical and geochemical anomalies of Late Permian coals from the Fusui coalfield, Guangxi province, Southern China: Influences of terrigenous materials and hydrothermal fluids. Int. J. Coal Geol. 105, 60 (2013)
27. T. Ge, A Research into the distribution characteristics & Natural existing condition of the mercury in the coal of Huainan city. Adv. Mater. Res. 599, 448 (2012)
28. S.L. Zhao, C. Sun, Y.Q. Zhang, T.T. Jiao, W.R. Zhang, P. Liang, H.W. Zhang, Determination of mercury occurrence and thermal stability in high ash bituminous coal based on sink-float and sequential chemical extraction method. Fuel. 253, 571 (2019)
29. J.H. Pan, C.C. Zhou, L.F. Cong, N.N. Zhang, C. Liu, C.B. Peng, C.H. Ouyang, Mercury in Chinese coals: Modes of occurrence and its removal statistical laws during coal separation. Energ. Fuel. 31, 986 (2016)
30. S.F. Dai, V.V. Seredin, C.R. Ward, J.C. Hower, Y.W. Xing, W.G. Zhang, W.J. Song, P.P. Wang, Enrichment of U-Se-Mo-Re-V in coals preserved within marine carbonate successions: Geochemical and mineralogical data from the Late Permian Guiding coalfield, Guizhou, China. Miner. Depos. 50, 159 (2015)
31. A. Kolker, C.L. Senior, J.C. Quick, Mercury in coal and the impact of coal quality on mercury emissions from combustion systems. Appl. Geochem. 21, 1821 (2006)
32. T. Dziok, A. Strugała, A. Włodek, Studies on mercury occurrence in inorganic constituents of Polish coking coals. Environ. Sci. Pollut. Res. 26, 1 (2018)
33. X.F. Bai, W.H. Li, W. Yue, D. Hua, The distribution and occurrence of mercury in Chinese coals. Int. J. Coal Sci.Tech. 4, 172 (2017)
34. G.J. Liu, L.G. Zheng, Y. Zhang, C.C. Qi, Y.W. Chen, Z.C. Peng, Distribution and mode of occurrence of As, Hg and Se and Sulfur in coal seam 3 of the Shanxi formation, Yanzhou coalfield, China. Int. J. Coal Geol. 71, 371 (2007)
35. I. Kostova, J.C. Hower, D. Apostolova, Q. Wei, Concentration, distribution, and mode of occurrence of mercury in Bulgarian high-sulphur coals. Earth. Env. Sci. 609, 12 (2020)

A New Method to Determine the Lower Limit of Reservoir Physical Properties—Corrected Minimum Flow Pore-throat Radius Method

Jinyou Dai[1,2,*], Lixin Lin[2], and Rui Wang[2]

[1]State Key Laboratory of Petroleum Resources and Prospecting, China University of Petroleum Beijing, Beijing, China
[2]School of Petroleum Engineering, China University of Petroleum Beijing, Beijing, China

Abstract. The lower limit of reservoir physical properties is an important parameter for identifying reservoirs and determining effective thickness in reserves evaluation, and is also an important basis for selecting perforated test intervals in oilfield exploration and development. There are many methods to determine the lower limit of reservoir physical properties, and the minimum flow pore throat radius method is one of the commonly used methods. The method uses 0.1μm as the minimum flow pore-throat radius, and uses this to calibrate the lower limit of reservoir physical properties. However, according to the water film theory, the minimum radius of the reservoir's flowing pore throat is not a definite value, but varies with the displacement dynamics. Therefore, there is no exact basis for using 0.1μm as the minimum flow pore-throat radius, so it needs to be corrected. To this end, a new method for determining the lower limit of reservoir physical properties—the corrected minimum flow pore-throat radius method is proposed. The correction method comprehensively considers the factors of oil and gas accumulation dynamics, and determines the lower limit of reservoir physical properties by obtaining the minimum flow pore-throat radius value suitable for oil and gas accumulation dynamics. A case study of Chang 6^3 reservoir in A Oilfield shows that the minimum flow pore radius of oil and gas determined by the correction method is 0.08 μm, and the lower limit of reservoir physical properties (porosity 9.1%, permeability 0.117×10^{-3} μm^2). The traditional method has a minimum flow pore-throat radius of 0.1 μm and a lower limit of reservoir physical properties (porosity of 9.8% and permeability of 0.133×10^{-3} μm^2). Due to full consideration of the impact of oil and gas accumulation dynamics, the minimum flow pore-throat radius determined by the correction method is more reliable than the traditional method, and the lower limit of the reservoir physical property calibrated by it has practical significance.

1 Introduction

The lower limit of reservoir physical properties is the minimum effective porosity and minimum permeability that can store and percolate fluid. It is usually expressed by a certain value of porosity or permeability [1]. The lower limit of reservoir physical properties is an important parameter for identifying reservoirs and determining effective thickness in reserves evaluation, and it is also an important basis for selecting perforated test intervals in oilfield exploration and development [2-3]. In view of the basicity and necessity of the lower limit of reservoir physical properties, research on the lower limit of reservoir physical properties is the focus of oil and gas reservoir geology and engineering. For a long time, many scientists and technicians have conducted a lot of research on the lower limit of reservoir physical properties, and put forward more methods to determine the lower limit of reservoir physical properties [4-31]. Among them, the minimum flow pore-throat radius method is one of the commonly used methods to determine the lower limit of reservoir physical properties [24-31]. The method believes that the pores and throats of rocks are spaces and channels for oil and gas storage and flow. Whether oil and gas can flow out of rocks under a certain pressure difference depends on the thickness of the throat. That is, the radius of the rock throat is the key factor that determines whether oil and gas can flow out of the rock under a certain pressure difference. This minimum throat radius that can store oil and gas and allow oil and gas to seep is the minimum flow pore-throat radius of oil and gas [24,25,32]. After determining the minimum flow pore-throat radius, the correlation curve of pore-throat radius and porosity and permeability can be drawn according to the principle of statistical analysis, and the corresponding porosity and permeability lower limit values can be calibrated according to the minimum flow pore-throat radius [26-27]. At present, the minimum flow pore-throat radius method generally adopts 0.1μm as the minimum flow pore-throat radius of oil and gas. It is also believed that 0.1μm is equivalent to the thickness

*Corresponding author e-mail: d772512281@qq.com

of the water film attached to the surface of the water-wet clastic rock. The pore-throat radius is less than this value. Oil and gas are not flowing, and the corresponding porosity and permeability values are the lower limits of reservoir physical properties [24-31].

The advantage of the minimum flow pore-throat radius method is that the lower limit of the reservoir physical properties and the minimum flow pore-throat radius are linked together, and the lower limit of the reservoir physical properties corresponding to the specific pore-throat radius (0.1μm) is calibrated. The process is simple and clear, easy to master. However, this method also has obvious shortcomings. The theoretical basis for determining the minimum flow pore-throat radius as a fixed value (0.1 μm) is not sufficient. According to previous studies on water film theory [33-34], the thickness of the water film attached to the rock surface is not fixed, and the thickness of the water film is a function of the displacement pressure. As the displacement power increases, the thickness of the water film gradually decreases, and the corresponding minimum flow pore-throat radius of oil and gas must also decrease, which shows that the minimum flow pore-throat radius should not be a certain value. From this point of view, it is not appropriate to use 0.1μm as the minimum flow pore-throat radius, and there is a certain deviation in the lower limit of the physical properties calibrated by this. Therefore, how to calculate the minimum flow pore-throat radius value more accurately? It is the key to objectively calibrate the lower limit of reservoir physical properties using the minimum flow pore-throat radius method. To this end, this paper proposes a new method to determine the lower limit of reservoir physical properties-the corrected minimum flow pore-throat radius method (hereinafter referred to as the correction method).

2 Method principle

Pore throats in reservoirs vary in size and geometry, and pore throats of different orders interweave to form an intricately interconnected system. The complexity of the reservoir pore-throat network, especially the order of pore-throat distribution, will inevitably have a direct and significant effect on the seepage response of the reservoir fluid. When the displacement power is constant, only the part of the pore-throat network with pore-throat size greater than a certain limit will seep, while other parts of the pore-throat network below this limit cannot seep. That is, the fluid seepage in the pore-throat network is selective. This phenomenon can be understood as, as the pore-throat size of the reservoir decreases, the liquid particles are continuously strengthened by the capillary force and the molecular force of the surrounding solid interface, and the seepage resistance increases accordingly. When the pore-throat radius reaches a certain limit, the displacement power and the seepage resistance are in equilibrium. When the pore-throat radius is greater than this limit, the displacement power is greater than the seepage resistance, and the liquid flows. When the pore-throat radius is smaller than this limit, the displacement power is less than the seepage resistance, and the liquid does not flow. Therefore, the pore-throat radius in the displacement equilibrium state is the minimum flow pore-throat radius.

As the displacement power increases, the minimum flow pore-throat radius of the reservoir gradually decreases. When the minimum flow pore-throat radius of the reservoir is reduced to just the full use of oil and gas in the reservoir, the corresponding minimum flow pore-throat radius is the minimum flow pore-throat radius of the oil and gas, and the lower limit of physical properties calibrated based on this is the effective lower limit. Because the production process is limited by many factors, it is difficult to accurately determine whether the oil and gas are fully utilized and the minimum flow pore-throat radius when the oil and gas is fully utilized. However, considering that the process of oil and gas accumulation and production is the reciprocal process of oil drainage and water flooding, the minimum flow pore-throat radius of oil and gas can be obtained according to the oil and gas accumulation dynamics.

The mercury intrusion curve is a curve reflecting the change of the saturation of the non-wet phase with the displacement pressure during the displacement of the wet-phase fluid by the non-wet phase fluid. The mercury intrusion process can be regarded as the accumulation and filling process of oil drainage, so the mercury intrusion curve actually reflects the correspondence between oil saturation and accumulation dynamics. In other words, the mercury inlet pressure corresponding to the original oil saturation read on the mercury pressure curve is the accumulation power, and the pore-throat radius corresponding to this accumulation power is the minimum pore-throat radius of oil and gas. Based on the determination of the minimum flow pore-throat radius, and then based on the principle of statistical analysis, the correlation curve of pore-throat radius and porosity and permeability is drawn, and the lower limit of reservoir physical properties can be calibrated according to the minimum flow pore-throat radius. Taking the Chang 6_3 reservoir in the A oilfield of the Ordos Basin as an example, the lower limit of the physical properties of the reservoir is determined by the correction method.

3 Calculation example

3.1 Geological overview

The structure of the A oil field belongs to the southwest of the Yishan slope in the Ordos Basin. It is located in Huachi and Qingyang in Gansu Province, with an area of 2600km^2. More than 300 exploration and evaluation wells have been drilled. The main production layer of the oil field is the Chang 6_3 oil group of the Upper Triassic Yanchang Formation in the Triassic system, which belongs to the gravity flow sedimentation of deep lake-semi-deep lake facies, with an average thickness of 47m and a sand-to-land ratio of 0.52. The surface of the area belongs to the loess plateau landform, the terrain is undulating, the ground elevation is about 1150-1650m, and the relative height difference is about 500m. The structure of Chang 6 period is relatively simple. The overall structure is a gentle west-dipping monoclinic with an inclination angle of less than 1 degree. Analysis of core physical property data of 8288 samples from 77 cored wells shows that the porosity of Chang 6_3 reservoir in A oilfield is distributed between 4 and 15%, with an average porosity of 9.1%. The permeability distribution is between 0.01 and 0.8 × $10^{-3}\mu m^2$, and the average permeability is 0.152 × $10^{-3}\mu m^2$. The reservoir belongs to ultra-low porosity-ultra-low permeability reservoir.

3.2 Characteristics of mercury intrusion curve

Based on core observation and sample collection, five rock samples were selected for mercury intrusion experiment. The characteristics of the rock samples are shown in Table 1. It can be seen that the porosity of the rock samples is distributed between 6.86 and 13.51%, with an average of 9.97%; the permeability is distributed between 0.035 and 0.203 × $10^{-3}\mu m^2$, with an average of 0.137 × $10^{-3}\mu m^2$; It can reflect the physical properties of Chang 6_3 low porosity and low permeability reservoir.

Table 1. Characteristics of rock samples.

Well no.	Well depth (m)	horizon	Core number	Length (cm)	Diameter (cm)	Porosity (%)	Permeability ($\times 10^{-3}\mu m^2$)
Shan 127	1951.25	Chang 6_3	1 #	6.45	2.53	12.06	0.203
Bai 221	2064.1	Chang 6_3	2 #	6.46	2.53	13.51	0.186
Bai 269	1936.27	Chang 6_3	3 #	6.1	2.53	9.21	0.123
Shan 156	2060.1	Chang 6_3	4 #	6.41	2.53	8.23	0.137
Wu 85	1991.79	Chang 6_3	5 #	6.67	2.53	6.86	0.035
Average				6.42	2.53	9.97	0.137

According to the experimental data of mercury intrusion, a mercury intrusion curve (Figure 1) was drawn and the characteristics of mercury intrusion parameters of rock samples were calculated (Table 2).

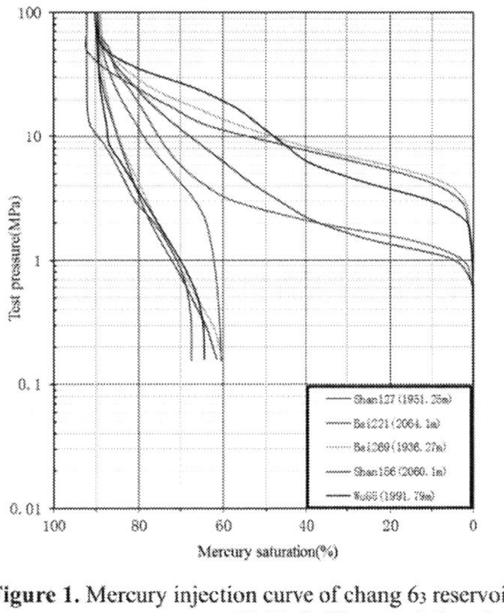

Figure 1. Mercury injection curve of chang 6_3 reservoir.

It can be seen that the displacement pressure of the five rock samples is distributed between 0.78 and 2.85MPa, with an average of 1.82MPa. The median pressure is distributed between 2.61 and 14.31MPa, with an average of 8.51MPa. The maximum pore-throat radius is distributed between 0.258 and 0.943μm, with an average of 0.524μm. The median throat radius is between 0.051 and 0.282μm, with an average of 0.135μm. The maximum mercury saturation is between 83.94 and 92.13%, with an average of 87.58%. The mercury withdrawal efficiency is distributed between 26.83 and 32.6%, with an average of 30.36%. Overall, the displacement pressure and median pressure of the Chang 63 reservoir are higher, the median throat radius is lower (average 0.135μm), the reservoir throat is small, the maximum mercury saturation is high (average 87.58%), and mercury is withdrawn Low efficiency (average 30.36%).

Table 2. Characteristics of mercury injection parameters of rock samples.

Well no.	Core number	Displacement pressure (MPa)	Median pressure (MPa)	Maximum pore-throat radius (μm)	Median radius of pore throat (μm)	Maximum mercury saturation (%)	Exit efficiency (%)
Shan 127	1 #	0.98	2.61	0.750	0.282	86.67	32.60
Bai 221	2 #	0.78	3.60	0.943	0.204	87.47	31.50
Bai 269	3 #	2.85	12.63	0.258	0.058	87.67	32.19
Shan 156	4 #	2.55	9.38	0.288	0.078	92.13	26.83
Wu 85	5 #	1.93	14.31	0.381	0.051	83.94	28.66
Average		1.82	8.51	0.524	0.135	87.58	30.36

3.3 Reservoir-forming power and minimum flow pore-throat radius

The original oil saturation is an important basis for determining the accumulation power. Since the rock samples taken are not in the original state, the original oil saturation cannot be measured. To this end, based on the logging interpretation results, the original oil saturation of the five rock samples is counted, and then combined with the mercury intrusion test curve to obtain the accumulation power and the minimum flow pore-throat radius (Table 3). It can be seen from Table 3 that the oil saturation of the five rock samples is distributed between 44.3 and 74.4%, with an average of 56.1%. The accumulation power is distributed between 8.97 and 9.31MPa, with an average of 9.18MPa. This shows that although the mercury intrusion curves of different rock samples are different in shape (Figure 1) and the oil saturations are also very different, the accumulation power is relatively stable (all around 9MPa). If factors such as thin Chang 6_3 reservoir, gentle underground structure, and interpretation error of oil saturation logging are considered, the reservoir-forming power of Chang 6_3 reservoir in A oilfield is a certain value. For this reason, based on various reservoir-forming dynamics values, the reservoir-forming power of Chang 6_3 reservoir calculated by arithmetic average method is 9.18 MPa, and the corresponding minimum flow pore radius is 0.08 μm.

Table 3. Characteristic table for determining the minimum flow pore-throat radius of rock samples.

Well no.	Core number	Original oil saturation (%)	Reservoir-foming pressure (MPa)	Minimum flow pore-throat radius (μm)
Shan 127	1 #	74.4	9.31	
Bai 221	2 #	66.1	9.19	
Bai 269	3 #	44.3	9.17	
Shan 156	4 #	49.4	9.26	
Wu 85	5 #	46.2	8.97	
Average		56.1	9.18	0.08

3.4 lower limit of reservoir physical properties

On the basis of the determination of the minimum flow pore-throat radius of oil and gas in Chang 6_3 oil layer, according to the principle of statistical analysis, the correlation curves of pore-throat radius and porosity and permeability are drawn (Figure 2-Figure 3). According to the minimum flow pore-throat radius, the corresponding lower limit of porosity and permeability is calculated. Figure 2 is a graph of the intersection of the median radius and porosity of the reservoir pore throats of five mercury-injected samples. It can be seen that the two are logarithmically related and the fitting relationship is as following,

$$POR = 3.219 \times \ln(R_{50}) + 17.19 \quad (1)$$

The correlation coefficient is: $R^2 = 0.810$.

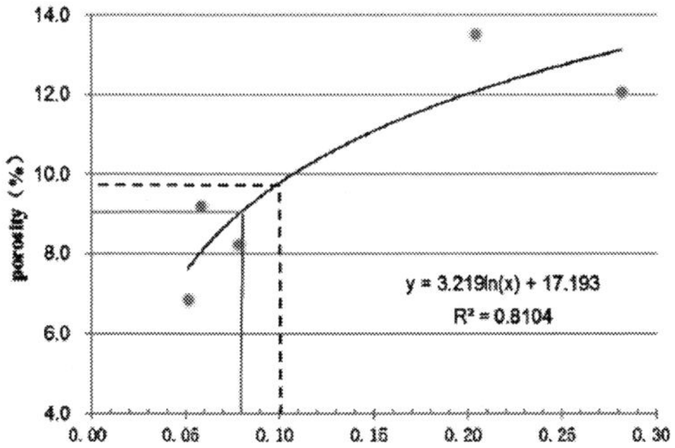

Figure 2. Crossplot of median radius of pore throat and porosity in chang 6_3 reservoir.

Figure 3 is the intersection graph of the median pore-throat radius and permeability of five mercury-injected samples. It can be seen that the two are logarithmically related, and the fitting relationship is as following,

$$PERM = 0.075 \times \ln(R_{50}) + 0.306 \quad (2)$$

The correlation coefficient is: $R^2 = 0.781$.

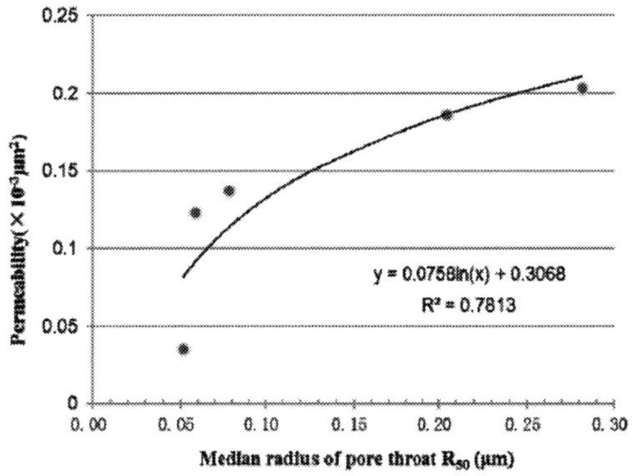

Figure 3. Crossplot of median radius of pore throat and permeability in chang 6_3 reservoir.

According to the minimum flow pore-throat radius of 0.08μm for oil and gas, the lower limit of physical properties of Chang 6_3 reservoir in A oilfield is calculated according to formula (1) and (2): porosity is 9.1%, permeability is 0.117 × 10^{-3}μm^2. If calculated according to the minimum flow pore-throat radius of 0.1μm for oil and gas, the lower limit of the physical properties of Chang 6_3 reservoir in A oilfield is: porosity 9.8%, permeability 0.133 × 10^{-3}μm^2. It can be seen from the comparison that the absolute error of the porosity is 0.7% and the relative error is 7.1% between the lower limit of the physical property of the reservoir calibrated according to 0.1 μm and the lower limit of the physical property of the reservoir calibrated by the correction method. The absolute error of permeability is 0.016 × 10^{-3}μm^2, and the relative error is 12%. Due to the full consideration of the influence of oil and gas accumulation dynamics, the minimum flow pore-throat radius determined by the correction method is more reliable than the traditional method, and the lower limit of the reservoir physical property calibrated by it is of guiding significance.

4 Conclusion and understanding

The minimum flow pore-throat radius of reservoir oil and gas is not a definite value, but a function of reservoir-forming power. As the reservoir-forming power increases, the minimum flow pore-throat radius of oil and gas decreases. The traditional method of using 0.1μm as the minimum flow pore-throat radius of oil and gas lacks a precise theoretical basis and needs to be corrected. The mercury intrusion curve reflects the accumulation process of oil drainage and the corresponding relationship between oil saturation and accumulation dynamics. The mercury intrusion curve can be used to determine the reservoir accumulation dynamics and the minimum flow pore-throat radius.

The mercury intrusion test and logging interpretation oil saturation analysis of 5 samples of Chang 6_3 reservoir in A oilfield show that the reservoir forming power of Chang 6_3 reservoir is 9.18 MPa, the corresponding minimum flow pore radius of oil and gas is 0.08 μm, and the lower limit of the reservoir physical properties is: porosity 9.1%, permeability $0.117 \times 10^{-3}\mu m^2$. While the traditional method takes the minimum flow pore-throat radius of 0.1μm, the lower limit of the calibration reservoir physical properties is: porosity 9.8%, permeability $0.133 \times 10^{-3}\mu m^2$. Because the influence of reservoir forming dynamics is fully considered, the minimum flow pore-throat radius determined by the correction method is more reliable than the traditional method, and the lower limit of reservoir physical properties calibrated by this method has practical significance.

Acknowledgments

This work was financially supported by the National Science and Technology Major Project "Effective Development Technology for Extra Low Permeability Reservoirs" of China (Grants No. 2011ZX05013—006).

References

1. Guo, R. (2004) Supplement to determining method of cut-off value of net pay. Petroleum Exploration and Development, 31: 140-143.
2. Liu, M.L., Feng, Z.P., Cai, Y.L., et al. (2014) Present situation and developmental trend of the research on methodology for determination of physical properties cut-off of an effective reservoir. Acta Geologica Sichuan, 34: 9-13.
3. Lu, Z.Y., Han, X.H., Zhang, X., et al. (2016) Research status and outlook for methods of determining petrophysical property cutoffs. Journal of China University of Petroleum(Edition of Natural Science), 40: 32-40.
4. Cong, L., Ma, S.Z., Li, W.L., et al. (2013) Study on physical property lower limit of effective reservoir of deep layer in Qingshui-Yuanyang Area. Advanced Materials Research, 2117: 104-108.
5. Zhang, C., Jiang, Y.Q., Guo, H.G., et al. (2010) Methods to determine the matrix petrophysical property low limits of the effective reservoir. Petroleum Geophysics, 8: 11-16.
6. Zhang, X.G. (2012) Application of empirical statistical method in the determination of lower limit of physical properties. World Well Logging Technology, 191: 17-18.
7. Geng, L.X., Cao, Y.S., Yi, Z.W., et al. (1999) A study on petrophysical property cutoffs of sandstone reservoirs in Pucheng oil field. Petroleum Exploration and Development, 26: 81-83.
8. Dong, L.Q. (2005) A research on reservoir physical texture lower-bound of the second zone of Xujiahe Formation of south area of west Sichuan, MS dissertation, Chengdu University of Technology, (April 2005).
9. Wan, N.M., Cao, Y.C., Zheng, L.J., et al. (2012) Lower limit studies of the physical properties for the effective reservoirs in sha no.4 member of Binnan-lijin area. Petroleum Geology & Oilfield Development in Daqing, 30: 50-56.
10. Wang, Y.Z., Cao, Y.C. (2010) Lower property limit and controls on deep effective clastic reservoirs of paleogene in Chezhen depression. Acta Sedimentologica Sinica, 28: 752-761.
11. Yu, W.Q., Ye, S.D., Lu, M.J. (2011) Study on lower limit of physical properties of E_1f_3 effective reservoir in Gaoyou Sag. Complex Hydrocarbon Reservoirs, 4: 5-9.
12. Jiao, C.H., Xia, D.D., Wang, J., et al. (2009) Methods for determining the petrophysical property cutoffs of extra-low porosity and permeability sandstone reservoirs——an example from the Xishanyao Formation reservoirs in Yongjin oilfield. Oil & Gas Geology, 30: 379-383.
13. Li, J., Yang, Y., Wang, S.F., et al. (2012) Determination of the lower limits of the physical properties of consolidated sandstone reservoirs in eastern Sulige gas field. Special Oil & Gas Reservoirs, 18: 52-56.
14. Jin, Y.Q., Wang, G.C., Zhang, T. (2010) A new study on the lower limit standard of effective reservoir thickness in Maliangou area of Zhiluo oilfield. Science and technology of western China, 9: 11-13.
15. Cui, Y.B. (2007) Method for determining the lower limit value of effective reservoir physical properties. World Well Logging Technology, 22: 32-35.
16. Li, X.Y., Guo, J.X., Zhang, Q.X., et al. (2008) Determining method for the lower limit of physical-property parameters in gas reservoirs. Natural Gas Exploration and Development, 31: 33-38.

17. Li, J., Zhao, F., Liu, P. (2012) Determination of lower limits of porosity and permeability of tight sand gas reservoirs in the eastern block of the Sulige Gas Field. Natural Gas Industry, 32: 31-35.
18. Han, X.H., Cui, Y.F., Li, F.B., et al. (2012) Classification and conventional logging identification of inhomogeneous reservoir in DB gas field. Progress in Geophysics, 27: 186-192.
19. Wang, Y.Z., Cao, Y.C., Song, G.Q., et al. (2008) Application of production test data to evaluation of the effective reservoir in deep clastic of Bonan Sag. Acta Petrolei Sinica, 29: 701-706, 710.
20. Wu, X.H., He, L., Fang, H.C. (2007) Analysis and application of specific productivity index. Petroleum Exploration and Development, 34: 745-749.
21. Shao, C.X., Wang, Y.Z., Cao, Y.C. (2008) Two new methods used to determine the low limits of effective reservoir physical properties and their applications—a case study on deep clastic reservoir of Palaeogene in Dongying depression. Journal of Oil and Gas Technology, 30: 414-416.
22. Wang, Y.Z., Cao, Y.C., Song, G.Q., et al. (2009) Determination of physical property lower limit of deep clastic effective reservoirs of Paleogene in Dongying depression. Journal of China University of Petroleum(Edition of Natural Science), 33: 16-21.
23. Li, M., Liu, J.Z., Wang, X.L., et al. (2010) Study on the lower limit standard of physical properties of Chang 6 ultra-low permeability reservoir in W South area of S oil field. Liaoning Chemical Industry, 39: 1065-1068.
24. Peng, B., Lv, G.X. (2008) Method for determining lower limit of physical properties of low porosity and low permeability sandstone reservoirs in Shan 23 member of Zizhou gas field. Petroleum Geophysics, 6: 11-13.
25. Gao, Y., Jiang, Y.Q., Yang, C.C., et al. (2011) Minimum flow pore throat radius for determination of the lower limits of parameters in low permeability reservoir. Science & Technology Review, 29: 34-38.
26. Wei, X.W., Xie, J.R., Tang, D.H., et al. (2008) Methods of determining the matrix petrophysical cutoffs of low porosity and low permeability sandstone reservoir—taking the J_2s1 oil reservoir in LA field in central Sichuan as an example. Natural Gas Industry, s1: 28-31.
27. Jiang, Y.Q., Gao, Y., Xu, H.W., et al. (2010) New method for determining cut-off parameter in hydrophilic and low permeable sandstone reservoir—case of gas reservoir. Petroleum Geology and Recovery Efficiency, 17: 58-60.
28. Fu, J.H., Luo, A.X., Zhang, N.N., et al. (2014) Determine lower limits of physical properties of effective reservoirs in Chang 7 Oil Formation in Ordos Basin. China Petroleum Exploration, 19: 82-87.
29. Shi, Y., Yang, Z.M., Huang, Y.Z. (2009) The study of cut-off pore-throat in reservoir. Journal of Southwest Petroleum University(Science & Technology Edition), 31: 42-44.
30. Duan, X.G., Wang, H.H., Hu, Y.Z., et al. (2003) Methods of determining the lower limits of different parameters of reservoir in Anpeng oilfield, Henan. Journal of Chengdu University of Technology(Science & Technology Edition), 30: 169-173.
31. Wang, W.M., Lu, S.F., Tian, W.C., et al. (2016) A new method to determine porosity and permeability cutoffs of tight oil reservoirs by using thickness of adsorption water film:A case study from the Damintun Sag,Liaohe oifield. Oil & Gas Geology, 37: 135-140.
32. Xiang, J.H., Yin, D.Y. (2018) The lower limit of the flowing pore throat radius in the extra-low permeability reservoir. IOP Conference Series: Earth and Environmental Science, 113.
33. He, C.Z., Hua, M.Q. (1998) Thickness of water film in oil and gas reservoirs. Petroleum Exploration and Development, 25: 91-93.
34. Liu, D.X., Yue, X.A., Hou, J.R., et al. (2005) Experimental study of adsorded water layer on solid particle surface. Acta Mineralogica Sinica, 25: 15-19.

Analysis and Solution of Condensation Phenomenon in Ring Main Unit

Kexin Shi[1*], JunjieYang[1]

[1] Shanghai Dianji University , Shanghai,201306, China

Abstract. The current condensation phenomenon inside the ring main unit will cause corrosion of components, short circuits, partial discharges and other problems, which will seriously affect the safety and durability of power supply. In response to the above problems, this paper designs and builds a condensation test platform, conducts an experimental analysis on the factors that cause condensation, and combines the conclusions drawn to propose a remote monitoring system for anti-condensation. Operation and maintenance personnel can use a small program to remotely monitor the internal environment of the ring network cabinet and deal with condensation in a timely manner. The application of this system provides great convenience for the operation and maintenance personnel, and also has a significant impact on the monitoring and maintenance of the anti-condensation of the ring network cabinet.

1 Introduction

With the introduction of the smart grid concept and the further increase in the scale of the demand for the distribution network, power companies have higher and higher requirements for the quality of power supply and the reliability of power supply [1]. An important means to improve the reliability of power supply is to adopt the ring network power supply technology. The ring network power supply can reasonably distribute the electricity load and shorten the power supply distance to ensure the power quality of the user side [2]. The key equipment of the ring network power supply is the ring network cabinet, which is a combined switchgear, which can combine multi-loop in and out lines, play a role in switching the operation mode of the distribution network, and ensure the power supply safety of the distribution network. It is widely distributed in substations, switch stations, factories and other areas.

Compared with other power equipment, the internal structure of the ring main unit is more complicated. Excessive temperature will generate a lot of water vapor in the closed ring main unit. When the temperature difference between the inside and outside of the cabinet is large, it is easy to cause condensation on the equipment. When the dust and liquid water vapor adhering inside the ring main unit are mixed, a conductive channel will be formed in the cabinet, which will affect the electrical insulation performance of the ring main unit equipment, so that the area that should not be conductive is transformed into an abnormally conductive area. Increased the probability of equipment failure. Once a failure occurs, it will inevitably cause local power supply interruption and even regional power grid paralysis accidents, causing huge economic losses, so the research on the analysis of the condensation phenomenon of the ring network cabinet and the study of solutions is very important.

2 Analysis of condensation phenomenon

2.1 Factors and hazards affecting condensation

The air under natural conditions is composed of a small amount of dust, water vapor and absolutely dry air[3], the water vapor that the air can hold is proportional to the ambient temperature, that is, the higher the ambient temperature, the air can hold a greater amount of water vapor[3], and the so-called dew point temperature refers to the highest temperature at which condensation occurs in the air with a specific humidity, and the water vapor contained in the air at a higher temperature [3]Due to the sudden temperature change, the water vapor in the air cannot be contained in the cabinet and can only be discharged in the form of liquid water vapor. Under normal circumstances, the moisture in the cable trench at the bottom of the ring network cabinet continues to rise with the passage of the cable room in the cabinet. When the humid air meets the top plate, terminals and other metal components in the cabinet, the surface temperature of the components is relatively high. Low, so that the humid air will be precipitated in the form of liquid water vapor in the collision of cold and heat, and condensation will occur at this time.

Once the condensation of mixed dust attached to the top plate of the cabinet exceeds a certain weight, it will drip on the wiring terminals in the cabinet, forming a

*Corresponding author's e-mail: yangjj@sudj.edu.cn

https://doi.org/10.1051/e3sconf/202129003005

conductive channel that does not exist originally, causing confusion in logic pulses[4], leading to electronic components The probability of failures such as failure and short circuit of the power supply is greatly increased, and it is more likely to cause serious accidents such as fires.

From the above analysis, it can be seen that by destroying the condensation conditions such as temperature difference and humidity, the occurrence of condensation can be eliminated, and destroying any of these conditions can reduce the occurrence of ring network cabinet failures.

2.2 Calculate the dew point value

When the water vapor content in the air is constant and the air pressure is kept constant, the temperature at which the air is cooled to saturation is called dew point temperature, referred to as dew point[4], which is actually the temperature at which water vapor and water reach an equilibrium state. According to the relevant information, there is a great relationship between the conditions of condensation and the dew point temperature. Therefore, we use the Magnus-Tetens approximation method to obtain the dew point for the next operation. The formula is:

$$T_d = \frac{b \bullet \gamma(T, RH)}{a - \gamma(T, RH)} \tag{1}$$

Among them, Td is the dew point temperature, T is the temperature, RH is the relative humidity, the constant a is equal to 17.27 degrees Celsius, the constant b is equal to 237.7 degrees Celsius, and γ is:

$$\gamma(T, RH) = \frac{aT}{b+T} + \ln(RH/100) \tag{2}$$

The dew point temperature is calculated according to the Magnus-Tetens approximation method formula, and the result of the dew point temperature calculation is shown in Figure 1. It can be seen from Figure 1 that when the air temperature in the cabinet does not change, the dew point temperature will continue to approach or exceed the air temperature in the cabinet as the relative humidity in the air in the cabinet continuously increases, and condensation will occur more easily. phenomenon. For example, when the air temperature in the cabinet is 40°C and the relative humidity is 10%RH, the dew point temperature in the cabinet is -1°C, which is much lower than the air temperature in the cabinet, so there is no condensation at this time. When the relative humidity is 70%RH under the same conditions, the dew point temperature in the cabinet is 37.4℃. It can be seen that the condensation phenomenon is caused by a temperature difference of 2.6°C. In the same way, when the relative humidity in the cabinet does not change, the dew point temperature will continue to approach or exceed the air temperature in the cabinet as the air temperature in the cabinet continuously increases, and condensation is more likely to occur. For example, when the air temperature in the cabinet is 30°C and the relative humidity is 70%RH, the surface dew point temperature of the objects in the cabinet is 26.9°C. When the relative humidity in the cabinet is 60%RH under the same conditions, the dew point temperature in the cabinet is 21.4°C.

Therefore, we conclude that the condensation phenomenon has a great relationship with the dew point temperature. When the dew point temperature is always lower than the temperature in the cabinet, it is not easy to produce condensation.

2.3 Test results

After calculating the dew point temperature in 2.2, the dew point temperature values under different temperatures and humidity are obtained, as shown in Figure 1. It can be seen from the figure that when the air temperature in the cabinet is constant, the dew point temperature will continue to approach or exceed the air temperature in the cabinet as the relative humidity in the air in the cabinet continuously increases, and condensation will occur more easily. For example, when the air temperature in the cabinet is 40°C and the relative humidity is 10%RH, the dew point temperature in the cabinet is -1°C, which is much lower than the air temperature in the cabinet, so there is no condensation at this time. When the relative humidity is 70%RH under the same conditions, the dew point temperature in the cabinet is 37.4℃. It can be seen that the condensation phenomenon is caused by a temperature difference of 2.6°C. In the same way, when the relative humidity in the cabinet does not change, the dew point temperature will continue to approach or exceed the air temperature in the cabinet as the air temperature in the cabinet continuously increases, and condensation is more likely to occur. For example, when the air temperature in the cabinet is 30°C and the relative humidity is 70%RH, the surface dew point temperature of the objects in the cabinet is 26.9°C. When the relative humidity in the cabinet is 60%RH under the same conditions, the dew point temperature in the cabinet is 21.4°C.

Therefore, we conclude that the condensation phenomenon has a great relationship with the dew point temperature. When the dew point temperature is always lower than the temperature in the cabinet, it is not easy to produce condensation.

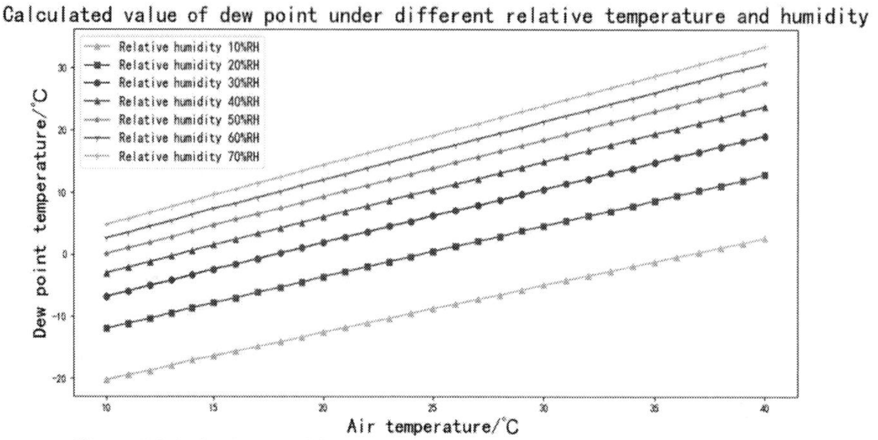

Figure 1 Calculated value of dew point under different relative temperature and humidity

At the same time, refer to related literature [5、6] and conduct experiments on the length of condensation time under different temperature and humidity conditions. The experiment was carried out when the temperature difference between the inside and outside of the cabinet was 5℃, 10℃, 20℃, the relative humidity was 70%-90%, and the temperature inside the cabinet was 20℃, 30℃, and 40℃. The experimental results are shown in Figure 2、Figure 3.

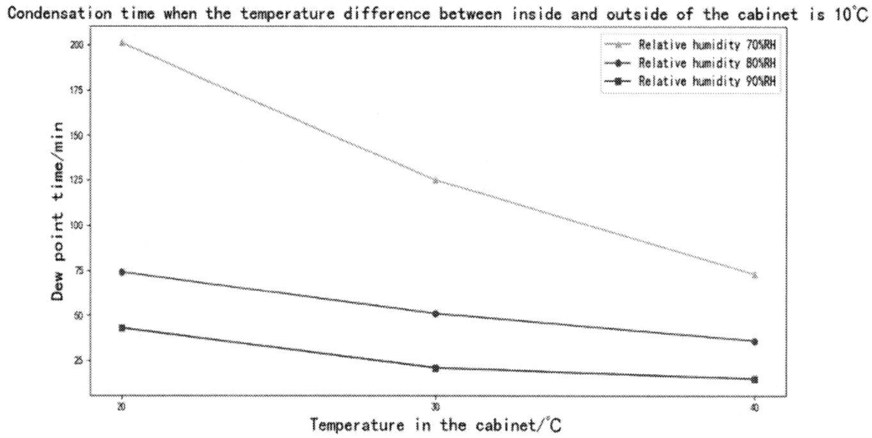

Figure 2 Comparison of condensation time under the same temperature difference between inside and outside the cabinet

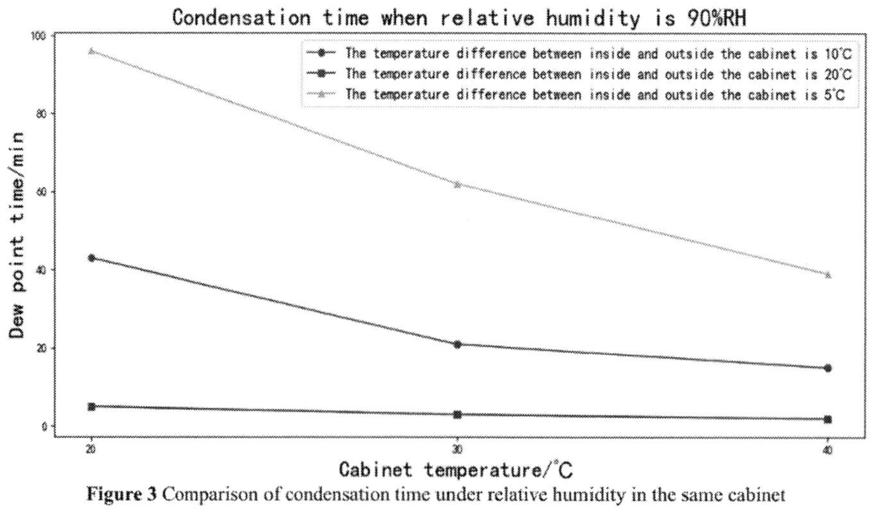

Figure 3 Comparison of condensation time under relative humidity in the same cabinet

2.4 Test summary

Through the design and construction of the temperature and humidity collection and condensation test system, the TH10S-B high-precision temperature and humidity transmitter is used to collect the temperature and humidity in the cabinet and calculate the dew point in the cabinet under different conditions, and record the condensation time under different conditions. The following conclusions are reached:

(1) The most important factor that causes condensation is the dew point temperature. When the air temperature in the cabinet is close to or higher than the dew point temperature, condensation is easy to occur.

(2) Sudden changes in any of the factors such as temperature difference inside and outside the cabinet, relative humidity inside the cabinet, and air temperature inside the cabinet will directly affect the occurrence of condensation. Similarly, the combined influence of the three will also cause condensation. As the temperature and humidity inside the cabinet continue to increase, the temperature difference between inside and outside the cabinet continues to increase, and the time for condensation to occur will become shorter and shorter.

3 Formatting the text

According to the conclusions drawn in Chapter 2, this paper designs and implements the anti-condensation smart applet of the ring network cabinet, which includes three parts: smart hardware terminal, cloud server and client. After collecting various important data running in real-time in the ring network cabinet, the collected data information is analyzed and transmitted to the monitoring center through the 4GDTU communication module to determine the current operating environment status of the ring network cabinet and determine whether there is an abnormality. If the sensor data measured inside the ring network cabinet deviates from the preset threshold of the sensor, an early warning message will be sent through the WeChat applet and operation and maintenance personnel will be arranged for rescue in time.

The modules designed and implemented are as follows:

Homepage map module: The homepage of the applet displays the map of the current area, displays the location information of all the ring network cabinets on the map and can reach the specific ring network cabinet address through the route navigation map.

Data information management module: Display the collected internal temperature and humidity data of the ring network cabinet in this module. Set up the Excel export function, you can save all the collected data in the form of a table locally, which is convenient for future viewing.

Abnormal data management module: Under this module, it is convenient for operation and maintenance personnel to view abnormal information about the ring network cabinet they manage on their mobile phones, such as viewing the internal operation of the ring network cabinet through video, and turning on the dehumidification device with one button if necessary.

Monitoring management module: Under this module, you can view the video of the abnormal time period online to determine whether a misjudgment has occurred. At the same time, the video download function is set up, and all monitoring videos can be saved locally for easy viewing in the future.

4 Formatting the text

Through the design and construction of the condensation test platform, the TH10S-B high-precision temperature and humidity transmitter is used to collect the temperature and humidity inside the cabinet, and the DHT11 temperature and humidity sensor is used to collect the temperature difference information inside and outside the cabinet. It is concluded that the most important factor for the condensation phenomenon is the dew point. Temperature, when the air temperature in the cabinet is close to or higher than the dew point temperature, condensation is easy to occur. In addition, the temperature and humidity inside the cabinet, and the temperature difference between inside and outside the cabinet, any sudden changes in any of them will affect the generation of condensation. As the temperature and humidity inside the cabinet continue to increase, the temperature difference between inside and outside the cabinet continues to increase, and the time for condensation to occur will become shorter and shorter.

According to the conclusions drawn from the experiment, we designed a remote monitoring system for anti-condensation, using the temperature and humidity information inside and outside the cabinet as the judgment standard, and can remotely monitor the appearance of condensation through the operation and maintenance personnel in the first time, so as to be able to timely Suppress condensation. This system will produce good results for practical applications.

References

1. Tong Zijing. Research on real-time temperature measurement method of T-connector of ring main unit[D]. North China Electric Power University (Beijing), 2018.
2. Lian Shuang. Analysis and Solution of Condensation Phenomenon in Outdoor Ring Network Cabinet [J]. Science and Technology Wind, 2020 (01): 184..
3. Yang Jie, Feng Hao, Yuan Jie, Xu Xian, Wang Hualiang, Lu Lijun, Chen Haifei. Design of a Concentrating Photovoltaic-Photothermal Coupled Seawater Desalination System[J]. Solar Energy, 2019(07): 60-63+ 18.Li Honglian, Bai Yu. Development of secondary anti-condensation device for switchgear[J]. Modern Business and Trade Industry, 2016, 37(01): 218-219.
4. Liu Ruoxi, Guo Qin, Geng Jianghai, Che Yao, Gui Yuan, Xu Xingquan, Guo Wei. Research on the anti-condensation measures of the secondary compartment of the ring network cabinet [J]. Journal

of Electric Power Science and Technology, 2017, 32(04): 154-158.

5. Guo Qin. Analysis and research on condensation mechanism and protective measures of 10kV ring network cabinet [D]. North China Electric Power University, 2017.

https://doi.org/10.1051/e3sconf/202129003006

Preparation of ring crosslinked poly (p-nitrile) nanoparticles and Study on adsorption behavior of uranium (VI)

Meixue Xu[1,*],Pei Tian[1], Jing Wen[1],Yang Yang[1],Yanfei Wang

[1]School of chemistry and chemical engineering, University of South China , 421000, PRC

Abstract. Cyclo crosslinked polyphosphazenes (PPNs) were prepared by Hexachlorocyclotriphosphazene (HCCP) with unique P = n structure and active p-cl group and green compound phloretin. The synthesis pro cess of polyphosphazene nanospheres with Hexachlorocyclotriphosphazene as the main body and phloretin was discussed. The effects of pH value and solid-liquid ratio of solution on the adsorption of uranium by pol yphosphazene nanospheres were studied. The results show that polyphosphazene nanospheres are off white powder in appearance and spherical in microstructure. When the pH value of the solution is 4 and the solid-l iquid ratio is 0.5g • L-1, the maximum adsorption rate of uranium is obtained.

1 INTRODUCTION

Adsorption method is a kind of wastewater treatment me thod that uses adsorbents with a large number of adsorpti on active groups to adsorb the radioactive substances in HLW to the surface. It is widely used in the treatment of HLW because of its advantages of simplicity, low cost a nd good effect. With the rapid development of social eco nomy, environmental protection has become one of the most concerned hot spots in today's society. Among the m, nuclear pollution is a widely concerned environmenta l problem, and uranium, the main resource of nuclear fue l, is one of the radioactive elements in waste liquid. Ther efore, the search for a green material that can efficiently remove uranium by adsorption has a far-reaching impact on the future environmental protection. At present, diyun Chen [2] Research Group has studied the adsorption beh avior of phosphate rock for uranium (VI) in mine sewage, the adsorption capacity is 0.1133mg • g-1, and the rem oval rate of uranium (VI) is 76.89%. M. K. sureshkumar [3] prepared chitosan tripolyphosphate (CTPP) particles by liquid internal solid method. The adsorption capacity of CTPP particles for uranium (VI) was 236.9mg • g-1. Xingcheng Ding [4] the research group prepared polyacr ylonitrile (cts-pao) grafted with amidoxime chitosan and used it to adsorb uranium (VI) from aqueous solution, wi th the removal rate of 86.02%. M. H. khani [5] used the s ynthesized polypyrrole to adsorb uranium (VI) in waste water, and the adsorption capacity was about 87.72 mg g-1. In addition, polyphosphazene microspheres have attracted the attention and application of some researche rs because of their superior characteristics to traditional a dsorption materials. However, there are few studies and l iterature reports on the application of polyphosphazene materials in HLW at home and abroad. In this paper, He xachlorocyclotriphosphazene co phloretin nanospheres w ere synthesized with green inorganic material Hexachlorocyclotriphosphazene as the main body and green organi c material phloretin, and characterized by SEM. Then, a series of single factor experiments were carried out to stu dy the optimal pH value and solid-liquid ratio of the synt hesized microspheres for uranium adsorption in aqueous solution.

2 Experimental methods

2.1 Experimental process

2.1.1 Preparation of cyclocrosslinked Hexachlorocy clotriphosphazene co phloretin nanospheres

Add 0.201 g Hexachlorocyclotriphosphazene and 0.4733 g phloretin into a 250 ml round bottom flask, pour in 10 0 ml acetonitrile solvent, disperse and dissolve in ultraso nic (100 W), add 4 ml triethylamine after dissolution, the n pass in nitrogen as protective gas, vacuum the flask thr ee times to remove oxygen, and react in ultrasonic (100 W) constant temperature water bath for 3 h, The device was transferred to an oil bath at about 60 ℃ for 12 h. Th e solution with the product was centrifuged to obtain soli d. Then the crude product was cross washed with ethanol and distilled water for three times, and finally dried in a n oven for 48 h to obtain crosslinked polyphosphazene m icrospheres. The synthesis route of cyclocrosslinked poly phosphazene phloretin nanospheres is shown in Fig. 1.

* Corresponding author: xmx_2902088301@163.com

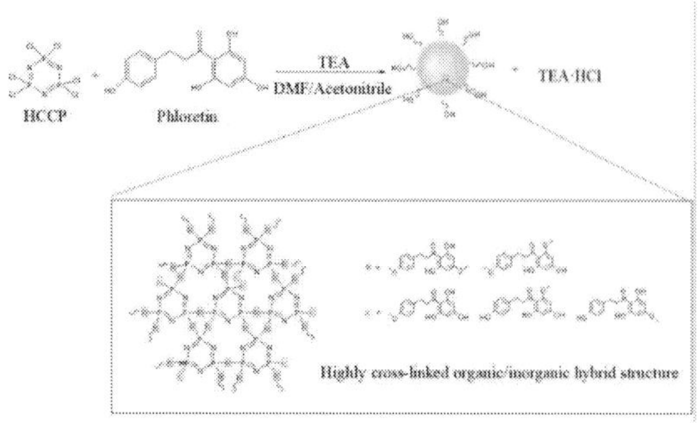

Fig. 1 Synthesis process of cyclic cross-linked polyphosphazen e-phloretin microspheres

2.1.2 Preparation of uranium (VI) standard solution

At first, 1g • L-1 of uranium (VI) reserve solution was prepared, and then the reserve solution was gradually dil uted to 60 mg • L-1, 50 mg • L-1, 40 mg • L-1, 30 m g • L-1, 20 mg • L-1 and 10 mg • L-1. Measure 1ml of each solution, add 1ml of arsenazo Tris, and prepare c hloroacetic acid sodium acetate buffer solution with pH of 2.5 to titrate the above solution to the scale line. Use d eionized water group as blank sample, measure three tim es in parallel on the visible spectrophotometer, record th e corresponding absorbance data at each concentration, a nd then use Origin Software for linear fitting, A linear cu rve and a linear fitting equation can be obtained, which i s the standard curve of uranium (VI) solution.

2.1.3 Effect of pH value on adsorption of uranium (Ⅵ) by cyclocrosslinked polyphosphazene nanosph eres

10ml of 30mg • L-1 uranium (Ⅵ) solution was accurat ely measured. The pH of the solution was adjusted by H NO3 and NaOH respectively to be 2,3.0,4.0,5.0,6.0 respe ctively. 20mg of HCRP co-pgm nanoparticles were adde d respectively. The reaction solution was separated by ul tra-fine filter membrane after the oscillation was complet ed, and 1ml of filtrate was transferred into 10ml volumet ric flask, and 1ml of arsenazo Ⅲ (0.05wt%) reagent was added into the volumetric flask with a pipette, Then, the buffer solution of chloroacetic acid sodium acetate with pH of 2.5 was added to the scale line by the drop of rubb er head dropper. The absorbance of filtrate was determin ed by uv-2100 spectrophotometer. The effect of pH on th e adsorption of uranium (Ⅵ) by the ring crosslinked poly phosphate root skin nanomicrospheres was analyzed. Th e calculation formula of adsorption quantity (q) and rem oval rate [6].

$$Q(mg\,g^{-1}) = \frac{(C_0 - C_e)V}{m} \quad (1.1)$$

$$R(\%) = \frac{(C_0 - C_e)}{C_0} \quad (1.2)$$

Where, q is the adsorption capacity (mg • g-1), C0 and CE represent the concentration of uranium (VI) ion befor e and after adsorption (mg • L-1), V is the volume of ur anium (VI) solution (L), and M is the amount of adsorbe nt (g).

2.1.4 Effect of solid-liquid ratio on adsorption of ura nium (Ⅵ) by cyclocrosslinked polyphosphazene na nospheres

The pH of 30mg • L-1 uranium (Ⅵ) solution was accur ately measured, and the pH of the solution was adjusted t o 3.5. 5mg, 10mg, 15mg, 20mg, 25mg and 30mg ring cr osslinked polyphosphate nanomicrospheres were added r espectively. The reaction solution was separated by ultra -fine filter membrane after the vibration was completed. 1ml of filtrate was transferred into 10ml volumetric flask, and 1ml of arsenazo Ⅲ (0.05wt%) reagent was added in to the volumetric flask with a pipette, Then, the buffer so lution of chloroacetic acid sodium acetate with pH of 2.5 was added to the scale line by the drop of rubber head dr opper. The absorbance of filtrate was determined by uv-2100 spectrophotometer. The effect of solid-liquid ratio on the adsorption of uranium (Ⅵ) by the ring crosslinked polyphosphate root skin nanosphere was analyzed.

3 Results and discussion

3.1 Characterization of cyclocrosslinked Hexachloroc yclotriphosphazene co phloretin nanospheres by SE M-EDS

Fig.2 Scanning electron microscopic study of pre adsorption ri ng crosslinked polyphosphazene phloretin nanospheres

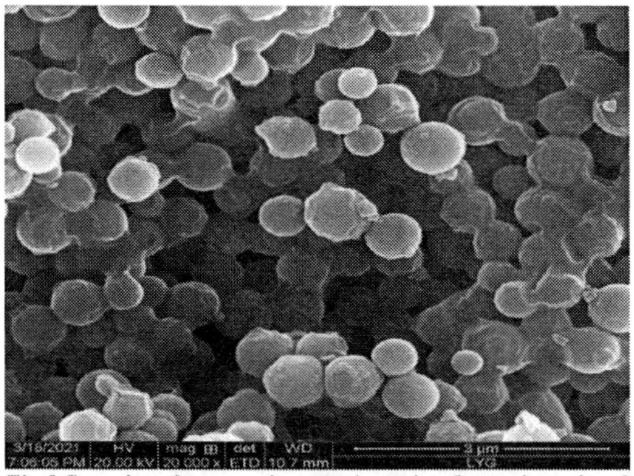

Fig.3 Scanning electron microscopic study of pre adsorption ri ng crosslinked polyphosphazene phloretin nanospheres

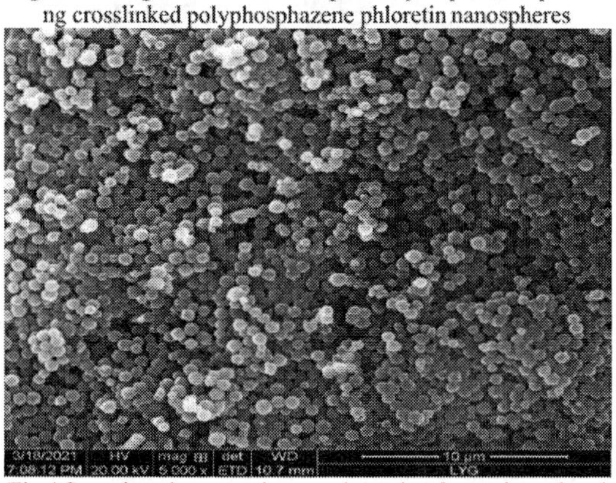

Fig.4 Scanning electron microscopic study of pre adsorption ri ng crosslinked polyphosphazene phloretin nanospheres

It can be seen from figures 2, 3 and 4 that the material pr esents regular and uniform spherical shape, which is in li ne with the initial conjecture on the structure of cyclocro sslinked Hexachlorocyclotriphosphazene co phloretin na nospheres.

3.2 Establishment of standard curve of uranium (VI) solution

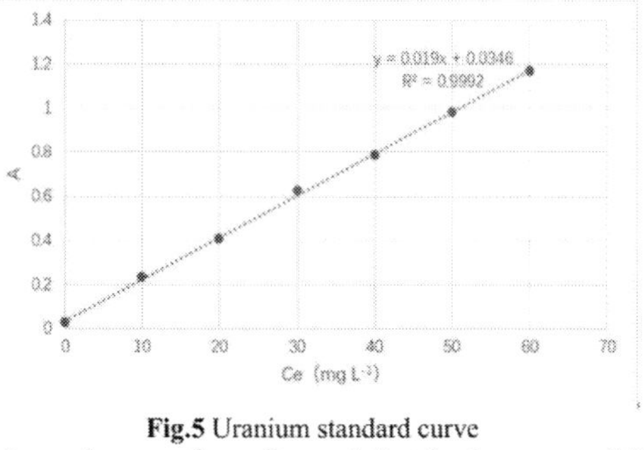

Fig.5 Uranium standard curve

It can be seen from figure 5 that in the range of 10-60mg • L- 1, there is a good linear relationship between the concentration of uranium (VI) and its absorbance (R2 = 0.9992), and the line ar equation can be obtained, as shown in formula 2.1

$$A = 0.019Ce + 0.0346 \qquad (2.1)$$

Where a is the absorbance of uranium (VI) and Ce (mg L-1) is the concentration of uranium (VI). When the absorbance of unknown uranium (VI) solution is known, the fitted equation can be used to calculate its concentrat ion.

3.3 Effect of pH value on adsorption of uranium (VI) by cyclophosphazene nanoparticles

One of the key factors for the adsorption of uranium (Ⅵ) by the cyclocrosslinked HCH co-pgm nanoparticles is th e pH value in the wastewater containing uranium, which has a great influence on the adsorption behavior. The pH value not only affects the binding sites on the adsorbent surface, but also affects the morphology of the adsorbate in the solution [1]. The relationship between the pH valu e of solution and the adsorption amount of uranium (VI) is shown in Figure 6, which has an effect on the adsorpti on of uranium (VI) by microspheres. It can be seen from the figure that when the pH is less than 2.0, the correspo nding adsorption amount is zero, indicating that when th e pH is lower than 2.0, the microspheres have no adsorpt ion effect on uranium containing wastewater. When the pH is less than 4 and greater than 2, the adsorption amou nt increases with the increase of pH value, The adsorptio n capacity of U (Ⅵ) by the cyclocrosslinked ppn-pgms i ncreases gradually. The above phenomena may be due to the presence of a large amount of hydrogen ions in urani um solution under acidic conditions, which may be easie r to bind with the adsorption sites in the cyclocrosslinked ppn-gpm nanoparticles, which leads to the particle of th e microspheres, which is due to the repelling of the same -sex charge, The particles will repel the positively charge d uranium (Ⅵ) ions, which is not conducive to the adsor ption of uranium (Ⅵ) [7] in the solution of the microsph eres, so this trend is presented. When the pH value of the solution is 4, the content of hydrogen ion in the solution decreases gradually. At this time, a large number of activ e adsorption sites are preferentially combined with urani um (Ⅵ) ions with positive charge, thus obtaining the ma ximum adsorption capacity. However, it can be seen fro m the figure that the adsorption of uranium by the micros pheres will decrease when the pH of the solution continu es to increase and is greater than 4. It is possible that ura nium (Ⅵ) is easy to hydrolyze in this pH range, and even form (UO2) 3 (OH) - 7 with negative charge. The appea rance of such ions may produce repulsion with the surfac e of the ring crosslinked hch-co-gpn nanoparticles, whic h also leads to the decrease of the adsorption strength of the adsorbent. Therefore, when the pH value of the soluti on is 4, the adsorption behavior of the cyclocrosslinked p pn-gpm nanoparticles to uranium (Ⅵ) is the best. At this time, the best adsorption capacity is 39.33mg • g-1, and the removal rate reaches 65.86%.

https://doi.org/10.1051/e3sconf/202129003006

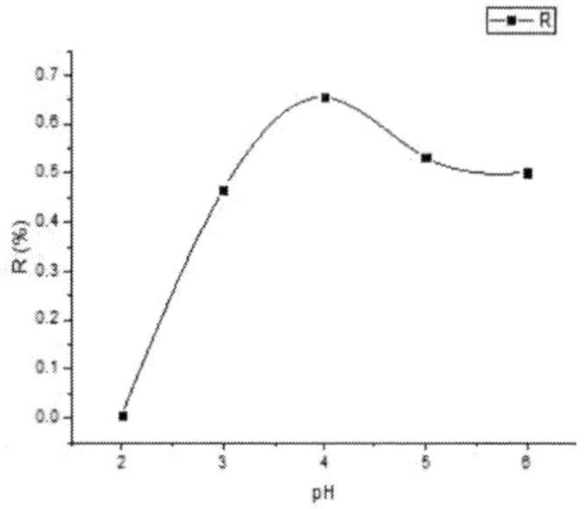

Fig.6 The effect of pH on adsorption of uranium

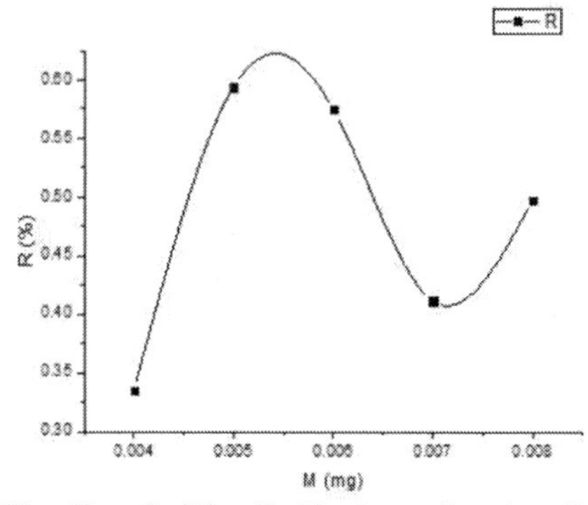

Fig.7 The effect of solid-to-liquid ratio on adsorption of uraniu m

3.4 Effect of solid-liquid ratio on adsorption of urani um (VI) by cyclocrosslinked Hexachlorocyclotriphosp hazene co phloretin nanospheres

The ratio of solid to liquid affects the economic benefit o f adsorption behavior. Too high or too low ratio of solid t o liquid is unfavorable to adsorption behavior. For this e xperiment, when the solid-liquid ratio is too large, not on ly the excess adsorbent will be wasted, but also the adsor ption capacity will be reduced [8], and the removal rate will not increase significantly, or even decline. When the solid-liquid ratio is low, the adsorption rate of the adsor bent for uranium (VI) will slow down, and even can not absorb the remaining uranium (VI) ions in the solution. I n order to study the effect of solid-liquid ratio on the ads orption of uranium (Ⅵ) by cyclohexachlorocyclotriphos phazene co phloretin nanospheres, the initial concentrati on of uranium (Ⅵ) was 3G • L-1, the adsorption time was 2 h, and the pH value was adjusted to 4.0. The volu me of uranium (Ⅵ) solution was 10 ml, and the mass of adsorbent was 4-8 mg. It can be seen from the influence of solid-liquid ratio on adsorption of uranium (VI) by mi crospheres in Fig. 7 that with the increase of solid-liquid ratio (M / V), the removal rate of uranium (VI) by cycloh exachlorocyclotriphosphazene co phloretin nanospheres gradually increases, and when the solid-liquid ratio is gre ater than 0.54g • L-1, it gradually tends to be flat, and i n a slight decline, the removal rate is 62.13%. This may be due to the existence of a certain "shielding effect [9]" on the surface of the adsorbent after reaching a certain ad sorption capacity, which has a certain repulsive effect on the remaining uranium (VI) ions. At this time, due to the decrease of the concentration of uranium (VI) ions, ther e is not enough driving force to offset the repulsive force. Therefore, after the concentration of 0.54 g • L-1, the r emoval rate gradually decreased. It is speculated that the best removal rate is 62.13% when the solid-liquid ratio is 0.54 g • L-1.

4 Conclusion

In this study, a novel adsorbent, cyclocrosslinked Hexac hlorocyclotriphosphazene co phloretin nanospheres, was synthesized and characterized by SEM. A series of singl e factor experiments were carried out to study the adsorp tion conditions of the synthesized microspheres for urani um in aqueous solution. The results showed that when th e solution pH value was 4 and the solid-liquid ratio was 0.5g · L-1, the adsorption behavior of cyclophosphazene phloretin nanospheres for uranium (VI) was the best.

References

1. Fu Ying Yi, Ding Xin Geng, Dou Tian Jun, et al. Ap plication of adsorption technology in HLW [J]. Mate rials guide, 2013, 27 (2): 255-258
2. Chen B, Jin W, Kong L, et al. Adsorption of uraniu m from uranium mine contaminated water using pho sphate rock apatite (PRA): Isotherm, kinetic and cha racterization studies [J]. Colloids & Surfaces A Physicochemical & Engineering Aspects, 2017, 520:612-621.
3. Sureshkumar M K, Das D, Mallia M B, et al. Adsor ption of uranium from aqueous solution using chitos an-tripolyphosphate (CTPP) beads [J]. Journal of Ha zardous Materials, 2010, 184(1): 65-72.
4. Xu C, Wang J, Yang T, et al. Adsorption of uranium by amidoximated chitosan-grafted polyacrylonitrile, using response surface methodology [J]. Carbohydr ate Polymers, 2015, 121:79-85.
5. Abdi S, Nasiri M, Mesbahi A, et al. Investigation of Uranium (VI) Adsorption by Polypyrrole [J]. Journa l of Hazardous Materials, 2017, 332: 132-139.
6. Ghalami Z, Ghoulipour V, Khanchi A. Hydrogen an d deuterium adsorption on uranium decorated graph ene nanosheets: A combined molecular dynamics an d density functional theory study [J]. Current Applie d Physics, 2019, 19(4): 536-541.
7. Huynh J, Palacio R, Safizadeh F, et al. Adsorpt ion of Uranium over NH2-Functionalized Ordered Silica in Aqueous Solutions [J]. Acs Appl Mater Int erfaces, 2018,22:35-38.

Practices and countermeasures for the treatment of black and odorous water bodies in the rural areas of Shandong Province

Shi Xiuhua [1,a*], Wu Qiong[2,b]

[1,2]School of Literature,law and Economics,Wuhan University of Science and Technology,WuHan, Hubei Province.

Abstract: In recent years, with the rapid development of the rural economy, while the living standards of rural residents have been greatly improved, the environmental conditions in the rural areas are not optimistic, and water pollution is one of the most important ones. Therefore, in order to improve the rural ecological environment and promote the task of water pollution control, Shandong Province has issued "the Governance Action Plan of black and odor water in rural areas of Shandong Province". This article mainly uses " the Action Plan" as a guide to analyze the current situation of rural black water bodies in Shandong Province and the current problems and related countermeasures, and follow the "comprehensiveness, applicability, economy, long-term effectiveness and safety. According to the black and odorous water problems caused by natural, technological and human factors, we will adopt comprehensive methods such as water purification, dredging, source control and pollution interception[1] to promote the treatment of black and odorous water bodies in rural areas, solve outstanding water environment problems in rural areas, and speed up construction of beautiful and livable countryside.

1 Analysis of the status quo of rural black and odor water body in Shandong Province

Driven by the requirements "Action Plan for Prevention and Control of Water Pollution" regarding the nationwide implementation of urban black and odorous water treatment work, the current urban black and odorous water bodies have been effectively treated[2]. However, because Shandong Province is a large agricultural province, there are more than 70,000 villages in the province, and the black and odor water bodies in rural areas have the characteristics of wide-ranging, concealed and repetitive, and obvious seasonal changes, which have a greater impact on the ecological environment. It is far-reaching and has not yet been effectively governed. From June to November 2020, Shandong Province organized the investigation of black and odorous water bodies in rural areas, and took the lead in using satellite remote sensing monitoring and manual investigation to investigate all administrative villages in the province. Specifically, the main gathering areas of villagers in each administrative village appropriately extend the water bodies within 1,000 meters of the area, as well as the black and odor water that the villagers reflect strongly. The investigation found a total of 1398 black and odor water bodies in rural areas.Therefore, in order to carry out rural black and odorous water treatment, the Provincial Department of Ecology and Environment, the Provincial Department of Water Resources, the Provincial Department of Agriculture and Rural Affairs and the Provincial Department of Finance jointly issued "the Action Plan" for rural black and odorous water treatment in Shandong Province, which provided a good policy basis for the rural black and odorous water treatment.

2 Causes of black and odorous water bodies in rural areas

Rural black and odor water bodies can be identified based on sensory characteristics such as the smell and color of the water body. If the water body has peculiar smell or the color is obviously abnormal, it is regarded as black and odor water body. At the same time, for suspicious water bodies with controversial sensory judgments, water quality monitoring can be used to determine whether they are black or smelly. The water quality monitoring indicators include three indicators of transparency, dissolved oxygen, and ammonia nitrogen. The specific indicator thresholds are as follows:

*Corresponding author: [a]Shi Xiuhua (1968-) , associate professor and Ph.D. in the School of Literature,law and Economics ,Wuhan University of Science and Technology, has been engaged in social governance research for many years. email:1716682261@qq.com
[b]Wu Qiong, a graduate student majoring in Public Administration in the School of Literature,law and Economics , Wuhan University of Science and Technology. email:2719160095@qq.com

Table 1. Water quality monitoring index.

Water quality monitoring index(unit)	Index threshold
transparency（cm）	<25
Dissolved oxygen(mg/L)	<2
Ammonia(mg/L)	>15

As far as the reasons for the formation of black and odor water bodies in rural areas are concerned, it can be divided into three aspects: natural, technological and human factors.

2.1 Natural factors

2.1.1 The lack of oxygen in the water causes black and odor problems[3].

With the rapid development of the global economy, the industrial sector has continuously expanded its production scale, and the global warming climate problem has become increasingly serious with the growth of development, which poses a great threat to the normal operation of river courses and aggravates the degree of oxidation of water resources.

2.1.2 The bottom sludge is deposited and pollutants are continuously released.

After the pollutants enter the water, they will continuously accumulate in the bottom sludge, causing the bottom sludge to be polluted. If the amount of pollutants adsorbed by the bottom sludge reaches saturation, it will continuously release pollutants into the water body, causing more serious water pollution.

2.2 Technical factors

2.2.1 The coverage of the sewage collection pipe network is poor[4].

A large amount of domestic sewage is discharged into rivers and other water bodies without being collected and treated. In addition, the misconnection of the pipe network is common, and the sewage will be directly discharged into the water body after the sewage pipe is wrongly connected to the rainwater pipe, causing the water body to become black and smelly.

2.2.2 The drainage system is not sound enough.

At present, the sewage pipes and drainage pipes in many villages and even cities in China are not separated. The sewage will be discharged into the water body along with the rainwater channel, which brings huge sewage treatment pressure to the sewage treatment plant. If the rainfall increases, the sewage treatment plant cannot treat the excess water and can only discharge it into the water body, causing pollution to the rural water environment.

2.3 Human factors

2.3.1 Random dumping of garbage leads to black and odor water bodies.

There are informal garbage dumps around some black and odorous water bodies; Some residents build kitchens and toilets along the river for convenience, and discharge pollutants directly without treatment[5]; In addition, the domestic sewage in some primary and secondary school canteens is directly discharged, and rainwater will carry pollutants into ditches and ponds continuously, thus causing the water body to be black and smelly.

2.3.2 The industrial sector is also an important factor that seriously damages the quality of urban rivers.

In order to effectively promote the production efficiency and scale of enterprises, a large number of toxic pollutants, such as waste gas and wastewater, are directly volatilized into the atmosphere and rivers without standardized treatment in the manufacturing process of some enterprises, causing large-scale water pollution to urban rivers.

2.3.3 Inadequate supervision and management will affect the urban water landscape.

In addition to the above reasons, inadequate supervision of the surrounding environment of the water body is also an important reason for the black and odor water body. In addition, due to the incomplete identification and incomplete treatment of black and odor water bodies, local inspections of new black and odor water bodies are incomplete, and some newly added black and odor water bodies require public participation and national supervision and discovery.

2.3.4 It is difficult to guarantee funds for the treatment of black and odor water bodies.

The comprehensive improvement of black and odor water bodies requires a lot of human and financial resources, and the financial capacity of local governments is difficult to support the advancement of the comprehensive improvement project of black and odor water bodies. Local governments have been unable to provide financial support. In addition to the impact of the COVID-19 in 2020, the funding guarantee for the normal operation of various departments is limited, which makes the treatment of black and odorous water bodies more difficult.

3 Goals and principles of harnessing black and odorous water body in rural areas of Shandong province

In order to implement the requirements of rural human settlements improvement, Shandong Province, guided by

Xi Jinping's ecological civilization thought, closely combined with the requirements of rural human settlements improvement and pollution prevention and control, gradually eliminated the rivers and ponds in front of and behind the houses and the black and odorous water bodies strongly reflected by the masses in rural areas, and planned to use 3 years to complete the existing 1,398 rural black and odorous water treatment projects by 2023. The specific plan is shown in the following table:

Table2. Shandong Province Rural black and odor Water Body Treatment Task Target Table

city	Now	2021	2022	2023
Jinan	103	25	25	53
Qingdao	82	20	20	42
Zibo	33	8	8	17
Zaozhuang	50	25	25	0
Dongying	118	29	29	60
Yantai	10	2	2	6
Weifang	24	5	5	14
Jining	317	159	158	0
Taian	24	12	12	0
Weihai	4	2	2	0
Rizhao	25	5	5	15
Linyi	94	23	23	48
Dezhou	103	25	25	53
Liaocheng	35	8	8	19
Binzhou	138	33	34	71
Heze	238	119	119	0
Sum	1398	500	500	398

In the treatment of black and odorous water bodies in rural areas, the principles of "comprehensiveness, applicability, economy, long-term effectiveness, and safety" must be followed. The so-called comprehensiveness means that when choosing a treatment technology, you can not only achieve water quality improvement from one aspect or rely on a single technology, but comprehensively consider the combination of various technologies to achieve the treatment of black and odorous water bodies. Adhere to adapting measures to local conditions, resolutely prevent "one size fits all", and fully integrate rural types, natural environment and economic development level and other factors to carry out governance in different areas. The so-called applicability means that different treatment technologies have different characteristics,and it is necessary to prioritize the selection of applicable technologies according to the degree of water pollution, pollution causes and different treatment stages. Economy, that is, when choosing treatment technology, a comprehensive comparison and selection of different treatment technology solutions should be carried out, and the economically feasible technology should be selected[6]. Long-term effectiveness means that when studying black and odorous water treatment countermeasures, not only the short-term effects after the implementation of the technical plan should be considered, but also the long-term water quality improvement effects and water quality stability should be paid attention to. Taking full account of the lagging rural environmental infrastructure construction in Shandong Province, a weak working foundation, and a wide range of people, a phased goal of promoting rural black and odorous water treatment is proposed. Safety is to consider the adverse effects and secondary pollution of the water environment after the implementation of the technology.

4 Relevant measures for the treatment of rural black and odorous water bodies in Shandong province

It is a very complicated systematic work to fundamentally treat black and odorous water bodies and prevent the black and odorous water bodies from turning back to black and repeated deterioration of water quality in rural areas. In the specific practice of black and odorous water bodies, physical, biological and chemical technologies should be organically integrated according to the specific conditions. In combination, develop high-efficiency and low-consumption treatment technologies suitable for the actual conditions of Shandong Province. According to the requirements "Action Plan", in view of the existing black and odorous water bodies in the rural areas of Shandong Province, the source control, pollution interception, silt dredging and dredging were taken on the basis of a thorough analysis of the source of pollution and the causes of the black and odor from the three aspects of nature, technology and man-made. 3 measures such as water purification and targeted treatment.

4.1 In response to the phenomenon of black and odorous water bodies in rural areas caused by human factors, adopt the method of source control and pollution interception.

First of all, rural black and odorous water bodies can be divided into rural domestic sewage, livestock and poultry breeding sewage, aquaculture, rural domestic garbage, planting non-point source pollution, industrial sewage, and other 7 types.Scientific selection and technology are feasible.Economic and reasonable treatment methods to eliminate external source pollution.

Secondly, the government needs to strengthen the investment in black and odorous water treatment[7], incorporate rural black and odorous water treatment into the important content of rural ecological revitalization work, encourage and guide social capital to participate in rural black and odorous water treatment, and actively seek financial support from higher authorities to ensure the source of funds for black and odorous water bodies.

Finally, in order to ensure the effectiveness of governance, relevant provincial departments and units need to strengthen inspections and regular inspections in the process of carrying out specific supervision work, and it is strictly forbidden for enterprises to surreptitiously discharge and leak discharge.

4.2 In view of the black and odorous water phenomenon in rural areas caused by technical problems, the method of removing the silt and then dredging.

In view of the current remediation path of black and odor water bodies in rural areas, we must start with the investigation of pollution sources and continuously strengthen the protection of water bodies. Therefore, from a feasible point of view, rationally construct river basin pollution interception projects, establish a pipe network connection function, run through the sewage interception work upstream and downstream of the river, and promote the centralized collection of rural domestic waste water, thereby promoting the efficiency of waste water treatment.

For water bodies that still have black and odor after adopting source control and interception measures to eliminate external source pollution, internal source treatment shall be carried out. Comprehensively evaluate the water quality and bottom mud conditions of rural black and odorous water bodies, and scientifically formulate dredging and dredging plans.

4.3 Because of the phenomenon of black and odor water bodies in rural areas caused by natural factors, water purification methods should be adopted.

In terms of water purification, areas with better economic conditions can carry out ecological care in time, and at the same time reconstruct the water ecosystem and improve the self-purification capacity of water bodies. For villages that are too scattered and ecologically fragile, the natural shoreline of rural water bodies should be maintained on the premise of meeting the requirements of flood control and drainage. Taking into account the actual conditions of the region's economic conditions, topography, recent planning, etc., adopt scientific methods to reconstruct the water ecosystem and enhance the self-purification capacity of water bodies.

Strictly control water transfer and pollution flushing on the grounds of restoring hydrodynamic force, so as to ensure the landing of treatment projects. All localities have formulated engineering schemes one by one, implemented engineering investment, formulated implementation plans, and established dispatching ledgers. In the use of funds, in strict accordance with the provisions of the measures for the administration of project funds, make overall plans for the construction and management, and bring into play economic, social and environmental benefits.

5 Conclusion

The treatment of black and odor water bodies in rural areas itself is a systematic and complicated process[8]. In addition, the pollution of rivers in Shandong Province is relatively serious and the distribution is relatively scattered. Therefore, the phenomenon of "to treat the symptoms rather than getting to the root of the problem" should be avoided. In the specific implementation process, it is necessary to follow the requirements "the Action Plan" and learn from the successful governance experience of other provinces and cities, and grasp the principles of adapting measures to local conditions and timeliness, so that the restoration of black and odorous water bodies can be carried out for a long time and reach the maximum Good results.

Acknowledgments

This article was supported by Hubei NPO Research Cent er (Project Number: HBNPO2020YB01)

References

1. Chen,J.,L.(2021).Design Points for Comprehensive Treatment of Urban Black and Odorous Water Bodies.Low Carbon World,11:73-74.
2. Zhao,Y.C.,Li,Y.C.,(2020)Study on the Causes and Treatment of Black and Odor Water in Rural Areas of Anhui Province.Environment and Sustainable Development,45:154-157.
3. Li,X.M.(2016)Study on the Cause of the Black Odor Water Body and Its Comprehensive Treatment.Chemical Intermediate,8:85-86.
4. Ma,T.,Liang,J.B.(2020)The Causes of Urban Black and Odorous Water Bodies and Treatment Methods.Technology and Application,12:109-111
5. Wu,S.X.,Lu,D.L.,Xia,Y.H.(2020)Current Situation and Counter Measure of Black and Odorous Water in Guizhou Province.Environment and Development,32:63-64+66.
6. Hu,H.Y.,Sun,Y.,Xi,J.Y.,Zhao,T.T.(2015)Treatment and Water Quality Improvement Technology of Black-and-Malodorous Water Body in Urban Area.Environmental protection,43:24-26.
7. Sun,Z.L.(2021)Comprehensive Treatment and Long-term Management of Urban Black and Odorous Water Bodies.Anhui Architecture,28:85-86.
8. Cheng,Q.L.,He,Y.,Huang,M.S.,Tong,M.(2011)The Discussion on the Treatment Methods of Urban Malodorous River.Shanghai Chemical Industry,36:25-31.

 https://doi.org/10.1051/e3sconf/202129003008

Environmental effects and monitoring system design of energy cooperation between China and Russia in the context of "One Belt, One Road"

Jie Gao[1]

[1]School of Foreign Languages,Northeast Petroleum University, Daqing, China.

Abstract. In the context of the Belt and Road, China-Russia energy cooperation has become closer, and energy cooperation industries are distributed in various energy fields. This paper firstly proposes that environmental monitoring of energy cooperation industries has positive significance for energy cooperation in the context of the Belt and Road, then elaborates the current situation and environmental effects of Sino-Russian energy cooperation, leads to the design of environmental effect monitoring system for energy cooperation industries, which is based on satellite and spectral estimation to analyze the changes of water and soil areas such as water and vegetation and the degree of atmospheric pollution, and derives the environmental benefits of industries. Finally, we conclude with a summary and outlook on the prospects and effects of Sino-Russian energy cooperation.

1 Introduce

Russia, as one of the most important countries in China's "One Belt, One Road" strategy, is a very important pivot point. Therefore, strengthening the cooperation between the Russian and Chinese governments in the energy industry is in line with the political interests of both countries and the demands of their economic development, as well as the promotion of the economic development of both sides. It is good for the stability of the current economic situation in Asia and Europe and the world as a whole. In this paper, we analyze the situation of energy industry cooperation between China and Russia under the "One Belt, One Road" strategy proposed by China, and design an environmental monitoring system based on energy industry cooperation by analyzing the impact and effect of energy extraction, transportation, and use on the environment, and through this system, we promote the smooth development of energy industry cooperation between the two countries. [1]This system will facilitate the smooth development of energy industry cooperation between the two countries, and develop corresponding environmental protection strategies and industrial standards for energy cooperation, promote energy security, make the cooperation develop in the direction of high quality, set a model for cooperation between countries in the region, and expand the influence of China's "One Belt, One Road".

2 Current Situation and Environmental Effects of China-Russia Energy Cooperation

2.1.Current Situation of Russian-Chinese Natural Gas Energy Cooperation

Chinese and Russian cooperation in natural gas is also in fact with the laying of natural gas pipelines. At the same time, Chinese domestic energy companies have invested in natural gas projects developed by Russia in the Arctic region. In 2015, Russia was hit by Western economic sanctions, which caused damage to the domestic economy and a significant devaluation of the currency, and put gas development projects on hold. At this time, China entered into cooperation agreements with domestic Russian gas development projects and purchased part of the project's equity. [2]To date, China's stake in the Russian Arctic gas project has risen to nearly 30%. This project is also the first natural gas trade between China and Russia since the "One Belt, One Road" economic strategy was introduced, which reflects the positive impact of the "One Belt, One Road" Silk Road on energy cooperation and strategic relations between China and Russia. The delivery of Russian natural gas will greatly alleviate China's domestic energy needs, improve energy reserves, and provide strategic energy security. And to a certain extent it will improve the environmental pollution in China. China's energy structure is shown in Figure 1.

* Corresponding author: gj_dqpi@126.com

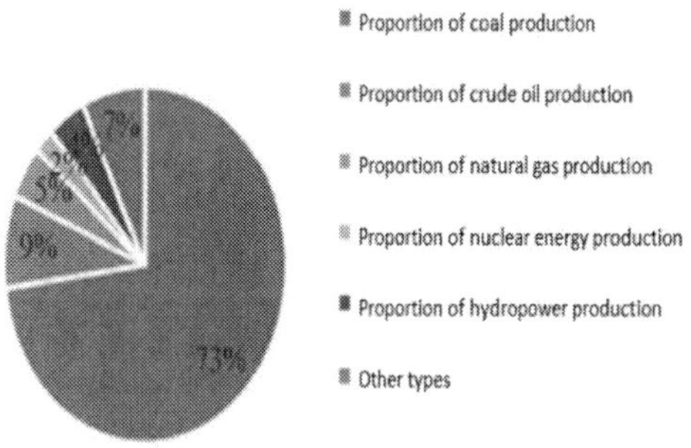

Fig.1 China's energy structure diagram

2.2 Current status of Russian-Chinese power and energy cooperation

In addition to the above-mentioned cooperation between China and Russia in the oil and gas energy industry, the electricity and coal industries are also important objects of cooperation between the two countries in the energy industry[3]. The cooperation between China and Russia in electric power industry cooperation has been laid out widely and for a long time, mainly through the sale and purchase of electric power, the construction of power generation equipment and the technology of power generation equipment to carry out this trade exchange and industrial cooperation. At present, the cooperation between the two countries is mainly focused on wind power generation, which has no impact on the environment. The largest wind power project in Russia is the Ulyanovsk wind power project, which is also a joint project between the two countries. The project was undertaken and assembled by China's Heilongjiang Thermal Power Company and is currently achieving good power transmission results. In addition, with the intention of cooperation between the two governments, power generation companies of both countries hope to develop hydroelectric power generation in the Russian Far East and Siberia, and promote development through the provision of appropriate sites and facilities. Yangtze River Hydropower Group has also signed a letter of intent with the Russian counterpart. At the same time, the Russian government has concentrated its main resources for hydroelectric power generation in the Far East, the Chinese counterpart of which is the northeastern region of China. The northeastern region is located close to the Far East and has a large population and a high demand for electricity. As China's old domestic industrial base, the northeast has a single energy structure, making the region's economic development at a backward level in the country. The cooperation agreement with Russia's Far East region for power transmission will enable the transfer of its hydroelectric power to the northeastern part of China, thus alleviating the environmental impact of the northeastern part of the country due to the burning of coal for power generation. In the future, hydroelectric power generation is likely to be a major focus of cooperation between China and Russia in the energy industry as a good project that does not pollute the environment and produces energy at the same time.

2.3 Environmental effects of cooperation

The impact of energy consumption on the environment is mainly manifested in the pollution of water, atmosphere, soil and biological vegetation. In particular, the polluting gases and pollutants released after energy consumption are easy to cause great damage to the environment and pose a threat to the production and life of human beings, which is also an urgent concern and a problem that needs to be solved by governments and people of all countries. For China, the coal-based energy consumption structure, including the consumption of other fossil energy sources, mainly causes a large amount of emissions of waste water, waste gas and waste residue, which easily causes damage to water, air and soil. The main pollution of coal-based energy consumption structure is the emission of harmful gases such as sulfur dioxide and waste residues from combustion, which are harmful to the environment. The impact of Sino-Russian oil and gas energy cooperation on China's environmental quality is in general an indirect one. The first thing that Russia and China can do through oil and gas energy cooperation is to change China's traditional coal energy-based energy consumption structure and start rational optimization of energy consumption. Secondly, when Russian oil and gas energy enter the Chinese energy consumption market, they can play a corresponding substitution role, replacing a part of coal energy consumption, which will correspondingly reduce a part of sulfur dioxide and other emissions and waste from coal combustion, thus contributing to the improvement of China's environmental quality.[4]

3 Design of environmental effect monitoring system for energy cooperation industry

3.1 Monitoring the environmental effects of water and soil in the energy cooperation industry

The main means of monitoring the soil and water environment is to extract the soil and water information from the original rivers and reservoirs through the remote sensing images of satellites, and to obtain the remote sensing images of the study area after the pre-processing of radiation calibration, atmospheric correction and band fusion, and then to calculate the area of the original rivers, reservoirs and green vegetation, which is used to assess the impact of the energy cooperative industry on the soil and water environment.[5]

The specific research method is based on satellite data, calculating the annual normalized difference vegetation index (NDVI) and annual normalized difference water body index (NDWI) for the years before and after the construction of the energy cooperative industry, inversion to obtain the vegetation cover and land use type data, and analyzing the impact of the industrial area on the vegetation growth and ecological environment through buffer zone analysis and spatial superposition analysis methods. The flow chart of the analysis is shown in Figure 2.

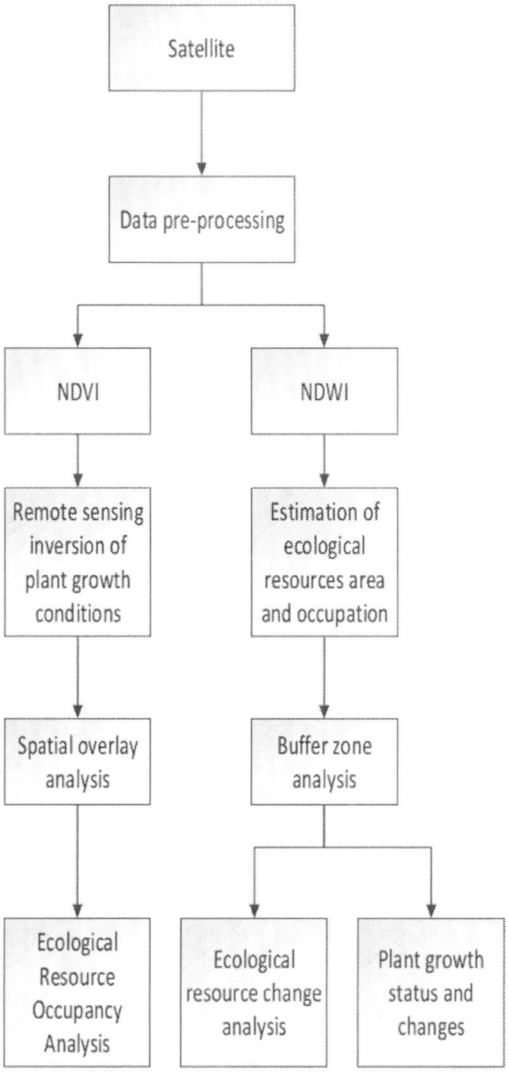

Fig.2 Flow chart of soil and water environment monitoring and analysis

The normalized differential water body index, calculated from the ratio of the difference between the apparent reflectance and the sum of the green light band and the near-infrared band, mainly takes advantage of the strong absorption of water bodies in the near-infrared band and the strong reflectance of vegetation to extract the water body information in the image by suppressing the vegetation and highlighting the water bodies, which has the feature of high accuracy of water body classification.

$$NDWI = \frac{\rho_{\text{green}} - \rho_{\text{nir}}}{\rho_{\text{green}} + \rho_{\text{nir}}} \qquad (1)$$

In equation (1), ρ_{nir} is the apparent reflectance in the near infrared; ρ_{green} is the apparent reflectance in the green wavelength band. After extracting the range, the set of overlap places are manually trimmed in combination with visual interpretation to obtain the final water range.[6]

Surface vegetation plays a key role in ground-air energy exchange and the Earth's carbon and nitrogen water cycle, so it is closely related to the ecological environment. The vegetation cover is the percentage of the vertical projection of vegetation on the ground to the total area of the study area, which can effectively reflect the characteristics of surface vegetation distribution and monitor the ecological environment changes, and is an important parameter index in ecological environment evaluation. The NDVI is the best indicator of vegetation growth, biomass and vegetation cover dynamics, and is calculated from the ratio of the difference between the apparent reflectance and the sum of the red wavelengths.

$$NDVI = \frac{\rho_{\text{nir}} - \rho_{\text{red}}}{\rho_{\text{nir}} + \rho_{\text{red}}} \qquad (2)$$

$$FVI = \frac{NDVI - NDVI_s}{NDVI_v - NDVI_s} \qquad (3)$$

In equation(2), ρ_{nir} is the apparent reflectance of NIR, and ρ_{red} is the apparent reflectance of red light band; in equation (3), $NDVI_s$ and $NDVI_v$ represent the NDVI completely covered by soil and vegetation in the region, respectively, and the larger the FVC, the more luxuriant the vegetation is and the better the ecological environment is.

3.2 Atmospheric impact monitoring of the energy cooperative industry

The main pollutants in the air are sulfur dioxide, nitrogen dioxide, and total suspended solids, and the use of differential absorption spectroscopy in the energy cooperative industry can effectively identify the type and concentration of pollutants and analyze the impact of the industry on the atmosphere.

Differential absorption spectroscopy by its nature is based on the absorption of light radiation by molecules. When a beam of light passes through the atmosphere, the light is selectively absorbed by the molecules in it, causing a change in the intensity and spectral structure of the light, and the absorption spectrum can be derived by comparing it with the spectrum of light emitted from a light source. By analyzing the absorption spectrum in a certain wavelength band, it is possible to determine not only the presence of certain components, but also the content of these substances and to obtain the concentration of the corresponding gas in the atmosphere. The absorption of light by substances is in accordance with the Lambert-Beer law.

$$I(\lambda) = I_0(\lambda)\exp\left[-L_\sigma(\lambda)c\right] \qquad (4)$$

In equation (4), I(λ) is the detected light intensity; I_0 (λ) is the light intensity of the light source; L is the light range length and c is the concentration; σ is the absorption cross section, which indicates the absorption capacity of the substance.[7]

In actual atmospheric measurements, the extinction factors mainly include Rayliegh scattering, Mie scattering, and the influence of other substances in the atmosphere. When these effects are considered, equation (4) is expanded as :

$$I(\lambda) = I_0(\lambda)\exp\{-L[\textstyle\sum_1 (\sigma_i(\lambda)c_1) + \varepsilon_R(\lambda) + \varepsilon_M(\lambda)]\}A(\lambda) \qquad (5)$$

where, $A(\lambda)$ denotes the transmission function of the system. The change in optical thickness caused by

Rayliegh scattering and Mie scattering varies slowly with wavelength, while the change in optical thickness caused by molecular absorption properties varies rapidly with wavelength. The spectral change caused by scattering is called "broadband" spectrum (low frequency part), and the spectral change caused by molecular absorption is called "narrowband" spectrum. The "narrow-band" spectrum, which varies rapidly with wavelength, can be separated out using a high-pass filter. The separated molecular absorption spectrum is fitted with a reference spectrum, and the concentration of light-absorbing substances in the atmosphere can be calculated. This is the analytical process of differential absorption spectrometry. The diagram of the atmospheric environment monitoring equipment based on differential absorption spectroscopy is shown in Figure 3.

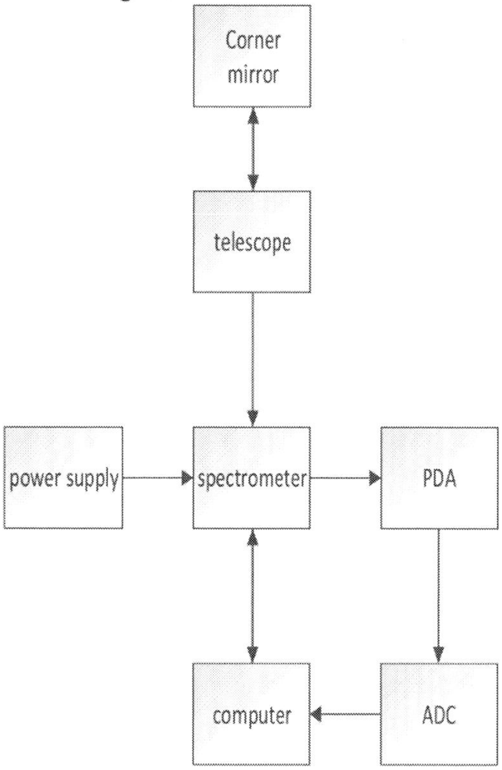

Fig.3 Atmospheric environment monitoring equipment diagram

4 Conclusion

In the context of "One Belt, One Road", based on theoretical and empirical analysis, it is concluded that the environmental impact of Sino-Russian oil and gas energy cooperation on China is positive. On the one hand, China's import of Russian oil and natural gas through Sino-Russian oil and gas energy cooperation can replace part of the consumption of inefficient energy sources such as coal, which can improve China's energy utilization efficiency and reduce pollutant emissions, and thus improve China's environmental pollution situation. Therefore, as long as China's economic growth reaches a certain critical value, pollutant emissions will be continuously reduced, and thus Sino-Russian oil and gas energy cooperation will reduce pollutant emissions after reaching a critical value when promoting China's economic growth, which will have a positive effect on China's environmental pollution control. By analyzing and monitoring the environmental impact effects of China-Russia energy cooperation industry, we can reasonably carry out industrial planning and guide the environmental management policy and strategy.

Acknowledgments

This paper is the result of the research project on the deep integration of Daqing into the construction of "One Belt and One Road" (No. DSGB2021069).

References

1. Mu Hang, He Chao, Ruan Qiuming, Wang Danlu, Zhou Xiaoyan, Yang Lu, Tu Peiyue, Wang Haiyang, Hong Song. Spatial and temporal distribution characteristics of PM_(2.5) pollution and population exposure risk in countries along the "Belt and Road"[J/OL]. Journal of Environmental Science:1-12[2021-06-05]
2. Gong Wei, Li Li, Liu Qinhuo, Xin Xiaozhou, Peng Zhiqing, Wu Mingquan, Niu Zheng, Tian Haifeng. Remote sensing monitoring of ecological and environmental impacts of hydropower station projects in the "Belt and Road" region[J]. Journal of Geoinformation Science,2020,22(07):1424-1436.
3. BobrukIgor. Research on the influencing factors and countermeasures of Sino-Russian energy industry cooperation[D]. Xi'an University of Petroleum, 2020.
4. Yang Zhongshan, Wei Xiaoxue. Total Factor Energy Efficiency in Key Regions of "Belt and Road"-Measurement, Decomposition and Analysis of Influencing Factors[J]. China Environmental Science,2018,38(11):4384-4392.
5. Ran H-L. Study of differential absorption spectral inversion method in environmental monitoring system[J]. Science and Technology Perspectives,2016(17):216.
6. Xu Chengliang. Research on the impact of China-Russia oil and gas energy cooperation on the environmental effects in China [D]. Hubei University,2015.
7. Cui Houxin,Qi Rubin,Zhang Wenjun,Xu Kexin. Design of online continuous monitoring system for atmospheric quality by differential absorption spectrometry[J].

Comparison of Two Different Calculation Methods of Fractional Vegetation Cover

Liao kaitao[1]

[1]Jiangxi Academy of Water Science and Engineering, 330029 Nanchang Jiangxi, China
Jiangxi Normal University, College of Geography and Environment, 330022 Nanchang Jiangxi, China

Abstract. Fractional Vegetation Cover (FVC) is an important parameter for soil erosion equation. The Fractional Vegetation Cover of typical soil erosion county in red soil region of South China was calculated by parameter revision method (PRM) and fusion method (FM).The results of the two methods were compared from two parts of FVC level and 24 half-months FVC. The results show that the FVC level calculated by PRM was concentrated in high level, and that calculated by FM is mainly in medium-high level and high level. The time variation of FVC in 24 half-months calculated by the two methods was the same, but the majority of FVC value calculated by PRM was higher than that calculated by FM. It is better to choose FM method to calculate FVC in estimating soil erosion.

1 Introduction

Fractional Vegetation Cover (FVC) is not only an indicator of surface vegetation status, but also an important index to evaluate regional hydrological, ecological and environmental changes [1-3]. FVC is one of the important factors affecting soil erosion, which can directly reflect the change of vegetation coverage or the ability of vegetation measures to control soil erosion [4], and it is an important parameter of Chinese Soil Loss Equation (CSLE) [5]. With the rapid development of remote sensing technology, using satellite image data to estimate regional FVC has become the main method [6-8]. The Technical Regulations on Dynamic Monitoring of Regional Soil Erosion (Trial), which promulgated by Monitoring Centre of Soil and Water Conservation, Ministry of Water Resources of the People's Republic of China in 2020, required local government to use Landsat satellite data and MODIS-NDVI data to calculate regional FVC, and provided two calculation methods, one is called fusion method (FM), the other is called parameter revision method (PRM).

At present, most regions of China used the parameter revision method to estimate FVC. However, the data parameter revision method used was from 2011, cannot accurately express the current situation, because the land cover had changed a lot. In order to compare the difference between the fusion method and the parameter revision method, Wuning County of Jiangxi Province was taken as the study area, two different methods were used to estimate FVC, and the results were compared and analyzed. This hopes to provide reliable basic parameters for improving the calculation accuracy of annual soil erosion dynamic monitoring in China and provide beneficial enlightenment for soil erosion research.

2 Materials and Methods

2.1 Study area

Wuning County is located in the northwest of Jiangxi Province, and in the middle and lower reaches of the Yangtze River, within a longitude of 114°28′E to 115°27′E and a latitude of 28°52′N to 29°35′N, occupying 3507 square kilometres. The study area is in the southern edge of the middle subtropical zone, belongs to the typical subtropical humid monsoon climate zone with an annual average of 1488mm, and the annual average temperature is about 16.6℃.

The main land use type of study area are forest land, agricultural land, water. The forest land area is 2687.14 km^2, 76.64% of the total land use area, the agricultural land area is 346.16km^2, 97% of the total land use area, and the water area is 279.67km km^2, 9.86% of the total land use area. The area of soil erosion is 222.04km^2, accounts for 6.33% of the total land use area, all of which are water erosion. Among them, the light erosion area is 198.05 km^2, 89.20% of the total area of soil erosion. The moderate erosion area is 16.08 km^2, 7.24% of the total area of soil erosion.

2.2 Remote image

The data used in the study include Landsat OLI (2017-2019), MODIS-NDVI (2017-2019), land use (2019) and correction coefficient image (2011).

The resolution of Landsat OLI image is 30m, and there are 9 bands in total, of which 4 and 5 bands are Red band and NIR band respectively for NDVI calculation, and 9 band is cirrus band for cloud detection. The resolution of

* Corresponding author: Liaokaitao@jxnu.edu.cn

https://doi.org/10.1051/e3sconf/202129003009

MODIS-NDVI image is 250m, revisit period is 16 day, 23 images per year. The radiometric calibration, geometric correction, cloud / snow / shadow mask, mosaic and NDVI calculation of Landsat OLI images are all completed on the Google Earth Engineer (GEE) cloud platform [9].

The land use data was based on visual interpretation of GF-2 image, which resolution is about 2m. Correction coefficient image was supplied by Ministry of Water Resources of the People's Republic of China.

2.3 Image fusion method （FM）

Based on land use data, the half-month frequency vegetation index series of all kinds of pure pixels were determined:

$$V_M(t) = \frac{1}{3\times N}\sum_{y=k}^{k+2}\sum_{n=1}^{N} NDVI(t,y,n) \qquad (1)$$

Where $V_M(t)$ is a certain category of Multi-year NDVI mean value, N is the number of pure pixels, y is the year, k is the starting year of the first three years of the monitoring year.

Then based on function 2, process Landsat OLI NDVI data with MODIS-NDVI data, obtained 23 NDVI images of 30m.

$$V_H(t_i) = V_M(t_i) + \frac{\sum_{j=1}^{n}\left[\omega(t_i,t_j)\left(V_T(t_j)-V_M(t_j)\right)\right]}{\sum_{j=1}^{n}\omega(t_i,t_j)} \qquad (2)$$

Where $V_H(t_i)$ is the result of image fusion, $V_M(t_i)$ is the MODIS-NDVI multi-year mean value sequence corresponding to each pixel, $V_T(t_j)$ is the Landsat NDVI value of each pixel at the certain time, n is the number of Landsat OLI images, $t_i$is the Julian day of MODIS-NDVI image, t_j is the Julian day of Landsat OLI image, $\omega(t_i,t_j)$is the weight of Landsat OLI image.

Then based on function 3, converting NDVI data to FVC :

$$FVC = \frac{NDVI - NDVI_{min}}{NDVI_{max} - NDVI_{min}} \qquad (3)$$

Where $NDVI_{max}$ is the NDVI value of pure vegetation coverage, and $NDVI_{min}$ is the NDVI value of pure bare soil. In this paper, according to the cumulative frequency table of NDVI gray distribution in the whole image, the upper and lower cumulative frequencies in the confidence interval of 5 % and 95 % were selected as $NDVI_{min}$ and $NDVI_{max}$, respectively.

2.4 Parameter revision method （PRM）

Firstly, used MODIS-NDVI data obtain 23FVC_L images by function 3. Then based on function 4, obtain 23 FVC_H images of 30m.

$$FVC_H = FVC_L \times a \qquad (4)$$

Where a is the correction coefficient image supplied by Ministry of Water Resources of the People's Republic of China.

In CSLE model, it need 24 FVC images to calculate soil erosion, but MODIS-NDVI only have 23 images a year, so take the average value of the 8th and 9th FVC image as the new 9th FVC image.

Use the same method, the FVC values of the 24 in 2017, 2018 and 2019 were calculated, then average the data of three years as the finally FVC results.

The FVC is divided into five levels: 0-30% for low level, 30-45% for medium-low level, 45-60% for medium level, 60-75% for medium-high level and 75-100% for high level.

3 Results and Discussion

FVC is related to vegetation, so we compare the FVC of forest land, grassland and orchard calculated by the two different methods. The vegetation in the study area grows best in September, so select the 17th period FVC for statistical analysis.

3.1 FVC Level analysis

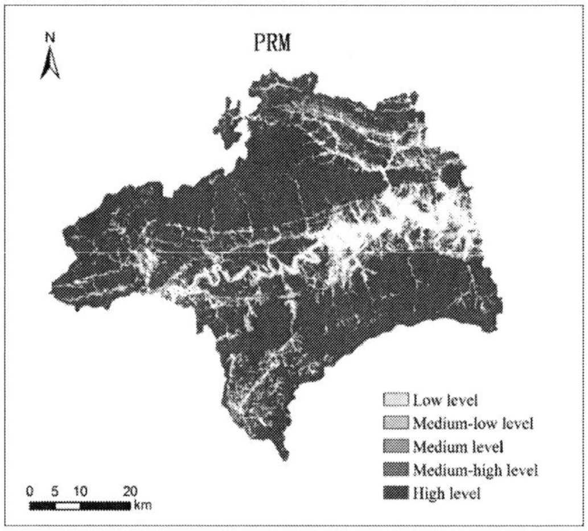

Fig. 1. The FVC level calculated by PRM

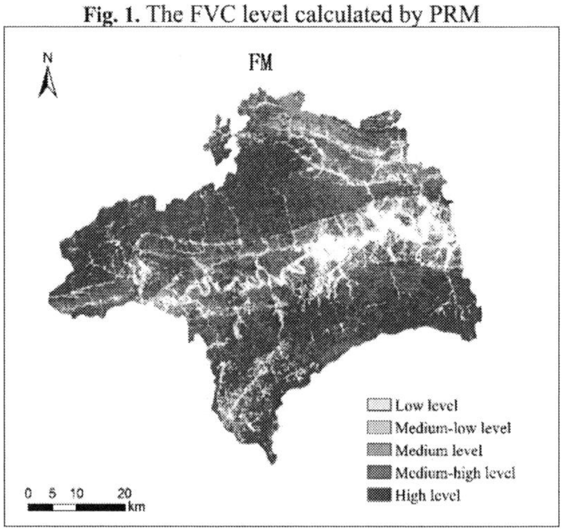

Fig. 2. The FVC level calculated by FM

Compared with figure 1 and Figure 2, it can be seen that the FVC level calculated by PRM was mainly concentrated in high level, and the other several levels of FVC were basically invisible. But the FVC level calculated by FM was mianly distributed alternately in high level and medium-high level.

Table 1 shows that the FVC of Wuning County calculated by the two different methods were all mainly

 https://doi.org/10.1051/e3sconf/202129003009

in high level. In the low level, the area calculated by PRM is 39.96km^2, and the area calculated by FM is 21.37km^2, which is 18.59km^2 smaller than that PRM. In the medium-low level, the area calculated by the PRM is 9.93km^2, the area calculated by FM is 29.73km^2, which is 19.80km^2 smaller than that PRM. In the medium level, the area calculated by PRM is 24.04km^2, the area calculated by the FM is 106.34km^2, which is 82.30km^2 larger than that PRM. In the medium-high level, the area calculated by PRM is 69.24km^2, the area calculated by the FM is 1099.78km^2, which is 1030.54km^2 larger than that PRM. In the high level, the area calculated by PRM is 2626.66km^2, and the area calculated by the FM is 1512.63km^2, which is 1114.03km^2 smaller than that PRM. Compared the results of FM and PRM, the FVC calculated by PRM is higher and concentrated in the high value area.

Table 1. The area of different FVC level

FVC Level		FM	PRM
Low	Area（km^2）	21.37	39.96
	Percentage（%）	0.77	1.44
Medium-low	Area（km^2）	29.73	9.93
	Percentage（%）	1.07	0.36
Medium	Area（km^2）	106.34	24.04
	Percentage（%）	3.84	0.87
Medium-high	Area（km^2）	1099.78	69.24
	Percentage（%）	39.71	2.50
High	Area（km^2）	1512.63	2626.66
	Percentage（%）	54.61	94.83
Total（km^2）		2769.85	2769.85

3.2 Analysis of FVC in 24 half months

Figure 3 shows that FVC of orchard in 24 half-months calculated by PRM is 0.33 to 1, the lowest is 0.33 in the 4th half-month, and the FVC reach the highest 1 from 10th to 20th half month. The FVC of orchard in 24 half-months calculated by FM ranged from 0.42 (the 5th half-month) to 0.93 (the 13th half-months). From the 6th half-month to the 23rd half-month, the vegetation coverage calculated by PRM is higher than that calculated by FM.

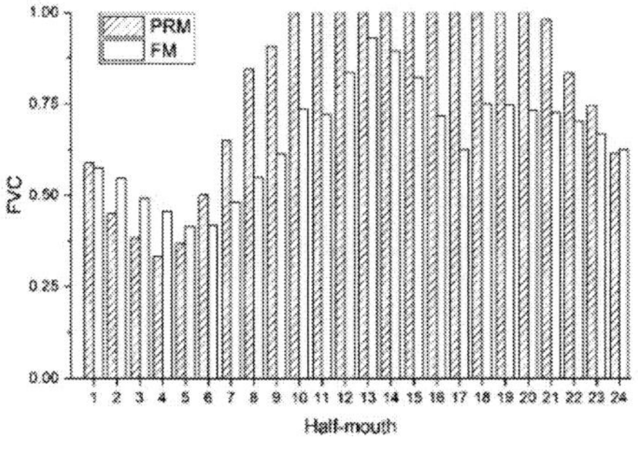

Fig. 3. FVC of orchard in 24 half-months

Figure 4 shows that FVC of forest land in 24 half-month calculated by PRM is 0.83 to 1, the lowest is in the second half-month, and the FVC from the 8th half-month to the 24th half-month is 1. The FVC calculated by FM ranged from 0.76 (the 9th half-month) to 0.98 (the 5th half-month). From the 8th half-month to the 24th half-month, the FVC calculated by PRM is higher than that calculated by FM.

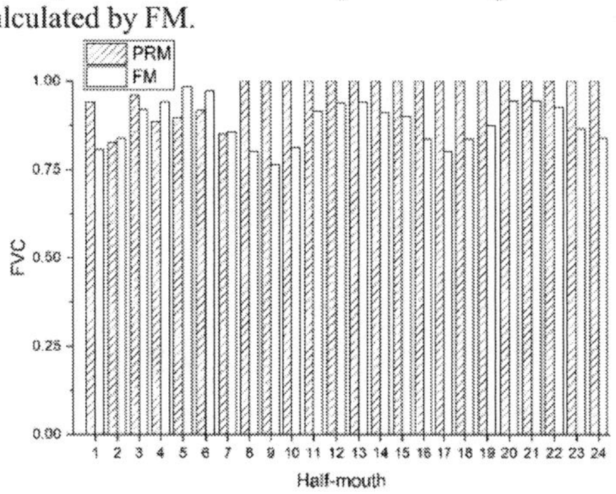

Fig. 4. FVC of forest land in 24 half-months

Figure 5 shows that the FVC of grassland calculated by PRM in 24 half-month is 0.34 to 1, the lowest in the 5th half-month, and the FVC from the 10th to the 19th half-month is 1. The FVC of grassland calculated by FM in 24 half-month is 0.36 (the 4th half-month) and 1 (12th half-month). From the 6th half-month to the 24th half-month, the FVC calculated by PRM is higher than that calculated by FM.

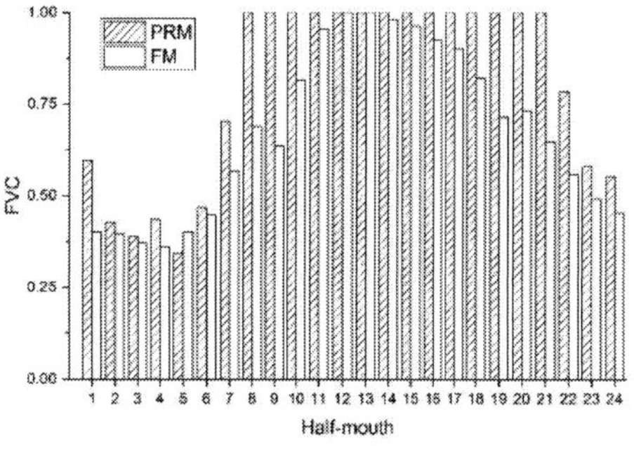

Fig. 5. FVC of grassland in 24 half-months

Generally speaking, the minimum value of the FVC calculated by PRM generally appears in the 4th or 5th half-month. The maximum value is 1, and it always begins to appear in the 8th or 10th half-month, and often lasts to about the 20th half-month. The maximum value of FVC calculated by FM is also 1, but only appear in few half-mouth. The maximum value usually occurs in summer or autumn, and the overall value is lower than that calculated by PRM. The overall trend of the two methods is consistent.

4 Conclusions

In this paper, a typical red soil erosion county in Jiangxi Province, South China was taken as the study area. The PRM method and FM method were used to calculate FVC

https://doi.org/10.1051/e3sconf/202129003009

of Wuning County. Compared the difference between the two methods in FVC classification level and 24 half-mouths. The FVC level calculated by PRM was mainly in high level, while FVC level calculated by FM was mainly distrubuted in medium-high level and high level. The time variation of FVC in 24 half-months calculated by the two methods was the same. The minimum value of the FVC often appeared in 4th or 5th half-mouth, and the maximum value always began to appear in the 8th or 10th half-month. The FVC in summer was the highest during the year. No matter the Vegetation types were, the FVC calculated by PRM method was higher than that calculated by ME method. It is better to choose FM method to calculate FVC in estimating soil erosion.

Acknowledgments

The study was supported by the Water Conservancy Science and Technology Project of Jiangxi Province (NO. 202124ZDKT25, 202022YBKT20, 201922ZDKT08, 202123YBKT16, 201821ZDKT18) and the National Natural Science Foundation of China(NO.41967011) .

References

1. Zhang J , Zhang Z , Chen J , et al. Estimating soil salinity with different fractional vegetation cover using remote sensing[J]. Land Degradation and Development, 2020.
2. Wu D , Hao W , Zhao X , et al. Evaluation of Spatiotemporal Variations of Global Fractional Vegetation Cover Based on GIMMS NDVI Data from 1982 to 2011[J]. Remote Sensing, 2014, 6(5):4217-4239.
3. Wang X , Jia K , Liang S , et al. Fractional Vegetation Cover Estimation Method Through Dynamic Bayesian Network Combining Radiative Transfer Model and Crop Growth Model[J]. IEEE Transactions on Geoscience & Remote Sensing, 2016, 54(12):7442-7450.
4. Yang, Xihua. Deriving RUSLE cover factor from time-series fractional vegetation cover for hillslope erosion modelling in New South Wales[J]. Soil Research, 2014, 52(3):253–261.
5. Meusburger K , Konz N , Schaub M , et al. Soil erosion modelled with USLE and PESERA using QuickBird derived vegetation parameters in an alpine catchment[J]. International Journal of Applied Earth Observation and Geoinformation, 2010, 12(3):208-215.
6. Zhan Y , Meng Q , Wang C , et al. Fractional vegetation cover estimation over large regions using GF-1 satellite data[J]. Proceedings of SPIE - The International Society for Optical Engineering, 2014, 9260.
7. Zheng Z , Yuan Z , Zhao Y , et al. Monitoring and dynamic analysis of fractional vegetation cover in southwestern China over the past 15 years based on MODIS data[J]. Remote Sensing for Land & Resources, 2017.
8. Zhou Z , Yang Y , Chen B . Estimating Spartina alterniflora fractional vegetation cover and aboveground biomass in a coastal wetland using SPOT6 satellite and UAV data[J]. Aquatic Botany, 2017, 144:38-45.
9. Vermeulen L M , Munch Z , Palmer A . Fractional vegetation cover estimation in southern African rangelands using spectral mixture analysis and Google Earth Engine[J]. Computers and Electronics in Agriculture, 2021, 182(12):105980.

https://doi.org/10.1051/e3sconf/202129003010

Study on comprehensive utilization technology of low concentration coal bed methane

Lu XIAO[1,2*]

[1]Chongqing Research Institute of China Coal Technology & Engineering Group Corporation, No.55, Shangqiao village, Shapingba District, Chongqing, China.

[2]National Key Laboratory of Gas Disaster Detecting Preventing and Emergency Controlling, No.55, Shangqiao village, Shapingba District, Chongqing, China.

Abstract: Aiming at the problems of low utilization rate and serious environmental pollution caused by low concentration coal bed methane emission in a coal mine, the utilization technology of low concentration coal bed methane was studied by means of field investigation and theoretical calculation, and the technical solution of comprehensive utilization of low concentration coal bed methane combining regenerative oxidation and cryogenic liquefaction was obtained, which realized the "coal and gas are co-mined and shared" and constructed the virtuous cycle development of " use to promote pumping, pumping to promote safety". The research shows that: the low concentration coal bed methane with concentration of about 5% is converted into high temperature flue gas by regenerative oxidation, and the heat energy is extracted to realize the heating in the mining area; the coal bed methane with concentration of more than 35% is purified and liquefied into LNG product by cryogenic liquefaction, so as to realize long-distance transportation. The technology improves the utilization rate of coal bed methane in the mining area and eliminates the burning of coal for heating. The annual utilization of pure gas is 53.38 million m^3, generating economic benefits of 211 million yuan and reducing CO_2 equivalent by 767,000 t. The safety, economic and environmental benefits are remarkable. This technology has practical significance to improve the utilization rate of gas and promote the realization of the goal of zero gas emission.

1 Introduction

With the vigorous development of industrial technology and the rapid increase of resource consumption [1], the situation of air pollution is grim, regional atmospheric environment problems are becoming increasingly prominent, and the pressure of pollution prevention and control is increasing, which damages people's health and affects social harmony and stability [2]. Environmental protection is one of the important supports for sustainable development. The main component of coal bed methane (i.e. coal mine gas) is methane [3]. Methane is a kind of greenhouse gas, which will produce greenhouse effect when discharged into the atmosphere [4]. The greenhouse effect produced by direct emission of methane is 20~30 times of that of carbon dioxide, which causes great environmental pollution [5]. In addition, coal mine gas has been a major hidden danger affecting the safety of coal mining [6].

Coal bed methane mainly comes from underground gas drainage, and the drainage pipe network is in a negative pressure state [7]. A large amount of air infiltrates into the pipe network, resulting in the low concentration of coal bed methane [8], which is not convenient to use and directly discharged [9]. The United Nations Framework Convention on Climate Change and the Kyoto Protocol require the world to control and reduce greenhouse gas emissions. Coal mine gas is fully burned and has little pollution. If it is used, the harm can be changed into treasure. It not only saves energy, but also reduces the air pollution, so as to achieve the purpose of environmental protection. The coal bed methane with concentration more than 35% is purified and liquefied into LNG product by means of deep cold liquefaction, so as to realize long-distance transportation.

In this context, combined with the technology of coal and gas co-mining and sharing [10-12] and according to the actual energy demand of coal mines, the emission of low-concentration coal bed methane can be reduced by utilizing the technology of comprehensive utilization of low-concentration coal bed methane. The coal bed methane with a concentration of about 5% is converted into high temperature flue gas by means of regenerative oxidation, and the heat energy is extracted from it [13], which can realize the heating in the mining area, thus replacing the coal-fired boiler. The coal bed methane with concentration more than 35% is purified and liquefied into LNG product by cryogenic liquefaction method, so as to realize long-distance transportation. Aiming at the problem of low utilization rate of low concentration coal

* E-mail: xiaolu8317@126.com

bed methane in coal mine, this paper analyzes and studies the comprehensive utilization technology of low concentration coal bed methane by means of field investigation and theoretical calculation in the background of coal bed methane utilization and energy utilization in this coal mine, which has a guiding significance for other coal mines to realize low concentration coal bed methane emission reduction.

2 Investigation of gas source in coal mine

A coal mine can produce 5 million tons of coal every year, which belongs to the coal and gas outburst mine, and there is a large amount of coal seam gas in the coal seam. According to the report of field supplement survey and the measurement data of coal seam gas parameters, the total CBM reserves of the coal mine are 24.36 billion m^3, of which the recoverable coal seam reserves are 12.9 billion m^3, the reserves of the non-recoverable adjacent coal seams within the mining impact range of the recoverable coal seam are 7.4 billion m^3, and the surrounding rock reserves are 4.06 billion m^3.

Coal mine as the central air shaft square pumping station adopts the comprehensive gas drainage method of this coal seam, adjacent strata and goaf drainage, and the high and low negative pressure drainage systems are independent. A total of 10 gas drainage pumps are installed in the pump station, 5 for use and 5 for standby. The drainage system of this coal seam is equipped with 6 sets of water ring vacuum pumps with a total extraction capacity of 2040 m^3/min (excluding standby); the drainage system of adjacent strata is equipped with 4 sets of vacuum pumps with a total extraction capacity of 1248.6 m^3/min (excluding standby).

According to the information provided by the coal mine ventilation department, in 2018, the low concentration coal bed methane extracted from the coal seam in the mine was 27.24 million m^3, with a concentration of about 5%, which was completely drained. The coal bed methane extracted from adjacent strata is 61.46 million m^3, the concentration is more than 35%, and the utilization is 20.77 million m^3. It is mainly used for heating in mining area, bathing, central air shaft heating, coal bed methane driven air conditioning unit, coal bed methane power generation, slime drying, canteen gas, and gas for nearby residents, etc. The remaining 40.69 million m^3 is not used, all of which are discharged, resulting in the waste of high-quality clean fuel in coal mine.

At present, the utilization of gas in the mining area is mainly high concentration gas, the utilization of low concentration gas is 12.5 million m^3/a (converted into pure gas), and the surplus is 27.5 million m^3/a (converted into pure gas), with the utilization rate of about 37.5%. In addition, the high concentration gas allowance fluctuates greatly with seasons. The allowance in non-heating season is 91200 m^3/d (pure), and that in high load period in heating season is only 19200 m^3/d (pure).

In addition, there are a lot of low concentration gas emptying and waste heat of power plant flue gas unused in the mining area, which is a serious waste. If the waste heat of low concentration gas and gas power plant flue gas is utilized, it can be used for slime drying, central wind wells wellbore heating, bathing and heating in living area.

In order to solve the problems existing in gas utilization, it is proposed to optimize the whole gas utilization system, use the remaining low concentration gas and waste heat of power plant to replace part of high concentration gas, and utilize the high concentration gas by cryogenic liquefaction way, so as to improve the utilization rate of gas and maximize the benefit.

3 Comprehensive utilization scheme of coal bed methane

The selection of comprehensive utilization scheme of coal bed methane should follow the principle of "zero emission of gas drainage, comprehensive utilization of heat energy and maximum benefit". According to the concentration and volume of coal bed methane drainage, the location of pumping station and the utilization location, utilization mode and concentration of downstream products, as well as the matching relationship of utilization with seasonal changes, the current situation and expected demand of energy in the mining area should be comprehensively considered to optimize the energy supply mode and the utilization scheme of coal bed methane.

As mentioned above, 27.24 million m^3 of low concentration coal bed methane with a concentration of about 5% is discharged every year. This part of coal bed methane can generate heat energy by regenerative oxidation. One part of the heat energy can be used for coal slime drying in the mining area to replace the hot air stove for coal slime drying, and the other part can replace the coal-fired boiler in the production area to produce high-temperature flue gas and heat the air into the shaft to guarantee the working face is free of ice in the winter, the surplus heat energy can provide steam or hot water for the mining area. Steam can be used for power generation, and hot water can be used for bathing and laundry. According to the heat consumption scale of the mining area and the gas supply capacity of low concentration coal bed methane, it is suggested to build a low concentration coal bed methane regenerative oxidation device with a feed gas treatment capacity of 200000 m^3/h, and the designed treatment capacity of methane concentration is 1.2%. Considering the oxidation efficiency and heat transfer efficiency of the device, the total heat output is 17745 kW, the low concentration coal bed methane consumption is 38.6 m^3/ min, and the operation scale of the device is 193000 m^3/ h. In the process of heat transfer, the heat efficiency of high-temperature flue gas heat exchanger and waste heat boiler is 85%, so the heat provided by high-temperature flue gas heat exchanger and waste heat boiler is 15083kW, which can fully meet the heat load demand of slime drying after being transported to the heating end through the pipeline network along the way, and the remaining unused heat can replace part of the heat load of gas steam boiler.

For the 40.69 million m^3 of unused and drained CBM, the concentration of which is more than 35%, the method of cryogenic liquefaction can be used to extract the methane in low-concentration CBM, and then the methane can be purified and liquefied into LNG product. According to the supply and concentration of CBM, a set of cryogenic liquefaction device with processing scale (mixing capacity) of 300000 m^3/d of feed gas can be built, and the concentration of CBM can be designed as 35%. It has been calculated that the device can produce 25,333t of LNG product per year.

Both the regenerative oxidation device and the cryogenic liquefaction and extraction LNG device can operate throughout the year with high utilization rate. Except for the maintenance period of the device, zero emission of coal mine gas resources has been basically realized, and considerable subsidy income of coal bed methane utilization can be obtained. In addition, since the heating season is the peak of LNG price, when the gas quantity in the heating season is insufficient, the gas supply of other utilization methods can be adjusted appropriately to ensure the operating load of the liquefaction plant as far as possible and improve economic benefits.

4 Cryogenic liquefaction

4.1. Technological process.

The process of extracting methane from low concentration coal bed methane by cryogenic liquefaction mainly includes: feed gas metering and compression, acid gas removal, dehydration and drying, coal bed methane liquefaction and distillation separation, LNG storage and loading, bog treatment and LNG gasification system. Figure 1 is the process flow chart of the device. In the process design, the explosion characteristics of the pressurization process should be theoretically calculated and experimentally demonstrated according to the specific mole fraction of methane and other combustible impurities in the feed gas, and the appropriate operating pressure, operating temperature and fluid flow rate should be selected to ensure that the methane concentration in the pressurization process of coal bed methane is in the safe area beyond the explosion limit.

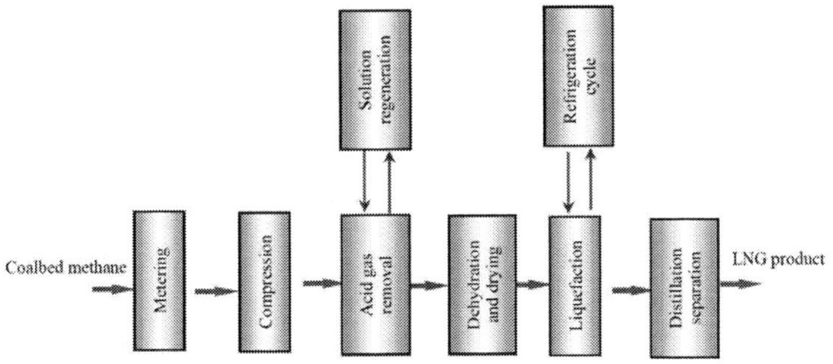

Figure 1. Cryogenic liquefaction process diagram of low concentration coal bed methane

There are two channels in the liquefied cold box, which are coal bed methane channel and refrigerant channel respectively. The liquefaction cold box is mainly composed of plate fin heat exchanger group and distillation column. The plate fin heat exchanger group mainly plays the role of precooling before separation of coal bed methane. The distillation column provides a place for liquefaction and purification of coal bed methane. LNG product mainly flow out after gathering at the bottom of the column. The cooling capacity required for liquefaction and separation is provided by the mixed refrigerant cycle, which is powered by the mixed refrigerant compressor.

After acid gas removal and dehydration, the feed gas can be regarded as a mixture of CH_4, O_2, N_2 and other components. The molar fraction of CH_4 is 35%. The mole fraction of is 13%; the mole fraction of N_2 is 52%. Feed gas flow rate F = 300 thousand m^3/d=12500 m^3/h. According to the separation task, the purity of LNG product at the bottom of the distillation column is required to be ≥99.5%; O_2 content ≤0.5%, CH_4 recovery ≥ 98.5%.

4.2. Material balance

According to the material balance before and after the separation of CBM, the methane recovery rate is calculated as follows:

$$\eta = \frac{W \times \varphi_W}{F \times \varphi_F} \qquad (1)$$

Where, η is the recovery rate of methane in LNG product; W is LNG production, m^3/h; φ_W is the purity of LNG, that is, the mole fraction of methane in the LNG product at the bottom of the tower; F is the feed gas flow, m^3/h; φ_F is methane content (mole fraction) of feed gas.

Substituting F=12500 m^3/h, φ_F=0.35, φ_W=0.995, η=0.985, it can be calculated that the gas production at the bottom of the tower is W = 4331 m^3/h; the nitrogen and oxygen tail gas flow at the top of the tower is $D = F - W = 8169$ m^3/h; the nitrogen and oxygen tail gas flow at the top of the tower is $\varphi_D = \frac{F \times 0.35 \times (1-0.985)}{D} = 0.803\%$; the nitrogen content at the top of the tower is $\varphi_{D,N} = \frac{F \times 0.52}{D} = 79.57\%$; the oxygen content at the top of the tower is $\varphi_{D,O} = 1 - 0.803\% - 79.57\% = 19.63\%$. Table 1 shows the material balance data of CBM channel.

 https://doi.org/10.1051/e3sconf/202129003010

Table 1. Material balance of coal bed methane in liquefied cold box

Gas name	rate of flow (m^3/h)	mole fraction		
		CH_4	O_2	N_2
feed gas	12500	0.35	0.13	0.52
tail gas at the top of the tower	8169	0.00803	0.1963	0.7957
product at the bottom of the tower	4331	0.995	0.005	0

4.3. Cooling capacity calculation

According to the energy balance calculation, the cooling capacity of the oxygenated coal bed methane liquefaction cold box is shown in Figure 2. According to the field test, the relevant process data are as follows.

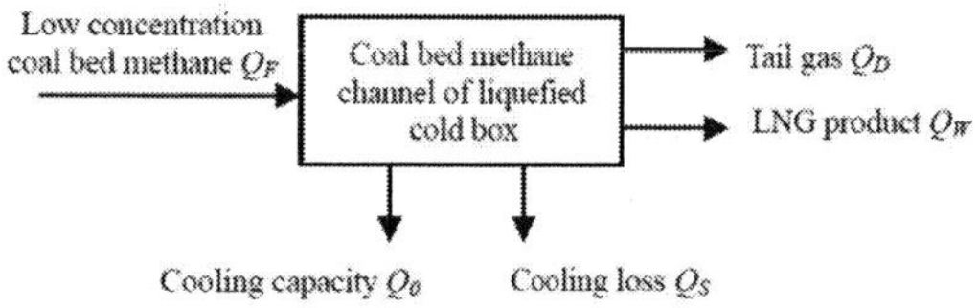

Figure 2. Energy balance diagram of coal bed methane channel in liquefied cold box

Before coal bed methane enters the plate fin heat exchanger, the pressure is 0.3 MPa (gauge pressure, the same below), the temperature is 30 ℃, and the mole fraction of each component is as follows: CH_4 is 35%; O_2 is 13%; N_2 is 52%. According to the above parameters, the specific enthalpy of feed gas entering the cold box is 1533.36kJ/kmol.

The pressure of nitrogen and oxygen tail gas is 0.2MPa, the temperature is 25 ℃, the mole fraction of each component is 0.8% for CH_4, 19.63% for O_2 and 79.57% for N_2. According to the above state parameters, the specific enthalpy is 105.1028 kJ/kmol.

The pressure of LNG product is 0.22 MPa, the temperature is -152℃, and the molar fraction of each component is as follows: CH_4 is 99.5%; O_2 is 0.5%. Specific enthalpy -10192.6956 kJ/k mol.

According to the above data, the heat balance calculation of CBM channel is carried out according to the following correlation formula:

$$Q_F + Q_S = Q_0 + Q_D + Q_W \qquad (2)$$

Where , the heat of oxygen-containing coal bed methane $Q_F = 1533.3615 \times \frac{12500}{22.4} = 855670.5$ kJ/h; the heat of oxygen-containing coal bed methane $Q_D = 105.1028 \times \frac{8169}{22.4} = 38329.7$ kJ/h; Heat of nitrogen and oxygen tail gas $Q_W = -10192.6956 \times \frac{4331}{22.4}$ $= -1970739.5$ kJ/h; Cooling capacity loss is 6.5kJ/ (m^3 feed gas) , that is $Q_S = 6.5 \times 12500 = 81250$ kJ/h. Therefore, the cooling capacity required for the coal bed methane to cool to low temperature is $Q_0 = 2.87 \times 106$ kJ/h= 797.2 kW.

4.4. Distillation tower parameters

Before CBM enters the distillation tower, the resistance in the plate-fin heat exchanger group is about 0.08MPa, so the operation pressure of the distillation tower is 0.22MPa. The state of CBM after precooling is as follows: the pressure is 0.22MPa, the temperature is -176℃, and the mole fraction is as follows: CH_4 is 35%, O_2 is 13%, N_2 is 52%; flow rate = 12500 m^3/h. Nitrogen and oxygen tail gas: 0.20MPa, flow rate = 8169 m^3/h. LNG product: 0.23MPa, methane molar fraction 99.5%.

According to the above data, the numerical simulation of the distillation column is carried out, and the calculation results are as follows: theoretical plate number is 15 (including reboiler and condenser), feed position is the fifth plate, reflux ratio is 0.1905, reboiler load is 429.58kW, condenser load is -120.72kW.

4.5. Refrigerant compressor parameters

Refrigerant compressor is the heart of refrigeration cycle, which provides power for refrigeration of cold box, and also provides energy for consumption "big household" of the whole project. The selection of parameters directly relates to the reliability and economy of the operation of the device. In the liquefaction process of natural gas, if mixed refrigerant is used for refrigeration, the refrigerant is generally selected from N_2, CH_4, C_2H_4 or C_2H_6, C_3H_8, C_4H_{10}, C_5H_{12} and other substances. The performance of C_2H_4 is slightly lower than C_2H_6, but the cost of C_2H_4 is much lower than C_2H_6. Considering the economic performance, C_2H_4 is selected for this device. According to the preliminary test and research, the mixture refrigerant is proposed to use the proportion shown in Table 2 with flow rate of 31250 m^3 / h.

Table 2. Mole fraction of each component of refrigerant mixture

material	N_2	CH_4	C_2H_4	C_3H_8	C_4H_{10}	C_5H_{12}
mole fraction	0.366	0.31	0.158	0.025	0.082	0.057

Before the mixed refrigerant enters the compressor, the pressure is 0.08 MPa, the temperature is 25 ℃; the pressure after compression is 4.38 MPa, 50 ℃. The ratio of absolute pressure before and after compression is 4.48 ÷ 0.18 = 24.89. In order to ensure safety and reduce the power consumption of the compressor, the three-stage compression and intermediate cooling are adopted, and the ratio of three-stage pressure is 2.92. According to the relevant data, the adiabatic efficiency is 80%.For the sake of insurance, the inlet water of cooling water is 0.2MPa, 30℃, and the outlet water is 50℃. After calculation, the total power consumption of the compressor is 4284KW, and the cooling water circulation capacity is 193t/h.

4.6. Safety measures

The low concentration CBM is inflammable and explosive. In order to operate safely, stably and continuously operation for a long period, corresponding safety measures are set up in three levels: safety system, process link and specific single equipment in the plant. The pressure vessel is equipped with safety valve and bursting disc. When the pressure of the vessel is greater than the rated unloading pressure of the safety valve, the safety valve opens and the drain main pipe is drained. In the case of safety valve failure, the bypass gate valve of the safety valve can be opened for pressure relief; the electrical equipment in the plant area is explosion-proof, the socket is equipped with leakage protection, and the electrical circuit is laid directly with armored cable; the whole equipment is explosion-proof, and the socket is equipped with leakage protection; the electrical circuit is directly buried with armored cable; The whole plant explosion-proof electrical design, installation and selection of electrical equipment are carried out in accordance with relevant standards and specifications; DCS and ESD system are adopted to realize the distributed control of the plant and emergency stop of accident state. In terms of lightning protection, class I lightning protection is adopted and independent lightning receiving device is set. In the aspect of anti-static, copper skin jumper and automatic resistance detection are adopted to ensure that the grounding resistance of the device is less than 1 Ω.

An automatic explosion suppression device is installed in front of the feed gas compressor, including an automatic explosion suppression valve, an automatic powder spraying explosion suppression device, and a water seal fire-proof and explosion venting device. Detection elements of safety assurance system are installed at the inlet end of key equipment, such as flame sensor and pressure sensor. In order to increase the reliability, the system is equipped with a separate control device, which operates independently and is not controlled by the DCS control system in the plant area. In case of flame or overpressure at any detection point, the corresponding signal will be transmitted to the control device through the sensor, the explosion-proof valve will immediately cut off the gas source, and the automatic powder spraying device will immediately spray powder to prevent the flame from spreading; in case of overpressure, the safety valve, bursting disc and bursting disc of water seal fire-proof and explosion relief device in the plant are operated to ensure the safety of equipment and operators in the plant.

5 Analysis of practice effect

By adopting the CBM comprehensive utilization technology, the coal mine can basically realize the full utilization of the gas extracted (except for the maintenance period of the device), and the annual additional gas consumption is about 53.38 million m^3, which is equivalent to 767,000 tons of CO_2 equivalent emission reduction. In addition, after the completion and operation of the project, it is estimated that the annual average income is 211 million yuan, and the annual average net profit is 84.98 million yuan. The financial risk of the investment and construction is small, and the economic effect is obvious.

6 Conclusion

(1) The low concentration coal bed methane extracted from the coal mine is comprehensively utilized according to the energy demand of the mining area. The low concentration coal bed methane with concentration of about 5% is transformed into high-temperature flue gas by regenerative oxidation, and the heat energy is extracted to realize heat supply in the mining area; the unused and empty coal bed methane with concentration of more than 35% is purified and liquefied into LNG product by cryogenic liquefaction, so as to realize long-distance transportation. The comprehensive utilization scheme of low concentration coal bed methane in coal mine follows the principle of "zero discharge of gas drainage, comprehensive utilization of heat energy and maximum benefit", which can basically realize the full utilization of gas drainage.

(2) The comprehensive utilization technology of low concentration coal bed methane can not only improve the air environment quality of the project site, but also alleviate the contradiction between supply and demand of gas shortage in the mining area, and realize the sustainable development of the enterprise.

(3) Low concentration coal bed methane comprehensive utilization is an important resource comprehensive utilization technology, which integrates the functions of realizing the energy demand of coal mining enterprises, reducing cost and increasing efficiency, promoting mine safety production, improving gas utilization rate, optimizing energy structure, promoting energy conservation and emission reduction. The gas source of the coal mine is stable and reliable, all construction conditions are available, the adopted technology is safe and feasible, and the project has remarkable energy saving, environmental protection, social and economic benefits.

Acknowledgements

This work was supported by the China National Science and Technology Major Project (2016ZX05045-006) and the technology innovation and entrepreneurship fund special project of Tiandi Technology Co., Ltd. (2019-TD-ZD004). The authors gratefully acknowledge the support provided by Technical Institute of Physics and Chemistry, CAS for the experimental study.

Funding

This research was funded by the China National Science and Technology Major Project (2016ZX05045-006) and the technology innovation and entrepreneurship fund special project of Tiandi Technology Co., Ltd. (2019-TD-

ZD004). The APC was funded by the technology innovation and entrepreneurship fund special project of Tiandi Technology Co., Ltd. (2019-TD-ZD004).

References

1. Zhou H. Air pollution control action plan [J]. Green China, 2017, No. 474 (08): 55-57.
2. Jiang S, Xue Z, Li W, et al. The environmental air quality and the impact of air pollution on health in China [J]. North China electric power technology, 2015, 000 (008): 7-13.
3. Ma H , Wang L , Jia H , et al. Experimental Study on Desorption Characteristics of Coal bed methane under Variable Loading and Temperature in Deep and High Geothermal Mine [J]. Advances in Civil Engineering, 2020, 2020(3):1-17.
4. Richardson A D , Hollinger D Y , Shoemaker J K , et al. Six years of ecosystem-atmosphere greenhouse gas fluxes measured in a sub-boreal forest [J]. Scientific Data, 2019, 6(1):117.
5. Ngnikam E , Tanawa E , Rousseaux P , et al. Evaluation of the potentialities to reduce greenhouse gases (GHG) emissions resulting from various treatments of municipal solid wastes (MSW) in moist tropical climates: application to Yaounde [J]. Waste Management & Research the Journal of the International Solid Wastes & Public Cleansing Association Iswa, 2002, 20(6):501.
6. Pu H , Zhang L , Dong X , et al. Simulation of the Extraction Efficiency of Coal bed methane under Water Injection: A Gas-Liquid-Solid Coupling Model [J]. Geofluids, 2020, 2020(3):1-14.
7. Cheng Y P , Dong J , Li W , et al. Effect of negative pressure on coal bed methane extraction and application in the utilization of methane resource [J]. Journal of China Coal Society, 2017, 42(6):1466-1474.
8. Cheng Y P , Dong J , Li W , et al. Effect of negative pressure on coal bed methane extraction and application in the utilization of methane resource [J]. Journal of China Coal Society, 2017, 42(6):1466-1474.
9. Kang J. Experimental study on wellbore heating technology of low concentration gas thermal storage oxidation [J]. Coal science and technology, 2018, 046 (006): 211-216.
10. Lian C. Research and practice effect of coal and gas CO mining and sharing technology in small high gas coal mine [J]. Mining research and development, 2019 (11): 63-66.
11. Yuan L, Xue J. Key technology of CO mining non pillar coal and gas in low permeability coal seam group [J]. Coal science and technology, 2013, 41 (1): 5-11.
12. Zhang X, Zhang Z, et al. Study on gas emission law and comprehensive treatment in Yangquan mining area [J]. Coal science and technology, 2014, 42 (9): 123-125.
13. Luo S. Design of low concentration gas thermal storage oxidation heating in Huoerxinhe coal mine [J]. Shaanxi coal, 2019, 038(004):10-13.
14. Huang K, Wang Y, Xiao Z. Application of ESD system in low concentration coal bed methane oxygen liquefaction device [J]. Mining safety and environmental protection, 2012, v.39; No.200 (001): 36-39.

Reservoir characteristics of Chang 2 Member of Yanchang Formation in Area A, Ordos Basin

Yajie Jing[1], Zhiwu Zhang[1], Peng Xu[1] and Shasha Yang[2,*]

[1]Zhidan Oil Production Plant, Shaanxi Yanchang Petroleum (Group) Co. Ltd., Yan'an, Shaanxi, 716000, China
[2]Shaanxi Key Laboratory of Safety and Durability of Concrete Structures, XiJing University, Xi'an, Shaanxi, 710123, China

Abstract. With the continuous enhancement of exploration and development in the Ordos Basin, in-depth research has been carried out on the petrological and reservoir characteristics of Chang 2 reservoir in Area A, which provides a geological basis for the efficient development of oil reservoirs. Comprehensive use of reservoir sandstone thin section identification casting analysis, mercury intrusion analysis, logging analysis and other methods to systematically study the petrological characteristics, pore characteristics and reservoir physical properties of Chang 2 reservoir in Area A. The results show that the reservoirs in the study area are dominated by fine-grained sandstones, with low component maturity and high structural maturity. They are all medium-low porosity, low-permeability and ultra-low permeability reservoirs. Primary intergranular pores and residual intergranular pores are developed, the reservoir drainage pressure is low, which is good-medium, and the mercury removal efficiency is high, indicating that the reservoir has good storage performance and seepage properties.

1 Introduction

Reservoir is one of the main controlling factors affecting the distribution of oil and gas, and it is the basis for finding remaining oil and developing oilfields [1-3]. Therefore, it is very important to conduct in-depth research on reservoir characteristics and comprehensive evaluation of reservoirs.

The Ordos Basin is a large-scale multi-cycle craton basin. The main oil-bearing layer of the Triassic Yanchang Formation is divided into 10 oil-bearing groups according to sedimentary cycles [4-6]. The Chang 2 reservoir in the study area mainly developed delta plain deposits, mainly sandstone deposits, with the characteristics of multi-stage channel superimposed development. Reservoir sand bodies are controlled by the distribution of sedimentary facies belts, with large oil layers and shallow burial. Chang 2 reservoir is the main production layers in Area A. This study used reservoir sandstone thin section identification, casting analysis, electron microscope scanning, mercury intrusion analysis methods to comprehensively evaluate the petrological characteristics and reservoir characteristics of Chang 2 reservoir in Area A area, and provide a geological basis for the efficient development of oil reservoirs.

2 Reservoir petrological characteristics

Through the identification and statistics of rock flakes, combined with observation under a scanning electron microscope, the petrological characteristics of the reservoirs in the study area were studied in detail, and the rock types were classified based on the three-terminal components of quartz, feldspar, and cuttings (figure 1).

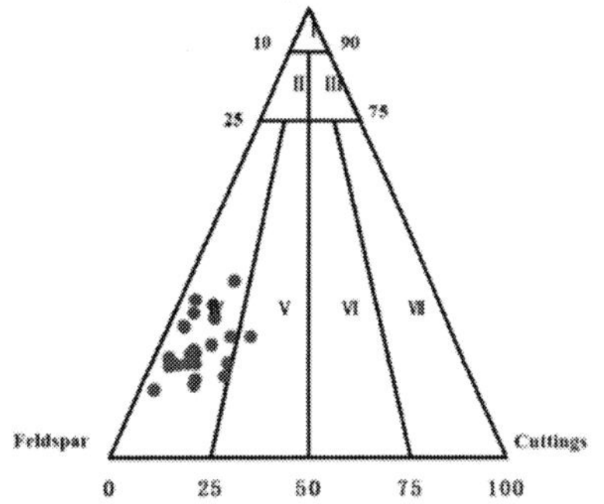

I quartz sandstone, II feldspar quartz sandstone, III lithic quartz sandstone, IV feldspar sandstone, V lithic feldspar sandstone, VI feldspar lithic sandstone, lithic sandstone.

Figure 1. Triangular diagram of Chang 2 rock types in Area A

2.1 Sandstone characteristics

The mineral composition maturity of the Chang 2 reservoir rock is relatively low. Statistics show that the clastic composition is dominated by feldspar, with a content of 46.3% ~ 48.3%, with an average of 47.2%. Followed by quartz, with a content of 28.6% ~ 29.6%, with an average of 29.1%. The content of biotite and cuttings is relatively low, with an average content of 7.3% and 3.1%, respectively. The composition of the cuttings is mainly acid extruded rock, granitic rock and dynamic metamorphic rock, and mica is mostly chlorite(figure 2).

* Corresponding author: 20190100@xijing.edu.cn

https://doi.org/10.1051/e3sconf/202129003011

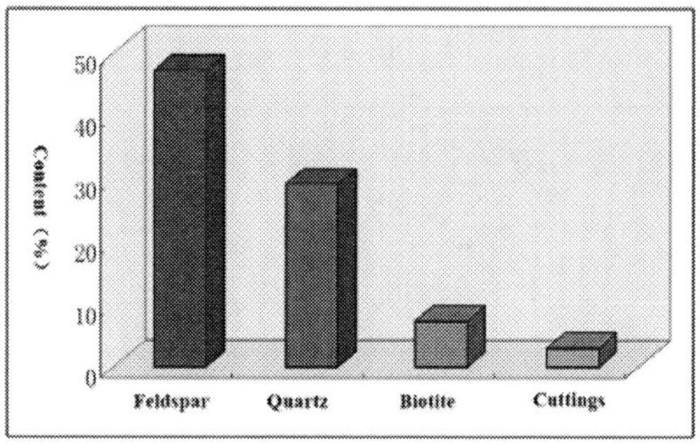

Figure 2. Distribution diagram of rock composition of Chang 2 in Area A

2.2 Clay mineral characteristics

The clay minerals in the Chang 2 reservoir in Area A can be divided into two types, the original cause and the autogenous cause. The microscopic characteristics of clay minerals are: primary clay minerals are mostly filled in the pores between the particles, and there is no fresh crystal form. The microscopic characteristics of authigenic clay minerals are more complex and the crystal shape is relatively complete. According to statistics, the absolute content of clay minerals in Chang 2 oil layer in this area is 4.4% to 10.8%, with an average of 8.4%. There are four main types of clay minerals, namely chlorite, illite, kaolinite, and illite/smectite layers. Among them, chlorite is the main type, with a relative content of 57.5% to 90.7% of the total clay minerals, with an average of 74.2 %. Observed under a scanning electron microscope, chlorite is mostly distributed vertically on the surface of detrital particles in needle-like or rose-like shapes, and is produced in a pore-filling or cushioning type (Figure 3). The second is kaolinite, with a content of 8.2% to 18.8%, with an average of 12.7%. It is mostly produced in the form of pages or worms in the reservoir, but its distribution varies greatly in different parts. The content of illite in the reservoir is relatively high, ranging from 4.1% to 11.7%, with an average of 7.6%. Most of the illite is produced in the pores in the form of filaments, bridges or flakes, blocking the throat. The content of clay minerals in the Iraqi/Mongolian mixed layer is the lowest, with a content of 3.6% to 6.7%.

Figure 3. Scanning electron microscope photo of chlorite clay minerals in Chang 2 oil layer in Area A

3 Reservoir pore type and pore structure

3.1 Pore types

Pore types can be divided into three categories: primary pores, secondary pores and micro-fractures according to their genesis. Primary pores mainly refer to the intergranular pores of detrital particles, and also include interlayer pores and pores. Secondary pores refer to the pores and fractures in the rock after the formation of sedimentary rocks due to leaching, dissolution, metasomatism, dissolution and recrystallization. During the diagenesis process, the primary pores will gradually decrease through compaction, cementation and pressure dissolution. At the same time, the soluble detrital particles and easily soluble formations will dissolve and metasomatize as the buried depth increases, thereby contributing to Development of secondary pores in clastic rocks.

The Chang 2 oil layer group of Yanchang Formation in Area Area A is shallower, and the compaction and pressure solution are weakened, and the primary pores are mainly developed. The primary intergranular pores remaining after the secondary enlargement of mechanical compaction of feldspar and quartz and a variety of cementation are also called residual intergranular pores. Including the residual intergranular pores after the early film-like chlorite and illite cementation, the residual intergranular pores after the secondary growth of quartz and feldspar, and the residual intergranular pores after the filling and cementation of calcite and clay minerals (Figure 4).

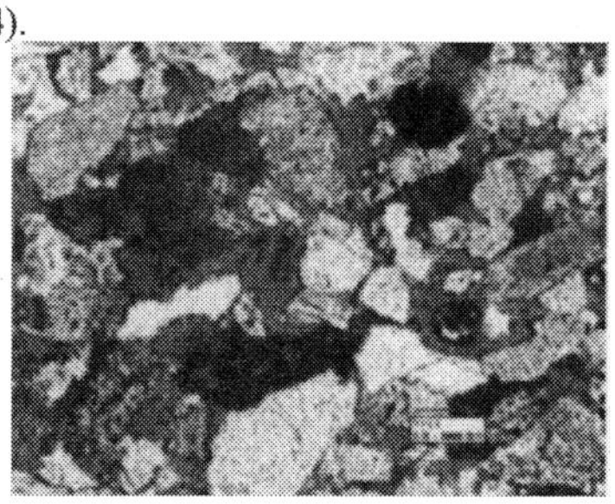

A.Primary intergranular pores

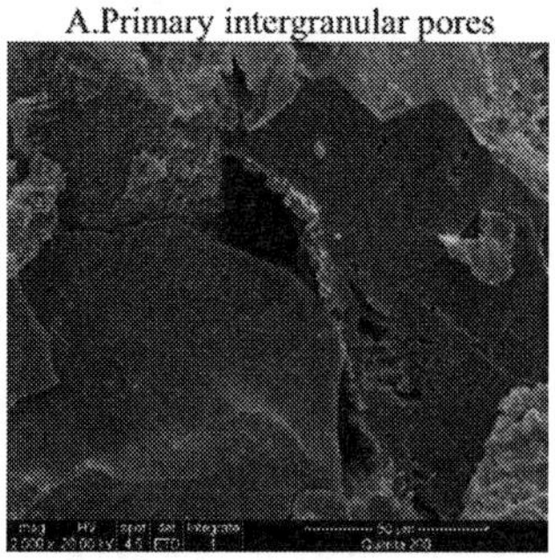

B.Residual intergranular pores

Figure 4. Primary pores in Chang 2 reservoir of Well B95 in Area A

https://doi.org/10.1051/e3sconf/202129003011

3.2 Pore structure characteristics

In this study, mercury intrusion method was selected to study the pore structure. The capillary pressure curve reflects the size of the pores through which mercury may enter the throat and the volume of pores connected by this throat under a certain displacement pressure. The capillary pressure curve is not only a function of the pore size distribution and pore volume, a function of the pore-throat connection method, but also a function of porosity, permeability and saturation [7,8]. The capillary pressure curve is used to characterize the pore structure, the length of the relatively flat section in the middle of the capillary pressure curve indicates the concentration of rock throat distribution and the quality of the sorting. The longer the flat section, the more concentrated the distribution of rock throats, the better the sorting, and the better the flat section. Lower, the larger the radius of the rock throat, the lower the displacement starting pressure, and the better the rock permeability. On the contrary, the higher the gentle section, the smaller the radius of the rock throat, the higher the displacement starting pressure, and the worse the rock permeability.

According to the capillary pressure curve and mercury injection curve data, it can be seen that the displacement pressure, the saturation median pressure, and the median throat radius of the Chang 2 reservoir in the study area have not changed much. The average displacement pressure is 0.19MPa, the average saturation median pressure is 3.41MPa, the average throat median radius is 0.24 μ m, the average mercury removal efficiency is 39.98%, and the average maximum mercury saturation is 93.46%. In general, the Chang 2 reservoir in Area A area has low displacement pressure and high mercury removal efficiency, indicating that the reservoir has good reservoir and permeability properties.

4 Reservoir physical properties

4.1 Classification standards for reservoir physical properties

According to the reservoir classification standard in the petroleum industry standard, the physical properties of sandstone reservoirs are divided into 5 types of reservoirs, as shown in Table 1.

Table 1. Classification standards for physical properties of sandstone reservoirs

Classification	Porosity Φ (%)	Permeability K ($\times 10^{-3}\mu m^2$)
Extra High	Φ≥30	K≥1000
High	25≤Φ < 30	500≤K < 1000
Medium	15≤Φ < 25	50≤K < 500
Low	10≤Φ < 15	5≤K < 50
Extra Low	Φ < 10	K < 5

4.2 Reservoir physical characteristics

According to the logging data, the porosity and permeability of the studied well sections are statistically classified. The results show that the overall physical properties of the Chang 2_1 layer are better than those of the Chang 2_2 layer, which may be related to the compaction of the formation. The maximum permeability varies greatly between each small layer, especially the difference between the maximum permeability of the Chang 2_2^1 and Chang 2_2^3 layers Nearly 10 times. On the one hand, there may be a large difference in the cored samples, and the Chang 2_2^3 layer has less physical property data, which cannot fully represent the real state of the physical properties of the layer. On the other hand, there may be deviations in logging data or calculation methods. Lead to the emergence of undue real data. However, there are more than 60 data points in the physical property statistics of each sub-layer, and more than 200 data points in individual layers. However, there are only 1 to 2 suspected under-real data obtained by statistics, which account for a small proportion in the overall sample. can be ignored. Therefore, the physical property data of each sublayer of Chang 2 in this statistics can basically reflect the physical property characteristics of the formation.

After sub-layer statistics of physical property data, it is found that the average porosity of each sub-layer is greater than 10%, the average permeability is less than $50\times 10^{-3}\mu m^2$, and the physical property classification of the entire Chang 2 formation has little difference. Therefore, in this study, all the physical property data of Chang 2 formation were calculated according to the physical property classification standard, and the frequency distribution histogram of porosity and permeability (Figure 5, Figure 6) was drawn to grasp the change of Chang 2 physical property from a macro perspective feature.

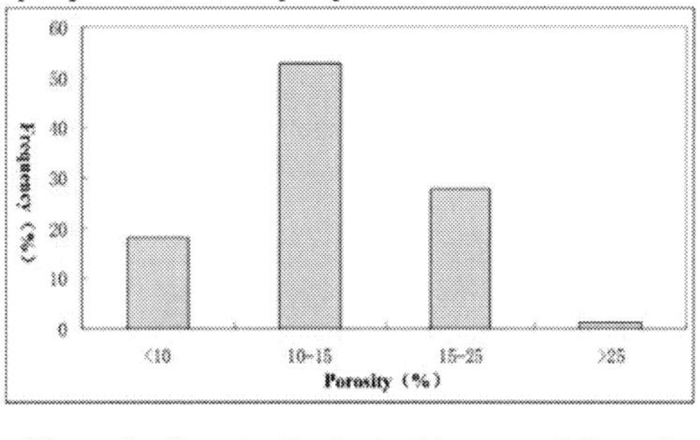

Figure 5. Porosity distribution histogram of Chang 2 reservoir in Area A

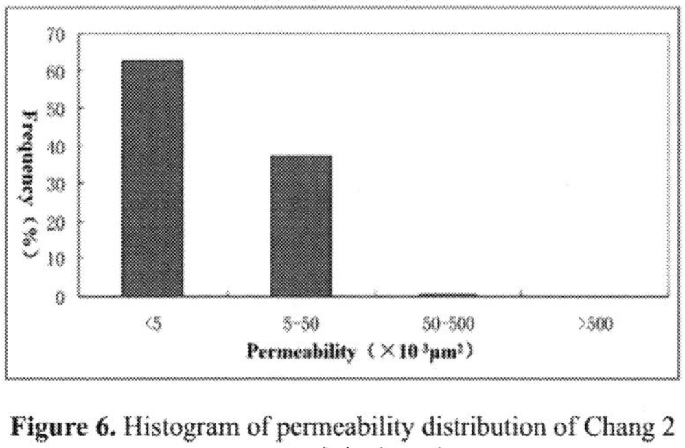

Figure 6. Histogram of permeability distribution of Chang 2 reservoir in Area A

From the above two physical property distribution histograms, it can be seen that the porosity of the Chang 2 formation in Area A is mainly 10-15%, followed by 15-25%, and the proportion of porosities less than 10% and greater than 25% is relatively small. Most of the permeability is less than $5\times10^{-3}\mu m^2$, followed by 5 ~ $50\times10^{-3}\mu m^2$, and there is almost no data with permeability greater than $50\times10^{-3}\mu m^2$. According to the reservoir physical property classification standard, the Chang 2 reservoir is a medium-low porosity, low-permeability, and ultra-low permeability sandstone reservoir.

5 Conclusion

1. Chang 2 in Area Area A is dominated by fine-grained sandstone, and the compositional maturity of the reservoir sandstone is low. The clastic composition is dominated by feldspar, followed by quartz, and the content of rock fragments is low. Reservoir sandstone has low component maturity and high structural maturity.

2. The Chang 2 oil layer of Yanchang Formation in the study block is shallower, the compaction and pressure solution are weakened, and the primary intergranular pores and residual intergranular pores are mainly developed. The reservoir drainage pressure is low, the mercury removal efficiency is high, indicating that the reservoir has good reservoir and permeability properties.

3.The porosity of Chang 2 formation is mainly 10-15%, and the permeability is mostly less than $5\times10^{-3}\mu m^2$. According to the classification standard of reservoir physical properties, the Chang 2 reservoir is a medium-low porosity, low permeability, and ultra-low permeability sandstone reservoir.

References

1. Zhao Jingzhou, Wang Yongdong, Meng Xiangzhen. Distribution law of Triassic Chang 2 reservoir in the eastern part of the northern slope of Shaanxi, Ordos Basin[J]. Petroleum Exploration and Development 2007(1):23-27.
2. Guo Yanqin, Study on the micro characteristics of Yanchang Formation reservoir in Fuxian exploration area[D]. Northwest University, 2006.
3. Liu Chunyan, Wang Yi. Sedimentary characteristics and physical properties of Yanchang Formation in Fuxian area, Ordos Basin[J]. World Geology, 2009,28(4):23-26.
4. Xu Jianhong.Geological characteristics of the Triassic Yanchang Formation reservoir in Chenglinsong Ordos Basin[J]. Journal of Southwest Petroleum University,2007,29(5)13-17.
5. Wu Fuli, Li Wenhou, Li Yuhong. Delta deposition and evolution of the Upper Triassic Yanchang Formation in the Ordos Basin[J]. Journal of Palaeogeography,2004(3):307-315.
6. Wang Kai, Wu Fuli, Wang Erhu, etc. Reservoir characteristics analysis of Chang 2 oil layer group in Zichang Oilfield[J]. Journal of Yangtze University, 2013, 10(8):29-30.
7. Guo Yanqin. Study on the micro characteristics of Yanchang Formation reservoir in Fuxian exploration area[D]. Northwest University, 2006.
8. Wang Feng. Study on the sedimentary, sequence evolution and lithologic reservoir characteristics of the Triassic Yanchang Formation in the Ordos Basin[D]. Chengdu University of Technology, 2007.

https://doi.org/10.1051/e3sconf/202129003012

Brief Introduction of Carbon Neutral International Organizations

Wang xin[1],*, Ma Cuifeng[1]

[1]National Geological Library of China, 100083 Beijing, China

Abstract. We searched the yearbook database of international organizations with "carbon" and "CO_2" as key words, and got six international organizations related to carbon neutrality on the basis of cleaning the data. This paper will briefly introduce six carbon neutral international organizations included in the Yearbook database of international organizations to provide references for researchers.

1 Introduction

At the general debate of the 75th UN General Assembly held on September 22, 2020, China announced to the world that it would enhance its national independent contribution. The emissions of the carbon dioxide will reach the peak by 2030 by taking more effective policies and measures, and the carbon neutrality will be realized by 2060. It has aroused wide discussion and attention of global researchers and policy makers on carbon neutrality after the goal was put forward[1]. This paper will briefly introduce six carbon neutral international organizations included in the Yearbook database of international organizations to provide references for researchers.

2 Methods and Materials

The ~66,000 international organizations maintained in Yearbook database of international organizations include Intergovernmental Organizations (IGOs) and International Nongovernmental Organizations (INGOs) in 300 countries and territories. Approximately 1,200 new organizations are added each year[2].

We searched the yearbook database of international organizations with "carbon" and "CO_2" as key words, and got six international organizations related to carbon neutrality on the basis of cleaning the data. We briefly introduced the establishment time, place, mission, member structure and activities of these organizations. See Table 1 for details.

Table 1. Six Carbon Neutral International Organizations.

Name	Founded Time	City HQ	Country/ Territory HQ
European Network of Excellence on the Geological Storage of CO_2	2004	Orléans	France
Global CCS Institute	2009	Melbourne VIC	Australia
International Institute for Carbon-Neutral Energy Research	2010	Fukuoka	Japan
Carbon Offsets To Alleviate Poverty	2011	Oakland CA	USA
International Sustainability and Carbon Certification	2011	Cologne	Germany
Carbon Neutral Cities Alliance	2014	Network	Network

* Corresponding author: 357941967@qq.com

3 International Organizations about Carbon Neutral

3.1. European Network of Excellence on the Geological Storage of CO_2 (CO_2GeoNet)

3.1.1 Founded

European Network of Excellence on the Geological Storage of CO_2 (CO_2GeoNet) was founded in Venice (Italy) in April, 2004[3].

3.1.2 Mission

The mission of CO_2GeoNet is to enable safe and efficient deployment of CO_2 Capture and Storage (CCS).

3.1.3 Membership

CO_2GeoNet is the European scientific body on CO_2 geological storage. The nonprofit scientific association currently is comprised by 29 research institutes from 17 European countries. See Figure 1 for details.

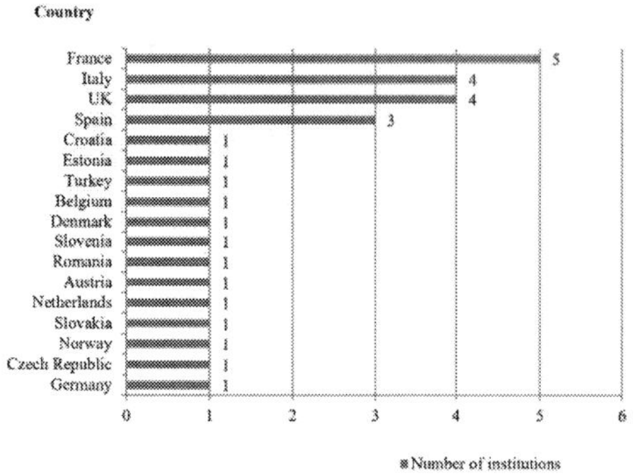

Fig. 1. Number of institutions in each country in CO_2GeoNet.

3.1.4 Activities

3.1.4.1 Joint Research

The strategic research aims of the CO_2GeoNet Association are to produce and disseminate integrated scientific research results, contribute to the identification of knowledge gaps and the formulation of new research targets, and to foster the interaction and exchange of information and views between CO_2 storage researchers and CCS stakeholders.

3.1.4.2 Scientific Advice

Scientific advice represents a major channel for CO_2GeoNet to position itself as a reference body for European science on CO_2 Geological Storage. The Association strategy includes the creation of an organized advice system to enable the Network's contribution to advice processes both based on external and own funds.

3.1.4.3 Capacity Building

The CO_2GeoNet strategy includes training and capacity building to foster the development of upcoming generations of scientists that will be needed for the widespread deployment of the technology.

3.1.4.4 Dissemination

The CO_2GeoNet strategy includes information and communication as a way to promote the Association and specifically identifies the provision and dissemination of integrated scientific research results and the fostering of the exchange of information and views between CO_2 storage researchers and CCS stakeholders.

3.2 Global CCS Institute（GCCSI)

3.2.1 Founded

Global CCS Institute was Founded in Canberra (Australia) in April, 2009. Officially founded on July 1, 2009. Also referred to as Global Carbon Capture and Storage Institute (GCCSI) of The Institute[4].

3.2.2 Mission

The mission of GCCSI is to accelerate the deployment of carbon capture and storage (CCS).

3.2.3 Membership

The diverse international membership includes governments, global corporations, private companies, research bodies and nongovernmental organizations; The Global CCS Institute is comprised by 98 research institutes from 19 countries. See Figure 2 for details.

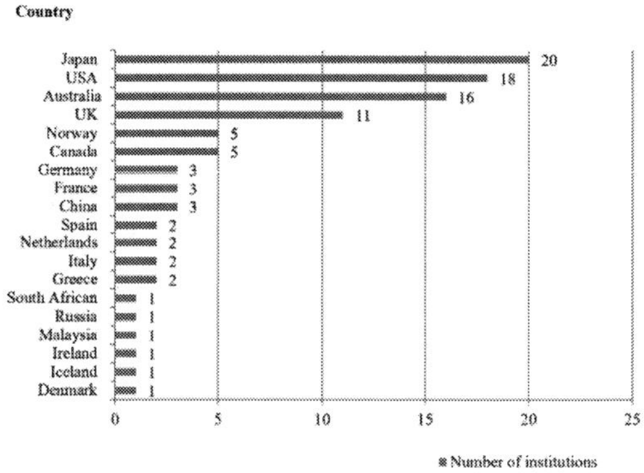

Fig. 2. Number of institutions in each country in GCCSI.

3.2.4 Activities

3.2.4.1 Build Knowledge

As the world's leading CCS think tank, it is an unparalleled source of research, data and information on all aspects of CCS deployment. The Global CCS Institute offers expertise and knowledge across every aspect of CCS, providing our members with accurate, insightful and impactful information and data about CCS, assisting them to build knowledge, connect with the right people and make informed business decisions. No other organization has the holistic capabilities or CCS expertise on a global scale. Our team includes geologists, scientists, engineers, economists, policy analysts, climate change experts, business developers and marketers, who provide expertise across the entire CCS spectrum.

3.2.4.2 Shift the Narrative

It promote and build awareness of CCS and its vital role in achieving energy security and emissions reductions targets to key media outlets, policymakers and important influencers around the world. This is critical given the

https://doi.org/10.1051/e3sconf/202129003012

proliferation of mis-information regarding the reliability, economics and applicability of CCS.

3.2.4.3 Enable Investment

It works with its members and other businesses and governments, leveraging our CCS knowledge, skills and expertise, to enable investment in the next wave of CCS facilities. As policy and legislative changes create renewed opportunities around the world, now is the time to drive this investment.

3.3 International Institute for Carbon-Neutral Energy Research (I²CNER)

3.3.1 Founded

International Institute for Carbon-Neutral Energy Research was founded in Fukuoka (Japan) in 2010. Also referred to as I²CNER[5].

3.3.2 Mission

The mission of I²CNER is to creation of a sustainable and environmentally friendly society by advancing fundamental science to reduce CO_2 emissions and establish a non-fossil based energy carrier system.

3.3.3 Membership

The Institute is a unique collaborative project between Kyushu University and the University of Illinois at Urbana-Champaign, with the main facility located on the Ito campus in Japan, and a Satellite Institute located in the United States of America.

3.3.4 Activities

3.3.4.1 Advanced Energy Materials Thrust

Develop molecular, nano, and bulk materials based on new science of surfaces, interfaces, and microstructures for applications involving H_2, H_2O, and CO_2.

3.3.4.2 Advanced Energy Conversion Systems Thrust

Development and evolution of systems which either lower carbon emission, increase energy efficiency, or both.

3.3.4.3 Multiscale Science and Engineering for Energy and the Environment Thrust

Pull together the range of challenges facing Japan's and the world's energy transition, namely the transition from largely fossil fueled energy technology to a carbon-neutral or a carbon-free energy supply.

3.4 Carbon Offsets To Alleviate Poverty (COTAP)

3.4.1 Founded

Carbon Offsets To Alleviate Poverty (COTAP) was Founded in 2011, Registered in the State of California (USA)[6].

3.4.2 Mission

The mission of COTAP is to empower individuals and organizations in developed countries to address both climate change and global poverty.

3.4.3 Membership

The Institute is an Internationally-oriented national organizations. American Tim Whitley launched COTAP. The COTAP is comprised by 6 research institutes from 2 countries, including five American institutions and one British institution.

3.4.4 Activities

3.4.4.1 Ecotrust's Trees for Global Benefits (TFGB) project

Ecotrust's Trees for Global Benefits (TFGB) project, located in the Albertine Rift valley in Uganda, is a small-scale, farmer-led agroforestry program which produces long-term, plan vivo-accredited emissions reductions while measurably improving farmer livelihoods and emphasizing sustainable land-use practices. Uganda, in sub-Saharan East Africa, is among the poorest countries in the world, ranked 163th of 187 countries in the UNDP's 2014 Human Development Index. The TFGB project began as a pilot in the Bushenyi district in 2003, and the pilot's success has led to expansion into the districts of Rubirizi, Mitooma, Kasese, Hoima, and Masindi, as well as into the Mount Elgon ecosystems of Rwenzori, Mbale, Manafwa, Bududa, Bulambuli, and Sironko.

3.4.4.2 Taking Root's CommuniTree Project

Taking Root's CommuniTree Project began near the rural municipality of San Juan de Limay in the province of Esteli, Nicaragua. Nicaragua is the second poorest country in the Americas, after Haiti. As of 2016, Taking Root has partnered with 65 different communities comprising 409 small-scale family farmers and spanning an area of 7,400 hectares. As of the end of the 2016 planting season, the project had cumulatively planted 2.2 million trees.

3.4.4.3 The Khasi Hills Community REDD+ Project

The Khasi Hills Community REDD+ Project is located in the remote northeast state of Meghalaya with Bangladesh to the south and Bhutan to the north.

 https://doi.org/10.1051/e3sconf/202129003012

Spanning 27,000 hectares, it deploys strategies for both forest protection (Reducing Emissions from Deforestation and Forest Degradation, or "REDD") and restoration (Assisted Natural Regeneration, or "ANR"). The project also provides detailed and long-term plans for improving the livelihoods of 4,400 households, 80% to 90% of which live below the poverty line. The region experienced a rapid 28% loss in forest cover between 2000 and 2005. Further, India ranks 136th of 186 countries in the UNDP's 2012 Human Development Index, and annual incomes in the region are $490 for a family of 5 to 6.

3.4.4.4 Envirotrade's Sofala Community Carbon Project

Envirotrade's Sofala Community Carbon Project is located in the buffer zones around Gorongosa and Marromeu National Parks in the central Sofala province of Mozambique. Mozambique, in sub-Saharan East Africa, is among the poorest countries in the world, ranked 184th of 187 countries in the UNDP's 2011 Human Development Index. Its 16-year civil war, which ended in 1992, left it one of the poorest countries in the world, cost the lives of an estimated 1 million of its citizens, and resulted in significant environmental degradation.To address these difficult challenges, the Sofala project was started as a pilot in 2002 by Envirotrade, the University of Edinburgh, and the European Union with 53 farmers in the wards of N'hambita and Munhanganha.

3.5 International Sustainability and Carbon Certification (ISCC)

3.5.1 Founded

Current statutes adopted in February, 2011. Registered in accordance with German law[7].

3.5.2 Mission

The mission of ISCC is to contribute to the implementation of environmentally, socially and economically sustainable production and use of all kinds of biomass in global supply chains.

3.5.3 Membership

The ISCC is comprised by 171 research institutes from 33 countries. Among them, there are 23 countries with two or more institutions, see Figure 3 for details. There are 10 countries with only one institution, including Australia, Colombia, Denmark, Greece, Hungary, Iceland, Ireland, Lithuania, Slovakia, and South Korea.

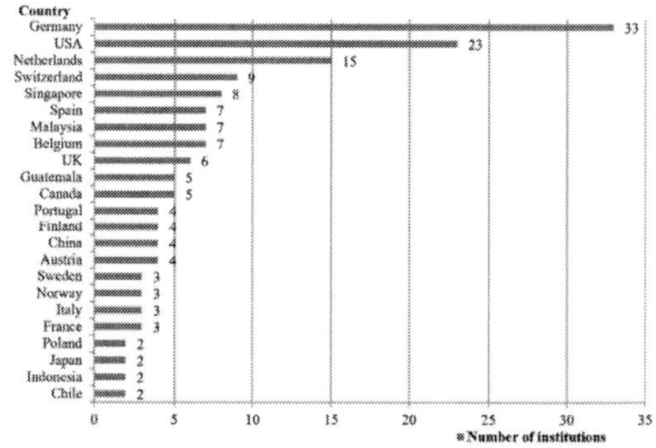

Fig. 3. Countries with 2 or more institutions in ISCC.

3.5.4 Activities

3.5.4.1 Governance & Transparency

ISCC has set up a working group to improve the transparency of ISCC by mandatory publication of summary audit reports. The set up of the working group was agreed at the General Assembly 2016 with the scope to elaborate a summary audit report that should be made publicly available for each issued certificate. The working group had to consider a trade-off between enhancing the transparency of ISCC and protecting sensitive company information.

3.5.4.2 Food Security

ISCC is partner of the Food Security Standard (FSS). By adopting the FSS, companies along the value chain can now demonstrate that they are committed to the Human Right to adequate Food and the UN Sustainable Development Goals (SDGs).

3.5.4.3 Sustainable Development Goals

ISCC supports the Sustainable Development Goals. The 2030 Agenda for Sustainable Development adopted by all UN Member States in 2015 aims to foster peace and prosperity for people and the planet, today and in the future. At its heart are the 17 Sustainable Development Goals (SDGs) which call for climate action, environmental conservation, an end to poverty, improved health and education, reduced inequality and economic growth. Especially the private sector is called upon to contribute to a sustainable development. Therefore, ISCC supports businesses to act more responsibly and secure sustainability throughout the entire supply chain. However, ISCC is aware of the challenges and seeks to continuously improve its contribution to the fulfilment of the SDGs.

3.5.4.4 Circular Economy and Bioeconomy

ISCC supports the transition to a circular economy and bioeconomy.

3.5.4.5 Benchmarking

SCC regularly participates in benchmarking processes comparing different systems for sustainability certification. ISCC uses the results of those processes and the feedback provided to learn and improve continuously.

3.5.4.6 Sustainable Aviation Fuels

As a well-experienced sustainability certification scheme, ISCC can guarantee compliance and enhance traceability through site-specific audits. ISCC's vast experience in road transportation is of high value when it comes to certifying sustainable aviation fuels. Since large volumes of certified feedstock are already available and many major fuel producers are ISCC members, SAF certification with ISCC is feasible and practical.

3.6 Carbon Neutral Cities Alliance (CNCA)

3.6.1 Founded

The Carbon Neutral Cities Alliance (CNCA or "Alliance") was born in Copenhagen in June 2014[8].

3.6.2 Mission

The mission of CNCA is to enable leading cities worldwide that are working aggressively toward a zero-carbon future to advance their own transformational efforts, collaborate with each other and key partners to overcome barriers, foster innovative approaches, and share lessons with other cities ready to pursue similar goals.

3.6.3 Membership

At present, the alliance has attracted 22 cities from 13 countries to participate in carbon neutrality. See Figure 4 for details.

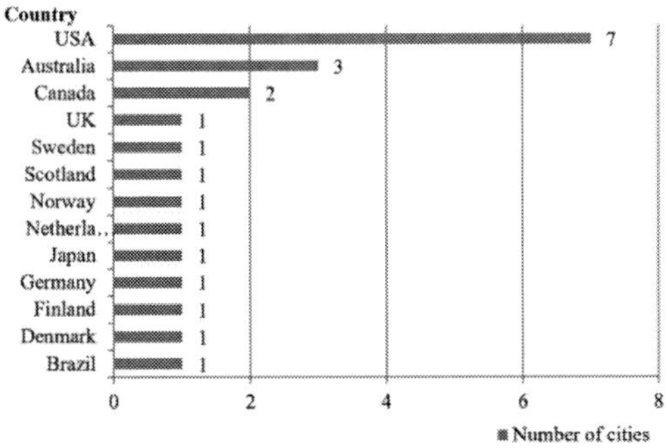

Fig. 4. Number of cities in each country in CNCA.

3.6.4 Activities

3.6.4.1 Developing Carbon Neutrality Planning Standards

Developing approaches, analysis, and tools to support carbon neutrality; standardizing measurement and verification methodologies for tracking progress.

3.6.4.2 Advancing "Transformative Change" in Key Urban Sectors

Sharing and implementing best practices for achieving "transformative" deep carbon reduction strategies in urban transportation, energy use, and waste systems.

3.6.4.3 Advocating for Policy Change

Identifying and advocating for policies at the state, regional, and federal levels to reduce emission sources not controlled directly by cities and engaging with other external stakeholders who are critical to cities' success.

3.6.4.4 Speaking with a Common Voice

Helping CNCA cities demonstrate their leadership and communicate with a common voice.

3.6.4.5 Creating a CNCA "Innovation Fund"

Investing in high-potential, city-led projects that develop, test, implement, and amplify deep decarbonization strategies and practices.

3.6.4.6 Increasing Alliance Impact

Sharing Alliance learnings with a broader audience to benefit the "next wave" of cities striving for carbon neutrality.

4 Conclusions

The goal of six carbon neutral international organizations is to offset carbon dioxide emissions through afforestation, energy conservation and emission reduction, so as to achieve "zero emission" of carbon dioxide and contribute to the creation of a sustainable and environment-friendly society.

Among the six carbon neutral international organizations, most of the institutions and cities are from the United States, Germany, Australia, Japan and the United Kingdom.

Acknowledgements

This paper is supported by China Geological Survey Project "Update and service of geoscience literature information" (DD20190413).

References

1. Xi J P. Speech at the general debate of the seventy-fifth United Nations General Assembly [N]. The People's Daily, 2020-09-23 (in Chinese)
2. https://ybio.brillonline.com/ybio/,2021-5-10.
3. http://www.co2geonet.com/ ,2021-5-12.
4. https://www.globalccsinstitute.com ,2021-5-12.
5. http://i2cner.illinois.edu/ ,2021-5-12.
6. http://cotap.org/ ,2021-5-12.
7. http://www.iscc-system.org/ ,2021-5-12.
8. https://carbonneutralcities.org/ ,2021-5-12.

Research on the detection and early warning technology of harmful marine organisms in the water intake of nuclear power plant by 3D image sonar

Jianfei Zhang [1,*], Zhongpeng Wu [1], Chong An [1]
[1] Dalian measurement and Control Technology Research Institute, NO. 16, BinhaiStreet,Zhongshan District, Dalian,Liaoning,116013, P. R. China Jianfei_zhang760@163.com

Abstract. In view of the frequent water intake blockage caused by marine biological invasion in many nuclear power plants, which affects the safety of cold source in nuclear power plants, a defense system in depth including marine biological identification, detection, early warning and hierarchical response is established through investigation and analysis of the causes of the event. Underwater acoustic high-resolution multi beam detection method is used to detect jellyfish, hairy shrimp and other marine organisms, The echo characteristics are obtained, the distribution density in sensitive sea area is analyzed, and the effective early warning judgment is given. Among them, detection and early warning is the front-end disposal measure, which can provide accurate and timely marine biological information for the subsequent hierarchical response. At the same time, it can replace the underwater inspection work of divers,reduce the risk of divers,reduce the risk of industrial safety,and improve the reliability of cold source.

1 Introduction

Some data show that with the change of climate, marine and river environmental conditions, the frequent occurrence of water intake blockage events indicates that the nuclear power plants have insufficient risk prediction of marine organisms in the water intake area, the monitoring and early warning mechanism of marine organisms is not perfect, and the response measures are not targeted. Since 2011, the safe and reliable operation of circulating water system has been affected by jellyfish, Phaeocystis, sediment, small fish and shrimp in many nuclear power units at home and abroad, which blocked the cooling water pipe from the turbine of the power plant, resulting in the power plant having to shut down the reactor, seriously affecting the safety of the unit. Since 2016, the national nuclear safety administration has reported that dozens of incidents occurred in domestic nuclear power plants, which affected the safety of water intake due to marine organisms or foreign matters blocking the water intake system. ([1]-[3])Ten typical incidents of marine organisms affecting the safety of water intake of nuclear power plants were disclosed. Under the risk of frequent marine biological invasion, the problem of cold source support of coastal nuclear power plants is becoming increasingly prominent. How to solve the cold source problem of nuclear power plants from the source has become an urgent task. Therefore, the establishment of marine biological detection and early warning mechanism has become the top priority.

2 Design and development of autonomo us monitoring and early warning syste m for marine organisms in nuclear pow er plant

The automatic monitoring and early warning system for marine organisms in nuclear power plant uses high-resolution multi beam sonar as sensing means, and uses the feature analysis and statistics module to analyze and process the obtained marine organism image information and identify the features, so as to obtain the density distribution of jellyfish, hairy shrimp and other marine organisms in the monitoring area in real time, and give an alarm when the threshold is exceeded.

2.1 Data processing and analysis of interception network + kelp echo

The autonomous monitoring and early warning system for marine organisms of nuclear power includes shore based monitoring center and underwater monitoring platform. The shore based detection center is composed of early warning processing unit and database; The underwater monitoring platform is composed of cruise monitoring platform and several fixed monitoring platforms (configured according to the monitoring water area width of the cold source water intake of the nuclear power plant). The fixed monitoring platform is mainly composed of controllable lifting platform, high-resolution multi beam sonar, underwater camera,

* Jianfei Zhang: Jianfei_zhang760@163,.com

https://doi.org/10.1051/e3sconf/202129003013

hydrological and water quality monitoring module and wireless transmission module; Cruise monitoring platform is mainly composed of autonomous vehicle, high-resolution multi beam sonar and wireless transmission module.([4])

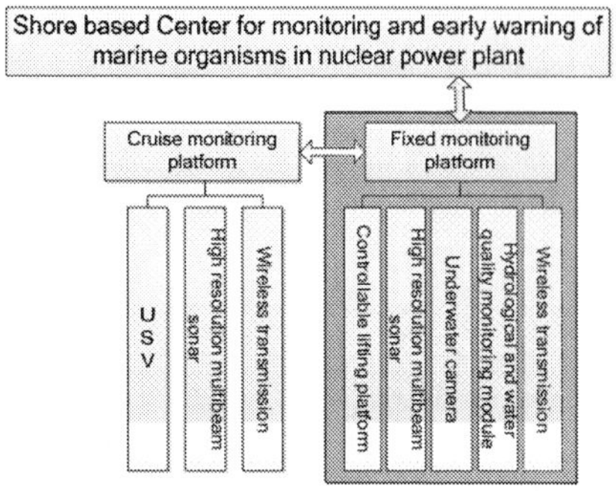

Fig.1. Test pond Block diagram of autonomous monitoring and early warning system for marine organisms in nuclear power plant

The data flow of the system is shown in Figure 2. At the front end of the detection, the acoustic signal is converted into electrical signal by the receiving sensor array, and xtf type three-dimensional acoustic image data and JPG format image file are directly processed at the pre-processing end. In the post-processing process, the image data is exported to the txt document or mat type document used by the algorithm program.([5])Finally, the density distribution is compared with the alarm threshold, the high and low levels are output, and the alarm device is controlled to start. In the initial stage of the project, the post-processing is carried out by manual processing. In the follow-up development of the project, professional software developers will be used to design the system architecture and realize the program of automatic processing.([6],[7])

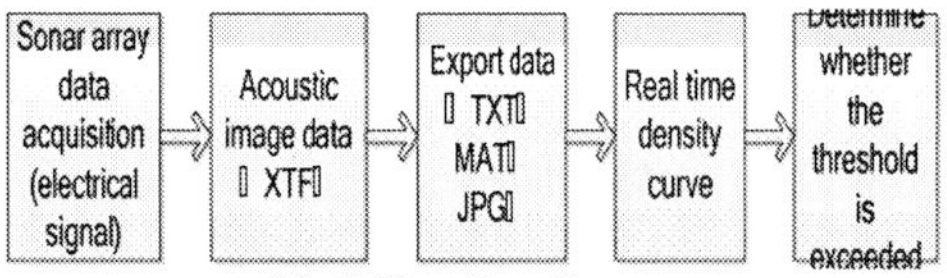

Fig.2. Data flow diagram

The input and output of the system after modularization. When the marine life automatic monitoring and early warning system is embedded into other parent systems as a subsystem, the input and output parameters of the subsystem are as follows: the input parameters include high frequency multi beam data (xtf format) collected by sonar array, GPS positioning data and control instructions; Taking jellyfish as an example, the output parameters include real-time (or quasi real-time) acoustic image data (jpg format), jellyfish density, maximum jellyfish size and maximum jellyfish echo intensity. As shown in Figure 3:

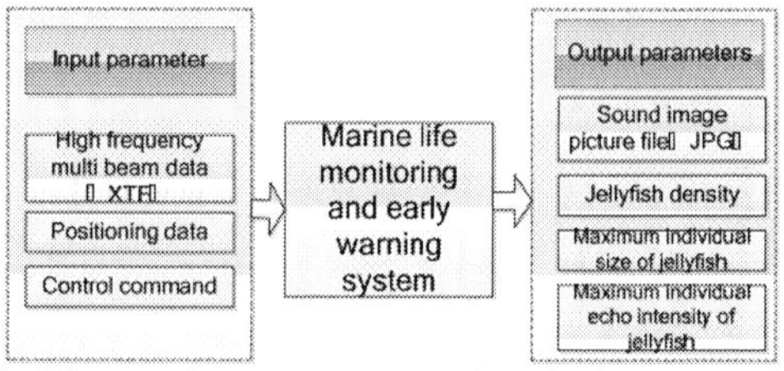

Fig.3. Input and output parameters of monitoring system after modularization

2.2 Design of fixed marine life monitoring and early warning network

Based on the high-resolution multi beam acoustic method to detect marine life, the marine life early warning system is constructed, as shown in Figure 4. At a certain distance from the cold source water intake of the power station, the monitoring line is established, the data link between the monitoring sonar and the communication buoy is established, and the data of the monitoring sonar is transmitted back to the shore monitoring room by the communication buoy.

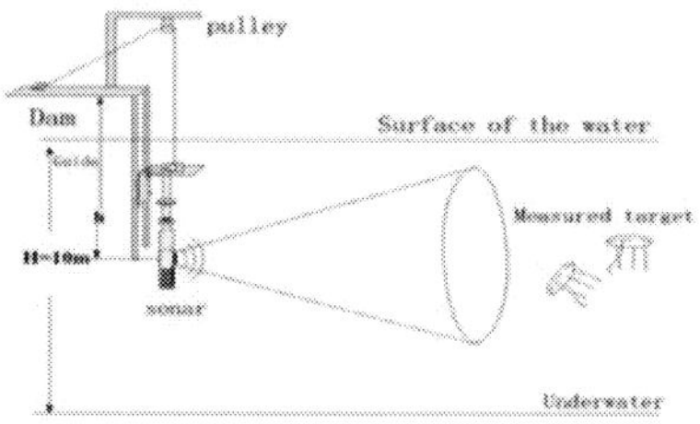

Fig.4. Working mode diagram of fixed monitoring node

2.3 Mobile monitoring and early warning scheme based on USV

USV (unmanned surface vehicle) is a low-cost and high-efficiency small surface self-propelled vehicle. Based on the USV platform, the sonar is placed at a depth of about 1m below the sea surface, and the acoustic beam is emitted obliquely downward to detect the jellyfish and other marine biological targets within the range of the sonar beam. The test system consists of USV, detection sonar, GPS positioning, wireless data transmission and shore control, as shown in Figure 5.

https://doi.org/10.1051/e3sconf/202129003013

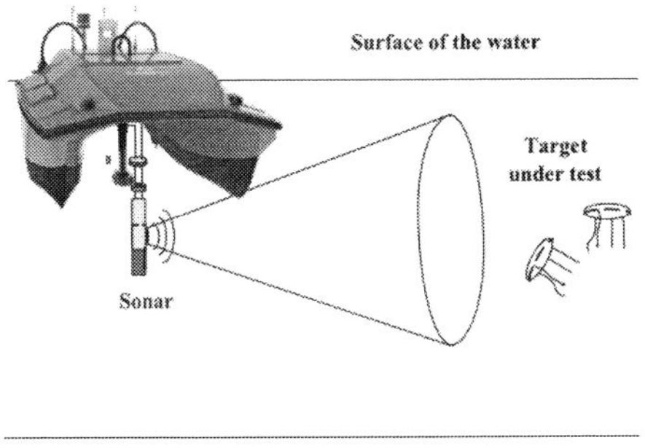

Fig.5. Schematic diagram of working mode based on USV

Different from the fixed monitoring platform, the mobile monitoring and early warning system based on USV makes full use of the mobility of USV platform, and can monitor the distribution of marine organisms on a large scale in cold source waters. In order to effectively monitor the distribution of marine organisms in the whole cold source sea area, it is necessary to reasonably design the mobile lines of USV platform, which must fully consider the specific conditions such as ocean current, sea terrain and coastline.

3 Research on marine life self monitorin g and early warning technology

3.1 Investigation on species, characteristics and habits of marine organisms in nuclear power plant

According to the types, characteristics and habits (hazard time period) of the harmful marine organisms in the sea area around the nuclear power plant, the systematic investigation and hazard assessment are carried out, and the knowledge base of the habits of the harmful marine organisms in nuclear power plant is established, so as to provide a priori perception information for the monitoring and early warning of the harmful marine organisms in nuclear power plant.

3.2 Study on the monitoring methods of typical marine organisms affecting nuclear power safety

Through the long-term detection of typical marine organisms individuals and clusters that affect the safety of nuclear power, the echo data of typical marine organisms are obtained, and the acoustic characteristics (relative echo intensity, acoustic scale, etc.) of typical marine organisms are analyzed and extracted, Finally, the real-time monitoring of typical marine organisms can be realized.

3.3 Research on the typical early warning mechanism of marine organisms affecting nuclear power safety

Using high-resolution multi beam sonar as sensing means, the echo data of marine organisms are analyzed, processed and identified by using the feature analysis and statistics module, and the density distribution of jellyfish, hairy shrimp and other marine organisms in the monitored sea area is obtained in real time. Combined with the threat degree of marine organisms invasion to the operation of nuclear power plant in the data information database, the alarm threshold is set, Real time warning of harmful marine organisms beyond the warning threshold.

4 Research on system implementation a nd field monitoring scheme

Based on the high-resolution multi beam acoustic detection technology of marine life, the marine life early warning system is constructed. It is proposed to arrange fixed monitoring stations on both sides of the water intake and at the center of the water intake, so as to solve the problem of rapid water area coverage monitoring capacity. Subsequently, the autonomous mobile monitoring platform will be built, the mobile route will be planned, and the whole area coverage monitoring of water intake width will be realized through cruise.

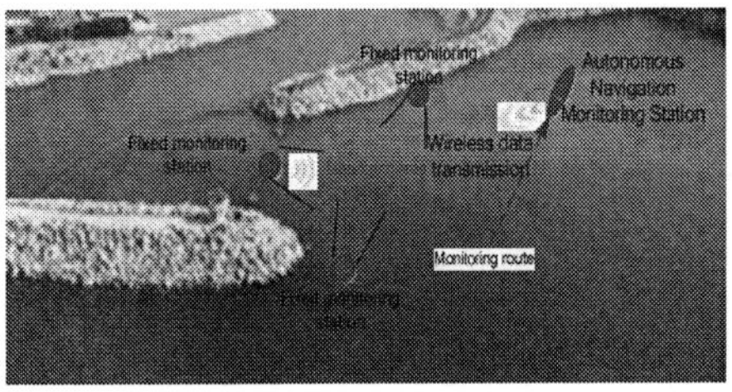

Fig.6. Layout of monitoring points of cold source water intake

Fig.7. Test platform of a project

In the process of the research on the detection method of marine organisms based on the high-resolution multi beam acoustic detection method, the acoustic characteristics of individual marine organisms such as jellyfish and fish swarm are analyzed and extracted, and the estimation method of the number and distribution density of marine organisms cluster is further studied.

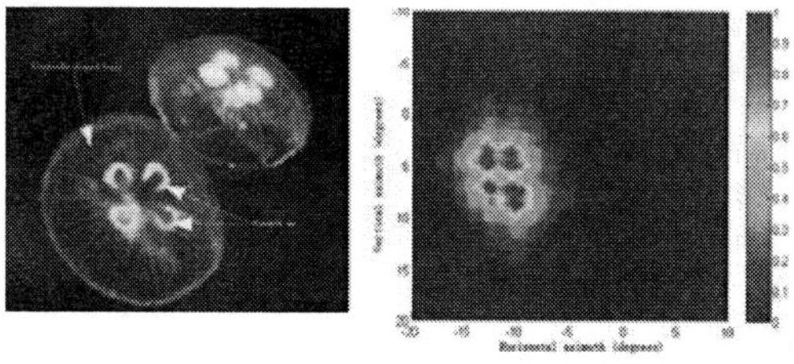

Fig.8. Real image and acoustic image of jellyfish

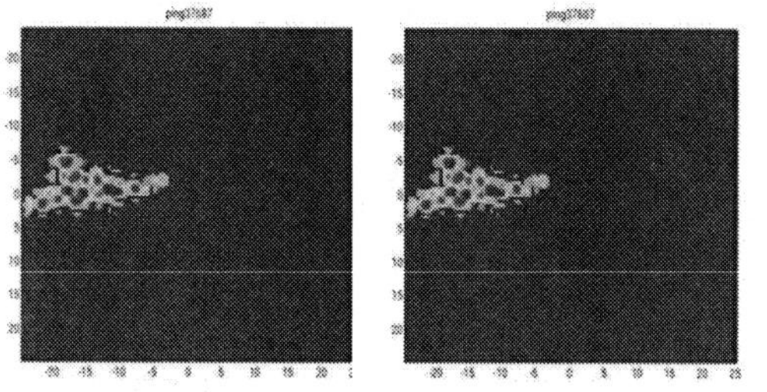

Fig.9. Fish echo image

Through the acoustic detection test of jellyfish and fish swarm in a certain project, we can clearly observe the echo image of jellyfish individual, the temporal spatial variation of jellyfish distribution density and the process of fish swarm appearing in the sonar field of vision, etc., which can provide technical support for the future research on monitoring and early warning technology of marine organisms harming nuclear power plants.

5 Conclusion

Based on the design and analysis of the detection and early warning scheme for harmful marine organisms at the water intake of nuclear power plant based on high-resolution multi beam acoustic detection method, an automatic monitoring and early warning system for marine organisms is constructed, and the image information of marine organisms obtained is analyzed, processed and identified by using the feature analysis and statistics module with high-resolution multi beam sonar as sensing means, The density distribution of jellyfish, fish and other marine organisms in the monitoring area is obtained in real time, and the alarm is given when the threshold is exceeded, which verifies the technical feasibility.

Therefore, in order to avoid the threat of jellyfish, fish and other marine organisms attacking the water intake of nuclear power plant and improve the reliability of cold source sea area, it is necessary to establish a defense in depth system of marine organisms identification, detection, early warning and hierarchical response, in which detection and early warning is the front-end disposal measure, providing accurate and timely marine organisms information for subsequent hierarchical response, and can replace the underwater inspection work of divers, Reduce the risk of industrial safety.

References

1. Fischell EM, Schmidt H Classification of underwater targets from autonomous underwater vehicle sampled bistatic acoustic scattered fields. J Acoust Soc Am 138(6) 357: (2015)
2. Morse SF, Marston PL, Kaduchak G High-frequency backscattering enhancements by thick finite cylindrical shells in water at oblique incidence: experiments, interpretation, and calculations. J Acoust Soc Am 103(2):785–794 (1998).
3. Ristic, B., Vo, B.N., Clark, D. and Vo, B.T.. A metric for performance evaluation of multi-target tracking algorithms, IEEE Transactions on Signal Processing, 59(7), 3452-3457(2011).
4. Raj, K.D. and Krishna, I.M. Kalman filter based target tracking for track while scan data processing, Proceeedings of the 2nd International Conference on Electronics and Communication Systems (ICECS), IEEE, Coimbatore, India, pp. 878-883(2015).
5. Tang, X., Tharmarasa, R., McDonald, M. and Kirubarajan, T.. Multiple detection-aided low-observable track initialization using ML-PDA, IEEE Transactions on Aerospace and Electronic Systems, 53(2), 722-735(2017).
6. Li XK, Meng XX, Xia Z. Characteristics of the geometrical scattering waves from underwater target in fractional Fourier transform domain . Acta Physica Sinica, 64(2015).
7. Fan J, Zhu BL, Tang WL. Modified geometrical highlight model of echoes from nonrigid surface sonar target. Aeta Acustica, 26(6), 545 (2001).

A study on the characteristics of diatoms and ecological environment pollution in the main water of Tianjin, China

Sugur Enh[1], Jing Fang[1*], Pengyi Ma[1], Bo Liu[1], Heng Zhang[1], Dexuan Lan[1], Ziyao Yang[1], PíngCuoDolka[1], Zhifei Wang[1]

[1]school of Geography and Environmental Science Tianjin Normal University Tianjin, China

Abstract. Twenty lake bottom surface layers mud and water samples in the Qilihai Wetland and the Beidagang Reservoir in Tianjin were taken to do diatoms analysis. More than 6000 diatoms were identified and divided into 117 species and 28 genera. Based on the calculation of DAIpo(Diatom Assemblage to organic water pollution), the water pollution level of the Beidagang Reservoir can be divided into: Xenosaprobic, β -oligosaprobic, α -oligosaprobic, β -mesosaprobic, from clear to slightly cloudy water. The water pollution level of Qilihai wetland can be divided into: extremely oligarchy dirt belt, β oligarchy dirt belt, α oligarchy dirt belt, β Medium dirt belt, that is, the water quality is clear to slightly turbid. The water quality of the Beidagang Reservoir is clearer than that of the Qilihai Wetland. The physical and chemical indexes such as EC, pH, TN, TP and COD_{Mn} were determined for the water samples, and the correlation between DAIpo and these measured values was analyzed. The results showed that the degree of water pollution detected was consistent with the conclusion obtained by diatom analysis. The results show that the water quality of the Beidagang Reservoir and the Qilihai Wetland was mildly polluted in the spring and summer of 2019, but the eutrophication is more serious.

1 Introduction

Many scholars analyzed the ecological changes of water environment by using the characteristics of diatom species and their community combinations, and combine the nutritional status index, sediment pollution degree, total phosphorus content, EC, pH and other water environment factors, as well as the main component analysis method, to analyze and evaluate the aquatic ecological environment. In the 1990s, Van Dam et al. revealed that the diatoms are very effective in indicating pH, salinity, organic nitrogen, oxygen saturation, humus, and eutrophication (Van Dam H,et aL 1994). Watanabe et al. (1986) proposed a calculation method, DAIpo, based on adhesion diatoms to evaluate water pollution. With DAIpo, the degree of organic pollution of fresh water can be calculated, and the water pollution situation can be compared with spatiotemporal analysis(Watanabe et aL 1986a). Since then, the effectiveness of DAIpo in different environments, such as rivers and lakes, had been validated (Watanabe et al 1988).

The Qilihai Wetland and the Beidagang Reservoir are important wetlands in the Tianjin Plain in China. In recent years, many scholars have made many achievements in the study of Tianjin's wetland environment by different methods.Diatom analysis has been widely used in the restoration of paleoenvironment of the coastal plains of China, but it has been applied less in the monitoring and evaluation of water quality, especially in North China.

Based on the sediment sampling of the Qilihai Wetland and the Beidagang Reservoir, diatom species were identified. By summarizing the genus of diatoms , DAIpo (Diatom Assemblage Index to organic water pollution) was calculated, and through the pollution index, the pollution degree of different water bodies in the same period is classified quantitatively.

2 Research areas and methodology

The Qilihai wetland is located in the northeast of Tianjin, the geographical coordinate is39°17′N, 117°47′E, with a total area of about 95km2 . It formed by the land and sea changes, originally part of the ancient Bohai Sea, and then retreated through the Bohai Sea. Because of low-lying terrain, today's wetland landscape was formed. It is also a transit point for migratory birds, inhabiting and breeding more than 200 species of birds, with very important ecological and conservation value (Wang et al.,2005).

The Beidagang wetland is located in the southeast of Tianjin Binhai New Area, the geographical coordinates is 38°36′N~38°50′N, 117°11′E~117°37′E, an area of 34887hm², and it is the largest area of *the Wetland Nature Reserve in Eastern Asia*, which is of great significance to the ecological protection of migratory birds.

On June 1 and 2, 2019, 20 different sites, including the surface mud and water samples of the lake, were collected at the edge of the water body and the estuary of

* Corresponding author: mdfangjing@163.com

the river, respectively, in the Qilihai Wetland, Northeast of Tianjin and the Beidagang Reservoir, Southeast of Tianjin(Fig.1).

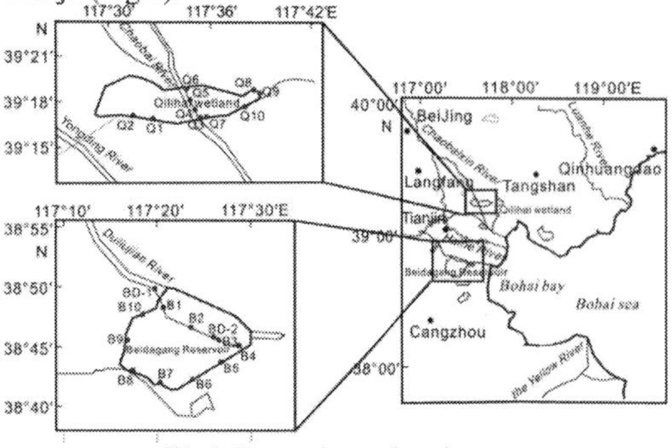

Fig.1. Research area location map

Based on the sediment sampling of the Qilihai Wetland and the Beidagang Reservoir, diatom species were identified. By summarizing the genus of diatoms , DAIpo (Diatom Assemblage Index to organic water pollution) was calculated, and through the pollution index, the pollution degree of different water bodies in the same period is classified quantitatively.

Diatom identification is based on Watanabe et al.(2005) to be named. Diatoms were divided into three groups, such as Saprophilous taxa ,Eurysaprobic taxa, and Saproxenous taxa, and the relative percentage content of 3 species was calculated, and the number of diatoms (DAIpo) based on the attached diatom population in Watanabe et al. (2005) was calculated according to its share(Watanabe et al.,2005).

The search for the value of DAIpo:

$$DAIpo=50+(A-B)/2 \qquad (1)$$

A: The relative frequency (%) of all occasional stoic species that appear at the survey site.

B: The relative frequency (%) of all Saprophilous taxa that appear at the survey site.

The physical and chemical indexes such as EC, pH, TN, TP and CODMn were determined for the water samples.

3 Results and discussion

3.1 Diatoms from the Beidagang reservoir and Qilihai wetland

20 surface mud samples (10 in the Qilihai wetland, 10 in Beigang reservoir) were collected on the edge of the Qilihai wetland and the Beidagang reservoir to do diatoms analysis and identification. More than 6000 diatoms were identified and divided into 117 species and 28 genera.

Among them, the diatoms' advantage synods in the Beidagang Reservoir are mainly Navicula, Nitzschia. Saproxenous taxa includes Achnanthes conspicua ,Achnanthes marginulata ,Navicula pseudoreinhardtii,Comphoneis pseudokunoi ,Diploneis interrupta ,Encyonopsis minuta ,Eunotia veneris ,Fragilaria capitellata ,Navicula cincta,Stephanodiscus pseudosuzukii and other 16 species. Saprophilous taxa includes Cyclotella delicatula,Cyclotella meneghiniana ,Navicula confervacea,Navicula pupula ,Nitzschia amphibia. Eurysaprobic taxa includes Aulacoseira islandica,Brachysira brebissonii ,Cyclotella ocellata ,Gyrosigma procerum ,Navicula aryptocephala ,Neidum ampliatum ,Peronia fibula ,Stephanodiscus hantzschii ,Tabularia fasciculate ,Synedra ulna and other 24 species.

The advantages of diatoms in the Qilihai Wetland are mainly:Navicula,Nitzschia,Achnanthes. Saproxenous taxa includes Achnanthes clevei,Diploneis interrupta,Encyonopsis minuta,Fragilaria capitellata,Gomphonema olivaceum,Hannaea arcus,Navicula contenta,Navicula cryptotenella ,Nitzschia tabellaria,Synedra inaequalis and other 28 species. Saprophilous taxa includes Cyclotella meneghiniana,Eunotia asterionelloides ,Navicula pupula,Navicula subminuscula,Nitzschia amphibia. Eurysaprobic taxa includes Aulacoseira Canadensis,Diatoma hyemalis,Gyrosigma procerum,Hantzschia amphioxys,Navicula amphiceropsis,Nitzschia filiformis,Rhopalodia gibba,Stephanodiscus alpinus,Surirella bifrons,Synedra ulna and other 19 species(Fig.2).

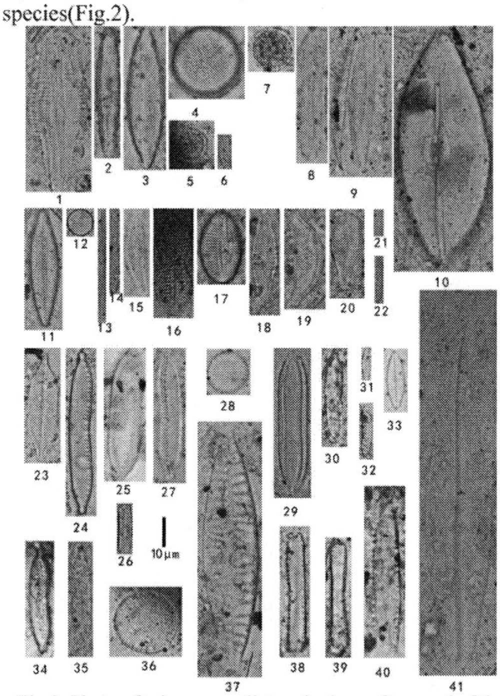

Fig.2. Photo of microscope diatom in the surface mud of the Beidagang Reservoir and the Qilihai Wetland.

(1)Rhopalodia gibba,(2)Achnanthes lanceolata,(3)Navicula delicatilineolata,(4)Orthoseira roeseana,(5)Cyclotella ocellata,(6)Achnanthes oestrupii,(7)Stephanodiscus pseudosuzukii, (8)Surirella andoi,(9)Pinnularia acidobionta,(10)Navicula confervacea,(11)Navicula cincta,(12)Cyclotella delicatula,(13)Peronia fibula,(14)Neidum ampliatum,(15)Eunotia veneris,(16)Diploneis interrupta,(17)Diploneis subovalis,(18)Navicu notha,(19)Epithemia sorex,(20)Gomphonema olivaceum,(21)Achnanthes clevei,(22)Nitzschia tabellaria,(23)Navicula amphiceropsis,(24)Fragilaria capitella,(25)Achnanthes biasolettiana,(26)Eunotia ueneris,(27)Synedra inaequalis,(28)Cyclotella meneghiniana,(29)Nitzschia constricta,(30)Encyonopsis minuta,(31)Fragilaria capucina,(32)Hannaea arcus,(33)Navicula subminuscula,(34)Hantzschia amphioxys,(35)Synedra ulna,(36)Stephanodiscus alpinus,(37)Surirella bifrons,(38)Navicula pupula,(39)Navicula contenta,(40)Nitzschia zvidoclinata,(41)Gyrosigma procerum.

The diatoms analysis results are shown in Fig. 3.

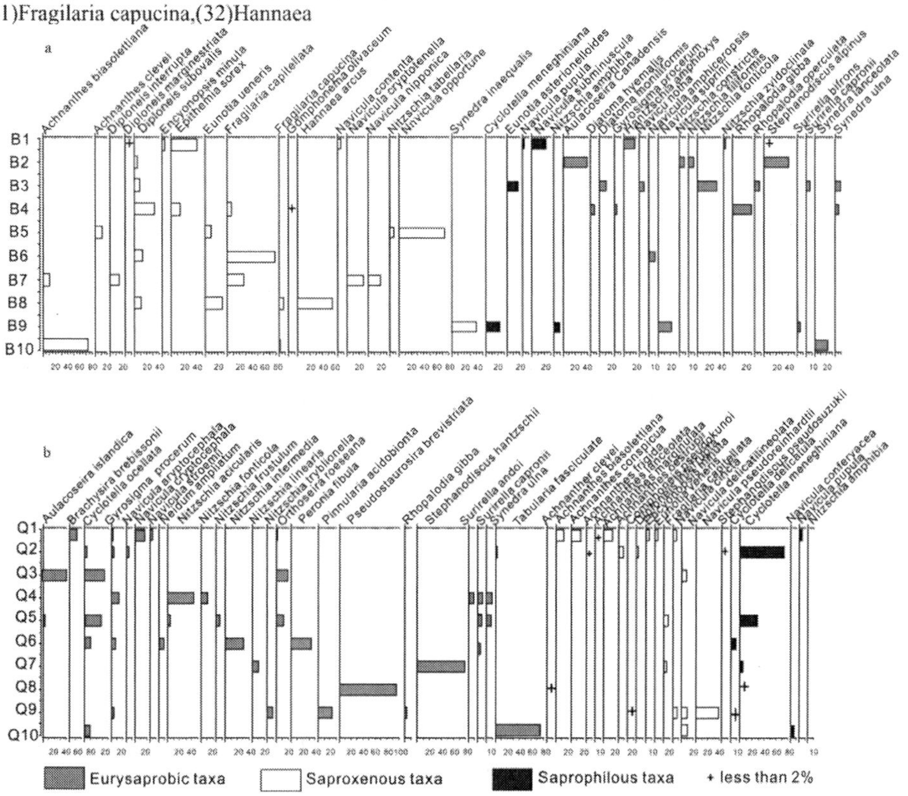

Fig.3. Results of diatoms in Beidagang reservoir(a) and Qilihai wetland (b)

3.2 EC、pH、DAIpo、TN、TP and COD_{Mn}

The diatom data of 20 samples of the Beidagang Reservoir and the Qilihai Wetland were substituted into the formula (1). The EC, pH, DAIpo, TN, TP and COD_{Mn} were measured for the samples of the Beidagang Reservoir and the Qilihai Wetland, shown in Fig. 4.

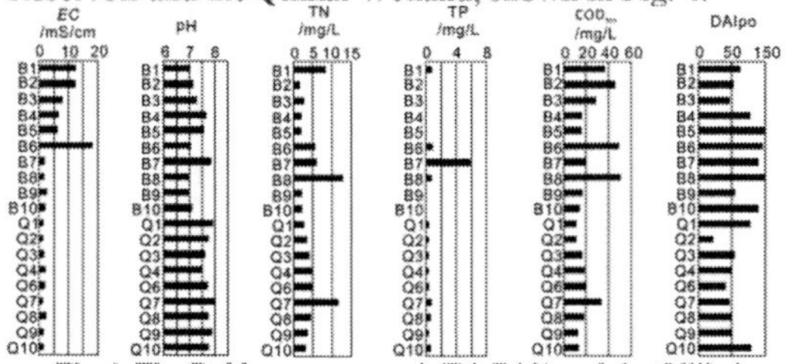

Fig.4. The Beidagang reservoir(B1-B10) and the Qilihai wetland(Q1-Q10) EC、pH、TN、TP、COD_{Mn} and DAIpo

3.3 Evaluation of water quality of the Beidagang reservoir and Qilihai wetland

3.3.1 Beidagang reservoir

Based on the DAIpo, the water quality pollution level of Beidagang reservoir can be divided into xenosaprobic, β-oligosaprobic, α-oligosaprobic, β-mesosaprobic, that is, the water quality is clear to slightly dirty. The EC value of Beidagang reservoir was higher, ranging from 1.61 to 18.18 mS/cm. The pH value was 6.95-7.81, which reached the weakly alkaline environment. Besides B7 exceeding the standard, the total phosphorus concentration in Beidagang water body was in the mesotrophic and meso-eutrophication state (50 - 1300 μg/L). According to the national evaluation criteria for lake eutrophication, the total nitrogen content of the three sampling points, B2, B4 and B5 in Beidagang reservoir, were eutrophication (less than 2.0mg/L), but all of them reached the state of hypereutrophication (more than 2.0mg/L). It was speculated that the main reason was that, with the development of industrial and agricultural in this area chemical substances discharged into the water body, causing an explosion of green algae, cyanobacteria and other algae and other plankton, resulting in the degradation of the self-purification ability of the water ecology and the deterioration of the water quality.

Overall, B10 in the northwest of the Beidagang Wetland and B5 in the southeast are very clear (the DAIpo index are both 100), indicating that the southern water body is clear (the DAIpo index is 100-89.93), the northeast and west side are mild-slightly polluted (DAIpo is 68.55-46.83) , with pollution levels from low to high.

3.3.2 Qilihai wetland

Based on the DAIpo pollution index, the water quality pollution levels of the Qilihai wetland can be divided into β-oligosaprobic, α-oligosaprobic, β-mesosaprobic, α-mesosaprobic, that is, the water quality is clear to dirty. The EC of the Qilihai wetland was relatively low, all of which were Marine sediments between 1.13 and 2.4 mS/cm. The pH value was 7.49~7.95, which reached the weakly alkaline environment. The total phosphorus mass concentration in the Qilihai land was rich. According to the national evaluation standard for lake eutrophication, the total nitrogen content of the Qilihai wetland was all greater than 2 mg/L, which is ultra-eutrophication(Wang et al.2019). The main reason was that, under the influence of fish pond culture and human activities, nitrogen, phosphorus and other nutrients enter the water in large quantities, leading to eutrophication.

The above analysis showed that the water in the southern and eastern part of the Qilihai wetland was clear - slightly dirty (DAIpo was 77.76-50), and the water in the central part of the road passing through the Qilihai wetland was slightly polluted (DAIpo was 39.32-54.03). Among them, Q2 water at the sampling point at the inlet of Yongding river was seriously polluted (DAIpo was 19.9), which was speculated to be caused by the deposition of pollutants carried by the river in the lake.

Comparing Beidagang reservoir and Qilihai wetland ecology, Beidagang DAIpo index is slightly higher than the Qilihai point DAIpo index. Therefore, compared with the latter, the former has better water quality, less water pollution and better ecological environment protection. It shows that the pollution degree of Beidagang and Qilihai in 2019 is not serious, but mild, and the water quality is relatively stable.

4 Conclusion

By analyzing the diatoms in the surface mud samples of the lake bottom, which were collected from the surface layer of the lake bottom from the edge of the Qilihai Wetland, the Beidagang Reservoir in Tianjin and the Duliujian River, and the physical and chemical index such as EC, PH, TN, TP and COD_{Mn} were measured.

(1) According to the tolerance of different diatoms to pollution, the DAIpo of the Beidagang Reservoir and the Qilihai Wetland was calculated. It was found that the DAIpo index of sample points was very different. The grade of water pollution in Beidagang reservoir can be divided into: xenosaprobic, β-oligosaprobic, α-oligosaprobic, β-mesosaprobic, from clear to slightly dirty water. The level of water pollution in Qilihai wetland can be divided into: β-oligosaprobic, α-oligosaprobic, β-mesosaprobic, α-mesosaprobic, that is, the water quality is clear to muddy. Compared with the Qilihai Wetland, the DAIpo index of the Beidagang Reservoir's sample points is slightly higher than that of the Qilihai Wetland's. Therefore, the water quality of the Beidagang Reservoir is better, the water pollution degree is lighter, and the ecological environment protection is better. The EC in the study area is higher, and the pH value reaches a slightly alkaline environment. The content of TP and TN is relatively high, showing a state of eutrophication. The results show that the degree of water pollution monitored by physical and chemical indicators is consistent with the conclusion obtained by diatom analysis.

(2) Based on the DAIpo index and other physical and chemical indexes, the comprehensive evaluation of the study area was conducted. The results showed that, in 2019, the water quality of the Beidagang Reservoir and the Qilihai Wetland was slightly polluted, but the eutrophication was serious.

Acknowledgements:

This research was financially supported by the Tianjin Natural Science Foundation (18JCYBJC91100) and Innovation Training Program for College Students (201910065373).

References

1. Van Dam,H.,Mertens,A.,Sinkeldam,J.(1994)A coded checklist and ecological indicator values of freshwater diatoms from the Netherlands.Netherlands Journal of Aquatic Ecology, 28(1) :117-133.
2. Watanabe,T.,Asai,K.and Houki,A.(1986) Numerical estimation to organic pollution of flowing water by using the epilitic diatom assemblage—Diatom Assemblage Index (DAIpo)—.The Science of the Total Environment, 55, 209-218.
3. Watanabe,T.,Asai,K.,Houki,A.,Tanaka,S., Hizuka,T.(1986a)Saprophilous and eurysaprobic diatom taxa to organic water pollution and diatom assemblage index (DAIpo). Diatom, 2:22–73.
4. Watanabe,T.,Asai, K. & Houki, A.(1988)Numerical water quality monitoring of organic pollution using diatom assemblages.In:Round,F.E.(ed.) Proceedings of the 9th International Diatom Symposium pp.123-141.
5. Wang,Z.,Lui,M.,Li,Z.,Lv,S.(2005)Degeneration Characteristics and Rehabilitation of Qilihai Wetland Environment Ecological System. Resea rch o f Soil a nd Wa ter Co nserv ation,12(5): 244-247.

https://doi.org/10.1051/e3sconf/202129003014

6. Watanabe,T,Ohtsuka,T.,Tuji,A.,Houki, A.(2005)Picture book and ecology of the freshwater diatoms. Uchida Rokakuho Publishing Company, Tokyo.,666.

Research on the Coordination of Environmental Pollution and Industrial Structure Upgrading from the Coupling Perspective

Wenjing Zhang[1]

[1] School of Management, Tianjin University of Technology, Tianjin, China

Abstract. Economic transformation and industrial structure upgrading are critical to achieving environmentally sustainable development. Based on the inaccurate data analysis of the VEC model and the logarithmic average exponential decomposition method, this paper proposes an analysis of the two-way coupling relationship between natural gas energy consumption and environmental pollution. Through the identification of conceptual models of the re-linking and decoupling relationship between energy consumption and environmental pollution, two-dimensional endogenous variables and exogenous variables are determined. The research results show that both energy consumption and CO2 emissions are on the actual data statistical line, with an error of 0. At the end, the article puts forward policy recommendations to promote the coordinated development of urban agglomeration environmental pollution and industrial structure upgrading.

1 Introduction

On October 26, 2020, the Fifth Plenary Session of the 19th Central Committee clearly emphasized the need to achieve the new requirements of the "ecological civilization construction to achieve new progress" and "economic development to achieve new results", which means that domestic industrial development must take into account economic transformation and environmental sustainability. Dual goals. Reasonable environmental regulations can not only achieve the goal of environmental improvement, but also achieve green economic development by optimizing the industrial structure [1]. It can be seen that attaching importance to environmental protection and upgrading the industrial structure has very important practical significance for realizing environmental protection and economic development. The use of energy, such as natural gas in the process of drilling, mining, gathering and transportation, and purification, will produce various tail gas and waste residues, which cause a lot of pollution to the environment and endanger human health [2]. In terms of drilling wastewater pollution, the treatment has achieved initial results, and the pollution area and scope have been initially controlled. However, large-scale and in-depth treatment cannot effectively control the pollution caused by desulfurization tail gas, gas field water, and operation wastewater. It is still difficult to solve the problem deeply troubles the country and the people [3].

Scholars have conducted a lot of related paradigm studies on carbon emissions. Zhou Bin et al. [4] believe that “by integrating the low-carbon-circular economy coupling development model into the planning development ideas, main goals, spatial layout, and key industries, the spatial distribution reflects the circulation of regional resources; in key industries Zhongjun runs through the low-carbon-circular economy concept design industry chain, especially the vein industry designed to lay the foundation for the smooth implementation of the circular economy industrial cluster." Yang Zhicheng [5] discussed in depth the changes in the total amount of carbon “sources” and economic development. The relationship between carbon emission structure and economic structure. It is believed that the distribution of carbon emissions reflects social and economic factors such as the total economic volume, structure, efficiency, and population of the country and region, and it is recommended to strengthen the research and monitoring of the dynamic changes of carbon "source/sink". Wang Mingyue et al. [6] believe that “low-carbon circular economy is an organic combination of low-carbon economic model and circular economy model. It is a 'two-dimensional' concept and a strong coupling model. Low-carbon circular economy has resource efficiency and carbon emissions. It is the driving force, government guidance is a reliable guarantee for strong coupling, remanufacturing is a new bright spot, industrial clusters and symbiosis support carriers in the park, and the necessary conditions for the industrial symbiosis coupling structure.” It can be seen that the carbon circular economy as an energy sector reduces CO2. Discharge methods have attracted the attention of experts and institutions. According to previous studies, the energy consumption of natural gas is determined through the use of the VEC model; the logarithmic average index decomposition method is used to analyze the energy consumption of natural gas based on environmental pollution driving factors [7]. However, it is extremely rare to analyze the relationship between the two under the perspective of coupling. Therefore, discuss the coupling relationship between natural gas energy consumption and

Corresponding author’s e-mail:18822281152@163.com

https://doi.org/10.1051/e3sconf/202129003015

environmental pollution, prove the potential of energy saving and emission reduction, for reference by the natural gas service industry, and provide valuable theoretical basis and data support for the low-carbon development of the natural gas service industry.

2 Quantitative analysis of the coupling relationship between energy consumption and environmental pollution

2.1 Analysis of the relationship between energy consumption and environmental pollution

Energy promotes economic development, but it also causes serious pollution to the human living environment [8]. The current energy production and use of natural gas produce a large amount of harmful gases, waste liquids and wastes, which not only pollute the atmosphere, rivers, and land, but even destroy ecological balance, food safety and other issues, and seriously threaten human health and the living environment. Energy consumption and environmental pollution problems refer to the environmental effects produced by human beings in the process of mining, transportation, processing, transformation and utilization of energy. How to solve the pollution of natural gas energy emissions is one of the biggest problems facing the modern environment. While energy consumption continues to improve human material life, it also deteriorates the human living environment. To achieve sustainable development, mankind must resolve this contradiction.

2.2 Metrological analysis

2.2.1 Energy consumption

According to the "General Principles of Comprehensive Energy Consumption Calculation", the comprehensive energy consumption of energy consumption mainly includes various energy consumed by the production system, auxiliary production system and auxiliary production system, energy conversion losses, and energy consumption of raw materials. Therefore, the "Statistical Yearbook" "Include selection. Therefore, the statistical limit of energy consumption is set as the entire coal-to-gas production process. The energy of the input device includes raw coal, fuel coal and other electricity consumption, and the energy of the output system includes by-products such as external air.

2.2.2 Environmental pollution

Environmental pollution indicators are industrial waste gas emissions in units of 100 million standard cubic meters, industrial wastewater emissions in units of 10,000 tons, and industrial waste solid emissions in units of 10,000 tons.

According to the statistical yearbook, the formula for estimating CO2 emissions from the carbon emissions data generated by fuel is as follows:

$$MCO_{2j}=qnet_j \times 10^{-9} \times ratio_{c\_j} \times 10^{-3} \quad (1)$$

In the calculation formula, $qnet_j$ represents the low calorific value generated by the use of natural gas; $ratio_{c\_j}$ represents the carbon oxidation rate. Obtain the following formula to calculate the CO2 emissions of province i in year t:

$$CO_{2i,t} = \left(\sum_{j}^{8} M\_CO_{2j}\right) \times q_{i,j,t} \quad (2)$$

In the calculation formula, $q_{i,j,t}$ represents the natural gas energy consumption degree of province i in year t.

3 Coupling model construction

Analyze the impulse disturbance response of environmental pollution and energy consumption through the impulse response function, and obtain the following impulse response function calculation formula:

$$x_i = b^0\mu_j + b^1\mu_{j-1} + b^2\mu_{j-2} \quad (3)$$

In the calculation formula, b^j represents the impulse function obtained by impulsively affecting the two-dimensional endogenous variables between natural gas energy consumption and environmental pollution.

Through the changes of endogenous variables in the two-way coupling model of natural gas energy consumption and environmental pollution, the importance of the contribution of different information shocks can be evaluated. According to the two-way coupling relationship between environmental pollution and energy consumption, the variance analysis is carried out, and the expression (4) of the variance decomposition is:

$$\mathrm{RVC} = \frac{\sum_{i=0}^{\infty}(k^t)^2\mu_j}{\sum_{j=1}^{k}[\sum_{i=0}^{\infty}(k^i)^2\mu_j]} \quad (4)$$

4 Coupling relationship analysis

Using the coupling relationship between natural gas energy consumption and environmental pollution to analyze the energy consumption decoupling elastic coefficients are all located in the first quadrant, distributed in the expansion negative decoupling zone, expansion connection zone and weak decoupling zone, most of which are located in the weak decoupling zone. The study found that the economic growth rate of the natural gas service industry during the observation period was higher than that of most years; although it did not reach the ideal strong decoupling state, the overall decoupling state was relatively good. Set 10-period impulse response, obtain the impulse response relationship between energy consumption and environmental pollution as shown in Table 1.

Table.1 Energy consumption and environmental pollution response value

Shock period	Impact periodCO2Emission response value	Solid emissions Response	Liquid discharge Response

		value	value
1	0	0	0
2	-0.15	-0.05	0
3	-0.22	-0.15	-0.05
4	-0.28	-0.14	0.1
5	-0.3	-0.13	0.12
6	-0.26	-0.10	-0.11
7	-0.23	-0.08	0.09
8	-0.18	-0.07	0.08
9	-0.15	-0.05	0.06
10	-0.1	-0.01	0.05

It can be seen from Table 1 that the energy consumption response values corresponding to CO2 emissions and solid emissions are all negative numbers. Therefore, the relationship between energy consumption and CO2 emissions and solid emissions is negative; while the relationship between liquid emissions is negative. The response value fluctuates on the zero value line, so it is concluded that the liquid discharge volume and the energy consumption volume are restricted.

5 Coupling mechanism design

5.1 Data collection

By querying the original data of China Statistical Yearbook and China Energy Statistical Yearbook from 2015 to 2020, draw the following table, and deeply explore the coupling relationship between energy consumption, CO2 emissions and the environment.

Table 2. Energy consumption and CO2 emissions statistics table

years	Energy consumption $/10^{11}$KJ	CO2 emissions $/10^4$t
2015	45978. 62	39850. 85
2016	48745. 62	42435. 75
2017	54112. 15	47533. 15
2018	63584. 65	55115. 30
2019	69551. 85	60525. 25
2020	85120. 15	68452. 01

5.2 Result analysis

This paper uses the VEC model, log-average exponential decomposition method and coupling model analysis, and finds that the energy consumption and CO2 emissions obtained using the VEC model have a large error; the energy consumption and CO2 emissions obtained using the log-average exponential decomposition method The emission error value is smaller than that used in the VEC model; the energy consumption and CO2 emission using the coupled model both reach the actual data statistical standard, and the error is 0. Based on this, it is concluded that the coupling model analysis effect is good.

6 Conclusions and recommendations

This paper constructs a coupling model based on the decoupling and double hooking theory to analyze the relationship between natural gas energy consumption and environmental pollution. Through experiments, it is known that the model can accurately determine the relationship between natural gas energy consumption and environmental pollution and has a good analytical effect. Under the background of the new normal, on the basis of ensuring the quality of economic development and the continuous expansion of economic scale, reducing the growth rate of energy consumption and achieving zero growth in energy consumption is an arduous and urgent task. Based on the analysis results of this article, we plan to propose a regulation path value to reduce the growth of energy consumption from the following three points:

First, in the path of new urbanization, the government should thoroughly implement the concept of energy-saving consumption, scientifically guide the population growth of the country, especially the central and western regions, and actively promote the construction of small and medium-sized cities, which will reduce the energy consumption and the impact of population growth. It is of great significance to coordinate the balance of regional development. Through the dual means of policy incentives, publicity, and training, people's traditional consumption outlook is changed.

Second, give full play to the leverage of energy prices and vigorously promote the reduction of industrial energy intensity. Continue to strengthen scientific and technological innovation, make full use of the advantages of energy price leverage, transform traditional industries, reduce the use of high-polluting energy sources such as crude oil, fuel oil, coal, gasoline, and promote emerging industries, and strive to tap and reduce energy sources in various industries, especially high-energy-consuming industries. Strengthen the potential, actively develop renewable energy and new energy, increase the proportion of renewable energy, and explore a diversified energy consumption structure.

Third, optimize the direction of industrial structure transformation, and earnestly exert the energy consumption reduction effect of industrial structure adjustment. Reduce dependence on foreign-funded enterprises, guide capital to flow to high-end industries, take extensive application of high-tech and advanced applicable technologies to transform and upgrade traditional superior industries as an important task, provide support from various aspects such as policies, funds, and talents, promote corporate restructuring and eliminate backward production capacity , to inject vitality and vitality into the transformation and upgrading of traditional industries, and promote the upgrading and transformation of industrial structure.

References

1. Yang Qian, Qin Wenjin, Liu Huajun. Does environmental regulation promote the optimization and upgrading of industrial structure? [J] . Shanghai

 https://doi.org/10.1051/e3sconf/202129003015

Economic Research, 2019, (6): 83-95.

2. Yang Xiaoye, Mi Yanxiang, Xu Donghai. Coupling analysis of environmental pollution and environmental governance—taking industrial waste gas emission and governance as an example [J]. Ecological Economy, 2019, 35(1): 191-194; 205.
3. Guan Xueling, Zhou Min. Research on the Coupling Development of Urbanization and Energy Consumption [J]. Journal of China University of Mining and Technology, 2019, 48; 231(6): 226-233.
4. Zhou Bin, Chen Xingpeng, Xue Bing, et al. Theoretical research and empirical research on the coupling development model of low carbon and circular economy [J]. Resources and Industry, 2010 (6): 19-27.
5. Yang Zhicheng. Carbon cycle and economic development [J]. Science and Technology Plaza, 2012 (6): 122-127.
6. Wang Mingyue, Zhang Qiqi, Shi Wenqiang. Research on the Connotation and Development Strategy of Low-Carbon Circular Economy [J]. Journal of Northeast Agricultural University (Social Science Edition), 2017 (5): 46-52.
7. Chen Yue, Wei Wei, Liu Feng, etc. Optimal decision-making of consumers in thermal-electricity market based on CES-type utility function [J]. Automation of Electric Power Systems, 2018, 42(13): 118-126.
8. Wang Renxiang, Lu Pengfei. The coupling relationship between technological innovation, green finance and industrial policy — — Based on the analysis of my country's energy conservation and environmental protection industry [J]. Journal of Beijing University of Posts and Telecommunications (Social Science Edition), 2019, 21(1): 30-41.

Fluid inclusions, H-O-S isotopic characteristics and genesis of the Chambishi copper deposit, Zambia

Hu Qiaofan[1,2,*], Feng Zuohai[1], Mo Jiangping[2], and Fang Ke[2]

[1]Guangxi Key Laboratory of Hidden Metallic Ore Deposits Exploration, Guilin University of Technology, 541004 Guilin, China
[2]China Nonferrous Metals (Guilin) Geology and Mining Co., Ltd., 541004 Guilin, China

Abstract. In this paper, fluid inclusions and H-O-S isotope geochemistry of the Chambishi copper deposit in Zambia are studied. According to the fluid inclusion in quartz and H-O-S isotope characteristics, it is concluded that ore-forming hydrothermal fluid is derived from mantle source and crust source magma mingling, the cause of copper precipitation, sedimentary type sulfur layered mineralization are mainly from diagenetic sulfides and seawater sulfate. Sulfate is mainly reduced by thermochemical method. The hydrothermal vein mineralization of Chambishi copper deposit is closely related to the magmatic activity in the middle Neoproterozoic, and the sedimentary stratified mineralization is mainly related to the large-scale orogeny and regional metamorphism in the late Neoproterozoic.

1 Introduction

Sedimentary copper deposits are an important source of copper, and the Central African copper belt is the highest grade and largest copper deposit in the world. As an important part of the Central African copper belt, Chambishi copper deposit in Zambia is a typical large-scale sand shale copper deposit. Previous studies have been carried out on the basic geological characteristics, ore-bearing structures, mineralization types, ore-controlling structures, prospecting prospects and prospecting marks of Chambishi copper deposit, and abundant results have been obtained.

Some researchers by geochemical anomalies delineated the ore-prospecting prospective area ([1]), puts forward the regional structure, strata, sedimentary equal in mineral zonality and the ore-prospecting criteria ([2]), at the same time also to the southeast of the copper deposit ore body, the main orebody and west ore body have found some molybdenite ([3]), Re-Os isotopic dating shows that molybdenite was formed in Late Cambrian (491 ~ 509 Ma, [4]).

In addition, some researchers have carried out comparative studies on the Chambishi deposit with large sedimentary and metamorphic deposits, such as Dongshengmiao deposit, Tanyaokou deposit and Huogeqi deposit in Inner Mongolia, China ([5]).

However, the study on fluid inclusions and H-O-S isotope geochemistry of Chambishi copper deposit is still weak, and the study on the relationship between fluid evolution and mineralization is seldom involved, which is not beneficial to the correct understanding of the ore-forming fluid properties and the source of ore-forming materials of Chambishi copper deposit. Therefore, in this paper, the analysis of the ore-bearing vein fluid inclusions and chalcopyrite/bornite sulfur isotopes in Chambishi copper deposit is helpful to reveal the properties of ore-forming fluids and trace the source of ore-forming materials, thus providing a new basis for the correct understanding of the origin of Chambishi copper deposit.

2 Methods

2.1 Temperature measurement of fluid inclusion

Eleven fluid inclusion samples were collected for the observation of the petrographic characteristics and temperature measurement of the inclusions. The samples were all from the ore-bearing vein in the mineralization stage of the hydrothermal period in the southeast mining area of Chambishi copper deposit.

The microthermometry of fluid inclusions was completed in Guilin Mineral Geology Analysis and Testing Center. The instrument is Linkam-THMS600 hot and cold platform (temperature range is below zero 196 ℃ ~ 600 ℃) made in UK.

2.2 H-O isotope

The hydrogen and oxygen isotopic compositions of 11 quartz fluid inclusions were analyzed in Beijing Institute of Geology of Nuclear Industry. The instrument MAT-253 mass spectrometer was used to obtain $\delta^{18}O$ and δD with SMOW as standard.

* Corresponding author: 2585752797@qq.com

2.3 S isotope

The sulfur isotope test was carried out on 25 chalcopyrite/bornite samples collected from the main west orebody and southeast orebody of the deposit. Single mineral separation was completed in Langfang Tuoxuan Rock-Mine Inspection Service Co., Ltd., and sulfur isotope test was completed in the State Key Laboratory of Mineral Deposit Geochemistry, Institute of Geochemistry, Chinese Academy of Sciences. After ultrasonic cleaning, the selected single mineral chalcopyrite/bornite was ground to 200 mesh or below. The sulfur isotope composition was tested by MAT253 stable isotope mass spectrometer. The test results were presented according to V-CDT standard.

3 Results

3.1 Petrographic characteristics and inclusion temperature measurement

Through the observation of the petrographic characteristics of fluid inclusions, it is found that the fluid inclusions of ore-bearing vein with hydrothermal filling vein mineralization are of different shapes and sizes, with the largest being up to 20 μm and the smallest being about 1 μm. They are mainly elliptic, oblate and irregular (Fig. 1), and are primary fluid inclusions. The temperature measurement results show that the homogenization temperature of the hydrothermal ore-bearing vein fluid inclusions varies widely from 97 to 445 °C, mainly from 100 to 350 °C, with a peak value of 200 to 250 °C (Fig. 2a), belonging to the medium and low temperature fluid. The range of salinity also varies greatly, ranging from 7.3 to 46 wt.% NaCl, mainly from 11 to 19 wt.% NaCl (Fig. 2b), with an average value of 14.9 wt.% NaCl, belonging to the medium and low salinity fluid.

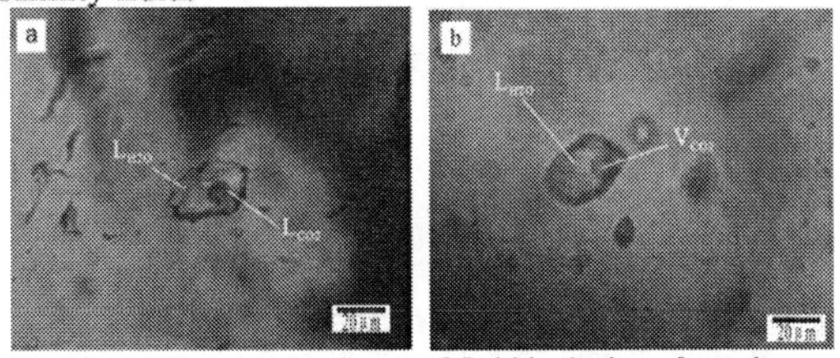

Fig.1. Microscopic photos of fluid inclusions from the Chambishi Cu deposit.

a: Type I-rich liquid and gas-liquid two-phase primary fluid inclusions; b: Type II - three-phase inclusion containing CO_2: L_{H2O}- liquid phase water

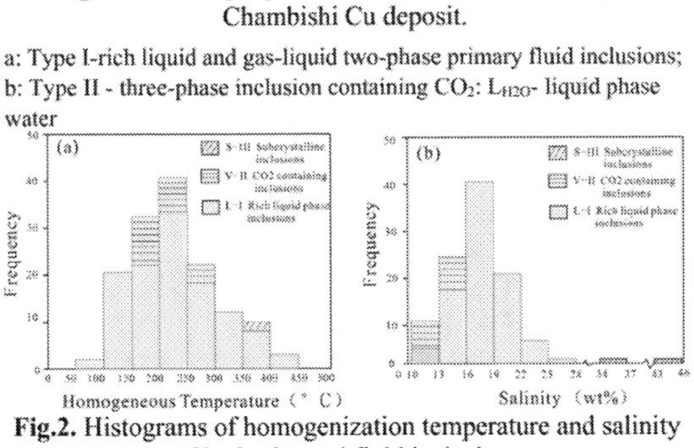

Fig.2. Histograms of homogenization temperature and salinity of hydrothermal fluid inclusions

3.2 H-O isotope

According to the test results (Table 1), the variation ranges of $\delta^{18}O_{V\text{-}SMOW}$ and $\delta D_{V\text{-}SMOW}$ are 11.5‰ to 13.4‰ and -64.0‰ to -52.6‰, and the average values are 12.5‰ and -58.7‰, respectively.

3.3 S isotope

The sulfur isotope test results are shown in Table 2. The results show that the ($^{34}S_{CDT}$ values of 18 sulfur isotope samples ranged from 6.0‰ to 21.0‰, with an average of 10.65‰ and a standard deviation of 4.33; The ($^{34}S_{CDT}$) values of seven sulfur isotope samples in the late vein copper-molybdenum orebody ranged from 5.5‰ to 12.1‰, with a mean value of 7.07‰ and a standard deviation of 2.43.

Table 1. H-O isotopic compositions of samples

Sample No.	Mineral	$\delta D_{V\text{-}SMOW}$(‰)	$\delta^{18}O_{V\text{-}SMOW}$(‰)	$\delta^{18}O_{H_2O}$(‰)
QES560-01	Quartz	-52.6	12.0	2.07
QES560-02	Quartz	-57.7	12.3	2.37
QES680-01	Quartz	-58.6	12.9	2.97
QES680-02	Quartz	-53.0	12.4	2.47
ES680C8-1	Quartz	-63.5	12.6	2.67
ES680C8-2	Quartz	-60.5	12.5	2.57
QES560-03	Quartz	-60.0	12.7	2.77
QES560-04	Quartz	-58.5	12.8	2.87
QES680-03	Quartz	-58.4	12.6	2.67
QES680-04	Quartz	-58.6	13.4	3.47
ES680C8-3	Quartz	-64.0	11.5	1.57

Table 2. S isotopic compositions of samples

Sample No.	Describe	Mineral	$\delta^{34}S_{CDT}$(‰)
w232m6-1	Layered copper bodies	Chalcopyrite	10.4
w232m6-2-H	Layered copper bodies	Chalcopyrite	8.9
w232m6-2-B	Layered copper bodies	Bornite	8.2
w184m4-1-H	Layered copper bodies	Chalcopyrite	7
w184m4-1-B	Layered copper bodies	Bornite	6.8
w184m3-2-H	Layered copper bodies	Chalcopyrite	7.3
w184m3-2-B	Layered copper bodies	Bornite	7.4
w184m4-2	Layered copper bodies	Chalcopyrite	7.2
w184m3-2-1	Layered copper bodies	Chalcopyrite	6.6
w184m4-h7-B	Layered copper bodies	Bornite	6
se980-9-H	Layered copper bodies	Chalcopyrite	19.2
se980-9-B	Layered copper bodies	Bornite	21
se980-16	Layered copper bodies	Chalcopyrite	11.5
se980-10-H	Layered copper bodies	Chalcopyrite	13.8

https://doi.org/10.1051/e3sconf/202129003016

Sample No.	Describe	Mineral	$\delta^{34}S_{CDT}$(‰)
se980-10-B	Layered copper bodies	Bornite	14.1
es98004-4	Layered copper bodies	Chalcopyrite	12.3
es98004-5	Layered copper bodies	Chalcopyrite	12.1
es98004-3	Layered copper bodies	Chalcopyrite	11.9
w184m4-h7-H	Vein-like copper - molybdenum ore body	Chalcopyrite	6.1
mob-1-H	Vein-like copper - molybdenum ore body	Chalcopyrite	8.4
mob-1-B	Vein-like copper - molybdenum ore body	Bornite	12.1
mob-2	Vein-like copper - molybdenum ore body	Chalcopyrite	6.1
mob-3-H	Vein-like copper - molybdenum ore body	Chalcopyrite	5.6
mob-3-B	Vein-like copper - molybdenum ore body	Bornite	5.5
mob-4	Vein-like copper - molybdenum ore body	Chalcopyrite	5.7

4 Discussion

4.1 Ore-forming fluid

The average $\delta^{18}O_{H_2O}$ value of hydrothermal filling vein mineralized fluid inclusions is 2.59‰ (Table 1), different from the $\delta^{18}O_{H_2O}$ value of magmatic water (5.5‰ ~ 9.5‰) and the $\delta^{18}O_{H_2O}$ value of meteoric water (-80‰ ~ -40‰) ([6]).

In the $\delta^{18}O_{H_2O}$-$\delta D_{V\text{-}SMOW}$ diagram (Fig. 3), the sample points mainly fall into the region near magmatic water/metamorphic water, suggesting that the type is mainly magmatic water/metamorphic water.

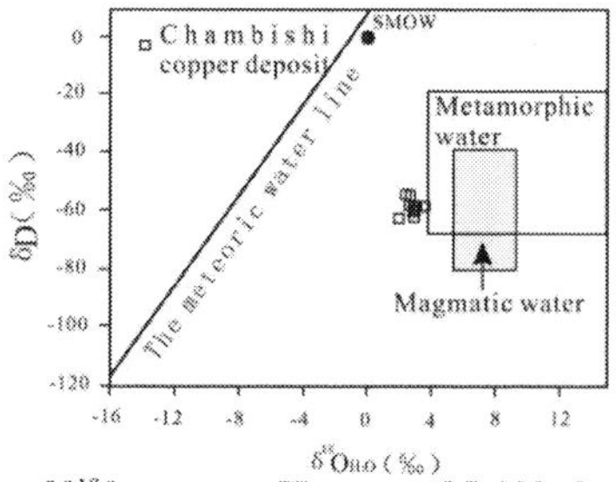

Fig.3. Plot of $\delta^{18}O_{H_2O}$ versus $\delta D_{V\text{-}SMOW}$ of fluid inclusions from the Chambishi Cu deposit (modified from Taylor, 1974)

Hydrothermal stage of microscopic characteristics of fluid inclusions shows that type I, II, III inclusions have no close coexist in the same horizon (Fig. 1), the micro measurement results show that the main metallogenic stage I type inclusions homogenization temperature are mainly distributed in the low temperature range, in the uniform temperature and salinity diagram, the salinity of the fluid inclusions and uniform temperature range is larger, these features also indicate fluid mixing. Therefore, the ore-forming fluids in the late hydrothermal period of Chambishi copper deposit are formed by the mixing of various fluids with different properties, and their sources are characterized by multi-source characteristics. The mixing of fluids is the main cause of copper precipitation.

4.2 Source of ore-forming material

4.2.1 Sedimentary stratified mineralization

The $\delta^{34}S_{CDT}$ values of chalcopyrite and bornite in sedimentary stratified deposits are positive, ranging from 6.0‰ to 21.0‰, with the highest $\delta^{34}S_{CDT}$ = 21.0‰. The S isotope composition is similar to that of Marine sulfate ($\delta^{34}S_{CDT}$ = 20‰), and it is presumed that part of S is derived from palaeo-seawater sulfate. It is suggested that the sulfur source of Chambishi copper deposit is mainly the mixed sulfur of diagenetic sulfide and seawater sulfate. Previous studies have suggested that the variation of sulfur isotopes in seawater is related to transgression and regression. In general, transgression leads to lighter $\delta^{34}S_{CDT}$ values, while regression leads to heavier $\delta^{34}S_{CDT}$ values ([7]). The high $\delta^{34}S_{CDT}$ value of sulfide in the lower orebody of sedimentary stratified ore (Table 2) may be caused by the decay of $\delta^{34}S_{CDT}$ in a large water body during the regressive stage, which reflects that the sedimentary environment has undergone a transition from regressive to transgressive.

4.2.2 Hydrothermal filling vein mineralization

The $\delta^{34}S_{CDT}$ values of seven sulfur isotope samples of the late hydrothermal filling vein copper body range from 5.5‰ to 12.1‰, with an average of 7.07‰, which is significantly lower than the $\delta^{34}S_{CDT}$ values of the sedimentary bedded copper body (6.0‰ to 21.0‰, with an average of 10.65‰), and close to the $\delta^{34}S_{CDT}$ values of mantle flow (0 ~ 3‰). It is suggested that part of S in the late ore-forming hydrothermal fluids originated from the late mantle-derived magmatism. Considering the chambishi region has experienced strong Lufrian orogeny in the late neoproterozoic (600 ~ 500 Ma) ([8]), It is speculated that the sulfur isotopes of the late hydrothermal filling vein mineralization show the characteristics of mantle flow, which may be related to the mantle-derived magmatism caused by the strong orogeny in the late Neoproterozoic.

The $\delta^{34}S_{CDT}$ values of the sedimentary and hydrothermal type ores from Chambishi copper deposit are positive (6.0‰ ~ 21.0‰ and 5.5‰ ~ 12.1‰, respectively, Table 2), suggesting that their sulfate reduction mechanism is mainly a thermochemical reduction mechanism, which causes SO_4^{2-} to H_2S more thoroughly.

In conclusion, the sulfur sources of Chambishi copper deposit may have multiple sources or multi-stage mineralization. The sulfur sources in the early sedimentary period were mainly derived from diagenetic sulfide and palaeo-seawater sulfate, and the S isotopic composition was also affected by transgression and regression. Late-stage hydrothermal sulfur source affected by late Neoproterozoic orogeny ([9]) and shows the characteristics of mantle flow.

It is worth noting that there is still a lack of detailed chronological data for the formation age of Chambishi copper deposit. According to the previous studies, the main mineralization age of Chambishi copper deposit is the middle Neoproterozoic (883 ~ 760 Ma, [10]), based on the fact that the ore bodies of Chambishi copper deposit mainly occurred in the Lower Ronan formation. It is mainly characterized by sedimentary stratified mineralization. Since then, the Chambishi region in the late Neoproterozoic (600 ~ 500 Ma) has experienced strong Lufrian orogeny and regional metamorphism, leading to mineralization in the area activated again, prompting copper, cobalt, molybdenum, such as enrichment of ore-forming elements occur again. It is mainly characterized by hydrothermal filling vein mineralization.

5 Conclusion

The sulfur of the Chambishi sedimentary stratified mineralization mainly comes from the diagenetic sulfide and the palaeo-seawater sulfate, and the S isotopic composition is also affected by transgression and regression. The sulfur isotopes of hydrothermal vein mineralization mainly exhibit the characteristics of mantle sulfur. The sedimentary stratified mineralization is closely related to the magmatic events in the middle Neoproterozoic (883 ~ 760 Ma), and the hydrothermal filling vein mineralization is mainly related to the large-scale orogeny and regional metamorphism in the late Neoproterozoic (600 ~ 500 Ma).

Acknowledgments

This study was supported by the research grant of Guangxi Natural Science Foundation (grant No. 2017GXNSFBA198203) and Guangxi Key Laboratory of Hidden Metallic Ore Deposits Exploration (No.20-065-17-K02).

Reference

1. Wei, Y. B., Liu, X. C., 2012. Geological Characteristics and Prospecting Potential of Kalulushi Copper Deposits in Copperbelt Province, Zambia. Journal of East China Institute of Technology (Natural Science Edition), 35 (4): 364–370.
2. Zhu, H. B., Zhang, D. H., Xiao, B., 2016. Geological characteristics and genesis of Chambishi copper deposit in Copperbelt of Zambia. Mineral Resources and Geology, 30 (1): 19–23.
3. Fang, K., Hu, Q. F., Li, M. J., et al., 2019. Characteristics of molybdenum mineralization in the south-east ore body of Chambishi copper deposit in Copperbelt Province of Zambia. Mineral Resources and Geology, 33 (6): 987–994.
4. Qin, P., Hao, B., Fang, K., 2019. The geological significance for Re-Os isotopic dating of Cu-Mo orein the lower footwall of Chambishi copper deposit within Copper Belt province of Zambia. Mineral Resources and Geology, 33 (3): 377–384.
5. Zhao, X. G., 2010. Geological characteristics of Chambishi copperdepsit in Zambia. Geology and Exploration, 46 (1): 183–190.
6. Taylor, H. P., 1974. The application of oxygen and hydrogen isotope studies to problem of hydrothermal alteration and ore deposition. Economic Geology, 69 (6): 843–883.
7. Ohmoto, H., Rye, R. O., 1979. Isotopes of sulfur and carbon. In: Geochemistry of hydrothermal ore deposits. New York: John Wiley and Sons, 509–567.
8. Kampunzu, A. B., Cailteux, J., 1999. Tectonic Evolution of the Lufilian Arc (Central Africa Copper Belt) During Neoproterozoic Pan African Orogenesis. Gondwana Research, 2: 401–421.
9. Ren, J. P., Wang, J., Zhang, D. H., et al., 2018. Reactivation of Lufilian Arc in Zambia: Zircon and Apatite Fission Track Chronology. Earth Sicence, 43 (6): 1850–1860.
10. Torrealday, H. L., Hitzman, M. W., Stein, H. J., et al., 2000. Re-Os and U-Pb dating of the vein-hosted mineralization at Kansanshi copper deposit, north Zambia. Economic Geology, 95: 1165–1170.

Study on Photocatalytic Degradation of Pollutants by Zinc Oxide Nanocomposites

Yuanyuan Du[1*], Xinjun Li[1*], Ying Fu[1], Xing Gao[1], Wen He[1] , Li Han[1] ,Weijuan Gong[1] and Pengwu Zheng[2]

[1]School of Chemical and Biological Engineering, Qilu Institute of Technology, Jinan Shandong. 250200. China
[2]School of Pharmacy, Jiangxi Science and Technology Normal University, Nanchang Jiangxi. 330013. China
Yuanyuan Du[1*] and Xinjun Li[1*] contributed equally to this work.

Abstract. In recent years,the research of zinc oxide nanocomposites has attracted much attention.Due to the high probability of photonized carrier recombination of a single nanometer zinc oxide material,and the low utilization of sunlight,its application in the degradation of pollutants is limited.Compounding zinc oxide with other materials can solve this problem.In this paper,through the study of the photocatalytic properties of zinc oxide nanocomposites,the synthesis methods and photocatalytic properties of TiO_2/ZnO and ZnO/graphene nanocomposites are reviewed.Nanocomposite photocatalytic degradation mechanism of pollutants and its application in the degradation of pollutants.

1 Introduction

In the contemporary era where the industrial level is rapidly developing, green and energy have become hard core issues in contemporary development. As a novel and effective green technology, photocatalysis has played a huge role in energy utilization. The current photocatalysts include TiO_2, ZnO, WO_3, ZnS, CdS, g-C_3N_4, etc., which are the most widely used semiconductor photocatalysts, including oxides , Sulfides, new photocatalysts, etc. Among the semiconductor photocatalytic materials, ZnO is cheap, non-toxic, stable, hot spots, high photosensitivity, controllable crystal shape and excellent photocatalytic activity, and has attracted wide attention in the degradation of various pollutants[1]. However, because of the unique performance structure of ZnO, it also encounters extremely severe problems in photocatalytic applications, because its electron-hole pair combination is more likely to be combined , resulting in its absorption and utilization of solar energy. The rate is relatively low. Since the discovery of TiO_2 and graphene, researchers in this field have begun to discover a composite photocatalyst of TiO_2, graphene and ZnO, which can effectively improve the utilization efficiency of photocatalysis. This article mainly introduces the research progress of TiO_2/ZnO and ZnO/graphene nanocomposites, their preparation methods, reaction mechanism and photocatalytic degradation of pollutants[2].

2 Nano materials

2.1 Nanomaterials and their basic properties

Nanomaterials refer to the substances stored in the nanometer. Its three-dimensional space size is at least one-dimensional, composed of nanoparticles, and its size is between atoms, molecules and macroscopic systems . Nanotechnology refers to new materials that are directly processed on the nanometer depository, and are composed of one atom or molecule[3]. Because nanomaterials are very small, the electronic activities inside them are not free, making them different from the electronic energy level structure of bulk solids, which specifically exhibits four characteristics.

(1) Surface effect

When the particle size is reduced to nanometer size, its specific surface area increases rapidly, the number of atoms on it increases, its surface energy increases, and the activity of the particle also increases.

(2) Small size effect

When the particle size is reduced to nanometer size, as its volume becomes smaller, the wavelength of the light source becomes shorter, the de Broglie wavelength becomes smaller, the size of the single magnetic domain becomes smaller, and the critical size condition of the particle lattice becomes worse. Out of an unstable state, there is a significant change in properties[4] .

(3) Quantum size effect

When the particle size is reduced to nanometer size, the continuous combination of electronic activities changes its electronic energy level structure, resulting in changes in physical properties.

(4) Macroscopic quantum tunneling effect

The ability of microscopic particles (electrons) to enter and penetrate an obstacle is called tunneling. The microscopic mechanism of the nanometer size will affect its macroscopic physical quantities, such as the magnetization and magnetic flux of microscopic particles. The macro-quantum tunneling effect shortens the length

*Corresponding author. duyuan01@126.com 928878031@qq.com

https://doi.org/10.1051/e3sconf/202129003017

of time the optical disc stores information, and is bound to become a hard core in the field of microelectronics development, driving the further development of science and technology to miniaturization[5].

2.2 ZnO photocatalytic nanomaterials

A prominent feature of nano-semiconductor materials is the photocatalytic feature. It can achieve high differentiation and degradation ability of pollutants under the condition of low energy consumption. This new technology material is called nano-semiconductor photocatalytic material, which plays a key role in the field of environmental sewage treatment. The principle of photocatalysis is that the electrons in the valence band of the semiconductors irradiated by sunlight are excited, and electron-hole pairs are formed in the conduction band[6]. The electrons have strong reductive properties and the holes have strong oxidizing properties. The compound on the surface of the catalyst undergoes a redox reaction with it upon stimulation .

ZnO photocatalytic reaction process: When ZnO is irradiated with high energy and strong light, electrons in the valence band are excited to the conduction band, and holes are generated in the valence band, and electron-hole pairs are generated inside the semiconductor. The holes oxidize the OH- and H_2O molecules on the surface, and generate OH•. The generated OH• can continue to oxidize adjacent organic substances, and OH• can also oxidize organic substances in the liquid phase. Various organic substances undergo a series of oxidation processes. In the end, they all become H_2O and CO_2, and the degradation process of organic matter is completed[7] .

As a new type of photocatalyst, ZnO differentiates and degrades organic pollutants. The main problems of ZnO itself are: first, the probability of electron-hole pair binding is relatively high, resulting in low catalytic reaction efficiency; second, the utilization rate of solar energy is low. , Limiting its application. This article will describe the composite material formed by TiO_2 or graphene combined with nano-zinc oxide, which prevents the combination of electron-hole pairs, increases the photocatalytic response range, and has excellent photocatalytic performance .

3 Study on the photocatalytic performance of TiO_2/ZnO nanocomposite

In the fields of environmental beautification, pollution control, and biological sciences, photocatalysts play a very important role. By combining TiO_2 and ZnO to increase the light absorption range, improve the absorption and utilization of solar energy, and prevent the combination of electron-hole pairs, In this way, the photocatalytic performance of Shanshan becomes better.

3.1 Nano TiO_2

Nano-TiO_2, green and environmentally friendly, is a heterogeneous semiconductor photocatalyst. It plays an important role in various application fields, such as qualified wastewater treatment and air purification quality standards. It is non-toxic and harmless, stable in chemical properties, low in energy efficiency, and can be used repeatedly for many times. Its high photocatalytic activity is a notable feature, and it is one of the most promising semiconductor materials[8].

3.1.1 Nano-TiO_2 photocatalytic mechanism

TiO_2 is an n-type semiconductor material. When sunlight is irradiated, electrons in the valence band are excited and transition to the conduction band, and corresponding holes (h^+) are generated in the valence band. Under the action of the electric field, the electrons and holes are separated and transferred to the particle surface[9]. A part of the electron-hole pairs recombine, and the O_2, H_2O and OH^- on its surface react with another part to form HO•, which then oxidizes various organic substances and is differentiated and degraded.

3.1.2 Application of nano-TiO_2 in the field of photocatalysis

(1) Removal of wastewater pollutants

Wastewater mainly contains various pollutants such as phenols and hydrocarbons. Among them, the most toxic is heterocyclic compounds. Because of the complex composition of wastewater, it is difficult to separate the pollutants from the wastewater[10].

(2) Air purification

Air purification has always been a key topic of green development today. At present, TiO_2 occupies an important position in air purification, because it can effectively decompose air pollutants. TiO_2 can be coated on the surface of various devices, can also be used in steel bars, and even directly used on pollution sources, which has played an important role in environmental governance.

(3) Sterilization, corrosion resistance and self-cleaning

TiO_2 photocatalysis is used in all aspects of life, because it can not only decompose bacteria, but also sterilize and disinfect. Because of the difference in energy level, it can protect the metal surface from corrosion. It is excellent in hygiene and health. If it is applied to materials that are not easy to clean, it shows good self-cleaning ability. It also has a good function of sterilization and disinfection, strong practicability, and outstanding performance in antivirus of daily necessities[11].

3.2 Nano TiO_2/ZnO composite material

3.2.1 Preparation of nano-TiO_2/ZnO composite materials

At present, there are three main research methods for nano-TiO_2/ZnO composite materials.

(1) Sol-gel method

When the organometallic compound or metal salt is processed, it will become a liquid colloidal substance after the hydrolysis reaction. After losing water, it will become a solid colloid. Apply it to the outermost layer of the substance[12]. Will form a hard protective film. We can also use other technologies to produce smaller particles. The main reason is that this method is simple to operate and low in cost. It can also be used for large-scale production of film products .

(2) Hydrothermal method

Using water as the reaction medium, heating the reaction kettle, insoluble or insoluble substances will become soluble under high temperature and high pressure conditions, and can be recrystallized. It is similar to the sol-gel method and also has the advantages of simple operation and low cost. Vendula et al. synthesized TiO_2/ZnS by melting and degrading $TiOSO_4$ and $ZnSO_4•7H_2O$, which was dried and then continuously calcined to generate a photoactive TiO_2/ZnO nanocomposite. Small-molecule spherical substances constitute the material, and the more TiO_2 content, the smaller the particles formed. Remi et al. used this method to obtain the TiO_2/ZnO nanocomposite, which is uniformly dispersed and has a small average particle size[13].

(3) Two-step method

First, TiO_2 nanostructures are generated, and the second hydrothermal method is used to synthesize TiO_2/ZnO nanocomposites. Its advantages are: (1) The shape of the substance can be well controlled to obtain a satisfactory shape of the substance; (2) The obtained substance can realize the classification of nano-layers and has ultra-high performance.

3.2.2 Photocatalytic degradation mechanism of nano-TiO2/ZnO composite

When the composite is irradiated by solar energy, the electrons in the valence band will be stimulated and migrate to the conduction band, and the valence band will produce holes (h+): TiO_2/ZnO (hv) → TiO_2-ZnO Holes will oxidize various organics to generate CO_2 and H_2O, which completes the degradation of organics. Because of the doping of other substances, the doped energy level will be generated, so that photons of small energy can excite the electrons and holes generated at this energy level, and the possibility of photons being absorbed and utilized is increased[14]. The more doping, the greater the free length of carriers, the longer the survival time of electrons and holes, and the restriction of e^-/h^+ combination, which increases the possibility of solar energy absorption and utilization of the composite material, and the photocatalytic effect Excellent. It is concluded that composite materials have a much better catalytic effect than pure nano-materials as photocatalysts.

3.2.3 Application of photocatalytic degradation of pollutants by nano-TiO2/ZnO

Under the conditions of 180 ℃ and 0.8 MPa, using low-temperature and low-pressure hydrothermal method, nano-TiO_2/ZnO composite materials can be synthesized. This material can absorb visible light well, of course, ultraviolet light is no exception, and it can fully differentiate and decompose acid red B And acid black 234 dye. It has one of the most significant advantages, it can directly use sunlight to optimize the space environment[15].

According to the research of Fang Yanju et al., after the two substances are combined, the absorption spectrum will produce a red shift phenomenon, which significantly improves the ability of the composite film to oxidize glucose, which is higher than that of a single raw material. Zhang Hao et al. Pointed out that for the composite formed by the Zn/Ti atomic ratio of 2:1, its photocatalytic performance is excellent, and the composite material has outstanding performance in degrading methylene blue, and its degradation ability is much higher than that of a single raw material.

Kostedt et al. studied the nanocomposite of the two substances, and made an objective judgment on the adsorption capacity of the red MX-5B dye and the photocatalytic oxidation capacity of the shoots. The study found that because these two raw materials have similar zeta potential and energy band energy, materials with a mixing ratio of almost 100% ZnO or TiO_2 exhibit excellent catalytic performance. Rad et al. believed that when the mass ratio of ZnO to TiO_2 in the composite is 1:1, it has a more prominent performance effect on the degradation of red dyes. The larger the specific surface area of the composite, the more the composite is decomposed. The greater the possibility, on the contrary, the larger the particle size, the smaller the possibility of its differentiation and decomposition. Liu Yue et al. found that the composite has much higher catalytic activity than P25 when photocatalytic oxidation of NO. Under sunlight, when the atomic fraction of Zn^{2+} is 0.5%, it prevents OH on the surface. The formation of nitroso groups makes NO easier to be oxidized, and the photocatalytic effect is the best.

SakaiA et al. found that the complex can be used to remove methyl mercaptan. TiO_2 coated with ZnO improves the adsorption capacity of TiO_2, especially for acidic substances. When the exposed part can absorb sunlight, photoions will be formed and CH_3SH will be oxidized and decomposed. Ndrea.E et al. studied the TiO_2/ZnO nanocomposite electrode and found that the more appropriate the amount of ZnO in the composite is controlled, the more breakthrough the photocatalysis problem is. Different materials have their own suitable composite ratio of TiO_2/ZnO. The photocatalytic reaction mechanism and the shape of the composite are also factors that influence the optimal composite ratio.

4 Study on the photocatalytic performance of ZnO/graphene nanocomposite

ZnO/graphene nanocomposite material helps to separate electron-hole pairs effectively. Because of its large specific surface area, it plays an important role in the diffusion and mass transfer of pollutants and free radicals. The hole of ZnO is on the high-energy conduction band. The electrons can be directly transferred to graphene, which greatly improves its photocatalytic performance under ultraviolet or visible light.

4.1 Graphene

4.1.1 The structure and properties of graphene

Graphene, a hexagonal single-layer carbon sheet, is the basic unit of carbon materials . It is formed by the way of hybridization of carbon atoms and presents a hexagonal crystal structure. C-C is connected by a σ bond, and a 120° bond is formed between C-C, and the distance between C-C is about 0.142 nm . Its single layer thickness is only 0.35 nm, which is extremely light and thin.

Graphene is a semi-metallic material with no band gap . Its ultraviolet penetration effect is very good. Each layer of graphene has an absorption rate of only 2.3% for ultraviolet and visible light . It also has extremely high transparency in the near and mid-infrared bands, and its mechanical strength has always been the first . With the same excellent thermal properties, BalandinAA et al. measured that single-layer graphene has a higher thermal conductivity of 5.2 kW. (m.K)-1 than carbon materials such as diamond and carbon nanotubes. The super large specific surface area makes it possess extremely strong adsorption capacity.

4.1.2 The production method of graphene

1. Peeling method, including mechanical peeling method [39] and chemical peeling method.

2. Reduction and oxidation method. The graphite is oxidized to obtain graphene oxide (GO), and GO is reduced by chemical reduction method, electrochemical reduction method, thermal expansion reduction method and hydrogen reduction method.

4.2 ZnO/graphene nanocomposite

ZnO nanoparticles are uniformly dispersed and bound to graphene, which can increase its specific surface area and charge separation more easily.

4.2.1 Preparation of ZnO/graphene nanocomposite

Using the hydrothermal method, the ZnO/GO composite material obtained by Pokharel P et al. is petal-shaped, which improves the possibility of combining photoelectrons and holes and improves the photocatalytic performance of the composite material. Using the uniform precipitation method, He Guangyu et al. prepared ZnO/GO composite materials. Because of the electron transfer effect between the two kinds of nanoparticles, the combination of photogenerated electron-hole pairs is limited, and the visible light catalytic performance is greatly improved. increase.

Using the one-step method, Sun Hongqi et al. produced a ZnO/CTAB/GO composite material with low band gap energy, fast electron transfer, and strong adsorption of organic dyes. Hou Zongwei et al. obtained a zinc oxide nanorod/graphene composite photocatalyst, and used a vacuum-assisted pressure induction method to input the reaction solution into the expanded graphite room to synthesize the material.

Using the microwave-assisted synthesis method, Liu Xinjuan et al. synthesized ZnO/GO composite materials with zinc nitrate and graphite oxide as reactants, which increased the absorption range of ultraviolet light, reduced the neutralization effect, and increased the electron transfer efficiency.

Using the one-pot method, Zou Changwu et al. produced ZnO/graphene nanocomposite microspheres, which accelerated the speed of electron transfer and had a good effect on the differentiation and decomposition of methyl blue under ultraviolet light.

Using the auxiliary method of ionic liquids, Zhang Kai et al. developed a ZnO/graphene composite material with a p/n heterojunction, which improved the possibility of photoelectrons and holes binding, and improved the ability to adsorb pollutants. improve.

4.2.2 The photocatalytic degradation mechanism of ZnO/graphene nanocomposite

Light absorption, generation and separation of electron-holes, electron-hole transfer, and oxidation-reduction reactions are the key steps for the material to differentiate and decompose pollutants. Mainly manifested in : First, because graphene is mixed, the material has a larger surface area, and the diffusion and mass transfer of pollutants and free radicals is further improved; second, the high-energy conduction band of ZnO is -4.05 eV , Its low-energy valence band full of electrons is -7.25 eV, while graphene is -4.42 eV. The electrons on the conduction band of ZnO can be directly transferred to graphene, and the photocatalytic performance of the composite material is improved . Third, graphene has a strong e-capturing ability, which can effectively separate e- and make O_2 react with it to produce O_2^-, and OH- captures the separated h^+ to produce OH, which makes it possible to combine electrons and holes. The performance decreases and the photocatalytic efficiency increases. Therefore, composite materials have a much better catalytic effect than pure nano-materials as photocatalysts.

4.2.3 Application of ZnO-graphene nanocomposite material for photocatalytic degradation of pollutants

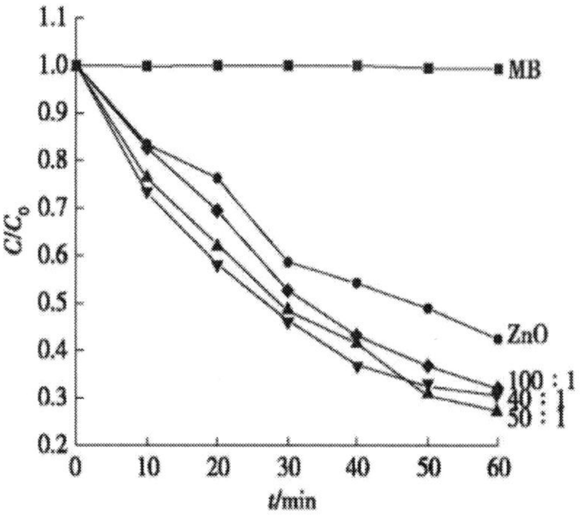

Figure 1 Degradation of methylene blue solution (without any catalyst (MB); pure ZnO; ZnO/graphene mass ratio 100:1, 50:1, 40:1) under ultraviolet light

Figure 1 shows the changes in the degradation rate of methylene blue over time when pure ZnO and ZnO/graphene are selected as catalysts. MB was used as a blank control, which did not add any catalyst, which can be observed under the irradiation of ultraviolet light. At this time, the nature of the methylene blue solution is very stable, and the concentration of the solution is basically unchanged. It can also be observed that the photocatalytic performance of all ZnO/graphene samples is better than pure ZnO. This is because in the composite material where zinc oxide is in the graphene sheet structure, they have very close band gaps, and the electrons on the valence band of zinc oxide can directly migrate into graphene. The energy required for this process is not high, so It makes the transmission and transfer of photo-generated electrons more convenient, and the photocatalytic performance is improved. When the mass ratio of graphene in the composite material changes from 100:1 to 40:1, its photocatalytic activity will show a peak and then decrease. This peak value is when the mass ratio is 50:1, the material exhibits the strongest activity When the photocatalytic time reaches 60min, using the formula $\mu=(1-C/C_0)\times100\%$, the efficiency of photocatalytic degradation of methylene blue solution at this time is calculated to be higher than 70%, and at this time, the mass ratio is 100 The photocatalytic decomposition efficiency of :1 is 67.9%, and the photocatalytic decomposition efficiency of 40:1 by mass is 69.4%. The results indicate that the highest efficiency of photocatalytic degradation is that the ZnO/graphene composite material has a mass ratio of When the ratio of 50:1 appears, the efficiency of photocatalytic degradation of methylene blue solution is slightly lower than that of other ratios . Under the irradiation of ultraviolet light, the mass ratio of ZnO/graphene changes from 100:1 to 50:1, the electron migration speed becomes faster, the possibility of electron-hole pair combination is lower, and the photocatalytic performance of the material is higher; The graphene content continues to increase to 40:1, but the photocatalytic performance decreases. The reason should be: when the mass ratio of graphene in the composite material continues to increase, after the mass ratio of ZnO/graphene exceeds 50:1, it will inevitably increase The thickness of graphene results in poor transparency, resulting in a decrease in the intensity of light that can reach ZnO, which in turn reduces the number of photo-generated electron-hole pairs in ZnO, which will reduce the photocatalytic performance . Therefore, the best quality ratio of ZnO/graphene should be around 50:1.

5 Summary

For the nano-ZnO raw material, because it has a small band gap width, it can be used in the photocatalytic decomposition of pollutants, but its shortcomings limit the improvement of its performance. Due to the low valence band potential and poor light stability, the raw material is light. The effect of catalytic differentiation and decomposition of pollutants is not ideal. In view of the advantages and disadvantages of a single raw material, nanocomposite TiO_2 and ZnO, and prepare a TiO_2/ZnO nanocomposite system. The greater the coupling between the nanoparticles, the greater the possibility of solar energy absorption and utilization, and the organic matter is differentiated The faster the rate of decomposition will make it more and more important in the differentiation and decomposition of pollutants. Nano-composite ZnO and graphene, combine ZnO particles on graphene through different experimental methods and synthesize composite materials. As a green advanced composite, it plays a very important role in photocatalytic applications, especially in Differentiating and decomposing pollutant dyes has both theoretical value and practical significance. As far as the current research status is concerned, the removal and mechanism of new refractory pollutants are not perfect enough, and we still need to continue to work hard. We also believe that with the continuous development of nanotechnology, people are concerned about the photocatalytic degradation of pollutants by zinc oxide nanocomposites. Research will also go to the next level.

Reference

1. Zhang Zhonghai, Yuan Yuan, Fang Yanju, et al. Preparation of photocatalytic nano-ZnO/TiO_2 film and application for determination of Chemical oxygen demand[J]. Talanta, 2007, 73: 523.
2. Chen Da, Zhang Hao, Hu Song,et al.Preparation and enhanced photoelectro-Chemical performance of coupled bicomponent ZnO-TiO_2 nanocomposites[J]. J Phys Chem C, 2008, 112: 117.
3. William L, Kostedt I V, Adel A Ismail, et al. Impact of heat treatment and composition of ZnO-TiO_2 nanoparticles for photocatalytic oxidation of an azo dye[J]. Ind Eng Chem Res, 2008, 47: 1483.
4. Asl S K, Sadrnezhaad K, Rad M Kianpoor. Photocatalytic decolorization of red dye in aqueous ZnO-TiO_2 suspensions[J]. Smart Mater, 2008, 55-57: 577.
5. Wang Haiqiang, Wu Zhongbiao, Liu Yue, et al. The characterization of ZnO-anatase-rutile three-component semiconductor and enhanced

photocatalytic activity of nitrogen oxides[J]. J Mol Catal A, Chem, 2008, 287: 176.

6. Sakai A, Kanehata M, Ogawa T, et al. Improvement of methyl mercaptan removal by ZnO-coated TiO_2[J]. Trans Mater Res Soc Japan, 2007, 32(4): 1079.
7. Indrea E, Dreve S, Silipas D T, et al. Semiconductor photoelectrodes for solar of splitting water[J]. J Optoelectron Adv Mater, 2008, 10(9): 2213.
8. Zhu Y, Murali S, Cai W, et al. Graphene and graphene oxide: synthesis, properties, and applications[J]. Advanced Materials, 2010, 22(35): 3906-3924.
9. King A, Johnson G, Engelberg D, et al. Observations of intergranular stress corrosion cracking in a grain-mapped polycrystal[J]. Science, 2008, 321(5887): 382-385.
10. Nair R R, Blake P, Grigorenko A N, et al. Fine structure constant defines visual transparency of graphene[J]. Science, 2008, 320(5881): 1308.
11. King A, Johnson G, Engelberg D, et al. Observations of intergranular stress corrosion cracking in a grain-mapped polycrystal[J]. Science, 2008, 321(5887): 382-385.
12. Balandin A A, Ghosh S,Bao W, et al. Superior thermal conductivity of single-layer graphene[J]. Nano Letters, 2008, 8(3): 902-907.
13. Novoselov K S, Jiang Z, Zhang Y, et al. Room-temperature quantum hall effect in graphene[J]. Science, 2007, 315(5817): 1379.
14. Geim A K, Morozov S V, et al. Electric fieldeffect in atomi[J]. Science, 2004, 306(5696): 666-669.
15. Si Y, Samulski E T. Synthesis of water soluble graphene[J]. Nano Letters, 2008, 8(6): 1679-1682.

https://doi.org/10.1051/e3sconf/202129003018

A Qualitative Study on the Influencing Factors of Public Participation in the Construction of "Waste-Free City" Exploratory Analysis Based on Grounded Theory

Mao Chai[1], Jing Jiao[1*], Yu Han[2]

[1]School of Public Administration, Xiangtan University Xiangtan, Hunan P.R. China
[2]School of Public Administration, Northwest University Xi 'an, Shaanxi P.R. China

Abstract. As a cellular project of beautiful China and a systematic project of the whole people, the construction of "Waste Free City" cannot be separated from the extensive participation of the public. This research in China's Henan province "no waste city" construction of public participation as the research object, using the grounded theory, first through open and spindle coding that affecting public participation "no waste city" of the construction of the six main factors: attitude, subjective norm, perceived behavioral control, driving factors and social environment and identity characteristics; Then a model of influencing factors of public participation is constructed through selective coding. The results show that participation attitude, subjective norms and perceived behavior control directly affect the willingness of the public to participate in the construction of "waste-free city", which are internal psychological factors. Government behavior, group influence, community factor and enterprise factor constitute the external environmental factors that affect public participation. The three categories of publicity and education, reward and punishment mechanism and laws and regulations form the driving factors of public participation in the construction of "waste-free city", which play a moderating role between internal psychological factors and external environmental factors. At the same time, identity characteristics also directly affect the public participation behavior in the construction of "waste-free city".

1 Problem Raising and Literature Review

With the continuous development of economy and society, and the accelerating process of urbanization and industrialization, China's annual increase of solid waste amounts to about 10 billion tons, and the trend of growth year by year, seriously affecting the sustainable development of urban economy. In order to reduce the negative impact of solid waste on the ecological environment, "Waste Free City" is a new urban development model guided by the five new development concepts and using green production development as the means. The Fifth Plenary Session of the 19th CPC Central Committee made major strategic arrangements for the construction of ecological civilization and ecological environmental protection. It puts forward the long-term goal of "basically realizing the goal of building a beautiful China" in 2035 and the new goal and task of "realizing new progress in ecological civilization construction" in the "14th Five-Year" period. On May 6, 2021, [1] the National Development and Reform Commission issued the "14th Five-Year Plan for the Development of Urban Domestic Waste Classification and Disposal Facilities", which provides another important institutional guarantee for the development of "Waste Free Cities". As a systematic project of the whole people, the steady progress of the construction of "waste-free city" cannot be separated from the extensive participation of the whole public.

Foreign research and exploration on the construction of "waste-free city" started earlier, and the concept of "waste-free" or "zero waste" began to appear in the 1990s. At the national level, At the national level, Japan, the United Kingdom, Singapore [2] have implemented the "zero waste" strategy earlier. At the urban level, New York and San Francisco in the US, Vancouver in Canada, and Stockholm in Sweden have all announced their zero-waste plans. In C40 City Group, an international organization of cities to jointly deal with climate change, 23 cities have also signed the Declaration toward Zero Waste [3]. Compared with foreign studies, the exploration of "waste-free city" in China started relatively late. China's "Zero Abandoned" Alliance was established in December 2011, and academic discussions on "waste-free city" were almost zero before 2018, but there has been a significant increase in recent years. Research topics can be roughly divided into solid waste management and resource utilization, the construction pattern and development path to explore, scheme evaluation and effectiveness evaluation three categories, there are also involved in construction waste recycling use of the recycled product [4], risk of hazardous wastes and their legal regulation [5],

Corresponding Author: jiaojing213@163.com

as well as China's provincial cities without waste "space-time evolution trend of the development level of study [6], etc.

In addition, many domestic scholars have conducted related studies on public participation in ecological civilization construction from different perspectives, covering natural environment governance, public environmental protection behavior, ecological function area construction, ecological city construction, etc. Among them, some research conclusions have important reference significance for this study. First, The first is the research on the influencing factors of public participation in the construction of ecological civilization. The academic discussion on this topic mainly includes subject factor [7], objective factor [8], external environment [9] and so on. Second, the study on the path and mechanism of public participation in ecological civilization construction. Specifically, it includes the cultivation path of ecological environmental awareness [10], the judicial path of public participation in environmental protection [11], the "empowerment, identification and cooperation" mechanism of public participation in environmental governance [12], and the community action and collective cooperation in environmental protection [13].

To sum up, the research results on public participation in ecological civilization construction have been quite fruitful, which has important enlightenment for the study of the influencing factors of public participation in the construction of "waste-free city". "No waste city" as a new urban development pattern, public participation "no waste city" construction is different from general environmental volunteering, uniqueness, and few of the current "no waste city" related research on public participation in the discussion, so it is necessary for the public to participate in the "no waste city" of the construction of the influencing factors and explore the mechanism of action, To provide theory and experience for the better development of "waste free city".

2 Study Design

2.1 The research methods

Grounded theory is a qualitative research method based on systematically collecting and processing data, and then extracting concepts and constructing models from the bottom up. [14] It develops a theoretical model through coding and refining data, which is very suitable for exploring the influencing factors of public participation. In addition, the theory of planned behavior has been widely used to explain and predict the decision-making process of individual behavior due to its outstanding explanatory and predictive ability. [15] Therefore, this study combined with the theory of planned behavior, processed and analyzed the data through the three-level coding of the grounded theory, constructed a theoretical model, and revealed the main factors affecting the public participation in the construction of "waste-free city".

2.2 Sample selection

This study selected of public participation in China's henan province "no waste city" as the research object on the construction of major for two reasons: one is as zhongyuan urban agglomeration in China only pilot "no waste city" construction, the authors formed the advanced experience of the solid waste disposal and characteristic pattern, won several authoritative media coverage in China. Second, Xuchang gave full play to the role of system guarantee, market drive, technical support and cultural maintenance to guide the active participation of the public, and its specific implementation effect is worth further exploration.

2.3 Data collection

Considering the actual situation of the pilot work of "Waste Free City" in Xuchang City, 36 residents of Weidu and Jian 'an two central urban areas were selected as the main interviewees. The first-hand interview data were obtained through on-site in-depth interviews with interviewees. In addition, secondary data such as news reports and official government documents were collected and cross-verified with interview data to improve the reliability and validity of this study. The following is a summary of the interview.

Table 1: The Interview Outline

Interview project	*Specific problems*
Personal information	Gender, age, occupation, type of residential area
The respondents' understanding and participation in the "waste-free city" and its driving factors	Do you know the construction status of Xuchang "Waste Free City"?
	What do you think is a "waste free city"?
	In what ways do you participate in the construction of "Waste Free City" in your daily life?
	What are your reasons for participating (not participating) in the construction of "Waste Free City"?
	What factors will you consider when you participate in activities related to "Waste Free City"?
	What are your suggestions for the construction of "waste-free city"?

3 The Research Process

In this study, software NVivo12.0 was used for three-level coding, followed by saturation test and qualitative results analysis. In this study, 32 of the 36 interview data were randomly selected for coding, and another 4 were reserved for testing theoretical saturation.

3.1 Open coding

Open coding is a process of word-for-word coding, labeling and conceptualization of the original data. In this study, through coding 32 randomly selected interview data, 15 categories were finally extracted.

3.2 Axial Coding

Axial Coding is to analyze and refine each category of open coding according to certain logic, and establish the logical relationship between concepts, so as to develop the category of spindle [16]. In this study, by combining and clustering 15 independent categories obtained from open coding, 6 main axis categories were obtained at the macro level.Axial coding and the corresponding category are shown in table 2.

Table 2: Axial Coding and its Corresponding Categories

The main categories	*Category connotation*	*Corresponding categories*	*Initial concepts*
Attitude to Participation	The public's cognition and judgment of "waste free city", the public's environmental responsibility and environmental values	"Waste - free city" cognition, environmental responsibility, environmental values	Negative cognition, environmental awareness, environmental psychology ……
The driving factor	Publicity and education, reward and punishment incentive and legal restraint and other measures can guide public behavior and form driving factors	Publicity and education, reward and punishment mechanism, laws and regulations	Laws and regulations, environmental education, Incentives and punishments ……
Perceptual behavioral control	The perception of public participation on the ability, energy, opportunity and guarantee level of infrastructure conditions of the construction of "waste-free city"	Participation opportunity cognition, participation behavior guarantee	Knowledge and ability, guarantee of participation in the process ……
The social environment	The government, enterprises, communities, social organizations and other subjects jointly influence the social environment of the construction of "waste-free city"	Government action, group influence, community factor, enterprise factor	The power of government, management of business, influence of voluntary groups ……
Subjective norms	The moral norms followed by individuals and the social group pressure perceived when the public participated in the construction of "Waste Free City"	Individual norms, social pressures	Value pursuit, community impact ……
Identity and Traits	The difference of public status, class structure and individual personality characteristics will lead to different participation behaviors in the construction of "waste-free city"	Identity and Traits	Status, personality traits ……

3.3 Selective Coding

Selective coding refers to the process of analyzing and refining the spindle coding again, and then digging out the "core category" which can play the role of summarizing, and then analyzing the relationship between the core category and non-core category and establishing the connection relationship. In this study, a model of influencing factors for public participation in the construction of "waste-free city" was constructed through selective coding, as shown in Figure 1. After that, the four interview data previously reserved were tested for saturation, and no new concepts and categories were shown. Therefore, it was determined that the model had achieved theoretical saturation

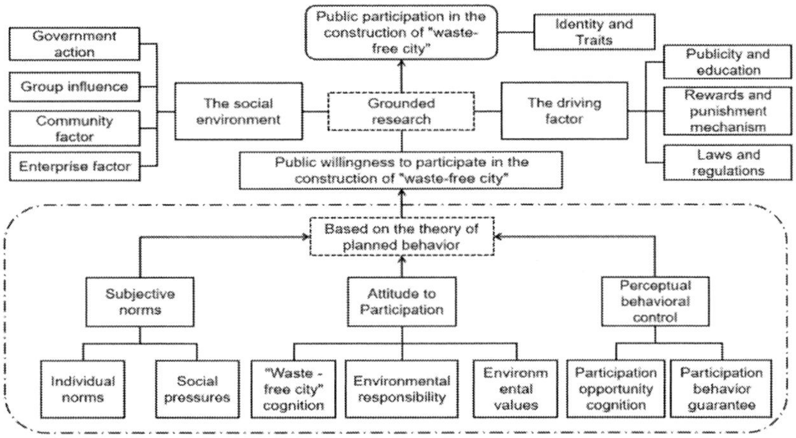

Fig. 1 : Influencing factor model of public participation in the construction of "waste-free city" from the perspective of grounded theory

3.4 Model to interpret

From the perspective of realizing the research purpose, the method of grounded theory is not only to construct the influencing factor model, but more importantly, to analyze the current situation and problems of public participation through the acquired theory. [17] Therefore, this study explains the internal mechanism of public participation in the construction of "waste-free city" from the following dimensions.

(1) Consistent with the theory of planned behavior, participation attitude, subjective norm and perceived behavioral control are the three main variables affecting behavioral intention. But its concrete connotation differs somewhat [18]. First, the attitude of participation refers to the attitude formed by the conceptualization of the positive or negative evaluation of a specific behavior. In the construction of "waste-free city", the participation attitude of the public is composed of the cognition of "waste-free city", the sense of environmental responsibility and the environmental values, which are the internal and pre-factors influencing the participation of the public in the construction of "waste-free city". Secondly, subjective norms refer to the moral norms that individuals follow and the social group pressure they feel when taking a particular action, including individual norms and social pressure. Individual norms mainly include personal value pursuit, living habits and the influence from family and relatives. Social pressure refers to the collective norms from communities, units and groups. This kind of group pressure or collective norm will make the public form a motive to comply with the universal social pressure and promote the public to participate in the construction of "waste-free city". Third, the perceived behavioral control reflects the public's past experience and expected obstacles, and refers to the difficulty of the public's perceived participation in the construction of "Waste Free City", which is composed of the perception of participation opportunity and the guarantee of participation behavior. Participation opportunity cognition is an individual's speculation and judgment on whether he or she has the ability to complete a certain behavior. Participation guarantee refers to the guarantee factors of public participation in the construction process and effect of "waste-free city".

(2) Social environment. Social environment includes government behavior, group influence, community factor and enterprise factor, which belongs to external environment factor. According to sociological theory, social environment determines social consciousness and has an important influence on public behavior intention and behavior motivation. Under the large social background, the public is easily affected by the environmental policies of the government, as well as the incentives or constraints of corporate rules and systems, codes of conduct of communities and social groups, etc., thus forming their own behavioral value norms for treating the ecological environment and influencing their willingness and behavior to participate in the construction of "Waste Free City".

 https://doi.org/10.1051/e3sconf/202129003018

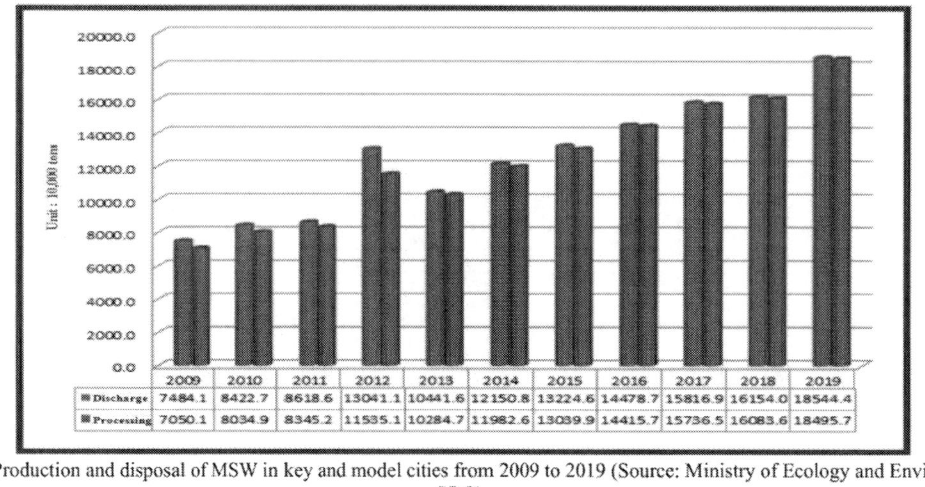

Fig. 2 : Production and disposal of MSW in key and model cities from 2009 to 2019 (Source: Ministry of Ecology and Environment, PRC)

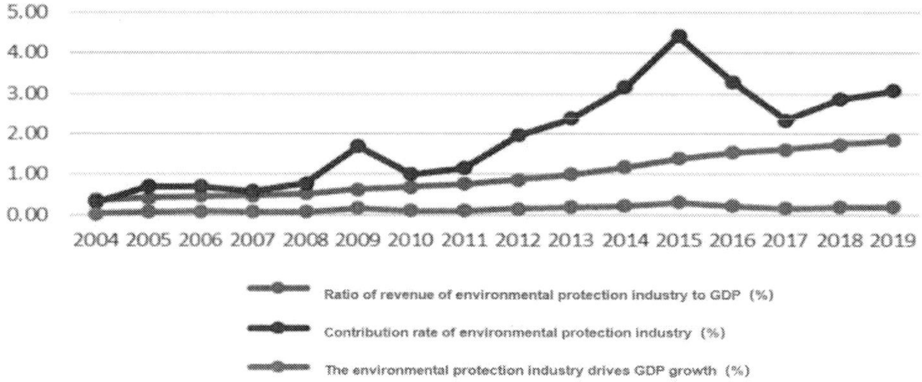

Fig. 3 : Contribution Rate of Environmental Protection Industry to National Economy (Source: China Environmental Protection Industry Association)

With the great attention and support from the government, the disposal capacity of MSW has been continuously improved, as shown in Figure 2. Figure 3 shows the contribution of the environmental protection industry to the national economy.

(3) Driving factors. Driving and promoting factors include publicity and education, reward and punishment mechanism, and laws and regulations. They are situational variables that affect public participation in the construction of "Waste Free City", and play a moderating role between psychological factors and external environmental factors. When the effect of this situational variable is weak, the public participation in the construction of "waste-free city" is mainly affected by internal psychological factors. When the effect of situational variables is significant, it may greatly promote or restrict the occurrence of public participation behavior. In the construction process of "waste-free city", the government mainly forms driving factors through strengthening publicity and education, establishing reward and punishment mechanism and improving laws and regulations, so as to improve public participation.

(4) Identity traits. Identity trait includes identity status and personality trait. Among them, status refers to the social role and class structure of the public. The public will make different considerations and judgments to participate in the construction of "waste-free city" based on the comprehensive consideration of social and economic status and social inequality. Personality traits reflect the individual's stable and unchangeable personality and psychological characteristics. According to the Big Five model of personality, there are about five traits that cover all aspects of personality description: openness, conscientiousness, extraversion, agreeableness, and neuroticism. Because the public have different personality traits, their participation in the construction of "waste free city" is also different.

4 Conclusion and Discussion

Taking the public participation in the construction of "waste-free city" in Xuchang as the sample, this study concluded through rooted analysis that six categories, including participation attitude, subjective norms, perceived behavior control, social environment, driving factors and identity traits, were the main factors affecting the public participation in the construction of "waste-free city". Specifically, the public's participation attitude, subjective norms and perceived behavior control directly affect the public's willingness to participate, which are internal psychological factors. The social environment constitutes the external environmental factors that affect public participation; The driving factors are situational variables, which play a moderating role between internal psychological factors and external environmental factors. The identity characteristics directly lead to the difference of public participation behavior.

The innovation and theoretical contribution of this study are as follows: First, it deeply excavates the connotation composition of the factors influencing public participation in the construction of "waste-free city", adds accumulation and material to enrich the system of public participation in ecological civilization construction, and makes up for the vacancy of the current research on public participation in the construction of "waste-free city". It has a certain reference significance for enhancing the enthusiasm of the public to participate in the construction of "waste-free city" and other ecological and environmental protection behaviors from a deep level. Second, it systematically sorted out the action mechanism of the factors influencing public participation in the construction of "waste-free city", combined subjective psychological activities with objective situational factors [19-23], which can more comprehensively analyze the influencing factors of public participation. Thirdly, it verifies and extends the theory of planned behavior, and well verifies the influence of behavioral attitude, subjective norms and perceived behavioral control on individual behavior. Based on the public participation in the construction of "Waste Free City" in Xuchang, the influencing factor model is built to expand the research direction and scope of the theory of planned behavior. Fourthly, taking the public participation in the construction of "waste-free city" in Xuchang as an example, primary materials are obtained based on field research and root analysis. The research conclusions are of reference significance for other pilot cities or regions to enhance the public participation effect.

Although this study is innovative in exploring the influencing factors of public participation in the construction of "waste-free city", it also has shortcomings and limitations, which need to be verified and improved through further empirical research. Specifically, it includes the following points: First, the data collection method of this study is mainly interview, and the open questionnaire and other forms are not used enough, and the diversity of methods needs to be improved. Second, the interview process is difficult to accurately and intuitively reflect the psychological activities of interviewees, and there may be some deviation in the way described by interviewees. Thirdly, different factors have different effects and intensities on public participation, so the government also needs differentiated implementation when taking specific measures to enhance public participation, which is not analyzed in depth in this study. Fourth, different identities and personality traits of the public will have a specific impact on their participation behaviors. Subfollow-up studies can be conducted to refine the study of citizens with different positions, fields and psychological characteristics, in order to more accurately enhance the effectiveness of public participation in the construction of "waste-free city".

Acknowledgment

This research is supported by the National Social Science Foundation of China (No.17CZZ022); Project of Social Science Achievements Evaluation Committee of Hunan Province (No.XSP17YBZZ111); Philosophy and Social Science Foundation of Hunan Province (No.16YBQ058)

References

1. Sohail, M. T., Mahfooz, Y., Azam, K., Yen, Y., Genfu, L., & Fahad, S. Impacts of urbanization and land cover dynamics on underground water in Islamabad, Pakistan. Des, and water treat, 159, 402-411(2019).
2. Qiu Qiwen, Wen Xuefeng. Research Report on the Construction of "Waste Free Cities" in Japan [J]. Environmental Protection,48(Z1):57-60 (2020).
3. Wen Zongguo. Waste Free City: Theory, Planning and Practice. Beijing: Science Press,2020.
4. Sohail, M. T., Mahfooz, Y., Aftab, R., Yen, Y., Talib, M. A., & Rasool, A. Water quality and health risk of public drinking water sources: a study of filtration plants installed in Rawalpindi and Islamabad, Pakistan. Des. and water treat., 181, 239-250 (2020).
5. Doe Xiaodong. Legal Regulation of Hazardous Waste Risks under the Construction of Waste Free Cities -- Interpretation of "Three-dimensional Mechanism" and Exploration of "Three-dimensional Path". Journal of Social Sciences of Hunan Normal University, 49(03):1-10 (2020).
6. Sohail, M. T., Xiuyuan, Y., Usman, A., Majeed, M. T., & Ullah, S. Renewable energy and non-renewable energy consumption: assessing the asymmetric role of monetary policy uncertainty in energy consumption. Environmental Science and Pollution Research, 1-10 (2021).
7. Zheng Junjun, Li Chengzhi, Liu Chunyan. Journal of Zhengzhou University (Philosophy and Social Sciences Edition),50(03):60-65+159 (2017).
8. Sohail, M. T., Aftab, R., Mahfooz, Y., Yasar, A., Yat, Y., Shaikh, S. A., & Irshad, S. Estimation of water quality, management and risk assessment in Khyber Pakhtunkhwa and Gilgit-Baltistan, Pakistan. Desalination and Water Treatment, 171, 105-114 (2019).
9. Lu Yanzhao, Wu Xiaojing, Yan Wenjing. Study on the influencing factors of public participation in urban solid waste management. Journal of Arid Land Resources and Environment,31(11):21-25 (2017).
10. Sohail, M. T., Delin, H., Talib, M. A., Xiaoqing, X., & Akhtar, M. M. An analysis of environmental law in Pakistan-policy and conditions of implementation. Research Journal of Applied Sciences, Engineering and Technology, 8(5), 644-653 (2014).
11. Yasara, A., Farooqa, T., Tabindaa, A. B., Sohailb, M. T., Mahfooza, Y., & Malika, A. Macrophytes as potential indicator of heavy metals in river water. DESALINATION AND WATER TREATMENT, 142, 272-278 (2019).

12. Sohail, M. T., Huang, D., Bailey, E., Akhtar, M. M., & Talib, M. A. Regulatory framework of mineral resources sector in Pakistan and investment proposal to Chinese companies in Pakistan. American Journal of Industrial and Business Management, 3(05), 514 (2013).
13. Yang Jiaying, Xu Chun. Promoting public participation in ecological civilization construction through community intervention. Guangdong Social Sciences, (05):79-84 (2018).
14. Glaser B,Strauss AL.The Discover of Grounded Theory : Strategies for Qualitative Research. Nursing Research,17(4):377-380 (1968).
15. AJZENI. The Theory of Planned Behavior. Organizational Behavior and Human Decision Processes,50(2):179-211 (1991).
16. Mahfooza, Y., Yasara, A., Guijianb, L., Yousafb, B., Sohailc, M. T., Khanb, S., ... & Khana, M. An assessment of wastewater pollution, treatment efficiency and management in a semi-arid urban area of Pakistan. DESALINATION AND WATER TREATMENT, 177, 167-175 (2020).
17. Yen, Y., Wang, Z., Shi, Y., Xu, F., Soeung, B., Sohail, M. T., ... & Juma, S. A. The predictors of the behavioral intention to the use of urban green spaces: The perspectives of young residents in Phnom Penh, Cambodia. Habitat International, 64, 98-108 (2017).
18. Ma Bojun, Kang Hongyan. A Review of Behavioral Consumption Theory. Journal of Social Sciences of Hunan Normal University,42(03):101-107 (2013).
19. Xie Pengxin, Feng Jiaojiao, Yang Fu. Mediation Strategy of Local Government Collective Disputes and Its Influencing Factors -- A Rooted Study Based on Beijing Case. Chinese Administrative Management (03):115-122 (2021).
20. Sohail, M. T., Delin, H., & Siddiq, A. Indus Basin Waters A Main Resource of Water in Pakistan: An Analytical Approach. *Current World Environment*, *9*(3), 670 (2014).
21. Muhammad, A. M., Zhonghua, T., Dawood, A. S., & Sohail, M. T. A Study to Investigate and Compare Groundwater Quality in Adjacent Areas of Landfill Sites in Lahore City. *Nature Environment & Pollution Technology*, *13*(1) (2014).
22. Chai, M., Deng, Y., & Sohail, M. T. Study on Synergistic Mechanism of Water Environment Governance in Dongting Lake Basin Based on Evolutionary Game. In *E3S Web of Conferences* (Vol. 257, p. 03075). EDP Sciences (2021).
23. Zhao, P., Yen, Y., Bailey, E., & Sohail, M. T. Analysis of urban drivable and walkable street networks of the ASEAN Smart Cities Network. *ISPRS International Journal of Geo-Information*, *8*(10), 459 (2019).

A Visual Study of Jiangxi Music Non-legacy Knowledge Based on VR Technology

Chuanli Liu[1] Chunbao Cai[1] Lizhi Wei[1]

[1]Nanchang Vocational University,Nanchang,Jiangxi,330004,China

Project No. GJJ206309 of Science and Technology Research Project of Jiangxi Education Department "Visualization of Music Non-legacy Knowledge Based on VR Technology

Abstract. As the development of society and the innovation of science and technology, VR technology, as a new product of the 21st century, brings us not only shock, but a fusion of innovation. As a hot spot in this era, VR technology is also called "spiritual technology ". This technology makes a virtual world by computer, makes the diversified information merge into a three-dimensional body information, through the equipment, can make people in a virtual real environment can not extricate themselves. The progress of science and technology has always surprised us ." VR technology "is a comprehensive innovation for all walks of life. How to make Jiangxi music non-legacy knowledge get attention again, VR technology should be a breakthrough.

1 Introduction

At this age of science and technology, the spread of culture needs to seek opportunities through new ways, VR technology as a hot spot in the 21st century, it brings a feast of science and technology. VR technology as a new medium of cultural protection, its mission is to make Chinese culture re-line the cultural stage. VR technology and Jiangxi kind of non-legacy knowledge culture fusion is not so much an innovation, but also a reform, it uses its most beautiful side to show in everybody's eyes, lets Jiangxi culture carry forward again.

2 VR Overview of Virtual Reality Technology

2.1 Development of VR technology

VR technology, commonly known as virtual reality, is a virtual world created by computer system, which is formed by computer graphics system and various realistic control devices. In this three-dimensional world, provide visual, auditory, tactile and a series of sensory simulation, so that you seem to live. VR as a new technology in the 21st century, the demand for VR technology is increasing in all walks of life. At present, VR technology has been widely used in industry, medical treatment, education and other fields, which has a good effect on people's perceptual experience. Nowadays, virtual reality technology has become more and more mature and developing. According to the project statistics of 805 VR technologies in the world, the results show that the proportion of entertainment, education and art is as high as 21.4, military aviation 12.7, medical 6.13, robot 6.21 and commerce 4.96. Below is a scale of market demand analysis for VR technologies across industries[1].

Table 1. Proportion of demand VR technology banks

Proportion of demand VR technology banks	
Entertainment, Education and Arts	21.4%
military aviation	12.4%
medical	6.13%
robot	6.21%
commerce	4.96%
others	48.9%

2.2 Basic functions of VR technology

As a hot spot in this era, VR technology is a multi-disciplinary technology, which includes a series of social disciplines, such as multimedia, network, sensor, simulation and three-dimensional display. Nowadays, virtual reality technology is recognized as the most influential technology in the world. Its development affects the progress of various industries. VR technology uses the principle of simulation, through computer information technology, multimedia technology, Internet technology and electronic technology to make devices form a perceptual world, can make people feel the same in this perceptual world. VR technology provides strong support for VR simulation system technology through lifelike and real-time interaction. In the whole virtual

*Corresponding author's e-mail:qgjxzbt@163.com

reality system process, the system must satisfy three characteristics, first, the user's immersion in the virtual world, in the virtual process to give the user an ability to distinguish between right and wrong, is the user in the whole body and mind into the virtual world, To watch and feel makes him feel that in this world, everything is true, as in his real life. Second, the controllability of the virtual world refers to the user in the virtual process, through the use of simulators in the virtual world sense things, when there is something in the hands of the feeling is, you can feel in the illusory world weight, Can easily capture the illusory world things, let you have a real feeling. Third, the user's autonomy to the virtual world, through a virtual multidimensional space, the user through their own perception ability, obtain multi-party information so that they have the initiative, seek more opportunities to solve problems[2].

3 VR technology for Jiangxi music non-legacy knowledge inheritance significance

3.1 VR technology will be an opportunity for the development of Jiangxi music non-legacy knowledge

Jiangxi music non-legacy knowledge is of great significance to the composition of non-legacy protection in our country. It has high value cultural information resources and protects non-legacy culture in the whole historical process. How to re-innovation is the most important part of Jiangxi music culture.

VR technology has played a key role in entertainment and art in recent years, making our lives more exciting. VR technology uses virtual reality to provide advanced protection measures for restoring Jiangxi music non-heritage cultural knowledge[3]. Cultural inheritance can not be separated from the innovation of science and technology. VR technology, as the innovation combination of Jiangxi music non-legacy knowledge, will lead to a new direction for the spread of culture. In the process of national deployment, the use of virtual reality technology to promote Jiangxi music non-legacy knowledge is the turning point of cultural innovation and development. VR technology and Jiangxi music integration, is an innovation, not only to make up for Jiangxi music non-heritage knowledge inheritance and dissemination process deficiencies, but also let people re-understand Jiangxi music non-heritage knowledge culture, This is of great significance to the whole Jiangxi music non-heritage knowledge.

3.2 VR technology is a valuable embodiment for the inheritance of Jiangxi music non-legacy knowledge

Value is the embodiment of a culture, Jiangxi music non-legacy knowledge in the whole history has a very important value. As a tool to break down barriers, VR technology breaks the constraints of time and space. In the illusory three-dimensional space, in the most intuitive way, Jiangxi music non-legacy knowledge can be re-emerged. When all non-legacy knowledge is recorded, the value of inheritance will become larger and find the inheritor for Jiangxi music non-legacy knowledge[4].

4 The Present Situation of Jiangxi Music Non-legacy Knowledge

4.1 The Rise of New Music and the Decline of Traditional Music

One is the development of the times, accompanied by collision, in this road of development, the two times collision process, one high and low is their sharp contrast. The rise of new music may be inevitable in the development of this world. Human beings always incorporate fresh elements and fresh affairs into music. With the passage of time, the way of music expression changes at any time, and the multi-element music becomes the current trend. And those traditional music is gradually forgotten and ignored, resulting in the inability to get fresh blood to pass on.

4.2 The lack of cultural interest among the new generation of young people inheritance difficult

In recent years, with the continuous changes in the economic market, Jiangxi Province, in the whole process, the economic system is constantly changing, class distribution has also changed significantly, people also have their own unique views on culture. There is a higher standard for the pursuit of organizational life. Therefore, Jiangxi non-culture has also suffered the impact of various cultures. As a new generation of young people, their interest in traditional intangible culture has gradually disappeared, especially the intangible cultural heritage of music has even appeared the situation of no one inheriting and losing culture. The attitude of the new generation of young people to ignore the traditional intangible cultural heritage is an important reason for the loss of most musical intangible cultural heritage[5].

4.3 A fault crisis

At present, the most severe situation of cultural heritage is that no inheritor, no one wants to learn, which leads to a lot of inheritance is about to appear fault crisis. These inheritance is more harsh for the selection of people, most of them are old inheritance, and aging inheritance is undoubtedly the most difficult in the current inheritance, the speed of inheritance will be fast, once there is no suitable inheritor, There will be a fault crisis in this culture.

4.4 Culture is spread in a single way

Jiangxi music non-legacy knowledge in each era has its own development characteristics, in the eyes of Jiangxi people also have a different position. But with the development of the times, Jiangxi music non-legacy knowledge culture also encountered unprecedented challenges, cultural knowledge and spirit slowly weakened, people's love for it can not return to the past. The main reasons are as follows: first, the art communication mode of Jiangxi music is relatively single, the lack of innovation to its own music, and the lack of culture that truly combines the regional characteristics of Jiangxi people. Second, blindly follow the trend, want to emulate other music culture but forget their own culture. Third, lack of new ways to open up the market, continue the old way to spread culture. It is these points that Jiangxi music non-legacy knowledge dissemination process has been in a dangerous state, need a new communication interface, break through this obstacle[6].

4.5 Inadequate local protection policies

The protection and inheritance of intangible cultural heritage can not be separated from the government's policy, and the process of cultural inheritance is accompanied by forgetting. If there are no good policies and measures, culture will slowly be forgotten and disappear in the vast crowd. Only by investing a lot of human, financial and material resources to protect the government can culture be preserved and continued.

5 Application of VR Technology to the Non-legacy Knowledge of Jiangxi Music

5.1 Into a completely new experience

For most people, listening to songs is simply listening to words and sounds, but the most valuable content ," artistic conception ", is often ignored. Jiangxi music, as a member of non-legacy knowledge, can only experience artistic conception by watching it in person, which is difficult to meet.

5.2 The transmission of culture will have a new development

Jiangxi music as a non-heritage knowledge protection object, its survival and development especially emphasizes the inheritance of tradition, Jiangxi music to the emotional description is more direct, can vividly show ups and downs, let the audience feel the same. But in the process of transmission, there is no opportunity to go to the stage of the world. And this opportunity is VR technology, for music performance, virtual reality technology[7], not only bring new ways of music dissemination, but also should be created new space. A multi-dimensional space of performance, VR technology simulation of real-action performance, the diversity of forms, can make art have a stable dissemination space.

5.3 Let music escape the bondage of time and space

In this changing world, things are shuffling, as if "change" has become the same theme of the world, including music. From traditional music to digital music, the form of music has changed a lot, people also have a new understanding of music. Throughout the various times of music it has its own form of expression, Jiangxi music non-legacy knowledge in its history of the river, it uses its unique rhythm and emotion, conquered many people. Now, it is in the sea of smoke can not reproduce the eyeball, there is a barrier, so that it can not cross. What opportunity can break this barrier[8], I think VR technology is its breakthrough, using VR technology to bring people into the scene of that era, relive the feelings of that era, let reality and the past reconnect, Let the new era young people understand Jiangxi local customs, music rhyme. Jiangxi music non-legacy knowledge can reflect its most fundamental value.

5.4 Review music and improve aesthetics

Music as a human emotional expression, its rhyme story at any time affect our emotions. A prosperous performance, every time can fluctuate our hearts. When technology becomes a performance, traditional music will be a review. As the medium of Jiangxi non-legacy music, VR virtual technology can make people appreciate the culture of traditional music and feel the cultural atmosphere of that era. As long as you feel the culture of the times, music will lead you to that world and sublimate your heart.

5.5 Cultural heritage needs to continue

In the rivers of the years, Jiangxi music has been destroyed many times, some have disappeared in the vast universe, some stick to the traditional heritage, imagine the future, Our descendants will never see the cultural heritage left by their ancestors again. What a sad thing it should be. If the use of VR technology to retain all non-heritage skills, Jiangxi music non-heritage knowledge will be permanently preserved, so that continued[9].

6 Strategies and Suggestions on the Protection of Jiangxi Cultural Non-legacy Knowledge Based on VR Technology

6.1 Based on VR technology to protect Jiangxi culture non-legacy knowledge implementation strategy

6.1.1 By using VR technology, the original preservation of all Jiangxi non-legacy knowledge. Jiangxi music non-posthumous knowledge of the greatest value lies in the historical, thousands of years of historical heritage of culture, with its own way of transmission, as long as not arbitrarily changed, this historical value can continue, VR technology as the media of Jiangxi music non-heritage knowledge, Its function lies in keeping this historic culture intact, recording all processes in all aspects, vividly and vividly displaying cultural heritage and breaking regional boundaries. No matter how the society develops, its inheritance is still there. As long as you enter the illusory world of preservation, you can understand the oldest memory skills.

6.1.2 By using VR technology, non-posthumous culture can be brought back to the stage. The most direct way to protect culture is to make culture go to the stage again, VR technology use scientific and technological performance to make more people attract attention, let more people watch Jiangxi music non-heritage cultural tradition. For example, the Beijing Museum's collections are digitally viewed and the Shaanxi Terracotta Army's Digital Museum, all of which are repackaged by VR technology to absorb sound in new forms. If Jiangxi music non-legacy knowledge seize this opportunity, can let their own culture, re-expression, re-attract people's attention, let the cultural road forward, re-entering the cultural stage.

6.1.3 Building an exhibition space of Jiangxi music non-legacy culture by using VR technology. With the increase of the national policy on the protection of cultural heritage, the design of cultural heritage exhibition hall has become an upsurge. If we use VR technology to create a cultural exhibition belonging to Jiangxi music class, it will be a new form of expression. Jiangxi music non-legacy knowledge itself has historical value and exploration, the integration of all non-legacy culture will become a cultural appreciation, it will have a strong attraction, more people to explore and learn[10].

6.1.4 By using VR technology, Jiangxi music will interact deeply with people. Virtual reality interaction refers to the computer simulation environment in the aspects of vision, hearing, touch, smell and taste, in which users interact to produce a new experience. Jiangxi music non-heritage cultural scene made by VR technology, multi-form performance can drive people's emotional emotion at any time, when the process of interaction with it, people will further understand Jiangxi music non-heritage knowledge. Interaction is a passive acceptance process that allows you to subconsciously accept it.

6.1.5 Building brand culture by using VR technology using VR technology. Building a Jiangxi music brand by VR technology is not only of great significance to the inheritance and dissemination of Jiangxi non-heritage knowledge culture, but also a continuation of the cultural spirit of Jiangxi music non-heritage knowledge. Jiangxi music non-legacy knowledge culture records the historical process, its tradition and knowledge, is the entire Jiangxi non-legacy culture has the important historical basis. At present, science and technology are the productive forces of social development, and the integration of VR and Jiangxi music non-legacy knowledge and culture has a new guide to the spread of Jiangxi culture. The brand culture will promote the development of the whole industry. The brand strength of Jiangxi music knowledge culture will give art a new way for all Jiangxi non-legacy culture, so that more people can accept her and understand it[11]. At the same time, it has a leading role in the inheritance of Jiangxi's non-heritage culture, which makes Jiangxi's non-heritage culture have new connotations and tastes, and makes the culture have a new cultural charm, so that Jiangxi's music non-heritage culture can speak and go out of China.

6.1.6 Dissemination of cultural exhibitions through VR technology. In the current era of Internet science and technology, whether it is cultural propaganda or transmission of information, most institutions will choose to carry out exhibitions, want to convey the information, so that their own publicity to achieve greater results. VR and Jiangxi music non-legacy knowledge culture fusion can be through the exhibition way to promote, so that more people know Jiangxi music non-legacy knowledge culture. The transmission of culture must open up a new way, the development of the exhibition is a new means for Jiangxi music knowledge culture, using the image of the exhibition to declare themselves, improve their own value has a very important role.

6.2 Based on VR technology to protect Jiangxi cultural non-legacy knowledge implementation recommendations

6.2.1 Stick to tradition and develop together. As the frontier of social development, VR technology has changed along this era, and Jiangxi music knowledge is fixed in that era. Although culture should keep pace with the times, the root of essence in the process of seeking development can not be changed. Once changed, tradition will not be that traditional culture. Jiangxi music non-legacy knowledge can use VR technology to board the ship, but must not rely solely on it ashore. Only to show their own light, is based on the fundamental. Steady progress and coordinated development is the survival of Jiangxi music non-legacy knowledge[12].

6.2.2 Follow the rules and strive for excellence. The law is moving and follows the natural development. The development of any thing must follow the thing itself, only conform to the law can be done, reverse law, will only be difficult, to destroy. VR technology should be applied to Jiangxi music knowledge, pay attention to culture itself, strive for excellence, find out their own law before moving forward.

7 Conclusion

With the rapid development of science and technology, the development of VR technology will become the most important way to inherit Jiangxi music culture. Based on the characteristics of virtual reality, VR technology has broken many barriers in protecting Jiangxi music non-legacy knowledge and solved the limitations of traditional culture communication. As the VR technology becomes more and more perfect in the future, virtual reality technology will perfectly fit Jiangxi music non-legacy knowledge, show its uniqueness, vitality and inheritance, and carry on the inheritance and dissemination in the multi-dimensional angle of time and space, so that Jiangxi music non-legacy culture can go to the world stage again.

References

1. Tian,Q.(2010)A Case Study on the Theory and Practice —— Protection of Music Non-legacy Projects[D].Anhui Literature and Art Publishing House.
2. Liu,J.Huang,J.M.Zhang,H.Y.(2019)The Development of Rural Folk Music in Shangrao[J].Arts Assessment.
3. Wang,Y.L.(2020)Application and Exploration of 5G Technology in Music Education[J].Media Forum3.
4. Fang,C.H.Yang,J.Ma,K.Zhang,X.Y.(2019)Study on the Communication Mode and Protection of Shifang Non-posthumous Culture in the New Media Context[J].New Media Research.
5. Ning,F.Hou,J.J.(2016)Study on the Communication Strategy of Intangible Cultural Heritage in the All Media Age[J].Journal of Nanchang Normal University.
6. Shi,S.M.(2020)The Innovation and Promotion of Non-posthumous Culture from the Perspective of the Whole Media[J].Chinese character culture.
7. Ji,Y.(2020)Research on the Application of Virtual Reality Technology in the Inheritance and Dissemination of Non-posthumous Culture[C]. University of Science and Technology Liaoning.
8. Zhang,Y.Y.(2020)Research on the Influence of Short Video in New Media Age on Non-posthumous Culture Communication[J]. Media Forum.
9. Fan,X.Y.(2020)An Inquiry into the Situation Teaching of Junior Middle School Music Course Based on VR Technology[C]. Inner Mongol Normal University.
10. Shi,L.(2015)Study on the Participation of Music Education in Intangible Cultural Heritage in Jiangxi University[J].Theatre House.
11. Ren,F.Jiang,X.T.(2017)Study on the Protection of Intangible Cultural Heritage in Jiangxi Province[J]. Journal of Qiqihar University (philosophy and Social Sciences Edition).
12. Liu,J.Huang,J.M.Zhang,H.Y.(2019)Study on the Protection of Intangible Cultural Heritage in Jiangxi Province[J].Theatre House.

Advance of modified ELISA and their application

Hejia Li[1,a,*,†], Yuhan Li [2,b,*,†] and Yihan Xia[3,c,*,†],

[1]Ulink College of Shanghai, China

[2]The Second High School attached to Beijing Normal University

[3]Biochemistry, Department of Chemistry, University of Washington, the United States

[†]These authors contributed equally

Abstract. ELISA (enzyme-linked immunosorbent assay) is a reliable and sensitive method of detection technique. It is commonly used in the daily research of scientific institutions and laboratories world widely. Although ELISA has an acceptable sensitivity and accuracy, when it comes to timely information about urgent detections, conventional ELISA lacks the ability to provide that. In this article, we introduce various types of modifications of ELISA and their applications in many fields, including the clinic diagnosis and the food detection, as well as the introduction of practical ELISA systems.

1 Introduction

As a common technique used for disease-associated biomarker detection, ELISA is widely adopted in the clinal field for diagnosis of disease and predicts treatment success, and ELISA is also used in non-clinical detection, like in food quality control, environmental and chemical field. In fact of the rapid spread of usage of ELISA in bio-analytical domain, numerous promising modifications of ELISA are introduced in terms of seven parameters of the bio-analytical platform that comes up. In addition to the modification of ELISA, the main goal of modifying ELISA is to apply ELISA for various fields. Clinical diagnosis and food safety inspection are the two main practical uses enumerated in this work. To extend the understanding of the modified ELISA, one specific example provided is the digital ELISA. This work describes and analyses the digital ELISA with its development status and some of its application.

2 ELISA modifications

2.1 sensitivity

Although ELISA is one of the most sensitive immunoassays available, considerable approaches appear to enhance the performance of ELISA. It is because of the limiting resources, and there is the urgency of improving the sensitivity of ELISA when a low concentration sample exists in order to obtain a wider usage of ELISA. Tons of modification was introduced to elevate its sensitivity, involving in addition of nanoparticles to antibodies in the basic indirect ELISA. For instance, the utilization of DNA to replace the enzyme used in standardized ELISA and the nucleic amplification performed by CRSPR/Cas 13a to achieve the amplification of the signal.

2.1.1 Nanoparticles-conjugated ELISA

The standard ELISA consists of the antigen from the sample, antibody binding to the antigen, so the elevation of the signal ability of antibody by attaching antibodies to nanoparticles helps to improve the sensitivity of ELISA. (Fig 1)

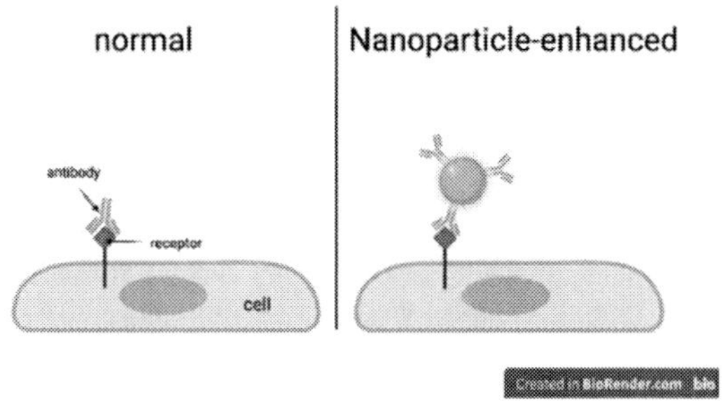

Fig 1. Normal vs. nanoparticle enhanced ELISAs.

In normal ELISA (left), primary antibodies flow freely and bind to their target receptor, which provide limited amount of binding sites for secondary antibodies. In nanoparticle-enhanced ELISA(right), nanoparticles are coated with primary antibodies to increase the number of binding sites for secondary antibodies.

Aiming to increase the sensitivity of ELISA, Margaret M. Billingsley et al. come up with the way of incorporating nanoparticles conjugated antibody into

* Corresponding author: [a]Yihanx2@uw.edu [b]libeccaz@163.com [c]liyuhan30@126.com

https://doi.org/10.1051/e3sconf/202129003020

standard ELISA. The primary antibody is decorated in nanoshells for detecting breast cancer cells. Because of the nanoparticles, more secondary antibodies could bind and show a higher resultant signal [1].

Additionally, the use of metal nanoparticles as an antibody carrier could also develop an effective way to improve its sensitivity. Adriano Ambrosi et al., utilizes gold nanoparticles (AuNPs) as carriers for anti-CA15-3-HRP (horseradish peroxidase). As a result, it shows an amplification in sensitivity as well as shorter assay time [2]. The novel probe is based on the colloidal gold nanoparticle (AuNP) that is modified with antigens in order to amplify the signal. Also, HRP synthesized used as signal amplification in many cases. Zhou Yu et al., have modified the probe with goat anti-mouse IgG and synthesize HRP. Based on the prob developed, the enhanced ELISA is developed. The newly synthesized probe bound with monoclonal antibody (McAb) which bound by coated BSA-ITCBE-Pb(II) on the plate and Pb(II) in the sample [3]. The HRP acts as a signal amplification to generate the optical signal to be measured. Due to the sphere shape of AuNP, more HRP conjugated comparing to IgG, result in a higher sensitivity. Both ways of similar application of AuNP indicates that metal nanoparticles, due to its large interaction area, enhance the basic ELISA that have flat and less-effective conjugation of signals, lead to an elevated sensitivity

2.1.2 Nucleic amplification ELISA

Another route to improve the sensitivity of ELISA is the use of DNA or RNA amplification method. Based on the knowledge that nucleic acid attached with optical signal it also plays a role in signaling response for detection. Qian Chen's group utilized the double-standard DNA containing T7 promoter sequence to replace the enzyme (horseradish peroxidase), T7 polymerase recognizes the promoter and proceeds transcription to produce single-standard RNA (ssRNA). After that, they designed CRISPR/Cas 13a system, which trans-cleaves ss-RNA. Short ssRNA reporter labeled with fluorophore and quencher group at both ends can be cleaved by trans-cleavage activity, leads to the amplification of signal [4].(Fig 2) This process is named CRISPR/ Cas 13a signal amplification linked immunosorbent assay (CLISA), which is confirmed at least 10^2-fold more sensitive than conventional ELISA in detecting inflammatory factors, tumor markers etc [4].

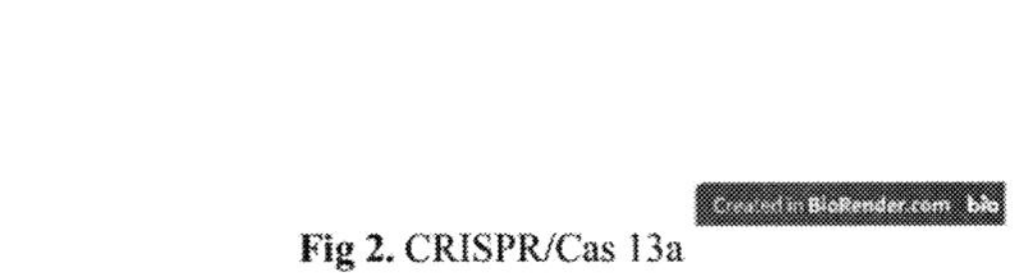

Fig 2. CRISPR/Cas 13a

T7 polymerase synthesizes ssRNA from dsDNA with NTP addition for modification in ssRNA. Then, CRISPR/Cas 13a cleaves ssRNA into short RNA fragment and add fluorophore and quencher group on both end as reporter signal.

Nucleic acid sequence-cased amplification (NASBA) is commonly used to continuously amplify nucleic acid bases on primer-dependent technology. Therefore, with the ability to amplify, nucleic acid sequence-based amplification with enzyme-linked immunosorbent assay (NASBA-ELISA) was developed for more sensitive ELISA. In order to amplify nucleic acid, a specific primer would be designed targeting RNA genome for amplification, producing amplicons. The amplicons were facilitated for probes then detected by a microplate reader. For instance, Gill P et al., performed this method for the detection of Mycobacterium tuberculosis. By introducing primer targeting on bacterial RNA, the amplification is made by isothermal digoxigenin (DIG)-labeling NASBA process. The accumulation of DIG-labeled RNA amplicons was hydrolyzed with a specific DNA probe that immobilized on the plate. The DNA-RNA hybrids were then detected [5]. The amplification by NASBA increases the sensitivity and specificity. Furthermore, for genotype II Grass carp reovirus (GCRV-II) detection in China, Zeng W et al., also comes up with NASBA-ELISA method for overcoming the problem of low sensitivity [6].

2.2 specificity

As one of the significant parameters for the clinical analytical platform, specificity contributes to validating the quality of ELISA outcomes. Different from sensitivity, which mainly concentrates on correctly identify disease, specificity measure the ability to correctly identify people without the disease, inhibiting non-specific binding. In order to improve the specificity of ELISA, numerous modifications including Ethanol vortex ELISA (EVELISA) and ZEDIII-based ELISA.

2.2.1 EVELISA

EVELISA is commonly used in the diagnosis of Johne's disease (JD) in dairy cattle. This test is objected to the detection of anti-MAP antibodies in milk. Due to the various environmental mycobacteria, more than 70% of the sample were identified as false positive [7]. Antibodies in the sample reacted in high activity with antigens of various environment mycobacteria, showing cross-reaction antibodies in the sample. The cross-reactive antibodies of MAP in the sample were highly inhibited by the use of Mycobacterium phlei antigen for antibody absorption [7]. Ashutosh Washwa et al., also choose EVELISA to test and optimize the concentration of antigens, milk and secondary antibody for segregating MAP-infection herds [8]. By using ethanol extract of MAP, EVELISA performs a basis for a higher sensitivity diagnostic test with a higher level of specificity than the current commercial ELISA test.

2.2.2 ZEDIII-based ELISA

In order to relieve the epidemic of Zika virus (ZIKV) and Dengue Virus (DENV), which are often co-epidemic, numerous amounts of exercise have been introduced by scientists. Their high protein-sequence homology of *flaviviruses*, ZIKV and DENV had highly cross-relative against each other's antigen(s). [9] Denis et al., observed a ZEDIII-based ELISA using Domain III of the ZIKV Envelope (ZEDIII) sequence. ZEDIII sequence is virus-specific, thus it is purified for specific detection of ZEDIII-induced IgG [9]. This kind of recombinant domain of ZEDIII-based ELISA distinguishes between past and current DENV and ZIKV infections. The challenge of serological detection of ZIKV, which has highly cross-reactivity against the related virus in detection. The ZEDIII-based ELISA assay showed 90% specificity and 92% sensitivity according to Jessica and her fellows.

2.3 Other parameters

2.3.1 Rapid

Aside from being sensitive and specific, the lengthy incubation of ELISA is also a key parameter for ELISA to improve for detecting disease. The faster result contributes a longer time for doctors to outline a treatment. Such techniques particularly serve as the early detection of fatal illness. Samira et al., manufactured a combination of microspheres and microfluidic disks to perform sandwich ELISA [10]. This mixing mechanism increases the molecular contact between target analyte molecule and microsphere, thus leads to more effective detection. Along with lower detection limits, the incubation time lengthy deduce from 2 hours to 5 minus. This huge success in enhancing the rapid of ELISA by modifying two mechanisms together contribute an earlier detection result, which gives a promising future in the early detection of fatal illness.

2.3.2 User-friendliness

Scientists and medical companies always longed for convenient equipment that releases the tasks of the scientist. Automation of equipment is one of the demands for ELISA in further development of modifying ELISA technique. Conventional ELISA is a series of mixing, washing, and incubation steps. The technology of micro-fluidic devices, the chance to release scientists' hands and promise the accuracy can be taken into consideration. The immobilization of biomarkers takes place in the channel, and due to the miniaturized size, the assay could perform in automation by the electrical or magnetic forces. Al-Faqheri et al., introduced a microfluidic compact disk (CD) platform that presents an easy to fabricate air and liquid check valves rather than a wide range of active and passive valving methods that is tested in other microfluidic platforms that is too complex to analyze [11].

2.3.3 Affordability

After the sensitivity and specificity improvement, in order to benefit the population with ELISA detection machine, the modification to provide a more affordable price is also one of the focuses. Enzyme-linked immunosorbent spot (ELISpot) contains the advantage of affordability. ELISpot will yield information about a number of cells secreting cytokine of interest. IFN-g-based assay is one of the most popular forms of ELISpot technique. IFN-g is secreted by activated T cells abundantly. The IFN-g assay could be widely used in vaccine discovery by the immune response quantity. High sensitivity, multiple target analytes, and affordable prices are some of the main advantages of this technique [12].

3 MODIFIED ELISA APPLICATION

Since the ELISA can be used to detect the presence of particular antigens or other substances, it is widely used in plenty of areas including clinical diagnosis and food safety inspection.

3.1. Clinical diagnosis

To diagnose the disease accurately, doctors need to analyze the elements collected from patients' bodies besides just asking their symptoms. In most cases, a specific antigen is a key to indicate the occurrence of one disease because of the specificity. However other substances such as proteins can also be used for detection. In this section, several clinical diagnosises using ELISA will be present.

3.1.1 Chikungunya(CHIK) diagnosis

Chikungunya (CHIK) fever is an acute arthropod-borne viral illness in which the pathogen is CHIK virus. The explosive epidemic of CHIK emerged major public health problem in India and Indian Ocean islands. It is a good example of ELISA application in clinical diagnosis. The original diagnosis was essentially based on virus isolation, IgM ELISA and reverse transcriptase polymerase chain reaction (RT-PCR). However, its high cost and long time for waiting the result led to an alternative way for early clinical diagnosis. The advanced sandwich ELISA employed rabbit and mouse anti-CHIK IgG antibodies as the capture and the detector antibodies, respectively.

The particular strain for the research was an India strain CHIKV ISW HYD06. To produce antibodies against CHIKV, purified CHIKV antigen mixed with equal volume of Freund's complete adjuvant (FCA) to immunize rabbit New Zealand strain). At the same time, a group of 3 to 4-week-old Swiss albino mice was also immunized subcutaneously with 100 μg of purified CHIK viral antigen and an equal volume of FCA. Thus, the production was subjected to titration by indirect microtiter plate ELISA using the CHIK viral antigen.

The purified CHIK viral protein was diluted to 300 ng/μL in 0.1 mol/L carbonate buffer, pH 9.6, and used for coating in 96-well microtiter plate (100 μL/well) and incubated at 4°Cfor overnight. The coated wells were washed once with PBS (1×PBS) and then blocked with 3% Bovine serum albumin (BSA) in 1×PBS followed by incubation at 37°C for 2 h. After washing with 1×PBS, 100 μL of rabbit and mouse HIS was titrated at an optimum dilution ranging from 1:1000 to 1:1 024 000, which were added in the wells and incubated at 37°C for 1 h. The plate was incubated with anti-rabbit and anti-mouse HRP conjugate at a dilution of 1:2000 made in 3% BSA followed by washing steps. The reaction was developed with 100 μL of tetramethylbenzi dine (TMB) substrate and kept at room temperature (RT) for 5 min. The peroxidase reaction was terminated with 100 L of 1 N H2SO4, and the absorbance was read at 490 nm in an ELISA reader.

To determine the optimum working concentration of both the rabbit and mice HIS, titration was performed through indirect double antibody sandwich microplate ELISA. The optimum titer of rabbit and mice HIS to be used as capture and detector antibodies in the antigen capture ELISA was found to be 1:32 000 and 1:128 000, respectively.

Combined with ELISA and real time PCR, the result showed an accordance of 96%. Moreover, it had high sensitivity (95%) and specificity (97%). This study was able to detect the presence of viral antigen as early as the second day of fever and, thus, it can be very useful for the early clinical diagnosis of CHIK with acute-phase patient serum and CSF samples [13].

3.1.2 Allergy diagnosis

Toshinori Ohashi and his team developed a practical and more sensitive micro-ELISA system for allergy diagnosis. Based on the previous method for allergy diagnosis, they modified the links of BSA-biotin-avidin so they were immobilized on the polystyrene surface by physisorption, ameliorated non-specific adsorption. Another improvement was the stopped-flow conditions of the enzymatic reaction to produce dye molecules.(Fig.3)

In their micro-ELISA system including fluid and optical components, the samples and reagent are stocked in tubes on an automated rotating stage. In the next automated stage, a specific volume of the sample and reagents would be sucked. The samples and reagents were then driven to a bead-packed area at a determined flow rate by the syringe pump. The length of the tube set shorter to 20 mm to prevent broadening of the dilute sample to decrease the sensitivity. Micro beads (45 mm) were introduced to the micro channel and a precise 5 mm-long bead-packed area was formed by introducing more beads and flushing the extra length with a syringe and valve. (Fig.3)

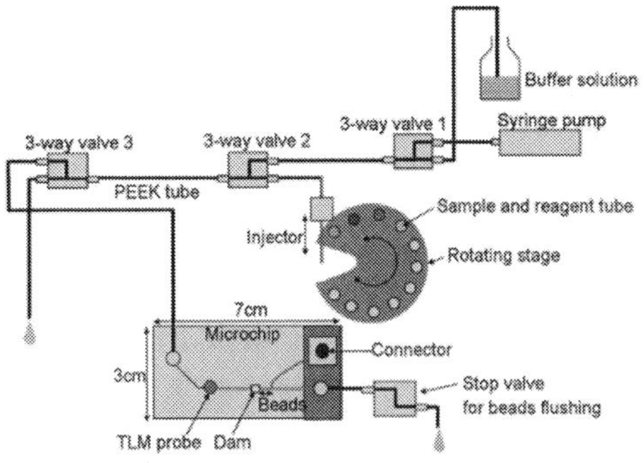

Fig.3 block diagram of micro-ELISA system.[14]

The measurement protocol for allergy was similar to the traditional way, including analyte aspiration, antibody aspiration, substrates aspiration and enzymatic reaction. The process from the introduction of the bead to bead removal only lasts 12 minutes. BSA buffer solution (0.2%) was used as the micro-ELISA system carrier solution. In their research, Cedar pollen allergen Cry j1, Cedar pollen allergen Cry j2, biotin-labeled cedar pollen allergen Cry j1 and biotin-labeled cedar pollen allergen Cry j2 were used for allergens.

This system was tested using human serum samples from 85 allergy patients. Signal intensity was measured by the micro-ELISA system, and the concentration was calculated by utilizing 4-parameter logistics for improving the accuracy. Total IgE and Cedar specific IgE were selected. For comparison, the same serum samples were measured by the CAP method (conventional method).

The advantage for this micro-ELISA system was more sensitive and faster. It was demonstrated using a smaller sample volume (1/10) and a shorter analysis time (1/20). All the processes were automatically applied with a micro-fluid system, extending the applicability to real clinical diagnostic situations [14].

3.2 Food safety inspection

In the last few decades, the rapid development of modified ELISA has become a widely used technique in food detection because of its specificity and sensitivity. Both monoclonal and polyclonal antibodies can be used in ELISA for food analysis. Polyclonal antibody is able to recognize a mixture of different epitopes of the antigens, more tolerance to small changes in the nature of antigen while monoclonal antibody has defined biological activity, consistent specificity and their production are not limited [16]. The most used immunoassays kits for food component identification are ELISA and lateral flow tests. The use of ELISA technique on food identification and detection would be illustrated by examples of application in the milk and meat industry [15].

3.2.1 Application in milk products

In the milk industry, health issues need to be considered from the production of the raw milk from cows to the storage of the final products. Bovine respiratory syncytial virus (BRSV) is one of the most important causes of respiratory disease in beef and dairy cattle. Fever, coughing, respiratory distress, and a marked drop in milk production are the main symptoms. M. Elvander, S. Edwards, IS. Näslund, N. Linde evaluated the NVI ELISA detection for BRSV in serum and milk. The results showed that there was a good correlation between the ability to detect antibodies in serum and the ability to detect them in milk. There was a clear distinction between antibody levels in diseased and healthy herds. The NVI ELISA is a rapid and reliable test for detecting antibodies to BRSV in milk, bulk milk, and serum samples [17].

Aflatoxins (AFs) are extremely toxic substances produced by certain species of Aspergillus, especially Aspergillus flavus, A. parasiticus, and A. nomius that contaminate plants and its products. AFM1 forms in the liver and it is secreted into milk in the mammary gland of dairy cows. AFM1 intake, even at low concentrations, causes a significant risk to human health, especially to children who are the major consumers of milk. A China research group evaluated the occurrence of milk and milk products contamination with AFM1 in Northeast of China by ELISA using monoclonal antibodies. Of 135 milk samples tested, 55 (41%) samples contained AFM1 at levels that ranged from 0.32–0.50 ng/ ml, 24 (18%) samples contained 0.16–0.32 ng/ml, and 18 (13%) samples contained 0–0.16 ng/ml; in 38 (28%) samples AFM1 was not detected. Therefore, precautions were necessary for minimizing the AFM1 contamination in milk and milk products in Chin [18].

The aminoglycoside antibiotics are used for the treatment of Gram-negative and Gram-positive infections. However, the presence of neomycin in animal-origin food is potentially hazardous to human health. For consumer protection, it is important to monitor neomycin residual levels in biological matrixes. The advanced ELISA was developed for the fast screening of neomycin in milk samples with high sensitivity, specificity and minimised matrix interference. The optimised ic ELISA could be finished in 55 minutes [19].

3.2.2 Application in meat

Two main purposes using ELISA in the meat industry is to identify the animal species so that to prevent the adulteration and detect the pathogen in meat.

For example, Martin, Chan, and Chiu (1998) carried out successfully a quantitative evaluation of pork adulteration in raw ground beef by ELISA [20]. It can be also applied for non-meat proteins, such as soybean proteins which is the potential health threat for Allergy sufferers. Macedo-Silva, Shimokomaki, Vaz, Yamamoto, and Tenuta-Filho discovered soy proteins in hamburgers containing beef, chicken and swine meat. Consequently, they obtained polyclonal antibodies against these proteins and quantified soy protein in these meat products by ELISA [21].

Other contaminants such as viruses, bacteria and antibacterial residues contributed to the concerns about food safety. A broad spectrum antimicrobial agent Chloramphenicol (CAP) was used against many Gram negative and Gram positive bacteria. Some undesirable effects in humans had been indicated so many countries have banned its use in food animals. H. Tajik, S.M. Razavi-Rouhani, M.R. Pajouhi, R. Mahmoudi, A. Haghnazari investigated CAP residue in 160 broiler chickens carcasses collected from three provinces of Iran by Four Plate Test (FTP), ELISA and HPLC. The ELISA assay showed that out of 28 positive samples in FPT, 22 liver, 21 kidney and 14 muscle samples were positive for CAP. ELISA analyses demonstrated that the minimum and maximum levels of 0.54 and 155.2 ng/g were detected in the kidney and liver, respectively. The data showed that despite the prohibition of CAP application in food animals including poultry, the CAP residue was detectable [22].

The results for detection can also contribute to some regulatory agencies and help pesticide residue analysis. Some companies (Neogen Corporation, Zeu Inmunotec, ELISA Systems and Tepnel, among others) commercialize immunochemical diagnostic kits for the rapid detection and quantification in food [15].

4 Practical ELISA system

4.1 Digital ELISA

4.1.1 Introduction

The sensitivity of protein assays still lags behind nucleic acid assays, such as polymerase chain reaction (PCR). Isolation and detection of single protein molecules offer a promising method for measuring very low concentrations of proteins

4.1.2 Research status

Single Molecule Array (SIMOA) is aside from known as digital ELISA. It uses a femtoliter-sized array of reaction chambers to segregate and ascertain distinct individual enzyme molecules. The point is to use the ability to capture and detect individual enzymes to detect different protein molecules labelled by enzymes. In the first step of this single-molecule immunoassay, the sandwich antibody complex is formed on a 2.7 μm diameter microsphere. The binding complex is labelled with an enzyme, just like a traditional bead-based ELISA. When testing samples containing very low concentrations of protein, the ratio of protein molecules to beads is small and usually less than 1:1. For example, if 3000 molecules in 0.1 mL of protein were captured and tagged on 200,000 beads, then 1.5% of the beads would contain protein molecules and 98.5% of the beads would have no protein molecules. These low numbers of enzyme markers cannot be detected using standard detection techniques because each enzyme produces a fluorophore that diffuses over a large number of measurements, typically 0.1-1 mL, requiring thousands of enzyme markers to produce a fluorescence signal higher than the background.

SIMOA needs to be able to detect enzyme markers at very low concentrations by limiting the fluorophore produced by a single enzyme to a very small volume, resulting in high local concentrations of fluorescent product molecules. To achieve this position in the immunoassay, in the second step, the beads are placed into a series of spike-sized holes. Approximately 50,000 Wells were drilled for the project, measuring 2 mm wide, 4.5 μm in diameter, and 3.25 μm in depth. The loaded array is then sealed on a rubber gasket in the presence of luciferase substrate droplets, isolating each bead in the ascending chamber. In a 50 fl reaction chamber, beads containing a single enzyme labelled immune complex produced a localized high concentration of the fluorescent product. By using standard microscope optics to acquire time-lapse fluorescence images of the array, beads associated with individual enzyme molecules can be distinguished from those associated with non-enzyme molecules. Protein concentration in the test sample is determined by calculating the number of pores containing beads and fluorescent products relative to the total number of pores containing beads. With SIMOA, the concentration is determined digitally rather than using the total analog signal; therefore, the use of SIMOA to detect a single immune complex is called a digital ELISA.

4.1.3 Comparison between Digital ELISA and early Elisa methods

The sensitivity of SIMOA assay was 0.62 pg/mL and ELISA assay was 78.0 pg/mL. So, we found that SIMOA was more sensitive than ordinary ELISA tests. The highly sensitive SIMOA technique merits further study in a larger cohort of patients.

The advantage of ELISA is that it only requires a simple procedure, and under the simple procedure, it is also provided with high specificity, high sensitivity and high efficiency. Because ELISA is just simply based on antigen-antibody reaction. Also, about the safety, because ELISA does not require the radioactive substances and numerous organic solvent, it is safe and eco-friendly. Last, the reagents are in a low cost, so one of the strengths of ELISA, compared with other assays, is cheap. [26]

The digital ELISA contains the same advantages and even better. Even though the whole technology seems more complicated, digital ELISA only needs to use a relatively simple set of equations based on biomolecular interactions for each step of the assay. Moreover, the limit of detection has been improved. The limit of detection of PSA is very low. Many of the current detection methods are not effective, but the digital ELISA exhibits a 3-log improvement in sensitivity compared to the most sensitive commercially available PSA assays. In addition, the accuracy of Digital ELISA can be ensured while the sensitivity is improved. Digital ELISA is a method that literally counts the number of molecules during detection, so provides an excellent way to compare the number of molecules predicted by theory and those detected in experiments. The theoretical results of Digital ELISA are basically the same as the actual results. And the automation of digital ELISA is also an advantage. [24]

4.1.4 Applications

The recent increase in global incidence and severity of Clostridium difficile infection (CDI) is a major concern. CDI currently consumes significant resources in diagnosis, treatment and infection control, and accurate diagnosis remains the cornerstone of effective management.

Toxin detection plays an important role in CDI diagnosis, but the current methods of toxin detection are not perfect. Traditional qualitative toxicity immunoassay is widely used, but its LODS is high and its sensitivity is poor. In view of these limitations, the field is poised to develop a simple toxin detection test that combines high analytical sensitivity with clinical specificity for toxin detection.

The ability to isolate and detect individual protein molecules greatly increases the sensitivity to detect the concentration of FG protein per milliliter, typically up to 1000 times more sensitive than traditional ELISA. Digital ELISA has demonstrated high accuracy and reproducibility in detecting low levels of proteins in highly complex matrices such as cell culture supernatants, serum, plasma and cerebrospinal fluid (CSF). The SIMOA method is suitable for the determination of protein in feces. The higher clinical specificity of SIMOA offers the potential to improve diagnostic and management approaches.

Also, Digital ELISA (SIMOA), within higher sensitivity, it is able to detect prostate cancer, Crohn's disease, and neurological disorders. It is able to measure neurological conditions at their very earliest stages. [25]

4.1.5 Frontiers

The antibody is difficult to produce and needs a huge amount of labor which causes it to be expensive. Also, the antibody is fragile that it has to be saved in a special temperature to be more specific, cold storage, and has a storage life because the antibody is a protein. Last but not least, insufficient blocking of immobilized antigen results in false consequences. There is a high possibility of false positive or negative. Also, for digital ELISA, because is a single molecule detection, it needs to eliminate the molecular interactions that come with numerous cells, so a lot of noise has to be eliminated.

4.2 Plasmonic ELISA

4.2.1 Research status

In recent years, plasmonic immunoassay has been developed by combining traditional enzyme-linked immunosorbent assay (ELISA) with nanomaterials. Compared to other immunoassays, plasmonic immunoassays are highly sensitive and allow open-eye reading without the need for the use of complex instruments. Metal nanoparticles, such as gold and silver nanomaterials, are widely used in plasmonic Nano-sensors due to their excellent local surface plasmonic resonance (LSPR) properties. In plasmonic immunoassay, an enzyme-labeled antibody catalyzes the formation of a product from its substrate, which triggers the aggregation or shape change of the nanomaterial. [23]

4.2.2 Applications

Plasmonic immunoassay has been widely used in clinical diagnosis, environmental pollution monitoring and food safety detection. [23]

4.2.3 Frontier

Robust, ultra-fast and highly stable AuNPs aggregation in colorimetric analysis remain a challenge due to the dynamic aggregation process of AuNPs. In addition, in order to quantitatively test molecular targets, relatively complex and large instruments are needed to measure the spectra of nanomaterials as well as the plasmonic immunoassay. The inconveniences associated with these instruments make them unsuitable for point-of-care analysis. Therefore, there is an urgent need to develop a portable, inexpensive and easy-to-use plasmonic immunoassay reader for immediate detection (POCT) diagnosis.

5 Conclusion

Modified ELISA is undoubtedly the best choice for detection small particles like antigens in clinical detection, bacteria in some food safety tests. In our research, the improvement of ELISA with the development of ELISA application in different regions indicate the advanced modified ELISA techniques have the characteristics of sensitive, accurate and practical.

ELISA is already a proven biosensor, but it is less accurate and cost effective than other methods such as PCR for a particular test. Therefore, the use of ELISA in conjunction with other technologies can be more effective, as in the case of Digital ELISA and Plasmonic ELISA, which have higher precision.

References

1. M.M. Billingsley, R.S. Riley, E.S. Day, Antibody-nanoparticle conjugates to enhance the sensitivity of ELISA-based detection methods, PLOS ONE 12(5) (2017) e0177592.
2. A. Ambrosi, F. Airò, A. Merkoçi, Enhanced Gold Nanoparticle Based ELISA for a Breast Cancer Biomarker, Analytical Chemistry 82(3) (2010) 1151-1156.
3. Y. Zhou, An enhanced ELISA based on modified colloidal gold nanoparticles for the detection of Pb(II), Biosensors & bioelectronics v. 26(no. 8) (2011) pp. 3700-3704-2011 v.26 no.8.
4. Q. Chen, T. Tian, E. Xiong, P. Wang, X. Zhou, CRISPR/Cas13a Signal Amplification Linked Immunosorbent Assay for Femtomolar Protein Detection, Analytical Chemistry 92(1) (2020) 573-577.
5. P. Gill, R. Ramezani, M.V.P. Amiri, A. Ghaemi, T. Hashempour, N. Eshraghi, M. Ghalami, H.A. Tehrani, Enzyme-linked immunosorbent assay of nucleic acid sequence-based amplification for molecular detection of M. tuberculosis, Biochem Biophys Res Commun 347(4) (2006) 1151-1157.
6. W. Zeng, W. Yao, Y. Wang, Y. Li, S.M. Bermann, Y. Ren, C. Shi, X. Song, Q. Huang, S. Zheng, Q. Wang, Molecular detection of genotype II grass carp reovirus based on nucleic acid sequence-based amplification combined with enzyme-linked immunosorbent assay (NASBA-ELISA), J Virol Methods 243 (2017) 92-97.
7. M.C. Scott, J.P. Bannantine, Y. Kaneko, A.J. Branscum, R.H. Whitlock, Y. Mori, C.A. Speer, S. Eda, Absorbed EVELISA: a diagnostic test with improved specificity for Johne's disease in cattle, Foodborne Pathog Dis 7(11) (2010) 1291-1296.
8. J.P.B. Ashutosh Wadhwa, Todd M. Byrem, Traci L. Stein, Arnold M. Saxton, C.A. Speer, and Shigetoshi Eda, Optimization of Serum EVELISA for Milk Testing of Johne's Disease, Foodborne Pathog Dis 9(8) (2012) 749-754.
9. J. Denis, S. Attoumani, P. Gravier, B. Tenebray, A. Garnier, S. Briolant, F. de Laval, V. Chastres, G. Grard, I. Leparc-Goffart, B. Coutard, C. Badaut, High specificity and sensitivity of Zika EDIII-based ELISA diagnosis highlighted by a large human reference panel, PLOS Neglected Tropical Diseases 13(9) (2019) e0007747.
10. S. Hosseini, M.M. Aeinehvand, S.M. Uddin, A. Benzina, H.A. Rothan, R. Yusof, L.H. Koole, M.J. Madou, I. Djordjevic, F. Ibrahim, Microsphere integrated microfluidic disk: synergy of two techniques for rapid and ultrasensitive dengue detection, Sci Rep 5 (2015) 16485.
11. W. Al-Faqheri, F. Ibrahim, T.H.G. Thio, M.M. Aeinehvand, H. Arof, M. Madou, Development of novel passive check valves for the microfluidic CD platform, Sensors and Actuators A: Physical 222 (2015) 245-254.
12. A.C. Karlsson, J.N. Martin, S.R. Younger, B.M. Bredt, L. Epling, R. Ronquillo, A. Varma, S.G. Deeks, J.M. McCune, D.F. Nixon, Comparison of the ELISPOT and cytokine flow cytometry assays for the enumeration of antigen-specific T cells, J Immunol Methods 283(1-2) (2003) 141-153.
13. Asensio L, González I, García T, et al. Determination of food authenticity by enzyme-linked immunosorbent assay (ELISA)[J]. Food control, 2008, 19(1): 1-8.
14. Ohashi T, Mawatari K, Sato K, et al. A micro-ELISA system for the rapid and sensitive measurement of total and specific immunoglobulin E and clinical application to allergy diagnosis[J]. Lab on a Chip, 2009, 9(7): 991-995.
15. Asensio L, González I, García T, et al. Determination of food authenticity by enzyme-linked

immunosorbent assay (ELISA)[J]. Food control, 2008, 19(1): 1-8.

16. Goding, J. W. (1996). Monoclonal antibodies: Principles and practice. London: Academic Press Inc. Limited.
17. Elvander M, Edwards S, Näslund I S, et al. Evaluation and application of an indirect ELISA for the detection of antibodies to bovine respiratory syncytial virus in milk, bulk milk, and serum[J]. Journal of Veterinary Diagnostic Investigation, 1995, 7(2): 177-182.
18. Pei S C, Zhang Y Y, Eremin S A, et al. Detection of aflatoxin M1 in milk products from China by ELISA using monoclonal antibodies[J]. Food control, 2009, 20(12): 1080-1085.
19. Xu N, Qu C, Ma W, et al. Development and application of one-step ELISA for the detection of neomycin in milk[J]. Food and agricultural immunology, 2011, 22(3): 259-269.
20. Berger, R. G., Mageau, R. P., Schwab, B., & Johnston, R. W. (1988). Detection of poultry and pork in cooked and canned meat foods by enzyme-linked immunosorbent assays. Journal of the Association of Official Analytical Chemists, 71, 406–409.
21. Macedo-Silva, A., Shimokomaki, M., Vaz, A. J., Yamamoto, Y. Y., & Tenuta-Filho, A. (2001). Textured soy protein quantification in commercial hamburger. Journal of Food Composition and Analysis, 14, 469–478.
22. Tajik H, Malekinejad H, Razavi-Rouhani S M, et al. Chloramphenicol residues in chicken liver, kidney and muscle: A comparison among the antibacterial residues monitoring methods of Four Plate Test, ELISA and HPLC[J]. Food and chemical toxicology, 2010, 48(8-9): 2464-2468.
23. Quanli Yang, Ruitian Cai, Wei Xiao, Zengfeng Wu, Xia Liu, Yan Xu, Miaomiao Xu, Hui Zhong, Guodong Sun, Qihui Liu, Qiangqiang Fu & Junjian Xiang. (December 05). Plasmonic ELISA for Sensitive Detection of Disease Biomarkers with a Smart Phone-Based Reader. *Spring Open*, 7.
24. Linan Song, D. Richard Lachno, David Hanlon, Adam Shepro, Andreas Jeromin, Dipika Gemani, Jayne A. Talbot, Margaret M. Racke, Jeffrey L. Dage & Robert A. Dean. (2016, December 15). A digital enzyme-linked immunosorbent assay for ultrasensitive measurement of amyloid-β 1–42 peptide in human plasma with utility for studies of Alzheimer's disease therapeutics. *Spring Open*, 7.
25. Linan Song, Mingwei Zhao, David C. Duffy, Joshua Hansen, Kelsey Shields, Manida Wungjiranirun, Xinhua Chen, Hua Xu, Daniel A. Leffler, Susan P. Sambol, Dale N. Gerding, Ciarán P. Kelly, Nira R. Pollock. (2015, July 15). Development and Validation of Digital Enzyme-Linked Immunosorbent Assays for Ultrasensitive Detection and Quantification of Clostridium difficile Toxins in Stool. *Journal of clinical microbiology*, 7.
26. Elena Pérez-Ruiz. (2018, July 26). Digital ELISA for the quantification of attomolar concentrations of Alzheimer's disease biomarker protein Tau in biological samples. *ScienceDirect*, 7.

Study on cyclic crosslinked polyphosphazene microspheres and its adsorption behavior for uranium (VI)

Meixue Xu[1,*],Kaifa Liao[1], Mouwu Liu[1],Yi Tan[1],Yanfei Wang[1]

[1]School of chemistry and chemical engineering, University of South China , 421000, PRC

Abstract. Poly (cyclotriphosphazene-co-4,4 '- diaminodiphenylsulfone) (PZD) microspheres were synthesi zed by precipitation polymerization of Hexachlorocyclotriphosphazene (HCCP) and polyfunctional organic monomers. The products were characterized by FTIR, SEM-EDS, XPS and bet. The adsorption behavior of PZD microspheres for uranium (VI) in aqueous solution and the influence of adsorption behavior were disc ussed. The results show that the PZD microspheres have a certain adsorption capacity for uranium (VI) in a queous solution. When pH = 3.5, adsorption time is 6h, solid-liquid ratio is 2.0g · L-1 and initial concentra tion of uranium (VI) is 30mg · L-1, the adsorption rate of uranium reaches the maximum.

1 INTRODUCTION

At present, many methods have been developed to remo ve uranium (VI) from aqueous solution. However, many traditional methods have their own limitations. For exam ple, although membrane separation technology can effect ively remove uranium, the cost of this method is too high [1]. Chemical precipitation method is easy to operate an d low cost, but it is easy to cause secondary pollution [2]. Compared with the above two methods, adsorption meth od is widely favored because of its easy implementation, low cost and few secondary pollutants. Hai Nguyen tran [3] synthesized a new adsorbent for uranium (VI) adsorp tion by impregnating br PADAP onto multi walled carbo n nanotubes (MWCNTs). Xiangke Wang [4] has synthes ized magnetic titanate nanosheets (Fe3O4 @ TNS) to ad sorb uranium (VI) in wastewater, and the maximum adso rption capacity is relatively large. Ramzi zarrougui [5] us ed nonfluorinated ionic liquids (ILS) to extract uranium (VI) from radioactive wastewater. Polyphosphazenes are a kind of new polymer materials with unique P = n struc tural units and active p-cl groups. Their physical and che mical properties can be adjusted by nucleophilic substitu tion of functional groups. A poly (cyclotriphosphazene c o-4,4 '- diaminodiphenylsulfone) microsphere (PZD) was synthesized by precipitation polymerization of Hexachlo rocyclotriphosphazene (HCCP) and 4,4-diaminodiphenyl sulfone (DDS) under the protection of nitrogen and trieth ylamine (tea). The microspheres were used to adsorb ura nium (VI) from radioactive waste water.

2 Experimental methods

2.1 Experimental process

2.1.1 Preparation of PZD microspheres

65g HCCP and 4,4-diaminodiphenylsulfone (DDS) were dissolved in 200ml acetonitrile, and then the mixed solut ion was poured into a three port flask. After the solid wa s completely dissolved, nitrogen was used as protective g as. The three port flask was placed in an ultrasonic instru ment (200W, 40KHz) for ultrasonic reaction, and the rea ction temperature was set at 50 ℃. After 5 minutes of ult rasound, 3 ml triethylamine was added dropwise into the three port flask. After ultrasonic treatment for 6 hours, th e reaction system was transferred to an oil bath, heated i n an oil bath at 50 ℃ and stirred by magnetic force for 6 hours.

At the end of the reaction, the reaction product was centr ifuged in a centrifuge (800r / min) for 5 minutes, the sup ernatant was removed, and the white precipitate was was hed several times with anhydrous ethanol and deionized water. Finally, the washed solid material was dried in a v acuum drying oven at 60 ℃ for 48 hours to obtain the sy nthetic product.

2.1.2 Preparation of uranium (VI) standard solution

1 g · L-1 of uranium (VI) reserve solution was prepare d, and then the reserve solution was gradually diluted to 10 mg · L-1, 20 mg · L-1, 30 mg · L-1, 40 mg · L-1, 50 mg · L-1 and 60 mg · L-1. Measure 1ml of the a bove solution respectively, add 1ml of arsenazo Tris, titr ate it with chloroacetic acid sodium acetate buffer solutio n with pH = 2.5 to the scale line, take deionized water as blank sample, measure it in parallel for three times on th e visible spectrophotometer, and obtain the correspondin g absorbance at each concentration, which is linearly fitt ed by origin [6]. The linear fitting equation obtained is th

* Corresponding author: xmx_2902088301@163.com

e standard curve of uranium (VI) solution.

2.1.3 Effect of pH value on adsorption of uranium (Ⅵ) by PZD microspheres

10 ml of 30 mg • L-1 uranium (VI) solution was accura tely measured. The pH of the solution was adjusted to 2, 3.0, 4.0, 5.0 and 6.0 with HNO3 and NaOH respectively. 20 mg PZD microspheres were added respectively, and t he adsorption time was 6 h under the condition of oscilla tion rate of 120 rpm and temperature of 25 ℃. After the oscillation, the reaction solution was separated by ultrafil tration membrane. 1 ml of filtrate was transferred into a 10 ml volumetric flask, and then 1 ml of arsenazo Ⅲ (0. 05 wt%) chromogenic agent was transferred into the vol umetric flask with a pipette gun, Then add chloroacetic a cid sodium acetate buffer solution with pH of 2.5 dropwi se to the scale line, and determine the absorbance of the f iltrate by uv-2100 spectrophotometer, so as to analyze th e effect of pH on the adsorption of uranium (VI) by cycl ocrosslinked polyphosphazene phloretin nanospheres. Ca lculation formula of adsorption capacity (q) and removal rate [6] are as follows:

$$Q(mg\ g^{-1}) = \frac{(C_o - C_e)V}{m} \quad (1)$$

$$R(\%) = \frac{(C_o - C_e)}{C_o} \quad (2)$$

Where, q is the adsorption capacity (mg • g-1), C0 and CE represent the concentration of uranium (VI) ion befor e and after adsorption (mg • L-1), V is the volume of ur anium (VI) solution (L), and M is the amount of adsorbe nt (g).

2.1.4 The effect of solid-liquid ratio on adsorption of uranium (Ⅵ) by PZD microspheres

Accurately measure 10ml of 30mg • L-1 uranium (VI) solution and adjust the pH of the solution to 3 5. 5mg, 10 mg, 15mg, 20mg, 25mg and 30mg PZD microspheres w ere added respectively. Under the condition of oscillation rate of 120rpm and temperature of 25 ℃, the adsorption was vibrated for 6h. After the oscillation, the solid-liquid separation of the reaction solution was carried out by ult rafiltration membrane. 1ml of the filtrate was transferred into a 10ml volumetric flask, and then 1ml of arsenazo Ⅲ (0.05wt%) chromogenic agent was transferred into th e volumetric flask by pipette gun. Then the pH 2.5 chlor oacetic acid sodium acetate buffer solution was added dr opwise to the scale line with a rubber tipped burette, and the absorbance of the filtrate was determined by uv-2100 spectrophotometer, so as to analyze the effect of solid-li quid ratio on the adsorption of uranium (Ⅵ) by cyclicall y crosslinked polyphosphazene phloretin nanospheres.

2.1.5 Effect of initial concentration of uranium (VI) o n adsorption of uranium (VI) by PZD microspheres

The concentration of 10 mg • L-1, 20 mg • L-1, 30m g • L-1, 40 mg • L-1 and 50 mg • L-1 of uranium (V I) solution was accurately measured. The pH of the soluti on was adjusted to 3.5 with HNO3 and NaOH respective ly. 20 mg PZD microspheres were added. The shaking ra te was 120 RPM and the adsorption time was 6 h at 25℃. After shaking, the reaction solution was separated b y ultrafiltration membrane. 1 ml of filtrate was put into a 10 ml volumetric flask, Then, 1 ml arsenazo Ⅲ (0.05w t%) chromogenic agent was transferred by pipette gun an d added into volumetric flask. Then, chloroacetic acid so dium acetate buffer solution with pH of 2.5 was added dr opwise with rubber tip dropper to make volume to the sc ale line. The absorbance of the filtrate was determined b y uv-2100 spectrophotometer to analyze the effect of init ial concentration on the adsorption of uranium (VI) by P ZD.

3 Results and discussion

3.1 Characterization of PZD microspheres

3.1.1 Infrared characterization of PZD microspheres

The absorption peaks at 3361cm - 1 and 3215cm - 1 are t he stretching vibration peaks of - NH2. This indicates tha t the polymerized PZD microspheres contain a certain a mount of - NH2 group. Obviously, the two strong absorp tion peaks of 1593 cm-1 and 1500 cm-1 are the absorptio n peaks of C = C in aromatics. 1300 cm-1 and 1103 cm- 1 are characteristic absorption peaks of o = s = O stretchi ng vibration of sulfodiphenol group. The characteristic a bsorption peak of P = n stretching vibration is at 1143cm -1. 880cm-1 is the stretching vibration peak of p-n and 9 31cm-1 is the stretching vibration peak of ar-n-p. due to t he formation of ar-p-n, it can be preliminarily inferred th at HCCP and DDS have polymerized to form a new subs tance.

3.1.2 XPS characterization of PZD microspheres

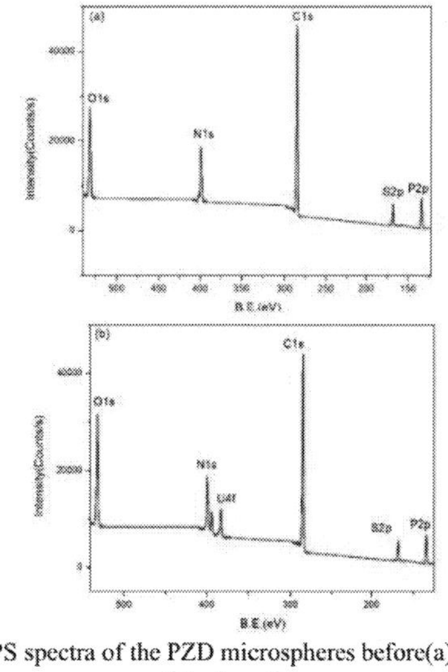

Fig.1XPS spectra of the PZD microspheres before(a) and after (b) adsorption

In order to determine the composition of the elements in PZD microspheres, XPS was used. The XPS spectrum of PZD microspheres is shown in Fig.1 It can be seen from figure a that the surface of PZD microspheres is compos ed of C, N, O, P and S elements, in which Cl, N, O, P an d S elements account for 63.93%, 11.33%, 11.52%, 6.9 4% and 6.27% respectively. It can be seen from figure B that the peak of uranium (VI) appears obviously after ads orption, and the proportion of uranium (VI) is 0.33%, wh ich indicates that PZD microspheres have a certain adsor ption effect on uranium (VI) in aqueous solution.

3.1.3 Bet characterization of PZD microspheres

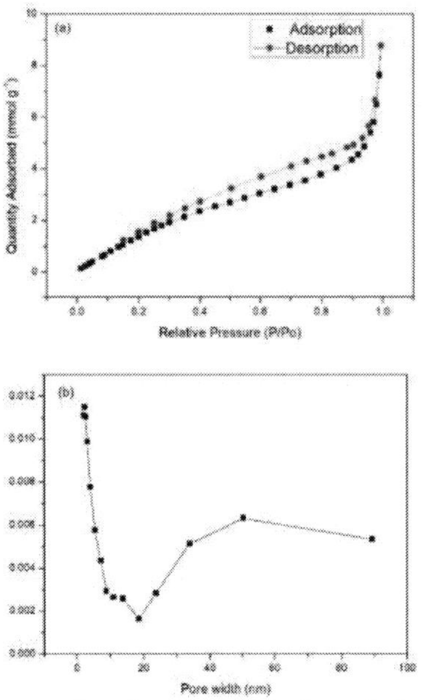

Fig.2 Nitrogen adsorption-desorption isotherm obtained at 77K for the PZD（b）The pore size distribution curve of the PZD

In order to determine the specific surface area and pore s ize of PZD microspheres, the PZD microspheres were ch aracterized by bet. As shown in Figure 2.4, the BET spec ific surface area of PZD microspheres is 10.42m2/g, the t otal pore volume is 0.012cm3/g, and the pore size is 5.21 nm, which is a mesoporous material. It can be seen that t he specific surface area of the adsorption material is goo d, which can provide a certain adsorption site for the ads orption of uranium (VI) [7].

3.1.4 Characterization of PZD microspheres by SE M-EDS

Accurately measure 10ml of 30mg • L-1 uranium (VI) solution and adjust the pH of the solution to 3.5. 5mg, 10.It can be seen from figure 6 that the material presents a r egular and uniform ball shape. However, after adsorption, the material becomes irregular and even flat. It can be se en from Fig.7 that there is one more uranium element on the PZD microspheres after adsorption than before, whic h further indicates that the PZD has successfully adsorbe d uranium (VI) on its surface.

3.2 Effect of pH value on adsorption of uranium (VI) by PZD microspheres

When the pH is less than 3.5, with the increase of pH val ue, the adsorption capacity of PZD microspheres for ura nium (VI) increases gradually. This may be due to the ex istence of a large number of positively charged H + in th e uranium solution under strong acid conditions, and thes e positively charged hydrogen ions may be easier to com bine with the adsorption sites in PZD microspheres, resul ting in protonation of PZD microspheres, The protonated microspheres will repel the positively charged uranium (VI) ions, which is not conducive to the adsorption of ur anium (VI) in solution by PZD microspheres [8]. Howev er, when the pH of the solution is higher than 4, the adso rption capacity of PZD microspheres for uranium present s a downward trend. This is because when the pH value e xceeds 4, uranium (VI) is easy to hydrolyze, even to for m negatively charged (UO2) 3 (OH) 7 -, and the appeara nce of this ion will produce repulsive force with the PZD surface, resulting in the decrease of the adsorption stren gth of the adsorbent. In conclusion, when the pH value o f the solution is 3.5, the adsorption behavior of PZD micr ospheres is the best, and the optimal adsorption capacity is about 12.79 mg • g-1.

3.3 Effect of solid-liquid ratio on adsorption of urani um (VI) by PZD microspheres

In order to study the effect of solid-liquid ratio on the ad sorption of uranium (Ⅵ) by PZD microspheres, experiments were carried out under the conditions of initial conc entration of uranium (Ⅵ) 30 mg • L-1, shaking adsorpti on for 6 h, pH = 3.5, volume of uranium (Ⅵ) solution 10 ml and adsorbent dosage (5-30 mg). It can be seen that with the increase of solid-liquid ratio (M / V), the remov al rate of uranium (VI) by PZD microspheres gradually i ncreases, and when the solid-liquid ratio is greater than 2. 0 g • L-1, it gradually tends to be flat, and the removal rate is 86%. This may be due to the existence of a certain "shielding effect [9]" on the surface of the adsorbent aft er reaching a certain amount of adsorption, which has a c ertain repulsion to the remaining uranium (VI) ions. At t his time, due to the decrease of the concentration of uran ium (VI) ions, there is not enough driving force to offset the repulsion force. Therefore, it can be inferred that the best adsorption condition is when the solid-liquid ratio is 2.0 g • L-1.

3.4 Effect of initial concentration of uranium (VI) on adsorption of uranium (VI)

When the initial concentration of uranium (VI) ion is in t he range of 10-30 mg • L-1, the adsorption capacity of PZD microspheres for uranium (VI) increases with the in crease of uranium (VI) ion concentration. When the initi al concentration of uranium (VI) ion in the solution is gr eater than 30 mg • L-1, the increase of adsorption capa city gradually slows down. It may be that when the conc entration of uranium (VI) ion is high, the number of effe ctive collisions between uranium (VI) ion and adsorbent is increased, and the driving force of mass transfer in soli d-liquid phase is improved, so more uranium (VI) ions ar e combined with PZD adsorbent. When the initial conce ntration of U (VI) in the solution is in the range of 10 ~ 5 0 mg • L-1, the removal rate of U (VI) by PZD microsp heres decreases slowly. It may be that at a higher concen tration of uranium (VI) ion, after uranium (VI) ion reach es a certain concentration, due to the increased competiti on of uranium (VI) ion in the adsorption sites, the adsorp tion sites on the adsorbent that can combine with metal i ons reach saturation [10].

4 Conclusion

In this experiment, PZD microspheres, a new adsorbent, were synthesized by precipitation polymerization. The P ZD microspheres were used to adsorb uranium (Ⅵ) in w astewater. The influence of pH value, solid-liquid ratio, i nitial solution concentration on adsorption of uranium (Ⅵ) by PZD microspheres was studied by single factor method. The results show that PZD microspheres have c ertain adsorption capacity for uranium (Ⅵ) in aqueous s olution. The removal rate was over 85%, and the adsorpt ion capacity was close to 12.19mg · g-1. The best conditi ons for adsorption of uranium (Ⅵ) in the waste liquid by PZD microspheres are: pH = 3.5, adsorption time is 6h, solid-liquid ratio is 2.0g · L-1, and initial concentration o f uranium (Ⅵ) is 30mg · L-1.

Acknowledgment

Thank you for the financial support of the innovation trai ning program of Nanhua University, project number: 200 xcx158

References

1. Bhalara P D, Punetha D, Balasubramanian K. A revi ew of potential remediation techniques for uranium
2. (VI) ion retrieval from contaminated aqueous enviro nment [J]. Journal of Environmental Chemical Engi neering, 2014, 2(3): 1621-1634.
3. Hu e Ming, Shao Er Yan, Zhao Jing et al. Research progress of uranium containing wastewater treatmen t technology [J]. Hunan Journal of Ecological Scienc es, 2016, 3 (1): 42-48
4. Khamirchi R, Hosseini-Bandegharaei A, Alahabadi A, et al. Adsorption property of Br-PADAP-impregn ated multiwall carbon nanotubes towards uranium a nd its performance in the selective separation and de termination of uranium in different environmental sa mples [J]. Ecotoxicology and Environment Safety, 2 017, 150:136-143.
5. Yin L, Song S, Wang X, et al. Rationally designed c ore-shell and yolk-shell magnetic titanate nanosheet s for efficient U(VI) adsorption performance[J]. Env ironmental Pollution, 2018, 238:725-738.
6. Zarrougui R, Mdimagh R, Raouafi N. Highly efficie nt extraction and selective separation of uranium (VI) from transition metals using new class of undiluted ionic liquids based on H-phosphonate anions [J]. Jou rnal of Hazardous Materials, 2017, 342:464.
7. Tan Juan. Adsorption properties of poly (cyclotripho sphazene-co-4,4 '- dihydroxydiphenylsulfone) for th orium (IV) and uranium (VI) [D]. 2018
8. Lu Z, Sun Y, Song L, et al. Dihydroxy bezladely der ivatives functionalized mesoporous silica SBA-15 fo r the sorption of U(VI) [J]. Journal of Radioanalytic al & Nuclear Chemistry, 2016, 310(1): 1-13.
9. Huynh J, Palacio R, Safizadeh F, et al. Adsorption o f Uranium over NH2-Functionalized Ordered Silica in Aqueous Solutions [J]. Acs Appl Mater Interfaces, 2018,22:35-38.
10. GIRI, A. K, PATEL, et al. Artificial neural network (ANN) approach for modelling of arsenic (III) bioso rption from aqueous solution by living cells of Bacil lus cereus biomass [J]. Chemical Engineering Journ al, 2011, 178(24): 15-25.
11. Fu J, Chen Z, Wu X, et al. Hollow poly(cyclotriphos phazene-co-phloroglucinol) microspheres: An effect ive and selective adsorbent for the removal of cation ic dyes from aqueous solution [J]. Chemical Enginee ring Journal, 2015, 281:42-52.

Big data method and its application in innovation education research

Yankui Song[1], Chuijiao Jie[2] and Zhijin Xu[2*]

[1]Department of College English, Capital Normal University, Beijing, 100048, China

[2]Department of Management, Capital Normal University, Beijing, 100048, China

Abstract. Big data technology is a new stage of information development. In recent years, it has been widely used in many fields, especially in social science research. This paper analyzes the development status and significance of the combination of big data technology and social science research, on the basis of summarizing and combing the concept of big data and its important role. Taking the application of big data method in the research of innovation education as an example, this paper makes a series of visualization analysis with Citespace software on the related literature with the theme of "big data and innovation education" collected by CNKI, such as annual analysis, literature source analysis, co-occurrence analysis of authors, organization analysis, keyword clustering analysis and keyword timing analysis. This paper also draws the corresponding knowledge mapping, clarifies its research status, hot spots and development trend, and provides scientific basis for the research of innovation education. Thus the paper believes that the research on big data and innovation education needs to strengthen interdisciplinary communication and cooperation, refine and deepen the research theme and content.

1 Introduction

With the rapid development of information technology, human beings have entered the "big data era" of data explosion. Big data technology is a new stage in the development of informatization and a "sharp tool" to boost economic development, improves social governance and people's lives. In recent years, big data technology and methods have attracted more and more attention in humanities and social sciences, especially in literature, history, economics and Marxist theory, which are in the ascendant and have achieved many important research results [1]. How to mine and apply massive data by big data technology has become one of the focuses of academia, especially in social science.

This paper analyzes the development status and significance of the combination of big data technology and social science research, on the basis of summarizing and combing the concept of big data and its important role. Taking the application of big data method in the research of innovation education as an example, this paper makes a series of visualization analysis with Citespace software on the related literature with the theme of "big data and innovation education" collected by CNKI, such as annual analysis, literature source analysis, co-occurrence analysis of authors, organization analysis, keyword clustering analysis and keyword sequence analysis.

2 Concept and function of big data technology

Big data is a major technological breakthrough in research. McKinsey, a well-known global consulting company, first proposes the arrival of the era of big data, in its report *Big data: the next frontier of innovation, competition and productivity* (2011), defines big data as a data set whose size exceeds the ability of conventional database tools to acquire, store, manage and analyze. At the same time, McKinsey emphasizes that the data set does not necessarily exceed a specific TB level in order to be regarded as big data. IDC (2011) defines big data technology as a new generation of architecture and technology that can economically extract value from high frequency, large capacity, different structures and types of data. As for its concept, people have put forward "3V", "4V" and "5V" successively, and its understanding is consistent [2]. Big data technology methods mainly include data clustering, data mining, spatial analysis, time series analysis, visualization technology and others [3].

Big data technology and methods play an important role in "explaining the past, predicting the future, and making present decisions" [4]. Compared with traditional data analysis methods, big data technology analysis has the following two characteristics or advantages: on the one hand, big data analysis transcends the limitations of sampling analysis and small amount of data. Big data analysis no longer adopts sampling analysis method, but

*Corresponding author's e-mail: xzjccc@sina.com

directly analyzes all samples with higher accuracy of analysis results instead; on the other hand, big data analysis can dynamically track data [5]. Social science problems often have the characteristics of real-time and evolution, so it is difficult for traditional data analysis methods to track and feedback in real time. Big data and its technology system provide technical feasibility for the complex adaptive characteristics of effective human social activities, so as to provide strong support for promoting social sciences to learn from natural science achievements and form a new model of social sciences research based on data driving.

Big data technology and methods have developed rapidly in recent years. It is widely used in many fields and industries, such as commerce, manufacturing, transportation, communication, medical treatment, education, finance, especially in social science research. Wal-Mart was the first to analyze sales by big data technology, and found the matching goods suitable for trading together, creating a classic business case of "beer and diapers" [6].

3 Situation and significance of the combination of big data and social science research

The combination of data science and humanities & social sciences, a new research is in the ascendant and has broad prospects. In 2015, the Fifth Plenary Session of the 18th CPC Central Committee adopted the "proposal of the CPC Central Committee on formulating the 13th five-year plan for national economic and social development", which first proposed the implementation of the national big data strategy and the network power strategy. In the same year, the State Council issued "the action outline for promoting the development of big data". The theme of the second collective learning conference of the Political Bureau of the CPC Central Committee after the 19th CPC National Congress is the national big data strategy. Since then, there has been a rapid emergence of application findings on the combination of big data technology, philosophy and social sciences in domestic academic circles. The year 2015 is also called "the first year of big data application" by many domestic industry insiders [7].

In recent years, big data technology and methods are more and more widely used in the fields of literature, history, anthropology, sociology and Marxist theoretical research in the multi-dimensional and multi-level aspects of research methods, research problems and research subjects. They have made many extraordinary scientific research achievements, formed increasingly important academic influence, and are profoundly changing people's production mode and consumption mode and the way of thinking. Some scholars believe that big data is the complementary knowledge of empiricism and rationalism, which not only develops the understanding of causality, but also emphasizes the tolerance of uncertain results, and deepens the cognition of uncertainty [8]. Based on the literature of science & technology and philosophy in the past 60 years, some scholars have introduced the theory and method of scientific knowledge mapping to conduct scientometrics research on the development of philosophy of science & technology. Through flexible use of various tools to conduct in-depth mining of literature data, they have described the macro, meso and micro prospects of the development of science & technology and philosophy in China [9]. In fact, the past form of "spoon feeding" education has far been unable to meet the humanized demand of people in the big data era nowadays. We must vigorously develop the "internet+ education" mode, make full use of the internet to understand teachers and students mutually and timely, and promote the personalized development of continuing education [10].

4 Application of big data method in innovation education research

In recent years, how to apply big data and give full play to the role of big data analysis methods authentically in the field of innovation education has aroused wide spread concern in academic circles. A series of quantitative and visual knowledge mapping analysis by Citespace software, such as annual literature analysis, literature source analysis, organization analysis, keyword clustering analysis and keyword sequence analysis, are carried out to grasp the research hotspots and development trends, so as to provide scientific reference for the research of big data and innovation education.

4.1 Data sources and research tools

On the CNKI database page, click "advanced search" and take "(subject = innovative education or title = innovative education or v_subject = Chinese and English extension (innovative education) or title = Chinese and English extension (innovative education)) and (subject = big data or title = big data or v_subject = Chinese English extension (big data) or title = Chinese English extension (big data))" as the search condition. After eliminating useless literatures, 378 eligible literatures are selected and exported in Refworks format recognized by Citespace, named as database 1. On the CNKI database page, click "advanced search" and take "(core journal = y or CSSCI journal = y) and ((subject = innovative education or title = innovative education) or v_subject = Chinese and English extension (innovative education) or title = Chinese and English extension (innovative education)) and (subject = big data or title = big data or v_subject = Chinese and English extension (big data) or title = Chinese and English extension (big data))" as the search condition. After eliminating useless literatures, 45 eligible literatures are selected and exported in Refworks format recognized by Citespace, as in database 2. The retrieved deadline is Mar. 12, 2021.

This paper selects the software of Citespace as research tools [11]. It is suitable for the quantitative and visual analysis of titles, authors, organizations, keywords, word frequency, co-occurrence and citation information of the literature, and then draws the corresponding knowledge mapping.

4.2 Visual analysis of research status

This paper analyzes the annual distribution of research literature by Citespace, the status quo of institutional cooperation and the research hotspots and development trends of big data and innovative education by institution and keyword co-occurrence function, and then draws a visual mapping.

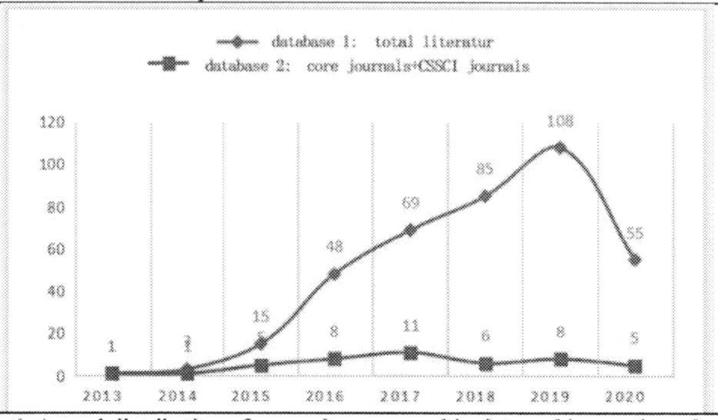

Figure 1. Annual distribution of research papers on big data and innovative education.

4.2.1 Annual analysis of research literature. According to the year of publication, the literatures in database 1 and database 2 are statistically analyzed, and the publishing trend chart is drawn in Figure 1. We can see that the total literature with the theme of "big data and innovative education" and the literature published in core journals or CSSCI journals began in 2013, only accounting for 1.18%, which indicates that the research quality is not high and it is in the exploratory period. The general trend has been increasing since 2013, a slow growth period from 2013 to 2015 and a fast growing period from 2015 to 2019. In early 2020, the outbreak of Covid-19 in China caused great concern in academic circles, which distracted some scholars' research efforts. Consequently, the number of documents on the theme of "big data and innovative education" decreased, but on the whole the research on that theme is still on the rise.

4.2.2 Analysis of literature sources. Based on the statistical analysis of literature sources in database 1, the top ten literature sources are ranked from high to low in Figure 2. The top ten literature sources are mainly in journals which focused on innovative education by big data. Among them there are 31 articles, the largest number, published in Age of Think Tanks. These journals provide academic exchange platform for the innovative education research by big data, and reflect its latest research trends.

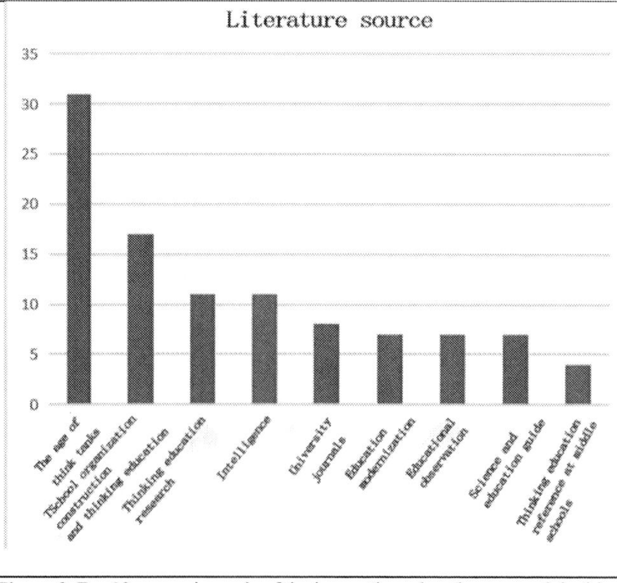

Figure 2. Top 10 source journals of the innovative education research by big data.

Table 1. Institutions of the innovative education research by big data.

Serial number	Name of organization	Number of papers	Serial number	Name of organization	Number of papers
1	Central China Normal University	12	6	Lanzhou University	4
2	University of Electronic Science & Technology	8	7	University of Defense Science and Technology	4
3	Northeast Normal University	4	8	China West Normal University	4
4	Nanjing Institute of Political Science	4	9	Renmin University of China	3
5	Southwest University	4	10	Harbin Normal University	3

4.2.3 Institutional Analysis. According to the distribution of institutions in CNKI group browsing, the top ten institutions are selected, and its paper number of big data and innovative education are drawn in Table 1. Among them Central China Normal University obtains the most scientific research achievements, published 12 articles in big data and innovative education.

4.2.4 Research hotspots and trend analysis. In the CX project of Citespace, Time Slicing is selected from Jan. 2013 to Dec. 2020, Years Per Slice for 1 year, Node Types for Keyword, and other parameters remain in the default state. Click Go! and K (Label clusters with indexing terms) to draw the keyword topic clustering of big data and innovative education research literature in Figure 3.

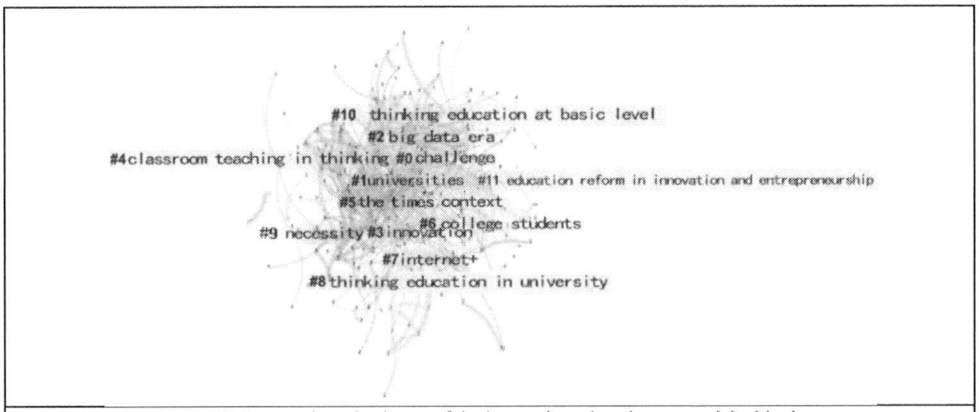

Figure 3. Keywords topic cluster of the innovative education research by big data.

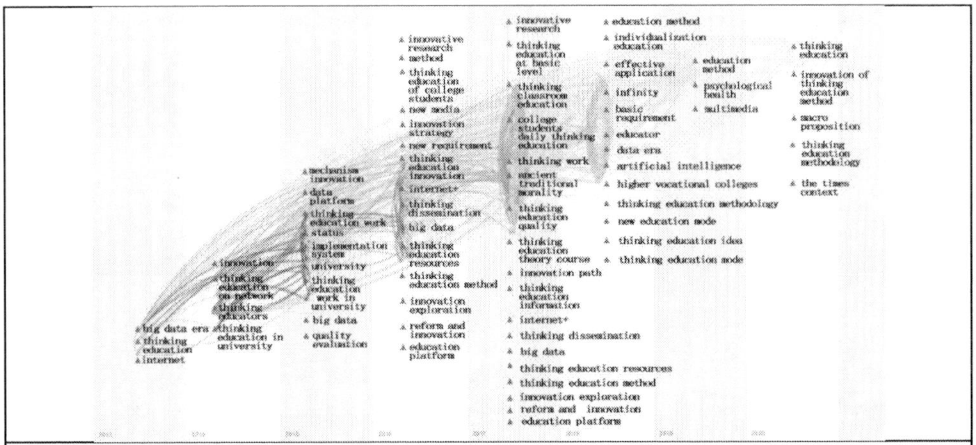

Figure 4. Keywords sequence of the innovative education research by big data.

According to Figure 3, 12 major clusters of "big challenge", "big university", "big data era", "innovation", "classroom teaching", "era context", "university student", "internet+" and "innovation & entrepreneurship education reform" are found in big data and innovative education research.

Each cluster represents a research topic of big data and innovative education. Combined with the top five tag words and the high-frequency literature in each topic, this paper describes the research topics of big data and innovative education. On the basis of Citespace keyword topic clustering, select Timezone, calculate Node Labels according to By Degree, and select 5 as Threshold. Nodes are adjusted appropriately to ensure the clear display, and the keywords sequence of big data and innovative education is drawn in Figure 4.

According to the sequence chart in Figure 4, big data and innovative education research can be divided into three stages, namely, the budding period in 2013, the development period from 2014 to 2018 and the deepening period from 2019 to 2020. Big data and innovative education research has gone through three stages of combination: "can or cannot - where - how to". At present, most research of big data and innovative education focuses on multiple topics, and discusses how to carry out innovative education from multiple themes and a macro perspective with big data method.

5 Conclusions

The paper analyzes 378 documents on the theme of big data and innovative education collected by CNKI, including a series of visual knowledge mapping of annual analysis, source analysis, organization analysis, keyword clustering analysis and keyword sequence analysis. It provides quantitative, visual and dynamic scientific support to clarify the research status of big data and innovative education in this field. Due to the short time of big data research, there is still a large space for the development of big data and innovative education research. There is less research on how to put the proposed method or model into practice, such as how to predict students' innovative behavior by analyzing students' past behavior habits and other specific issues through big data. We should identify the combination of big data and innovative education, strengthen interdisciplinary exchanges and refine the research content in the future.

In short, big data technology and methods will be more and more widely used in more fields, especially in social science research with the development of economy, technology and society. The research on big data and innovation education needs to strengthen interdisciplinary communication and cooperation, refine and deepen the research theme and content, in order to show broader and stronger application value in the process of "explaining the past, predicting the future, and making present decisions" in big data technology development. This interdisciplinary research has prominent characteristics, obvious advantages, great significance and broad development prospects.

Acknowledgement

This paper was sponsored by 2020's construction project of university think tanks and social service ability in Capital Normal University "Big data mining and analysis of Marxist symbiosis thought" (SK2020ZK03).

References

1. Zhang, S., Jiao, D. (2019) 2nd symposium of big data, philosophy and social sciences: let data open the door of good governance and link to a better life. Guangming Daily, June 18.
2. Yin, M.Y. (2019) Big data application and analysis. Technology and Innovation, 04: 31–39.
3. Manyika, J., Chui, M., Brown, B., et al. (2018) Big data: the next frontier for innovation, competition and productivity. http://www.mckinsey.com/insights/business_technology/bid data the next frontier for innovation.
4. Yu, L. (2013) Big data and big data economics. China Soft Science, 07: 23–31.
5. Luo, X.Y., Huang, X.R. (2017) Big data method in social science research. Journal of Systems Science, 04: 57–45.
6. Gao, Y. (2008) Beer and diaper. Tsinghua University Press, Beijing.
7. Du, X.Y. (2018) How can big data become a national strategy? Financial Economy, 11: 17–25.
8. Pan, P., Zheng, H., Lan, L.S. (2015) Essential characteristics of big data system and its philosophical reflection. Journal of Systems Science, 03: 13–21.
9. Zhang, D.J. (2017) Big data analysis of the development trend of Chinese philosophy of science and technology. University of Science and Technology of China, Beijing.
10. Yu, H.Y. (2019) Application of management information system in continuing education. E-commerce, 04: 51–59.
11. Zhan, P.W., Xie, X.J. (2018) Overview of big data system and key technologies and tools. Network Security Technology and Application, 08: 29–37.

Study on the Spatial Distribution of Gansu Longshen Cult in Qing Dynasty Based on ArcGIS

Zhang keze*[1]

[1]Collaborative Innovation Center for Security and Development of Western Frontier China, Sichuan University, Chengdu, Sichuan ,610065, China

Abstract. Influenced by the limitations of the times, the agricultural belief and sacrificial system are developed and widely used in the traditional society of China, and the Longshen cult is a typical one. Taking the Gansu Longshen cult in Qing Dynasty as a case, this paper considered that the Longshen temple throughout Gansu are mainly the result of the correspondence between the rain function of Longshen and the characteristics of climate drought. Based on the statistics of the number of Longshen temple, the spatial distribution map of Gansu Longshen cult in Qing Dynasty was drawn by ArcGIS software.

1.Introduction

In the agriculture-oriented society of ancient China, the Longshen were widely distributed as part of the agricultural sacrificial system. The folk society of Gansu in the Qing Dynasty generally worshiped the Longshen, which was closely related to the role of the Longshen in the traditional Chinese folk beliefs and the natural environment of Gansu. On the one hand, the Longshen temple throughout Gansu were due to the correspondence between the rainfall function of the Longshen and the characteristics of climate drought. On the other hand, there were differences due to different agricultural forms.

2. The background of Gansu Longshen Cult in Qing Dynasty

2.1. Rainfall: the main function of Longshen

Water resources play an important role in the development of agriculture. In a society of "building a country by agriculture", water resources are the source of the survival and development of agriculture, farmers and even state power. Ancient ancestors generally believed that precipitation was mainly dominated by gods. On the basis of this understanding, the relationship between precipitation and Longshen was linked for a long time, and a set of quite complete rain seeking ceremony was gradually developed. With the popularity of the Buddhist concept of Longwang, the cult of Longshen has been widely spread in civil society, and gradually from the folk to the temple, completed the institutionalized leap. As the god in charge of Rain Water's apologies in the ancient Chinese agricultural belief system, Rainfall is its most basic function, and it is also the fundamental reason why it can be widely worshipped in an agricultural society. It is nothing more than a kind of expectation for the rain to be worshipped by the people.

In the ancient Chinese ancestors' perception, the dragon, as a deity specializing in rainfall, was the main object chosen by civil society when seeking rain in every drought. The association between dragon and rain was established as early as the pre-Qin period. Although the association between dragon and rain was established very early, it was only during the Han Dynasty that a set of strict rituals for seeking rain took shape. For example, Dong Zhongshu made a detailed account of the ritual of seeking rain in the four seasons of spring, summer, autumn and winter when there was a drought. With the improvement of rain rituals in civil society and the widespread spread of the concept of the Longwang in Buddhism, the institutionalization of the Longshen was further promoted, and gradually evolved from a spontaneous act in civil society to a formal system at the national level. Since the Tang Dynasty, the belief in Longshen has gradually entered the formal system of the state and become an important part of the state ritual system. Through the continuous recognition and strengthening of the official system since the Tang and Song dynasties, the Longshen were gradually promoted and developed and gradually extended to the whole country.

2.2. Drought: climatic characteristics of Gansu Province

Gansu is deeply inland in northwest China, and its topography and geomorphology are complex. It is the rendezvous area of Qinghai-Xizang Plateau, Loess Plateau and Mongolian Plateau. Most of the areas in Gansu are arid areas with dry climate and less rain, the annual precipitation is $42 \leq 760$ mm [1], and the spatial distribution of precipitation is also very uneven, showing the law of

*Corresponding author's e-mail: 357243039@qq.com

 https://doi.org/10.1051/e3sconf/202129003023

decreasing from southeast to northwest on the whole. The annual precipitation in the southeast of Longnan is 600 ≤ 900 mm, and the climate is relatively humid. The annual precipitation in southeastern Qinghai, Hedong of Gansu and southern Ningxia is 300 ≤ 600 mm, mostly semi-arid and semi-humid climate, while the annual precipitation in Hexi Corridor area and northern Ningxia is 15 ≤ 300 mm, which is arid and extremely arid climate area[2], and drought and less rain is the background of Gansu climate. This also determines that drought has become a key factor perplexing the development of agriculture in Gansu Province. In the traditional society where the level of productive forces is backward, the nearly extinct caused by drought has occurred from time to time. The number of droughts occurred in the history of Gansu is very frequent. If the number of droughts occurred between the Western Han Dynasty and the Republic of China, the average number of droughts occurred about once in three years. If the number of droughts from the Sui Dynasty to the Republic of China is counted, the average drought occurs once in two years. It can be said that the overall characteristics of Gansu climate can be vividly explained by two droughts in three years [3]. At the same time, affected by geographical location and other regional factors, the drought suffered by different areas in Gansu will be different. In terms of the administrative scope of Gansu in the Qing dynasty, according to the different topography, climate and other areas under its jurisdiction can be divided into Longdong (Qingyang, Pingliang, Jingzhou, Guyuan and Huapingchuan), Longnan (Qinzhou and Jiezhou), Longzhong (Lanzhou and Gongchang), Hexi (Liangzhou, Ganzhou, Suzhou and Anxi), Haidong (Xining), Ningbei (Ningxia) different geographical units. According to the relevant statistics in the Qing dynasty archives, 290 droughts occurred in the Longdong region; 77 droughts occurred in the Longnan region; 408 droughts occurred in the Longzhong region; 154 droughts occurred in the Hexi region; 47 droughts occurred in the Haidong region; 107 droughts occurred in the Ningbei region [4].

In traditional societies with low productivity levels, the effective means to face frequent droughts were extremely limited, and the power of the gods to counteract them became one of the options. The high degree of correlation between the Longshen role in traditional Chinese agricultural rituals and Gansu's climate, which is characterized by dryness and low rainfall, created an innate condition for the widespread distribution of Longshen cult in Gansu, satisfying the spiritual need of the people to pray for a good harvest in a drought-stricken agricultural society, which explains the socio-cultural phenomenon of Longshen temple all over Gansu during the Qing Dynasty.

3. The Spatial Distribution of Gansu Longshen Cult in Qing Dynasty

As an encyclopedia documenting the political, economic and social conditions of a place, local chronicles have rich historical data value and are an important entrance to the study of regional social history[5]. In order to master the actual distribution of Gansu Longshen cult in Qing Dynasty as much as possible, this paper starts with *Gansu Tongzhi* during Qianlong period and *Gansu xin Tongzhi* during Guang Xu's years, supplemented by the local chronicles of Gansu counties, using the relevant contents of the chronicles, on the basis of mutual reference and supplement, outlines the complete schema of Gansu local social Longshen cult in Qing Dynasty. It should be explained that due to the regional differences, in the local historical records of Gansu Province, in addition to the Longwang, there are Qiushen, Quanshen, Chaonaqiu and so on, and even some personification of the Longshen. Because it has the function of praying for rain in a general sense, it is also regarded as the Longshen cult in a wide sense and included in the statistical scope in the process of investigating its distribution. In addition, the administrative division of Gansu in Qing Dynasty is slightly different from that at present, which roughly includes some regions in Ningxia and Qinghai today. In order to fully reflect the distribution of Gansu Longshen cult in Qing Dynasty, some areas belonging to the above provinces and autonomous regions will still be included in the statistical scope.

With the help of the relevant records in the local historical records, it can be roughly outlined that the Gansu Longshen cult in Qing Dynasty is a common cultural phenomenon. Temples worshiped to the Longshen are widely distributed in Jiezhou, where the climate is relatively humid, and in Anxi, where the climate is relatively dry. The distribution characteristics of Longshen cult in Gansu Province are not only wide in scope, but also different among regions (see Figure 1). The narrow geographical characteristics of Gansu Province have created great differences in the natural environment between the eastern and western regions, on the basis of which different agricultural areas have been formed. Generally speaking, the Yellow River has been used as the boundary to form Hedong area (Lanzhou, Gongchang, Pingliang, Qingyang, Jiezhou, Qinzhou, Jingzhou, Guyuan, Huapingchuan) and Hexi (Liangzhou, Ganzhou, Ningxia, Xining, Anxi, Suzhou). Hedong is a rain-fed agricultural area, Hexi is an irrigated agricultural area. The emergence of Longshen cult is mainly related to agriculture [6], so when investigating the distribution difference between Longshen cult regions, we can choose different agricultural areas to observe. From the statistical results of the belief distribution of Gansu Longshen cult in Qing Dynasty, it can be seen that the number of Longshen temples in the rain-fed agricultural area in the east of the river is slightly lower than that in the irrigated agricultural area in Hexi.

https://doi.org/10.1051/e3sconf/202129003023

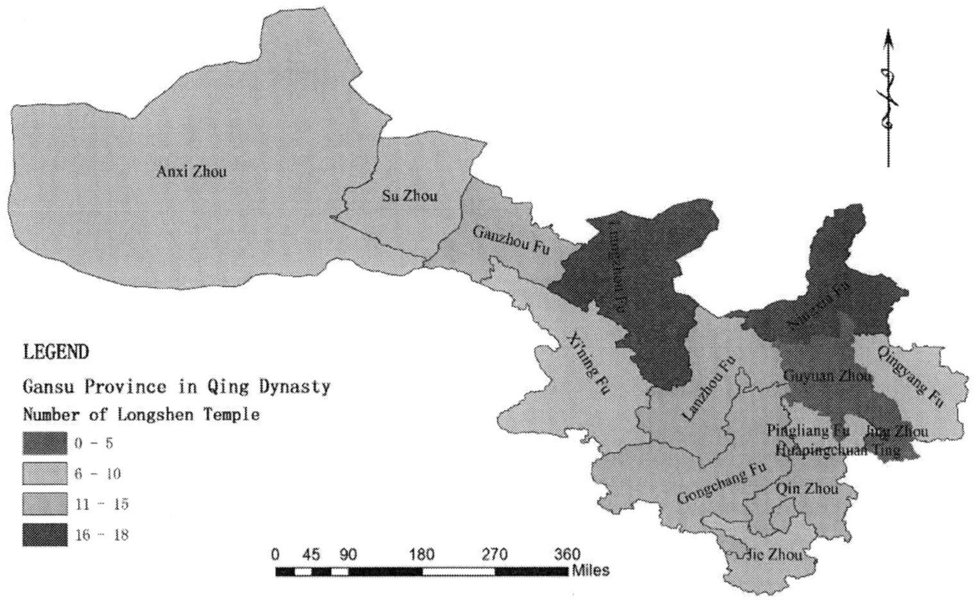

Figure 1. Spatial distribution map of Gansu Longshen temple in Qing Dynasty

4.Conclusions

After investigating the distribution of Gansu Longshen cult in Qing Dynasty, it is found that the widely distributed Longshen temple in arid and rainy Gansu reflects a general psychological state of people 's desire for harvest in the traditional agricultural society with low productivity. As a symbolic symbol of carrying agricultural civilization, Longshen cult not only expresses a hope for rich life, but also is affected by climate and agricultural form in the process of its historical transmission and evolution. In fact, it is a cultural phenomenon shaped by multiple factors. Therefore, when examining the Longshen Cult during the traditional period, especially under the perspective of modern science, we cannot simply attribute the Longshen Cult as a product of ignorance under the backward productivity, but as a long-standing social and cultural phenomenon, engraved with a deep brand of the times and rich in economic, political and social cultural connotations.

References

1. Chen Peixuan,Lu Guoyang,Bai Huzhi.(2020) Analysis of the Main Characteristics and Change Trend of Arid Climate—Taking Gansu Province as an Example. Anhui agricultural sciences ,18:210-219.
2. Zhang Qiang, Wang Runyuan and Deng Zhenyong. (2012) Effects of arid climate change on agriculture and ecology in northwest China. Meteorological Press, Beijing.
3. Yuan Lin. (1994) A Study on the Law of Drought Occurrence in the Historical Gansu,Ningxia and Qinghai. Journal of Lanzhou University, 02: 111-116.
4. Tan Xuming. (2013) Qing dynasty drought archives historical materials. China Book Publishing House, Beijing.
5. Chang Jianhua. (2006) Value of Chinese Local Chronicles as Social History Data. Chinese Social and Historical Review, 00: 61-73.
6. Zhao Shiyu. (2002) Revelry and Ordinary Life—— The Temple Fair and the Folk Society Since Ming and Qing Dynasties. SDX Joint Publishing Company, Beijing.

Analysis of Application Scenarios of Electronic Decision Theater in the Military Field

Xian Zhang[1], Qi Xue[2*]

[1] Center for Assessment and Demonstration, Academic of Military Sciences, Beijing, 100091, China
[2] Center for Assessment and Demonstration, Academic of Military Sciences, Beijing, 100091, China

Abstract. The Decision Theater(DTs) is an immersive collaborative decision-making visualization solution, which plays a very important role in improving the scientific level of decision making. Based on the analysis of the technical origins, design concepts, application fields, and technical characteristics of the DTs, a general system architecture for the DTs for the military field is proposed, four typical application scenarios were discussed, including drafting and argumentation of major military decisions, simulation of military action plans, training of senior commanders' professional skills, and emergency response decision-making, which is of great value for improving the scientific nature of military decision-making.

1 Introduction

The "Recommendations on Formulating the Fourteenth Five-Year Plan for National Economic and Social Development and the Long-Term Goals for 2035" mentioned that it is necessary to improve the pre-evaluation and post-evaluation systems for major policies, unblock the channels for participating in policy formulation, and improve decision-making level of socialization, democratization, and rule of law. For the field of military full of confrontation and competition, whether it is the formation of various major military decision-making processes or the organization and implementation of major military combat exercises, it is necessary to handle scientific, democratic, and efficient decision-making issues. It is necessary to compare and analyse different decision-making schemes, and perform and simulate the main program methods. Based on the study of the DTs. we conduct a targeted analysis in combination with the main characteristics of the military field, which is of great value for improving the scientific nature of military decision-making.

2 Current Research Status of Electronic Decision Theater

2.1 The origin of electronic decision theater

The DTs develops custom software solutions to build scenarios, visualize tradeoffs and discover solutions across a wide range of social and organizational systems [1]. DTs can be viewed as meeting rooms characterized by specific input/display technologies and seating arrangement that allow a group of people to interact with each other and with the data in order to contextualize a decision-making situation, evaluate the impacts of decisions, and find a common solution.

The world's first electronic Decision Theater was built by the Public Decision Laboratory of Arizona State University in 2004[2]. which is suitable for specific decision-making problems, by making some assumptions around a hypothesis and predicting or simulating possible options, the decision makers can collectively solve the problem. The core component of the DTs is called "The Drum". The Drum can accommodate 25 people at the same time, with various tools for collecting input from participants, the three-dimensional animation demonstrations can be simultaneously used. Participants can watch the behaviour through three-dimensional glasses and evaluate various scenarios for future scenarios. Huazhong University of Science and Technology built the world's largest DTs in 2010 by absorbing the technology of DTs of Arizona State University(ASU)[3]. The DTs consist of an eight-channel 240-degree ring screen system, a professional graphics workstation cluster, a virtual scene resource configuration management system, computer cluster with supercomputing capabilities, was able to interactive simulation and remote decision-making information transmission and exchange, and has super integration capabilities with computing power, it can provide decision-makers with an interactive three-dimensional intuitive decision-making environment. The DTs of Harbin Institute of Technology built at the same time includes design planning system, high-performance computer cluster, video image distribution processing system, audio distribution processing system, central integrated control system, meeting room/decision-making discussion system, etc, which could complete various kinds of decision analysis and process deduction analysis.

*Corresponding author's e-mail: scarebuff@163.com

https://doi.org/10.1051/e3sconf/202129003024

2.2 Frontiers of Research on Electronic Decision Theater

Since the world's first DTs was built at ASU in 2004, The Global Electronic Decision Theater Alliance was established. The alliance consists of Huazhong University of Science and Technology, Wuhan University [4], Arizona State University, Massachusetts Institute of Technology, and George City University, Renmin University of China, Harbin Institute of Technology and more than ten well-known universities at home and abroad [5]. A major joint R&D team composed of different universities has been formed. The research fields cover socioeconomics, mathematics, computer software and hardware, and virtual simulation, many scientific decision-making system platforms was developed.

The ASU DTs has completed more than 70 cases, including the Phoenix Water Supply Project, Forest Planning, Urban Climate Change, Urban Development Planning, and the Scientific Design of School Curriculum, covering urban planning, intelligent transportation, Environmental resources, smart buildings, social economy and many other public management decision-making fields. The DTs of Huazhong University of Science and Technology has been used in government public decision-making, education services and supervision, urban planning, construction, supervision and traffic modelling. The CITIC Electronic DTs has formed a DTs for scientific and technological policy evaluation. It has been used in the evaluation of innovative cities, the perception of user needs of Chinese scientific and technological information resources, the evaluation of listed companies, and the management of emergency warnings.

2.3 The Core technology of DTs

The DTs integrates multiple disciplines and forms of expertise to leverage and manage the tension between differences, rather than compartmentalize them. The working process of DTs can be abstracted as a process from simulation, to three-dimensional visualization, and finally to group decision-making. The following six technologies have been comprehensively used:

Firstly, scenario-based gaming and simulations. Participants role play what-if scenarios by engaging with DTs' game and simulation-based learning techniques. An environment was created which allow participants to explore and understand the consequences and impacts of different levels of risk[6].

Secondly, Software. DTs uses an in-house, open source complex systems framework software platform. The platform powers multi-screen model interaction to visually follow the effects and consequences of decision alternatives as they ripple throughout an integrated system. DTs constantly reduces cycle time and costs through increasingly growing capabilities and a mature technology shelf.

Thirdly, data analytics. DTs data analytics capabilities leverage staff expertise in geographical information systems and spatial optimization/modelling, social network analysis, statistical analysis, data mining, agent based modelling, machine learning, textual analysis, social media analysis, to customize solutions for a variety of innovative projects across a wide variety of topic areas.

Fourthly, user experience and user interaction. DTs models are designed to communicate the intricacies of a problem as well as the broader landscape of consequential action. DTs builds custom graphic dashboards for users to engage with databases, analytics and computations in real-time. Projects developed with DTs empower non-experts to engage in research, data exploration and computational modelling [6].

Fifthly, rapid data visualization. DTs can be utilized as a final product, as a way to explore data, or as a step in the iterative process for a larger project. Rapid data visualizations can take the form of interactive dashboards, exploratory data maps, timelines, or a combination of them and many others.

Sixthly, high-performance computing. The use of high-performance computing supports the integration of a wide range of computationally intensive models in various fields, including genetics, weather forecasting and climate research, enabling near-real-time decision-making.

3 The Analytical Framework in the Military Field

Compared with the public policy field of local governments, the military field has wider coverage, stronger antagonism, and greater risk in decision-making. Especially in the process of drafting and making major military decisions and organizing and implementing major exercises, it faces complex decision-making issues. It is necessary to reach a consensus on the content of the main decision-making matters in a short time, and make a scientific decision from a variety of different decision-making options. So we can design the following analysis framework.

3.1 General Architecture of DTs

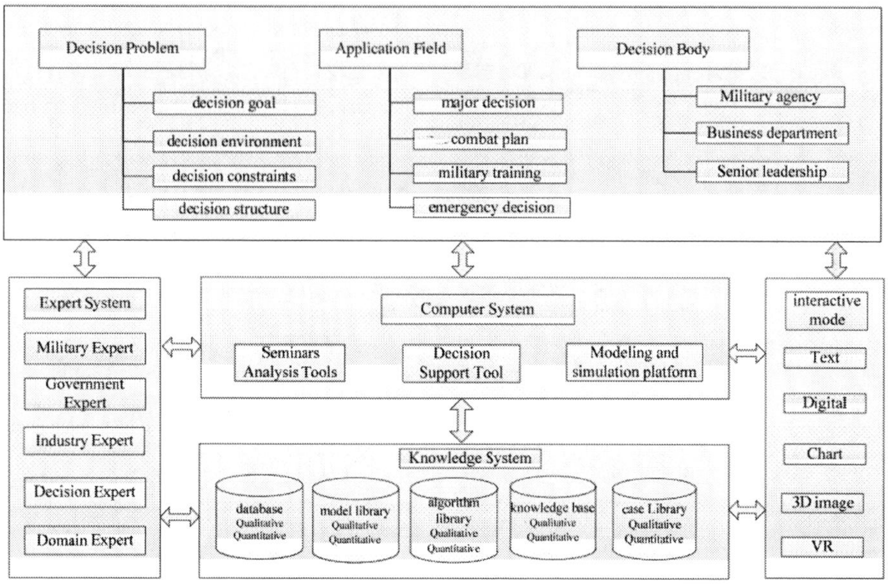

Figure 1. General Architecture of DTs

The DTs mainly includes three components: The Expert System, The Knowledge System, and The Computer System[7]. The Knowledge System refers to the sum of various intelligence information and fact data related to decision-making activities, including previous experience, domain knowledge, problem solving knowledge. The four types of knowledge are specifically embodied as databases, model libraries, algorithmic paths, knowledge bases, and case libraries that combine qualitative analysis and quantitative analysis. The Expert System, is aimed at the participants of the seminar, including various experts directly related to decision-making, including military commissioners, department leaders, industry experts, decision-making experts, and domain experts. The Computer Service System refers to the construction of computer supporting conditions in the integrated seminar environment. Specifically, it refers to the use of computer data storage capabilities, information processing capabilities, and knowledge presentation capabilities to collect, analyse, and process the opinions of seminar participants, so that many participants Opinions can be integrated and fed back into the system, and the ability to obtain new results through the system.

3.2 Application Scenarios of DTs

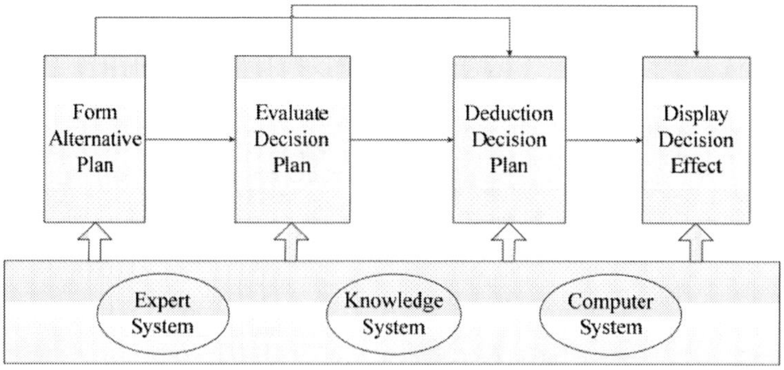

Figure 2. Application Scenarios of DTs

The DTs is consisted of The Expert System, The Knowledge System and The Computer System. According to the decision-making process, there are mainly the following four application scenarios[8-10]: The first is the analysis of alternative decision-making schemes, based on the expert experience system, the idea of forming alternative schemes through the analysis of existing cases, and the use of scientific modelling analysis tools to form alternative schemes on this basis. The second is the evaluation of alternatives, through the design of decision-making models and the use of technical indicators to analyse the pros and cons of multiple alternatives. The third is to perform deductions on the decision-making plans, through model adjustment, parameter adjustment and human-computer interaction deduces the implementation of alternative plans. Fourth, the effect of decision-making plans is displayed. Virtual simulation technology is used to comprehensively display the effects of a single set of decision-making plans and multiple sets of decision-making plans.

3.3 The basic analysis process of DTs

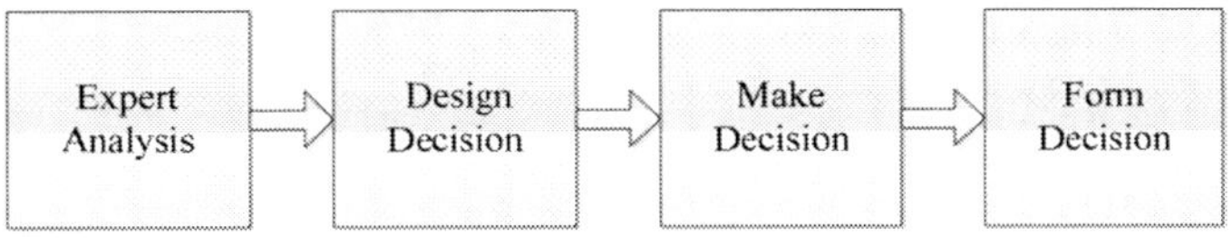

Figure 3. The basic analysis process of DTs

For general decision-making problems, decisions can be made by the four stages.

In the qualitative analysis stage, the relevant decision-making issues are first raised, experts are grouped to discuss the issues, and the issues are structured and an evaluation index system is created.

In the plan design stage, after creating a basic scenario based on the data, and construct an index system to solve the scenario, the initial data was entered to generate multiple sets of plans.

In the decision-making stage, firstly organize relevant personnel to hold a decision-making meeting, pour into multiple sets of plans for comparison and analysis, and then modify the impact factor based on expert opinions, generate the latest calculation results according to the calculation model, use the DTs to display the decision results, use the voting system to make the final decision.

In the plan formation stage, the generated decision plan is stored in the knowledge base, the storage model is imported into the model library, the final results and query opinions are released through the information release system.

4 The possible applications in the military field

The unique immersive visual experience of the DTs, the front-end and back-end cooperative operation mode, the completion of diversified decision-making tasks, the advantages of effectively stimulating the image thinking, determines its great application scenarios in the military field.

4.1. Drafting for major military decisions

In the process of drafting and demonstrating major military decisions, the main decision-making texts are detailed and decomposed to form a list of relevant decision-making matters. The selection part was focused on and conducted according to the responsible parties involved in the specific decision-making clauses, the human-in-the-loop mode was used to judge the pros and cons of the decision-making plan. Through the immersive interactive experience of experts on the main decision-making issues, the decision-making risks are reduced and the decision-making quality was improved.

4.2. Simulation of military action plan

In the deduction of major military action plans, especially for the main links and content of the combat plan, the DTs can transform the decision model into a decision script, adopt modelling and simulation technology to model and analyse the combat space, combat forces, and combat actions, adopt the virtual reality technology to construct a virtual battlefield environment, integrate the commanders into the virtual reality environment, and complete the deduction simulation of the combat action plan based on the simulation deduction engine.

4.3. Professional skills training for senior commanders

During the professional training of senior commanders, the ability and quality required by senior commanders are transformed into specific index content, and a list of optional operating procedures can design and deduct in the DTs. The DTs can not only complete the assessment of the comprehensive ability and quality of the senior commanders, but also complete the assessment of the individual professional ability of the senior commanders, and improve the professional ability level of the senior commanders through interactive assessment.

4.4. Emergency response decision-making process

In the process of emergency response decision-making, different emergency decision-making and disposal are transformed into different decision-making scripts. According to the main body involved in the decision-making relationship, occurrence probability and time sequence connection of various decision-making activities, by designing a multi-attribute decision model, According to user interaction, the pros and cons of various decision-making schemes are sorted.

5 Conclusion

Based on the analysis of the origin, design idea, application field and technical characteristics of DTs, we propose a general system architecture of DTs suitable for the military field, and discusses four different application scenarios. It is of great reference value to enhance the in-depth integration of decision-making in the military field.

References

1. Shutters, S.T.; Kandala, S.S.; Wei, F.; Kinzig, A.P (2021). Resilience of Urban Economic Structures Following the Great Recession. Sustainability, 13, 2374.
2. Miller, J., Salla, R., Amresh, A., Smith, H., Kandala, S., Hinrichs, M., Gorantla, R., Sokteva, E., Wei, F., Hirsch, K., & others. (2019). The Decision Theater: Collaborative Research Methodology.
3. Mao Zijun, Xu Xiaolin, Xu Xiaodong (2018). Electronic Decision Theater for Scientific and Democratic Decision Navigation, Chinese Administration, 2011, Issue 9,120-122.
4. White, D. D., Larson, K. L., & Wutich, A. (2018). Boundary Organizations and Objects Supporting Stakeholders for Decision Making on Sustainable Water Management in Phoenix, Arizona USA. In Transformations of Social-Ecological Systems (pp. 333-352). Springer, Singapore.
5. White, D., Jones, J., Maciejewski, R., Aggarwal, R., & Mascaro, G. (2017). Stakeholder analysis for the food-energy-water nexus in phoenix, Arizona: Implications for nexus governance. Sustainability, 9(12), 2204.
6. https://dt.asu.edu/technology.
7. Wang Xinglin (2008). The Decision-making mechanism of Decision Theater under the sciences of public policy. Harbin Institute of Technology.
8. Boukherroub, T., D'AMOURS, S., & Rönnqvist, M. (2016). Toward decision theater design for community forest management & planning: The case of Québec. In 11th International Conference on Modeling, Optimization and Simulation, Montréal, Québec, Canada.
9. Gober, P. (2018). Social Learning for Water Sector Resilience. In Building Resilience for Uncertain Water Futures (pp. 163-190).
10. Amresh, A., Salla, R., Sinha, M., & Birr, R. (2016). Design, implementation and evaluation of a game-based intervention targeting latino children for improving obesity outcomes. In 2016 IEEE International Conference on Serious Games and Applications for health (SeGAH) (pp. 1-8). IEEE.

Research progress and optimization prospect of constant boiling distillation technology

Heng Ma[1,a]

[1]School of Chemical Engineering, Qingdao University of Science and Technology, Qingdao, China

Abstract. Coboiling distillation is a special distillation method of separating the azeotropic system mixture, which is widely used in petroleum refining and chemical production. This paper reviewed the selection and characteristics of azeoliling agents, and introduced the research progress of home and abroad in detail. The system compares the advantages and disadvantages of the traditional constant boiling distillation and the new heat pump, compare the advantages, and presents the prospect of constant boiling distillation.

1 Introduction

Coboiling distillation is the mass transfer and heat transfer to separate the coboiling system mixture, which is commonly used in the actual chemical production, widely used in petroleum refining and chemical product refining. When there are two cozeolite in the components or system with close relative volatility, it is difficult to separate the ordinary distillation method. For such systems, a large return ratio must be set by the conventional distillation method, which increases the equipment cost and operation cost in the actual production process and the economy is reduced.

Cozeoiling distillation is to add a third component with a low boiling point, called an azeoiling agent or binding agent, the new added component can form a low boiling point with one or two components of the original system, destroy the original boiling system, so that the dilemma separation component is effectively separated, the method becomes azeotropic (constant boiling) distillation [1-3].

Cozeolite distillation is mainly suitable for the separation process of various organic compounds and hydrocarbon oxides. At present, there is more research on coboiling distillation technology in [4-6], and the new heat pump and next tower technology are relatively reported. This paper summarizes the technical methods of traditional distillation process, new heat pump and heat pump, compares the advantages and disadvantages of the next process of new heat pump, obtains the advanced nature of the distillation, and provides theoretical reference and basis for the improvement ofindustrial production separation technology.

[a] Corresponding author: [a]1106938964@qq.com

2 Brief introduction of cozeoiling distillation

2.1 Selection of the azeotherite

In the separation process of cozeolite distillation, the selection of azeolite agent plays a decisive role in the separation effect. The currently reported selection principles of coboiling agent [7-8] mainly include:

(1) The selected azeozeolite should preferably be inhomogeneous to making the separation process simpler.

(2) Cozeolite can significantly affect the vapor-liquid equilibrium relationship of the key components to be separated.

(3) The cozeolite forms with at least one or two components in the original system, destroying the cozeolite system in the original component. The newly formed azeolite should be more than 10 ℃ lower than the coboiling point of the coboiling ite with the original component.

(4) In the new formation of cozeolite, the lower the cozeolite content, can reduce the dosage and circulation.

(5) The azeolizer shall have small latent heat for vaporation to save energy consumption, and the azeolizer should have excellent chemical properties, with the characteristics of cheap price, rich raw materials, good thermal stability and small corrosion ability, non-toxic and environmental protection.

2.2 Characteristics of azeoboiling distillation

If ordinary distillation methods cannot meet the separation requirements, special distillation methods such as coboiling, extraction or variable pressure distillation can be used. Coboiling distillation has the following characteristics [9] relative to other separation methods:

(1) The co-boiling distillation process is highly

applicable and applicable for both continuous and intermittent operation.

(2) The boiling point of the azeotherite is high and must form a minimum azeolite with the components in the original component system and the selection range is limited.

(3)entrainer is often steamed from the top of the tower, large energy consumption, only the content of azeotropic agent is small in the new azeolite is more economical.

(4) The operating temperature of coboiling distillation at fixed pressure is relatively low and more suitable for separating thermosensitive materials.

3. Progress in cozeoiling distillation

3.1 Progress of cozeolite distillation process

3.1.1 process of coboiling distillation

Yao[10]took the separation of A + B mixture as an example and introduced the process of azeolboiling distillation. A is volatile component, low boiling point and light key component; B embarrassed volatile component, high boiling point and heavy key component.

In the process of coboiling distillation, the azeolite is soluble in component A and forms the lowest azeolite with A. Due to the low new coboiling point, the top material is heterogeneous mixture, and one phase contains more A and enters the component A recovery tower for further refining; the other is a mixture of azeolite and A (containing a small amount of A component), whose top backflow flows into the azeolboiling distillation tower. The specific flow chart of the coboiling distillation process is shown in Figure 1.

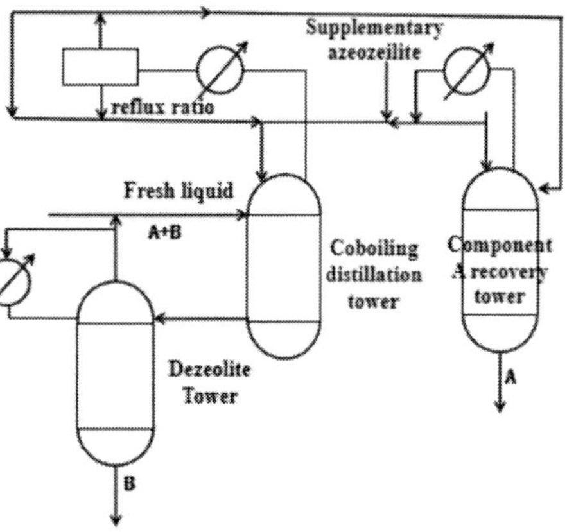

Figure 1 Distillation process with a two-component inhomogeneous azeotherite

From the above process flow chart, the fresh liquid mixture A and B enters the coboiling distillation tower, the top extracts the coboiling agent and the mixture of component A as reflux, the other part enters the component A recycling tower, which can finally be recovered at the bottom of the tower component A. The material of the coboiling distillation tower enters the dezeolizer tower, where the top part of the B, tower returns, and the other part is mixed with the fresh material liquid and added into the azeolize distillation tower. In order to maintain the concentration of the feed or reflux, the process is the distillation process of heterogeneous cozeolite. In the case of ternary azeolite, the azeolizing agent can be added with fresh material liquid, cycle the ternary component azeolite at the top of the demazeolite agent tower, the pure B component distilled at the bottom of the tower, the upper layer of azeolite reflux at the top of the main tower, and the lower phase rich in component A enters the component A recovery tower for refining.

3.1.2 Optimization direction of constant boiling distillation

The biggest defect of the separation process is the large energy consumption. Many new energy-saving processes have been developed to efficiently and fully utilize energy and save [11]. It includes thermal integration technology, is a kind of heat logistics heat to heat cold logistics, reduce utility heating as far as possible, achieve heat integration and full utilization; in addition, thermal coupling technology, by improving the internal structure of the distillation tower, form a more complex tower structure, reduce the operation cost and equipment investment cost. Common energy-saving processes mainly include heat pump distillation, multi-effect distillation and adjacent tower distillation. In addition, middle condensation, intermediate reboiling and side extraction are also an energy-saving process through the middle condensate is the middle heat exchange, change the operation line of the distillation section, evenly distribute the heat transfer driving force in the tower, reduce the irreversible degree of heat transfer process, keep the fluid in steady flow state, and improve the separation efficiency.

3.2 Constant boiling distillation process

3.2.1 Traditional constant-boiling distillation process

Zhu Liuliu [12] et al. produced 99.5% hydrless alcohol as an example, introduced the process of traditional constant boiling distillation process, Figure 2 is a process flow chart of production purity of 99.5% hydrless alcohol from conventional constant boiling distillation.

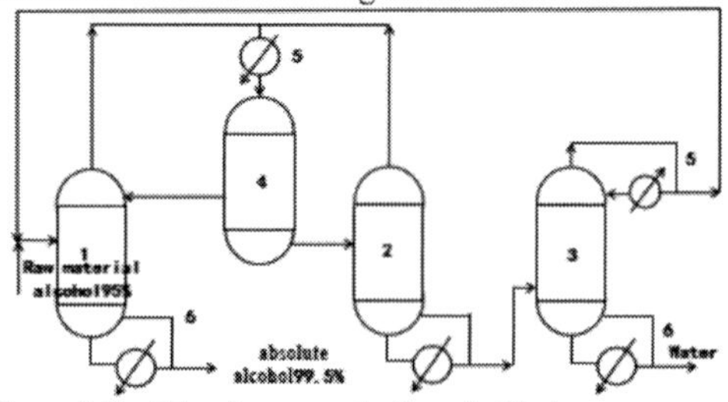

Figure 2 Traditional constant boiling distillation process raw material alcohol

Because it is more difficult to produce anhydrous alcohol under normal pressure conditions, the traditional constant boiling distillation process adds the azeolite that can form the lowest azeolite with ethanol (such as pentane, benzene), the azeolite can form ternary azeolite with water and ethanol, and the boiling point of the original solution, to produce high purity alcohol through 95% of the raw material alcohol.

The traditional constant boiling distillation method produces anhydrous alcohol after long development and improvement. The process has been mature and is widely used in anhydrous alcohol production enterprises. Its advantages lie in high product quality, large production, relatively stable production and relatively low operating cost, but its energy consumption compared with other advanced processes is the biggest problem, in addition, the azeoiling agent is high and may cause environmental pollution problems in the actual production.

3.2.2 New process of Alcohol by constant boiling distillation of heat pump

As a new energy-saving process, heat pump distillation is to pressure the steam on the top of the tower heat up operation, recycle the tower top steam condensation latent heat, this part of the heat as the heat source of the tower kettle reboiler, so as to reduce the tower bottom reboiler, to achieve the effect of energy saving and consumption reduction. Heat pump distillation has achieved good energy saving and consumption reduction effect in the separation of alkane and alcohol components. It has been studied to combine the heat pump technology with constant boiling distillation. This process is not only applicable to the difficult separation system, but also has a good energy saving effect. In the actual production, the system operates in the fully closed system, producing no harmful by-products, which is relatively environmental friendly.

Considering the small temperature difference of the dehydration tower under constant pressure (10-15 ℃) Conditions for heat pump distillation. Liu Zongkuan et al. developed a new azeotropic distillation heat pump to pro duce anhydrous ethanol [13]. The ternary azeotrope on the top of the dehydrating tower (64.6 ℃) is heated to about 100℃ directly through the heat pump, without the conde nsation process of the condenser. The energy produced b y this process can be used as heat at the bottom of the de hydrating tower, eliminating a reboiler. After the separati on of the condensate, the upper benzene-rich phase is used as the top reflux, and the lower water-rich phase flows into the entrainer recovery column. The ethanol and water mixture at the bottom of the recovery c olumn flows into the alcohol recovery column for further refining. The cozeolite distilled at the top of the alcohol recovery tower does not undergo the condensation process at 78 ℃, and is pressurized to the heat pump to 120 ℃ to provide heat to the alcohol recovery tower, once again saving a rezeiler and saving energy consumption.The list of process parameters of anhydrous ethanol developed by Liu Zongkuan et al. [13] is as follows:

Table 1. Process parameters of anhydrous ethanol produced by constant boiling distillation heat pump

name	numerical value
Temperature difference between tower top and tower kettle	13℃
Alcohol dehydration tower tower top steam temperature	64.6℃
Heat pump compressor heat up 1	100℃
Alcohol recovery tower top vapor temperature	78℃
Heat pump compressor heat up 2	120℃
Alcohol recovery tower and kettle temperature	103℃

Heat pump constant boiling distillation has many advantages. First of all, to meet the requirements of technical and economic indicators, the product anhydrous ethanol is not vaporized, the distillation tower does not need a large return ratio, reduce operating costs and equipment costs. In addition, through the combination of heat pump technology and constant boiling distillation, the tower top materials are not condensing, and the heat pump compressor can be used as the heat source of the tower kettle reboiler, reducing the use of the reboiler and meeting the requirements of energy saving and consumption reduction. Through this process product high yield and no toxic substances during operation, is a clean production process.

3.2.3 Technology of co-boiling distillation of heat pump

Next door tower distillation [14] is a higher separation requirements by improving the internal structure of the tower, and can also be combined with a heat pump to further reduce energy consumption. The next tower is divided into four parts, the feed first enters the preseparation section, the A-B, B-C component mixture is separated in advance, the light component A-B enters the distillation section, the heavier B-C enters the distillation section, the lighter A component of the distillation tower is extracted from the top of the tower, the heavy C component is extracted from the tower kettle, the B component enters is the lateral extraction section, and the material is the more pure B component. Steam in the tower flows upward, the liquid phase flows downward, the gas and liquid of each tower plate are in the phase balance process, the operation is relatively stable, and the quality transfer heat transfer effect is good.

The process of the next tower of the heat pump is the intermediate heat exchange, which can achieve efficient separation of the ternary complex system, and expand the heat pump technology to the wide boiling series separation[15]. Constant boiling distillation of the adjacent tower is a typical example of heat-coupled distillation, improving the structure of the two distillation towers into one distillation tower.

Chen Mengqi [16] et al. have developed a new process

of co-boiling distillation of the next tower. The steam on the top of the main tower is compressed as the heat pump compressor heat source at the bottom of the tower, and conducts heat exchange with the tower kettle logistics. The feed and organic phase obtained from the phase divider enter the top of the main tower, the tower top pressure is set to 101.35KPa, the partition door top material enters the coboiler recovery tower, and the pressure is also set to 101.35KPa. The heat load of the tower in the coboiler recovery tower is 1208KW, temperature is 102 ℃, and the final EDA purity reaches 99.7%. This process has high yield, high purity and energy saving, providing new ideas for the improvement of coboiling distillation process and useful reference for the next tower technology of heat pump and the combination of coboiling distillation process.The process parameters of heat pump co-boiling distillation next tower developed by Chen Mengqi [16] et al. are as follows:

Table 2. Process parameter value of adjacent tower with coboiling distillation of heat pump

name	numerical value
Number of main tower(MC)	71 boards
Main tower top pressure	1.013bar
Top pressure of the partition door	1.013bar
Heat load of coboiler recovery tower kettle reboiler	1207.6kW
Temperature of cozeolite recovery tower kettle	103℃
Pressure of azeotropic agent recovery tower kettle	1.063bar
EDA Moore purity	99.7%

4 Conclusions

Distillation is an important separation and purification method in the chemical field. For some difficult separation or azeoliling systems, it can be efficiently separated by destroying the lowest azeolite system, and is widely used in chemical product refining. At present, there has been a substantial theoretical foundation in fluid mechanics, process control and gas-liquid heat and mass transfer of distillation tower. Modern chemical industry has put forward higher requirements for separation requirements, and separation technologies such as coboiling distillation are also constantly optimized and improved. Firstly, the research status of constant boiling distillation and tower distillation of the heat pump introduced the three processes in detail. Secondly, the advantages and disadvantages of the three processes are analyzed, highlighting the advantages of constant boiling distillation next to the heat pump, which not only realizes high purity, high yield, greatly reduces energy consumption, makes full use of steam energy, and has good heat transfer quality transfer effect, which is an effective combination of the heat pump tower technology and constant boiling distillation. Finally, the development prospect of coboiling distillation process is looked forward and provides theoretical reference and design basis for the improvement of industrial production separation technology.

References

1. Liu Tong, Ye Qing, Chen Jingxing, Cen Hao. Simulation and Optimization of the Separation of Bioalcohol-Water Mix [J]. Petroleum Journal (Petroleum Processing), 2019,35 (05): 938-947.
2. Li Zhizhuo, Jiang Zhankun, Zhang Xianghe. Research and application progress of azeotropic distillation technology [J]. Shandong Chemical Industry, 2015,44 (03): 37-39.
3. Hu Song, Zhang Bingyan, Yu Zhihong, et al. Application and Research of Cozeiling Distillation in Chemical Production [J]. Chemical progress, 2010,29 (12): 2207-2219.
4. Zhou Qiang, Wu Gang, Zhuang Lizhen. Application of cozeiling distillation in tetrachloroethylene dehydration [J]. Chloralkali, 2019 (09): 39-41.
5. Sun Guoxiu, Cui Xianbao. Study on the heterogeneous intermittent coboiling distillation process [D]. Tianjin: Tianjin University, 2007.
6. Xu Bin. Application and optimization of energy-saving technology of constant boiling distillation [J]. Chemical Management, 2020 (09): 58-60.
7. Ren Helen, Andenchao, Zhu Taoyue, Li Hailong, Li Xingang, Rectification Technology Research Progress and Industrial Application [J]. Chemical Progress, 2016,35 (06): 1606-1626.
8. Lyu X Y, Zhao L, Wang W C, et al. Heat-pump azeotropic distillation for ternary azeotrope ethanol-toluene water separation [J]. Journal of Changzhou University, 2017, 29(6): 26-31.
9. Li Fangsheng, Wang Xuyun, Qi Yunguo. Study on butanone-water heterogeneous intermittent coboiling distillation [J]. Chemical Engineer, 2020,34 (05): 25-29.
10. Yao cang. Simulation Study of Phase Balance and Separation Process of Sparse Acetic Acid Extraction-Coboiling Distillation System [D]. Yangzhou University, 2017.
11. Tian Peng. Process Design and Control of Ethonitrile-n-propanol Cozeolite System [D]. Qingdao University of Science and Technology, 2016.
12. Zhu Liuliu. Simulation study on strengthening of distillation of fuel ethanol [D]. China University of Petroleum (East China), 2017.
13. Liu Zongkuan, Gu Zhaolin, He Yanling, Feng Xiao. Research on constant boiling distillation of fuel ethanol heat pump [J]. Chemical progress, 2003 (11): 1147-1149.
14. Xia Ming. Design and control of energy saving adjacent tower for separating cozeolite [D]. Tianjin University, 2014.

15. Li Murong, Xu Liang, Xin Chun. Simulation of Wide BoSystem in Next Tower of Heat Pump distillation [J]. Chemical Journal, 2017,68 (5): 1906-1912.
16. Chen Mengqi. Control of conventional adjacent tower and heat pump [D]. China University of Petroleum (East China), 2018.

Effect of cadmium and arsenic contamination on element content and effectiveness in rhizosphere soils of different forages

Na Jiang[1,2], Jingmin Yang[1], Zhihong Huang[3], Yuan Li[1] and Yanqun Zu[1,a]

[1]College of Resources and Environment, Yunnan Agricultural University, 650201 Kunming, Yunnan, China

[2]College of Animal Science and Technology, Yunnan Agricultural University, 650201 Kunming, Yunnan, China

[3]Shengqing environmental protection Co., Ltd, Kunming, China

Abstract. To explore the effects of Cd and As composite contaminated soils on the element contents and effectiveness in rhizosphere soils of different forages, the root morphology, rhizosphere soil elements and their availability of 21 forages were studied and analyzed. The results showed: The 21 cultivars showed significant differences in root morphological characteristics. Root length is an index of root morphology that has the highest correlation with other elements and availability. And the correlation between the availability of each element and root morphology was higher than that of element content. Among them, Fe availability was significantly correlated with root morphological characteristics (P<0.05). The contents of different elements and availability varied greatly among the different cultivars. Among the six rhizosphere soil elements, the difference of toxic metal Cd was the largest, while the difference of Al availability was the largest. According to the correlation analysis between the elements and availability of each cultivar, the interaction of the availability of each element is higher than that of the soil element under the composite contaminated of Cd and As. Toxic metals Cd and As affected each other, and there was a very significant positive correlation between the content and availability. Fe, Mn and P had higher correlation with Cd and As.

1 Introduction

Toxic metal pollution is one of the important research problems in soil ecological remediation, which is shown not only by single element pollution, but also by two or even multi-element compound pollution[1]. Among them, the problem of cultivated land caused by Cd and As combined pollution is causing more and more economic losses to China. These two elements have a wide range of pollution and great toxicity. However, there are complex additions, synergism and antagonism between soil and organisms, which makes the impact of their combined pollution on the environment more complex than that of single pollution. The interaction between different metal elements also provides a new idea for biological detoxification and metal pollution remediation[2-3]. The interaction between metal and other elements is very complex under the stress of metal compound pollution, which will show different forms of action due to the existence of different elements. At present, the research on Cd and As compound polluted soil mainly focuses on the different absorption of toxic elements by cultivated varieties[4-5]. There are relatively few studies on the effects of metal pollution elements in soil on root morphology and the content of other elements in soil, which need to be studied.

Gejiu City, Yunnan Province is an industrial city rich in mineral resources. The problems left by the early mining and metallurgy are the main sources of Cd and As pollution in the local soil. According to the investigation, the local characteristic dairy industry has developed into a regular model, forages is the main forage crop. Therefore, 21 forages cultivars planted locally were used as test materials on the farmland soil polluted by Cd and As. The root morphological characteristics, different element contents and effective statuses of the forages cultivars were investigated to explore the interaction between each element under Cd and As composite contaminated soils. It is hoped to provide a basis for the future regulation of toxic metal elements in contaminated soils and the safe prevention and control during forages production in agricultural fields.

[a] Corresponding author: zuyanqun@ynau.edu.com

2 Materials and methods

2.1 Experimental site location and forage plants

The experimental area was located Zhadian District, Jijie Town, Gejiu City, Yunnan Province, with an altitude of 1428 m, 103 ° 15 ′ E and 23 ° 46 ′ N. Soil physical and chemical properties: The pH between 7.64-8.19, 40.26 g/kg of organic content, 2.54 g/kg of total nitrogen content, 1.36 g/kg of total phosphorus content, 2.26 g/kg of total potassium content, 142.23 mg/kg of alkali-hydrolyzable nitrogen content, 179.5 mg/kg of available phosphorus content, 178.91 mg/kg of available potassium, 5.09 mg/kg Cd and 226.13 mg/kg As, respectively. The contents of Cd and As in the experimental area exceeded the control values of soil pollution risk of agricultural land in *"Soil environmental quality risk control standard for soil contamination of agricultural land"* (GB 15618-2018) by 1.27 times and 2.26 times, which belonged to severe pollution.

The tested materials were 21 cultivars of forages mainly comprising 3 families (*Gramineae*, *Leguminosae*, and *Compositae*), including 18 species and 21 cultivars (Table 1.).

Table 1. Species, families and genera of forages materials.

Number	Cultivar	Species	Family
1	Farn	*Festuca arundinacea* Schreb.	*Gramineae*
2	Denata	*Dactylis glomerata* L.	*Gramineae*
3	Reyan 4	*Pennisetum purpureum* Schumach. × *Pennisetum glaucum* (L.) R. Br.	*Gramineae*
4	Pearl millet	*Pennisetum alopecuroides* (L.) Spreng.	*Gramineae*
5	Ye Mingzhu	*Phleum pratense* L.	*Gramineae*
6	Dongmu-70	*Secale cereale* L.	*Gramineae*
7	Munzhou I	*Pennisetum giganteum* z.x.lin	*Gramineae*
8	VNS	*Coronilla varia* L.	*Leguminosae*
9	Bondi	*Lolium Multiflorum* L.	*Gramineae*
10	Maddy	*Lolium perenne* L.	*Gramineae*
11	Super King Tang	*Sorghum bicolor* × *S. Sudanense*	*Gramineae*
12	12SU9001	*Sorghum bicolor* × *S. Sudanense*	*Gramineae*
13	12SU9003	*Sorghum sudanense* (Piper) Stapf.	*Gramineae*
14	12SU9004	*Sorghum sudanense* (Piper) Stapf.	*Gramineae*
15	13FB7001	*Sorghum bicolor* (L.) Moench	*Gramineae*
16	12FS9003	*Sorghum bicolor* (L.) Moench	*Gramineae*
17	WL525HQ	*Medicago Sativa* L.	*Leguminosae*
18	Grand Slam	*Cichorium pumilum* Jacq.	*Compositae*
19	You-12	*Zea mays* ssp. *mexicana*	*Gramineae*
20	Rainbow	*Pennisetum purpureum* Schum cv.Guiminyin	*Gramineae*
21	Chaosheng	*Lactuca indica* L.	*Compositae*

2.2 Experimental design

The test materials were sown in October 2019. In the in-situ field test, each material was set up with three replicates. The plot area was 3m × 5m, and the materials were randomly distributed. In the in-situ field test, each material was set up with three groups of repetition, the plot area was 3m × 5m, and the materials were randomly distributed. During the growth period, insecticidal, weeding, watering and topdressing were carried out according to the growth status. Samples were collected in June 2020. Five sampling sites were sampled evenly and randomly by "X" method, and the plant root and rhizosphere soil samples were collected. The aboveground and underground parts of the plant were opened with scissors and put into polyethylene bags.

2.3 Determination of root morphology

The plant root scanner EPSON perfection V700 photo was used for scanning and analysis.

2.4 Determination of element content and available state in rhizosphere soil

The content of cadmium was determined by flame atomic absorption spectrometry (Thermo ICE 3000 SERIES) and recorded it in mg/kg[6-7]. The content of arsenic was determined by atomic fluorescence spectrometer (AFS-9710) and recorded it in mg/kg[8-9]. Determination of total phosphorus and available phosphorus in soil by molybdenum antimony anti colorimetric Spectrophoto-metry (Metash UV-5800)[6]. The content and effective state of Fe, Mn and Al were determined by ICP-OES (ThermoScientific iCAP6300) [6, 10].

2.5 Statistical analyses

The statistical significance analyzed used one-way analysis of variance (ANOVA) and Duncan test the difference of the average value of different treatments at the 0.05 level by SPSS 20.0 (n=3). And the correlation was analyzed by Spearman method.

3 Results

3.1 Differences in rhizosphere soil pH and root morphology

The rhizosphere soil pH of the 21 cultivars ranged from 7.33 to 8.70, and the differences were not significant ($P > 0.05$), whereas obvious differences were observed in root morphology among the different cultivars (Table 2.). The maximum difference of total root length was 13.41 times, the maximum value was 'Denata' of *Gramineae*, and the minimum value was 'Grand slam' of *Compositae*, and the maximum differences of root length, root surface area, root volume and root diameter among cultivars were 1.97, 6.26, 30.82 and 179.3 times, respectively. Except for root length, the maximum value is *Compositae*.

 https://doi.org/10.1051/e3sconf/202129003026

Table 2. Soil pH and root morphology in the Rhizosphere of 21 cultivars.

Cultivar	pH	total root length (cm)	root length (cm)	root surface area (cm^2)	root diameter (mm)	root volume (cm^3)
Farn	7.73±0.07a	210.94±52.21efg	8.68±2.09d	28.96±8.83h	0.43±0.04d	0.3±0.13c
Denata	7.71±0.04a	681.71±160.46a	11.63±2.52cd	125.67±28.13abcde	0.48±0.02d	1.49±0.3c
Reyan 4	8.27±0.02a	619.69±112.16ab	17.85±0.88a	113.94±27.1abcdef	0.58±0.04cd	1.68±0.51c
Pearl millet	8.02±0.09a	397.92±36.45bcde	11.89±2.11bcd	74.03±10.86defgh	0.59±0.04cd	1.11±0.23c
Ye Mingzhu	8.00±0.12a	370.36±78.34de	11.79±1.02bcd	41.66±5.32fgh	0.38±0.06d	0.39±0.09c
Dongmu-70	7.46±0.10a	301.13±21.73def	13.5±0.6abcd	52.4±14.97efgh	0.55±0.16cd	0.84±0.45c
Munzhou I	8.13±0.03a	611.69±170.39abc	15.31±1.61abc	151.24±13.62abc	0.62±0.11cd	1.9±0.6c
VNS	8.07±0.05a	411.66±66.17bcde	15.14±1.28abc	107.97±9.56abcdefg	0.89±0.17cd	2.41±0.51c
Bondi	8.00±0.00a	273.13±9.89defg	12.48±1.09bcd	33.22±3.67gh	0.39±0.05d	0.33±0.08c
Maddy	8.11±0.03a	431.44±99.5bcde	8.87±0.32d	53.42±12.3efgh	0.39±0.01d	0.53±0.12c
Super King Tang	8.01±0.01a	350.36±6.32de	14.29±1.49abc	114.47±1.2abcdef	1.04±0.01cd	2.98±0.05c
12SU9001	7.90±0.03a	279.56±28.65defg	14.29±1.49abc	78.79±6.54cdefgh	0.90±0.05cd	1.78±0.16c
12SU9003	7.62±0.05a	200.82±37.54efg	12.38±1.57bcd	51.44±8.64efgh	0.84±0.09cd	1.07±0.2c
12SU9004	8.13±0.04a	279.49±14.29defg	12.4±2.22bcd	95.41±8.74bcdefgh	1.09±0.1cd	2.64±0.45c
13FB7001	7.82±0.19a	467.84±38.72abcd	14.41±0.39abc	134.12±18.27abcd	0.91±0.08cd	3.11±0.65c
12FS9003	7.48±0.03a	384.67±53.46cde	16.31±0.52abc	153.14±29.23abc	1.25±0.1cd	4.9±1.19c
WL525HQ	7.68±0.05a	281.75±39.49defg	17.14±0.89ab	148.21±53.75abcd	1.60±0.43c	3.73±1.03c
Grand Slam	8.03±0.03a	50.84±11.02g	11.51±3.28cd	158.32±55.72ab	11.71±0.6a	46.98±17.68b
You-12	7.58±0.02a	246.25±26.03defg	15.05±1.57abc	117.32±10abcdef	1.10±0.21cd	2.62±0.97c
Rainbow	7.87±0.12a	466.04±71.62abcd	16.09±1abc	144.36±15.32abcd	0.53±0.04cd	1.48±0.42c
Chaosheng	7.73±0.05a	72.24±1.1fg	14.12±0.84abc	181.37±23.41abcd	9.08±1.16b	53.79±15.24a

Note: Data are the mean ± standard errors of three replicates. Different lowercase letters represent significant differences between data in the same column ($P < 0.05$).

3.2 Difference of element content in rhizosphere soil

There were differences in the contents of different elements in the rhizosphere soil of different forage grasses (Figure 1.). The Cd and P contents in soil of 'Denata' was the highest. The highest contents of As, Fe, Mn and Al were 'You-12', 'Reyan 4', 'Faren' and 'Grand Slam', respectively, with significant difference ($P < 0.05$). After planting 'Reyan 4', the contents of Cd, As and Mn in rhizosphere soil were the lowest, the lowest contents of Fe were 'denata' and 'Dongmu-70', the lowest contents of Al were made in China, and the lowest contents of P were 'Munzhou I'. Among the six rhizosphere soil elements, the biggest difference was the Cd content, and the difference reached 4.76 times.

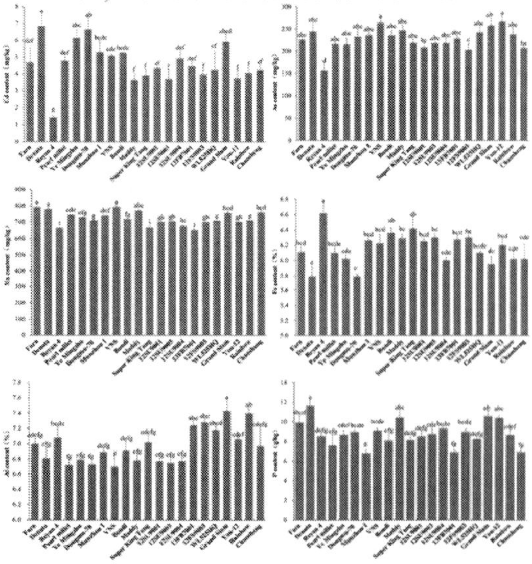

Figure 1. Elements in the rhizosphere soil of 21 cultivars.

3.3 Difference of effective content of elements in rhizosphere soils

The highest availability of As and P was found in 'Denata' (Figure 2.), the lowest in 'Super King Tang' and '12SU9004', respectively. And 'Yemingzhu', 'Dongmu 70' and 'Super King Tang' had the highest availability of Cd, Mn and Al, while 'You-12', 'Reyan 4' and 'VNS' had the lowest. Moreover, The highest availability of Fe was 'Reyan 4', the lowest was 'denata' and 'Dongmu-70'. There were significant differences in rhizosphere soil availability among different cultivars, and the biggest difference among these available states was Al availa- bility, which was 4.22 times higher than that in rhizosphere soil of different cultivars.

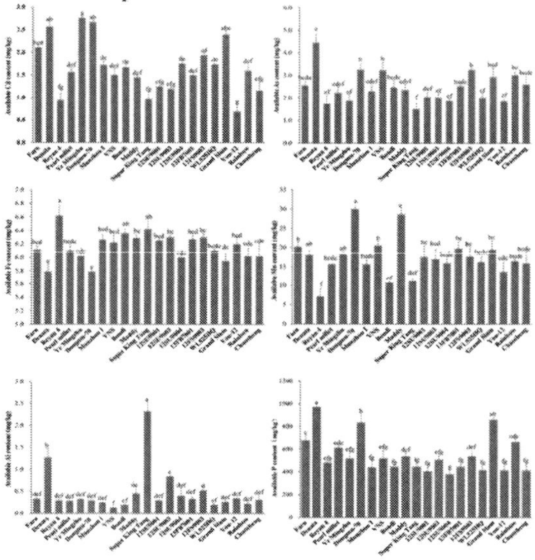

Figure 2. Availability of elements in the rhizosphere soil of 21 cultivars.

3.4 Correlation analysis

The root morphology of forages was not significantly correlated with the content of Cd and As, the same as the soil pH (Table 3-5.). But the pH was significantly correlated with the As availability, and the root length was negatively correlated with the available state of Cd, Fe, Mn and P ($P < 0.05$). The contents of Cd and As were significantly positively correlated with the available state of Cd and As, and both of them were significantly correlated with Fe, Mn content and availability. ($P < 0.05$). There was extremely significant positive correl-ation between Cd, As availability and P availability ($P < 0.01$). The correlation of the available state of each element to the root morphology in the soil polluted by Cd and As was higher than that of the element content, and the interaction of the available state of each element was higher than that of the soil element.

https://doi.org/10.1051/e3sconf/202129003026

Table 3. Correlation analysis between root morphology and soil elements among 21 cultivars.

	pH	Total root length	Root length	Root surface area	Root diameter	Root volume
Cd content	-0.171	-0.051	-0.222	-0.002	0.090	0.096
As content	-0.157	-0.104	-0.004	0.120	0.081	0.085
Fe content	0.305*	0.090	0.154	-0.098	-0.276*	-0.297*
Mn content	0.012	-0.065	-0.359**	-0.095	0.203	0.189
Al content	-0.052	-0.137	0.151	0.373**	0.303*	0.175
P content	-0.088	-0.015	-.0278*	-0.111	0.036	0.013
Cd availability	-0.231	-0.019	-0.294*	-0.092	0.084	0.047
As availability	-0.303*	0.153	-0.134	0.130	0.127	0.134
Fe availability	-0.090	0.071	-0.264*	-0.291*	-0.288*	-0.258*
Mn availability	-0.233	-0.088	-0.298*	-0.172	0.028	0.072
Al availability	-0.047	0.086	-0.123	0.010	-0.100	-0.085
P availability	-0.237	0.066	-0.291*	-0.033	0.177	0.120

Note: *At the 0.05 level, correlations were significant.
** At the 0.01 level the correlation was extremely significant. The same below.

Table 4. Correlation analysis of the content of elements in rhizosphere soil of 21 cultivars.

	Cd	As	Fe	Mn	Al	P
Cd	1					
As	0.358**	1				
Fe	-0.632**	-0.257*	1			
Mn	0.375**	0.351**	-0.320*	1		
Al	-0.207	0.039	0.284*	-0.212	1	
P	0.102	0.205	-0.264*	0.297*	-0.002	1

Table 5. Correlation analysis of availability in rhizosphere soil of 21 cultivars.

	Cd	As	Fe	Mn	Al	P
Cd	1					
As	0.469**	1				
Fe	0.463**	0.306*	1			
Mn	0.415**	0.380**	0.521**	1		
Al	-0.085	-0.040	-0.145	-0.159	1	
P	0.628**	0.593**	0.325**	0.409**	0.115	1

4 Discussions

Roots showed a high degree of plasticity to changes of soil environmental factors during the growth process. Difference in genotype and growing environment can lead to changes in morphological characteristics of part or the whole root system[11]. Metal ions on the root surface can be absorbed actively and passively through cation exchange, diffusion and metabolic energy, and show a high selectivity to some metal elements[12]. Larger root morphological characteristics, such as root length, root surface area, root volume and root tip number, not only help to increase the contact area between plant roots and soil, and promote roots to absorb trace elements in soil. It may also increase the absorption and accumulation of metal elements[13]. In this experiment, the effect of pH on the rhizosphere soil of all forage cultivars before and after planting was small. There were significant differences in total root length, root surface area, root diameter and root volume among different forages, and the degree of difference varied with cultivars, which was consistent with the results of Gao et al.[14]. In root morphology, the total root length and root length of *Gramineae* were generally higher than those of *Leguminosae* and *Compositae*, while the root surface area, root diameter and root volume of *Compositae* were higher than those of *Gramineae* and *Leguminosae*, which was mainly related to the growth characters and interspecific differences of each material. Li et al.[15] also shows that the change in root form structure is an important embodiment of plant adaptation to stress. In addition, correlation analysis showed that there was a significant correlation between Fe availability and root morphology, and it was found that root length had more correlation with the content of Mn, P and the available state of Cd, Fe, Mn, which fully indicated that root length was a very important index in Cd and As contaminated soil, which may be more affected by some trace elements and Cd.

There are complex additive, synergistic and antagonistic effects between soil and organisms, which makes the impact of metal compound pollution on the environment more complex than that of single pollution[16]. Cd and As exist in different forms in soil with different migration characteristics and bioavailability. After Cd and As enter the soil environment, the interaction between soil pH, and hazardous metals is an important influencing factor[17]. Studies have shown that toxic elements tend to affect mineral elements in soils[18-19]. However, the contents of both toxic metals including Cd and As showed extremely significant positive correlation in this experiment ($P < 0.01$), and it indicated that the interaction had a large effect and showed a synergistic relationship. Guo et al.[20] reported that the coexistence of Cd and As in soil could additive to alfalfa, and Cd promote the accumulation of As. At the same time, the toxic elements were positively correlated with Fe content and negatively correlated with Mn content, but the effectiveness was positively correlated with the available state of Fe, Mn and P, which had obvious synergistic effect. This result suggests that increased or accumulated nutrients such as Fe and P in soils may exacerbate the risk of Cd or As contamination in cultivated forages.The interactive interaction between Cd, As and each element is an important factor affecting the absorption and accumulation of the pollutants toxic metals in soils by forages. It is similar to that of Cheng et al.[21].

5 Conclusions

There were significant differences in root morphology among different cultivars, and the root length was more affected by some trace elements and Cd in the composite contaminated soils. The correlation between the available state of each element and root morphology was higher than that of element content, and the available state of rhizosphere soil elements had a greater effect on plant root morphology. There was a significant positive correlation between Cd and As, and the interaction was synergistic. At the same time, in the element content of rhizosphere soil, the difference of Cd content was the biggest, while in the available state, the difference of Al content was the biggest. The elements with high correlation with metal Cd and As are Fe, Mn and P. The results can provide a certain scientific basis for the selection of agricultural cultivars of farmland polluted by hazardous metals in mining area and the appropriate measures to control the pollution of Cd and As in farmland soil, and realize the effective control and safe utilization.

Acknowledgments

This work was supported by the Key R&D Projects of Yunnan Province (grant No. 2019BC001-04).

References

1. Yao, B., Yang, A.P., Chen, H.Y., et al. (2020) Ssil heavy metal pollution and risk assessment of agricultural soils in the Yunnan-Guizhouarea, Upper Pearl River Basin. Journal of Agro-Environment Science. 39(10): 2259–2266.
2. Li, Y., Shang, J.Y., Huang, Y.Z., et al. (2020) Research progress of passivation materials in cadmium-arsenic contaminated soil. Acta Pedologica sinica.1–12.
3. He, Y.L. (2020) Speciation and bioavailability of cadmium and arsenic in soils. Journal of Green Science and Technology. 10: 108–109+112.
4. Shan, T.Y., Liu, Q.X., Yan, X.L., et al. (2017) Cd and As absorption and transport characteristics of rice in a paddy field co-contaminated by Cd and As. Journal of Agro-Envirornment Science. 36(10): 1938–1945.
5. Zhao, D.B., Cao, Z., She, W., et al. (2015) Effects of Cd, As stress on growth and Cd, As uptake of Ramie (Boehmeria nivea L.). Plant fiber sciences in China., 37(04): 183–188.
6. Bao, S.D. (2005) Agricultural soil analysis. Chinese Agricultural Press. Beijing.
7. National environmental protection standards of the people's Republic of China. (2009) GBT23739-2009GB Soil quality-Analysis of available lead and cadmium contents in soils-Atomic absorption spectrometry. Standardization Administration of China.
8. National environmental protection standards of the people's Republic of China. (2008) GBT22105.2-2008 Soil quality-Analysis of total mercury , arsenic and lead contents-Atomic fluorescence spectrometry. Standardization Administration of China.
9. Yang, Y.P., Wang, P., Yan, H.J., et al. (2019) NH4H2PO4-extractable arsenic provides a reliable predictor for arsenic accurmulation and speciation in pepper futs(Capsicum annum L.). Environmental Pollution. 251:651–658.
10. National environmental protection standards of the people's Republic of China. (2016) HJ804-201Soil - Determination of bioavailable form of eight elements -Extraction withbuffered DTPA solution/Inductively coupled plasma optical emission spectrometry. China Environmental Science Press: issued by Ministry of environmental protection.
11. Ostonen, I., Pittsepp, U., Biel, C., et al. (2007) Specific root length as an indicator of environmental change. Plant Biosystems-An International Journal Dealing with all Aspects of Plant Biology. 141(3): 426–442.
12. Wang, X.J., Wang, W.B.,Yang, L., et al. (2015) Transport pathways of cadmium (Cd) and its regulatory mechanisms in plant. Acta Ecologica Sinica. 35(23): 7921–7929.
13. Lu, Z.W., Zhang, Z., Su, Y., et al. (2013) Cultivar variation in morphological response of peanut roots to cadmium stress and its relation to cadmium accumulation. Ecotoxicology and Environmental Safety. 91(4): 147. DOI: 10.1016/j.ecoenv.2013.01.017.
14. Gao, Q.L., Zhen, R.L., Li, H.F. (2010) Effects of transpiration rate and root character on cadmium absorption by pakchoi cultivars. Chinese Journal of Ecology. 29(09): 1794–1798.
15. Li, J.H., Xu, C.Y., Zhu, J.Y., et al. (2019) Phenotypic Adaptation Strategy of Cotinus coggygria Seedlings in Continuous Drought Environments. Journal of Northwest Forestry University. 34(2): 28–34.
16. Gu, J.F., Zhou, H., Yang, W.T., et al. (2016) Effect of Combined Soil Amendment Regulating Chemical Forms of Cadmium and Arsenic in Paddy Soil and Their Bioaccumulation and Translocation in Rice. Acta Pedologica sinica. 53(6): 1576–1585.
17. Zhou, L., Zhen, X.Q., Ding, Y.Z., et al. (2017) Probes of porevention and contrml of famland poluation by cadinium & arsenic and crmp prowhuctonsafety. Journal of Agro-Environment Science. 36(4): 613–619.
18. Norton, G.J., Dasgupta, T.M., Islam, M.R., et al. (2010) Arsenic influence on genetic variation in grain trace-element nutrient content in Bengal Delta grown rice.Environmental science & Technology. 44: 8284–8288.
19. Li, H.F., Zhen, Z.Y., Zhang, F.S., et al. (1999) Effect of iron on the uptake of Cd from different compounds by wheat plants. Acta Ecologica Sinica. 19(2): 170–173.
20. Yang, Z.M., Zhen, S.J., Hu, A.T., et al. (1999) Research Progress on interaction between phosphorus and heavy metal elements zinc and cadmium in plants. Plant Nutrition and Fertilizer Science. 5(4): 366–376.
21. Ye, W.L., Li, L.L., Lu H.J., et al. (2015) Effet of zinc fertilization on cadmium and arsen ic accumulation in wheat. J Environm en talScience & Technology. 38(7): 17–20.

22. Guo, G.L., Zhou, Q.X., Li, X.Y. (2005) Advances in research on in situ chemo-immobilization of heavy metals in contaminated soils. Chinese Journal of Applied Ecology. 16(10): 1990–1996.
23. Cheng, D.W., Zhang, G.P., Yao H.G., et al. (2006) Genotypic and environmental variation and their stability of As, Cr, Cd, Ni and Pb Concentrations in the Grains of Japonica Rice. Acta Agronomica Sinica, 32(4): 573–579.

Effect of Ultrasound Treatment on the Sterilization and Quality of *Cucurbita pepo*

Suo Guanwen[1], Cai Meijuan[1], Zhou Chunli[a1], Su Wei[1], Hu Xueyan[1]

[1](School of Life Science, Jiangxi Science &Technology Normal University (Nanchang 330013)

Abstract. The influence of ultrasonic technology on the sterilization effect and quality of Cucurbita pepo juice were studied in this paper. The sterilization rate under conditions of control 100W, 200W, 300W ultrasonic, color difference, total chlorophyll content, pH changes on juice Cucurbita pepo were determined. The results showed that the total color difference, browning index, whiteness index and total chlorophyll content of the juice had significant changes, while the pH had no significant changes. Ultrasonic has a certain influence on the sterilization effect of Cucurbita pepo, and ultrasonic treatment is helpful to maintain the quality of pumpkin juice in a certain range. In the future, it is hoped that ultrasonic technology will combine with other technologies to achieve commercial sterility while minimizing the loss of quality in Cucurbita pepo.

1 Introduction

Cucurbita pepo, commonly known as zucchini, is a common vegetable on the family table, rich in nutritional value, deeply loved by people. Pumpkin pulp sweet and mild, are rich in polysaccharides, carotenoids, pectin, vitamins (C, E, K, thiamine riboflavin and pyridoxine), amino acids, minerals (Fe, Mg, K, P, Se), dietary fiber, phenolic compounds (flavonoids, phenolic acids) and other substances beneficial to human health, so which has high nutritional and medicinal value [1-7]. *Cucurbita pepo* in the past use of stacking, cellaring, storage and other ways, there are certain requirements for space, temperature, humidity, general requirements need ventilation, temperature is not too high or too low, higher than the set temperature pumpkin easy to rot, less than 0 ℃ prone to cold damage.

Maintain asepsis can extend the storage period of *Cucurbita pepo* in the processing, storage, transportation. The high temperature of the traditional thermal processing sterilization method will have a certain impact on the quality of pumpkin, while the Ultrasonic treatment can reduce the loss of nutritional quality of Cucurbita pepo. Ultrasonic processing is one of the non-thermal processing technologies to alternative to thermal pasteurization of food products and a new low-cost technology. The mechanism of ultrasonic processing is simple, no chemical residue, high safety, low energy consumption, economic and environmental protection, and product quality deterioration can be ignored [8]. Ultrasonic treatment reduces the pathogenic microorganisms of fruit juice, retains the bioactive compounds of fruit juice, and has effects on the color, antioxidant, vitamin C, polysaccharides, pectin, total phenol content and enzyme activity of fruit juice. It has been found in mango juice [9], cranberry juice [10], grapefruit juice [11], orange juice [12], and other relevant literatures have reported. However, there are few studies to evaluate the effect of ultrasonic treatment on pumpkin juice. The aim of this study is to analyze the effects of ultrasonic treatment on pumpkin juice, and to provide a theoretical basis for ultrasonic processing and production of pumpkin juice and development of fruit juice-pumpkin juice combination products.

2 Materials and methods

2.1 Experimental materials

2.1.1 Material

Pumpkin (bought from Lemaijia Supermarket)

2.1.2 Drugs and reagents

Drug: NaCl, LB Nutrition Agar, Ethanol, Acetone Reagent:

①LB medium, 40 g nutrient AGAR powder was dissolved in 1000 mL distilled water, and autoclaved for 15 minutes after packing;

②The volume fraction was 80% acetone, and 80 mL acetone (analytical pure) was added with distilled water to a constant volume of 100 ml;

③0.9% NaCl solution, 0.9g NaCl was dissolved in a small amount of distilled water, and then the volume was constant to 3 100 ml;

④75% ethanol solution.

[a] Corresponding author: zhoucl6212@126.com

https://doi.org/10.1051/e3sconf/202129003027

2.1.3 Experimental equipment

15 ml plastic centrifuge tube, 25 ml with calibration test tubes, glass rod, a beaker, taper bottle and a cork, wipe mirror paper, gauze, funnel, sterile liquid gun, 1 ml spear, plate count AGAR, constant temperature incubator, Ultrasonic cell crusher SCIENTZ- Ⅱ D (Ningbo Xinzhi Biotechnology Co., Ltd.), super clean workbench, juicer (philips), Colori7 color difference meter, ultraviolet spectrophotometer (Shanghai Fine China Technology Co., Ltd.), Shanghai Lei Magi PHS-3C pH tester.

2.2 Experimental method

2.2.1 Pre-treatment of pumpkin

Clean the Cucurbita pepo, remove the tail, pedicle and fruit pulp, cut into small pieces, and press the juice with a juicer. Divide the pumpkin juice into four parts and put it in a 100 ml brown conical flask, and store it in a refrigerator at 4 ℃ for later use.

2.2.2 Ultrasonic treatment

Ultrasonic treatment conditions: frequency: 25 K Hz; time: 5 min; power: 100 W, 200 W, 300 W; temperature: 25 ℃.

2.2.3 Measurement of chlorophyll

The contents of total chlorophyll in Cucurbita pepo were calculated according to method of Arnon [13].

X=(20.0D663 nm+8.02D645 nm) 25/V (1)

In the formula, X: total chlorophyll content, μ g/ mL fresh weight; V: the volume of pumpkin juice, ml; D663 nm and D645 nm are respectively 663 nm and light absorption value at 645 nm.

2.2.4 Determination of sterilizing rate

Test method reference GB 4789.2—2016. The bactericidal effect of the Cucurbita pepo treated by ultrasonic is expressed by the bactericidal rate, which is calculated by the formula:

$$\text{Sterilizing rate (\%)} = \frac{X_0 - X}{X_0} \times 100 \qquad (2)$$

in the formula, X_0: total number of bacteria before treatment; X: total number of bacteria after treatment

2.2.5 Determination of color

For color determination of the treated pumpkin juice, a colorimeter (Color i7) was used. For measure the color changes, the following parameters were determined: L*(brightness/darkness),a*(redness/greenness),b*(yellowness/blueness), chroma (C), and hue angle (h), total color difference (TCD) and browning index (BI) were expressed according to the following equations [14].

$$C= (a^{*2}+ b^{*2})^{1/2} \qquad (3)$$

$$h=\tan^{-1}(a^*/ b^*) \qquad (4)$$

$$TCD = [(\triangle L)^2 + (\triangle a)^2 + (\triangle b)^2]^{1/2} \qquad (5)$$

$$BI = 100(X-0.31) \qquad (5)$$ in the formula,

$$X=(a^*+1.75L^*)/(5.645L^*+a^*-3.012b^*) \qquad (6)$$

2.2.6 Determination of pH

The pH value of Cucurbita pepo juice was measured by PHS-3C pH tester.

2.3 Statistical Analysis

For the three samples of pumpkin juice in each group. All treatments were carried out in triplicate for each condition and all measurements were done in triplicate. The results were analyzed by Origin 2019 and SPSS 21.

3 Results and analysis

3.1 Influence of ultrasonic technology on the sterilization effect of Cucurbita pepo

As can be seen from Figure 1, with the increase of ultrasonic power, the sterilization rate of Cucurbita pepo gradually increased, but the sterilization rate was low. In the experiment, the ultrasonic power was low, and the time of 5min was short, and the ultrasonic temperature was controlled below 25 ℃, which led to the insignificant sterilization effect and failed to reach the state of commercial sterility. Ultrasonic sterilization mechanism is due to its cavitation can have a certain impact on the structure of microbial cells, cavitation makes cells and organelles within the cell, all kinds of substances shock, cell is destroyed and then achieve a sterilization effect [15]. It has been found that ultrasound treatment alone is effective in controlling the microbiome (thermophilic bacteria, cryophilic bacteria, and total coliforms), but the combination of heat and ultrasound is more suitable for controlling pathogenic bacteria and spoilage bacteria [15]. However, the combination of heat and ultrasound is less effective in controlling pathogenic bacteria and spoilage bacteria. In the future research, it is necessary to adjust the ultrasonic treatment parameters to further explore its sterilization effect.

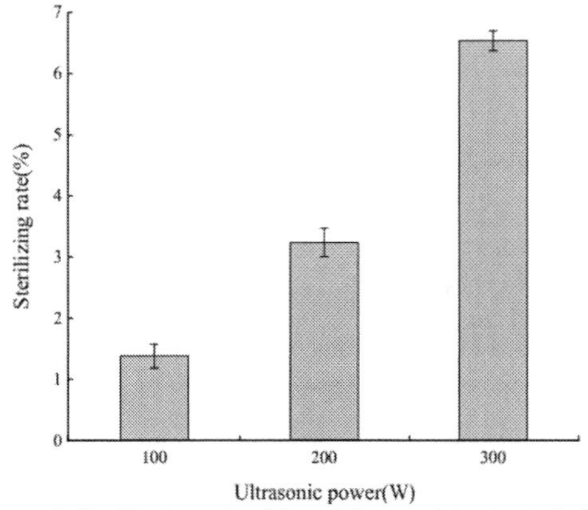

Figure 1. Sterilization rate of Cucurbita pepo juice treated with ultrasound

3.2 Influence of ultrasonic technology on color aberration and PH of Cucurbita pepo

Table 1. Effects of ultrasound treatment on the color attributes and PH in the pumpkin juice

	Control	100W	200W	300W
C	63.41±0.024	61.90±0.047	59.18±0.038	62.31±0.007
h	94.10±0.009	94.62±0.006	95.21±0.003	94.74±0.006
TCD	—	7.81±1.110	30.37±1.790	9.45±1.200
BI	127.37±4.613	193.33±0.650	257.22±2.174	230.33±0.826
PH	6.17±0.120	6.12 ±0.170	6.11± 0.230	6.14±0.150

In this study, Effects of ultrasound treatment on the color attributes in the pumpkin juice was evaluated (Table1). As shown in Table 1, The main results suggest that ultrasound did not induce significant changes in C and h values, but did induce significant changes in TCD, BI, Wang et al, by studying the application of ultrasound in strawberry juice, observed that the technique was responsible for an increase in the parameter TCD [16]. Similarly, in Souza et al study, sonication significantly affected all color parameters evaluated in cape gooseberry - Physalis peruviana juice [17]. This may be due to the inhibition or activation of chromatic aberration-related enzymes caused by the cavitation effect during the ultrasonic process under extreme physical conditions, thus measuring this result [18,19,20]. At the same time, the use of ultrasound can cause greater cell disruption, leading to the release of intracellular compounds, which can serve as substrates for oxidative processes and thereby affect the color of the final product [21].

Acidity is an important parameter of juice quality. It affects not only the taste of the juice but also the possibility of the development of microorganisms [22]. It can be seen from the figure that there is no significant change in the pH of untreated pumpkin juice and ultrasonic treated Cucurbita pepo, which is consistent with the result of Starek et al. 's study on the influence of ultrasonic treatment on the pH of tomato juice [22]. The effect of sonication on pH of juice have been reported also by other authors—Bhat et al and Abid et al [23,24]. These results show that the ultrasound treatment does not change the pH quality of pumpkin juice.

3.3 Influence of ultrasonic technology on chlorophyll content of Cucurbita pepo

It can be seen from Figure 2. that the total chlorophyll content in Cucurbita pepo decreases with the increase of ultrasonic power. With the increase of ultrasonic power, the total chlorophyll content in Cucurbita pepo decreases, which may be due to the energy and cavitation effect in ultrasonic, which destroys the cellular structure and decomposes unsteadily after the release of chlorophyll, leading to the decrease of chlorophyll content. If you want to better retain chlorophyll, ultrasonic power should not be too large. This result is related to Ordnonez-Santos et al. had the opposite effect on the carotenoid content of Cape Gooseberry Juice after ultrasonic treatment [25]. This study results indicated that ultrasonic wave could inhibit the loss of fruit and vegetable quality to a certain extent. The specific influencing mechanism needs further study and proof.

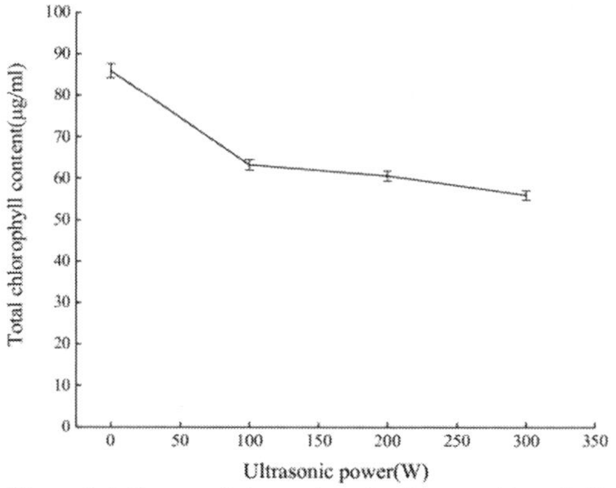

Figure 2. Influence of ultrasonic technology on chlorophyll content of Cucurbita pepo

4 Conclusion

In this study, the bactericidal effect, visual color, total chlorophyll content and pH value of *Cucurbita pepo* juice were evaluated by ultrasonic treatment. After ultrasonic treatment, C, h and pH values did not change significantly, TCD and BI increased significantly, but total chlorophyll content decreased significantly. The results show that ultrasonic treatment is helpful to improve the quality of pumpkin juice to a certain extent. In the further study, to improve the technical parameters, it is expected to replace the hot processing technology.

Reference

1. Zdunić, Gordana, Menković, Nebojša, Jadranin, Milka, Novaković, Miroslav, Savikin, Katarina P, Živković, Jelena Č. Phenolic compounds and carotenoids in pumpkin fruit and related traditional products[J]. Institute of Chemistry, Technology and Metallurgy, 70(2016):429-433.
2. S Sharma, T V R Rao. Nutritional quality characteristics of pumpkin fruit as revealed by its biochemical analysis[J]. Int Food Res J, 20(2013):2309–2316.
3. C H Azevedo-Meleiro, D B Rodriguez-Amaya. Qualitative and quantitative differences in carotenoid composition among Cucurbita moschata, Cucurbita maxima, and Cucurbita pepo[J]. J Agric Food Chemistry, 55(2007):4027–4033.
4. C Kurz, R Carle, A Schieber. HPLC-DAD-MSn characterisation of carotenoids from apricots and pumpkins for the evaluation of fruit product authenticity[J]. Food Chemistry, 110(2008):522–530.
5. A Nawirska-Olszanska, A Biesiada, A, Sokol-Letowska, A Z Kucharska. Characteristics of organic acids in the fruit of different pumpkin species[J].

Food Chemistry, 148(2014):415–419.

6. C L Zhou, W Liu, J Zhao, C Yuan, Y Song, D Chen, Y Y Ni, Q H Li. The effect of high hydrostatic pressure on the microbiological quality and physical–chemical characteristics of pumpkin (Cucurbita maxima Duch.) during refrigerated storage[J]. Innov Food Sci Emerg Technol. 21(2014):24–34.
7. M Y Kim, E J Kim, Y N Kim, C Choi, B H Lee. Comparison of the chemical compositions and nutritive values of various pumpkin (Cucurbitaceae) species and parts[J]. Nutr Res Pract, 6 (2012):21–27.
8. X Li, L Zhang, Z Peng. The Impact of Ultrasonic Treatment on Blueberry Wine Anthocyanin Color and its In-vitro Anti-Oxidant Capacity[J]. Food Chemistry, 333(2020):127455.
9. V Santhirasegaram, Z Razali, C Somasundram. Effects of thermal treatment and sonication on quality attributes of Chokanan mango (Mangifera indica L.) juice[J]. Ultrasonics Sonochemistry, 20(2013):1276-1282.
10. W F Gomes, B K Tiwari, Ó Rodriguez, E S de Brito. Effect of ultrasound followed by high pressure processing on prebiotic cranberry juice[J]. Food chemistry, 218(2017):261-268.
11. R M Aadil, X A Zeng, Z H Zhang, M S Wang. Thermosonication: A potential technique that influences the quality of grapefruit juice[J]. International Journal of Food Science & Technology, 50(2015):1275–1282.
12. P Khandpur, P R Gogate. Understanding the effect of novel approaches based on ultrasound on sensory profile of orange juice[J]. Ultrasonics Sonochemistry, 27(2015):87–95.
13. D I Arnon. Copper enzymes in isolated chloroplasts. polyphenoloxidase in Beta vulgaris[J]. Plant Physiology, 24(1949):1-15.
14. [14] L E Ordóñez-Santos, J Martínez-Girón, M EArias-Jaramillo. Effect of ultrasound treatment on visual color, vitamin C, total phenols, and carotenoids content in Cape gooseberry juice[J]. Food Chemistry, 233(2017):96–100.
15. L M Carrillo-Lopez, I A Garcia-Galicia, J M Tirado-Gallegos et al. Recent advances in the application of ultrasound in dairy products: Effect on functional, physical, chemical, microbiological and sensory properties[J]. Ultrasonics Sonochemistry, 73(2021):105467.
16. J Wang, SK Vanga, V Raghavan. High-intensity ultrasound processing of Kiwifruit juice: Effects on the ascorbic acid, total phenolics, flavonoids and antioxidant capacity[J]. LWT, 107(2019):299–307.
17. Francisca das Chagas do Amaral Souza a, A Lgsm, A Kdob et al. Thermosonication applied on camu–camu nectars processing: Effect on bioactive compounds and quality parameters[J]. Food and Bioproducts Processing, 116(2019):212-218.
18. T J Mason, E Joyce, S S Phull, J P Lorimer. Potential uses of ultrasound in thebiological decontamination of water[J]. Ultrason Sonochem, 10(2003):319–323.
19. T G Leighton. What is ultrasound? [J]. Prog Biophys Mol Biol, 93(2007):3–83.
20. A R Jambrak, V Lelas, Z Herceg, M Badanjak, Z Werner. Application of high-power ultrasound in drying of fruits and vegetables | Primjenaultrazvukavisokesnage u sušenjuvoc´aipovrc´a, Kem. U Ind, Chem. Chem. Eng. 59(2010):169–177.
21. J Wang, S K Vanga, V Raghavan. High-intensity ultrasound processing of kiwifruit juice: Effects on the microstructure, pectin, carbohydrates and rheological properties. Food Chemistry, 313(2020)126121.
22. A Starek, Z Kobus, A Sagan. Influence of ultrasound on selected microorganisms, chemical and structural changes in fresh tomato juice[J]. Scientific Reports, 11(1) 2021.
23. R Bhat, N S B C Kamaruddin, L Min-Tze, A A Karim. Sonication improves kasturi lime (Citrus microcarpa) juice quality. Ultrason. Sonochem, 18(2011):1295–1300.
24. M Abid et al. Effect of ultrasound on different quality parameters of apple juice. Ultrason Sonochem, 20(2013):1182–1187.
25. L Eduardo Ordonez-Santos, J Martinez-Giron, M Enith Arias-Jaramillo. Effect of ultrasound treatment on visual color, vitamin C, total phenols, and carotenoids content in Cape gooseberry juice[J]. Food Chemistry, 233(2017):96-100.

https://doi.org/10.1051/e3sconf/202129003028

The Review of Biosensor and its Application in the Diagnosis of COVID-19

Jiyun Wu[1,*,†] and Qiuyao Wu[2,†]

[1]School of Chemistry, Faculty of Science, UNSW
[2]Hangzhou Foreign Language School CAL Center
[†]These authors contribute equally.

Abstract. The objective of this article is to summarize the available technologies for biosensing applications in COVID-19. The article is divided into three parts, an introduction to biosensing technologies, applications of mainstream biosensing technologies and a review of biosensing applications in COVID-19. The introduction of biosensors presents the history of inventing the biosensing technology, which refers to the ISFET. The resonant biosensor with the example of MEMS. the principle of optical biosensor, and the thermal biosensor. In the second part, the main use of biosensing techniques, it was discussed the field of the food industry, environmental monitoring, and the medical industry. In the part of biosensor application in COVID-19, it was mentioned that the technique of POCT, the use of RT-LAMP-NBS in the early detection in China, and the use in gRT-PCR for the detection of the DNA code to determine the presence of pathogen of COVLD-19 in the human body.

1 Introduction

According to recent reports from WHO, by 6:02pm CEST, 22 April 2021, there have been 143,445,675 confirmed cases of COVID-19, including 3,051,736 deaths [1]. Moreover, the number of new cases has been increasing for 8 weeks continuously, the number of new deaths increased for 5 weeks continuously by 18 April 2021, 10 am CET [2]. Based on the current situation, there is an increasing demand for the ability to diagnosing COVID-19 to stop spreading. One possible method is the biosensor. It is not the first time that biosensor is used to test the virus: type A influenza virus can be detected by Surface Plasmon Resonance biosensor (SPR) [3]. This makes SPR a possible choice. Another choice may be an electrochemical biosensor (SPR is an optical biosensor) [4]. In fact, not only the two types of biosensor but also many other kinds of biosensors are currently or are going to playing increasingly important roles in preventing the spread of the virus, like point-of-care biosensors (POC) and the requirement have developed from detecting the virus inside the human body through the biological process to various requirements: Chip-based biosensor can produce results instantaneously [5], paper-based electrochemical biosensor has advantages in low cost [6], both 2 biosensors belong to POC [5]. And the target of detection maybe some environment that the virus could survive and spread, like sewage rather than suspected cases [7]. All those requirements ask for research and innovation toward biosensors. In This article, the definition, structure, classification, and application of biosensor would be introduced, and we will analyze and discuss how biosensors would help in the diagnosis of COVID-19.

2 Biosensor introduction

Biosensors are analytical devices that convert a biological response into an electrical signal [8]. In 1962, Clark invented the Clark oxygen electrode, which can be considered as the first biosensor. In general, biosensors are combinations of two parts: a bio element and a sensor element (shown in figure 1). Bio element can be something like enzyme, which interact with specific analyte; sensor element then converts the change of bio element to an electric signal (shown in figure 2). Depending on the transferring mechanism, the biosensors can be separated to many types such as: resonant biosensors, optical-detection biosensors, thermal-detection biosensors, ion-sensitive field-effect transistor (ISFET) biosensors, and electrochemical biosensors [9]. The first biosensor, the Clark electrode, is an electrochemical biosensor using Amperometry to determine the concentration of glucose.

[a] Corresponding author: z5221565@ad.unsw.edu.au

https://doi.org/10.1051/e3sconf/202129003028

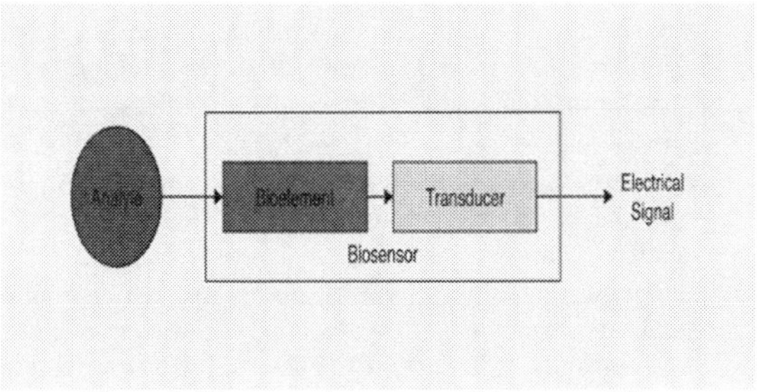

Figure 1. The general structure of biosensors [9].

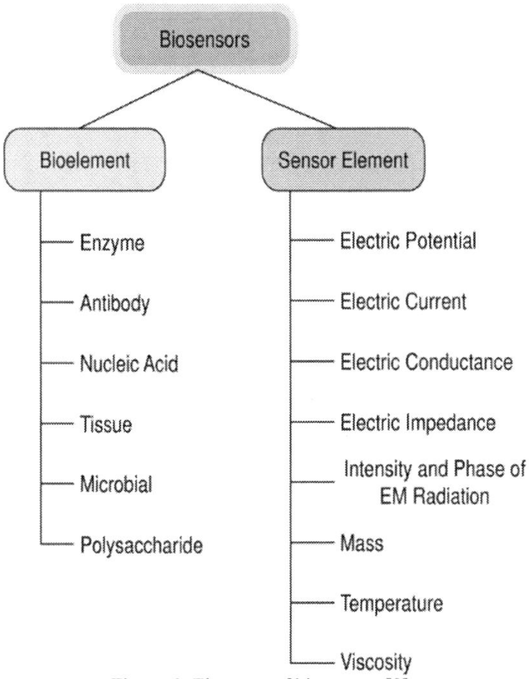

Figure 2. Elements of biosensors [9].

For resonant biosensor, the principle is to let the device coupling with the analyte (antibody or another bio element), then measuring the difference in physical property under the acoustic wave, like resonant frequency, the difference is related to the amount of attached analyte [9, 10]. An advantage of this biosensor is low cost, as it may use quartz in the acoustic sensor. An example is Micro/nano-electromechanical systems (MEMS), which have been used to detect Escherichia coli and vaccinia virus [10-12].

Generally, the principle of the optical biosensor is exploiting the interaction of the optical field with a biorecognition element, generating a signal which is related to the concentration of the analyte. [13]. An example is the Surface plasmon resonance biosensor (SPR), this biosensor was first constructed in research in 1983 [14]. The main advantages of the optical biosensor are that the test is real-time and sensible.

The thermal biosensor is based on detecting and analyzing the heat released during biochemical reactions [15]. The principle is to let the analyte contract the enzyme, causing heat reaction, and measure the change in temperature with temperature sensors, which is related to the amount of analyte that takes part in the reaction [9]. An example is the microelectromechanical systems (MEMS) differential thermal biosensor, which has advantages in thermal detection sensitivities and time responses [15].

Ion-sensitive field-effect transistor (ISFET) biosensor are semiconductor FETs having an ion-sensitive surface. When ions and the semiconductor interact at the surface, the electrical potential would change, which can be measured and is related to the ions (analyte) [9]. An application is using ISFET to monitor the PH change inside the human body, recording the PH change in different position of the digestive system [16].

As mentioned previously, the electrochemical biosensor is the earliest. The general principle of this type of biosensor is to make the analyte a part of an electrochemical cell and analyze its change through the change in some parameters of the cell, like potential or current. Based on that, electrochemical biosensors can be classified as potentiometry, voltammetry, Amperometry, coulometry, conductometry etc. A more current example is research in 2016, using voltammetry to analyze the DNA binding molecule, which is rapid, sensitive and low-cost [17].

3 Biosensor application

Biosensor could be used in many areas. In this paper, we will discuss its application mainly in terms of the food industry, environmental monitoring, and medical industry.

3.1 Food Industry

Food security and quality control are really important in the food industry, and biosensor is a reliable and efficient way. The biosensor can be used to measure many important nutrients in food, like the Tea polyphenols in black tea, protein in pure milk or dairy products, ethanol in alcoholic beverages etc. Or, on the other side, test the possible harmful substances, like bacteria in fresh meat [18]. An example of this is an electrochemical biosensor used to monitor the presence of urea in milk (structure shown in figure 3). This urea biosensor used urease to hydrolyze urea, forming ammonium ion and interact with an ion-selective electrode, then detected by potentiometric transducer [19].

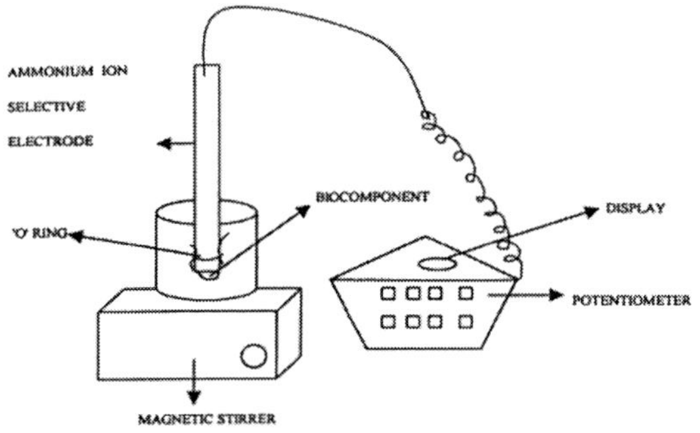

Figure 3. Structure of urea biosensor [19].

3.2 Environmental monitoring

Similar to the food industry, biosensor also has an advantage in terms of cost and efficiency in environmental monitoring, compared to the traditional analytical chemistry method like chromatography [20]. An example of this is the Biochemical Oxygen Demand (BOD) biosensor. BOD is used to measure the oxygen demand of wastewater, indicate the amount of biodegradable organic material (carbonaceous demand) and other important quantitative or semi-quantitative analytical information. The biosensor (as shown in figure 4) makes dissolved oxygen transport through the dialysis membrane and get consumed by the immobilized microorganisms. The remaining oxygen then diffuses through the gas-permeable Teflon membrane and get detected by the oxygen electrode [21].

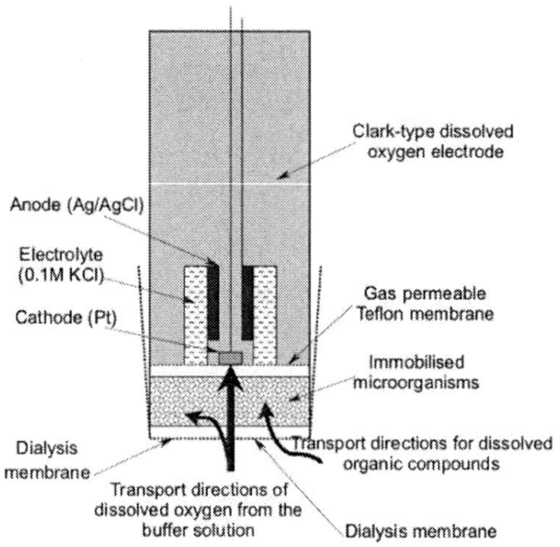

Figure 4. BOD biosensor [21].

3.3 Medical industry

The advantages in cost, like the previous two, also makes biosensor important in medial industry, especially the clinical diagnostics industry. The most common example of this is a single-use glucose strip for diabetic patients. Similarly, the biosensor can be used to monitor important analyte related to diseases, like lactate, urea, creatinine, and cholesterol [22]. Another medical application is the detection of biological agents, which may be not far from people's daily life, like Salmonella, Escherichia coli, Ricin and even some viral pathogens. These may be tested by antibody or some specific DNA fragments [23]. Further discussion will be made relating to COVID-19 in the later part.

4 Biosensor application in COVID-19

COVID-19, which originated in Wuhan, China, in 2019, is extremely widespread and has now been identified by experts as having the potential to become a cyclical epidemic like influenza. Because of the high infection rate, the low number of initial symptoms and the fact that effective agents and vaccines have only been developed in some regions, there is a need for early development of technologies that can easily detect the virus for early medical treatment of patients and to detect the presence of items carrying the virus in important transport channels.

In the early days, before COVID-19 had expanded globally, various inventions were already being developed in China to better detect the virus and to contain its expansion. Blood lactate biosensor used in the diagnosis and treatment of COVID-19. It is proposed that blood lactate level or lactate clearance is an important indicator for COVID-19 stratification, assessment of treatment efficacy and prognosis. The POCT (instant test) blood lactate biosensor is fast, economical, and reliable, enabling rapid and continuous blood lactate testing in clinical patients or individuals (families), facilitating COVID-19 disease severity stratification and assessment of treatment efficacy, as well as enabling patients to identify their condition promptly, relieving psychological concerns and promoting physical recovery [24].

Another invention that played the role in the early detection of COVID-19, is the invention of reverse transcription loop-mediated isothermal amplification combined with gold nano-biosensing in Hainan, China. This invention uses RT-LAMP-NBS for the detection of 2019-nCoV by designing two sets of LAMP amplification primers for the ORF1ab and N genes of 2019-nCoV and labelling the 5' end of the loop primer with a semi-antigen or the two sets of LAMP amplification primers were designed for the ORF1ab and N genes of 2019-nCoV, and the 5' ends of the loop primers were labeled with semi-antigen or biotin to achieve simultaneous detection of both ORF1ab and N genes. The detection system established by the present invention for 2019-nCoV has the advantages of high specificity, high sensitivity, and simple and rapid operation [25].

About a month later, in mid-April, COVID-19 broke out all over the world and became a global infectious disease. More and more comprehensive experiments have been realized, the trend of biosensor for large-scale detection of COVID-19 and the gene detection with qRT-PCR as the diagnostic standard of COVID-19 mentioned in the innovation (shown in figure 5). In RT-PCR, RNA is transformed into complementary DNA, and then the DNA signal is amplified by real-time PCR. In the extension stage of the PCR cycle, when more DNA copies are produced, the fluorescence signal will increase. When it exceeds the set threshold, the test result is positive. If there is no virus in the sample, the fluorescence threshold cannot be achieved, and the detection result is negative. The report also shows Isothermal amplification and point-of-care molecular diagnostics, CRISPR-based diagnosis, and DNA sequencing-based diagnosis of COVID-19 which are based on the bio-sensing technology to detect the antigens in the human bodies [26].

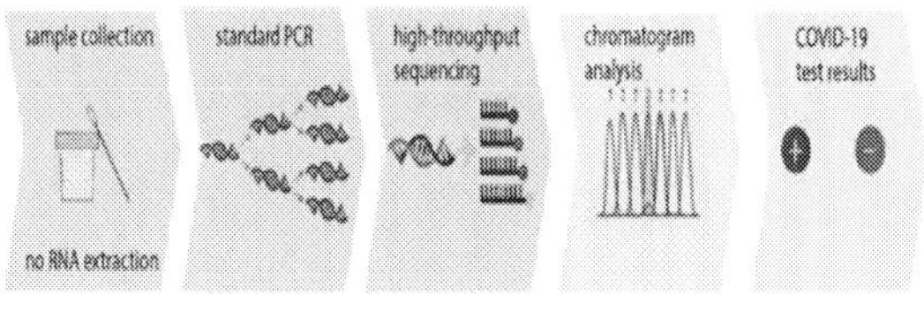

Figure 5. COVID-19 diagnostic whole-genome sequencing workflow [26].

5 Conclusion

To conclude, the biosensing technique becomes more popular in different kinds of fields in this century. There is a trend that the biosensing technique will be used more in the medical aspect and also in the industry aspect which detects the microelements that may affect human health. It will play an important role in the diagnosis, and probably other aspect, of COVID-19. We expect that this article can be helpful to researchers who are interested in this area, give them an basic understanding of biosensors.

Reference

1. "WHO Coronavirus (COVID-19) Dashboard," World Health Organization, accessed April 23, 2021, https://covid19.who.int/
2. "COVID-19 Weekly Epidemiological Update," World Health Organization, last modified April 20, 2021, https://www.who.int/publications/m/item/weekly-epidemiological-update-on-covid-19---20-april-2021
3. Amano, Yoshihisa, and Quan Cheng. "Detection of influenza virus: traditional approaches and development of biosensors." Analytical and bioanalytical chemistry 381.1 (2005): 156-164.
4. Abid, Suhad Abbas, et al. "Biosensors as a future diagnostic approach for COVID-19." Life Sciences (2021): 119117.
5. Choi, Jane Ru. "Development of point-of-care biosensors for COVID-19." Frontiers in chemistry 8 (2020): 517.
6. Yakoh, Abdulhadee, et al. "based electrochemical biosensor for diagnosing COVID-19: Detection of SARS-CoV-2 antibodies and antigen." Biosensors and Bioelectronics 176 (2021): 112912.
7. Mao, Kang, Hua Zhang, and Zhugen Yang. "An integrated biosensor system with mobile health and wastewater-based epidemiology (iBMW) for COVID-19 pandemic." Biosensors and Bioelectronics 169 (2020): 112617.
8. Mehrotra, Parikha. "Biosensors and their applications–A review." Journal of oral biology and craniofacial research 6.2 (2016): 153-159.
9. Mohanty, Saraju P., and Elias Kougianos. "Biosensors: a tutorial review." Ieee Potentials 25.2 (2006): 35-40.
10. Fogel, Ronen, Janice Limson, and Ashwin A. Seshia. "Acoustic biosensors." Essays in biochemistry 60.1 (2016): 101-110.
11. Gfeller, Karin Y., Natalia Nugaeva, and Martin Hegner. "Micromechanical oscillators as rapid biosensor for the detection of active growth of Escherichia coli." Biosensors and Bioelectronics 21.3 (2005): 528-533.
12. Johnson, Luke, et al. "Characterization of vaccinia virus particles using microscale silicon cantilever resonators and atomic force microscopy." Sensors and Actuators B: Chemical 115.1 (2006): 189-197.
13. Damborský, Pavel, Juraj Švitel, and Jaroslav Katrlík. "Optical biosensors." Essays in biochemistry 60.1 (2016): 91-100.
14. Liedberg, Bo, Claes Nylander, and Ingemar Lunström. "Surface plasmon resonance for gas detection and biosensing." Sensors and actuators 4 (1983): 299-304.
15. Wang, Li, et al. "A MEMS thermal biosensor for metabolic monitoring applications." Journal of microelectromechanical systems 17.2 (2008): 318-327.
16. Duroux, Ph, et al. "The ion sensitive field effect transistor (ISFET) pH electrode: a new sensor for long term ambulatory pH monitoring." Gut 32.3 (1991): 240-245.
17. Diculescu, Victor Constantin, Ana-Maria Chiorcea-Paquim, and Ana Maria Oliveira-Brett. "Applications of a DNA-electrochemical biosensor." TrAC Trends in Analytical Chemistry 79 (2016): 23-36.
18. Meshram, B. D., et al. "Biosensor and its application in food and dairy industry: a review." Int. J. Curr. Microbiol. Appl. Sci 7 (2018): 3305-3324.
19. Verma, Neelam, and Minni Singh. "A disposable microbial based biosensor for quality control in milk." Biosensors and Bioelectronics 18.10 (2003): 1219-1224.
20. Rodriguez-Mozaz, Sara, et al. "Biosensors for environmental monitoring: A global perspective." Talanta 65.2 (2005): 291-297.
21. Liu, Jing, and Bo Mattiasson. "Microbial BOD sensors for wastewater analysis." Water Research 36.15 (2002): 3786-3802.
22. Malhotra, Bansi D., and Asha Chaubey. "Biosensors for clinical diagnostics industry." Sensors and Actuators B: Chemical 91.1-3 (2003): 117-127.
23. Kirsch, Jeffrey, et al. "Biosensor technology: recent advances in threat agent detection and medicine." Chemical Society Reviews 42.22 (2013): 8733-8768.
24. H. Sun, G. Gao, Q. Liu, Y. Ma, J. Li, J. Shi, Exploring the use of blood lactate biosensors in the diagnosis and treatment of COVID-19, 2019
25. X. Wang, X. Zhu, L. Wang, H. Li, X. Fu, Reverse transcription loop-mediated isothermal amplification combined with gold nano-biosensing for the detection of COVID-19, 2019

https://doi.org/10.1051/e3sconf/202129003028

26. Santiago, Ibon. "Trends and innovations in biosensors for COVID-19 mass testing." ChemBioChem 21.20 (2020): 2880.

https://doi.org/10.1051/e3sconf/202129003029

Study on Coupling Effect of Additives on NOx Control in Coal Pyrolysis-Combustion

Xianglu Kong[1,3], SongYang[2,3], Shoujun Liu[2,3],Kaixia Zhang[1,3], Tingting Jiao[1,3] and Ju Shangguan[1,3*]

[1]State Key Laboratory of Clean and Efficient Coal Utilization, Taiyuan University of Technology, Taiyuan 030024, Shanxi, China
[2]College of Chemistry and Chemical Engineering, Taiyuan University of Technology, Taiyuan 030024, Shanxi, China
[3]Shanxi Engineering Center of Civil Clean Fuel, Taiyuan University of Technology, Taiyuan 030024, Shanxi, China

Abstract. A large amount of nitrogen oxides produced by loose coal combustion has an important impact on the ecological environment. To solve this problem, it is proposed to prepare clean coke instead of loose coal combustion to reduce the emission of nitrogen oxides from civil coal combustion. Clean coke is prepared by pyrolysis raw coal adding additives, and the gas generated by pyrolysis is collected uniformly, thus avoiding the emission problem in the process of loose burning. In addition, the clean fuel catalyzes the reduction of nitrogen oxides to produce N_2 in the combustion process, thus reducing the emission of combustion nitrogen oxides. In this paper, the additives were investigated, and finally it was found that loading 1 wt.% Fe and Ni had a better effect of decrease nitrogen content in coke through pyrolysis of coal and denitrification during the combustion of coke, and had a coupling effect on nitrogen oxide control in the pyrolysis-combustion process.

1 Introduction

Coal occupies a large proportion in China's energy consumption, and its status as the main body of energy in China will not change in a short time. But the nitrogen oxides produced by coal combustion will cause environmental problems and other social problems. Therefore, it is very important to study clean fuels for energy reduction. Before coal combustion, a part of volatile nitrogen will be released through a pyrolysis process. If the volatile nitrogen can be effectively controlled, it is helpful to reduce the emission of nitrogen oxides in coal combustion. Sowa et al [1] made relevant research and found that iron additives can really change the law of nitrogen migration, improve the yield of volatile matter, and play an important role in reducing the nitrogen content of coke. He et al [2] carried out swelling treatment on coal, and then carried out pyrolysis with additives Ca, Co and Cu, found that the better nitrogen removal ratio was Co and then Cu. In addition, Deng et al [3] found that nitric oxide is always the main component of nitrogen oxides during coal combustion, and its release amount increases sharply with the increase of combustion temperature. Zhu et al [4] found that Ce, Co and Cu additives have better catalytic activity for nitrogen oxides, and iron-cerium composite additives have great synergistic effect on nitrogen oxides control. Lei et al [5] found that the reduction ratio of NO supported by Mn-Ce bimetallic promoter was higher than that of single metal. Many scholars have done a lot of work on the control of nitrogen oxides during pyrolysis and combustion. Some additives can really play a good role in reducing the solid nitrogen content during pyrolysis, but the effect during combustion is not good. In the combustion process, it is studied that some additives can reduce NOx in selective catalytic reduction, so as to reduce NOx emission in the combustion process, but they are not suitable for the combustion of loose coal due to limited conditions and other harsh conditions. In view of this, we hope to find a kind of auxiliary agent for combustion of loose coal, which can reduce nitrogen to prepare clean coke during pyrolysis and produce selective catalytic reduction during combustion, so as to play a role in coupling control between coal pyrolysis and combustion, so as to study the coupling control effect of nitrogen oxides and reduce the overall nitrogen oxide emission from coal combustion.

2 Experimental part

2.1 Coal sample treatment

Firstly, the raw coal is pretreated, crushed and sieved, and the coal sample with a particle size of about 0.2mm is dried in an oven at 65℃ for about 12 hours and taken out for later use, hereinafter abbreviated as CYM. After that, the raw coal was treated by ethanol swelling to expand the molecular network structure of coal and enhance the pyrolysis reaction performance. The swelling treatment method is as follows: 100g coal sample with particle size larger than 80 meshes is put in a beaker, mixed evenly according to the ratio of coal to ethanol of 2:1, covered with plastic wrap, stirred in a magnetic stirrer for 6 hours, then stood for 12 hours, and dried in an oven at 50℃ to

*Corresponding author:shanggj62@163.com

https://doi.org/10.1051/e3sconf/202129003029

obtain the swelling coal sample named RZM.

Table 1. Industrial analysis and elemental analysis of raw coal (wt. /%)

Name	Industrial analysis				Elemental analysis				
	M_{ad}	A_d	V_{daf}	FC_{ad*}	C_{ad}	H_{ad}	N_{ad}	O_{ad*}	$S_{t,ad}$
Jet coal	10.74	11.38	35.55	50.98	61.83	4.43	0.75	10.41	0.46
Swelling coal	5.47	11.33	36.63	53.12	67.22	4.61	1.23	18.30	0.48

Note:d,dry base; ad,air drying base; daf ,dry and ashless; * ,calculated by difference subtraction

2.2 Additive loading

In CYM pyrolysis experiment, $FeCl_3$, $CaCl_2$ and NaCl additives with 1wt.% raw coal were loaded into raw coal by mechanical mixing and stirred evenly. After that, the pyrolysis experiment of RZM adopts the loading mode of equal volume impregnation, and the chloride of metal auxiliary agent is dissolved in proper deionized water, and then coal sample can be added for complete absorption, which is dried in an oven at 65℃ for 2h and taken out for later use.

2.3 Experimental process

Pyrolysis experiment: Accurately weigh 4.0000g of coal sample and put it into a quartz tube. Under the protection of high purity argon, the temperature is raised from room temperature to 1000℃ at a rate of 25℃/min. After being kept constant for 30min, it is naturally lowered to room temperature. The solid product coke is collected, the liquid tar is collected from the condenser, and the gaseous products such as HCN, NH_3 and N_2 are collected respectively. Wherein nitrogen is collected by an air bag; NH_3 and HCN produced by pyrolysis were absorbed by 1000mL, 0.02mol/L methanesulfonic acid solution and 800mL, 0.1mol/L NaOH solution respectively. Among them, CYM-1000 refers to coke produced by direct pyrolysis of long flame coal at 1000℃; CYM-Fe-1000 refers to coke made from long flame coal loaded with 1 wt.%Fe at 1000℃.

Coke combustion experiment is also carried out in a small tube furnace. Every time, weigh 0.5000g of coal (coke) sample and put it into a quartz tube,Both sides of the sample are equipped with high-temperature resistant Shi Ying cotton to ensure the uniform passage of gas. First, introduce argon gas to purge the quartz tube to ensure the replacement of air. After the temperature is programmed to the specified temperature and stabilized, switch the gas to O_2/Ar, push the constant temperature zone of the furnace to the charging zone to start the combustion experiment, and use the flue gas analyzer to detect NO in the combustion flue gas online. Coal ash is collected after the experiment is completed for subsequent testing. Each experiment was repeated three times. When the error of parallel experimental data was less than 2%, the average value was taken.

2.4 Sample analysis and characterization

Tar nitrogen was determined by chemiluminescence nitrogen analyzer, the concentration of NH^{4+} and CN^- in the absorption solution was determined by solution absorption and detection method. N_2 is detected by gas chromatograph. The chromatographic carrier gas is Ar, and the nitrogen content of the collected gas is determined by external standard method. NOx in flue gas is detected online by flue gas analyzer. The components in coke and combustion ash were characterized by XRD.

Calculation formula of coke nitrogen yield and nitrogen removal ratio: coke nitrogen yield (%) = [nitrogen content in coke (mg)]/[nitrogen content in raw coal (mg)] × 100%; Nitrogen removal ratio (%)=[1-(nitrogen content of coke/nitrogen content of raw coal) ×100%]. Volatile nitrogen such as NH_3, HCN, N_2 and tar nitrogen are calculated based on the nitrogen content in raw coal. Taking HCN as an example, the formula is: HCN generation ratio = [N content in generated HCN (mg)]/[N content in raw coal (mg)]×100%. The amount of nitric oxide produced is shown by the following formula:

$$m_{NO}=Q\times\frac{M_{NO}}{22.4}\times10^{-6}\int_0^t C_{NO}\ (t)\ dt$$

m_{NO} -The amount of NO produced by coke combustion,mg

Q-Volume flow of air,mL/min

C_{NO}(t)-Volume concentration of NO in flue gas at time t,ppm

M_{NO}-The molar mass of NO,30g/mol

3 Results & Discussion

3.1 Effect of additives on nitrogen distribution in coal pyrolysis

In order to investigate the influence of the interaction between minerals in raw coal long-flame coal and additives on nitrogen distribution during pyrolysis, pyrolysis experiments were carried out on long-flame coal without and with 1 wt.% additives in a tube furnace at 1000℃, and pyrolysis coke, tar and pyrolysis gas were collected and their nitrogen content was calculated, in which N_2 is calculated by difference subtraction to achieve nitrogen balance, as shown in Fig. 1.It can be seen from the Fig. 2 that iron, calcium and sodium additives loaded in raw coal will reduce the nitrogen content of coke, which shows that the interaction between additives and minerals in coal shows a positive effect on the nitrogen removal from coal, catalyzing the transformation of solid nitrogen in coal to other phases. The nitrogen removal ratio of raw coal and after adding calcium, sodium and iron are 38%, 38.75%, 40.07% and 42.89% respectively. The best additives for nitrogen removal are iron, then sodium and calcium. It was observed that iron loading caused more coke nitrogen to be converted into N_2.

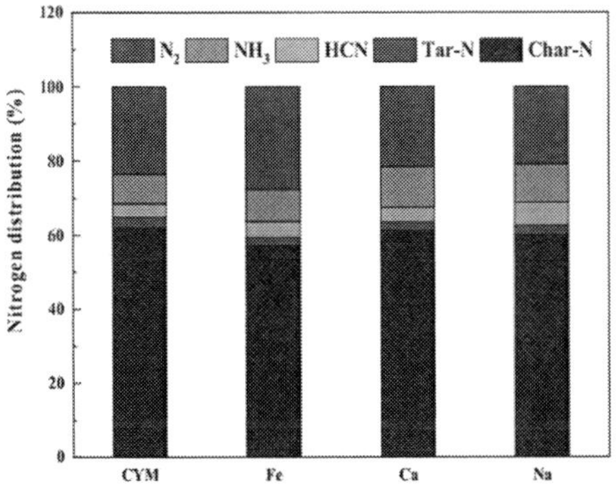

Fig. 1. Nitrogen balance of pyrolysis of raw coal and supported additives at 1000℃.

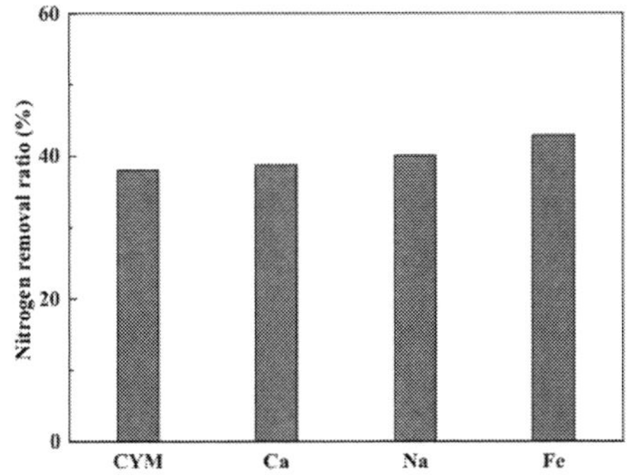

Fig. 2. Nitrogen removal ratio of raw coal pyrolysis with auxiliary agent at 1000℃.

3.2 Effect of additives on nitrogen reduction in pyrolysis and denitrification in combustion

In order to further explore an assistant with better effect in reducing nitrogen in coal pyrolysis and denitration by combustion, experiments were carried out with swelling treated coal as raw coal, and pyrolysis experiments were carried out at 1000℃ under Ar conditions. Gas-solid-liquid three phases were collected to detect their nitrogen content, the overall nitrogen content of tar and HCN is calculated by difference subtraction to balance nitrogen as shown in Fig. 3. It is observed that when Ti, Fe and Ni are loaded, the nitrogen content of coke is low, and more nitrogen migrates into gas phase N_2, and the nitrogen removal ratio are 47.98%, 30.72% and 30.02%, respectively. When loading two metals at the same time for pyrolysis, it is found that the effect of loading iron and cerium is relatively good under this condition, but the overall effect is still not ideal, this shows that the loading of promoter and the loading ratio of composite promoter play an important role in the catalytic activity, and it is necessary to study the catalytic effect of composite promoter separately in the later research. Fig. 4 shows that the coke collected by pyrolysis burns at 1000℃, and the NO production is detected and collected online and compared. It is found that the NO production of loaded Ni-Co-Cu-Fe-Ti compared with swelling raw coal decreases to different degrees, which are 0.2608mg, 0.05886mg, 0.0676mg, 0.1560mg, 0.1927mg and 0.2467mg respectively. In the combustion process, it is found that Ni, Co and Fe have better effects on inhibiting NO generation.

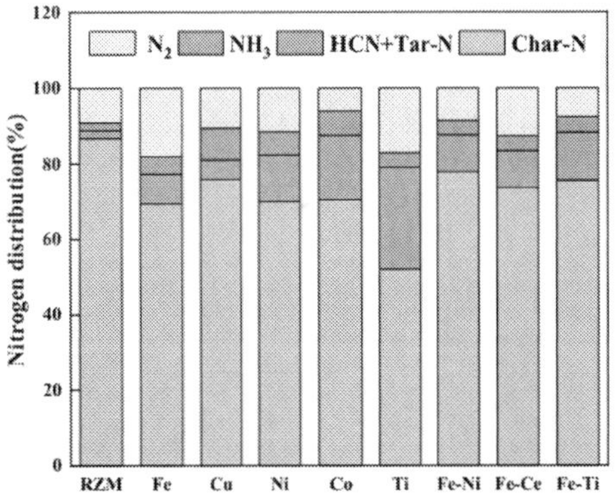

Fig. 3. Nitrogen balance of swelling coal and supported additives during pyrolysis at 1000℃.

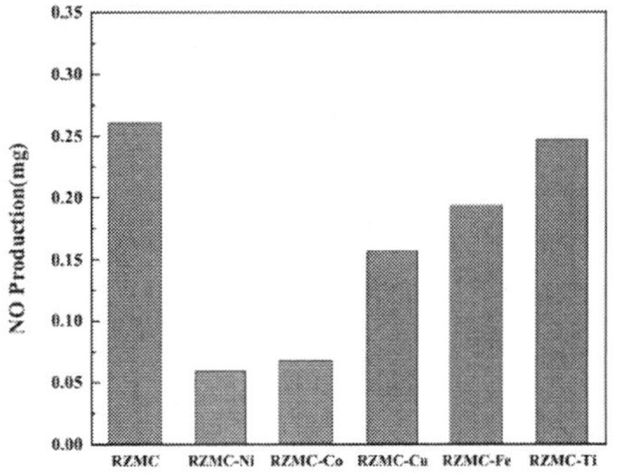

Fig. 4. NO production of swelling coal and supported additives during combustion at 1000℃.

3.3 Coupling effect of additives on coal pyrolysis-combustion process

In order to explore the coupling control effect of additives on nitrogen reduction in pyrolysis and NOx removal in combustion, the coupling control effect of pyrolysis on combustion was explored by adding Ni, Fe and Ti additives with better effect in pyrolysis and combustion. Coke combustion is also carried out at 1000℃, and other conditions are completely consistent with pyrolysis and combustion conditions except for different ways of adding additives. RZMC-M indicates that the coke prepared by swelling coal loaded with auxiliary agent M burns at 1000 ℃ , and RZMC+M indicates that swelling coal pyrolysis coke burns at 1000℃ with auxiliary agent M. It can be clearly seen from Fig. 5 that nickel, iron and

https://doi.org/10.1051/e3sconf/202129003029

titanium additives loaded in the combustion of the original coal char can inhibit the release of NO to a certain extent. This shows that the supported additives play a certain role in reducing nitrogen and denitration during pyrolysis or combustion. It can be seen from the above figure that purple represents coke combustion after pyrolysis of raw coal loaded with additives, and it is found that the NO released by combustion of three additives under the same conditions is less than that by adding additives to pyrolysis coke represented by yellow, which shows that the coupling effect of nitrogen oxide control does exist in the pyrolysis-combustion process. This is because the additives added before pyrolysis combine with coal to form solid solution in coal pores, and they contact more closely after pyrolysis gas is discharged, which can inhibit the production of nitrogen oxides. Adding additives to coke is only a mixture of solid and solid phases, which can only catalyze the reduction of nitrogen oxides on the surface layer and promote the reduction of nitrogen oxides by coke and CO in local anoxic environment.

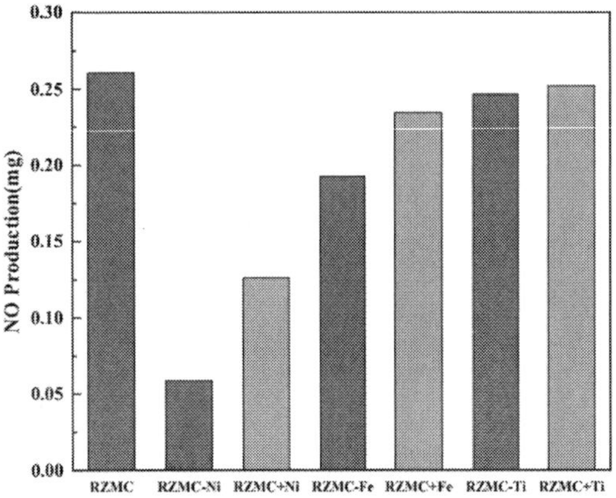

Fig. 5. NO production of swelling coal supported auxiliary coke and swelling coke supported auxiliary combustion at 1000℃.

3.4 Phase morphology analysis before and after combustion of coke supported additives

The coke and combustion ash with good catalytic effect in the above experiment process were characterized by XRD, and their auxiliary forms were detected, as shown in the following Fig. 6 and Fig. 7. XRD characterization analysis was carried out on the prepared coke, and the phase morphology of additives after pyrolysis was determined, the analysis results are shown in Fig. 6. Fe-doped coke has obvious Fe_2O_3 and α-Fe phases, which indicates that $FeCl_3$ introduced before pyrolysis changes with the increase of pyrolysis temperature, and local oxidation-reduction produces a new iron phase, compared with XRD characterization of ash components after combustion, the results are shown in Fig. 7. According to the analysis of ash, it is found that the iron-containing phase of ash after combustion is only oxide, which may be related to the strong oxidation system of combustion reaction, which indicates that iron oxide in coke participates in the reduction of nitrogen oxides and constitutes the balance of Fe, NO and Fe_2O_3.

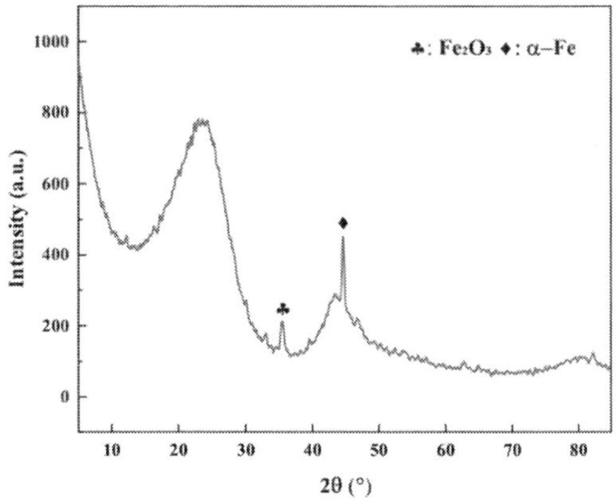

Fig. 6. Iron-containing phase of coke before combustion.

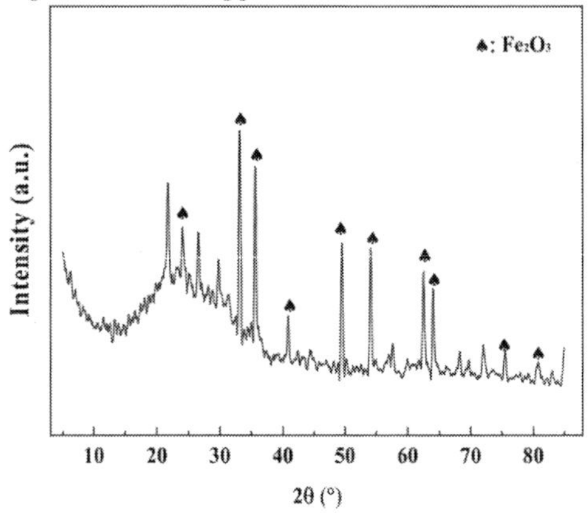

Fig. 7. Iron-containing phase of ash after combustion.

4 Conclusions

Coals loaded with $FeCl_3$ or NiCl can enhance the coupling effect of NOx control during pyrolysis and combustion. When the additive amount is 1 wt.% and the pyrolysis temperature is 1000 ℃, the effect is better, and the denitrification ratio are increased to 30.72% and 30.01%, respectively. Under these conditions, the combustion of coke loaded with additives has a better effect on reducing NOx emission. On the one hand, the combustion produces less NOx, on the other hand, the catalytic reduction produces NOx into N_2, and the lowest NO produced when loaded with nickel is 0.05886mg. This shows that the iron and nickel additives introduced in one step have coupling effect on NOx control in coal pyrolysis-combustion process, which promotes the conversion of nitrogen in coal into N_2 in pyrolysis process, catalyzes NOx reduction to N_2 in combustion process, and achieves ultra-low NOx emission from clean fuel combustion.

Acknowledgments

This work was supported by National Natural Science Foundation of China (21878210), Shanxi "1331" Civil Clean Fuel Engineering Research Center, Scientific and Technological Innovation Programs of Higher Education

https://doi.org/10.1051/e3sconf/202129003029

Institutions in Shanxi (2019L0313), Patent Promotion and implementation in Shanxi Province (20200719) and sponsored by Taiyuan Green Coke Energy Co., Ltd. (China).

References

1. S. John, M. Thomas. Fuel Process. Technol, 92, 2211-2218 (2011)
2. C. He, X. Min, H. Zheng. Energy Fuels, 12, 13558-13571 (2017)
3. L. Deng, X. Jin, Y. Zhang. Fuel. 175, 217-224 (2016)
4. Z. Lin, Z. Zhong, Y. Han. Korean J. Chem. Eng, 34, 1-9 (2017)
5. Z. Lei, S. Hao, J. Yang. Chemosphere, 270, 128646 (2020)

AUTHOR INDEX